Sixth Edition

Geology
of National Parks

Ann G. Harris
Youngstown State University

Esther Tuttle
Science Editor

Sherwood D. Tuttle
Professor Emeritus
University of Iowa

KENDALL/HUNT PUBLISHING COMPANY
4050 Westmark Drive   Dubuque, Iowa 52002

Book Team

Chairman and Chief Executive Officer   Mark C. Falb
Vice President, Director of National Book Program   Alfred C. Grisanti
Assistant Director of National Book Program   Paul B. Carty
Editorial Developmental Manager   Georgia Botsford
Developmental Editor   Tina Bower
Assistant Vice President, Production Services   Christine E. O'Brien
Prepress Project Coordinator   Sheri Hosek
Prepress Editor   Charmayne McMurray
Permissions Editor   Renae Heacock
Design Manager   Jodi Splinter
Designer   Deb Howes
Senior Vice President, College Division   Thomas W. Gantz

Cover photo of Kilauea Volcano, Hawaii, by Marli B. Miller, University of Oregon.

Library of Congress Catalog Card Number: 2003108444

ISBN 978-0-7872-9971–2 perfect bound version
ISBN 978-0-7872-9970–5 loose leaf version

Printed in the United States of America
10 9 8 7 6

Contents

Part III Landscapes Shaped by Continental or Alpine Glaciation 289

The Theory of Glacial Ages 290
Types of Glaciers 291
How Glaciers Form and Move 291
The Work of Glaciers 292
Glacial Advances and Retreats 292
Carbon-14 Dating 293
Is Another Ice Age Coming? 293

Preface

The National Park System

The 1871 government-sponsored expedition led by F. V. Hayden into the Yellowstone region and his report, combined with the photographs of W. H. Jackson and paintings by Thomas Moran, convinced Congress to withdraw this public land from sale. On March 1, 1872, President Grant signed the bill authorizing the creation of Yellowstone National Park. The proclamations stated that this first national park was to be "a pleasuring ground for the benefit and enjoyment of the people."

Because of the dedication of many individuals, there are now 57 national parks, numerous national monuments, scenic trails, rivers, lake shores, and wilderness areas that have been set aside. Areas of historic importance, such as battlefields and historic sites, are also administered by the National Park Service, which was established in 1916.

National parks are established to protect areas that have great scenic and scientific importance, with boundaries that provide for enough land and water to protect the resources, including plant and animal life. National monuments, in general, are smaller and preserve only one or a few unique resources, but lack the diversity and breadth of a national park. Another difference is that an Act of Congress must create a national park, but a national monument can be set aside by a presidential proclamation. Over time, a number of national monuments have been upgraded to national park status.

The National Park Service owns all the national parks except American Samoa. In some instances, when funding has not been available, land has been purchased by an organization, such as the Nature Conservancy, and later turned over to the National Park Service. Channel Islands National Park came into the park system in this way.

Some of our national parks that are close to metropolitan areas, such as Yosemite National Park in California, have become overused and overcrowded. Automobile traffic has had to be restricted, but shuttle buses transport visitors into Yosemite Valley. Shuttle bus service is also available at Zion National Park in order to relieve traffic congestion.

There are also other concerns. Lack of funding of the parks has caused maintenance and staffing problems. There are pressures from groups that want to exploit the natural resources within the parks. Some park visitors do not take into account that they may be disturbing the wildlife or damaging vegetation when they are snowmobiling or riding ATVs throughout the parks. There is even a problem with poaching, such as killing the bears in Great Smoky Mountains National Park. Animals have also been hunted in Yellowstone National Park.

The National Parks as a Means of Learning about Geology

Geology is a dynamic science that is changing as new discoveries are made. With a few dramatic exceptions, the natural features in the national parks have not changed appreciably; but explanations of how the landscapes developed have undergone many changes. Sometimes two apparently opposing concepts are combined into one comprehensive theory as more information has become available. The design of the sixth edition of **Geology of National Parks**, as in previous editions, has been to provide scientific explanations that are current, accurate, and understandable to readers who may or may not have a background in geology. Within the national parks, examples of interesting geologic features abound—large and small, old and new, in a grand array of natural environments. **Geology of National Parks** uses these examples to present basic elements of physical and historical geology in a nontraditional format, with terms and concepts defined or explained as they are introduced. The general plan of the book has been to proceed from simpler geologic themes—as shown in parks where flat-lying sedimentary rock predominate—to areas of complex mountains where the geology is exceedingly challenging.

Adding the e-mail address and web site for each park in this edition will be helpful to students and visitors. Three new chapters have been added: Great Sand Dunes and the Black Canyon of the Gunnison, both in Colorado, and Cuyahoga Valley in Ohio. Several chapters have been rewritten, and all have been updated. The photographs and diagrams are now in color, which will help readers to gain a better understanding of the landscapes. A CD-ROM is included, which contains additional photographs of all the parks, some with geologic features labeled.

Acknowledgments

Photographs of geologic features for the 57 national parks for the sixth edition of *Geology of National Parks*. have been derived from individuals, the National Park Service, and other sources, such as public domain material on the Internet. We are grateful to Ted Williams, who volunteered his time and whose help has been invaluable. Gratitude is also due to David B. Hacker, who not only wrote the new chapter of Cuyahoga Valley National Park, but also is the creator of the CD-ROM accompanying this book. Many of the photographs in the book are either his or his father's—Ronald F. Hacker. Our profound thanks also to Phil Brease, a ranger in Denali National Park, who revised and updated the Denali chapter.

Again we thank the contributing authors of the earlier editions, Richard A. Davis, College of Mount Saint Joseph; D. D. Trent, Citrus College; Lauren A. Wright, Pennsylvanian State University; Martin Miller, University of Oregon; Ann F. Budd, University of Iowa; Rodney M. Feldman, Kent State University; Donald S. Fellows, Anchorage, Alaska; Katherine A. Giles, New Mexico State University; Arthur N. Palmer, State University of New York, Oneonta; Monte Wilson, University of Idaho; and Lisa A. Rossbacher, Southern Polytechnic State University.

Special thanks also go to colleagues who reviewed chapters, made suggestions, corrected errors, lent illustrations, and gave support during the preparation of earlier editions. They include Robert Badger, State University of New York; Richard Baker, University of Iowa; Robert F. Biek, North Dakota Geological Survey; Robert R. Churchill, Middlebury College; Julie A. Dumoulin, U.S. Geological Survey; J. Thomas Dutro, U.S. Geological Survey; Brian F. Glenister, University of Iowa; Mary L. Griffitts, Boulder, Colorado; Lee Grim, Rainy River Community College; W.K. Hamblin, Brigham Young University; Robert D. Hatcher, Jr., University of Tennessee; George Haselton, Clemson University; Carol A. Hill, Cave Research Foundation; Richard A. Hoppin, University of Iowa; Patricia Humberson, Youngstown State University; Richard W. Jones, Wyoming State Geological Survey; Gwendolyn W. Luttrell, U.S. Geological Survey; Walter L. Manger, University of Arkansas; Robert B. Neuman, U.S. Geological Survey; Willard H. Parsons, Wayne State University; Charles C. Plummer, University of California, Sacramento; Weldon W. Rau, Department of Natural Resources, State of Washington; Mark Reagan, University of Iowa; John C. Reed, Jr., U.S. Geological Survey; Jeffrey P. Schaffer, Napa Valley College; Holmes A. Semken, University of Iowa; Keene Swett, University of Iowa; R.W. Tabor, U.S. Geological Survey; the late Paul Tychsen, University of Wisconsin, Superior; Robert E. Wallace, U.S. Geological Survey.

A heartfelt thanks goes to all the individuals who have generously lent me their photographs: the late Myron Agnew, Raymond Beiersdorfer, Anthony Belfast, James Ellashek, Robert Fletcher, Douglas Fowler, David B. Hacker, Ronald F. Hacker, Earl Harris, Alan Jacobs, Carol Kriner, David W. Little, Michael Little, Dianne Ludwig, Jan McGuire, Martin Miller, Edward Mooney, Gerald Morrison, Lois Petticord, Richard Pirko, Douglas Price, Sherry Schmidt, Thomas Serenko, Charles R. Singler, Robert K. Smith, D.D. Trent, and Kelli B. Weibel.

We express appreciation and thanks to the staff of Kendall/Hunt Publishing Company for their continuing help and encouragement through the many years of our association.

And, finally, we cannot adequately express our gratitude to all of the persons in the National Park Service and the affiliated Natural History Associations who through the years have patiently answered our questions, offered assistance, given us time, corrected our mistakes, helped us locate things needed, and demonstrated again and again their concern for the natural environments in their care and their dedication to the ideals of the National Park Service. All of us owe them support, now and in the days ahead, so that they and we together can continue to protect these national treasures for generations to come.

Ann G. Harris
Esther Tuttle
Sherwood D. Tuttle

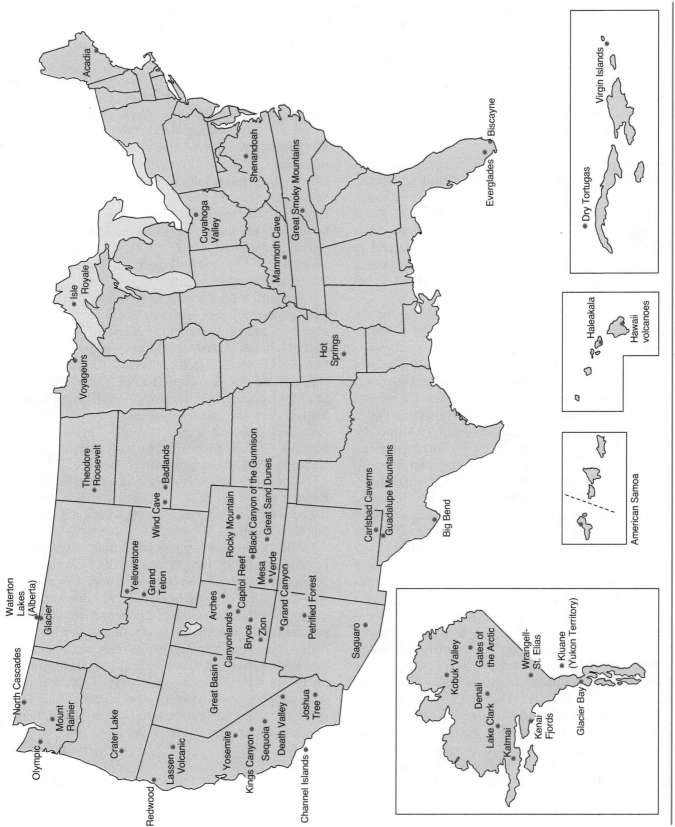

Locations of the U.S. National Parks.

PART I

Scenery Developed by Weathering and Erosion on Flat-Lying Rocks

FIGURE PI.1   Grand Canyon National Park as photographed by NASA satellite. This photograph has been computer enhanced to give the 3D effect and to alter the color. Photo by NASA.

Spectacular cliffs, canyons, arroyos, pinnacles, arches, pedestals, and dunes are landforms produced by weathering and erosion that we see when we visit the national parks in this group. Most are located in drier areas of the United States and all but five are in the Colorado Plateaus physiographic province. (Portions of the earth's surface that display a similar array of landforms and scenery are recognized as *physiographic provinces*.) The high plateaus of western Colorado, Utah, northern Arizona, and northwestern New Mexico make up the Colorado Plateaus province. Eight national parks and more than a dozen national monuments have been set aside in this remarkable and beautiful region. We'll vist these first in Part I. We next describe two national parks in the northern part of the Great Plains province, Badlands National Park (South Dakota) and Theodore Roosevelt National Park (North Dakota). The next national park in Part I is Kobuk Valley National Park in northwest Alaska, where some 500 square miles are covered by either active or stabilized dunefields. Great Sand Dunes is located in south central Colorado, while Cuyahoga Valley is found in northern Ohio.

Despite the dry climate of these areas, with the exception of northeastern Ohio, running water is the dominant agent of erosion shaping the land surface. Weathering processes, which prepare rock for erosion, operate wherever water is present, even in small amounts. Stream erosion—the most important external geologic process here—sculptures the slopes and valleys. Scarcely a landscape exists anywhere that does not have erosional and depositional features produced by the action of streams.

Landforms are the individual features we see, such as a hill or a slope; *landscapes* are combinations of landforms making up an expanse of natural scenery— what the eye sees in a view. *Geologic agents* and *geologic processes* continually form and alter landforms and landscapes. Running water and wind, for example, are geologic agents; stream erosion and wind erosion are the geologic processes. *Weathering* of rock at the earth's surface, which includes both chemical decomposition and physical disintegration; *mass wasting,* the downslope movement of material under the influence of gravity; *entrainment,* the picking up of particles by running water and wind; transportation by a variety of means; and eventual *deposition* are some of the natural processes that recycle rocks and modify landscapes over time. Human activities are also a significant factor in bringing about environmental changes.

The generally low precipitation (rainfall and snow) in the national parks of the Colorado Plateaus and most of the other parks in Part I supports a sparse or spotty cover of vegetation. We can see the geologic framework, the rock structures, and the dynamic interaction of erosional processes more easily and dramatically here than in humid areas where forests tend to obscure the geology. The bedrock of the Colorado Plateaus has been elevated so that much of the present land surface stands well over a mile above sea level. The elevation gives streams so much erosive power that they have cut deeply into the rocky crust, and the resulting landforms are unmatched for size and scale in most areas of the world. Depictions of these landscapes on film, postcards, and paintings seldom capture the grandeur of the views as seen by the human eye.

How does the sculpturing of the land take place? Wherever moisture exists, weathering processes soften the rocks, and loosen the constituent particles. Water enters minute cracks, soaks into spaces between rock or mineral grains, and percolates slowly downward. Atmospheric gases and organic acids from decaying plants combine with water to form a weak acid solution that slowly decomposes rock. When temperatures drop below freezing, trapped water freezes and expands, shattering rocks and breaking grains apart. Fragments, individual grains, clay, and soluble compounds are shed from rocks of all sizes and shapes. Loose sediment accumulates on rock surfaces, around the edges of blocks, at the base of cliffs, and at the foot of slopes. Impelled by gravity, the processes of mass wasting (sheetwash during rainstorms, creep of soil, slumping, sliding, and avalanching) move vast amounts of rock material downslope toward the valley bottoms.

Wind, a powerful geologic agent in dry climates, picks up and drops loose sediment. The size and the amount of the particles moved depend on weather conditions and how rock fragments are exposed. Wind also scours and abrades rock and piles up sand in dunes. Ancient or "fossil" dunes form prominent cliffs in Zion National Park in Utah (Chapter 2), while modern, active dunes still accumulate and move in Kobuk Valley National Park in Alaska (Chapter 11).

Running water, in trickles and in streams, picks up material and moves its load by rolling or jumping fragments, and by carrying fine particles in *suspension,* and

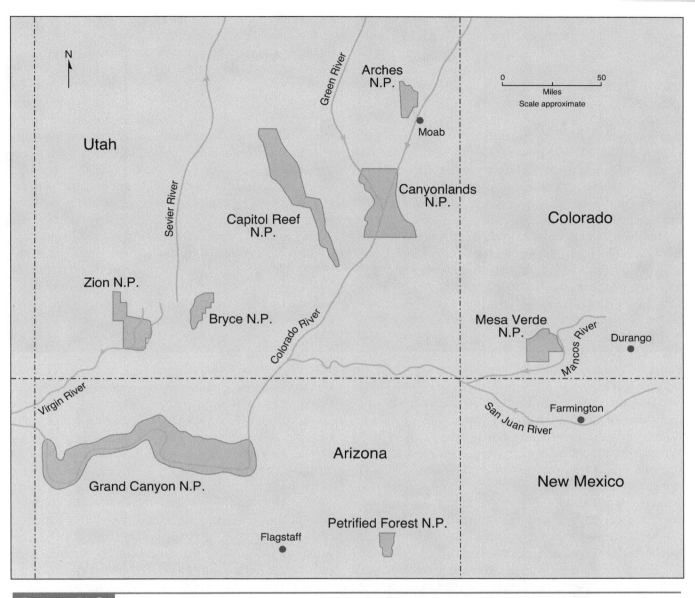

FIGURE PI.2 Location on the Colorado Plateaus of the first eight national parks described in Part I. This large physiographic province is about 130,000 square miles in area. The region's distinctive characteristics are: (1) flat-lying sedimentary rocks, with red the dominant color; (2) relatively high elevations, between 5000 and 11,000 feet; (3) deep canyons dissecting plateau surfaces; (4) retreating escarpments produced by erosion of rock layers of varying resistance; (5) arches, natural bridges, pinnacles, fins, and many other erosional features; (6) semiarid climate.

A

B

FIGURE PI.3   Two views of erosional features in Canyonlands National Park, Utah. *A.* Along the upper Colorado River, near Shaffer Trail, a distinctive butte seen from river level. The butte has been carved by erosion out of the Triassic aged Wingate Sandstone Member of the Glen Canyon Group, that is, a cliff-former. The first bench below it consists of the Permian-aged White Rim Sandstone. Photo by Ann G. Harris. *B.* A view of Canyonlands National Park, taken from Island in the Sky, shows the differences in the weathering of resistant and less resistant beds forming cliffs and slopes. Photo by David B. Hacker.

soluble matter in *solution.* Sediments are continuously eroded, deposited, and re-eroded. Fragments and particles become rounded, abraded, sorted, and mixed. They may lie undisturbed in a channel bottom or a river bar for many years or be spread across a valley floor during a flood. But inevitably, sediments move downhill and downstream and away from their source area. Estimates of rates of erosion suggest that even high regions may be reduced to lowlands in a few million years. Erosion rates in the national parks in Part I are comparatively high because downcutting goes rapidly in regions of sparse vegetation and high elevation. The Grand Canyon, deep as it is, was eroded only within the last one or two million years.

The Geologic View of Time

Geology draws on other sciences, both physical and biological, for many of its concepts; but its unique intellectual aspect is its historical approach and its view of time. Geologists think in terms of sequences of events and develop explanatory patterns that closely follow cause and effect within a framework that's called "geologic time." A geologist observes running water gullying a slope and carrying off soil and postulates that in the past, running water eroded slopes, transported sediment, and deposited it in a similar way. This fundamental principle is called *uniformitarianism,* which means that natural processes have operated in the past (recent or ancient) in much the same way and at about the same rate as now. "The present is the key to the past" is another way of expressing this idea. (Occasionally, a catastrophic event, such as an asteroid impact, may interrupt or alter usual patterns of geologic processes.)

The time ordering, or sequencing, of events in the geologic history of a feature or region becomes very important in studying and understanding modern landscapes. By going farther back, a geologist attempts to fit the events into the overall time framework of earth history. For example, we look at a glaciated mountain range and assume that glaciation occurred after the range was uplifted. Streams began dissecting the glacial deposits while the glaciers melted.

By various techniques, the rocks in the mountain range might be dated Jurassic and Cretaceous in age with the glaciation occurring late in Pleistocene time. These are *relative* time terms (shown in the Geologic Time Scale, page xiv). Relative time units are like days

of the week or cultural and historical eras of human history, except for their much greater expanse of earth time. Relative dating techniques, which are based on paleontology (study of ancient life forms) and stratigraphy (interpretation of rock layers), are mostly logical applications of careful observations and tests in order to build a sequence of geologic history. Prior to the discovery of radioactivity, these relative dating methods, which are still necessary and important, were the only means of determining when rocks formed.

The determination of *absolute ages* of rocks became possible with radiometric techniques that measured the rates of radioactive decay, or disintegration. *Radioactive decay* is the spontaneous nuclear disintegration of certain isotopes of radioactive elements, a process that begins when some types of rock are formed and that continues at a constant rate. The "radioactive clock" in such a rock can be read by laboratory techniques that compare the amount of radioactive elements with the amount of decay products. This determines how much time has elapsed since the rock was formed because radioactive elements in rocks have characteristic and constant rates of decay. The time units are expressed as "years before the present," or y.b.p. An absolute age, for example, might be 265 million years, plus or minus 30,000 years.

The Flat-Lying Rocks

In the national parks discussed in Part I, most of the solid bedrock, whether exposed or underlying the soil cover, is relatively undeformed, flat-lying sedimentary rock. In many places in the Colorado Plateaus, a sequence of sedimentary rock layers, one to two miles thick overall, is exposed. You can stand on the floor of a canyon or before the face of a cliff and count the beds (*strata*) from bottom to top. (This is called "layer-cake geology.") Many varieties of sedimentary rock, formed in different periods of geologic time and by different ways are found in these national parks, but all resulted from the consolidation of loose sediment (particles, grains, shells, etc.) that accumulated in layers. Sand grains, compacted and cemented, become *sandstone.* Finer quartz grains, clays, and muds may lithify (form rock) as *shales. Limestones* may precipitate from seawater (an inorganic process) or accumulate from the skeletal remains of marine organisms. By careful examination of the layered rocks, geologists produce data and make interpretations about the compositions of the rocks, their ages, the environments in which the sediments accumulated, where the sediments came from, and the remains of ancient life (fossils) preserved in the rocks.

We describe the sedimentary rocks in the Colorado Plateaus, Ohio and the Dakota badlands as "undeformed" and "flat-lying," but this is true only in a general sense. Great Sand Dunes are essentially horizontal but are in a valley of complex structure. Some of the layered rocks have been subjected to stress and in places have yielded to bending *(folding)* or breaking *(faulting)*. As a park visitor, you may recognize broad open folds, or you may see faults where rock has moved several hundred feet. However, in comparison to deformation that rocks have undergone elsewhere, mild bends and breaks in layers can be regarded as reflections of local circumstances rather than as a regional characteristic.

Red-Colored Rocks

In the plateau country, you can see in the landscapes reds of every hue and intensity—pinks light and dark, purples, orange-yellow reds, even brownish reds. The reddish tints of the rocks change as sunlight and shadow shift over the land from dawn to dusk. The remarkable colors are due to the presence of minute amounts of iron compounds. Chemically, the tints are similar to iron rusts resulting from oxidation and hydration. The colors are usually not in the sediment grains but in the intergranular cement that fills the pore spaces. Less common are rocks stained black, brown, yellow, gray, or green because of traces of carbon, manganese, or copper. The coloring elements, which are added during processes of deposition, are concentrated from ground water by evaporation and precipitation in a semiarid continental environment.

The scenery of the national parks in Part I—except for Kobuk Valley, Great Sand Dunes, and Cuyahoga Valley National Parks—developed because of deep dissection of brightly colored sedimentary rocks in areas of low rainfall where vegetation does not blanket the landscape. The difference in the Kobuk Valley, which is also dry and sparsely vegetated, is that the sediments are comparatively young and are unconsolidated. In Great Sand Dunes there are huge dunes that are up to 700 feet high. Cuyahoga Valley is a park located in Ohio where there is heavy vegetation, stream erosion and continental glaciation. But in all these national parks the interactions of complex weathering, downslope movements, and wind and stream erosion have produced magnificent landscapes.

CHAPTER 1

Grand Canyon National Park

N O R T H E R N   A R I Z O N A

Area: 1,217,158 acres; 1902 square miles
Proclaimed a National Monument: January 11, 1908
Established as a National Park: February 26, 1919
Designated a World Heritage Site: 1979

Address: P.O. Box 129, Grand Canyon, AZ 86023
Phone: 928-638-7888
E-mail: deanna_prather@nps.gov
Website: www.nps.gov/grca/index.htm

The incomparable Grand Canyon of the Colorado River is ten miles wide and over a mile deep. The national park encompasses 277 miles of river, plus adjacent uplands. Weathering and erosion have produced colorful cliffs, slopes, spires, buttes, and mesas. Exposed in these features are rock formations that illustrate a vast span of geologic history. Two sequences of Precambrian rocks record early mountain-building episodes. A classic Paleozoic section includes carbonates, shales, and sandstones deposited in five geologic periods.

FIGURE 1.1   The Grand Canyon is so large that it can be seen from outer space. The canyon was formed around two million years ago as the Colorado River eroded down as rapidly as the Colorado Plateau was uplifted. Erosion exposed one of the most complete geologic section of rocks in the world, from the Precambrian to the Cenozoic. The river developed along a fault zone where rock was weaker. The canyon rims are from 9 to 18 miles apart. Photo by John Van Hosen.

The Indians of the Southwest had legends to explain the origin of the Grand Canyon. The stories reveal an awareness of natural causes and effects. The Navajos, for example, believed that rain for many days and nights caused a great flood that covered the land with water that rose higher and higher. Finally an outlet formed, and as the rushing waters drained away, the Grand Canyon was cut deeply into the earth.

More than 500 Indian sites in the park, both on the rim and within the canyon, indicate that as recently as 600 years ago large numbers of Indians lived in the Grand Canyon vicinity. Havasupai Indians still live in a branch canyon watered by Havasu Creek in the western section of the park.

In 1540, a band of Coronado's conquistadores were the first white men to see the Grand Canyon, after being guided by Hopi Indians to the rim. More than 300 years later, Lt. Joseph C. Ives, a U.S. Army surveyor, came upriver from the southwest and reached the lower Grand Canyon. After an unsuccessful attempt to reach the rim, he concluded that surveying the Grand Canyon of the Colorado River was impractical.

John S. Newberry, a member of the Ives party and a geologist, thought otherwise. He convinced Major John Wesley Powell, a fellow geologist (who had lost a hand in the Civil War), that an expedition by boat down the Colorado River through the Grand Canyon would be worth the risk in order to complete the survey. With scientific help from the Smithsonian Institution, the Powell expedition of four boats and nine men left Green River, Wyoming, in May 1869, and began their hazardous journey downstream. One boat was smashed; scientific instruments and food were lost. Most of the time the party was wet, tired and hungry, and some were convinced that they would never get out of the canyon alive. Three men climbed out of the canyon but were killed on the plateau by hostile Indians or Mormons. The men who stayed with Powell survived and reached the settlements near the western end of the Grand Canyon. Two years later, Powell went down the Colorado River again in order to collect more data for his geographical and geological report on the region.

Theodore Roosevelt, after a trip to the Grand Canyon in 1903, was determined that it should be preserved. He was able to have part of the area placed under government protection in 1908, but opponents blocked the establishment of a national park until 1919.

The Colorado River carries tons of sediment past any one point each day. The rate of erosion is not constant throughout the course of the river, but it is estimated that every thousand years the drainage basin of the Colorado River system is lowered by an average of 6 1/2 inches. It was his observation of the sediment load that convinced Powell that he would not be washed over high falls (as some people had predicted) on his initial expedition. He reasoned correctly that a river carrying so much sediment would have scoured its bed down to a more or less even grade, or slope.

The Colorado River no longer rampages every spring or falls to lowest low-water stage in summer. The Colorado has been a regulated stream since 1966, when Glen Canyon Dam, just upstream from Lees Ferry, impounded river water to fill Lake Powell in Glen Canyon National Recreation Area. The tremendous sediment loads that used to scour the canyon during spring floods now settle to the bottom of Lake Powell, and the pre-1966 average load of half a million tons per day has been greatly reduced. The curtailment of the sediments has resulted in the removal of sand bars and beaches, to the detriment of the environment. A "test flood" in 1996 lasted a week, and another in 1997 lasted two days. Both deposited sand bars and beaches, but they disappeared within a year once the source of the sediments was removed. The river below the Grand Canyon has been controlled since 1935 by Hoover Dam, which backed up water in Lake Mead to Grand Wash, where the river leaves the national park.

Grand Canyon National Park extends from the mouth of the Paria River, which joins the Colorado River just below Glen Canyon Dam, to the eastern boundary of Lake Mead National Recreation Area, a river distance of 277 miles. The Grand Canyon is over a mile deep, with the distance from rim to rim ranging from 9 to 18 miles. The elevation of the North Rim (8900 feet) averages some 1200 feet higher than the South Rim (6900 feet).

Dimensions of such magnitude have biological as well as geological significance. Life forms on the two canyon rims evolved differently because of physical separation over time. At the higher altitude of the North Rim, under conditions of cooler average temperatures and greater annual precipitation, plants and animals characteristic of Canadian life zones became established. Trees and shrubs such as blue spruce, ponderosa pine, fir, and aspen, that are darker in color, took over. A typical North Rim animal is the Kaibab squirrel, which has a black body, black stomach, and white tail.

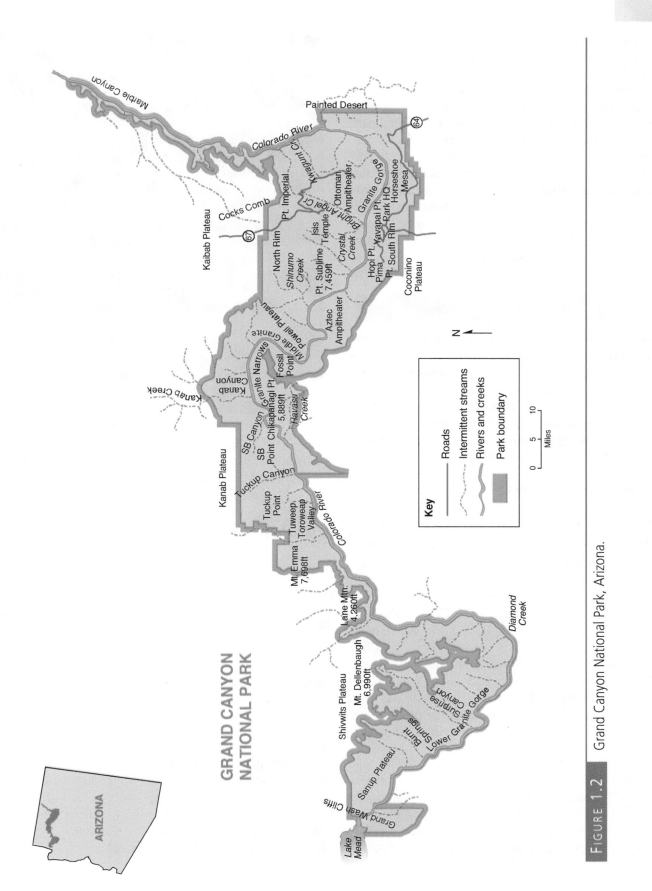

FIGURE 1.2 Grand Canyon National Park, Arizona.

On the South Rim, vegetation that is lighter in color and typical of the Upper Sonoran life zone (junipers, pinon, etc.) predominates. The Abert squirrel is found here, with its grizzled gray body, white stomach, and gray tail. The common ancestor of these two modern species was a black and white squirrel with tufted ears that lived on the plateau before the Grand Canyon was incised. When the squirrels were no longer able to migrate back and forth across the physical barrier of the canyon, and as differences in climate became pronounced between the North Rim and South Rim, new squirrel species developed due to evolutionary adaptation—the darker Kaibab squirrel on the North Rim and the lighter Abert on the South Rim.

Geologic Features

In terms of geologic time, the Grand Canyon of the Colorado River is surprisingly young. The river accomplished the major part of the downcutting during the last two million years. It is the rocks that are ancient, not the canyon itself. Exposed by this downcutting is one of the most nearly complete geologic columns on earth, encompassing some two billion years of geologic history. The layers of sedimentary rock seen in the Grand Canyon are essentially flat-lying; many of these beds extend throughout the Colorado Plateaus region.

How Does a River System Evolve?

The Colorado River is a major through-flowing stream with a total length of 1450 miles from the Rocky Mountains to the Gulf of California. Like other major streams, the Colorado developed in response to environmental variables over time; in the Colorado's case, its history has been more complicated. For over a hundred years, geologists have argued about possible courses and directions of flow of tributaries and "ancestral" streams that may have had some part in the Colorado's evolution. So far, no hypothesis has met with overall acceptance. The details of the problem are fairly technical, but some general principles that apply to stream development are outlined below.

Base level. The limiting surface to which a stream can cut down is called *base level*. Sea level is the ultimate base level for through-going streams like the Colorado, but sometimes local base level, such as a lake, a reservoir, or an interior basin without an outlet, controls a

FIGURE 1.3   Natural cave and a flowing spring in the Redwall Limestone. Major Powell named the area Vasey's Paradise after the botanist in the 1869 expedition. Ground water, seeping down through rock layers, over time dissolves parts of the limestone and enlarges openings in the rock. Prehistoric animals and humans found shelter in the caves in the canyon walls. Photo by Michael Little.

stream's downcutting ability. The Colorado's base level has been affected by uplift and also by sea level fluctuations, especially during the Pleistocene Ice Ages. Moreover, segments of the river, probably before it became through-flowing, may have emptied into inland lakes that functioned as temporary base levels.

Gradient. A stream whose *gradient* (rate of descent, or slope) is low does not have much erosive power. But if an area is uplifted, the gradient steepens, increasing the velocity of stream flow. When this happens, a stream downcuts rapidly, especially if its volume of water also increases. Sometimes, during uplift, a stream is able to maintain its original course and cut a narrow trench down through bedrock.

Headward erosion and stream capture. While a trunk stream is downcutting or entrenching itself, its tributaries lengthen themselves by *headward erosion,* which means that they cut back into a plateau or upland at the head of each valley. A drainage system that is favorably situated, perhaps flowing on weaker rock, for example, is able to erode headward faster and eventually intersect and *capture* the headwaters of a neighboring stream or drainage system. Geologists are sure that stream capture occurred numerous times while the Colorado was evolving into a major stream; but the details of how, when, and where these events happened is not clear.

The key factors in the Colorado River's evolution are probably these:

1. The river's source area in the Rocky Mountains, in addition to the Colorado Plateau, has undergone a long series of relatively rapid uplifts since the end of Mesozoic time (65 million years ago). Overall, the Colorado Plateaus rose 5000 to 10,000 feet. Stream gradients became steeper, increasing stream velocity and erosion capability.

2. More rain and snow fell in the Rocky Mountains as the elevation increased. In the colder Pleistocene climates, glaciers formed and grew. As more meltwater and runoff became available, streams gained volume. The streams that drained the Rocky Mountains were powerful forces in erosion as they poured out on the plateaus.

3. The climate of the plateaus was (and is) drier than the mountain climate. Therefore, the high-volume, swift rivers cut narrow, deep canyons. Because rainfall is low on the uplifted plateaus, runoff is not ample enough to widen the valley side slopes.

4. When the Gulf of California opened more than 6 million years ago, this lowered base level for the Grand Canyon region and caused rapid headward erosion. The modern course of the through-flowing Colorado River was probably established after this time. Previously to the opening of this arm of the sea, the Colorado—or an ancestral stream that drained the Rockies—flowed into the ocean at a different location.

5. The east-west Grand Canyon segment of the Colorado's course across the Kaibab Plateau is asymmetrical; that is, the river flows closer to the south side due to a slight tilting that occurred during uplift. The fact that the North Rim receives more moisture and sheds more sediment also tends to push the river toward the south side of the canyon. The rate of erosion on the Kaibab Plateau is greater because all of the precipitation falling on the north side of the river drains to the south and into Grand Canyon. On the South Rim, rain and meltwater also drain to the south; since the runoff goes away from the canyon, it has less erosional effect.

6. How the Colorado River is regulated will determine how it evolves in this century and the next. Lake Powell became filled to capacity in 1980. How long will it be before silt accumulating in the lake bottom will cause the river to run over the dam? If such overflow is allowed to happen, the dam will inevitably wash out.

The controlled average release of water by Glen Canyon Dam is 12,200 cubic feet per second. Contrast this with the range of flow before dams were built above the Grand Canyon: a low of 1000 cubic feet per second during the driest time of year and a recorded flood high of 325,000 cubic foot per second. The Paria River, which joins the Colorado just below Glen Canyon Dam, brings some sediment, seasonal high-water flow, and occasional flash-flooding to the Grand Canyon. To a lesser extent, the Little Colorado River, with its confluence just above the big bend where the main river turns westward, does the same. But these contributions, added to sediment-free water released by the dam, are not powerful enough to keep the canyon bottom free of debris. Meanwhile, beaches on bank areas that used to be replenished by sediment during seasonal floods are being eroded due to fluctuating river levels when large amounts of water are released from the power station at Glen Canyon Dam.

Restoring the river's normal flow is not a practical answer because the Colorado River is the only significant source of surface water in the entire Southwest. More of its water is legally allocated to users than the river normally carries. How national parks and recreational areas are affected by the regulation of the Colorado River is only one aspect of a difficult problem of crucial importance to millions of people.

Igneous and Metamorphic Rocks in the Grand Canyon

As the Colorado River sliced its way deeper into the upwarped rocks of the plateau, it lets itself down onto ancient rocks and structures beneath the flat-lying

Paleozoic beds. These igneous and metamorphic rocks, exposed in the Inner Gorge of the Grand Canyon, are Precambrian, and over 600 million years old. The oldest is the Vishnu Schist at the very bottom of the canyon. The rock originally accumulated as silts, muds, and volcanic material and lithified about 2 billion years ago. During mountain-building episodes, the layered rocks were highly deformed and metamorphosed. On dark-colored, contorted Vishnu outcrops alongside the river, little, rounded garnets stick out from the rock surface like a scattering of seeds.

A *pluton* (a body of cooled magma), the Zoroaster Granite, intruded the metamorphosed Vishnu rocks deep underground and gradually cooled. The granite shows up as light-colored, irregular bands and blotches in the Vishnu Schist.

Later in Precambrian time, many layers of shales, sandstones, and other sedimentary rocks—interbedded with countless lava flows—covered the region. These rocks, which comprise the Grand Canyon Supergroup, were also metamorphosed by heat and pressure during tectonic activity. Some of the lavas show up as dark ribbons or bands between or cutting through the old sedimentary beds. Geologists call such features *dikes,* when they cut through older rocks, and *sills,* if they are squeezed between rock layers (fig. 1.4). A long period of erosion removed a great portion of the Precambrian rocks.

In recent geologic time, around a million years ago, dark (basaltic) lava erupted from fissures and flowed down into the Grand Canyon in the western section of the park. Volcanic ash, thrown out from small cinder cones on the canyon rim, fell on the land. These flows and ash falls made the youngest rocks in the park.

Geologic Structures

Unconformities. Before erosion planed off the igneous and metamorphic rocks of the Inner Gorge, the rock units had been broken into large blocks and tilted about 15° by mountain-building. But the surface was relatively flat by the time an ocean covered the region and the sedimentary beds of Paleozoic time began to accumulate. The surface or contact between Paleozoic and Precambrian rocks, which is called an *unconformity,* represents a gap in the geologic record. In this case, it is an *angular unconformity* because the bedding planes of the rock units above and below the erosion surface are *not* parallel; specifically, the younger over-

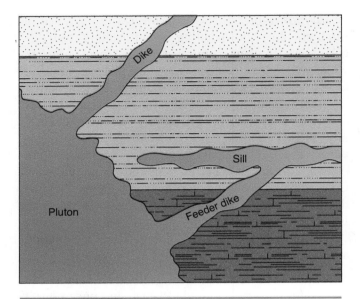

FIGURE 1.4   Magma from a small pluton intrudes sedimentary rock beds. Dikes cut across rock layers, while a sill squeezes between bedding planes.

lying sediments rest upon the eroded surface of tilted or folded older rocks (fig. 1.7).

What causes gaps in the rock record? The marine sedimentary rocks in the walls of the Grand Canyon cover a large portion of geologic history, but some geologic periods are not represented. For example, a major erosional unconformity exists between horizontal Cambrian and Devonian strata in the canyon sequence. Were the marine beds of the missing geologic periods completely eroded away? Or did marine sedimentation not occur during those periods?

Tectonic forces generated within the earth have resulted in uplift, movement, and deformation of portions of the earth's crust. Continents have stood at times higher and at times lower during geologic history. Ocean levels have fluctuated over millions of years.

When sea level rises, the oceans *transgress* onto the land; and when sea level falls, they *regress* from the land. During the long and slow process of transgression or regression, streams carrying sediment down to the ocean continue to erode. If the ocean is regressing, the streams may strip and redeposit (at lower elevations) most or all of the sediment layers that were put down during an earlier period of transgression. When the seas finally come back and again cover an area that has been

Box 1.1   How Rocks Are Classified

How a rock originated determines its category. The three main divisions are *igneous, sedimentary,* and *metamorphic.* Igneous rocks are: (1) those that cool from molten rock material, called *magma,* deep in the earth; or (2) those that cool from *lava,* derived from magma that rises to the earth's surface. Sedimentary and metamorphic rocks are formed from preexisting rocks. Loose particles of rock (sediment) are consolidated, or *lithified,* to make up sedimentary rocks. Metamorphic rocks, which originally may have been igneous, sedimentary, or other metamorphic rocks, have been so altered by burial, heat, chemical fluids, and pressure that they acquire a new rock identity.

When you look at a hand specimen of a rock, you may or may not be able to tell how the rock originated; but you can probably see grains or particles. Either with the naked eye or with the aid of a hand lens, you may be able to recognize some of the mineral constituents by their physical characteristics.

Rocks are made up of grains of one or more varieties of minerals. In the geological sense, a mineral is a naturally occurring, inorganic, crystalline solid with a definite chemical composition and characteristic physical properties that are either uniform or variable within definite limits. Therefore, because the grains in rock represent the constituent minerals and because each mineral has a specific crystalline structure, the mineral grains, their size, and how they combine provide the basis for identifying and classifying rocks.

Rock texture, a means of describing the grain size of component minerals (or sediments, in the case of sedimentary rocks), ranges from extremely fine to extremely coarse. For example, most *extrusive igneous rocks* (those erupted at the earth's surface) have a fine-grained texture with crystals too small to be seen by an unaided eye. This is because at surface temperatures and pressures, lava cools too fast for crystals to attain much size while the rock hardens. A rock that cools so quickly that no crystals have time to form becomes a natural glass, called *obsidian.* (In the plateau country, weathered pieces of black obsidian are called "Apache tears.")

Intrusive igneous rocks (which harden far below the earth's surface) tend to have a coarse-grained texture because magma cools at great depth under confining pressure over thousands, and perhaps millions, of years. As the mineral crystals separate from the melt, they grow, interlocking and filling all the available space.

When igneous rocks and any other rocks are exposed at the earth's surface, they are subjected to processes of weathering that bring about chemical changes (*decomposition*) and physical changes (*disintegration*). The rock particles produced by weathering are usually transported and then deposited, commonly in layers, and may eventually become sedimentary rock; that is, rock made up of consolidated sediment.

Sediment is a collective term for loose, solid particles of rock that can originate in several ways. *Clastic* sediment refers to any accumulation of unconsolidated fragments freed by weathering and erosion of preexisting rock. Sizes of fragments range from large boulders to microscopic clay particles. During transportation by running water, glaciers, or other agents, these rock particles become rounded and broken further by processes of *abrasion,* which is the wearing away of rock by friction.

Decomposition of rock involves chemical alteration by the addition or subtraction of various substances. Some minerals, for example, go into *solution* when exposed to water and are transported in dissolved form. Later, the minerals may be *precipitated out* by evaporation or by organic secretion (by marine organisms, etc.). Sediment formed by chemical precipitation or by organic secretion is referred to as *nonclastic.*

A sedimentary rock made up of grains or fragments of clastic sediment is described as having *clastic texture.* A *conglomerate,* for example, is coarse-textured rock made up of consolidated rounded gravel. *Shale* is fine-grained rock made of silt or wet mud that has been lithified by compaction and cementation.

Cementation is the process by which clastic sediments are converted into sedimentary rock by

(continued)

Box 1.1	How Rocks Are Classified (continued)

Table 1.1	Classification of Common Sedimentary Rocks	

Source	Sediments	Sedimentary Rocks
Clastic — Coarse-textured	Boulders, cobbles, pebbles, granules	Conglomerate (rounded fragments) or breccia (angular fragments)
	Quartz grains; other mineral and rock particles of sand size	Sandstone (graywacke, arkose)
Fine-textured	Particles of silt and/or clay size	Shale (thin-bedded), mudstone, siltstone
Nonclastic: Chemical Precipitation and/or Evaporation	Calcium carbonate CaCO$_3$ (calcite)	Limestone
	Calcium magnesium carbonate CaMg(CO$_3$)$_2$ (dolomite)	Dolomite/dolostone
	Silicon dioxide SiO$_2$ (silica)	Chert/flint
	Sodium chloride NaCl (halite)	Rock salt
	Calcium sulfate CaSO$_4$ (gypsum)	Rock gypsum
Organic remains plus biochemical secretions, etc.	Shells, shell fragments, skeletal material (bones, teeth, etc.) calcareous or silicious ooze	Limestone, some chert
	Buried vegetable matter	Peat, coal

Grade sizes of clastic sediments:

Boulders	
	10.08 in.
Cobbles	
	2.52 in.
Pebbles	
	0.157 in.
Granules	
	0.079 in.
Sand	
	0.0025 in.
Silt	
	0.00015 in.
Clay	

precipitation of a mineral cement (such as silica or calcite) between the sediment grains. As mineral cement binds the sediment grains together, it becomes an integral part of the rock.

The most common sedimentary rock having a *nonclastic texture* (i.e., made up of nonclastic sediment) is *limestone,* which is composed mainly of interlocking crystals of calcite (calcium carbonate) that have been precipitated by inorganic or organic means.

Table 1.1 shows how the more common sedimentary rocks are classified. Exposures of these rocks in cliffs, buttes, and mesas form the distinctive landscapes of the Colorado Plateaus.

The third major rock category, metamorphic rocks, is the least abundant group in the Colorado Plateaus; but examples are found in some localities, such as in the deeper parts of the Grand Canyon. Generally (but not always), a metamorphic rock has a texture and composition different from those properties in its "parent" rock. For example, an erodible sandstone may be changed into highly resistant *quartzite;* a soft shale into hard *slate;* and a dull limestone into gleaming *marble.* If metamorphism persists or intensifies, slate may change into *phyllite,* which, in turn, may become a *schist,* and then a *gneiss.* During metamorphism, if the melting point of a rock is exceeded, the rock becomes molten. The resulting magma may eventually be "recycled" as an igneous rock.

This continual reworking of the earth's crust over geologic time can be depicted graphically by a theoretical *rock cycle* (fig. 1.5). A rock cycle model shows the interrelationships among weathering, erosion, sedimentation, deposition, burial, tectonic forces, melting, volcanism, and the major rock types. With the passage of time, new rocks are formed from old rocks under many different conditions, and minerals combine and recombine to produce the infinite variety of rocks found in nature.

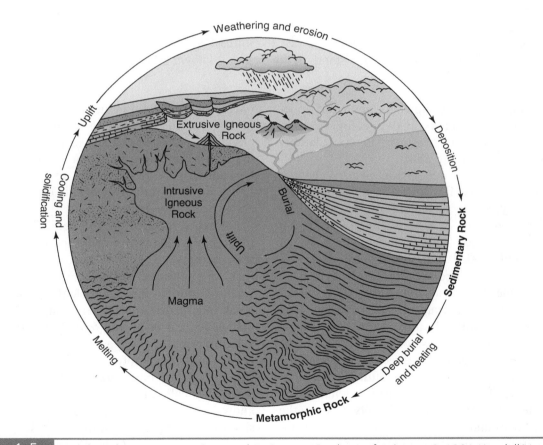

FIGURE 1.5   Rock cycle. From D. Nations and E. Stump, *Geology of Arizona.* © 1981 Kendall/Hunt Publishing Company. Used by permission.

FIGURE 1.6   The early Precambrian Vishnu Schist and intrusions of the Zoroaster Granite in the Inner Gorge. A remnant of lava that flowed into the canyon stands in the river; it is known as Vulcans Forge. Photo by Michael Little.

stripped of sediment, a gap, shown as an unconformity in a geologic column, occurs in the geologic record. The rock above an unconformity is considerably younger than the underlying rock.

Geologists who try to reconstruct ancient changes in elevation and sea level find unconformities to be useful indicators. Typically, an unconformity can be recognized by the presence of a *basal conglomerate* that rests on the old erosion surface and is composed of fragments of old eroded sediment incorporated with younger sediment deposited by a transgressing sea.

As a shoreline shifts landward or seaward (depending on whether the sea is transgressing or regressing), beach deposits accumulate in this narrow depositional environment. A *time-transgressive rock unit,* with the rocks becoming younger in the direction in which the sea was moving, may show the shifting location of the old shoreline. The Tapeats Sandstone, which is on the upper side of the angular unconformity between the Precambrian rocks and the Paleozoic beds, is an example of a time transgressive rock unit (fig. 1.8).

Folds and faults. Layered sedimentary rock units, such as those exposed in the canyon walls above the igneous and metamorphic rocks of the Inner Gorge,

have other interesting clues about geologic conditions in the past. The presence of marine fossils leads to the obvious conclusion that these rock layers have been uplifted more than a mile from their original position beneath an ancient sea. Uplift must have been essentially uniform over a large area in order to keep the layers so nearly horizontal.

What happens to layered bedrock when the uplifting forces are not uniform? Keep in mind that most tectonic activity takes place at considerable depth when rocks are subjected to great heat and pressure. Under such conditions, layered rock that would fracture or break in a surface environment may become *plastic,* or malleable, and bend into *folds* rather than break. The sides, or arms, of a fold are called *limbs*. Most folds are either *anticlines* (upfolds) or *synclines* (downfolds) (fig. 1.9). If the limbs are parallel (as with a hairpin), the fold is called an *isocline* or *isoclinal fold,* and we know it must have been formed by intensely compressive forces (i.e., squeezed together). Folds of this type are found in some of the Precambrian metamorphic rocks in the Grand Canyon.

A type of fold more characteristic of the Colorado Plateaus is the monocline, which usually shows up in flat or slightly tilted beds. *Monoclines, or monoclinal*

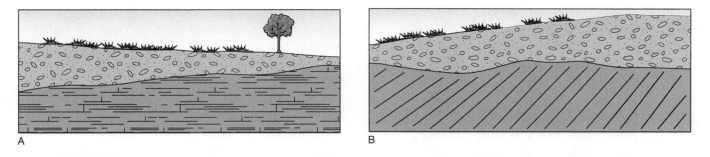

A   B

FIGURE 1.7 Unconformities represent a gap in the geologic record. They are old erosional surfaces separating younger rock units from older rocks exposed to erosion before being covered over by younger rock layers. *A.* This **unconformity** may be due to an interruption in sedimentation or to uplift and erosion. Deformation did not occur while processes of weathering and erosion operated because beds above the unconformity are approximately parallel to layers below the contact. *B.* This **angular unconformity** shows that the older rocks below the contact were folded ("deformed") and subsequently eroded. Younger beds were deposited on the old erosional surface.

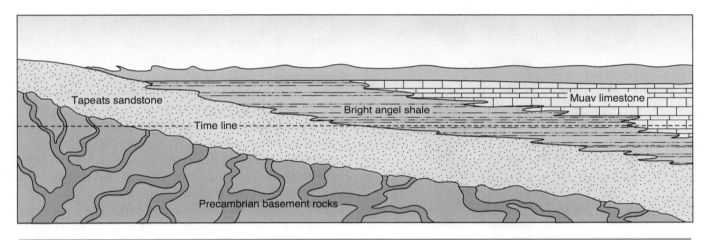

Tapeats sandstone   Muav limestone

Bright angel shale

Time line

Precambrian basement rocks

FIGURE 1.8 Time-transgressive rock layers may be deposited when a shoreline shifts laterally, causing the age of a formation to become progressively younger across an area. Beach deposits of the Tapeats Sandstone accumulated as an ocean advanced over Precambrian rocks; mud and silt settled in shallow water offshore and became the Bright Angel Shale; the Muav Limestone formed in deeper water. The time line in the cross section represents a theoretical equal age across the formations shown. From D. Nations and E. Stump, *Geology of Arizona.* © 1981 Kendall/Hunt Publishing Company. Used by permission.

folds, have one limb and a gentle bend, or flexure, connecting rock layers at one level (elevation) with the same layers at another level (fig. 1.9). The East Kaibab monocline, which outlines the eastern edge of the Kaibab Plateau, is an example.

A *fault* is the result of movement along a break or fracture in bedrock causing *displacement.* Faults are named and identified according to the type of movement or displacement that occurred. Faulting may or may not be related to the topography of an area. In

some regions the surface features give no indication of the geologic structure of the bedrock below, but in the Basin and Range Province, just west of the Grand Canyon, great mountains and valleys display evidence of faulting on a grand scale.

Normal faults separate blocks that have been either raised or dropped, mainly as a result of tensional and vertical stresses affecting the earth's crust. Tectonic forces that stretch the crust cause normal faults. The normal faults in the Precambrian rocks in the Grand

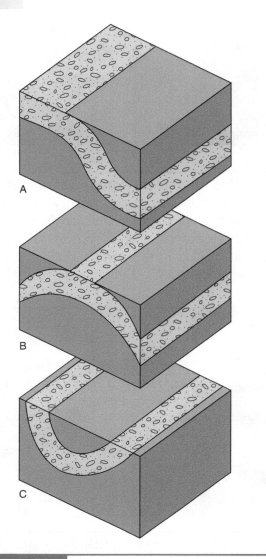

| FIGURE 1.9 | Block diagrams of simple folds. |

A. **Monocline.** Local steepening of flat-lying beds produces a bend in the rock unit. *B.* **Anticline.** Compression of the layered rocks creates an upfold, opening downward, with older beds in the center of the fold. *C.* **Syncline.** A downfold, also the result of compression, opens upward, with younger beds in the center of the fold.

Canyon indicate block-faulting during ancient geologic times and are unrelated to present surface features.

Faults involve relative movement between two blocks, one of which is called the footwall block and the other the hanging wall block. If a person were able to stand on a fault surface within the earth (as if in a mine), his feet would be supported by the *footwall*,

while the *hanging wall* would be the rock surface over his/her head. In a normal fault, the hanging wall has moved *down* in relation to the footwall.

If the hanging wall has moved *up* in relation to the footwall, it is called a *reverse fault*. A special type of reverse fault, in which the fault plane has a low angle (about 10° or less), is referred to as a *thrust fault* (fig. 1.10). Reverse faults are produced when tectonic forces compress crustal rocks, causing them to break vertically or at an angle. In the Grand Canyon, some of the Precambrian normal faults (produced by tensional forces) were changed millions of years later to reverse faults by compressive forces.

Sometimes compressive forces cause blocks to slide past each other, producing a *strike-slip fault* (fig. 1.10) that shows lateral displacement.

Sedimentary Environments

Studying the sizes, arrangement, and composition of sediments can reveal a great deal about the origin of sedimentary rocks and the environmental conditions that prevailed at the time the sediments accumulated. In the Grand Canyon many interesting examples of weathering processes can be seen. *Brecciated* deposits of rock that had been broken up and crushed into angular fragments are found where a cave ceiling has fallen or in fault zones where movement along fault surfaces has crushed rock. Rock broken by faulting weathers more rapidly because more surfaces are exposed to water and the atmosphere. Tributaries of the Colorado River tended to form along fault zones where the rock is susceptible to weathering and erosion. An example is Bright Angel Canyon.

Some of the Grand Canyon limestones have *karst* features, which are the result of slightly acidified underground water dissolving soluble rock. Because limestone consists principally of the mineral calcite (calcium carbonate), which is soluble under humid conditions, water trickling down through cracks and crevices in rocks tends to enlarge any voids into hollows or caverns. The caves in the Red Wall Limestone developed during an earlier time when the climate was more humid than now.

In the semiarid climate of the present, the limestones of the Colorado Plateaus tend to be resistant to weathering. The Kaibab Limestone, which forms the caprock on the rim of the Grand Canyon, is an example. However, visitors approaching the North Rim can see circular pits and hollows, some containing water,

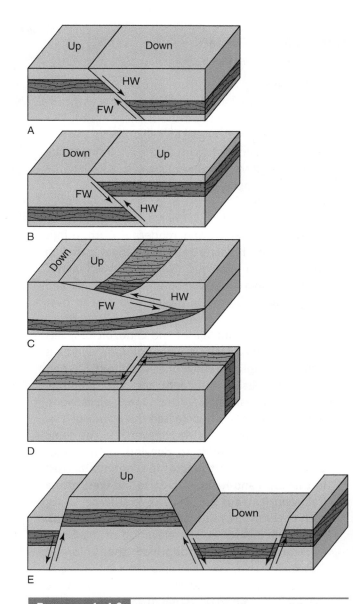

Block diagrams of types of faults. Stresses that create faults may cause one block to move up, down, obliquely, horizontally, or even to rotate, in relation to another block. *A.* **Normal fault.** The hanging wall side of the fault moves *down* in relation to the footwall side. *B.* **Reverse fault.** The hanging wall side moves *up* in relation to the footwall side. *C.* **Thrust fault.** A reverse fault in which the fault plane dips at a low angle. *D.* **Strike-slip fault.** One side of the fault moves horizontally by the other side. *E.* **Block-fault structure.** The upthrown block and the downdropped block are bounded by normal faults. Extensional forces stretching the crust produce normal faults. Reverse faults are the result of crustal compression.

beside the highway. These are *sinkholes,* which are depressions dissolved beneath the surface in the Kaibab Limestone. The Kaibab Plateau is flat enough so that rain and meltwater tend to collect in pools that over time eat away the soluble rock.

Precipitation that falls on the Kaibab Plateau also percolates down through the sedimentary beds of the North Rim (fig. 1.3). When the water reaches an impervious rock unit, it emerges from the canyon wall as seeps or natural springs. Roaring Springs, one of the larger springs supplied in this way, gushes from a chasm high on the north wall. Roaring Springs provides water (with the aid of a pumping station) for the Park Service facilities on the South Rim.

When water containing dissolved calcium carbonate from the limestone evaporates, deposits of finely crystalline, pale *travertine* are precipitated. Along Havasu Creek, travertine deposits have encrusted the canyon walls and formed low, sinuous dams in the creek bed.

Sandstones may be formed from old dunes, such as the *eolian* (windblown) deposits that make up the Coconino Sandstone. Other sandstones were formed from water-laid deposits, either in freshwater lakes or in marine environments. The degree of sorting and rounding of sand grains, the presence or absence of fossils, the type of bedding, and the kind of cementing (or matrix) holding the grains together can all provide clues as to the geologic history of an area.

Geologic History

1. **Accumulation of sediments early in Precambrian time.**

Around 2 billion years ago, sands, silts, muds, and volcanic ash were deposited in a shallow marine basin associated with an orogenic belt. Fossil evidence of life has not been found in these marine beds.

2. **Ancient mountain-building and igneous intrusions.**

About 1.7 billion years ago, tectonic activity, resulting from plate collision, converted the accumulation of sedimentary beds and volcanic material into the metamorphic Vishnu Schist. Then molten rock pushed into the schist, slowly cooling to form Zoroaster Granite. Some of the rock became metamorphosed to gneiss. In Trinity Chasm near Isis Temple, light-colored gneiss

Box 1.2	An Outline of the Geologic History of the Colorado Plateaus

1. Precambrian Time

The Precambrian rock units exposed in the Grand Canyon do not crop out in the other national parks of the Colorado Plateaus, presumably because erosion and stream downcutting have not reached low enough depths.

2. Lower Paleozoic: Cambrian to Devonian Periods

The Colorado Plateaus were a tectonically stable area for millions of years, probably as the trailing, subsiding edge of a *continental plate* (a segment of continental crust). It is believed that continental rifting (the splitting up of a continent) had occurred, possibly before the end of Precambrian time. Shallow seas spread over a vast area of the continental plate that would eventually become North America. In these seas, thousands of feet of sediment were deposited.

3. Upper Paleozoic and Mesozoic: Mississippian through Cretaceous Periods

A series of *orogenies* (mountain-building episodes) deformed parts of western North America surrounding the Colorado Plateaus. Instead of breaking up or pulling apart, some land masses were coming together and closing up former oceans. Sand dunes overspread vast plains. The abundance of quartz-rich sands that built the great Jurassic dune belts in coastal deserts were derived from the weathering and recycling of quartz grains from older rocks.

In Cretaceous time, the last great inundation by ocean waters spread over western North America. Thick units of richly fossiliferous, marine rocks were laid down. The Mesozoic Era drew to a dramatic close as accelerated orogenic activity lifted the land and drained the shallow ocean. This great event, which began 65 million years ago and lasted for millions of years, drastically changed the face of western North America. It is called the Laramide orogeny.

4. Cenozoic Era

Minor uplifts and adjustments, accompanied by gentle deformation and moderate outpourings of dark lava, characterized early Cenozoic time in the Colorado Plateaus. Nonmarine sediments, washed down from surrounding highlands and ranges, accumulated in basins, and spread over plains.

Toward the middle of the Cenozoic Era, tectonic activity resumed. The Colorado Plateaus region was uplifted still higher and tilted. Streams began to dissect the landscape into plateaus, benches, mesas, buttes, arches, and deep canyons. Each successive uplift increased the power of erosional processes. Uplift was uneven, causing some faults and open folds to develop in the rocks. Upwarping and stretching of the continental crust in the Basin-and-Range region to the west of the Colorado Plateaus produced several large faults, such as Hurricane fault, between the two regions.

In the meantime, the elevation of the Sierra Nevada to the west deprived the Colorado Plateaus of rain, and the climate became increasingly arid. (When moisture-laden clouds from the ocean are forced to drop rain or snow when rising

---

(containing appreciable amounts of quartz, feldspar, and micas) shows up as veins in the ancient rocks.

3. Uplift and erosion of the early Precambrian rocks.

We know little about the Mazatzal Mountains that were constructed; only their roots are left. For millions of years, sediments were stripped from highlands and deposited in lowlands or oceans.

4. Deposition of the Grand Canyon Supergroup (1.2 billion to 800 million years ago).

The rock units of the *supergroup* (an assemblage of vertically related formations and groups) consist of sedimentary beds that were later subjected to varying degrees of metamorphism. (Since they have retained the layering and certain other characteristics of sedimentary rocks, most have been given sedimentary rock names.) The Unkar Group, the older section of the

over a mountain range, the far side of the range is said to be in a *rain-shadow zone*.)

The Pleistocene Epoch was a time of colder climates, increased precipitation, and intensified erosion. Mountain glaciers formed at higher elevations, and ice and frost broke up rocks every-where on the plateaus. The results of colder temperatures, and more rain show up in the present Colorado Plateaus landscapes as dry valleys that once held streams, abandoned lakebeds, land-forms eroded by fluvial sculpturing, and stream deposits unrelated to today's creeks and rivers.

FIGURE 1.11   The area known as The Box on the floor of the Inner Gorge. The bedrock exposed is the Precambrian Vishnu Schist. Note the size of the debris on the flood plain that indicates the great velocity of the floodwater. © Lois Peddicord.

supergroup, was deposited in an offshore environment. Waves worked over and smoothed out the early Precambrian land surface. Larger fragments accumulated as gravel, creating a basal conglomerate known as the Hotauta Member of the Bass Limestone, a formation. (A *group* consists of two or more geographically associated formations with notable common features. A *formation* is a rock unit containing one or more beds with distinctive physical and chemical characteristics. A *member* is a minor unit within a formation.)

Ranging from 120 to 340 feet in thickness, the Bass Limestone accumulated in a shallow sea as a mixture of limestone, sandstone, and shale.

The Hakatai Shale, the next younger formation of the Unkar Group, is composed of units of thin-bedded shales, mudstones, and sandstones. Since most of the shale is nonmarine, the formation represents a temporary retreat of the sea.

The Shinumo Quartzite, a marine sandstone before it was metamorphosed, was so resistant to erosion that it juts up as "fossil islands" into overlying Cambrian beds. Apparently cliffs of Shinumo Quartzite withstood the wave action of the Cambrian ocean until the islands were completely engulfed. The Dox Sandstone, mostly sandstone but containing some shale layers, is another accumulation of shallow ocean deposits.

The flows of dark (basaltic) Cardenas Lava, the youngest rock unit of the Unkar Group, intruded zones of weakness in the sedimentary rock. Lava flows up to 1000 feet thick erupted from fissures and spread over older rocks.

The beds of the Nankoweap Formation, which are shallow-ocean deposits, lie unconformably over the

eroded surface of Cardenas Lava and separate the Unkar Group from the younger rocks of the Chuar Group. Three formations—Galeros, Kwagunt, and Sixty Mile—make up the Chuar Group. Evidence of algae has been found in Kwagunt rocks, but fossils are rare in rocks this ancient.

5.   Block-faulting of the Grand Canyon Supergroup.

Remnants of the Grand Canyon Supergroup crop out in scattered sections of the Inner Gorge and the deeper tributary canyons. The pockets of rock are isolated from each other because when tilting and block-faulting occurred, some rocks were elevated and some

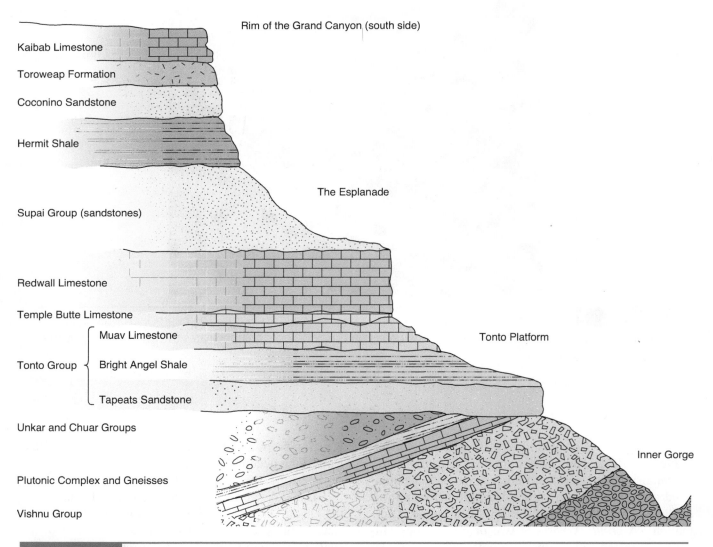

Kaibab Limestone

Toroweap Formation

Coconino Sandstone

Hermit Shale

Supai Group (sandstones)

Redwall Limestone

Temple Butte Limestone

Tonto Group { Muav Limestone

Bright Angel Shale

Tapeats Sandstone

Unkar and Chuar Groups

Plutonic Complex and Gneisses

Vishnu Group

Rim of the Grand Canyon (south side)

The Esplanade

Tonto Platform

Inner Gorge

FIGURE 1.12   Schematic profile of rock units exposed on the south wall of the Grand Canyon. Relative thicknesses of formations are approximate, but slope angles are greatly exaggerated. In the semiarid climate, limestones tend to be cliff-formers, while shales form gentle slopes. The resistance of sandstones depends on their purity and cementation. Adapted from Billingsley and Breed 1980.

Table 1.2	Geologic Column, Grand Canyon National Park

Time Units			Rock Units		Geologic Events

Era	Period	Group & Supergroup	Formation	Geologic Events
Cenozoic	Quaternary Tertiary		Lavas Slump and slide deposits, terrace gravels, river deposits, travertine.	Minor lava flows Development of Grand Canyon by uplifting and tilting, accompanied by down-cutting Uplift, erosion, minor deformation Erosion
Mesozoic	Cretaceous Jurassic Triassic		Moenkopi	Flood-plain deposition Withdrawal of seas Erosion
Paleozoic	Permian		Kaibab Limestone	Transgression of shallow seas; regression Erosion
			Toroweap	Transgression of shallow seas, flattening sand dunes
			Coconino Sandstone	Flood-plain deposits covered by migrating sand dunes Erosion
			Hermit Shale	Deposition on flood plain, lagoons Erosion
	Pennsylvanian	Supai	Esplanade Sandstone Wescogame Manakacha Watahomigi	Flood-plain deposition Retreat of seas Marine sedimentation Erosion
	Mississippian		Surprise Canyon	Channel deposits; uplift Erosion
			Redwall Limestone	Karst development Marine sedimentation Erosion
	Devonian		Temple Butte Limestone	Marine deposits in channels Erosion
	Cambrian	Tonto	Muav Limestone Bright Angel Shale Tapeats Sandstone	Deposition in a transgressing sea, forming a facies change in these formations Erosion
Late & Middle Precambrian time (Proterozoic)		Grand Canyon — Chuar	Sixty Mile	Stream deposition Erosion
			Kwagunt Galeros	Volcanic activity; Grand Canyon orogeny, block-faulting, thrust-faulting, folding Erosion
			Nankoweap	Deposition in a shallow sea Erosion
		Grand Canyon — Unkar	Cardenas Lava Dox Sandstone Shinumo Quartzite Hakatai Shale Bass Limestone (including Hotauta Conglomerate)	Faulting, igneous intrusions, lava flows Shallow sea deposits Deposits as a marine sandstone Temporary retreat of the sea Deposition in warm, shallow sea Transgression of seas; major unconformity Erosion
Early Precambrian time (Archean)			Trinity and Elves Chasm Gneisses	Block-faulting; uplift. Igneous intrusions; metamorphism Erosion
			Zoroaster Granite	Uplift of Mazatzal Mountains; orogeny, folding, faulting, intrusions, metamorphism
			Vishnu Schist	Deposition of marine sediment; volcanic activity

Source: Modified after Huntoon et al., 1981, 1982, 1986.

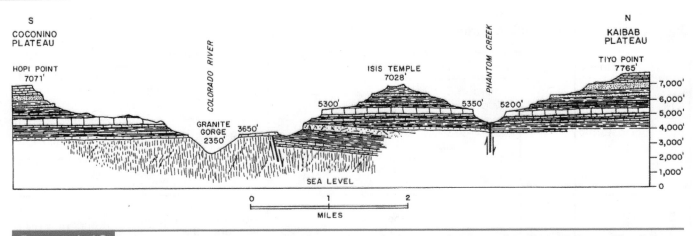

FIGURE 1.13   South Rim to North Rim cross section through the Grand Canyon from Hopi Point to Tiyo Point. Canyon features are shown in their relative sizes and correct proportions. Uplifting of the Colorado Plateau enabled the Colorado River, with the help of mass wasting, to remove immense amounts of rock and carve the Grand Canyon in a few million years. Adapted from F. Matthes, U.S. Geological Survey Bright Angel quadrangle, 1906.

were dropped down (fig. 1.10E). Tensional tectonic forces were stretching and rifting crustal segments at that time. Movement was along fault lines, such as the Bright Angel fault, creating north-south trending fault-block mountain ranges.

6. "The Great Unconformity."

During the long erosional period that followed the uplifting and tilting of the fault-block mountains, most of the Chuar beds and some of the Unkar Group were stripped from the mountain slopes. This exposed some of the more resistant beds, such as the Shinumo Quartzite, that stood as islands in the encroaching ocean of Cambrian time. Most of what is left of the Grand Canyon Supergroup is preserved in the down-faulted blocks. The horizontal contact between the Precambrian and Paleozoic beds is the Great Unconformity. In 1869, when John Wesley Powell recognized the angular relationship between the older tilted rocks and the flat-lying formations above, he named the contact the Great Unconformity.

7. Deposition of Cambrian rock units.

About 550 million years ago, a western ocean began to spread over the eroded coastal lowlands. Three formations were deposited, each representing a different environment. The way in which the sediments were laid down illustrates the principle of time transgression (fig. 1.8).

Imagine yourself standing on a sandy beach that would eventually become the Tapeats Sandstone. Pebbles, cobbles, and a few boulders are on the sand after falling from the cliffs along the shore. This is a turbulent coast. Pounding storm waves attack the cliffs; cobbles and sand are rolled back and forth in the surf and tossed up on the shore. These are the constituents of the basal conglomerate at the bottom of the Tapeats. Also preserved in the Tapeats are the shapes of ancient coastal sand dunes.

Farther out in the water, the sea bottom becomes muddy. These muds, which contain the remains of primitive brachiopods, trilobites, and worms, formed the Bright Angel Shale. Offshore, calcium carbonate precipitates from seawater and accumulates as a limy ooze on the bottom. From this ooze was formed the Muav Limestone.

Because sea level is gradually rising, the shoreline and the beach are shifting eastward, and the pattern of sediment deposition is also moving eastward without a break in continuity. Thus when the Tapeats Sandstone and the Bright Angel Shale were first laid down, it was early in Cambrian time; but by the time the last grains were being deposited some distance to the east, the time was Middle Cambrian, about 30 million years later. Evidence that the sea was transgressing is shown by finer-grained shale deposited over coarser-grained sandstone. Had the sea been regressing, coarser sediment would have overlaid the finer.

The Tapeats Sandstone, Bright Angel Shale, and Muav Limestone belong to the Tonto Group and are the beds that make up the Tonto Platform. These beds lie in their original horizontal position, virtually undisturbed since they were deposited on top of the Precambrian erosional surface. The Tapeats Sandstone and the Muav Limestone are consistent cliff-formers in the walls of the Grand Canyon, while Bright Angel Shale is a slope-former (box 1.3).

8. Unconformity between the Cambrian and Devonian beds.

Beds from the Ordovician and Silurian periods are missing from the Grand Canyon geologic column. There is no evidence to indicate whether beds were deposited in the region during these time periods and subsequently removed by erosion, or whether deposition did not occur. However, deep channels were carved by streams (or possibly by marine scour) on top of the Muav. Devonian Temple Butte Limestone was deposited in these depressions. The channel deposits are especially well displayed in Marble Canyon in the eastern section of the park. Toward the western end of the park, near Grand Wash Cliffs, Temple Butte is a cliff-former, in some places a thousand feet thick. Because the limestone in Temple Butte has largely been replaced by dolomite, only a few fossil fish teeth remain in the formation. Erosion, producing an unconformity, brought the Devonian Period to a close.

9. Deposition of the Mississippian Redwall Limestone.

The visually striking cliffs of the Redwall in the Grand Canyon today are stained red by rainwater dripping down from the overlying Supai and Hermit Shale redbeds. The Redwall is actually a bluish gray limestone containing chert nodules. It was deposited in a shallow tropical sea in early to middle Mississippian time, some 330 million years ago. Thickness averages 500 feet. Fossil remains of crinoids, brachiopods, bryozoans, and other marine creatures are abundant in the formation. Added to the list of fossils are the chambered nautiloid fossils that were discovered in the streambed of a branch of Marble Canyon. Following the deposition of the Redwall Limestone, the region was gradually uplifted and the beds underwent erosion late in Mississippian time. Because limestone is highly susceptible to solution by ground water, some natural caves developed.

10. A geological "Surprise" in the Grand Canyon.

In very latest Mississippian time (and possibly very earliest Pennsylvanian time), a singular depositional event occurred in what is now the Grand Canyon region (Beus 1986). A sedimentary rock, heretofore unrecognized, was first discovered by McKee and Gutschick, but was not considered as a separate formation until 1985 by Billingsley and Beus. This "new" unit has been appropriately named the Surprise Canyon Formation. The exposures, which are in discontinuous lenses a few hundred feet thick at the top of the Redwall Limestone, record the evolution of tidal estuaries on a low coastal plain that gradually subsided, drowning the valleys. At the end, a short period of erosion was followed by deposition of the lowermost beds of the Supai Group. Fossils in the Surprise Canyon Formation—from abundant logs and plant materials mixed with coarse sands in the bottom beds to marine shells in beds of sand and lime deposits—reflect the changing environments that produced this unusual sequence.

11. Deposition of the Supai Group during Pennsylvanian and early Permian time.

The Supai Group, a prominent ledge- and cliff-former in the canyon walls, is 600 to 700 feet thick. Mostly nonmarine, red siltstones and sandstones accumulated in a swampy environment. Distinctive crossbedding may be an indication of deltas and coastal dunes. Fossils include footprints of amphibians and an abundance of plant materials. The Esplanade Sandstone of early Permian age has been oxidized to a bright red color.

12. Early and middle Permian deposition.

The Hermit Shale, a red slope-former, accumulated in an environment similar to the underlying Supai Group. The beds contain ripple marks, mud cracks, footprints, plant fossils, and a few insect remains.

Migrating sand dunes that buried the swampy flood plain formed the Coconino Sandstone. This formation is the whitish cliff-former near the top of Grand Canyon. The uniformly sized, perfectly rounded, frosted quartz grains, arranged in a cross-bedded pattern characteristic of dunes, testify to a wind-blown (eolian) origin. The dunes must have been large because the Coconino is 400 feet thick. Reptile tracks are preserved in the sandstone.

The Toroweap Formation, with its nearly 200 feet of sandstones and limestones interbedded with gypsum,

Box 1.3 — The Terraced Walls of the Grand Canyon

Observers who look down into the Grand Canyon from either the South Rim or the North Rim are usually fascinated by the "stairstep" appearance of the canyon walls (figs. 1.12, 1.14). The upper part of the Grand Canyon is terraced while the lower part, or Inner Gorge, is narrower and V-shaped.

Four geological rock units are responsible for the most prominent cliffs or terrace risers. They are, from the top, the Permian Kaibab Limestone and the Supai Group, the Mississippian Redwall Limestone, and the Cambrian Tapeats Sandstone. Each cliff has at its base blankets of loose rock fragments, or *talus,* overlying gentle slopes. On top of the Tapeats Sandstone, a broad terrace top, the Tonto Platform, has been created by removal of less resistant rock units. A similar platform, called the Esplanade, in the western part of the park, has developed on the resistant Esplanade Sandstone. These slope segments, formed as the Grand Canyon was eroded down through successive geological formations, illustrate how some rock units are more resistant to weathering and erosion than others. This phenomenon is called *differential erosion.* The narrow Inner Gorge is not terraced or stepped because the metamorphic and intrusive igneous rocks have fairly uniform hardness and are more resistant to weathering and mass wasting than the Paleozoic formations.

Because the canyon is more than 10 miles wide from rim to rim, it becomes evident that the processes of valley widening and cliff recession are also significant. Except on the faces of the steepest cliffs, great sheets of broken rock on the valley sides prove the effectiveness of weathering, which produces the fragments, and mass wasting, which moves loose rock downslope.

North of Grand Canyon National Park, stepped topography on a larger scale is produced by the retreat of the margins of Mesozoic and Cenozoic formations in the form of the Chocolate Cliffs, Vermilion Cliffs, White Cliffs, and Pink Cliffs. These cliffs, which make up "the Grand Staircase" of the Colorado Plateaus, trend generally east-west, and step up toward the north. The most prominent features in Zion National Park (Chapter 2) are formed in the White Cliffs. The Pink Cliffs are strikingly displayed in Bryce Canyon National Park (Chapter 3).

FIGURE 1.14   Differences in resistance to erosion of sedimentary rock layers produced the stairstep topography of the Grand Canyon. The Tonto Platform (flat surface, center of the photograph) was eroded from weak Bright Angel Shale, but is upheld by the underlying, resistant Tapeats Sandstone. Muav Limestone, a resistant formation, makes up the first tier of cliffs that partially ring the Tonto Platform. Photo by Ronald F. Hacker.

records marine advances and retreats. Brachiopods, mollusks, corals, and other organisms lived in the ancient ocean. The Toroweap is a ledge- and cliff-former.

Forming the rim of the Grand Canyon and the edge of the Kaibab Plateau is the massive, thick-bedded Kaibab Limestone of middle Permian age. More than 300 feet thick, the Kaibab Limestone is a prominent cliff-former. Some zones in the Kaibab have sand grains and chert nodules. Fossils include organisms such as crinoids and brachiopods that lived in warm, shallow seawater.

13. Withdrawal of the Permian Sea; Mesozoic deposition.

As the Mesozoic Era began, the continent was rising. Streams carved wide, low valleys. Sediment was transported from higher land to lower slopes and basins. These deposits became the Moenkopi Formation of early Triassic age, a mixture of brightly colored sandstones and shales with interbedded gypsum layers. The number and variety of fossils indicate an abundance of life forms.

Presumably the early Triassic rock units, such as those now exposed along the Colorado River above Marble Canyon, may have overspread the uplands of the Grand Canyon region. The soft Triassic shales, less resistant to erosion than the Permian Kaibab and Toroweap rock units, were probably stripped off long ago. Younger Mesozoic and early Cenozoic beds, identified in nearby areas, may also have been removed by erosion.

14. Mesozoic uplift and mountain-building.

The compressive forces generated by the Laramide orogeny had profound effects on western North America at the end of the Mesozoic and in the early Cenozoic. Nevertheless, the horizontal beds of the Colorado Plateaus remained stable and relatively undeformed while still being elevated like a great platform. All around the Colorado Plateaus, rock units were folded, faulted, and uplifted, and great mountain ranges were being pushed up. The East Kaibab monocline (and other monoclinal folds) developed along the edge of the

Kaibab Plateau during this time. (Presumably a south-flowing segment of the Colorado River began to erode Marble Canyon along the lower flank of the monocline, long before the stream cut across the Kaibab Plateau.)

15. Middle Tertiary uplifting and faulting.

About 20 million years ago, as the Colorado Plateaus were uplifted and warped by tensional forces, local faulting occurred—sometimes along old fault lines—along with moderate volcanic activity. The stress of crustal stretching that was going on to the west in the Basin and Range province created several major north-south trending faults between the Colorado Plateaus and the Basin and Range. The Grand Wash fault, which intersects the Colorado River at the western edge of the park, is on the boundary between the two provinces. The Hurricane fault extends north into Utah. By late in Tertiary time, the Colorado Plateaus region had attained most of its present elevation.

16. The connecting of Colorado River segments and establishment of its present course.

The uplift and tilting of the plateaus accelerated the development of the drainage system. With its gradient increased and base level lowered by the opening of the Gulf of California, the young Colorado River downcut

FIGURE 1.15   View from Bright Angel Point on the North Rim. The Coconino Sandstone stands out as a cliff-former near the top of the cliff wall on the opposite side, overlying the Toroweap Formation. Both are Permian in age. Note the resistant dike (center bottom) that stands out in stark relief as it cuts across the horizontal beds. Photo by Ann G. Harris.

rapidly, captured the headwaters of other streams, and carved out the configuration of the Grand Canyon.

17. Intensification of weathering and erosion during Pleistocene glacial episodes.

The colder, wetter climates of Pleistocene time increased erosional rates in and around the Grand Canyon. Most of the work was done by mighty floods of meltwater in the spring and by sudden flash floods from violent summer storms. In the last two million years, most of the downcutting of the Grand Canyon was accomplished.

18. Late Cenozoic volcanism.

Only a million years ago, flows of fluid basaltic lava covered parts of the plateau. Some poured down side canyons and dammed the Colorado River in the western section of the Grand Canyon. These lava flows are still being eroded by the Colorado River. The remains of small cinder cones on the rim mark the sites of ash eruptions.

19. Continued deepening of the Grand Canyon by downcutting; widening of the Canyon by mass wasting.

The climatic patterns of spring and summer floods, along with the age-old processes of weathering and erosion, continued into Holocene time, although at a somewhat slower pace. Since the Colorado River has become a regulated stream system from its source area to its outlet, its power to incise its channel has been reduced. However, as weathering loosens particles, mass wasting—powered by gravitational energy—moves materials downslope. In an arid climate these processes operate slowly except when quantities of loose sediment are transported rapidly during flash floods or landslides.

Geologic Map and Cross Section

Billingsley, G.H., and Breed, W.J. 1980. Geologic Cross Section Along Interstate 40, Kingman to Flagstaff, Arizona. Petrified Forest National Park: Petrified Forest Museum Association.

Breed, W.J., Stefanic, V., and Billingsley, G.H. 1986. Geologic Guide to Bright Angel Trail, Grand Canyon, Arizona. Tulsa, Oklahoma: American Association of Petroleum Geologists.

Huntoon, P.W., Billingsley, G.H., and Clark, M.D., 1981. Geologic Map of the Hurricane Fault Zone and Vicinity, Western Grand Canyon, Arizona. Grand Canyon Natural History Association.

——— 1982. Geologic Map of the Lower Granite Gorge and Vicinity, Western Grand Canyon, Arizona. Grand Canyon Natural History Association.

Huntoon, P.W., Billingsley, G.H., Breed, W.J., Sears, J.W, Ford, T.D., Clark, M., Babcock, R.S., and Brown, E.H. 1986. Geologic Map of the Eastern Portion of the Grand Canyon National Park, Arizona. Grand Canyon Natural History Association.

Maxson, J.H. 1966. Geologic Map of Bright Angel Quadrangle, Arizona. Grand Canyon Natural History Association.

Sources

Baars, D.L. 1983. *The Colorado Plateau, a Geologic History.* Albuquerque, New Mexico: University of New Mexico Press. p. 3–48.

Beus, S.S. 1986. A Geologic Surprise in the Grand Canyon. *Fieldnotes* 16 (3): 1–4. Tucson, Arizona: Arizona Bureau of Geology and Mineral Technology, University of Arizona.

———, and Morales, M. (eds.) 1990. *Grand Canyon Geology.* Oxford University Press, 582 p.

Collier, M. 1980. *An Introduction to Grand Canyon Geology.* Grand Canyon Natural History Association.

Hamblin, W.K. 1994. *Late Cenozoic Lava Dams in the Western Grand Canyon.* Geological Society of America, 144 p.

———, and Murphy, J.R. 1980 (revised edition). *Grand Canyon Perspectives.* Provo, Utah: Brigham Young University Geology Studies, Special Publication No. 1, 48 p.

Hunt, C.B. 1969. Geologic History of the Colorado River, in *Colorado River Region: John Wesley Powell.* U.S. Geological Survey Professional Paper 669, pp. 59–130.

McKee, E.D. 1969. Stratified Rocks of the Grand Canyon, in *Colorado River Region: John Wesley Powell.* U.S. Geological Survey Professional Paper 669, pp. 23–58.

Nations, D. and Stump, E. 1981. *Geology of Arizona.* Dubuque, Iowa: Kendall/Hunt Publishing Company.

Powell, J.W. 1981 reprint (1875 *Scribner's Monthly* articles) W.R. Jones, editor. *The Canyons of the Colorado.* Golden, Colorado: Outbooks. 64 p.

Rigby, J.K. 1977. *Southern Colorado Plateau*, K/H Geology Field Guide Series. Dubuque, Iowa: Kendall/Hunt Publishing Company.

Note: To convert English measurements to metric, go to www.helpwithdiy.com/metric_conversion_calculator.html

Zion National Park

SOUTHWEST UTAH

Area: 146,598 acres; 229 square miles

Proclaimed a National Monument: July 31, 1909

Established as a National Park: November 19, 1919

Address: Zion National Park, SR9, Springfield, UT 84767-1099

Phone: 435-772-3256

E-mail: ZION_park_information@nps.gov

Website: www.nps.gov/zion/index.htm

Near-vertical canyon walls and great towers on a grand scale display hues of bright red, salmon pink, orange, yellow, and off-white. Faults, fractures, and ancient dunes have created spectacular landforms and landscapes. Zion Canyon, 3000 feet deep, is a comparatively young erosional feature, cut by the Virgin River through Mesozoic clastic sedimentary rocks. Sandstone Kolob Arch, with a span of 310 feet, is the largest of its type in the world.

FIGURE 2.1   Downcutting by the North Fork of the Virgin River, weathering, and mass wasting have sculptured Zion Canyon in Zion National Park. Photo by David B. Hacker.

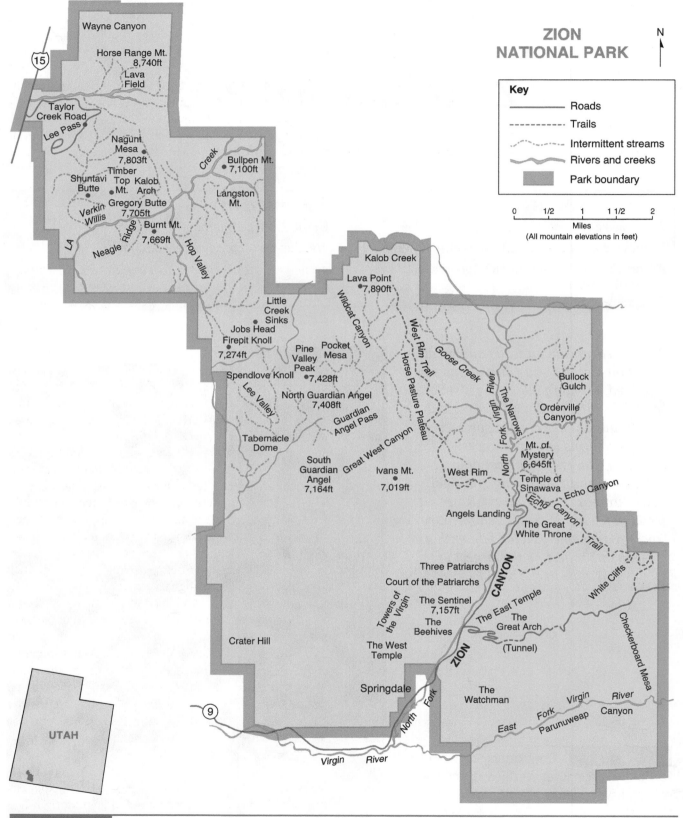

FIGURE 2.2   Zion National Park, Utah.

In 1869, Major John Wesley Powell and his fellow explorers were in desperate straits because of their loss of boats and supplies during their first trip through the Grand Canyon. They knew that they would be safe if they reached the Mormon settlements below the canyon, near the confluence of the Colorado and Virgin Rivers.

Mormons had started settling the Virgin River region in 1847 with plans to farm and grow cotton. Their search for farmland led them to Zion Canyon in 1858, about 75 miles up the Virgin River from the older settlements. They farmed on tillable lands scattered in narrow fields along the river banks and on an old lakebed in the canyon until 1909, when a national monument was first set aside.

After his arduous passage through the Grand Canyon, Powell was determined to complete the task of surveying the whole Colorado River region. His exploration party included geologists Clarence Dutton and Grove Karl Gilbert, photographer Jack Hillers, and artist William H. Holmes—all famous names in the history of scientific exploration of the American West. Over the next two decades, these hardy scientists put together a picture of time and territory on a giant scale: how through periods of geologic history the Colorado Plateau was uplifted while river systems steadily cut down, layer by layer, through sedimentary rocks.

Powell and Gilbert made a geological reconnaissance of Zion Canyon in 1872 and named several features. Later Dutton carried on extensive mapping of the region. It was their glowing reports, plus Hillers' photographs and Holmes' drawings, that first created interest in Zion. In 1903, artist Frederick S. Dellenbaugh painted scenes of the canyon and exhibited his canvases at the St. Louis World's Fair. Spectators were doubtful that such strange and remarkable landscapes could exist, but the attention led to the presidential proclamation of a national monument. Congress established Zion National Park in 1919; additional areas were included in the park in 1937 and 1956.

The Zion-Mt. Carmel highway (Utah 9), an engineering marvel of its time, was opened in 1930 to provide access to Zion. From the park's east entrance, the highway leads to the famous Pine Creek Tunnel, which is over one mile long and has six large windows cut into the rock for views. From the tunnel, switchbacks go down to the scenic drive on the floor of Zion Canyon. The Kolob section of Zion National Park has a visitor center and scenic drive and is accessible from Interstate 15.

In Zion National Park, climatic zones, determined by elevation, affect weathering and erosion as well as the distribution of plant and animal life. In the high desert (Upper Sonoran) zone of the canyons (elevation, 3500 to 5000 feet), summers are hot and dry and winters are mild. On the canyon rims (elevation, 5500 to 7500 feet) in the transition zone, snow may last all winter and summer temperatures are more moderate. On the higher parts of the plateau (around 8000 feet), in the montane (Canadian) zone, stands of fir and aspen thrive under conditions of increased moisture and colder winters.

Zone boundaries are not sharply defined. Within the desert zone, for example, protected tributary canyons and shaded slopes with springs support plants and animals characteristic of higher elevations. Wet seasons are normally in early spring and late summer. Most of the scouring in the canyons is done by spring floods or by flash floods during intense summer storms.

Geologic Features

Zion Canyon, a deep chasm with nearly vertical walls, was cut by the North Fork of the Virgin River, a stream with great erosive power whose gradient, or slope, ranges between 50 and 80 feet per mile. In addition to high velocity because of the steep gradient, the Virgin River has a higher volume during wet seasons and runs rampant when flooding. At such times, the river transports the greater part of the three million tons of rock and sediment it carries downstream annually.

Since the Virgin River is downcutting more rapidly than it is widening, its tributaries with smaller volumes of water are left hanging because they are unable to downcut as fast as the main stream. Their valley floors are at a higher level than the floor of the main stream, making them *hanging valleys*. Waterfalls connect the hanging valleys with the river below. The tributaries have smaller drainage basins and some of them are *intermittent;* that is, they flow only during wet seasons. Angels Landing is a typical hanging valley, as are the valleys between Twin Brothers peaks and Mountain of the Sun (located on the eastern side of the canyon, north of Pine Creek).

How much more can the Virgin River cut down before it "hits bottom," or reaches base level? Geologists estimate that the Virgin can cut down another thousand feet and still have enough energy to transport its sediment to the Colorado River.

The erosive power of sediment-laden water running at high velocity is well shown in the famous Narrows at the head of Zion Canyon. Here the stream runs through an erosion-enlarged fracture about 12 miles long. The floor is only a few feet wide but the enclosing walls rise about 2000 feet, nearly straight up. Sunlight reaches the streambed only when the sun is directly overhead. When the water level is low, visitors are permitted to hike through the Narrows. However, the canyon walls are so close to each other that the water level may rise 25 feet in 15 minutes during a heavy rain or flash flood. This explains why the rock walls show evidence of scouring high above normal stream levels (fig. 2.3).

Mass wasting in Zion Canyon is gradually widening the valley, and in many places the canyon rims are developing a scalloped look as they retreat unevenly. Sapping by springs and the undermining of sandstone cliffs as the softer underlying shales give way has had a twofold effect. First the sandstone cliffs become overhanging; then secondly, due to lack of support, blocks of rock break off and fall to the canyon floor. Occasionally a block weighing several tons crashes down. Sometimes a "blind arch," or *inset arch,* forms on a cliff face when the upper rock layers remain intact after the lower part has fallen (fig. 2.4A). A prominent example is Great Arch on a cliff that can be seen from tunnel windows on the way into Zion. Great Arch measures 600 feet across its base and is 400 feet high. Some arches, such as Kolob Arch in the Kolob section of the park, have become freestanding instead of inset. With a span of 310 feet, Kolob is the largest freestanding arch in the world.

Slump blocks, large masses of material that slide downslope as a coherent unit, are another example of mass wasting in Zion National Park. Mudstone layers in the Kayenta Formation, underlying the Navajo Sandstone, become slippery when saturated. Sometimes when the mudstones give way, slump blocks of the Navajo break loose along vertical joints (fractures) and slide down to the canyon floor. About 4000 years ago, a great slump block in Zion Canyon dammed the Virgin River and created a lake that backed up in the canyon as far as Weeping Rock. Lake-bottom deposits accumulated on the canyon floor until the Virgin River was able to cut through the slump and drain the lake. Eventually, all traces of the former lake will be

FIGURE 2.3   The Narrows near the head of Zion Canyon. Turbulent waters of the North Fork of the Virgin River have cut a slot canyon through resistant Moenave beds of purplish sandstone and siltstone. The canyon walls are fluted and polished by stream abrasion. A smooth coating of calcium (tufa), precipitated out from the flowing water, partly covers the rock surface. Photo by Douglas Fowler.

removed and the river will resume downcutting bedrock in that part of the canyon.

Recent earthquakes in the Colorado Plateau region have triggered rockfalls and landslides. In 1993, a moderate earthquake (magnitude 5.6) was the cause of a large landslide at the west entrance of Zion National Park. The landslide continued to move for nine hours after the initial shaking, possibly because of water pressure in the landslide subsurface. An unusual aspect of these earthquakes is that the rockfalls and landslides have occurred at fairly large distances from epicenters. The reason for these patterns is not clear.

The great variety of landform shapes in Zion that astonished the early explorers are due to a combination of factors. Resistant and nonresistant types of bedrock are exposed to the ravages of weathering; and shales tend to be less resistant than sandstones. Providing access to the interior of the layered rocks are many

FIGURE 2.4A   Processes of weathering and mass wasting attack the rocks in Zion National Park. The Great Arch, an inset (blind) arch, near the tunnel on the Zion-Mt. Carmel Highway. Photo by David B. Hacker.

FIGURE 2.4B   Weathering in the Navajo Sandstone. Weathering along the joints has created pinnacles on top. Ground water percolating down through the cliff dissolves cementing compounds; loosened sand grains then fall or are washed away. Photo by Ann G. Harris.

bedding planes, joints, and faults, some of which are curved, others horizontal or vertical, some close together, and still others widely separated.

Frost action has significant effects, mainly at higher elevations on the canyon rims and plateau, where there are more freeze-thaw days and more moisture. Rainwater, seepage, and tree roots are also effective agents of weathering, even in the arid to semiarid climate of southern Utah. The massive Navajo Sandstone is sufficiently porous so that water percolates through

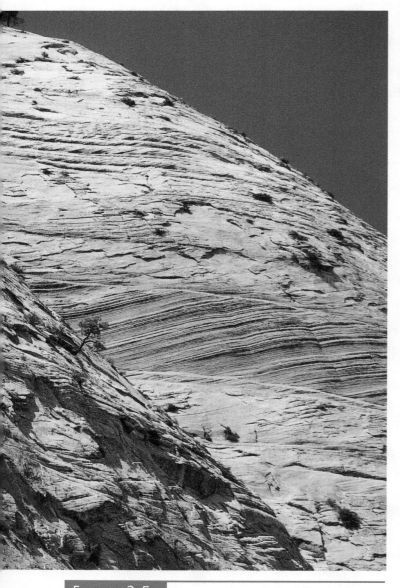

FIGURE 2.5   Large scale cross-bedding in the Jurassic Navajo Sandstone. Photo by Ann G. Harris.

it. On its downward path, some water seeps out along fractures and bedding planes and drips off the cliff face; still more water collects along the top of an impermeable rock layer and flows out in a spring. On Weeping Rock in Zion Canyon, hanging gardens of wildflowers cling to dripping crevices in the wall above a spring near the base. Weeping Rock is the loveliest of many springs that issue from the "spring line" between the Navajo and Kayenta Formations. The Kayenta is an impermeable shale.

The rectangular stream pattern is the result of closely spaced *joints* (fractures in bedrock without displacement) and vertical faults that permit broken or exposed rock to weather and erode more easily to form canyons (fig. 2.4B). The "temples" are the result of resistant, unfractured rock standing as columns or pyramids after the highly fractured rock between has been eroded and transported away. Examples are East Temple, Checkerboard Mesa, Sentinel Mountain, Twin Brothers, Temple of Sinawava, West Temple, the Three Patriarchs, Towers of the Virgin, and the Beehives.

The rock unit that dominates Zion National Park is Navajo Sandstone. This massive formation is 1500 to 2000 feet thick and is famed for its *cross-bedding* that developed as sand dunes migrated on the shores of an ancient shallow sea (fig. 2.5). Some lower parts of the formation were deposited in shallow water with the cross-bedding caused by currents. The eolian cross-bedding formed as shore dunes were impelled by the prevailing winds. In this process, wind picks up sand grains on the gentle upwind slope of a dune and deposits them on the (steeper) downwind side. The loose grains cascade down the face of the dune, which tends to keep the slope at about 34°, the "angle of repose." As sand underneath becomes compacted and the top of the dune becomes oversteepened, the face of the dune shifts direction slightly as more sand grains keep tumbling down. Gradually the entire dune is overridden as it moves in a downwind direction. Any shift in wind direction intensifies the cross-bedding effect (fig. 2.6). The most striking cross-bed sets are displayed in the upper parts of the Navajo, as found in Checkerboard Mesa and other cliffs along the Zion–Mt. Carmel highway.

The quartz sand grains in the Navajo are well rounded and frosted due to long exposure to abrasion by water and wind. The sand was washed into the ocean from the continental interior in an earlier time.

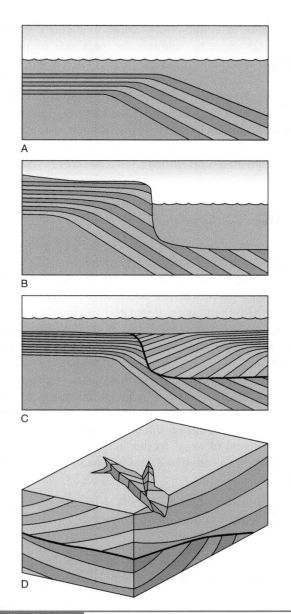

FIGURE 2.6   Cross-bedding in sedimentary rocks may result from changes in direction or velocity of wind currents or water currents that deposit sediment. In this example, showing "cut and fill," new sedimentary layers were deposited in a channel after earlier layers had been partially washed away. *A.* Dipping beds are deposited. *B.* Parts of some beds are eroded. *C.* New deposits are laid down by currents coming from a different direction. *D.* Over time, after consolidation and lithification have occurred, the sedimentary rock is exposed in an outcrop. The cross-bedding appears to go in several directions because of the nature of the cut-and-fill process when the sediment layers were deposited.

Waves and currents piled the sand onto the emerging coastal plain, and shifting westerly winds shaped the great coastal dune belt that became the Navajo Sandstone.

The Great White Throne, a natural feature sculptured by erosion of the Navajo Sandstone, is a large monolith rising approximately 2450 feet from the canyon floor. The color of the throne—the upper part of the monolith—is actually a light tan; but from a distance, particularly if the sunlight is striking it, the rock appears to be a brilliant white. The base of the throne is stained a deep red. This is the result of iron oxide leaching out from higher rock layers as they were weathered. The Great White Throne is also an example of how the lower beds of the Navajo Sandstone tend to form vertical cliffs, while the upper portion tends to weather as rounded domes.

The highest point in the park is the summit of Horse Ranch Mountain, elevation 8726 feet, which is located in the Kolob section, north of Kolob Finger Canyons. The lowest point, near South Campground, is about 3900 feet above sea level. Thus the *relief* in the park (i.e., the difference in elevation between the highest and lowest points) is a little over 4800 feet.

The colorful purple and gray Chinle Formation is exposed near Cougar Mountain and near Taylor Creek trailhead in the Kolob section. The sediments composing the Chinle are a mixture of sandstone, shale, limestone, and volcanic ash. The volcanic ash is interesting because it supplied the silica necessary for wood petrification in the Chinle Formation (Chapter 8). Pieces of petrified wood weather out of Chinle beds in the Zion region and other parts of the Colorado Plateaus.

Volcanic Rocks

The only igneous rocks in the park are young basaltic lavas that broke through to the surface along north-south trending normal faults. One flow, about one half million years ago, flooded and dammed the Coalpits Wash drainageway in the southwest corner of the park. After the lava cooled, a large lake filled the wash. Crater Hill, a cinder cone, represents the last eruption in the Coalpits Wash. Eventually, as the lake filled with sediment, a spillway developed that gullied and drained off the water. The Coalpits Wash is now the driest, most desolate area in the park. Lava Point, near the northern edge of the park, and Horse Ranch Mountain in the Kolob section are also locations accessible by trails where lava flows can be seen.

Geologic History

The geologic history of Zion Canyon begins where the history of the Grand Canyon ends. The rock sequence in Grand Canyon National Park is primarily Precambrian and Paleozoic and covers about two billion years, whereas Zion formations are Mesozoic and represent only about 150 million years of geologic time.

1. Deposition of the Permian Kaibab Limestone.

The Kaibab, which is the same formation that caps the rim of Grand Canyon, is a yellowish gray marine limestone that was deposited in a shallow sea. In Zion National Park, the Kaibab Limestone is exposed along the Hurricane Cliffs, behind and to the north of the Kolob Visitor Center.

2. The varied sediments of the Lower Triassic Moenkopi Formation.

The shales, siltstones, sandstones, gypsum, mudstones, and limestones of the Moenkopi Formation crop out on pink, white, and chocolate-brown slopes along the scenic drive in the Kolob section of Zion National Park. When these beds were deposited, the vast Kaibab sea had withdrawn, leaving a low coastal plain that was covered from time to time as sea levels rose or fell. The gypsum beds formed when seawater left in lagoons evaporated. The limestones and limey mudstones contain marine fossils. The three members of the Moenkopi, from oldest to youngest, are the Virgin Limestone, Shnabkaib Shale, and Upper Red Member.

3. Deposition of the Shinarump Conglomerate at the bottom of the Chinle Formation.

By the middle of Triassic time, the land had risen and the Moenkopi surface was being eroded. Rivers draining highlands picked up sands and gravels and washed down trees that became buried in stream bars during floods. Later these channel deposits were strongly cemented by silica dissolved in ground water and the wood became petrified. The resistant Shinarump is a prominent cliff-former in Zion and is up to 200 feet thick. The name *Shinarump* means "wolf's rump." Indians in the region where the Shinarump crops out thought the gray beds, which have a tendency to weather to rounded hills, looked like the back end of a wolf.

4. Deposition of the Chinle Formation in a warm, humid environment.

Fossils of amphibians, pond-dwellers, and other freshwater creatures suggest the presence of tropical swamps when Chinle beds were laid down. Fine-grained sediments, slowly accumulating on the flat plain, became the purple and gray claystones, mudstones, and thin-bedded limestones of the Chinle Formation in Zion. The volcanic ash that supplied the silica, which percolated down into the Shinarump, was wind-deposited and probably came from volcanoes some distance away.

5. Uplift in late Triassic time; deposition of the Moenave Formation.

In Zion National Park, a reddish, slope-forming siltstone called the Dinosaur Canyon Member makes up the older beds in the Moenave (mo-en-ah-vee) Formation. Ripple marks and cross-bedding suggest that the sediments were deposited in sluggish streams and ponds. The cross-bedding was due to shifting currents and the building of sandbars. The upper member, the Springdale Sandstone, was deposited by streams with greater velocity and volume. Gray mudstone and shale, which accumulated in lakes, make up the thin-bedded Whitmore Point Member between the upper and lower members of the Moenave Formation. The fossil remains of large freshwater fish similar to sturgeon have been found in Zion Canyon in the Springdale Sandstone.

6. Increasing aridity in early Jurassic time as the Kayenta Formation accumulated.

The climate became like today's northern tropical belt, with wet summers and dry winters. Streams deposited silt and sand in channels and on the flood plain. Dinosaurs that were three-toed and walked upright, came down to the stream for water and left footprints in the damp sediment along the banks. The Kayenta Formation, a slope-former underlying the Navajo Sandstone, consists of red and mauve siltstones and sandstones.

7. Migrating dunes preserved in Navajo Sandstone.

A vast coastal desert, similar to the modern Sahara, spread over thousands of square miles of western North America. The thickest accumulation of the dune sands (about 2200 feet) was in the Zion region. Strong dry winds picked up sand grains and piled them in dunes that migrated because no vegetation could grow to anchor them. Some cross-bedded sands were transported by currents and accumulated in the shallow waters

| Box 2.1 | **Plate Tectonics: An All-Encompassing Explanation** |

The spectacular scenery of western North America is largely the result of the past 65 million years of tectonic pulses and continuous erosion. We have suggested that tectonic forces generated within the earth elevated the Colorado Plateaus and built mountain ranges all around the plateaus. Moreover, we have seen that at times tectonic forces compressed the rocks of the crust and at other times have pulled segments of the crust apart. How does this happen?

A Composite, Unifying Hypothesis

Historians of science regard the plate tectonics hypothesis, including sea-floor spreading, to be the most significant idea developed in the physical sciences in the twentieth century. Plate tectonics is now the dominant geological concept in planning research strategies, framing hypotheses, and synthesizing geologic history.

The essential idea of plate tectonics states that the earth's outer shell consists of twelve or more major crustal *plates,* or slabs, plus about a dozen smaller ones. Continents and ocean floors occupy the larger plates; some smaller plates are the size of islands. The plates range in thickness from 10 to 100 miles and move slowly over the global surface. They interact mostly along their edges because they tend to be internally rigid. Each plate consists of three zones: a shallow upper part deforms by brittle breaking or by elastic bending; a deeper part yields plastically; and the lowest part is viscous. It is on this viscous layer that a plate slides, or "drifts."

Plates are pushed together in some places because they are being pulled apart at other places on the global surface, mainly at *spreading* centers (rifts) in the deep ocean basins along the crest of the great mid-oceanic ridge system. A nearly continuous series of submarine mountain ranges makes up the *mid-oceanic ridge.* Along its crest, upwelling magma continuously flows from the spreading center and adds new material to the earth's crust while the plates move away from the ridge in either direction.

Sometimes rifting occurs on a continental plate, breaking it apart. When this happens, an arm of the sea may move into the gap. The opening of the Gulf of California is an example of continental rifting.

Since the earth is not expanding, new crust added at a spreading center, or *diverging plate boundaries,* must be consumed elsewhere. This happens in *subduction zones,* at *converging plate boundaries,* where plates collide and some crustal material plunges several hundred miles to depths where rock material is melted and recycled. Spreading centers and converging zones, where plate boundaries interact, are where the most intense tectonic activity takes place. Earthquakes are frequent; volcanic belts may develop; and mountain ranges may rise.

The plates that are composed mainly of continental crust tend to ride higher on the global surface because they are less dense than plates of oceanic crust. That is why continental plates tend to override oceanic plates when they converge. When an oceanic plate tilts downward, or is subducted, the bend creates an *oceanic trench* in the subduction zone.

Continental rocks and oceanic rocks get jammed together in a subduction zone and may get pulled down into the oceanic trench as a jumble of mashed and broken rock material. Other continental rocks are folded and faulted by the compression of colliding plates, and mountains are pushed up. Deep below the surface, rocks subjected to intense heat and pressure are metamorphosed or melted. The less dense molten rock gradually works its way toward the surface. Some of the melt collects in *magma chambers* that may eventually cool as great *batholiths* beneath mountains. Some magma chambers are reservoirs from which lava erupts in volcanoes.

In a third type of plate interaction, the converging plates may scrape past each other as they collide, instead of one being subducted. This is called a *transform plate boundary.* The San Andreas fault in California marks such a boundary at the surface between the Pacific plate and the North American plate (fig. 2.7). California earthquakes are caused by shifting of crustal segments along the fault zone. Channel Islands National Park (Chapter 50) is on the Pacific side of the plate boundary.

(continued)

Box 2.1	**Plate Tectonics: An All-Encompassing Explanation (continued)**

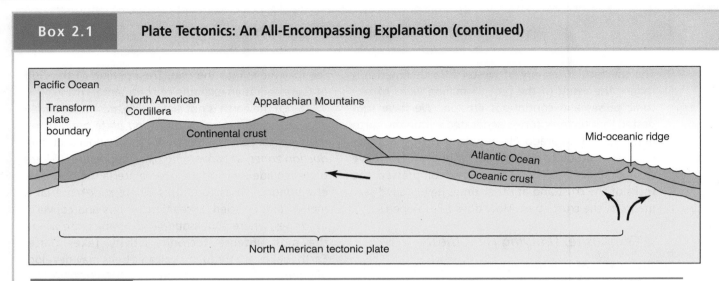

FIGURE 2.7   The North American plate from the mid-oceanic ridge in the Atlantic Ocean to the transform plate boundary in western California. The mid-oceanic ridge is a spreading center where the North American and European plates are pushed apart as new oceanic crust is being formed. At the transform plate boundary, the Pacific plate is rotating slowly in a counterclockwise direction as it slides past the westward-moving North American plate. At the present time, the North American plate moves at a rate of approximately an inch per year.

Plate Tectonic Activity and the Colorado Plateaus

Much of the geologic history of the Colorado Plateaus is based on the principle of *uniformitarianism;* that is, that geologic processes in the past operated in much the same way as they do in the present. For example, the "fossil" sand dunes that we see preserved in the cliff faces of Zion National Park probably accumulated and migrated in the same way as the modern dunes in Kobuk Valley National Park (Chapter 11).

Geologists examine the ancient, contorted, crystalline rocks in the bottom of the Grand Canyon and assume that plate interactions were going on in Precambrian time that were not unlike the behavior of colliding plates and opening ocean basins that are being investigated today. Evidence suggests that for

millions of years, periods of intense tectonic activity occurred involving the deformation of rocks, mountain-building, and volcanism. Interspersed with this activity were long spans of quiet erosion.

The land mass that eventually became North America was probably in the Southern Hemisphere in Precambrian time. Throughout Paleozoic time, the North American continental plate slowly drifted northwest, crossing several global climatic belts. Some rotation occurred, which means that the plate has had different orientations in the past. Part of the time the North American plate was joined with other plates in a supercontinent (called Pangea). And for long periods of geologic time, vast shallow oceans covered the stable interior of the plate. Many of the rock layers of the Colorado Plateaus accumulated in these oceans. The

of the nearshore environment, but most of the sand was in coastal dunes that spread gradually inland. At first the dunes interfingered with the Kayenta stream deposits, but eventually desert sands covered the entire area. Because of differences in the types and amounts of cement, the Navajo Sandstone varies in color from pale tan to red.

8. **Early mid-Jurassic deposition of the Temple Cap Formation.**

The caprock at the top of East Temple and West Temple records a marked change in climate at about the midpoint of Jurassic time. Streams loaded with red mud flooded the dunes and partially leveled them. After thin

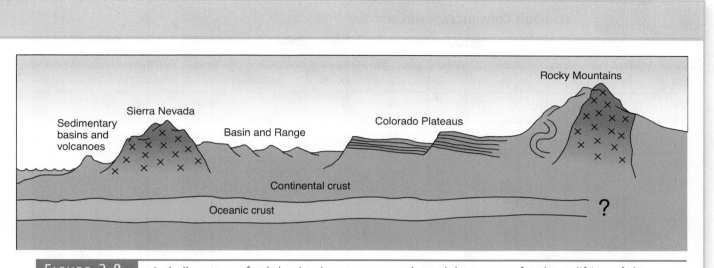

FIGURE 2.8   A shallow type of subduction is one concept that might account for the uplifting of the Colorado Plateaus and the development of the Rocky Mountains. Along the western continental margin, sediments accumulated in basins while volcanic arcs (volcanic chains, offshore or on continental crust near the coast) erupted quantities of pyroclastics and lava. Inland, an unyielding block of plateaus rose, buoyed up by the thrust of a huge slab of oceanic crust sliding beneath continental crust. Along the eastern edge of the plateau block, intense compression resulted in folding, thrust faulting, and the pushing up of a great mountain range. Pendick 1997.

sediments and structures of the rocks record times of emergence, submergence, and reemergence during the drifting of the plate over degrees of latitude and many miles of distance.

In the latter part of Mesozoic time, the lengthy and complicated period of Laramide mountain-building began when the westward-moving North American crust began to be affected by the eastward-moving Pacific Ocean crust. The ocean floor slipped under the western edge of the continent and broke pieces off the continental shelf, dragging down sediments that thickened the continental crust and increased its buoyancy. The far western mountain ranges developed due to the compression and crumpling of the parts of North America closer to the con-

verging zone. The Colorado Plateaus were a solid, unyielding, discrete block that resisted the compression. The plateaus rose, partly from pressure underneath and partly from buoyancy, and rotated slightly as the continent moved northward. The Rocky Mountain area was crumpled and pushed up due to intense compression along the eastern boundary of the plateaus (fig. 2.8).

In Cenozoic time, toward the end of the Tertiary Period, the Colorado Plateaus again underwent uplift with minor warping and deformation. This time the tectonic forces were extensional as crust in the Basin and Range to the west thinned and stretched, causing extensive faulting and tilting in that neighboring area.

beds of clay and silt were deposited, desert conditions resumed, but not for long.

9. **Deposition of the Carmel Limestone in a nearshore marine environment from middle to late Jurassic time; erosional unconformity.**

About 150 million years ago, a shallow, warm sea gradually spread over the area. Wave action flattened the sand dunes as calcareous silt buried them. The Carmel Limestone accumulated in massive beds of compact limestone that were one to four feet thick and separated by thin, sandy limestone beds containing

Table 2.1	Geologic Column, Zion National Park

Time Units			Rock Units			
Era	Period	Epoch	Group	Formation	Member	Geologic Events
Cenozoic	Quaternary	Holocene Pleistocene		Basaltic lavas		Weathering, canyon-cutting, cliff retreat, slump blocks Volcanic eruptions Earthquakes
	Tertiary — Neo-gene			Pliocene		Gravel deposits
				Miocene		Block faulting; uplift, erosion
	Tertiary — Paleo-gene	Oligocene		Lava flows		Episodes of volcanic eruption; uplift
Mesozoic	Cretaceous			Dakota Sandstone		Stream deposits
						Erosion
	Jurassic	Upper	San Rafael	Carmel		Silts deposited in shallow sea Erosion
		Middle and Lower		Temple Cap		Flooding followed by desert conditions
	?	?				
	?	?	Glen Canyon	Navajo Sandstone		White Cliffs Area buried by migrating sand dunes; increasing aridity Erosion
				Kayenta		Intermittent stream deposits
				Moenave	Springdale Sandstone	Stream deposits
					Whitmore Point	Lake deposits
	Triassic				Dinosaur Canyon	Stream deposits Erosion
				Chinle	Shinarump Conglom-erate	Stream, swamp deposits, plus volcanic ash; wood fossils Basal conglomerate Erosion
					Upper Red	Inland stream deposits
				Moenkopi	Shnabkaib Shale	Coastal plain; gradually changing from marine deposition
					Virgin Limestone	
Paleozoic	Permian			Kaibab Limestone		Deposition in shallow sea
			Older units not exposed.			

Source: Hamilton, 1984.

FIGURE 2.9   The Three Patriarchs have been created by the weathering of the Navajo Sandstone in Zion Canyon. © Robert Fletcher.

fossils of shallow-water organisms. Exposures of Carmel Limestone overlie the Navajo Sandstone on Horse Ranch Mountain in the Kolob section of the park.

Late Jurassic beds are not present in Zion and were probably removed by uplift and a long period of erosion that brought the Mesozoic Era to a close in this region. Nevertheless, a small remnant of the Dakota Formation, probably deposited late in Cretaceous time, crops out at the top of Horse Ranch Mountain (Hamilton 1984). The beds consist of a basal conglomerate and a fossiliferous sandstone.

During the latter part of Mesozoic time, the layered rocks of Zion were mildly affected by the compressive effects of tectonic activity to the west. The results of folding and subsequent thrust faulting can be seen in beds in the Taylor Creek area in the Kolob section of the park.

10. Normal faulting, uplift and erosion; lava flows.

Late in Tertiary time (about 20 million years ago), the Zion rocks began to "feel" the tensional forces that were stretching and breaking up the crust in the region to the west that eventually became the Basin and Range.

Two major north-south-trending vertical faults, which form the margins of the Markagunt Plateau, developed on either side of Zion National Park: Sevier fault on the east and Hurricane fault on the west (fig. 2.10). Several normal faults developed in the park, including the Cougar Mountain faults and the Beartrap fault. It was after the uplifting and tilting of the Markagunt Plateau that the Virgin River began downcutting.

Much later and after considerable erosion had taken place, flows of basaltic lava reached the surface by pushing up through the fractured rocks in the fault zones. The oldest dated lava flows, at Lava Point, are 1.4 million years old; the youngest, near the lower end of Cave Valley, are only a little more than 250,000 years old (Grater and Hamilton, 1978). Several cinder cones in the southwest corner of the park are much younger, possibly Holocene in age.

11. Dissection of the Markagunt Plateau by the Virgin River and its tributaries.

As renewed faulting and tilting increased stream gradients, downcutting and mass wasting sculptured the Zion landscape. The differences in the lithology of the rock layers, their resistance or lack of resistance to

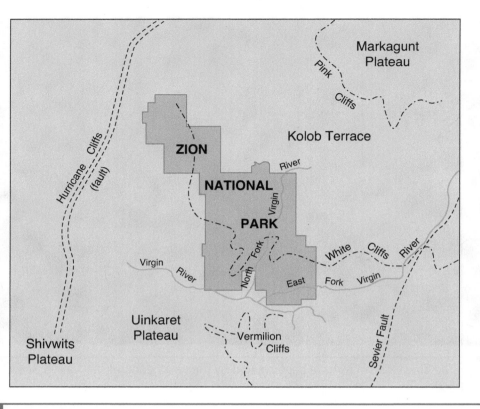

Major plateaus and faults in Zion National Park and vicinity, southwestern Utah. Hurricane fault extends south past the western edge of Grand Canyon National Park.

weathering, and the location of the faults and joints (fractures) produced the features that we now view.

Geologic Map and Cross Section

Hamilton, W.L. 1978. Geological Map of Zion National Park. Springdale, Utah: Zion Natural History Association.

Sources

Goetzmann, W.H. 1982. Explorer, Mountain Man, and Scientist, in National Park Service Handbook 116: *Exploring the American West, 1803–1879*. p. 17–95.

Grater, R.K. and Hamilton, W.L. 1978. Zion. Unicorn Associates, Denver, Colorado, in Cooperation with Zion Natural History Association.

Gregory, H.E. 1950. Geology and Geography of the Zion Park Region, Utah and Arizona. U.S. Geological Survey Professional Paper 220.

———— 1956. *Geological and Geographic Sketches of Zion and Bryce Canyon National Parks*. Springdale, Utah: Zion-Bryce Natural History Association.

Hagood, A. 1985 (2nd edition). *This Is Zion*. Zion Natural History Association and National Park Service. 73 p.

Hamilton, W.L. 1984. *The Sculpturing of Zion*. Springdale, Utah: Zion Natural History Association.

Humphreys, E.D. 1995. Post-Laramide Removal of the Farallon Slab, Western United States. *Geology*, November, v. 23, no. 11, p. 987–990.

Pendick, D. 1997. Rocky Mountain Why, Mountain Mystery. *Earth* v. 6, no. 3 (June), p. 26–33.

Rigby, J.K. 1976. *Northern Colorado Plateau*, K/H Geology Field Guide Series. Dubuque, Iowa: Kendall/Hunt Publishing Company.

Woodbury, A.M. 1950. *A History of Southern Utah and Its National Parks*. Originally published by Utah State Historical Society (1944, v. XII, n.3,4); revised and reprinted by A.M. Woodbury, Salt Lake City.

Note: To convert English measurements to metric, go to www.helpwithdiy.com/metric_conversion_calculator.html

Bryce Canyon National Park

S O U T H E R N   U T A H

Area: 35,835 acres; 56 square miles

Proclaimed a National Monument: June 8, 1923

Established as a National Park: February 25, 1928

Address: P.O. Box 170001, Bryce Canyon, UT 84717

Phone: 435-259-8161

E-mail: brca_reception_area@nps.gov

Website: www.nps.gov/brca/index.htm

Brightly colored pinnacles, windowed walls, spires, and pedestals stand in horseshoe-shaped amphitheaters eroded from the Pink Cliffs, along the edge of the Paunsaugunt Plateau. The erosional sculptures developed in the red, pink, mauve, buff, and off-white beds of the Tertiary Claron Formation.

FIGURE 3.1 The Grottos created by the weathering and erosion of the horizontal beds of the Eocene-aged Claron Formation have produced many interesting shapes. Photo by Ann G. Harris.

Geographic Setting

Bryce Canyon National Park is about 50 miles to the northeast in a direct line from Zion National Park, but the two parks have striking differences in geology, topography, climate, and scenery. Because Bryce Canyon is about one thousand feet higher in elevation, the climate is cooler and moister. The two canyons have undergone intense erosion for about the same length of time, but the rocks in Bryce Canyon are around a hundred million years younger than those found in Zion Canyon. When you arrive in Zion Canyon, you stand on the canyon floor and look up; but at Bryce Canyon, you come in at the top of the Paunsaugunt Plateau and look down over the rim (fig. 3.2). The cliffs and towers in Zion Canyon National Park are bold and colossal. The amphitheaters of Bryce Canyon National Park are immense, and within them the standing walls, fins, and pinnacles display intricate, delicately formed spheres, knobs, windows, and pendants in an array of colors.

Cedar Breaks National Monument

An interesting way to go from Zion National Park to Bryce Canyon National Park is by way of Cedar Breaks National Monument, which is about 25 highway miles north of Zion's Kolob section. The monument is on the western rim of the Markagunt Plateau at an elevation of over 10,000 feet. The rocks and erosional features of the amphitheater at Cedar Breaks are similar in age and appearance to those in Bryce Canyon National Park, located about 35 miles to the east on the Paunsaugunt Plateau. In an earlier time, one continuous, unbroken deposit covered the two areas. Now the two plateaus stand at different elevations, separated by the north-south trending Sevier fault and the Sevier River valley.

The western explorers and pioneers used the term "breaks" to describe places like Cedar Breaks and Bryce Canyon, where the edge of an upland descends abruptly or breaks down by eroding to a lower level. Thus Bryce Canyon is not a true canyon, but a series of

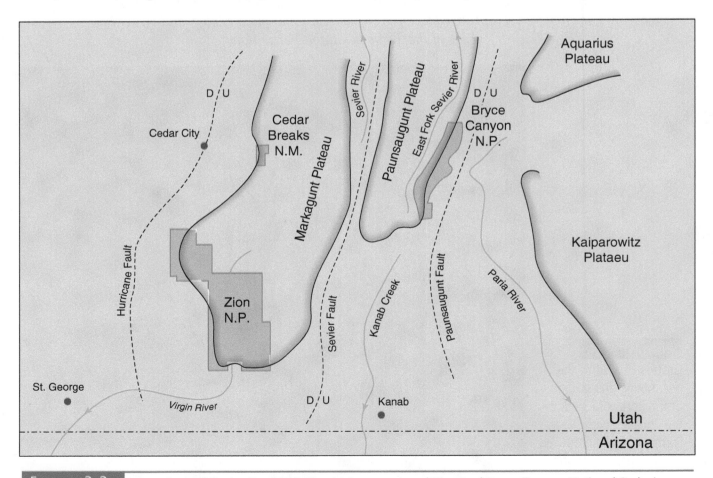

FIGURE 3.2   Locations of Cedar Breaks National Monument and Zion and Bryce Canyon National Parks in relation to major faults, fault blocks (the plateaus), and river valleys of southwestern Utah.

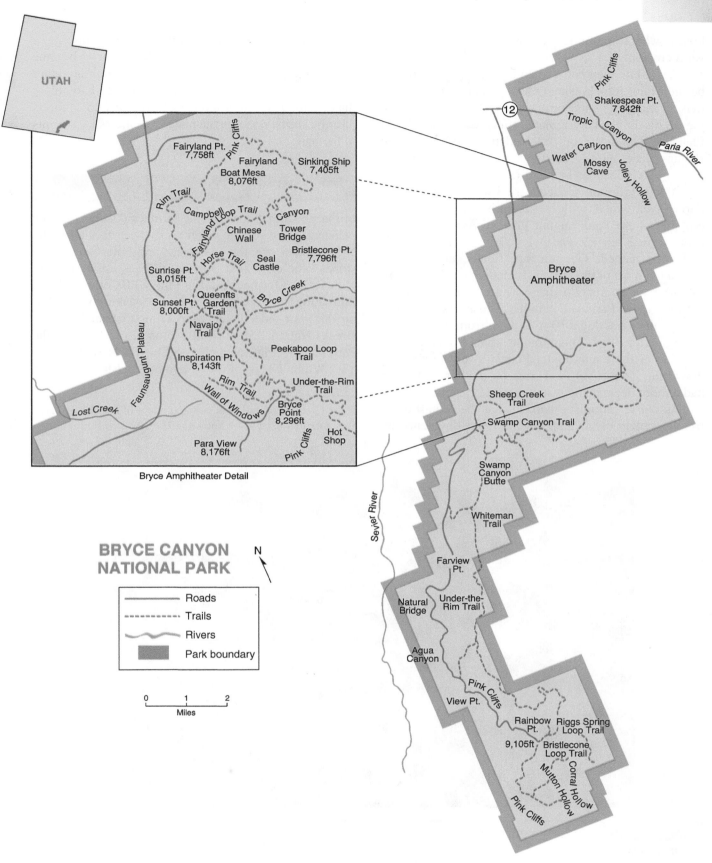

UTAH

Fairyland Pt.
7,758ft

Pink Cliffs

Fairyland

Boat Mesa
8,076ft

Sinking Ship
7,405ft

Rim Trail

Campbell Loop Trail

Fairyland Loop Trail

Canyon

Chinese
Wall

Tower
Bridge

Horse Trail

Seal
Castle

Bristlecone Pt.
7,796ft

Sunrise Pt.
8,015ft

Queenfts
Garden
Trail

Bryce Creek

Sunset Pt.
8,000ft

Navajo
Trail

Faunsaugunt Plateau

Inspiration Pt.
8,143ft

Peekaboo Loop
Trail

Rim Trail

Under-the-Rim
Trail

Lost Creek

Wall of Windows

Bryce
Point
8,296ft

Para View
8,176ft

Pink Cliffs

Hot
Shop

Bryce Amphitheater Detail

Pink Cliffs

Shakespear Pt.
7,842ft

Tropic Canyon

Water Canyon

Mossy
Cave

Jolley Hollow

Paria River

Bryce
Amphitheater

Sheep Creek
Trail

Swamp Canyon Trail

Swamp
Canyon
Butte

Sevier River

Whiteman
Trail

Farview
Pt.

Natural
Bridge

Under-the-
Rim Trail

Agua
Canyon

Pink Cliffs

View Pt.

Rainbow
Pt.

Riggs Spring
Loop Trail

9,105ft

Bristlecone
Loop Trail

Corral Hollow

Mutton Hollow

Pink Cliffs

BRYCE CANYON NATIONAL PARK

N

——— Roads

- - - - - Trails

〜〜〜 Rivers

▨ Park boundary

0   1   2
Miles

FIGURE 3.3   Bryce Canyon National Park, Utah.

breaks where amphitheaters are developing by headward erosion.

Few white men had visited the Bryce Canyon area before Major John Wesley Powell and his team of mapmakers and geologists began to survey southern Utah in the 1870s in their systematic study of the Colorado Plateaus. The surveyors retained many of the Paiute Indian place names on their maps. For example, *Paunsaugunt,* means "home of the beaver."

A few Mormon pioneers tried to settle along the Paria River, which, with its tributaries, drains the eastern side of the Paunsaugunt Plateau. Among the pioneers were Ebenezer and Mary Bryce, for whom the park was named. Overgrazing, floods, and drought forced out most of the settlers, including the Bryce family, after an attempt to divert water from the Sevier drainage to the Paria Valley failed. Those who were left, a determined lot, dug a 10-mile ditch from the east fork of the Sevier River. This brought water from the western side of the Paunsaugunt Plateau over the rim and down into Tropic Valley. Farms on the east side then had a more dependable water supply.

As conservationists realized that the fantastic but fragile scenery of the amphitheaters was threatened by overgrazing and lumbering on the plateau, a movement was begun to protect the area, which eventually became a national park. Bryce Amphitheater and the other amphitheaters along the plateau rim extend for more than 20 miles through the park from north to south. A 35-mile scenic drive on the plateau connects viewpoints in the park.

Geologic Features

In the prevailing climate, the terraces ("steps") on the plateaus formed on the softer rocks, such as shales. The cliff-formers are the resistant sandstones and limestones. The cliffs that rim the plateaus are associated with various formations. The Pink Cliffs of the Claron Formation*, for example, are most strikingly exposed on the west side of the Markagunt Plateau in Cedar Breaks National Monument and on the east side of the Paunsaugunt Plateau in Bryce Canyon National Park. The White Cliffs of the Navajo Sandstone rim the

*The Claron Formation is now the name of rock units in this area that used to be known as the Wasatch Formation.

FIGURE 3.4   Looking north over Bryce Amphitheater from Bryce Point. The pink and red hues of the pinnacles and towers standing in the vast amphitheater appear to change from hour to hour as the sun passes overhead. Looking at the landscape, it is easy to understand the Paiute name for Bryce, which is "red rocks standing like men in a bowl-shaped canyon." Photo by David B. Hacker.

FIGURE 3.5   The "Silent City" area near Sunset Point; turreted vertical columns stand above gullies several hundred feet deep. Joint systems determine the locations of columns and gullies. Photo by David B. Hacker.

southern side of the Markagunt Plateau in Zion National Park.

The Pink Cliffs developed differently from the White Cliffs because of their lithologies; that is, the physical characteristics of the rocks. Instead of the imposing monoliths of the White Cliffs, the Pink Cliffs display barren badlands with very little soil or vegetation. However, these are not typical badlands made up of stream-dissected slopes. W.L. Stokes (1986) describes the scenery of Bryce Canyon and Cedar Breaks as displaying "a combination of slopes and cliffs, a compromise between the erosion of material too soft to form uniform cliffs and too hard to form uniform slopes."

In Bryce Canyon National Park, the processes of weathering, *slope wash* (downward trickling or running water, not in channels), and mass wasting have been mainly responsible for the remarkable scenery that visitors see. Downcutting by streams has been less significant. Bryce Amphitheater, the largest and most spectacular scenic feature, is a horseshoe-shaped basin about 12 miles wide, 3 miles long, and 800 feet deep. The arches and windows in Bryce, its sculptured terraces, and its intricate pinnacles have all been weathered and eroded out of the pink and white Claron Formation—a mixture of limestone, limy siltstone, shale, sandstone, and a few remnants of conglomerate, laid down in an inland lake of vast extent. The pink hues of the beds are due to the presence of iron oxide and manganese in the sediments.

The inland lake, which existed early in Eocene time, spread over low-lying, shallow basins that covered several thousand square miles. The size and shape of the lake system varied as it expanded and contracted in response to changes in climate and environment over millions of years. The differences in the Claron beds reflect the different depositional environments in which they accumulated.

Subsequent uplift and faulting formed blocks that streams carved into plateaus. The joint systems and cracks produced by uplift control the location of walls, arches, windows, and natural bridges. The pinnacles are the result of differential weathering acting along vertical joints (fractures) on beds with varying degrees of hardness. The lime-rich beds are more resistant to weathering in the Bryce climate than the softer, lime-poor beds. The difference in weathering rates produces the characteristic parallel grooving on the features; protrusions follow lime-rich layers and indentations appear along weaker, lime-poor beds (fig. 3.5).

A feature called Queen Victoria, located in the Queen's Garden area, is one of many notable examples of differential weathering. Some beds have been completely removed. For example, the conglomerate capping two buttes at Bryce Point is about all that remains of that particular layer in the entire park.

Most arches in the park, such as the famous Natural Bridge, are carved from sandstone beds in the Claron Formation. Despite their appearance of solidity, they are not permanent features. In the Fairyland area, an arch known as Oastler's Castle collapsed under its own weight in 1964. However, as old ones are destroyed, new ones are created by the unremitting processes of weathering (table 3.1).

Weathering Processes

The fragmentation or *disintegration* of rock is a *physical* process that is inextricably linked to *decomposition* of rock, which involves *chemical* processes. In other words, rock weathers physically and chemically at the same time. In Bryce Canyon National Park, the climatic zonation strongly influences how the various weathering processes interact. Moreover, the intermittent pattern of precipitation throughout the year creates cycles of wetting and drying that affect how the cliffs retreat.

At the higher elevations along the plateau (roughly 8500 to over 9000 feet) in the montane, or Canadian, life zone, precipitation averages 25 inches annually, much of it in the form of snow. Winters are relatively severe, and late spring snow squalls are not infrequent. Spruce and fir thrive in the cool, moist conditions. Water from rain and melting snow running off the edge erodes the rim of the plateau. Moisture seeping through the soil and transporting organic acids down into bedrock increases chemical weathering.

In the Upper Sonoran (high desert) life zone at the base of the plateau (elevation around 6500 feet), the average annual precipitation is 13 inches, with summers tending to be hot and dry. Only hardy trees and shrubs can survive on the semiarid eastern slopes. Rain falls mostly in late winter and in late summer. Thunderstorms may cause rapid erosion of loose sediments on dried out or weakened slopes.

In the transition zone (7000 to 8000 feet), ponderosa pines are the dominant trees and precipitation averages 15 inches per year. Average temperatures range from lows in the 30°s (Fahrenheit) to highs in the 80°s.

On the steep, unstable slopes within the park amphitheaters, very little vegetation is available to hold

Table 3.1	Weathering Processes

A. Mainly physical (mechanical) weathering-disintegration

Type	Process
1. Granular disintegration	Separation of coarse mineral grains due to temperature extremes
2. Frost wedging	Grains and particles forced apart by freezing water
3. Frost heaving	Rock fragments raised up by ice pressure
4. Exfoliation	Stripping of concentric rock slabs from outer surface of a resistant rock mass (principally in granite, sandstone and quartzite; usually large-scale)
5. Spheroidal weathering	Rounding by weathering (partly chemical) from an initial blocky shape; forms boulders
6. Organic weathering	Root pry, animal burrows, human activity, etc.
7. Fire	May cause rocks to crack or explode

B. Mainly chemical weathering—decomposition

Type	Process
8. Oxidation	Reaction of mineral grains with atmospheric oxygen
9. Carbonation	Reaction of carbonic acid (from water and carbon dioxide) with mineral grains; frequently in combination with organic acids
10. Hydration	Minerals in rock combining with water
11. Solution	Soluble minerals go into solution; insoluble mineral grains transported as sediment

soil and sediment and soak up rainwater. Moreover, the high clay content in many of the beds prevents water from penetrating deeply. That is why a sudden, intense downpour can sometimes change the appearance of features within minutes.

Zone boundaries throughout the park are not distinct. Southwest-facing slopes and erosional features, for example, are exposed more to sunlight, to greater

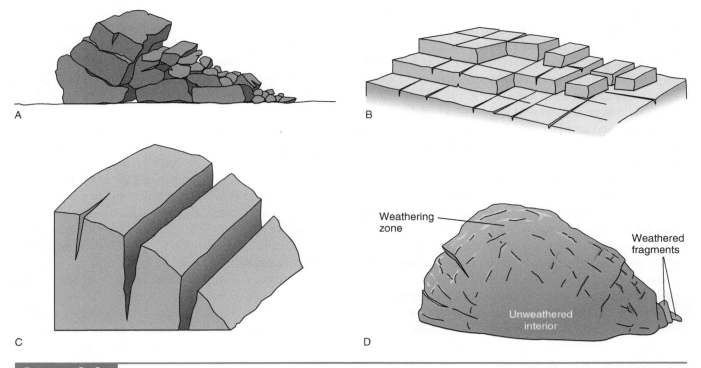

FIGURE 3.6   Examples of weathering processes that break up rocks. *A.* **Granular disintegration.** Massive rock (with few bedding planes or joints) tend to break into irregular blocks and fragments. As more surfaces are exposed, mineral grains loosen and weather out. *B.* **Separation along joint blocks.** Weathering penetrates fractures, causing a rock to come apart. Once loosened, blocks break into smaller pieces. *C.* **Frost wedging.** Pressure of freezing water, expanding in cracks and fractures, enlarges openings in rock. Many cycles of freezing and thawing cause slabs or blacks to split off. *D.* **Spheroidal weathering.** While processes of physical and chemical weathering operate on exposed edges and corners of outcrops, cracks tend to form just beneath a rock's exterior. Eventually concentric shells or layers loosen and spall off.

evaporation, and to more freeze-thaw days. Sometimes when the morning sun warms a southern exposure after a freezing night, rocks come crashing down. The north-facing slopes that stay in the shadow remain covered with snow longer and have fewer freeze-thaw days. As a result, erosional processes operate a little more slowly.

Frost action is especially effective in attacking bedrock on the exposed and wind-swept promontories at higher elevations. Fairyland View, for example, is 7586 feet above sea level; Bryce Point, 8294 feet; and Rainbow Point, the highest spot on the rim, 9105 feet. Rim erosion is intensified at locations that are subject to greater extremes of temperature.

Along with freezing and thawing, extremes of temperature also tend to cause *granular disintegration* of coarse-grained rock masses such as conglomerate and coarse sandstone, both of which are found in the Claron Formation. The mineral grains tend to separate from one another to produce coarse sand or gravel (table 3.1).

Frost action attacks rock primarily by physical (mechanical) weathering. Moisture in cracks and joints freezes and expands, causing *frost wedging* (fig. 3.6C). Cracks enlarge and rock fragments break off. Subsurface freezing of water forces a portion of rock upward and causes more fragmentation by a process called *frost heaving*. As cracks deepen, moisture penetrates farther into the rock and enables chemical weathering to attack the minerals by loosening and altering the grains.

Spheroidal weathering is a physical and chemical process that has the effect of rounding off pinnacles and knobs in tile amphitheaters. Most of the features tend to be angular; they break up along vertical joints (fig. 3.6B). A sharp corner has more surface area exposed than a smooth, flat rock face, so weathering works faster on corners. Angular shapes become rounded as corners retreat. If enough moisture is present, rounded layers may spall off as concentric shells (fig. 3.6D).

A similar process on a more massive scale is called *exfoliation.* Large curved sheets or slabs of rock peel off, *or exfoliate,* from a bedrock surface. Igneous rocks and resistant sandstones that have widely spaced fractures or joints tend to be subject to this type of weathering.

Plants and animals (including humans) cause fragmentation of rock by both physical and chemical processes. As plant roots pry open cracks, plant acids react chemically with rock. Animals dig burrows, and engineers blast out rock when building roads. All of these activities expose fresh rock surfaces to moisture and organic and atmospheric acids.

The effects of chemical weathering are evident, even in the drier sections of the park where evaporation rates are high. *Oxidation,* which involves the reaction of certain mineral constituents in rock with the oxygen in the air, is the process responsible for the many shades of color characteristic of Bryce Canyon. The reds, pinks, and browns are produced by the iron oxide hematite (Fe_2O_3); the yellows by another iron oxide, limonite ($Fe_2O_3 \cdot nH_2$); and the purple hues by manganese oxide (MnO_2). Most of the white or light-colored beds were deposited as relatively pure limestones lacking the minerals that produce the colors. In a number of places in the amphitheaters, a natural stucco has given white layers an orange coating.

Hydration is a weathering reaction involving the combining of water with minerals in rock. It causes the grains to expand and the rock to be weakened. (Limonite is an oxide that has been hydrated.)

The *carbonation* reaction begins with the combination of carbon dioxide in air and soil with water to form carbonic acid, which plays an important role in the weathering of many kinds of rock. Some feldspars, for example, weather by carbonation to produce soluble calcite ($CaCO_3$) that becomes the main constituent of limestone and also forms a cementing medium in clastic sedimentary rocks.

Examples of *solution weathering* (i.e., the dissolving of the soluble minerals in rock) can be seen in such features as the Wall of Windows along Peekaboo Trail. Here again, physical and chemical processes interact as spheroidal weathering and exfoliation also aid in carving out the windows (fig. 3.7).

Many of the odd and unusual rock shapes in the park, such as the standing pillars, or *hoodoos,* are

FIGURE 3.7   The Wall of Windows. Chemical and spheroidal weathering plus exfoliation attack the less resistant parts of the rock, producing windows in the wall. Photo by Ronald F. Hacker.

Thor's Hammer was developed by weathering and erosion attacking beds of Eocene sedimentary rock, forming a pillar capped by a large knob. Photo by David B. Hacker.

The Natural Bridge is a combination of a cap rock of resistant limestone to form the "Bridge" and exfoliation below it to form the window. © Robert Fletcher.

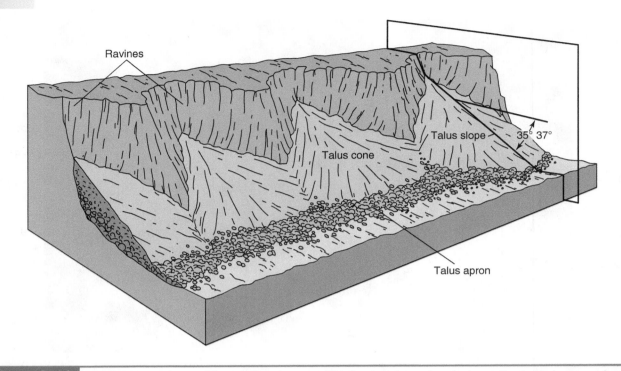

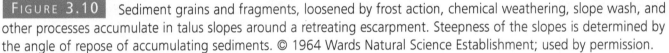

FIGURE 3.10   Sediment grains and fragments, loosened by frost action, chemical weathering, slope wash, and other processes accumulate in talus slopes around a retreating escarpment. Steepness of the slopes is determined by the angle of repose of accumulating sediments. © 1964 Wards Natural Science Establishment; used by permission.

due in part to solution weathering of less resistant portions of the original bedrock. Hoodoos form where gullying has cut through resistant rock and exposed softer layers beneath. Many of the pillars are capped by remnants of resistant rock. In the Hat Shop on the Under-the-Rim Trail, some hoodoos are topped by boulders that have fallen from the cliffs above. The boulders act like umbrellas by protecting the weak, silty or clayey beds from erosion by pelting raindrops.

Frost action on jointed rocks in a region of high relief accounts for the effectiveness of mass wasting in Bryce Canyon National Park. Evidence for this can be seen in the extensive accumulations of talus that overspread the base of slopes, pile up against the cliffs, and surround the columns (fig. 3.10). Typically the steepness of a talus slope or cone is at the angle of repose (roughly 35°). Loose fragments continually roll downslope or shift position as they move downward.

The Paria River and its headwaters, which drain Bryce Canyon, carry away sediment and debris plus rock material in solution. The Paria flows southeastward and empties into the Colorado River at Lees Ferry, just above the Grand Canyon.

Geologic History

1. **Deposition of the Dakota Sandstone in a shallow Cretaceous sea.**

A few remnants of thin Dakota Sandstone, the oldest formation exposed in the vicinity of the park, have been found well below the plateau in the Paria Valley. The Dakota Sandstone lithified from a *blanket sand;* that is, a sedimentary layer thinly covering a large geographic area. In this region the deposit represents the beach area, lagoons, and coal swamps of a transgressing sea (fig. 3.11).

2. **Deposition of the Tropic Shale in a deepening seaway.**

At the lowest elevations in the park are scattered outcrops of the Tropic Shale that accumulated in a relatively calm sea and contains fossil ammonites (Cephalopods).

3. **Erosion of the Late Cretaceous formations, creating unconformities.**

Most of the sandstones and shales that were deposited in the Bryce Canyon vicinity late in

Table 3.2			Geologic Column, Bryce Canyon National Park	
Time Units			**Rock Units**	
Era	Period	Epoch	Formation	Geologic Events
Cenozoic	Quaternary	Holocene		Weathering and erosion of Claron beds to form present landscape
		Pleistocene	Sevier River	Sandstone with patches of conglomerate
				∼Uplift and erosion∼
	Tertiary	Pliocene	Boat Mesa Conglomerate	Coarse deposits on land
		Miocene Oligocene		
		Eocene	Claron	Streams depositing in vast, inland lakes
				∼Erosion∼
Mesozoic	Upper Cretaceous		Kaiparowits	Flood-plain deposition
				∼Erosion∼
			Wahweap Sandstone	Deposition along a shifting shoreline
			Straight Cliffs Sandstone	Deposition in a shallow sea and on beaches and flood plains
				∼Erosion∼
			Tropic Shale	Deposition in a deepening seaway
			Dakota Sandstone	Beach deposits as a blanket sand in a transgressing sea; coal swamps
(Older Units not exposed in parks.)				

Source: Gregory, 1956; Rigby, 1976; Stokes, 1986; Decourten, 1994.

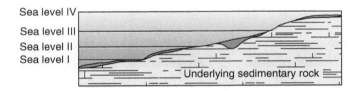

Sea level IV
Sea level III
Sea level II
Sea level I

Underlying sedimentary rock

FIGURE 3.11 Beach sand, accumulating on a shoreline migrating inland, may form a blanket sandstone. The rock formation shown here becomes progressively younger toward the right as the sea encroaches on the land.

Cretaceous time were stripped away by erosion as the period came to a close. A few outcrops of the Straight Cliffs and Wahweap Sandstones and the Kaiparowits Formation are found at low elevations in the southern portion of the park.

4. Effects of plate convergence late in Cretaceous and early in Paleocene time.

The Laramide orogeny, a mountain-building episode that profoundly affected western North America (box 2.1), began as the westward-moving North American plate slid over the Pacific oceanic plate. The compressive tectonic forces produced by this convergence caused low uplands to be uplifted in what is now southern Utah. Between the uplands, low-lying basins gently subsided.

5. Deposition of the Claron Formation in inland lakes in Eocene time.

A broad, shallow lake system that spread over the basins was fed by rivers that sorted and transported vast quantities of clastic sediment. Over the millions of years that the Claron Lake system existed, it expanded and contracted many times; as it did so, alternating beds built up, some thick, some thin, some coarse, some fine, some in water, and others along shorelines. The continually changing depositional environments created a formation lacking uniformity in texture: nearly pure limestones, laid down in deep water, graded into calcium-rich mudstones, which were often interspersed with lenses of standstone and conglomerate. The colorful lower member of the Claron Formation (called by Clarence Dutton the "Pink Cliffs Series") is the rock unit from which nearly all the hoodoos and other spectacular landforms developed.

Fossils are not abundant in the Claron beds, but some subtropical, aquatic snails, turtles, reptiles, and other species have been identified.

6. Deposition as the land rose.

Remnants of rock units younger than Claron beds occur in a few places along the plateau rim and in the northern part of the park. The Boat Mesa Conglomerate (50 to 100 feet thick) of Oligocene age consists of gravel and fragments of limestone derived from Claron rocks. The Sevier River Formation, which was deposited in Pliocene and early Pleistocene time, is a brownish-gray sandstone with interbedded conglomerates.

7. Mid-Tertiary (Miocene) uplift and blockfaulting, creating the plateaus.

From Rainbow Point (elevation 9105 feet), you can look out over a broad expanse of southern Utah and see most of Bryce Amphitheater, the Table Cliffs, the Aquarius Plateau, White Cliffs, Vermilion Cliffs, and the Henry Mountains. The culminating uplift of the high plateaus of southwest Utah that raised these features (and the whole Colorado Plateaus region) several thousand feet from near sea level occurred in Miocene time, some 16 million years ago.

During this episode, the tectonic forces were tensional rather than compressional. A portion of the western continent, especially in Nevada and southwest Utah, was being pulled apart as well as elevated. As a result, deformation was greater on the western margins of the Colorado Plateaus than toward the east. The stress and strain of the crustal disturbance created the great north-south normal faults (such as the Sevier and Paunsaugunt faults) that broke up and separated crustal blocks (fig. 3.2). Some blocks dropped, but the blocks that formed the southern Utah plateaus rose and tilted. Occasional earthquakes in the region are an indication that tectonic adjustments are still occurring far below the surface.

8. Erosion of the Paunsaugunt Plateau and development of the amphitheaters in late Tertiary and Quaternary time.

The river systems that transported the Claron sediments were completely disrupted by the Miocene upheavals. Moreover, as the North American plate continued to drift toward northern latitudes, while moving generally westward, the climate cooled, then became more moist. During the times of greatest cold in the Pleistocene Epoch, the low temperatures maintained snow year-round at higher elevations. For a time glaciers existed at the top of windswept Cedar Breaks to the west.

The steeper gradients of the reorganized drainages and the increased volume from rain and snowmelt made the streams of the region swift and powerful. The headwaters of the Paria River (a tributary of the Colorado River) cut rapidly headward into the eastern side of the Paunsaugunt Plateau. The more erosion broke through protecting soil and vegetation, the more the bedrock layers were laid bare, exposed in all their colors and structures, to the endless processes of weathering.

Geologic Map and Cross Section

Bowers, W.E. 1991. Geologic Map of Bryce Canyon National Park and Vicinity, Southwestern Utah. Map I-2108, U.S. Geological Survey Miscellaneous Investigations.

Gregory, H.E. 1951. The Geology and Geography of the Paunsaugunt Region, Utah (with geologic map). U.S. Geological Survey Professional Paper 226.

Sources

Decourten, F. 1994. *Shadows of Time, the Geology of Bryce Canyon National Park.* Bryce Canyon Natural History Association, 128 p.

Rigby, J.K. 1976. *Northern Colorado Plateau,* K/H Geology Field Guide Series. Dubuque, Iowa: Kendall/Hunt Publishing Company.

Stokes, W.L. 1986. *Geology of Utah.* Salt Lake City: Utah Museum of Natural History and Utah Geological and Mineral Survey. p. 161–163.

Note: To convert English measurements to metric, go to www.helpwithdiy.com/metric_conversion_calculator.html

Capitol Reef National Park

SOUTH CENTRAL UTAH

Area: 241,904 acres; 378 square miles

Proclaimed a National Monument: August 2, 1937

Established as a National Park: December 18, 1971

Address: HC 70 Box 15, Torrey, UT 84775

Phone: 435-425-3791

E-mail: CARE_interpretation@nps.gov

Website: www.nps.gov/care/index.htm

". . . a maze of cliffs and terraces, red and white domes, rock platforms gashed with profound canyons, burning plains barren even of sagebrush—all glowing with bright color and flooded with blazing sunlight."

Clarence E. Dutton, 1880

The dome-shaped, white Navajo Sandstone rimrock is uplifted and tilted by the Water Pocket Fold, a monocline. Narrow, high-walled gorges, cut through the uplifted fold, reveal vividly colored beds of Mesozoic sedimentary rocks.

FIGURE 4.1   View looking along the Water Pocket Fold. The Navajo Sandstone is the oldest formation exposed and forms the higher ridge; the Carmel Formation is a narrow band along the base. On the valley floor the Entrada Sandstone, Summerville, and Morrison formations crop out. The Dakota Sandstone forms a low sinuous ridge. On the left, beds of the Mesa Verde Group cap a butte of Mancos Shale, youngest rocks in the sequence. Photo by Ronald F. Hacker.

Geographic Setting

Capitol Reef National Park lies along a north-south-trending monocline, called the Water Pocket Fold, about halfway (in a direct line) between Bryce Canyon National Park and Canyonlands National Park. Only about seven inches of precipitation a year falls in the Capitol Reef region, some of it in brief, intense storms that may cause flash flooding. Capitol Reef National Park is in desert country, except for the oasis along the Fremont River, a year-round stream. The Fremont River cuts through the Water Pocket Fold, from west to east, on its way to the Dirty Devil River, a tributary of the Colorado.

Indians of the Fremont culture, who raised corn, beans, and squash by irrigation, lived along the river around a thousand years ago. They drew pictographs on canyon walls. The stone granaries they built for storing corn were called "moki huts" by the Paiute Indians, who came into the region many years later. The Paiutes thought that a race of tiny people, or moki, lived in the huts, some of which are preserved in sheltered locations in the park.

Almon H. Thompson, one of Major J.W. Powell's surveyors, explored the formidable region to the east of Bryce Canyon and crossed the Water Pocket Fold in 1872. Subsequently, geologist Clarence Dutton spent several summers in the area engaged in field work. His findings are reported in his classic monograph on the High Plateaus geology (1880).

Mormon settlers moved into the Fremont River valley in the 1880s. They lived in and around a community called Fruita. Their orchards and lawns are now a picnic area and campground within the national park. The old Fruita Schoolhouse and a settler's cabin have been restored by the National Park Service. Old lime kilns in Sulfur Creek and near the campgrounds on Scenic Drive are all that remain of an early industry. The settlers used lime from limestone beds in Capitol Reef for whitewash, fertilizer, and cement. In the 20th century several small uranium mines were worked.

The name Capitol Reef was applied to a line of white domes and cliffs of Navajo Sandstone that extend along part of the Waterpocket Fold from the Fremont River south to Pleasant Creek. Early visitors were reminded of the white, domed buildings of Washington, D.C. The term "reef" meant any rocky barrier or escarpment that made traveling difficult. (Reef in this sense has nothing to do with the rocky masses built by marine organisms.) A national monument that was proclaimed in 1937 comprised roughly the Capitol Reef area only. Its expansion in 1971 to a national park encompassed the greater part of the Water Pocket Fold.

Utah highway 24, which crosses the park from east to west, was built along the Fremont River in 1962. Before that time, a frequently washed-out "highway" went through Capitol Gorge, about ten miles to the south. Abandoned now except for travel by foot, the old wagon road is strewn with flood boulders and hummocks of reworked gravel. In some places, Capitol Gorge, with towering cliffs on either side, was barely wide enough for a wagon or car to pass through (fig. 4.4).

Geologic Features

The Water Pocket Fold, a large monoclinal fold dipping eastward and trending north-south for over 90 miles, forms the geologic framework of Capitol Reef National Park (fig. 4.5). A *monocline* is a fold creating a local steepening of dip in layered rocks in an area where the bedding is otherwise generally flat or very gently dipping. All but about 30 miles of this dominant structure lies within park boundaries (fig. 4.5). Along the east side of the fold, erosion sculpted the escarpment, or so-called reef, that was for many years a formidable barrier to travelers and settlers going west.

Capitol Reef National Park is an ideal place to see large-scale examples of differential erosion of harder and softer beds because of the way they are exposed in the monoclinal fold. The more resistant beds in the structure form imposing ridges, cliffs, domes, mesas, and buttes, as well as smaller features, such as pinnacles, towers, and arches.

Major cliff faces in the park are formed from Wingate Sandstone, while the Navajo Sandstone and the Carmel Formation form the summit and much of the eastern slope of Capitol Reef and the Water Pocket Fold. The eroded turrets of the Castle, for example, are carved from the orange and red Wingate Sandstone, which juts up in westward-facing rimrock along the "spine" of the fold. The Wingate Sandstone also forms the Circle Cliffs, which lie just outside the west boundary of the park's south section. The Circle Cliffs are an eroded remnant of an uplift that flanks the west side of the Water Pocket Fold.

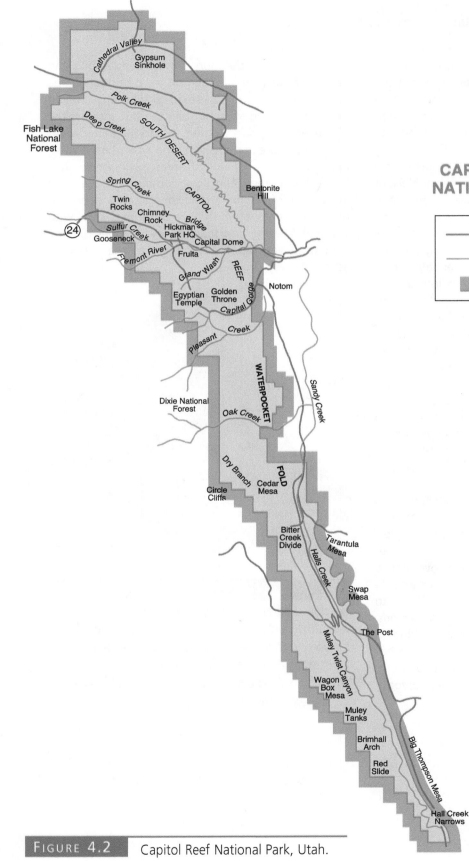
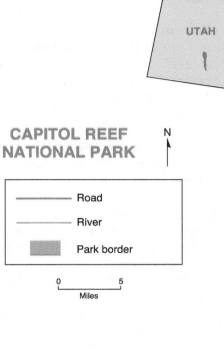

FIGURE 4.2   Capitol Reef National Park, Utah.

| Box 4.1 | **How Mass Wasting Operates in Arid Regions** |

The effects of mass wasting on landforms in the national parks of the Colorado Plateaus region are plain to see. Loose pieces or chunks of sedimentary rock that fall from cliffs accumulate in piles of *talus* and talus slopes that surround the base of a mesa or butte. Roadcuts made through bedrock may bring about *rockslides* that block a highway. On some steep slopes, scars of old slides can be seen above a block or mass of hillside material that has separated and slipped down along a curved surface, slumping at the bottom so that the base bulges out as a hummocky toe. Sudden torrential rains, perhaps accompanied by pelting hail, may loosen, wash down, or saturate so much debris that a desert stream turns into a thick *mudflow* that continues down a canyon picking up or burying everything in its path. When the rain stops, momentum keeps the mudflow going until the water dissipates, leaving the channel or the canyon mouth choked with boulders, uprooted trees, gravel, and silt.

With the driving force of gravity supplying the necessary energy, loosened rock or debris may fall, slide, slip, slump, or flow from higher locations to lower locations. Controlling the rate and type of mass wasting processes are such factors as climate, degree of slope, relief, characteristics of the bedrock or sediments, and vegetation cover.

Relief and slope angle are important because steep slopes are more susceptible to mass wasting than gentle slopes. Slopes in areas of high relief tend to be less stable than slopes in areas of low relief. Climate influences what types of weathering occur and governs how much and what type of vegetation grows in an area. In cold climates, freezing and thawing contribute to downslope movement in both humid and arid regions. In desert country, infrequent but heavy rain pelting on slopes unprotected by vegetation aids the processes of mass wasting. Depending upon the amount of runoff, debris may wash down or may become saturated, forming a debris slide or a mudflow.

In contrast to the more gently rounded hills and valleys typical of humid regions, the desert landscape looks angular with its steep slopes and walled canyons. This kind of topography tends to increase the effectiveness of mass wasting. Due to the thin soil and lack of moisture in the desert, one type of mass wasting characteristic of humid areas is extremely slow here. This is *creep*, the slow, continuous downslope movement of soil or any unconsolidated debris. In temperate regions, which have water in the soil with seasonal freezing and thawing, even slopes covered with thick vegetation are subject to the continual action of creep (fig. 4.3).

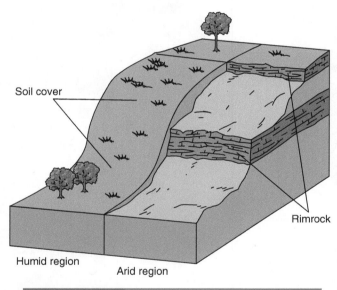

Soil cover

Humid region

Arid region

Rimrock

FIGURE 4.3   Slopes in humid regions tend to develop differently from slopes in arid regions, even when the bedrock is similar. Slopes are rounded and mantled with soil in humid regions because more moisture is available for weathering processes. In arid regions, soil cover is sparse and angular rock faces are common.

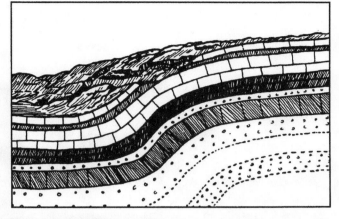

FIGURE 4.5   John Wesley Powell made this drawing of a typical Colorado Plateau monocline in 1875 for his survey report.

FIGURE 4.4   Capitol Gorge is strewn with rocks left by flash floods that fill the normally dry canyon with water 10 to 20 feet after a sudden thundershower. Photo by David B. Hacker.

The Hickman Natural Bridge, 133 feet wide and 125 feet above the streambed, is a true natural bridge that was formed over a watercourse as weathering and stream erosion wore a hole through a narrow "fin" of sandstone in the Kayenta Formation. Successive floods deepened the hole and the canyon, leaving the bridge high and dry most of the time. Hickman Bridge can be reached by a trail from the north side of Highway 24.

The flat-topped hills in the park, capped by resistant rock layers, are erosional remnants carved from the nearly horizontal sedimentary rock layers. These landforms, called *mesas,* have steeply sloping erosion scarps on all sides. *Buttes* are very similar but are smaller in size with less summit area than mesas. Most buttes were probably mesas originally. Examples of easily accessible buttes in the park are Chimney Rock, Twin Rocks, and Egyptian Temple. In these buttes, soft Moenkopi beds are capped by the resistant Shinarump Member of the Chinle Formation (fig. 4.6).

The "waterpockets," for which the monocline is named, are really *potholes* that have been carved in bedrock by the abrasive action of fine and coarse sediment whirling in turbulent water and deepening a hole that may have started as a small depression. Some may owe their origins to the meltwater torrents of waning Pleistocene time, but many are still being actively eroded whenever flash floods fill narrow gorges and canyons in the park with rushing water. Boulders and cobbles, trapped in low places and swirled around by the water, may act like a drill, boring holes in the rock. Some waterpockets are found at the foot of a cliff in the bed of an *intermittent stream,* that is, a temporary or seasonal stream. The action of falling water when the stream is in flood may create bowl-shaped potholes. The water stored in these natural pockets is an important source of drinking water for park wildlife (fig. 4.7).

Rockhounds find many examples of rock layers in the park and outside its boundaries that exhibit *primary structural features,* such as ripple marks and footprints of dinosaurs and amphibians, which were put there during deposition but before consolidation. Slabs of ripple-marked stone from red Moenkopi beds are quarried outside the park and sold as "ripple rock" for decorative stonework.

Igneous Rocks

Dark-colored basaltic magma squeezed between some sedimentary beds in Tertiary time, while they were still deeply buried. The intrusions hardened and formed sills. Subsequently, more magma pushed up and cut

FIGURE 4.6   The Shinarump Conglomerate, which caps Chimney Rock, protects softer, underlying beds of the Moenkopi Formation. Photo by David B. Hacker.

through both the sedimentary layers and the sills and created dikes. (By definition, *sills* are intrusions that harden parallel to bedding while *dikes* cut across structures.) Because the basalt is more resistant than the sedimentary rocks, some sills and dikes are now exposed and standing in relief at the northern end of the Water Pocket Fold.

The black basaltic boulders scattered about the park on top of Mesozoic bedrock are extrusive in origin, rather than intrusive. They came from lava flows capping Boulder and Thousand Lake Mountains west of the park and were transported by rockslides, glacial action, and floods of glacial meltwater. The Fremont people used some of the boulders for the foundations of their pit-house dwellings.

Patterns of Stream Erosion

Stream erosion has been significant in the shaping of the Capitol Reef even though in today's arid climate most of the streams are intermittent. The Fremont River and a few others are permanent streams. Sulphur Creek enters the park from the west, cuts through the Goosenecks, and joins the Fremont east of the Park Headquarters and Visitor Center. (The Goosenecks are an example of "incised meanders," which are explained in Chapter 5.) Another year-round stream, Pleasant Creek, which flows northeast, crosses the park about ten miles south of the Fremont. The only other permanent stream in the park is Oak Creek, which crosses the Waterpocket Fold in the area south of the domes.

It is interesting to note the effect of structure on the stream patterns of the area. All the present year-round streams had enough erosive power to cut across the monocline from west to east while it was being uplifted. However, some canyons, such as Capitol Gorge, which were carved out by east-flowing streams, are now dry or intermittent. These streams lost out to more favorably located drainages.

FIGURE 4.7 In dry seasons, potholes are the main source of water. Abrasive action of the sediments in the streambed enlarge depressions in the bedrock to form these "rock tanks." © Robert Fletcher.

Several intermittent streams follow the trend or strike of the monoclinal structure. (The *strike* is the direction, or trend, of the bedding. In this case, it's generally north-south.) A tributary that follows the strike of the beds has adjusted its course to the underlying rock structure. For example, Hall Creek in the southern part of the park flows south, parallel to the east flank of the Water Pocket Fold. The Hall follows a course of less resistance along the base of Navajo Sandstone cliffs, cutting down its channel in softer beds of the Carmel Formation and the Entrada Sandstone. However, just before Hall Creek crosses the park's southern boundary, it cuts the Narrows, a three-mile defile through Navajo Sandstone, and then empties into the Colorado River. In a sense, Hall Creek "let itself down" in order to reach the Colorado's lower level.

The Fremont River flows east *across the strike* of the beds, following the regional slope toward the Colorado River. The regional slope coincides approximately with the tilt, or dip, of the beds. (*Dip* is defined as the vertical angle that beds form with a horizontal surface.) The Fremont River is a swift-flowing stream that rampages through its valley when augmented by even a moderate increase in volume. A drop of roughly 2000 feet between the west side of the monocline (near Torrey) and the junction with the Dirty Devil (near Hanksville) gives the Fremont River enough gradient to generate considerable erosive power. The Mormon farmers finally gave up their orchards and fields in the Fremont valley after years of trying to maintain dams and irrigation ditches on an uncontrollable river that alternated between extremes of too little and too much flow.

Alluvial fans have been built up where streams loaded with sediment come out of steep canyons onto an open, flat floor, basin, or valley. The sudden loss of velocity causes the streams to unload as water spreads out and dissipates.

Sudden flash floods that come with little warning are dangerous to persons who may become trapped in narrow, steep-walled gorges and canyons. Grand Wash and Capitol Gorge are well known examples of areas subject to flash flooding. Park visitors must be especially cautious if rain threatens.

Geologic History

1. **Deposition of the Cutler Formation and Kaibab Limestone during Permian time.**

Toward the end of the Paleozoic Era, the Cutler Formation and the Kaibab Limestone were deposited along the margin of a shallow, transgressing sea. Only the White Rim Sandstone, the youngest member of the Cutler Formation, is exposed in Capitol Reef National Park. The White Rim, a cross-bedded dune deposit, crops out in the bottom of Sulphur Creek and in the center of the Circle Cliffs outside the western edge of the park. However, the White Rim Sandstone and the other members of the Cutler Formation are prominently displayed in Canyonlands National Park (Chapter 5), which is roughly 60 miles in a direct line to the east. The Kaibab Limestone is well known as the rim of Grand Canyon (Chapter 1). In the Capitol Reef area, where the Kaibab Limestone is exposed in the deeper parts of some canyons, the formation consists of dolomite with interbedded sandstone. In dolomitic rocks, magnesium has replaced much of the calcium of the original limestone.

2. **Deposition of the Triassic Moenkopi and Chinle Formations.**

Mesozoic rocks (about 10,000 feet of sedimentary layers) dominate the landscape in Capitol Reef

| Table 4.1 | Geologic Column, Capitol Reef National Park |

Time Units		Rock Units			Geologic Events
Era	Period	Group	Formation	Member	Geologic Events
Cenozoic	Quaternary	Sands, gravel, glacial erratics, boulders			Mass wasting, stream erosion Canyon cutting Glacial climate
Cenozoic	Tertiary	Dikes, sills, mostly basalt Volcanic plugs, lavas			Extrusive and intrusive activity Uplift and deep fluvial erosion
Mesozoic	Cretaceous	Mesaverde			Waterpocket Fold formed during Laramide orogeny
Mesozoic	Cretaceous		Mancos Shale		Sandy shales deposited in oscillating sea
Mesozoic	Cretaceous		Dakota Sandstone		Nonmarine, then shore deposits in transgressing sea Erosion
Mesozoic	Jurassic		Morrison	Brushy Basin	Gray-green and maroon clays, dinosaur bones in flood-plain deposits
Mesozoic	Jurassic		Morrison	Saltwash	Conglomerates, siltstones; contains uranium Erosion
Mesozoic	Jurassic	San Rafael	Summerville		Sand dunes; shales on tidal flats and in shallow lakes
Mesozoic	Jurassic	San Rafael	Curtis		Green shales, tidal-flat deposits Erosion
Mesozoic	Jurassic	San Rafael	Entrada Sandstone		Reddish brown sandstone, shallow-water deposits
Mesozoic	Jurassic	San Rafael	Carmel		Shales, limestone, gypsum deposited by advancing and retreating seas Erosion
Mesozoic	?	Glen Canyon	Navajo Sandstone		Accumulation of white, well-rounded sand grains in cross-bedded dunes; cliff-former
Mesozoic	Triassic	Glen Canyon	Kayenta		Fluvial, interbedded red shale and sandstone; slope-former
Mesozoic	Triassic	Glen Canyon	Wingate Sandstone		Bright-red sandstone, cross-bedded dunes; cliff-former Erosion
Mesozoic	Triassic		Chinle		Shale, sandstone, siltstone; slope-former; contains petrified wood
Mesozoic	Triassic		Chinle	Shinarump	Conglomerate; contains uranium Erosion
Mesozoic	Triassic		Moenkopi		Brownish-red shales, gypsum, sandstones in tidal-flat deposits; "ripple rock"
Paleozoic	Permian		Kaibab Limestone		Dolomitic limestone, siltstone
Paleozoic	Permian		Cutler	White Rim Sandstone	Marine sandstone
		Older rocks not exposed in park			

Sources: Baars, 1985; Billinglsey, Breed, Huntoon, 1981; Collier, 1987; Stokes, 1986.

National Park. First laid down were the reddish brown, Triassic Moenkopi beds of siltstone. Many of them bear ripple marks that indicate their deposition by streams flowing toward a marine embayment to the north.

After uplift, erosion of the Moenkopi Formation produced an unconformity, on which discontinuous beds of basal conglomerate of the Shinarump Member of the Chinle Formation were deposited. Uranium salts that accumulated in this sedimentary unit were taken out as ore from the Oyler Mine in Grand Wash. The first claim was filed in 1904, many years before the uranium boom of the 1950s.

Sediments accumulating in the Chinle beds above the Shinarump tended to be finer as the rivers slowed and deposited irregular layers of sand and mud over the subsiding basin. Logs washed down, were buried, and eventually petrified, probably aided in this process by volcanic ash blown in from distant eruptions. The soft,

red and pink shales of the Chinle Formation tend to weather into rounded hills and slopes.

3. **Accumulation of the Glen Canyon Group of sandstones in late Triassic and early Jurassic time.**

The most visible and prominent rocks in Capitol Reef National Park are the three formations of the Glen Canyon Group that make up the spine of the Water Pocket Fold, namely, the Wingate Sandstone, the Kayenta Formation, and the Navajo Sandstone. These three formations, especially the Navajo, "star" in the majority of well known landmarks of the Colorado Plateaus.

The Triassic Period ended in a time of increasing aridity. Migrating sand dunes on the shore of an ancient sea advanced over the Chinle surface and buried it with 350 feet of a cross-bedded, orange sandstone that became the cliff-forming Wingate Sandstone.

FIGURE 4.8   Dipping beds at the north end of the Water Pocket Fold, which is a large monoclinal fold; that is, a fold that bends or flexes from the horizontal in only one direction and has only one limb. Photo by Ronald F. Hacker.

As the Triassic Period came to a close and the Jurassic Period began, sluggish streams deposited the limey, thin-bedded sandstones of the Kayenta Formation in channels and on low-lying plains. At places where dinosaurs and tritylodonts (crocodile-like animals) came to drink on the banks of streams, their footprints have been preserved in the rock. The Kayenta tends to form ledgy slopes where it crops out between the massive Wingate Sandstone below and the Navajo Sandstone above.

Again, desert sands overspread the land, encompassing a vast area for millions of years. Migrating dunes accumulated to a thickness of 1000 feet. These remarkably clean cross-bedded sands became the white Navajo Sandstone that is famous for its beautiful canyons, gleaming domes, smooth, unbroken cliffs, and striking monoliths. Examples are the Golden Throne in Capitol Reef National Park and the Great White Throne in Zion National Park (Chapter 2). The cross-bedding of the Navajo Sandstone causes it to weather into curving canyons and rounded domes.

4. Deposition of the San Rafael Group unconformably over the Glen Canyon Group.

The four formations of the San Rafael Group—Carmel Formation, Entrada Sandstone, Curtis Formation, and Summerville Formation—are exposed on the dipping east side of the Waterpocket Fold. Around the middle of Jurassic time, after an erosional period had planed off the Navajo dunes, the limey silts, sands, and gypsum beds of the Carmel Formation accumulated in what may have been a down-faulted basin that was periodically invaded by seawater. At the north end of the park, in Cathedral Valley, some of the Carmel gypsum dissolved, causing a surface area to collapse, which formed a sinkhole. A more resistant part of the Carmel Formation caps the Golden Throne, located south of Grand Wash.

The environment in which the overlying Entrada Sandstone accumulated consisted of tidal flats, low shores, offshore bars and barrier islands. The reddish brown color of the Entrada Sandstone is the same here as in Arches National Park (Chapter 6), but since the joint systems in the rock are different, arches did not form in the Capitol Reef exposures of the Entrada. The oddly shaped monuments and "goblins" (hoodoos) that stand in Cathedral Valley are typical of the weathering of Entrada Sandstone in this park.

The Curtis Formation, which is light gray and only 50 feet thick, is resistant to erosion and has a basal conglomerate overlain by sandstone and shale beds. The presence of glauconite, a green iron potassium silicate mineral, indicates that the beds were deposited in a

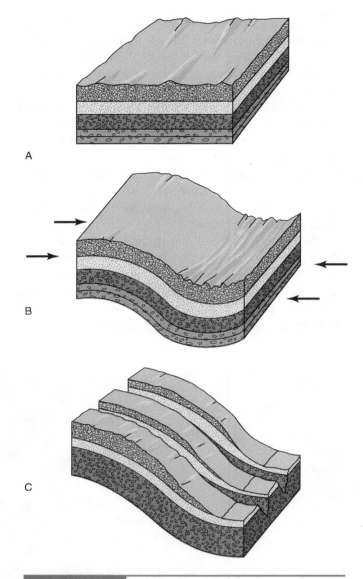

A

B

C

FIGURE 4.9   How the Water Pocket Fold developed. *A.* Sandstones, shales, and limestones, both marine and nonmarine, accumulated for millions of years in flat-lying beds. *B.* Compressive forces, generated during the Laramide orogeny, produced a great monocline, the Water Pocket Fold. *C.* As the region was uplifted, streams began to dissect the monocline, cutting narrow canyons through the Navajo Sandstone and wider canyons in the Chinle Formation. From M. Collier, *The Geology of Capitol Reef National Park.* © 1987 Capitol Reef Natural History Association. Used by permission.

shallow sea. The formation crops out only in the northern section of the park, where numerous features in Cathedral Valley are capped by Curtis bedrock.

As the Curtis sea retreated, the reddish brown mudstones and white sandstones of the Summerville

Formation accumulated on tidal flats. The Summerville tends to form ledgy cliffs.

5. Deposition of the nonmarine Morrison Formation, bringing the Jurassic Period to a close.

The Morrison Formation, overlying the San Rafael Group, consists of stream-deposited sandstones and mudstones laid down in lakes on swampy plains. The lower Salt Wash Member was the source of most of the uranium that was mined in the park area in the 1950s. The upper Brushy Basin Member is best known for its dinosaur remains. Because the bones are scattered, however, the dinosaurs have been difficult to assemble and identify.

6. Cretaceous deposition of Dakota Sandstone and Mancos Shale.

The lower units of the Dakota Sandstone are nonmarine and resemble the Jurassic beds on which they were deposited; but eventually, a vast sea overspread the Colorado Plateaus and much of the region beyond. Exposures of Dakota Sandstone, some containing fossil oyster shells, are found in the southern section of the park.

Units of Mancos Shale were deposited during times of advancing and retreating shorelines. Within the park's southern section, some of the mesas and buttes are eroded from Mancos Shale beds. A few mesas of Mancos Shale on the park's eastern boundary are capped by beds of the more resistant Mesaverde Group, the youngest sedimentary rocks in the park. East of the park, some soft Mancos units have formed badlands.

7. Folding and uplifting of the monocline at the end of Cretaceous time.

Like the other monoclines of the Colorado Plateaus, the Waterpocket Fold has a north-south alignment with the steep side facing east. The structure is the result of the intense compression exerted on the Colorado Plateaus during the Laramide orogeny. Perhaps the fold formed here because of a weak zone caused by a basement fault in Precambrian rocks far below the surface. Minor earthquakes in 1979 that centered below the Water Pocket Fold may indicate shifting of a basement fault.

The uplifting of the Capitol Reef region was part of the overall elevation of the Colorado Plateaus as the land was raised several thousand feet from being near sea level at the beginning of Cretaceous time.

8. Tertiary igneous activity.

The compression, uplifting, and fracturing that accompanied the Laramide orogeny were followed by relatively gentle stretching and adjusting of the crust. Magma began to rise and pushed into weakened crustal

FIGURE 4.10 Petroglyphs chipped out of the desert vanish by the Fremont Indians cover the sandstone. The desert varnish is a dark pantina of iron and manganese oxides, with some silica left behind as the water is drawn to the surface and evaporates. Photo by Ronald F. Hacker.

FIGURE 4.11   Hickman Bridge is a true natural bridge that formed over a watercourse as weathering and stream erosion wore a hole through a narrow "fin" of sandstone in the Kayenta Formation. Photo by David B. Hacker.

zones. In the South Desert and Cathedral Valley, dikes and sills were intruded along joints and bedding planes in the sedimentary beds. Some magma reached the surface and erupted as basaltic lava through fissures. Along with dikes and sills, volcanic plugs have been stripped of their sedimentary covers and now stand exposed above the desert sands. The igneous activity probably began in Oligocene time and continued sporadically. A relatively young dike that cuts the Curtis Formation has been dated at just under 4 million years B.P. or Pliocene time (Collier, 1987).

9. **Development of the Fremont River and other Colorado River tributaries; effects of Pleistocene glaciation.**

The uplifting of the Water Pocket Fold caused rearrangement of the drainage in the area. While the Fremont River and several other streams continued to carve their courses across the fold during uplift,

streams such as Sandy Creek and Halls Creek began to flow parallel to the strike, or trend, of the monocline, downcutting their channels along less resistant beds. Some streams carved steep-sided canyons and later abandoned these courses. The Fremont River and its tributaries continue to shape the landscape in the park, but the processes of weathering and mass wasting play a significant part as well.

Erosional processes were more intensive during the Pleistocene Epoch and during moister climatic cycles of the Holocene. Pleistocene mountain glaciers (outside the park area) and the accompanying cooler, wetter climate kept streams well supplied with water. Glacial ice plucked basaltic boulders from Boulder Mountain; meltwater streams washed them downslope and distributed them over much of the park area.

The present landscape is remarkably varied with dry canyons, plateaus, mesas, buttes, domed cliffs, arches, and colorful rocks—a place of magnificent scenery and few people.

Geologic Map and Cross Section

Billingsley, G.H., Breed, W.J., and Huntoon, P.W. 1987. Geologic Map of Capitol Reef National Park and Vicinity, Utah (4 sheets). Utah Geological Survey.

Billingsley, G.H., and Breed, W.J. 1985. Geologic Cross Section from Cedar Breaks National Monument through Bryce Canyon National Park to Escalante, Capitol Reef National Park, and Canyonlands National Park, Utah. Capitol Reef Natural History Association.

Sources

Baars, D.L. 1983. *The Colorado Plateau, a Geologic History.* Albuquerque: University of New Mexico Press.

Collier, M. 1987. *The Geology of Capitol Reef National Park.* Capitol Reef Natural History Association.

Davidson, E.S. 1967. *Geology of the Circle Cliffs Area, Garfield and Kane Counties, Utah.* U.S. Geological Survey Bulletin 1229.

Dutton, C.E. 1880. *Geology of the High Plateaus of Utah.* U.S. Geographical and Geological Survey of the Rocky Mountain Region.

Olson, V.J., and Olson H. 1979. *Capitol Reef—The Story Behind the Scenery.* Las Vegas, Nevada: KC Publications.

Rigby, J.K. 1976. *Northern Colorado Plateau,* K/H Geology Field Guide Series. Dubuque, Iowa: Kendall/Hunt Publishing.

Stokes, W.L. 1986. *Geology of Utah.* Salt Lake City: Utah Museum of Natural History and Utah Geological and Mineral Survey. p. 111–112.

C H A P T E R   5

Canyonlands National Park

S O U T H E A S T   U T A H

Area: 337,570 acres; 559 square miles

Established as a National Park: September 12, 1964

Address: 2282 S. West Resource Blvd., Moab, UT 84532-3298

Phone: 435-719-2313

E-mail: canyinfo@nps.gov

Website: www.nps.gov/cany/index.htm

Utah's largest national park is a geological wonderland of spires and mesas rising to more than 7800 feet. A large triangular plateau, called Island in the Sky, stands high above the confluence of the Green and Colorado Rivers. The two rivers come together through deep, sheer-walled, meandering canyons; below their confluence is Cataract Canyon, a 14-mile-long stretch of white-water rapids. The Maze, an almost inaccessible jumble of canyons, lies west of the rivers; beyond are the red towers and walls of the Land of Standing Rocks. An eroded salt dome, Upheaval Dome, exposes rings of colorful Mesozoic rocks. On the Colorado River's southeast side is the Needles district, a natural exhibit area of arches, fins, spires, grabens, canyons, potholes, and Indian ruins.

FIGURE 5.1   Angel Arch, one of several arches in Canyonlands National Park, is located in the southeast section. The arch is part of the Cutler Formation of Permian age. Photo by Edward Mooney.

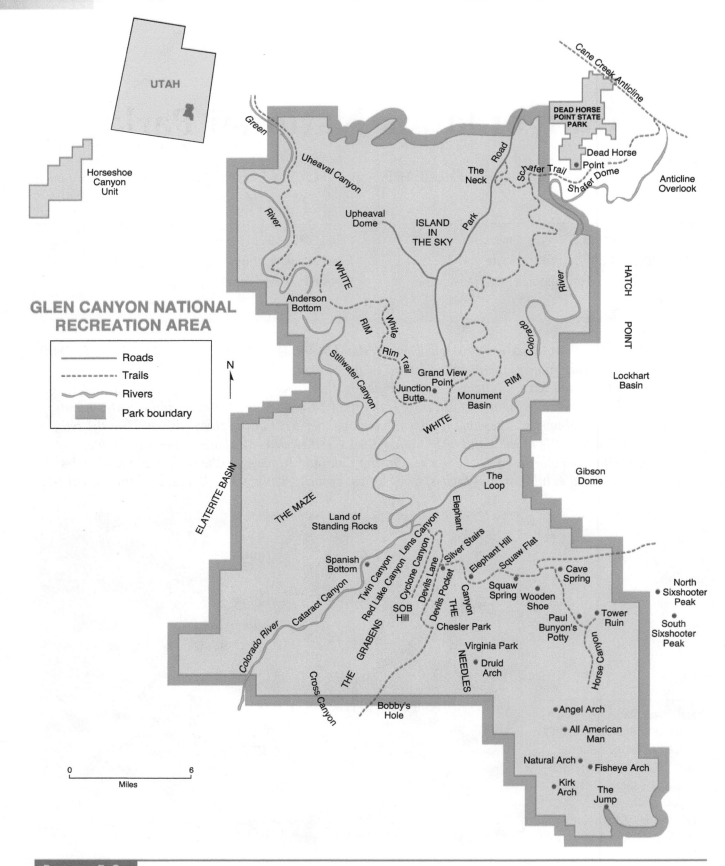

FIGURE 5.2   Canyonlands National Park, Utah.

Canyonlands National Park is a wild and primitive desert region between Capitol Reef and Arches National Parks in the heart of the Colorado Plateaus. The Green and Colorado Rivers, which join in the park, and much of the surrounding wilderness, are little changed from when Major John W. Powell first explored the canyons in 1869. He entered what is now the park from the north, through the Green River's Labyrinth Canyon. He described the landscape between Labyrinth Canyon and the confluence of the Green and Colorado Rivers as

> . . . naked rock with giant forms carved on it, cathedral-shaped buttes towering thousands of feet, cliffs that cannot be scaled, and canyon walls that make the river shrink into significance, with vast hollow domes, tall pinnacles, and shafts set on the verge overhead, and all the rocks, tinted with buff, gray, red, brown, and chocolate, never lichened, never moss-covered, but bare, and sometimes even polished. (Powell, 1981 reprint.)

After resting several days at the confluence in order to dry out gear and provisions, the intrepid explorers started out again through a stretch of white water that Powell named "Cataract Canyon," because of "bad rapids in close succession." (In parts of the canyon, huge blocks of rock that have fallen from the cliffs impede the water, producing turbulence.) One of Powell's boats swamped, three oars were lost, and all the boats were damaged and leaked badly from banging against rocks. After a narrow escape from being trapped in the gorge near the foot of Cataract Canyon by a flash flood, the party emerged safely in an open area of the river beside deserted Indian dwellings.

Being as interested in archaeology as geology, Powell carefully described the ruins, flint chips, arrowheads, broken pottery, and what he called "etchings" (pictographs) on the cliffs. He was excited to find stairsteps cut into a cliff leading up to a watchtower. "I stood," he wrote, "where a lost people had lived centuries ago, and looked over the same strange country."

FIGURE 5.3   The confluence of the Colorado and Green Rivers, where Major Powell and his men camped in 1869. In this view, looking downstream (south), the Green River comes from the right. The mesa top in the right foreground is the south end of the Island in the Sky. Photo by Myron Agnew.

Powell's "lost people" may have been Indians of the Fremont Culture, who lived in this area until A.D. 900. Or they could have been a later group, the Anasazis, who arrived around A.D. 1075 and left during the twelfth century. The pictographs and camp sites of these early people can be seen in numerous places in the park. The Indians made pictures on the rock by scratching through the coating of desert varnish to the lighter-colored, unweathered rock beneath. *Desert varnish* is a thin dark film of iron and manganese oxides that forms on the surface of rocks.

Since this is desert country (only 5 to 9 inches of precipitation per year), water is precious and water holes are few and far between. Settlers were not attracted to the region, but in the late 1800s and early 1900s, cowboys and sheepherders passed through the area from time to time. Butch Cassidy and his "Wild Bunch" knew the region well and used their knowledge to evade posses of lawmen. The area was ideal for outlaws and cattle rustlers who knew the location of water holes and could come and go without getting lost in the jumble of canyons, cliffs, and rough terrain.

Canyonlands National Park was established in 1964 in order to preserve its outstanding scenic, scientific, and archeological resources. As the Colorado River leaves Cataract Canyon and crosses the park's southern boundary, the river flows into the head of Lake Powell at the northern end of Glen Canyon National Recreational Area. A strip of Glen Canyon National Recreational Area also extends along the western boundary of Canyonlands National Park (fig. 5.2).

Geologic Features

Island in the Sky

Island in the Sky is the best observation tower from which to gain perspective on the geological processes that have created the "architecture" of Canyonlands. This broad, level, roughly triangular mesa is wedged between the Green and Colorado Rivers and stands over 2000 feet above the two streams that come together at the foot of the mesa point. Views from the overlooks encompass canyon after canyon, stretching off to mountains ringing the horizon a hundred miles distant. Within the park, the Maze is to the west and the Needles is to the south. Closer to the mesa edge, about 1200 feet down, is the White Rim Sandstone bench, the habitat of desert bighorn sheep. The rivers are another 1000 feet down (fig. 5.3).

Red sandstones are the dominant rock of the Canyonlands landscape; but a rich variety of sedimentary rocks in hues ranging from near-white through orange, brown, or near-black provides contrast and diversity to individual landforms, as well as the overall panorama. The sandstones consist mainly of uniformly sized, clear quartz grains. The red rocks are cemented with iron oxide, which gives them their color. During and after the uplifting of these sedimentary rocks, differential weathering along joints and erosion by wind and water dissected the flat layers into hundreds of brightly colored canyons, mesas, buttes, fins, arches, spires, and thousands of smaller features.

Upheaval Dome

Upheaval Dome (fig. 5.4), located in the northwest corner of Island in the Sky, is a circular feature about three miles across. From the air it looks like a bull's-eye. The view from the top of Whale Rock (elevation 5820 feet), a smooth, whale-shaped monolith of Navajo Sandstone, gives an idea of Upheaval Dome's enormous size. From the Wingate Sandstone rim (5700 feet), you can look down into Upheaval Dome's core more than 1000 feet below. Concentric rings of sedimentary beds are exposed between the core and the rim. The more resistant layers jut up in ridges and the less resistant layers form valleys. Upheaval Canyon breaches the western wall and provides a flume for runoff to carry away erosional debris. The canyon drains to the Green River five miles downslope (fig. 5.5 A, B).

Geophysical studies indicate that a *salt dome* forcing its way through sedimentary beds pushed up Upheaval Dome. The salt is from the Paradox Formation (a unit of the Hermosa Group) that was deposited during the Pennsylvanian Period. The Paradox salt beds evaporated from the brine of a landlocked ancient sea, trapped by a rising mountain block (the Uncompahgre Uplift). In this particular location, a rise in the granite floor deep below the sea bottom caused a slight thinning of salt layers above the crest, creating a weak zone.

Eventually, many feet of sediment washed down from highlands, filled the basin, and covered the salt beds. The pressure of about a mile of sediment weighing down the salt caused it to become *plastic* (i.e., able to flow without rupture). Salt flowing upward over the hump as a huge plug (not unlike toothpaste oozing from a tube) arched the overlying beds into a dome, or circular anticline. As the beds were pushed up almost 1000 feet and turned almost on edge, they became

FIGURE 5.4   The intrusion and subsequent collapse of a salt dome formed a view of Upheaval Dome in the northwestern part of Canyonlands National Park. The feature is about three miles across and almost 1600 feet deep. Photo by David B. Hacker.

cracked and severely deformed. This made them more susceptible to weathering and erosion and the top beds were stripped off. Since the beds exposed in the center of Upheaval Dome are softer than those forming the rims, the less resistant rock eroded more rapidly, creating the depression that resembles an open-pit mine.

The Maze, the Land of Standing Rocks, and the Fins

West of the Colorado and Green Rivers is the Maze, the wildest section of Canyonlands, described in the park brochure as a "30-square-mile puzzle in sandstone." Outstanding examples of prehistoric rock art are hidden in some of the deepest canyons. Beyond the jumbled canyons of the Maze is the Land of Standing Rocks, where weirdly shaped pinnacles, buttes, and mesas display sandstone layers.

The Fins are narrow sandstone ridges, or walls, that are eroding along parallel joints. Cracks between closely spaced vertical joints become enlarged by chemical and physical weathering as water seeps down, dissolving iron oxide cement and loosening sand grains. Frost action, with its alternate freezing, expanding, and thawing, also breaks up the sandstone. Loose sediment is removed by wind and runoff. The joint spaces gradual-

ly enlarge into narrow valleys that separate the narrow walls, called *fins*.

The Needles, Arches, and Grabens

The east side of the Colorado River is an area of great diversity. The Needles themselves are similar to fins, but have more crisscross fractures. Needles, like fins, are the result of weathering and erosion along joint lines and fault lines, some of which are quite close together, a feature characteristic of the Cedar Mesa Sandstone. The Joint Trail, which goes through a crack in this formation, is only a couple of feet wide. The distinctive red-and-white banding in the Needles (and elsewhere in the park) is due to the interfingering of two units of the Cutler Group: the red Organ Rock Shale and the white Cedar Mesa Sandstone.

The *needles* and *pillars* develop when two intersecting joint sets are spaced closely together. The second set of joints, which is at an angle to the first set, prevents fins from developing. Instead, pointed needles or rounded pillars evolve as weathering and erosion sculpture the rock. Some pillars are capped by resistant material that makes balancing rocks. Where fins have developed, they tend to evolve into a line of pillars as weathering attacks cross joints.

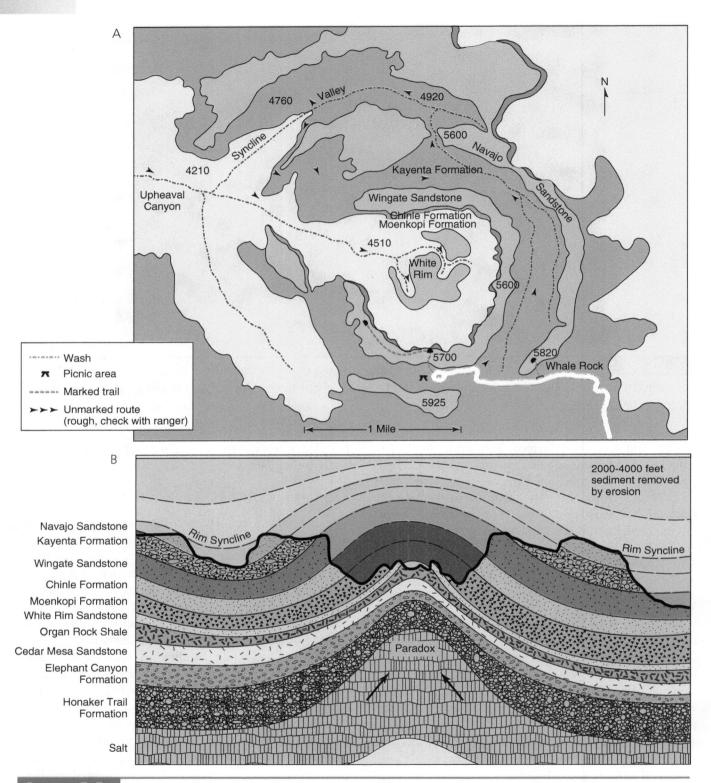

A

N

Valley

4760   4920

Syncline   5600

Navajo

4210   Kayenta Formation

Upheaval   Wingate Sandstone
Canyon   Chinle Formation   Sandstone
               Moenkopi Formation

               4510

               White
               Rim   5600

               5700   5820
                                  Whale Rock

               5925

Legend:
- Wash
- π Picnic area
- Marked trail
- ►►► Unmarked route (rough, check with ranger)

|← 1 Mile →|

B   2000–4000 feet
          sediment removed
          by erosion

Navajo Sandstone
Kayenta Formation   Rim Syncline   Rim Syncline

Wingate Sandstone

Chinle Formation

Moenkopi Formation
White Rim Sandstone

Organ Rock Shale

Cedar Mesa Sandstone   Paradox

Elephant Canyon
Formation

Honaker Trail
Formation

Salt

FIGURE 5.5   *A.* Upheaval Dome in map view. The topography is directly related to the rock formations, with the more resistant formations, such as the Wingate Sandstone, forming concentric walls. *B.* Cross section of Upheaval Dome, showing the salt dome at the core and the sedimentary layers that were arched up by pressure of the rising salt plug. Adapted from "Upheaval Dome," National Park Service and Canyonlands Natural History Association, 1981.

FIGURE 5.6   The Maze, located west of the Green and Colorado Rivers, is the wildest section. The park brochure describes it as a "78 square kilometer (30 square-mile) puzzle in sandstone." Uplift at the same rate as downcutting has allowed the rivers to maintain the dendritic stream pattern producing the Maze. © Robert Fletcher.

Natural arches, some of the distinctive landmarks in the Needles district, may form from fins. As the joint spaces widen and deepen, beds at the base of the fins are exposed. If the lower beds are slightly softer, they are removed more rapidly than the top beds. Because the fins are narrow, a hole, or *window,* appears where the rock is worn completely through. Sand blown by wind (sandblasting) polishes and rounds off the window edges. Snow, rain, and ice continue to attack the rock, dislodge fragments, and cause rockfalls (a process of mass wasting) until a window is enlarged to form a natural arch.

Because of their composition, type of cement, and structure, certain rock formations have a tendency to form arches. Canyonlands has 25 arches, mostly sculpted in Permian Cutler Group sandstones, which are more likely to form standing rocks than arches. In nearby Arches National Park (Chapter 6), more than 300 arches have developed in Jurassic Entrada Sandstone.

"Potholes" form in Cedar Mesa Sandstone, a member of the Cutler Group, because the calcium carbonate that cements its sand grains is soluble in natural carbonic acid (carbon dioxide plus water). Carbonic acid attacks any weak spot or zone in the sandstone. Loosened sand grains fall, are washed, or blown away. Once a slight depression is started, more and more of the sandstone is exposed to the processes of weathering and erosion. Potholes, caves, and sometimes arches develop in this way. Paul Bunyan's Potty is a well known example.

The name Chesler Park was given to an expanse of a thousand acres of flat, grassy land surrounded by pinnacles of the Cedar Mesa Sandstone. Chesler Park is east of the Grabens.

The Grabens is an area of elongated collapsed troughs parallel to Cataract Canyon. As was the case in Upheaval Dome, the underlying salt deposits of the Paradox Formation acted as an unstable platform for the millions of tons of overlying shales and sandstones. Great thicknesses of salt accumulated in subsiding, elongated valleys parallel to the rising, ranges of the Uncompahgre Uplift. As the weight of sedimentary overburden (washed down from the mountains) became ever greater, the salt flowed as a plastic mass toward the southwest, away from the source and greatest weight of sediment. When the flowing salt came against the ridges, it was deflected upward, forming salt walls and plugs that penetrated the overlying beds. In this way, the salt took the path of least resistance to reach areas where the pressure of overburden was lower.

When the Colorado Plateaus region was uplifted millions of years later, enough of the sedimentary cover was stripped off so that ground water seeped into the salt, dissolving some of the upper parts and causing the overlying sandstones and shales to collapse into the salt-valley grabens.

The Rivers

The Colorado River flows south through Canyonlands in a spectacular, deep, narrow-walled gorge. The Colorado is joined in the park by its largest tributary, the Green River, which also has a deep, narrow canyon (fig. 5.3). Together they flow quietly for about three miles until they enter Cataract Canyon at Spanish Bottom. The turbulent water in Cataract Canyon, which rushes through at a rate of around 40 million gallons per minute, is constantly boiling up as "white water" because of changes in the bedrock of the stream channel, slumping of the lower beds in canyon walls, and blocks of rock falling from the cliffs into the stream.

The Colorado River appears to follow a course through Cataract Canyon along the edge of the block-faulted area. The large faults on the east wall of the canyon do not appear on the west wall—nor do the intricate fracture patterns in the rocks on the southeast side of the Colorado, such as the lines of needles and their intervening, north-south, "racetrack" valleys, cross the river. Presumably, the path of salt-dissolving ground water terminates where it is diverted into the Colorado River drainage.

Both the Green River and the Colorado River have abrasive bedloads, a significant factor in their ability to cut down their channels to form *incised meanders* (fig. 5.7). These two rivers may have meandered back and forth on a fairly low, flat surface in the geologic past, and may have retained their meandering pattern as the region was raised. However, some geologists believe that during uplift, as base level was lowered, the streams began downcutting their flood plains so rapidly by headward erosion that they could do very little lateral eroding. In soft alluvial sediments, meanders migrate over a flood plain, but when meanders are trenched into resistant rock layers, as in Canyonlands, the meandering pattern becomes "trapped." The stream's energy, supplied

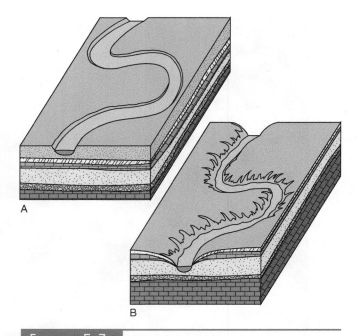

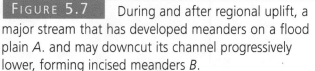

During and after regional uplift, a major stream that has developed meanders on a flood plain *A.* and may downcut its channel progressively lower, forming incised meanders *B.*

by increased velocity, cuts its channel ever deeper but cannot widen it unless it encounters a zone of weak or easily erodible bedrock, as at Spanish Bottom, where gypsum beds have been eroded.

The water in the Green and Colorado Rivers is not potable because of the saline ground water that seeps into the rivers from the salt beds. People and animals in the park use water in potholes or rock depressions in the White Rim Sandstone and Cedar Mesa Sandstone, or depend on spring water that seeps from canyon walls along contacts between the Organ Rock Shale and the White Rim Sandstone. The hollows where water collects—locally referred to as "tanks" or "charcos"—are formed by the combined process of solution of calcium carbonate cement and removal of loose rock grains by wind (deflation). Tanks are found by Junction Butte near Grand View Point in the central portion of the park and also in the Needles district near Chesler Park.

Geologic History

1. **The development in Pennsylvanian time of the Paradox Basin and the Uncompahgre Uplift.**

The oldest rocks exposed in Canyonlands National Park are beds of the Paradox Formation of Pennsylvanian age in the bottom of Cataract Canyon. These exposures are of rock gypsum and interbedded black shale that are less soluble than the halite salt beds underneath Upheaval Dome and the Needles section. A gypsum plug, similar to a salt plug but more resistant, has been exposed by erosion at the mouth of Lower Red Lake Canyon, near the head of Cataract Canyon. The plug was pushed up into the overlying Honaker Trail Formation, the uppermost unit deposited in Pennsylvanian time. Both formations are units of the Hermosa Group.

The thousands of feet of evaporites that make up the Paradox Formation began to accumulate in Middle Pennsylvanian time when blockfaulting caused the Paradox Basin to subside and the Uncompahgre ranges to be uplifted.

Water from a shallow sea that had covered the area early in the Pennsylvanian Period became landlocked in the sinking basin, and a system of interior drainage developed as mountains rose along the margin. In the *closed basin,* water escaped only by evaporation; and since the climate was hot and dry, water evaporated rapidly, concentrating salts and mineral matter in shallow lagoons.

When the salinity of a brine reaches a ratio of 10 parts salt per 1000 parts water (3 times the ratio of salt in seawater), the minerals anhydrite and gypsum begin to precipitate as crystals. If evaporation continues and no water is added to dilute the brine, the ratio of salt to water may reach 35 parts per 1000 parts water, or 10 times that of normal seawater. Under these conditions, halite (rock salt) precipitates. The heavier salt crystals tend to sink, accumulating over the bottom of the lagoon. This is what happened in the Paradox Basin.

From time to time, storms washed sediment down from the mountains. The coarser gravels accumulated in wedges along the base of the range, and the silts and muds spread over the salt flats. Occasionally fresh seawater breached the lagoons and moved into the basin over a shallow threshold in the margin. The salty brines were not flushed out because the fresh seawater acted as a barrier that prevented the escape of the dense bottom water. Thin black shale layers between some salt beds mark these incursions of seawater. For thousands of years, the layers of salt and clastic sediment accumulated, becoming ever thicker—up to 5000 feet in some places. The gypsum layers in the bottom of Cataract Canyon represent the last phases of the evaporite deposits.

Presumably, the weight of the accumulating overburden caused the salt beds to begin flowing plastically before the close of the Pennsylvanian Period, and this process continued throughout the Permian and well into Mesozoic time. The migration of the salt beds had probably stopped by the end of the Jurassic Period.

Toward the end of Pennsylvanian time, a warm tropical sea spread over the region. About 1500 feet of fossiliferous limestones, shales, and sandstones blanketed the salt basin. These are the gray beds of the Honaker Trail Formation that crop out at lower elevations in the deep canyons of the park, especially along the Colorado River.

2. **Permian deposition of the extensive, varied formations of the Cutler Group.**

In some places the red, terrestrial Permian beds lie unconformably over gray, marine Pennsylvanian beds; but elsewhere the transition is more gradual. Early in Permian time, advancing seas deposited the Halgaito Shale, which grades into the Elephant Canyon Formation, a rock unit deposited in coastal lowlands. In Canyonlands, beds of these formations are exposed in Cataract Canyon, Elephant Canyon, and upriver from the confluence.

All during this time, the Uncompahgre Mountains to the east were being severely eroded. Streams stripped sediment off the slopes and dropped it on enlarging alluvial fans that filled the basin at the foot of the range. The material in the fans lithified to iron-rich, red, arkose sandstones that are mapped as "undivided" rock units of the Cutler Group.

Interfingering with Cutler red beds are white Cedar Mesa Sandstone beds that were deposited as submarine bars and coastal dunes. The zone of facies change between these continental and marine rock units occurs as a four- to five-mile-wide belt across Canyonlands National Park from south of the Needles through the Maze and into the Elaterite Basin along the park's western boundary. Many of Canyonlands' most dramatic and colorful rock sculptures are in this zone where the two types of sedimentation shifted back and forth.

Muds that oxidized and became the *Organ Rock Shale* were laid down over the white sands of the Cedar Mesa. The Organ Rock beds make up brightly colored red, orange, and brown slopes that separate cliff-forming units above and below (fig. 5.8). Some of the most interesting exposures of the Organ Rock Shale are in the Land of Standing Rocks.

The "type locality" of the *White Rim Sandstone* is the topographic bench, mentioned earlier, about 1200 feet down from the surface of Island in the Sky. (A *type locality* is the location from which a stratigraphic unit takes its name.) The White Rim Sandstone that forms striking cliffs between Cutler and Organ Rock red beds

Table 5.1	Geologic Column, Canyonlands National Park			
Time Units		**Rock Units**		
Era	**Period**	**Group**	**Formation**	**Geologic Events**
Cenozoic	Quaternary Tertiary	Alluvium Dune sand Landslide deposits		Arches and needles formed Entrenching of Green and Colorado Rivers Continuing uplift and block-faulting Forming of salt valleys by solution and collapse
Mesozoic	Cretaceous			Uplift and warping due to compression during Laramide orogeny Erosion
Mesozoic	Jurassic	San Rafael	Entrada Sandstone	Scattered patches exposed in park vicinity
Mesozoic			Carmel	Erosion
Mesozoic	- - - - ? - - - -	Glen Canyon	Navajo Sandstone	Cross-bedded, white sands deposited by migrating dunes; forms cliffs, knobs, hummocks
Mesozoic	Triassic		Kayenta	Deposited on flood plains and in lakes; slope-former
Mesozoic			Wingate Sandstone	Cross-bedded sands deposited by migrating dunes; forms red rimrock
Mesozoic			Chinle	Bright-colored shales; slope-former; contains petrified wood Erosion
Mesozoic			Moenkopi	Tidal flat deposits; shales with ripple marks, mud cracks Erosion
Paleozoic	Permian	Cutler	White Rim Sandstone Organ Rock Shale Cedar Mesa Sandstone Elephant Canyon Halgaito Shale	White, cross-bedded sandstone deposited in coastal dunes Reddish brown siltstones, sandy shales and sandstones deposited on oscillating shoreline and interfingered with deposits on beach dunes, in sand bars, and in arkosic alluvial fans (Because formations thin and the facies change from west to east, rocks are mapped as "Undivided Cutler" along the eastern margin) Erosion
Paleozoic	Pennsylvanian	Hermosa	Honaker Trail	Shallow marine deposits
Paleozoic			Paradox	Gypsum and salt evaporites interbedded with shale and black limestone

Source: Huntoon, Billingsley, and Breed, 1982.

FIGURE 5.8   Chesler Park is a natural meadow, walled in by the Needles and the Grabens area. Photo by Ann G. Harris.

includes ancient sand dunes as well as large marine sand bars. Interesting exposures of this rock unit can be examined along the White Rim Trail.

In the Elaterite Basin along the park's western boundary, the White Rim Sandstone displays a "fossil" offshore bar that may have been a barrier island late in Permian time. This structure contains petroleum that seeps out as a dark-brown, tarry substance called "elaterite." The oil probably migrated into the sandstone from a source rock and became trapped by the impermeable red shale beds along the east side of the sandbar. Because the structure has been exhumed by erosion and the gas has escaped, the oil is not recoverable.

3. Mesozoic Moenkopi and Chinle Formations.

An unconformity between Paleozoic and Mesozoic beds represents withdrawal of seas and a lengthy period of erosion when the continent was high. The Triassic units in the Canyonlands area are red beds consisting of shallow-water clastic sediments deposited by streams on flood plains throughout a broad lowland that sloped gently toward a western ocean. The mudstones of the Moenkopi Formation, which accumulated on tidal flats,

are exposed in the northern and western sections of the park. Mudcracks and ripple marks are visible on many of the outcrops.

An unconformity separates Moenkopi beds from brightly colored, overlying Chinle beds, which are mostly slope-forming shales. A geologist describes Chinle beds as showing up on canyon walls in the park like "a brilliant ring of fire" (Baars 1983, p. 167). In some places, pieces of petrified wood from the Petrified Forest Member of the Chinle Formation have weathered out of cliffs and collected around the base.

4. Sandstones of the Glen Canyon Group.

The prominent cliffs of Wingate Sandstone, the cliffs and slopes of the Kayenta Formation, and the cliffs, knobs, and rounded monoliths of the Navajo Sandstone characterize Glen Canyon Group rock units in Canyonlands. Imposing outcrops of these strata encircle the western and northern sections of the park; and the cliffs, especially the red Wingate cliffs, form an effective barrier to travel across many parts of Canyonlands and vicinity. Much of the beauty and uniqueness of the Colorado Plateaus is due to the units of the Glen Canyon Group, which are well exposed in Canyonlands and other national parks and national monuments in the region.

As the Triassic climate became drier, sand dunes migrated across the region, burying the rivers and their flood plains. The dune sediment became the Wingate Sandstone. The Wingate cliffs rise several hundred feet throughout Canyonlands and run for hundreds of miles with very few breaks in their semicircular wall. Separating the Wingate dunes and the windblown desert deposits of the Jurassic Navajo Sandstone are the stream-deposited beds of the Kayenta Formation. Apparently a cycle of somewhat moister climate prevailed between the windy desert environments of the Wingate and Navajo. The Kayenta Formation, consisting of reddish brown to lavender sandstones with interbedded siltstones and shales, forms more gentle, ledgy slopes. The Navajo Sandstone, the top formation of the Glen Canyon Group, has well developed cross-bedding, and forms light-colored (buff to pale orange), steep, rounded cliffs, knobs, and hummocks. Some of the more striking ones resemble beehives, turbans, and temples in shape. Canyon View Point Arch and Millard Canyon Arch are also examples of Navajo Sandstone features.

5. San Rafael Group deposition, ending the Jurassic Period.

Rock units of the San Rafael Group unconformably overlie the Glen Canyon Group in the Canyonlands vicinity, mainly in scattered patches east and west of the park. In this area, the Carmel Formation beds accumulated on mud flats. The Entrada Sandstone, a massive sandstone; forms vertical cliffs and hummocky knobs that can be seen along some of the approaches to Canyonlands National Park.

Erosion stripped off most of the San Rafael beds in this area, along with any Cretaceous rock units that may have overlaid them.

6. The relationship of the Laramide orogeny to the development of Canyonlands.

The Laramide orogeny, which brought the Mesozoic Era to a close and lasted into Tertiary time, elevated the Canyonlands region thousands of feet, but left the sedimentary rock units essentially horizontal. The dip of the beds is so slight that they appear flat. However, the tectonic compressive forces that pushed up the surrounding mountains also produced joint systems and fractures that strongly influenced patterns of landform development in Canyonlands. Erosional

processes, intensified by uplift, tended to follow the joint patterns while dissecting the flat-lying rocks.

The presence of the salt beds of the Paradox Formation, as explained earlier, complicated the fracturing and jointing. Sometime after uplift began, and as erosion stripped off the younger beds, ground water reached the buried salt beds. Some of the upper salt layers were dissolved, but the less soluble gypsum was left in place.

In places where salt near the surface was reduced in volume—as it was under the grabens—the overlying beds collapsed.

7. Quaternary erosion and deposition.

The periods of cold and increased moisture that characterized Pleistocene climates increased the dissection and deepening of the canyons, especially the incising of the Green and Colorado River canyons. The volume of the two rivers was augmented by increases in runoff from glaciers in the Rocky Mountains and other highlands drained by these streams. Some of the dune deposits, landslide deposits, alluvial fans, and other unconsolidated sediments in the park accumulated—or at least began to form—in Pleistocene time. These processes continue today at a somewhat slower rate in the drier climate.

Geologic Map and Cross Section

Huntoon, P.W., Billingsley, G.H., and Breed, W.J. 1982. Geologic Map of Canyonlands National Park and Vicinity, Utah. Moab, Utah: Canyonlands Natural History Association.

Sources

Baars, D.L. 1983. *The Colorado Plateau, a Geologic History*. Albuquerque: University of New Mexico Press. 279 p.

Campbell, J.A. 1987. *Geology of Cataract Canyon and Vicinity*, a field symposium. Guidebook, Four Corners Geological Society, 10th Field Conference.

Huntoon, P.W. 1982. The Meander Anticline, Canyonlands, Utah: An Unloading Structure Resulting from Horizontal Gliding on Salt. *Geological Society of America Bulletin* 93:941–950, October.

Lohman, S.W. 1974. *The Geologic Story of Canyonlands National Park*. U.S. Geological Survey Bulletin 1327.

Powell, J.W. 1981 reprint. *The Canyons of the Colorado*. Originally published in *Scribner's Monthly*, 1875. Golden, Colorado: Outbooks. 64 p.

Rigby, J.K. 1976. *Northern Colorado Plateau*. K/H Geology Field Guide Series. Dubuque, Iowa: Kendall/Hunt Publishing Company.

CHAPTER 6

Arches National Park

SOUTHEAST UTAH

Area: 73,379 acres; 115 square miles
Proclaimed a National Monument: April 12, 1929
Established as a National Park: November 12, 1971
Address: P.O. Box 907, Moab, UT 84532-0907

Phone: 435-719-2299
E-mail: archinfo@nps.gov
Website: www.nps.gov/arch/index.htm

"These are natural arches, holes in the rock, windows in stone, no two alike, as varied in form as dimension. They range in size from holes just big enough to walk through to openings large enough to contain the dome of the Capitol in Washington, D.C. Some resemble jug handles or flying buttresses . . . . In color they shade from off-white through buff, pink, brown, and red, tones that change with the time of day and the moods of the light, the weather, the sky."

Desert Solitaire, Edward Abbey, 1968.

The greatest density of natural arches in the world is in Arches National Park. More than 700 arches, in all stages of development, are carved in the salmon-colored Entrada Sandstone. Joint sets and faults controlling the size, location, and orientation of fins and arches are related to salt beds in the Paradox Formation that forms an unstable base beneath the sandstones.

FIGURE 6.1 Landscape Arch, 291 feet across, is the longest arch in Arches National Park and one of the longest natural arches in the world. Photo by Edward Mooney.

Arches National Park is about 25 miles northeast of Canyonlands National Park (Chapter 5) by road, and approximately the same distance up the Colorado River. A few miles upstream from Canyonlands, the river's enclosing walls become farther apart, the flood plain widens, and the water has less turbulence. Just beyond Moab, the Colorado's course becomes the southern boundary of Arches National Park. The finest panorama of natural arches, fins, windows, pinnacles, and balanced rocks to be seen on the Colorado Plateaus—and probably the finest in the world—is in Arches National Park.

Yet, for all its strange beauty, the lack of water and the barren appearance of the terrain discouraged settlers. The climate of the park area is harsh. Temperatures range from 110°F in summer to –20°F in winter. Precipitation averages five to nine inches per year, and shelter from the strong winds is hard to find.

Prehistoric Indians were attracted to the area around Delicate Arch by the presence of chalcedony, a variety of quartz that can be chipped to make a sharp edge. Probably the Indians spent winters making tools, arrowheads, and spear points at their campsites. They left many pictographs on rock walls throughout the park, and believed that the Great Spirit had made the arches. Some of the early settlers, however, thought that the Indians had handcrafted the arches.

Geologic Features

Since Arches National Park is only a short distance up the Colorado River from Canyonlands National Park, the geology of the two parks is similar in many ways; but significant differences are obvious. Canyonlands and Arches have several of the same sandstones—the Cutler sandstones of Permian age, the Triassic Wingate, and the Jurassic Navajo. Some Cutler red beds produced huge fins and some natural arches in Canyonlands but not in Arches National Park. In Arches most of the formation is missing or covered by younger beds.

Several sandstones with very minor exposures in Canyonlands are present in Arches National Park and vicinity. The Jurassic Entrada Sandstone is the most prominent and the most important because most of the arches are in this rock unit. With all of these sandstones in the same arid climate, why are most of the arches concentrated in Arches National Park? And why are natural bridges not present in Arches?

Arches and Natural Bridges

By definition, a *natural bridge*, which is chiefly the result of erosion by running water, spans a watercourse (that may now be a dry stream bed). A natural *arch,* on the other hand, is formed by physical and chemical weathering plus mass wasting, but stream erosion is not involved. Once a natural bridge is begun, however, it is enlarged and shaped by the same forces that mature natural arches.

Two national monuments in southern Utah, south of Arches and Canyonlands National Parks, protect outstanding examples of natural bridges: **Natural Bridges National Monument,** with three spans in White and Armstrong Canyons, carved by tributaries (now intermittent) of the Colorado River; and **Rainbow Bridge National Monument,** farther downstream and also located in a tributary canyon. These features are formed in massive cross-bedded sandstones. The spans in Natural Bridges National Monument are in Permian Cedar Mesa Sandstone; Rainbow Bridge is in Jurassic Navajo Sandstone.

To find answers to why arches but not natural bridges formed from the Entrada Sandstone in Arches National Park, we look at the processes by which the features are made, the differences in the origin of the sandstones, variations in joint sets and composition, differing reactions to uplift and weathering, and—most important of all—the local geologic structure. Although stream erosion is not an important process in arch formation in Arches National Park, streams do have a clean-up function. During flash floods or times of snow melt, sediments accumulated from weathering processes are swept down dry washes by intermittent streams that drain to the Colorado River.

How the Arches Form

Like other massive cross-bedded sandstones, the Entrada is highly susceptible to weathering by *exfoliation;* that is, the peeling off from bedrock of slabs or shells usually in a series of concentric layers. (This process was discussed in the examples of weathering in Bryce Canyon National Park, Chapter 3.) Massive rocks, such as quartzite, granite, and some sandstones, have a tendency to weather in this manner.

Several weathering processes help exfoliation get started. The vertical joints in the Entrada Sandstone, 10 to 20 feet apart, are zones of weakness through which weathering agents, such as water, ice, and frost, attack

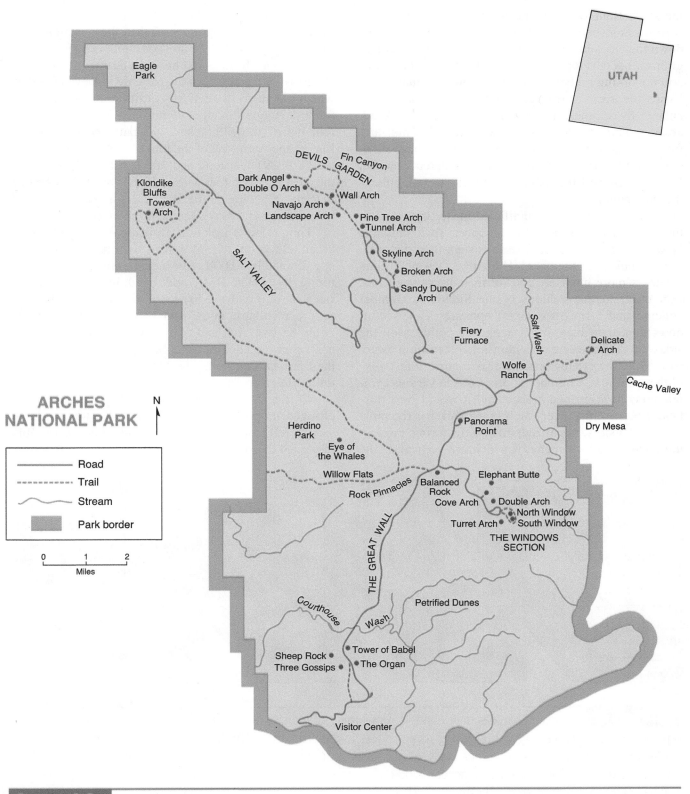

FIGURE 6.2   Arches National Park, Utah.

the rock. Rain water, being slightly acidic from dissolved atmospheric carbon dioxide, slowly breaks down the calcium carbonate cement, releasing the sand grains. *Fins* develop in this way. The process of exfoliation, working on both sides of the fin, eventually perforate the rock, forming windows. Water and frost enlarge these holes, and the wind, carrying fine sand, abrades them. At the base of the fins, added moisture from seepage evaporates at a slower rate, so that beds at the base weather more rapidly. This undercuts the fin, causing loosened fragments and blocks to fall, enlarging the opening.

These processes are significant in starting many of the Entrada arches because the lower rock unit, the Dewey Bridge Member, is less resistant than the overlying Slickrock Member. Thus the Dewey Bridge weathers more intensely and forms hollows or niches. Left unsupported, sections of the Slickrock Member collapse and form (or enlarge) openings. A few windows appear to have been started as potholes on the sides of cliffs, where the abrasion of rushing water formed hollows. This process also enlarges the "sills" of the windows. A number of processes in various combinations have formed arches, and it would be a mistake to assume that every arch in the park has the same erosional history. Localized, intense fracturing, producing zones of joints, many of which are vertical, is the

factor that appears to be most significant in arch formation. What is important is that these features are impermanent, that they are constantly changing, and that new ones form as old ones are destroyed (fig. 6.3).

Most of the arches in the park are in Devils Garden, which has them in all stages of development from the faintest suggestion of a depression to the collapsed remains of those that have all but weathered away. Landscape Arch, one of the longest natural spans in the world, is 291 feet long and 105 feet high. One part of the span is only 6 feet wide, which is an indication that in terms of geologic time, the arch is in the last stage of its existence and may soon be no more (fig. 6.1).

Unique in its isolation, Delicate Arch stands alone at an elevation of 4829 feet. Estimations are that approximately 70,000 years were required for its sculpturing by water, frost, and wind in a dry climate. Because of a narrowed segment on one side, this arch may also collapse in the near geologic future. Delicate Arch, 85 feet high and 65 feet wide, is formed at the contact of the Slickrock and Moab Tongue Members of the Entrada Sandstone, the Moab being the uppermost unit of the formation.

In the Devils Garden area, Skyline Arch last changed its profile in November 1940, when a large section fell from the arch to the ground and noticeably increased the size of the opening.

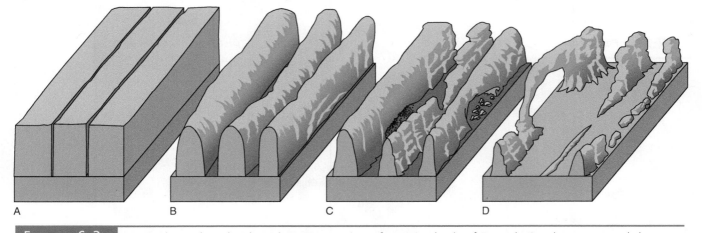

FIGURE 6.3   How the arches developed. *A.* Upwarping of massive beds of Entrada Sandstone caused deep vertical, parallel fractures to form, exposing rock surfaces to the elements. *B.* As weathering and erosion enlarge cracks and fractures, narrow sandstone walls, or fins, develop. *C.* Frost action, exfoliation, and crumbling round off the tops of the fins. Some fins are breached; blind arches and windows form in others. *D.* Rockfalls and weathering gradually enlarge holes to arches. Eventually arches wear through and collapse, leaving buttresses that will, in time, succumb to erosion. Adapted from Arches National Park brochure, National Park Service, 1985.

FIGURE 6.4   Double Arch in the Windows' section has been produced by the exfoliation of the Jurassic Entrada Sandstone. Processes of weathering, and not stream erosion, form a natural arch. Photo by Ann G. Harris.

Balanced Rocks

Balanced Rock, near the Windows, is a striking example of a feature shaped by wind erosion as well as by solution from seepage (fig. 6.5). A resistant block of Slickrock Sandstone is balanced on top of a pedestal of crinkly Dewey Bridge beds, which, in turn, rest on a Navajo Sandstone base. (A *balanced* rock is formed by weathering and erosion in place and rests more or less precariously on its base.) Normally the rock on top is a cap rock that is more resistant than the rock below. The dissolving of intergranular cement by moisture plays an important role in shaping the feature. Helping in the development of balanced rocks in Arches National Park (and in some desert regions) is *sandblasting,* or abrasion by wind-blown sand. Wind also removes the small fragments loosened by weathering. Wind velocity is at its greatest less than 20 inches above the ground, so the most intense erosion occurs at this level. First a pedestal is formed. As it becomes thinner, the weight of the cap rock causes the top part to tilt very slowly to one side, thus compressing the rock on that side and making it more difficult to erode. The opposite side, not under as much compression, weathers away slightly faster until this causes the rock to tilt

back and compress *that* side, and then the process starts all over again. The rock keeps tilting very slightly back and forth until the pedestal gives way or the rock falls from its own weight.

The Controlling Influence of an Ancient Structure

Many feet below the surface in Arches National Park there are layers of rock salt and gypsum considerably older than the rocks we see exposed. These beds belong to the same Pennsylvanian-age Paradox Formation responsible for so much of the unique scenery in Canyonlands National Park. Because the relationship of rocks and structure is somewhat different in the Arches area, however, the resulting expression of scenery is not the same.

Arches National Park lies over a deeper part of the Paradox Basin, closer to the northwest-trending Uncompahgre fault and the upthrown mountain block beyond, which shed sediment in vast quantities into the basin. In a landlocked arm of the sea, held in this closed drainage basin, layers of salt and interbedded sediments accumulated for millions of years (as explained in Chapter 5).

FIGURE 6.5   Balanced Rock, 55 feet high, weighs 3500 tons. The huge rock, eroded from the Slickrock Member of the Entrada Sandstone, is perched on crinkly beds of the Dewey Bridge Member. The pedestal, in turn, rests on the Navajo Sandstone. Photo by David B. Hacker.

The increasing pressure of the overburden on the salt beds caused the rock salt to flow plastically, pushing upward through old fault lines, which were zones of weakness. In the Arches area, the subterranean flowage of salt created northwest-trending "salt anticlines." Much later, dissolution of the upper beds by ground water caused collapse, creating sunken valleys (like grabens), bounded by escarpments. Salt Valley and Cache Valley are examples inside the park. The Moab-Spanish Valley (with the town of Moab at its northern end) is a sunken valley mostly outside the park. Nearly all of the arches and fins in the park stand on the cliffs rising above Salt and Cache Valleys.

So much silt flowed into the salt anticlines that the thickness under Salt Valley in Arches National Park is at least 10,000 feet (Doelling, 1985). Rock salt is not exposed at the surface. As fresh water seeping down through fractures dissolves the buried salt, a residue of impurities like gypsum and shale is left behind,

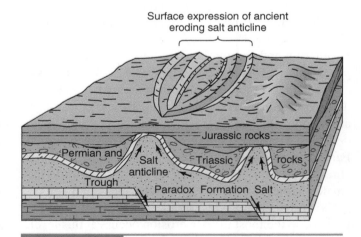

FIGURE 6.6   Block diagram showing how the pattern of surface erosion is controlled by an underlying salt anticline so that some formations are locally thick, thin, or missing. From Doelling 1985.

Table 6.1	Geologic Column, Arches National Park

Time Units		Rock Units			
Era	**Period**	**Group**	**Formation**	**Member**	**Geologic Events**
Cenozoic	Quaternary		Alluvium Terrace gravels Sand dunes		Weathering and erosion forming arches, etc. Solution along faults, forming salt valleys
Cenozoic	Tertiary				Block faulting Uplift; Laramide orogeny
Mesozoic	Cretaceous		Mancos Shale	Upper	Marine shales
Mesozoic	Cretaceous		Mancos Shale	Ferron Sandstone	Lenticular, deltaic, and barrier island sands
Mesozoic	Cretaceous		Mancos Shale	Lower	Deposited as marginal marine beds, some coal; regressing sea
Mesozoic	Cretaceous		Dakota Sandstone		Beach deposits along the edge of a transgressing sea
Mesozoic	Cretaceous		Cedar Mountain		Stream deposits
Mesozoic	Jurassic				Erosion
Mesozoic	Jurassic		Morrison	Brushy Basin	Greenish gray, maroon gray mudstones; contains dinosaur bones
Mesozoic	Jurassic		Morrison	Saltwash	Channel sandstones, siltstones; contains petrified wood
Mesozoic	Jurassic		Morrison	Tidwell	Thin, dark-red siltstone; contains concretions Erosion
Mesozoic	Jurassic	San Rafael	Entrada Sandstone	Moab Tongue	White sandstone; dune deposits
Mesozoic	Jurassic	San Rafael	Entrada Sandstone	Slickrock	Orange-red, resistant, marine sandstone most arches eroded from this member
Mesozoic	Jurassic	San Rafael	Entrada Sandstone	Dewey Bridge	Red and gray beds deposited in shallow sea; has crinkly appearance Erosion
Mesozoic	?	Glen Canyon	Navajo Sandstone		Yellow and gray sandstone deposited in dunes—cliff-former
Mesozoic	Triassic	Glen Canyon	Kayenta Sandstone		Deposited by streams and in lakes; produces slabby fragments
Mesozoic	Triassic	Glen Canyon	Wingate Sandstone		Sandstone (accumulated in dunes) cliff-former
Mesozoic	Triassic		Chinle		Reddish brown, silty sandstone Stream, beach, and lake deposits Erosion
Mesozoic	Triassic		Moenkopi		Reddish brown sandstone and shale, accumulated on tidal flat Erosion
Paleozoic	Permian		Cutler		Red sediment deposited in fans Erosion
Paleozoic	Pennsylvanian		Honaker Trail		Limestone, sandstone, and shale deposited in shallow sea
Paleozoic	Pennsylvanian		Paradox		Gypsum, salt, and shale deposited in lagoons

Source: Doelling, 1985.

slumped down, or contorted in masses. Gypsum exposures of this type occur in Salt Valley beside the jeep trail to Klondike Bluffs.

By the time the Entrada Sandstone was deposited on tidal flats, beaches, and coastal dunes, the flowage of salt had slowed and may have stopped. The Entrada beds do not slope over the salt anticlines as much as the older beds, which were much thinner—or missing—over the tops and thicker on the sides. The arching and flexure of beds over the anticlines produced parallel, northwest-trending fractures. When compressive tectonic forces later caused gentle warping of the rocks in the region, additional closely spaced, parallel joints developed. Several thousand feet of regional uplift, which accelerated erosion, set the stage for the development of the present landscape.

Products of weathering. Most of the loose sediment that accumulates in shifting dunes or piles up in joints between fins and arches consists of fine sand grains that are clean, well rounded, and very smooth. In fact, they are so spherical that they do not cling together well, even when wet. When the sand grains in stream beds become saturated with water, they may form *quicksand.* Such a deposit is found in Salt Wash in the eastern part of Arches National Park. Water flowing upward due to hydrostatic pressure beneath the surface keeps the grains separate and neutralizes their frictional strength. The soft, semiliquid mass of quicksand instantly shifts under the weight of an object, an animal, or a person and has a tendency to suck anything down. However, few of the quicksand beds are more than waist-deep to an adult. The important thing to remember if you find yourself in quicksand is not to struggle, because the sand is reluctant to release anything. The best method for extricating yourself is to move slowly in a breast-stroke motion, so as to distribute body weight more evenly. Then gradually work your way out of the quicksand. If you have to cross a quicksand area, walk slowly and steadily without stopping or standing in one place.

The *chalcedony,* from which the Indians made stone tools, is a cryptocrystalline variety of quartz (i.e., having no visible crystals) that is sometimes found in chert or as concretions that are weathered out from sedimentary rocks. Because chalcedony can be chipped or made to break with a smooth, curved surface, the Indians used the stone for making tools and points with sharp, scalloped edges. This property or characteristic of some minerals to break along curved surfaces is called *conchoidal fracture.*

For "rockhounds," this region yields chalcedony, apache tears, almandine garnet, quartz crystals, geodes, flint, petrified wood, and other kinds of fossils. **Specimens must not be collected in the park; it is illegal to take or disturb natural objects or Indian relics in a national park.**

Geologic History

1. **Deposition of Paleozoic beds; erosion.**

Paleozoic sedimentary beds exposed in nearby Canyonlands were also deposited in the Arches National Park area, but underwent more extensive erosion. Although the Paradox salt beds are not exposed in the park, a few nonsalt, contorted, gypsiferous masses of the Paradox Formation crop out in Salt Valley. Nongypsiferous, marine beds of the Honaker Trail Formation, which were laid down over the Paradox beds late in Pennsylvanian time, have mostly been eroded away or covered, although some are exposed near the Arches National Park Visitor Center and Highway 191. Honaker Trail limestones in these outcrops contain fossils of corals, bryozoans, crinoid stems, and other organisms. Permian Cutler red beds, so well displayed in Canyonlands, are missing in Arches except in the cliffs bordering the southwest corner of the park.

2. **Deposition of the Moenkopi and Chinle Formations and the Glen Canyon Group in Triassic/Jurassic time.**

Remnants of Moenkopi and Chinle Formations are best exposed along the canyon of the Colorado River where it forms the southern boundary of Arches National Park. The Chinle also underlies the Wingate Sandstone in the cliffs along Salt Valley. The sandstones of the Glen Canyon Group—the Wingate, Kayenta, and Navajo—make up the cliffs and the plateau tops throughout much of the park. By the time these sandstones were deposited, the flowage of salt beds had slowed down so that sedimentary beds completely covered the salt anticlines, although some thinning occurred over the top. In Arches National Park, the Wingate, Kayenta, and Navajo look much as they do in the other Colorado Plateau parks. The Wingate and Navajo are cliff-formers and the Kayenta in between forms ledgy slopes. The "petrified" sand dunes of the Navajo beside Scenic Drive in Arches look as if a slight increase in the wind's velocity might set them moving again!

3. How the Entrada Sandstone, with its three members, accumulated in Jurassic time.

When an arm of the sea moved from the north over central Utah in middle Jurassic time, the Arches region became an area of tidal flats, beaches, low islands, and coastal dunes. The sediments deposited in this environment became the three distinctive members of the Entrada Sandstone, the formation in which most of the arches are found. The bottom member is the dark, reddish brown Dewey Bridge, a finegrained, silty sandstone. It's easy to recognize because it tends to be "crinkly" or wavy. This waviness, or contorted bedding, affects some of the lower parts of the overlying Slickrock Member. Slickrock, an orange-red, smooth, more resistant unit, is the main body of the Entrada Sandstone. The gleaming red cliffs, fins, arches, and windows of Slickrock are a memorable sight as they stand against the blue desert sky (fig. 6.7, 6.8).

The uppermost member of the Entrada, the Moab Tongue, is a lighter-colored (yellowish, pink, or buff) resistant sandstone that crops out on the flanks of the Salt Valley anticline and in some of the erosional features.

4. Deposition of the Morrison Formation late in Jurassic time.

Three members of the Morrison Formation are exposed in Arches National Park. The Tidwell Member (formerly mapped in this area as part of the Summerville Formation) is an interesting unit that crops out in tilted blocks in Cache Valley and Salt Valley. Tidwell is a thin, dark red siltstone that contains large, white, siliceous concretions. (This unit was probably a source of the Indians' chalcedony.) Good exposures occur beside the trail to Delicate Arch.

The middle Salt Wash Member and the upper Brushy Basin Member are interbedded channel sandstones, siltstones, mudstones, and conglomerates that were deposited under varying conditions on broad flood plains. The channel sandstones in the Salt Wash Member contain fossil wood fragments and a few dinosaur bones. The colors of the Brushy Basin slope-forming mudstones range from purplish outcrops just north of the park to greenish exposures in the southern part of the park like those found near Delicate Arch. In the park vicinity some of the thicker sandstone beds of

FIGURE 6.7 The Jurassic-aged Entrada Sandstone is divided into three members. The oldest member is the Dewey Bridge, made of red and green beds, that has a wavy, crinkly appearance. It weathers more intensely than the overlying Slickrock Member and forms hollows and niches. Left unsupported, sections will collapse and enlarge the openings, with the overlying Moab Tongue forming the top of the arch. Delicate Arch is an excellent example; it is formed at the contact of the Slickrock and Moab Tongue Members. Delicate Arch is 85 feet high and 65 feet wide. Photo by Ronald F. Hacker.

Box 6.1   **Colorado National Monument and Dinosaur National Monument**

Colorado National Monument is about 50 miles east of Arches National Park and a few miles off I-70, near Grand Junction, Colorado. The monument is small (barely 32 square miles in extent), but it is an ideal place to get acquainted with the geology of the Colorado Plateaus. The sheer walls of the deep canyons expose many of the colorful formations displayed in other parts of plateau country, and the views from Rim Rock Drive are breathtaking. This 23-mile drive climbs from the Colorado River valley to the top of the Uncompahgre Plateau and then winds along the plateau rim. Interpretive plaques at overlooks explain the origin and development of the geological features. The canyon bottoms are floored by ancient Precambrian rocks. Lying unconformably over them are bright red Chinle (Triassic) beds that are, in turn, topped by the cliff-forming Wingate Formation, the Kayenta Formation, Entrada Sandstone, Summerville Formation, and the Morrison Formation—which contains dinosaur fossils. Three Cretaceous formations are exposed on top of the plateau. A well known landmark is Independence Monument, a free-standing monolith that rises 450 feet from the canyon floor.

Dinosaur National Monument, famous as a dinosaur quarry, straddles the Utah-Colorado boundary, about a hundred miles north of Arches National Park. This interesting national monument is actually in the same physiographic province as the Middle Rocky Mountains, but its geologic history and many of its rock formations are closely related to the Colorado Plateaus. The Jurassic Morrison Formation is the rock unit holding the concentration of dinosaur remains in Dinosaur National Monument. From this "graveyard," once a large sandbar in a powerful stream, have come hundreds of skeletons that have been assembled and shipped to museums all over the world.

The Morrison Formation, which is exposed in many localities in western North America—as well as in several national parks of the Colorado Plateaus—has been one of the most thoroughly studied formations on the continent, both for its unique fossil assemblages (dinosaurs, other animals, plants, etc.) and for its mineralization, especially uranium deposits. The entire formation consists of terrestrial deposits that accumulated in and alongside streams and in shallow lakes. Judging by the variety and abundance of life forms in the fossil record, the late Jurassic climate must have been unusually benign. However, in no place is the abundance and variety of dinosaur fossils greater than in Dinosaur National Monument. Apparently season after season dead animals were washed downstream by flood waters, stranded on the sandbar, and buried by sediment and debris.

Long after the lithification and burial of the rocks of the Morrison Formation, the Laramide orogeny began. The monument area, which underwent more severe deformation and uplift than did the Colorado Plateaus rocks to the south, became part of the Uinta Mountains anticline. Erosion by the Green and Yampa Rivers exposed the dinosaur boneyard.

the Salt Wash Member contain uranium, vanadium, and copper ores. The minerals have been transported as solutes in ground water and concentrated or precipitated by organic material that was present in ancient stream channels, sandbars, and lakeshores.

5. Cretaceous deposits, nonmarine and marine.

During a long period of erosion, thin beds of the Cedar Mountain Formation were deposited by streams. The last terrestrial deposits were thin, discontinuous beds of Dakota Sandstone that accumulated in stream valleys in coastal regions. The Dakota Sandstone now forms a rimrock over the shaley Cedar Mountain strata in parts of Salt Valley.

In Late Cretaceous time, the last great sea spread over the Colorado Plateaus. In this region, several thousand feet of gray mud, which became the Mancos Shale, covered everything. Between upper and lower Mancos Shale members, the Ferron Sandstone was deposited. Within the park, all Mancos beds have been removed by erosion except for a few remnants that are exposed in collapse features in Salt Valley (fig. 6.9).

6. Laramide orogeny and uplift.

At the end of Cretaceous time and the beginning of Tertiary time, stress and strain accompanying mountain-building around the Colorado Plateaus and the subse-

FIGURE 6.8   The fins are eroded along vertical, parallel fractures of the Entrada Sandstone in Arches National Park. Photo by David B. Hacker.

quent elevation of the region caused minor warping and faulting, along with fracturing of the rock layers. Deformation during Tertiary time produced gentle anticlines and synclines that generally followed the same northwest trends that were there before. The Courthouse syncline, which follows the Courthouse Wash, is an example.

The Moab fault, which skirts the southwest corner of the Arches National Park near the Visitor Center, displays the effects of relatively recent movement. The beds on either side of U.S. Highway 191 do not match. The Navajo Sandstone and the Entrada Sandstone, which were dropped down, can be seen on the northeast side of the road. Across the highway, in the upthrust block, Paleozoic beds of the Hermosa Group and the Cutler Formation are exposed at the bottom of the canyon. Prominent in the high cliff above are beds of

the Moenkopi and Chinle Formations; at the top is the Wingate Sandstone. Vertical movement along the fault amounted to 2600 feet.

7. **Development of a young landscape in Quaternary time.**

Continuing uplift, accompanied by enlargement of the joint spaces and intensified erosion in Tertiary time, initiated the development of the unique landforms in Arches National Park. The most rapid period of development was during the colder, wetter episodes of the Pleistocene. In the drier environment of the present, the construction and destruction of arches and other features continues at a slower rate.

Modern sediments accumulate mostly in the form of talus along the base of cliffs, as landslide deposits, in dunes and sheets of wind-blown sand, and terrace grav-

FIGURE 6.9   The Salt Valleys are actually grabens created when the underlying salt beds below are either dissolved or migrate because of the weight of the overlying beds. With the support removed, the beds drop, creating the parallel depressions. Photo by David B. Hacker.

els along the Colorado River, which brings sediments into the park area as well as carries them away.

Between floods, rock fragments and soil accumulate in intermittent stream beds, a feature called a *dry wash* in plateau country.

Geologic Map and Cross Section

Doelling, H.H. 1985. Geologic Map of Arches National Park and Vicinity, Grand County, Utah. Map 74. Utah Geological and Mineral Survey, Department of Natural Resources.

Sources

Abbey, E.F. 1968. *Desert Solitaire, a Season in the Wilderness*. New York: Ballantine Books. 303 p.

Baars, D.L. 1983. The Colorado Plateau, a Geologic History. Albuquerque, New Mexico: University of New Mexico Press.

Cruikshank, K.M. and Aydin, A. 1994. *Role of Fracture Localization in Arch Formation, Arches National Park,* *Utah*. Geological Society of America Bulletin, v. 106, p. 879–891 (July).

Doelling, H.H. 1985. *Geology of Arches National Park*. Utah Geological and Mineral Survey, Department of Natural Resources. 15 p.

Hagood, A. 1972. *Dinosaur, the Story behind the Scenery*. Las Vegas, Nevada: KC Publications. 30 p.

Lohman, S.W. 1975. *The Geologic Story of Arches National Park*. U.S. Geological Survey Bulletin 1393.

Lohman, S.W. 1981. *The Geologic Story of Colorado National Monument*. U.S. Geological Survey Bulletin 1508.

Rigby, J.K. 1976. *Northern Colorado Plateau*, K/H Geology Field Guide Series. Dubuque, Iowa: Kendall/Hunt Publishing Company.

Stokes, W.L. 1986. *Geology of Utah*. Salt Lake City: Utah Museum of Natural History and Utah Geological and Mineral Survey. p. 120–121.

Note: To convert English measurements to metric, go to www.helpwithdiy.com/metric_conversion_calculator.html

Mesa Verde National Park

S O U T H W E S T   C O L O R A D O

Area: 52,122 acres; 81 square miles
Established as a National Park: June 29, 1906
Designated a World Heritage Site: September 6, 1978
Address: P.O. Box 8, Mesa Verde National Park,
CO 81330-0008

Phone: 970-529-4465
E-mail: meve_General_Information@nps.gov
Website: www.nps.gov/meve/index.htm

*Pre-Columbian cliff dwellings, multi-room pueblos, and irrigation systems of Anasazi
people are the best preserved in the United States. They built cliff houses in natural alcoves
produced by weathering and erosion of Late Cretaceous Sandstone and shale beds.*

FIGURE 7.1   Mesa Verde is a cuesta that slopes gently to the south. An outcropping of resistant sandstone, which is a remnant of a pediment that once connected the Mesa Verde Plateau with the La Plata and San Juan Mountains, forms a steep 2000-foot high escarpment on its north side. © Robert Fletcher.

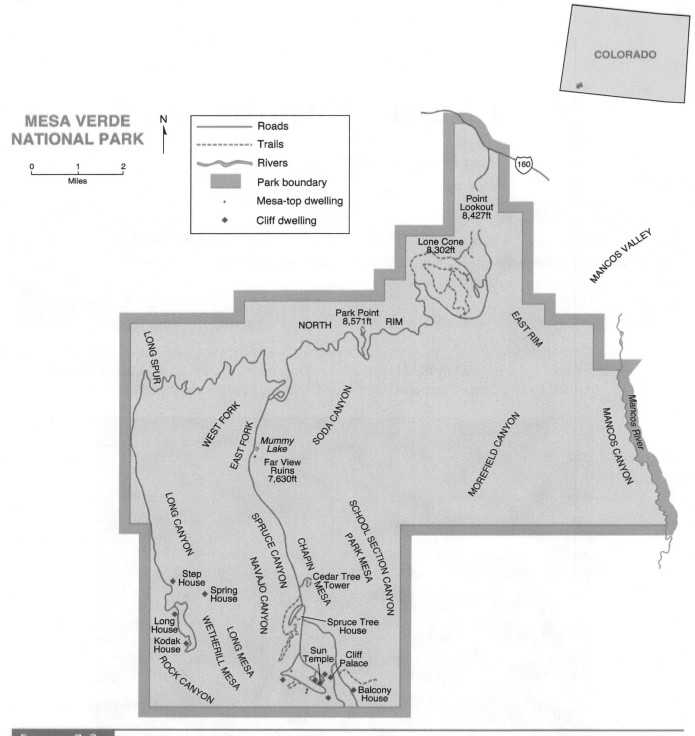

FIGURE 7.2   Mesa Verde National Park, Colorado.

FIGURE 7.3   Anasazi occupied Far View Ruins, located near the center of the mesa top, around 800 years ago. Skilled masons built the multi-room dwellings and also constructed extensive stone-lined irrigation systems. Photo by Edward Mooney.

Mesa Verde National Park, west of Durango, is in the Colorado corner of the Four Corners area. The park lies on the top of a high mesa that rises from the canyon of the Mancos River, a tributary of the Colorado River. Early Spanish explorers, who camped along the Mancos, named the feature *La Mesa Verde,* meaning "the Green Table."

The first American to make an official report on the area was William H. Jackson, the famous photographer with the Hayden Survey, who photographed cliff dwellings in Mancos Canyon in 1874. Virginia McClurg, a New York journalist, was sent to the Four Corners area in 1882 to investigate rumors of "lost cities." Because of Indian uprisings, she was unable to do much exploring until 1886, at which time her expedition went into several lower canyons that drain into the Mancos River. Her most important discovery was Balcony House (near the park's southern boundary),

which was one of the last cliff dwellings occupied by the Anasazi people. She was determined that the archaeological resources of Mesa Verde should be protected and preserved, and for the next 20 years, she campaigned to gain support for this effort.

The existence of the pueblo ruins on the mesa top and the larger, hidden cliff houses in the alcoves of inner canyons was unsuspected by non-Indians until 1888, when ranchers, Richard Wetherill and Charles Mason, happened on Cliff Palace and Spruce Tree House when they were searching for cattle. Fascinated by their discovery of abandoned communities, they set about exploring the dwellings during the winter when ranch chores were less demanding. Later, after word of the discoveries had spread, Wetherill helped Gustaf Nordenskiold (the young son of a Swedish nobleman) to excavate several of the cliff dwellings. Nordenskiold published his findings in 1893 in a book entitled *The Cliff Dwellings of Mesa*

Verde. His collection of artifacts from Mesa Verde is displayed in the Finnish National Museum in Helsinki.

The Colorado Cliff Dwellings Association, founded by Virginia McClurg, rallied support to stop the removal and sale of artifacts and worked on arrangements to preserve the ruins. Due to their efforts, Mesa Verde National Park was established in 1906 in order to protect the archeological sites.

The Natural World of the Mesa Verde Anasazi

The Anasazi (from a Navajo word meaning "Ancient Ones") were a Stone Age people who were skillful at making tools and furnishings from natural materials, such as stone, bone, wood, clay, feathers, skins, plant fibers, and whatever was at hand. They also learned how to use the local geology to their utmost advantage in securing shelter and dependable food and water supplies. Management of water resources for agricultural purposes and daily life was vitally important in Anasazi

culture. The task demanded foresight and a high degree of community cooperation.

The climate of the region is semiarid, but more rain falls on the high plateaus and mesas than in the parched lowlands of the Montezuma and Mancos Valleys, some 2000 feet below the mesa. The heaviest snowfalls are in the area of the north rim of Mesa Verde where elevations are 8000 feet or more. Runoff from melting snow drains south due to the slight tilt of the mesa surface. Furthermore, because of prevailing storm patterns, the heaviest average annual precipitation, 18 inches per year, falls in the middle section of the mesa, where elevations are around 7000 feet. The most favorable growing conditions are, therefore, in this middle area that receives more moisture from rain and snowmelt. Not surprisingly, the villages of Far View, which had the densest population, were located near the central part of the mesa. With no permanent streams (except the Mancos), the Indians depended on springs and seeps for drinking water. Summer thunder-

FIGURE 7.4   Cliff Palace, the largest cliff dwelling in the park, had over 200 living rooms, 23 kivas, and many storage recesses. As many as 400 people at one time lived in the complex, which had at least eight floor levels. Photo by Michael Little.

storms and melting snow in spring replenished the springs and provided water for crops. In addition to higher precipitation, the growing season on the mesa is longer than in the valleys at the foot of the cuesta because temperatures are more moderate in both winter and summer on the mesa.

The presence of fertile soil on the mesa top is another favorable geologic factor. While some of this soil may have been formed by weathering of underlying bedrock, most of it accumulated as wind-deposited silt (*loess*), blown from the southwest during the Pleistocene Epoch and early in Holocene time.

Much of what we know about the Anasazi who lived on the mesa top and in the cliff alcoves is due to excavations and restorations begun in 1916 by Jesse Walter Fewkes, a Smithsonian Institution archaeologist. His work is carried on by Park Service archaeologists and other scientists.

About 550 A.D., the first Anasazi, called by archaeologists, the "Modified Basketmakers," settled in Mesa Verde. They lived in pithouses on the mesa top, cultivated crops, and made handsome baskets. The Anasazi descended to the alcoves in the cliffs for shelter in winter, made crude dwellings there, and stored food in hollows in the sandstone.

During the evolving of the Pueblo culture (750 A.D. to 1100 A.D.), the Anasazi built rows of houses with poles and mud above ground on the mesa. The oblong excavations for pithouses gradually acquired the round shape of kivas, while the crude row houses developed into stone-walled pueblos. By 1000 A.D., the Anasazi had become skilled stone masons. They fashioned sandstone into rectangular building blocks and used mud and water for mortar. Some buildings were three stories high with thick, double-coursed walls, joined in units of 50 or more rooms. Ceremonial activities were carried on in the underground kivas.

Scattered over the mesa are the remains of round stone towers that may have been watch or signal towers. Perhaps the towers were for keeping track of the sun's position so that solar observers could tell the people when to plant and harvest crops.

The "Classic" or most culturally advanced period was between 1100 and 1300. Probably several thousand people lived in the stone-walled villages, and most of the mesa top was under cultivation. Irrigation ditches channeled available water into the fields, and check dams in stream courses collected silt and helped to prevent erosion of precious topsoil.

A water supply system developed for the Far View area (south of the visitor center) was remarkable. A man-made reservoir, Mummy Lake (now dry), was constructed to trap and hold runoff from rain and snow in order to supply water for crops and household use. The Anasazi hollowed out a large circular depression that was then lined with sandstone blocks and rimmed with a low retaining wall. A stone-lined intake channel, where silt could settle out, allowed clear water to go into the reservoir. A ditch above Mummy Lake may have diverted water for irrigating.

Around 1200 the Anasazi began to move back into the cliff alcoves where their ancestors had sheltered centuries before. Perhaps the soil on the mesa had become depleted from overuse, or perhaps increasing aridity and a harsher climate forced the mesa dwellers to move closer to springs in the protected environment of the canyons. Most likely a combination of reasons led to construction of cliff dwellings in the middle decades of the thirteenth century. There are more than 600 dwellings in the park. Most of them are from 1–5 rooms. They range in size from one- or two-room houses built into an overhang to cliff alcoves, such as Cliff Palace, which holds over 200 rooms. Spruce Tree House, which is the third largest in the park, the best preserved, and the one easiest to get to, was constructed in an alcove measuring 216 feet wide and 89 feet deep. A natural spring at the head of the canyon supports lush vegetation and is protected from evaporation by shade most of the day. The canyon springs provided the Indians with a reliable supply of water close at hand, except in times of extreme drought. The narrow canyons provided protection from possible enemies as well as from the elements. The alcoves that face south-southwest were the best for dwellings. They were cooler in the summer and warmer in the winter.

The cliff houses were inhabited less than one hundred years. By the beginning of the fourteenth century, after a series of droughts and crop failures, the Anasazi left Mesa Verde. They may have moved south into Arizona and New Mexico. Some of the Pueblo Indians of today are believed to be descendants of the people of Mesa Verde.

Geologic Features

Mesa Verde is a *cuesta* that slopes gently to the south. Its steep, 2000-foot high escarpment on the north side is formed by an outcropping of resistant sandstone (fig. 7.1). The dip of the sedimentary beds southward from the escarpment is slightly steeper than the slope of the mesa surface. The Mesa Verde cuesta is a remnant of a pediment that once connected Mesa Verde with the La

Plata and San Juan Mountains. A *pediment* is a broad, flat (or gently sloping) bedrock erosional surface that typically develops at the base of a mountain front in arid regions.

When the Mesa Verde region was uplifted, streams incised themselves into the pediment and divided it into separate components. McElmo Creek, which flows west outside the park, captured the headwaters of streams that used to flow south along the north and west sides of the cuesta. The *stream capture* occurred because the south-flowing streams had lower gradients even though they were at higher elevations than McElmo Creek. When headward erosion by the McElmo tributaries cut back far enough, they encountered the south-flowing streams and captured their headwaters, thus changing their direction of flow from south to west.

The Mancos River, which drains the cuesta, has cut a deep, broad valley along the mesa's eastern and southern edge. Intermittent streams, fed by rain and meltwater, dissected the mesa by headward erosion into 15 long, parallel, steep-sided canyons, all of which drain into the Mancos. Narrow uplands separate the fingerlike canyons. The Anasazi built their dwellings in alcoves in the canyon walls and on the intervening mesa surface (fig.7.5).

How the Alcoves Formed

The shallow, cavelike alcoves that the Anasazi enlarged and adapted for cliff dwellings developed along a shale-

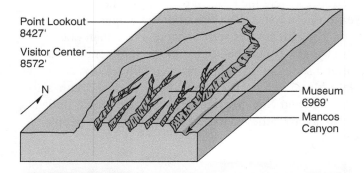

FIGURE 7.5 The Mesa Verde cuesta slopes gently to the south, as shown in the diagram. Along the east side of the mesa, the Mancos River has eroded a deep canyon. Tributary streams have cut deep, narrow side canyons into the mesa top. The cliff dwellings were constructed in alcoves in the canyon walls.

sandstone contact where massive, cliff-forming sandstone is underlain by soft, erodible shale beds. Some of the cliff dwellings in the park are at the base of the top member (the Barker Dome Tongue) of the Cliff House Sandstone, which contains a discontinuous layer of shale (fig. 7.6).

Because the Cliff House Sandstone has many pore spaces (openings) and can hold a considerable quantity of ground water, it has high *porosity*. It is also *permeable,* which means that water can pass through it. On the mesa top, melting snow and rain soaks in and readily percolates down through sandstone until reaching shale beds, which are porous, but not permeable because the pore spaces are so small that water is trapped and cannot pass through. The ground water is forced to move laterally along the contact until it seeps out on the face of the canyon wall or at the back of an alcove, creating a spring.

The percolating ground water dissolves cement along joints and bedding planes and loosens sheets and blocks of sandstone. Seasonal freezing and thawing of the ground water breaks up the bedrock. Exfoliation, mass wasting, and trickling water remove the loosened material, enlarging hollows into alcoves.

Geologic History

1. **Deposition of the Mancos Shale over the Dakota Sandstone late in Cretaceous time.**

The Dakota Sandstone, which underlies most of Colorado, crops out west of Mesa Verde National Park, but is not exposed in the park. The Mancos Shale (above the Dakota) forms the long talus slope of the north escarpment of Mesa Verde. The highway up to the mesa top from the park entrance has to climb unstable slopes on weak Mancos Shale in order to reach the sandstone rimrock. Because the Mancos has a high clay content, it expands when wet and causes slumping and sliding so that the road has to be repaired frequently.

The Mancos beds in this area are fine-grained sandstones, mudstones, and shales that accumulated in deep water, far enough from shore so that coarser sediments had settled out. The Mancos members that crop out at the surface in the park are the highly fossiliferous Carlile Shales; the Niobrara beds of sandy, limey shale; and a Transition Member, containing sandy shale and shaley sandstone.

Table 7.1		Geologic Column, Mesa Verde National Park			
Time Units		**Rock Units**			
Era	**Period**	**Group**	**Formation**	**Member**	**Geologic Events**
Cenozoic	Quaternary	Stream deposits, talus Loess soil Pediment gravels, cobbles			Weathering of alcoves in cliff walls Headward erosion of canyons Uplift and tilting; isolation of Mesa Verde cuesta by erosion Development of pediment Uplift of La Plata and San Juan Mountains Uplift of Colorado Plateaus during Laramide orogeny
	Tertiary				
					Erosion
Mesozoic	Upper Cretaceous	Mesaverde	Cliff House Sandstone	Barker Dome Tongue	Cliff-former; thick accumulations of fine, cross-bedded sand and thin shale beds on oscillating shoreline
				Lower Member	
			Menefee		Backshore and lagoonal deposits; silts, muds, sand, coal beds
			Point Lookout Sandstone		Cliff-former; massive, fine, cross-bedded sands
			Mancos Shale		Slope-former; deep-water deposits, muds, silts, fine sands
			Dakota Sandstone		Not exposed in park

Source: M. Griffitts (personal communication); Wenger, 1980.

2. Shallow-water deposits of the Point Lookout Sandstone, Mesaverde Group.

The three formations comprising the Mesaverde Group—Point Lookout Sandstone, Menefee Formation, and Cliff House Sandstone—reflect the changes in depositional environment that were occurring toward the end of Cretaceous time when the area that is now the park lay along an oscillating shoreline. The type locality of the Mesaverde Group is Mesa Verde National Park.

The Point Lookout Sandstone, which forms the rim of the escarpment above the entrance to the park, is named for Point Lookout (elevation, 8427 feet) (fig. 7.1). The sediments making up the formation, which is about 400 feet thick, accumulated as the shoreline of the Cretaceous sea was shifting toward the northeast.

The upper layers of the sandstone contain fossils of invertebrates. The Point Lookout Sandstone is massive (because it has no bedding planes), fine-grained, cross-bedded, and very resistant. Since the formation is virtually all marine, the cross bedding is probably the result of wave action and currents in shallow water.

3. Nonmarine deposits of the Menefee Formation.

The middle formation of the Mesaverde Group consists of interbedded carbonaceous shales, siltstones, and sandstones that were deposited in brackish water in bays, swamps, and lagoons. In areas outside the park so much organic material accumulated that coal beds of commercial grade were produced. A few thin coal seams are exposed in canyon walls in the park. The Upper Member of the Menefee Formation interfingers with the overlying Cliff House Sandstone.

FIGURE 7.6   Anasazi took advantage of the smaller recesses above the main recess where the living dwellings were located by using them as storage areas. The recesses in the cliffs were the result of the interbedding of the Menefee Shale and the Point Lookout Sandstone. Weathering removed the less resistant Menefee Shale from the Point Lookout Sandstone, leaving alcoves where the dwellings were located. Photo by Edward Mooney.

FIGURE 7.7   Remnants of Anasazi dwellings can be seen as outlined on the floor of the alcove with remnants of some of the walls. These dwellings can be seen along Petroglyph Point Trail. © Robert Fletcher.

4. **A shifting shoreline during deposition of the Cliff House Sandstone.**

Great quantities of fine-grained sand accumulated in sand bars, beaches, and shore dunes, with clastic materials overlapping and interbedding as the shoreline kept shifting back and forth. The sediments became the massive, cross-bedded Cliff House Sandstone, the youngest bedrock in the park. The formation averages 400 feet in thickness and forms the bold, yellow-orange cliffs in the canyons. Fossil beds, containing clam shells, snail shells, fish teeth, etc., occur in lenses in the upper 100 feet of the Cliff House. Concretions lacking nuclei that erode out of the Cliff House Sandstone appear to be the result of variations in porosity. The shale zones in the Cliff House Sandstone determine the location of the alcoves in which the cliff dwellings were constructed (figs. 7.6, 7.7).

5. **Beginning of successive uplifts at the close of the Cretaceous, continuing through Tertiary time.**

Although marine sediments were deposited over the formations of the Mesaverde Group, the youngest

 FIGURE 7.8   Not all the buildings are square or rectangular; some are round. © Robert Fletcher.

Cretaceous beds were stripped off by erosion while the Colorado Plateaus were uplifted during the Laramide orogeny.

The uplifting of the San Juan and La Plata Mountains in Tertiary time and the accompanying erosion created the Mesa Verde pediment, which gently slopes southward. On this erosional surface, meandering streams flowed toward the southwest depositing gravels brought down from the mountains. Scattered patches of the gravels remain in the Mesa Verde area.

These waterworn pebbles and cobbles, generally of greater hardness than the sandstones, were prized by the Indians for toolmaking and other uses.

6. **Dissection of the pediment plateau and isolation of Mesa Verde from the La Plata and San Juan highlands.**

A final uplift and southward tilting late in Tertiary time caused the streams on the pediment to downcut rapidly, removing sediment, creating canyons and broad

FIGURE 7.9   The interior of a kiva, a subterranean ceremonial building. In the center of the photograph is the firepit; just above it the upright stone is a deflector to distribute the heat. Behind the deflector is a ventilator shaft. The small hole at the bottom of the photograph is called a sipapa (the hole for the gods to enter). Photo by Edward Mooney.

FIGURE 7.10   Row houses typical of Pueblo culture (A.D. 750 to 1100). Photo by Rochelle Beiersdorfer.

valleys, and isolating erosional remnants, such as the Mesa Verde cuesta. A slight difference in steepness between the slope of the mesa surface and the dip of the sedimentary formations is shown by the fact that Point Lookout Sandstone forms the north rim of the mesa but is buried under younger beds of the Mesaverde Group in the center of the mesa. Only the Menefee Formation and the Cliff House Sandstone are exposed in the canyons.

Headward erosion of the canyons in the cuesta, developing a parallel drainage pattern, was especially intense during the cooler, more humid climates of Pleistocene time. Also, because the bedrock was wetter, older alcoves formed, deepened, and collapsed in the canyon walls, while new alcoves were being created. In the drier climate of the present, the erosional processes continue at a slower rate.

Geologic Map and Cross Section

Tweto, O. 1979. Geologic Map of Colorado. U.S. Geological Survey.

Sources

Brown, K.A. 1995. *Four Corners, History, Land, and People of the Desert Southwest.* New York: Harper Collins, p. 123–162.

Griffins, M.L. 1990. *Guidebook to the Geology of Mesa Verde National Park, Colorado.* Mesa Verde Museum Association, Mesa Verde National Park.

Hunt, C.B. 1969. *Geologic History of the Colorado River.* U.S. Geological Survey Professional Paper 669.

National Park Service. n.d. Geology Information Sheet, Mesa Verde National Park.

Rigby, J.K. 1977. *Southern Colorado Plateau,* K/H Geology Field Guide Series. Dubuque, Iowa: Kendall/Hunt Publishing Company.

Wanek, A.A. 1959. *Geology of Fuel Resources of the Mesa Verde Area, Montezuma, and LaMata Counties, Colorado.* U.S. Geological Survey Bulletin 1072M.

Wenger, G.R. 1980. *The Story of Mesa Verde National Park.* Mesa Verde Museum Association, Mesa Verde National Park.

Note: To convert English measurements to metric, go to www.helpwithdiy.com/metric_conversion_calculator.html

Petrified Forest National Park

EAST CENTRAL ARIZONA

Area: 93,533 acres; 146 square miles
Proclaimed a National Monument: December 8, 1906
Established as a National Park: December 9, 1962
Address: P.O. Box 2217 Petrified Forest National Park, AZ 86028

Phone: 928-524-6228
E-mail: PEFO_Superintendent@nps.gov
Website: www.nps.gov/pefo/index.htm

"I sit silent and alone . . . in the deeper silence of the enchanted old, old, forests. . ."
John Muir, 1906

A profusion of logs and parts of trees that have changed to multicolored stone. Indian ruins and petroglyphs. Brightly colored badlands in the Painted Desert.

FIGURE 8.1   Agate
Bridge, a petrified log over 100 feet long, spans an eroded beam. The National Park Service has placed supports beneath Agate Bridge to prevent its collapse. Photo by Edward Mooney.

Geographic Setting

Petrified Forest National Park is located near the southern edge of the Colorado Plateaus about 120 miles (air distance) southeast of Grand Canyon National Park. The geology of Petrified Forest is similar to other national parks of the Colorado Plateaus. Erosional landforms, characteristic of arid regions and sculpted from flat-lying rocks, dominate the scenery. However, scattered over the desert surface in Petrified Forest National Park are thousands of jewel-like logs, broken branches, segments of tree trunks, and sparkling wood chips. All have been changed from wood to stone. Here is an array that is unsurpassed anywhere in the world for variety and profusion.

The petrified trees have been eroded from a sequence of Triassic rocks that are a part of the vast Painted Desert of Arizona. Although the American Southwest has many areas of brightly colored rocks referred to as "painted deserts," the geographic entity officially designated as the Painted Desert curves southwestward across Arizona from Lees Ferry, at the northern tip of Grand Canyon National Park, to Holbrook, then east across Petrified Forest National Park, and on to the New Mexico border.

The Little Colorado River, which, in a general way, skirts the southern edge of the Painted Desert, flows northwest to join the Colorado River, collecting the drainage of the area as it goes. The intermittent Puerco River, which cuts across the middle of the park in a wide, flat-bottomed arroyo, meets the Little Colorado west of the park at Holbrook.

Evidence of stream erosion by the Puerco and its tributaries is everywhere about the landscape, but water is seldom seen on the surface. The arid plateau on which the park is located (elevation 5500 feet) is smooth in some places, but elsewhere is intricately dissected. Mesas, buttes, gullies, arroyos, sinkholes, and other features show that even in the desert, water is the dominant agent of erosion.

The semiarid climate is variable, with wide fluctuations in rainfall and sudden changes in temperature. Mean annual precipitation is 9 to 10 inches. Most of it comes in brief, violent thunderstorms during the summer months. Annual temperature readings may range as much as 127 Fahrenheit degrees between maximum highs and lows! Occasionally snow falls in winter, but does not remain long on the ground. In times of drought, many months may pass without precipitation.

In this harsh climate, rocks are well exposed, but vegetation is sparse.

Since prehistoric times, a well traveled, east-west trade route has followed the valley of the Puerco River through this part of Arizona. Indians, and later explorers and pioneers used the trade route to get to Baja, California and the Pacific Coast. A transcontinental railroad and later a transcontinental highway followed the same route. Today, Interstate 40 goes through Petrified Forest National Park and provides convenient access to its features.

The oldest Indian ruins in the park are on The Flattops, an area of sandstone-capped mesas near the southern entrance. Ancestors of the Pueblo people lived in a village of pit houses that are partially dug into the earth and lined with sandstone slabs. The inhabitants raised crops on the flat mesa tops, when the climate was less arid than now. Nearby, south of Rainbow Forest Museum, is Agate House, a partially restored pueblo built entirely of petrified wood.

A village at Twin Buttes (west of the park) was occupied by people of a later culture. Pit houses were excavated more deeply and had more rooms. The inhabitants built long, low structures of stone that served as windbreaks to protect crops and reduce the movement of sand over the land surface. The Hopi Indians of today still build such walls. Pottery and tools found at the site include artifacts made from Pacific seashells and other materials foreign to this region that were probably traded for articles made from petrified wood.

Droughts that began in the thirteenth century brought about changes in the Indians' way of life. They built villages on the flood plains where the water supply was more dependable. The Puerco Indian Ruin, which may have had a population of more than 200 people, is a large, rectangular pueblo of uncemented masonry construction on a terrace overlooking the Puerco River. The Indians grew crops on the level terrace tops.

Around the beginning of the fourteenth century, precipitation increased; but instead of being evenly distributed throughout the growing season, the rain came in heavy bursts and thunderstorms. Time and time again, the river flooded, washing out fields, eroding topsoil, and ruining crops. According to Hopi tradition, their ancestors moved from the Puerco region many generations ago to what is now the Hopi Indian Reservation located about 80 miles to the northwest.

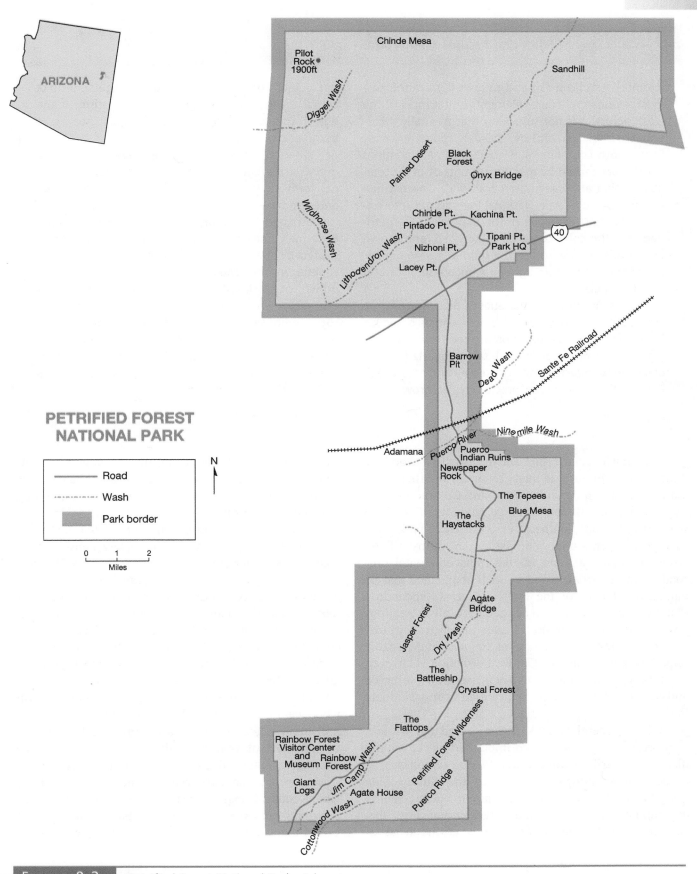

ARIZONA

PETRIFIED FOREST NATIONAL PARK

N

— Road

–·– Wash

▨ Park border

0   1   2
Miles

Chinde Mesa

Pilot Rock ● 1900ft

Sandhill

Digger Wash

Painted Desert

Black Forest

Onyx Bridge

Windhorse Wash

Chinde Pt.

Kachina Pt.

Pintado Pt.

Lithodendron Wash

Tipani Pt.
Park HQ

Nizhoni Pt.

40

Lacey Pt.

Barrow Pit

Dead Wash

Sante Fe Railroad

Nine mile Wash

Adamana

Puerco River

Puerco Indian Ruins

Newspaper Rock

The Tepees

Blue Mesa

The Haystacks

Agate Bridge

Jasper Forest

Dry Wash

The Battleship

Crystal Forest

The Flattops

Petrified Forest Wilderness

Rainbow Forest Visitor Center and Museum

Rainbow Forest

Jim Camp Wash

Giant Logs

Agate House

Puerco Ridge

Cottonwood Wash

FIGURE 8.2 Petrified Forest National Park, Arizona.

Box 8.1	Silica Can Assume Many Forms

Inasmuch as silicon (Si) and oxygen (O) are the two most abundant chemical elements in the rocks of the earth's crust, it is not surprising that naturally occurring silica (silicon dioxide, SiO_2) is found in many forms, both crystalline and noncrystalline. Variations in color are caused by minute amounts of other compounds that are precipitated with silica. The silicified logs in Petrified Forest National Park display interesting and unique examples of this mineral. In the petrified logs the color tones of yellow, red, and brown were produced by iron in the ground water. Copper added tints of blue and green; manganese or carbon blended in black tones.

All of the crystalline varieties of precipitated silica are quartz, but they have different names based on their color. The most common are rock crystal (clear), amethyst (lavender or purple), smoky quartz (gray to black), milky quartz (white or cream), and citrine (yellow). If the mineral can precipitate, or grow, in an unoccupied space, it may develop its regular, distinct crystal form as a six-sided pyramid or prism. If the quartz grows in very small or irregular spaces, it becomes crystalline grains.

When the internal crystalline arrangement of the quartz is so irregular that it has no characteristic external form, or when its internal structure cannot be detected by a petrographic microscope, the specimen is described as amorphous or *cryptocrystalline.* (Extremely fine crystallinity may, however, be demonstrated by the use of an electron microscope.) Such varieties include: *chalcedony*—white, pale blue, gray, brown, or black with a waxlike luster and uniform tint; *jasper*—red chalcedony or chert (a sedimentary rock composed primarily of silica); *carnelian*—a translucent red or orange-red variety of chalcedony; *agate*—a banded or variegated chalcedony, characterized by colors arranged in alternating stripes or bands but occurring in other forms as well.

The first official report on the petrified wood by American explorers was made by an army officer who led an expedition down the Zuni and Colorado Rivers in 1851. He told of finding large amounts of petrified wood a few miles southeast of the present park area. Five years later, Amiel Whipple, a U.S. Army officer, while surveying a route for the transcontinental railroad, found petrified wood in such abundance that he named a tributary of the Puerco River Lithodendron ("stone-tree") Creek. Later geologic investigations throughout the Colorado Plateau revealed that fossilized tree remains are not uncommon in Mesozoic rocks, especially in the Chinle Formation. However, nowhere else are specimens found so well preserved and in such abundance as in Petrified Forest National Park.

When General W.T. Sherman toured the western territories in 1879, he was so impressed by the natural display of petrified wood in the Puerco region that he asked that specimens be sent to the Smithsonian Institution in Washington, D.C. An Army detail, sent to accomplish the task, found huge buried logs that seemed suitable near Bear Springs at the head of Lithodendron Wash. They dug up two large segments and transported them by wagon to Fort Wingate, New Mexico. From there, the larger piece was shipped by rail to the Smithsonian.

When the transcontinental railroad was extended through Arizona, the railroad company constructed a coaling and water stop in the Puerco River Valley that became the town of Adamana. Train passengers were encouraged to take excursions to "Chalcedony Park," as the area was then called, to see the "petrified forests." Word spread, and from then on, souvenir hunters, gem collectors, and commercial jewelers began hauling away quantities of petrified wood, quartz and amethyst crystals, and Indian relics. Some collectors, looking for crystals, used dynamite to break up the logs. Fossil wood was carted away by the wagonload. When a stamp mill was erected at Adamana to crush the logs for the manufacture of abrasives, the local people became alarmed and intensified their efforts to preserve the petrified forests. The territorial legislature petitioned the United States Congress to set aside Chalcedony Park as a national preserve. Lester Ward, a paleobotanist of the U.S. Geological Survey, investigated the area in 1899 and recommended that the land be withdrawn from homesteading and placed under government protection. John Muir, in his old age, studied the fossils in Petrified Forest. Horrified at seeing petrified wood carried away by the boxcar load, he persuaded the Congress and President Theodore Roosevelt to preserve the treasures of Petrified Forest in a national monument. Later, part of the Painted Desert was added when the national park was established. **Removal of petrified wood or any unauthorized materials from Petrified Forest National Park is prohibited.**

Geologic Features

The Petrified Wood

In Petrified Forest National Park, fossil logs and chunks of all sizes lie scattered on the ground or stick out of the soft, shaly bedrock. Examination of the logs shows that some have roots, but many are bare trunks without roots, limbs, or twigs. Much of the bark is scarred and abraded. The silica that infiltrated the wood made the logs heavy as well as brittle. Some are split lengthwise and some are broken clean across their width as if sawed. Probably some logs were broken after petrification during uplift, possibly by earthquakes, or perhaps when tumbling down slopes as they were uncovered and loosened by erosion (fig. 8.3).

Most of the trees appear to have been uprooted, transported, tumbled, and swirled about by raging streams and then dumped in a jumble of woody trash and sediment far downstream from where they originally grew. To account for so much wood being washed out, transported, and then buried, we have to visualize large rivers that overspread flood plains and deltas during times of high water. Perhaps the climate was mon-

FIGURE 8.3   Differential weathering and erosion uncovered the petrified logs, which are more resistant than the Chinle Formation beds in which they were buried. Photo by Ronald F. Hacker.

Box 8.2	How Does Wood Turn to Stone?

Like a number of other geochemical processes, petrifaction is not completely understood. How helpful it would be if wood could be silicified and the process studied in a laboratory under controlled conditions. So far no one has successfully accomplished this, although some investigators have tried. A human lifetime is too short for the completion of experiments to duplicate processes that operate on geologic time in nature.

The logs in Petrified Forest demonstrate that processes of petrifaction are infinitely varied. In some logs, the internal woody structure has been preserved with amazing fidelity. Others are *pseudomorphs*; that is, they are recognizable as logs, but the internal woody structure was not retained in the precipitated jasper, chalcedony, or agate. The wood has been entirely replaced by silica. Moreover, in some logs both replacement of woody tissue by silica and preservation of cellular structure have occurred in close proximity.

The process by which the details of cellular structure are preserved is called *permineralization*; it is a type of fossilization whereby mineral matter is deposited within and around the cells or pores without distorting or destroying them. Wood that has been silicified in this way can be immersed in hydrofluoric acid so that the silica is chemically removed, which leaves a soft mass of woody tissue from which slides can be made for microscopic study, as with any modern wood.

What probably helps to make this process of fossilization possible under certain natural conditions is the fact that inorganic molecules are much smaller than organic molecules. In other words, the preservation of wood structure in faithful detail by petrifac-

tion is not accomplished by replacement of wood cells by silica ("molecule for molecule," as was once thought to be the case), because there is considerable discrepancy in size and geometry between organic and inorganic molecules. What does happen is that the woody components are penetrated and surrounded by "microcrystalline mineral matter often without visible disturbance of either the tissue or the interpenetrating microscopic crystal patterns" (Schopf 1971, p. 523).

Natural conditions under which permineralization of wood may take place involve ample moisture as well as time. In the swampy lowlands in which the wood was deposited, the water table was close to the surface. Ground water circulated very slowly and the plant material was constantly saturated. From time to time, fine volcanic ash from volcanic eruptions to the west fell from the atmosphere or was washed down from the highlands. The ash was the main source of the silica. Ground water, somewhat acidic from decaying plants, began dissolving silica and traces of other mineral matter from the ash. Thus enriched, the ground water solutions gradually permeated the woody tissue, infiltrating the cells, filling the intercellular cavities, and coating the cell walls. We can only speculate as to what chemical or environmental conditions caused some of the precipitating silica to *replace* cells and some to *preserve* cells. We cannot know whether the processes operated concurrently or successively.

The well developed crystals of quartz and amethyst (some of gem quality) that are found in some of the petrified logs grew in hollows or open spaces that probably existed in the wood of the original tree.

soonal, with winds bringing in seasonal downpours from a coast that lay to the west.

About 40 different species of fossil plants (most of them extinct) have been identified in the Chinle Formation, the principal rock unit exposed in the park. In addition to the petrified logs, careful searches have uncovered roots, leaves, seeds, and other

plant debris preserved in fine grained shale beds. Included in the Triassic flora are rushes, and horsetails, large and small ferns, and large fernlike cycads. Such plants thrive on warm, swampy flood plains and in marshes.

Most of the logs are conifers belonging to a group named *Araucarioxylon*. Distant living relatives of these conifers are the monkey puzzle tree (native to Chile) and the Norfolk Island pine (native to Norfolk Island in the South Pacific). Some of the petrified trees were more than 200 feet tall and 10 feet in diameter. Evidently the conifers grew on highland slopes upstream before they were buried in the lowlands by sediment-laden flood waters. Some trees were healthy specimens, perhaps growing on riverbanks undercut during high water; but other trees were dead or dying of natural causes when they toppled down or were blown down. Evidences of insect borings, fungus rot, and charring by forest fires have all been preserved in the petrified remains of the fossil trees. Burial was rapid once the logs came to rest. Not many logs rotted before being covered. The water-saturated muds and sands sealed out air and prevented decomposition.

Fossils

Although plant fossils are by far the most abundant remains in the park, teeth, scales, and bones of extinct species of fish, clams, amphibians, and dinosaurs record the presence of animal life in the Triassic swamps. Many of these fossils were petrified by permineralization or pseudomorphic replacement. In 1985, the bones of a small, very early species of dinosaur, perhaps the oldest dinosaur skeleton yet discovered, was found near Chinle Point in the northern part of the park. The dinosaur, the size of a large dog, lived early in Triassic time. Remains of crocodiles, phytosaurs (crocodile-like animals), and archosaurs (advanced reptiles) have been found. During late Triassic the dinosaur plateosaurus lived here.

Petroglyphs and Desert Varnish

Carving petroglyphs in desert varnish on rocks seems to have been a favorite pastime of the prehistoric people who lived in the American Southwest. Thousands of their mysterious inscriptions are in Petrified Forest National Park. Some of the petroglyphs are drawings of human and animal figures, and others are abstract, geometric patterns, similar to the designs on Indian pottery. Newspaper Rock, near the Puerco Indian Ruin, is a massive sandstone block with many examples of rock art. The meaning of the petroglyphs is not clear, although some of them are clan symbols and many probably have ceremonial significance. Astronomers have discovered solar markers, carved on rock in Petrified Forest National Park, that correlate with the sun's position as the seasons change. Ancient sun-watchers used the petroglyphs as a calendar to keep track of the sun's progress across the sky so that they would know when to plant crops.

Archeologists and anthropologists who have studied the rock art assume that the drawings were done at different times and by individuals from different cultures, but no satisfactory method of dating the petroglyphs has been discovered. The Indians made the drawings by using a sharp tool, probably a piece of petrified wood, to scratch through the dark desert varnish, exposing fresh, light-colored rock underneath.

As sandstone weathers in a desert climate, a thin, dark brownish red or black patina, or *desert varnish,* develops on the rock surface. The varnish consists of iron and manganese oxides with some silica and is probably caused by the repeated evaporation of water drops from the rock surface. As the moisture dries, dissolved mineral matter exuded from the rock is left. Some authorities believe that acid-secreting lichens help to fix the varnish coating to the rock. Because the rate at which desert varnish accumulates is highly variable, it cannot be used for dating inscriptions.

Caliche and Other Desert Features

Tubes and nodules of caliche are exposed in ravines and on the flanks of mesas. *Caliche* is a general term for concentrations of soluble calcium salts (such as calcium carbonate) that precipitate from soil moisture at or near the surface. Under arid conditions caliche accumulates because rainfall is insufficient to wash calcium carbonate away or dissolve it. Sometimes caliche forms an impermeable "hardpan" layer beneath desert soil, as well as nodules and tubes. The latter represent decayed roots that have been coated.

Desert pavement tends to form on flat surfaces, such as on the flood plain of the Puerco River. Wind has removed soil and fine particles and left a closely packed surface of wind-polished pebbles and rock fragments, usually somewhat cemented together by calcium carbonate. Loose rocks, many of them fragments of petrified wood, have been so abraded by wind-blown sand that they have smooth, polished faces and sharp edges. These are *ventifacts.*

Badland Topography and the Painted Desert

To the north along Interstate 40 in this part of Arizona is the panorama of the Painted Desert—a landscape of brilliant, improbable colors against a deep blue sky. The multicolored rocks of the Chinle Formation are exposed in the myriad shapes of the badlands—gullies, ridges, arroyos, washes, mounds, escarpments, and mesas. Like the colors of Badlands National Park in South Dakota (Chapter 9), but much more brilliant, the hues of the Painted Desert change with the hours of the day. Colors are most brilliant in the morning and late afternoon but least intense at midday. Colors change when the surface is wet; and on a partly cloudy day, shaded areas and sun-lit slopes contrast beautifully. Thus the natural variations in daylight act on the sedimentary rocks to produce spectacular chromatic effects. The colors of the various shales, marls, sandstones, and limestones of the Chinle are purple, pink, gray, yellow, ash gray, lavender, rose, maroon, sienna, lilac, cream, chocolate, and various other shades of red and brown. Patches of blue, white, and even black are seen. Dominating overall are the reds, pinks, and purples.

Although annual precipitation is low, the nonresistant Triassic rocks are being eroded. Landforms change with every cloudburst and thunderstorm. Much of the badlands topography, however, was formed when rain was more plentiful than in today's climate. Interbedded with the thin layers of poorly cemented layers of gravel, mud, and sand are water-laid units of volcanic ash. Weathering has altered the ash to *bentonite,* a type of soft clay that absorbs water and swells as it becomes wet. When saturated, the clay slumps or may even flow as a fine muck. After it dries, the clay hardens again. Local slopes that develop on the bentonite clay units are rounded and bulbous because of the clay's plasticity (fig. 8.4). *Piping* causes sinkholes to develop in the tops of mounds and in gully bottoms. During downpours, water drains through tunnels, or "pipes," inside the mounds and the gullies, washing out loose material and undermining slopes that later collapse as sinks and slumps. New slope profiles and sinkholes may develop in a few hours when flash floods occur. Steeper slopes and escarpments develop on the sandstone and conglomerate beds of the Chinle. Some of the more resistant units form cap rocks on the buttes and mesas of the landscape.

Geologic History

1. Subsurface Paleozoic rocks.

Paleozoic rock units that are well exposed in many other parts of the Colorado Plateau are not seen in the region of Petrified Forest National Park. Some of the Paleozoic rocks were probably removed by erosion; the rest are buried by younger rocks.

2. Deposition of the Moenkopi Formation early in Triassic time.

At the beginning of Triassic time, the landmass that eventually became North America was still part of the supercontinent Pangea, and the Colorado Plateaus region was close to the equator. On a low coastal plain that sloped gently westward, sandy and muddy beds

FIGURE 8.4   Badlands topography in the Painted Desert portion of Petrified Forest National Park. Bentonite clay and shale beds of brightly colored Chinle Formation erode into bulbous slopes and slumps. A sandstone bed capping the ridge in the background provides some protection to the underlying clay and shale. Photo by David B. Hacker.

accumulated, 300 to 600 feet in thickness. Marine and calcareous units interfingered with continental beds, indicating that at times the sea encroached on the coastal plain. In some places gypsiferous beds formed in shallow lagoons between layers of mud. Sandfilled channels trace the courses of ancient streams. The climate was warm and alternately dry or humid. Only a few beds of the Moenkopi Formation remain in the Petrified Forest area because most of the formation was eroded as the plain was upwarped and the sea withdrew.

3. Deposition of the Shinarump Member of the Chinle Formation.

In a subsiding basin in north-central and northeastern Arizona, a typical basal conglomerate, containing many Moenkopi pebbles, was deposited unconformably over the eroded Moenkopi Formation. The gravelly, sandy, lower layers are indicative of high source areas and (presumably) streams that were able to transport the clastic material. The coarse sediment grades upward into finer material. This conglomerate, averaging 35 to 50 feet in thickness, is the Shinarump Member of the Chinle Formation.

4. Deposition of Chinle beds later in Triassic time.

The sea to the west continued to regress, leaving an immense alluvial plain in the Colorado Plateaus area. Sediment transported by the rivers became finer. The climate, still warm, became more humid, and tropical grasslands and marshes developed. Aprons of sand, mud, and silt spread over the northern Arizona lowlands, covering old flood plains, deltas, and backswamp areas. Localized units of gypsiferous clay represent evaporite deposits in lagoons or playa lakes, perhaps due to seasonal wet and dry periods.

Several hundred feet of sandy, shaly material accumulated during this time to form the Lower Petrified Forest Member and the Upper Petrified Forest Member of the Chinle Formation. These two members arc separated in some places by a discontinuous unit, the Sonsela Sandstone Member, which contains more petrified wood than any of the other rock units in the park. The Owl Rock Member at the top of the section completes the Chinle Formation in Petrified Forest National Park (Billingsley and Breed 1980).

Tectonic activities toward the end of the Triassic affected the northern Arizona basin. In the western sea, a chain of volcanoes (about where the California-Nevada boundary is now) began erupting large quantities of ash that were carried eastward by winds and dropped on the plains, basins, and highlands of the interior. Heavy rains washed quantities of ash, mixed with mud and other debris, into the Petrified Forest area. The volcanic ash is the source of the silica in the petrified wood.

5. The wood in the Chinle Formation.

The Chinle Formation is famous in the western states for its petrified wood. Although the finest and

Table 8.1			Geologic Column, Petrified Forest National Park		

Time Units			Rock Units		Geologic Events
Era	Periods	Epochs	Formations	Members	
Cenozoic	Quaternary	Holocene	Stream, wind, slope wash, and slump deposits		Fluctuating climate, becoming more arid; development of badland topography. Intensified erosion during pluvials
		Pleistocene			
	Tertiary	Pliocene	Bidahochi		Coarse sediment interbedded with and topped by pyroclastics from explosive eruptions northwest of park area
					──── Extensive erosion ────
		Paleocene			Upwarping, faulting. Laramide orogeny
Mesozoic	Cretaceous Jurassic		(rock units eroded from park area)		Extensive marine and nonmarine sedimentation
					──── Erosion ────
	Triassic		Chinle	Owl Rock	Gypsiferous clays accumulating in playas and lagoons; ashfalls
				Upper Petrified Forest	
				Sonsela Sandstone	Contains largest accumulation of trees and plant debris
				Lower Petrified Forest	Extensive deposition in interior basin; shales, sandstones
				Shinarump	Basal conglomerate containing Moenkopi fragments
					──── Erosion ────
			Moenkopi		Shallow marine and nonmarine sedimentation
			(older rock units not exposed in park)		

Source: Modified after Billingsley and Breed, 1980.

most abundant specimens are in the Petrified Forest area, the same species of conifers have been found in Chinle beds in other places where the formation crops out. Obviously the forests were widespread. A late Triassic highland extending from central Colorado to northern New Mexico may have been the source of the large conifers washed into the northern Arizona basin. It was there in the swampy lowlands that the ferns and cycads flourished. The fact that their fossil remains are found together, showing little evidence of decay, suggests that the uprooting, transportation, and swamp burial of the conifers occurred in relatively short spans of time, possibly during monsoonal floods of considerable magnitude.

Just when and for how long the processes of petrifaction and permineralization went on is not known.

Perhaps the elevation of the Colorado Plateau region, beginning as the Mesozoic Era closed, terminated the silicifying of the buried wood.

6. Accumulation of Jurassic and Cretaceous sediments.

Several thousand feet of Jurassic and Cretaceous sediments were deposited on top of the Chinle Formation in the Colorado Plateaus region; but in the Petrified Forest area, these beds were subsequently eroded. The weight of the rock units helped to compact the Chinle Formation, while groundwater, percolating through the sediments, probably affected the wood buried in the Chinle.

7. Tertiary uplift, warping, and erosion.

Through the last of the Mesozoic and the beginning of the Cenozoic Eras, the Laramide orogeny, with its pulses of tectonic activity, shifting from west to east, profoundly affected western North America. Although the Colorado Plateaus were uplifted and slightly tilted during Laramide mountain-building, the flat-lying rocks remained relatively undeformed. Around the margins of the plateau, some broad folding and upwarping occurred. In northeastern Arizona, the upwarping (Defiance uplift) resulted in the shaping of the park landscape that we see today. Erosion stripped off Jurassic and Cretaceous rocks and some of the Chinle beds from the Petrified Forest region. (Some of these units are preserved in downwarped basins only about 50 miles away from the park.) Meanwhile the climate became cooler and increasingly arid. The mountain barriers rising to the west cut off moisture-laden winds from the Pacific.

8. Late Cenozoic sedimentation and volcanism.

In Pliocene time, an irregular blanket (in places several hundred feet thick) of coarse sediment and volcanic debris was spread unconformably across the eroded surface of Mesozoic rocks in this part of Arizona. This is the Bidahochi Formation. Only a few remnants are left in the park, mainly as cap rock on

FIGURE 8.6   In the Petrified Forest, prehistoric Indians left pictographs (picture writings) scratched out in the desert varnish, a thin dark patina, consisting of iron and manganese oxides, that develops on rocks in arid regions. Photo by Ronald F. Hacker.

mesas and divides in the Painted Desert. Overlying the Bidahochi sediments (and interbedded with them) are units of basaltic debris derived from volcanic vents northwest of the park. Large *volcanic bombs* (ellipsoidal fragments that were ejected in a molten state and then hardened in the air) can be seen in roadcuts in the northern part of the park. The bombs, as well as large angular boulders of basalt, show that the eruptions were nearby and highly explosive.

9. Pleistocene and Holocene erosion.

During the waxing and waning of Pleistocene ice sheets to the north and northeast, regions outside glaciated areas such as Petrified Forest had intervals of *pluvial climate,* that is, periods of high precipitation. In general, the pluvials correspond to transitional or glacial periods in the glaciated regions. Erosion accelerated during the pluvial intervals, and probably more rain fell throughout the year instead of being largely concentrated in summer as at the present time. The nearly complete removal of the Bidahochi sediment and volcanic debris indicates that the intense erosion of Tertiary time continued through the Pleistocene but that erosion is somewhat slower today due to the increasing aridity of the region. By before Quaternary time, the Little Colorado River and its tributary, the Puerco, had cut down their valleys into the soft Triassic shales of the Chinle Formation. In the semiarid lands of the Southwest, slight changes in weather patterns and annual precipitation that become climatic trends or persist for more than a few years have profound effects on plant, animal, and human life.

Sources

Ash, S.R.; and May, D.D. 1969. *Petrified Forest: The Story Behind the Scenery.* Petrified Forest National Park, Arizona: Petrified Forest Museum Association.

Bezy, J.W.; and Trevena, A.S. 1975. *Guide to Twenty Geological Features at Petrified Forest National Park.* Petrified Forest National Park, Arizona: Petrified Forest Museum Association.

Billingsley, G.H. and Breed, W.J. 1980. *Geologic Cross Section along Interstate 40 from Kingman to the New Mexico-Arizona State Line.* Petrified Forest National Park, Arizona: Petrified Forest Museum Association.

Gregory, H.E. 1917. *Geology of the Navajo Country.* U.S. Geological Survey Professional Paper 93.

Long, R.A. and Houk, R. 1988. *Dawn of the Dinosaurs, the Triassic in Petrified Forest.* Petrified Forest Museum Association, Petrified Forest National Park, 96 p.

Mears, B. Jr. 1968. Piping. In Encyclopedia of Geomorphology, ed. R.W. Fairbridge, p. *849.* New York: Reinhold Book Corporation.

Nations, D. and Stump, E. 1981. *Geology of Arizona.* Dubuque, Iowa: Kendall/Hunt Publishing Company.

Rigby, J.K. 1977. *Southern Colorado Plateau,* K/H Geology Field Guide Series. Dubuque, Iowa: Kendall/Hunt Publishing Company.

Schopf, J.M. 1971. Notes on Plant Tissue Preservation and Mineralization in a Permian Deposit of Peat from Antarctica. *American Journal of Science* 271:522–543 (December).

———1975. Modes of Fossil Preservation. *Review of Paleobotany and Palynology* 20:27–53.

Trimble, S. 1984. *Earth Journey, a Road Guide to Petrified Forest.* Petrified Forest Museum Association, Petrified Forest National Park.

Note: To convert English measurements to metric, go to www.helpwithdiy.com/metric_conversion_calculator.html

Badlands National Park

WEST CENTRAL SOUTH DAKOTA

Rodney M. Feldmann   Kent State University

Area: 242,751 acres; 379 square miles
Proclaimed a National Monument: January 25, 1939
Redesignated a National Park: November 10, 1978
Designated a World Heritage Site: 1979

Address: P.O. Box 6, Interior, SD, 57750
Phone: 605-433-5361
E-mail: badl_information@nps.gov
Website: www.nps.gov/badl/index.htm

An intricately sculptured landscape created by rain, ground water, and frost, in layered rocks, nearly barren of vegetation.

Badlands slopes, pinnacles, and buttes are eroding back into remnants of the shortgrass prairie on high tablelands. Ancient soil horizons and a rich variety of fossils depict a Tertiary time of verdant flood plains, grasslands, and forests, teeming with life. Also exposed are concretions, geodes, clastic dikes, fissures, and faults.

FIGURE 9.1 The Brule Formation is present in much of Badlands National Park. It consists of fine-grained shales, siltstones, and volcanic ash. Photo by Sherry Schmidt.

The Badlands terrain is beautiful and desolate; fascinating, yet formidable. Early settlers avoided this land because it was poor for farming and a barrier to travel.

Yet, from this hostile terrain has come much of what we know about North America's "Golden Age of Mammals" in the Oligocene Epoch. The strange bones in the walls of the canyons and cliffs and along stream banks aroused the curiosity of nineteenth-century naturalists. As it turned out, this region was probably the richest storehouse of vertebrate fossils in all of North America. Among the first scientists were paleontologists Fielding B. Meek and Ferdinand V. Hayden, who were with an expedition of the Corps of Topographical Engineers sent out to explore the Upper Missouri River and the Dakota Badlands in the 1850s. Meek and Hayden brought back many specimens from the great caches of extinct mammal fossils so that these bones could be studied and identified by scholars. Dr. Joseph Leidy of Philadelphia, who found in the collection bones of tiny primitive horses, wrote a paper on the evolution of the horse, the first significant study of the subject.

Paleontologists came to the area in increasing numbers. The U.S. Geological Survey, the great natural history museums, and universities of America and Europe sent expeditions and established "digs." One can open practically any textbook on historical geology or vertebrate paleontology and find pictures of fauna from the Brule and Chadron Formations of the White River Group—famous names, these, in American geology. An excellent collection of Badlands fossils is displayed in the Museum of Geology of South Dakota School of Mines and Technology in Rapid City.

Unfortunately, private collectors and amateurs—perhaps more enthusiastic than discriminating—began collecting in alarming numbers. Congress decided in 1929 that this record of ancient life forms, set in remarkable and extensive badlands, should be preserved. Ten years later the national monument was proclaimed. After several additions to the protected area, primarily to conserve the shortgrass prairie, Badlands was redesignated a national park.

Today a 28-mile scenic highway with numerous overlooks winds through the park's North Unit. The best times to take this drive are early in the morning and late in the afternoon. At those times of day, the sun's rays slant into the deep gorges and gullies and light up hues of pink, yellow, green, maroon, cream, gray, and red on the spires and pinnacles. Buttresses and cliffs stand out in relief. After every rainstorm—the short, infrequent, heavy rains typical of a dry climate—the colors and shapes are different.

In the park's South Unit (which lies within the Pine Ridge Indian Reservation) is Stronghold Table, a large butte where "ghost dances" were performed by Oglala Sioux Indians late in the nineteenth century. They believed that the ceremony would bring back the buffalo and cause the white men to disappear from Indian lands. In December 1890, Chief Big Foot tried to elude the 7th Cavalry by leading his people through the Badlands toward the Pine Ridge Reservation. However, cavalry troops caught up with the Indians on December 28 and escorted them to Wounded Knee Creek (south of the park) where they camped for the night. In the morning, when the troops were disarming the Indians, a rifle discharged. Indiscriminate shooting followed until more than 200 Sioux—men, women, and children—were dead. Twenty-five soldiers died and 39 were wounded, some of them hit by their own fire. In bitter cold as a blizzard approached, the wounded soldiers and about 50 wounded Indians were taken in wagons to Pine Ridge. The bodies of the dead Sioux were later interred in a mass grave above the creek. This was the last major event of the Indian Wars.

Geologic Features

Prehistoric Animals of the Badlands

During the Oligocene Epoch, between 38 and 24 million years ago, the region that is now the Badlands supported many kinds of animals. The climate was especially benign early in the Cenozoic Era when it was warmer and wetter than later in the era. Plants and animals flourished on the flood plains. The remains of successive generations of animals were covered over by relatively rapid sedimentation, and their bones and teeth were preserved in the rocks. Sometimes animals perished in seasonal floods and were quickly buried; thus whole skeletons became fossilized.

Some Oligocene species were the direct ancestors of modern horses and camels. Strangely enough, although most of the evolution of the camel and the horse took place in North America, both had become extinct on this continent before the arrival of the white men. What caused their disappearance is a mystery. Nevertheless, horses survived and multiplied on the western plains after the Spaniards reintroduced them to

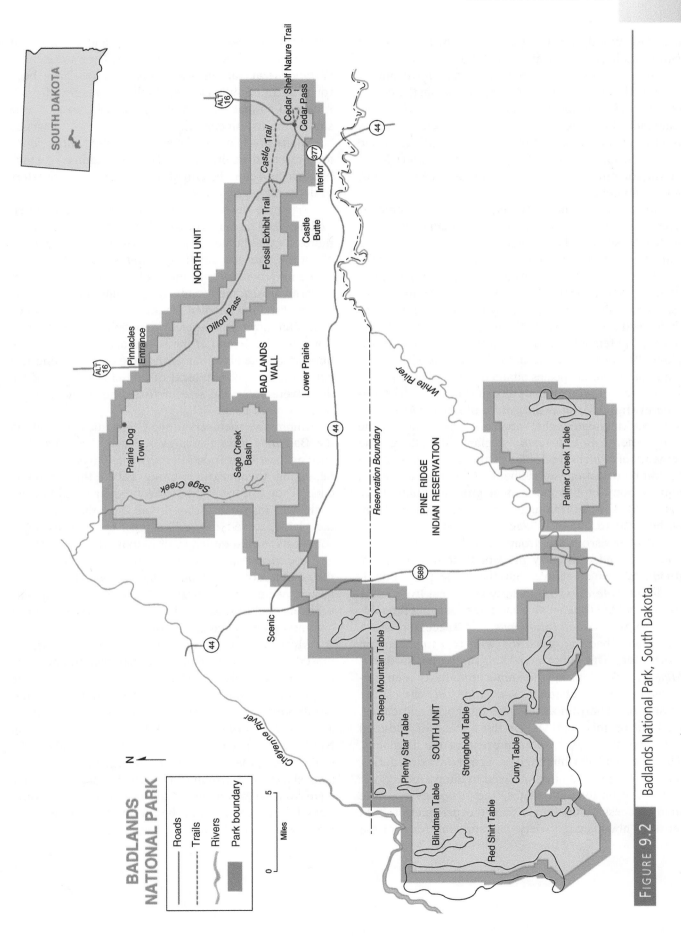

FIGURE 9.2 Badlands National Park, South Dakota.

the New World. Clearly, the North American habitat was suitable for horses after the 1500s.

Some groups, the saber-toothed cats, for example, have died out completely. Ancient turtle shells, on the other hand, the most common fossil found in the Badlands, look just about the same as the shells of modern species. Other reptiles, such as alligators and lizards, lived in this lush, well watered land. Many groups of birds were present, including eagles, owls, gulls, and pelicans.

Most significant, however, were the mammals. They began to take over after the dinosaurs (the rulers of the Mesozoic) disappeared from the earth. By the time the Oligocene Epoch began, due to the mammals' rapid expansion into habitats left by the dinosaurs, most of the modern mammalian orders had become established. Carnivores (flesh-eaters) diversified rapidly as their main source of food, the ungulates (hoofed mammals), proliferated in forests and marshes and on flood plains. Not only was this a time of great diversity among mammals; several species attained impressive size.

Largest of all mammals at that time were the *titanotheres* (fig. 9.3), an extinct order of *perissodactyls,* or odd-toed ungulates, that attained their peak in the Oligocene. *Brontotherium,* a giant titanothere, the remains of which are found in the sandstones of the Chadron Formation, stood about 12 feet high and had a pair of bony protuberances that grew hornlike above the snout. Like other ungulates, the titanotheres were herbivorous (eaters of foliage and grasses).

Most perissodactyl groups had three toes to begin with and tended to carry most of their weight on the middle toe, which evolved into the single hoof of modern horses. *Mesohippus,* an early horse that lived in the forests and ate foliage, still had three toes, but the middle one was large. *Pliohippus,* his successor and a grass-eater, had one great toe with two tiny ones on either side. The bones and teeth of the perissodactyl *Metamynodon,* an aquatic rhinoceros, have been collected from a channel sandstone in the Brule Formation. The rhinoceros, like the horse and camel, died out on this continent. A third group of odd-toed ungulates, the tapirs, were present in the Badlands. Their descendants migrated to South America during the Pleistocene and have survived to the present.

The even-toed ungulates, the *artiodactyls,* were more successful mammals than the perissodactyls in terms of abundance, variety, and survival. Most of the

Oligocene artiodactyls had four toes and carried their weight on the third and fourth digits. Bones of camels (*Tylopoda*) at varying evolutionary stages have been found in the South Dakota Badlands, ranging from a four-toed animal the size of a jack rabbit to a species similar to modern camels.

Many bones of piglike animals, called *entelodonts,* are preserved in the Badlands. These animals, which became extinct in the Oligocene, were not the ancestors of domestic pigs.

Primitive flesh-eaters, called *creodonts,* are classified as "archaic carnivores." They were forerunners of more efficient hunters, the true carnivores. *Hyaenodon,* about the size of a small bear, was an Oligocene, carrion-eating creodont. In the evolutionary development of carnivores, some became doglike and others became catlike; but they retained features in common for some time after they differentiated from their shared ancestor. For example, several of the Badlands dogs had curved, hooded, catlike claws rather than the curved cylindrical cones that are more typical of dog claws. Most of the dogs were about the size of foxes, but some were as large as black bears.

Impressive members of the Oligocene cat family in the Badlands were the large saber-toothed cats. Two species have been found in the White River beds, both belonging to an extinct group called the "stabbing cats." They had the ability to swing their lower jaw out of the way (as some snakes do) and stab their prey with their large, curved, daggerlike upper fangs. These awesome cats preyed on large, slow-moving mammals.

Insectivores are represented in the fossil record by the golden mole, hedgehogs, and shrews. Other rodent groups were ancestors of modern beavers, rabbits, squirrels, and rats. Squirrels and rabbits were the most common rodent ancestors in the Badlands.

Most of the modern ruminants (cud-chewers), the most diverse and abundant of all mammal orders today, evolved in Eurasia and Africa. However, a large group of primitive ruminants, unrelated to modern species, did thrive in the Oligocene. Fossilized bones of these animals, called *oreodonts* (fig. 9.3), are found only in North America. The cud-chewing oreodonts, about the size of foxes, roamed the plains in large herds. Many of their skeletons are found in huddled groups as if they were overcome by a natural disaster such as a blizzard. They became extinct well before the beginning of the Ice Age.

FIGURE 9.3   Excavation of bones has revealed the lower jaw of a titanothere, an extinct group of animals that first appeared in the Eocene. Many animal skeletons have been found in the park, such as pigs, horses, rhinoceros, titanotheres, "stabbing cats," and rodents. Photo by David B. Hacker.

The Animals' Environment

The Great Plains are part of a vast midcontinent area overlying the North American craton. (A *craton is* a portion of the earth's crust that has been stable for a very long time.) Extending eastward from the Rocky Mountains to the Mississippi valley, the Great Plains supported grasslands from southern Canada nearly to Mexico before the settlers·came.

Although they remained generally flat-lying, the sedimentary rocks of the Great Plains were strongly affected by the intense tectonic activities that took place to the west at the end of the Mesozoic. During the Laramide orogeny, the regional tilt of the Great Plains was increased from west to east as the land rose gradually.

In an early Tertiary uplift, the dome of the Black Hills bulged upward, along what is now the South Dakota-Wyoming border, only about 70 miles west of the Badlands. As erosion removed Cretaceous marine shales and sandstones from the Black Hills, and even part of the Precambrian crystalline rocks, rivers spread the great quantities of sediment over the Badlands in Oligocene time. Late in the epoch, clouds of volcanic ash, borne by winds from eruptions far to the west, were a main source of sediment. All these materials became consolidated into the conglomerates, sandstones, mudstones, and associated *paleosols* (ancient buried soil horizons) that formed the tablelands and badlands of the Dakota Great Plains.

Interpreting Paleosols in the Badlands

Periods of uplift alternated with quieter times and resulted in alternating intervals of stream erosion and deposition. During periods of stability, well developed soils formed over wide areas. The earlier soils, which formed under warm, moist conditions, resemble soils that are presently found in southeastern United States. Although the vegetation is poorly known, it was probably a dense, canopied forest of warm-temperate and subtropical angiosperm trees (tree species that bear seeds enclosed in seed cases).

Strongly renewed erosive activity by streams early in Oligocene time, resulting from an episode of upwarp and tilting to the west, deepened the valleys temporarily in the Badlands region. However, as the valleys filled with sediment, the landscape became flatter. The paleosols of this time exhibit calcium carbonate horizons and other features that indicate a progressively drier, cooler climate. Forests became increasingly confined to flood plains. The uplands changed first to open savannas, and finally, in latest Oligocene time, to semiarid steppes. (A *savanna* is a subtropical grassland having scattered trees and a pronounced dry season; a *steppe* is a broad, flat, treeless, dry region of scattered bushes and short-lived grasses.)

Recent careful study of Tertiary sedimentary layers in the Badlands has revealed that the entire sequence contains abundant evidence of paleosols (Rettalack, 1983). At least 87 different paleosols were examined, and each one provides evidence of soil-forming factors, such as climate, organisms, parent material, and topography that prevailed at the time. Characteristic features—including fossil root traces, fossil burrows, mammal bones, mollusk shells, typical soil structures, color banding, and calcareous layers—help to identify the fossil layers and interpret their environment of formation. Many of the reddish bands that are seen throughout the Badlands represent paleosols.

How the Badlands Developed

The *badlands topography* of the park is characterized by complex, intricately dissected steep slopes and sharp ridges. Fine drainage networks are developed on soft,

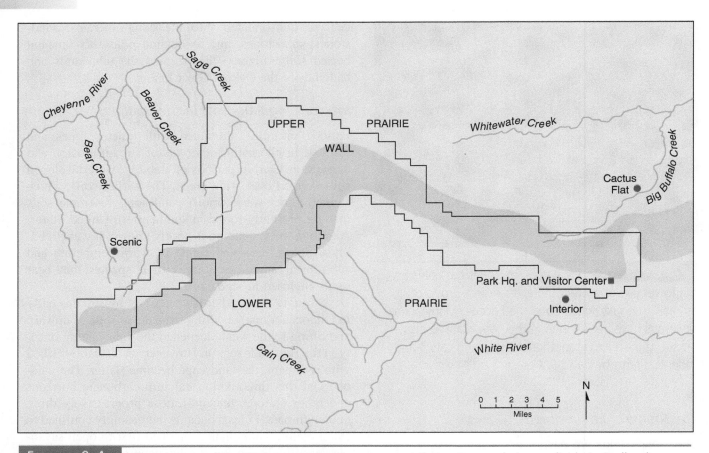

FIGURE 9.4   Sketch map of the Wall, an erosional escarpment that acts as a drainage divide in Badlands National Park. Adapted from R.R. Churchill 1979.

erodible, flat-lying sedimentary rocks that lack a vegetation cover. Local relief is about 100 to 400 feet. Conditions favorable for the sculpturing of the Badlands landscape came about as follows: (1) erosive activity became more intense when renewed uplift occurred late in Tertiary time; (2) the climate became semiarid; (3) precipitation occurs mainly as infrequent heavy rains or cloudbursts; (4) vegetation did not become well established even during moister episodes in Pleistocene time; and (5) the high bentonite clay content in the sedimentary rocks causes rapid surficial erosion when they are exposed to moisture.

Bentonite is a weak sedimentary rock, containing a high proportion of clay minerals (belonging to the montmorillonite group), which are derived from weathered volcanic ash. Because the clay minerals swell in contact with moisture, bentonite can absorb large quantities of water and may increase its volume about eight times. When bentonite dries out after a soaking by rain or melt-

ing snow, it "slakes," which means that the surface material crumbles and disintegrates so that fine sediment is either blown away or washed away by the next rain.

The South Dakota Badlands are being eroded into the Wall, a south-facing, retreating escarpment, 60 miles long and three to five miles wide (figs. 9.4, 9.5). From the top of the Wall, sloping gently to the north is the Upper Prairie, a nearly flat grassland, slightly incised by north-flowing tributaries of the Bad River. In some places, the Wall juts up above the Upper Prairie and is being eroded from both sides. The deeply dissected badlands terrain of the Wall is drained by tributaries that flow southward into the White River. The top of the Wall is, therefore, a drainage divide. The Cheyenne River, west of the park, also receives some Badlands drainage. The water of all three rivers eventually discharges into the Upper Missouri.

Below the Wall is an abrupt transition to the Lower Prairie, a partly grassy pediment area that slopes down to the flood plain of the White River. Between the top of the Wall and the distinct boundary where the Lower Prairie begins, relief averages about 150 feet but rises to more than 200 feet in some places. Incised, intermittent stream channels and "sod tables" (erosional remnants) break up the surface of the Lower Prairie. Park Headquarters is on the Lower Prairie, which is about 3 miles in width. The White River flood plain, below the Lower Prairie, is about a half mile wide.

Once much closer to the White River, the Wall is steadily eroding northward as the river widens its flood plain and the escarpment is dissected into badlands topography. The White River has cut all the way down through the Cenozoic rock units into the Mesozoic formations below. It is this downcutting and escarpment retreat that has exposed the fossils and paleosols.

Faults

A series of linear normal faults (associated with a structural trough) crosses the Badlands area. The amount of displacement, a few feet in most cases, shows up in the alternating bands of light- and dark-colored sediment that can be seen on either side of a fault. The fault at Dillon Pass, however, shows that movement there was as much as 49 feet. The faults are related to movement in the Precambrian-Paleozoic basement complex far below the sedimentary rocks of the Badlands.

Erosional Features of the Badlands

Erosion rates in the Badlands are among the highest known. When photographs of Badlands landforms taken early in the century are compared with recent photographs of the same features, it is apparent that marked changes in form and height have taken place. During World War II, planes from a U.S. Army Air Force base near Rapid City sometimes discharged surplus ammunition in the Badlands on the way back to base after target runs. Around 1960 a park ranger found a projectile in the ground, its top flush with the surface on a sloping hillside. Today the projectile sticks up more than a foot above the ground.

FIGURE 9.5   Looking northward at the Wall, which is cutting back into the Upper Prairie. The Lower Prairie grassland is in the lower two-thirds of the photograph. Photo by Ann G. Harris.

The rate of erosion varies with location and rock type. Spires in loosely consolidated ash may lose half a foot of height per year. Tops of mudstone mounds may be lowered about one inch per year. Resistant sandstone caprock, on the other hand, may show an erosion rate of an inch in 500 years.

The dominant geologic agent in the Badlands is running water. During cloudbursts, pelting raindrops loosen grains and particles, and running water washes loose material downslope. Miniature gullies, eroding headward, enlarge rapidly. When infiltration occurs, the slope material may become saturated and move downward by mass wasting processes, slowly as creep or faster by sliding and slumping. Sometimes an entire slope face moves at one time. Cliff Shelf, a large slump area near Cedar Pass, continues to slip down, especially during wet seasons.

The lithology of the rocks influences erosional processes. The high clay content and fine-grained texture of the shale, siltstone, and ash of the Brule rocks limits infiltration of rain water, making runoff the dominant process. Thus many spires and pinnacles have formed. In the Chadron rocks, which have more sandstone and weaker cementation, a higher rate of infiltration makes mass wasting processes and downwastage more significant.

Frost action, which can be severe in the South Dakota climate, speeds the breaking up of the rocks. Repeated wetting and drying also causes cracks to form, with each tiny fissure providing more access for water.

Topographic position is a factor in slope development of tables, or buttes. The north-facing slopes, being protected to some extent from the sun's radiation, retain moisture longer than south-facing slopes. Water has more time to react with calcium carbonate cement that holds the rock together, so that chemical weathering penetrates farther into the rock. Thus creep or slump of weathered material is more likely to occur on north-facing slopes; and when rain comes, the loose, weathered material is easily washed downslope. North-facing slopes tend to have more risers or slope segments and are usually not so steep as south-facing slopes.

South-facing slopes, on the other hand, tend to have high, steep slopes and angular dropoffs. Weathering on

FIGURE 9.6   Resistant sandstone caprock forms a "mushroom" after most of the surrounding mudstones and siltstones have eroded away. In time, as undermining continues, the caprock will tumble down the slope. Photo by Douglas Fowler.

south-facing slopes is extremely limited because they dry very quickly when storms end and the sun comes out. Once dried out, any loosened fragments or blocks fall rapidly, collecting as debris at the bottom of the slope face (Churchill 1979).

Some of the strange-looking erosional landforms in the Badlands are caused by layers or patches of resistant caprock. The rounded tops of the "toadstools" (rock pedestals or hoodoos) found in the park are made up of channel sandstone boulders that have broken off and dropped down during slumping. The clay pedestals supporting the sandstone caps erode faster and become thinner until the structure topples over (fig. 9.6). Arches have formed where weakly consolidated beds have washed out faster than overlying beds that are more firmly consolidated. These arches have a short life and soon cave in. Meanwhile, new ones are carved out by the never-ending processes of erosion.

A ground water process called *piping* (a term from engineering geology) speeds mass wasting in the Badlands, particularly in the weakly cemented sandstones of the Chadron Formation. Percolating ground water, infiltrating soil and rock, and dissolving cement in sandstone, initiates the formation of little tunnels, or pipes, in sediment and rock. These conduits enlarge as rivulets drain into them or as storm waters rush into them and scour out tunnels. Eventually a valley slope, an arch, or a slump body becomes undermined by the pipes and may collapse suddenly during or right after a hard rain. Landslide masses that have temporarily dammed tributary valleys may give way or wash out after being undermined by water flowing in pipes or tunnels beneath the surface.

Clastic dikes stand out in relief against adjacent rocks that have been eroded back. These features were formed by inwash of clastic sediment (i.e., sand, silt) that filled cracks and fissures in the sedimentary rock. Chalcedony (silica) precipitated by circulating groundwater cemented the sandy sediment into hard, resistant rock. Little veinlets of chalcedony also filled joints and fractures in the rock. Once the less resistant enclosing rock has been removed by erosion, the dikes and veinlets are left as thin, resistant ridges, walls, and "shingles." Some of them are a foot or more wide and others only an inch or two wide (fig. 9.7).

Why the Water Is "White"

When streams are running in the Badlands, they are always loaded with sediment. However, even when

FIGURE 9.7 Clastic dikes formed by volcanic ash that washed away into a fissure or fracture in sedimentary rock and then hardened. Vertical clastic dikes crisscross many areas of the Badlands. Photo by Ann G. Harris.

streams slow down or water stands so that material in suspension drops, the water remains cloudy and light-colored. This is because submicroscopic particles of clay are in *colloidal dispersion;* that is, the particles carry a slight charge of electricity that causes them to repel each other and prevents them from settling to the bottom. The White River was given its name by early visitors to the region because of its cloudy water.

Pioneers in the Badlands region discovered that pouring a pail of "alkali water" (taken from a waterhole in shale) into a barrel of white water would cause it to settle so that it could be used. When a railroad was built through the Badlands in 1907, providing drinking water for the teams of horses used in construction work was a serious problem until someone discovered that throwing slices of cactus pads into a horse trough or waterhole would clear the water enough so that the animals could drink it. Obtaining a supply of clear water in the Badlands region can be a problem even today; however, the National Park Service has drilled shallow wells in gravel deposits near the White River that furnish enough clear water for park facilities.

Concretions and Geodes

Like the fossils, these curious objects are to be seen but *not collected* within the park. The calcareous concretions, found in all the formations of the Badlands, are more numerous and more varied in size, shape, and type than those ordinarily found in sedimentary beds. The nodular zones of the Brule Formation have the most concretions. These zones, which are probably floodplain deposits, show up in the sedimentary sequence as red beds with a lumpy appearance. The nodules (concretions) that have weathered out tend to be rusty brown due to oxidation. Many of the concretions contain a fossil fragment as a nucleus. Around the nucleus are tightly cemented layers of clay particles.

Like the calcareous concretions, geodes are present in all of the park's formations and are most numerous in the nodular zones of the Brule.* The nodules that weather out are apt not to show by their outward appearance whether they are calcareous concretions or geodes. But geode nodules that have been broken can be identified as such by small crystals in hollow interiors. By definition, *geodes* are hollow (or partly hollow), globular-shaped bodies that occur in sedimentary rock and that weather out as discrete nodules. (Geodes are a type of concretion.) A geode's outer layer is made of dense chalcedony, and crystals project from its lining into a central cavity. The crystals, most commonly quartz or calcite, are usually perfectly formed and are not the same minerals as those of the enclosing rock (box 9.1). Sometimes the chalcedony shell is further enclosed by coatings of cemented clay.

Unusual specimens of calcareous concretions and geodes, as well as fossils and minerals, are on display in the Museum of Geology at the South Dakota School of Mines and Technology in Rapid City and in the Badlands National Park Museum.

Geologic History

1. Deposition of Pierre Shale, over 1200 feet in thickness.

The geologic record exposed in the White River Badlands began in Late Cretaceous time, between 80 and 65 million years ago, with the deposition of silts and muds in the shallow Pierre Sea, which stretched from the area of the Gulf of Mexico north into Canada. The skeletal remains of marine animals, including sea lizards, fish, and turtles, have been preserved as fossils in the shale beds. When the land rose, due to Laramide mountain-building in the west at the end of the Cretaceous Period, the sea retreated.

2. Development of Yellow Mounds paleosols.

Millions of years passed while sun, rain, and vegetation weathered the upper layers of the black Pierre Shale to a bright yellow and purple soil called the Yellow Mounds paleosol. Erosion has exposed these layers in the park at the Yellow Mounds Overlook and in the Sage Creek area.

3. Severe erosion and extensive deposition due to the doming of the Black Hills; Chadron Formation.

By the beginning of the Oligocene Epoch, uplift had brought the Black Hills to their maximum elevation. Rivers draining the mountains carved a rolling terrain of hills and valleys. In the Badlands region, where relief was not great, streams loaded with sand and gravel dumped the sediments as they slowed down. Layers of clastics covered the old weathered and eroded surface. These deposits became the Chadron Formation, the oldest formation of the White River Group.

Three members, Ahearn, Crazy Johnson, and Peanut Peak, make up the Chadron Formation. The Ahearn was deposited as channel fill in an ancient drainage network that antedated the drainage patterns of today. Because the river system's conditions changed frequently, crossbedding and cut-and-fill deposits are common. The streams that deposited the Crazy Johnson and Peanut Peak Members* had lost some volume and erosive energy. Crazy Johnson Member is a mixture of mudstone and channel fill, its color ranging from green-gray conglomerate to gray-blue or green mudstone. Peanut Peak Member is similar to Crazy Johnson but thinner and lighter in color. It has more tan and orange mudstone and less of the green

---

*Dr. Philip Bjork, Director, Museum of Geology, South Dakota School of Mines and Technology. Personal communication.

*The names of these Chadron members refer to a Badlands settler, dubbed "Crazy Johnson" by his neighbors, because he tried to grow peanuts on a large, flat-topped sod table, still called "Peanut Peak."

Box 9.1   How Are Geodes Formed?

Geodes are like snowflakes; no two are exactly alike. They may be large, small, hollow, or nearly solid. Some, called "rattle rocks," contain loose crystals; some are filled with connate water (water trapped in sediment). In the Badlands, geodes have even been found clustered together *inside a* calcareous concretion or—even more surprising—encased in the long bone of a mammal fossil.

Geode occurrences are not very common worldwide. They are found most typically in limestone, but occasionally in marine shales. The origin of geodes is not clear. No one theory seems to account satisfactorily for all varieties of geodes and their different sedimentary environments. The puzzle involves getting what was inside *out* and putting what is now inside *in.*

A widely accepted theory that appears to fit many geodes suggests that they began with the burial of a shelled animal or other object in the limy ooze of an ocean bottom. The shell or bone provided the original cavity; then silica gel accumulated in the void, surrounding and trapping a small amount of sea water. Later, after the sedimentary unit had been uplifted, groundwater began circulating through the rock. Osmotic pressure between the trapped sea water inside the silica gel and fresh water outside caused the membrane of silica to expand and eventually to dry out and become the chalcedony layer. Minute cracks in the chalcedony permitted mineral ions in circulating groundwater to enter the geode and gradually crystallize out of solution. Layers of clay particles may have become cemented to the exterior because they could not pass through.

Obviously, some modifications in this theory must be made in order to account for the Badlands geodes. The only Badlands marine sequence is the Cretaceous Pierre Shale. All the overlying beds formed in a terrestrial environment. Most of the rock layers are shales, siltstones, and sandstones; there is no marine limestone, and there are only a few lenses of freshwater limestones here and there. Yet geodes are found in all formations (but not in every layer).

But what was happening during and after lithification of the sediment to make such a varied assortment of both calcareous concretions and geodes in the Badlands formations? Ground water must have been plentiful and it was probably close to the surface (high water table). But did it just percolate downward, or was it also forced upward by hydrostatic pressure that brought mineral ions from rocks farther down? Silica in soluble form must have been readily available over a long period of time. Was the highly siliceous volcanic ash that settled on and washed into the region the source? Why are Badlands geodes sometimes associated with calcareous concretions and sometimes not? Rockhounds, who regard geodes as fascinating natural objects, can doubtlessly add more questions.

and gray. As rainfall decreased, vegetation changed, bringing about changes in faunal habitat. The titanotheres were approaching the culmination of their development, and many of their bones are found in the Chadron Formation, along with bones of other mammal orders. Chadron rocks tend to weather to low, rounded "haystack hills."

4. Interior paleosols.

Transitional between the Yellow Mounds paleosols and the Oligocene formations are the red, weathered clays of the Interior soils, named for the nearby town of Interior. These red paleosol layers, which overlie the Yellow Mounds, also developed on flood-plain deposits and channel fills in the Chadron and the Brule Formations. The paleosols represent stable times during which soil horizons formed.

5. Deposition of the Brule Formation.

A thin, discontinuous nonmarine limestone marks the boundary between the Chadron and Brule Formations, suggesting the existence of shallow ponds at that time. Overlying this limestone sheet is the Lower Nodular Zone that contains, in addition to concretions, many fossils of turtles, oreodonts and titanothere (fig. 9.3). Although the overall trend during Brule time was toward a cooler and more arid climate, cycles of warm, wet climate with widespread flooding did occur. The flooding laid down sheets of thick, gooey mud that preserved fossil bones as it dried.

Table 9.1			Geologic Column, Badlands National Park			
Time Units			**Rock Units**			
Era	Period	Epoch	Group	Formation	Member	Geologic Events
Cenozoic	Quaternary	Holocene	Alluvium			Deposits in streambeds
			Landslide deposits			Slumped masses of shale, silt, and ash
		Pleistocene				Stabilized dunes
			Windblown sand			Erosion of badland topography during last 500,000 years
			Sod tables			Channel fills
			River gravels			Pediment deposits
			Older alluvium			
						Erosion ————————————————
	Tertiary	Oligocene	White River	Sharps (Arikaree)	Rockyford Ash	Volcanic ash, silt, pebbles, concretions, sands, poorly consolidated beds
				Brule	Poleslide	Arid climate, sheet flows, mudflows, flash floods; ashfalls, concretions
					Scenic	
				Forming of red Interior paleosols		
				Chadron	Peanut Peak	Interfluve deposition; sandstone, clay, conglomerate
					Crazy Johnson	Climate changing from humid to semiarid
					Ahearn	Flood-plain deposition; channel fill
	?	?				Erosion and Uplift———————————————
Mesozoic	Upper Cretaceous			Yellow Mounds paleosols formed by deep weathering in warm, wet climate		
				Pierre Shale		Black marine shale deposited in Pierre Sea

Sources: Raymond and King 1976, Retallack 1983, National Park Service, 1988.

Above the Lower Nodular Zone is the Upper Nodular Zone. Since both of these are continuous across the Badlands, as well as distinctive in appearance, they serve as excellent markers (fig. 9.8).

The Scenic and the Poleslide are the two members of the Brule Formation. The upper member, the Poleslide, is thicker. The Scenic Member, which includes the nodular zones, consists of interbedded clays, silts, sand, and volcanic ash, as well as channel fillings. In the Poleslide Member, increasing amounts of volcanic ash are found, some of it fluvially reworked and mixed with silts, fine sands, and clays. The Poleslide channel-fill deposits are called "Protoceras sandstones" because they contain bones from this extinct, distant relative of the camel (fig. 9.8).

The Brule rocks, which make up the major part of the Wall, tend to form steep cliffs, knife-edge ridges, and pinnacles. On Brule slopes are many steps, overhangs, and flutings that reflect various responses to weathering processes. Some Brule beds are made up of loose clay; others of firm clay. Some are finer, some coarser. These differences suggest rapidly changing conditions of deposition and consolidation that result in variations in slope as the rock weathers and erodes. The distinctive, reddish, horizontal banding in the Brule Formation is due to the presence of beds having a high concentration of oxidized iron. Lighter bands contain less iron, not so heavily oxidized.

6. Deposition of the Sharps Formation at the close of Oligocene time.

The highest and youngest rocks in the Badlands are those of the Sharps (Arikaree) Formation, which contains fluvial deposits, as well as large amounts of volcanic airborne and water-laid ash. The Rockyford Ash Member in the Sharps is a layer of nearly pure, stream-deposited, white volcanic ash. Capping the Rockyford Ash are resistant sandstone beds that form some of the ridge tops in the Badlands.

7. Development of Badlands topography.

A long quiet span with little erosion or deposition followed the Oligocene. About half a million years ago, during Pleistocene time, erosion of the White River Group began. Large deposits of river gravels are associated with the general disruption of drainage that went on during the severe climates of Pleistocene time. The Badlands were not covered by continental glaciers, but were close to glacial margins. In post-Pleistocene time, high winds created sand dunes, most of which later stabilized. Even during times of greater moisture, vegetation never became established well enough to control the erosion of the poorly consolidated beds of ash and clastics. As long as these conditions continue, the Badlands will continue to erode.

Animals that have adapted to the Great Plains environment thrive in the park, as did ancient species who lived in the region millions of years ago. American bison (buffalo) and bighorn sheep have been successfully reintroduced on the open plains and slopes. Prairie dogs busily dig their "towns" on the prairie. Pronghorns, mule deer, and many species of small animals find their favorite habitats in this beautiful and fascinating national park.

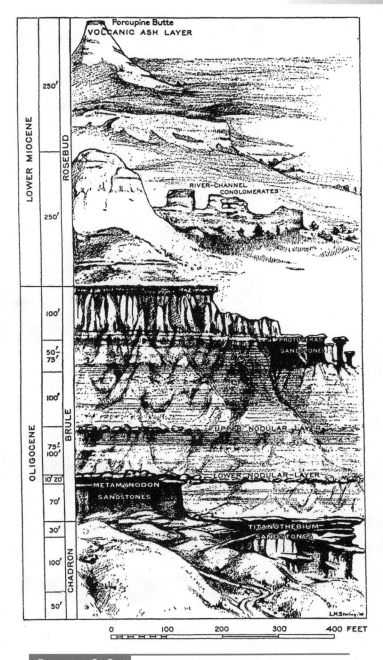

FIGURE 9.8   Composite diagram of Oligocene rock units in the South Dakota Badlands. Adapted from H.F. Osborn 1909.

FIGURE 9.9   During Upper Cretaceous time the upper layers of the black Pierre Shale underwent extensive weathering and changed to a bright yellow paleosol called the Yellow Mounds paleosol. Paleosols are soils of past geologic time buried by younger sediments. Photo by Ann G. Harris.

Geologic Map and Cross Section

Raymond, W.H., and King, R.U. 1976. Geologic Map of the Badlands National Monument and Vicinity, West-Central South Dakota. U.S. Geological Survey, Miscellaneous Investigations Series Map I-934.

Sources

Churchill, R.R. 1979. Topoclimate as a Controlling Factor in Badlands Hillslope Development. Ph.D. thesis, University of Iowa.

———— 1979a. The Importance of Mass Movement and Piping in Badlands Slope Development. In *Proceedings, Iowa Academy of Science* 8 6 (1): 10–14.

Clark, J., Beerbower, J.R., and Kietzke, K.K. 1967. Oligocene Sedimentation, Stratigraphy, Paleoecology, and Paleoclimatology in the Big Badlands of South Dakota. *Fieldiana, Geology Memoirs,* vol. 5. Chicago Museum of Natural History.

Goetzmann, W.H. 1982. Explorer, Mountain Man, and Scientist, in *Exploring the American West,* 1803–1879, Handbook 116, National Park Service Division of Publications. p. 81.

Graham, R.W., Semken, H.A. Jr., and Graham, M.A. (editors). 1987. *Late Quaternary Mammalian Biogeography and Environments of the Great Plains and Prairies.* Illinois State Museum Scientific Papers, v XXII, 491 p.

Harksen, J.C., and Macdonald, J.R. 1969. Type Sections for the Chadron and Brule Formations of the White River Oligocene in the Big Badlands, South Dakota. South Dakota Geological Survey, Report of Investigation 99.

Hauk, J.K. 1969. *Badlands, Its Life and Landscape.* Bulletin no. 2. Interior, South Dakota: Badlands Natural History Association.

Kurten, Bjorn. 1972. *The Age of Mammals.* New York: Columbia University Press.

National Park Service. 1988. Geologic History, Badlands National Park. Information sheet.

Osborn, H.F. 1909. *Cenozoic Mammal Horizons of Western North America.* U.S. Geological Survey Bulletin No. 361.

Retallack, G.J. 1983. Late Eocene and Oligocene Paleosols from Badlands National Park, South Dakota. Special Paper 193, Geological Society of America.

Schumm, S.A. 1956. The Role of Creep and Rainwash on Badland Slopes. *American Journal of Science* 254: 693–706.

Smith, K.G. 1958. Erosional Processes and Landforms in Badlands National Monument, South Dakota. *Geological Society of America Bulletin* 69: 975–1008.

Wanless, H.R. 1923. The Stratigraphy of the White River Badlands, South Dakota. *American Philosophical Society* 62:190–269.

CHAPTER 10

Theodore Roosevelt National Park

WESTERN NORTH DAKOTA

Rodney M. Feldmann   Kent State University

Area: 70,447 acres; 110 square miles

Established as a Memorial Park: April 25, 1947

Redesignated as a National Park: November 10, 1978

Designated a World Heritage Site: 1979

Address: P.O. Box 7, Medora, ND 58645-0007

Phone: 701-623-4466

E-mail: Victoria_mates@nps.gov

Website: www.nps.gov/thro/index.htm

Scenic badlands along the Little Missouri River. Huge slump blocks on the sides of the valley. Flat-lying sedimentary rocks; interbedded coal seams; burned-out lignite beds; "scoria." Quantities of petrified wood; "cannonball" concretions. Continental glacier boundary.

FIGURE 10.1   Badlands along the Little Missouri River in Theodore Roosevelt National Park. Photo by David B. Hacker.

In 1883, when he was 24 years old, Theodore Roosevelt traveled west on the newly constructed Northern Pacific Railroad. This was his first visit to Dakota Territory. His purpose was to hunt buffalo, but he was so impressed with the badlands along the Little Missouri River that he purchased the Maltese Cross Ranch, located six miles south of Medora, and began a 15-year career as a rancher. He expanded his enterprise by developing the Elkhorn Ranch site in the badlands along the river 35 miles north of Medora. (These badlands are some 250 miles north of Badlands National Park, South Dakota, Chapter 9.)

In an article written by Roosevelt and illustrated by Frederic Remington for *Century* magazine in February 1888, Roosevelt observed that the success of a ranching venture in the northern Great Plains was dependent upon the essential elements of food, water, and shelter. During the hot summer months, food and water were the two essentials and, because the Little Missouri River was one of the few perennial streams in the area, proximity to it assured a continuous, abundant water supply for cattle. Additionally, the badlands region of the Little Missouri River supported adequate vegetation to permit grazing. As Roosevelt pointed out, during the long harsh winters, the most important element for survival of cattle was shelter. Availability of water was less essential because cattle could derive adequate moisture from the snow cover. The canyons and gulleys of the region, as well as the cottonwood thickets along the Little Missouri River, provided the animals with protection against the winter winds that severely limited ranching enterprises in the bleak, exposed country of the Missouri Plateau.

Thus, the badlands were a paradox; they were both help and hindrance to the development of the West. Wagon travel across the area was extremely difficult. Wagons had to be disassembled and handcarried across the more rugged parts of the terrain. On the other hand, that very ruggedness provided shelter that permitted development of a ranching industry in a climate that was otherwise too severe to permit anything but summer grazing.

Even at its best, the badlands did not support a large number of animals. By the mid-1880s, the number of cattle in the area had increased to the point where overgrazing of the range land resulted. The winter of 1886-1887 was extremely severe and dealt an almost killing blow to the ranching industry along the Little Missouri. The ranchers did survive, however, and

Roosevelt maintained his ranching interests in the area until 1898.

Before the ranchers brought cattle into the northern Great Plains, herds of buffalo (bison) had thrived on the grazing lands for thousands of years. Hardier than cattle, the buffalo were able to survive periods of drought and the most severe winters. The buffalo were a mainstay of the Indians, who used the meat for food and hides for clothing and shelter. Hostilities between Indians and ranchers were the result of Indian resistance to the loss of their hunting lands and the slaughtering of the buffalo herds. Organized resistance by the Indians had ended only a few years before Theodore Roosevelt came to the region.

Upon being elected President of the United States, Theodore Roosevelt initiated and supported numerous policies to assure conservation of lands in the National Park System. His interest in protecting these areas was influenced by his years as a cattleman in the badlands. Two large separate tracts of land (North Unit and South Unit) along the Little Missouri River were established as the Theodore Roosevelt National Memorial Park in 1947. A part of Elkhorn Ranch, between the two units, is also under park jurisdiction. In 1978, the area was redesignated Theodore Roosevelt National Park.

Geologic Features

The topographic expression of any area is a function of structure, stratigraphy, and climate. In the badlands region of Theodore Roosevelt National Park, the rocks were elevated after deposition and burial but have not been substantially deformed tectonically. Therefore, with the exception of local mass wasting features, the beds are essentially horizontal (fig. 10.1). As stream erosion dissects the area, the drainage pattern tends to be dendritic, and no obvious "grain" related to structural deformation has developed. Much of the park's land lies within the deeply dissected valley walls of the Little Missouri River, which meanders across the southwestern corner of North Dakota, and minor ephemeral tributaries that feed it. (*Ephemeral streams flow only after heavy rains or snow melt.*) The valley is bounded by an upland surface, the Missouri Plateau, which represents a stripped structural surface.

Much of the interest in this particular badlands region lies in the variety of exposed sediments. The sedimentary types include sandstone, siltstone, shale,

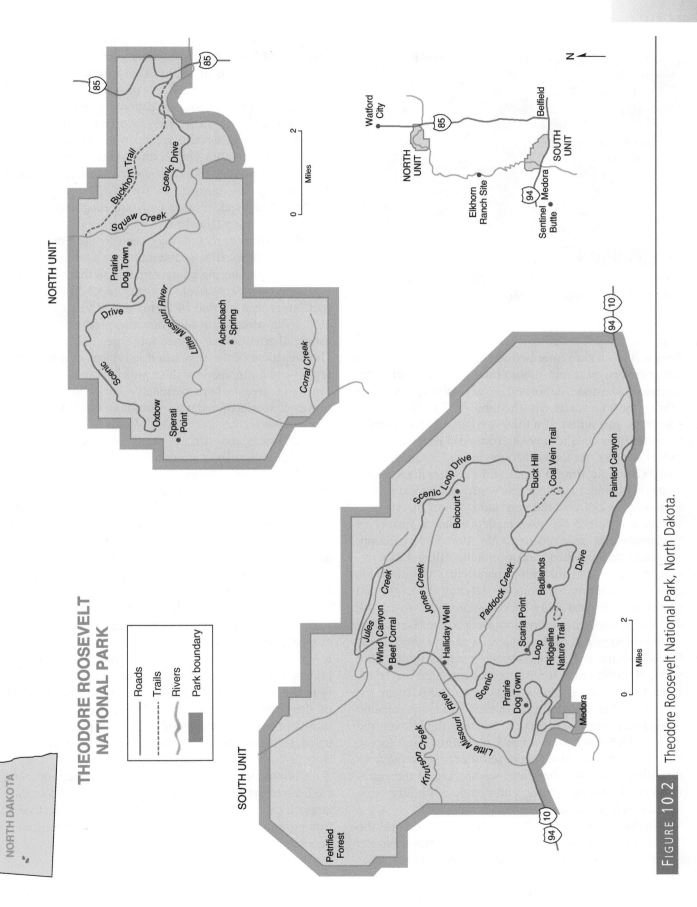

NORTH DAKOTA

THEODORE ROOSEVELT NATIONAL PARK

Roads
Trails
Rivers
Park boundary

NORTH UNIT

85
85

Buckhorn Trail
Scenic Drive
Squaw Creek
Prairie Dog Town
Scenic
Drive
Little Missouri River
Achenbach Spring
Scenic
Drive
Oxbow
Sperati Point
Corral Creek

Miles
0   2

Watford City
85
NORTH UNIT
Elkhorn Ranch Site
SOUTH UNIT
Belfield
94
Medora
Sentinel Butte

N

SOUTH UNIT

94
10

Scenic Loop Drive
Boicourt
Buck Hill
Coal Vein Trail
Painted Canyon
Jules
Wind Canyon
Beef Corral
Creek
Jones Creek
Halliday Well
Paddock Creek
Badlands
Drive
Scaria Point
Loop
Ridgeline Nature Trail
Scenic
Prairie Dog Town
Little Missouri River
Knutson Creek
Petrified Forest
Medora
10
94

Miles
0   2

FIGURE 10.2   Theodore Roosevelt National Park, North Dakota.

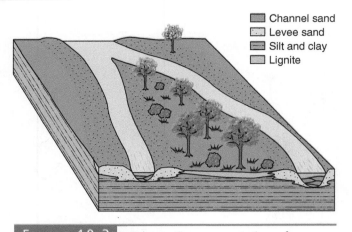

Channel sand
Levee sand
Silt and clay
Lignite

FIGURE 10.3   Schematic representation of environments in which the more common Fort Union sediments were deposited. Modified from Jacob 1976.

lignite (brown coal intermediate in degree of coalification between peat and subbituminous coal), and occasionally limestone deposited in a rhythmic sequence during the Paleocene Epoch (table 10.1). These sediments are just a part of a thick wedge of fluvial, lacustrine, and floodplain deposits formed during the uplift and erosion of the Rocky Mountains in Cretaceous and early Cenozoic time. Faunal and floral evidence suggests that the climate then was warm and humid so that lush vegetation covered the land. Sluggish streams meandered across the region, depositing a variety of sediments now referred to as the Fort Union Group. The streams deposited sand as channel fill and in *natural levees,* or embankments, paralleling the channels (fig. 10.3). During episodes of flooding, finer-grained material, silt and clay, accumulated as overbank deposits. Low-lying regions adjacent to the river valley were swampy, and in these areas organic material was buried, compacted, and eventually became lignite. Thin beds of nonmarine limestone accumulated in lakes. Finally, occasional bursts of volcanic activity in the mountain ranges to the west yielded sufficient volcanic ash to deposit layers several inches thick over the region. These volcanic ashes have either been mixed with fine-grained floodplain sediments or remain as discrete layers. In either case, weathering of the material has yielded bentonitic clays with a lithology distinct from other units in the Fort Union.

Perhaps the most distinctive rock unit in the Theodore Roosevelt National Park, as contrasted to the sediments of Badlands National Park, South Dakota, is

lignite. The badlands in South Dakota do not contain lignitic beds and, therefore, the general appearance of the terrain is quite different. The lignite in Theodore Roosevelt National Park is prominently exposed throughout the area and, additionally, has produced another interesting rock type, locally called "scoria." (The term *scoria* usually refers to vesicular rocks of volcanic origin only.) The North Dakota scoria is baked clay with a pottery-like consistency that has resulted from the *in situ* burning of lignite. The incinerated bright red to black clinker beds are diagnostic of the Fort Union Group (fig. 10.4).

The variation in lithology seen in the Fort Union Group is expressed as sediments with a wide range of colors and weathering characteristics so that the terrain is studded with erosional remnants capped by resistant rock units in a myriad of forms and sizes (figs. 10.5, 10.6). The variation in lithology also produces an interesting distribution of vegetation types. In the present climate, most of the vegetation is typical of a semiarid region and dominated by short prairie grasses and sagebrush. In areas where permeable rocks are exposed, however, stands of cedar are common. On lowland surfaces adjacent to the Little Missouri River, where ground water and surface water are plentiful, thick groves of cottonwood are dominant.

The regional climate is best described as severe. Annual temperatures fluctuate over a range of 150° F; and rainfall, which averages about 15 inches per year, typically occurs in short, violent bursts. This climatic setting results in a relatively sparse vegetal cover. The absence of dense vegetation, coupled with periodic heavy rainfalls, makes an ideal setting for development of the high drainage density of badlands topography (fig. 10.7).

Petrified wood, which is common throughout the park, is mostly preserved as kindling-sized and smaller pieces that weather out from slopes or accumulate in low spots; but nearly complete stumps (fig. 10.8) and logs of petrified wood can be found in place, especially in the lower part of Sentinel Butte Formation. Most of the petrified wood occurs in the silty and clayey sediments representative of the flood plain environment during the time of deposition of the Fort Union Group.

Concretions of various sizes also weather out from the beds of the Fort Union Group. Some are remarkably large, especially in North Unit. Near Squaw Creek Campground alongside the Scenic Drive, large, round concretions called "cannonballs" have accumulated at the base of the bluffs (fig. 10.6).

FIGURE 10.4   Small knolls capped by "scoria," baked clay resulting from burning lignite. Photo by David B. Hacker.

FIGURE 10.5   Erosional remnants capped by resistant sandstone in the Petrified Forest Plateau. Photo by David B. Hacker.

FIGURE 10.6 Sandy concretions protect pedestals from being eroded away. Photo by David B. Hacker.

The occurrence of lignite in the Fort Union Group has yielded another interesting combination of geological events. The lignite beds exposed throughout the region are subject to spontaneous combustion. If beds begin to burn, they not only bake the overlying sediments, but also affect the topography of the area. Coal seams tend to burn from the area of exposure back into the hillside, and this results in collapse of the overlying material and development of a hummocky terrain. Burning continues until such time as the lignite has burned so far back into the hillside that cracking and slumping do not provide access to oxygen for further combustion. A major coal seam that smoldered and burned until 1978 is in the area of Buck Hill in the South Unit of the park. The effect on the topography is still evident. Areas in which scoria has been produced lack the lignite beds that should occupy that stratigraphic position.

Although the area is structurally simple, that is, the beds typically dip at angles of less than 1°, there are several places where much greater dips, on the order of 45°, can be observed. Many of these tilted blocks are tens of yards in length and width. Invariably they occur along slopes, and investigation of the rock on the slopes suggests that the dips are of local extent and the result of mass wasting processes. Huge slump blocks (fig. 10.9) have moved downslope and rotated under the influence of gravity.

Geologic History

1. **Pre-Paleocene deposition of sediments in the Williston Basin.**

Throughout the Paleozoic Era and most of the Mesozoic, the area comprising western North Dakota, eastern Montana, and southern Saskatchewan and Manitoba was one of continued, slow subsidence and accumulation of mostly marine sediments. Although none of the Paleozoic or Mesozoic rock sequence is exposed in the vicinity of Theodore Roosevelt National Park, the sequence is of interest. The Williston Basin is a significant oil-producing region, and the area surrounding the park is dotted with pumps and drill rigs extracting petroleum from the Paleozoic rock sequence. Until near the close of the Mesozoic Era, the entire area was either slightly below sea level or about at sea level.

2. **Uplift of the Rocky Mountains and the end of marine conditions.**

Beginning late in the Cretaceous Period and continuing into Paleogene time, the Laramide orogeny resulted in the development of the Rocky Mountain chain. One of the effects of the uplifting of the Rocky Mountains, hundreds of miles west of the park, was that erosion of these young mountains produced hundreds of cubic miles of sediment that was carried eastward and deposited in great clastic wedges. This event is recorded in a sequence of terrestrial sediments ranging in age from Late Cretaceous through Oligocene time in many parts of the Dakotas and Montana. The Fort Union Group represents the Paleocene element of that clastic wedge. During the time of deposition of the Fort Union Group in the badlands area, the last vestige of the midcontinent seaway in North America was rapidly being filled in east of the park in central North Dakota. These rocks, comprising the Cannonball Formation, are

FIGURE 10.7   View of Painted Canyon, illustrating the high-density drainage pattern. The Missouri Plateau is the flat surface on the horizon. Photo by David B. Hacker.

FIGURE 10.8   Petrified stump preserved in a standing position in the Petrified Forest Plateau. Photo by Diana Ludwig and Douglas Fowler.

Table 10.1	Geologic Column, Theodore Roosevelt National Park

Time Units			Rock Units			Geologic Events
Era	Period	Epoch	Group	Formation	Marker Bed	
Cenozoic	Quaternary	Holocene	Alluvial deposits, landslide features, slump features, "scoria" formation			Development of present badlands topography
		Pleistocene	Cut-and-fill structures and terraces along Little Missouri River and tributaries, glacial erratics, slump features			Glaciation reroutes Little Missouri Missouri River; possible early Pleistocene glaciation
		?	/////////////			
	Tertiary	Paleogene — Paleocene	Fort Union Group	Sentinel Butte Formation	Upper Sand	Grayish to brownish siltstone, clay, and sandstone deposited in habitats similar to Tongue River sediments; lignites less common; "Blue" bed represents weathered volcanic ash marker
					Bullion Butte lignite	
					Upper "yellow" bed	
					Lower "yellow" bed	
					"Blue" bed	
					Basal sand	
				Tongue River Formation	HT Lignite	Yellowish, pastel siltstone, clay, and sandstone deposited in fluvial, lacustrine, and overbank environments with numerous lignitic beds formed in swampy, flood plain setting
					Meyer Lignite	
					Garner Creek Lignite	
					Harmon Lignite	
					Hanson Lignite	
					H Lignite	
					Basal sandstone bed	

Source: Modified from Royse, 1970.

FIGURE 10.9   Slump blocks (dipping beds) are masses of rock material that slide downward, usually with a backward tilting relative to the slope from which it descends. Photo by David B.Hacker.

FIGURE 10.10   The Maltese Cabin is located in the South Unit. Theodore Roosevelt used the Maltese Cross as a brand for his cattle. When he used it, it was an Indian symbol; its orientation is in the reverse direction of the German swastika. Photo by David B. Hacker.

well exposed along the Missouri River and some of its tributaries.

Scattered remnants of post-Paleocene sediments are found throughout western North Dakota in isolated remnants capping butte tops. These units, the Golden Valley Formation and the White River Formation, span Eocene and Oligocene time.

3. Post-Oligocene uplift and erosion of sediments.

No rock units of any extent in western North Dakota are younger than the Oligocene White River beds. The area had apparently been uplifted and was in the process of being dissected by rivers ancestral to the modern Little Missouri River system. The only evidence of deposition subsequent to Oligocene time is in the form of gravel and sand accumulations that are difficult to date.

4. Pleistocene glaciation and development of the modern drainage system.

Pleistocene continental glaciers advanced southward from Manitoba and Saskatchewan at least as far as the North Unit of Theodore Roosevelt National Park. Evidence of glaciation may be seen at "Edge of Glacier" pullout in the North Unit.

In preglacial time, the Little Missouri River flowed north toward Hudson Bay; but with the onset of glaciation, north-flowing streams were blocked. The drainage of the Little Missouri, along with that of the Yellowstone River to the west, was diverted eastward and southward along the edge of the ice front. This ice-marginal drainage eventually developed into today's Missouri River, to which the Little Missouri is tributary. The rate of erosion and downcutting of the Little Missouri River was not constant, and as erosion and deposition continued, the stream cut a series of terraces, remnants of which can still be seen in several places in the park.

Geologic Map and Cross Section

Clayton, L., Moran, S.R.; Bluemle, J.P.; and Carlson, C.G. 1980. Geologic Map of North Dakota: U.S. Geological Survey.

Sources

Bluemle, J.P., and Jacob, A.F. 1981. Auto Tour Guide along the South Loop Road. North Dakota Geological Survey Educational Series 4.

Chronic, H. 1984. *Pages of Stone, Geology of Western National Parks and Monuments,* v. 1, Rocky Mountains and Western Great Plains. Seattle, Washington: The Mountaineers, p. 128-136.

Clayton, L., Carlson, C.G., Moore, W.L., Groenewold, G., Holland, F.D., Jr., and Moran, S.R. 1977. The Slope (Paleocene) and Bullion Creek (Paleocene) Formations of North Dakota. North Dakota Geological Survey Report of Investigation 59.

———— and Freers, T.F. 1967. *Glacial Geology of the Missouri Plateau and Adjacent Areas.* North Dakota Geological Survey Miscellaneous series 30.

Holtzman, R.C. 1978. Late Paleocene Mammals of the Tongue River Formation, Western North Dakota. North Dakota Geological Survey Report of Investigation 65.

Jacob, A.F. 1976. Geology of the Upper Part of the Fort Union Group (Paleocene), Williston Basin, with Reference to Uranium. North Dakota Geological Survey Report of Investigation 58.

Laird, W.M. 1950. *The Geology of the South Unit Theodore Roosevelt National Memorial Park.* North Dakota Geological Survey Bulletin 25.

————. 1956. *Geology of the North Unit Theodore Roosevelt National Memorial Park.* North Dakota Geological Survey Bulletin 32.

McCullough, D. 1981. *Mornings on Horseback.* New York: Simon & Schuster.

Royse, C.F., Jr. 1967. Tongue River-Sentinel Butte Contact. North Dakota Geological Survey Report of Investigation 45.

————. 1970. A Sedimentologic Analysis of the Tongue River-Sentinel Butte Interval (Paleocene) of the Williston Basin, Western North Dakota. *Sedimentary Geology* 4:19-80.

Schoch, H.A. 1974. *Theodore Roosevelt; The Story Behind the Scenery.* Las Vegas, Nevada: KC Publications.

Ting, F.T.C., ed. 1972. Depositional Environments of the Lignite-Bearing Strata in Western North Dakota. North Dakota Geological Survey Miscellaneous Series 50.

Note: To convert English measurements to metric, go to www.helpwithdiy.com/metric_conversion_calculator.html

Kobuk Valley National Park

NORTHWESTERN ALASKA

Area: 1,750 acres; 2735 square miles

Proclaimed a National Monument: December 1, 1978

Established as a National Park: December 2, 1980

Address: P.O. Box 1029, Kotzebue, AK 99752

Phone: 907-442-3890

E-mail: wear_webmail@nps.gov

Website: www.nps.gov/kova/index.htm

Located entirely north of the Arctic Circle, the park includes geological, biological, and cultural resources, such as Great Kobuk and Little Kobuk Sand Dunes; permafrost, glacial, and fluvial features; border zone between boreal forests and tundra; prehistoric migration route for humans and animals; Onion Portage Archeological District; Salmon Wild River.

FIGURE 11.1   A meandering tributary stream empties into the main stream in Kobuk National Park. National Park Service photo.

Geographic Setting

Kobuk Valley National Park lies in a mountain-rimmed basin, 270 miles south of the Arctic Ocean. Although this large wilderness park—larger than the state of Delaware—has no roads or facilities, the Kobuk River has been a major transportation route for humans and animals for thousands of years. For native hunters and fishermen, the Kobuk River has been an east-west highway between the ocean and the Alaskan interior ever since people came to this continent. Every autumn caribou herds ford the Kobuk on their way south to the shelter of the subarctic woodlands; every spring they return north to the tundra to graze. The arctic tundra is a treeless, gently undulating, marshy plain that is covered with mosses and low shrubs and underlain by mucky soil and permafrost.

The size of Kobuk National Park, 2735 square miles, reflects the philosophy of the planners to set aside a hydrologic basin and self-sustaining ecosystem within park boundaries. In arctic latitudes, biological support systems are spread thin over large areas. It takes more than one hundred square miles to sustain a single brown or grizzly bear and many times that size habitat for a caribou herd. Kobuk National Park, the Salmon Wild River within the park's boundaries, the adjoining Noatak National Preserve on the north, and the Selawik National Wildlife Refuge abutting on the south are all part of the Alaska National Interest Lands that were withdrawn by Act of Congress from "all forms of appropriation" and classified as wilderness.

The Kobuk River, which rises in the central Brooks Range, flows south from the Arctic Divide and then nearly due west to the Chukchi Sea. (In its upper reach-

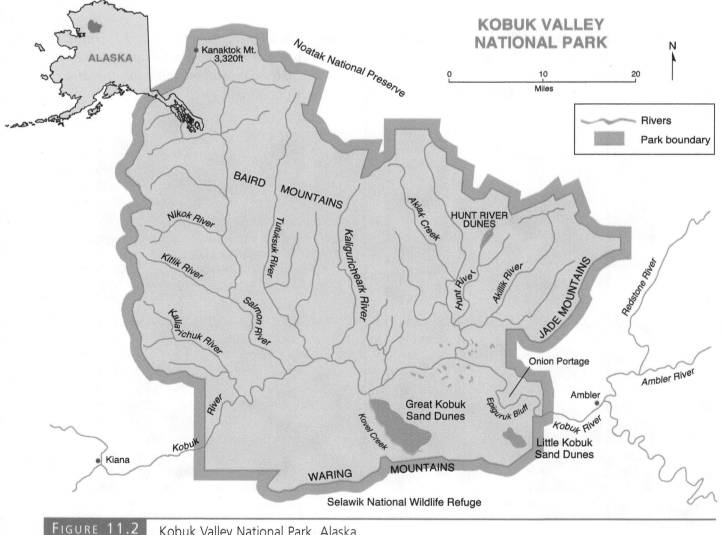

FIGURE 11.2　Kobuk Valley National Park, Alaska.

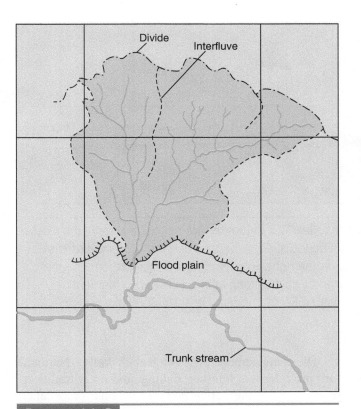

FIGURE 11.3 Small drainage basin, part of a larger hydrologic system drained by a trunk stream. Interfluves (ridges) separate tributary streams; divides along crests define drainage basins. The drainage pattern is *dendritic;* that is, it resembles veins in a leaf. This kind of drainage pattern develops on a land surface that is fairly uniformly erodible. In Kobuk Valley National Park, where drainage patterns are predominantly dendritic, the bedrock in the highlands on both sides of the valley is largely metamorphic and not severely fractured.

es, within Gates of the Arctic National Park and Preserve, the Kobuk has been designated a part of the Wild and Scenic Rivers System.) The northern and southern boundaries of Kobuk National Park follow drainage divides: along the crest of the Baird Mountains (3500 to 5000 feet high) on the north; and on the south side, along the Waring Mountains that rise to elevations of 1000 to 2000 feet. The larger tributaries, including the Salmon Wild River, enter the Kobuk River from the north side. The tributaries draining slopes on the south side are small creeks. All the streams in the park run clear with low sediment loads because permafrost keeps soil from eroding; at present,

the streams do not carry glacial meltwater. Two Eskimo villages outside the park are the only settlements— Kiana, 10 miles downriver from the western boundary, and Ambler, a few miles upstream beyond the eastern boundary.

Whalers and traders visited the lower Kobuk River in the eighteenth century, but a party in small boats from the U.S. Revenue Service in 1884 were probably the first Americans to explore the river's upper reaches. In 1903, a U.S. Geological Survey geologist, W.C. Mendenhall, made a reconnaisance of the Kobuk River and its tributaries by canoe. By 1910 geologists were using pack strings in the area, searching for mineral localities. (The Nome gold rush had begun around the turn of the century, and prospectors were looking for "pay dirt" along the Arctic rivers and tributaries.) In the 1950s, Arthur Fernald of the U.S. Geological Survey mapped the surficial geology using aerial photographs, float planes, and canoes. Today, outboard motors and snowmobiles are the principal means of transportation in the region.

The Kobuk Valley has a long and fascinating prehistoric record. When archeologists began to investigate the Kobuk Valley, looking for traces of "Early Man" on this continent, they were astonished to find at Onion Portage, now listed in the National Register of Historic Places, a record of human occupation longer than any other that has been discovered in North America. At this one site (which has become the standard against which all other Alaskan sites are measured), some 30 layers of human habitation document changes of tools and other adaptations indicating eight cultures. Dr. Louis J. Giddings of Brown University began to excavate Onion Portage and other sites in the Kobuk Valley in the 1960s. His investigations and those that followed show that humans have lived in the valley and used its resources for at least 12,500 years (fig 11.7).

The significance of Onion Portage (named for a wild chive that flourishes there) is that it is a caribou crossing, which has made it a magnet for hunters ever since the earliest people of the Paleoarctic culture discovered it. They were nomadic hunters from the Asian mainland, who made their way across the Bering Land Bridge that connected northeast Siberia and northwest Alaska when sea levels were lower during the Ice Age. These Stone Age people left weapons and tools in the Kobuk Valley. Fashioned of chert and flint, they are similar to artifacts found in Siberia and Japan. Some groups of immigrants stayed on in the north and adapted to arctic and subarctic conditions.

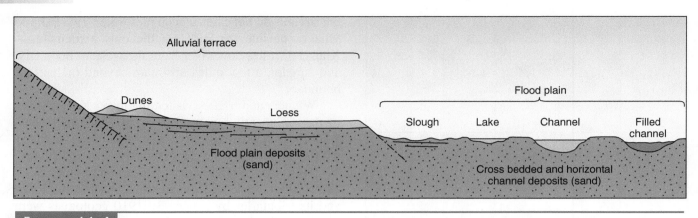

FIGURE 11.4 Simplified cross section of the Kobuk River Valley. The alluvial terrace on the left, topped by loess deposits and dunes, consists of many layers of older flood-plain deposits. Cross-bedded and horizontal deposits of alluvial material (mainly sand) underlie the modern flood plain on the right. High bluffs beside the present channel expose layers of muck, paleosols, and sand (many beds deformed by solifluction), along with wood fragments, artifacts, bones, tusks, and the like. Modified after Hamilton, Ashley, Reed, and Van Elten 1982.

In that earlier time, the Kobuk Valley was an ice-free corridor although glaciers still covered large areas of North America. As the Ice Age mammals drifted southward, hunting clans followed. One route may have been an ice-free corridor that presumably opened along the eastern side of the Canadian Rockies. Gradually the hunters and the hunted spread over North America. Several species that were hunted, such as mastodons, mammoths, large-horned bison, and horses, later became extinct on this continent.

A gap in the cultural record (around 6000 B.C.) was followed by a wave of migrants. They were Northern Archaic people, who came about 6000 years ago from forested lands to the south and east. Fish were apparently their main food.

Asian people of the Arctic Small Tool tradition, who were well adapted to arctic conditions, arrived in the Kobuk Valley around 2000 B.C. They were part of the last great migration from Asia, that gradually spread eastward as far as Greenland. Their habit was to make seasonal journeys to coastal regions from inland locations, and they never lost their dependence on coastal and marine resources.

By about A.D. 1200, people of the Arctic Woodland culture were living in the Kobuk Valley. They were probably coastal Eskimos who adapted to inland living, and their subsistence activities, such as salmon fishing, caribou hunting, and berry gathering, continued into historic times.

The management plan for Kobuk Valley National Park provides that hunting, fishing, and other traditional activities of the native people will take precedence over tourism. No accommodations are available in the park and none are planned, although campsites exist for primitive camping.

Climatic Conditions

With mean annual precipitation only 16 inches, the Kobuk Valley is semiarid. Most of the rain falls during the short, cool, moist summer seasons. Winters are long, cold, and dry, with an average snowfall of 72 inches. The January average temperature is –10° F, with –61° F being the coldest value on record. With a mean annual temperature of 22° F, the permafrost (permanently frozen ground) has little chance of melting, even though the average reading for July is 57° F. Runoff is high in summer, and the valley is awash because the permafrost keeps water from soaking into the soil.

The Kobuk Valley's east-west orientation makes it an open funnel for the dry polar easterlies that blow from September to May. Return winds blowing in from the ocean bring the summer rains to the valley. The polar easterlies are strong and blow for long periods without letup because of intensified atmospheric circulation and steep pressure gradients between ice-covered and ice-free regions. Conditions such as these prevailed over a large part of North America during the Ice Age.

That is why the climate of the Kobuk Valley is regarded as *relict,* or a remnant, of Pleistocene climates.

The park is in a transition zone between boreal forests that can exist in the northern climate and the treeless arctic tundra. On the south flanks of the Baird Mountains, trees finger up the tributary valleys, thin out, and then stop. Where the tree line ends, the tundra begins. Over the last 10,000 to 15,000 years the tree line has shifted north and south as the climate has changed. Studies of sediment layers and pollens show that when the earliest people came, the climate was colder and drier and most of northwest Alaska was thinly vegetated. The Kobuk Valley was a sheltered, ice-free refuge for both animals and humans.

FIGURE 11.5   The crescent-shaped barchan sand dunes have formed the farthest from the source area and are the youngest, most active sand dunes in Kobuk National Park. National Park Service photo.

Geologic Features

The Flood Plain

The broad floor of the Kobuk River Valley is filled with fluvial landforms: sandbars, cutbanks, point bars, oxbows, and sloughs on the flood plain; terraces and fans against the slopes of the encircling hills (box 11.1). Active and stationary sand dunes and dunefields rising south of the flood plain are spread over many square miles and give evidence of the strong winds that sweep down through the valley. Below the surface, at depths of one to five feet lies the permafrost, the continuously frozen ground of the Arctic. Permafrost features, such as ice-wedge polygons and pingoes, roughen the floodplain surface. Both are the result of summer freezing and thawing that makes the ground expand and contract. Water fills cracks in the permafrost and refreezes forming *ice wedges.* Intersecting ice wedges form *polygons.* Some are bordered by ridges, and some by troughs. *Pingoes* are conical frost mounds pushed up by ice.

The base of the permafrost has been located at a depth of 390 feet below the flood plain surface and at more than 600 feet in the Waring Mountains. (The arctic permafrost probably froze before the beginning of the Pleistocene Epoch.) Because water cannot sink into the impermeable permafrost, summer meltwater collects on the surface in pools, marshes, and thaw lakes, making the ground wet and soggy. Conditions are ideal for the annual spring invasion of mosquitoes and black flies that swarm in dense clouds about people and animals.

The Sand Dunes

As a geologic agent, wind is generally regarded as having only minor influence in the shaping of landscapes; but here in the Kobuk Valley the wind's erosional and depositional activities are spectacular. Ideal conditions exist: a plentiful supply of sand to be picked up and moved, winds of intensity and high velocity, and a place for the sand to be deposited.

At least 500 square miles in Kobuk Valley National Park are covered with dune sand. The greater amount of this sand is now stable; some is partly stable but occasionally active; and about 30 to 40 square miles of the valley are occupied by active dunes as barren of vegetation as a dunefield in a desert. The two major areas of active dunes, both on the downwind or south side of the river, are Great Kobuk Sand Dunes and Little Kobuk

Box 11.1	Landforms in River Valleys

We can see unusual and unique natural features in the valley of the Kobuk River, but we also find riverine features that are characteristic of any well developed river system. Onion Portage, for example, is located on a meander loop where canoeists paddling upstream can save five or six miles by portaging across the neck of the meander.

The Kobuk River has built up the relatively flat floor of its *flood plain,* season after season, by over-bank flow. Meltwater roars down from the mountains in the spring, carrying a fresh load of sediment. The river boils up over its banks and spreads out over the valley floor. Branching channels, called *sloughs,* cut across the flood plain, distributing water and sediment. Then the flood waters slow down, become sluggish, and drop sand and mud. Sloughs become stagnant during the low-flow seasons. *Sandbars* accumulate in the river channel near a riverbank in places where the current slows down enough to drop some of its sediment load. Sandbars build up to water level or to just below the water surface. Oftentimes they are washed away or relocated by the next flood.

Because of their natural hydrologic characteristics and helical (screwlike) flow patterns, rivers tend to create sweeping curves, bends, and loops as they swing from side to side over a flood plain. These sinuous windings are called *meanders.* A stream erodes flood-plain alluvium on the outside of a meander, forming a *cutbank,* and then deposits sand and silt on a *point bar* on the inside of a meander curve. The layers of sediment exposed in an eroded part of a flood-plain, such as a cutbank or a bluff, sometimes record a river's history. Geologists have been able to interpret thousands of years of climatic changes in Kobuk Valley by studying deposits exposed in Epiguruk Bluff, upstream from Onion Portage.

The universal habit that streams have of eroding on the outside of a meander curve and depositing on the inside causes the channel to migrate laterally on the floodplain. During a flood, lateral movement may amount to many feet. Sometimes meander loops are brought close together in such a way that the narrow neck of one of them may be cut through—or the river may flow over a neck and develop a new channel. When this happens, the meander is cut off and the entrance to the loop fills in with sediment that separates the water in the loop from the river. This is how *oxbow lakes* form. (The name describes the U shape.) Oxbow lakes tend to get filled in by sediment from overbank flow and vegetation and eventually become oxbow-shaped swamps, or *meander scars,* on the flood plain.

Natural levees, which are parallel to the river-banks, are low ridges of sand several feet high but usually many feet wide. They are deposited by flood waters that are slowing down. The size and width of natural levees depend on the river's volume and velocity and the amount of sediment it carries.

A pronounced change in the regimen of a stream, such as an increase in precipitation or a decrease in sediment load, may cause a river to incise its channel into the soft material of its flood plain and then begin to build a new flood plain at the lower level. The remnants of the old flood plain, stretching away toward the valley-side slopes, become *alluvial terraces.* Apparently this has happened more than once in the Kobuk Valley. At present, the sediment load of the Kobuk River system is comparatively low, mainly because permafrost stabilizes the soil.

Alluvial fans are low cones or fan-shaped mounds of sediment that accumulate where a tributary coming down from a steeper side valley reaches the flood plain and abruptly slows and spreads out. The sediment load that is dropped at the edge of the flood plain builds up an alluvial fan.

Sand Dunes. The Great Kobuk Sand Dunes have an elevation of about 100 feet near the river and are about 500 feet high where they bump up against the base of the Waring Mountains. NASA scientists have compared these dunes and their characteristics to areas on the surface of Mars that are cold and dry and have strong winds.

The abundance of sand comes from reworked glacial drift and alluvium in sandbars, terraces, and in the dunes themselves. The wind picks up loose grains, a process called *deflation,* and rolls or bounces them along by *traction* and *saltation.* This is similar to the way sediment grains are transported by stream action. *Blowout* and *parabolic dunes* are erosional forms. Their centers are eaten out by the wind so that they become U-shaped with their arms pointing upwind. Dunes of this type are more common on the upwind sides of the dunefields. Toward the center the dune shapes become irregular. In the extreme downwind section, farthest from the source area, the youngest, most active dunes are accumulating. These are the *barchans.* They are half-moon shaped, in map view, with arms pointing downwind. Sand is swept up over the top or carried along the base, thus lengthening the arms as the middle of the dune moves along. Where there is room

to migrate, some dune forms move several miles a year. Internally, the dunes show extensive cross-bedding.

The grains of dune sand are well rounded, "frosted" (opaque), and non-uniform in size. These characteristics indicate long-continued abrasion during numerous cycles of erosion and deposition. Much of the work of abrasion was done by glacial scouring of bedrock in the surrounding highlands. The sand fragments consist of resistant quartz, feldspar, and minor amounts of other minerals.

Bedrock

Except for rocky slopes that rim the upwind side and a few knolls of rock that jut up through the dunes on the downwind side, the bedrock of Kobuk Valley is deeply buried. Outcrops are found in the foothills and slopes of the Baird and Waring Mountains within the park's boundaries. Some reconnaissance work on the hard-rock geology has been done, but the area has not been studied in detail. In the Baird Mountains, Paleozoic sedimentary rocks and older metamorphosed rocks (phyllites, schists) crop out, as well as Tertiary conglomerates, cross-bedded sandstones, and shales with thin coal beds. To the south, in the Waring Mountains,

FIGURE 11.6   Great Kobuk Sand Dunes are encroaching on the boreal forest. National Park Service photo.

there are Cretaceous conglomerates, sandstones, arkoses, and thick beds of shale. In the Jade Mountains, just outside the eastern park boundary, are deposits of nephrite jade of gem quality. During the winter the native people move jade boulders by sleds down to the riverbank. In spring, after the ice breaks up, the jade blocks are barged to Kotzebue on the coast, where native artisans make jade jewelry and carvings.

Geologic History

1. Paleozoic and Mesozoic sedimentation.

The marine sediments making up the older rocks in the park region may have accumulated in another part of the globe, along continental margins and shelves far to the south. The Kobuk area is a part of Alaska that probably drifted north for millions of years and eventually collided with and became attached to the North American continent. Associated mountain-building, accompanied by volcanic activity and emplacement of

igneous rocks in plutons, deformed the rocks, and some were metamorphosed. The jade in the Jade Mountains, for example, is an indication of high-pressure metamorphism of igneous rocks, probably in Jurassic time. In chapters on Alaskan national parks in Part III, this tectonic history is explained more fully.

2. Tertiary sedimentation.

The Tertiary sedimentary rocks in the Baird Mountains accumulated on deltas, tidal flats, and along coastal margins. Interlayered beds of soft coal indicate proximity to sea level and a warm, moist climate. Mountain-building activities that deformed and elevated the rocks probably began early in Tertiary time.

3. Quaternary glaciation.

As the climate cooled late in Tertiary time, glaciers formed on the Brooks Range and grew until a large icecap covered the region. The permafrost has been frozen for several million years, probably since late in Tertiary time. Sea level lowered as more water became locked

Table 11.1			Geologic Column, Kobuk Valley National Park	
Time Units			**Rock Units**	**Geologic Events**
Era	Period	Epoch		
Cenozoic	Quaternary	Holocene	Alluvial deposits Dunes, loess Glacial drift	Reworking of glacial and fluvial sediments by wind and streams Continued dune building

		(Wisconsinan
		    stage)
		
		Pleistocene
		    (Illinoian stage)
			Glacial drift, outwash
			Dune deposits
			Peat deposits
			
			Till and outwash
				Several episodes of alpine glaciation in
				    nearby mountains
				Intense wind erosion and deposition
				Kobuk glaciation; entire region
				    blanketed by icecap

	Tertiary	(late)
		
		(early)
			Conglomerates, soft shales and
			    cross-bedded sandstones with
			    interbedded coal
				Freezing of permafrost
				Uplifting of Baird and Brooks Ranges
				Accumulation of sediments in coastal
				    areas and on deltas

Mesozoic	Cretaceous – –(?)– – Jurassic (?)	(late)	Conglomerates, arkose, shales Jade (nephrite)	Continental shelf, near-shore, and shore deposition, marine and nonmarine High-pressure metamorphism
Paleozoic	(late)		Granite rocks, volcanic rocks Marine sedimentary rocks	Tectonic activity Extensive deposition

Sources: Smith, 1939, Fernald, 1964; Wahrhaftig, 1965.

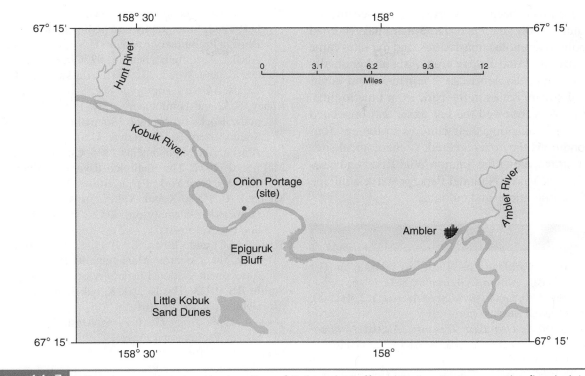

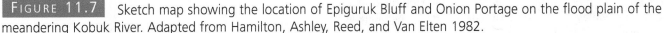

0   3.1   6.2   9.3   12
Miles

FIGURE 11.7   Sketch map showing the location of Epiguruk Bluff and Onion Portage on the flood plain of the meandering Kobuk River. Adapted from Hamilton, Ashley, Reed, and Van Elten 1982.

up in glacial ice. This episode, named Kobuk glaciation, took place during the Illinoian stage of the Pleistocene Epoch, around 200,000 years ago. The deposits correlate with Illinoian glacial material at **Cape Krusenstern National Monument** about 150 miles to the west.

In Wisconsinan time (the most recent major glacial stage), alpine glaciers advanced and retreated several times. They moved down the valleys of the Baird Mountains and the Brooks Range but did not completely override the Kobuk Valley. Peaty material older than the Wisconsinan glacial outwash in the Kobuk Valley has been dated by radiometric methods as 38,000 to 33,000 years before the present. More recently, in Holocene time, ice from the east that moved toward the park sent floods of meltwater and glacial drift downstream. Small glaciers still occupy cirques in mountains to the east and northeast of the park.

4. The development of the present landscape.

The extensive areas of stabilized dunes in the Kobuk Valley indicate that eolian activities in the Pleistocene were even more intense than those in recent times, remarkable as the presently active dunes are. Dune-building probably began after the retreat of Kobuk glaciation. Along with the dunes are smaller deposits of silt (loess) laid down by wind.

Epiguruk Bluff, an eroded river terrace a few miles upstream from Onion Portage, has revealed in its sediment layers an unusually complete and detailed record of changing environments and life forms in the Kobuk Valley throughout a long segment of Quaternary time. A paleosol (buried soil horizon) at the base of the bluff, which has been dated as 35,000 B.P., contains peat and muck deposited on a flood plain. A younger paleosol (24,000 B.P.), containing loess and alluvium, buried the older layer. Overlying the soils are deposits of crossbedded channel sands, horizontal sand layers, wind-blown sands, and loess (wind-blown silt). Plant and animal fossils found in some of the deposits help to document climatic changes, as for example, a change from tundra vegetation to spruce vegetation. At times in the past, the river was sometimes higher or lower than present. Average annual temperatures have been somewhat warmer at times and at other times much colder. Precipitation rates have likewise varied in range between wetter and drier (fig. 11.7).

As the climate began to warm and the ice melted back around 14,000 years ago, large mammals (horned bison, woolly mammoths, mastodons, and the like) came across the Bering Land Bridge from Asia and gradually spread throughout the Americas. Their tusks and bones wash out of stream banks in the park from time to time. Bands of hunters followed the big game and populated the New World, adapting their cultures to the environmental conditions they encountered. Archeological sites at Onion Portage, along the Salmon Wild River and elsewhere in Kobuk Valley National Park, preserve a unique record of the migrations and cultures.

Sources

Belous, R. 1981. Parklands in Northwest Alaska, in The Kotzebue Basin, *Alaska Geographic,* 8(3): 1–16.

Beikman, H.M. 1980. *Geologic Map of Alaska,* 1:2,500,000, U.S. Geological Survey.

Brown, W.E. 1982. *This Last Treasure, Alaska National Parklands.* Anchorage, Alaska: Alaska Natural History Association, p. 106–113.

Fernald, A.T. 1964. Surficial Geology of the Central Kobuk River Valley, Northwestern Alaska, U.S. Geological Survey Bulletin 1181K, p. 1–31.

Galloway, J.P. and Koster, E. 1981. Comparison of Grain Size Statistics from Two Northern Alaska Dunefields. U.S. Geological Survey in Alaska, Accomplishments 1981. Circular 868, p. 20-21.

Hamilton, T.D., Ashley, G.M., Reed, K.M., and Van Etten, D.P. 1982. Stratigraphy and Sedimentology of Epiguruk Bluff, a Preliminary Account. U.S. Geological Survey in Alaska, Accomplishments 1982. Circular 939, p. 12–14.

Hunter, C. and Wood, G. 1981. Alaska National Interest Lands, *Alaska Geographic* 8(4):8-16, 76–83.

Loney, R.A. and Himmelberg, E.R. 1984. Ophiolitic Ultramafic Rocks of the Jade Mountain-Cosmos Hills Area, Southwestern Brooks Range. U.S. Geological Survey in Alaska, Accomplishments 1984. Circular 967, p. 13–15.

McPhee, J. 1977. The Encircled River, in *Coming into the Country,* New York: Farrar, Straus, Giroux, p. 5–91.

Melham, T. 1984. Kobuk Valley, in *Alaska's Magnificent Parklands,* Washington, D.C.: National Geographic Society, p. 155–167.

Shaver, M.C. and Gilbert, C. 1985. Kobuk Valley, National Park, in General Management Plan/Environmental Assessment, National Park Service, p. 22–54.

Smith, P.S. 1913. The Noatak-Kobuk Region, Alaska, U.S. Geological Survey Bulletin 536.

———. 1939. Areal Geology of Alaska, U.S. Geological Survey Professional Paper 192.

Wahrhaftig, C. 1965. Physiographic Divisions of Alaska, U.S. Geological Survey Professional Paper 482, p. 27–28.

Note: To convert English measurements to metric, go to www.helpwithdiy.com/metric_conversion_calculator.html

Great Sand Dunes National Park and Preserve

SOUTH CENTRAL COLORADO

Area: Park: 107,265 acres; 168 square miles
　Preserve: 41,686 acres; 65 square miles

Proclaimed a National Monument: March 17, 1932

Established as a National Park and Preserve: November 22, 2000

Established: Wilderness on October 20, 1976
　(86 percent of original monument)
　Sangre de Cristo Wilderness Area: August 13, 1993

Address: Great Sand Dunes National Park and
　Preserve, 11500 Highway 150, Mosca,
　CO 81146-9798

Phone: 719-378-2312

E-mail: GRSA_Interpretation@nps.gov

Website: www.nps.gov/grsa/index.htm

The sand dunes here are the tallest in North America, attaining heights of 700 feet above the surrounding area in southern Colorado. The dune field began to form as Pleistocene glaciers retreated, leaving behind huge quantities of sandy material to be worked over by prevailing winds and running water. As the climate changed, vegetation has largely stabilized the dunes. Within the Park and Preserve, beside dunes, are alpine lakes; six mountain peaks (over 13,000 feet in elevation); forests of pine, spruce, cottonwood, and aspen; and grassland areas.

FIGURE 12.1　The dune field rises as much as 750 feet above the surrounding area. The people walking on the dunes (left central) can barely be seen, showing how large the dunes really are. Photo by Jeffrey Dick.

Geographic Setting

Great Sand Dunes National Park and Preserve is located in the San Luis Valley between the Tertiary-aged volcanic San Juan Mountains to the west and Precambrian-aged granitic rocks of the Sangre de Cristo Mountains to the east in south-central Colorado. A fault runs along the western base of the Sangre de Cristo Mountains. The Rio Grande River flows through the valley.

The valley has an average elevation of over 7500 feet and extends for over 100 miles south into New Mexico. The area is part of the desert ecosystem of the Colorado Rockies. Average rainfall is less than 10 inches per year.

The dune field rises as much as 750 feet above the surrounding area. The prevailing wind is usually from the southwest and blows almost perpendicular to the trend of the Sangre de Cristo Mountains. This produces dunes with the sand moving up the gentle windward slope and cascading down the steeper leeward slope. However, during the spring when there may be strong storms, the winds may change direction, and reverse as the wind goes through the passes of the San Juan Mountains instead of the passes of the Sangre de Cristo Mountains. This produces a counter dune that has a steep slope on the west and gentle slope on the east. Three low passes are in the Sangre de Cristo Mountains: the Mosca, Music and Medano (see fig. 12.3A, 12.3B, 12.3C). They funnel the wind into the area that is protected by the projection of Blanca Peak (14,363 feet) causing the dunes in the park to form where they have. Strong winds, especially in the spring and fall pick up the sand from the valley floor, transport it across the valley, and deposit in the area of the park. The oval-shaped dune field is about six miles across and contains 39 square miles of dunes.

A complicated system of wind, erosion, and sand deposition has produced the dunes. The building materials for the dunes, the sand and water, are derived from areas above and outside the original 60 acres of the park. The water erodes the dunes and deposits it on a sand sheet that is located below the original boundaries. This is the main reason for the expansion of the park to protect these resources.

Proposals had been made to tap the ground water supply and divert it to urban development along the Colorado Front Range. If this were to happen, the dunes could no longer form.

The climate ranges from 70° to 80° F during the day and may drop down to 40° F at night during the summer. Because of the presence of dark minerals in the sand, the temperature of the sand can reach 140° F. The annual precipitation is only 10 inches per year. The warmest month is July. The temperature drops below freezing in the winter with an average of 38 inches of snow. The heaviest snow is in March. It is a dry snow averaging only 10 percent moisture. Frost penetration has been measured as deep as four feet below the surface of the sand. The coolest month is January. The lowest temperature recorded in the park was in 1963 at 25° F; the highest temperature was 98° F in 1982.

The strongest winds are in April, May, and early June. The strongest winds are from the northeast and have been clocked as high as 40 miles per hour. When they come from the southwest they can be anywhere from violent to mild. It is sunny most of the time, with only 30 to 40 cloudy days. The fall usually has an Indian summer.

The vegetation includes blowout and Indian rice grasses, scurf pea and prairie sunflower (grows up to four feet). A total of 20 species of plants grow directly on the dunes. A large stand of ponderosa pines, the largest known in the United States, has ancient tribal markings. Animals include rabbits, mice, kangaroo rats, ground squirrels, and mule deer. Also found are birds such as western tanagers, red-tailed hawks, mountain bluebirds, and golden eagles, among others. Rare plants and animals (e.g., six species of insects, including the Great Sand Dunes tiger beetle) are confined to San Luis Valley.

Archaeological sites indicate that roaming bands of Indians visited the area 11,000 years ago, according to the Clovis/Folsom artifacts, especially broken spear points. The hunters came into the San Luis Valley during the spring, summer, and fall months hunting mammoth, bison, and other animals. The Utes visited this area in the late 18th to early 20th century, but never established any permanent camps.

The first European to see the dunes was probably Don Diego de Vargas during his search for gold and other precious materials in the Sangre de Cristo Mountains in 1694.

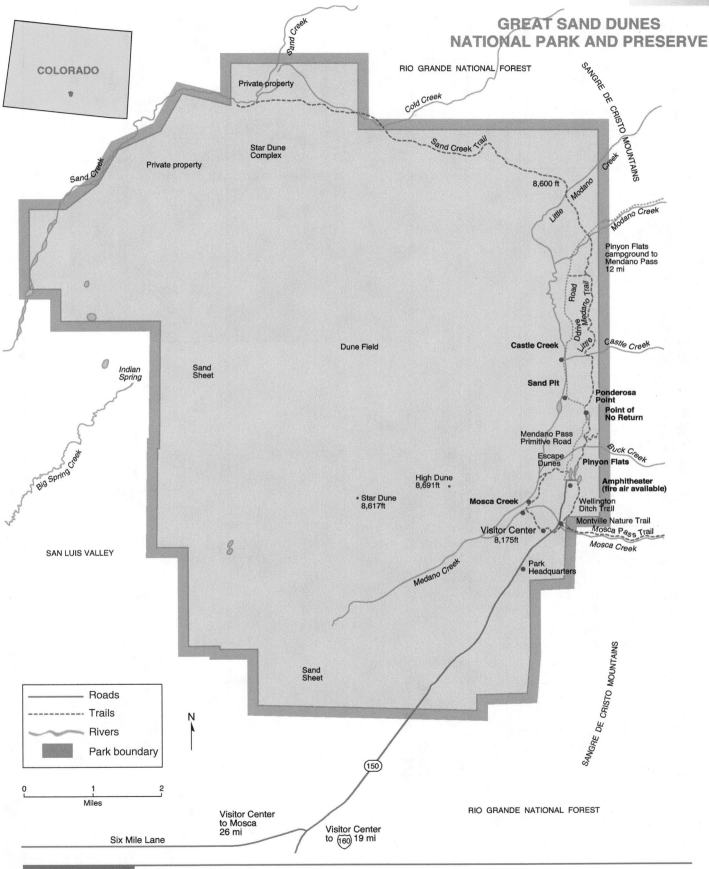

FIGURE 12.2A   Great Sand Dunes National Park and Preserve, Colorado.

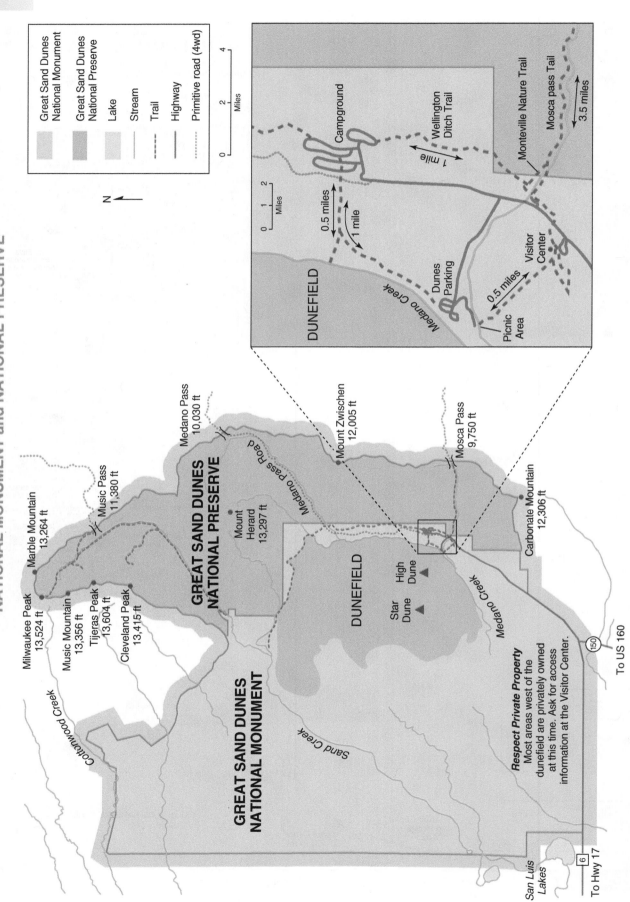

GREAT SAND DUNES
NATIONAL MONUMENT and NATIONAL PRESERVE

FIGURE 12.2ʙ   Great Sand Dunes National Monument and National Preserve, Colorado.

A

B

FIGURE 12.3   The Santa de Cristo Mountains have three low passes: *A*. Mosca, *B*. Music, and *C*. Medano. They funnel the wind into the area that is protected by the projection of Blanca Peak (14,363 feet), causing the dunes in the park to form at that location. National Park Service Photo.

C

Apparently there were attempts by other individuals to extract minerals from the dunes. However, they had no luck.

Early explorers, such as Zebulon Pike (1807), Captain John Fremont (1848–9), and Captain John Gunnison (1853) used Mosca and Medano Passes to get into the valley. There is a description of the dunes in Pike's journal. The Utes signed a series of treaties starting in 1855 that forever gave up the valley in exchange for the west slope of Colorado and their traditional hunting grounds; but the treaties were broken time after time. In 1883, the Ute Indians were forced to a small reservation in southwestern Colorado.

After gold was discovered in the San Juan Mountains to the west of the dunes, Mosca Pass was the main access to the San Luis Valley by many of the settlers. Frank Hastings obtained a charter from the Territory of Colorado in 1871 and operated a store near an old trading post. He improved the trail and established a tollgate.

The settlers, ranchers, and homesteaders raised fruits and vegetables for themselves, using irrigation canals. Range wars broke out in the 1880s because of the competition for grazing lands. These wars were to lead to the establishment of timber preserves and eventually to the first national forests.

The chief industries by 1900 in the San Luis Valley were ranching, farming, and mining. In 1927, Frank Wellington dug by hand a ditch from the mountains to his homestead as a water supply for his orchards, gardens, pastures, and livestock. Other people built extensive canals near the Great Sand Dunes for both irrigation and mineral development.

In June of 1930 Elizabeth Spencer of Monte Vista presented a program on the unique sand dunes habitat and asked for local support for a national monument designation, which is created by a presidential declaration. The superintendent of Yellowstone National Park visited the dunes in 1931. In 1932 President Herbert Hoover set aside 39 square miles of the dune area as a national monument.

Geologic Features

When most people think of a desert, they think of sand dunes everywhere. In this case it is true, although most deserts are just barren land with scattered vegetation. The region is classified as a desert because it receives less than 10 inches of rainfall per year.

Three types of sand deposits are within the park: an *active dunefield,* a *dry lake area,* and a *sand sheet.*

The active dunefield is an area where the dunes migrate because of wind action. The height of the closely spaced dunes is considerable: 700 feet.

A *sand dune* is a low mound capable of movement that is created when strong winds pick up sand-sized grains from a valley floor and transport the grains by various methods. Those methods include *deflation,* the wind removal of material; *surface creep,* the slow downwind advance by impact of smaller grains; *saltation,* the skipping downwind by impact and rebounding along a surface; and *suspension,* the fine sediments held above the ground being carried by a body of air, until there is a reduction in velocity and the grains are dropped. About 95 percent of the sand grains are transported by saltation. The sand dune retains its characteristic shape as it moves forward because of the wind. The sand is blown up the shallow dipping face, the *windward face,* and is deposited on the steep *leeward face* or *slip face.* The sand avalanches downslope when the angle of repose is exceeded, which, in the case of sand dunes, is 34°.

Different physical and geologic conditions form different types of dunes. Two types of dunes form when the wind is blowing in one direction. They are the

Figure 12.4A   Transverse dunes are strongly asymmetrical and are elongated in a direction perpendicular to the prevailing wind. The windward face is gently sloping and the slip face has a steep dip at or near the angle of repose. This type of dune usually forms in an area of sparse vegetation. National Park Service photo.

FIGURE 12.4B   Barchan dunes are isolated, crescent-shaped dunes that lie transverse (across) the direction of prevailing winds. The side facing the wind is gently sloping and convex in shape (curved outward). The horns of the crescent point downwind and are located on either side of the concave (curved inward) leeward side that is steep. This type of dune usually forms on a hard, flat surface where there is a limited supply of sand and the wind, with a moderate velocity and blows in the prevailing direction. National Park Service photo.

transverse dune (fig. 12.4a) and the *barchan dune* (fig. 12.4b). The transverse dunes are strongly asymmetrical and are elongated in a perpendicular direction to the prevailing winds. The windward face is gently sloping, and the slip face has a steep dip at or near the angle of repose. This type of dune generally forms in an area of sparse vegetation. The barchan (Arabic for "ram's horn") dunes are isolated crescent-shaped dunes that lie transverse (across) the direction of the prevailing wind. The side facing the wind is gently sloping and convex in shape (curved outward). The horns of the crescent point downwind and are located on either side of the concave (curved inward) leeward side that is steep. This type of dune usually forms on a hard, flat surface where there is a limited supply of sand, the wind has a moderate velocity and blows in a prevailing direction. These dunes move horizontally since the sand is blown from the back to the steep slip face.

If there are two opposing directions of wind, a *reversing dune* will form with material from the steep slip face being blown back onto the top of the gentle slope adjacent to the slip face. Hence it has a slip face on either side (fig. 12.4c).

Star dunes are isolated hills of sand with the base of the dune resembling a star. The ridges diverge from a central peak having three or more slip faces. They tend to remain fixed in one location but have a tendency to grow vertically. They develop because of the wind blowing in all directions.

The northwest lobe of the sand dune field has a complex of star dunes. These dunes are located at the base of Mount Herard (13,297 feet), which separates Medano Pass (10,030 feet) and Music Pass (11,380 feet). The winds become complex because they have to flow around the sides of Mount Herard and toward either Medano or Music Pass.

The term *sand* actually indicates the size of the individual grains of a substance and not the composition. The sand-sized grains range between 0.00248 to 0.08 inches and can have many compositions. The sand for Great Dunes National Park and Preserve was derived from the igneous and metamorphic rocks of the

FIGURE 12.4c   Star dunes are named because of their shape. They are isolated hills of sand that have three or more slip faces and have ridges radiating from a central peak. They tend to remain at the same location and have a tendency to grow vertically. They form because the wind is blowing in all directions. National Park Service photo by Sarah Andrews.

surrounding mountains. The average composition of the sand is 51.7 percent volcanic rock, 29.1 percent quartz, 8.9 percent feldspar, 3.7 percent other rocks, 3.4 percent other minerals, 2.5 percent sandstone, and 0.7 percent magnetite. The size of the grains in this park range in size from 0.008 to 0.011 inches in diameter.

The presence of magnetite usually indicates the presence of gold; hence in the early 1900s some of the prospectors tried mining the sand dunes for gold called "flour gold." But there wasn't enough gold to make a profit so the ventures were abandoned.

Surging flows (fig. 12.5) can be seen in Mendano Creek and Sand Creek. Ridges of sand build up in the creek beds and hold the water back. More and more water builds up as the ridges get higher. Eventually enough pressure builds up to break the ridges and the water surges downstream. Not every stream will have surging flows because there are three requirements that must be met first: a smooth channel, swift water velocity, and shallow water depth.

Medano Creek has an interesting feature called *antidunes*. They develop when sand ridges form in water that is flowing fast. The ridges are a temporary form of a ripple on the floor of the stream similar to a sand dune. It is in phase with the surface water waves; in other words, the flow of the water mirrors, the form of the ridges. On occasion they may migrate upstream.

Braided streams (fig. 12.6) are those that are divided into an interlacing network with the individual streamlets being separated by channel bars, having a braided appearance. They form in areas of low gradient and a high sediment load.

Medano Creek flows around the southeastern boundary of the dune field. The creek acts as a barrier to the overall movement of the dunes. The creek is known for its surging flow during the spring when the runoff is higher. The dunes cannot migrate toward the Sangre de Cristo Mountains because the creek erodes the sand from the dune and carries the material downstream before it disappears into the sand.

A *sand sheet* (fig. 12.7) is a thin accumulation of fine gravel or coarse sand formed of grains that are too large to be transported by the process of saltation. It is a flat or plain-like surface broken only by small sand

FIGURE 12.5   Surging flows are caused by ridges of sand that build up in creek beds and hold the water back. As the ridges get higher, more and more water builds up until pressure causes the ridge to break and the water surges downstream. This type of flow has a smooth channel, swift water velocity, and shallow water depth. National Park Service photo.

ripples or small hills that were once active sand dunes. The small hills are now covered by vegetation and can no longer migrate because they are fixed by the root system of the plants (fig. 12.7).

The sand sheet in the geologic past used to be the main source for the dune field. This happened during the drier periods when there was not so much vegetation and a lot of sand could be transferred into the dune field.

Dry lake areas are basins that formerly contained water. They are located in regions that have the groundwater table near the surface. Water is brought up to the surface by *capillary action*. The capillaries are produced by interconnecting spaces of sand grains that are so small that the water is pulled to the surface against the pull of gravity. When the water reaches the surface, only the water can evaporate, leaving behind any dis-

solved mineral matter it contains. As the minerals precipitate out of the evaporating water, they cement the sand grains together.

The *active dune field* is the area where nothing impedes the sand grains by cementation or vegetation. It is divided into three distinct areas because of the complexity of the wind, the nearby mountains, and the streams.

Most of the active dune field contains north-trending reversing dunes that are separated from one another by large troughs. At the present time these dunes lack a source of sand because they are adjacent to the dunes that have been stabilized by vegetation.

The complex network of the star dunes make up the northwest lobe of the dunefield. The location is at the base of Mount Herard, which separates the Music and Medano passes. As the winds flow toward the passes

FIGURE 12.6   Medano Stream has an interlacing network with individual streamlets being separated into channel bars. This is a "braided" stream pattern. The intertwining pattern forms in streams that have a low gradient and a high sediment load. National Park Service photo.

FIGURE 12.7   A sand sheet is a thin accumulation of fine gravel or coarse sand formed of grains that are too large to be transported by the process of saltation. It is a flat or plane-like surface broken only by small sand ripples or small hills that were once active sand dunes. However, the area is covered by vegetation, thus preventing the migration of the dunes. In the geologic past, the sand sheet was the main source for the dune field. National Park Service photo.

they have to flow around the sides of Mount Herard, producing a complex movement pattern.

To create sand dunes, three requirements must be met: wind, sand, and a natural trap. The bend in the Sangre de Cristo Mountains has produced the trap and prevented the sand from moving farther eastward. Weathering of the San Juan Mountains, to the west, provided the sand after the glaciers, during the Pleistocene Epoch, removed the mountaintops by eroding them into sand-sized particles. The meandering Rio Grande River supplied additional material as it eroded the local bedrock. The wind currents supplied the energy to transport the material. It had to take many thousands of years for the dune field to develop.

Today's environment is not conducive to the formation of sand dunes because of the San Luis Lakes in the center of the San Luis Valley. The surface of the land is cemented together by the precipitation of salt from the shallow ground water. The wind is not strong enough to carry the sand from the alluvial fan of the Rio Grande River (derived from the San Juan Mountains) across this cemented land. Analysis of the sand found in the park reveals that it is a mixture of fine-grained quartz, lava, ash and pumice. The sand was derived from the San Juan Mountains located to the west of the park rather than from the Sangre de Cristo to the east. Because these mountains are a mixture of rock types, the sand formed from them is coarser and cannot travel as far as finer-grained sand. Another possible source of sand would be the sand sheet area, but since it is stabilized with vegetation, the wind cannot pick up sand from this site. Hence there is a limited amount of sand for the dunes today, with no chance of material being added to the site. The dunes of today are actually a relict landform of previous climatic conditions during the mid-Holocene Epoch when the climate was drier and hotter.

Not all of the water of Medano Creek flows around the dunes. Much of the creek on the east side of the dunes actually flows beneath the dunes. This happens because either it sinks into the gravel that makes up the floor of San Luis Valley or it finally emerges as springs on the west side of the dunes. Therefore, the sand is constantly recycled.

Geologic History

1. **Formation of Precambrian rocks about 1.7 to 1.8 billion years ago. Erosion of the bedrock.**

Metamorphic and igneous rocks not only form the basement of the San Luis Valley, but also the uplifts on either side. The Sangre de Cristo rocks have been exposed in the upper plates of the Laramide uplift on its western side. The Precambrian rocks have been divided into four groups. The oldest are 1.7-1.8 billion years old and are a quartz monzonite (an igneous rock with an intermediate composition); leucogneiss (a light-colored banded metamorphic rock); and gneiss (a light and dark banded metamorphic rock). The age of these rocks is Lower Proterozoic. The Upper Proterozoic formation, 1.4 billion years old, is a quartz monzonite. There were many intrusions and other deformations, such as faulting. Erosion occurred before the deposition of sedimentary rocks took place.

2. **From Late Proterozoic time through Middle Paleozoic, shallow seas occupied the area, surrounded by low-lying land in the region of the Sangre de Cristo Mountains. Periodic deposition and erosion occurred.**

In the vicinity of today's Sangre de Cristo Mountains, there was deposition of Lower Paleozoic sedimentary rocks from shallow seas onto a stable continent. The Ordovician rocks were primarily fine-grained. An unconformity resulted when the seas retreated during the Silurian Period. The seas then transgressed back onto the land during the Devonian Period. This was followed by deposition of Mississippian limestones with chert nodules in the warm shallow seas. At this time, the area was near the equator. All these beds were removed from this region by erosion, but can be seen elsewhere.

3. **During the Upper Paleozoic Era in the Ouachita mobile belt region, mountain building produced the Ancestral Rockies. There was continued uplift, rapid erosion, and cyclothemic deposition of the newly built highlands during the Pennsylvanian and Permian Periods. Erosion removed most of the Ancestral Rockies.**

The deposition of the Pennsylvanian-aged sandstone and coaly shale reflect the uplifting of the Ancestral Rockies and the retreat of the seas. Continued uplift, rapid erosion of the rising land, and rapid deposition took place. Also during the Pennsylvanian Period was cyclic deposition of a basal arkosic conglomerate that gradually graded to fine sandstones with nodular limestones. By the end of the

Table 12.1	Geologic Column, Great Sand Dunes National Park

Time Units					Rock Units		Geologic Events
Eon	Era	Period	Sub-Period	Epoch	Group	Formation	
Phanerozoic	Cenozoic	Quaternary		Holocene		Alluvium	Formation of sand dunes about 10,000 years ago; most of the sand is derived from the weathering of the San Juan Mountains deposited at the foot of the Sangre De Cristo Mountains
				Holocene		Alamos Formation	Closed basin deposition, silts, sand, lake, wind, and river deposits, alluvial sand deposition at the base of the Sangre De Cristo Mountains by the Rio Grande River; the material was derived from the erosion of the terminal moraines about 12,000 years ago; the filling of the San Luis Valley with 40,000 feet + of lava, ash, sand, clay, and gravel
				Pleistocene	Pinedale Durango Cerro Glacial Stages		Growth of glaciers on the San Juan and Sangre De Cristo Mountains over 1 million years ago, forming outwash, moraines, kettle lakes, etc., on the valley floor during the various stages of glaciation
		Tertiary	Neogene	Pliocene	Santa Fe Group	Servilleta Formation	Sandy sediments grading to fanglomerates near the Sangre De Cristo mountains; deposition of sediments and eruption of occasional lava flows
						Heinsdale Formation	Deposition after the formation of the Rio Grande rift/graben of basalt and local lava flows
				Miocene		Las Pinos Gravel	Rapid erosion of the San Juan Mountains and deposition of an alluvial gravel derived from the volcanic mountains; long interval of erosion accompanied by the warping and tilting of the eastern San Juan Mountains, the formation of the Rio Grande rift zone
			Paleogene	Oligocene		Conejos Formation	A mixture of reworked volcanic debris, breccias, and lavas; volcanic rocks and lava flows created the San Juan Mountains 30–35 million years ago; composition was intermediate. Second set of San Juan Mountains
				Eocene		Blanco Basin Formation	The beginning of volcanic eruptions about 40 million years ago continued formation of the San Juan Mountains as a cluster of volcanoes; erosion
				Paleocene		Laramide Orogeny	Arching of the Paleozoic and Mesozoic bedrock; formation of anticlinal structures; the Sangre De Cristo Mountains became an elongated, narrow, linear fold; the San Juan Mountains became an almost cricular dome. Erosion of Mesozoic and Paleozioc beds down to the Precambrian cores
Precambrian	Proterozioc	Proterozoic Y					Faulting about 1.4 billion years ago. Intrusion of a quartz monzonite into metamorphic rocks; erosion
		Proterozoic X					Between 1.7–1.8 billion years ago the formation of gneiss, leucogneiss, and quartz monzonite; erosion

Source: After Chronic, Larkin, Gregory, Peters et. al.

Permian Period, the Ancestral Rockies were eroded to low hills and plains.

4. **Deposition of Mesozoic Era sediments in the vicinity of the Sangre de Cristo Mountains. Later these sediments were removed by erosion during the Laramide orogeny, which created the Rocky Mountains, and the rift valley where the dunes are located.**

On the West Coast of North America, there was eastward subduction of the Pacific Plate beneath the North American Plate. Formation of alluvial (stream) and aeolian (wind) deposits resulted. A shallow transgressive sea eventually covered the area. The Laramide orogeny that took place from Late Cretaceous to Middle Eocene time (80–50 million years ago) caused uplifting and tilting of beds, and thrust and reverse faulting throughout western North America by compression. The orogeny produced west-dipping reverse faults that are exposed in the Sangre de Cristo Mountains. The Sangre de Cristo normal fault trends in a northwest-southeast direction and runs through Great Sand Dunes National Park. Seismic and gravity studies show the normal fault zone is a narrow belt of step faults that was first covered by alluvial fan facies about 19 million years ago during the Miocene Epoch. The fault zone was later covered by the sediments of the Basin Fill Facies in the Rio Grande Rift. The study showed that there was rapid movement along the fault, with the Sangre de Cristo Mountains providing most of the clastic material. The fault scarp indicates that a large earthquake of a magnitude of 7+ happened from 10,000 to 14,000 years ago. The San Luis Valley occupies the site of the former San Luis-Brazos uplift. This was a broad area of crustal warping and high-angle reverse faulting. About 40 million years ago (Eocene Epoch), the San Juan Mountains formed again as a cluster of volcanoes. This uplift was eroded to low relief and created the unconformity separating the Eocene Blanco Basin Formation from the overlying Oligocene-aged Conejos Formation.

5. **During the Cenozoic Era, cessation of uplift, subsidence of marginal basins, and creation of the San Juan Mountains by volcanic activity. Volcanic activity continued during the Oligocene Epoch and created the Conejos Formation.**

For 38 to 20 million years (until Miocene time) there were repeated eruptions of volcanic ash and lava. Lava flows, tuffs, and the formation of a large caldera characterize the San Juan volcanic field. A series of intermediate volcaniclastic ash flows, called the Conejos Formation, occurred from 26 to 30 million years ago (Oligocene Epoch). The deposits were derived from the volcanic field located west of the San Luis Valley. They predated the formation of the rift. There were successive periods of volcanism with each episode followed by subsidence. During the late Oligocene Epoch, ring dikes formed, accompanied by faulting and hydrothermal fluids moving through the fractures.

6. **Mid-continental regional uplift in the Miocene Epoch from 28 to 10 million years ago.**

Much of the western United States, including Colorado and adjacent states, underwent uplifting. The central section was uplifted 5,000 feet. The San Luis Valley was not uplifted, as it is part of the Rio Grande rift, which is a *graben,* that is, a down-dropped section of land bounded on either side by faults, hence the adjacent land is at a higher elevation. The San Juan Luis Valley stayed at essentially the same elevation as the surrounding area rose. After the formation of the Rio Grande rift during the end of the Miocene Epoch and beginning of the Pliocene Epoch, the rift area was filled with lava flows and basalt of the Heinsdale Formation. There was continued deposition of the sediments and lava flows during the Pliocene-aged Servilleta Formation. Included were sandy sediments grading to a fanglomerate (sediments mixed in size that were first deposited as part of an alluvial fan and then cemented together). Both formations are part of the Santa Fe Group.

7. **Increased precipitation and runoff. Rapid downcutting and valley filling.**

Today the faults on the western side of the valley are buried by lava flows and river deposits. But the eastern side is exposed as the Sangre de Cristo fault. The Sange de Cristo (rampart) has a steep rise along the fault line. During the last 10 or 20 million years (Miocene Epoch) the Rio Grande graben was filled. It was much deeper than the San Luis Valley is today, but lava and ash from the San Juan Mountains, debris from the Sange de Cristo Mountains, and river deposits eventually filled it. Periodically, lakes formed in the valley.

8. **Infilling of San Luis Valley during the Pleistocene Epoch; formation of the dunes after the glaciers retreated.**

During the Pleistocene Epoch, valley glaciers formed in both the Sangre de Cristo and San Juan Mountains. There were at least three episodes of glaciation. From the oldest to the youngest, they were the Cerro, Durango, and Pinedale. Glaciers transported large amounts of rock debris into the San Luis Valley, but began to retreat about 12,000 years ago in this region. The Rio Grande River began to deposit material in this closed basin environment with silts and sands that were derived from the glacial material by wind and river and deposited at the base of the Sangre de Cristo Mountains. Lakes periodically formed. These beds are called the Alamos Formation. They accumulated and filled the San Luis Valley with over 10,000 feet of material. Some sections of the valley have over 30,000 feet.

The last of the glaciers disappeared about 10,000 years ago. With their disappearance, the climate became warmer and drier. As the vegetation gradually disappeared, the bare soil was exposed. The winds picked up, removed the finer particles, and picked up sand grains from the valley floor. They were moved to the northeast through the low areas in the Sangre de Cristo Mountains, and accumulated as the sand dunes that we see today.

A fire April 18 to 21, 2000, burned 5,400 acres, destroyed three buildings, and closed the park for two days. It started as a grass fire near a highway. Eventually 70 mph winds produced flames 10 to 15 feet high and blew them into the park.

Geologic Map and Cross Section

Colorado Geologic Highway Map. 1991

Kluth, Charles F. & Schaftenaar, Carl H. 1994. *"Depth and Geometry of the Northern Rio Grande Rift in the San Luis Basin, South-Central Colorado,"* in Basins of the Rio Grand Rift: Structure, Stratigraphy and Tectonic Setting: Geological Society of America Special Paper 291, pgs. 27–37

McCalpin, James P. n.d. *"General Geology of the Northern San Luis Valley, Colorado"* GEO-HAZ Consulting, P.O. Box 1377, Estes Park, Colorado, 80517 or 600 East Galena Avenue P.O. Box 837, Creststone, Colorado; website http://www.geohaz.com./TRIP2021.PDF

Sources

Aber, James S., n.d., *"Rocky Mountain Geology South Central Colorado"* Emporia State University; Website <http://www.academic.emporia.edu/aberjame/field/rocky_mtn/rocky.htm

Chronic, Halka. 1984 *"Pages of Stone: Geology of Western National Parks and Monuments"* 1: Rocky Mountains and Western Great Plains. Seattle: The Mountaineers.

Clayton, Sam 2000-2001, *Thesis: Study Area in Southwestern Colorado";* WebSite http://www.samclayton.co.UK/thesis/toc.htm

Connors, Tim. 1998 National Park Service *"Park Geology"* WebSite <http://www.agd.nps.gov/grd/parks/grsa/

Cyberwest Magazine, Inc. n.d. *"Science West Geology: Great Sand Dunes National Park Inches Closer to Reality with Baca Purchase"* & *"Western Geology, Plate Tectonics"* http://www.cyberwest.com/cu08/v8scwst.html

Desert USA. 1998 *"Great Sand Dunes National Monument"* Digital West Media Incorporated; Website <wysyg://AnswerFrame.82/http://www.desertusa.com/sand/index.html

Digital West Media. 1996-2002, Desert USA: *"Great Sand Dunes National Monument: Description";* Website http://www.desert.com/sand/sand_desc.html

Kiever, Eugene P. & Harris, David V. 1999, *Geology of Western Parklands* John Wiley & Sons, Inc., New York p. 644–652

Kluth, Charles F. & Schaftenaar, Carl H. 1994 *"Depth and Geometry of the Northern Rio Grande Rift in the San Luis Basin, South Central Colorado"* in Keller, G.R. and Cather, S.M., eds., Basins of the Rio Grande Rift: Structure, Stratigraphy, and Tectonic Setting: Geological Society of America Special Paper 291, p. 27-37.

Larkin, Robert P., Greggor, Paul K. & Peters, Gary L. 1980 *"Southern Rocky Mountains"* Geology Field Guide Series, Kendall/Hunt Publishing, Dubuque, Iowa.

McCalpin, James P. n.d. *"General Geology of the Northern San Luis Valley, Colorado"* GEO-HAZ Consulting, P.O. Box 1377, Estes Park, Colorado or 600 East Galena Avenue, P.O. Box 837, Crestone, Colorado; Website http://www.geohaz.com/TRIP2021.PDF

National Park Service "Great Sand Dunes National Monument Flyer" Reprint 1999

National Park Service Handout n.d. "Discovering the Dunes"

National Park Service Handout n.d. "Geology of Sand and Dunes"

National Park Service Handout, 2001. "Hey Ranger: Questions Most Asked by Visitors"; Website NPSHandout\\INP-GRSA-SERVER\Shared\interp\Handout

National Park Service Map "Great Sand Dunes National Monument and National Preserve." n.d.

Note: To convert English measurements to metric, go to www.helpwithdiy.com/metric_conversion_calculator.html

C H A P T E R   1 3

Cuyahoga Valley National Park

N O R T H E A S T   O H I O

David B. Hacker   Kent State University

Area: 33,000 acres; 52 square miles

Established as a National Recreation Area: December 27, 1974

Designated a National Park: October 11, 2000

Address: Cuyahoga Valley National Park, 15610 Vaughn Road, Brecksville, OH 44141-3018

Phone: 216-524-1497

E-mail: cova_canal_visitor_center@nps.gov

Website: www.nps.gov/cova/index.htm

Sculpted over time by running water, glaciers, and mass wasting, the valley's diverse landscape includes meandering streams, deep gorges, a glacially buried valley, steep valley walls, ravines, waterfalls, and towering rock ledges. Examples of ancient streams are preserved in the exposed Mississippian and Pennsylvanian portions of the bedrock and retreating Pleistocene glaciers deposited a complex assemblage of drift.

FIGURE 13.1   The meandering (crooked) Cuyahoga River. National Park Service photo.

Geographic Setting

The Cuyahoga River is the backbone and central natural feature of Cuyahoga Valley National Park. Known as "Ka-ih-ogh-ha" or "crooked river" to Native Americans (fig. 13.1), the Cuyahoga River begins about 30 miles east of its mouth in Cleveland, and follows a roughly V-shaped course for approximately 100 miles before emptying into Lake Erie. Flowing lazily on its relatively flat flood plain past lush forests, fields, and towns, the river provides excellent habitat for park wildlife within this densely populated urban setting of northeast Ohio. The Cuyahoga River is at an elevation of about 730 feet above sea level as it enters Cuyahoga Valley National Park from the south, and at 600 feet as it leaves the park to the north. Above the Cuyahoga's flood plain the landscape is rugged, with steep (6 to 70 degrees) valley walls backed by high, narrow hills. The rugged slopes give way to upland areas at elevations of about 900 feet to as much as 1170 feet. The uplands are relatively flat, with gentle slopes of less than 6 degrees. It is here that the tributaries to the Cuyahoga originate. Many of them are funneled through closely spaced ravines as they flow toward the crooked river. Some drop up to 600 feet in a distance of only a few miles through steep gorges. Many are gently flowing while others are more aggressive, flowing rapidly toward the Cuyahoga and sometimes dropping suddenly over scenic waterfalls. The geometry or pattern of the tributaries is still presently evolving in Cuyahoga Valley through headward, downward, and lateral erosion.

The Cuyahoga River has long attracted people and wildlife; therefore the valley has a long history of human habitation and use. People have lived here for nearly 12,000 years, and it is no surprise that they left a legacy of archeological sites and disturbed lands throughout the valley. Lands now within the park had many different uses in the past, including: agriculture, quarries, mining (sand, gravel, and topsoil), industry, dumps, canals, and residential development.

The valley was an important transportation route for Native Americans, who traveled south from the cold waters of the Great Lakes to the short portage across the divide near present day Akron. Then they traveled along the Tuscarawas River, which drains into the Ohio River and the warmer waters to the south. European explorers and trappers arrived in the 17th century, followed by traders and settlers who appreciated the river's potential as a source of livelihood (hunting and agriculture) as the Native Americans had done before them. In 1786, Connecticut reserved 3,500 acres in northeastern Ohio, known as the "Western Reserve," for settlement by its citizens. Some of the oldest homes in the valley, such as Frazee House and Hale Farm Homestead (both circa 1826), preserve a distinct New England look in the valley.

The Ohio & Erie Canal (from Lake Erie to the Ohio River) opened in 1827 between Akron and Cleveland and played a large part in the early economic development of the area (fig. 13.3). The canal paralleled and was partly watered by the Cuyahoga River, which it replaced as the primary transportation artery. The canal was a means of opening up and developing inland Ohio, and provided a boon to the development of commerce in the Midwest. It helped the economic base of the Valley change to manufacturing. The historic towns of Boston and Peninsula in the park boomed with canal-related industry. Many of the locks, canal houses, and mills associated with the canal industry are preserved within the national park. The park is also a part of the Ohio & Erie Canal National Heritage Corridor that extends 110 miles from Cleveland to New Philadelphia, uniting communities that share a common history and influence from the canal.

By the 1860s, railroads had become prevalent and led to the eventual demise of the canal in the valley, but contributed to the growth of the cities of Akron and Cleveland. The Jaite paper mill was one of the significant industries in the valley in the early 1900s. The area around the mill developed into a true company town, complete with a railroad depot, a company store, and company-owned houses. Natural gas from five local wells powered the mill instead of water. Restored buildings of the town now serve as the Cuyahoga Valley National Park Headquarters. Because of the valley's long and diverse history, many structures in the park are listed on the National Register of Historic Structures.

Nearly 15 percent of Ohio's population lives within the Cuyahoga River watershed. The major metropolitan centers of Akron and Cleveland, at the southern and northern ends of the Cuyahoga Valley National Park, contain most of the watershed's population. This high concentration of urban and heavy industrialized areas along the river's banks environmentally stresses the lower Cuyahoga, which had been troubled by pollution since development in the 1800s. In 1969 the Cuyahoga became internationally known as the "river

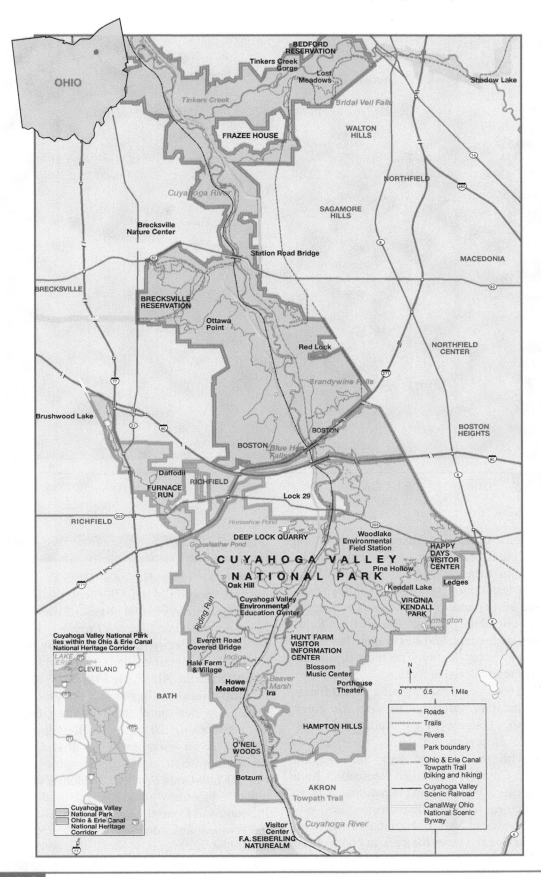

FIGURE 13.2 Cuyahoga Valley National Park, Ohio. National Park Map.

FIGURE 13.3   Remnants of the Ohio & Erie Canal are preserved throughout the park. Photo by Edward G. Soldo.

that burned" because of the fire caused by the pollution in the river. This incident received a great deal of media attention, which in turn spurred the passage of the Clean Water Act less than a year later. Although substantial clean-up efforts have progressed since then, the Cuyahoga is not completely healed. Runoff from fields and parking lots, sediments, and sewer overflows continue to impair the water quality. The river is still not clean enough for recreational activities.

In the early 1960s concerned citizens joined forces with state and local governments to save the green space and historic features in the valley from the onslaught of development. In 1974, their efforts crossed paths with the National Park Service, and Congress created Cuyahoga National Recreation Area as an urban park as a way to bring a national park to people living in cities. In 2000, the recreation area was redesignated Cuyahoga Valley National Park. The National Park Service manages the park in cooperation with others who own property within its boundaries, including Metro Parks of Summit County and Cleveland Metroparks.

Geologic Features

Cuyahoga Valley National Park is located on the glaciated portion of the Allegheny Plateau (fig. 13.4), a physiographic region in the northwestern part of the larger Appalachian Plateaus Province. Relatively undeformed sedimentary rocks dissected by steep-sided valleys with intervening ridges and gently rolling uplands characterize the Allegheny Plateau. With the exception of local mass wasting slump features, the rocks are essentially horizontal, having a regional dip to the southeast of only 20 feet/mile. Bounding the Allegheny Plateau is an escarpment of a steeper topography descending to adjacent geomorphic regions of the Central Lowlands Province, which is characterized by older sedimentary rocks. Where both the Allegheny Plateaus and Central Lowlands have been modified by the comings and goings of multiple continental glaciations, the sedimentary rocks are covered by thin to thick glacial drift deposits. Bedrock, glacial deposits, fluvial processes, and mass wasting have joined here to create a diverse and interesting landscape.

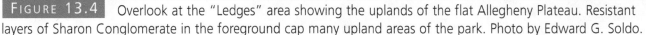

FIGURE 13.4 Overlook at the "Ledges" area showing the uplands of the flat Allegheny Plateau. Resistant layers of Sharon Conglomerate in the foreground cap many upland areas of the park. Photo by Edward G. Soldo.

The Sedimentary Bedrock

The bedrock exposed in Cuyahoga Valley National Park, as well as throughout Ohio, consists of relatively flat-lying, layered Paleozoic sedimentary rocks. The bedrock within the park consists of a sequence of clastic units that record transgressional and progradational (or regressional) events during Late Devonian, Mississippian, and Early Pennsylvanian Periods. The Devonian system is represented by dark gray to black shales (Ohio Shale) that accumulated in a tropical, stagnant sea. The Mississippian system is represented by a sequence of shale and interbedded siltstone (Bedford Shale), overlain by sandstone (Berea Sandstone), and silty, clayey, sandy shales (Cuyahoga Formation). This sequence was deposited in changing environments from offshore marine, to coastal, to fluvial, and back to offshore marine. The Pennsylvanian system is represented by sandstone and conglomerate (Sharon Conglomerate) deposited by fluvial braided streams. During this span of time, Ohio lay on the western flank of the Appalachian Basin (a structural basin) and

received clastic sediment from the east from the eroding mountains of the Appalachian Mobile Belt. This was produced by continental collisional events that dominated the Paleozoic Era in eastern North America and led to the eventual formation of the supercontinent of Pangaea by the end of the era.

Within the Cuyahoga Valley, the less resistant shale units form soil-covered slopes and are best exposed in tributary stream valleys such as the 200 feet deep Tinkers Creek Gorge, which has one of the finest exposed sections in northeast Ohio. The best known and most easily recognizable rock units in the park are the more resistant Berea Sandstone and Sharon Conglomerate and conglomerates deposited in a fluvial environment. The Berea is a fine- to medium-grained sandstone that weathers to a light or dark brown color. Two distinct phases are recognized: an upper thinly bedded phase of uniform sheet sandstone exhibiting surface oscillation (symmetrical) ripple marks, and a lower more massive phase exhibiting large-scale trough and planar cross-bedding and surface current (asymmetrical) ripple marks. Because of its resistance to erosion, the Berea frequently forms the lip of scenic water-falls such as Brandywine, Chippewa Creek, and Blue Hen Falls (fig. 13.5) located in the side tributary streams to the Cuyahoga River. Brandywine Falls (fig. 13.6) is the most spectacular of the waterfalls in the park, falling 63 feet into a large plunge pool. The waterfall is capped by 15 feet of the massive, cross-bedded Berea Sandstone that overlies the softer gray Bedford Shale that erodes out and eventually undercuts the sandstone until it breaks off under its own weight. Evidence of this process can be seen in the huge boulders of Berea Sandstone found at the base of the falls. The top surface of the Berea, over which the stream flows, exhibits potholes and current fluting (shallow gutter-like channels worn in the rock surface by water action). The contact between the Berea Sandstone and Bedford Shale is made prominent by rust or orange iron staining from springs issuing along the contact. The Brandywine Falls also shows the relationship between geology and early industry found throughout the park. Beginning in 1810, the falls was used to power various kinds of milling

operations, including grist, lumber, distillery, woolen, and cider industries. Foundations and other remains of some of these mills still remain.

The fine, uniform, angular quartz grains of the Berea Sandstone, as well as its ability to be easily sawed and shaped, made it well suited as a structural and decorative stone. Therefore, Berea has long been quarried as a dimension stone for foundations, curbing, flagstone sidewalks, and grindstones at numerous places outside as well as in the park. One of these places within the park is at Deep Lock Quarry (fig. 13.7), named after the nearby lock #28 on the Ohio and Erie Canal, which is the deepest one at 17 feet along the entire canal system. This is the best place in northeast Ohio to see the thick lower massive phase of the Berea sandstone in three dimensions. The quarry operated in the 1800s and supplied grindstones and building stones that were shipped via the canal. Several grindstones (more accurately hulling stones, which are smaller and thicker for hulling grains) can be found along the trail leading to the quarry (fig. 13.8). The top of the Berea is covered with gla-

FIGURE 13.5   Blue Hen Falls capped with resistant Berea Sandstone overlying shales and siltstones of the Meadville Shale Member of the Cuyahoga Formation. Reddish-orange iron staining from springs marks the contact. Photo by Edward G. Soldo.

FIGURE 13.6   Brandywine Falls, capped with resistant Berea Sandstone overlying shales and siltstones of the Meadville Shale Member of the Cuyahoga Formation. Photo by David B. Hacker.

FIGURE 13.7   Massive phase of the Berea Sandstone exposed in the historic Deep Lock Quarry. Photo by David B. Hacker.

cial till, but this has been partially removed during the quarrying operations to reveal glacial striations on the bedrock surface.

The Sharon Conglomerate is the youngest rock formation in the park and can be found capping the highest hills and ridges and forming scenic rock ledges (fig. 13.9). The name is misleading since the sandstone fraction exceeds the conglomerate fraction in most outcrops. The Sharon is a clean, white to yellow, medium-grained, friable (loosely cemented), quartz sandstone (fig. 13.10). Conglomerate layers and lenses are common and composed of well rounded milky white quartz and some quartzite. The purity of silica (up to 98%) has made the Sharon an important economic source of silica for foundry and glass sands in Ohio. The most prominent and striking features of the Sharon are sedimentary structures that include abundant *trough* and *planar cross-bedding* (fig. 13.10 and 13.11) and large

FIGURE 13.8   Abandoned hulling stones from late 1800s quarried from Deep Lock Quarry. Photo by David B. Hacker.

FIGURE 13.9   Ledges of Sharon Conglomerate exposed at the "Ledges" area (Ritchie Ledges or Kendall Ledges). Photo by David B. Hacker.

FIGURE 13.10 Typical outcrop of Sharon Conglomerate showing planar cross-bedded sandstone with white quartz pebbles exposed at the "Ledges" area. Photo by Edward G. Soldo.

FIGURE 13.11 Planar and deformed (overturned) cross-bedding of the Sharon Conglomerate exposed along joints at the "Ledges" near Ice Box Cave. Photo by David B. Hacker.

FIGURE 13.12   Vertical jointing in Sharon Conglomerate at the entrance to Ice Box Cave at the "Ledges." Photo by David B. Hacker.

cut-and-fill channel features. Planar cross-bedding is abundant in tabular sets throughout the formation while trough cross-bedding is found mainly in channels. Several layers of planar cross-beds show various degrees of overturning of their tops (fig. 13.11). These bedding types are characteristic of a *braided stream*, one in which wide channels were choked by sediment and formed numerous mid-channel sand and gravel bars that streamwaters flowed around in an intricate and ever-changing network of interconnected channels. These features are found in modern braided stream environments like the Platte River in Nebraska.

Jointing is also an obvious feature in the Sharon and forms two prominent sets (fig. 13.12) trending

N80°W and N20°E. Additionally, some of the rock surfaces show well developed *honeycomb weathering* (fig. 13.13) produced by differential weathering of the less tightly cemented areas, creating a pitted or Swiss cheese-like appearance.

Glacial Features

Overlying and mostly burying the bedrock within the park are glacial deposits (glacial drift) laid down directly or indirectly by Pleistocene continental glaciers. A number of glacial features are recognized within Cuyahoga Valley, including ground moraine, end moraine, erratics, kames, kame terraces, kame deltas, outwash valley trains, and lake sediments (fig. 13.14).

Ground moraine and end moraine are composed of unstratified drift known as *till*, which is characterized as being unsorted and deposited (or carried) by a glacier. *Ground moraine* is deposited directly by a receding glacier and forms a fairly thin blanket of till that tends to mimic the contours of the underlying bedrock. Bedrock may be exposed where the relief is pronounced. *End moraine* forms at the margins of stationary glaciers, producing thicker till deposits. *Erratics* are cobbles or boulders that have been transported from their place of origin by glaciers and do not have the same composition as the underlying bedrock. Pink granites and banded gneiss are the most conspicuous erratic compositions and can be seen frequently in streambeds. These igneous and metamorphic rocks have been transported from the Precambrian Canadian shield, where they are exposed on the surface. Erratics commonly display beveled edges or striations (scratches) on their surfaces.

Kames, kame terraces, kame deltas, and valley trains are composed of sorted, stratified drift deposited by glacial meltwater streams. *Kames* are knolls or elongate hills of sand and gravel deposited along the edge of the melting glacier. A *kame terrace* is composed of clay, silt, and sand deposited by streams flowing along and between a stagnant glacier and its valley wall (fig. 13.15). Coarser material was deposited where the current was swift and the finer where it was slow moving. A *kame delta* is composed of silty clay, sand, and gravel formed at the glacier margin, where meltwaters flowed into ponded water or lakes. The large Yellow Creek scarp is an example of stratified silts and sands deposited as part of a kame delta. A *valley train* or outwash is composed of gravel, sand, and silt dropped and spread within the valley from a meltwater stream flowing out from the glacier.

FIGURE 13.13   Honeycomb weathering of the Sharon Conglomerate caused by differential weathering of weakly cemented areas (pitted or hollowed out areas) of the sandstone exposed at the "Ledges" area. Photo by David B. Hacker.

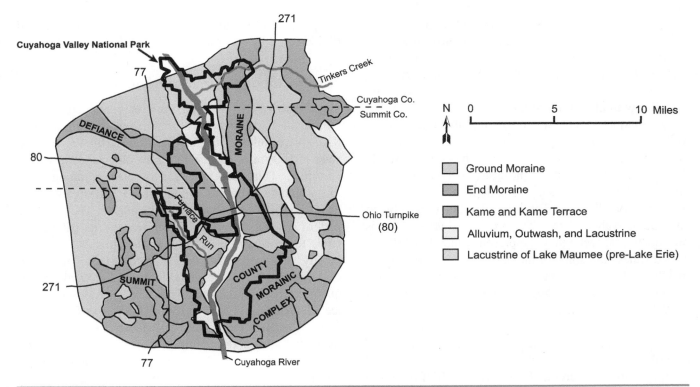

FIGURE 13.14   Map of typical glacial features found in the lower Cuyahoga Valley area. After Corbett and Manner 1988.

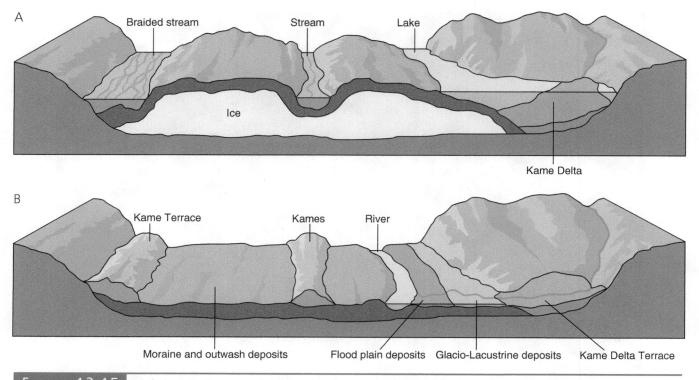

FIGURE 13.15 Schematic diagrams showing development of glacial features found in Cuyahoga Valley. *A.* Depicts valley during initial phase of glacial retreat when stagnant ice fills most of the valley. *B.* Shows the valley following melting of the ice, leaving various types of drift deposits. Modified from Pringle 1982.

As glacial ice melted in the valley, a series of ephemeral lakes were formed as meltwaters were trapped between the receding ice and end moraines. Lake sediments consist of clays and silts, with some exhibiting varve-like laminations. Lake plain areas are very flat with almost no relief.

A unique aspect worth noting about Cuyahoga Valley is the fact that it is a classic example of a *buried glacial valley*, which is a valley that has been partially filled in with glacial drift. In the case of Cuyahoga Valley, all of the above-mentioned glacial deposits contributed to the burial (or filling) of a pre-glacial valley carved by an ancestral Cuyahoga River. The pre-glacial valley floor is now about 500 feet below the present valley floor. These buried valleys are some of the best groundwater aquifers in Ohio when they contain thick deposits of outwash gravels. Because Cuyahoga Valley's deposits contain more fine-grained drift than coarse, the buried valley of the park has not been utilized much as an aquifer,

The glacial features in the Cuyahoga Valley are irregular in their distribution and internal structure compared to the bedrock. Therefore, much of the present topography is controlled by the distribution of these glacial deposits, even though the present valleys over-lie pre-glacially carved valleys and the ridges, and divides tend to occur above bedrock highs and ridges.

The Cuyahoga River Watershed

Cuyahoga Valley National Park is contained entirely within the Cuyahoga River watershed that drains over 810 square miles of northeastern Ohio (fig 13.16). The river is 100 miles long with an average gradient of 7.2 feet/mile. Thirty-seven major tributaries, along with many unnamed streams totaling over 1,100 stream miles, enter the Cuyahoga River along its course. The drainage pattern developed by the Cuyahoga and its tributaries, tends to be dendritic in form, as the streams flow over mostly uniformly erodible glacial drift. The preglacial drainage exercised considerable control over all subsequent drainage in the area. The drainage following each ice advance would tend to follow the debris-filled older valleys because these were still lower topographically than the surrounding thinly drift-veneered bedrock uplands. This was also because the unconsolidated material in the valleys would erode more easily.

Following a roughly V-shaped course, the Cuyahoga River begins within 15 miles of the Lake Erie

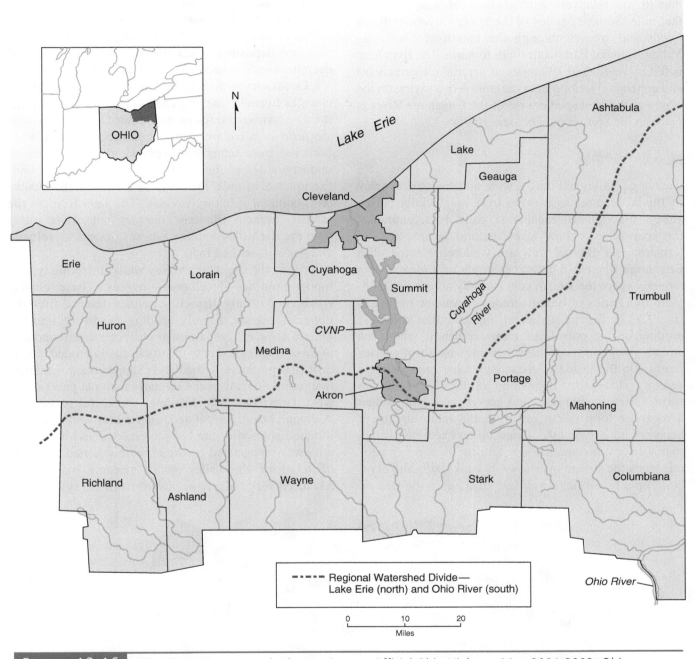

FIGURE 13.16   Cuyahoga River watershed map. Source: Official Ohio Highway Map 2001-2002, Ohio Department of Transportation.

shore as two bubbling springs. The river first flows 60 miles to the southwest in a shallow and uneven channel through thick forests and rich farm fields, until it reaches the populated urban areas near Akron. At this point, the river hits an east-west continental divide and turns sharply west, forming the short bottom of the "V." Here the river falls 225 feet into a gorge, known as the "Gorge of the Cuyahoga," 1.5 miles long and about 240 feet deep. The river then flows 40 miles northwestward along a lower tortuous course until it reaches its terminus in Lake Erie at Cleveland. The actual river length along this lower section (nearly twice the air distance) is due to the relatively low gradient of less than two feet/mile. Some 22 miles of the lower Cuyahoga River winds and weaves through the length of Cuyahoga Valley National Park from south to north. The river here is fed by more than 190 miles of perennial (permanent) and ephemeral (temporary) streams. A discussion on the origin of the V-shaped course of the Cuyahoga River is given in the section on Geologic History.

Fluvial Features

Cuyahoga Valley exhibits a wide and marvelous array of fluvial features as it flows over its glacially buried valley. Floodplains, meanders, point bars, cutbanks, meander scars, oxbow lakes, natural levees, alluvial terraces, and alluvial fans are well represented. The Cuyahoga River is a classic example of a *meandering stream*, where the stream cuts laterally across its floodplain in a series of broad sinuous curves, or meanders (fig. 13.17). During its lateral migration the stream is eroding on the outside of meander bends, where the stream's velocity is greater, forming steep-walled *cutbanks*. On the inside of meanders, where the stream's velocity slows, the deposition of bedload sediment forms characteristic sand and gravel *point bars*. When a meander loop is cut off from the main stream, it becomes an *oxbow lake*. Dried up and sediment-filled oxbows form *meander scars*. During high water periods when the stream overflows its banks, silts and clays build up a broad flat *floodplain* adjacent to the stream.

Because of high flood potential, development of the Cuyahoga floodplain has been minimal. *Natural levees* also form along the upper edge of stream channels during flood periods. These are linear ridges of coarser sediment that slope gradually away from the stream; they are deposited when the velocity of the stream decreases once it leaves the confines of its banks.

Occasionally the level at which a stream is eroding changes (usually due to a change in its base level) and it cuts downward through its former floodplain to create *alluvial terraces*. Alluvial terraces are flat areas bordered by a scarp and occur singly or in pairs on opposite sides of the valley. The valley also contains examples of *alluvial fans* that are formed at the mouths of ephemeral tributary ravines. The fans form as the steeper gradient tributary emerges out of its valley onto the flat valley bottom, where it looses its velocity and deposits its bed load.

Along the Cuyahoga Valley walls, a striking type of topography has developed—*ravines*. These closely spaced and deeply dissecting ravines descend from the uplands and are actively eroding headward into the upland surface. The particular character and abundance of these ravines is due to the thick, easily eroded glacial drift that produces a "badlands" appearance similar to that found in many of the semiarid national parks of the west (see Badlands National Park, for example). Although heavily vegetated, the ravines were probably initiated soon after the last ice receded and before the area was revegetated as the climate warmed. Because of relief of the valley walls, streams most likely advanced very quickly headward from the valley floor.

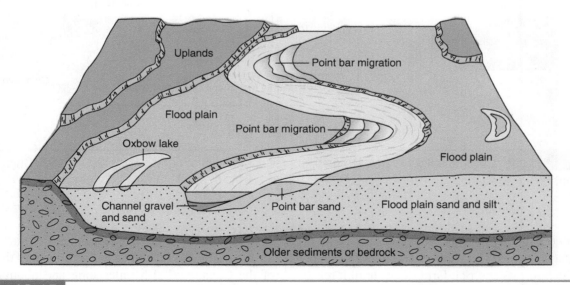

FIGURE 13.17   Schematic diagram of a stream meandering across a flood plain, showing common fluvial features and migration of cutbanks and point bars. After Coogan and others 1974.

Thus erosion would have been rapid, developing the badlands topography which is preserved today beneath the deciduous forests.

Mass Wasting Features

Mass wasting processes are continually moving masses of material downslope within the park, often contributing to geologic hazards. Driven by gravity and aided by Ohio's moist climate and freezing and thawing conditions, slide and flow type structures are abundant, especially within the thick glacial drift and clay lake sediments. North-facing slopes also tend to hold more moisture and exhibit less stability than the south-facing slopes. The speed and form material exhibits when moving downslope varies. Slump, debris slides, mudflows, earthflows, and creep are common forms of mass wasting found on the valley walls in Cuyahoga Valley.

Slump is a slow to rapid rotational movement of surficial material downslope as a single mass or multiple masses along a curved rupture surface. Slumping occurs in Cuyahoga Valley in the thicker unconsolidated debris (glacial drift and soil) on the valley slopes, where the material ruptures along a spoon-shaped surface (concave upward or outward), and the block's upper surface is sometimes tilted backward. A *debris slide* is similar to a slump in which slow to rapid movement of surficial material moves downslope as a single mass or multiple masses along a more or less straight rupture surface angled parallel to the slope. A *mudflow* is a rapid type of movement involving water-laden debris moving downslope as a viscous fluid, mostly within channels and ravines. An *earthflow* is similar to a mudflow (but contains less water) in which water-saturated debris moves downslope on the hillsides as a viscous fluid during times of heavy precipitation or snowmelt. In the spring of 1983, a large earthflow about 150 feet long and 50 feet wide moved downslope and blocked a portion of Boston Mills Road. Scars left behind by earthflows and slumps are abundant on the slopes throughout the park.

Creep conditions are much slower (imperceptibly slow), but still an efficient process of mass movement, involving the gradual downslope movement of material. Signs of this slow movement can be seen in the abundance of curved tree trunks of trees growing on slopes when creep occurs in the top layers of the soil and drift. Creep also occurs in the movement of large bedrock blocks of Sharon Conglomerate. The jointing of the Sharon Conglomerate produced a series of huge rectilinear blocks that slide and slump slowly downslope on the underlying Meadville Shale Member of the Cuyahoga Formation (fig. 13.18). Ground water passes easily downward through the porous sandstones and conglomerates of the Sharon but cannot pass through the impermeable shales below. It can be seen flowing out as springs at the base of the Sharon. The water helps to make the shales slippery and allows the jointed Sharon to move more freely over it. This creep process has created a series of narrow passageways along the joint surfaces (fig. 13.19), producing the characteristics of the "ledges" area of the park. Ice Box Cave (fig. 13.12) received its name from the cool air that constantly blows out from its 50 feet recesses.

FIGURE 13.18 Huge rectilinear blocks produced by jointing of the Sharon Conglomerate that slowly slide and slump downslope (to the left) on the underlying Meadville Shale Member of the Cuyahoga Formation.

FIGURE 13.19 Narrow passageways created by creep along joint surfaces in the Sharon Conglomerate.

Geologic History

1. Subsurface Precambrian and early Paleozoic Rocks.

The earliest record of the geologic history in northeast Ohio is preserved in the igneous and metamorphic Precambrian basement rocks that underlie Paleozoic sedimentary rocks at a depth of over 5,000 feet in Cuyahoga Valley National Park. These rocks are known in Ohio only from deep-well drilling and have been dated at more than one billion years old. In central and eastern Ohio, these subsurface rocks are part of the Grenville Province, and are the remains of a Proterozoic plate that collided with an older portion of the North American continent, the Archean Superior Province to the west. This formed the one to two billion year-old Grenville Mountains.

Following the Grenville mountain building episode, erosion leveled the ancient mountains and formed a great plain over the region, upon which, starting in the Cambrian Period (about 570 million years ago) and continuing through the rest of the Paleozoic Era, several shallow seas transgressed and regressed over the craton, depositing sands, clays, and carbonate sediments. Beginning in the Ordovician Period, collision between the North American and European continents formed a series of island arcs and mountains east of Ohio. Known as the Taconic Orogeny, this was the first in a series of collisional disturbances within the Appalachian Mobile Belt that produced the eventual suturing of the continents together. They formed the supercontinent of Pangaea by the end of the Paleozoic Era. Although Ordovician and later rocks in Ohio were not involved in this or future collisional events, they record these activities in the sediments that accumulated in the subsiding Appalachian Basin that formed throughout the Paleozoic Era in the foreland area west of the Appalachian Mobile Belt. Ohio was located on the northwestern margin of the Appalachian Basin during the Paleozoic Era events and had its western flank occupied by the Cincinnati Arch. Toward the end of the Silurian Period, the seas gradually receded and evaporation of seawater formed thick salt beds in subsiding restricted areas of northeast Ohio. These salt beds are actively mined more than 1639 feet beneath Lake Erie at Cleveland.

Although none of the Precambrian or lower Paleozoic rocks are exposed in or near Cuyahoga Valley National Park, the sequence is of interest to the casual park observer. The Appalachian Basin is a significant oil and natural gas producing region, and the area inside and surrounding the park is dotted with pumps extracting petroleum from older Paleozoic rocks. The most important producing zone is the Early Silurian "Clinton" sands, deposited in deltas and offshore bars as a result of erosion of the Taconic highlands to the east. Farther to the north in Canada, the same Precambrian Ohio basement rocks of the Grenville and Superior Provinces are exposed in the Canadian Shield area of Ontario, where later Pleistocene glaciers plucked and transported them south to Ohio and left them after melting. The larger glacial erratics, of mostly granite and gneiss, form the resistant boulders scattered on the uplands or in the streambeds within the park.

2. Deposition of Devonian Marine Ohio Shale.

Rocks of Late Devonian time are the oldest exposed in northeast Ohio and in Cuyahoga Valley

| Table 13.1 | | | Geologic Column, Cuyahoga Valley National Park | | | |

Time Units			Rock Units			
Era	**Period**	**Epoch**	**Group**	**Formation**	**Member**	**Geologic Events**
Cenozoic	Quaternary	Holocene	Alluvium			Stream erosion, weathering, and mass wasting
		Pleistocene	Ground moraine, end moraine, kames, outwash, lake deposits			Continental glaciation
	Tertiary		~~~~~~			Major erosion interval of 245 million years
Mesozoic			~~~~~~~~~~~~~ Unconformity ~~~~~~~~~~~~~			
Paleozoic	Pennsylvanian		Pottsville	Sharon Conglomerate and sandstone		Sandstone and minor conglomerate of up to 98% quartz, braided streams
			~~ Unconformity ~~~~~~~~~~~~~~~~~			
	Mississippian		Cuyahoga	Meadville Shale		Gray shale, siltstone, and sandstone, shallow marine shelf deposits and nearshore marine depositions
				Sharpsville Sandstone		
				Orangeville Shale		
				Berea Sandstone		Delta distributary stream channels and coastal sands
				Bedford Shale		Gray shale and siltstones, prodelta marine deposits
	Devonian			Ohio Shale	Cleveland Shale	Dark gray and black shales, stagnant anoxic seas; fish fossils
					Chagrin Shale	Deposition in offshore marine basin arthopods

Sources: Coogan, 1996; Corbett and Manner, 1988; Hansen, 1999, 2001.

National Park. In Early Devonian time, another plate collision along the northeastern margin of North America resulted in the second burst of mountain-building activity within the Appalachian Mobile Belt. Known as the Acadian orogeny, red clastics that formed the Catskill Delta spread westward toward Ohio across New York and Pennsylvania as sediments were being eroded from the older mountains. However, this part of Ohio was seaward beyond the delta complex and received only dark muds rich in the organic plant and animal material that became the great marine shale deposit known as the Ohio Shale. This rock unit is one of the most unusual in the state and records an episode of unique environmental conditions during Late Devonian time.

During the Devonian Period, Ohio was located just south of the equator, and the Ohio Shale accumulated in an extensive tropical sea that developed a stagnant, stratified water column where there was no continual or periodic mixing of bottom and surface waters. The reasons for the stagnation of the sea are poorly understood, and geologists have argued for the sea being either shallow or extremely deep. What is known is that the bottom waters became very oxygen poor (anoxic) while the surface waters remained oxygenated (aerobic). In this environment, bottom-dwelling (benthic) animals and bacteria were absent and abundant organic matter from above was allowed to settle into the stagnant bottom sediment and not be destroyed. Fossils are rare in the shale, but those found include terrestrial plants or animals that floated out to sea or those that lived in the upper waters. A wide variety of fish lived in the oxygenated surface waters, and upon death, their remains settled into the anoxic bottom mud and became exquisitely preserved fossils. Bones of heavily armored, jawed fishes (arthrodires) such as *Dunkleosteus* can be found within large concretions in the shale. Twenty-two species of arthrodires have been reported from the Cleveland Shale Member; however *Dunkleosteus* was by far the largest predator fish of the time, attaining lengths of more than 39.37 feet. Remains of the earliest well known shark, *Cladoselache*, are also found with soft parts preserved to the finest detail. The Cleveland Museum of Natural History possesses and displays the world's largest collection of these fishes.

The Ohio Shale is rich in hydrocarbons, with nearly one-third organic matter by volume, and has been a minor source of natural gas and is considered an oil shale containing about 20 to 25 gallons of crude oil per ton. These darker black shales, such as those of the Cleveland Shale Member, are easily identified because they give off a distinct petroliferous odor when broken. In the Cuyahoga Valley, the Cleveland Shale is believed to mark the upper boundary of the Devonian Period and lies, for the most part, in conformable contact with the overlying Bedford Shale of Mississippian Age.

3. Deposition of Mississippian marine and non-marine sediments.

The Mississippian-age rocks consist of a somewhat monotonous sequence of poorly fossiliferous shales, siltstones, and sandstones that record a quiescent phase in Late Paleozoic continental collision and mountain-building in the Appalachian Mobile belt to the east. During the Mississippian Period, beginning about 360 million years ago, the stagnant waters of late Devonian time began to clear and the Acadian highlands and Catskill Delta complex eroded during a long, relaxational phase of Late Paleozoic continental collision. Streams carried these sediments into the Ohio portion of the Appalachian basin, creating muddy to sandy seas. Fluvial, deltaic, and marginal marine clastic sediments of the Bedford Shale, Berea Sandstone, and Cuyahoga Formation in northeast Ohio reflect this sharp change in depositional environments from the dark Devonian shales that they buried.

The Bedford Shale consists of about 95 feet of gray, sandy shale that is reddish in the middle part of central and southern Ohio. The reddish facies can be found in the Cuyahoga Valley National Park on the west side of the valley in the gorge of Chippewa Creek, where it overlies the gray facies. The Bedford Shale defines the fine-grained (clay and silt) offshore (or prodelta) phase of a delta. The Berea Sandstone is a fine- to medium-grained sandstone that represents stream deposits or sandbars associated with delta development. The upper part has a blanket type geometry while the lower has thick, channel-form features of controversial origin. The lower sands are interpreted to be a constructional phase of a delta where distributary streams or lobes of different deltas deposited sand bodies that slumped and deformed into the underlying soft shales of the Bedford, which shows deformed and contorted bedding near the thicker Berea sands. The upper sands represent a more transgressive destructive phase of the delta. The overlying Cuyahoga Formation consists of interlayered shales, siltstones, and sandstones, deposited in a shallow shelf type of environment. The Cuyahoga is moderately fossiliferous, containing brachiopod- and mollusk-rich faunas as well as some excellent crinoid assemblages. These deltaic and nearshore marine environments would dominated Ohio for the remainder of the Paleozoic Era.

4. Regression, Erosion, Pennsylvanian nonmarine deposition.

At the close of the Mississippian Period, a major regression and subsequent erosional event ended Mississippian deposition in Ohio and removed younger Mississippian rock units (overlying the Cuyahoga Formation) in northeastern Ohio. During the early part of the Pennsylvanian Period, beginning about 330 million years ago, a blanket of coarse sand and gravel was spread over this erosional surface by braided streams from the north. This deposit of very pure quartz sands with lenses of quartz pebbles became the Sharon Conglomerate of the Pottsville Group. The Sharon represents the onset of the third and final phase of continental collision and mountain-building (the Alleghenian orogeny) in the Appalachian Mobile Belt. Warm tropical seas transgressed back over Ohio at this time, as streams carried great volumes of sand and mud eroded from the rising Appalachian Mountains to the southeast and in Canada to the north. These streams formed great deltas covered by vast peat (coal) swamps where they entered the sea and represented the final episodes of Paleozoic deposition in Pennsylvanian and Early Permian times in Ohio. Within Cuyahoga Valley National Park, the Sharon Sandstone is the youngest rock unit and the only representative of the Pennsylvanian system.

5. Erosion and development of pre-glacial topography.

During the Mesozoic Era and the Tertiary Period of the Cenozoic Era, a total span of about 245 million years, Ohio was undergoing extensive erosion, which removed vast quantities of deposits of these ages (if indeed there were any) as well as a significant portion of Devonian through Permian rocks that may have been west of their present outcrops. Extensive river systems, such as the Teays, developed at this time, dissecting the entire surface of Ohio and producing large and deep river valleys, including the ancestral Cuyahoga Valley, prior to the onset of the Pleistocene Ice Age. The Teays River system formed the main westward drainage from the newly formed Appalachian Mountains and crossed western Ohio, northern Indiana, and central Illinois on its way to the ancestral Mississippi River. Other rivers flowed north and discharged into the ancestral St. Lawrence River, including the ancient Dover River that carved part of the ancestral Cuyahoga Valley into the bedrock.

6. Pleistocene Glaciation.

During the last 2 million years, several periods of glaciation occurred when great continental ice sheets accumulated in Canada and repeatedly spread southward to cover about two-thirds of Ohio. There were at least four major continental ice age periods, known as stages, which are widely recognized. From oldest to youngest, they are: Nebraskan, Kansan, Illinoian, and Wisconsinan. Deposits from the latter three have been identified in Ohio but only those of the Wisconsinan (70,000 to 14,000 years ago) have been identified on the surface in Cuyahoga Valley National Park. Wisconsinan glacial drift deposits cover nearly all the bedrock in Cuyahoga Valley, as well as evidence of the previous glaciations. The glacier dropped unsorted drift sediments as till deposits, while meltwater from the glacier deposited stratified drift as outwash, and ice-contact features such as kame terraces. A series of ephemeral lakes developed in the valley when glacial meltwaters became trapped between a dam of receding ice to the north and moraines to the south. Within the park, as well as the rest of glaciated Ohio, weathering of the drift deposits (especially the tills) since the last retreat of the glaciers produced thick, rich soils that support lush deciduous forests and agricultural crops.

7. Evolution of the present Cuyahoga River.

The drainage systems in northeast Ohio, including the Cuyahoga, were greatly altered by Pleistocene glaciations. The Cuyahoga did not acquire its curious V-shaped (southward to northward flowing) course until about 10,000 years ago, following the retreat of Wisconsinan glaciers. Prior to glaciation, the lower Cuyahoga Valley was part of a larger river system known as the Dover River system, which had its headwaters farther south in Tuscarawas County (fig. 13.20A). The Dover River flowed north, carving a deep valley (the future Cuyahoga Valley) into the Allegheny Plateau, and joined a larger stream to the north that flowed eastward in the general vicinity of present Lake Erie. Between 14,000 and 25,000 years ago, glacial ice blocked the northward flow of the Dover River, causing it to reverse its course southward through the Canton Massillon area (fig. 13.20B). Meanwhile, the upper ancestral Cuyahoga River section was being supplied water from the melting of two glacial lobes (the Cuyahoga and Grand River lobes). It was between these two lobes that the headwaters of the Cuyahoga River developed and were forced (as a result of being blocked by moraines or ice to the north) to flow southward as part of the ancestral Tuscarawas River.

As the glacier began to retreat northward, it left behind a large recessional moraine complex (Summit County moraine complex) near Akron that produced a drainage divide between the future Lake Erie basin and the southward flowing ancestral Tuscarawas River. A minor glacial advance later produced the Defiance Moraine across the present lower Cuyahoga Valley. These moraines acted as dams trapping water in the valley and forming lakes depositing lacustrine silts and clays. Streams flowing down the valley walls from the uplands to the lakes deposited deltas and alluvial fans of gravels and sand. Thus the pre-Cuyahoga Valley (ancient Dover River valley) north of Akron was filled (buried) with as much as 500 feet of till, outwash, lacustrine, delta, and alluvial deposits of varying thickness, extent, and composition.

About 10,000 years ago (fig. 13.20C), following the retreat of ice from Ohio, the northward flowing segment of the Cuyahoga River developed in the lower basin of the Cuyahoga Valley and emptied into Lake Erie. Cutting to the south by headward erosion through the buried valley drift, the north-flowing river segment breached the moraine complexes and captured and pirated the upper basin of the southward-flowing segment near Cuyahoga Falls (fig. 13.20D). Thus the upper and lower Cuyahoga were united to form the V-shaped course of the present Cuyahoga River. The remainder of the ancestral Cuyahoga River still flows south (south of Akron) as the present Tuscarawas River.

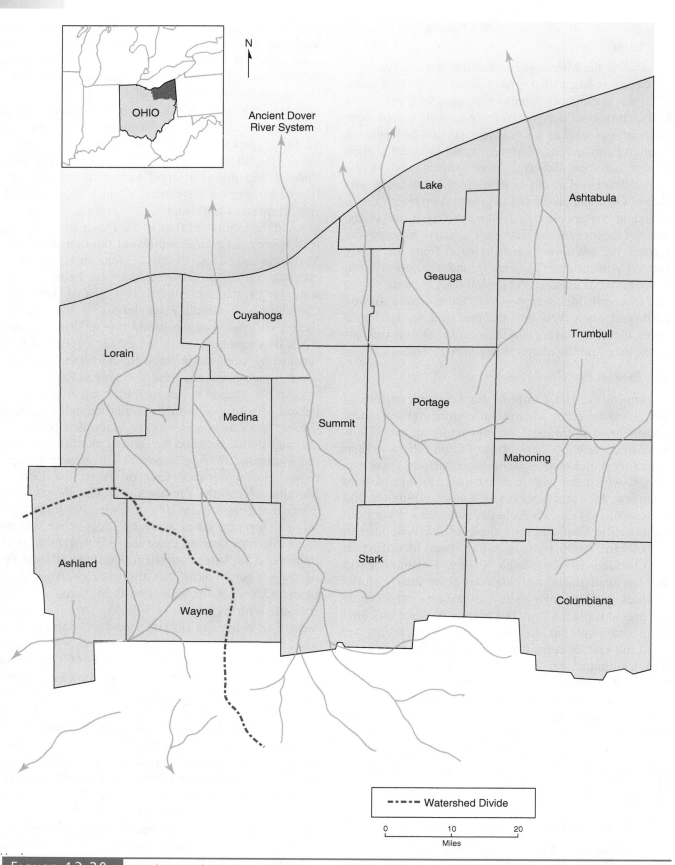

FIGURE 13.20A Evolution of Cuyahoga River maps. *A.* Showing preglaciation (pre-Wisconsinan, 70,000 years ago) streams and flow directions. Sources: White 1984 and Layton 1980.

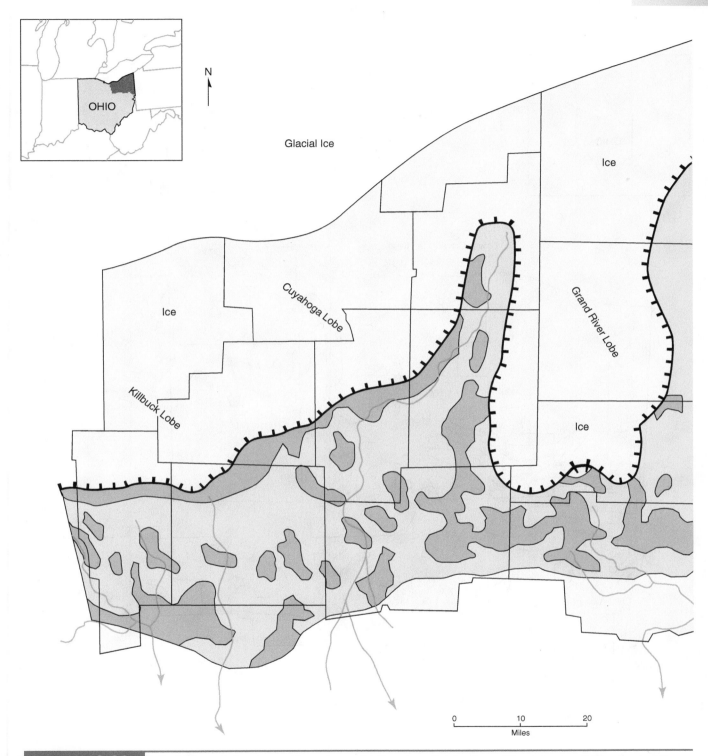

OHIO

N

Glacial Ice

Ice

Cuyahoga Lobe

Grand River Lobe

Killbuck Lobe

Ice

Ice

Ice

0   10   20
Miles

FIGURE 13.20B Evolution of Cuyahoga River maps. *B.* Showing distribution of ice lobes and tills of retreating glaciers about 14,000 years ago and southerly flow of ancestral Cuyahoga River. Sources: White 1984 and Layton 1980.

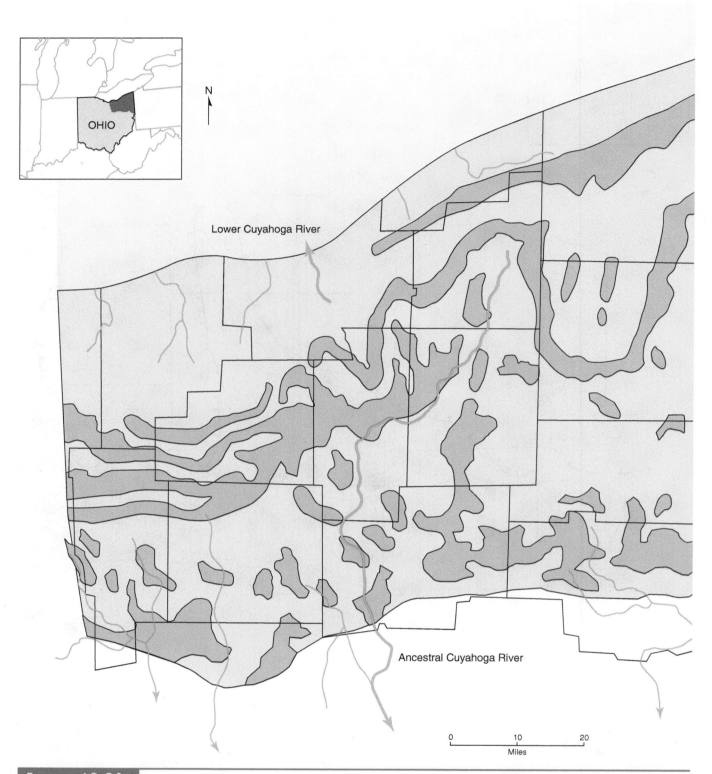

FIGURE 13.20C   Evolution of Cuyahoga River maps. C. Showing postglacial streams and headward erosion of lower Cuyahoga River about 10,000 years ago. Sources: White 1984 and Layton 1980.

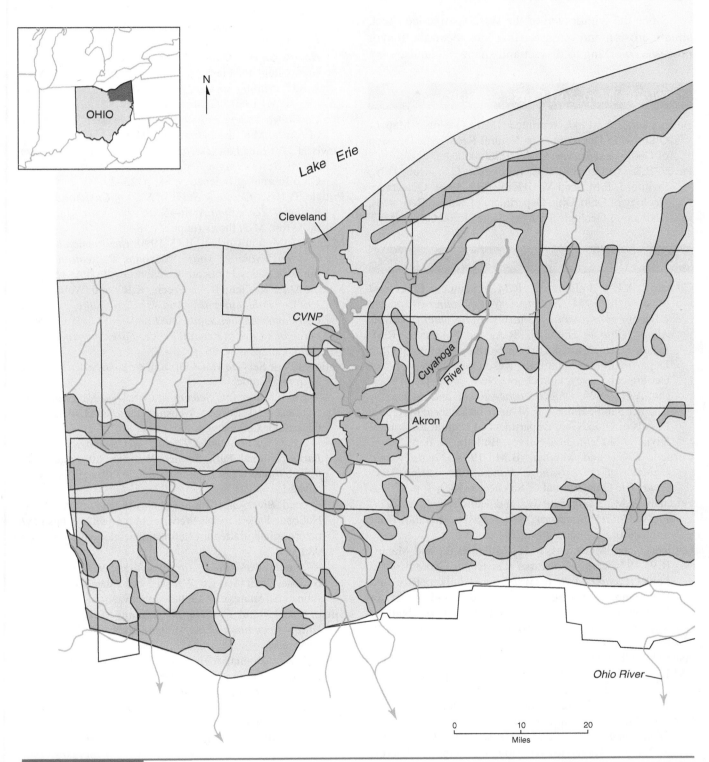

OHIO

N

Lake Erie

Cleveland

CVNP

Cuyahoga River

Akron

Ohio River

0   10   20
Miles

FIGURE 13.20D Evolution of Cuyahoga River maps. *D.* Showing present-day streams and stream capture and piracy of lower Cuyahoga River with upper Cuyahoga River to form V-shaped course of modern Cuyahoga River. Sources: White 1984 and Layton 1980.

8. Recent surficial processes.

Since the withdrawal of the Wisconsinan ice sheet, stream erosion and mass wasting have been the mayor processes working to dissect and shape the landscape.

Geologic Map and Cross Section

Bownocker, J.A. 1947, reprinted 1981. Geologic Map of Ohio. Ohio Department of Natural Resources, Division of Geological Survey. Scale: 1:500,000.

Pavey, R.R., Goldthwait, R.P., Brockman, C.S., Hull, D.N., Swinford, E.M., and Van Horn, R.G. 1999. Quaternary Geology of Ohio. Ohio Department of Natural Resources, Division of Geological Survey Scale: 1:500,000.

Sources

Coogan, A.H., Feldmann, R.M., Szmuc, E.J., and Mrakovich. 1974. *Sedimentary Environments of the Lower Pennsylvanian Sharon Conglomerate Near Akron, Ohio* in Heimlich, R.A. and Feldmann, R.M. (eds.), Selected Field Trips in Northeastern Ohio: Ohio Department of Natural Resources, Division of Geological Survey, Guidebook, No.2, p. 19-41.

Coogan, A.H., 1996. *Ohio's Surface Rocks and Sediments.* Chapter 3 in Feldmann, R.M. and Hackathorn, M. (eds.), Fossils of Ohio. Ohio Department of Natural Resources, Division of Geological Survey, Bulletin 70, p. 31-50.

Corbett, R.G., and Manner, B.M. 1988. *Geology and Habitats of the Cuyahoga National Recreation Area, Ohio.* The Ohio Journal of Science, v. 88, n.1, p. 40–47.

Feldman, R.M., Coogan, A.H., and Heimlich, R.A. 1977. *Field Guide: Southern Great Lakes.* Kendall/Hunt Publishing, Dubuque, Iowa, 241 p.

Gardner, G.W., McComas, M.R., Miller, B.B., and Manus, R.W. 1974. *Engineering and Pleistocene Geology of the Lower Cuyahoga River Valley* in Heimlich, R.A. and Feldmann, R.M. (eds.), Selected Field Trips in Northeastern Ohio: Ohio Department of Natural Resources, Division of Geological Survey, Guidebook, No. 2, p. 45–59.

Hannibal, J.T. and Feldmann, R.M. 1987. *The Cuyahoga Valley NRA, Ohio Devonian and Carboniferous Clastic Rocks.* Geological Society of America Field Guide–North Central Section, p. 403–406.

Hansen, M.C. 1993. *Guide to the Geology along Interstate 77 between Marietta and Cleveland.* Ohio Department of Natural Resources, Division of Geological Survey, Educational Leaflet No. 15.

———. 1999. *The Geology of Ohio—The Devonian.* Ohio Geology: A Quarterly Publication of the Ohio Department of Natural Resources, Division of Geological Survey, 1999, No. 1.

———. 2001. *The Geology of Ohio—The Mississippian.* Ohio Geology: A Quarterly Publication of the Ohio Department of Natural Resources, Division of Geological Survey, 2001, No. 2.

Kent State Geological Society. 1969. *Geology of the Cuyahoga Gorge and Chippewa Creek Sections Northeastern, Ohio.* in Frank, G.W. (ed.), Ohio Intercollegiate Field Trip Guides 1950-51–1969-70, K.S.U. Printing Service, p. 1-1–1-15.

Layton, A.W. 1980. *Geological Study of the Northern Cuyahoga Valley National Recreation Area.* University of Akron, M.S. thesis (unpub.), 94 p.

Lewis, T.L. 1988. *Late Devonian and Early Mississippian Distal Basin-Margin Sedimentation of Northern Ohio.* Ohio Journal of Science, v. 88, p. 23–39.

Pringle, P. 1982. *Geologic Study of Sites of Cuyahoga Valley National Recreation Area—Southern Sector.* University of Akron, M.S. thesis (unpub.), 181 p.

Manner, B.M. and Corbett, R.G. 1990. *Environmental Atlas of the Cuyahoga Valley National Recreation Area.* Surprise Valley Publications, Monroeville, PA, 150 p.

Mullet, D.J., Kurlich, R.A., Frech, K.R., and Wells, N.A. 1990. *Paleocurrent Analysis of the Sharon Conglomerate and Sandstone Lithosomes in the Vicinity of Ice Box Cave at Kendall Ledges Park, Summit County, Ohio.* Compass, v. 68, p. 21–32.

National Park Service Handout. n.d. *Cuyahoga Valley.* Park Map and Guide.

———. n.d. *Nature and Science at Cuyahoga National Park.* National Park Service Website. http://data2.itc.nps.gov/nature/index.cfm?alphacode=cuva

———. n.d. *Environmental Factors at Cuyahoga National Park.* National Park Service Website. http://data2.itc.nps.gov/nature/EnvironmentalFactors.cfm?alphacode=cuva&loc=3

———. n.d. *Rivers and Streams at Cuyahoga National Park.* National Park Service Website. http://data2.itc.nps.gov/nature/subnaturalfeatures.cfm?alphacode=cuva&topic=20&loc=4

———. n.d. *Watersheds at Cuyahoga National Park.* National Park Service Website. http://data2.itc.nps.gov/subnaturalfeatures.cfm?alphacode=cuva&topic=18&loc=4

Sifritt, S.K. 1983. *A Guide to the Geology of the Cuyahoga Valley Recreation Area,* unpublished, National Park Service.

Wells, N.A., Coogan, A.H., and Majoras, J.J. 1991. *Field Guide to Berea Sandstone Outcrops in Black River Valley at Elyria, Ohio—Slumps, Slides, Mud Diapirs, and Associated Fracturing in Mississippian Delta Deposits.* Ohio Journal of Science, v. 91, p. 35-48.

_____, Richards, S.S., Peng, S., Keattch, S.E., Hudson, J.A., and Copsey, C.J. 1993. *Fluvial Processes and Recumbently Folded Crossbeds in the Pennsylvanian Sharon Conglomerate in Summit County, Ohio, U.S.A.* Sedimentary Geology, v. 85, p. 63-83.

White, G.W. 1984. *Glacial Geology of Summit County, Ohio.* Ohio Department of Natural Resources, Division of Geological Survey, Report of Investigations No.123, 25 p., map, scale: 1:62,500.

PART II

Caves and Reefs

FIGURE PII.1   In the Blue Grotto, which is in the lower part of Wind Cave, the boxwork is thick and durable. Uplift severely fractured the Pahasapa Limestone and calcite was deposited in the fractures as secondary filling by ground water. At a later time the original limestone was removed by solution leaving the crisscrossing thin fins behind. Boxwork is found in only a few caves in the world, including Wind Cave and Jewell Cave.
Photo by David Little.

"Soft rocks" (an informal term geologists use, meaning sedimentary rocks) are dominant in the eight national parks in Part II. In these parks, as in Part I, most of the stratified bedrock is relatively flat-lying or undeformed. A significant difference is that in the caves and reefs of the national parks included in Part II, carbonate rocks are overwhelmingly the dominant rock type. Most natural *caves,* or *caverns,* and the features in them, are the result of erosion, solution, and deposition by ground water (the geologic agent) in carbonate rocks. *Reefs* are rigid, wave-resistant structures built in nearshore environments by marine organisms (corals, calcareous algae, and the like) and infilled by broken corals, shell debris, and sediment. Encrusting calcareous algae, foraminifera, and calcium carbonate (precipitated from seawater in voids and spaces) bind the reef structures together.

Caves and Reefs Protected in the National Park Systems

Mammoth Cave, Kentucky; Wind Cave, South Dakota; and Carlsbad Cavern, New Mexico (Chapters 14–16), are not the only caves protected in the National Park System, but they are without question the largest and most notable. At least two other national parks, Great Basin (Chapter 45) and Sequoia (Chapter 49) have caves with interesting and unique features, but they are located in mountain areas where rocks have undergone deformation, metamorphism, plutonism, and other activities associated with mountain-building. National monuments set aside to preserve caves are **Jewel Cave** (near Wind Cave National Park), South Dakota; **Oregon Caves,** Oregon; **Timpanogos Cave** in Utah, and **Russell Cave** in Alabama.

Ancient reefs and modern reefs are dominant features in the five remaining national parks in Part II. Guadalupe Mountains National Park in west Texas, which has ancient reefs, is just south of Carlsbad Caverns National Park, New Mexico, and the geology of these two national parks is closely related. In Virgin Islands National Park in the Caribbean, modern reefs are growing on and around "fossil" reefs. **Buck Island Reef National Monument,** also in the Virgin Islands, is noted for its underwater trail among "coral gardens." Everglades, Biscayne, and Dry Tortugas National Parks, all in Florida, display a great variety of living and fossil reef features, some underwater and some at or above present sea level. Carbonate rocks of various ages in other national parks (for example, Glacier

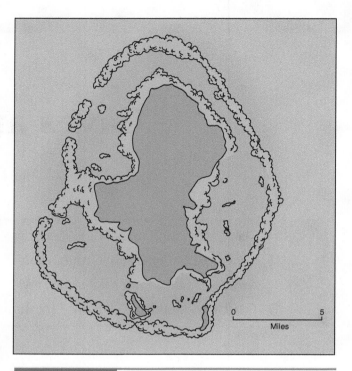

FIGURE PII.2   A small tropical island ringed by an outer coral reef, or barrier reef, with an inner fringing reef rimming the shore. A lagoon lies between the outer and inner reefs.

National Park, Montana, Chapter 26) preserve the remains of ancient reefs that help geologists reconstruct the geologic history of the past.

Limestones and Dolomites

By definition, carbonate rocks are composed of more than 50 percent by weight of carbonate minerals, such as *calcite* ($CaCO_3$) and *dolomite* ($CaMg(CO_3)_2$). Also present in carbonate rocks are varying amounts of clastic material (clay, quartz grains, etc.) plus iron oxide, iron carbonate, and traces of other compounds. Geologists use the word "dolomite" for both the mineral dolomite and the sedimentary rock called *dolomite,* the latter being composed of more than 50 percent dolomite (the mineral) and most commonly in excess of 90 percent. Similarly, *limestones* contain more than 50 percent of the mineral calcite and usually more than 95 percent.

Limestones and dolomites are often found together in reefs and caves. Sometimes they are interbedded or laterally gradational. Most dolomites were originally

Most of the carbonate rock layers in the national parks in Part II have remained relatively undeformed since they were laid down. Some have been deeply buried, elevated, and subsequently uncovered and reburied, as ancient oceans opened and closed and highlands were pushed up and then eroded. Many of the fossils and sedimentary structures within these rocks are remarkably well preserved. The geologic structures in the areas described in Part II are, in general, not complicated. With a good eye and only a moderate amount of geologic knowledge, you can "read" or surmise much of the geologic history when you visit caves and reefs. The Virgin Islands (Chapter 18) have been most affected by tectonic activities, but even here, the reefs themselves, since they are relatively young, have not been deformed.

The Role of Ground Water

Learning about caves (and visiting a cave, if possible) can help us to better understand and appreciate our *most important geologic resource;* namely, ground water, an indispensable component of Planet Earth's water budget. *Ground water* is the subterranean water that occupies the crevices, pores, and voids in rocks and sediments beneath the surface. More ground water is stored in the rocks of this continent than all the surface water held in rivers, reservoirs, and lakes—including the Great Lakes. Nevertheless, our vital supplies of ground water are presently seriously threatened from overuse (excessive drawdown) and from pollutants that seep underground and spread throughout ground water systems. Problems are especially acute in regions underlaid by carbonate rocks because their high permeability and fairly high porosity make them susceptible to pollution. National park rangers and interpreters try to alert park visitors to the need for protection of ground water supplies.

Surface water (from rain, snow, and runoff) is the source of virtually all ground water. Under the influence of gravity, water sinks into soil and percolates downward through cracks, joints, and along bedding planes of bedrock, passing through an unsaturated zone, called the *zone of aeration* or the *vadose zone.* When ground water has filled all the pores and open spaces (the depth is variable), it has reached the *saturated zone,* known also as the *phreatic zone.* An irregular surface separating these two zones is termed the *water table,* which rises in times of plentiful rain and

FIGURE PII.3   Carlsbad Caverns, located in New Mexico, has spelothems that are found in many of the caves. There are stalactites hanging from the ceiling, stalagmites on the floor, and popcorn coating on many of the surfaces. This is the Big Room, which covers 14 acres. Photo by David B. Hacker.

deposited as limestones, but magnesium ions replaced calcium ions during recrystallization and caused the limestone to be partially or completely "dolomitized."

Carbonate rocks occur in many different textures ranging from massive and very fine-grained to coarse and detrital (i.e., made up of rock fragments and organic debris). Limestones are of several types. Biohermal limestones are reefs or reef-like accumulations of organic origin; inorganic (or chemical) limestones are those formed by precipitation in seawater; and clastic limestones are accumulations of transported fragments of largely carbonate composition (fossil remains, bits of rock, etc.).

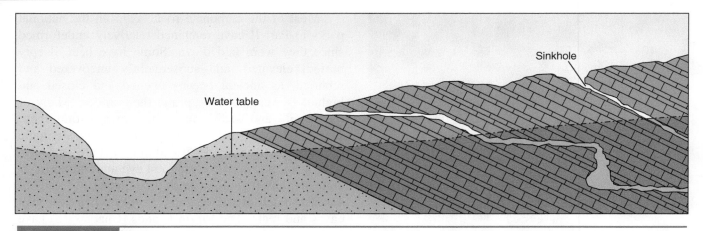

A cave system may develop in carbonate rock as ground water, moving along bedding planes, into joints, or finding weak spots, gradually dissolves rock. The water table moves higher or lower over time. Surface water may drain into the passages through sinkholes, such as the one in the upper right.

falls during droughts. Sometimes uplift or subsidence may raise or lower the water table (and the base level) of a region over the passage of time.

Carbon dioxide in the atmosphere and from decaying organic matter in soils makes ground water slightly acidic and thus able to slowly dissolve carbonate rocks above and below the water table. Openings form in fractures and along bedding planes that may eventually develop into caves. In general, the sequence of events is as follows: (1) As ground water moves through carbonate rocks, solution creates openings, tunnels, chambers, pits, or galleries. (2) If the water table drops, due to changes in climate or lowering of base level, air enters the part of the cave system that is in the vadose (unsaturated) zone. Because water that seeps into air-filled caves is very close to saturation with dissolved carbonate, the water loses carbon dioxide to the cave atmosphere and deposits much of its dissolved load as *travertine,* a dense, finely crystalline type of limestone that occurs in an infinite variety of shapes and cave "formations." *Speleothem* is an inclusive term for any secondary mineral deposit formed in caves.

Abrasion by underground stream action (where water acts under conditions of open channel flow), as well as deposition of alluvial fill, may take place in caves intermittently with solutional erosion.

Carbonate rocks in reefs are affected by circulating ground water when they become exposed to erosion by changing sea levels or tectonic activity. Ancient reefs in sedimentary sequences may become buried and subsequently undergo erosion and deposition by solution, as in Guadalupe Mountains and Carlsbad Caverns National Parks. Living and recently constructed reefs, such as those in the Virgin Islands and Florida, are also modified by ground water processes.

How Organisms Are Named

"Binomial nomenclature" is the term for the modern system of naming and classifying life forms, including fossil species, that was established by the Swedish naturalist Carl Linnaeus in the mid-18th century. The International Commission of Zoological and Botanical Nomenclature now governs the system. Every species has a unique scientific name, or "binomen," that consists of two words: a generic name and a specific name. The commission requires that binomens be latinized, to avoid language confusion, and the scientific names are customarily printed in italics. The zoological name of a genus is always capitalized, and the species name is not. One of the examples introduced in Part II is *Acropora palmata,* the scientific name of the elkhorn coral. Botanical nomenclature follows the same rules except that a plant species named for a person may be capitalized. An example is *Rhizophora Mangle,* which is the scientific name of the red mangrove.

CHAPTER 14

Mammoth Cave National Park

WESTERN KENTUCKY

Arthur N. Palmer   State University of New York

Area: 52,708 acres; 82 square miles	**Address:** P.O. Box 7, Mammoth Cave, KY 42259
Established as a National Park: July 1, 1941	**Phone:** 270-758-2251
Designated a World Heritage Site: 1981	**E-mail:** maca_park_information@nps.gov
Designated a Biosphere Reserve: 1990	**Website:** www.nps.gov/maca/index.htm

Longest known cave in the world, with more than 300 miles of explored and mapped passageways. Formed in nearly flat-lying Mississippian limestones and dolomites by circulation of ground water, fed by infiltration from the surface. Cave enlargement continues today; the lowest levels still contain streams. Great variety of passage shapes, including tubular and canyonlike passageways and vertical shafts. Numerous passage levels show relationships between cave development and the evolution of the landscape on the surface. The karst topography of the region is the most extensive in the United States.

FIGURE 14.1 Looking down from the top of Mammoth Dome, as seen on the Historic Tour. The eddying and coring of descending water formed this immense vertical shaft. Photo by David B. Hacker.

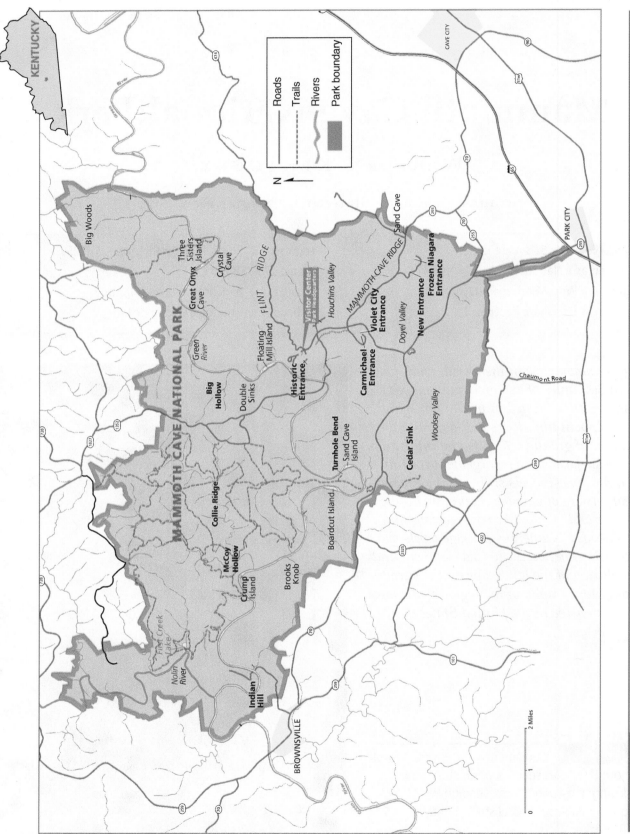

FIGURE 14.2 Map of surface features of Mammoth Cave National Park, Kentucky.

Regional Setting

Mammoth Cave National Park is located in west-central Kentucky, midway between the cities of Louisville, Kentucky, and Nashville, Tennessee, in the heart of America's most extensive limestone region. Solution of the limestone by underground water has created the longest known series of caves in the world. Mammoth Cave, the largest of these, contains more than 345 miles of surveyed passageways. About 12 miles of the passageways are open to the public.

The park is in a plateau of limestone bedrock capped by insoluble clastic rocks, mainly sandstones, at the southeastern edge of a regional downwarp, the Illinois Basin (a structural basin). The sedimentary rock layers in the plateau dip gently northwestward. Steep-walled valleys have breached the insoluble rocks, exposing 300 feet of cavernous limestone in irregular ridges. This rugged, hilly region is called the Chester Upland. To the southeast, the insoluble rocks have been removed by erosion, and the limestone is exposed at the surface in an uninterrupted plain, covering thousands of square miles called the Pennyroyal Plateau. Together, the Chester Upland and the Pennyroyal Plateau comprise one of the world's finest examples of *karst topography,* a landscape dominated by solutional features such as caves, closed surface depressions, sinking streams, and large springs. (fig. 14.2).

Guided cave tours of varied length and degree of difficulty are available to visitors in Mammoth Cave National Park. Year-round tours include: (1) the Historic Tour, which emphasizes pre-Columbian archaeology of the cave and saltpeter mining during the early nineteenth century, and also gives the clearest view of the limestone stratigraphy and the numerous levels at which the cave has formed; (2) the Frozen Niagara Tour, which enters the cave through a spectacular series of vertical shafts and extends through the area most profusely decorated with travertine speleothems; (3) the Travertine Tour, which consists of only the well-decorated sections of the Frozen Niagara Tour; (4) the Grand Avenue Tour, which is a 4.5 hour hike through passages of various types; and (5) the Violet City Lantern Tour, which visits passages beyond the Historic Route that have only primitive trails, with illumination from hand-held lanterns. Several other tours are offered on a limited basis, including an Introduction to Caving, which involves a strenuous 3.5 hour trip through passages off the maintained trails, requiring some crawling through narrow spaces.

Tours offered only in summer include: (1) "The Making of Mammoth," which visits the stream passages

FIGURE 14.3   In the early days of exploration and tours before Mammoth Cave became a National Park people would leave their names on the walls of the cave either by using soot from the torches or scratching it on the walls. Photo by Kelli Weibel.

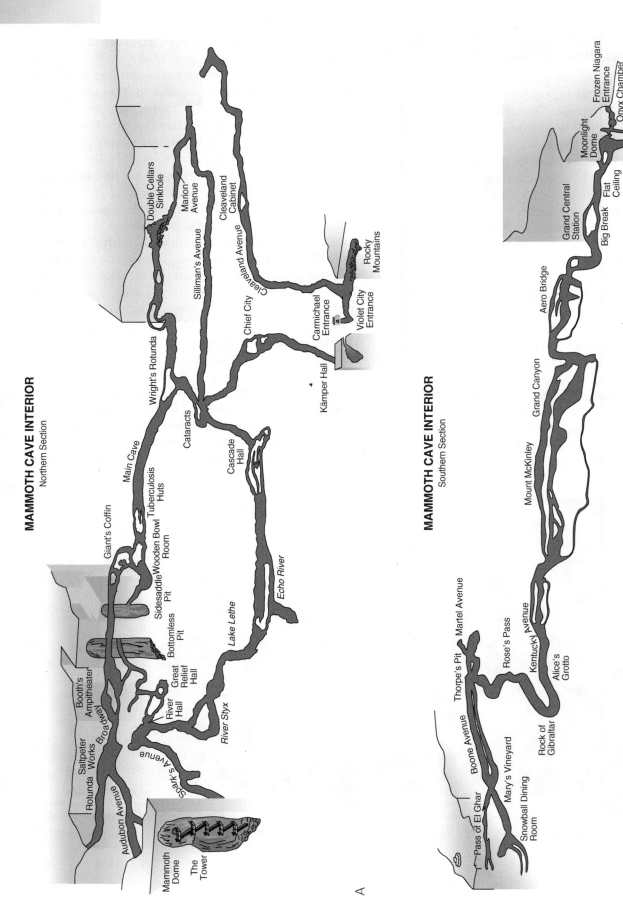

FIGURE 14.4   Public areas of Mammoth Cave. *A.* Underground passages in the Northern section. *B.* Underground passages in the Southern section. Adapted from *The Underground World of Mammoth Cave National Park,* R. Schlecht and A.N. Palmer © 1984 National Geographic Society. Used by permission.

at the lowest levels below the Historic Routes; (2) the Cleaveland Avenue Tour, which visits a passage well decorated with gypsum speleothems; and (3) the Wild Cave Tour, a strenuous six-hour trip for cave explorers to off-trail parts of the cave. Hiking shoes and a light jacket or sweater are recommended for all tours. The cave temperature is about 55° to 57° F, corresponding to the mean annual temperature of the region.

Prehistoric Indians of the Woodland Culture, who lived in the area roughly 3000 to 4000 years ago, used the larger cave entrances for shelter and explored several miles of passages in their quest for minerals used as medicines and flint for making tools. For light they made torches of cane, reeds, and twigs, remnants of which have been found in many of the dry upper-level passages of the cave. Other archaeological finds include bare footprints in mud and dust, woven baskets, fiber slippers, and two Indian "mummies" whose desiccated bodies were preserved naturally in dry parts of the cave.

Mammoth Cave became known to white settlers in the years just prior to 1800. Under private ownership, some of its passages were mined for saltpeter (nitrates) to make gunpowder, particularly during the War of 1812. Well preserved remnants of the saltpeter works are shown on the Historic Tour.

The saltpeter operation closed down after the War of 1812, but in its place grew an increasing tourist trade, spurred by many published articles about the vast cave. One of the early guides, a black slave named Stephen Bishop, became America's first truly great cave explorer. He discovered sections of Mammoth Cave far beyond the previous limits of exploration, including Echo River and some of the passages shown today on the Half-Day Tour.

In 1908 the great size of the cave was verified for the first time when a young German visitor, an engineer named Max Kaemper, surveyed 45 miles of passageways. Although he used only a hand-held compass and measured distances by pacing, his map was remarkably accurate.

At that time the known parts of the cave did not extend beyond the flanks of the five-mile-long Mammoth Cave Ridge, one of many flat-topped limestone ridges in the Chester Upland capped by insoluble rocks and separated from neighboring ridges by dry karst valleys. In the late nineteenth and early twentieth century many new caves were discovered and explored in nearby ridges. Some of these were opened for tourists, and commercial rivalry developed among the various cave owners. Although Mammoth Cave was still by far the most popular attraction of the region, its tourist trade suffered a blow when new entrances into the same cave were excavated beneath neighboring property and its southeastern end was opened to tours in competition with the "historic" northwestern end.

One of the residents of nearby Flint Ridge, to the northeast of Mammoth Cave Ridge, was an enthusiastic cave explorer named Floyd Collins. He discovered a large cave on his family's property, which he called Great Crystal Cave, and opened it for tourists. Of the many tourist caves in the area, his was one of the least accessible by road, so Floyd spent a great deal of time searching for a more conveniently located cave. In 1925, while exploring Sand Cave, a tight, unstable cave, he dislodged a loose rock that fell on his ankle, trapping him. Despite efforts by many people to free him, he died of exposure. Floyd Collins' fate received enormous publicity and brought even more visitors to the Mammoth Cave area.

Mammoth Cave was authorized as a national park in 1926, and by the time it was officially dedicated in 1941, the park boundaries encompassed many of the neighboring caves that had been open to the public earlier. Several of the former operators and explorers of Floyd Collins' Great Crystal Cave formed the Cave Research Foundation in 1957 that was dedicated to the exploration and scientific study of these and other caves within the park. By 1961 the group had discovered connections between all of the major caves in Flint Ridge, revealing a vast cave system that rivaled Mammoth Cave in length. In 1972 a connection was found between the Flint Ridge Cave and Mammoth Cave through a narrow, wet passageway that extended beneath the dry karst valley that separates the two ridges. This joined the two caves into a single enormous system with a combined length of more than twice that of any other known cave in the world. Continued explorations will almost certainly reveal connections with other large caves nearby.

Geologic Features

Stratigraphy and Geologic Structure

The rocks of Mammoth Cave National Park originated as sediments on the floor of a continental sea that covered most of southern North America during the

Box 14.1	**How Caves Are Formed**

Virtually all major caves are formed by the dissolving action of underground water within soluble bedrock. Limestone, dolomite, marble, and gypsum are the most common rocks of this type. All but gypsum require that the underground water be at least slightly acidic for the solutional activity to be rapid enough to form sizable caves. Limestone, by far the most common cave-forming rock, is attacked mainly by water containing dissolved carbon dioxide, a natural gas present in the atmosphere and in soil, that combines with water to form weak carbonic acid. The carbonic acid breaks down calcite, the principal mineral in limestone, into soluble components that are carried away by water. The chemical reactions that take place are shown below.

Water that enters the ground through openings, such as fissures in bedrock or cave entrances, is able to dissolve limestone throughout its entire underground flow path (though at a decreasing rate with time), provided that the water does not lose its solutional potential by releasing carbon dioxide gas (CO_2). Discrete flows of water, such as those that enter the ground through major depressions, or karst features, at the surface, generally gain carbon dioxide from the subsurface openings rather than losing it, so their ability to dissolve may be sustained for a great length of time.

The water that infiltrates through soil picks up a great deal of carbon dioxide from the decay of vegetation and other organic processes. Therefore, a given volume of this ground water is able to dissolve much more limestone than water from most other sources.

However, if the ground water that has escaped down through soil emerges into an air-filled cave, some of the dissolved carbon dioxide is released into the cave atmosphere. When this happens, the solution process is reversed, causing calcium carbonate ($CaCO_3$) to be precipitated in the form of calcite (limestone). This is how many of the cave decorations, or *speleothems,* are formed.

Gravity and capillary forces also control the movement of infiltrating water. Except temporarily and locally, water does not entirely fill the available cracks and pores in the ground. Wherever the water flow is substantial enough to overcome the capillary forces, gravity pulls it downward via the steepest available openings until it reaches a zone where all available openings in the ground are filled with water. In this zone the water flows by a combination of gravitational force and hydrostatic pressure, eventually emerging as springs at lower elevations such as river valleys. The zone of descending gravitational water is called the *vadose zone,* and the zone in which all openings are filled with water is called the *phreatic zone.* The top of the phreatic zone is known as the *water table.*

Cave-forming processes can operate in both the vadose and phreatic zones. Vadose cave features show a gravitational orientation (e.g., stalactites and stalagmites, vertical solution pits, and cave passages with consistent downward trends), whereas phreatic features have no consistent gravitational orientation (e.g., irregular, vaulted rooms).

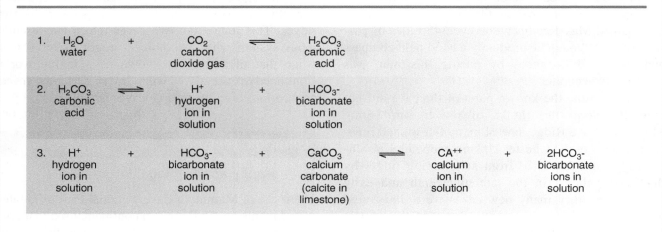

Mississippian Period, approximately 350 to 325 million years ago. Limestone was the major rock type deposited in this vast sea, although sand, silt, and clay were carried in from the land area that lay to the north. The water was shallow, despite its broad extent, with average depths of a few tens of feet. The limestones deposited at this time are highly varied in character and possess rather thin and prominent bedding.

Mammoth Cave is located within three different limestone formations. From oldest to youngest, these are the St. Louis Limestone, the Ste. Genevieve Limestone, and the Girkin Formation. The St. Louis is about 200 feet thick, although Mammoth Cave extends into only the upper half. The cave occupies the entire thickness of the Ste. Genevieve (100–120 feet) and the Girkin (135–140 feet). The climate was arid while the St. Louis Limestone was forming. As a result, gypsum beds are common at a depth below the surface, although the gypsum has been almost entirely removed from the near-surface parts of the formation by ground water solution. Chert beds and lenses are common, and it is thought that many of these were formed by the replacement of carbonates by silica (SiO_2). The climate grew more humid while the Ste. Genevieve Limestone was accumulating, so this formation contains little or no gypsum. Chert is limited to large rounded nodules and dike-shaped bodies. The upper half of the Ste. Genevieve Limestone contains several thin, impure, silty beds that represent the first stages of encroachment of detrital sediment from the land into the Mammoth Cave area. The Girkin Formation consists of thick limestone members, each about 20 feet thick, interspersed with thin silty limestones and shales.

Above the Girkin Formation lie the younger rock units of the Chester Series, in which the relative proportion of limestone and detrital sediment is reversed (table 14.1). Thick formations of sandstone and shale are interspersed with limestones only 10 to 40 feet thick. The Chester Series forms the resistant cap rock on the limestone ridges of the Chester Upland. This series is truncated by a widespread disconformity with local relief of several hundred feet in places. (The surface of a *disconformity,* representing missing rock layers, is parallel to beds above and below it.) Pennsylvanian conglomerates and other detrital rocks overlie the disconformity in the northern part of Mammoth Cave National Park, but in the southern ridges of the park, including Mammoth Cave Ridge, the Pennsylvanian rocks have been removed by erosion.

Downwarping of the Illinois Basin occurred while the Paleozoic sedimentary rocks were being deposited, causing the beds to dip gently toward the center of the basin (located in southeastern Illinois). In Mammoth Cave National Park, the dip averages one-third to one-half degree toward the northwest, although the direction and amount of dip vary a great deal from place to place because of irregularities in the beds.

Joints are rather small and inconspicuous in the cavernous limestones. Faults with up to 50 feet of displacement occur within the park, but the largest that have been observed in the cave have displacements of less than a foot. Because the fracturing of the limestone is so slight, individual beds and bedding-plane partings exert the main control over the directions of passages in Mammoth Cave.

Surface Features

Unlike Carlsbad Caverns (Chapter 16), Mammoth Cave is intimately related to the hydrology and geomorphic evolution of its surrounding landscape. Almost all the runoff in this humid region sinks underground through karst features in the land surface. Such features include *sinkholes,* which are funnel-shaped depressions where soil has subsided into solutionally enlarged cracks in the limestone. Collapse of bedrock into the underlying openings also plays an important role in forming large sinkholes. In the Mammoth Cave area these features number in the hundreds of thousands and range in size up to a thousand feet in diameter and more than a hundred feet deep. Some have open holes in their bottoms or sides that lead directly to caves. Throughout much of the Pennyroyal Plateau the density of sinkholes is so great that there is virtually no surface runoff. Surface streams form where water collects on adjacent insoluble rocks, but where they encounter limestone they disappear underground into sinkholes. These are called *sinking streams.* Despite the large number of karst features, the overall surface of the Pennyroyal Plateau is of remarkably low relief, appearing almost flat when viewed from an overlook. Actually the land slopes gently toward the few entrenched surface rivers that cross the area.

The Chester Upland, in which Mammoth Cave is located, rises abruptly 200 to 300 feet above the Pennyroyal Plateau surface. This steep rise separating the two regions is called the Chester Escarpment. The highly irregular edge of the Chester Upland has many erosional embayments and isolated remnants called *knobs,* which are outliers of limestone capped by insoluble rocks. The Chester Upland is a rugged, hilly region rising to altitudes of nearly 1000 feet. It is extensively dissected by steep-walled valleys, many of which have lost their streams to underground piracy and are now dry, sinkhole-floored karst valleys. The only surface streams in the Chester Upland and Pennyroyal Plateau are major rivers into which ground water emerges through karst springs. The master stream in the Mammoth Cave area is the Green River, which passes westward through both the Pennyroyal Plateau and the Chester Upland.

The Pennyroyal Plateau is lower than the Chester Upland because the erosional removal of the late Mississippian detrital rocks exposed the less resistant limestones beneath. Rocks like sandstone, which make up most of the cap rock of the Chester Upland, are much more resistant in this humid environment than limestone, which is attacked by rapid solution as well as by mechanical weathering. Sandstone is affected very little by solution. The very low relief of the Pennyroyal Plateau resulted from a prolonged period of weathering at or near base level when erosional dissection of the region slowed or stopped. Although the limestones were weathered and eroded almost to a flat surface at that time, the resistant cap rock prevented all but a modest widening of stream valleys in the Chester Upland. Subsequently, renewed dissection of the region

has caused major rivers to entrench about 200 feet into the Pennyroyal Plateau and to deepen the valleys in the Chester Upland (fig. 14.5).

Underground water in Mammoth Cave National Park emerges at dozens of springs in the Green River valley. Their flow varies greatly with the seasons and responds quickly to rainfall. Most of the largest springs are fed by cave passages that formed when the Green River valley was deeper than it is now. These passages were flooded when river sediment filled the bottom of the valley to depths of at least 30 feet late in Pleistocene time. At present, in order to reach the river, underground water from these passages is forced to rise upward through funnel-shaped openings in the sediment covering the floor of the Green River valley. In addition, many small springs emerge *above* the Green River level in the walls of valleys, particularly in the thin limestone formations sandwiched between detrital rocks in the cap rock overlying the main cavernous limestones.

Streams in the main part of Mammoth Cave discharge through sediment in the floor of the Green River valley at Echo River Spring. During wet periods the underground water overflows through other passages to River Styx Spring, which is located exactly at the level of the Green River. Both of these springs can be viewed by following short surface trails.

The cavernous limestones have such an enormous permeability that ground water is filtered hardly at all.

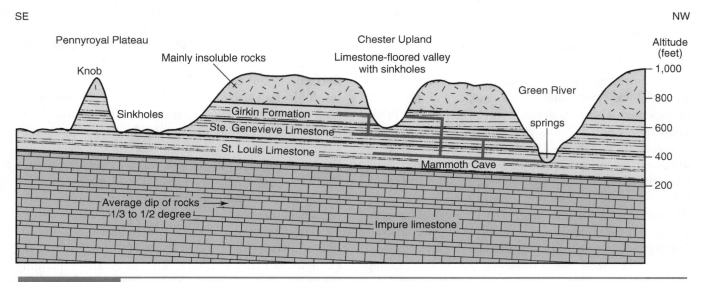

FIGURE 14.5 Geologic profile through Mammoth Cave National Park, showing the relationship of the cave to local topography and rock units.

Contamination of ground water supplies can be a serious problem. Recent dye tests by the National Park Service show that water passes through Mammoth Cave from sources as far as 10 miles away in the Pennyroyal Plateau. With so much water flowing through open conduits in the limestone, great care must be taken to avoid contamination of water supplies in the region by chemical spills and improper waste disposal.

Some of the larger streams in Mammoth Cave are occasionally contaminated when wastes from populated areas to the east and south seep into ground water. Although none of the passages open to the public are affected, National Park Service staff are concerned about the impact on the delicate cave ecology. As a result, they have worked closely with local officials to reduce the threat of ground water pollution in the park and the surrounding area by controlling waste disposal and by planning a regional waste-water drainage system.

Cave Passages

Mammoth Cave consists mostly of long, winding, nearly horizontal tunnels, many of which can be followed for several miles without interruption. These passages were dissolved by running water that has infiltrated from the land surface mainly through sinkholes. The various passages join each other as tributaries, forming a crude dendritic pattern. This pattern is not at all apparent from a map of the cave for several reasons:

1. Subsurface water is strongly controlled by the local geologic structure, as will be shown later. As a result, cave passages have much more erratic patterns than those of dendritic surface streams.
2. Many different levels of cave development are superimposed upon each other. As the Green River erodes downward, the water in the cave passages diverts to progressively lower routes and forms new passages. The abandoned passages are left high and dry. However, at any given time during the evolutionary history of the cave, the passages that are actively forming possess a dendritic pattern.
3. Most passages cannot be followed the entire distance from water input to spring, because they either become too small to explore or are terminated by collapse or sediment fill.
4. Overflow passages can form where periodic flooding of the lowest cave levels causes water to be forced into alternate routes through the limestone,

particularly where a stream passage is partly blocked by collapse material or sediment.
5. Many passage junctions have resulted from the random intersection of passages, such as by collapse of bedrock separating two passages where they cross one another on different levels.

As a result of these factors, Mammoth Cave appears to the visitors as a confusing three-dimensional maze, rather than an orderly dendritic pattern of stream channels.

All of the passages in Mammoth Cave are well adjusted to the patterns of past or present ground water circulation in the region. Passages emanate from areas of ground water recharge, such as the Pennyroyal Plateau and karst valleys in the Chester Upland, and terminate at or near the Green River valley, the only significant area of outflow. The chemical character of this water indicates that Mammoth Cave formed by solution of limestone by carbonic acid in water. Measurements in active stream passages show that the limestone is dissolving away at a maximum rate of slightly less than a millimeter each year.

Although this process is responsible for nearly all major caves and karst features, Carlsbad Caverns National Park in New Mexico (Chapter 16) is a uniquely interesting exception. Carlsbad Caverns, which formed by the solutional action of sulfuric acid generated by the oxidation of sulfides, does not possess the close association with the land surface and the present patterns of ground water circulation that Mammoth Cave shows so well; nor does Carlsbad Caverns contain distinctly stream-carved tunnels.

Passages in Mammoth Cave exhibit a variety of shapes determined by the conditions of water flow that formed them. Infiltrating water flows downward through cracks and pores in the limestone through the *vadose zone* and is pulled down mainly by the force of gravity. This water eventually reaches the *phreatic zone,* where all openings in the rock are filled with water. In the phreatic zone, the water moves under the combined influence of gravity and pressure along sub-horizontal paths to springs in the nearest available river valley.

A misconception among some geologists is that caves form in the phreatic zone, at or beneath the water table, and are exposed to vadose conditions only when the water table drops with time. This is not true, because vadose water that enters the ground through sinkholes and sinking streams is able to form caves as readily as phreatic water. In fact, the majority of the passages in Mammoth Cave formed in the vadose zone above the water table. Many vadose passages are still being enlarged by underground streams. The sinkholes that feed these passages evolved along with the cave, and, except in rare cases, do not represent points where surface water was diverted underground into preexisting phreatic caves.

Vadose water descends along the steepest-available paths, preferably through vertical joints or faults. However, in the Mammoth Cave area, faults are sparse, and most joints are narrow, inconspicuous, and extend through only a single bed. If no vertical opening is present, vadose water follows the next steepest path. Thus the most common passage types in Mammoth Cave are those formed by vadose water that initially flowed not vertically downward, but instead followed the dip of the limestone along bedding-plane partings or favorably jointed beds.

When enough solution occurred to allow a discrete stream of water to flow through, such a passage would enlarge by the downward solution and erosion of the floor, thus creating a narrow, sinuous *canyon passage*. A canyon passage looks like the canyon of an arid landscape, except that it is roofed by limestone. An exam-

ple is Fat Man's Misery on the Historic Tour. Most canyon passages in Mammoth Cave are less than 10 feet wide, but they range up to 100 feet high. However, some of the largest passages (fig. 14.6) in the cave are broad canyons as much as 80 feet wide, such as Kentucky Avenue, Broadway, and Audubon Avenue. The pattern of a canyon passage is intimately controlled by the local direction of dip, and minor irregularities in dip cause a marked sinuosity in the passage. Because the direction of dip is so variable, canyon passages are not all oriented in the direction of the northwesterly regional dip.

Where water is able to flow vertically downward along a fracture, the fracture is widened by solution wherever the flow is greatest, and a *vertical shaft* is formed. These features are shaped like the inside of a silo or well with vertical, cylindrical walls. From the bottom, a vertical shaft may look like the inside of a high, narrow dome, but from the top it looks like a pit. For this reason, they are sometimes called *domepits*. An example is Mammoth Dome on the Historic Tour (fig. 14.1). Water generally enters laterally through a passage at ceiling level (usually a canyon passage) and exits through a similar passage that leads out through the shaft wall at floor level. In the poorly jointed lime-

FIGURE 14.6   One of the large passageways that is nearly filled with detrital sediments. Photo by David B. Hacker.

stones of the area, most vertical shafts owe part of their vertical extent to the coring action of the descending water upon the limestone floor. Abandoned drain passages, representing former floor levels, are common in shaft walls. Vertical shafts in Mammoth Cave range in size up to 30 feet in diameter and 200 feet from top to bottom. They may intersect other passages as they grow, thus providing connections between different levels in the cave. Most shafts are still actively forming and contain water flow ranging from small trickles to major waterfalls.

Tubular passages are tunnels with elliptical or lenticular cross sections formed in the phreatic zone at or below the water table. The long dimension of their cross sections is controlled by bedding planes or, less commonly in Mammoth Cave, by joints or faults. They vary in size up to 30 feet high and 100 feet wide. An example is Cleaveland Avenue on the Cleaveland Tour (named for a nineteenth century geologist; the spelling is correct). Tubular passages either form while completely filled by the water flowing through them, or are filled with water only during periods of high flow. In either case the solutional enlargement is not only downward but lateral and upward as well, so that broad, arched ceilings are formed. The overall gradients of these passages are very low, but in places there may be abrupt jogs upward or downward across the bedding for as much as 30 feet. As the Green River erodes its valley downward, the underground water seeks lower paths through the limestone, abandoning the older phreatic passages as relict features within the vadose zone.

Tubular passages show no consistent relationship to the dip of the strata, because phreatic water has no inherent tendency to flow straight down the dip. Instead they develop along the most efficient flow path to the nearest available spring outlet. Because of the prominent bedding and the decreasing openness of partings and fractures with depth, many tubular passages follow a single stratigraphic horizon for thousands of feet, or even miles, commonly with a trend nearly parallel to the local strike of the beds. This trend represents the intersection between the low-relief water table and the dipping beds, where phreatic solution is most intense.

Most large passages in Mammoth Cave have vadose sections in their upstream ends and phreatic sections in their downstream ends. In an actively forming passage, the transition between the two sections is obvious, but the transition is also rather clear in a dry, abandoned passage. The passage may change from a relatively steep canyon to a nearly horizontal tube, and from a predominantly dip orientation (i.e., controlled by dip of beds) to a direction other than down the dip. A fine example of this transition is where the upper level of Boone Avenue, a canyon, changes to the tubular Cleaveland Avenue.

The major cave passages cluster at several significant levels: 600 feet, 550 feet, and 500 feet above sea level. The Green River is presently at 425 feet, and any cave passages below this elevation are completely flooded. The upper level (600 feet) consists of wide canyons and tubes partly filled with sediment. These are the largest and oldest passages in the cave. Audubon Avenue is an example of a wide canyon at this level that is filled with stream-deposited sand and gravel to more than half its height, so its canyonlike shape has been obscured. Apparently the filling took place soon after the passages at this level formed, while they still carried water. These passages correlate in elevation and probably also in age with the surface of the Pennyroyal Plateau. Their large size and sediment fill indicate slow entrenchment by the major rivers in the area, interrupted by at least one period of aggradation (deposition). The two lower levels are formed mainly by large tubular passages that represent periods of rather static regional base level. It is important to note that the cave levels are not the result of favorable zones caused by differences in bedrock solubility. Each level consists of passages occupying different sedimentary layers, depending on the dip of the rocks.

By examining abandoned cave passages, it is possible to reconstruct former patterns of ground water flow in the region. Although specific flow routes have undergone a great amount of capture (piracy) and diversion throughout the history of Mammoth Cave, the overall pattern of flow has remained constant.

Solutional Features in the Cave

In addition to the cave passages themselves, several minor solutional features provide evidence of past flow conditions.

Scallops are asymmetrical solution hollows formed by turbulent water flow. The bedrock surfaces of most canyons and many tubular passages are covered with a continuous dimpling of scallops. The scallops vary in length from roughly one-half inch to several feet, with the length inversely proportional to the velocity of the water that formed them. Their asymmetry also indicates the direction of flow, as with sand ripples or dunes; the steep faces of the scallops are oriented in the downstream direction. Scallops are easily visible in River Hall on the Historic Tour.

Flutes are parallel grooves formed in cave walls by water dripping vertically or flowing down a steep slope. Aligned in the direction of flow, they resemble the fluting of columns in classical architecture. Flutes are common in vertical shafts, such as Mammoth Dome on the Historic Tour, and in the shafts at the beginning of the Frozen Niagara tour.

Anastomoses are small, winding tubes that interconnect with each other in a braided pattern, normally along bedding-plane partings. The tubes have semicircular to circular cross sections and are dissolved upward into the base of the overlying bed. They range in diameter from about one-fourth inch to a foot. Anastomoses open into the walls of many tubular passages, and in some places they are exposed in the cave ceiling where the underlying bed has dropped or dissolved away. They are usually interpreted as remnants of the initial solution channels from which the cave passages developed. However, recent evidence suggests that they are more readily formed during periodic floods late in the solutional history of a passage, when solutionally active floodwaters fill the passage under pressure, forcing water into partings in the cave walls. Remnants of anastomoses can be seen in Kentucky Avenue on the Frozen Niagara Tour and in Little Bat Avenue on the Historic Tour.

Cave Deposits

Various minerals precipitate in the cave as "cave formations," more properly called *speleothems*. The great majority of speleothems are precipitated by vadose water seeping into the cave. Where water runs over a cave wall, sheets or draperies called *flowstone* are formed (fig. 14.7). Icicle-shaped *stalactites* are deposited by water dripping from the ceiling, and where the drops hit the floor they deposit mounds shaped like totem poles called *stalagmites*. Where a stalactite and stalagmite are produced by the same source of dripping water and grow large enough to meet, a *column* is formed. If the amount of water entering the cave is too small to form distinct drips, it is held to the growing speleothems by surface tension, and the calcium carbonate is deposited in irregular, twisted fingers called *helictites*.

Nearly all of the speleothems are still actively growing. Various forms of travertine ($CaCO_3$) can be seen on the Frozen Niagara Tour and in a few scattered places throughout the other tours. These include flowstone, stalactites, stalagmites, columns, rimstone, and popcorn (explained in Chapter 15). The largest of these features is the 75-foot-high flowstone drapery called Frozen Niagara.

Travertine is deposited in caves only when the water infiltrating from the surface is able to lose carbon dioxide to the cave air. This requires that the water first pass through an overlying soil in which the carbon dioxide content is higher than that of the cave. Vadose seepage passes through the soil, picks up a large

FIGURE 14.7 Flowstone draperies in Mammoth Cave. These travertine deposits are formed where water high in dissolved limestone content enters the cave and loses carbon dioxide to the cave air, precipitating calcium carbonate. Photo by David B. Hacker.

amount of carbon dioxide, and readily dissolves the underlying limestone. If the seepage takes place in small quantities, it quickly approaches saturation with respect to the limestone. When seepage water loses carbon dioxide to the cave air, it becomes supersaturated and precipitates some of its dissolved carbonate as travertine. Large discrete flows of vadose water, particularly those that enter the ground through sinkholes, do not possess such a high carbon dioxide content and may even gain, rather than lose, carbon dioxide in the cave. They remain undersaturated and account for the solutional development of cave passages.

The insoluble cap rock that overlies most of the ridge in which Mammoth Cave is located severely limits the amount of water infiltrating into the cave. Only around the perimeter of the ridge, where the cap rock has been eroded away, is it possible for large amounts of seepage to enter the cave through soil-covered limestone and to deposit large amounts of travertine.

Beneath the insoluble cap rock, passages that have been abandoned by the water that formed them are generally dry. Capillary water is drawn toward them in small quantities from the moist surrounding limestone and evaporates when it emerges into the cave. Many varieties of evaporite minerals are deposited by this water. The most common is gypsum ($CaSO_4 \cdot 2H_2O$), which is derived in part from the oxidation of pyrite (FeS_2) in the limestone and overlying sandstone. The process is similar to the sulfuric-acid reaction and gypsum deposition described in Chapter 15, except that in Mammoth Cave the capillary nature of the water produces speleothems quite different from the bedded gypsum of Carlsbad Caverns. In general, the gypsum in Mammoth Cave forms a thin white or golden crust on the cave walls up to several inches thick (fig. 14.8). Where the deposition of gypsum is particularly intense, delicate *gypsum flowers* grow from the walls as curved fronds and form clusters that resemble lilies

FIGURE 14.8   Snowballs on the ceiling of the cave. The gypsum is derived in part from the oxidation of pyrite (FeS_2) in the overlying limestone and sandstone. In Mammoth Cave the gypsum forms a thin white or golden crust on the ceiling and cave walls up to several inches thick. Where the process is very active *gypsum flowers* grow as curved fronds and clusters. Photo by David B. Hacker.

FIGURE 14.9 Flowstone and stalagmites. The stalagmites form on the floor as the result of the water seeping through the bedrock and dripping onto the floor from the ceiling of the cave. Photo by Michael Little.

or cotton. Many gypsum deposits are seen on the Cleaveland Avenue Tour.

Mirabilite ($Na_2SO_4 \cdot 10H_2O$) crystallizes as clear needles or stalactite-like forms; *epsomite* ($MgSO_4 \cdot 7H_2O$) forms white, hairlike tufts. Both are deposited in areas of high evaporation where the relative humidity is less than about 80 to 90 percent. They can form remarkably fast, sometimes overnight, but when the humidity rises, they become unstable and may redissolve.

Nitrates, especially calcium nitrate, are abundant in the cave sediment of the Historic Route. In the saltpeter mining operations of the early nineteenth century, nitrates were leached out of the sediment with water. The resulting solution, when passed through wood ashes, produced potassium nitrate, one of the essential ingredients of gunpowder. It appears that most of the nitrates were derived from the guano of bats, which were once numerous in this part of the cave.

Deposits in Mammoth Cave resulting from mechanical weathering include sediment carried in by cave streams and breakdown blocks that have fallen from the ceiling and walls. The sediment consists of gravel, sand, silt, and clay derived mainly from weathering of the overlying insoluble rocks, plus a small insoluble residue from solution of the limestone. *Breakdown* is collapse material in the form of blocks, slabs, and chips. It is produced in several ways: (1) by solution along the partings and joints that bound a block of limestone; (2) by the loss of buoyant force on the ceiling when a passage is drained of its water by lowering of the water table; (3) by sagging and peeling of limestone beds in the cave ceiling after the underlying beds have been removed by solution; and (4) by the wedging effect of minerals such as gypsum that grow within fractures in the limestone. Breakdown of a passage ceiling usually progresses upward to the top of a weak rock unit such as shaly limestone, and afterward the passage is relatively stable. Where one passage crosses over another, breakdown may occur between them and produce a junction room. Several entrances to the cave have been formed by the intersection of a passage by surface erosion, but in most such places the incipient entrance is totally obscured by breakdown.

Geologic History

1. **Deposition of limestones during the Mississippian Period, 350 to 325 million years ago.**

The shallow continental sea, in which the St. Louis Limestone, the Ste. Genevieve Limestone, and the Girkin Formation were deposited, had covered most of the southern part of the North American craton since early in Paleozoic time. Many beds of older siltstones and impure limestones (not exposed in Mammoth Cave) underlie the 400 to 500 feet of uninterrupted Mississippian limestones. As the limestones accumulated, the climate gradually changed from relatively arid to humid with progressively greater amounts of detrital sediment being carried in from land areas to the north. Downwarping of the Illinois Basin occurred simultaneously with this deposition, imparting a gentle northwesterly dip to the rocks.

Table 14.1			Geologic Column, Mammoth Cave National Park	

Time Units			Rock Units	Geologic Events
Era	Period	Epoch		
Cenozoic	Quaternary	Holocene	Alluvium	Alluviation; drowning of lowest levels of Mammoth Cave
		Pleistocene		Rapid entrenchment alternating with minor aggradation; lower levels of Mammoth Cave formed
	Tertiary (Neogene)	Pliocene		Upper levels of Mammoth Cave formed; nearly flat surface of Pennyroyal Plateau formed during period of relatively static base level
		Miocene		
	Tertiary (Paleogene)	Oligocene		Slow erosion, probably interspersed with periods of aggradation
		Eocene		
		Palocene		
Mesozoic				
Paleozoic	Permian			
	Pennsylvanian	Morrowan	Caseyville Formation (conglomerate and sandstone)	Insoluble rocks deposited; gradual emergence of continent above sea level
			⟨ Disconformity ⟩	⟨ Erosion ⟩
	Mississippian	Chesterian	Leitchfield Formation (shale and sandstone)	Mainly insoluble rocks deposited in continental sea
			Glen Dean Limestone	
			Hardinsburg Formation (sandstone)	
			Haney Limestone	
			Big Clifty Formation (sandstone and shale)	
			Girkin Formation (limestone)	
		Meramecian	Ste. Genevieve Limestone	Relatively pure limestones deposited in continental sea
			St. Louis Limestone	
			Salem Limestone	Impure limestones deposited in continental sea
			Harrodsburg Limestone	

Source: Palmer, 1981.

FIGURE 14.10 The Styx River flowing out of Mammoth Cave. Photo by Earl Harris.

2. Deposition of predominantly detrital rocks late in Mississippian time and in the Pennsylvanian Period.

Increasing amounts of detrital sediment accumulated in the continental sea and produced the varied rocks of the Chester Series, which today (with the exception of the limestones of the Girkin) form the largely insoluble cap rock overlying the cavernous limestone of Mammoth Cave National Park. A drop in sea level relative to the surface of the continent allowed extensive stream erosion of the Mississippian rocks. With a subsequent rise in sea level early in the Pennsylvanian Period, conglomerate and sandstone were deposited over the erosion surface.

3. Slow subaerial weathering and erosion, beginning in late Paleozoic time.

The Pennsylvanian rocks are the last to have been deposited in the Mammoth Cave region. Since then, several hundred feet of Pennsylvanian and late Mississippian rocks have been removed by erosion, exposing Mississippian limestones over broad areas. Karst features began to develop wherever limestone was exposed at the surface. Mammoth Cave belongs to the last stages of this erosional history.

4. The forming of Mammoth Cave, late in Tertiary time, more than 10 million years ago.

The earliest passages to form in Mammoth Cave were large tubular and canyon passages now located 570 to 690 feet above sea level. The highest of these (in Crystal Cave, not open to the public) lies above the level of the Pennyroyal Plateau and has an irregular ceiling profile, indicating that it originated below the contemporary elevation of the Green River. Therefore, this passage predates the Pennyroyal surface. Later aggradation filled the passage nearly to the ceiling with stratified silt, sand, and gravel.

Recently the quartz-rich sediment in the various levels of Mammoth Cave have been dated with cosmogenic nuclides. In sunlight, cosmic radiation produces tiny amounts of radioactive aluminum-26 and beryllium-10 in quartz sandstone and conglomerate. When the weathered material is carried underground, the ^{26}Al and ^{10}Be decay at different rates. By measuring their ratio, dates up to 5 million years can be determined. In the upper levels of Mammoth Cave the sediment is 2.3 to 4 million years old. The passages themselves are still older. The highest level contains sediment dated at 2.3 million years, which represents a widespread period of aggradation that obscured the original sediment. It is likely that the highest level is much older than its sediment date.

The major late Tertiary passages originated during the period of rather slow fluvial entrenchment that allowed the nearly flat Pennyroyal Plateau surface to form. Erosion was interrupted by at least one period of aggradation. Cave passages were formed by sinking streams draining from the Pennyroyal Plateau and from karst valleys in the Chester Upland. Audubon Avenue, Broadway, Main Cave, and Kentucky Avenue were formed at this time as a single enormous passage that has since been segmented by breakdown. Analysis of aluminum and beryllium isotopes in quartz sediment in these passages gives ages of 2.3 to 3.4 million years. Since these dates apply to the fill material, the passages are considerably older. A major period of sediment accumulations to as much as 100 feet filled all upper-level passages 2.3 million years ago, representing widespread aggradation over the entire region. This was probably the result of climatic change just prior to widespread continental glaciation.

5. Development of lower cave levels during the Pleistocene Epoch.

Pleistocene glacial advances fell short of the Mammoth Cave region, but their influence on the erosional history of the region was profound. Prior to continental glaciation, the Ohio River was not nearly so large as it is today. Drainage from most of the Appalachian Mountains flowed through what is now northern Indiana into the Mississippi River, bypassing the Ohio River entirely. Early Pleistocene glaciers blocked this drainage and diverted it to the Ohio River, which was located approximately along the farthest advance of the ice. The greatly increased discharge caused rapid entrenchment of the Ohio. The Green River, which controls the levels of passage development in Mammoth Cave, is a tributary of the Ohio River, so it also began to entrench rapidly. As a result, the regional water table dropped and streams in the caves were diverted to lower routes. All but the largest surface streams on the Pennyroyal Plateau responded in a similar way by flowing underground as sinking streams.

Several lengthy pauses in fluvial entrenchment that occurred throughout the Pleistocene were accompanied by minor interludes of aggradation. Each pause allowed the cave streams to remain stable at the same level for a long period of time, thus enlarging the cave passages at that level to comparatively great size. The longest pauses resulted in major tubular passage levels at the present altitudes of 550 and 500 feet. Cleaveland Avenue is the best example at the 550-foot level. It reaches widths of 50 to 60 feet and a height of 15 to 20 feet. Great Relief Hall is part of the 500-foot level. As the water level dropped in the cave, travertine and evaporite speleothems began to accumulate in dry passages. Dating of quartz sediment by analysis of Al/Be isotopes gives ages of about 1.5 million years for the 550-foot level and 1.2 million years for the 500-foot level. It appears that the rapid entrenchment between stable periods was caused by adjustments in pattern of surface rivers. It is possible that the diversion of Appalachian runoff into the Ohio accounts for entrenchment below the 500-foot level, and that the intervals between higher cave levels represent earlier changes in drainage pattern with the region.

Terraces formed in the Green River valley during the periods of relatively static base level, and they correlate in elevation with the cave levels of the same age. Only the 500-foot-level terrace is relatively intact today, as the higher terraces have been severely eroded.

Sediment dates for the 550-foot level are about 1.5 million years, which suggests that the passages themselves are about 1.5 to 2 million years old. Sediment dates for the 500-foot level are about 1 million years old, and so the passages are probably about 1 to 1.5 million years old.

A more recent geomorphic event has been the filling of the Green River valley with 30 to 50 feet of alluvium late in the Pleistocene Epoch, probably during Wisconsinan time when the last major ice advance occurred. As a result, the lowest passages in Mammoth Cave are now completely water filled.

Natural entrances to Mammoth Cave have probably existed at springs and at points where sinking streams entered the ground since the cave first began forming. However, the present entrances are mainly secondary features created by the random intersection of cave passages by surface erosion. The Historic Entrance is an example of an entrance formed where an upper-level passage was intersected by a minor tributary valley of the Green River valley.

The processes that formed Mammoth Cave continue to operate today. For this reason it is an ideal "underground laboratory" for studying the ways in which caves and their related karst features evolve in a humid environment.

Geologic Map and Cross Section

Haynes, D.D. 1962. *Geology of the Park City Quadrangle, Kentucky.* U.S. Geological Survey Map GQ-183.

_____. 1964. *Geology of the Mammoth Cave Quadrangle, Kentucky.* U.S. Geological Survey Map GQ-351.

_____. 1966. *Geology of the Horse Cave Quadrangle, Kentucky.* U.S. Geological survey Map GQ-558.

Klemic, H. 1963. *Geology of the Rhoda Quadrangle, Kentucky.* U.S. Geological Survey Map GQ-219.

Sources

Brucker, R.W., and Watson, R.A. 1976. *The Longest Cave.* New York: Alfred A. Knopf.

Granger, D.E., Fabel, D., and A.N. Palmer, 2001. Pliocene-Pleistocene Incision of the Green River, Kentucky, Determined from Radioactive Decay of ^{26}AL and ^{10}Be in Mammoth Cave Sediments. *Geological Society of America Bulletin* 113(7):825–836.

Harmon, R.S., Thompson, P., Schwarcz, H.P., and Ford, D.C. 1975. Uranium-series Dating of Speleothems. *National Speleological Society Bulletin* 37:21–33

Hess, J.W. 1976. A Review of the Hydrology of the Central Kentucky Karst. *National Speleological Society Bulletin* 38:99–102.

_____, and White, W.B., 1993. Ground Water Geochemistry of the Carbonate Karst Aquifer, South-central Kentucky, U.S.A. *Applied Geochemistry* 8:189–204.

Hill, C.A. 1981. Origin of Cave Saltpeter. *National Speleological Society Bulletin* 43:110–126.

Klimchouk, A., Ford, D.C., Palmer A.N., and Dreybrodt, W. (editors), 2000. *Speleogenesis-evolution of Karst Aquifers.* Huntsville, AL, National Speleological Society, 527 p.

McGrain, P., and Walker, F.H. 1954. Geology of the Mammoth Cave Region, Barren, Edmonson, and Hart Counties, Kentucky. Kentucky Geological Survey Field Trip Log.

Palmer, A.N. 1981. *A Geological Guide to Mammoth Cave National Park.* Teaneck, N.J.: Zephyrus Press.

_____, 1985. The Mammoth Cave Region and Pennyroyal Plateau, in P.H. Dougherty, editor, *Caves and Karst of Kentucky,* Kentucky Geological Survey, Special Publication 12, Series XI, p. 97–118.

_____, 1991. Origin and Morphology of Limestone Caves. *Geological Society of America Bulletin* 103:1–21.

Pohl, E.R. 1970. Upper Mississippian Deposits in Southcentral Kentucky. Kentucky Academy of Sciences, *Transactions* 31:1–15.

_____, and White, W.B. 1965. Sulfate Minerals: Their Origin in the Central Kentucky Karst. *American Mineralogist* 50: 1462–1465.

Quinlan, J.F., and Rowe, D.R. 1977. Hydrology and Water Quality in the Central Kentucky Karst: Phase I. University of Kentucky, Water Resources Research Institute, *Research Report* No. 101.

Schmidt, V.A. 1982. Magnetostratigraphy of Sediments in Mammoth Cave, Kentucky. *Science* 217(4562): 827–829, August 27.

Watson, P.J., ed. 1974. *The Archaeology of the Mammoth Cave Area.* New York: Academic Press.

White, W.B., Watson, R.A., Pohl, E.R., and Brucker, R.W. 1970. The Central Kentucky Karst. *Geographical review* 60:88–115.

_____, and White, E.L. (eds.) 1989. *Karst Hydrology: Concepts from the Mammoth Cave Area.* New York: Oxford University Press, 346 p.

Note: To convert English measurements to metric, go to www.helpwithdiy.com/metric_conversion_calculator.html

C H A P T E R   1 5

Wind Cave National Park

SOUTHWEST DAKOTA

Rodney M. Feldmann   Kent State University

Area: 28,295 acres; 44 square miles

Established as a National Park: January 9, 1903

Established as a National Game Preserve: 1912;
  addition to park: 1835

Address: R.R. 1, Box 190, Hot Springs, SD 57747-9430

Phone: 605-745-4600

E-mail: phyllis_cremonini@nps.gov

Website: www.nps.gov/wica/index.htm

*Eighth longest limestone cavern in the world, famous for its "boxwork"
and calcite crystals. The park's location is in the beautiful Black Hills.*

FIGURE 15.1   In Wind Cave National Park there are actually two parks; the park on the surface, and the park below. Photo by David B. Hacker.

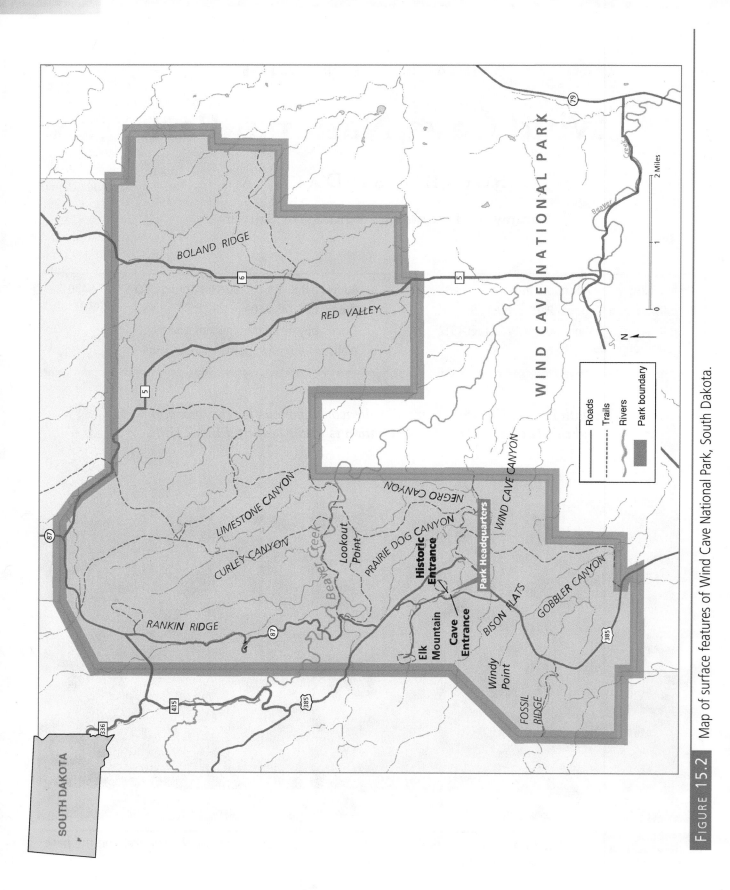

FIGURE 15.2 Map of surface features of Wind Cave National Park, South Dakota.

The Sioux Indians (or Dakotas) told legends about the "Sacred Cave of the Winds." They believed it was the cave through which their ancestors arrived at the Black Hills. Another myth was that Wakan Tanka ("Great Mystery") sent the buffalo out into the Sioux hunting grounds from the cave.

Legends persist in the Black Hills region to this day, as to how Wind Cave was discovered. One story claims a cowpuncher was riding by when the wind blew his hat off and into a small hole. When he went to look for it he discovered the cave. Since the opening was very small, he could not retrieve the hat. The next day he rode that way again and the wind brought his hat back to him, placing it on his head.

The official discoverer of Wind Cave was Tom Bingham, an early settler in the Black Hills. As he was hunting one day, he heard a strange whistling sound and began to search for the source. He found a small hole, through which the wind was rushing and making noise (fig. 15.3). As the natural opening was too small to permit entrance, another was dug nearby to a depth of six feet that made access to the cave possible.

The region around the cave was open to claims. In 1890 the South Dakota Mining Company filed a claim on the cave and hired Jesse McDonald and his sons, Alvin and Elmer, to take care of the property. Alvin McDonald did most of the exploration. He named many of the features in the cave and kept a diary of all his mapping. Alvin was buried near the cave; a plaque marks his grave site. In 1892, Jesse McDonald and several partners formed the Wonder Wind Cave Improvement Company. They took over the property, developed some passageways, put in stairs, and opened the cave to the public. Magazine stories printed about the cave in 1900 and 1901 brought it to the attention of the government. President Theodore Roosevelt signed a bill to create the national park in 1903.

Wind Cave National Game Preserve, established in 1912, was added to the park in 1935. On their way to the cave entrance, park visitors may see elk, deer,

FIGURE 15.3 The natural opening to Wind Cave. It is only 8 by 12 inches in size. The noise that was made by air passing in and out of the original opening (cave breathing) is what gave the cave its name. Photo by Michael Little.

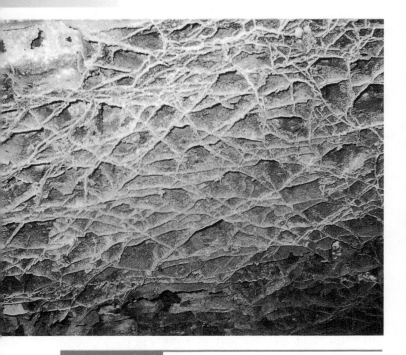

FIGURE 15.4 Boxwork results when a fracture system is filled by secondary calcite that is left standing in relief, following solution of surrounding original bedrock. Wind cave is noted for this unusual feature. This is the boxwork found on the ceiling at Grand Central Station. Photo by David Little.

pronghorn, prairie dogs, and bison, as well as a rich variety of birds and plants. In its beautiful setting in the Black Hills, Wind Cave National Park is truly two parks—one aboveground and one below.

Geologic Features

Wind Cave is a remarkably ancient cave with many unmapped and still unexplored passageways. In recent years hydrologic investigations and surveying of previously unknown passages have added to our knowledge of the park's subterranean levels Dr. James Pisarowicz, a Park Service naturalist, and his colleagues have discovered and surveyed many feet of passageways, including an alternate route between the eastern and western parts of Wind Cave. Among the cave features found by the team are five remarkable underground lakes, one of them 150 feet long and 60 feet wide.

With more than 100 miles of mapped rooms, tunnels, and passages, Wind Cave is now ranked third longest in the United States and eighth longest in the world. Spelunkers (cave explorers) predict that eventually a connection may be found between the subterranean passages of Wind Cave National Park and those of **Jewel Cave National Monument** that lies (via the land surface) 18 miles west-northwest of Wind Cave. Jewel Cave, which has 73 miles of mapped passageways, is the second largest cavern in the nation and ranks fourth longest worldwide.

Origin of Caves

Caves in limestone are formed when joints and fractures in soluble rock permit passage of slightly acid ground water along selective pathways. These pathways are then enlarged by solution of adjacent bedrock and subsequent precipitation of material as some form of cave deposit. Variations in rock type, jointing, pattern, and ground water chemistry result in a wide diversity of forms and a broad spectrum of caves and cave deposits. Each of the caves preserved by the National Park Service illustrates features that are unique. In Wind Cave the dominant precipitate is the unusual boxwork that formed early in the cave's development (fig. 15.4). Less common in Wind Cave are the stalagmites, stalactites, and other dripstone features that dominate most other caves.

Primary and Secondary Features

The development of caves involves an initial period of growth and enlargement, primarily by solution, followed by a period of infilling by precipitation of minerals from ground water. The "formations" observable in caves can be classified on the basis of whether they were formed during the enlargement stage, *primary features,* or during the stage of infilling, *secondary features.*

Primary features. The most important feature developed in Wind Cave is the *boxwork.* In addition to the regional joint pattern that dictated the orientation of the major passageways in the cave, the Pahasapa Limestone became highly fractured sometime during its history. The orientation of the fractures differs somewhat from the regional joint pattern and is expressed on an entirely different scale. These primary fractures, in which boxwork ultimately formed, are separated from one another by inches or fractions of inches. Calcite was deposited in the fractures as a secondary filling by ground water. Later, the primary limestone was removed by solution, leaving the secondary material behind as thin fins that crisscross one another. Nowhere

WIND CAVE INTERIOR

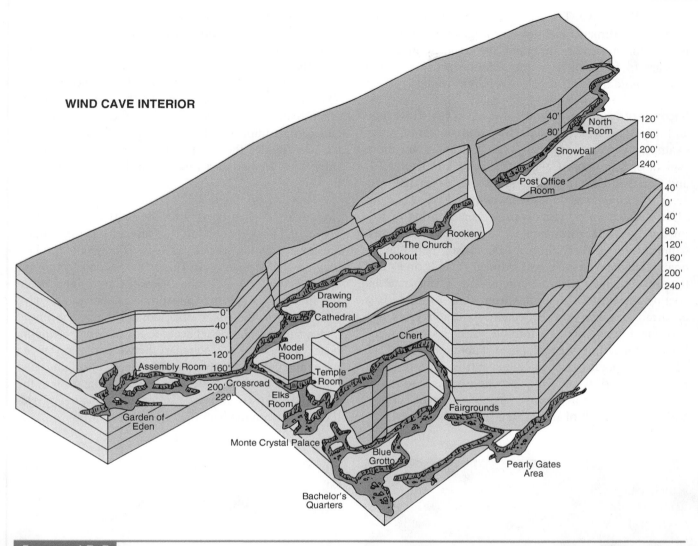

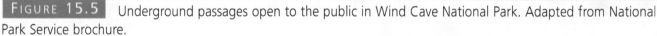

FIGURE 15.5   Underground passages open to the public in Wind Cave National Park. Adapted from National Park Service brochure.

is boxwork developed in greater volume or in wider variety than in Wind Cave.

Cave earth, a residue of clay particles and other insoluble materials, is found in all caves. In Wind Cave, cave earth ranges from a thin veneer to layers several feet thick. It accumulated either as the insoluble residue from solution of limestone or as material washed into the cave from sources at the surface. The Pahasapa Limestone consists of about 98 percent soluble material, so that, given the large volume of cave earth in Wind Cave, it is likely that most of the material infiltrated the cave during weathering and erosion of overlying rocks.

Collapse blocks are portions of the ceiling and walls that have detached themselves from the surface of the cave and are now lying on the floor. Probably the blocks loosened and fell when the cave was flooded because boxwork on the underside of blocks is undamaged. Had the cave been dry, the impact of the fall would have shattered the delicate structures.

Collapse breccia is found in parts of the Pahasapa Limestone where the cave ceiling collapsed. This caused blocks of rock from the overlying Minnelusa Formation to fall, along with the Pahasapa, as a plug into the cavern.

Spongework, a maze of openings, or pores, ranging in size from a few inches to several feet across, is common in Wind Cave. The passages form a randomly oriented, three-dimensional network large enough in some places for a person to crawl through. Spongework appears to form early in the solution process as the most soluble material is selectively removed.

Secondary features. Wind Cave has a few flowstone and dripstone features (stalactites and stalagmites, for example) but not very many. Such features are more typical of Mammoth Cave National Park (Chapter 14) and Carlsbad Caverns National Park (Chapter 16).

Cave lining was deposited when the cave was completely under water. The deposit formed during or following boxwork formation and coated all the walls. Composition of the lining can be calcite, hematite, or opaline silica. The cave was drained several times and then resubmerged with the composition of the material coating the boxwork varying from time to time. This occurred at some time between 50 million years ago (Eocene Epoch) and 10 million years ago (Pliocene Epoch). Final draining took place at about the end of the Pliocene Epoch (two or three million years ago). Erosional and weathering processes have removed much of the lining in the cave passages near the surface, perhaps as a result of fluctuating relative humidity.

Cave ice features, or calcite rafts, form in bodies of standing water in a cave. They are common in Wind Cave and are also found in Carlsbad Caverns. They are roughly rectangular pieces of calcite so thin that they are supported on top of the water by surface tension. They range in size from one to three inches across. Disturbance of any type causes them to sink. New ones can form within a year. In an area referred to as the Calcite Jungle, the cave floor is covered with these rafts; they formed at a time when the water level was higher.

Popcorn (globulites) is a series of layered nodules (less than 2/5 inch or .039 to .39 inch) supported on a very small stalk (about 4/100 of an inch). If the nodules are covered with a bristlelike cluster of aragonite needles, they are called *frostwork*. This is the second feature for which Wind Cave is famous. (Jewel Cave is also noted for frostwork.) At first, popcorn was thought to have formed below the water table, but at the present time it appears to be forming adjacent to a wall where dripping water splashes at the boundary between the top of the water and the air. Another possible theory of origin is that the features form where water seeps from the walls at the water line. Frostwork may also form as a coating of needles directly on the wall (fig. 15.6 A,B,C,D).

FIGURE 15.6A   Speleothems in Wind Cave. Flowstone sometimes forms unusual features such as one named "Bleeding Heart." The flowstone formed over a knob of rock in a stream in the shape of a heart. Photo by Michael Little.

FIGURE 15.6B   Cave popcorn is nodular, grapelike protrusions formed where water seeps out of pores in the rock wall of caves. Photo by Michael Little.

FIGURE 15.6c Frostwork is bristle-like clusters of aragonite needles deposited on cave popcorn. They tend to form where air currents move through a passageway. Photo by Michael Little.

The needles may be formed of either aragonite or selenite. *Aragonite* is chemically the same as calcite but belongs to a different crystal system; calcite is hexagonal, whereas aragonite is orthorhombic. *Selenite* is a clear, cleavable form of gypsum. Temperature seems to determine which forms develop. For example, aragonite has a tendency to form in areas where the mean annual temperature is between 56° F and 63° F. Presently, the mean annual temperature of the region is about 50° F and the temperature of the cave is only 47° F; so no aragonite is forming. However, Wind Cave is situated in an area of the Black Hills with known geothermal activity. Thermal springs are currently active just south of the park in the city of Hot Springs, and evidence in surrounding areas suggests that in earlier times, activity was more extensive. Ground water may have ranged through much higher temperatures in the past. A sinkhole in Hot Springs, called Mammoth Site, is preserved as a National Historical Landmark because of a large assemblage of Pleistocene mammals and other fossils that have been excavated there.

Several other secondary deposits can be identified in the cave, such as *dogtooth spar* (fig. 15.7), which is well formed calcite crystals, and *hydromagnesite* or "moon milk," a white, earthy magnesium mineral, but none of these forms occurs in Wind Cave in significant quantity.

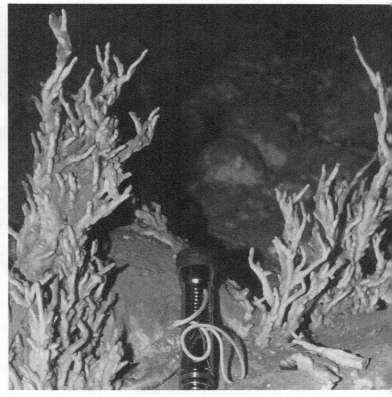

FIGURE 15.6d Helictites or helictite bushes are contorted cave deposits (speleothems) that angle or twist erratically. Photo by Mike Little.

Geologic History

1. **Deposition of Paleozoic and Mesozoic beds on Precambrian basement.**

During much of the Paleozoic and Mesozoic Eras, the Black Hills region was either slightly below sea level and receiving marine sediments, or was elevated slightly above sea level. Nearly the entire sequence of Paleozoic and Mesozoic rock units found elsewhere in the Black Hills is present in Wind Cave National Park. The Pahasapa Limestone, in which the cave developed, accumulated as a combination of organic fragments and calcareous mud in warm, shallow marine habitats during the Mississippian Period (360 to 320 million years ago). Following deposition of the Pahasapa beds, some erosion and development of karst features resulted in a hummocky surface in the region. The overlying Minnelusa Formation was, therefore, deposited on an irregular surface. Preliminary investigation of the upper

FIGURE 15.7 Dogtooth spar are well formed calcite crystals frequently found in caves. National Park Service photo.

levels of the cave suggests that the Pahasapa surface may have had a few hundred feet of relief, but this has not been confirmed.

2. Uplift of Black Hills from about 70 to 50 million years ago.

During Cretaceous and Paleogene time, the Black Hills region was subjected to forces that elevated two crustal blocks of the Precambrian basement and the overlying sedimentary rock sequence. The result was an ovoid dome about 150 miles long and 60 miles wide (fig. 15.8). Had the uplift occurred all at once with no accompanying erosion, the Black Hills would have been some 7000 feet higher than they are today. Instead, erosion has stripped sediments from the central part of the Black Hills and redeposited them elsewhere. Some of the sediments were deposited on the flanks of the uplift during Oligocene and Miocene time and are now exposed in the nearby badlands.

3. Formation of Wind Cave in the Mississippian Pahasapa Limestone.

The cave developed in the upper 100 to 150 feet of the Pahasapa Limestone Formation, which has a total unit thickness of 300 to 600 feet in the Black Hills area. Perhaps cave development began as early as middle Eocene time (50 million years ago), but as late as Pliocene time (10 million years ago) is more likely. In either case, Wind Cave is one of the oldest in the world. Most caves have formed only in the last million years.

As is typical of any cave, Wind Cave developed along the regional joint system and passed through a solutional phase (when it was below the ground water level) and a depositional phase, generally above the ground water level. The regional joint system, developed during the Black Hills uplift, trends northwest-southeast and northeast-southwest. Examination of the plan of the cave shows that the northwest-south-east direction has been the primary compass direction

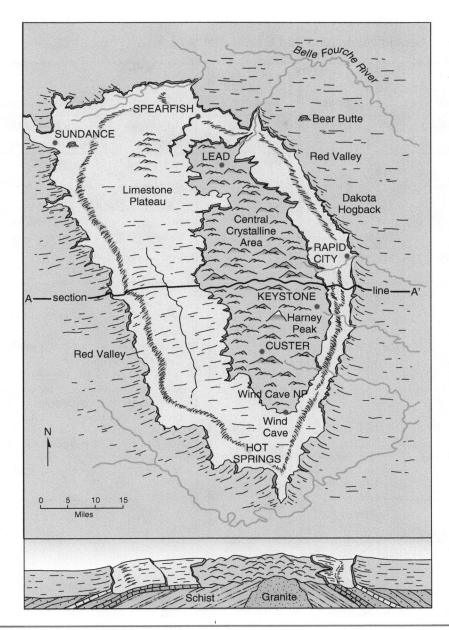

FIGURE 15.8   Physiographic diagram and cross section of the Black Hills showing how the domal structure controls erosion and landscape development. From *The Black Hills,* R.M. Feldmann and R.A. Heimlich. © 1980 Kendall/Hunt Publishing Company. Used by permission.

of cave development. A third direction, parallel to the bedding in the Pahasapa, is also significant. Because of a combination of structural and stratigraphic conditions, the cave can enlarge only along the strike (trend) of the beds; up or down the dip, pathways intersect either the land surface or the water table. Much of the growth of Wind Cave has been lateral; this is reflected in the observation that the cave is confined to the upper portion of the formation, even though jointing extends throughout this and other sedimentary units in the area. This suggests that the special combination of conditions necessary to cave development (i.e., opening of the joint system, position relative to the ground water level, and water

Table 15.1		Geologic Column, Wind Cave National Park		
Time Units		**Rock Units**		
Era	**Period**	**Formation**	**Member**	**Geologic Events**
Cenozoic	Quaternary	Alluvium		Development of surface landscape; solution and precipitation of features in Wind Cave
	Tertiary			Cave formed
				Erosion
Mesozoic	Upper	Several formations		Uplift of Black Hills associated with Laramide orogeny; deposition of marine and nonmarine sediments throughout most of Mesozoic Era
	Lower	Spearfish		
Paleozoic	Permian	Minnekahta Limestone		Deposition of marine and some nonmarine sediments
	Pennsylvanian	Minnelusa Sandstone		Development of karst terrain on Pahasapa surface
	Mississippian	Pahasapa Limestone	Upper Unit	Brecciated, debris that accumulated in sinkholes. Contains spongework.
			Middle Unit	Gray or red-brown fossiliferous cliff-former, organic fragments, and calcareous muds; well developed boxwork and cave lining; has longest caverns
			Lower Unit	
				Controlled by joint system, cave cut dogtooth spar
		Englewood Limestone		Pink, thin-bedded carbonate rocks
				Erosion
	Cambrian	Deadwood Sandstone		Clastic rocks

Sources: Feldmann and Heimlich, 1980; Palmer, 1988.

chemistry) were met only in the upper portion of the Pahasapa.

4. Development of cave zones.

Since the growth of Wind Cave has been dominantly lateral, with the cave floor sloping gently, total relief is only 450 feet. Initial solution of the cave passageways may have occurred simultaneously as rising ground water dissolved limestone along joints and on bedding surfaces. Over time, three "levels," or zones, developed in the cave that are controlled by differences in lithology within the Pahasapa Limestone. The upper zone formed in breccia, the angular debris that accu-mulated in ancient sinkholes. Crawlways of small and mazelike spongework are characteristic of this zone. Neither boxwork nor cave lining exist. The middle zone contains the biggest caverns, well developed boxwork, and cave lining. In the lowest zones, passages tend to be narrow and strongly controlled by the joint system. The Calcite Jungle with its cave rafts or "ice" is located here, as well as displays of dogtooth spar crystals.

5. Development of boxwork and other features.

The small-scale fracture system that developed parallel to bedding was filled with secondary calcite. Subsequently, the more soluble limestone was dis-

FIGURE 15.9   An outcropping of the Permian-aged Minnekahta limestone on the roadcut on Gobbler Pass, located in the southwest portion of the park. The distinctive color and banding makes it easy to identify. Note the miniature faults in the banding. Photo by Michael Little.

solved, leaving calcite fins or boxes. The colors are various shades of yellow, brown, pink, and blue. In some chambers (e.g., Fairground) the calcite is fluorescent and glows blue, green, red, and violet under "black light."

The boxwork has preferred directions of N 70° to 80°E and N 10° to 20°W. The boxwork even cuts through geodes formed in the limestone, which shows that the geodes developed first and the fracture systems later.

6. Erosion and formation of topography.

The climate is semiarid today, as is shown by Wind Cave Canyon Creek, an *intermittent stream* (i.e., does not flow year-round) that drains through Wind Cave Canyon and Red Valley and may flow through the cave at times when it has sufficient water. Evidence of stream flow through the cave can be found near the Garden of Eden. Beaver Creek, the only permanent stream in the park, cuts across the ridges or hogbacks in the region. The landscape is a combination of steep canyons, hills, and flat upland areas. Origin of these flats (e.g., Bison Flats) is uncertain. They could be structural or erosional.

Geologic Map and Cross Section

Robertson, F. *Geologic Map Portfolio—A Laboratory Study for Historical Geology.* Williams & Heintz Pub. Map no. 7.

Sources

Bakalowicz, M.J., Ford, D.C., Miller, T.E., Palmer, A.N., and Palmer, M.V. 1987. Thermal Genesis of Dissolution Caves in the Black Hills, South Dakota. *Geological Society of America Bulletin* 99(6): 729–738.

Darton, N.H. 1909. Geology and Water Resources of the Northern Portion of the Black Hills and Adjoining Regions in South Dakota and Wyoming. U.S. Geological Survey Professional Paper 65.

Feldmann, R.M., and Heimlich, R.A. 1980. *The Black Hills,* K/H Field Guide Series. Dubuque, Iowa: Kendall/Hunt Publishing Co.

Harpster, J. 1987. Exploring the Underworld: Seasonal Naturalist Makes Discovery at Wind Cave. *Courier,* March, p. 14.

National Park Service. 1979. *Wind Cave National Park.* National Park Service Handbook no. 104. 145 p.

Palmer, A.N. 1988. *Wind Cave—An Ancient World beneath the Hills.* Wind Cave/Jewel Cave Natural History Association.

———— and Palmer, M.V. 1986. Outline of Geologic Research in Wind and Jewel Caves. National Park Service.

Note: To convert English measurements to metric, go to www.helpwithdiy.com/metric_conversion_calculator.html

Carlsbad Caverns National Park

SOUTHEAST NEW MEXICO

Arthur N. Palmer   State University of New York, Oneonta

Area: 46,766 acres; 73 square miles

Proclaimed a National Monument: October 25, 1923

Established as a National Park: May 14, 1930

Address: 3225 National Parks Highway, Carlsbad, NM 88220

Phone: 505-785-2232

E-mail: cave_park_information@nps.gov

Website: www.nps.gov/cave/index.htm

Largest underground rooms of any cave in the United States. Great diversity in cave "formations," particularly dripstone. Formed in Permian limestone by sulfuric acid derived from hydrogen sulfide gas ascending from adjacent oil fields. Gypsum beds in cave are a byproduct of the reaction between sulfuric acid and limestone. Passages extend through massive reef limestone, bedded back-reef limestone, and fore-reef breccia. Carlsbad Caverns is no longer being enlarged by solution.

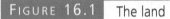

FIGURE 16.1 The land surface of Carlsbad Caverns National Park is a semiarid plateau, made up of reef limestone deposited in Permian time around the perimeter of the Delaware Basin. The caverns are located in several formations of the reef system, primarily in the reef talus, Capital Limestone, Yates and Tansill Formations. National Park Service photo by Raymond E. Beiersdorfer.

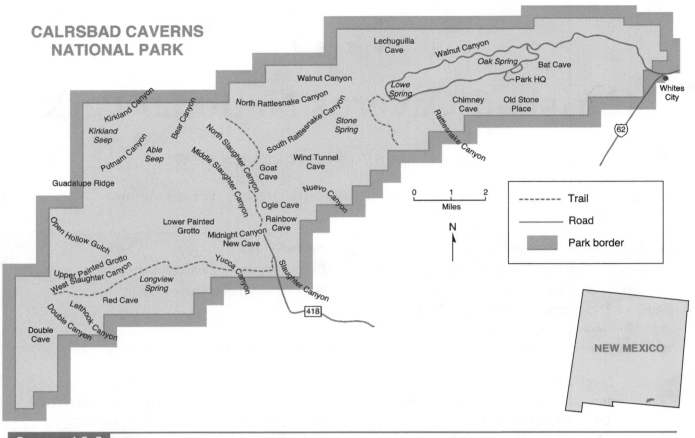

FIGURE 16.2 Map of surface features of Carlsbad Caverns National Park, New Mexico.

Regional Setting

Carlsbad Caverns National Park is located in the southeastern corner of New Mexico along the prominent southeast-facing Guadalupe Escarpment that extends northeast with diminishing height from Guadalupe Mountains National Park (Chapter 17) in Texas. The two parks share nearly the same geologic setting of massive reef and bedded back-reef limestones of Permian age. This is a region of deeply dissected plateaus having an average altitude of 4000 to 6000 feet that overlooks the arid Delaware Basin 1000 feet below.

Carlsbad Cavern,* one of the largest caves in the park, matches the popular image of how a cave ought to look. It has huge, interconnected rooms with shadowy, vaulted openings leading off in every direction;

massive and profuse decorations of stalactites and stalagmites; and prodigious length and depth. The cave's largest room, the Big Room, one of the world's largest, covers 14 acres (equivalent to 14 football fields). Its maximum floor-to-ceiling height is 370 feet at the so-called "Bottomless Pit." In the past, geologists treated the cave as a grand example of how limestone caves in general must form. It is surprising, therefore, to learn that in its geology and origin, Carlsbad Cavern is actually quite different from most other caves.

Public access to the cave is facilitated by well lighted trails. The tour is self-guided with an interpretive program keyed to various locations in the cave through portable radio receivers with headphones. Park Service guides are stationed throughout the cave to answer questions. Exit is provided directly to the surface by a 750-foot elevator that also affords entry for those who wish a shorter and less strenuous tour. The cave temperature is approximately 56° F year-round, equivalent to that of the mean surface temperature of the region.

---

*The park's major cave is now referred to as Carlsbad Cavern (without the "s"). The name of the national park is unchanged.

The Guadalupe Mountains were inhabited by humans as early as 12,000 years ago, but Indians apparently did not explore Carlsbad Cavern because of a sharp drop-off a short distance inside the cave entrance. By the late nineteenth century, the entrance was well known to local ranchers because of the dense cloud of bats that exits each evening for nocturnal feeding on insects, and then returns to the cave near daybreak. These useful animals are Mexican free-tailed bats and number in the hundreds of thousands. They inhabit a room called Bat Cave (not open to the public), which is just inside the cavern. Each night the bat colony consumes more than a ton of insects.

The first person to enter the cave, lowered on a rope by his father in 1883, was apparently 12-year-old Rolph Sublett, but he did not explore beyond the limit of daylight. The Bat Cave contains enormous deposits of bat droppings (guano), and the economic potential of the material as a fertilizer did not escape the attention of local speculators. In 1903 Abijah Long filed a claim for the area around the mouth of the cave and established a guano-mining operation. A shaft was constructed from the surface directly into the Bat Cave. Guano was removed in large cable-driven buckets, which also served as the primary means of entry and exit for the miners. The enterprise proved to be an economic failure, although over a 20-year period, six different companies tried unsuccessfully to run the operation. More than 100 million pounds of guano were excavated, lowering the cave floor as much as 50 feet.

A young guano miner named Jim White was fascinated by the huge, steeply descending underground corridor that led from Bat Cave past the natural entrance into the unknown. He made frequent explorations, often alone, deep into the cave and brought back wondrous tales. When the final mining company went bankrupt after World War 1, Jim White stayed in the area and led private tours into the cave. At first his enthusiastic reports were scoffed at by most people. However, it was not long before others joined in his praise, including several officials of the federal government. Willis T. Lee of the U.S. Geological Survey led a six-month National Geographic Society expedition to study and map the cave with White serving as a guide. With the size and beauty of the cave substantiated, President Coolidge initiated legislation in 1923 to create Carlsbad Cave National Monument. In 1930 it was established as Carlsbad Caverns National Park.

Today Carlsbad Cavern is the most popular cave in the country and has close to a million visitors each year. More than 80 other caves are known within the park boundaries. Cave explorers have found spectacular crystalline gypsum speleothems in Lechuguilla Cave (fig. 16.8), which is longer than Carlsbad Cavern although the latter has larger rooms (fig. 16.9). Its currently

FIGURE 16.3   Remnants of a ladder that earlier visitors used to enter the cave. Photo by Raymond E. Beiersdorfer.

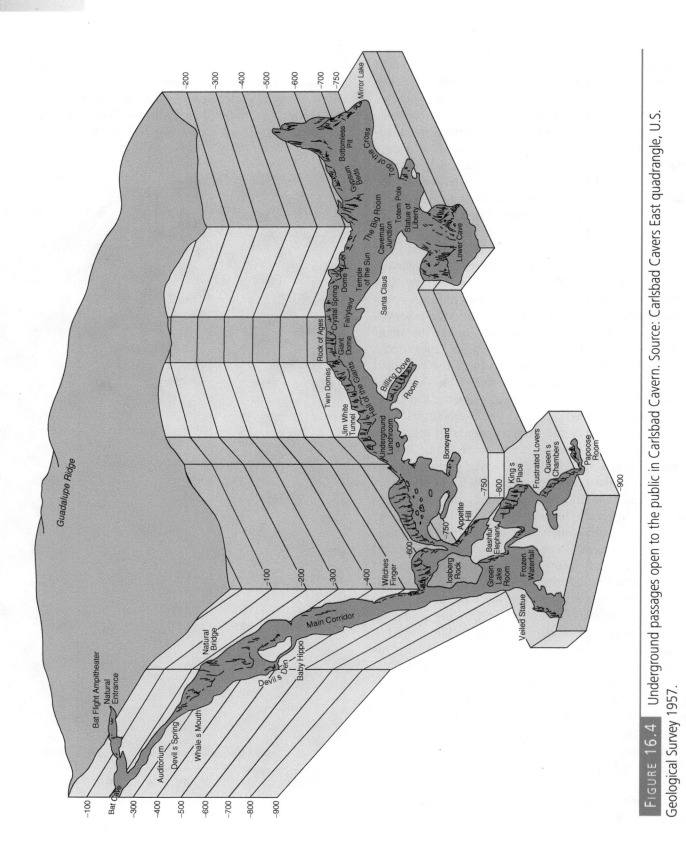

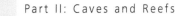

FIGURE 16.4   Underground passages open to the public in Carlsbad Cavern. Source: Carlsbad Cavers East quadrangle, U.S. Geological Survey 1957.

mapped length is more than 107 miles, third longest in the U.S.; and at 1593 feet deep, it is the deepest cave in the country. Another large cave, Slaughter Canyon Cave, which contains travertine formations every bit as grand and bizarre as those of its larger neighbors, is open to the public on an unimproved basis. Park Service guides lead groups through with hand-held lanterns.

Geologic Features

Origin of the Cave

Unlike most caves, which are formed by carbonic acid from surface sources, Carlsbad Cavern was dissolved mainly by sulfuric acid created by the oxidation of sulfides. This acid is uncommon in ground water except where hydrogen sulfide gas is available from deep sources or where iron sulfides, such as pyrite, are exposed to weathering. Recent evidence suggests that Carlsbad Cavern formed when hydrogen sulfide rising from depth reacted with oxygen at or near the water table and formed sulfuric acid, which readily dissolves limestone. The reactions are as follows:

1. Oxidation of hydrogen sulfide:

$$H_2S + 2O_2 \longrightarrow 2H^+ + SO_4^-$$
$$\text{sulfuric acid}$$

2. Solution of limestone:

$$H^+ + CaCO_3 \rightleftharpoons Ca^{++} + HCO_3$$

Additional information about the origin of the cave is explained in later paragraphs.

Bedrock and Geologic Structure

Details of the geologic setting are similar to those of nearby Guadalupe Mountains National Park in Texas (Chapter 17), so only a brief overview is given here. Three rock sequences, all about the same age but of different lithologies, occur in Carlsbad Caverns National Park. The Guadalupe Escarpment is formed by an exhumed barrier reef that developed around the perimeter of the Delaware Basin during the Permian Period. This reef, called the Capitan Limestone, contains a framework of bryozoan, sponge, and algae fossils in a matrix of unbedded, fine-grained limestone. To the north-northwest of the reef are the back-reef limestones (Artesia Group), which are prominently bedded. They contain a large percentage of silt eroded from nearby land

areas, as well as gypsum that accumulated in areas of restricted water circulation. To the southeast, the Capitan Reef is bordered by an apron of reef talus that merges with dark, fine-grained limestones deposited in the deeper waters of the Delaware Basin, contemporaneously with the other rock units. The basinal limestones are overlain by gypsum and evaporites, mainly of the Castile Formation, which are also Permian in age. Figure 16.5 is a diagrammatic representation of the sedimentary facies that make up the bedrock of the Carlsbad region.

As a result of Mesozoic and Cenozoic uplift, processes of solution and erosion have removed much of the gypsum. The more resistant limestones now stand out in relief, with nearly the same topographic expression they had while they were being deposited in the inland sea 200 to 250 million years ago. Southeasterly tilting and broad folding accompanied uplift. Faults with displacements up to several tens of feet are scattered throughout the park, particularly along the border zone between the reef facies and basinal facies. Normal faults along the western edge of the Guadalupe Mountains (outside the park) attain a displacement of several thousand feet. Prominent joints in the vicinity of Carlsbad Cavern are oriented in two sets, parallel to and perpendicular to the reef front. Many of the major joints parallel to the reef front have been filled with shallow-water deposits of land-derived sediment. These deposits now show up in the limestone as clastic dikes of sandstone and siltstone.

Cavern Features

A landscape dominated by solutional features is referred to as *karst topography*. A typical karst region contains many solutional fissures, pits, and funnel-shaped depressions called sinkholes where surface water collects and flows underground. Caves show evidence of subsurface solution by underground water. In the semiarid climate of the Guadalupe Mountains, surface karst features are almost entirely absent, so the region is not considered true karst topography, despite its many caves. Karst topography is best developed in a humid climate, such as that of Mammoth Cave National Park in Kentucky (Chapter 14).

Most caves consist of long, conduitlike passageways dissolved by underground streams. In such a cave it is easy to distinguish the points through which ground water recharge took place, as well as the springs through which water emerged at the surface, particular-

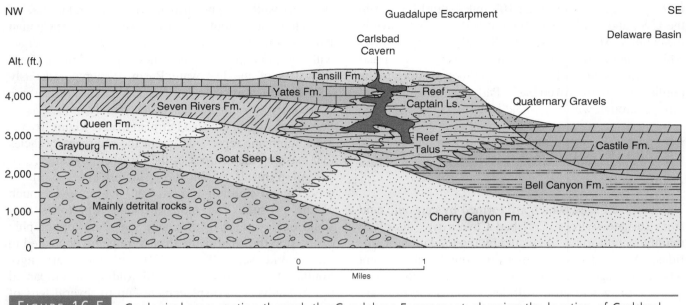

NW   Guadalupe Escarpment   SE

Delaware Basin

Carlsbad
Cavern

Alt. (ft.)

4,000 —   Yates Fm.   Tansill Fm.   Reef
                                                                      Captain Ls.   Quaternary Gravels
         Queen Fm.   Seven Rivers Fm.

3,000 —   Reef   Castile Fm.
         Grayburg Fm.   Talus
                            Goat Seep Ls.

2,000 —
                                                                                                   Bell Canyon Fm.

1,000 — Mainly detrital rocks
                                                        Cherry Canyon Fm.

    0 —

                              0   1
                                        Miles

FIGURE 16.5   Geological cross section through the Guadalupe Escarpment, showing the location of Carlsbad Cavern.

ly since many caves are still actively forming today. In contrast, Carlsbad Cavern and most other caves in the Guadalupe Mountains consist of large, interconnected rooms with highly irregular patterns and profiles. Nearly all of these caves are inactive today after being abandoned by the water that formed them. Since clues to the original water sources, patterns of flow, and springs are virtually absent, attempts to describe Carlsbad Cavern in terms of "normal" circulation of ground water and the carbonic-acid reaction do not explain the unusual character of the cave complex.

Carlsbad Cavern is located mainly in the reef limestone, but also extends into the adjacent back reef beds (in the northwestern part of the cave) and the fore-reef talus (at the south end of the Big Room). Nevertheless, the pattern and character of the cave are remarkably uniform, despite the great contrast in rock type. The main chambers of the cave are broad, arched rooms and linear, fissurelike passages in which major trends are controlled by the prominent jointing of the Capitan Limestone. In places, the rooms are bordered by intricate zones of *spongework,* where the original pores in the limestone have been enlarged by solution to form an interconnected maze of openings similar to that of Swiss cheese. The Boneyard is one of the best examples. The rooms, fissures, and spongework are the result of solution in the phreatic zone (below the water table). Vadose solution features are limited to a few minor rills and drip pockets.

Many of the largest rooms and passages are grouped at three different depths below the entrance: Bat Cave at –160 feet; the Big Room and Lunch Room at –660 feet; and the Lower Cave, King's Palace, and neighboring rooms at –740 feet. These levels are not controlled by differences in rock type and may represent separate stages of cave development. Several small lakes are scattered throughout the lowest rooms, including the lowest point in the cave, Lake of the Clouds, at a depth of 1037 feet below the entrance. These are merely local zones of ponding in the vadose zone and do not represent the water table. The original shape of the cave was modified in various places by the dropping of large rocks from the ceiling. The enormous Iceberg Rock is the largest of these *collapse blocks.* The presence of vadose travertine on some of the blocks shows that the collapse took place late in the developmental history of the cave; yet the ceilings are quite stable today. No rocks have dropped in the entire time that people have known about the cave.

The deposits within Carlsbad Cavern are unusually extensive and varied. The most spectacular are composed of travertine, which is calcium carbonate pre-

FIGURE 16.6   A stalagmite in the Hall of Giants. The rate of growth of each layer of a stalagmite is the equivalent of a thin coat of paint every 65 years. Photo by Richard Pirko.

FIGURE 16.7   Stalactites hang from the ceiling and are icicle-shaped, dripstone pendants. They form when water seeping through the bedrock drips so slowly from the ceiling that it evaporates and deposits the calcium carbonate as a ring on the ceiling. Photo by Carol Kriner.

cipitated when the carbon dioxide in groundwater is released into the cave atmosphere. The rate at which these features grow can vary a great deal, but maximum rates of about one centimeter per hundred years have been observed in other caves. Stalactites, stalagmites, and helictites are more abundant in Carlsbad Cavern than in Mammoth Cave or Wind Cave (Chapters 14, 15). Some of the stalagmites reach heights of 60 feet (fig. 16.6, 16.7).

Where water seeps out of pores in the rock, nodular, grapelike protrusions called *cave popcorn* are precipitated. These deposits may form on other speleothems in the same way. Cave popcorn is most abundant in areas where capillary water is drawn from the moist rock by gentle, dry air currents that flow through the cave, as shown by the fact that popcorn growing on stalactites and stalagmites is generally much more profuse on the windward side. However, warm, moist air rising from lakes deeper in the cave has caused water to condense on the colder rock surfaces above. The windward sides of speleothems in such an area are partly dissolved, whereas popcorn is deposited only on the lee sides.

Where water drips into a shallow basin containing small particles such as sand grains, the particles may become coated with calcium carbonate. If the agitation is sufficient to keep the growing particles from being cemented to the floor, rounded *cave pearls* are formed. The largest cave pearls in Carlsbad Cavern are more than an inch in diameter.

Certain speleothems are deposited in ponded water. These include calcite crystals such as dogtooth spar, travertine deposits called *rimstone* around the edges of pools, and bulbous, mammillate wall coatings that look like clouds (hence the name "Lake of the Clouds").

Most of the large speleothems are not actively growing today and are in the process of drying out. Their profusion shows that the climate must have been much more humid in the past, presumably during pluvial episodes of the Pleistocene Epoch. It was discovered recently that a great deal of moisture was being lost as water vapor through the elevator shaft, so air locks have been placed at the elevators to prevent this accelerated drying of the cave.

FIGURE 16.8 Pearlsian Gulf is a room in the Lechugilla Cave. This cave is located about five miles northwest of the Park Headquarters. It has been closed to the public since its accidental discovery in 1986 when cavers decided to investigate the bottom of Misery Hole, a 90 feet deep pit. They had to dig their way through the bottom and discovered the entrance to the cave. It is opened only to approved scientific researchers with caving experience and exploration and survey teams and National Park Service management-related trips. It is considered the deepest cave in the United States. National Park Service photo by Dave Bunnell.

Sediments in the Cavern

A variety of cave sediments of both mechanical and chemical origin provide evidence for the processes that formed Carlsbad Cavern. Most prominent of these are extensive bodies of gypsum deposited in the middle levels of the caverns. The gypsum deposits, which reach a maximum thickness of 20 feet, have been partly removed by solution, so only isolated blocks remain today. Water dripping from the cave ceiling has dissolved vertical "drill holes" through the gypsum in many places.

Most of the gypsum precipitated on the cave floor from sulfate-rich brine that once filled the cave; but in some areas the gypsum has actually replaced the limestone wall rock to a depth of several feet. Recent analysis of the gypsum shows that it is highly enriched in the light sulphur isotope S^{32}, which indicates a biological influence upon its origin. The gypsum was not derived from the Castile Formation, at least directly, because the Castile is not enriched in S^{32}.

The presence of the clay mineral endellite, overlain by montmorillonite, gives a further clue to the origin of Carlsbad Cavern. *Endellite* (hydrated halloysite) is formed by the alteration of other clay minerals in an acidic, sulfate-rich environment, whereas the *montmorillonite* (also a clay mineral) is a typical insoluble residue from the solution of limestone. The combined presence of gypsum and endellite suggests that sulfuric acid played a more significant role than carbonic acid in the origin of the cave. Sulfuric acid can be produced by the oxidation of pyrite, which is abundant in the back-reef limestones; but except for a few rare examples, the cave patterns show no relationship to the distribution of pyrite. Oxidation of sulfides in rising oilfield brines seems to be a more likely source of the sulfuric acid. Hydrogen sulfide gas (H_2S) is dissolved in water within rocks of the Delaware Basin that are rich in petroleum. Hydrogen sulfide oxidizes to sulfuric acid, usually in several steps that include sulfur, although with the aid of certain sulfur-oxidizing bacteria, oxidation to sulfuric acid can be achieved in a single step.

Remnants of pure sulfur have been found in Carlsbad Cavern and other caves nearby; this supports the view that hydrogen sulfide played a role in the cave's origin. The lack of relationship between the way the cave evolved and the usual patterns of groundwater flow can also be explained. The three levels in the cave may represent average positions of the water table at various times, inasmuch as the greatest rate of oxidation of sulfides occurs at and above the water table but diminishes downward within the phreatic zone.

The bedded gypsum overlies silt that is mainly insoluble material released during the solution of the limestone. Other sediments include sparse, rounded limestone pebbles carried in from the surface during the developmental history of the cave, but they represent only a minor event that has little bearing on the cave's origin.

FIGURE 16.9 Aragonite in Lechugilla Cave located in Carlsbad Caverns National Park. Aragonite has the same chemical formula ($CaCO_3$) as calcite but belongs to a different crystal system; calcite is hexagonal while aragonite is orthorhombic. National Park Service photo.

Geologic History

1. Deposition of limestones in the Delaware Basin during the Permian Period.

The Delaware Basin was a small inland sea 250 to 200 million years ago during the Permian Period. The Capitan Reef was one of many that developed around the perimeter of the basin, with impure, bedded limestones and gypsum accumulating on the landward side of the reef and fine-grained limestones in the center of the basin. Details of the sedimentary history are given in Chapter 17, Guadalupe Mountains National Park.

2. First stage of karst development took place late in Permian time.

Late Permian glaciation in the polar regions caused a drop in sea level and exposed the recently deposited limestones. Solution took place by surface runoff and by circulating ground water. Karst features formed at this time have been largely destroyed by later erosion or filled by detrital sediment.

3. Filling of the Delaware Basin with gypsum.

Waning of late Permian glaciers caused sea level to rise. During and after the polar glaciation, the Delaware Basin was nearly isolated from the ocean, and a high rate of evaporation caused gypsum and other evaporites to fill the basin at least to the height of the Capitan Reef.

4. Deposition of detrital sediment in a continental environment near the end of the Permian Period.

Silt and sand, eroded from nearby highlands, covered the area, creating the Salado and Rustler Formations. Some of the fill in the earlier karst features may date from this episode.

5. Tectonic uplift and deformation late in Cretaceous and early in Tertiary time; early stages of cave development.

As part of the Laramide orogeny, the Guadalupe Mountains were formed by a local uplift, which was accompanied by faulting and southeastward tilting. Erosion and karst development began in the higher regions, while sediment derived from these processes may have filled any remaining karst features of Permian age at lower elevations. Mobilization of hydrogen sulfide gas and sulfide-rich brines by the tec-

Table 16.1	Geologic Column, Carlsbad Caverns National Park

Time Units			Rock Units	Geologic Events
Era	**Period**	**Epoch**		
Cenozoic	Quaternary	Holocene	surficial gravels	Drying of climate; deposition of cave popcorn
				Deposition of most speleothems in cave
				Draining of cave
		Pleistocene		Erosion of Castile Formation exposing reef front; erosional dissection of limestones
	Tertiary (Neogene)	Pliocene	surficial gravels?	
		Miocene		Renewed uplift, faulting, tilting; major solutional development of Carlsbad Caverns; deposition of gypsum in cave
	Tertiary (Paleogene)	Oligocene		Slow erosion
		Eocene		Early stages of origin of Carlsbad Caverns
		Paleocene	surficial gravels and sand?	Laramide orogeny—uplift, faulting, tilting toward northeast
Mesozoic	Cretaceous			
	Jurassic			
	Triassic			Minor erosion, low relief
Paleozoic	Permian	Ochoan	NW — Rustler Formation — Salado Formation — SE	Thin, widespread, terrigenous deposits
			Castile Formation	Deposition of evaporites in Delaware Basin
		Guadalupian	Tansill Fm. / Yates Fm. / Seven Rivers Fm. / Queen Fm. / Artesia Group / Grayburg Fm. — Capitan Limestone (reef) — Bell Canyon Formation	Possible first stage of karst development
			Goat Seep Dolomite (reef) — Cherry Canyon Formation	Limestone deposited in subsiding Delaware Basin

Sources: Modified after Hayes, 1957; Hayes and Koogle, 1958; Jagnow, 1979; and Queen, 1981.

tonic activity may have begun in mid-Tertiary time, producing sulfuric acid by mixing with oxygenated water that infiltrated from the surface. Some acidity could also have been produced by the mixing of descending water, rich in CO_2, with sulfate brines derived from the Castile Formation. Carlsbad Cavern began to form within the phreatic zone, but the extent of its development during this time is not yet known.

6. Renewed uplift of the Guadalupe Mountains during the late Tertiary Period; major phase of cave development.

The major phase of solutional enlargement dates from this time of uplift. Argon dating of cave deposits of the mineral alunite, a rare potassium-aluminum sulfate derived from the sulfuric-acid breakdown of clay, gives a good estimate of the actual times of cave development. The highest cave levels in the Guadalupe Mountains are approximately 12 million years old. The main levels of Carlsbad Cavern are about 4 million years old. Extensive gypsum deposits in the Guadalupe caves indicate that the dissolution was caused by sulfuric acid derived from the oxidation of dissolved hydrogen sulfide released from deeper rocks.

Dissolution was concentrated at the contemporary water table. The various cave levels were produced as the water table dropped as the result of continued uplift and erosional dissection. Passage levels correlate poorly in elevations between caves in the Guadalupes, which indicates that major passages were produced not by pauses in erosional dissection, but by the location and timing with which H_2S-rich water was released into the limestone.

7. Draining of caves during the Pleistocence Epoch; deposition of speleothems.

Carlsbad Cavern was no longer actively enlarging during the Pleistocence Epoch, because the influx of water rich in hydrogen sulfide had ceased. Further uplift and erosion allowed the water table to drop below the cave. Erosion and dissolution of the soft Castile gypsum exposed the Capitan Reef, which forms the Guadalupe Escarpment today.

Infiltration of CO_2-rich water from the surface deposited travertine speleothems, particularly during times of humid climate in the Pleistocene. Electron-spin resonance dating of speleothems in the Big Room indicate that Carlsbad Cavern was drained at that level more than 500,000 to 600,000 years ago.

Intersection of the upper cave level by erosion formed the entrance of Carlsbad Cavern. Gradual diminution of infiltrating water and drying of the cave in the semiarid climate since the last glaciation has reduced the rate of speleothem growth. This change has favored the deposition of nodular travertine, such as popcorn, rather than stalactites and stalagmites.

Today Carlsbad Caverns National Park is not only one of the greatest tourist attractions of the Southwest, but it is also the subject of considerable research in the fields of sedimentary processes, climatic history, and the unusual mode of cave origin that has occurred in the region.

Geologic Map and Cross Section

Hayes, R.T. 1957. *Geology of the Carlsbad Caverns East Quadrangle, New Mexico.* U.S. Geological Survey Map GQ-98.

Hayes, R.T., and Koogle, R.L. 1958. *Geology of the Carlsbad Caverns West Quadrangle, New Mexico.* U.S. Geological Survey Map GQ-112.

Sources

Bretz, J.H. 1949. Carlsbad Caverns and Other Caves of the Guadalupe Block, New Mexico. *Journal of Geology* 57: 447–463.

Cunningham, K.I., Northup, D.E., Pollastro, R.M., Wright, W.G., and LaRock, E.J. 1995. Bacteria, Fungi, and Biokarst in Lechuguilla Cave, Carlsbad Caverns National Park, New Mexico. *Environmental Geology* 25:2–8.

Davis, D.G. 1980. Cave Development in the Guadalupe Mountains, a Critical Review of Recent Hypotheses. *National Speleological Society Bulletin* 42: 42–48.

DuChene, H.R., and Hill, C.A. (editors), 2000. The Caves of the Guadalupe Mountains. *Journal of Cave and Karst Studies,* 62 (2), 107 p.

Dunham, R.J. 1972. Capitan Reef, New Mexico and Texas: Facts and Questions to Aid Interpretation and Group Discussion. Permian Basin Section, Society of Economic Paleontologists and Mineralogists, Publication 72–14.

Egemeier, S.J. 1981. Cavern Development by Thermal Waters. *National Speleological Society Bulletin* 43:31.

Halliday, W.R. 1966. *Depths of the Earth.* New York: Harper and Row.

Harmon, R.S., and Curl, R.L. 1978. Preliminary Results on Growth Rate and Paleoclimate Studies of a Stalagmite from Ogle Cave, New Mexico. *National Speleological Society Bulletin* 40: 25–26.

Hill, C.A., 1996. *Geology of the Delaware Basin— Guadalupe, Apache, and Glass Mountains, New Mexico and Texas.* Society of Economic Paleontologiests and Mineralogists, Permian Basin Section, Publication 96–39, 480 p.

——— Hill, C.A. 1987. *Geology of Carlsbad Cavern and Other Caves of the Guadalupe Mountains, New Mexico and Texas.* New Mexico Bureau of Mines and Mineral Resources, Bulletin 117, 150 p.

————, and Forti, P. 1986. *Cave Minerals of the World.* Huntsville, Alabama: National Speleological Society. 238 p.

———— 1990. Sulfuric Acid Speleogenesis of Carlsbad Cavern and Its Relationship to Hydrocarbons, Delaware Basin, New Mexico and Texas. American Association of Petroleum Geologists Bulletin 74 (11): 1685–1694.

Jagnow, D.H. 1979. *Cavern Development in the Guadalupe Mountains.* Columbus, Ohio: Cave Research Foundation.

———— and Jagnow, R.R. 1992. Stories from Stones: The Geology of the Guadalupe Mountains. Carlsbad, New Mexico: Carlsbad Caverns-Guadalupe Mountains Association, 41 p.

Newell, N.D., Rigby, J.K., Fischer, A.G., Whiteman, A.J., Hickox, J.E., and Bradley, J.S. 1953. *The Permian Reef Complex of the Guadalupe Mountains Region, Texas and New Mexico—a Study in Paleoecology.* San Francisco, California: W.H. Freeman and Co.

Polyak, V.J., McIntosh, W.C., Guven, N., and Provencio, P., 1998. Age and Origin of Carlsbad Cavern and Related Caves from ^{40}Ar/^{39}Ar of Alunite. *Science* 279: 1919–1922.

Taylor, M.R., and Widmer, U. (eds). Lechuguilla: Jewel of the Underground. Basel, Switzerland: Speleo Projects, 144 p. (distributed by Speleo Books, Box 10, Schoharie, New York 12157-0010.)

Thrailkill, J.V. 1971. Carbonate Deposition in Carlsbad Caverns. *Journal of Geology* 79: 683–695.

> Note: To convert English measurements to metric, go to www.helpwithdiy.com/metric_conversion_calculator.html

Guadalupe Mountains National Park

W E S T   T E X A S

Ann F. Budd, University of Iowa, and
Katherine A. Giles, New Mexico State University

Area: 86,416 acres; 135 square miles	**Phone:** 915-828-3251
Established as a National Park: September 30, 1972	**E-mail:** GUMO_Superintendent@nps.gov
Address: H.C. 60, Box 400, Salt Flat, TX 79847-9400	**Website:** www.nps.gov/gumo/index.htm

This fault-block range has spectacular exposures of an extensive Permian reef complex. A major normal fault marks the western side of the mountain mass; the Capitan Reef escarpment rims its eastern edge. The cave features of Carlsbad Caverns National Park, New Mexico (Chapter 16), formed in a less elevated portion of the Capitan Reef.

Guadalupe Peak (8749 feet), highest point in Texas, overlooks the Chihuahuan Desert. McKittrick Canyon offers a geologic cross section of the reef, showing types of deposits, many fossils, and small caves.

FIGURE 17.1 The Capitan Limestone that is exposed in the upper part of El Capitan represents the reef. The base is the Bell Canyon Formation, representing the fore-reef facies. The escarpment exposes part of the ancient reefs that encircled the Delaware Basin in Permian time. Photo by Ann G. Harris.

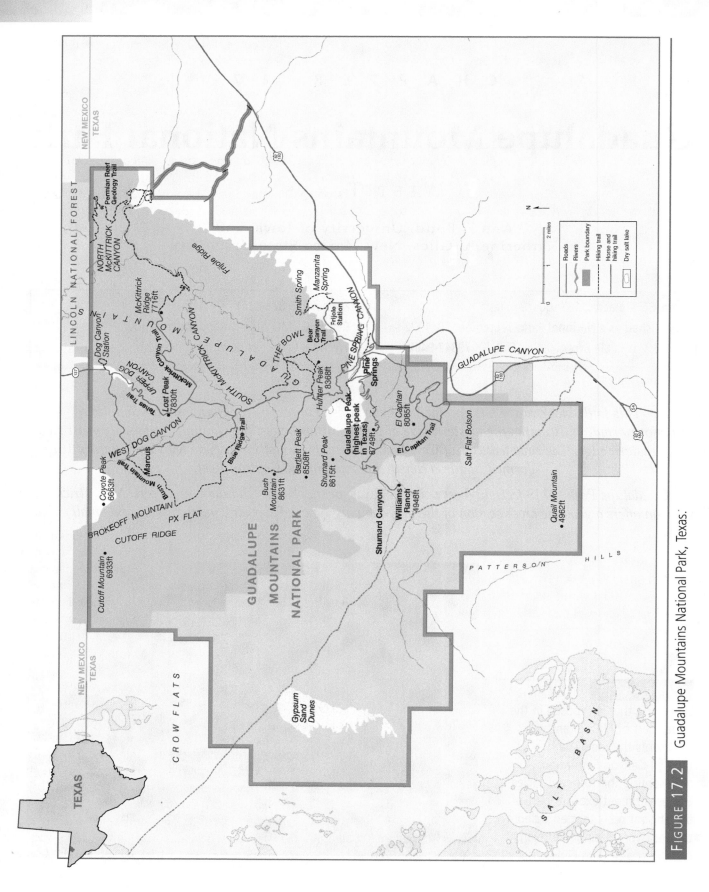

FIGURE 17.2 Guadalupe Mountains National Park, Texas.

Studies of pictographs in caves and other archeological finds suggest that humans inhabited caves in Guadalupe Mountains National Park about 10,000 years ago. They moved with the seasons from the canyon floors to the highlands while collecting plants and hunting animals. Later arrivals in the Guadalupe region were the fierce Mescalero Apache Indians, who discouraged travel or settlement in their domain. In the mid-1800s a series of surveys investigated possible wagon routes. The Butterfield Overland Stage set up a station at the Pinery in Guadalupe Pass on the first transcontinental mail route from St. Louis to San Francisco. The station was later abandoned because of conflicts with the Mescalero tribe; but its ruins are within the park, just off the main highway. Decades of hostilities followed before the Mescalero were subdued and moved to a reservation in 1879. Skirmishing also went on between the United States and Mexico, mainly over the salt flats, until 1878. The population remained sparse although a few ranchers homesteaded in the Guadalupe region early in this century. Bat guano was mined from caves for use as fertilizer for some years. Copper mining on a small scale was carried on in Dog Canyon near the park's northern boundary.

Wallace Pratt, a petroleum geologist and vice president of Standard Oil Company of New Jersey, donated the North McKittrick Canyon area to the federal government for use as a national park (a total of 5632 acres) in 1966. An additional 72,000 acres was purchased by the federal government and the park established in 1972.

Weather can be extreme in the Guadalupe Mountains. Frequent high winds, intense summer thunderstorms, occasional snow in winter, and a wide range of temperatures are to be expected. Precipitation is sparse. Rigorous conservation practices are preserving the fragile terrain and desert vegetation.

Geologic Features

The Guadalupe Mountains offer a rare opportunity for geologists to study a completely exposed sequence of evaporite, carbonate, and sandstone transitions from an ancient shelf down into a marine basin. Superbly exposed here is a portion of what is termed the "Permian Reef Complex," which contains one of the largest fossil reefs in the world, the Capitan Reef.

After being buried for millions of years, these reef deposits are now exposed in Guadalupe Mountains National Park (in an uplifted block), in karstic features in Carlsbad Caverns National Park (chapter 16) (in a down-faulted block), and to the south in the Glass Mountains and Apache Mountains. Subsurface drilling shows that the reef extends to the southeast for a distance of approximately 300 miles. It forms the shape of a horseshoe around what was in Permian time a deeper offshore marine basin, called the Delaware Basin (fig. 17.3). The most striking feature of Guadalupe Mountains National Park is the thousand-foot-high El Capitan Cliff, composed entirely of reef limestone (fig. 17.1).

The Capitan Reef

Unlike modern reefs that have been built mainly by a rigid framework of corals, the Capitan deposits consist of calcareous sponges and encrusting algae combined with other lime-secreting marine organisms. A great deal of lime mud is present that precipitated directly from seawater in a vast tropical ocean that covered the region in Permian time. (In general, reefs of Permian age are unusual because they were not built primarily by corals, the principal reef-formers of the middle Paleozoic Era and of the Mesozoic and Cenozoic Eras.)

By definition, a *reef* is a submerged resistant mound or ridge formed by the accumulation of plant and animal skeletons. Although reefs can form in deeper, cooler marine environments, the largest living reefs occur in shallow (not over 425 feet) tropical seas (temperatures over 68° F) where deep sea upwellings provide plentiful nutrients for organic growth. The seawater must also be clear, have normal marine salt content (27–40 parts per thousand), and be free of siliciclastic (noncarbonate) sediment or freshwater carried in by streams. Reefs generally form in areas with heavy wave activity so that the upper surface of the reef, the reef crest, is subjected to the constant pounding of breaking waves.

The sediments associated with reefs are quite different from those found in many other depositional settings. Reef sediments, largely composed of calcium carbonate ("lime") consist of an organic framework and broken fragments of the skeletons of plants and animals that live on or near the reef. These fragments are irregular in size and shape and poorly sorted. They show little evidence of being transported great distances.

Different portions of a reef are subjected to different environmental factors during construction. Relatively high wave action on the seaward-facing section, or *fore*

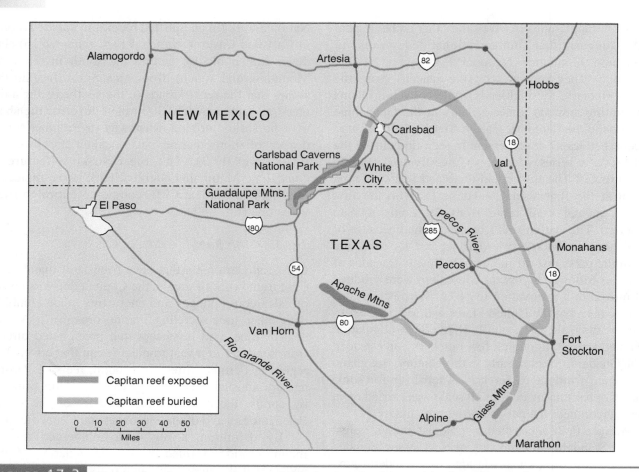

FIGURE 17.3 Exposed and unexposed parts of the Capitan Reef in Carlsbad Caverns and Guadalupe Mountains National Parks and vicinity. From D. Murphy, *The Guadalupes.* © 1984 Carlsbad Caverns/Guadalupe Mountains Natural History Association. Used by permission.

reef, causes both large and small blocks of living and dead skeletal material to detach and roll down the fore reef slope into the deep sea basin. This material accumulates as debris or talus at the base of the slope. Before they fall, these blocks are severely weakened by bioeroders (organisms) that rasp and bore through the dead skeletal material. Relatively high light intensity on the top of the reef, the *reef crest,* fosters a rapid upward growth of plants and animals that form the solid framework of the reef. Guadalupe reef crests were sedimentary shoals that received energy from currents. There is relatively low wave or current activity on the shoreward-facing section, or *back reef;* therefore, only fine sediment is carried back into this area, and the water itself is often stagnant and muddy. Salinity is especially high in areas lacking inflowing streams.

With each part of a reef having a characteristic marine environment, it is not surprising that the *biota* (i.e., the living organisms) of the reef area are highly diverse and also different from one zone of the reef to the next. For instance, plants and animals with more robust skeletons tend to inhabit the reef crest and fore reef areas, whereas those that burrow or thrive in muddy places tend to live in the back reef. The fastest growing plants and animals usually occupy the shallow fore reef where competition for space is especially intense. Because of the differences in the physical environment and in the constituent biota whose skeletons form the reef, distinctive sediment types are found in the fore reef, reef crest, and back reef areas. Each of these sediment types is called a *facies* (different rock types deposited in different environments at the same time). The sediments accumulate gradually as more plants and animals live and die, and their skeletons are washed over the lagoon and fore reef basins. As they are buried, the skeletal remains and the sediments created from them are compacted and recrystallized, becoming lithified as limestones. Sediment deposited in stagnant back reef or lagoon waters often contains high amounts of magnesium. The magnesium combines with limestone to form the rock dolomite. In especially salty water, intense evaporation causes precipitation of minerals and forms evaporites. Three types of evaporite rocks are associated with the Permian West Texas reefs: anhydrite, gypsum, and halite. As the salt content of the water increases during evaporation, first *anhydrite* (calcium sulfate, commonly as nodules or nodular layers) forms, then *gypsum* (hydrous calcium sulphate that tends to form long slender crystals), and last, *halite* (rock salt, sodium chloride).

During late Permian time (280–230 million years ago) in West Texas and New Mexico, three reef facies formed synchronously along the south-central portion of the North American craton in an inland sea comparable in size to today's Black Sea (fig. 17.4). To the east in the Appalachian region, orogeny took place as the African and North American plates collided. To the west along the edge of the craton through Nevada and Idaho, subduction of an oceanic plate was producing volcanoes. The reef facies formed in an arid tropical region south of the Permian equator. They consisted of: (1) a fore reef talus facies and deeper sea basin facies formed in the so-called Delaware Basin; (2) a reef crest and shallow fore reef facies formed in the surrounding El Capitan Reef Complex; and (3) a back reef and lagoon facies formed in the northeasterly Midland Basin. The rocks of the Delaware Basin consist largely of black limestones and fine-grained siliciclastic rocks. The black color of the limestone is believed to have been caused by high organic content deposited in a reducing environment and thus preserved. The process is similar to that of restricted marine environments of today where the bodies of floating microscopic animals fall to the bottom and are mixed with sediment. Such basin bottoms presumably contain little or no oxygen and thus few bacteria to consume the dead organic material.

The El Capitan Reef Complex consists of light colored, massive fossiliferous limestones; massive primarily because they were formed by growth in place and

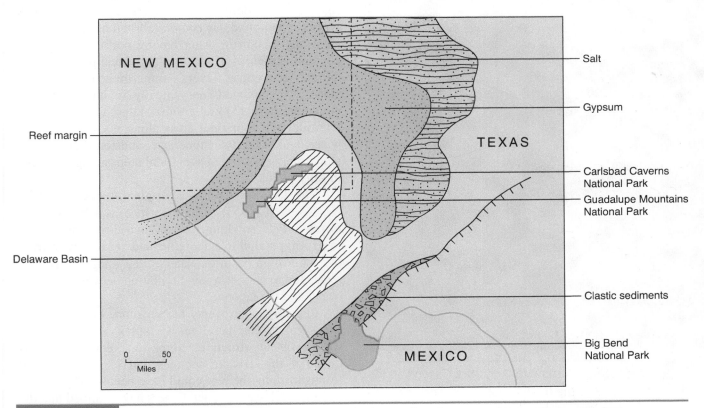

FIGURE 17.4 Delaware Basin and surrounding deposits in Permian time, superimposed on a modern map showing New Mexico, Texas, and Mexico. During Permian time, the region lay in the equatorial belt, far to the southwest of its present position.

accumulation of skeletons rather than by deposition of sediment from moving water. The skeletons were stabilized by encrusting organisms that grew over and thereby cemented the solid reef rock. The Permian organisms accumulated predominantly in shallow water where ample supplies of light existed to provide the energy necessary for algal photosynthesis.

A walk through McKittrick Canyon (fig. 17.5) along the Permian Reef Geology Trail proceeds

FIGURE 17.5   A trip into McKittrick Canyon shows exposures of the three major facies: deep basin (fore reef), reef crest, and back reef facies. McKittrick Canyon is located in the northeast corner of the park. National Park Service photo.

through all three major reef facies. The walk begins in the deep basin facies (the Bell Canyon Formation) and moves upward into the talus slope at the base of the fore reef (the Capitan talus). The Bell Canyon Formation contains numerous beds of *sedimentary breccia* (a rock composed of nonsorted, angular, and often large rock fragments) surrounded by fine-grained basin deposits. The breccias accumulated when submarine slides rolled down the steep face of the fore reef and talus slope. The Capitan talus consists of steeply sloping, thick beds of this breccia. On top of the steeply dipping Capitan talus beds lie more massive beds of the shallow fore reef and reef facies (the Capitan Formation), which are, in turn, overlain by thin, horizontal beds of the back reef/lagoon facies (the Carlsbad Group). The reef limestone itself is composed largely of sponges, algae, and bryozoans (tiny colonial animals). Corals make up a tiny percentage of the reef. Brachiopods (shelled animals), crinoids (sea lilies), and fusulinids (large, single-celled animals) are common in the fore reef and back reef facies. The back reef facies show a progressive change from carbonates (limestones and dolomites) through evaporites to clastic terrigenous (land-derived) rocks. Several specialized rock types can be found: (1) *coquina* ("shell hash"), deposited in the lagoon in water having normal marine salt content; (2) lenses of *pisolites* (spherical particles containing concentric layers, sometimes called "peastone"), precipitated inorganically in intertidal areas along the shore; (3) fine-grained dolomite, deposited in intertidal and supratidal areas along the shore in salty water; and (4) red siltstone and sandstone mixed with anhydrite, deposited in supratidal areas along the shore in salty ponds.

The Capitan Reef was originally interpreted by N.D. Newell (1953) to have formed as a barrier reef. *Barrier reefs* are long, narrow, wave-resistant ridges that form parallel to a shoreline and are separated from the shore by a basin of water called a lagoon. Barrier reefs tend to restrict water circulation between the open ocean and the lagoon.

Recently, controversy has developed over whether the Permian Capitan Reef Complex was actually a wave-resistant structure. Some workers have questioned if the delicate sponges and intertwined algal fronds forming the reef could have been strong enough to withstand wave impact. Others have claimed that the large volume of detrital (i.e., broken-up fragments) sediments in all three major facies suggests that the reef and the organisms forming the reef played only a minor role in controlling the sedimentation patterns of the area. Many geologists, therefore, have reinterpreted the Capitan Reef to represent merely a bank of skeletal debris below wave base (Achauer 1969). Despite this re-evaluation, most workers agree that the Delaware Basin sediments were deposited in an offshore enclosed marine basin; that the Capitan Reef Complex was formed on the edge of the basin in shallow marine conditions; and that the Midland Basin sediments were formed shoreward of the skeletal accumulations on an intertidal-supratidal, hypersaline (excessively salty) platform, or shelf.

Geologic History

1. **Development of the Delaware Basin (West Texas) early in Permian time (Wolfcampian Epoch).**

The Delaware Basin formed by earliest Permian time as an inland basin with a narrow outlet to the open ocean (fig. 17.6). More or less oval in shape, the body of water covered an area of around 10,000 square miles. Although the basin was largely enclosed, some ocean water entered periodically to replace water lost by evaporation. The circulation patterns within the basin are believed to have been similar to those in the present-day Mediterranean Sea. During this time, 1600 to 2200 feet of limestones and interbedded dark shale were deposited in the shallow, subsiding basin.

2. **Forming of small banks along the basin margin (Leonardian Epoch).**

By the beginning of mid-Permian time, the basin temporarily ceased to subside. Alternating beds of gypsum, sandstone, and dolomitic limestone of the Yeso Formation were deposited in nearshore areas and graded basinward into carbonate bank facies, the Victoria Peak Formation. Thin-bedded black limestones (the Bone Spring Limestone) accumulated in the stagnant deeper portion of the basin. Small, discontinuous patch reefs formed in shallow water along the rim of the basin.

3. **Continued patch reef development early in Upper Permian time (Guadalupian Epoch).**

By mid-Permian time, the basin began to subside again and many larger patch reefs formed along the basin margin. Cherty dolomites (San Andres Formation) built up close to shore, while quartz sandstones (Brushy Canyon Formation), containing scattered patch reefs, formed in and around the basin.

4. **Forming of the first barrierlike reef during the Upper Permian (Middle Guadalupian Epoch).**

As the patch reefs developed along the basin margins, the Delaware Basin itself began to subside rapidly. (This subsidence may have been related to compressive folding that resulted from the colliding of North American, South American, and African plates.) As the basin subsided, reef growth intensified, predominantly in an upward direction, forming the Goat Seep reef. Three distinct facies developed: (1) a lagoon and back reef facies composed of sandstones and dolomites (Queen and Grayburg Formations); (2) a reef crest and shallow fore reef facies containing dolomitized sponge

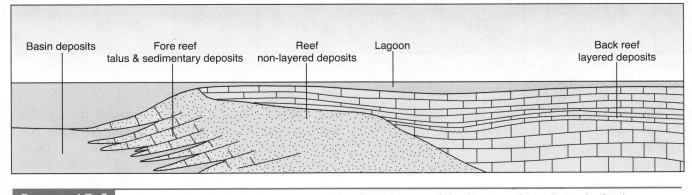

FIGURE 17.6   Cross section through the western side of Capitan Reef, looking southward. As the basin subsided, the reef grew upward and outward toward the basin center (left). Late in Permian time, as the equatorial sea dried up, the basin filled with evaporites. From D. Murphy, *The Guadalupes.* © 1984 Carlsbad Caverns/Guadalupe Mountains National History Association. Used by permission.

| Table 17.1 | | | Geologic Column, Guadalupe Mountains National Park | | | |

Time Units			Rock Units			Geologic Events
Era	Period	Epoch	Black reef lagoon	Reef crest, fore reef	Offshore basin	
Paleozoic	Permian	Upper — Ochoan?	Rustler Formation			Progradation of terrigenous red sandstones and siltstones across evaporites
			Salado Formation		Salado Formation	Widespread deposition of evaporites filling up the lagoon and back reef areas
					Castile Formation	
		Guadalupian	Carlsbad Group (Seven Rivers, Yates, Tansill Formations)	Capitan "Reef"	Bell Canyon Formation	Reduced basin subsidence and tectonic stabilization / Rapid reef growth outward over basin sediments
			Queen Formation	Goat Seep "reef"	Cherry Canyon Formation	Increased subsidence of deeper offshore basin during which three distinct reef facies (nearshore dolomites, small banks, offshore basin deposits), developed; reef increased in size but most growth was directed upward
			Grayburg Formation	Getaway "Bank"		
			San Andres Formation	Brushy Canyon Formation		Slow subsidence of offshore basin; reef development continued
	Lower	Leonardian	Yeso Formation	Victoria Peak Formation	Bone Spring Limestone	Infilling of the basin with carbonates and shale lenses to form a gently sloping basin and shelf margin; development of small patch reefs

Source: Modified after Newell et al., 1953.

and algae skeletons (the Goat Seep reef and the Getaway Bank); and (3) an offshore basin facies composed largely of quartz sandstones (the Cherry Canyon Formation).

5. **Development of an extensive reef complex during the Upper Permian (Late Guadalupian Epoch).**

By the end of mid-Permian time, the basin ceased to subside and achieved a stable position. The largest reef, the Capitan Formation, expanded rapidly 350 miles around the rim of the basin and gradually prograded toward the center of the basin. The three facies differentiated further. Limestones and sandstones (the Bell Canyon Formation) were deposited offshore in the basin; massive skeletal limestones (the Capitan Formation) accumulated in shallow areas encircling the basin; and dolomites and fine-grained sandstones (the Carlsbad Group) formed in nearshore areas.

FIGURE 17.7   In front and about 3000 feet below El Capitan is the Salt Flat Basin. Depending upon the amount of precipitation, it is a barren salt or a temporary shallow playa lake. Photo by Ann G. Harris.

6. **Infilling of the basin with evaporites during the Upper Permian (Ochoan Epoch).**

Continued sedimentation in the stable basin and possibly a drop in sea level caused shallowing and ultimate drying up of the basin late in Permian time. Extensive accumulations of deep-water evaporite deposits—the Castile Formation—covered the lagoon and reef and eventually the basin itself. Antiestuarine circulation may have caused the deposits to form. *Laminae* (bands of layers 1/16 inch in thickness) of gray anhydrite/gypsum, brown calcite, and halite precipitated first in the basin. Halite and potassium-rich salts (the Salado Formation) formed synchronously in nearshore areas and around the basin margin. Eventually the salts were deposited throughout the entire area.

7. **Migration of terrigenous red beds over evaporite deposits during the Upper Permian (Ochoan Epoch).**

At the very end of Permian time, terrigenous red siltstones and sandstone (Rustler Formation and Dewey Lake Formation) were deposited on the evaporites as rivers began to migrate across the area.

8. **Uplift and tilting of the area during late Mesozoic and early Cenozoic time.**

A series of tectonic movements created a major fault on the west side of the Guadalupe Mountains (the upthrown block), separating the range from a down-faulted block, the Salt Flat Bolson. The Guadalupe block was tilted, exposing the Capitan Limestone in the great rim escarpment with El Capitan at the "prow." The fault line along the Salt Flat Bolson (west side of the park) brings Cretaceous beds in the valley in contact with Permian beds that form the highlands.

9. **Erosion during the Cenozoic Era.**

Stream erosion has removed softer sediment and lowered the region to its present level. Ground-water erosion formed caves throughout the limestone units; the largest are those in Carlsbad Caverns National Park (Chapter 16). Because of the steep relief and lack of

FIGURE 17.8 The Castile Formation is an example of an evaporite. There are thin bands of alternating layers of gray anhydrite/gypsum, brown calcite, and halite that were precipitated in a basin. Photo by Ann G. Harris.

vegetation, landslides have eroded some hillsides, a process of mass wasting that continues in the present.

Geologic Map and Cross Section

Roswell Geological Society. 1964. *Geology of the Capitan Reef Complex of the Guadalupe Mountains* (Field Trip Guidebook). Roswell, New Mexico: Roswell Geological Society (map and text).

Sources

Achauer, C.W. 1969. Origin of Capitan Formation, Guadalupe Mountains, New Mexico and Texas. *American Association of Petroleum Geologists Bulletin* 53:(11) 2314–2323.

Bebout, D.G., and Kerans, C. (eds.) 1993. Guide to the Permian Reef Geology Trail, McKittrick Canyon, Guadalupe Mountains National Park, West Texas. Bureau of Economic Geology, University of Texas at Austin. Guidebook 26, 48 p.

Heckel, P.H. 1974. Carbonate Buildups in the Geologic Record: A Review. In *Reefs in Time and Space,* ed. L. F. Laporte, pp. 90–154, Special Publication 18. Tulsa, Oklahoma: Society of Economic Paleontologists and Mineralogists.

Hill, C.A. 1987. Geology of Carlsbad Cavern and Other Caves in the Guadalupe Mountains, New Mexico and Texas. New Mexico Bureau of Mines and Mineral Resources, Bulletin 117, 150 p.

Kendall, A.C. 1979. Continental and Supratidal (sabkha) Evaporites (#13) and Subaqueous Evaporites (#14). In *Facies Models,* ed. R.G. Walker, pp. 145–74. Ontario: Geological Association of Canada Publications.

Murphy, D. 1984. *The Guadalupes.* Carlsbad Caverns/ Guadalupe Mountains Natural History Association.

Newell, N.D.; Rigby, J.K.; Fischer, A.G.; Whiteman, A.J.; Hickox, J. E.; and Bradley, J.S. 1953. *The Permian Reef Complex of the Guadalupe Mountains Region, Texas and New Mexico.* San Francisco: W.H. Freeman & Co.

Selley, R.C. 1978. *Ancient Sedimentary Environments.* 2nd ed. Ithaca, New York: Cornell University Press.

West Texas Geological Society. 1960. *Geology of the Delaware Basin and Field Trip Guidebook.*

Wilson, J.L. 1975. *Carbonate Facies in Geologic History.* New York: Springer-Verlag.

CHAPTER 18

Virgin Islands National Park

ST. JOHN AND SURROUNDING WATERS
CARIBBEAN SEA, EAST OF PUERTO RICO

Ann F. Budd   University of Iowa

Area: 14,689 acres; water area, 5,650 acres; 23 square miles

Established as a National Park: August 2, 1956

Designated a Biosphere Reserve: 1976

Address: 1300 Cruz Bay Creek, Saint John, Virgin Islands 00830

Phone: 340-776-6201

E-mail: viis_superintendent@nps.gov

Website: www.nps.gov/viis/index.htm

Of volcanic origin, the Virgin Islands are overlaid by limestone. Tectonic uplift raised the volcanic islands to their present level. Both onshore and offshore features were modified by changing sea levels during the Pleistocene Epoch. Coral reefs began to develop in Pleistocence time.

FIGURE 18.1 Trunk Bay on the north side of Saint John Island. Tropical vegetation (breadfruit, mahogany palm, mangrove and kapok) masks the bedrock of the Louisenhoj Formation, which weathers to a reddish color, where it is exposed. National Park Service photo.

Geographic Setting

The four major island chains, or archipelagoes, in the Caribbean Sea (fig. 18.2) are: (1) the Bahama Islands (running northwest to southeast, southeast of Miami); (2) the Greater Antilles, including Cuba, Jamaica, Hispaniola, and Puerto Rico (running west to east, south of the Bahamas); (3) the Lesser Antilles, including Antigua, Guadeloupe, Martinique, and Barbados (running north to south, southeast of Puerto Rico); and (4) the Netherlands Antilles, including Curacao and Aruba (running west to east along the north coast of Venezuela). The Virgin Islands (about 100 in all) lie at the extreme eastern end of the Greater Antilles between that archipelago and the Lesser Antilles bending southward. The British Virgin Islands, in the northeast part of the archipelago, comprise about 40 islands, including Tortola, Virgin Gorda, and Anegada. The U.S. Virgin Islands, on the west side of the group, include St. Thomas, St. John, St. Croix, and smaller islands.

St. John Island, where the park is located (fig. 18.3), is only five miles wide by nine miles long; most of the topography is mountainous, with local relief being 500 to 1000 feet. Bordeaux Mountain, the highest peak, rises 1277 feet above sea level. Weathered, broken rocks—basalt, limestone, and chert—that mantle slopes are slowly being moved downhill. Along the shore are beautiful bayhead beaches, with fine-grained, white coral sands. The climate is pleasantly mild year round, except for hot, humid spells in summer and occasional hurricanes. Summer temperatures seldom reach the high 90s because easterly winds and the ocean moderate the tropical heat. Rainfall, which varies considerably from year to year, is supplied by the northeast trade winds and is collected in cisterns that are the chief source of fresh water.

In prehistoric times, Taino Indians camped on St. John shores and gathered food (fish, shellfish, iguanas, snakes, birds, sea turtles, and so forth) but did not establish villages because the land was too steep and rocky for the cultivation of manioc (a starchy root). Ancient petroglyphs have been found on the walls of a natural pool in basalt bedrock, as well as a few artifacts. Archeological digs at Lameshur Bay and a few other sites have yielded dates as early as 700 B.C. In those days, the island supported a dense rainforest of great hardwood trees with interlocking branches that formed a canopy. Freshwater collected in rock basins under the trees, and the trees made the air less hot and dry than now. Only a few patches of the original forest are left on St. John and the other Virgin Islands.

Christopher Columbus discovered the Virgin Islands in 1493 during his second voyage to America (box 18.2). He called the archipelago *Las Virgenes,* or "The Virgins," in honor of a martyred saint and her followers. After Columbus's voyages, naval vessels from the Netherlands, England, France, Denmark, and Spain visited the islands, as well as ships of pirates and traders.

The Danes officially claimed the Virgin Islands in 1687, but St. John was not settled by Europeans until 1716 when some Danish and Dutch planters moved from St. Thomas to St. John, divided the island into estates, and built plantations for the growing of sugar cane and cotton. The planters brought in African slaves, who cleared the steep hillsides, built miles of stone retaining walls for terraces to hold the thin, loose soil, constructed roads, and built stone mills with wood-burning boilers for processing sugar cane. The stonework consisted of fieldstone, chunks of brain coral, and ballast stone, held together with mortar that is now badly decayed. By 1726, all usable land was being farmed.

Although the plantations prospered for a time, intensive cultivation depleted soil nutrients. Despite the stonewalls, erosion washed soil and broken rocks downslope. Supplies of fresh water ran short. Several hurricanes, followed by insect plagues, discouraged the planters. Severe droughts triggered slave uprisings that were put down with heavy loss of life. Finally, in 1848, the Danish governor emancipated the slaves. Many planters and freed slaves left St. John because farming was no longer profitable. Remains of Danish plantations, the stone mills, and crumbling terraces can still be seen although native trees and bushes have regrown on the slopes.

During World War I, the United States bought the Virgin Islands from Denmark for $25 million because of their strategic position alongside the approaches to the Panama Canal. In order to protect the wildlife and natural resources, Laurance Rockefeller bought up nearly half the land on St. John Island and presented 5000 acres to the United States for a national park. Additional land and offshore areas (including Hassel Island in St. Thomas harbor) were incorporated into the park to protect the coral reefs from collection and commercial exploitation.

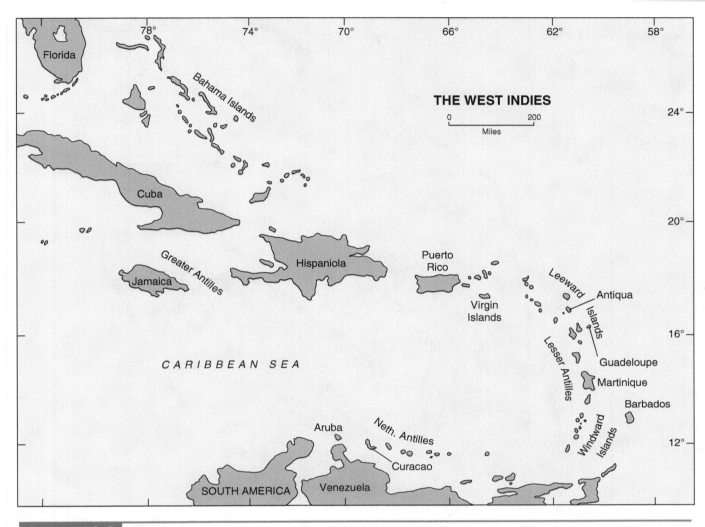

THE WEST INDIES

0   200
Miles

FIGURE 18.2   Major islands of the West Indies.

Development of Coral Reefs

Much of the modern offshore topography of St. John and other Caribbean Islands developed during the Pleistocene and Holocene Epochs as fringing reefs grew upward and outward along the windward coast on the eastern side of the islands. Such reefs range today from southern Florida (Chapters 19, 20, 21) to the southern margin of the Caribbean Sea. In these waters, temperatures are relatively high and sedimentation of fine particles low. Reefs in the Virgin Islands are mostly fringing or patch reefs, with an occasional barrier reef, and are constructed primarily of coral skeletons.

A *coral* is a cylindrical-shaped marine invertebrate that lives either singly or with others in a colony. Corals belong to class Anthozoa, phylum Cnidaria (formerly referred to as coelenterates). Coral larvae float free, then attach their cup-shaped bases to a hard rock substrate and begin to deposit calcium carbonate. Mature corals remain fixed for their entire lives (sometimes several centuries). An individual *polyp* may divide as it grows upward by continuously secreting an external lime skeleton around the lower cup-shaped portion of its body. The polyps occupy only the outer portion of a sometimes large skeleton that may be more than five to six feet in diameter. Corals feed by capturing microorganisms, small crustaceans, and other zooplankton by means of tentacles that encircle the mouth, which is located on the upper surface of each polyp. Some kinds of corals contain algae (zooxanthellae) within their liv-

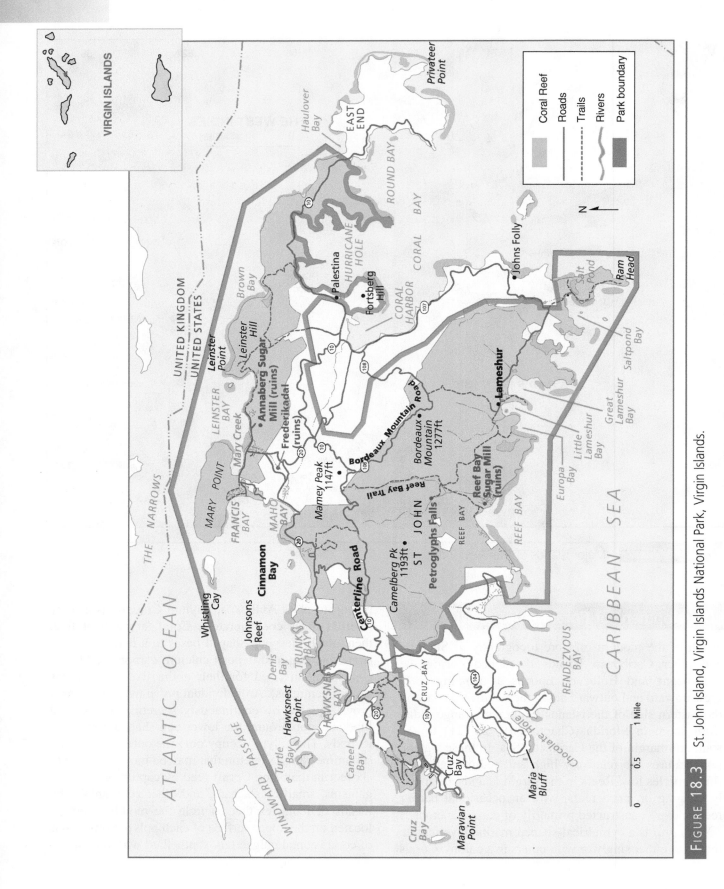

| Box 18.1 | **Tectonic Activity in the Caribbean Region** |

The Caribbean Sea formed during the early part of the Cretaceous Period as Europe and Africa moved away from North and South America (fig. 18.4-I). In the area now occupied by the Greater Antilles, a system of transform faults (explained below) and spreading centers (A) formed the Caribbean plate and separated North and South America. A subduction zone (B) (in central Venezuela) was active on the boundary between the Caribbean plate and the South American plate (fig. 18.2). Shortly thereafter, the spreading center system (A) was replaced by a subduction zone (A), and a new spreading center (C) became active in the area now between the Greater Antilles and the northern Venezuela coast (fig. 18.4-II). The core of the Greater Antilles consists of rocks extruded from volca-

noes in an island arc that developed along the newly formed subduction zone during early to middle Cretaceous time. The Netherlands Antilles, which were probably part of the Greater Antilles island arc, lay on the other side of the new spreading center (C) and therefore, "drifted" away from the Greater Antilles during middle to late Cretaceous time. The older subduction zone (B) through central Venezuela became inactive during the middle of Cretaceous time.

The newer spreading center (C) and subduction zone (A) ceased activity during late Cretaceous or early Tertiary time, and a (transform fault) (A) formed in the location of the subduction zone (A) (fig. 18.4-III). A transform fault is an extensive strike-slip (i.e.,

(continued)

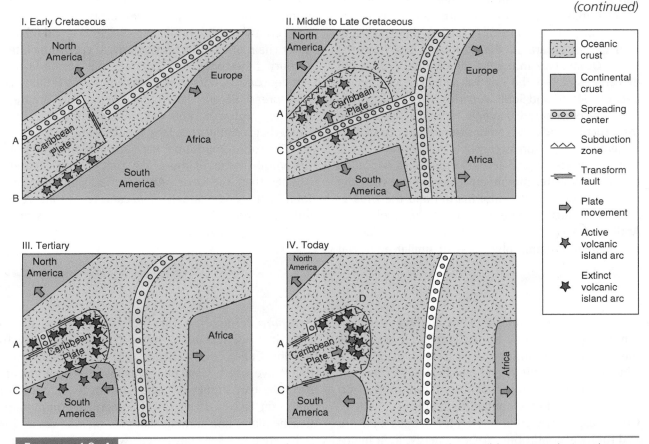

| FIGURE 18.4 | Schematic representation of major plate movements in the Caribbean Sea during the past 130 million years. Letters A through D indicate zones of significant tectonic activity. Adapted from Mattson 1977.

Box 18.1   Tectonic Activity in the Caribbean Region (continued)

having only horizontal movement) fault that occurs at a boundary between two tectonic plates. In this type of plate boundary, no crust is consumed (as in a subduction zone) or created (as in a spreading center); therefore, the North American and Caribbean plates merely slid by one another. A subduction zone (C) and associated volcanic island arc also formed at the same time near the spreading center (C) along the northern coast of Venezuela. A second subduction zone (D) and island arc became active during the early Tertiary along the line of the modern Lesser Antilles. Subduction along the north coast of Venezuela ceased during late Tertiary time and a transform fault (C) formed along its former position. Subduction of the North American plate beneath the eastern Caribbean plate margin continues today near the Lesser Antilles (fig. 18.4 II-IV). Throughout the last of the Mesozoic Era and in early Tertiary time, no major land barrier separated the Caribbean Sea from the Pacific Ocean. Late in the Pliocene Epoch, the Isthmus of Panama closed the Central American connection between North and South America.

This complicated geologic history is reflected in the structure and rock composition of various islands. Generally, the Bahamas consist of more than 10,000 feet of Cretaceous and Tertiary limestones deposited in a shallow-water marine environment between the North American continental margin and the Greater Antilles subduction zone (A) (fig. 18.4-II). The Greater Antilles are composed largely of highly deformed igneous and metamorphic rocks of similar ages that

formed in the volcanic portion of the boundary zone between continental and oceanic plates. Volcanic and seismic activity in the Greater Antilles has largely subsided, and the region is thought to be relatively stable tectonically.

Some scientists believe that the Greater Antilles are related to the North American Cordillera (on the western side of the North American plate) because the Greater Antilles are structurally more similar to northern Central America than to southern Central America (i.e., the Isthmus). Certainly both the North American Cordillera and the Greater Antilles experienced major deformation during Cretaceous time.

The Lesser Antilles, however, consist primarily of Tertiary and younger volcanics. The islands are arranged in two arcs: (1) an inner chain of volcanoes (some dormant and some active) consisting largely of late Tertiary and Quaternary volcanic rocks, and (2) an outer chain of early Tertiary volcanoes capped by late Tertiary limestones. In contrast, the Netherlands Antilles consist of Cretaceous volcanic rock overlaid by widespread Eocene limestones. As in the Greater Antilles, volcanic and seismic activity in the Netherlands Antilles has largely subsided.

The rock composition and deformation history of the Virgin Islands is more like that of the Greater Antilles than the Lesser Antilles. Rocks forming St. John Island consist largely of volcanic rocks that were intruded and deformed during late Cretaceous and early Tertiary time.

ing tissue; the photosynthesis of these algae produces oxygen and nutrients used by the corals. Hence corals containing numerous algae can secrete more skeletal material in well lit habitats and build reefs. Corals containing zooxanthellae grow much more rapidly than do types of coral that do not contain such algae.

Corals assume a great variety of shapes that are adapted for local conditions of wave energy, light intensity, and sedimentation rate. In offshore areas of St. John where wave energy is high, at Denis and Coral Bays, for example, the reef crest is dominated by the thick-branched elkhorn coral *Acropora palmata* and the mound-shaped brain coral *Diploria strigosa*. On either side of the crest are the thin-branched staghorn corals

Acropora cervicornis (fig. 18.5). Still farther from the crest in back reef and fore reef areas, one may find knobby finger corals *Porites porites* and other groups that thrive where wave energy is lower. Although corals form the reef framework, they do not live alone. Coral reefs consist of a diverse, well integrated community of marine plants and animals that may include sea urchins, fish, shrimp, sea fans, sea whips, snails, worms, sponges, and other species.

Reef Terraces

During the Pleistocene Epoch, several *reef terraces* (horizontal topographic plains bounded on either side

FIGURE 18.5 Elkhorn coral (*Acropora palmata*) is one of the dominant corals that make up the reef crest. Many animals live among the coral. National Park Service photo.

by steep slopes) formed along the coasts of all the Caribbean Islands. Most of the St. John reef terraces are now submerged; but on some islands, several terraces lie above sea level. Reef terraces are composed of lithi-fied skeletons of corals that lived and formed reefs immediately prior to terrace formation. It is believed that throughout the Caribbean, reef development flour-ished at times when sea level rose in the Pleistocene Epoch. Erosion occurred, forming a terrace each time sea level was lowered. Terraces formed repeatedly dur-ing the Pleistocene Epoch because sea level changed frequently. The ocean rose as much as 60 feet above today's sea level when the glaciers melted during inter-glacial periods, and dropped as much as 300 feet below today's sea level during periods when glaciers were growing. The exact number and timing of Pleistocene terrace formation throughout the Caribbean Islands is not clearly understood because of complications result-ing from vertical tectonic movements and isostatic adjustments that affected all the islands during this time. Recently, sea level has risen on St. John, causing upward reef growth to intensify and drowning some of the ancient reef terraces. One such terrace is in fairly deep water off the south shore.

Wave action has formed the beaches of St. John, with the reefs and submerged terraces being the main sources of sand. Because no permanent streams exist on the island, runoff does not supply much sediment to the lagoons and beaches except during heavy tropical storms or occasional hurricanes. However, storm waves and "swells" (large waves that travel across many miles of open ocean before breaking on a shore) erode both reefs and shore rocks, thus producing sand and a few cobbles. Some of the beaches on the island, such as Europa Bay on the southeast shore, are almost entirely made up of coral rubble; but nearby Little Lameshur is mainly fine sand. Cobbles, well rounded by wave action, cover much of Greater Lameshur, an adjacent beach. Each beach on the island's shore has been shaped by local conditions of the coastal environment, which include bottom topography offshore, character of the reef at that location, tidal currents, and wind and wave dynamics.

The Park Service's intention is to preserve this unique ecosystem of shore, ocean, and reef as much as possible in its natural state. However, reefs have been damaged by boats, divers, overfishing, and increased runoff due to development on shore. In 1989, a cruise ship dragged bottom off Francis Bay, ripping a jagged swath of some 300 square yards on the reef. Also in 1989, Hurricane Hugo severely damaged some of the reefs. Recovery of the reefs has been very slow.

Table 18.1			**Geologic Column, Virgin Islands National Park**	
Time Units				
Era	Period	Epoch	**Rock Units**	**Geologic Events**
Cenozoic	Quaternary		Alluvium	Erosion by streams and ground water; formation of submarine limestone terraces as sea level periodically rose and fell during glacial and interglacial stages
	Tertiary	Eocene	Scattered small dikes and plugs	Emplacement of dikes Widespread faulting; emplacement of dioritic plutons; localized metamorphism
		?	Hans Lollik Formation	Renewed volcanic activity; folding and strike-slip faulting
Mesozoic	Cretaceous	Albian	Tutu Formation	Renewed fault movement deposition of terrigenous material primarily as turbidites
			Outer Bass Limestone	Deposition of deep water limestones during a period of volcanic quiescence and sea bottom subsidence
			Louisenhoj Formation	Development of red soil horizon and subaerial volcanic activity; initial formation of deep ocean trench north and east of the Lesser Antilles
			-- -- -- unconformity -- -- --	-- subaerial erosion -- --
		Neocomian?	Water Island Formation	Submarine extrusion of volcanic material along an east-west high-angle fault onto a relatively flat sea bottom; gradual drop in sea level and/or tectonic uplift

Geologic History

1. Submarine extrusion of lava flows and subsequent uplift early in Cretaceous time (Neocomian Epoch).

The rocks forming the core of the Virgin Islands (the Water Island Formation) consist of more than 15,000 feet of *keratophyre* (a lightcolored igneous rock composed predominantly of plagioclase feldspar and quartz) and *spilite* (a mafic green igneous rock composed predominantly of plagioclase feldspar and chlorite). The lavas were extruded over deep sea clay sediments by quiet underwater volcanic flows onto a deep (more than 15,000 feet), relatively flat sea bottom. These rocks are found throughout the southeastern half of the island and can be best seen near Centerline Road and Bordeaux Mountain Road, near Chocolate Hole, and near Reef Bay trail. The lack of pumice and other land-derived sediment in these rocks, as well as their chemical composition, confirms their deep sea origin. The volcanoes that produced the lavas were part of the island arc chain that built up the Greater Antilles

Islands. During the Cretaceous Period, these islands lay in the western portion of the Tethyan seaway, an ancient ocean that extended around the central tropics from northern India through the Mediterranean Sea across the then-narrow Atlantic Ocean into the Caribbean Sea.

2. Subaerial volcanic eruption of ash and coarse debris and erosion in Lower Cretaceous time (Albian Epoch).

The Water Island Formation is immediately overlain by a 7000-foot-thick series of blue andesite ash beds, pyroclastic breccias (containing blocks as large as four feet in diameter), and brick-red soil layers (the Louisenhoj Formation). These rocks, which cover the western half of St. John, can be seen near Cruz Bay and Rendezvous Bay. The actual volcanic vent through which the lavas and ash were extruded is probably located in Pillsbury Sound between St. John and St. Thomas Islands. The ash beds and soil layers were formed subaerially after the growing volcanoes of the island arc emerged above sea level. The pyroclastics are separated from the Water Island Formation by a pronounced

| Box 18.2 | **Buck Island Reef National Monument and Salt River Bay National Historical Park and Ecological Preserve** |

The island of St. Croix, which is about 35 miles south of St. Thomas, is where these two national park areas are located. At Buck Island Reef, proclaimed a national monument in 1961, a superlative barrier reef surrounds much of the island as a submerged, fortress-like wall, rising over 30 feet from the sea floor. Some of the most massive specimens of elkhorn corals in the world have developed on the reef. Clusters of patch reefs grow both inside and outside the barrier reef, and a "pasture" of seagrass, where sea turtles graze, flourishes off the island's south shore. An underwater trail for snorkelers, laid out through a "marine garden," is in the lagoon between the island and the barrier reef (fig. 18.6).

The estuary and land area at Salt River Bay, which was added to the National Park system in 1992, is the only known site where Columbus landed on what is now U.S. territory. The first documented Native American resistance to European encroachment occurred here when an unequal skirmish between Spaniards and Tainos resulted in fatalities on both sides.

Of interest to geologists is the well developed submarine canyon that lies off the estuary and cuts into the St. Croix shelf. The canyon is ideal for study of oceanic processes and sedimentation. Over a distance of less than a mile, scientists can examine a transition from the terrestrial environment of the upper estuary to open marine conditions within the canyon.

The Very Long Baseline Array (VLBA)

The easternmost antenna of the VLBA system, which began operating in 1993, is located on St. Croix, with the westernmost dish located on Mauna Kea in Hawaii. This radio telescope, with its ten dish-shaped antennas, is spread across 5000 miles of sea and land. The antennas are synchronized by a control center in the National Radio Astronomy Observatory in Socorro, New Mexico. The VLBA, in addition to increasing our knowledge of outer space, will serve as a powerful tool for measuring geologic changes on earth, such as revealing movement of tectonic plates, faults, rifts, and other geologic features.

FIGURE 18.6   Brain corals (*Diploria stringosa*) grow on the reef crest in shallow water. The Park Service has a problem with snorkelers standing on them and crushing the polyps. National Park Service photo.

unconformity, which indicates that the Water Island Formation was exposed to air and eroded by running water. Ash and volcanic debris then exploded violently from volcanic craters rising above sea level or from shallow, underwater vents. Much of the pyroclastic material was later deposited below sea level by submarine slides and slumps on the steep slopes of volcano flanks.

3. Sea floor subsidence, volcanic quiescence, and limestone deposition (Albian Epoch).

Reduced volcanism and gradual sea-floor subsidence followed the eruptive activity. About 600 feet of thin-bedded, dark, silica-rich limestones (the Outer Brass Limestone) were deposited on a moderate slope in water depths of a few hundred feet. The limestones, which can be seen near Mary Creek, are silica-rich for two reasons: (1) the abundance of silica tests (shells) of microscopic floating animals, and (2) the occurrence of thin, interbedded layers of tuff.

4. Faulting and deposition of submarine slides and turbidites (Albian Epoch).

Toward the end of the limestone deposition, tectonic activity intensified, causing vertical fault movement that, in turn, created steep submarine slopes and caused the emergence of portions of the Louisenhoj Formation. Submarine slides and turbidity currents moved material eroded from the Louisenhoj Formation down the steep sea-floor slopes. During this time 6000 feet of flyschlike (widespread sandy and calcareous shales containing sandstone and conglomerate beds) rock suites composed of volcanic wacke (poorly mixed sediment consisting of volcanic particles) and limestone megabreccia (large angular blocks cemented together) were deposited, creating the Tutu Formation. The Tutu Formation is exposed from Maho Bay to Leinster Point. The limestone megabreccia occurs at Mary Point, Leinster Point, and in the Annaberg area.

5. Renewed volcanic activity, faulting, and plutonism from late in Cretaceous time through the Eocene Epoch.

Although they crop out only on islands northwest of St. John Island, at least 10,000 feet of water-laid andesite pyroclastic beds (the Hans Lollik Formation, similar to the Louisenhoj Formation) overlie the Tutu Formation. This suggests that a third period of intense volcanic activity occurred, possibly as late as the Eocene Epoch. This volcanic activity was accompanied by the intrusion of dioritic stocks and dikes and by contact metamorphism. Much of the British Virgin Islands and three islands south of St. John (Buck Island

National Monument, Frenchman Cap, and Flanagan Island) were formed entirely by these dioritic rocks. Outcrops on St. John can be found at Leinster Hill and near Mary Point. Extensive vertical faulting occurred concurrently with the dioritic intrusions. Later in the Eocene Epoch, dikes were intruded, including one exposed near Caneel Bay on St. John. By mid-Eocene time, the Caribbean region had assumed its modern geographic configuration, distinctly separated from the Mediterranean Sea by the widening Atlantic Ocean. Subduction of the North American plate below the Caribbean plate was especially intense during Eocene time, but thereafter movement along the northern margin of the Caribbean was largely horizontal along an east-west transform fault.

6. Alternations in sea level and reef development during the Quaternary Period (Pleistocene and Holocene Epochs).

As seawater warmed and sea level periodically rose in the Pleistocene Epoch, reefs developed around St. John but were later eroded to form terraces when sea level dropped. The cycle was repeated during major advances and retreats of continental glaciers as seawater was withdrawn from global oceans and then released. Over the past 15,000 years, sea level has slowly risen and submerged many of the older terraces. Living reefs have become established on the Pleistocene terraces and have grown upward during this last sea level rise.

Sources

Colin, T.J. 1981. The U.S. Virgin Islands. *National Geographic* 159(2):225–43 (February). Map Supplement.

Donnelly, T. W. 1966. *Geology of St. Thomas and St. John, U.S. Virgin Islands.* Geological Society of America Memoir 98, pp. 85–176 (also in Mattson 1977).

———. 1968. Caribbean Island Arcs in Light of the Sea Floor Spreading Hypothesis. *New York Academy of Science Transactions,* Series 2, 30:745–750 (also in Mattson 1977).

Greenberg, J., and Greenberg, I. 1972. *The Living Reef.* Miami: Seahawk Press.

Hubbard, D.K. (editor). 1989. *Terrestrial and Marine Geology of St. Croix, U.S. Virgin Islands.* Special Publication No. 8, West Indies Laboratory, Teague Bay, St. Croix, U.S.V.I.

Khadoley, K.M., and A.A. Meyerhoff. 1971. *Paleogeography and Geological History of Greater Antilles.* Geological Society of America Memoir 129.

Mattson, P.H., ed. 1977. *West Indies Island Arcs.* Stroudsburg, Pennsylvania: Dowden, Hutchinson, and Ross, Inc.

C H A P T E R   1 9

Everglades National Park

SOUTHWEST FLORIDA

Area: 1,506,539 acres; 2354 square miles

Established as a National Park:
December 6, 1947

Designated a Biosphere Reserve: 1976

Designated a World Heritage Site: 1979

Address: 40001 State Road 9336, Homestead,
FL 33034-6733

Phone: 305-242-7700

E-mail: EVER_information@nps.gov

Website: www.nps.gov/ever/index.htm

Largest remaining subtropical wilderness in the United States. Included in the park are freshwater and saltwater areas; offshore islands; mangrove forests; swamps; sawgrass prairies; sinkholes; relict coral reefs; and abundant wildlife. Park wetlands help to protect Florida's threatened water resources.

FIGURE 19.1   Sawgrass covers an expanse of freshwater marl prairie in Everglades National Park. A few clumps of red mangrove trees grow on slightly elevated knobs of limestone, called hammocks. National Park Service photo.

The Wetlands Environment— Past and Present

Although Everglades National Park and Big *Cypress National Preserve* (a large adjoining tract north of the park) occupy most of southwestern Florida, the Everglades wetlands area is larger still, extending as far north as *Lake Okeechobee* and originally as far east as the coastal ridge along the eastern shore.

Moreover, the Everglades are ecologically linked with a series of lakes, streams, and swamps that reach north from Lake Okeechobee almost to Orlando in the center of the state. The south Florida ecosystem, the whole lower half of the Florida peninsula, is an inter-

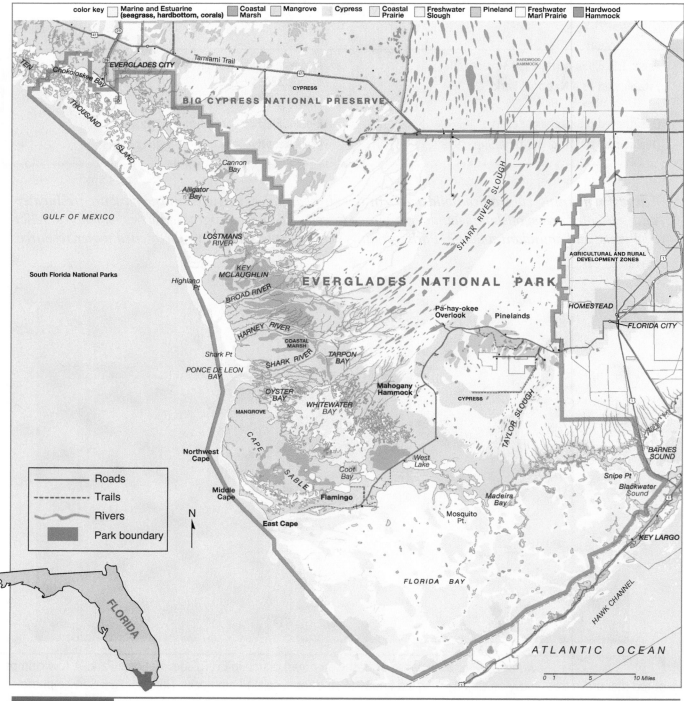

FIGURE 19.2   Everglades National Park, Florida.

dependent hydrologic system with the Everglades wetlands, both inside and outside park boundaries, acting as a delicate and vitally necessary balancing mechanism. Only recently has this function of the wetlands system been recognized, and much is still to be learned about its intricate workings that protect and preserve the state's water resources. The fact that Everglades National Park has been designated a Biosphere Reserve and a World Heritage Site is a significant recognition of the wetlands' importance. The Everglades wetlands are a unique, natural treasure as necessary to the enjoyment of life in southern Florida as the ocean, the beaches, and the benign climate.

The natural rhythms of the Everglades have been evolving for at least 5000 years, during which time climate and sea level have remained relatively unchanged. The water supply of the wetlands is renewed and restored annually by a rising water table and increased circulation of water during the hot, wet summer season. Most of the year's precipitation comes in brief, intense showers and in torrential rains between May and October. The mild winter season is drier; little rain falls between November and April. Most Everglades flora and fauna are not only adapted to, but dependent upon these wet and dry seasons. Their reproductive cycles have become synchronized with times of increasing and decreasing moisture.

In the past, disruptions of the system by periods of drought, lightning-caused fires, and hurricanes were accommodated without much difficulty; in fact, such natural disasters became part of the ecological processes of renewal. Today, controlled burning is a part of the National Park Service management policy in the Everglades because natural communities like the piney woods are restored by periodic burnovers.

The ecological crisis that the Everglades wetlands system has not been able to absorb and recover from is the loss and diversion of a great deal of its fresh water due to increasing residential, commercial, and agricultural development in surrounding areas. Until about a hundred years ago, water from Lake Okeechobee flowed very slowly, but unhindered, through sloughs and swamps toward the south, emptying at last into Florida Bay and the Gulf of Mexico. The interior of the Florida peninsula south of Orlando is a broad shallow basin, tilted just enough toward the south to permit natural drainage to move across the surface as sheetwash and flow sluggishly through irregular channels and stretches of open water (fig. 19.3). Decaying vegetation and fine sediment built up layers of peat and muck on

Figure 19.3 Air view of tree-lined waterways winding over a freshwater marl prairie in the Everglades. National Park Service photo.

the limestone bedrock. The Everglades vegetation thrives in these muck soils, which were enriched annually by saturation and flooding before the network of canals was excavated.

For several centuries after Spanish conquistadors discovered the Florida peninsula, the Everglades remained a little-known wilderness, uninhabited except for a few Seminole Indians and runaway slaves who took refuge in the swamps and secluded islands in the nineteenth century. Through the years a few hunters and trappers came to know the myriad waterways of the Everglades, but settlers seldom penetrated this formidable and mysterious wilderness. They were discouraged not only by its vastness and the difficulties of traversing its depths, but also by the swarms of ravenous mosquitoes and by several varieties of toxic plants.

The opening up of the Everglades began in this century. Drainage of the wetlands started around Lake Okeechobee with the digging of the first canal in 1906. Muck lands south and east of the lake were cleared for the growing of fruits and vegetables. This set off a land boom that reached its peak in the 1920s. Levees and dikes around the southeast side of Lake Okeechobee confined the water. Drainage ditches kept the fields

well watered but not saturated, and bountiful crops were harvested from the rich muck soils. During several drier than normal years in the early 1920s, land promotion schemes brought thousands of people into southern Florida. Several years of abundant rainfall ensued, including disastrous hurricanes in 1926 and 1928 that caused Lake Okeechobee to overflow its dikes. Reclaimed land was flooded, crops were ruined, and several thousand people perished. The land boom of the 1920s collapsed.

Population pressures around the Everglades slowed during the depression years of the 1930s, but picked up again with the influx of military personnel and defense establishments in the 1940s. More canals and drainage ditches were dug without adequate regional controls, and the cycle of disastrous floods, droughts, and fires began again. Finally, after the havoc wrought by the hurricane of 1947, a regional flood control district was set up by local authorities, working with the U.S. Corps of Engineers and the U.S. Geological Survey, to deal with the problems of too much water and too little water. Recognizing that Lake Okeechobee, with an area of 700 square miles (about one-third the size of Lake Michigan), was the key to

FIGURE 19.4 The canal along the Anhinga Trail acts as a refuge during times of drought. As much as 2500 gallons per minute are pumped into it so that the fish and other animals can survive, while the birds congregate around to feed. Photo by Alan Jacobs.

control of the system, engineers set about making the lake a huge storage reservoir.

Canals had been dug, both to the Atlantic and the Gulf, to improve drainage of excess water to the ocean. These were integrated into an elaborate network with pumps, gates, and levees designed to retain water needed during droughts (fig. 9.4). During heavy storms, excess water could be back-pumped from the reclaimed farm lands into the lake. Later, when it was needed, the water would be drawn back for irrigation. Another project involved dredging and channelizing the Kissimmee River, which collects water from the north and carries it into Lake Okeechobee. This shortened the river's length and increased the rate of flow into the lake.

The water management plan worked well for a number of years, but serious problems have recurred because population growth since the 1960s has put enormous stresses on the system. The threat of saltwater encroachment, which began when the first canals were dredged to the Atlantic Ocean—thus disrupting the natural drainage to the south—became more serious. During dry periods, saltwater moved into the canals and into the porous bedrock, displacing freshwater. A critical factor here is the level of the water table along the coast, and this level has permanently dropped. Freshwater wells near the coast became brackish, and new wells had to be dug farther inland. Attempts have been made to stabilize the water table by maintaining freshwater pressure high enough to prevent further encroachment of salt water; but whenever rainfall is less than abundant, the water supply problem in coastal areas becomes acute.

In the agricultural lands reclaimed from the Everglades, the muck soils are drying out, oxidizing, and blowing away, the loss being at the rate of about an inch per year. The only way of renewing and regenerating the muck is to allow tracts of land to be flooded for at least half of the year so that more organic material can decay and fine sediment can be deposited. The fact that about 1200 square miles of the Everglades agricultural area has already sunk 6 to 7 feet means that the loss of muck is approaching a critical point.

The channelization of the Kissimmee River lowered lake levels in the northern part of the system and caused the drying out of bottom lands exposed to air. This resulted in more soil loss. When the Kissimmee meandered over its mile-wide floodplain, more water soaked in, the water table was higher, and during rainy seasons the bedrock aquifer was recharged. As a

demonstration project, focusing on Taylor Slough, part of the canal system has been blocked so that water flows into old oxbows, refloods the floodplain, and reestablishes part of the wetlands.

The reduced flow of freshwater into the southern Everglades, including Everglades National Park, has been detrimental to the habitat of a number of plant, animal, and bird species, some of which are endangered. Sometimes water has been released too late in the breeding season or too early or too much all at once, instead of flowing slowly and continuously through the Everglades during the dry season as in the past. The most serious effect of reduced freshwater on life forms has been the stress on nurseries of shrimp and freshwater and saltwater fish in the southern Everglades and Florida Bay. The brackish estuaries open to Florida Bay have become more salty, greatly reducing the fish and shrimp population and resulting in losses to the fishing industry.

Most of the East Everglades wetlands, adjoining the park, is still an unspoiled area, but it, too, is threatened by residential development. Any further draining of this area will not only be detrimental to the park's ecosystem, but will also deprive the Biscayne Aquifer (box 19.1) of the freshwater it needs for recharge. This is a vital consideration, because the aquifer is the chief source of water for more than four million people.

Years of severe droughts were followed by Hurricane Andrew in 1992 and damaging record high water levels in 1994. These events have alarmed many residents of the Florida peninsula and have given renewed impetus to the conservation movement. The Everglades is now seen as the ecological base for repairs to the hydrologic system. Restoration is impossible; but with prudent and farsighted management, the water resources of south Florida can be maintained, used carefully, and not wasted.

Decades of effort by devoted conservationists, led by crusader Marjory Stoneman Douglas, went into the creation of Everglades National Park. The disastrous hurricane of 1947 provided the spur to action that finally brought enough land together by donation and purchase for its establishment. The second big step in the preservation of the wetlands came with the setting aside of Big Cypress Swamp as a National Preserve. The federal government continues to acquire wetlands acreage by purchase and donation; the National Audubon Society is creating a wet prairie sanctuary on the former flood plain of the Kissimmee River; and some tracts of land in East Everglades are being withdrawn from development. These measures and more are necessary for the protection of the Everglades wetlands and the preservation of water resources.

A National Park Service brochure that is given to visitors to Everglades National Park explains the wetlands problem:

"Water, fresh water, the lifeblood of the Everglades, appears to be everywhere, but... despite an apparent lush richness, water supplies are critical. Porous limestone underlies the entire park, and rooting plants have only a thin mantle of marl and peat atop this limestone for their support. If not protected, the Everglades' fragile richness would quickly vanish."

The Ecosystems of the Wetlands

Within Everglades National Park are eight zones, or ecosystems, that are dependent upon a balance between wet seasons and dry seasons, between freshwater and saltwater, and, in some cases, flood and fire. Primarily the zones are defined by climatic conditions; elevation in relation to sea level and to the water table; and the geologic aspects of the surface and subsurface materials.

The marine and estuarine zones. Most of Florida Bay is within the boundaries of Everglades National Park. This is a shallow body of water, roughly triangular in shape, lying at the southern end of the peninsula. Of special interest to geologists, Florida Bay is studied as a modern analog for ancient epicontinental seas in which deposition of fine sediment and organic material took place in a brackish environment with little circulation. Sandbars and banks covered by a dense growth of mangroves shut off the eastern end of the bay from tidal interchange with Barnes Sound; and the line of Florida Keys from Key Largo to Key West—along the edge of the continental shelf—keeps out most of the water from the ocean. Between the mainland and the Florida Keys, a lacelike pattern of winding, elongated mudbanks and bars, dotted with hundreds of tiny mangrove islands, spreads over the shallow bottom. Most of the bay is four or five feet deep; maximum depth is only about nine feet. Even on the west side of the bay, which is open to the Gulf of Mexico, daily tides have little effect since they penetrate the bay for only a short distance. It takes a full-fledged hurricane to flush out Florida Bay.

Florida Bay is part of the great "carbonate rock factory" (Chapter 20) that produces different kinds of lime-stone in the various marine environments off the Florida coast. By studying the chemical, biological, and environmental conditions under which carbonate rock is forming in Florida Bay, geologists seek to better understand the processes that formed ancient sediments, which are sources of petroleum. Much of the research being done is supported by petroleum companies.

The fine-grained calcareous mud that is accumulating in Florida Bay (and that may eventually be consolidated into limestone) is less than 5000 years old and lies on top of the oolitic facies of the Miami Limestone, which is Pleistocene in age. The southern Florida Keys (outside the park), which rim the southeast side of the bay, are also made up of oolitic limestone, which covers an older coral reef. Ooids have not formed in this area since Pleistocene time. (For an explanation of ooid formation, see Chapter 20.) The age relationship of the bedrock of the bay and the overlying mud suggests that Florida Bay has been under water a relatively short time.

The estuarine environments of Everglades National Park can best be observed along the course of the Wilderness Waterway, a boat and canoe route that goes from Flamingo on Cape Sable northwestward to the Gulf Coast Ranger Station (near Everglades City) in the northwest corner of the park. This winding, 100-mile inland waterway goes through long stretches of coastal swamp along an irregular, island-dotted shoreline. The underlying bedrock (rarely exposed because of the thick vegetation, fig. 19.5) is the bryozoan facies of the Miami Limestone in the southwest and the Tamiami Limestone in the northwest. Saltwater extends several miles up the estuaries.

Both the marine and estuarine zones serve as breeding grounds, spawning grounds, and nurseries for life forms of many kinds, from microscopic organisms to large species, such as crocodiles and manatees, both of which are in the endangered category. Crocodiles, which have a shorter, more tapered snout than alligators and are much less numerous, live mainly in the eastern end of Florida Bay and build their mounded nests on the Upper Key Largo shore and along the Bay shore. Manatees, or "sea cows," are large aquatic mammals that feed on vegetation in mangrove-lined creeks.

The mangrove zones. Mangrove trees, which grow in shallow, brackish water, establish themselves in the tidal zone of the marine areas and in the transition zone between freshwater glades and salty water in the estuaries. The mangroves extend the land as they spread

FIGURE 19.5 Bedrock is hard to find in the park, but occasionally the limestone is exposed. The bird seen is an anhinga or "water turkey." Photo by Alan Jacobs.

along the coastline and protect it from storm erosion. In the estuaries, mangroves line the banks and hold soil in islands (box 19.1).

In the Ten Thousand Islands region, part of which is within northwestern Everglades National Park, mangroves have created elongate islands from oyster beds in shallow waters offshore. When the upward growth of oyster beds and accumulated sediment reaches the tidal zone, mangrove seedlings are able to gain a foothold on top of the beds; eventually the trees displace living oysters, which reestablish beds in new locations farther offshore. Generations of mangroves have built up thick layers of peat over old oyster beds. These layers, representing successive tidal zones, provide a record of rising sea level along the coast (fig. 19.6).

The cypress zones. A cypress stand develops in solution pits and sinkholes that have been filled with muds and marls. Dwarf cypress grow in brackish areas inland from the southern coast and near the edge of the sawgrass glades. The larger pond cypress grow where more freshwater is available near the sloughs. Some build up slightly elevated "domes" of mud and vegetation on top of sinkholes. Mixed with other species, pond cypress thrive on tree islands (bayheads) in the marshes and sloughs.

The coastal prairie zone. Salt-tolerant plants, such as cactus and yucca, grow among grasses in the thin salty soils of the coastal prairie in the southern tip of the park on Cape Sable and at the lower end of Taylor Slough. North of West Lake a stand of hardwoods struggles for survival on an old shell mound left by prehistoric coastal Indians. The shell mound does not get the tree roots high enough out of the salt for the trees to do very well. Patches of coastal prairie are also found between the mangrove-lined shore and the coastal swamps on the west side of the park. Ancient shell mounds mark the sites of camps where Indians feasted on shellfish. One of these at Chokoloskee near Everglades City covers 135 acres. Twenty feet high, this man-made "hill" is the highest point in Everglades National Park. Pots made from marl clay, shark-tooth knives, and other artifacts have been excavated from the archaeological sites.

The freshwater sloughs and the freshwater marl prairie. Shark River Slough and Taylor Slough are the two main channels that move freshwater slowly southward through the Everglades. Both sloughs occupy broad, shallow depressions in the limestone bedrock and were probably free-flowing rivers at one time. Seen from the air, the sloughs are linear complexes of interconnected waterways and ponds, swamps, and saw-

FIGURE 19.6   Mangrove trees growing on tree islands, called "bayheads" or "hammocks," in a freshwater slough. National Park Service photo.

grass glades. The sloughs blend into the freshwater marl prairie on either side, and both prairie and sloughs are dotted with innumerable tree islands, called "bayheads" or "hammocks" (fig. 19.6).

Shark River Slough, the larger, central slough, is more than 40 miles wide during the rainy season but dwindles to a few narrow streams during winter droughts. At its lower end, the slough empties into the myriad waterways and lakes of the Shark River estuary at the southwestern end of the peninsula.

Between Shark River Slough and Taylor Slough, the southern tip of the Atlantic coastal ridge forms an interfluve, not more than three feet high, that divides the drainages of the southern Everglades. Before drainage canals were dug in the East Everglades area, Taylor Slough cleaned and stored water for a much larger part of southeast Florida. At the present time, state and federal authorities are attempting to preserve the wetlands between the park's eastern boundary and the metropolitan area spreading southwest from Miami in order to protect the region's water supply and also keep the feeding and breeding grounds for wildlife in the park from being further imperiled. In recent years heavily populated Miami and its environs have drawn so much water from the Biscayne Aquifer (box 19.1)

that the water table has dropped rapidly, permitting saltwater to seep into the aquifer in increasing amounts.

The freshwater marl prairies, which support many of the more than 100 species of grass growing in the park, have by far the greatest extent of the Everglades ecosystems. The endless expanse of grassy glades around the horizon are what inspired the term "everglades," and this became the name for the whole region. Small ponds and "gator holes" are scattered over the glades. These pits and solution holes in the limestone bedrock have been cleaned out by alligators during the dry seasons. They thrash their tails around, clearing out mud and vegetation so that water can collect in the depressions. Such waterholes enable Everglades species to survive droughts.

The sawgrass of the glades and sloughs is a type of sedge that grows as high as 12 feet and is hardier and tougher than most grasses. (Sedges have solid stems; true grasses have hollow stems.) Sawgrass stems have sharp, toothed edges, hence the name. The Everglades is the largest sawgrass marsh in the world (fig. 19.1).

The pinelands and hardwood hammocks. The pinelands in Everglades National Park are on the slightly elevated limestone at the end of the ridge between

Taylor Slough and Shark River Slough. Here on mostly bare limestone rock, the pines survive by sending their shallow roots down into pits and crevices in the oolitic rock. The stands of slash pine in the park are about all that is left of the pine forests that once covered the Atlantic Coastal Ridge for some 50 miles from north of Fort Lauderdale southward to the end of the ridge in the Everglades. The plant species in the pinelands have adapted to the periodic natural fires that sweep over the ridge during severe droughts. The slash pine, for example, germinates best after a ground fire has burned off low vegetation, releasing ash that enriches the scanty soil. The thick bark of mature trees protects them from the fire. Controlled burning of the pinelands is now an established part of the park's management.

The hammocks of the ridge area are slightly elevated, isolated clumps of tropical trees growing only in spots not burned over by fires. Like the pinelands, the hardwood hammocks were once more extensive on the Atlantic Coastal Ridge. Now they are found only in the park and in a few protected areas. The trees and shrubs of the tropical hammocks in south Florida are of West Indian origin. The taller, broad-leaved trees form a dense canopy that retains moisture for the forest floor. The hammocks also tend to grow on the edges of sinkholes or other sites where the roots can draw more water. The extra moisture has apparently enabled the groves to withstand the ground fires that have raced through the neighboring pinelands.

The hammocks of dense trees and shrubs that make up the tree islands (bayheads) of the sloughs and freshwater prairie may have less than a foot of elevation above the water surface. Many of them are teardrop-shaped and face into the current, which, although slow and weak, is able to gently mold the island banks. Some hammocks occupy rocky knobs or slight irregularities that jut up above the general flatness of the bedrock surface.

Geologic History

1. Cretaceous and Tertiary carbonate rocks of the Coastal Plain.

The surface bedrock of the Everglades is underlain by a three-mile thick sequence of marine sedimentary rocks, mainly carbonates, that were deposited in Cretaceous and Tertiary time and have undergone little deformation. None of these beds crop out at the surface in south Florida.

2. Miocene deposition of the Tamiami Formation.

In Miocene time, a blanket of cream, white, and greenish-gray clayey marl, silt, sand, and sandy marl—up to 100 feet in thickness—accumulated in the south Florida area. These shallow-water sediments were consolidated and cemented into the limestones of the Tamiami Formation, which makes up the bedrock surface of the northwestern part of Everglades National Park.

3. Pliocene emergence and slight erosion.

Pliocene rocks are not present in Everglades National Park, but sandy and shelly marls in the Okeechobee area to the north suggest that the land there may have been at or slightly above sea level.

FIGURE 19.7 Mangrove trees are the only trees that can tolerate salt water. The trees can grow in shallow water because the breathing roots hold the tree above the water level. The roots also act as a nursery for many forms of marine and freshwater life. Photo by Alan Jacobs.

| Box 19.1 | The Biscayne Aquifer |

This important water-bearing rock series is a nonartesian aquifer underlying all of southeast Florida. (An *aquifer* is an rock unit—or units—of moderate to high permeability from which ground water can be pumped.) The Biscayne Aquifer is the principal source of freshwater for the densely populated region from Fort Lauderdale to Key Largo.

The aquifer is supplied by rainwater and surface runoff that seep into the ground and fill the pore spaces in the bedrock. The ground water cannot penetrate deeply, because the impermeable lower part of the Tamiami Formation holds the water accumulating in the rocks above. The total section of bedrock saturated with water is approximately 10 feet in thickness in western Broward and Dade counties (in the center of the Florida peninsula), and increases to a maximum of about 200 feet along the southeast coast. Throughout most of the area, wells need only to be drilled below the water table in the aquifer in order to tap a supply of potable water.

The best water-bearing unit in the aquifer is the Fort Thompson Formation because of its thickness and high permeability. A good deal of the Fort Thompson is made up of shell hash and contains numerous cavities. The permeable Miami Limestone overlying the Fort Thompson is also part of the aquifer. Because of the solution-riddled characteristics of these limestones, the Biscayne is regarded as one of the most permeable aquifers in the world. What this means is that the aquifer benefits from fairly rapid recharge during abundant rains, but it is also subject to pollution and to encroachment by salt water.

The Biscayne Aquifer has always lost water due to natural causes such as evapotranspiration and ground water outflow into Biscayne Bay and the ocean. In recent years the amounts of available water have been greatly reduced by (1) the loss of recharge area to industrial, residential, and agricultural development; (2) the extensive system of drainage canals (many of which are polluted); and (3) increasing drawdown by pumping from wells.

The pressure of freshwater in the aquifer is what keeps the saltwater out (fig. 19.8). At the beginning of this century, the peninsular water table was so high and the aquifer was so full that freshwater bubbled up in Biscayne Bay, and freshwater wells could be dug a few feet back from the shore. Now saline water has seeped in along the base of the aquifer and extends inland underneath most of Dade and Broward Counties, which depend on the aquifer for most of their freshwater. Dams have been constructed in canals to maintain high water levels and prevent loss of freshwater. Other control measures have been put into effect; but the water supply situation remains critical.

4. Pleistocene deposition.

The Lower Pleistocene Fort Thompson Formation does not crop out in the park, but this very permeable unit of marine and freshwater marls, sandstones, and shelly limestones is the main water-bearing component of the Biscayne Aquifer (box 19.1).

When sea level was high during the last Pleistocene interglacial episode (preceding Wisconsinan glaciation), several rock units—deposited under different environmental conditions but all about the same age—accumulated in the south Florida area. These rock units make up the surface bedrock of most of south Florida and the Florida Keys, and underlie Biscayne Bay and Florida Bay. The Miami Limestone has two distinct facies. The oolitic makes up the Atlantic Coastal Ridge, the East Everglades region, parts of the southern coast, Florida Bay, and the southern Florida Keys. The oolitic facies interfingers with the bryozoan facies, which makes up the bedrock of southwestern and south central Florida. (The Key Largo Limestone of the same age, which does not crop out in Everglades National Park, is described in Chapters 20 and 21.)

Examination of a specimen of the Miami Limestone from the oolitic facies shows that its layers consist of billions of tiny spherical grains, or *ooids,* cemented together to make a rocky mass. Ooids are formed by precipitation of calcium carbonate, and the grains accumulate in a considerable thickness of limey oolitic ooze. Eventually they may be cemented by calcite.

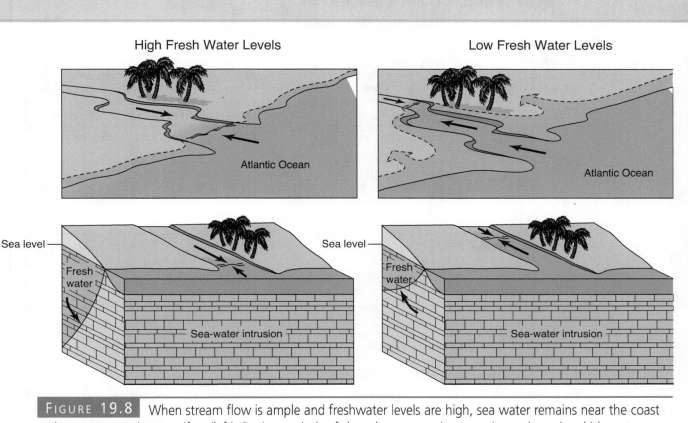

High Fresh Water Levels

Atlantic Ocean

Sea level

Fresh water

Sea-water intrusion

Low Fresh Water Levels

Atlantic Ocean

Sea level

Fresh water

Sea-water intrusion

FIGURE 19.8 When stream flow is ample and freshwater levels are high, sea water remains near the coast and cannot move into aquifers (left). During periods of drought or excessive pumping, salty or brackish water moves up tidal inlets and invades aquifers, accumulating beneath fresh ground water (right). From *The Face of Florida,* 2nd edition, C.M. Head and R.B. Marcus. © 1987 Kendall/Hunt Publishing Company. Used by permission.

The ooids that make up the Miami Limestone in the Atlantic Coastal Ridge were piled up along a late Pleistocene shoreline by ocean waves. The higher parts of the ridge toward the north, 20 to 40 feet above present coastal elevation, have been exposed to the atmosphere much longer than the southern part of the ridge, which is only about two feet high at its terminus in the southeast part of Everglades National Park.

In the central part of the park, the bedrock surface of the Miami Limestone is made up of bryozoan remains as well as ooids. Bryozoans, which are small marine invertebrates living in colonies, secrete calcium to form the tiny cuplike cells in which the individual organisms live. Bryozoan remains, accumulated over many generations, make up about 70 percent of the rock volume of the Miami Limestone. In Florida Bay modern bryozoans continue to grow in knoblike clusters encrusted with seaweed. These small animals have lived in marine environments and produced carbonate material since late in Cambrian time.

5. Holocene mucks, soils, and sediments.

As the great ice sheets of Wisconsinan glaciation melted, sea level rose again fairly rapidly, the trend continuing into early Holocene time. Then the rate decreased, and sea level has risen more slowly in the past 4000 to 5000 years. Unconsolidated sediments have accumulated on the land surface of southern Florida. Southward-flowing waters spread weathered and eroded material from the Tamiami and Miami

Table 19.1			Geologic Column, Everglades National Park

Time Units			Rock Units	Geologic Events
Era	Period	Epoch		
Cenozoic	Quaternary	Holocene	Carbonaceous sands in beach ridges; limy muds, mucks, peat, soil; limy marine muds in bays	Nonmarine accumulations of unconsolidated sediment in shallow stream channels and on bedrock surfaces
		Pleistocene	Miami Limestone (oolitic and bryozoan facies)	Accumulation of ooids and bryozoan masses in shallow seas
			Fort Thompson Formation (in subsurface of Biscayne Aquifer)	Marine and freshwater marls, limestones, sandstones, shell hash, etc.
	Tertiary / Neogene	Pliocene	/////////////////	Slight erosion
		Miocene	Tamiami Formation	Marine limestones deposited in shallow water; silts, shelly sands
	Tertiary / Paleogene		/////////////////	Disconformity
Mesozoic	Cretaceous		Underlying marine carbonates of the Coastal Plain (approx. 3 miles in thickness)	Only slight deformation

Source: Modified after Hoffmeister, 1974.

Limestones cover the saucer-like interior basin now occupied by the Everglades. Mixed with the sediment are abundant organic remains of decayed vegetation. The resulting peat and muck soils are immature but highly fertile, especially under conditions of high moisture and humidity. On land protected by natural vegetation, muck continues to accumulate, but on drained land the surface material becomes powdery during times of drought and is easily eroded.

State and federal authorities, including the National Park Service, continue their efforts to improve the hydrology of South Florida and protect the area's environmental resources. Programs include a combination of canal filling, modification of water control structures, restoration of flow in old sloughs and channels, and land acquisition.

Geologic Map and Cross Section

Purl, H.S., and Vernon, R.O. 1964. Geologic Map of Florida, plate 2. *Summary of the Geology of Florida and a Guidebook to the Classic Exposures.* Florida Geological Survey Special Publication No. 5.

Sources

Davis, R.A., Jr. 1995. Geologic Impact of Hurricane Andrew on Everglades Coast of Southwest Florida. *Environmental Geology* 25: 143–148.

Dawes, C.J., Bell, S.S., Davis, R.A. Jr., McCoy, E.D., Mushinsky, H.R., and Simon, J.L. 1995. Initial Effects of Hurricane Andrew on the Shoreline Habitats of Southwestern Florida. *Journal of Coastal Research* 21:103–110.

De Golia, Jack. 1978. *Everglades, the Story Behind the Scenery.* Las Vegas, Nevada: KC Publications.

George, J.C. 1972. *Everglades Wildguide.* Natural History Series, National Park Service. 106 p.

Head, C.M. and Marcus, R.B. 1987, 2nd edition. *The Face of Florida.* Dubuque, Iowa: Kendall/Hunt. 197 p.

Hiaasen, C. 1995. The Last Days of Florida Bay. *Sports Illustrated* 83 (12):76-87, September 18.

Hoffmeister, J. E. 1974. *Land From the Sea, the Geologic Story of South Florida.* Coral Gables, Florida: University of Miami Press.

Multer, H.G. 1977. *Field Guide to Some Carbonate Rock Environments, Florida Keys and Western Bahamas.* Dubuque, Iowa: Kendall/Hunt.

National Park Service. 1985. Everglades National Park. Map and brochure.

CHAPTER 20

Biscayne National Park

S O U T H E A S T E R N   F L O R I D A

Area: 172,925 acres land and water; land only, 4373 acres

Proclaimed a National Monument: October 18, 1968

Redesignated a National Park: June 28, 1980

Address: 9700 SW 328 St., Homestead, FL 33033-5634

Phone: 305-230-7275

E-mail: BISC_Information@nps.gov

Website: www.nps.gov/bisc/index.htm

A park of reef and water, with 33 keys (islands) forming a north-south chain between Biscayne Bay and the Atlantic Ocean plus a narrow strip along Biscayne Bay's western shore.

FIGURE 20.1   Aerial photograph of islands and channels in Biscayne National Park. Built up as coral reefs during Pleistocene times of higher sea level, the islands are now covered with vegetation. National Park Service photo.

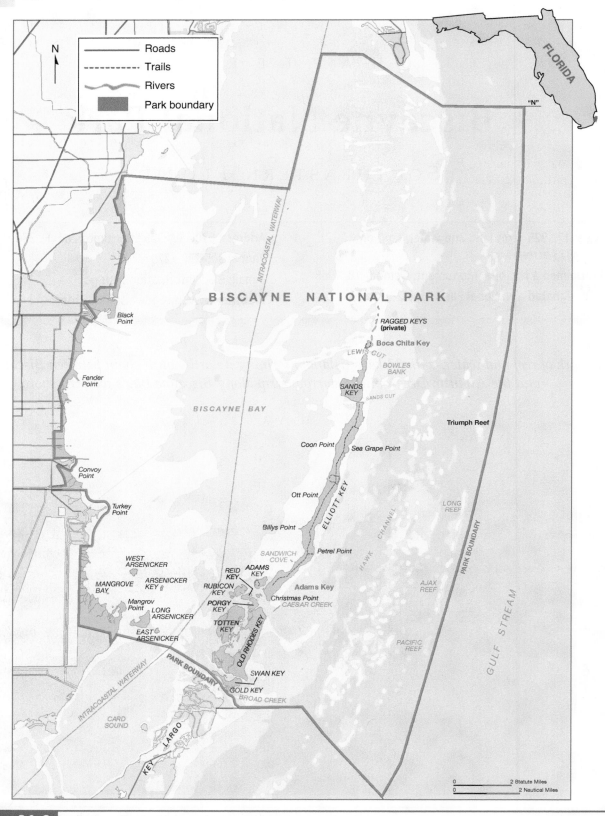

FIGURE 20.2   Land and water areas of Biscayne National Park, Florida.

Geographic Setting

Biscayne National Park is a large marine preserve south of Miami and offshore from the mainland of southeastern Florida. Within park boundaries lie most of shallow Biscayne Bay and a large reef tract, comprising a back reef area and the northern part of an outer reef. Hawk Channel, a seaway between the islands and the outer reef, is also in park waters. Biscayne National Park was established to protect three natural ecosystems:

1. the nursery grounds for many fish and shellfish species in Biscayne Bay
2. the native vegetation of the upper Florida keys
3. the marine plants and animals of the coral reefs.

The term "keys," derived from the Spanish word *cayo* meaning "islet," is given to the coral islands that form a line of fossil patch reefs off the southern Florida coast. A topographic and geologic distinction is made between the lower Florida Keys, which extend southwestward from Big Pine Key to Key West, and the upper keys, which run from Big Pine to Biscayne National Park's northern boundary. The park's southern boundary is between Old Rhodes Key and Key Largo. A causeway connects Key Largo and the rest of the southern keys with the mainland, but access to the keys in Biscayne National Park is by boat only. The Visitor Center at Convoy Point was destroyed by Hurricane Andrew in 1992, but new facilities have been constructed at that location (box 20.1).

Strong currents, high winds, hurricanes, and sudden violent storms, combined with the hidden reefs and shoals, make these waters dangerous for sailing vessels. In the seventeenth and eighteenth centuries pirates lay in wait in the Florida Keys for ships sailing through the Florida Straits between Cuba and the mainland. They attacked ships that were crippled by storms or blown off course onto the shoals and reefs, where they were easy prey. Hawk Channel is marked with the locations of some of the sunken wrecks.

The earliest inhabitants of the Florida Keys were Indians who left huge shell mounds that enlarged some of the islands. In the 1830s bands of Seminoles, driven from their homes in northern and central Florida by white settlers, took refuge in the Everglades and in the Florida Keys, where some of their descendants still live.

In recent years the coastal waters, shorelines, and islands of Florida have suffered severe environmental damage from the pressures of increasing population and from commercial and residential overdevelopment. Spurred by conservationists and alarmed by growing problems of water pollution, loss of fisheries, and destruction of wildlife habitat, the Florida legislature passed a Beach and Shore Preservation Act in 1965. This established coastal construction setback lines that have saved some of the state's beaches and inlets from being urbanized.

Geologic Features

The narrow strip of shoreline, the keys, the coral reefs, and the ocean floor in Biscayne National Park are all made of carbonate rock, most of it of organic origin. The marine platform on which the keys and the reefs were built is part of the Atlantic coastal shelf, which is at its narrowest here along the southeast coast of Florida. A few miles offshore, just east of the outer reef,

FIGURE 20.3   An aerial view south along Elliott Key. On the right hand side is Biscayne Bay and on the left side is Hawk Channel and the Atlantic Ocean. On the Atlantic side, the coral rock of the old fore reef drops off steeply. On the Biscayne Bay side, a wide beach slopes gently to the mainland across the bay. National Park Service photo by Claudine Laabs.

Box 20.1   The Geologic Role of Mangrove Swamps

In Biscayne National Park, and elsewhere on the Florida coast, mangroves have had an important role in the development of the shoreline. Scientists estimate that despite occasional losses of ground due to hurricanes, the mangrove swamps along the keys, Biscayne Bay, and Florida Bay have in this century added more than 1500 acres of forests where formerly open shoals existed. In the park, fringes of mangrove-built shoreline can be seen on both sides of the keys and at Mangrove Point on the mainland.

Among the salt-tolerant trees of the shore zone are the dominant red mangroves *(Rhizophora mangle)* (fig. 20.5B), the only true mangroves; the black mangroves *(Avicennia germinans)* (fig. 20.5C); and

white mangroves *(Laguncularia racemosa)* (fig. 20.5D), which belong to two other species although they are called "mangroves"; and the buttonwoods *(Conocarpus erectus)*. The buttonwoods grow farther inland than the mangroves and are often found mixed with mahogany *(Swietenia mahogani)* and other tropical hardwoods. The mangroves grow at the edge of the ocean in the zone between high tide and low tide. In addition to being salt-tolerant, these remarkable trees have an amazing ability to cling to unstable ground. When storms beat in on the coast and hurricanes bring devastating waves and winds, the mangroves usually survive. Only the most disastrous hurricanes can uproot them. Even then, some of the older, stronger trees that are flattened by the wind manage to send out new anchoring roots; then their trunks bend back upward and start growing toward the sun again. Before the days of advance warning for hurricanes, old-timers often took refuge in firmly anchored small boats in narrow creeks among the mangroves. The protection afforded by the mangroves enabled people to ride out the storm.

A full-fledged hurricane always produces coastal changes, but where mangroves grow, such changes are minimized. In fact, if it were not for the mangroves protecting the tidal environment, the Florida coastline would be reshaped by every big storm. Stretches of coastal swamps were drained, filled, and developed as waterfront real estate before the protective function of the mangrove swamps was understood. When hurricanes struck, as, for example, Donna in 1960 and Betsy in 1965, the damage was appalling.

Another important adaptation of red mangroves is their ability to colonize. Seeds germinate on the mature trees and then fall as seedlings. Some take root in water beneath the parent trees, but other seedlings are spread by tides and currents along the coast and sometimes travel considerable distances before stranding and sending out roots (fig. 20.4). The young trees anchor themselves by strong roots that sprout both from the trunk and from the lower branches. The roots (1) form a complex network of props and braces that hold up the crown; (2) bring up sap; (3) raise the trunk above the water; (4) furnish oxygen, which is unavailable in the watery muck; and

FIGURE 20.4   A mangrove sapling roots itself in the tidal zone. National Park Service photo.

FIGURE 20.5A   A tangle of old mangrove roots holds sediment, helping to protect the shoreline. National Park Service photo.

FIGURE 20.5B   The red mangrove is one of the salt tolerant trees found in Biscayne National Park. The dominant trees along the shoreline are the only true mangrove (*Rhizophora mangle*). National Park Service photo.

(continued)

| Box 20.1 | The Geologic Role of Mangrove Swamps (continued) |

FIGURE 20.5c   The black mangrove (*Avicennia germinans*) is not a mangrove but belongs to another species, although it is called a "mangrove." National Park Service photo.

(5) catch and hold silt and debris. It is this last function that extends the coastline into the water, creating new land.

For several thousand years, sea level has been slowly rising along Florida shores. Geologists have found mangrove peat accumulations to be a useful indicator of approximate rates of sea level change because mangroves grow only in the tidal zone. In tropical and semitropical latitudes, tidal ranges are low; on Florida coasts the average range between high and low tide levels is only about three feet.

Therefore, if sea level were static, mangrove peat accumulations would be about three feet in thickness. However, in protected locations on the shore, peat accumulations have been found to be 12 to 15 feet in depth, which means that sea level must have

the ocean floor drops off steeply into the deep water of the Florida Straits. The Gulf Stream flows through the 90-mile-wide straits, between Florida and Cuba to the south and the Great Bahama Banks to the east.

Landward of the outer reef, the water is shallow and the relief of the keys and the mainland is low. None of the keys rises more than 18 feet above sea level and some are barely above high tide level. The keys and all of southern Florida were under water during the last Pleistocene interglacial episode, when sea level was higher than now. The patch reefs that make up the Florida Keys were built by corals that lived and grew on the marine platform during times of high sea level in the Pleistocene Epoch. In the geologic sense, all the rocks, soil, and sediment of the region are very young, and the processes that are significant in the geologic story of the park are still going on.

The "Carbonate Rock Factory"

The marine preserve of Biscayne National Park is part of a large carbonate-producing region that includes the waters around the western Bahamas as well as the Florida Keys. Here in a variety of carbonate environ-

been rising while successive generations of mangroves grew and peat accumulated (Hoffmeister 1974).

The importance of the mangroves to the nutrient cycles of the coastal environment can hardly be overstated. Besides supporting a large number of rookeries, the mangrove swamps and bordering waters are the habitat of alligators, snakes, turtles, manatees, raccoons, and many other animals. Even the mosquitoes that breed by the millions in the mangrove swamps are a main source of food for many birds, frogs, reptiles, and fish. The tidal inflow and outflow circulate nutrients for the breeding grounds of fish and shellfish in the bays. Nutrients carried out to sea help to supply food for the corals and other organisms living in the reefs.

FIGURE 20.5D   The white mangrove (*Laguncularia racemosa*) is technically not a mangrove tree in spite of its common name. The tree is in bloom. National Park Service photo by F. Twitchell.

ments, a wide range of limestones are accumulating, both organic and inorganic. In the "labor force" are many species of marine plants and animals. The corals and calcareous algae are the main producers. "Manufacturing" processes include photosynthesis (corals and algae cannot work in darkness), precipitation, evaporation, and supersaturation. By these processes the chemical constituents, or "raw materials," are extracted from seawater and transformed into skeletal parts, crusts, and ooids (or ooliths).

Most *ooids* are tiny spheres of inorganically precipitated calcium carbonate. (Noncalcareous ooids, which are less common, are generally formed by replacement processes.) At the present time, ooids are forming only in the western Bahamas part of the carbonate-producing factory. Strong ocean currents there keep the waters agitated and sloshing back and forth over shallow and deep bottom areas. Intense sunlight raises the temperature of the tidal pools and shallows, and evaporation increases salinity, concentrating the calcite and causing it to precipitate around any fragment (a speck of shell, quartz sand grain, etc.) that acts as a nucleus. Precipitation of successive carbonate layers around such nuclei forms ooids.

Although conditions on the Florida side of the Florida Straits do not now permit ooid production, they accumulated in vast quantities in the past in the waters that covered the lower Florida Keys and southern Florida. Ooids were compacted and cemented to form oolitic limestone, some of which has been eroded to produce sand-sized grains and other sediment. Ooids in the form of sediment, either loose or in sedimentary rock, are referred to as *ooliths* or *oolites*. The small beaches on the Florida Keys and the larger beaches along the extreme southern tip of the Florida mainland consist principally of oolitic and coralline sand rather than quartz sand.

Geologists, marine biologists, and other scientists come from all over the world to study the carbonate-producing environments of the coastal waters of Florida and the Bahamas. Much of the research consists of detailed study of the processes of the present in order to better understand how the vast accumulations of carbonate rock on the earth's surface were produced in the geologic past. Because carbonate rocks are an important source of petroleum, a good deal of the research being carried on has significant economic implications.

The Living Reefs

The corals of the reef tract in Biscayne National Park are building reefs in the same way their ancestors built the older reefs that make up the keys. The fossil corals of the islands are not different from their living counterparts in the surrounding waters. The same coral species, about 50 in all, live in their chosen habitats throughout the Florida reef tract and the Bahamas. The coral groups are also closely related to those living in the reefs off the Virgin Islands (Chapter 18). A few species in the Florida reef tract seem able to thrive in many environments conducive to coral growth; but most of them prefer either (1) the outer reef along the open-ocean side of the marine platform; (2) the numerous patch reefs in the back reef zone (between the keys and the outer reef); or (3) the shoals on either side of the keys.

Each of the three habitats is dominated by a characteristic set of corals (fig. 20.6). The outer reef crest is

FIGURE 20.6   The coral shoreline on the east side of Elliot Key. National Park Service photo by F. Twitchell.

Table 20.1			Geologic Column, Biscayne National Park	
Time Units				
Era	Period	Epoch	**Rock Units**	**Geologic Events**
Cenozoic	Quaternary	Holocene	Accumulations of thin soils, marls, peat on land. Accumulations of fine bottom sediments, mainly carbonates, in seawater	Development of fringing mangrove swamps Generally rising sea level Growth of coral reefs
			Erosion	Erosion
		Pleistocene	Miami Limestone (oolitic facies) } Key Largo Limestone	Fluctuating sea levels, with coral reef growth and oolitic deposits during high sea stands; erosion during low sea stands

Source: Modified after Hoffmeister, 1974.

covered largely by the elkhorn coral, *Acropora palmata,* and below a depth of 10 meters by the staghorn coral, *Acropora cervicornis.* These two corals grow rapidly (.4–.98 in./year) in well-lit, agitated environments to form dense thickets of branches that shade and thereby inhibit the growth of other corals. The back reef is composed mostly of mound-shaped corals, including the star coral, *Montastraea annularis* (fig. 20.7), and the brain coral, *Diploria strigosa* (fig. 20.8). Single mounds of these corals can attain enormous sizes (more than 3 meters in diameter) after several centuries of growth. The corals found in the inshore shoals include small (4 cm diameter), almost round star corals, such as *Siderastrea radians,* and knob-shaped finger corals, such as *Porites divaricata* (fig. 20.9). These corals grow slowly but are able to remove sediment easily and to resist extreme or fluctuating temperatures and salinities.

Geologic History

1. **Pleistocene reef-building; Key Largo Limestone.**

During the last interglacial episode of Pleistocene time, when sea levels worldwide were high, a long, arc-shaped coral reef was built up on the marine platform. Data from drilling show that when it was living and growing, the reef had a total length of at least 220 miles and extended from Virginia Bay (part of Greater Miami) to Dry Tortugas National Park in the Gulf of Mexico (Chapter 21). However, the Key Largo Limestone is exposed at the surface only from Soldier

Key (northernmost key in the park) to Big Pine Key (about 35 miles east of Key West), a distance of 110 miles (Hoffmeister 1974). The distinction between the upper keys and lower keys is based on the fact that in the lower keys, from Big Pine Key to Key West, the Key Largo Limestone is overlain by the oolitic facies of the Miami Limestone (Chapter 19). The differences between these two types of limestone account for most of the erosional and topographic differences between the upper and lower keys.

Apparently the upper keys were constructed as a series of patch reefs in a back reef environment similar to the one that exists today. The principal framebuilders of both the keys and the living patch reefs are the star corals *(Montastrea annularis),* which usually grow in a moderately protected marine environment with limited wave energy. Species such as the elkhorn coral *(Acropora palmata),* which thrive in the surf of the outer reef, have not been found either as fossils in the Key Largo Limestone or as living corals in the back reef area. What seems likely is that an outer fringing reef was growing during Pleistocene time at the same location as that of the present outer reef.

2. **Erosion during the last major Pleistocene glaciation.**

During Wisconsinan time, as the last advance of the great continental glaciers occurred, sea level dropped more than 300 feet, exposing nearly all of the continental shelf off southern Florida. At the seaward edge of the marine platform, wave erosion lowered the

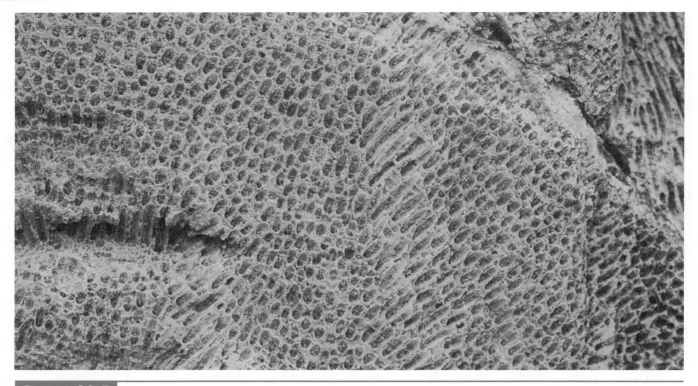

FIGURE 20.7   A fossil star coral (*Montastraea annularis*). The coral grows in the inshore shoals and is small—1.5 inches in diameter. Star coral forms almost round colonies. National Park Service photo.

FIGURE 20.8   Diver examining brain coral (*Diploria strigosa*) in Biscayne National Park. Biscayne has a problem with snorkelers standing on the brain coral and crushing the polyps, thus killing off the reef. National Park Service photo.

FIGURE 20.9 A finger coral (*Porites divaricata*) colony. These are found on the inshore shoals. National Park Service photo.

FIGURE 20.10 Boaters cause extensive damage to the sea grass meadows by running aground in them. The diver offers a sense of scale of the damage. National Park Service photo by Mark Nicholas.

| Box 20.2 | **Hurricanes** | By Richard A. Davis |

Day after day, year after year, the waves beat the shore in a ceaseless rhythm of erosion and deposition. Slowly, but surely, rock is reduced to sand, and sand is moved where the water takes it.

In the pre-dawn hours of August 24, 1992, just about at high tide, Hurricane Andrew came ashore—directly over Biscayne National Park. Sustained winds were about 150 mph, with gusts measured in excess of 175 mph.

It turns out that it is the occasional big storms that do the real work along the seashore—wreaking more changes in a few, or a few dozen, hours than ordinary wave action does in perhaps ten thousand of its days.

A storm is designated a hurricane once its winds reach 74 mph. Hurricanes are rated on what is called the Saffir/Simpson Scale on the basis of the velocity of the winds that comprise them (table 20.2). The winds in a hurricane may top 200 mph (for example, the Labor Day storm of 1935). To the owner of a beach-front cottage, that is indeed an attack by Huracan, Mayan god of the tempest.

Just how much work or damage a storm does is related to a number of factors: (1) the velocity of the wind (which is, in turn, related to barometric pressure); (2) storm surge (a function of wind velocity; (3) the state of the tide; (4) the direction of movement of the whole storm; and (5) how long the storm stays in the area.

It is not, in fact, the winds that do the most geologic work; it is the water driven by those winds. The storm surge produced by Hurricane Andrew, added to the tide, resulted in water 16.9 feet above normal. (The average elevation in the Florida Keys is three feet; the maximum elevation in the whole chain is only about 18 feet, with the highest point in Key West only some 17 feet above sea level!) Such masses of water can result in intense bottom scouring, major deposition of sediment, and extensive damage to corals, sponges, and other marine life.

Table 20.2	**Saffir/Simpson Scale with Ranges of Pressure, Wind Velocity, and Storm Surge**			
Saffir/ Simpson number	Central pressure		Wind velocity (mph)	Storm surge (ft.)
	Millibars	Inches		
1	980	28.9	74–95	4–5
2	965–979	28.5–28.9	96–110	6–8
3	945–964	27.9–28.5	111–130	9–12
4	920–944	27.2–27.9	131–155	13–18
5	< 920	< 27.2	> 155	> 18s

Source: Data from Pilkey et al., 1984, Table 2.1.

surface of the old outer reef by more than 50 feet. Erosion, first by waves and then by streams that began to drain off newly exposed land, also stripped off the tops of the patch reefs and the upper parts of the present keys. Channels between the keys, such as Caesar Creek, were widened and deepened. Hawk Channel, which parallels the upper keys, was scoured, and ridges or banks of sedimentary debris were deposited along its sides. (In the back reef area of the present time, tidal currents continue to scour the channels and deposit sediment in shoals and banks.)

Rain and ground water flushed salt from the reef rock of the keys and dissolved some of the calcite. Small pits, sinkholes, and other solution features developed in the rock. Gradually a thin soil developed and the upper keys became thickly wooded.

Although the damage caused by Hurricane Andrew on land was devastating, ironically, the underwater changes were relatively minor. This is because it moved through the area so rapidly. In November of 1994, tropical storm Gordon (with winds of only just over 50 mph) generated four days of 12-foot waves that did much more damage to the staghorn corals, for example, than did Andrew.

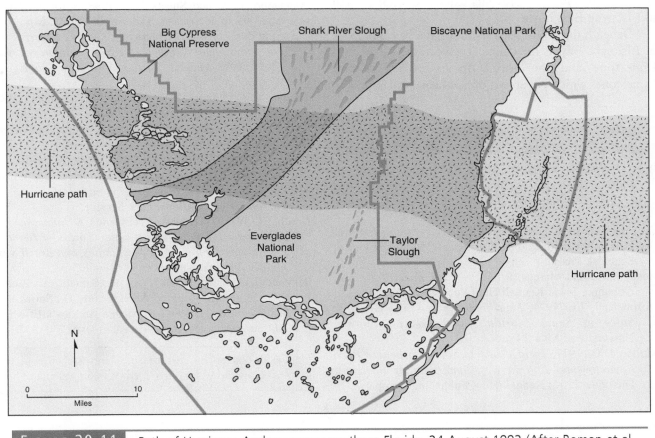

FIGURE 20.11   Path of Hurricane Andrew across southern Florida, 24 August 1992 (After Roman et al. 1994).

3. The development of Biscayne Bay.

Biscayne Bay, which lies between the northern part of the upper keys and the mainland, is only 6 feet deep near Miami Beach but deepens toward the south. The shallow bottom next to the mainland slopes off toward the keys to a 12- to 15-foot depth. When the keys were being built up by coral reefs during the high sea levels in the last Pleistocene interglacial episode, Biscayne Bay was a large coastal lagoon between the Atlantic Coastal Ridge rimming the Everglades (Chapter 19) and the reef complex offshore. When sea level was lowered due to Wisconsinan glaciation, the lagoon drained off, exposing the limestone bedrock of the ocean floor. Streams cut valleys in the sides and floor of the troughlike depression. Solution pits developed in the limestone. Freshwater calcitic muds that were deposited below present sea level indicate that for a time a freshwater lake occupied part of the Biscayne basin.

As sea level began to rise again, brackish water entered the lagoon and mangrove swamps developed. Eventually seawater filled the lagoon to its present depth, creating Biscayne Bay. Meanwhile, in the shallows next to the keys, small inshore corals began to grow again. The presence of sediment, constantly stirred up by currents, prevents extensive coral growth in Biscayne Bay, however.

Beneath the bottom sediment, on the bedrock floor of Biscayne Bay, is the contact between the Key Largo Limestone and the oolitic facies of the Miami Limestone, which covers all of southeastern Florida.

Sources

Bond, P.A. 1986. Carbonate Rock Environments of South Florida. *Geological Society of America Centennial Field Guide—Southeastern Section.* Boulder, Colorado: Geological Society of America. p. 345-349.

Greenberg, I. 1986. *Guide to Corals and Fishes of Florida, the Bahamas and the Caribbean.* Miami, Florida: Seahawk Press.

Head, C.M.; and Marcus, R.B. 1987. *The Face of Florida.* Dubuque, Iowa: Kendall/Hunt Publishing Company.

Hoffmeister, J.E. 1974. *Land from the Sea. The Geologic Story of South Florida.* Coral Gables, Florida: University of Miami Press.

Multer, H.G. 1977. *Field Guide to Some Carbonate Rock Environments—Florida Keys and Western Bahamas.* Dubuque, Iowa: Kendall/Hunt Publishing Company.

Pilkey, O.H., Jr., Sharma, D.C., Wanless, H.R., Doyle, L.J.; Pilkey, O.H., Sr.; Neal, W.J., and Gruver, B.L. 1984. *Living with the East Florida Shore.* Durham, North Carolina: Duke University Press.

Pimm, S.L., Davis, G.E., Loope, L., Roman, C.T., Smith, T.J., III; and Tilmant, J.T. 1994. Hurricane Andrew. *BioScience* 44(4):p. 224–229.

Purdue University Department of Earth and Atmospheric Sciences. 19 January 1996. http://thunder.atms.purdue.edu/hurricane/1994.

Rappaport, E.N., and Sheets, R.C. 1993. A Meteorological Analysis of Hurricane Andrew. In Tait, Lawrence S., compiler. *Lessons of Hurricane Andrew. Excerpts from The 15th Annual National Hurricane Conference.* April 13–16, 1993. Federal Emergency Management Agency, Washington, D.C. p. 1–18.

Roman, C.T., Aumen, N.G., Trexler, J.C., Fennema, R.J., Loftus, W.F., and Soukup, M.A. 1994. Hurricane Andrew's Impact on Freshwater Resources. *Bioscience* 44(4):p. 247–255.

Roy, C.J., et al. 1994, *Hurricane Andrew, 1992. The National Park Service Response in South Florida. Big Cypress National Preserve. Biscayne National Park. Everglades National Park.* National Park Service.

Smith, F.G.W. 1971. Atlantic Reef Corals. *A Handbook of the Common Reef and Shallow-water Corals of Bermuda, the Bahamas, Florida, the West Indies, and Brazil.* Coral Gables, Florida: University of Miami Press.

Tilmant, J.T., Curry, R.W., Jones, R., Szmant, A., Zieman, J.C., Flora, M., Robblee, M.B., Smith, D., Snow, R.W., and Wanless, H. 1994. Hurricane Andrew's Effects on Marine Resources. *BioScience* 44 (4): p. 230–237.

Note: To convert English measurements to metric, go to www.helpwithdiy.com/metric_conversion_calculator.html

Dry Tortugas National Park

SOUTHERNMOST FLORIDA

Richard Arnold Davis   College of Mount Saint Joseph, Cincinnati, Ohio

Area: 64,700 acres; 101 square miles; land area, 39.28 acres; 0.08 square miles

Coordinates: 24°34'–24°44' N. Lat., 82°46'–82°58' W. Long.

Proclaimed a National Monument: January 4, 1935

Established as a National Park: October 26, 1992

Mailing Address: P.O. Box 6208, Key West, FL 33041-6208.

Phone: 305-242-7700

E-mail: DRTO_information@nps.gov

Website: www.nps.gov/drto/index.htm

This handful of small islands and patch of ocean lying at the end of the Florida Keys include the least disturbed coral-reef system in the continental United States and the largest coastal fort ever built in the country.

FIGURE 21.1 Fort Jefferson, a part of Dry Tortugas National Park, as it is seen from the air. Note the dark shadows in the water; they are the coral reefs that surround the island. National Park Service photo.

Geographic Setting

Nestled in the azure and turquoise ocean some 75 miles west of Key West are the specks that comprise the land portion of Dry Tortugas National Park. The total area of the park is some 101 square miles, but dry land amounts to less than one-thousandth of that. The number of islands included depends on the state of the tide and on how low an island one is willing to count; generally seven are enumerated: Bush Key, East Key, Garden Key, Hospital Key, Loggerhead Key, Long Key, and Middle Key, although names have varied over the years. When the area originally was mapped, there were four additional islands: North, Sandy, and Southwest Keys were no longer above sea level by the early 1900s, and Bird Key washed away in the 1930s, following denudation by hurricanes in 1910 and 1919. In the last decade, however, sediment has accumulated at Bird Key Reef to the extent that its surface now is above sea level.

The Dry Tortugas constitute the western end of that arc of islands known as the Florida Keys. The Keys as a whole commonly are divided into the Upper Keys, at the north end (near Miami), the Middle Keys, and the Lower Keys, which include Key West. These three differ with respect to environmental conditions; for example, the trend of the Upper Keys is nearly north-south, facing almost directly into the prevailing east-southeast winds, whereas the Lower Keys lie in a line that nearly parallels those winds. (Recall that wind direction is stated with respect to the direction from which the wind comes, rather than the direction toward which the wind blows.)

Although this volume is devoted to the geology of the national parks, it is only fair to point out that the Dry Tortugas and environs are protected primarily for biological and historical reasons, rather than for geologic ones. The area was designated a wildlife refuge in 1908 to protect the nesting area of sooty terns (*Sterna fuscata*) and noddy terns (*Anous stolidus*) on Bush Key from human molestation, especially by egg collectors; in fact, to this day, Bush Key is off limits to visitors during the March-to-September nesting season, when upward of 100,000 birds gather there.

Due to its remoteness and hence, relatively unspoilt nature, the park area has been an important center for the study of coral reefs, their inhabitants, and other marine life. Beginning in 1904 and continuing for some four decades, the Carnegie Institution of Washington maintained a marine laboratory on Loggerhead Key where some of the most significant scientific work on these subjects was undertaken. (For more information on biologic work in the Dry Tortugas, see Tilmant's recent survey in Murphy, 1993, and the references cited therein.)

Coral reefs, such as those of the Florida Keys, are noted for their incredible biological diversity. For example, 442 species of fish have been identified in Dry Tortugas National Park; and, in a 15-minute period, an observer might see 40 to 70 species of fish within about 15 feet. It is no wonder that the Florida Keys Marine Sanctuary was established in 1990.

Logistics

Dry Tortugas National Park can be reached only by boat or seaplane. Commercial carriers operate out of Key West and other Florida locations; information on those approved by the National Park Service can be obtained by contacting park headquarters in Everglades National Park.

The park can be visited by private boat, too, but, because of the remoteness of the site, facilities are extremely limited. Visitors must bring everything they need, including drinking water and food, and must remove everything they bring, including their own waste.

This all sounds decidedly forbidding. However, the park is well worth the cost and planning necessary for a visit. In fact, over 100 individuals visit the park on an average day.

History

The Dry Tortugas were "discovered" in 1513 by Ponce de Leon, who called the islands "Las Tortugas," presumably from the sea turtles that abound in the adjacent waters and nest on the islands. The lack of fresh water soon caused the adjective "Dry" to be appended to the name.

During the first third of the nineteenth century, American military planners realized that by fortifying the Dry Tortugas they could control and protect sea traffic from the Gulf of Mexico through the 90-mile-wide Florida Straits that separate the islands from Cuba and provide the principal sea lane linking the Gulf of Mexico with the Atlantic. Moreover, the islands ring a major deep-water harbor that could have accommodated the entire U.S. fleet at the time. Hence, in 1846, construction of a fort that would be named after the country's third president was begun on Garden Key. Fort

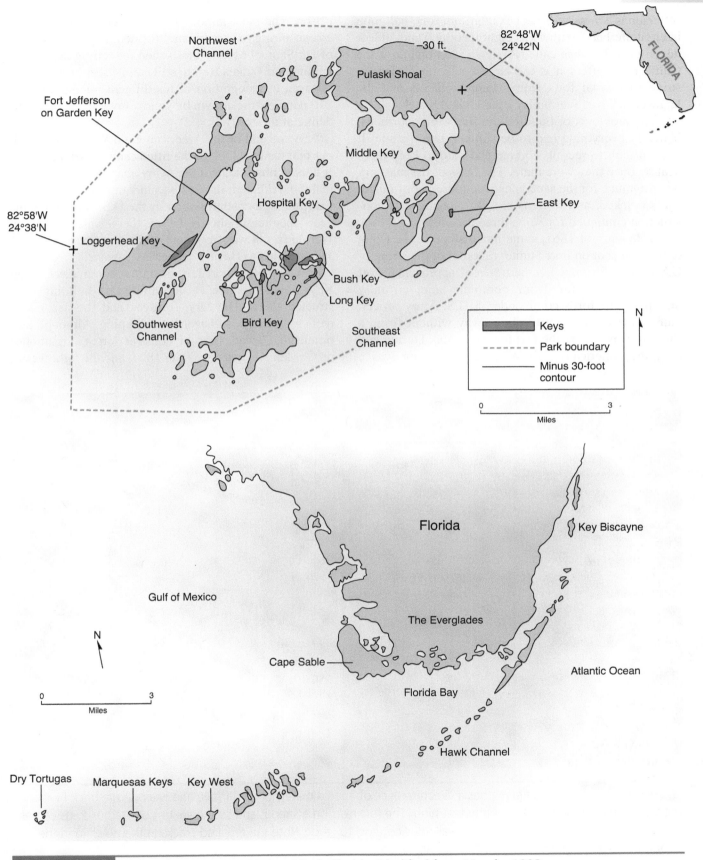

FIGURE 21.2 Map of Dry Tortugas National Park, Florida. Modified from Murphy 1993.

Jefferson is a massive brick hexagon some one-half mile in circumference, with walls eight feet thick and 50 feet tall (fig. 21.3). Once called America's "Gibraltar of the Gulf," Fort Jefferson is the largest of a once planned string of coastal forts from Maine to Texas and the largest brick fort ever built by the United States.

The problems of building the fort were enormous. Virtually everything had to be shipped into the site. It was difficult to recruit workers and hard to keep them healthy once they were there. The mass of the masonry was too much for the sand island, so foundations failed, walls cracked, and cisterns leaked. Nonetheless, construction continued for some three decades.

In January of 1861, some months before the firing of the first shot on Fort Sumter (another of the string of U.S coastal forts), a Southern vessel approached Fort Jefferson, then far from complete and virtually unarmed. The fort's commander, a U.S. Army officer, bluffed the intruder into sailing away without firing a shot, thereby saving the fort for the Union. During and after the American Civil War, Fort Jefferson was a prison. The most famous inmates were men convicted of conspiring to assassinate President Lincoln. The cell of Dr. Samuel Mudd, imprisoned for setting the broken leg of John Wilkes Booth, still can be seen. (Mudd was released in recognition of his diligent ministrations to 270 people struck down by yellow fever in an 1867 epidemic at the fort.)

By the end of the American Civil War, it was obvious that the invention of the rifled cannon had made Fort Jefferson obsolete, and the Army abandoned the project in 1874. Although the fort was incomplete, Garden Key was used as a coaling station by the U.S. Navy for some years, beginning shortly before the Spanish-American War; in fact, it was from the Dry Tortugas that the *U.S.S. Maine* sailed to Havana on her last voyage.

Even though Fort Jefferson now dominates Garden Key, it was not the earliest important building constructed there. The Dry Tortugas and the associated reefs long had been hazards to shipping. Although recommended decades earlier, construction of a lighthouse on Garden Key was begun in 1825, and the light was lit

FIGURE 21.3   Fort Jefferson has a circumference of approximately one-half mile and is a massive brick hexagon. The walls are eight feet thick and fifty feet high. The fort rests on a sand island so the weight of the buildings causes the sand to settle and foundations and wall (plus cisterns) to crack. Note the repaired cracks in the building in the center of the photograph. National Park Service photo.

for the first time on July 4, 1826 (ironically, this was the very day that Thomas Jefferson died). A second lighthouse in the Dry Tortugas, on Loggerhead Key, was first lit in 1858 and still is in use today. The lighthouse that now projects above Bastion C of Fort Jefferson replaced the earlier facility on Garden Key in 1876, but no longer functions as an aid to navigation.*

Geologic and Biologic Features

Note: Deep drilling would have to be done in the coral reefs of Dry Tortugas to determine the geologic column. Because of this, one has not been completed.

The islands called the Dry Tortugas and the associated banks and reefs together comprise an elliptical complex some 17 miles long by seven miles wide, with the major axis oriented northeast-southwest. There are three major bioherm systems: Pulaski Shoal to the northeast, Loggerhead Shoal to the west, and Southeast Reef (Long Key/Bird Key Shoal) to the southeast. In map view, the complex is reminiscent of an atoll. Unlike the Pacific atolls, however, there is no volcanic topographic high on the sea floor, upon which the bioherms grew. The term *bioherm* refers to a feature that is elevated above the sea floor and was produced by organisms; for example, by corals. A *reef,* on the other hand, is any feature elevated above the sea floor, the top of which comes close enough to the surface of the water so as to constitute a hazard to shipping; a reef may be of organic or inorganic origin. Thus, a coral reef is a bioherm, but not all bioherms, strictly speaking, are reefs. Having said that, it is only fair to point out that the terms reef and bioherm are used almost interchangeably in ordinary speech and writing. As discussed in earlier chapters, bioherms can be important sources of sediment; moreover, they, themselves, can be preserved in the rock record.

---

*As amply demonstrated in the fine volume edited by Larry E. Murphy [1993], the Dry Tortugas have a tremendous potential for historical and archaeological research. Unfortunately, it must be added that Fort Jefferson is in need of extensive conservation and restoration work—as evidenced by numerous signs that warn the visitor to be careful of crumbling masonry. One can only hope that sufficient labor and money will be made available for both the preservational and scholarly work due this important piece of our nation's history.

Three main channels through the discontinuous fringing reefs of the Dry Tortugas allow good circulation between the sea and the lagoon. Currents commonly measure 7.87 to 23.62 inches (0.3–1.2 knots), although tidal currents may be twice that.

The generally recognized lithologic units in south Florida are the Key Largo Limestone and the Miami Limestone (sometimes called the Key Largo Formation and the Miami Formation, respectively). The Key Largo Limestone is coralline limestone that formed during the most recent interglacial stage of the Pleistocene Epoch, some 100,000 to 125,000 years ago. Rock of the Miami Limestone, which underlies the mainland of southeastern Florida, is generally a cross-bedded, oolitic limestone and also is late Pleistocene in age. In the Lower Keys an oolitic limestone that has been called the Key West Oolite is present; it apparently is the same rock unit as the Miami Limestone, although the oolitic limestone of the mainland and the Upper Keys is at a somewhat higher elevation than is the oolitic limestone of the Lower Keys. On the basis of cores bored in the Lower Keys, it seems that the Key Largo Limestone was deposited contemporaneously with the Key West Oolite.

The Key Largo Limestone is not exposed in the Lower Keys. The oolitic-limestone unit present there extends westward from beneath Key West and lies under the Marquesas Keys and the Quicksands, which are between Key West and the Dry Tortugas. However, the present-day coral reefs of the Upper, Middle, and many of the Lower Keys are underlain by what appears to be Key Largo Limestone; for example, cores drilled in two Holocene coral reefs north and northwest of the Marquesas Keys recovered rock of the Key Largo Limestone beneath the two 26-foot-thick coral reefs. The positions of these coral reefs seem to be coincident with underlying topographic highs consisting of Key Largo Limestone.

Between the Marquesas-Keys/Quicksands area and the Dry Tortugas is a stretch of deeper water for which there is no information on the bedrock of the sea floor. Coring in the Dry Tortugas, however, has revealed Holocene coral reefs and sediments resting on coralline limestone, presumably of the Key Largo Limestone. The largest and most complete coral reef is Southeast Reef (Bird Key Reef), which consists of some 45 feet of biohermal material. Near the present reef base, in 79 feet of water, the Holocene deposits are only 26 feet thick.

The Holocene coral reefs of the Dry Tortugas and the Pleistocene coralline limestone there (which were once coral reefs, of course) have the same coral fauna (Shinn *et al.,* 1989; Tilmant, *in* Murphy, 1993). The scleractinian coral *Acropora palmata* (Elkhorn) (Lamarck) long has been considered the principal Holocene Caribbean reef-builder. Hence, it might be expected that corals of this species would be present in some abundance in both the Pleistocene and the Holocene coral reefs of the Dry Tortugas. It came as quite a surprise when none was found in the cores bored there in 1976. Moreover, only a few colonies could be found living on the present-day coral reef at that time. What's more, individuals of this species transplanted into the Dry Tortugas grew for about a year, but died when an unusually severe cold front passed through the area in 1977.

The dominant reef-building corals in the Dry Tortugas today belong to the genera *Diploria* (brain), *Montastraea* (star coral), and *Siderastraea* (round star coral), all of which belong in the order Scleractinia.[*]

Within the atoll lagoon, there were thriving thickets of the scleractinian coral *Acropora cervicornis* (staghorn) (Lamarck), but 90 to 95 percent of the individuals of this species were killed in late January of 1977. Subsequent recovery from this "freeze of 1977" had begun, but, recently, a coral disease has devastated them once again. This so-called "black-band disease" has become rampant in some areas of the Florida Keys in recent years. The cause is not clear; it may be related to the Caribbean-wide epidemic that has severely affected the algae-eating echinoderm *Diadema*. These sea urchins were known to have fed on algae that infested diseased areas of corals; thus, they may have served to control the spread of black-band disease.

Shinn et al. (1989) speculated that cold water that routinely enters from the Gulf of Mexico during the winter months may have prevented acroporid corals from becoming the major reef-builders in the Dry Tortugas. On the other hand, at the time of the visits of Louis and Alexander Agassiz in the second half of the nineteenth century, *Acropora palmata* (Elkhorn) was dominant on the crest of Long Key Reef (Bird Key Reef). As pointed out by Tilmant (in Murphy, 1993), it may be that although the Dry Tortugas reefs seem to be relatively stable, geologically speaking, the biological communities may be decidedly less so.

---

[*]For aid in identifying corals encountered in the park, see Greenberg (1986), Smith (1971), and appendix B in Shinn *et al.* (1989).

Adjacent to the coral reefs of the Dry Tortugas are several sand islands. Loggerhead Key, for example, consists of a 45-foot thickness of unconsolidated Holocene material. Fort Jefferson, on Garden Key, rests on a similar accumulation. Such islands are composed of present-day debris of coral and coralline algae, not *in situ* coral. In some areas, calcareous sand and debris have been welded into beachrock by calcium carbonate cementation. For example, the most extensive deposit of such material is located at Loggerhead Key, in south Florida. The age of these islands is unknown; nor is it known whether similar islands preceded the present ones.

The coral-reef community, *per se,* makes up less than four percent of the sea floor shallower than 60 feet; more than half is bare sand and rubble. In addition, between 15 and 20 percent of the area is underlain by a thin, partly lithified layer called a hardground; growing on the resulting hardened surface is a community of octocorals (these so-called "soft corals" are relatives of true corals; included here are "sea pens," "sea fans," etc.). A small area (less than one percent of the sea floor) is occupied by a community dominated by algae, but some 30 percent consists of a sea-grass community. For example, within the lagoon west of Southeast Reef, in water less than 6 1/2 feet deep, a sea-grass-covered flat is underlain by 40 to 50 feet of calcareous sediment. Although these are biological entities, the organisms present greatly affect the rate of production of calcium carbonate, whether sediment accumulates, and, if so, to what extent.

Although not a part of Dry Tortugas National Park, the area known as the Quicksands is worthy of mention. Lying some 25 miles east of the park is a deposit of current-swept sand as much as 40 feet thick that covers an area about 17 by 2 1/2 miles in size. If one approaches the park from the east by airplane, one can see spectacular subaqueous sand waves up to 16 feet high. Coral reefs cannot grow here, because of the shifting sand. On the other hand, the lack of coral allows extensive production of carbonate sand by algae, chiefly members of the genus *Halimeda*.

Geologic History

1. Formation of the Atlantic Ocean.

The basin of the Atlantic Ocean, as we know it, began forming in the early part of the Mesozoic Era. It is likely that faulting associated with that rifting result-ed in the submarine topography that became the foun-dation for the subsequent deposition of which we have direct evidence.

2. Cenozoic carbonate deposition.

Beginning in Mesozoic time and extending to today, what is now southern Florida, including the area of the present-day Florida Keys, has been predominantly an area of limestone deposition. A well southwest of the Marquesas Keys was bored some 15,000 feet deep; the lowest material encountered was limestone and dolomite of Cretaceous age. Similarly, a well drilled on Key Largo bottomed out at a depth of 12,057 feet in the Lower Cretaceous time.

3. Tilting to the west.

Evidence for gentle westward tilting, beginning in the Pliocene Epoch, includes: (a) the higher topography in the eastern part of Florida, (b) the lower elevation of the Key West Oolite, on the west, *versus* that of the contemporaneous Miami Limestone, on the east, (c) a known, once-horizontal linear Pleistocene beach deposit that slants downward to the west; and (d) the thickening of the Pleistocene section to the west. Tilting, although slight, may be continuing to this day.

4. Pleistocene glaciations and sea-level fluctuations.

Every one of the several major glacial advances of the Pleistocene Epoch involved a drop in world-wide temperature. In addition, each was accompanied by a lowering of sea level, as water removed from the sea by evaporation fell elsewhere as snow and eventually became incorporated in continental glaciers. Every major glacial retreat involved rises in both temperature and sea level. The subsurface Pleistocene rock record of Florida shows evidence of five major sea-level falls, each characterized by features diagnostic of a time interval of subaerial erosion.

During the most recent interglacial stage, some 125,000 years ago, sea level was at least 20 feet higher than it is today. During that time, the coral reefs now preserved as the Key Largo Limestone grew in the shal-low, warm sea. Simultaneously, in other portions of the sea floor, where conditions favored the production of oolites, the sediments that were to become the Miami Limestone/Key West Oolite were deposited.

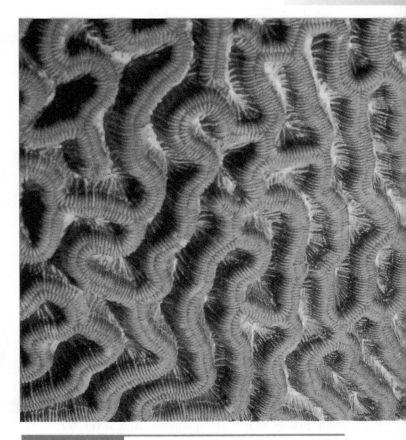

FIGURE 21.4   The coral reefs of Dry Tortugas National Park are the least disturbed of those in the continental United States. The brain coral (*Diploria stringosa*) is one of the main reef-building corals in this park. National Park Service photo.

5. Glaciation, and a drop in sea level.

At the height of the Wisconsinan glaciation, some 18,000 to 20,000 years ago, world-wide sea level was some 325 feet or more lower than it is today. As the water receded toward that low, and as temperatures dropped, the coral reefs that had grown up on the con-tinental shelf died and were subject to subaerial ero-sion, along with the rest of the now-emergent portions of the sea floor. What had been islands and coral reefs projecting up from the sea floor became topographic highs on the surface of the land.

6. Submergence, again.

As the continental glaciers of that portion of the Pleistocene Epoch called the Wisconsinan Glacial

Stage melted, the water that had been tied up in ice was released to return to the sea. (The time since the retreat of the Wisconsinan glacier is called by some the Holocene Epoch, or the Recent Epoch; however, we just may be in another interglacial stage within the Pleistocene. In either case, the chain of events is the same.)

Sea level began to rise rapidly 16,000 to 15,000 years ago. As it rose toward its present height, gradually what had been the pre-Wisconsinan continental shelf again was flooded; this began 7,000 to 6,000 years ago. Topographic highs became islands, and lower areas became Hawke Channel, Florida Bay, Biscayne Bay, and various channels between islands. The area now occupied by the deep waters east of the Dry Tortugas would have flooded relatively early, thereby isolating them from the mainland early in the marine transgression. Eventually temperatures became high enough to be conducive to the growth of hermatypic corals (*i.e.*, those that form bioherms). Many of the present-day coral reefs seem to be located on what had been Pleistocene topographic highs; ironically, these commonly were the remnants of Pleistocene coral reefs. Hence, many of today's coral reefs are mimicking older bioherms in shape and location. Sea level seems to have risen at a slow enough rate that the corals were able to grow upward rapidly enough to keep pace. In Bird Key Reef, where biohermal material accumulated to a thickness of 45 feet, the oldest such rock has been dated at 6019 years before present, with a possible error of 90 years either way (Tilmant, in Murphy, 1993).

Some present-day islands in the Florida Keys are old topographic highs that still protrude above sea level; these include ancient coral reefs. Others, of course, are the result of marine processes that operate today even as they always have operated in coastal areas of our planet's oceans.

7. The future?

If sea level were to continue to rise much more, all of the Florida Keys, including the islands of Dry Tortugas National Park, would become submerged (as would large portions of the mainland). This would have a significant effect on current and weather patterns. If the sea level rise weren't too rapid, the growth of the bioherms might be able to keep pace, and coral reefs could survive in the area, depending on what happened to the temperature, clarity, and chemical composition of the water—and here humans may be the deciding factor.

Geologic Materials Used in Fort Jefferson

The vast majority of the bulk of Fort Jefferson is composed of bricks, some 16,000,000 of them. Because the building of the fort stretched over three decades, including the American Civil War, and because of the vast quantities required, not all of the bricks came from the same place. They can be grouped into two general categories: "Southern bricks" (also called "Pensacola bricks") and "Northern bricks." The name Pensacola bricks notwithstanding, Southern bricks were obtained from brickyards not only in Pensacola, but also in Mobile, New Orleans, Charleston, Savannah, and Jacksonville. According to Anderson (1988), the main component of the fort are bricks produced by two different manufacturers in Baldwin County, Alabama, using Escambia clay. Although both Northern and Southern bricks were used in the fort as early as the first decade of its construction, as the American Civil War approached, Southern suppliers hesitated to provide bricks for the construction of a Federal fort, so Northern bricks came to dominate. These were brought in from Danvers, Massachusetts, and from Brewer, Maine; they are darker in color and have proved to be somewhat more resistant to weathering than have the Pensacola bricks. The two kinds of bricks also can be differentiated on the basis of size: the Pensacola bricks are 3 1/8 × 4 3/4 × 9 1/2", whereas the Northern bricks are 2 1/4 × 3 1/2 × 7 7/8" (Anderson, 1988) (fig. 21.5).

In places where larger sizes and greater strength were necessary, stone was used. Granite was utilized for steps, for lintel stones, and for traverse circles in gun emplacements; it was obtained from New York and Vermont (Murphy, 1993). What Anderson (1988) called "bluestone slate" was used for floors of casemates, for manhole covers, and elsewhere. ("Bluestone" is a commercial name for a variety of fine-grained sandstone that splits easily into slabs; rock of this type is quarried extensively along the Hudson River in New York State.)

A considerable amount of cement was used in the fort. It was shipped from New York and, presumably, was made there (Murphy, 1993). Coral rubble was obtained from outlying reefs and seems to have been mixed in with the cement in many places.

FIGURE 21.5   The building of the fort extended over a thirty-year period; hence not all of the bricks came from the same place. There are two general categories: "Southern Bricks" or "Pensacola bricks" and "Northern Bricks." The Northern bricks are darker in color, smaller, and more resistant to weathering. Granite was used for lintel stones and traverse circles in gun emplacements. National Park Service photo.

Geologic Map and Cross Section

Dry Tortugas, Florida, 1 x 2 degree quadrangle. U.S. Geological Survey and National Ocean Service. 1:250,000-Scale Metric Topographic-bathymetric Map.

Florida Keys. U.S. Geological Survey. International Map of the World Series. Sheet NG-17. 1:1,000,000.

U.S. National Oceanic and Atmospheric Administration, National Ocean Service. Nautical Charts 11434 and 11438.

Sources

Anderson, L. 1988. *Fort Jefferson National Monument (Florida). Historic Structure Report.* National Park Service, Architectural Data Section. xviii + 199 p.

Bond, P. A. 1986. Carbonate Rock Environments of South Florida. *Geological Society of America Centennial Field Guide—Southeastern Section.* p. 345–349.

Greenberg, I. 1986. *Guide to Corals and Fishes of Florida, the Bahamas and the Caribbean.* Seahawk Press, Miami, Florida.

Hoffmeister, J. E. 1974. *Land from the Sea. The Geologic Story of South Florida.* University of Miami Press, Coral Gables, Florida.

Hurley, N. E. 1994. *Lighthouses of the Dry Tortugas. An Illustrated History.* Historic Lighthouse Publishers, Aiea, Hawaii.

Logan, W. B., and Muse, V. 1989. *The Smithsonian Guide to Historic America. The Deep South.* Stewart, Tabori & Chang, New York.

Melham, T. 1994. Dry Tortugas National Park. In *Our Inviting Eastern Parklands. From Acadia to the*

Everglades. National Geographic Society, Washington, D.C., p. 150–157.

Multer, H. Gr. 1977. *Field Guide to Some Carbonate Rock Environments—Florida Keys and Western Bahamas.* Kendall/Hunt, Dubuque, Iowa.

Murphy, L. E., editor, 1993. *Dry Tortugas National Park. Submerged Cultural Resources Assessment.* National Park Service, Southwest Region, Submerged Cultural Resources Unit, Southwest Cultural Resources Center Professional Paper 45. xxviii + 434 p.

Shinn, E. A., Lidz, B. H., Kindinger, J. L., Hudson, J. H., and Halley, R. B. 1989. *Reefs of Florida and the Dry Tortugas. A Guide to the Modern Carbonate Environments of the Florida Keys and the Dry Tortugas.* U.S. Geological Survey, St. Petersburg, Florida. iii + 53 p.

Smith, F. G. W. 1971. *Atlantic Reef Corals. A Handbook of the Common Reef and Shallow-Water Corals of Bermuda, the Bahamas, Florida, the West Indies, and Brazil.* University of Miami Press, Coral Gables, Florida.

Stone, P. A. 1993. Dry Tortugas and South Florida Geological Development and Environmental Succession in the Human Era. In Murphy, L. E., editor. *Dry Tortugas National Park. Submerged Cultural Resources Assessment.* National Park Service, Southwest Region, Submerged Cultural Resources Unit, Southwest Cultural Resources Center Professional Paper 45. p. 5–26.

Tilmant, J. T. 1993. Relationship of Dry Tortugas Natural Resources to Submerged Archaeological Sites. In Murphy, L. E., editor. *Dry Tortugas National Park. Submerged Cultural Resources Assessment.* National Park Service, Southwest Region, Submerged Cultural Resources Unit, Southwest Cultural Resources Center Professional Paper 45. p. 51–62.

Voss, G. L. 1988. *Coral Reefs of Florida.* Pineapple Press, Sarasota, Florida.

Acknowledgments

The following people graciously provided advice, leads, references, publications, illustrations, contacts, and, most of all, their time; I am grateful to them all: Dr. Robert Brock and R. Cook (Dry Tortugas National Park headquarters at Everglades National Park), Dr. Robert B. Halley (of the U.S. Geological Survey's Center for Coastal Geology and Regional Marine Studies in St. Petersburg, Florida), and Dr. Larry E. Murphy (of the National Park Service's Submerged Cultural Resources Unit in Sante Fe, New Mexico).

This chapter is a compilation of material derived from the works of others, especially those listed under Sources. Because this is not a technical book, very few specific citations for the conclusions and items of information are included in the chapter. Even though sources are not specifically cited in the text, this in no way lessens the respect I have for these individuals.

Note: To convert English measurements to metric, go to www.helpwithdiy.com/metric_conversion_calculator.html

Landscapes Shaped by Continental or Alpine Glaciation

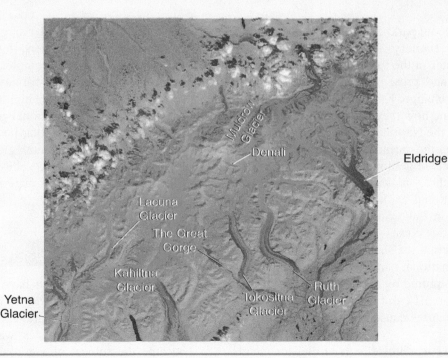

FIGURE PIII.1 The Western two-thirds of Denali National Park as photographed by Landsat5's Thermographic Mapper on June 15, 1986. This is a false-color image based on various wavelengths, such as infrared, near-infrared and several green wavelengths. The snow and ice are light blue, water is dark blue, green represents healthy vegetation, and bare land is pink. If you look carefully you can see both peaks of Denali. Photograph by NASA.

FIGURE PIII.2 Yosemite Valley and Yosemite Falls. The view into the valley caused John Muir to speculate that in the past glaciers had covered the uplands, widened and oversteepened the valley walls, and scoured the spurs along the lower part of the cliffs. Photo by Ronald R. Hacker.

Glaciers in national parks are more than fascinating elements of the scenery. Besides storing significant amounts of water in the form of snow and ice, glaciers are also past and present indicators of climatic and environmental change. Knowing something about how glaciers form and how they behave are concerns of vital importance. In the National Park System we can recognize features on the surface of the land caused by glacial action in recent geologic time.

The first three national parks in this section—Voyageurs National Park in Minnesota, Isle Royale National Park in Lake Superior, and Acadia National Park on the Maine coast—are in regions that were covered by continental ice sheets in Pleistocene time. The remaining ten national parks in Part III, all in mountain areas, were sculptured by alpine, or valley, glaciers. All the parks in Part III exhibit a variety of erosional and depositional features caused by the mighty glaciers of the Ice Age.

So pervasive were the effects of Pleistocene glaciation on the North American continent that Ice Age indicators of various sorts are discernible in all the national parks of the system. Even the offshore parks, such as Virgin Islands National Park and both national parks in Hawaii, were affected by the cooler temperatures,

increased moisture, and rising and falling sea levels characteristic of Pleistocene regimens. Several national parks in volcanic and complex mountain areas (Parts IV and V of this book) still have active valley glaciers or small remnant glaciers high on northern slopes or in sheltered hollows. A national park in Part I, Theodore Roosevelt in North Dakota, lies on the edge of an area that underwent continental glaciation by Pleistocene ice sheets; and in Kobuk Valley National Park, Alaska, Pleistocene glaciofluvial materials are the main source of the dune sands.

The Theory of Glacial Ages

For many years geologists have found studies of glacier characteristics and behavior to be both challenging and rewarding. Conclusions drawn from such studies before the theory of glacial ages was universally accepted became the subject of vigorous debates. When in the 1870s the renowned John Muir proposed, after months of climbing around Yosemite, that glacial erosion had sculptured that famous landscape, he was challenging the belief of the California state geologist that Yosemite Valley (Chapter 28) was the result of a violent cataclysm.

It was a Swiss geologist, Louis Agassiz, in the middle years of the nineteenth century, who advocated the theory that great ice sheets had moved down from the Arctic and spread over northern and central Europe in prehistoric times. He based the hypothesis on his observations of alpine glaciers and processes of glacial erosion and deposition in his native Switzerland. European naturalists had tried for years to explain such oddities as granite boulders completely unrelated to local bedrock that were scattered on the land surface. (*Erratics* is the geologic term for them.) Also strewn over the surface were deposits of sand, gravel, and cobbles—sorted or unsorted, thick and thin—in landforms of many sizes and shapes. Some thought these materials (called "drift") were left behind as Noachian floods receded. Or, could the erratics have been dropped by melting icebergs tossed up on the lowlands in storms? Another puzzling question was the origin of long scratches and grooves on exposed ledges. The scratches were nearly parallel and, more often than not, trended in a north-south direction. The explanation of the Noachian flood proponents that the scratches were made by flood-swept boulders was not convincing.

Having observed quantities of gravelly material left in front of retreating glaciers in the Alps, Agassiz concluded that the drift deposits and erratics of north-central Europe could have been pushed and carried there by a huge glacier or ice sheet. And since he had seen scratches and grooves (called *striations*) on bedrock in glacial valleys—scratches oriented in the direction from which the ice had come—he theorized that the ice sheet that had advanced over Europe had come from the polar regions.

Despite some initial opposition, the beautiful logic of Agassiz's ideas gained acceptance. The theory was soon extended to the British Isles and to North America, when Agassiz became a professor of geology at Harvard. Geologists came to recognize glacial ice as a powerful and dynamic agent of erosion and deposition and to understand how glaciers had dramatically created new landscapes over large areas.

Types of Glaciers

During Pleistocene time, *continental glaciers* repeatedly advanced and retreated over vast areas of northern Europe and northern North America. These great ice sheets were miles thick and flowed from a central high point downward and outward.

A *valley,* or *alpine, glacier* is confined to a valley (almost always a former stream valley) and flows from higher to lower elevations. Waterton-Glacier International Peace Park (Chapter 26), in Alberta, Canada, and Montana, is noted for its many valley glaciers on both sides of the border.

Examples of cirque glaciers can be seen in Rocky Mountain National Park in Colorado (Chapter 25). A *cirque glacier* occupies a hollow or basin (called a *cirque*) at the head of a mountain valley and is usually the remnant of a much larger glacier.

When several descending valley glaciers join together at lower elevations and spread out in a wide, bulbous ice mass at the foot of the range, this is called a *piedmont glacier.* The largest piedmont glacier in North America is the Malaspina in Wrangell-St. Elias National Park and Preserve, Alaska (Chapter 32).

How Glaciers Form and Move

Glaciers form in regions where more snow falls than melts each year. As more and more snow accumulates, a perennial snowfield builds up, while at the same time compacting and settling under its own weight. Individual snowflakes, formed from water vapor by direct crystallization, lose their characteristic six-pointed shape and become granules due to being compressed closer together. At the stage of transition from snowflakes to glacial ice, the granular snow is called *firen.* Continued compaction expels air and forces the granules more tightly together. Individual grains melt slightly and then recrystallize, filling voids and transforming the whole mass into ice. As soon as the ice mass becomes large enough and heavy enough so that gravity causes it to move or flow downward or outward, it is considered, by definition, a *glacier.*

The perennial snowfield that nourishes a glacier (due to a net gain of snow) is called the *zone of accumulation.* Downglacier, where melting and loss of snow and ice exceeds inflow (i.e., a net loss), a *zone of wastage* begins. Separating the two zones is an irregular, shifting band called the *firn line* that marks the highest level of retreat of winter snow cover during summer.

A glacier's motion is not uniform throughout its mass as it moves downslope. The top ice is brittle and tends to fracture and crack, forming vertical fissures, or crevasses, that may be a hundred or more feet in depth. This outer part of a glacier is the *rigid zone.* Underneath the rigid zone is the *plastic flow zone,* where ice deforms without fracturing (plastic behavior), due to pressure from overlying ice. This causes the ice nearer the "sole" of the glacier to move downslope a little faster than the brittle ice at the top (fig. III.3).

The Work of Glaciers

Glaciers are excavators. As they move, they expose outcrops that geologists find useful in investigating geologic structures and ore minerals. The great weight of a moving ice mass cracks and loosens bedrock. Most of this bedrock consists of igneous and metamorphic rocks of varying degrees of resistance because glaciers tend to accumulate at higher elevations where some or all of the sedimentary cover may have been stripped off. Because the national parks described in this section are in "hard rock country," the chapters include additional material about igneous and metamorphic rocks.

The work of glaciers is effective in mountains, even on resistant bedrock. Chunks of loose rock are *plucked* or *quarried* by glaciers and moved downslope. Large and small pieces of rock are frozen into the ice and scraped and dragged over underlying rock, scratching, grinding, scouring, pulverizing, and polishing. Glacial striations on bedrock are made in this way. The processes of glacial *plucking* and *abrasion* produce quantities of sediment of all sizes from *rock flour* (fine-

ly pulverized rock debris) to huge boulders. This mostly unsorted rock material, called *till*, is picked up and pushed or carried along on top of, beneath, beside, and in front of a glacier. Ridges of till that accumulate alongside and in front of glaciers are called *moraines*. (Till that is compacted underneath a moving glacier eventually becomes what is called *ground moraine* after the ice melts back.)

In summer seasons, meltwater drains off, through, and beneath glaciers, working over and sorting loose till and other debris into coarse and fine fractions, then redepositing them as sediments. *Glacial outwash* is the geologic term for such deposits. A broader term that includes all sorted and unsorted glacial deposits (till, outwash, etc.) is *glacial drift*.

The erosional and depositional processes of active alpine and continental glaciers produce distinctive landforms and other features characteristic of glacial scenery. However, the key indicators of former glacial action on a landscape are the presence of striations, erratics, and glacial drift.

Glacial Advances and Retreats

In Pleistocene time, which lasted about two and a half million years, numerous episodes of continental glaciation occurred, separated by ice-free interglacial periods. What brought on a glacial episode was a climatic trend of lower average summer temperatures, possibly for not very many years, and usually accompanied by increased precipitation. The onset of glacial episodes apparently took much less time than the melting back and decline of the ice sheets.

In the broadest sense, the episodes of global cooling and glacial advance may have been worldwide, but worldwide correlation of glacial features has proved not to be practical because local variations were so great. Intensive study and the use of radioactive dating has shown that advances, standstills, and retreats of glacial ice in different areas probably occurred at different times in different regions. Glaciers did not reach their maximum extent in Puget Sound (Washington), the Southern Rocky Mountains, or the coast of Maine all at the same time, nor did they retreat at the same rate. However, as a general time term, the Wisconsinan stage is still used in North America in referring to the last major glacial advances of the Pleistocene Epoch (fig. III.4).

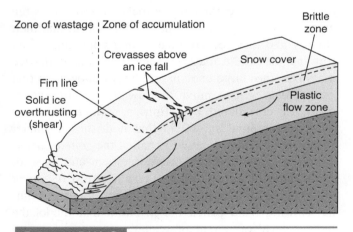

FIGURE PIII.3   Glacial ice moves downslope differently in different sections of a glacier, as shown in the diagram. Note that the firn line marks the transition between the zone of accumulation and the zone of wastage.

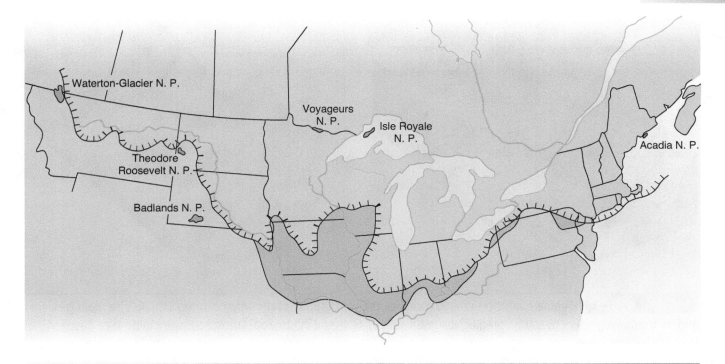

FIGURE PIII.4   Extent of the Pleistocene glaciation. The hachured line that goes from Waterton-Glacier International Peace Park to the Atlantic Ocean off southern New England marks the approximate location of Wisconsinan terminal moraines, the last major advance of continental glaciers. The darker colored areas were covered by earlier Pleistocene ice advances.

Carbon-14 Dating

Some glacial deposits of the later phases of Wisconsinan glaciation can be assigned absolute ages, mainly within the last 55,000 years, by means of carbon-14 (C_{14}) dating techniques. This method uses short-lived radioactive atoms of carbon-14 produced by cosmic ray activity in the earth's atmosphere. Minute amounts of carbon-14 combine with stable carbon atoms in atmospheric carbon dioxide. As living things incorporate carbon into their tissues, the ratio of carbon-14 to other carbon atoms remains the same as the ratio in the atmosphere. After an organism dies, the carbon-14 contained in its tissues decays radioactively at a fixed rate. When the ratio of carbon-14 to other carbon atoms in the organic remains is compared, a radiometric date can be calculated that establishes the length of time that has elapsed since the death of the organism. The technique is useful in dating recent glacial and volcanic events, young sedimentary accumulations, and archeological sites.

Is Another Ice Age Coming?

Scientists have speculated for many years on the ultimate cause (or causes) of global cooling that initiates an ice age and the subsequent warming that brings such an event to a close. No widely accepted explanation has been put forth that would account for all the phenomena observed, but most of those proposed deal with factors affecting precipitation; that is, increases or decreases in mean annual snowfall over time. Among the ideas that have been considered are the following: (1) "wobbling" of the earth's axis and variations in the earth's orbit, causing shifts in the amount of solar radiation received on the earth's surface; (2) changes in the concentration of atmospheric gases (such as carbon dioxide) that may affect retention of solar radiation; (3) increases in the size and number of volcanic eruptions, creating dust clouds that block the sun's rays; (4) tectonic "drifting" of large land masses toward polar latitudes; (5) changes in the size and extent of the polar icepack, affecting

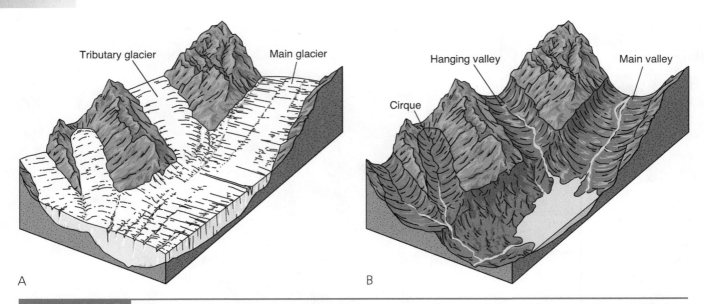

FIGURE PIII.5   Alpine glaciation greatly modifies the topography of mountainous regions. *A.* The main glacier and its tributary glaciers nearly fill preglacial stream valleys. *B.* In the postglacial landscape, the main valley has become a broad, U-shaped trough, streams tumble from hanging valleys, horns and cols sharpen the skyline, and a cirque is cut in the head of a tributary valley. From *Geology along Going-to-the-Sun Road, Glacier National Park, Montana* by O.B. Raup et al. © 1983 Glacier National History Association. Used by permission.

evaporation of moisture by polar winds; (6) and changes in ocean currents, especially those circulating warm sea water into colder regions, resulting in increases or decreases in the amounts of water vapor evaporated from oceans.

The Holocene Epoch, the geologic time unit we live in, is regarded as an interglacial stage that followed the Wisconsinan stage of the Pleistocene Epoch. Since the Wisconsinan ended, roughly 12,500 years ago, global temperatures have shifted back and forth, and modern glaciers have made minor advances and retreats. It seems likely that at some time in the future, great ice sheets will again form and spread over the northern part of the Northern Hemisphere.

CHAPTER 22

Voyageurs National Park

NORTHEAST MINNESOTA

Area: 218,035 acres; land area, 138,266 acres; water, 83,789 acres; 341 square miles

Established as a National Park: April 8, 1975

Address: 3131 Highway 53, South International Falls, MN 56649-8904

Phone: 218-283-9821

E-mail: VOYA_Superintendent@nps.gov

Website: www.nps.gov/voya/index.htm

Interconnected northern lakes, dotted with islands, once the route of French-Canadian voyageurs, are surrounded by boreal forest. A glacially scoured landscape exposes Precambrian rocks of the Canadian Shield. Patches of moraine, erratics, lakebed deposits, and peat bogs are scattered over bedrock. Pits of small, worked-out, gold mines are in a "greenstone belt" in the park.

FIGURE 22.1 A glacial erratic is a large rock fragment that has been transported by ice from its place of origin. National Park Service photo by Larry Edwards.

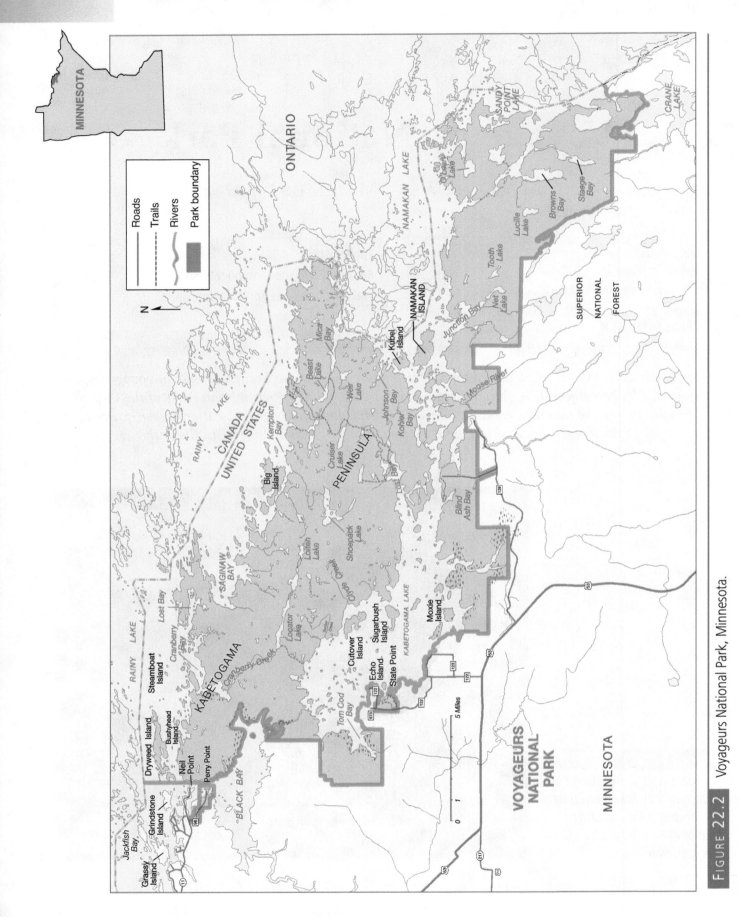

FIGURE 22.2   Voyageurs National Park, Minnesota.

Geographic Setting

Roads end a short distance beyond the entry points of Voyageurs National Park, which is located in the borderland of islands and waterways between Minnesota and Ontario, Canada. The main means of traveling among the islands is by canoe or boat. This is an area of water, woods, wildlife, and well exposed, very old igneous and metamorphic rocks, thinly mantled by Pleistocene glacial deposits. A southern boreal forest is protected in the park; that is, a subarctic forest characteristic of the transition zone between the Northern Temperate region and northern boreal forests.

The park is about 40 miles long and extends from International Falls on the west to the Boundary Waters Canoe Area on the east. The width of the park varies from about three to 15 miles. During the Pleistocene Epoch, continental glaciers shaped the landscape. As ice sheets advanced and retreated, they scraped, gouged, and pushed away trees, rock, and soil, converting a region of huge forests and a few rivers to a barren land of sparse soil, bogs, grassy meadows, and numerous lakes interconnected by a network of streams. The boreal forests that covered the land when the Indians, voyageurs, and fur traders were here grew in the thin soil left by melting glaciers or deposited in temporary lakes after the last ice sheet melted back about 10,000 years ago. Most of the virgin timber was logged off in the years before and after the turn of the century, but some tracts of the boreal forest remain in Voyageurs National Park, the Boundary Waters Canoe Area, and Quetico Provincial Park in Ontario.

FIGURE 22.3   Precambrian bedrock that was originally a mixture of sedimentary, igneous, and metamorphic rock dominates the landscape of Voyageurs National Park. National Park Service photo.

Winters are long and cold in the north woods, and daylight hours are few. Snow may cover the ground from October until May, and ice in the lakes and streams seldom breaks up before the middle of May. Although the summers are relatively brief, temperatures are comfortable and the days are long. Spring and fall are times of intense activity in the wilderness; migratory birds return to the northern breeding grounds as days lengthen, and then head south again as winter approaches.

"Red Paint People" came into the north woods about 8000 years ago. A few pictographs painted on rock cliffs with red ocher and artifacts, such as slate knives and arrowheads, give clues about their Stone Age culture. Chippewa Indians lived in the park area when the French Canadian trappers and voyageurs traveled along the waterways. The white men learned Indian tongues and Indian ways of adapting to the wilderness. Indian guides taught the voyageurs the safest routes among the myriad streams and lakes.

It was the voyageurs for whom the park was named, and their well established waterway between Lake Superior and Lake of the Woods was made the international boundary by the 1783 treaty ending the American Revolution. Fifty-six miles of that boundary, those passing through the waters north of the Kabetogama Peninsula, also mark the northern limits of Voyageurs National Park.

For 200 years, beginning in the 1600s, the voyageurs plied their trade by paddling and portaging thousands of miles through the wilderness waterways and bringing out many tons of furs. The voyageurs' 3000-mile route between Montreal and Lake Athabasca (northeastern Alberta) was divided into eastern and western stages. Each stage had its own type of birchbark canoes and group of men. The exchange point was first at Grand Portage, Minnesota, on the western shore of Lake Superior; but later at Thunder Bay in Canadian territory.

The larger, 36-foot Montreal canoes were paddled by ten men and transported trading goods and supplies out and furs back, between Montreal and Grand Portage or Thunder Bay. At the exchange station, cargoes were split up and reloaded into the lighter, 25-foot north canoes, each with a crew of five to six men, who traveled between Grand Portage and the western end of the line at Lake Athabasca. As soon as the ice began to break up in spring, the voyageurs loaded their canoes with the winter's accumulation of furs that had been traded for goods in remote Indian villages. The whole trip—getting the furs to the rendezvous, picking up the trade goods from Montreal, and returning to Lake Athabasca—had to be completed between spring thaw and autumn freeze-up. In making the 2000-mile journey, these brave, hardy men paddled and portaged 15 to 18 hours a day.

After the fur trade declined, lumbering and mining assumed more importance. Large-scale logging of the virgin forests went on for about 50 years. The Kabetogama Peninsula was the scene of a gold mining boom between 1893 and 1898. The boom town of Rainy Lake City (northwest corner of the park) filled up with prospectors and miners but then decreased from a settlement of 500 people to a ghost town in five years. Getting supplies and equipment to the mines and hauling out ore was difficult, with the closest transportation center 100 miles away.

The chief producer was the Little American Mine, located on the island of the same name near the south shore of Rainy Lake. About 500 tons of ore were extracted and crushed, yielding gold that was worth over $4600 in 1894 and 1895. The gold (of hydrothermal origin) was disseminated in a quartz vein 4 to 6 feet wide in a belt of severely metamorphosed rock (sheared chlorite and biotite schist). The open cut of the mine (a shaft 10 feet wide and 44 feet deep) was flooded when the water level of Rainy Lake was raised for the purpose of generating hydroelectric power.

The Big American Mine, the first mine dug in the area, was on Dryweed Island. The pit, five feet square and only five feet deep, was sunk on a quartz vein. Lyle Mine, on the north side of Dryweed Island, had a shaft 22 feet deep, but Old Soldier Mine, on a very small island off Dryweed, was only a shallow pit. Two other abandoned mine shafts are on the east end of Dryweed. On Bushyhead Island, an adit (tunnel) penetrated a shear zone of massive pyrite and quartz just above water level. Gold Harbor and Holman Mines, on a blunt point of the mainland, are shallow pits in a quartz-pyrite vein in a schistose area.

The mining and lumbering industries—although they brought a measure of prosperity to the region—drastically altered the natural environment. However, on both sides of the international boundary, people who loved the northern lake country felt that the forests and waterways should be preserved and protected. As early as 1891, the Minnesota legislature sought to have a national park established on the Kabetogama Peninsula.

FIGURE 22.4   Migmatitic erratic is sitting on chlorite schist of the greenstone belt. Note: canteen is for scale. National Park Service photo by Larry Edwards.

Geologic Features

The Precambrian Rocks

Through the years, similar efforts were being made in Ontario, resulting in the establishment of Quetico Provincial Park. The author-naturalist Sigurd Olson led the movement for the establishment of Voyageurs National Park. Both Canada and the United States have designated the region of boundary lands and waters "to be kept in a state of nature as far as possible."

All of the bedrock exposed in Voyageurs National Park and the surrounding region is Precambrian in age. These ancient rocks are part of a large area of basement rocks called the Canadian Shield, which is, in turn, a part of the North American craton. A *craton* is a part of the earth's crust that has been stable (without orogenic activity) for at least 1500 million years. The oldest cratonic rocks are thought to be the core or nucleus of early continental crust.

Exposures of the oldest rocks, classified as Archean, have been identified in shields in other parts of the world, such as the Fenno-Scandinavian region of northern Europe and in parts of Asia, Australia, South America, Antarctica, and Africa. The fact that the shield rocks are surrounded by platforms composed of somewhat less ancient Precambrian rocks (and early Paleozoic rocks) suggests that early continents may have grown by accretion. The platforms have some characteristics of once-active tectonic belts. Most Precambrian rocks are igneous or metamorphic. Erosion for millions of years brought about the characteristic low relief of the Precambrian shields.

Clusters of radiometric dates show that each Precambrian shield area is divisible into structural

Box 22.1	**Dating Precambrian Rocks**

Precambrian rocks, wherever they are found, represent geologic activities that took place during more than three-fourths of the entire time of the earth's geologic history. The rocks are "samples of time" that provide tantalizing clues about the origin of continents, and early tectonics (fig. 22.5).

The usual methods that geologists use in dating younger rocks are not reliable for Precambrian exposures. However, techniques of radiometric dating have yielded a number of absolute age determinations for Precambrian rocks. Such a value may be based on one or more methods of *radiometric dating;* that is, laboratory procedures for measuring how much time has elapsed in the process of radioactive "decay" since a particular rock formed. *Radioactive decay* is the spontaneous disintegration of certain unstable atoms, called the parent material, to *daughter products* through gain or loss of isotopes of the radioactive elements present. By carefully determining the ratio of daughter products to parent material in a given sample, the radiometric age of the rock can be calculated because the rates of decay are constant.

Igneous rocks containing mineral grains of radioactive elements tend to yield more precise radiometric values because, having crystallized from a melt, an igneous rock is less likely to be contaminated by previously decayed daughter products. For example, samples taken from a mafic dike on an island in Rainy Lake yielded K-Ar dates of 2130 m.y. and 2040 m.y. (Proterozoic X). K-Ar values are based on the process by which an isotope of potassium, K-40, converts to argon, Ar-40. (*Mafic* refers to igneous rocks with a high proportion of dark minerals that contain iron and magnesium compounds.)

In dating metamorphic rocks, several techniques may be combined to yield reasonably accurate age determinations. Sometimes radiometric dates for the mineral zircon in metamorphic rocks can be derived through the use of the rubidium-strontium method; that is, ascertaining the proportion of the rubidium radioactive isotope, Rb-87, to its daughter product, the strontium isotope, Sr-87. Age determinations for

various rocks sampled in the Rainy Lake area and analyzed by this technique have ranged between 2690 m.y. and 2750 m.y. (Archean).

Deeply weathered rocks are not suitable for radiometric dating because much of the original parent material has been lost or altered.

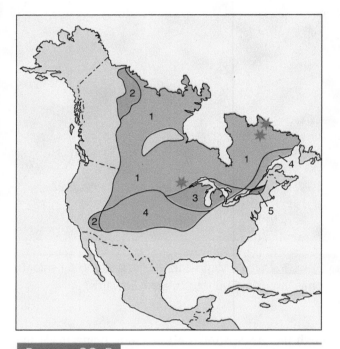

FIGURE 22.5   On this map are areas of North American Precambrian rocks that have a similar *age of origin* but that may or may not have been accreted to the North American craton at the same time. The stars mark the locations of crust over 3.5 bybp (billion years before present). Ages of rocks within the numbered areas are as follows: (1) older than 2.5 bybp; (2) 2.3 to 2.1 bybp in age; (3) 2.0 to 1.8 bybp; (4) 1.8 to 1.6 bybp; (5) 1.3 to 1.2 bybp. The entire North American craton was consolidated by 1.65 bybp. Modified after Hoffman 1989 and other sources.

Geologists have not found common factors other than radiometric age that can be used consistently to correlate Precambrian rocks from one region to another. Outcrops are scattered; index fossils are lacking; some rock units are too deeply buried for cores to be obtained; all have been subjected to repeated metamorphism and extensive erosion; and each geographic district has its own geologic history with different events having occurred in various sequences. Unconformities in Precambrian chronology show up on every continent, but where these interruptions occur varies so much from place to place that using them for purposes of correlation is not feasible.

Geologists are in general agreement that Precambrian time can be divided into two eons, Archean and Proterozoic. An *eon* is the longest geologic-time unit, the next in magnitude above an *era.* Because the Archean was a time of high heat flow, the earth's primitive continental crust probably developed rapidly. In Proterozoic time, the amount of heat generated by radioactive decay decreased, continents grew more gradually, tectonic mechanisms similar to those of the present began, and primitive life forms developed.

The U.S. Geological Survey's approach to Precambrian classification is to use the common factor that exists; that is, radiometric dating. The Archean-Proterozoic boundary has been set at 2500 million years ago, and the letters X, Y, and Z have been assigned to time units in the Proterozoic Eon. Although these time terms are regarded as being "without specific rank," they are useful for comparing and classifying groups of Precambrian rocks that yield dates falling within these informal divisions.

Table 22.1 compares the U.S. Geological Survey's Precambrian chronology with the Canadian classification. Although time boundaries in the two systems differ slightly, both systems place the end of Precambrian time at about 570 million years ago.

Research underway on Precambrian life forms may yield information that will be helpful in correlating Precambrian rocks in different parts of the world. In 1954 the first discovery of Precambrian microfossils was made near Thunder Bay on Lake Superior. The Gunflint Chert in which the fossils were found is Proterozoic X in age. Microfossils have not been found in rock units in Voyageurs National Park. However, about 60 miles northeast of the park, near Atikokan, Ontario, algae fossils, 2.7 billion years old (Archean or early Proterozoic), have been found in marine sedimentary rocks of the Steep Rock Formation of the Timiskaming group.

Table 22.1	Comparison of Precambrian Subdivisions used in the United States and Canada (Geologic Names Committee, U.S.G.S., 1980)				
Canadian Subdivisions		**Canadian Orogeniee**			**U.S.G.S. Subdivisions**
Proterozoic	Hadrynian				Proterozoic Z
	Helikian	Grenville-1000 m.y.	800 my – –		Proterozoic Y
	Aphebian	Hudsonian-1800 m.y.	1600 my – –		Proterozoic X
		Kenoran-2500 m.y.	2500 my – –		
Archean					Archean

FIGURE 22.6   Ancient crystalline rock scoured by glaciers. Hardy jack pines cling to crevices and grooves of the bedrock fringing Rainy Lake. National Park Service photo by Larry Edwards.

provinces, based on rock ages. Within each province, the trend of tectonic folds and the style of deformation are distinctive from those of adjacent areas. Thus metamorphic boundaries and fault boundaries, as well as radiometric dates, delineate the various provinces. Both Voyageurs National Park and Isle Royale National Park (Chapter 23) are in the Superior Province, which is a large Precambrian subregion with many bedrock exposures. (The Precambrian rocks in Isle Royale National Park in Michigan are much younger than those in Voyageurs.) Superior Province extends over a large part of south central Canada and dips slightly into the United States in northern Minnesota, Wisconsin, and Michigan. Because of its mineral resources, which include deposits of iron, gold, copper, and nickel, the Superior Province has been studied by Canadian and American geologists for many years. Much of the pioneer work on radiometric dating was done on Precambrian rock specimens from this area.

The Archean rocks of the Voyageurs region are associated with the Kenoran (Algoman) orogeny, which is regarded as the earliest datable orogeny in North America. Two contrasting assemblages are present in these ancient rocks—the granitoid gneisses, which are highly metamorphosed, and the greenstone belts, which are less highly metamorphosed. The granitoid gneisses were derived from high-silica rocks, possibly of volcanic origin, and sandstones. Because some greenstones are "host rocks" for valuable metallic ore deposits—in various parts of the world, as well as in the Superior Province—the greenstone belts are of special interest to economic geologists.

The greenstone belts occur in large, elongated downwarps between the more extensive granitoid gneiss accumulations. Dark greenish in color, the greenstones owe their name and appearance to the constituent minerals—chlorite, epidote, actinolite, etc.—that were produced by metamorphism. The presence of pillow lava structures in the greenstone belts indicates that the lavas were extruded in the waters of an ancient sea. Although most of the pillow structures in the Voyageurs area have been destroyed by metamorphism, a few "pillows" up to three feet long have been found on a small islet off Steamboat Island (north of Cranberry Bay). Some poorly sorted mixtures of clay, mica, and quartz grains (graywackes) that were derived from the decomposition of pyroclastic debris (tuff, ash, etc.) suggest that some of the volcanic activity was explosive. Some of the clastic material also has characteristics of rapid deposition by marine currents, possibly in an ocean trench.

The greenstone belt in Voyageurs National Park consists of schistose (i.e., finely foliated) and sheared lavas that have undergone intense metamorphism. In general, the schistosity is so extreme that the composition and characteristics of the original lavas are masked. Major islands in and near the northwest corner of the park, such as Dryweed, Grassy, and Grindstone Islands, are made up of greenstone, as are sections of Neil Point (across the Narrows) and part of Rainy Lake's south shore between Jackfish Bay and Black Bay.

Between the greenstone tracts are large areas of metamorphosed biotite and mica schists (originally sedimentary and granitic rocks). The main portion of the Kabetogama Peninsula and the Black Bay area are in part of the schist belt, and a contact zone between the

FIGURE 22.7   A close-up of one of the dikes shows the rectangular-shaped pinkish feldspar orthoclase, the dark gray quartz and black hornblende crystals. Any gold would be disseminated in extremely small grains throughout the rock and would not be visible to the naked eye. National Park Service photo.

greenstone belt and the schist belt can be traced in the Rainy Lake area.

The Vermilion batholith, which is associated with the Kenoran (Algoman) orogeny, underlies the Sullivan Bay and Blind Ash Bay area on the south side of Kabetogama Lake. Some of the coarse-grained, pink and gray granites show up in striking exposures along the shore. This large intrusion of granitic magma may have been the source of the gold-bearing quartz solutions that crystallized in veins along an east-west fault zone extending to Neil Point (just outside the western edge of the park).

Glacial Features

Glacial deposits. In the Voyageurs National Park area, most glacial deposits are associated with Wisconsin-aged glaciation and consist of four types: lakebed sediments, outwash gravels, peat, and till. The lake deposits accumulated in shallow margins of Glacial Lake Agassiz, a vast, inland body of water that for several thousand years covered hundreds of square miles north and south of the international boundary. The lake formed as water began to collect in the Red River valley in western Minnesota about 12,000 years ago and then spread over eastern North Dakota and adjacent Canadian Provinces. During this time of glacial retreat, meltwater draining the last continental glacier ponded because the north-flowing rivers headed for Hudson Bay were blocked by ice. As climatic conditions changed, lake levels were higher and lower a number of times. Recognizable beach lines recorded stillstands; that is, times when a lake level was constant for a long enough interval for waves to form a notch in the shore or a beach. Deposits of outwash gravels suggest that several outwash channels developed and were later abandoned.

The peat deposits, which have been dated as Holocene in age, formed in hollows or depressions, such as old lake basins, bogs, and swamps.

Till deposits, made up of unsorted glacial debris, occur as ground moraine and end moraine. Patches of *ground moraine,* deposited under the ice sheet, are interspersed with lakebed sediments to form an undulating surface called "swell-and-swale" topography. Till in *end moraines* was left in hummocky ridges dumped along the front of the ice sheet. The sections of end moraine that can be seen in places along the southwest shore of Rainy Lake were left by the last active glacier to occupy the area.

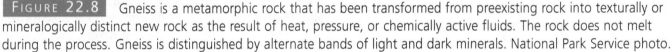

FIGURE 22.8 Gneiss is a metamorphic rock that has been transformed from preexisting rock into texturally or mineralogically distinct new rock as the result of heat, pressure, or chemically active fluids. The rock does not melt during the process. Gneiss is distinguished by alternate bands of light and dark minerals. National Park Service photo.

Erratics. Ranging in size from pebbles to blocks larger than automobiles, erratics abound in the Voyageurs area (figs. 22.1, 22.4). Some were carried by the ice a few miles, while others traveled long distances. Erratics from Precambrian gneiss and greenstone in the Boundary Waters region are scattered over the landscape of southern Minnesota and are even found in Iowa.

Striations and glacial polish. The parallel, linear scratches that show up on outcrops and erratics in the park are glacial *striations*, formed when quartz sand grains and hard, sharp rock fragments, frozen into the bottom of the ice sheet, scratched across rock surfaces as the glacier moved along. Striations show local directions of ice movement. In Voyageurs National Park, striations point south, south by southwest, and southwest. On some outcrops, the presence of cross-striations indicates more than one direction of ice movement.

Repeated abrasion by ice advances has smoothed and polished many rock exposures. *Rock flour*, consisting of fine particles held to the bottom of the ice sheet, put a high gloss on such rock surfaces. Glacial polish

FIGURE 22.9 As the granitic Vermilion batholith intruded into the local bedrock, hot liquids were injected into fractures and cooled into gold-bearing quartz and granitic dikes. The gold was probably derived from the rising magma. National Park Service photo.

on highly resistant outcrops in Voyageurs still looks fresh or only slightly weathered.

Effects of Glacial Rebound

The weight of huge ice sheets moving southward depressed the crust several hundred feet. After the ice retreated, lakes filled the basins and spread over the lowlands. Relieved of the weight of ice, the crust probably regained its normal elevation after major episodes of glaciation. This postglacial upwarp, called *glacial rebound,* is still going on, although probably at a slower rate in the last few thousand years. As the land has risen, some lakes have drained away; in others, water levels have dropped. Some lakes have become bogs and swamps, filled in by vegetation and sediment. The

lakes of the Boundary Waters region occupy deeper parts of the basins once filled by Glacial Lake Agassiz. Since rebound of the crust has been uneven, old beach lines, representing earlier lake levels in northwestern Minnesota, are now tilted slightly to the south.

Geologic History

1. **Early Precambrian (Archean) basement complex.**

These intensely metamorphosed rocks of sedimentary, volcanic, and intrusive igneous origin, dated as 2600 million years old, or older, are commonly referred to as "pre-Kenoran" (or "pre-Algoman"). The Coutchiching rocks (which may or may not be the oldest formation) are foliated gneisses and schists rich in mica that may

Table 22.2			Geologic Column, Voyageurs National Park	
Time Units			**Rock Units**	**Geologic Events**
Era	**Period**	**Epoch**		
Cenozoic	Quaternary	Holocene	Alluvium in streams, lakes, bogs, etc.	Continuing glacial rebound; some lakes drained or filled in Minor morainal and outwash deposition
		Pleistocene	Glacial drift	Glaciation by many ice advances removed weathered debris and smoothed bedrock
Paleozoic/ Mesozoic			/////////	Prolonged erosion of the shield
Proterozoic X (Middle Precambrian time)			Mafic intrusive rock	Intrusion of mafic dike
Archean (Early Precambrian time)			Kenoran (Algoman) granitic intrusives	Kenoran (Algoman) orogeny; mountain-building and uplift; metamorphism
			Pre-Kenoran (pre-Algoman) formations (position in column uncertain)	
			Timiskaming (Windigokan)	Rapid deposition of coarse conglomerates, graywackes, silt, etc.
			Laurentian granites	Small granitic intrusions
			Keewatin lavas	Submarine extrusion of mafic lavas; Occasional pyroclastic eruptions
			Coutchiching	Marine deposition of coarse and fine sands, silts, etc.

Source: Modified from Pye, 1968.

originally have been sandstones. A group of similar rocks of sedimentary origin, designated as Timiskaming (or Windigokan), are thought to be younger, mainly because the conglomerates contain large cobbles of granitic material that suggest rapid erosion and deposition. The volcanic greenstones, or Keewatin lavas, are largely basaltic but include some clastic material.

2. Kenoran (Algoman) orogeny at the end of Archean time and the beginning of Proterozoic X time.

The emplacement of large granitic batholiths in the Superior Province brought about intense folding and uplift of the earlier rocks and the forming of what were probably mountain ranges of considerable bulk and height. The deformation that accompanied the orogeny caused partial remelting of rocks as well as metamorphism. Subsequent erosion of the mountain ranges has exposed a number of ore-bearing structures in the region, such as the gold veins found in the Kabetogama Peninsula area.

3. Erosion of the mountains; minor intrusive activity.

The youngest bedrock in Voyageurs National Park is exposed in mafic dikes, about 2100 million years old (Proterozoic X), that crop out along the lake shore and on some of the islands.

4. Prolonged erosion throughout Paleozoic, Mesozoic, and most of Cenozoic time.

The end of Precambrian time brought to a close the long series of tectonic events that created the basement complex of the Canadian Shield. After many millions of years of erosion, only the roots of the ancient mountains remained. Relief was low. Probably a fairly uniform soil cover, supporting extensive forests, blanketed the ancient bedrock when continental glaciers began to form in Pleistocene time.

5. Pleistocene glaciation.

Ice sheets covered the Boundary Waters region many times, pushing and scraping away the old soil cover and gouging bedrock. The first glacial advance, probably early in Pleistocene time, came from the north-northwest. After an interval, ice from the northeast moved over the area, followed by the advance of a continental glacier from the west. Later, a large lobe of glacial ice came from the northeast, but may not have extended much beyond Rainy Lake. The later ice advances, associated with Wisconsinan glaciation, obliterated traces of the earliest glaciations.

6. The postglacial, constructional landscape.

Unconsolidated materials, Pleistocene and Holocene in age, have been superimposed on some of the oldest bedrock in North America. Because relief in the area is low, streams have been slow in reworking and sorting the glacial debris. Lakes occupy the large basins; and small ponds, bogs, and wetlands, interconnected by waterways, are all about. These are indications that drainage in the area is still disorganized.

Geologic Map and Cross Section

Day, W.C. 1985. Bedrock Geologic Map of the Rainy Lake Area, Northern Minnesota. U.S. Geological Survey Open File Report 85-0246.

Goebel, J.C. and Walton, M. 1979. Geologic Map of Minnesota, Quaternary Geology. Minnesota Geological Survey, University of Minnesota.

Sources

Day, W.C. and Weiblen, P.W. 1986. Geochemical Evidence from the Vermilion Granitic Complex of Northern Minnesota. *Contributions to Mineralogy and Petrology* 93(3):283–296.

Harris, F.R. 1974. Geology of the Rainy Lake Area, District of Rainy River. Geological Report 115. 94 p., two maps in pocket. Ontario Division of Mines, Ministry of Natural Resources.

Hoffmann, P.F. 1989. Precambrian Geology and Tectonic History of North America in Bally, A.W., and Palmer, A.R., eds. *The Geology of North America—An Overview*, v A, p. 447–512. Boulder, Colorado; Geological Society of America.

Minnesota Geological Survey. 1969. An Evaluation of the Mineral Potential of the Proposed Voyageurs National Park, with map. University of Minnesota.

Ojakangas, R.W. 1972. Rainy Lake Area. In *Geology of Minnesota: A Centennial Volume*, ed. P.K. Sims and G.B. Morey, pp. 163–70. Minnesota Geological Survey.

Ojakangas, R.W. and Matsch, C. L. 1982. *Minnesota's Geology*. Minneapolis: University of Minnesota Press. 255 p.

Pye, E.G. 1968. *Geology and Scenery, Rainy Lake and East to Lake Superior*. Geological Guide Book no. 1. Ontario Department of Mines.

Southwick, D.L. 1972. Vermilion Granite-Migmatite Massif. In *Geology of Minnesota: A Centennial Volume*, ed. P.K. Sims and G.B. Morey, pp. 108–118. Minnesota Geological Survey.

Wright, H.E., Jr. 1972. Physiography of Minnesota. In *Geology of Minnesota: A Centennial Volume*, ed. P.K. Sims and G.B. Morey, pp. 561–78. Minnesota Geological Survey.

Isle Royale National Park

NORTHERN MICHIGAN, IN LAKE SUPERIOR

Area: 571,790 acres; 893 square miles	**Phone:** 906-482-0984
Established as a National Park: April 3, 1940	**E-mail:** ISRO_Parkinfo@nps.gov
Designated a Biosphere Reserve: 1980	**Website:** www.nps.gov/isro/index.htm
Address: 800 East Lakeshore Drive, Houghton, MI 49931-1895	

A forested island, largest in Lake Superior, with long parallel ridges, lakes, and bogs. Plucking, abrasion, and scouring by Pleistocene continental glaciers shaped the landscape. Left by the last glaciation are drumlins, erratics, ground moraine, and end moraines. High shorelines, old beaches, and wave-erosion features mark levels of postglacial lakes. In late Precambrian time, a thick sequence of basaltic lavas with interbedded and covering sediments subsided to form the Lake Superior syncline.

Copper mines were worked by Indians several thousand years ago and by Americans in the nineteenth century.

FIGURE 23.1   An aerial photograph taken of the northeastern end of the island at Five-finger Bay. The view is along the strike of the major rock units; the beds are dipping to the left. Differing resistance of rock units to long subaerial erosion and glaciation produced the narrow bays, peninsulas, and islands of this Lake Superior Park. National Park Service photo.

Geographic Setting

In northern Lake Superior, the international boundary follows a water passage between Ontario, Canada, and Isle Royale, a long narrow island belonging to the state of Michigan. This wilderness island is 45 miles long and five to eight miles wide. Isle Royale National Park, which comprises the main island, along with its 200 surrounding islets and the waters 4.5 miles out from the shoreline, is accessible only by boat or floatplane. The park is about 50 miles from the shore of Michigan's Upper Peninsula and 15 to 20 miles from the Canadian shore. The island has no roads.

Native copper—the metallic element, uncombined and in a natural state—has long been known on Isle Royale. Indians called the island *Minong,* which means "a good place to get copper." However, Chippewas, who lived in the region when the French explorers came, knew nothing of mining copper or fashioning copper tools. They valued the pieces of "float copper" that they found in loose gravel or on beaches, and they used raw copper fragments for charms or crude cutting and scraping tools. The ancient pit mines that prospectors found on Isle Royale and the Keweenaw Peninsula on the Michigan mainland were a mystery to the Chippewas. Nor did tribal legends preserve the memory of the pre-Columbian people who for many centuries excavated and removed Lake Superior copper. Archaeologists investigating mine pits found charcoal that yielded radiometric dates of 3800 to 1500 years B.P. Ornaments of delicate design made from Isle Royale copper have been discovered in Hopewell mounds in southern Ohio and at other sites, proving that the people of the Copper Culture knew how to soften and shape the nearly pure copper. Artisans developed methods of heating and hammering the copper (pure copper is malleable and ductile) in order to make arrowheads, knives, and other artifacts.

Copper Culture mining was time-consuming and arduous. The Indians built a fire on top of a copper vein. When the rock was hot enough, they dashed cold water on it, causing the rock to shatter so that they could pound out chunks of copper with stone mauls. Thousands of the rounded beach cobbles used for this purpose have been found around the mines. Some of the stone mauls are grooved for attaching handles. Although Indian mine pits were left on all parts of the island, the highest concentration is along McCargo Cove on the north side. Although Isle Royale was given

to the United States by the 1783 treaty with England, the English remained in control until after the War of 1812. Meanwhile, the Chippewas considered the island to be their territory. Not until after the Chippewas ceded Isle Royale to the United States government in 1843 was the island overrun by prospectors and miners. The famous Ontonagon copper boulder, a glacial erratic long owned by the Chippewas, was the object of a complicated legal wrangle. At last the War Department confiscated the huge specimen, which weighed 3708 pounds, and removed it to Washington where it remains in the Smithsonian Institution.

A report by Douglass Houghton, Michigan's first state geologist, set off a copper boom in the 1840s, both on the mainland and the island. Because of the severity of the winters, the miners could work on the island only during the short summer season. In order to find ore outcrops more easily, the prospectors and miners burned the forests or cut them down for building settlements. When a promising deposit was found—a contact vein, a dike, or a fissure filled with copper— the miners blasted and excavated. If the site proved out, a shaft was sunk. As a result, in addition to worked-out mines and ancient Indian pits, hundreds of exploration pits are scattered over the island. Many old mines and pits are overgrown with blueberry thickets or have been filled in with dirt.

The nineteenth century miners found three classes of ore on Isle Royale and the Keweenaw Peninsula, which they called "moss," "barrelwork," and "stamp." Moss was large sheets of pure copper that weighed anywhere from a few hundred pounds to several tons. Barrelwork consisted of smaller pieces of copper enclosed in amygdules found at the top of brecciated (rough-textured) flows. An *amygdule* is a filled *vesicle,* or gas cavity, in igneous rock. Bubbles of trapped gas or steam in cooling lava may form vesicles. Later these small cavities may be filled with a secondary mineral (copper in this case) deposited by hot solutions that percolated upward through the rocks. Stamp copper consisted of pebble-sized pieces bound in a rock matrix. The copper was freed by pulverizing the ore in a stamp mill. Occasionally the miners came across masses of native copper weighing several thousand pounds. One mass exhibited at the 1876 Centennial Exposition in Philadelphia weighed 5720 pounds.

The principal mines on Isle Royale (operated at different times between 1843 and 1899) were the Smithwick Mine at Rock Harbor, Siskowit (which pro-

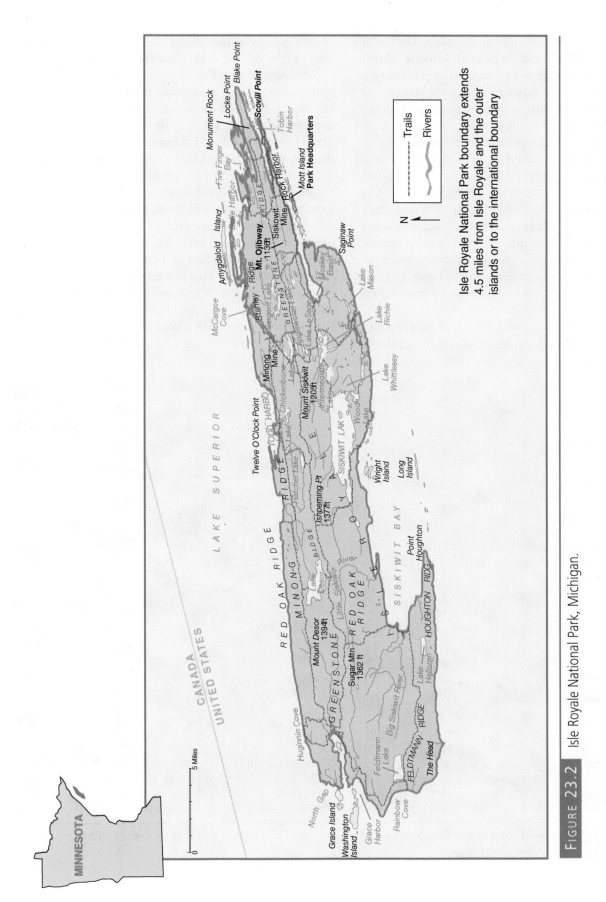

FIGURE 23.2   Isle Royale National Park, Michigan.

duced 190,000 pounds of copper between 1847 and 1855), Minong Mine, and Windigo. Minong Mine, located on Minong Ridge in the midst of ancient mines, was the largest. The Isle Royale mines were abandoned when the ore was exhausted and excavations became unprofitable. By the end of the century, only a few fishermen and their families were left on the island.

A new era began for Isle Royale when it became a summer resort in the early 1900s and gained a reputation as a haven for hay fever sufferers. The park was established in 1940. In 1980 Isle Royale National Park was designated a United Nations International Biosphere Reserve, which makes it part of a worldwide data-gathering system that monitors the impact of global pollution. The recent detection in Lake Siskiwit of airborne PCBs, toxaphene, and dioxins has caused some concern.

Isle Royale is both biologically and geologically interesting due to its isolated location and the geologic events that gave the island its unique character. The long ridges on the island are the result of upfaulting and erosion of late Precambrian volcanics. After prolonged erosion had worn down the surface, Pleistocene ice sheets overrode the whole region and their weight depressed the land. Isle Royale was under water during glacial retreats. Glacial rebound and the lowering of water levels of postglacial lakes exposed the present island surface late in Pleistocene time.

Most of the island was scraped bare of soil and weathered rock, although some glacial till and outwash was left on the western part of Isle Royale. How did plants and animals reestablish themselves on the island? In particular, how did small animals get across the miles of open water between Isle Royale and the Canadian mainland? These are speculative questions to which biologists do not have definitive answers. However, they do know about ecological changes that have come about during historic time. Caribou and

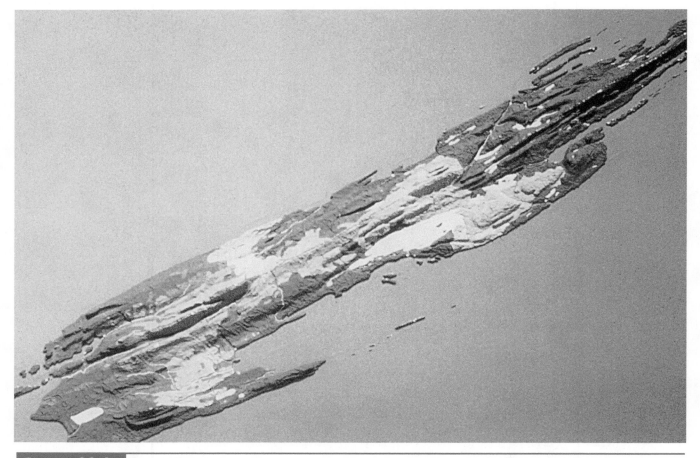

FIGURE 23.3A   A relief map of Isle Royale. The high ridges running down the center of the island are made up of fine-grained lava flows, called "trap rock." National Park Service photo.

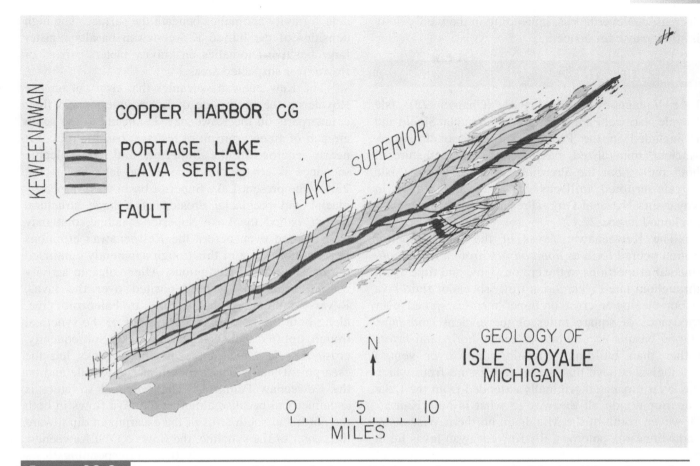

FIGURE 23.3B   A geological map of Isle Royale shows that the northwestern part of the island is made up of the Precambrian Portage Lake Lava Series, while the southeast portion of the island consists of the Precambrian Copper Harbor Conglomerate. The dark black lines represent faults. On the southeastern end of the island the faults run NE-SW; on the other end of the island they trend more or less NW-SE. National Park Service diagram.

lynx formerly lived on the island but left after the primeval forests were cut off or burned. The trees and undergrowth that replaced the original stand continue to evolve as a mixture of transitional hardwoods and subarctic conifers.

Sometime early in this century, moose arrived on Isle Royale. Presumably, they swam over from Canada or came when the lake was frozen. No predators were on the island, food was plentiful, and the moose population zoomed to several thousand animals, far more than the island could support. When the water plants, twigs, and leaves on which moose browse were depleted, hundreds of moose starved.

In the extremely cold winter of 1948-49, Lake Superior froze over, a rare event that occurs only a few

times in a century. A pack of eastern gray wolves from Canada crossed over on the ice and established themselves on Isle Royale. For the most part, the wolves killed moose that were very young, old, or weak. This culling process kept the moose population healthy and stable for a number of years. Beavers on the island also helped to maintain the balance of wolves and moose by building dams that create ponds and beaver meadows that supply browse for moose and food for other animals. In the last decade, however, the wolf population declined abruptly, while the moose increased in number. Biologists theorize that generations of inbreeding may have caused the wolves to become infertile. In the late nineties, the wolf population began to recover somewhat, while the number of moose on the island decreased. Nevertheless, inbreeding in the wolf colony is still a cause for concern.

Geologic Features

Like Voyageurs National Park (Chapter 22), Isle Royale National Park is part of the Canadian Shield and is included in the Precambrian Superior Province. Ancient, mineralized, basaltic lavas are a feature of both parks; but the Precambrian lava flows at Isle Royale erupted millions of years after those in Voyageurs National Park. (Precambrian chronology is explained in box 22.1.)

The Keweenawan lavas of the Lake Superior region poured forth as *flood basalts* from a rift of continental proportions in the craton. Time and time again throughout late Precambrian time, sheets of fluid lava (from the lower crust or upper mantle) spread over thousands of square miles of the ancient landscape. (Flood basalts accumulate in nearly horizontal layers rather than building volcanic cones over vents.) Geologists believe that the crustal fissure from which the lavas emanated originally extended from the Lake Superior region all the way to what is now Kansas. However, south of the Michigan, northern Wisconsin, and Minnesota outcrops, the Keweenawan lavas lie in a subsurface belt under many hundreds of feet of younger rock and glacial drift. The trace of the ancient rift has been mapped by geophysical techniques that detect gravity anomalies beneath the surface. The high densities of the buried Keweenawan basalts register large positive anomalies on gravity meters carried or flown over suspected areas.

The Keweenawan volcanics that crop out on Isle Royale on the north side of Lake Superior and their counterparts on the Keweenaw Peninsula to the south are part of an escarpment of resistant basaltic rock that nearly encloses Lake Superior, and the volcanic sequence is continuous beneath the lake basin (fig. 23.4). The present Lake Superior basin was formed by glacial and preglacial erosion of a large structural trough, called the Lake Superior syncline, that may have existed even before the Keweenawan eruptions began. Subsidence of this trough apparently continued during and after the eruptions. After volcanic activity ceased, sediments were deposited over the lavas. Reverse faulting, probably early in Paleozoic time, along both the north and south limbs of the synclinal trough, uplifted the Keweenawan rocks. Subsequently, erosion of less resistant sedimentary rocks left the escarpment that is now exposed at Isle Royale and on the Keweenaw Peninsula. The synclinal structure is asymmetrical because although the lava flows in both the north and south arms of the escarpment dip toward the center of the syncline, the flows on the Keweenaw Peninsula dip more steeply than corresponding flows on Isle Royale. The dips of the lavas on Isle Royale are less than 20°.

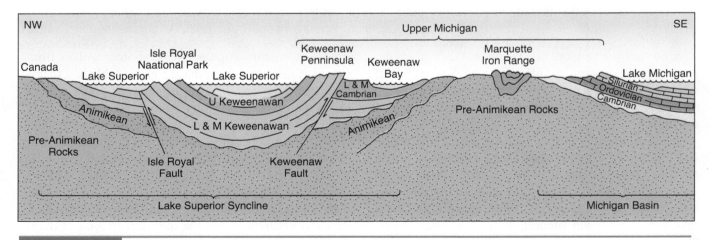

FIGURE 23.4 Schematic cross section showing how Isle Royale, the Keweenawan lavas, Keweenawan Peninsula, and major faults are related to regional structural features. From R.K. and R.A. Paull, *Geology of Wisconsin and Upper Michigan.* © 1977 Kendall/Hunt Publishing Company. Used by permission.

The Precambrian Rocks

Visitors to Isle Royale National Park can find exposures of volcanic and sedimentary rocks in a rich variety of textures and types, despite a dense ground cover of bushes and trees. The lava flows that were originally stacked more or less horizontally have been tilted by subsidence and faulting so that they are now exposed in strips or layers on ridge tops and along the shore. Thin sedimentary layers between some of the flows represent quiet intervals during which weathering of the flow surfaces and minor erosion and deposition could occur. Individual flows within the volcanic sequence have characteristic textures reflecting different rates of cooling and varying mineral compositions. Since most flows are basaltic, they are of *mafic* composition; that is, composed chiefly of dark-colored, ferromagnesian minerals. A few flows have an *intermediate* or *felsic* composition; that is, andesites containing an approximately equal amount of light- and dark-colored minerals or dacite, with a higher proportion of light-colored minerals.

The dark, massive, fine-grained flows, called "trap rock," tend to make up the resistant ridges, such as the Minong Ridge. Coarse-grained volcanic rocks, which solidified more slowly, are easily distinguished by their *ophitic texture;* that is, pyroxene crystals surround and enclose laths of plagioclase, giving the rock a mottled appearance. Weathered outcrops often have a knobby surface. The flows with *porphyritic texture* have well defined plagioclase crystals (slow cooling) scattered throughout a fine-grained groundmass (more rapid cooling).

The native copper in the Isle Royale volcanic sequences may have ascended with the basaltic lavas or via channelways between flows. The three classes of deposits in which the native copper was found correspond to the categories that the miners used: (1) copper in a matrix of conglomerate interbedded with lava flows, or what miners called "stamp;" (2) amygdules of copper filling vesicles, the "barrelwork;" (3) and "moss," the copper found in veins, fissure fillings, and fractures.

Michigan's state gem, a mottled-green, semi-precious stone called *chlorastrolite* (meaning "green star") is found in amygdules in amygdaloidal basalt on Isle Royale. This small, unusual gemstone, which fades in sunlight, is seldom found outside the Lake Superior region. Chlorastrolite crystals can be seen in bedrock at Scoville Point, Rock Harbor, and Todd Harbor, *but may not be collected within the park.*

Similar in composition to chlorastrolite is prehnite, another secondary mineral that occurs in crosscutting veins as well as amygdule fillings. Some of the small pink pebbles on the beaches are prehnite amygdules that have weathered out of lava flows.

Pink Lake Superior *agates,* a banded variety of chalcedony (quartz), formed in amygdules, most notably in the Amygdaloid Island flow. Agates also formed in cavities in volcanic ash (tuff) that occurs on top of the Greenstone flow. Some agates weathered out of the hardened flows and were redeposited with sediments interbedded between flows. Since the agates are resistant, many of them went through a second weathering cycle and now show up on Isle Royale beaches or in glacial till.

Sedimentary units in the volcanic sequences are mainly conglomerates and sandstones that appear to be fluvially deposited. Pebbles in the conglomerates consist of volcanic rock, mostly mafic but some felsic, reflecting the composition of the source rock. Accessible outcrops are in the Chippewa Harbor area, on the south side of Siskiwit Lake, and on the north side of Conglomerate Bay. Old mine dumps, such as those at Island Mine, also have accumulations of broken sedimentary rock.

The Copper Harbor Conglomerate.

On the south side of Isle Royale National Park, the Portage Lake Volcanics dip conformably and gently beneath the coarse sandstones and conglomerates of the youngest Precambrian formation, the Copper Harbor Conglomerate (fig. 23.5). About a fifth of the bedrock of the Isle Royale archipelago is made up of this sedimentary unit; the other four-fifths consists of the Portage Lake Volcanics. Exposures of the Copper Harbor Conglomerate crop out along the southwest shore of the main island, on the small islands offshore, and on the higher parts of Feldtmann Ridge and Houghton Ridge, which parallel the southwest shoreline of the main island. Surficial deposits, mainly of glacial origin, cover a good deal of the Copper Harbor Conglomerate. On the Michigan shore, the formation crops out again, lapping up on the Portage Lake Volcanics in the Keweenaw Peninsula. The beds may not be continuous beneath the lake due to erosive activity since their deposition.

FIGURE 23.5 The Precambrian Copper Harbor Conglomerate is exposed along the shore on the southeastern portion of the island. National Park Service photo.

Effects of Glaciation on the Preglacial Topography

The topographic "grain" of the Isle Royale archipelago reflects the structures of the bedrock. The main island and nearly all of the small islands are elongate with a northeast–southwest trend. The parallel ridges with valleys or water-filled troughs in between are the result of differential erosion on the uplifted fault-block. In general, massive basalt flows make up the resistant ridges; less resistant amygdaloidal basalts underlie the valleys, producing a "corrugated" preglacial topography that the continental ice sheets delineated even more sharply as they moved over the region (figs. 23.1, 23.2).

Fractures and minor faults in the bedrock also affected preglacial and glacial erosion. Ravines developed along fault zones that cut across the main trend of the ridges. McCargo Cove, a nearly north–south-trending inlet on the north side of the island, is an example of a linear topographic feature that was eroded from shattered rock in a fault zone that cuts across ridges.

The preglacial topography was accentuated by glacial erosion because glacial quarrying and plucking were more effective on the less resistant bedrock underlying the stream valleys. Elongate lakes, such as Angleworm Lake, Sargent Lake, and Hatchet Lake, occupy basins excavated by glacial ice that followed valley trends. Quarrying also deepened fiordlike inlets along the shore and some of the channels between the elongate islands. Some of the crosscutting ravines, especially those on the north side of the island, were gouged out by moving ice. On the ridge tops glacial abrasion smoothed and polished the resistant massive basalts and removed weathered bedrock from all but protected locations. Glacial striations are evident on many outcrops.

The last ice advance (which largely obliterated traces of previous ice sheets) moved southwestward over the eastern half of the island, following the trend of the parallel ridges and valleys. However, the ice veered more to the west over the western end of the island and flowed across the ridges. The change in direction of ice movement is probably the reason for the fact that bedrock on the island's eastern half is more severely eroded, while a number of depositional (constructional) glacial features were left on the western half of the island. Glacial deposits are lacking on the eastern part of Isle Royale except in the bottoms of valleys or small depressions. Glacial erratics are few in number. On the west end of the island, till mantles bedrock, and erratics are abundant.

Constructional features, such as drumlins and ice-margin deposits, are characteristic of continental glaciation. A *drumlin* is an elongate hill or ridge made up of compact glacial till that was carried in the basal part of the glacier and plastered onto the ground surface. The Isle Royale drumlins, on the western end of the main island, are of the "crag-and-tail" variety. The stoss end (facing the direction from which the ice advanced) is a steep, ice-smoothed knob of rock (the crag) that obstructed the flow of ice. The *lee* end (the tail), consisting of till, is streamlined and tapering. The alignment of the drumlins shows that on this end of the island the ice moved from east to west.

Morainal ridges (ice-margin deposits) are of irregular sizes and shapes and consist of coarse sand, gravel, and glacial debris. They are strung out across the island, west of Lake Desor, in a more or less northerly direction. They mark the stillstands of an ice front that paused in its retreat.

Abandoned shorelines and old beaches can be discerned in glacial drift on the western and southwestern

FIGURE 23.6 Rainbow Cove, located at the southwestern end of the island, represents an old shoreline that was created when the water level of Lake Superior was higher during the Pleistocene. National Park Service photo.

parts of the main island. Such features record former lake levels when the water was higher (or the land was depressed) as the last ice sheet retreated. On the east end of Isle Royale, where glacial drift is lacking, former shorelines are marked by wave-cut benches in bedrock, and by stacks, arches, and cliffs. Monument Rock (about two miles west of Blake Point) is a stack carved by wave erosion during the existence of Lake Minong, when the water level was about 80 feet higher than it is now. A wave-cut arch on Amygdaloid Island, now high and dry above the shore, was eroded by the waves of Lake Nipissing, a postglacial lake younger than Minong.

Geologic History

1. Middle Keweenawan eruptions during late Precambrian time.

Radiometric and paleomagnetic data collected from the Isle Royale lava flows places them with the Portage Lake Volcanics in the Middle Keweenawan rocks of Proterozoic Y time. All the Middle Keweenawan rocks have *normal polarity;* that is, when the lavas solidified, the earth's magnetic field was the same as it is today. Older lavas (Lower Keweenawan) that crop out in the Lake Superior region (and which may be buried beneath the Isle Royale flows) have *reverse polarity,* as do some younger intrusive rocks (Upper Keweenawan) outside park boundaries. During times of reverse polarity, the direction of the earth's magnetic field was *opposite* to what it is now. (Natural remanent magnetism in rocks and global polarity are explained in box 30.1 in Chapter 30.) The changes in polarity recorded in the rocks of the region are another indication of the length of time during which the rift in the craton was releasing tremendous volumes of mafic lava.

Twelve of the Portage Lake lava flows exposed on Isle Royale have been named and designated as marker flows (table 23.1), which means that they have distinctive characteristics and can be traced from place to place. Between the marker flows are unnamed lavas and sedimentary units that are less readily identifiable. Six of the marker flows, from oldest to youngest, are described briefly below because of their geologic significance and because outcrops are accessible from trails or along the shoreline (Huber 1973a).

Amygdaloid Island flow. This trap rock, which is more felsic in composition than the other volcanic rocks of the group, forms the resistant ridge that runs the length of Amygdaloid island.

Hill Point flow. The thickness and coarse grain of this ophitic rock make it a useful marker. Between Huginnin Cove (west end of the island) and McCargo Cove along the north shore are imposing cliffs eroded from the flow. The type locality outcrop is at Hill Point near the eastern end of Isle Royale. (A *type locality* is a place where a stratigraphic unit is situated and from which it takes its name.)

Huginnin flow. Porphyritic texture with large, tabular, plagioclase crystals scattered throughout a fine-grained groundmass makes this flow distinctive. The rock is easily recognized in beach pebbles and cobbles as well as in bedrock exposures, such as those near Huginnin Cove.

Minong flow. This flow, consisting of dark, fine-grained trap rock, is massive, thick, and resistant. It runs the entire length of the main island, forming the Minong Ridge.

Greenstone flow. Referred to as the "backbone" of Isle Royale, the Greenstone Flow forms the high resistant ridge that runs from east to west through the interior of Isle Royale. Mount Desor, the highest point (1394 feet) on the island, is on Greenstone Ridge, as are Sugar Mountain (1362 feet), Ishpeming Point (1377 feet), and Mount Siskiwit (1205 feet). The thickest part of the Greenstone Flow, about 800 feet, is in the middle of the island. The rock is mainly ophitic in texture.

Scoville Point flow. Scoville Point, where the flow is about 100 feet thick, is the type locality of this porphyritic rock. Fine plagioclase crystals uniformly distributed in a fine-grained matrix give the rock its characteristic appearance.

2. Mineralization in the Portage Lake Volcanics.

Mineral-bearing solutions may have risen concurrently with the lavas and saturated the rocks of the volcanic sequence during the long cooling intervals. Some metamorphism occurred but not to any great extent.

| Table 23.1 | | | Geologic Column, Isle Royale National Park | |

Time Units			Rock Units	Geologic Events
Era	Period	Epoch		
Cenozoic	Quaternary	Holocene	Beach deposits Alluvium	Development of Lake Superior Glacial rebound (Lake Nipissing shore features Stream erosion; wave action
		Pleistocene	Beach deposits Glacial outwash and debris Till	Shorelines formed at various lake levels by wave action (Glacial Lake Minong shore features) Drumlins, ice-margin deposits left by the last Wisconsin advance Erosion by repeated glaciations
Mesozoic Paleozoic			/////////	Prolonged erosional interval
Proterozoic Y (Late Precambrian)	Middle Keweenawan		Copper Harbor Conglomerate Portage Lake Volcanics *Marker flows:* Scoville Point (prophyry) Edwards Is. (trap) Middle Point (porphyry) Long Is. (trap) Tobin Harbor (porphyry) Washington Is. (ophite) Greenstone (ophite) Grace Is. (porphyry) Minong (trap) Huginnin (porphyry) Hill Point (ophite) Amygdaloid Is. (trap)	Isle Royale fault; uplift Deposition of volcanic and clastic sediments Emplacement of native copper and other minerals; mild metamorphism Continuing subsidence of Lake Superior syncline Extrusion of mainly basaltic lavas interbedded with sedimentary deposits

Source: Modified after Huber, 1973, 1973; Wolff and Huber, 1973.

3. **Deposition of the Copper Harbor Conglomerate, completing the Middle Keweenawan sequence.**

As volcanism in the Lake Superior region diminished, streams stripped off slopes and deposited sediments over the subsiding basin. The direction of sediment transport is assumed to have been generally eastward because of indications in the sedimentary structures, such as (1) the orientation of current directions in ripple marks and deflections; (2) thickening of the formation and decreasing coarseness eastward; and (3)

the composition of boulders and cobbles that appear to have come from older Keweenawan volcanics on the Minnesota shore. Nearly all of the coarser sediments in the Copper Harbor Conglomerate are of volcanic origin.

4. **Faulting and uplift, probably in early Paleozoic time.**

During the long subsidence of the Lake Superior synclinal trough, more than 40,000 feet of lavas, sandstones, and conglomerates accumulated in its center (fig. 23.4). Then a series of fault movements, possibly beginning before the end of Precambrian time, raised the Keweenawan rocks. The Isle Royale and Keweenaw faults, which are associated with the escarpment around Lake Superior, are two of the four major faults that developed along the north and south edges of the syncline during this time.

5. The forming of the lake basin.

Tectonic activity ceased, followed by many millions of years of erosion. The more resistant volcanic rocks tended to form escarpments, while softer sedimentary rocks were removed from the center of the syncline by streams that probably flowed across the area in a northeasterly direction. Pleistocene glaciers scoured and enlarged the preglacial drainage ways, creating the basin now occupied by Lake Superior, the deepest of the Great Lakes. From Isle Royale, on the

northwest rim of the lake, the bottom slopes off to a depth of more than 1000 feet.

6. Pleistocene glaciation.

Although Isle Royale was glaciated repeatedly during Pleistocene time, most evidence of earlier ice sheets was destroyed by a late phase of Wisconsinan glaciation about 11,000 years ago. This last advance brought ice back over the Lake Superior basin and removed surficial material that had previously been deposited on Isle Royale. As the ice began to retreat, the west end of the island was uncovered first. A pause in the retreat allowed accumulation of ice-margin deposits around tongues of stagnant ice, forming irregular morainal ridges. Shoreline features associated with a postglacial lake that existed at that time developed on the western end of the main island. Water level of the lake was about 200 feet higher than the present level of Lake Superior. During the long deglaciation process, a series

FIGURE 23.7 Sea cave at Todd Harbor located on the northern side of the northeastern end of Isle Royale. Wave action has produced the cave. National Park Service photo.

FIGURE 23.8   McCargo Cove on the northern side of the island is a trench fault. When faulting occurs, much of the rock is ground up at the fault surface (rock gouge). The softer material is then more easily eroded. National Park Service photo.

of lakes occupied the Lake Superior basin. They had different outlets and different elevations. When some outlets were blocked by glacial ice, other outlets opened. The final retreat of the ice margin was rapid. Glacial Lake Minong formed about 10,500 years ago, when the ice margin had melted back to what is now the northern shore of Lake Superior. Lake Minong, which filled the whole basin, existed long enough for well developed beaches to form on the island, especially in the glacial debris on the southwestern shore.

7. **Glacial rebound and the development of the present lake.**

Water levels continued to lower for a time but then began to rise again as glacial rebound elevated the outlets so that lake water backed up. Because of the tilting that accompanied rebound, abandoned shorelines are difficult to correlate, especially in the Isle Royale area

and the northern shore of the basin where upwarping was uneven. During Lake Nipissing's existence, about 4500 to 3500 years ago, strong wave action produced a number of well defined shoreline features, among them the wave-cut arch that stands in the middle of Amygdaloid Island. Nipissing was the largest of the postglacial lakes.

Continued uplift of the land to the north and the downcutting of the Port Huron outlet to the south brought the Nipissing stage to a close. Then the uplift of Lake Superior's outlet separated that basin from Lake Michigan and Lake Huron, forming the Lake Superior of today. Because glacial rebound is still causing gradual upwarping and tilting of the lake basin slightly toward the south, the shorelines in the vicinity of Isle Royale are rising, while those on the Keweenaw Peninsula on the opposite shore are being very slowly submerged.

Geologic Map and Cross Section

Huber, N.K. 1973. Geologic Map of Isle Royale National Park, Keweenaw County, Michigan. U.S. Geological Survey, Miscellaneous Geologic Investigations, Map I-796.

Sources

Books, K.L. 1972. Paleomagnetism of Some Lake Superior Keweenawan Rocks. U.S. Geological Survey Professional Paper 760.

Dorr, J.A., Jr.; and Eschman, D.F. 1970. *Geology of Michigan.* Ann Arbor, Michigan: University of Michigan Press.

Farrand, W.R. 1969. The Quaternary History of Lake Superior. Proceedings, International Association for Great Lakes Research, 12th conference, 1969. p. 181–197.

Holte, I. 1984. *Ingeborg's Isle Royale.* Grand Marais, Minnesota: Women's Times Publishing. 97 p. (with map)

Huber, N.K. 1973. Glacial and Postglacial Geologic History of Isle Royale National Park, Michigan. U.S. Geological Survey Professional Paper 754A.

———. 1973. The Portage Lake Volcanics (Middle Keweenawan) on Isle Royale, Michigan. U.S. Geological Survey Professional Paper 754C.

———. 1983. *The Geologic Story of Isle Royale National Park* (formerly U.S. Geological Survey Bulletin 1309). Isle Royale Natural History Association. 66 p.

Leskinen, C.; and Leskinen L. 1980. *Copper Country History.* Park Falls, Wisconsin: F.A. Weber & Sons, Inc.

Paull, R.K.; and Paull, R.A. 1977. *Geology of Wisconsin and Upper Michigan.* Dubuque, Iowa: Kendall/Hunt Publishing Company.

———. 1980. *Wisconsin and Upper Michigan*, K/H Geology Field Guide Series. Dubuque, Iowa: Kendall/Hunt Publishing Company.

Rakestraw, L. 1965. *Historic Mining on Isle Royale.* Isle Royale Natural History Association.

Wolff, R.G., and Huber, N.K. 1973. The Copper Harbor Conglomerate (Middle Keweenawan) on Isle Royale, Michigan. U.S. Geological Survey Professional Paper 754B.

Note: To convert English measurements to metric, go to www.helpwithdiy.com/metric_conversion_calculator.html

Acadia National Park

EASTERN MAINE COASTAL AREA

Proclaimed a National Monument: July 8, 1916	**Address:** P.O. Box 177, Eagle Lake Rd., Bar Harbor, ME 04609-0177
Established as Lafayette National Park: February 26, 1919	**Phone:** 207-288-3338
	E-mail: Acadia_information@nps.gov
Redesignated a National Park: January 19, 1929	**Website:** www.nps.gov/acad/index.htm

A rockbound coast, scoured by continental glaciers. Waves beat against a shoreline of cliffs, boulder beaches, tidal pools, bays, a fiord, many islands, and headlands. Bare, rounded, granite summits, above the forests, overlook the sea.

Paleozoic plutonic rocks emplaced in older rocks during the Acadian orogeny.

FIGURE 24.1   Mount Desert Island, with Sutton Island in the foreground. Somes Sound, a fiord, is on the far left. During Pleistocene time, when continental glaciers overrode the entire island, the ridge tops were smoothed over and the intervening valleys were scoured out. National Park Service photo.

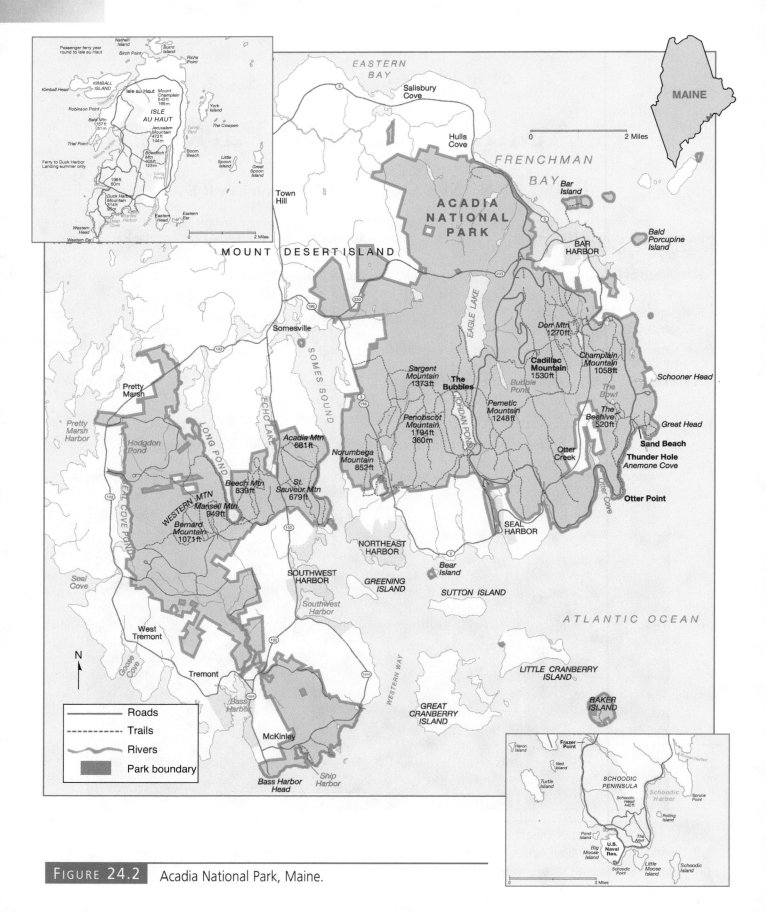

FIGURE 24.2   Acadia National Park, Maine.

Geographic Setting

Ledges, forests, beautiful harbors, and picturesque villages give the Maine coast great beauty and charm. Acadia National Park, located on this coast, is near the international boundary and about 200 miles "down east" from the port of Boston. Over 2,000,000 people visit the park each year.

Acadia National Park comprises much of Mount Desert Island, parts of the Schoodic Peninsula on the mainland and Isle au Haut offshore, and 12 smaller islands in the Gulf of Maine Archipelago. This is a deeply indented, rocky coast, with estuaries and scattered islands. The water temperature is cold, and summers are cool, the average air temperature for July and August being 69° F. Precipitation averages 40 inches annually, with some rain and fog in the summer months. A tidal range of around 10 feet (between high and low tide) unveils four different life zones, each populated by specifically adapted marine plants and animals, such as sea stars, urchins, periwinkles, anemones, and crabs.

The Red Paint People, so-named because they buried their dead with red ocher, were a prehistoric group who inhabited these shores several thousand years before Europeans arrived. Explorers and settlers encountered a later group of Indians who had developed a birchbark culture. They camped in the Somes Sound area to fish, dig for clams, and trap mink. Deep shell heaps indicate that they camped here for hundreds of years.

Perhaps the Norsemen and certainly the Portuguese explored this coast. A 1529 Portuguese map shows Mount Desert Island and the main estuaries of the region. Since Cadillac Mountain (1530 feet) is the highest point on the Atlantic coast, it was a landmark for sailing ships for the next several centuries.

Samuel Champlain, the father of New France, was the official discoverer of Mount Desert Island in 1604. Sieur de Monts, Champlain's patron, held a grant from the French king that comprised the entire French claim to North America (all the land between latitudes 40° N and 60° N). The territory was called *La Cadie* (later anglicized to "Acadia"). After establishing a trading post, de Monts sent Champlain to explore and chart the coast and encourage the local Indians to do business with the French. In his journal of the voyage, Champlain described what is now Acadia National Park as "an island, very high, and cleft into seven or eight mountains, all in a line. The summits of most of them are bare of trees, nothing but rock. I named it *l'Isle des Monts-deserts* ("the Isle of Bare Mountains"). Attracted by the smoke from an Indian camp, Champlain put in to shore near Otter Cliffs and, guided by friendly Indians, made a reconnaissance of the locality. His chart of the coastline shows fairly accurately the islands and estuaries.

In the 1840s and 1850s, the beauty of the island scenery was "discovered" by Thomas Cole, Frederick Church, and other American artists. College professors, wealthy summer visitors, and cruising yachtsmen began coming to the island. Coastal steamers served the area from the 1850s until the advent of the automobile. Although the island was mainly a resort for the very rich, an increasing number of tourists who appreciated the uniqueness of the region came to visit. By 1901, the wealthy people who owned most of the land began to fear that lumbering and commercial development would ruin the island's natural beauty. They formed a corporation to protect the area and acquired land on Mount Desert Island. Charles Eliot, president of Harvard, John D. Rockefeller, and George Bucknam Dorr were among the leaders in the effort to transfer the land to the federal government for a national park.

Geologic Features

Terranes That "Drifted"

The role of water and ice in the shaping of the dramatic Acadian scenery is significant, but the origin of the rocks that have resisted these forces is equally dramatic. Coarse-grained and medium-grained granites form the mountainous core and glacially smoothed summits of Mount Desert Island. Older rocks surround the plutons in a roughly circular pattern. The older rocks, originally sedimentary, were long ago deformed and metamorphosed. Later, as part of a minicontinent, they "drifted" northwestward for hundreds of miles, and eventually became terranes attached to the North American plate.

A *terrane* in an orogenic belt can be described as an "internally homogeneous geologic province with features that contrast sharply with those of nearby provinces. They are recognized by contrasts in any or all of the following: stratigraphy, geologic structure, metamorphic and plutonic histories, fossil faunas, mineral deposits, and paleomagnetic signatures. Their

Box 24.1	**More about Metamorphic Rocks**

The metamorphic rocks in Acadia National Park have "character." If you observe their distinctive appearance in outcrops along the shore, it's not hard to believe that long ago they were subjected to great heat and pressure far below the earth's surface. Contorted beds of different colors seem to coil and swirl as waves wash over the rock. Dark, banded layers, some thick, some thin, are cut by white quartz veins. Even the more massive rocks that stand in cliffs have subtle gradations in color, inclusions that look like patch pockets, and some have a sparkly texture.

Most of the Acadian metamorphic rocks were originally sedimentary, although a few were first formed as igneous rocks. All metamorphic rocks are formed from preexisting rocks, or "parent rocks." The composition of a parent rock has a great deal to do with its response to various metamorphic processes. The other important factors are the degrees of heat and pressure and the amount of water that is present. During metamorphism, parent rocks are transformed in a solid state—without being melted—into new rocks that are texturally and mineralogically distinct from the preexisting rocks.

Nearly all metamorphic rocks form deep within the earth's crust, usually as a result of *regional metamorphism,* which involves a general increase of temperature and pressure over a large area. Tectonic activities, such as subduction and orogeny, are associated with regional metamorphism. The most intensely deformed portions of mountain ranges are where metamorphic rocks are found—after the buried mountain cores have been exposed by erosion.

Regional metamorphism is usually *progressive*; that is, different conditions of temperature (ranging between 400° C and 800° C) and pressure (encountered at depths from 10 to 35 km) alter rocks in a

sequence that may vary from lightly metamorphosed to intensely metamorphosed. Shale, for example, becomes slate when temperature-pressure conditions are low. With an increase in temperature and pressure, implying deeper burial, a phyllite may be formed from the parent shale.

Some rocks, especially when intense, directed pressure is a factor, develop *foliated texture* because the pressure forces the minerals in a rock to become parallel to each other. *Foliation* refers to the parallel alignment of a rock's textural and structural features. Slate, a rock with foliated texture, splits easily along foliation planes because its pre-existing, microscopic, platy minerals were pushed into alignment during metamorphism.

The presence of water (from seawater and groundwater trapped in crustal rocks) is important in metamorphic processes because it aids chemical reactions that produce new rocks. Water under high pressure forces its way between mineral grains and dissolves and transports ions between minerals so that recrystallization can occur.

In *contact metamorphism,* high temperature is the dominant factor, with less pressure being involved. Magma intruding country rock is usually the source of the intense heat that metamorphoses, or "cooks," country rock. The zone of contact metamorphism around an intrusive rock mass is ordinarily narrow. Rocks produced by contact metamorphism tend to be fine-grained but do not have foliated texture because directed pressure is not involved.

Table 24.1 shows how common metamorphic rocks are classified. Note that nonfoliated rocks are classified on the basis of their composition, whereas foliated rocks are recognized by the type or degree of foliation.

boundaries are "sharp structural junctions marking discontinuities that cannot be explained by normal gradations . . . conventional facies changes, or standard unconformities." (Williams and Hatcher 1983, p. 34) A more formal term, used in geological literature, is *tectonostratigraphic terrane.*

The Metamorphic Rocks of Acadia

The Ellsworth Schist. This formation is the oldest of the three distinct units that make up the older rocks in Acadia National Park. Several hundred to several thousand feet thick, the Ellsworth Schist consists of mica

Table 24.1   Classification of Common Metamorphic Rocks

Nonfoliated Texture (Rock names based on mineral content)

Usual parent rock	Rock name	Predominant minerals
limestone dolomite	MARBLE	calcite dolomite
shale basalt	HORNFELS	plagioclase feldspars micas ferromagnesians clays
quartzose sandstone	QUARTZITE	quartz

Foliated Texture (Rock names based on nature of foliation)

Closely foliated	SLATE Splits into thin flat sheets, foliation finer than phyllites or schists	

Schistose (platy)	PHYLLITE
	   Crinkly, lustrous surfaces
	SCHIST
	   Well developed, parallel orientation of minerals	

Coarsely banded or layered	GNEISS Banded or layered foliation coarser than phyllites and schists	

schist and dense gneiss (fig. 24.3). Originally, the rock accumulated in thinly laminated beds of fine clastic sediments, intermixed with fluvially worked volcanic ash and tuff. The mica-rich schists in the Ellsworth are closely foliated and split readily along surfaces of foliation, but the gneisses, which resist splitting, have poorly developed foliation. Recrystallization by repeated metamorphism has increased the strength and density of the rock. Prominent exposures of the Ellsworth Schist are along the shore at the northern end of Mount Desert Island.

The Bar Harbor Formation. This formation accumulated as beds of gravel, sand, and silt on the floor of a shallow sea. The siltstone, which splits along bedding planes, is the source of much of the shingle on the beaches, especially around Bar Harbor. (*Shingle* is composed of smoothed, spheroidal, or flattened water-worn pebbles and cobbles.) Thin, white layers in the formation consist of volcanic ash that fell into the ancient sea and was washed by waves and currents.

Cranberry Island Series. These rocks were originally composed of interbedded volcanic and sedimentary layers, including pyroclastics and brecciated fragments. Because many of the beds are uniform and well stratified, the assumption is made that the volcanic materials were transported by streams and reworked by waves. The Cranberry Island units were sharply folded and squeezed not long after deposition. Great Cranberry and Little Cranberry Islands are made up of their "namesake" formation. The rocks also crop out on the southern part of Mount Desert Island in the vicinity of Bass Harbor.

The Mount Desert Intrusives

The light-colored, barren summits that so impressed Champlain when he sighted them from his ship are made up of massive, coarse-grained granite that is highly resistant to erosion. This is the granite, pinkish in color, that makes up the Mount Desert range and all the high points of the island (fig. 24.4). Wedged into the northwest section of the pluton is a somewhat younger body of gray, medium-grained granite that is speckled with biotite (black mica). From quarries in this granite (such as those between Somes Sound and Echo Lake), building stone of fine quality was shipped to cities along the eastern seaboard.

Along the northwest side of the island, next to the granites, is an *intrusive sheet* of diorite, the third most abundant rock type of the park. Long before the emplacement of the plutons, dioritic magma of intermediate composition came up very hot from great depths and worked its way through layers of crustal rock. Then the magma spread out horizontally between sedimentary bedding planes, forming sills in some places. The intrusive sheet, a large, somewhat irregular body, emanated from an extensive fracture and is sandwiched between the Ellsworth Schist and the Bar Harbor Formation. The presence of the metamorphic rock hornfels in the contact zones indicates that the magma was extremely hot when it injected the stratified rocks.

A fine-grained granite that crops out between Southwest Harbor and Goose Cove (southwest part of the island) was emplaced after the diorite but before the coarse- and medium-grained granites. This fine-grained granite may be part of a large body of similar rock of the same age that makes up nearly all of the Schoodic Peninsula and also crops out on Isle au Haut.

The Shatter Zone, Pegmatites, Veins, and Dikes

A zone of severely shattered, fractured, and brecciated rock, up to a mile or more in width, lies between the coarse-grained granite and the metamorphosed country rock. Called the *shatter zone*, this band of rock that partially encircles the island (along the eastern shore and across the southern part) marks the edge of an old collapsed area. Shatter zone rock, which is tough and coherent, is easy to spot because of its brecciated appearance (fig. 24.5).

Angular blocks of diorite, chunks of old sedimentary rock, coarse fragments of various colors, veins, small dikes, pegmatites, and partially blended rubble are all congealed in a matrix of granite.

The pegmatites, quartz veins, and light-colored dikes that fanned out from the granitic magma are most numerous in the shatter zone, but can also be found where cracks, fractures, or joints in the rock gave access to hot hydrothermal solutions and fingers of molten rock forcing their way upward. Small amounts of minerals (epidote, hematite, etc.) are present in the quartz veins and pegmatites, but none are of economic value.

FIGURE 24.3   The Ellsworth Schist, one of the oldest rocks on the island, consists of a mica schist and dense gneiss. It has undergone repeated metamorphism. National Park Service photo.

FIGURE 24.4   The frost-riven bedrock on the summit of Cadillac Mountain is the highest point (1530 feet) on the North American Atlantic coastline. The bare top is the result of exfoliation of the coarse-grained granite. Photo by Ronald F. Hacker.

FIGURE 24.5   The shatter zone is an area of randomly fissured or cracked rock; it may have a network of veins filled with mineral deposits. On Mount Desert Island the shatter zone lies between the coarse-grained granite and the metamorphic rocks. The zone partly encircles the island. National Park Service photo.

More numerous and more widely distributed than the granitic dikes are the basaltic (diabase) dikes that range in width from less than an inch to more than 50 feet. Some dike outcrops extend straight over mountaintops for thousands of feet. On bare ledges, the dark color of the basalt shows up against the lighter granites (fig 24.6). Contacts tend to be clear and sharp, which makes the dikes easy to follow. The Schoodic Point area (due east from Mount Desert Island, across Frenchman Bay) is the best place in the park to see basalt dikes. Here large, dark dikes cut through light, fine-grained granite. Where dikes come down to the shore, wave action has eroded straight-walled chasms. The less resistant basalt has been removed, leaving granite on either side (fig. 24.7). The basaltic intrusions apparently came from a very deep source. Radiometric dating techniques have yielded several sets of distinctly different ages for the dikes, indicating that from time to time magma was released and moved up toward the surface.

The Acadian Orogeny

These "native" and "foreign" rocks, so well exposed in Acadia National Park, became intermingled during a long, intensive series of tectonic episodes associated with the mid-Paleozoic Acadian orogeny. The ancient metamorphic rocks were part of continental crustal material carried by drifting tectonic plates from an area on the globe far to the east and south. When the drifting terranes reached the North American plate, a continent-continent collision ensued, squeezing and crumpling crustal rocks. Eventually, sizable sections of rock were subducted and became molten due to heat and pressure at depth. Accumulated magmas of varying composition began to work their way toward the surface. Some formed plutons and intrusive sheets; some erupted as volcanoes. Again the ancient rocks underwent metamorphism. Some of the intrusives were also altered by the tectonic events. The scenery that we

FIGURE 24.6   A basaltic dike is an igneous intrusion that cuts across the bedding planes of the rock. Mount Desert Island abounds with them. National Park Service photo.

enjoy today exposes the roots of what must have been grand and imposing Acadian mountains in middle and late Paleozoic time.

Glacial and Wave-Action Features

Before the arrival of Pleistocene glaciers, the coarse-grained granite of Mount Desert Island formed a highly resistant rock ridge, trending east and west and slightly dissected by the processes of weathering and stream erosion. Possibly the summits of the present range could be distinguished, although reconstructing the preglacial drainage patterns is speculative because most of the evidence has been destroyed. What we see today are six huge troughs, going north and south, cut through this tough, resistant ridge. The amount of debris that was excavated and the force required to scrape it out and transport it are awesome. Not much of the glacial debris is left on the island. Most of it was pushed or dragged out over what is now ocean bottom.

The Mount Desert ridge must have been a temporary barrier to the great continental glacier moving down from the north, but eventually the mass of ice that was banked up against the north side of the ridge overrode the summits, scouring and rounding them off. At shallow gaps in the ridge or areas of weakness, the erosive effect of the ice was intensified and channeled as flowing ice currents probed for access to the south. As time went on and more and more ice pushed down from the north, troughs became deeper and broader. The deepest trough, Somes Sound, which nearly bisects the island, is a fiord, the only true fiord on the U.S. Atlantic coast. A *fiord* is a long, deep arm of the sea, U-shaped, steep-walled, and always found on a mountainous coast that has been heavily glaciated. A fiord represents the seaward end of a glacial trough, partially submerged after melting of the ice. Most of the embayments on the coast of Maine are drowned valleys, and many of them have been modified by glacial action. Somes Sound qualifies as a fiord because notches and grooves were cut into the rock walls by glacial ice, and because the deepest part of the sound is some distance back from the submerged mouth. Characteristically, a fiord becomes shallower near the mouth where glacial scouring was less severe.

Trough lakes, such as Seal Cove Pond, Long Pond, Echo Lake, Eagle Lake, and Jordan Pond, filled the low areas in the floors of the U-shaped valleys after the ice melted. In the valleys occupied by Long Pond and

FIGURE 24.7   Thunder Hole, an eroded basaltic dike at Schoodic Point. National Park Service photo.

Jordan Pond, the water backed up to somewhat higher levels because of morainal dams at the south end.

The erosive power of the mile-thick ice sheet altered the shape of the mountains. Northern (up-ice) slopes became longer, smoother, and rounded; while southern (down-ice) slopes became steep and foreshortened, their surfaces rough and jagged from glacial plucking. The ridges dividing the up-ice and down-ice slopes tend to be perpendicular to the general flow of the ice mass. Two large *roches moutonnées* that resulted from this type of glacial action are The Bubbles that stand between Cadillac Mountain and Penobscot Mountain (fig 24.8). Smaller roches moutonnées with the characteristic smoothed, up-ice top and plucked, down-ice face, jut out from hillsides in numerous places in the park.

Glacial grooves and striations on bedrock have a generally north-south orientation, indicating the direction of ice flow. *Chattermarks* (a series of small, close, curved cracks made by vibratory chipping of ice-embedded rock fragments and boulders bearing down against a bedrock surface) show where some of the greatest pressures were exerted. Many glacial erratics, large and small, are scattered over the island.

Jointing in the bedrock influenced both the glacial and wave action features. On the summits and in road-cuts at high locations, particularly around Cadillac Mountain, sheet-jointing, or exfoliation, of the granite is noticeable. The joints that run parallel to the surface are the result of unloading when the rocks covering the pluton had been eroded, long after the granite had cooled. On some of the long-exposed surfaces, *weather pits* have developed in the coarse-textured rock because less-resistant mineral grains weather out first.

The geometric pattern of the intersecting vertical joints in the granites is visible wherever bedrock is bare of vegetation. On the summits, weathering and frost action have widened the joints and separated the rock into large blocks. On the south-facing slopes of the

FIGURE 24.8   The Bubbles, located between Cadillac Mountain and Penobscot Mountain are *roche moutonnées*, elongate hills of bedrock, smoothed and scoured by moving ice. On the upglacier side, the rock is gently inclined and rounded. On the downglacier side, the rock is steep and jagged, due to glacial plucking or quarrying. National Park Service photo.

FIGURE 24.9   Metamorphic rock, seen here south of Sand Beach, tends to break up into flat or shingle-shaped rocks. Eroded granitic rocks form rounded cobbles. Shingle and cobble beaches are typical of Mount Desert Island. A few small "pocket beaches" are in protected coves and inlets. Photo by © Robert Fletcher.

island, especially where high cliffs stand on the shoreline, the effect of the glacial plucking along joint planes is striking. Frost action and mass wasting of loosened blocks have steepened jagged slopes and in some places undercut cliffs. Cliffs exposed to wave action are further undercut as the sea continues the attack following the glacier's retreat. On the wave-cut bench formed at the base of a cliff, at and above the water level, rock fragments are rolled around, pounded by waves, and banged up against the cliff face during storms. Intensifying this mechanical action are the salt in the water, the spray, and the mist, which increase the rate of chemical weathering.

One of the many interesting features of the shoreline is the contrast between the effects of wave action on the granites, the shatter zone, and on the older metamorphic (originally sedimentary) rocks. At Bar Harbor, on the eastern side of Mount Desert Island, the rock along the harbor shore is in the Bar Harbor Formation, formed from gravel deposits, sand, silt, and some volcanic ash. Beds of purplish gray rock dip toward the water, while the risers lap up in sequence against the shore. Loose pebbles and cobbles, broken off from the beds, tend to be flattened, or shingle-shaped (fig 24.9). Most of the angles and corners of the beds have been smoothed and rounded by abrasion. Near the high-tide line is a small stack (a free-standing rock column, an erosional remnant, fig 24.10), consisting of siltstone layers. Wave erosion (since the retreat of the ice and the rise of sea level) appears to be moderately effective on these rocks.

To the south of Bar Harbor as far as Great Head, the shoreline is mostly in the shatter zone. Along this stretch of jumbled ledges and cliffs, there are few places where one could safely land a boat at either high or low tide. Most of the rock is fairly resistant, but

FIGURE 24.10 An elevated sea stack, a free-standing rock column, was eroded when sea level was higher; at its base is a granitic cobble beach. Photo by Ronald F. Hacker.

where weak spots exist, the waves have been able to erode. Anemone Cave is a large sea cave, facing the open ocean, that has been formed by wave action and frost action in a zone of fractured rock.

A short distance farther south, ledges of coarse-grained, resistant granite line the shore. Here the erosive action of the waves has had little effect since the Pleistocene except along joint planes at exposed locations. Thunder Hole, which emits loud booming sounds when waves rush in and out, is a chasm eroded by wave action along large joints in the coarse granite. Frost

action, by loosening slabs of rock, helps to enlarge the chasm. At low tide, between waves, one can see large, egg-shaped cobbles swirling and grinding away in the water at the head of the chasm.

One of the few places in the park where post-Pleistocene wave erosion of granite is obvious is on the unprotected ledges and cliffs of the headlands at the extreme southern tip of Mount Desert Island near Bass Harbor Head Lighthouse. Here waves from the open ocean, beating against the rock with full force, have eroded a wave-cut platform of considerable size. A *wave-cut platform,* which includes the lower part of the wave-cut bench, slopes gently seaward from the base of a cliff. At low tide, a wave-cut platform is partially exposed, but is generally covered at high tide. On the Maine coast, the tidal range is high, and varies from 10 to 20 feet. The average tidal range at Acadia is 10 feet.

Wave-Depositional Features

Most of the glacial drift and rock material scraped off from Mount Desert Island is strewn over the ocean bottom out in the Gulf of Maine. Not much sand has accumulated since sea level rose after Pleistocene ice sheets melted back. Wave-depositional features consisting of sand-sized material are minor except for small pocket beaches in coves and a few sandbars in protected areas, such as the one at Bar Harbor that connects the harbor shore with Bar Island at low tide. Most of the pocket beaches have more cobbles and shingles than sand. (Cobbles are generally spherical, while shingles tend to be disk-shaped.) Sand Beach (on the eastern side of the island between Great Head and Otter Cliff) was built across the head of a small bay and receives some sand from a small stream. However, Sand Beach is remarkable for the fact that the "sand" contains a high proportion of finely ground organic material (shells, bones, etc.) that has been washed up from the ocean and reworked by the waves.

Storm waves have built *sea walls* (natural beach ridges) of cobbles and shingle along the high-water line on some of the less rugged parts of the shoreline (where the preglacial slopes were gentle). The sea walls were the source of smooth, even-sized paving stones (called "popplestones" in Maine) that were shipped down to cities on the eastern seaboard for street construction in the nineteenth century.

FIGURE 24.11   The Porcupine Islands were cut off from the mainland and became islands with the rise in sea level due to the melting of continental glaciers. Photo by Charles R. Singler.

Geologic History

1. Origin of the Ellsworth Schist, Bar Harbor Formation, and Cranberry Island Series.

These metamorphosed (originally sedimentary) rock units are part of the Avalon terrane, or microplate, one of nearly a dozen such blocks that broke off from a Eur-African landmass, drifted north and west, and eventually became welded to eastern North America. Although some of the Avalon rocks appear to have close affinities with certain European and African rock units, the original sites of deposition and lithification of Avalon rocks are unknown, but the time during which they formed was late Precambrian through Ordovician. The Avalon rocks were deformed, meta-morphosed, and eroded before they reached North America.

2. Arrival of the Avalon terrane.

In mid-Devonian time, the Avalon terrane "docked" along the eastern edge of North America. Several other terranes arrived ahead of the Avalon, most of them reaching North America in middle and late Ordovician time during the Taconic orogeny. At least three more ter-ranes followed Avalon later in Paleozoic time. Estimates of the velocities of these movements indicate that ter-ranes drift two or three times faster than the large plates, such as the North American plate, which is moving two to three centimeters westward per year. The difference in velocities enables the microplates to overtake a large plate drifting in approximately the same direction.

3. Acadian orogeny in Devonian time.

Folding, faulting, and regional metamorphism accompanied plate convergence from mid-Devonian time on through most of the period. Tectonic activity was concentrated in New England and the Canadian Maritime Provinces.

4. Magmatic and plutonic activity accompanying the Acadian orogeny.

A large igneous complex was emplaced in a series of pulses. The earlier magma was dioritic (intermediate in composition) and the later magmas were more felsic, or granitic. Great quantities of diorite intruded the layers of the Bar Harbor Formation as sills, dikes, and an intrusive sheet. Some of the magma probably reached the surface and erupted in lava flows. Contact metamorphism, resulting from the magma's intense heat, baked the shales and the siltstones of the country rock into hornfels. Most of the diorite is fine-grained; but in the thicker intrusions, slower cooling in the center produced a coarse texture. Some of the sills in the Bar Harbor Formation are 100 feet thick. The largest body, the intrusive sheet, is about 3000 feet thick.

Table 24.2		Geologic Column, Acadia National Park		
Time Units				
Eras	**Periods**	**Rock Units**		**Geologic Events**
Cenozoic	Quaternary	Minor wave and stream deposits Minor glacial deposits		Minor wave erosion Crustal rebound Postglacial submergence Pleistocene glaciation: scouring, abrasion and excavation of troughs
	Tertiary			Minor uplift, renewed erosion
Mesozoic		Unconformity representing uplift, prolonged weathering and erosion		Unconformity representing uplift, prolonged weathering and erosion
Paleozoic	Mississippian ? Devonian	Miscellaneous granites Medium-grained granites Coarse-grained granites Fine-grained granites Diorite (intrusive sheets and sills)		Recrystallization Shattering, contact metamorphism Hydrothermal activity, pegmatites Uplift and erosion Docking of Avalon terrane Dikes and veins throughout this and succeeding intrusions Metamorphism, volcanism Beginning of Acadian orogeny
		Avalon terrane	Cranberry Island Series	Uplift and erosion Ash, tuff, felsite, interbedded sediments
	Silurian			Deformation, metamorphism, erosion, volcanism
			Bay Harbor Formation	Siltstones, shales, conglomerates—mainly marine
	?			Deformation, uplift, metamorphism, erosion
	Ordovician		Ellsworth Schist	Marine sedimentation; volcanic ash

Source: After Gilman and Chapman, 1988; Lowell and Borns, 1988.

Granitic magma that rose after the diorite had solidified apparently hardened near the surface because the texture is fine-grained. Heat and pressure from the magma shattered the country rock. Many fragments of rock from the Bar Harbor Formation and the diorite fell into the liquid magma and became inclusions or blended with the granite. Cracks and fractures in the country rock filled with magma that hardened as dikes, veins, and long stringers (fingers) within the older rock. Uplift of the land and long erosion followed.

5. Climax of Acadian orogeny; intrusions of coarse-grained and medium-grained granites.

In late Devonian time and perhaps extending into the Mississippian Period, crustal disturbances and pulses of intrusive and extrusive activity continued as a great mountain system was built parallel to the eastern coast of North America. Along the Maine coast, the dikes, sills, plutons of coarse- and medium-grained granite, and the patches of recrystallized granite are the roots of what must have been a lofty range. A long period of erosion followed while the mountains were being worn down.

6. Crustal extension in Mesozoic time.

The repeated plate collisions that brought about the Acadian orogeny finally ceased. (A later Allegheny orogeny had less effect in this area.) The impinging plates relaxed and slowly began to separate. This caused the development of several Triassic fault basins, one of which lies under the Bay of Fundy (between New Brunswick and Nova Scotia), about 150 miles northeast of Mount Desert Island. By Jurassic time, the plates had separated enough to open an ocean of significant size, which became the North Atlantic. When the eastern landmass drifted away, parts of the microplates were left behind as terranes joined to North America.

7. Mesozoic and Tertiary uplift and renewed erosion.

During plate collision and convergence, the piling up of continental crust, along with shallow subduction, caused the area that is now coastal Maine to became isostatically out of balance, which forced the lighter, less dense crustal material to rise. As uplift continued, erosion removed a mile or more of solid roof rocks that overlay the granite plutons.

In Tertiary time the Mount Desert range was a ridge, trending east and west. The park region, including the offshore islands, was part of the mainland. A minor uplift that tilted the region slightly seaward was due to continuing isostatic adjustment. Renewed erosion, followed by minor subsidence with encroachment of the sea, produced an irregular shoreline, perhaps similar in outline to what we see today. Shoreline fluctuations since that time are associated with glacial episodes that began in the Pleistocene Epoch.

8. Pleistocene glaciation.

The continental glaciers that covered New England and the Maritime Provinces were more than a mile thick. Ice extended southeastward over the Gulf of Maine, as far as the terminal moraine on Georges Banks. Sea level was then much lower because water had been drawn from the oceans to build the enormous ice sheets, and their great weight was depressing the land. When the ice melted, sea level rose rapidly (fig. 24.11). The ocean came in over the land, drowning the shoreline and filling the estuaries. Georges Banks, which had been a group of islands, was submerged and is now a famous offshore fishing ground. On Mount Desert Island, sea water was probably 300 feet higher than it is today. Old beaches and deltas on the island mark the locations of earlier sea levels. Not long after the unloading of ice, however, the coast of Maine began to rise slowly, a process called *glacial* or *crustal rebound*. Sea level continues to fluctuate slightly and has risen a few feet in the last several thousand years. Recent data also suggest slight tectonic subsidence along the east coast.

Geologic Map and Cross Section

Gilman, R.A. and Chapman, C.A. 1988. Bedrock Geology of Mount Desert Island, Bulletin 38, Department of Conservation, Maine Geological Survey.

Lowell, T.V. and Borns, H.W. Jr. 1988. Surficial Geology of Mount Desert Island, Bulletin 38, Department of Conservation, Maine Geological Survey.

Sources

Chapman, C.A. 1968. A Comparison of the Maine Coastal Plutons and the Magmatic Central Complexes of New Hampshire. In *Studies of Appalachian Geology,*

Northern and Maritime. Editors: E. Zen, W.S. White, J.B. Hadley, and J.B. Thompson, p. 385–396. New York: Interscience Publishers.

Gilman; R.A.; Chapman, C.A.; Lowell, T.V.; and Borns, H.W. Jr. 1988. *The Geology of Mount Desert Island.* Department of Conservation, Maine Geological Survey. 50 p.

Morison, S.E. 1960. The Story of Mount Desert Island. Boston: Little, Brown and Company.

Osberg, P.H.; Tull, J.F.; Robinson, P.; Hon, R.; and Butler, J.R. 1989. The Acadian Orogen in Hatcher, R.D., Jr.; Thomas, W.A.; and Viele, G.W. (eds.). *The Appalachian-Ouachita Orogen in the United States.* Boulder, Colorado: Geological Society of America, the Geology of North America v. F-2, p. 179–232.

Shaler, N.S. 1889. The Geology of the Island of Mount Desert, Maine. U.S. Geological Survey Annual Report 8:987–1061.

Williams, H. and Hatcher, R.D. 1983. Appalachian Suspect Terranes, In R.D. Hatcher, H. Williams, and I. Zeitz, editors. *Contributions to Tectonics and Geophysics of Mountain Chains.* Geological Society of America Memoir 158, p. 33–53.

Rocky Mountain National Park

NORTHERN COLORADO

Area: 265,727 acres; 415 square miles
Established as a National Park: January 26, 1915
Designated a Biosphere Reserve: 1976
Address: 1000 Highway 36, Estes Park, CO 80517-8397

Phone: 970-586-1206
E-mail: ROMO_Information@nps.gov
Website: www.nps.gov/romo/index.htm

*"Roof of the Rockies," a relatively flat upland, over 11,000 feet high, is crossed
by the Continental Divide.*

*The park encompasses the highest peaks of the Front Range, several more than 13,000 feet high.
Small, living glaciers, rock glaciers, and icefields survive at higher elevations. Glacial erosional
features were sculptured by alpine glaciers of the Pleistocene past.
Moraines document glacial advances and retreats.*

*Igneous and metamorphic rocks, mostly Precambrian but some much younger,
have been uplifted and exposed by faulting and erosion.*

FIGURE 25.1 The U-shaped glacial trough indicates that the Rocky Mountains were glaciated in the geologic past. Photo by Ronald Hacker.

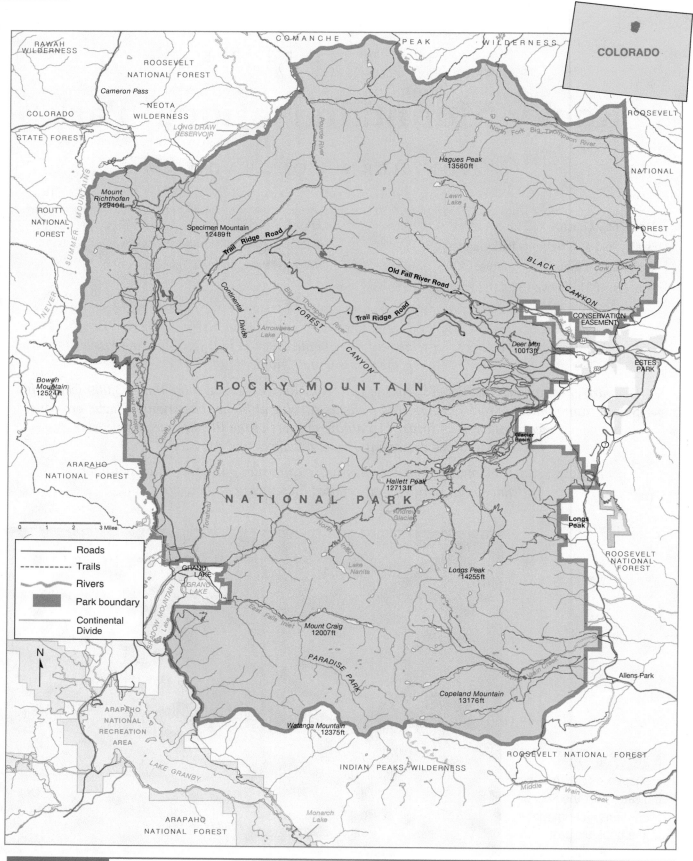

FIGURE 25.2   Rocky Mountain National Park, Northern Colorado.

Box 25.1	**More about Igneous Rocks**

In previous chapters, you have become familiar with terms referring to plutonic bodies of various sizes and shapes, such as dikes, sills, batholiths, and stocks, which formed from the solidification of magma, or molten rock, that originated under conditions of intense pressure and heat far below the earth's surface. *Intrusive igneous rocks* tend to be medium- or coarse-grained because they solidify from magma that cools slowly at depth over a long period of time. Magma that rises to the surface, erupting as lava, forms *extrusive igneous rocks.* They normally cool so rapidly that individual mineral grains in most specimens cannot be identified without the aid of a microscope.

In classifying or identifying igneous rocks, determining the texture, or grain size, is an important step (fig. 25.3). Are individual mineral grains in the rock visible to the unaided eye (larger than 0.0024 inches). Or are the grains too small to be seen without the aid of a microscope (generally smaller than 0.0024 inches)? If the grains can be seen, is the texture coarse, medium, or fine?

Sometimes changes in a magma's environment during cooling result in the formation of mineral crystals of distinctly different sizes within a rock, creating *porphyritic texture,* an effect that is sometimes described as similar to rice pudding with raisins. *Pegmatites,* which have very coarse texture, crystallize during the last cooling phase of a magma and are typically found around the margins of a large pluton. Sometimes large sheets of muscovite mica and deposits of rare minerals, such as beryl, can be mined from pegmatite bodies. (Other textures, mostly associated with volcanic activity, are defined in table 25.1.)

Igneous rocks are also classified on the basis of their mineral composition. The chemical composition of the parent magma determines this aspect of igneous rock classification. From *felsic* magmas, rich in aluminum, silicon, sodium, and potassium, come the light-colored rocks that mainly contain the minerals orthoclase feldspar, sodium-plagioclase feldspar, and quartz. At the other extreme are *mafic,* or iron-rich magmas, that are the source of the dark-colored rocks, made up primarily of the ferromagnesian minerals. Table 25.1 shows the classification of common igneous rocks by color (from light to dark), mineral composition, and texture.

Observe that a coarse-textured granite may have essentially the same mineralogical composition as a very fine-textured rhyolite, with both being light in color. Similarly, a dark, coarse gabbro may contain the same minerals as a dark, fine-grained basalt. What is surprising, however, is that the intrusive granite and the extrusive basalt are the most abundant igneous rocks on earth, while rhyolite (an extrusive rock) and gabbro (intrusive), the respective mineralogical equivalents, are relatively rare in the crust. For this paradox of nature, involving the most abundant extrusive and intrusive igneous rocks, geologists have as yet no wholly satisfactory explanation. However, in many respects, the distribution patterns of igneous rocks and their great variety can be accounted for by plate tectonic interpretations, based on field data; studies of composition and texture of rocks; and experimental work on artificial magmas.

Many of the igneous rock specimens that you see in Rocky Mountain National Park and other national

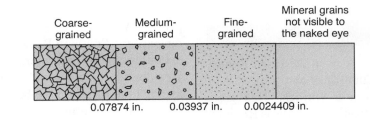

Coarse-grained	Medium-grained	Fine-grained	Mineral grains not visible to the naked eye
0.07874 in.	0.03937 in.	0.0024409 in.	

FIGURE 25.3	Comparative grain sizes in igneous rocks.

(continued)

Box 25.1	**More about Igneous Rocks (continued)**

parks have mineral grains large enough to be identified by eye using the characteristic physical properties of the constituent minerals. For convenience, diagnostic physical properties—grain shape, color, and cleavage—of common rock-forming minerals are

given in table 25.2. *Cleavage* is the breaking of a mineral along specific planes that reflect crystal structure. Not all minerals have cleavage, but minerals that do, break or "cleave" in a characteristic direction or set of directions.

Table 25.1	**Classification of Common Igneous Rocks by Color, Composition, and Texture**			
Characteristic rock color	**Light gray "salt-and-pepper"**	**Medium gray and medium green**	**Dark gray to black**	**Very dark green to black**
Mineral content	Quartz; feldspars (white, light gray or pink); a few ferromagnesian minerals	Feldspars (white or gray) and about 50% ferromagnesium minerals	Predominantly ferromagnesian minerals with some plagioclase feldspar	Entirely ferromagnesian minerals, usually olivine and pyroxene
Coarse-grained texture	GRANITE	DIORITE GRANODIORITE MONZONITE	GABBRO	ULTRAMAFIC ROCKS
Fine-grained texture	RHYOLITE	ANDESITE DACITE	BASALT	

Other varities of Igneous rocks classified by texture:
Obsidian: volcanic glass (usually of rhyolitic composition)
Tuff: pyroclastic rock consisting of consolidated volcanic ash (usually rhyolitic or andesitic)
Pumice: light-colored, highly vesicular, pyroclastic material; a form of frothy volcanic glass (commonly rhyolitic)
Breccia: pyroclastic rock consisting of coarse volcanic fragments
Pegmatite: very coarse-grained rock (usually granitic)
Porphyry: large crystals in a ground mass of smaller crystals
Vesicular: rocks that have holes created by gas trapped in cooling lavas

Geographic Setting

Rocky Mountain National Park is on the Front Range, a towering mountain mass that rises abruptly west of Denver, Boulder, and Colorado Springs. This great topographic barrier extends from the Arkansas River on the south to the north end of the Laramie Range in southern Wyoming. Between the park's northern and southern boundaries, about 21 miles, air distance, is the highest part of the Front Range, including the highest

peaks. The Front Range is the easternmost mountain range of the Southern Rockies and the first major range encountered by travelers coming from the east. Westbound pioneers, with their wagons and animals, avoided the Front Range because it was too much of a barrier. Wagon trains went north across Wyoming or south along the Sante Fe trail through New Mexico.

If you approach Rocky Mountain National Park from Estes Park, a community in the Big Thompson River valley, outside the park's eastern boundary, the

Table 25.2	Characteristic Appearance in Igneous Rocks of Common Rock-Forming Minerals
Mineral	**Diagnostic Physical Properties**
Orthoclase feldspar	1. Grains generally of uniform size 2. Usually white or pink 3. Two directions of cleavage nearly at right angles
Plagioclase feldspar	1. Lath-shaped grains 2. White, gray, or slightly greenish 3. Two directions of cleavage nearly at right angles; very fine parallel markings on cleavage faces
Quartz	1. Irregularly shaped grains formed between other grains 2. Colorless, glassy appearance 3. No cleavage
Olivine	1. Small, round grains 2. Pale green, glassy appearance 3. No cleavage
Amphibole group: example, hornblende	1. Lath-shaped grains 2. Black or dark green 3. Two directions of cleavage not at right angles
Pyroxene group: example, augite	1. Lath-shaped or blocky grains 2. Black or dark green 3. Two directions of cleavage nearly at right angles
Muscovite mica	1. Flat, flaky grains 2. Silvery white or pale green 3. One direction of cleavage
Biotite mica	1. Flat, flaky grains 2. Black or dark brown 3. One direction of cleavage

highway is at an elevation of 7500 feet. Upturned Paleozoic sedimentary rocks are exposed in roadcuts. Then if you drive through the park on Trail Ridge Road, you cross the highway's highest point (between Fall River Pass and Lava Cliffs) at an elevation of 12,183 feet above sea level. As you make the climb, you see displayed on cliffs, peaks, and precipitous valley slopes some of the oldest rocks in North America—contorted and twisted gneisses and schists and speckled granites. Several other highways and many trails lead visitors into the rich scenery and massive grandeur of this mountain park. Some of the trails follow paths that Indians used when crossing the mountains.

Lieutenant Zebulon M. Pike (for whom Pikes Peak, near Colorado Springs, is named) explored the region that is now Rocky Mountain National Park in 1806-07. He referred to the tallest summit in the range as "Great Peak" but it was renamed Longs Peak in 1820 by Major Stephen H. Long, whose expedition was sent out to map the region of the Great Plains and the Southern Rockies.

FIGURE 25.4   Tilted and contorted sedimentary beds exposed along Trail Ridge Road. Photo by Michael Little.

The first person to climb Longs Peak was William N. Beyer, who reached the summit on August 23, 1868. Among those in his party was Major John Wesley Powell, the geologist who explored the Grand Canyon in 1869. The Front Range has 42 peaks exceeding 12,000 feet in elevation; among them are Longs Peak (14,256), Mount Meeker (13,911), Ypsilon Mountain (13,314), Taylor Peak (13,153), Flattop Mountain (12,342), Snowdrift Peak (12,274), Mount Adams (12,121), and Battle Mountain (12,044).

The person who was most involved in the effort to preserve the region as a national park was Enos Mills. He built a home in Longs Peak Valley in 1866 when he was only 16 years old. After helping a survey party map Yellowstone in 1891, he was convinced that his home-site should be preserved like Yellowstone. A skilled mountain climber, Mills was the first to go up the east face of Longs Peak, a very difficult route.

The famous Trail Ridge Road, built in 1932, starts near Estes Park and climbs to the Roof of the Rockies, a broad, relatively flat upland extending over most of the central area of Rocky Mountain National Park. In a 40-mile trip over Trail Ridge Road, you traverse life zones equivalent to a round trip to the Arctic Circle. The drive may take you from hot summer weather to wintry conditions with even a snow squall. Above the treeline, the upland is a rolling terrain covered by frost-shattered rock. The fragile vegetation of alpine tundra struggles to survive among the rocks. Nearly continual high winds sweep over the upland, even during the short summer season. At Milner Pass (elevation 10,758 feet) near the park's western boundary, Trail Ridge Road crosses the Continental Divide (fig. 25.5) and then follows the Colorado River south to Grand Lake.

Geologic Features

The Crystalline Bedrock

The ancient igneous and metamorphic rocks exposed in Rocky Mountain National Park are *crystalline;* that is, consisting of crystals (or fragments of crystals), most of them large enough to be seen without a lens. The metamorphic rocks, which originally may have been sedimentary, have been recrystallized by the heat and pressure of intense and prolonged metamorphism. Igneous rocks, emplaced in plutons, crystallized from a melt. Both igneous and metamorphic rocks have been altered in varying degrees by several episodes of *regional metamorphism* (metamorphism caused by high temperature and pressure affecting rocks over a large area). Some of the rocks that were intruded by several gener-

FIGURE 25.5 The continental divide, off Trail Ridge Road, determines whether runoff drains to the east or the west. Photo by Michael Little.

ations of granitic magmas show the effects of *contact metamorphism* (metamorphism where high temperature is the dominant factor, usually in a narrow zone between intruding magma and country rock).

The central part of the Front Range has a complex of small plutons fitted between wedges of gneisses, schists, and quartzites. Large batholiths, such as the Pikes Peak Granite in the southern part of the range and the Sherman Granite at the north end, flank the smaller plutons.

Within the complexly faulted, anticlinal arch that makes up the Front Range, there are numerous cross folds and faults. (A *cross fold* is one that intersects a previous fold of different orientation.) The faults and shear zones, which trend generally northwest-southeast, are mainly of Precambrian age. Most of them were reactivated during the Laramide orogeny at the end of Cretaceous time.

The Development of the Rocky Mountains

Although geologists have studied the Front Range and the Rocky Mountains for many years, the relationships

of plate movements and episodes of orogeny have not been entirely worked out. One indication of a very long history of tectonic activity in this region is the elevation of the basement rocks—the Precambrian gneisses, schists, and granites—that are now exposed in the Front Range. You can see them thousands of feet above sea level on the Continental Divide and the peaks above the upland.

The present topography of the Front Range is due to a broad upwarping that took place during the Cenozoic Era, when vertical uplift brought Precambrian basement rocks to heights far above their former surroundings. However, the mountains that we see today are but the most recent manifestation of tectonic activities that have gone on in this part of the earth's crust for many millions of years.

In Paleozoic and Mesozoic time, the deeply eroded, crystalline Precambrian rocks were submerged by successive invasions of shallow seas in which many units of marine sedimentary rocks were deposited. Intermittent tectonic activity during the Pennsylvanian Period deformed Paleozoic rocks in central Colorado in an orogenic belt, trending northwest-southeast, that elevated the rocks into a range called the "Ancestral Rockies." These Paleozoic mountains, subsequently eroded away, may have been a westward extension of the Ouchita Mountains. The Ancestral Rockies are in no way related to the present-day Rocky Mountains, but their former presence can be inferred from remnants exposed around the edges of the high mountain ranges.

The Laramide orogeny formed the gross structures of the Rocky Mountains. The mountain-building began as a result of interaction between the North American plate moving westward and the Pacific plate drifting slowly eastward. Compressive forces, generated when the plates converged, intensified from west to east, culminating late in Cretaceous and early Tertiary time. Major orogenic activity, which occurred in the region that is now New Mexico, Colorado, northern Utah, Wyoming, Idaho, and Montana, formed the Northern and Southern Rockies.

Eastward and northeastward thrust faulting, folding, and crumpling of the crust produced a number of elongated mountain ranges where the crystalline basement rock was pushed up to form the cores of very large, faulted, anticlinal structures. The Colorado Front Range illustrates how this happened. Paleozoic and Mesozoic sedimentary rocks lap up against the eastern face of the Front Range (just to the east of the park boundary). High-angle thrust faults pushed the crystalline core thousands of feet upward and outward over the margin of the upfolded Paleozoic rocks.

The uplifting of the Front Range was but one aspect of the Laramide orogeny. Compressive forces deformed and uplifted rocks from Central America to Alaska, creating the North American Cordillera. In order to reach the Colorado area, subducted slabs of oceanic crust must have moved way inland from the western edge of the continent and must have been descending at a relatively low angle. The massive cores of the Rocky Mountain ranges transmitted stresses by large-scale faulting along their eastern margins but without intense deformation of their own structures. (See fig. 2.8 in Chapter 2.)

As the mountains were raised higher and higher, erosion intensified. Sediments from summits and mountain slopes filled the valleys and lowlands. A period of crustal adjustment, called the isostatic phase, ensued as the orogenic pressures subsided. The principle of *isostasy* proposes a condition of equilibrium—comparable to floating—involving units of brittle crust on the plastic zone of the *mantle* (the thick rock shell of the earth's interior between the crust and the core). Blocks of crust that are not in equilibrium, due to tectonic forces raising them up or the weight of accumulated sediment pushing them down, tend to compensate by moving up or down in a direction toward equilibrium. Block-faulting and other vertical adjustments of crustal units can sometimes be explained in terms of isostatic compensation. However, this principle cannot account for the much more powerful mechanisms of plate subduction and plate convergence that clearly involve forces of much greater intensity.

Volcanic Activity, Pegmatites, and Hydrothermal Veins

The major faults on both sides of the Front Range and the other faults and folds in the park region were the result of Laramide tectonic activity. In faulted areas, eruptions and intrusions of various kinds occurred repeatedly from the Cretaceous Period to early in the Pleistocene Epoch, with the greatest intensity being in Tertiary time.

Plate movement and increasing volcanic activity during the Tertiary Period was characteristic not only of the Rocky Mountain areas; it was widespread throughout much of western North America. In the Southern Rockies, more violent eruptions produced great quantities of pyroclastic dust, ash, and other debris that consolidated as tuff. Lava flows were mainly andesitic (of intermediate composition), but some were basaltic or rhyolitic (table 25.1). Some of the lava solidified as obsidian, or natural glass. Where lava flows overran streams, the cold water shattered the volcanic glass so that it consolidated as breccia. Stream gravels mixed with hardening lava and formed andesitic conglomerate. Remnants of these lavas can be seen in the Never Summer Range along the park's western border and in the northwest corner of the park, near where Mt. Richthofen stands. Lava Cliffs and Specimen Mountain, which were formed by volcanic eruptions of the same age, can be viewed from Trail Ridge Road.

Some of the igneous material solidified below the surface in dikes, sills, and stocks that were later exposed by erosion and uplift. Many of the smaller intrusions are Precambrian in age, but some are much younger. Among these bodies are the pegmatite dikes (fig. 25.6) and pods (small lenses). The coarse-grained pegmatites, made up mainly of large quartz and

FIGURE 25.6 Pegmatite dikes intruding the metamorphic rocks of the core of the mountains. The dikes were formed during an igneous intrusion. Photo by David B. Hacker.

feldspar crystals, hardened in faults and folds in the country rock around the edges of the main batholiths.

Hydrothermal veins of quartz, some of them associated with pegmatites, are the source of much of the mineral wealth of the Front Range. These veins formed when hydrothermal solutions containing mineral ions escaped from the hot magma and forced their way upward through cracks in the rock. As the liquid cooled, silica precipitated as quartz and other minerals were deposited in the quartz matrix. Most of the mineral-bearing pegmatites and veins are located in the Colorado Mineral Belt, south of Rocky Mountain National Park.

Old Erosional Surfaces

The early Rocky Mountain ranges, uplifted by Laramide mountain-building, were separated by intermontane valleys that rapidly filled with sediment. Interbedded with fluvial deposits washed down from the uplands were deposits of volcanic ash and lava from extensive eruptions. The whole region underwent a lowering of relief during early Tertiary time, and a sloping erosional surface with low hills and wide shallow valleys only a few thousand feet above sea level was

produced. The relatively flat summits of Deer Mountain and Flattop Mountain are remnants of this old erosional surface.

The broad upwarping that began in the Cenozoic Era was apparently related to the continuing subduction of the Pacific plate and/or isostatic adjustments. This affected a wide area that today stretches from Denver to Reno. Episodes of strong vertical uplift were particularly intense in the Front Range. As a result, streams began intensive downcutting, stripping away sedimentary beds and incising canyons in the resistant granitic rocks.

A second erosional surface of moderate relief developed late in Tertiary time as uplift slowed. Remnants of this surface are found on Giant Track Mountain, Old Man Mountain, Castle Mountain, Needles, and Twin Owls. The highest peaks stood as isolated knobs, or *monadnocks,* above the erosional surface. Again, alluvial deposits built fans, filled basins, and spread over the plains to the east. The increasing height of the mountain ranges to the west cut off some of the moisture supply from the Pacific and the climate became more arid.

Inferences that can be drawn from the old erosional surfaces suggest that (1) thousands of feet of rock and sediment were removed from the high areas during and after uplift, and (2) periods of uplift can be related to the erosional history of the region.

A last stage of uplift, near the end of the Pliocene Epoch, brought the mountains of the Front Range to their present height and grandeur. Streams cut deep canyons in the resistant rocks. Along the eastern side of the Front Range, much of the weaker sedimentary rock was eroded away, thus accenting the barrier appearance of the range. This was differential erosion on a grand scale.

For all their toughness, however, the basement rocks that make up the "roof" of the park are not impervious to erosive forces. The extensive joint systems, intensified by up-arching, provide conduits for weathering processes. The Needles and Twin Owls are examples of joint weathering. Exfoliation domes, such as McGregor Mountain, show the weathering of massive granite. The results of frost action at high altitudes are clearly evident along mountain trails and highways.

Glacial Features

A worldwide cooling trend that began in Tertiary time culminated in the severe glaciation of the Pleistocene Epoch. In the Rocky Mountains, the increasing eleva-

tion of the region brought about conditions favorable for glacier development early in the Pleistocene Epoch. Glacial ice displaced running water as the dominant agent of erosion. Mountain glaciers shaped the alpine scenery that we see in Rocky Mountain National Park today. On the high peaks are prominent cirques, cols, and horns. The crest of the Front Range itself is a serrated ridge. At their maximum extent, the glaciers of the Front Range coalesced to form a small icecap with the high peaks jutting up through the perennial snowfields and glacial ice.

Successive episodes of glaciation left great quantities of *glacial drift* (glacial deposits, debris, outwash, etc.) in the basins, valleys, and foothills. Morainal segments of various sizes, ages, and types are scattered over the eastern and western sides of the park in mountain meadows and glacial troughs. Some of the trails and roads in the park follow moraines. A section of the road west of the park entrance in Fall River valley was constructed on an end moraine. Most of the moraines are forested. Since the lodgepole pine prefers to grow on glacial drift, a good stand of these trees usually indicates the location of a moraine.

Rock glaciers, left by retreating ice glaciers, still contain enough ice so that freezing and thawing enables them to creep downslope. They form continuous, slow-

FIGURE 25.7   The core of the Rocky Mountains is made up of Precambrian granite that has a tendency to exfoliate, rounding off the mountaintops. Photo by David W. Little.

ly moving strips of talus for thousands of yards along the walls of high valleys. Rock glaciers that formed on some cirque floors enlarged as they pushed out over the rim and downward.

Glacial lakes appeared on the landscape as the glaciers departed. Grand Lake, just outside the southwest corner of the park, is dammed by an end moraine. Lakes in morainal kettle holes are Burstadt, Copeland, Bear, Dream, and Sheep Lakes. Cirque Lakes are Blue, Black, Shelf, Lawn, Crystal, Fern, Odessa, Tourmaline, and Chasm Lakes, as well as Lake Mills. Chains of *rock-basin lakes* are found on the "stair treads" of several rocky gorges that rise like staircases. Examples are the Gorge, Loch Vale, Cub Lake, East Inlet (including Lake Verna), and Ouzel Creek Lake. Some lakes, such as Nanita, Nokoni, and Bench Lakes, lie in hanging valleys. Streams that occupy hanging valleys are Roaring River, Chiquita Creek, Sundance Creek, and Fern Creek.

Roches moutonnées are common in the park. A *roche moutonnée* is an elongate knob or hillock of bedrock that has been smoothed and scoured by moving ice on the upglacier (stoss) side so that the rock is gently inclined and rounded. On the downglacier (lee) side, the rock is steep and hackly from glacial plucking or quarrying. Glacial Knobs is a well known area of roches moutonnées in the park.

Many rock fragments and exposed bedrock surfaces show glacial *striations*. These are furrows or lines that were inscribed on rock surfaces as rock fragments embedded in ice were dragged along with great force by a moving glacier. Striations on outcrops are usually oriented in the direction of glacier movement.

A bedrock surface that has been planed down, striated, and polished by glacial abrasion until it is relatively smooth is called *glacial pavement.* The area around North Inlet has particularly good examples of glacial pavement.

Glacial erratics are abundant at Boulder Field and on the sides of Fern Creek canyon. An *erratic* is a relatively large rock fragment that has been transported by ice from its place of origin that may be a considerable distance away. Boulder Field is also a *felsenmeer* area where in postglacial time, frost-shattered, angular blocks of rock have covered the well jointed bedrock.

Examples of V-shaped stream valleys in the park can be compared with glacially sculptured U-shaped valleys, all of which were originally V-shaped. Fox Creek, Red Creek, and Cow Creek have characteristic V-shapes. Fall River valley, Forest Canyon, Fern Lake valley, and North St. Vrain valley are U-shaped and display other evidence of glacial erosion and deposition.

Mass Wasting Features

Periglacial conditions in the park brought about some modification of glacial topography due to mass wasting on both steep and gentle slopes. On Sundance Mountain and along Trail Ridge Road, *solifluction,* or flowage of soil, has caused slumping and development of hummocky lobes on gentle upland slopes above the treeline and above glacial valleys. Because water cannot percolate down through the frozen ground that exists at this altitude, the soil becomes saturated during the melt season and begins to slide on even a slight slope. On steeper slopes, especially at lower elevations, landsliding and avalanching may occur seasonally. The abundance of talus debris, talus cones, and rock glaciers also indicates active mass wasting.

Where there are loose rock fragments or soil cover on high, flat surfaces above the treeline, various forms of *patterned ground,* such as stone rings, polygons, stripes, and nets, have developed. This is due to periglacial conditions, intense frost action, and very slow mass movement, which, in combination, tend to assemble stones in symmetrical patterns or designs.

Geologic History

1. **Accumulation of sediments during Precambrian time.**

The oldest rocks in the park, the schists and gneisses, were deposited over 1800 million years ago. Many thousands of feet of sediment formed shales, sandstones, and carbonate rocks interbedded with volcanic rocks (fig. 25.8).

2. **Folding, intrusion of igneous rock, and metamorphism.**

Tectonic activity compressed, intensely folded, metamorphosed, and uplifted the sedimentary beds. Intrusions of magma rose from a chamber far below the surface (fig. 25.9). As the orogeny continued, increased heat and pressure intensified the metamorphic processes, changing the rocks further into schists and gneisses. Prominent exposures of these ancient rocks are in outcrops along Trail Ridge Road and in cliffs along valley walls in the western and central parts of the park. Schists and gneisses can also be seen along the Continental Divide in the cirque headwalls (fig. 25.10).

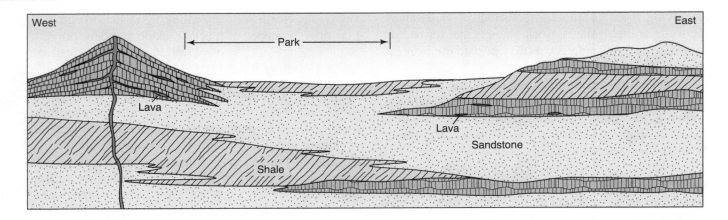

FIGURE 25.8 Great quantities of volcanic and clastic sediments accumulated throughout early Precambrian time.

FIGURE 25.9 Long after the sedimentary layers were buried and lithified, they were folded, metamorphosed, and intruded by granitic magma during Precambrian mountain-building.

FIGURE 25.10 The alternate light and dark bands indicate that this is the metamorphic rock gneiss. The zig-zag pattern shows that after the gneiss formed it was folded. Photo by David B. Hacker.

3. Precambrian uplift and intrusions.

While the earlier magma was cooling, a larger intrusion pushed its way up into the schists and gneisses, eventually creating a granitic mass about 30 miles across. Intensely hot hydrothermal solutions penetrated the country rock surrounding the batholith, and pods and lenses parallel to the banding of the schists and gneisses were formed. In some places the confining pressure was so great that pegmatites formed during the long, slow cooling. Radioactive minerals in the intruded rocks have provided an absolute age figure that the event took place about 1450 million years ago (fig. 25.11).

These granites crop out in the western part of Rocky Mountain National Park; in the area east of Shadow Mountain and Grand Lake; and also on the east side of the park, north and south of Estes Park. Additional exposures are on Hagues Peak (near the northern boundary) and on the east face of Longs Peak (near the southeast boundary). The contact between the older schists and gneisses and the granitic intrusions shows up on canyon walls east of the Continental Divide.

4. Erosion of the Precambrian highlands.

The mountains that were formed in the Precambrian orogenies were worn down during a span of about a thousand million years until the tops of the granite batholiths were uncovered. The sands and other clastics that were the products of erosion were lithified as sandstones blanketing the gneisses, schists, and granites. A few remnants of these ancient sandstones have been found overlying granite in the Colorado Springs area southeast of the park, but none within the park itself (fig. 25.12).

A basaltic dike, called "Iron Dike," suggests that another intrusion occurred in late Precambrian time. The dike crops out near Mount Chapin and Storm Pass north of Trail Ridge Road.

5. Uplifting of the Ancestral Rockies.

During early and middle Paleozoic time, shallow inland seas advanced and withdrew, at times covering most of the area that is now Rocky Mountain National Park. For 200 million years Paleozoic gravels, sands, muds, and oozes accumulated and were lithified. None of the sedimentary rocks deposited in this period remain in the park today.

Gradual uplift in the Pennsylvanian Period, about 300 million years ago, created the Ancestral Rockies in the region west of the Front Range. At their maximum the mountains may have been about 2000 feet high. Along with the uplift came increasing erosion and downcutting by streams, processes that stripped away the previously deposited marine sediments, along with some of the Precambrian igneous and metamorphic rocks (fig. 25.13).

6. Mesozoic reduction of relief and transgressions of seas.

By the end of the Paleozoic Era, the Front Range region was a low, swampy area with only low hills remaining where the Ancestral Rockies had stood. Amphibians, reptiles, and dinosaurs roamed over the land. Their footprints and fossilized remains are preserved in sedimentary rocks not far from the park.

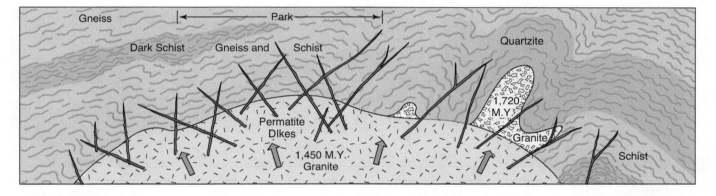

FIGURE 25.11   Late in Precambrian time, during a long period of intense mountain-building and regional metamorphism, a large granitic batholith was emplaced.

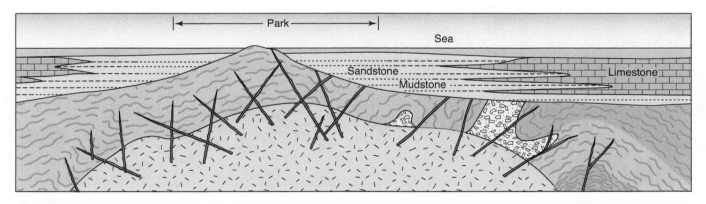

FIGURE 25.12   For millions of years, as erosion wore down the uplands, Paleozoic seas spread over the land. Many layers of sedimentary rock covered the crystalline Precambrian basement.

Jurassic coal beds formed in swamps, and coastal dunes migrated along the shores of an inland sea. Remnants of the dunes can be found in the sedimentary rocks east of the park along the mountain front.

7. Maximum advance of seas in Cretaceous time, followed by uplift.

About 100 million years ago, seawater completely covered the region (fig. 25.14). Many feet of sand and clay accumulated on the shallow sea floor. Remains of some of the shale beds that were deposited at this time are in the northwest part of the park, east of Lead Mountain.

About 70 million years ago, the region that eventually became the Front Range began to rise again, forming a low island in the sea. Uplift continued until, by the end of the Cretaceous, all of Colorado was above sea level. As the land rose, stream systems developed, eroding the marine rocks and redepositing the sediment in new layers over the surrounding areas.

8. Laramide orogeny; uplift and faulting; stripping of sedimentary beds.

At the end of the Cretaceous Period and the beginning of Tertiary time, a major event in the geologic history of North America took place. During the Laramide orogeny, the North American Cordillera, which extends from Mexico northward into Canada and Alaska, was created. In the Rocky Mountain National Park region, large blocks were uplifted along deep fractures in the Precambrian basement rocks (figs. 25.15, 25.16).

Periods of volcanism ensued. The lava flows at Mt. Richthofen mark an episode. Near Specimen Mountain, mudflows, lava flows, and obsidian are indications of a later and different type of volcanic activity of Oligocene age. Evidence of volcanic ash can be seen near Iceberg Lake along Trail Ridge Road (fig. 25.17).

During and between the episodes of volcanism and uplift, the processes of erosion went on, lowering relief, stripping away sedimentary rocks, and cutting into the Precambrian schists, gneisses, and granites

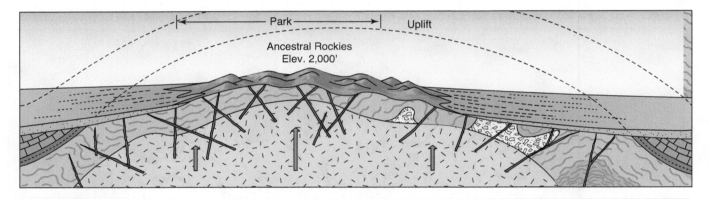

FIGURE 25.13   The region was uplifted in middle to late Paleozoic time, forming the "Ancestral Rockies." Erosion gradually lowered the highlands.

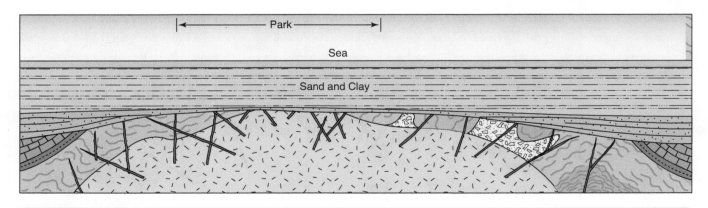

FIGURE 25.14   During the Cretaceous period, the sea advanced over the region for the last time. Marine sediments were laid down on the old surface.

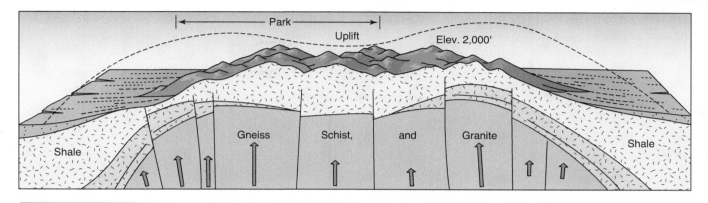

FIGURE 25.15 At the end of Cretaceous time, differential uplifting, due to Laramide mountain-building, produced fault blocks. Compressive forces caused thrust-faulting to occur.

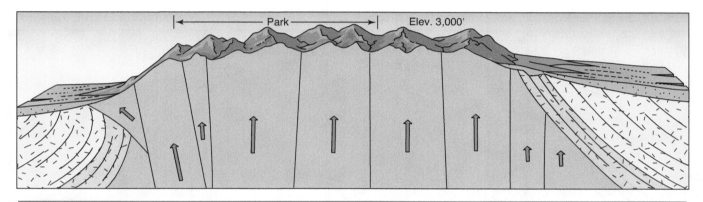

FIGURE 25.16 As uplift continued into Tertiary time, erosion stripped off sedimentary layers, exposing the rocks of the Precambrian crystalline basement.

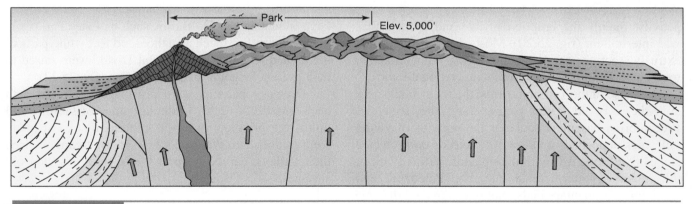

FIGURE 25.17 Volcanism accompanied uplift and erosion.

(fig. 25.18). In all, about 5000 feet of sedimentary rocks were removed.

9. Late Tertiary upwarping and erosion.

In late Tertiary time, successive uplifts that were part of a broad regional up-arching (probably due to iso-static compensation) brought the mountains of the park to their present dramatic height. The uplifts occurred primarily along the old faults and increased elevations as much as 4000 to 6000 feet (fig. 25.19). The Front Range appeared to rise as a series of steps, particularly on the east side of the Continental Divide. Volcanic

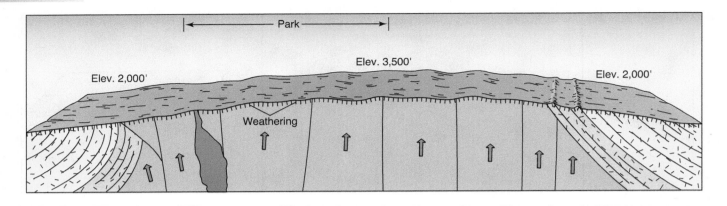

FIGURE 25.18 By middle Tertiary time, erosion had removed so much rock that summits were low and relief was subdued.

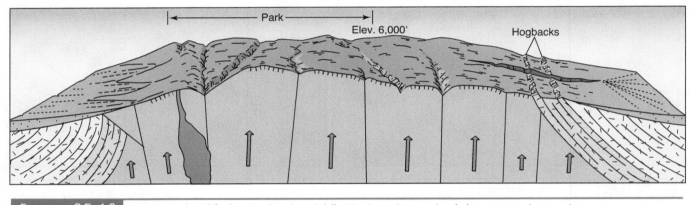

FIGURE 25.19 Renewed uplift, beginning in middle Tertiary time, raised the mountains again.

activity, mainly the eruption of ash and pyroclastics, was intermittent (fig. 25.20). Most of the intrusive and extrusive activity and accompanying mineralization took place south of Rocky Mountain National Park.

Because of the Tertiary uplifts, the rocks in the park are either very old or very young; that is, they are either Precambrian basement rocks or the remains of young volcanics deposited in valley-fill areas or on protected slopes that "rode up" with the uplands. Virtually all of the rocks of intermediate age, the Paleozoic and Mesozoic sedimentary beds, that once covered the area were stripped away. Erosional remnants of these rocks form the *hogbacks* of the foothill ranges that make up a narrow transitional belt east of Rocky Mountain National Park, between the plains and the Front Range (fig. 25.21).

By the end of the Tertiary, parts of the land surface had been markedly displaced by vertical movement of the large mountain blocks. Basins such as Estes Park were down-dropped as much as a thousand feet while parts of the upland along the Continental Divide were raised to their present height of over 12,000 feet (fig. 25.22).

Drainage patterns were disrupted or altered with each episode of uplift. In the last stage of faulting and uplift, the principal drainage patterns that exist today were established. Waterfalls formed as streams incised their valleys, carved deep canyons, and dissected old surfaces.

10. Pleistocene glaciation.

Glacial ice advanced and retreated many times in Rocky Mountain National Park; however, the major episodes can be grouped as follows:

1. Early glaciation (pre-Bull Lake), which involved at least two major advances, began around 1,200,000 years ago.

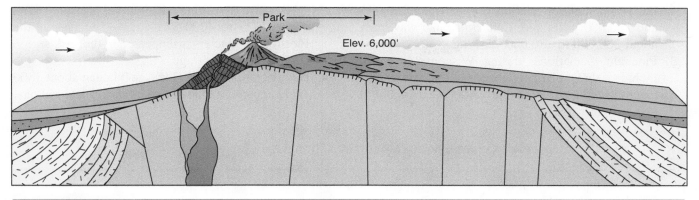

FIGURE 25.20   Volcanoes were built up by eruptions as lavas rose to the surface.

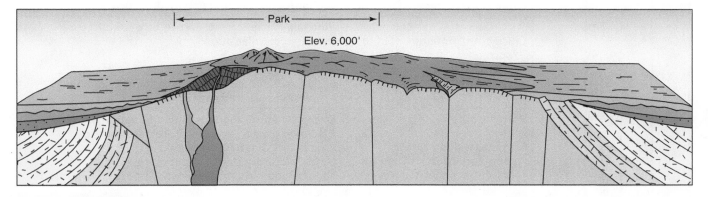

FIGURE 25.21   Increased erosion bared the uplifted crystalline summits. Remnants of Paleozoic and Mesozoic sedimentary rocks formed hogback ridges in the foothill ranges to the east.

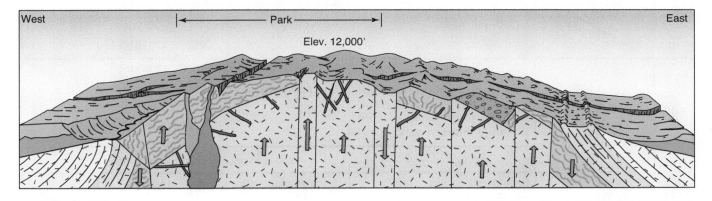

FIGURE 25.22   Vertical fault movements raised some blocks and dropped others, causing disruptions of drainage and some displacement of the land surface. The Front Range had attained roughly its present elevation by the beginning of Quarternary time.

2. Early Bull Lake glaciation was between 300,000 and 200,000 years ago; Late Bull Lake was from about 200,000 to 130,000 years ago.
3. Pinedale glaciations, during Wisconsinan time, included three distinct episodes. The maximum extent lasted from 20,000 to 18,000 years ago.

A warm (hypsithermal) interval prevailed from about 9000 to 4000 years ago. The last of the Pleistocene glaciers disappeared in the park region about 10,000 years ago.
4. Neoglaciation ("Little Ice Age") began about 3800 years ago and included two advances and retreats.

Table 25.3			Geologic Column, Rocky Mountain National Park	
Time Units				
Era	**Period**	**Epoch**	**Rock Units**	**Geologic Events**
Cenozoic	Quaternary	Holocene	Neoglacial deposits, alluvium, talus	Development of modern glaciers; active mass wasting; periglacial processes; stream erosion about 3800 ybp Melting away of earlier glaciers during post-Pleistocene warming about 10,000 ybp

		?
		Pleistocene
			Glacial tills, moraines, etc., of successive ages; early deposits largely buried or destroyed by later advances	Pinedale glaciation (at least three episodes)
				Intermediate glaciation
				    Late Bull Lake
				    Early Bull Lake
				Early pre-Bull Lake glaciation (at least two major advances and retreats)

	Tertiary	Paleogene/ Neogene ?	Volcanic ash Alluvial deposits Rhyolite, welded tuff, obsidian, pyroclastic debris, mudflow deposits	Development of modern drainage patterns Extensive erosion exposing basement rocks on summits, hogbacks on flanks Block-faulting forming basins Uplift along old faults raising Front Range to present height Volcanism (mainly to west and south of park) Mudflows, lava flows, ash falls
?				Laramide orogeny Intermittent uplift, faulting, tilting of sedimentary beds; stripping of beds producing erosional surface
Mesozoic	Cretaceous		Shales	Last transgression and withdrawal of seas Reduction of relief
	Jurassic Triassic			
Paleozoic	(Cambrian to Permian)		Remnants of sedimentary rock units exposed on eastern and western flanks of Front Range (none in park)	Uplift and erosion of Ancestral Rockies Long deposition, first marine, then terrestrial (due to uplift)
			/////////////////	Long erosion exposing granite batholiths
Late Precambrian (Proterozoic Y)			Basaltic dikes, granites, pegmatites	Successive intrusions and emplacement of batholiths; uplift
Middle Precambrian (Proterozoic X)			Granites, schists, gneisses (basement complex)	Uplift, orogeny, metamorphisms, intrusions Long accumulation of sediment interbedded with volcanics

Source: After Larkin, Grogger, and Peters, 1980; Richmond, 1974.

Traces of the earliest glaciations within the park were obliterated by later ice advances. The oldest glacial deposits in or adjacent to the park are found near the eastern boundary. Some are in roadcuts along Colorado Highway 7 in the southern Tahosa Valley and some are near the Fall River entrance to the park.

Since Bull Lake glaciations were extensive and severe, evidence of their erosion and deposition is distributed throughout the park and adjacent to park boundaries. Both Early and Late Bull Lake advances deepened the same cirque basins, enlarged canyons, joined together tributary glaciers, and left prominent end moraines, usually not far apart. Since the moraines tend to be similar in size, shape, and type of material, only the amount and depth of weathering of pebbles in the deposits distinguish them. Examples of Bull Lake end moraines can be seen in Aspenglen Campground and near the Beaver Meadows entrance. On the west side of the park, Bull Lake glaciers that occupied the Colorado River valley left end moraines near Shadow Mountain Lake.

Pinedale glaciation is notable for the sculpturing of the park's alpine scenery. The cirque headwalls along the Continental Divide and the nearly vertical east face of Longs Peak were steepened by Pinedale glaciation, even though the ice in the valleys did not attain the thickness of the Bull Lake glaciers.

On the east side of the park, Pinedale glaciers began in the perennial snowfields and cirques of the Mummy Range and then flowed down east-trending valleys. One glacier, 13 miles in length, occupying Big Thompson valley, flowed down from the eastern side of the Continental Divide. Another group of glaciers, which originated in the cirque basins of the Never Summer Mountains, flowed down to the Colorado River valley and joined westward-flowing glaciers from the Continental Divide upland to form the park's longest glacier. It extended for 20 miles from an area west of La Poudre Pass southward to the morainal islands in Shadow Mountain Lake.

By scouring preexisting valleys, the Pinedale glaciers gouged out chains of rock-basin lakes, such as the Gorge Lakes and those in Tyndall Gorge. Glacial snouts dug out basins, such as Horseshoe Park excavated by Falls River Glacier, Moraine Park by Thompson Glacier, and Bartholf Park by Bartholf Glacier. As the glaciers melted, they left drift in the basins.

Moraines left by Pinedale glaciers show less evidence of erosion and weathering than older glacial drift. Pinedale moraines tend to be prominent ridges with numerous boulders scattered over the surface. The end moraines west of the Fall River valley entrance are examples.

11. **Post-Pleistocene glaciers and erosion.**

Although all the Pleistocene glaciers melted away during the post-Pleistocene warming period, new glaciers reformed in the old cirques several thousand years ago during Neoglacial advances. The modern glaciers have shrunk to little more than perennial snowfields and *glacierets* (miniature alpine glaciers). In recent years, a few, such as Sprague, Andrews, Tyndall, and Taylor Glaciers, have increased slightly in volume. The moraines built by the post-Pleistocene glaciers are the easiest to identify because they are the least weathered and dissected. The best places to see Neoglacial moraines are near the cirques because the glaciers that built them did not extend far down the valleys.

With glaciers restricted to small areas at high altitudes, the processes of stream erosion and deposition have again become dominant in the shaping and modification of park landscapes. Processes of mass wasting, particularly those associated with periglacial conditions, also continue to play a significant role in landform evolution in Rocky Mountain National Park.

Geologic Map and Cross Section

Braddock, W.A., and Cole, J.C. 1990. Geological Map of Rocky Mountain National Park and Vicinity, Colorado. U.S. Geological Survey Miscellaneous Investigation series I-1973.

Sources

Chronic, H. 1984. *Time, Rocks, and the Rockies: Geologic Guide to Rocks and Trails of Rocky Mountain National Park.* Missoula, Montana: Mountain Press Publishing Company. 120 p.

——— 1984, *Pages of Stone: Geology of Western National Parks and Monuments.* I. *Rocky Mountains and Western Great Plains.* Seattle: The Mountaineers. p. 115–127.

Dickinson, W.R., and Snyder, W.S. 1978. Plate Tectonics of the Laramide Orogeny. In *Laramide Folding Associated with Basement Block-faulting in Western United States*, ed. V. Matthews, III, Geological Society of America Memoir 151, pp. 255–56.

Larkin, R.P.; Grogger, P.K.; and Peters, G.L. 1980. *Southern Rockies*, K/H Geology Field Guide Series. Dubuque, Iowa: Kendall/Hunt Publishing Company.

Pendick, D. 1997. Rocky Mountain Why, *Earth* v. 6, no. 3 (June), p. 26–33.

Richmond, G.M. 1965. Glaciation of the Rocky Mountains, H.E. Wright Jr. and D.G. Frey, editors. *The Quaternary*

of the United States. VII Congress of the International Association for Quaternary Research. Princeton University Press. p. 217–230.

———— 1974. *Raising the Roof of the Rockies.* Estes Park, Colorado: Rocky Mountain Nature Association, Inc.

Note: To convert English measurements to metric, go to www.helpwithdiy.com/metric_conversion_calculator.html

C H A P T E R   2 6

Waterton-Glacier International Peace Park

GLACIER NATIONAL PARK, NORTHWEST MONTANA
WATERTON LAKES NATIONAL PARK, SOUTHWEST ALBERTA

NORTHWEST MONTANA

Area: 1,013,572 acres; 1584 square miles

Established as a National Park: May 11, 1910

Proclaimed Part of Waterton-Glacier International Peace Park: June 30, 1932

Designated a Biosphere National Reserve: 1976

Address: Park Headquarters, West Glacier, MT 59936

Phone: 406-888-7800

E-mail: glac_information@nps.gov

Website: www.nps.gov/glac/index.htm

SOUTHWEST ALBERTA

Established as a National Park: 1895

Designated a Biosphere Reserve: 1976

Address: Waterton, Alberta, Canada T0K 2M0

Phone: 403-859-2445

Precipitous peaks, glacial troughs, lakes, and the Continental Divide, extend over both sides of the international boundary. Modern alpine glaciers occupy outsize cirques carved by Pleistocene glaciers; waterfalls leap from hanging valleys. Ancient, undeformed, sedimentary beds are displayed on summits and steep slopes. Along mountain trails and Going-to-the-Sun Highway are ripple marks, mud cracks, and fossils of primitive Precambrian plants. Horizontal displacement by the Lewis Overthrust pushed Precambrian strata miles eastward over much younger rocks.

FIGURE 26.1 Going-to-the-Sun Mountain near Siyeh Bend. The peak of the mountain is the Snowslip Formation, the bulk of the mountain front is the Helena (formerly the Siyeh) Formation. The Helena is easy to distinguish because of the black diorite sill that intruded the reddish beds; the heat from the intrusion baked the dolomitic Helena into a white marble on either side of the intrusion. It acts as a marker bed throughout the park. Photo by Douglas Fowler.

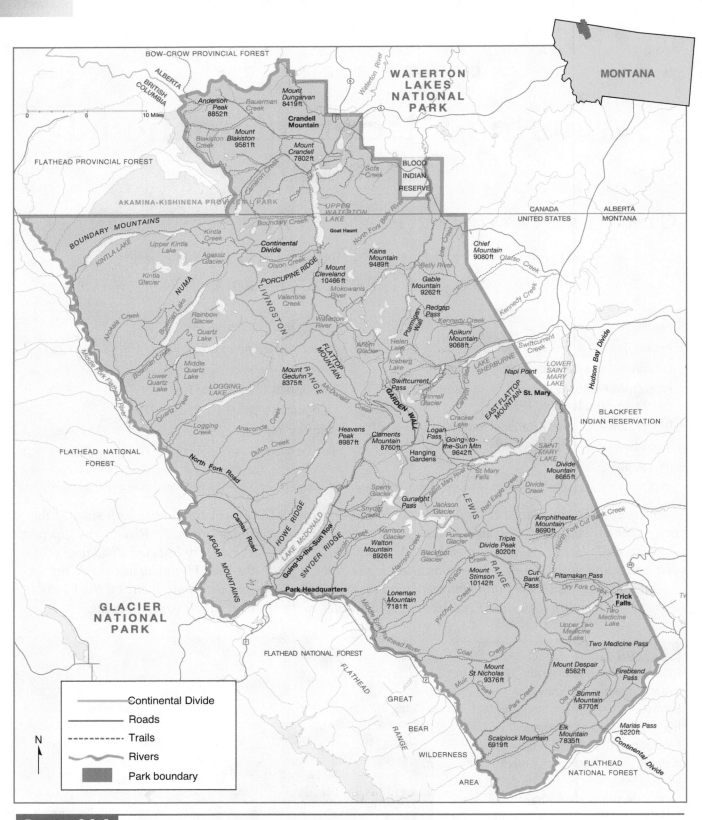

FIGURE 26.2 Waterton-Glacier International Peace Park, Alberta (Canada) and Montana.

Geographic Setting

Although Glacier National Park and Waterton Lakes National Park were set aside separately, Canada and the United States have shared this mountain wilderness for nearly 60 years. The joining together of the two parks to form Waterton-Glacier International Peace Park came about through the grass-roots efforts of Rotary International. The two governments cooperate in keeping the wilderness pristine and largely undisturbed by human activities while each jurisdiction administers the parkland on its side of the border. People move between the two parks almost as freely as the animals. Access roads are limited, but several trails cross the border and an excursion boat on Upper Waterton Lake makes frequent trips in summer between the village of Waterton, Alberta, and a visitor center on the U.S. side at the southern end of the lake.

The map of Waterton-Glacier shows the Continental Divide running from north to south along the western boundary of the Canadian park and through the U.S. park. Streams that drain the western side of the Continental Divide carry water that may eventually reach the Pacific Ocean. Water falling on the eastern side enters streams that end up in either Hudson Bay or the Gulf of Mexico. Triple Divide Peak, south of St. Mary Lake, is unique because water drains in three directions instead of two; namely, west toward the Pacific, northeast toward Hudson Bay, and southeast toward the Gulf of Mexico.

Because of its location straddling the Continental Divide, Waterton-Glacier makes its own weather. The dominant weather pattern is the flow of warm, moist air from the Pacific, which cools and drops rain or snow as air masses rise over the west side of the Continental Divide. The cooled air then plunges down the east slopes, becoming warmer as it descends. Periodically, cold fronts from the Arctic sweep down along the eastern side of the mountains, slide under warmer Pacific air from the west, and create fog, clouds, and gusty winds. Temperature ranges are extreme and may change rapidly. Snow may occur any month of the year. Because of the rain-shadow effect, winter accumulations of snow are much thicker on the western slopes than along the eastern edge of the park.

Indians have lived in the region for thousands of years. The Blackfeet Indian Reservation borders the eastern side of Glacier National Park. St. Mary Lake and Lake McDonald, the two largest of the more than 200 glacier-fed lakes in the park, were favorite areas of Indian habitation.

Government surveyors marked the 49th parallel as the international boundary in the 1860s. Today only a narrow swath where trees have been cut down marks the border. When the U.S. Geological Survey made a geological reconnaissance along the 49th parallel early in this century, Bailey Willis carried out the first geologic investigation of the Glacier National Park area.

In 1885, George Bird Grinnell, a naturalist and conservationist, vacationed in the region. He fell in love with the place, campaigned for preservation of the area for many years, and was influential in the legislation that established Glacier National Park.

Railroad surveyors came into the Northern Rockies after the Civil War looking for mountain passes suitable for potential rail routes. John Stevens discovered Marias Pass in 1889 and surveyed it for the Great Northern Railroad, which constructed its line through the region in 1892. The railroad now skirts the southern boundary of the park. After Glacier National Park was established, the railroad built hotels and mountain chalets to attract tourists.

Going-to-the-Sun Highway, named after a legendary Blackfeet chief, was opened in 1933 after a decade of costly and difficult construction. The 36-mile trip between the east and west sides of Glacier National Park is scenically majestic and geologically fascinating.

Geologic Features

The glaciation that produced the spectacular, ice-sculptured landforms in Waterton-Glacier was at its maximum extent during the Pleistocene Epoch. However, the small cirque and cliff glaciers now existing in the park developed only a few thousand years ago during the "Little Ice Age."

Examples of Glacial Erosion

At the head of a valley glacier, the scouring and plucking done by the ice mass produces a cirque basin. Many of the lakes in Glacier National Park, such as Avalanche, Cracker, Ellen Wilson, Helen, Hidden, Iceberg, Kennedy, Pocket, Ptarmigan, and Upper Two Medicine Lakes, are *cirque lakes* that lie in bowl-shaped depressions from which glacial ice retreated. Some of the higher lakes have a turquoise color enhancing their beauty.

FIGURE 26.3   Grinnell Glacier as seen in the early 1970s with its 165 foot high waterfall and lake. Since 1850 it has lost 90 percent of its volume and presently is only 10 percent of its original size. Scientists have estimated that around the year 3032 there will be no glaciers left in the park. More than a century ago there were about 150 glaciers; today only 35. The diorite dike is about 50 feet wide. Note the white bands above and below the sill, where the dolomite has been metamorphosed into marble by the heat of the diorite intrusion. Photo by Robert K. Smith.

Where two cirques have eroded headward until they are back to back or adjacent, the thin, blade-shaped wall of rock separating them is an *arête*. Garden Wall and Ptarmigan Wall, which can be viewed from Going-to-the-Sun Highway, are examples. A saddle-shaped depression, termed a *col,* results when glacial ice breaks down an arete between cirques. Jackson and Blackfoot Glaciers are separated by a col, and another can be found behind Haystack Butte. Some cols in the park are mountain passes, such as Logan Pass, Piegan Pass, Marias Pass, and Gunsight Pass. A glacial *horn,* a feature having a pyramidal shape, is produced when cirques cut back into a mountain summit from three or four sides. Some of the horns in Glacier National Park are Clements Mountain, Kinnerly Peak, Little Matterhorn, Reynolds

Mountain, Sinopah Mountain, Split Mountain, Mount Wilbur, St. Nicholas, and Haystack Butte.

Valleys that have been heavily glaciated have a characteristic U-shape. This is because valley glaciers tend to straighten bends or curves cut by preglacial streams, smooth and widen the side slopes, and deepen the valley floors. Typical U-shaped glacial valleys in the park are McDonald Valley, St. Mary Valley, Waterton Lake valley, Swiftcurrent and Bowman Lake Valley, Belly River Valley, and Cataract Creek Valley.

Many beautiful hanging valleys, some with waterfalls, are in Waterton-Glacier. The mouth of a *hanging valley* opens out high on the steep side of a larger, glacial valley, many feet above the valley floor. The larger valley held a trunk glacier, which had greater erosive power than its smaller tributary glaciers and thus was able to erode its valley more deeply. When the glaciers melted back, the tributary valleys were left hanging far above the main valley. Avalanche Falls is an example of a waterfall joining a stream in a hanging valley to a main stream. Other waterfalls, such as St. Mary, Rockwell, Baring, and Florence Falls, cascade down over glacial steps in bedrock walls. Trick Falls are unique because at the top the falls flow over resistant beds; when stream volume is low, the bottom falls (normally hidden from view beneath the top falls) issue from a cave formed by the enlargement of joints in limestone.

Some glacial valleys have a string of *rock-basin lakes* connected by a single stream. The lakes occupy depressions scoured out of highly fractured or weaker sections of bedrock. An example of this glacial feature can be seen between Many Glacier Hotel and Swiftcurrent Pass. Bullhead Lake and Swiftcurrent Lake are two of the rock-basin lakes in the string.

In late summer natural *ice caves* become accessible at the base of Grinnell and Sperry Glaciers. Glaciologists obtain data on ice pressures and temperatures from instruments placed at the ice-rock interface and left year-round in the cave beneath Grinnell Glacier. Some summers when scientists enter the cave, they find that instruments have been destroyed by the grinding of ice against the rock. Streams drain from the caves beneath the glacial snouts. The characteristic milky white color of the water is due to the heavy concentration of *rock flour* (finely pulverized rock debris) being carried off in suspension.

Glacial Deposition

Glacial erosion and mass wasting produce great quantities of sediment in all sizes—rock flour, sand, gravel, cobbles, huge boulders. Debris is picked up, flattened down, pushed, or carried along on top of, beneath, beside, and in front of a glacier. This largely unsorted material, called *till,* accumulates in *moraines* of various types and sizes. Lakes dammed by moraines are typical of glaciated topography. Examples in the park are Bowman, Josephine, and Kintla Lakes.

Streams work over and sort till and other debris, redepositing the sediment as *glacial outwash*. In the bottom of glacial troughs, glacial outwash usually spreads over till that was brought down by the ice. Eventually the scoured bedrock downvalley from a retreating glacier may be buried by many feet of sedi-

FIGURE 26.4 Bird Women Falls is a typical waterfall formed in a hanging valley. Photo by Thomas Serenko.

FIGURE 26.5   Because of the dip of the beds, water entering one side of Haystack Butte comes out as springs on the other side. Photo by Ronald F. Hacker.

ment. Sometimes streams back up, forming lakes held in by natural dams built up by a mixture of outwash and morainal material. Both St. Mary Lake and Lake McDonald, which lie in glacial troughs, were dammed in this way.

Erosion by sediment-laden streams draining off glaciers has developed large, deep *potholes* in some of the less resistant rocks in streambeds. Rock debris that collects in any slight depression or weak spot in bedrock or boulders acts as a drill, swirling around in the water and enlarging the pothole.

Avalanche Chutes

Visible on many of the steepest mountain slopes are straight, groovelike features, called *avalanche chutes,* that run hundreds of feet down the face of a slope. These are the tracks of avalanches that have been worn smooth by repeated slides of snow, ice, trees, and rock debris that roar down the slopes at high speed. In the spring of the year, when conditions cause snow packs to become unstable on steep slopes, an avalanche may be triggered by a sudden vibration or sharp noise. Snow, ice, and debris rushing down the mountainside—with a wall of air in front of the slide—can build up such force that trees are torn out by the roots. From Going-to-the-Sun Highway, you can see avalanche chutes on Mt. Cannon on the western side of the Continental Divide.

Glacier Recession

Grinnell and Sperry Glaciers, the two largest glaciers in the park, have shrunk considerably in the last several centuries. Grinnell Glacier, which is now about 275 acres, was twice as large when George Grinnell discovered it a hundred years ago. Precipitation probably was higher, and temperatures were lower when these glaciers were at their maximum extent several thousand years ago, but weather data and measurements of the glaciers have not been collected long enough for current trends to be determined with any certainty. Moreover, glacial studies have shown that normal "lag time" is such that it may take years before changes in average annual precipitation and temperature affect the size and thickness of a glacier.

In Waterton-Glacier International Peace Park, location and elevation have a significant relation to snowfall, temperature, and glacial advance or retreat. For example, at Babb, a village outside the park's eastern boundary and south of the international boundary, average annual precipitation is only about 20 inches, due to the rain-shadow effect. But over Grinnell glacier, 15 miles to the west and about 1700 feet higher in elevation, yearly precipitation is estimated at about 150 inches.

Data collected from a stream gage on the Kootenai River at Libby, Montana, about 75 miles west of the park, shows a downward trend in stream discharge from 1920 to the mid-1940s and an upward trend thereafter. Since the Kootenai drains much of the park area, one might tentatively conclude that for the short term, the trend of decreasing rainfall and rising temperatures has slowed. Measurements of glacier shrinkage and climate data collected in other areas have shown a similar pattern. However, long-term predictions of glacial advance or retreat due to climatic variations are extremely tentative.

The Belt Supergroup

Glaciers as agents of erosion may modify landscapes and landforms more rapidly and drastically than streams can. However, glaciers work on the preexisting topography of a region. The valleys and divides have already been molded by streams, which, in turn, are mainly controlled by the underlying rock structures. The glaciated alpine topography in Waterton-Glacier International Peace Park has a unique "architectural" character that is not seen in the Rockies farther south. That is because the angled, banded slopes and summits of the mountains in Waterton-Glacier are made up of

Waterton-Glacier International Peace Park

layered sedimentary and lightly metamorphosed rocks, rather than the granitic and metamorphosed rocks displayed in Rocky Mountain National Park (Chapter 25). What is surprising is that the dominant rocks in both parks are Precambrian in age.

The layered rocks in Waterton-Glacier are part of a thick series of sedimentary rocks called "the Belt rocks" or, more formally, the Belt Supergroup. (In Canada, the rocks are known as the Purcell Supergroup.) What makes these rocks so unusual is that they are relatively undeformed and only lightly metamorphosed despite the fact that they were deposited fairly late in Precambrian time, between 1600 million and 800 million years ago (Proterozoic Y).

The Belt rocks are spread over several states and Canadian provinces. In the park area, this great sedimentary wedge is over 18,000 feet thick, and probably totals 45,000 to 50,000 feet some miles to the west. The sands, silts, and carbonates that made up the rocks accumulated in a broad, subsiding basin, which was invaded by shallow seas that spread over a rift between continental landmasses. The rift was probably a spreading center that opened in a Precambrian continent. Eventually, due to deep burial, most of the sandstones, shales, and limestones that formed were lightly metamorphosed to quartzites, argillites, marbles, and dolomitic marbles.

On Going-to-the-Sun Highway, approaching from the east, you go "up-section" to the highest point at Logan Pass (elevation 6680 feet) and then you go "down-section" to the west entrance. This is the geologists' way of saying that the layers of rock get younger toward the top and older toward the bottom. Like a layer cake, the rock units are laterally continuous, so that you see the same formations going down as were exposed on the way up. The outcrops are not at the same elevation, however, because beneath the Continental Divide these Belt rocks are slightly deformed into a broad, open syncline.

Igneous Sill and Lavas

Visible in glacially undercut cliffs and steep slopes is a horizontal dark band that is conspicuous within the light-colored beds of the Helena (Siyeh) Formation. This band is a diorite sill, formed when very hot magma from deep in the crust was forcibly injected into Helena bedding planes near the top of the formation. Contact metamorphism "cooked" and bleached the limestone beds above and below the sill. The diorite along the sill margins is fine-grained due to rapid cooling, but the interior of the sill, which cooled slowly, has a fairly coarse texture. Because of their distinctive appearance, the sill and its enclosing beds are easily traced all over the park, and outcrops are marked on both the east slope and west slope of Going-to-the-Sun Highway. The age of the sill has not been determined, but similar sills in the Helena south of the park have yielded radiometric dates of 775 to 725 million years B.P. (Raup et al. 1983).

On several exposures in the park, *pillow structures* in basalt lava flows indicate that magma broke through

 FIGURE 26.6   The V-shaped profile of this valley indicates that it was eroded by streams. Photo by Thomas Serenko.

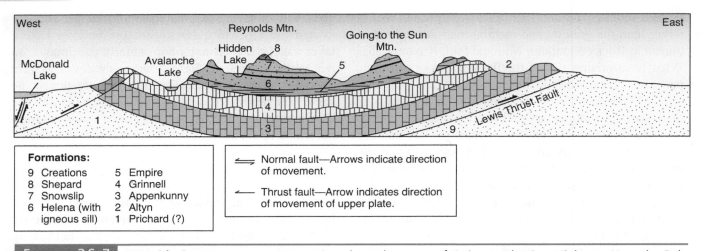

Formations:

9 Creations   5 Empire
8 Shepard   4 Grinnell
7 Snowslip   3 Appekunny
6 Helena (with   2 Altyn
  igneous sill)   1 Prichard (?)

⇐⇒ Normal fault—Arrows indicate direction of movement.

←— Thrust fault—Arrow indicates direction of movement of upper plate.

FIGURE 26.7 Simplified, west-to-east cross section along the route of Going-to-the-Sun Highway. Here the Belt Supergroup formations have been deformed into an open syncline, and the entire mountain mass has been thrust upward and eastward many miles by the Lewis thrust fault. The compressive stresses also produced the reverse fault on the lower part of the western slope. Adapted from O.B. Raup et al. *Geology Along Going-to-the-Sun Road, Glacier National Park, Montana.* © 1983 Glacier National History Association. Used by permission.

the sedimentary beds and erupted as "pillow basalt" on the floor of the Precambrian sea. The "pillows," which form when lava flows into water or erupts from a vent beneath the water surface, are close-fitting and tend to form masses or piles. Because the lava cooled suddenly in the water, it is fine-grained and has bubble holes, called *vesicles*. At a later time in geologic history, when the rocks had been elevated above sea level, percolating groundwater filled the holes with calcite and formed *amygdules* (similar to those in Isle Royale National Park, Chapter 23).

At Granite Park, north of Logan Pass, exposures of pillow basalt are well displayed. Years ago, this dark igneous rock was misidentified as "granite," and the name stuck, although there is no granite in, or close to Glacier National Park.

The Lewis Overthrust

The Northern Rockies in the Waterton-Glacier region are described as "mountains without roots" because they consist of ancient Precambrian rocks resting on top of much younger sedimentary beds.

Preserved on bedding surfaces in many of these rocks are clues to the ancient environments in which they formed. *Ripple marks* tell the direction of ancient waves and currents. *Mud cracks,* which formed when

wet sediments dried out in the sun, show that a particular bed was above water level for a time. *Rain-drop impressions* indicate that exposed soft mud was pelted by rain from a cloudburst.

The remains of several species of *calcareous algae* record the presence of colonies of single-celled plants in the Precambrian seas. These blue-green algae, collectively known as *stromatolites* in their fossil form, thrived in shallow water where they received abundant sunlight. By removing carbon dioxide from the seawater in which they lived, they released oxygen and secreted or deposited calcium carbonate around their single-celled bodies, forming colonies of various sizes and shapes. Some algae built roughly circular, bulbous masses, called "cabbage heads." Others *(Conophyton)* grew in columns of nested cones. The Proterozoic blue-green algae (cyanobacteria) functioned in the same way as modern, living algae and are essentially the same kind of organisms (fig. 26.6).

Belt Rocks exposed in Glacier National Park. A geologic guide, prepared by U.S. Geological Survey scientists (Raup et al. 1983), identifies eight formations within the Belt Supergroup that are keyed to stops along the cross-mountain, Going-to-the-Sun road (fig. 26.7). These formations, from oldest to youngest and from east toward the west, are Altyn, Appekunny,

Grinnell, Empire, Helena (Siyeh), Snowslip, and Shepard. Prichard, the eighth formation and the oldest one on the west side of the park, lies below the Appekunny, where it is thrust up against younger formations. The Altyn is the oldest formation exposed on the east side of the park. Altyn and Prichard are equivalent in age, but were deposited in different marine environments. Because the ancient ocean bottom sloped to the west, Altyn was deposited in shallow water near shore, while Prichard consists of deep-water deposits. Shepard Formation, youngest of the Proterozoic units described, does not crop out along the highway, but can be seen on the highest peaks in the view from the Logan Pass Visitor Center.

The reason for this is that a huge, low-angle thrust fault, known as the Lewis Overthrust, transported the sequence of Precambrian rocks eastward for many miles over the younger rocks. The first geologists to work in the area recognized the structure when they observed and mapped the ancient, tan-and-cream-colored beds of the Altyn Formation lying on top of dark-gray Cretaceous shales of the Montana Group. The trace of the fault surface, which separates the older beds on top from the younger beds beneath, can easily be seen on the face of Wynn Mountain and at other places in the park. The fault surface also crops out along the base of the vertical face of Appekunny Mountain.

Chief Mountain, a peak that stands alone astride the park's eastern boundary, is an erosional outlier, or *klippe* (i.e., a remnant) of the thrust fault, which was of much greater size and extent before being eroded (fig. 26.9). Making up the top third of Chief Mountain are the horizontal limestone beds of the Upper Member of the Altyn Formation. The lower two-thirds of the mountain consist of the more dolomitic beds of Altyn's Lower Member, which have been slightly tilted and show minor thrusting. Underneath the mountain are the

FIGURE 26.8 Just over Logan's Pass on the west side. The arete that was formed by glaciers eroding on both sides of a divide is known as Garden Wall. Calcareous algae called stromatolites are found here. Photo by Thomas Serenko.

FIGURE 26.9 The mountain on the left is Mahtotopah Mountain. From top to bottom are the Helena (Siyeh), Empire, and Grinnell Formations. To the right of Mahtotopah Mountain is Dusty Star and behind it is Fusillade. For both of these mountains the peaks are the Snowslip Formation and the bases consist of the Helena (Siyeh). The knife-edge shapes of these mountains are called arêtes and glacial U-shaped valleys separate them. Photo by Tom Serenko.

young Cretaceous beds separated from the Precambrian layers by the fault surface of the Lewis Overthrust, which is visible at the mountain's base (fig. 26.10).

How did this come about? The metasedimentary Precambrian beds must have been undisturbed for many millions of years when they began to be uplifted and upwarped by the intense compressive forces of the Laramide orogeny. The huge fault slab, several miles thick, from which the mountain ranges of Waterton-Glacier were carved out, broke loose and very slowly slid or was pushed eastward for over 50 miles.* The rel-

*Ongoing studies of Northern Rockies tectonics suggest that the displacement may have been of longer duration and greater magnitude than had been previously thought.

atively strong Precambrian rocks were able to move as a great thrust sheet, with only very slight deformation, over the soft Cretaceous shales.

Along the western boundary of Waterton-Glacier International Peace Park, the North Fork of the Flathead River meanders through a broad, north-south-trending valley, underlain by Cretaceous rocks. However, the Whitefish Range, which rises on the west side of North Fork Valley, is made up of Precambrian strata like those in Waterton-Glacier. The explanation seems to be that the thrust slab cracked or broke over North Fork Valley, like a slab of snow separating into two sections when sliding down a roof.

Erosion, first by streams and then by Pleistocene glaciers removed great quantities of rock from the thrust sheet. The fault trace on the west side of the park

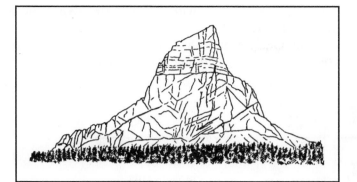

FIGURE 26.10   Chief Mountain, located on the eastern boundary of the park, is a klippe, formed by the erosion of a thrust sheet. The trace of the Lewis thrust fault, almost horizontal here, lies about at the top of the line of trees. Rocks of the Precambrian Belt Supergroup are above the fault; those below are Cretaceous in age. Drawing by Bailey Willis in *Geological Society of America Bulletin,* 1902, 13:334.

is still deeply buried, but in places on the east side (as around Chief Mountain), Mesozoic rocks beneath the fault surface are exposed.

Geologic History

1. **Deposition of the Belt Supergroup during late Precambrian time, Proterozoic Y.**

The basin in which the Belt Supergroup rock units were deposited held a sea that oscillated over the ancient landmass for millions of years during the span of Precambrian time called Proterozoic Y (explained in Chapter 22). The size, shape, and outline of the basin, plus the information derived from the sediment types and facies changes all indicate multiple source areas. The extent of the basin can be determined only approximately because the boundaries have been obscured by younger rocks and by deformation. Brief descriptions of the significant formations follow:

Waterton Formation of Canada. The dolomitic Waterton beds are exposed in Waterton Lakes National Park but not on the U.S. side of the border.

Altyn Formation. Sandy dolomite and limestone beds deposited in shallow water make up this formation, the oldest on the east side of Glacier National Park. Sedimentary features such as ripple marks, mud cracks, and stromatolites (cabbage-like fossil algae) can be easily recognized in many of the outcrops. Thin layers of quartz that filled in minute fractures have weathered out in sharp relief. Altyn rocks form the St. Mary Lake Narrows—the resistant ledge over which Swiftcurrent Falls tumble—and the rim upon which the Many Glacier Hotel is built.

Prichard Formation. This oldest rock unit on the west side of Glacier National Park consists of thin layers of black argillite and siltstone. The beds were deposited in deeper water than any other formations in the park. Cataracts and waterfalls along McDonald Creek above the Lake McDonald inlet flow over resistant ledges of the Prichard Formation. Large potholes have been carved out in the stream bed.

Appekunny Formation. The green mudstones of the Appekunny are darker on the west side than the east side because the beds on the west side were deposited in deeper water. Almost all of the green rocks that crop out in the park are Appekunny. Upstream, from the Prichard Formation, McDonald Creek flows over green Appekunny beds. Aster Creek carved its gorge in them. Soil weathered from Appekunny rocks supports a heavy forest cover.

Grinnell Formation. Near the top of the Appekunny Formation, layers of red mudstone are interbedded with green mudstone. The contact between Appekunny and the overlying Grinnell Formation is where red mudstone becomes dominant over the green. The presence of the mineral hematite (iron oxide) gives Grinnell outcrops their distinctive, bright red color. The beds were deposited on tidal mudflats. With repeated wetting and drying, mud cracks (fig 26.11) and ripple marks formed on the beds. A few striking beds of white quartzite in the Grinnell form prominent ledges between red mudstone layers.

Empire Formation. The argillites, siltstones, and carbonate beds in this rock unit are transitional between the rocks of the Grinnell Formation and those of the overlying Helena (Siyeh) Formation. The depositional environment was changing as the basin subsided and seawater became deeper. Some green carbonate rock layers of the Empire Formation contain "anomalous copper"; that is, higher-than-normal concentrations of copper minerals, such as chalcopyrite, chalcocite, and bornite. Anomalous copper (at least 100 parts per million, and much higher in some places) has been found in nearly all parts of the Belt basin. Silver and mercury, also in anomalous amounts, are sometimes associated with the copper. Exploration geologists refer to such copper deposits as "stratabound" because the copper minerals tend to be concentrated along bedding planes and in sedimentary

Table 26.1			Geologic Column, Glacier National Park		
Time Units			**Rock Units**		
Era	**Period**	**Epoch**	**Group**	**Formation**	**Geologic Events**
Cenozoic	Quaternary	Holocene		Alluvium Glacial deposits	Erosion and deposition by streams. Return of glaciers in Neoglacial time. Present glaciers between 6000- and 9000-foot levels on north and east slopes.
		Pleistocene		Glacial deposits	Valley and piedmont glaciations with many advances and retreats; intensive weathering and erosion
	Tertiary		//////////		Major erosional interval
Mesozoic			Montana		Return of seas depositing Cretaceous beds; Laramide orogeny—folding, uplift, thrust faulting, vertical faulting.
Paleo-zoic	//////////				Major erosional interval
Late Precambrian (Proterozoic Y)			Belt Supergroup	Kintla (Mt. Shields)	Red argillite, increasing aridity
				Shepard	Red to yellowish brown argillite, stromatolites
				Snowslip	Pale red to green argillite, pillow lava, ripple marks, mudrocks
				Helena (Siyeh)	Tan dolomite, diorite sill, oolite zone, stromatolites
				Empire	Buff to tan argillite and quartzite; copper
				Grinnell	Red to purplish red argillite and quartzite; ripple marks, mud cracks
				Appekunny	Gray to greenish argillite and quartzite; ripple marks, tidal mud flats
				Altyn (east) / Prichard (west)	Tan- to cream-colored limestone and dolomite, stromatolites, ripple marks (shallow water) / Dark gray to black argillite (deep water)
				Waterton (Alberta)	Red brown to gray dolomite

Sources: Alt and Hyndman, 1986; Harrison, 1972; Raup et al, 1983.

Mud cracks in the Grinnell Formation, a member of the Precambrian Belt Series. These beds were deposited on the tidal mud flats, which underwent repeated wetting and drying, forming mud cracks and ripple marks. Photo by Thomas Serenko.

structural features (e.g., mud cracks). Some of the copper in the Belt basin was deposited with the sediments (syngenetic), but probably some copper migrated or was reconcentrated subsequent to deposition (by diagenetic processes). Concentrations of ore grade copper are outside the park.

Helena (Siyeh) Formation.

Many algal zones accumulated in the limestones and dolomites of the Helena (Siyeh) Formation. Fossiliferous beds on the Garden Wall and along Going-to-the-Sun Highway have excellent examples of large and small "cabbage heads" and stacked, inverted cones *(Conophyton)*. The *Conophyton*

colonies formed a zone 100 feet thick in the Helena that is exposed at many places throughout the park.

An oolite zone occupies transitional layers between the carbonates of the Helena and the argillites and siltstones of the overlying Snowslip Formation. The *oolites*, which are rocks composed of small, round objects *(ooliths)* resembling fish eggs, are similar to oolites in Everglades and Biscayne National Parks (Chapters 19, 20) and probably formed in the same way.

The igneous sill, described earlier, that is sandwiched between limestone beds, near the top of the Helena, ranges in thickness from 10 to 200 feet. The sill and the bleached carbonate beds, altered to marble by contact metamorphism, are exposed in several roadcuts near the top of Logan Pass.

Snowslip Formation.

The colorful red and green argillites, purplish siltstones, and pale yellow quartzites of the Snowslip Formation were deposited on tidal mudflats. Land areas along the shore had been elevated enough to increase the supply of sediment being washed by erosion into the sea. Again, one can find ripple marks and mud cracks on outcrops. Some multicolored stromatolite beds are exposed below the Loop beside Going-to-the-Sun road. Iron- and copper-bearing minerals in the argillites have colored the algal heads.

Shepard Formation.

The red and tan rocks of the Shepard, which become more clastic toward the east, are a mixture of limestone and siltstone with some dolomite. Well-preserved fossil algae and undisturbed ripple marks and mud cracks suggest that no digging, crawling, or swimming organisms lived in the inland sea. No traces of organisms other than the calcareous algae have been found in these Precambrian rocks.

The climate became more arid, as shown in the increasing redness of the argillites. Oxidation occurring when the beds were exposed to air for longer intervals caused this. The Shepard beds blend into the red argillites of the Kintla (Mount Shields) Formation. Shepard and Kintla rocks form red caps on some of the highest peaks and passes in the northern and middle part of the park.

2. **Moderate uplift, erosion of some of the Belt rocks, and probable deposition of Paleozoic rocks.**

As the region of Waterton-Glacier International Peace Park was gradually uplifted, the seas withdrew. Stream systems developed and rapidly stripped the land

FIGURE 26.12   St. Mary Lake is a glacial lake that formed in a U-shaped glacial trough that was dammed up by glacial moraine. It is surrounded by mountains that have been carved into aretes and glacial horns. Photo by Thomas Serenko.

of sediments that were soft, unconsolidated, and easily eroded. Geologists surmise that in time the seas returned to this area because mountains both to the north and south of the park are composed of Paleozoic beds deposited in shallow water. Paleozoic rocks are not exposed at the surface in the park. They probably were eroded during and after the later uplift associated with the Laramide orogeny.

3. Cretaceous deposition.

A gap in the geologic record for most of Mesozoic time suggests that this part of North America remained above sea level during that era. Then a shallow inland sea extended into the Northern Rockies region as the Cretaceous Period was drawing to a close. Because the Cretaceous beds were never deeply buried, they are not so strongly lithified as the

Precambrian rocks and thus are more susceptible to weathering. Erosion of Cretaceous rocks east of the park has formed a gently rolling landscape in contrast to the rugged mountains and steep slopes eroded on the rocks of the Precambrian Belt Supergroup. The Cretaceous beds cropping out along the eastern side of the park contain oil, and those in North Fork Valley have natural oil seeps. Where the beds extend under the High Plains of Montana and Alberta, commercial oil fields are being worked. The site of the first oil well in Alberta is now a historical exhibit in Waterton Lakes National Park.

4. The Laramide orogeny and thrust faulting.

As the Cretaceous Period was ending, the upheavals that built the Northern Rockies tilted the flat-lying Belt rocks eastward. As the uplifting increased stream gradients, large stream valleys were eroded. Late in Cretaceous and early in Eocene time, tectonic compression caused massive thrust faulting. Deformation was especially severe in the Northern Rockies where Precambrian rocks in the huge slabs, moved by the Lewis thrust fault, overrode Cretaceous shales. North Fork Valley on the park's western margin formed where a fault slab split apart during its eastward sliding.

5. Cenozoic erosion and deposition; Miocene uplift.

The inland climate was desertlike and vegetation was sparse. Coarse sediments were dumped in valleys by slumps and flash floods. Erosion isolated sections of the overthrust sheet and created klippes, such as Chief Mountain, which is completely surrounded by younger beds.

A renewal of uplift in Miocene time (presumably due to isostatic adjustments) brought the Northern Rockies to their present elevation. The climate became cooler and moister, thus setting the stage for the development of glaciers, which may have begun to form in the high ranges by Pliocene time.

6. Pleistocene glaciation.

In Glacier National Park, evidence indicates extensive glaciation during the middle Pleistocene, two glacial substages during the Wisconsinan glacial stage, and at least two episodes of ice accumulation and advance during post-Pleistocene time. During the most extensive advances, glacial erosion and deposition in the park was done by alpine glaciers that filled all the valleys and piedmont glaciers that coalesced along the eastern mountain front. On the west side, North Fork Valley was covered by a belt of moving ice 10 miles across. Continental ice sheets (Laurentide) from Canada overlapped mountain glaciers on the northern and eastern sides of the park and joined the piedmont glaciers, but never overrode the high ranges. The high peaks in the park that stuck up above the ice show a line marking the ice level. Above the line, the cliffs are very rough; below the ice line, the bedrock surfaces are smooth due to the grinding and scouring of glacial ice.

In Glacier National Park (as throughout most of North America) the initial Pleistocene glacial stage was more effective than later glaciations. One indication of this can be seen on the crests of Two Medicine Ridge, Cut Bank Ridge, and Milk River Ridge, all of which

have deeply weathered till, obviously older than the blankets of till laid down on the floors of the intervening valleys by later glacial stages. Evidence also shows that two large piedmont glaciers—Two Medicine Glacier, that flowed east, and St. Mary Glacier, which flowed northeast—filled valleys with ice 500 to 1000 feet higher than occurred in succeeding glacial stages and advanced farther eastward onto the plains of Montana. The location of moraines and glacial outwash show that at its maximum, Two Medicine Glacier was 20 miles wide and extended some 10 miles out from the mouth of its valley.

7. Final retreat of Pleistocene glaciers about 10,000 years ago; evolving of the present landscape.

Near the end of the Pleistocene Epoch, the climate gradually became warmer. The result was that the park was denuded of glacial ice—except for a few small glaciers and snowfields that persisted at high altitudes and on north-facing slopes. No major changes in the landscape have come about since the ice disappeared, but the effects of weathering and erosion continue. Small gorges cut by streams after ice retreat are Sunrise Gorge and the gorge at Hidden Falls. Both were formed as the result of enlargement of the weak zones in the bedrock joint system. Mass wasting is active under the prevalent freeze-and-thaw conditions, especially on steep slopes. Talus slopes are building up at the foot of cliffs and along the base of the mountain ranges. Avalanches, rockfalls, mud flows, and soil creep all continue to modify the landforms.

Since the melting back of the glaciers took place over a much shorter time than their growth, enormous quantities of debris and sediment were released when the climate warmed. Overloaded streams responded, as streams will, by dumping the excess in alluvial fans and outwash features, on top of till beds, moraines, and bedrock. The valleys holding Waterton Lake, St. Mary Lake, Lower Two Medicine Lake, and other lakes in the park remain clogged because the water is dammed by glacial and landslide debris.

8. Origin of glaciers that exist today.

The postglacial climatic optimum that brought about the disappearance of glaciers was a warm interval that peaked about 7000 years ago. Average temperatures were warmer and rainfall was ample. The warm interval was followed by a cooling trend, but several climatic fluctuations have occurred. A period cooler than now that prevailed about 4000 years ago was the cause of renewed glacial activity that resulted in the development of the alpine glaciers that presently exist in the park. That the climate will continue to fluctuate seems certain.

The park's active glaciers are found between elevations of 6000 and 9000 feet. The perennial snowfields lie in protected areas, usually on slopes that face north or east. Wind-drifted snow from over the peaks is a main source of their nourishment. Movement of glacial ice may range from 6 to 8 feet per year to as much as 12 to 50 feet per year, depending upon conditions of precipitation and temperature. These velocities are slow in comparison with other valley glaciers that are monitored, especially some of the Alaska glaciers.

Geologic Map and Cross Section

Carrara, P. E. 1990. *Surficial Geology of Glacier National Park, Montana*. U.S. Geological Survey Miscellaneous Investigations series 1-508-D.

Earhart, R.L.; Raup, O.B.; Whipple, J.W.; Isom, A.L.; and Davis, G.A. 1989. *Geologic Maps, Cross Section, and Photographs of the Central Part of Glacier National Park, Montana*. U.S. Geological Survey Miscellaneous Investigations series I-1508-B.

Sources

Alt, D.D. 1983. Glaciated Park: Geology, in *Glacier Country, Montana's Glacier National Park*, no. 4, Montana Geographic Series, Montana Magazine, Inc. p. 6–25.

Alt, D.D., and Hyndman, D.W. 1973. *Rocks, Ice, and Water: the Geology of Waterton-Glacier Park*. Missoula, Montana: Mountain Press Publishing Co. 104 p.

———. 1986. *Roadside Geology of Montana*. Missoula, Montana: Mountain Press Publishing Co.

Beaumont, G. 1978. *Many-storied Mountains: The Life of Glacier National Park*. Natural History Series, National Park Service. 138 p.

Harrison, J.E. 1972. Precambrian Belt Basin of Northwestern United States: Its Geometry, Sedimentation, and Copper Occurrences. *Geological Society of America Bulletin* 83:1215–1240.

Johnson, A. 1980. Grinnell and Sperry Glaciers, Glacier National Park, Montana—A Record of Vanishing Ice. U.S. Geological Survey Professional Paper 1180, 29 p.

Karlstrom, E.T. 1987. Zone of Interaction Between Laurentide and Rocky Mountain Glaciers East of Waterton-Glacier Park, Northwestern Montana and Southwestern Alberta. *Centennial Field Guide*, v. 2, Rocky Mountain Section, Geological Society of America. p. 19–24.

Mudge, M.R. and Earhart, R.L. 1980. The Lewis Thrust Fault and Related Structures in the Disturbed Belt, Northwestern Montana. U.S. Geological Survey Professional Paper 1174, 18 p.

Raup, O.B., Earhart, R.L., Whipple, J.W., and Carrara, P. E. 1983. *Geology Along Going-to-the-Sun Road, Glacier National Park, Montana*. Prepared by U.S. Geological Survey and National Park Service; published by Glacier Natural History Association. 62 p.

Ross, C.P. 1959. Geology of Glacier National Park and the Flathead Region, Northwestern Montana. U.S. Geological Survey Professional Paper 296.

———. 1962. The Precambrian of the Northwestern United States—the Belt Series. In *The Precambrian*, K. Rankama, editor. 4:145–251. New York: Interscience.

Willis, B. 1902. Stratigraphy and Structure, Lewis and Livingston Ranges, Montana. *Geological Society of America Bulletin* 13:305–352.

Note: To convert English measurements to metric, go to www.helpwithdiy.com/metric_conversion_calculator.html

Gates of the Arctic National Park and Preserve

NORTHERN ALASKA

Area: 8,472,517 acres; 13,238 square miles

Proclaimed a National Monument: December 1, 1978

Established as a National Park and Preserve:
December 2, 1980

Designated a Biosphere Reserve (Portion): 1984

Address: 201 First Ave., Fairbanks, AK 99701

Phone: 907-692-5494

E-mail: GAAR_Visitor_Information@nps.gov

Website: www.nps.gov/gaar/index.htm

A wilderness park entirely north of the Arctic Circle, four times the size of Yellowstone National Park. Straddles the Arctic Divide and the Central Brooks Range.

Jagged peaks, glaciers, broad glaciated valleys. Forested southern slopes and treeless northern reaches.

Six streams are in the National Wild Rivers System.

A complicated tectonic history partly revealed in complex mountain structures.

FIGURE 27.1   Walker Lake which drains the highest part of the Brooks Range, occupies a large moraine-dammed lake valley. National Park Service photo.

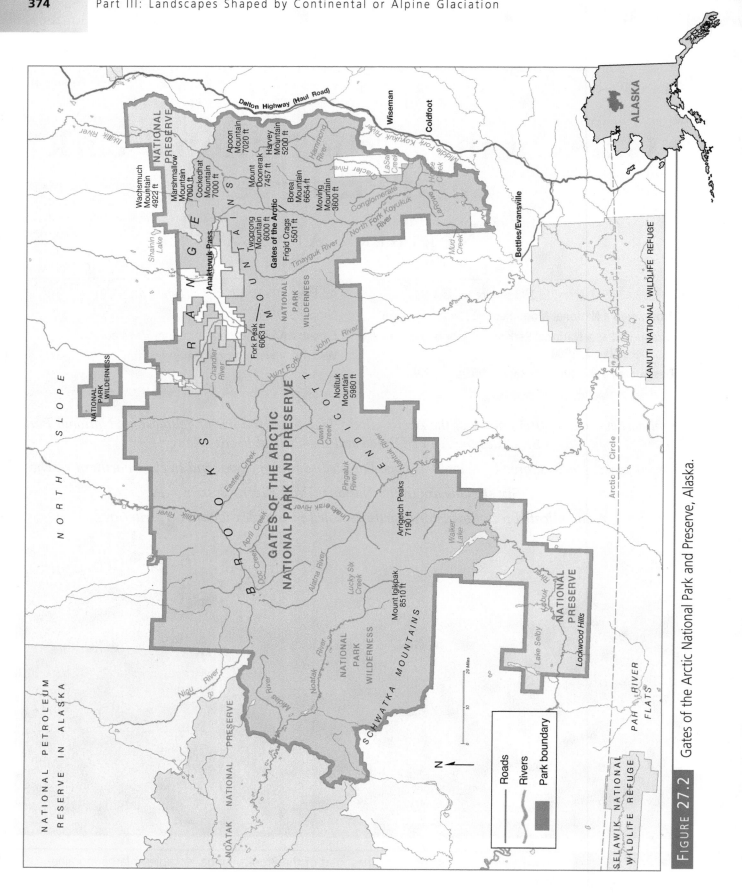

FIGURE 27.2   Gates of the Arctic National Park and Preserve, Alaska.

Geographic Setting

Gates of the Arctic National Park and Preserve is a vast wilderness extending east to west for some 200 miles along the crest and slopes of the central and highest part of the Brooks Range in Arctic Alaska. The Brooks Range, named for Alfred H. Brooks, a U.S. Geological Survey geologist who explored Alaska early in the 20th century, stretches like a great barricade from the Canadian border on the east to almost the Chukchi Sea on the west.

From south to north, Gates of the Arctic National Park and Preserve ascends from long, broad, glaciated valleys and rounded foothills to the sharp *arêtes* (narrow, rock ridges) and turrets along the crest of the range, then it goes down the steeper northern slopes to the edge of the Arctic coastal plain. Robert Marshall of the U.S. Forest Service extensively explored the region in the 1920s and 1930s. One day in 1929, as he was fol-

lowing up the North Fork of the Koyukuk Wild River, he saw against the sky two great facing peaks framing the valley he was ascending. He named this magnificent landscape "the Gates of the Arctic." He christened the peak of Boreal Mountain, the "east portal," and the peak of Frigid Crags the "west portal." Marshall's work and his writings about the region were instrumental in the later establishment of the park/preserve.

When the North Fork of the Koyukuk was placed in the National Wild Rivers System, the Tinayguk, its largest tributary, was also designated. Both rivers drain the south side of the Arctic Divide and the eastern part of the Endicott Mountains. The Endicott Mountains, which are part of the Brooks Range, extend from east to west over most of the park/preserve. The Schwatka Mountains, which have the highest summits in the Brooks Range, occupy the southwest corner of the park/preserve.

FIGURE 27.3   Ice-sculptured Arrigetch Peaks in the western part of Gates of the Arctic National Park. National Park Service photo.

The John Wild River, a major stream next west from the Tinayguk and the North Fork of the Koyukuk, rises in the vicinity of Anaktuvuk Pass on the crest of the range and also flows south. The river valley and the pass are a migration route for great herds of caribou that move each spring from the shelter of the forested valleys over the range and down to their summer pastures on the tundra of the North Slope. The Nunamuit Eskimo people, who have hunted caribou in the high valleys for many centuries, live in Anaktuvuk Pass, which is the only year-round settlement in the park/preserve.

Farther to the west, the Alatna Wild River, one of the best float streams in Alaska, bends through rugged country and then flows south along the eastern side of the lofty Arrigetch Peaks. Exposed in this river corridor are some of the complex geologic structures and metamorphic assemblages that challenge field geologists who have worked in the area. South of the park/pre-

serve, the Alatna and the John merge with the Koyukuk, which, in turn, flows into the Yukon. The last two Wild Rivers, the Noatak and the Kobuk, drain the glaciers in the Schwatka Mountains and Arrigetch Peaks. Both rivers flow west and empty into Kotzebue Sound (Chukchi Sea); but the Noatak flows out through Noatak National Preserve on the north side of the Baird Mountains, while the Kobuk passes through Kobuk Valley National Park (Chapter 11), south of the Baird Mountains. These three national parklands, Gates of the Arctic National Park and Preserve, Noatak National Preserve, and Kobuk Valley National Park, which encompass most of the Brooks Range, protect this country's last great pristine wilderness.

In addition to the six Wild Rivers, Gates of the Arctic includes two National Natural Landmarks, Arrigetch Peaks and Walker Lake. The glacially carved Arrigetch, a spectacular mountain area, was

FIGURE 27.4   The 83-mile long Alatna River, a component of the Wild and Scenic Rivers System, lies wholly within the Gates of the Arctic National Park and Preserve. The stream winds south from the high mountains of the Central Brooks Range and joins the Koyukuk River. National Park Service photo.

given its name by the Eskimos. The English equivalent is "fingers of the hand extended." The sharp, upthrust peaks, knife-edged arêtes, and undercut valleys of the Arrigetch have been eroded from a dark, massive, granitic intrusion—in contrast to the predominantly marine sedimentary rocks exposed in most of the park. Mt. Igikpak (8510 feet), highest peak in the park, is some 20 miles (air distance) west of the Arrigetch. A granitic pluton, similar to that of Arrigetch, is exposed on Mt. Igikpak. Walker Lake, lying below the Arrigetch and slightly to the southwest, is a long, moraine-dammed lake occupying a huge glacial trough in the headwaters area of the Kobuk Wild River (fig. 27.1).

Also rising in this high headwaters area of marshes and glacial lakes between Igikpak and Arrigetch, but draining to the north, is Noatak Wild River. Anthropologists believe that early hunters used this route to travel back and forth between the Noatak basin and the Kobuk Valley (Chapter 11) and that they set up winter camps in the deep, sheltered, glacial valleys.

Present-day Climate

Winters are long and cold in Gates of the Arctic; summers are short and mild. On the southern slopes and in the lowlands, midsummer temperatures occasionally rise to 80° F during the almost continuous sunlight; lows are near 50° F. On the northern slopes, summer temperatures range between 40° and 70° F. August is apt to be rainy. June and July are the worst times for gnats and mosquitoes. Freezing temperatures may occur at any time in summer, but are most likely after mid-August. Snow falls for eight or nine months of the year, with accumulations averaging 60 to 80 inches. January temperatures range from average highs of –3° to lows of –30° F. Ice in the Koyukuk River in the vicinity of Bettles, a town south of the park/preserve, usually breaks up around May 6 and freezes around October 21.

The ecology and environment of Gates of the Arctic are fragile because of the arctic and subarctic climate of the region. *Permafrost,* the permanently frozen ground of the far north, is at or near the surface throughout the park/preserve. Most of the permafrost is "dry"; that is, the frozen ground holds little moisture below what collects on the surface. The soil is very thin. Under these conditions, plants are slow-growing and easily damaged, especially north of the treeline, despite the apparent lushness of the arctic spring. The

northern limit of wooded country (the boundary of the boreal forest zone) passes well below the crest of the Brooks Range along the southern slopes. Above the treeline are barren rocks, varicolored and stark, like those in arid regions of southwestern United States. Because of low annual precipitation, only the higher summits have year-round snow and only the highest areas have glaciers. The northern slopes and the eastern part of the park/preserve are especially deficient in moisture. Average precipitation in the Mt. Doonerak area, near the park's eastern boundary, is only 8 to 12 inches annually, while Mt. Igikpak and the southwest corner of Gates of the Arctic receive from 12 to 18 inches per year. In the polar desert on the north side of the Brooks Range, only lowgrowing tundra vegetation survives as stunted shrubs, lichens, mosses, and herbs.

Visiting the Park/Preserve

Access to Gates of the Arctic is by scheduled flights to Fairbanks (200 miles southeast of the park/preserve) and to Bettles, the closest town. Arrangements for charter aircraft and gear can be made at Bettles. No roads, trails, or campgrounds are maintained in the park/preserve, but visitors may backpack or canoe into the wilderness. (The Trans-Alaska pipeline highway that parallels the park's eastern boundary is open during the summer as far north as Dietrich camp.) Attractions for visitors include the long glacial valleys; alpine lakes; float trips on the wild, clear-water rivers; and fishing (permitted in the preserve areas of Gates of the Arctic). Another plus is the experience of being in a wilderness where many of the scenic features are still unnamed.

The early history of the Gates of the Arctic region is similar to Kobuk Valley National Park (Chapter 11). The animals that migrated from Asia to this continent more than 12,000 years ago (when lowered sea levels exposed the Bering land bridge), and the human hunters who followed, spread over what is now northern Alaska and Canada. The numerous archeological sites that have been found within the boundaries of Gates of the Arctic National Park and Preserve mark the routes of the early people and give us clues as to their way of life. Most were nomadic hunters and gatherers. Artifacts of successive Indian and Eskimo cultures have been found at Anaktuvuk Pass, along the North Fork of the Koyukuk Wild River, among the lakes of the headwaters region between the Noatak and the Kobuk Wild Rivers, and in other areas. The Nunamiut Eskimos and Athabascan Indians who now live in the park/preserve retain ways of life similar to their ancestors. They hunt, fish, and gather berries in the park/preserve for their own subsistence.

Within historic time, the first explorers were American military officers who followed major rivers from the coast into the Arctic interior during the 1880s. Besides tracing the streams back to their sources, they scouted the country for possible mineral locations. One of these intrepid officers, S.B. McLenegan, a navy engineer, set out in a canoe in the summer of 1885, accompanied only by a seaman, to find the headwaters of the Noatak. He recorded in a journal their adventures, hardships, and dangers on the way to and back from what is now the western part of Gates of the Arctic National Park and Preserve. At one point, he wrote (after an "icy bath" when the canoe capsized):

". . . the difficulties encountered only seemed to awaken the stubborn elements of our natures, and with a determination not to be baffled, we prepared ourselves to meet anything short of utter annihilation." (Brown et al, 1982, p. 104.)

In the 1890s the search for gold lured prospectors and adventurers to the rivers of the far north, and in 1893 gold was discovered in the Middle Fork of the Koyukuk, east of the present park/preserve boundary. However, Alaska's main gold rush began in 1898 after gold in significant amounts was found in the Klondike. Mining camps sprang up at Bettles and along the Middle Fork of the Koyukuk. Some prospectors tried to find gold along the Kobuk and the Noatak Rivers and in the headwaters area between the drainage systems. Their high but unfulfilled hopes are shown by names such as Midas Creek and Lucky Six Creek, and by the ruins of a few cabins in the park/preserve.

FIGURE 27.5   The lineated pattern of the metasedimentary rocks strongly influence weathering and mass wasting. National Park Service photo.

During these years, the U.S. Geological Survey began sending expeditions into the region to do reconnaissance field work and mapping. Geologic investigations continue to this day, but a number of scientific questions about the area are unresolved.

Geologic Features

Effects of Glaciation

Erosion and deposition by a large system of Pleistocene valley glaciers produced most of the landscape features of the Brooks Range in Gates of the Arctic National Park. Horns, cols, cirques, cirque lakes, and other alpine landforms are characteristic of the higher parts of the range. Below the summits, the glacial troughs deepen and widen as they descend; among the foothills, tongues of glacial outwash extend in valley trains or spread over lowlands. Till and stratified drift make up morainal ridges and kame terraces. Some of the sediment is reworked and deposited as alluvial fans and lake fill. Many of these glacial deposits and valley fills have been prospected as possible sources of placer gold, but "shows" of significant size have not been found.

Most of the glaciers have disappeared because snowfalls of the present drier climate cannot support large active glaciers, like those in southern Alaska. The few glaciers that remain in Gates of the Arctic National Park and Preserve do not produce enough pulverized rock, or *rock flour,* to make the streams cloudy or "milky" during the melt season. Because most of the water comes from rain and clean snow, and runs down over permanently frozen ground, the arctic rivers are remarkably clear.

The Rocks

Many different kinds of rocks are exposed in outcrops in Gates of the Arctic. Some are carbonates and fine-grained clastics that accumulated in ancient marine shelf or deltaic environments. Some are volcanic, mostly basaltic, and others are plutonic and granitic. Most of the rocks have undergone moderate regional metamorphism, but some have been altered by local or contact metamorphism. Sedimentary rocks that indicate their tectonic origins include conglomerates, flysch sequences, and graywackes. (A *flysch* is a thinly bed-

ded, rapidly deposited marine sedimentary facies consisting of different sizes and types of sediments.) Original thicknesses of the sedimentary sequences exceeded thousands of feet but are difficult to determine because great thrust sheets have shortened, overridden, and compressed the beds.

Fossil correlations have yielded ages of sedimentary rocks ranging from Cambrian through Cretaceous. Radiogenic dates derived chiefly from plutonic rocks indicate Precambrian through early Tertiary igneous activity.

Geologic Structures

Structures of all sizes, from minute foliation to great regional thrust sheets, suggest that for a considerable span of geologic time a converging zone was active in this northern region between the North American continental plate and one or more oceanic plates. Intense folding occurred as plates collided. East-west trending structures were overturned to the north, and reverse faults were upthrust from south toward the north. Thrust faults moved rocks many miles. So intense was this thrusting that thrust sheets lie one over another, stacked like pancakes.

Oceanic crust was overridden and mangled as some of the marine sedimentary rocks were scraped off. Then the plate dipped down, carrying the remaining sediments as it sank. Under heat and pressure at depth, the mixture began to melt, allowing globs of molten rock to work up toward the surface. This process produced the plutons, stocks, dikes, and sills in the Brooks Range, along with some volcanic rocks. Meanwhile, large-scale regional metamorphism was altering most of the rocks undergoing compression at various depths.

As tectonism continued, the mountains were uplifted, causing erosional processes to intensify. Nonmarine sedimentary rocks accumulated in mountain basins. Complicating the picture are the blocks of *exotic terranes (tectonostratigraphic terranes)* that were "rafted" to the northwestern continental margin, presumably from south of the equator. These pieces of continental crust ("mini-continents") that rode on an oceanic plate like floating islands, were involved with the younger nonmarine rocks during later compression, faulting, igneous activity, and uplift that produced the Brooks Range (fig. 27.6).

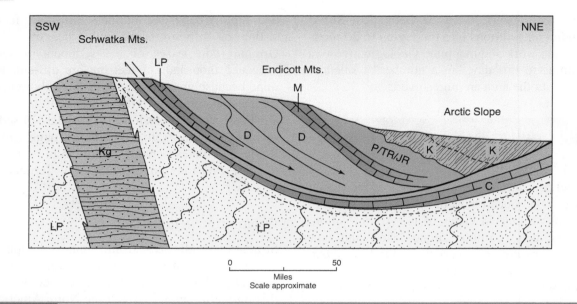

FIGURE 27.6   Generalized cross section through the Brooks Range, from southwest to northeast, showing geologic relationships. Key: LP=Lower Paleozoic; D=Devonian; M=Mississippian; C=Mississippian/Pennsylvanian; P/TR/JR=Permian/Lower Mesozoic; K=Upper Cretaceous; Kg=Cretaceous granite. The rocks above the thrust fault are mostly exotic (Endicott Terrane); northward-moving subduction not shown. Upper Cretaceous sediments (right) deposited on terrane during and after thrusting. The intrusive mass and folding are probably younger than the thrust faulting. Based on Mull et al., Connor and O'Haire 1988.

Geologic History

Because of the remoteness of the area and the complexity of the rocks and structures, the geologic history of the Brooks Range and Gates of the Arctic National Park and Preserve is still being pieced together. In general, the patterns and sequence are similar to those of other complex mountains discussed in this section and in Part V.

1. **Deposition of island arc and coastal plain sediments during Precambrian (?) to late Paleozoic time.**

Conformable, partially metamorphosed, sedimentary rocks at the bottom of exposed Paleozoic sequences in the "Doonerak anticlinorium" (a very large composite anticlinal structure, regional in extent) in the eastern part of the Brooks Range may be Precambrian or early Paleozoic in age. Their age is inferred from the fact that rock units above these beds have yielded fossils (trilobite fragments and a few brachiopods) correlated as middle Cambrian in age. Other

Paleozoic rocks include the Kanayuk Conglomerate (about 7500 to 8000 feet thick and part of the Endicott Group), which is one of the most extensive fluvial units in North America. The Kanayuk Conglomerate, made up of river flood-plain deposits, Late Devonian and Early Mississippian in age, crops out over an area about 30 miles north/south and 450 miles east/west. Geologists interpret the formation as being part of a large, compound delta, consisting mainly of coarse sediments washed down from the North American landmass. This delta prograded (built out its front) southwestward late in Devonian time but subsequently retreated, probably because the sea was advancing, early in Mississippian time.

Deformed coastal-shelf and deltaic rocks related to the North American plate have been cut by intrusive bodies dated as 470 and 350 million years old. Therefore, we can assume that compression and some subduction occurred early in Paleozoic time.

In the heart of the Endicott Mountains, rocks of the Lisburne Group include limestones and marbles of Mississippian and Pennsylvanian age. The Kanayuk

Table 27.1			Geologic Column, Gates of the Arctic National Park and Preserve	
Time Units **Era/Period/Epoch**			**Rock Units**	**Geologic Events**
Cenozoic	Quaternary	Holocene	Glacial drift Alluvium	Neoglacial episode Periglacial activity
		Pleistocene	Talus Glacial drift	Many glacial stages
	Tertiary			Continuing uplift Intense erosion
Mesozoic			Nonmarine rocks	Uplift, accretion of terranes, compression, thrusting, subduction, plutonism
Paleozoic			Nuka Formation Lisburne Group Kanayuk Conglomerate Marine sedimentary rocks of various ages Cambrian sedimentary rocks	Compression, thrusting, subduction, plutonism, metamorphism Deltaic sedimentation, metamorphism folding, plutonism
Precambrian	?		Plutonic rocks	Igneous activity

Sources: Dutro et al., 1981; Hamilton et al., 1986; Mull et al., 1982.

Conglomerate and the Lisburne Limestone have been used in ongoing paleomagnetic studies.

2. **Sedimentation, intrusions, and thrusting in late Paleozoic and early Mesozoic time.**

In this interval, more rocks were deposited locally, while larger and more complex geologic structures developed in the Brooks Range. Granitic plutons were emplaced under Mt. Igikpak and the Arrigetch Peaks in middle Devonian time. The late Paleozoic Nuka Formation was thrust north toward what is now the Arctic Coastal Plain.

3. **Major orogenic activity from late Mesozoic to early Tertiary time.**

Geologic activity speeded up toward the end of the Mesozoic Era. Events that took place in the Brooks Range are related to tectonic activity to the south in the Canadian and U.S. Rocky Mountains. Here in the north, several continental, oceanic, and island-arc crustal blocks reached the Arctic-Alaska region and were accreted. These tectonostratigraphic units probably rotated counterclockwise in varying amounts as the Arctic-Alaska block moved southward.

Early in Cretaceous time, the Lisburne Group was pushed some 50 miles northward. Granitic intrusions, dated as Late Cretaceous, uplifted the Schwatka Mountains. Widespread thrusting resulted in crustal shortening on a grand scale as rock units were carried from the southern foothills to the northern foothills. One field geologist has described five major thrust sheets made up of many thousands of feet of metasedimentary rocks, with lower Paleozoic units now on top, upper Paleozoic beds in the middle, and upper

Paleozoic and Cretaceous strata at the bottom! Each sequence has been telescoped northward in relation to the underlying sequence (Mull 1979). Intermixed with the thrust sheets are rocks from exotic terranes and from the North American plate.

Subduction with accompanying plutonism (dated at 92 to 86 m.y. by the potassium-argon method) and volcanism was ongoing. At first, subduction was apparently due to north-moving rocks dipping downward, but this activity was succeeded by the plunging of south-moving rocks under the complexly distorted beds. Injection of mineral-bearing solutions was widespread. Although traces of gold, lead, zinc, copper, antimony, and tungsten have been found, their exploitation has not been economically profitable. Of the several tectonostratigraphic terranes that were "sutured" to the Brooks Range during these orogenic spasms, some apparently came from the North American continent (perhaps from western Canada or from the "Lower Forty-eight") while others have been traced to possible source areas in the Southern Hemisphere, perhaps even the southwestern Pacific. The marbles, schists, and quartzites in the Brooks Range were regionally metamorphosed by compression, folding, and thrusting during the mountain-building activities.

4. Tertiary uplift and erosion.

Isostatic uplift that followed the orogenic activities elevated the Brooks Range still more and accelerated the processes of weathering and erosion. Sheets and wedges of sediment were shed from the mountains, accumulating on the Arctic Coastal Plain and spreading over the central Alaska lowlands. Many of these deposits have been buried by younger glacial and fluvial sediments.

5. Glaciation in late Tertiary and Quaternary time.

The processes of glacial and fluvial erosion and deposition dominate the more recent geologic history of the Gates of the Arctic region. In these northern latitudes frost action is intense, especially at the moderately high elevations of the Brooks Range. Frost-shattered rocks move downslope, breaking up more and more as they fall, roll, and slide. Eventually, streams haul the rock debris away.

The Brooks Range was thoroughly glaciated in the Pleistocene Epoch by at least four episodes of glaciation-two in late Pleistocene, one in middle Pleistocene, and one in early Pleistocene (Hamilton, 1977). These glaciations produced the alpine topography of the summits, the large troughs at mid-elevations, and the innumerable glaciofluvial deposits of the lowlands and foothills. The small glaciers on the higher mountains are remnants of Neoglacial ice that has retreated during the last 200 years.

Geologic Map and Cross-Section

Beckman, Helen M. 1980. Geologic Map of Alaska, U.S. Geologic Survey.

Sources

Brown, W.E. et al. 1982. Far North Parklands, in *Alaska National Parklands, This Last Treasure*. Anchorage: Alaska Natural History Association. p. 89–120.

Connor, C. and O'Haire, D. 1988. *Roadside Geology of Alaska*. Missoula: Mountain Press Publishing Co., 250 p.

Dutro, J.T. Jr., Palmer, A.R., Repetski, J.E. Jr., and Brosge, W.P. 1981. The Doonerak Anticlinorium Revisited. *U.S. Geological Survey in Alaska: Accomplishments During 1981*. Circular 868, p. 17–19.

Grybeck, D., and Nokleberg, W.J. 1978. Metallogeny of the Brooks Range, Alaska. U.S. *Geological Survey in Alaska: Accomplishments During 1978*. Circular 804B, p. B19–B21.

———, Cathrall, J.B., LeCompte, J.R., and Cody, J.W. 1983. Buried Felsic Plutons in Upper Devonian Redbeds, Central Brooks Range. *U.S. Geological Survey in Alaska: Accomplishments During 1983*. Circular 945, p. 8–10.

Hamilton, T.D. 1977. Late Cenozoic Stratigraphy of the South Central Brooks Range. *U.S. Geological Survey in Alaska: Accomplishments During 1977*. Circular 772B, p. B36–B38.

———, Reed, K.M., and Thorson, R.M. 1986. *Glaciation in Alaska, the Geologic Record*. Anchorage: Alaska Geological Society. 265 p.

Hillhouse, J.W., and Gromme, S. 1980. Cretaceous Overprint Revealed by Paleomagnetic Study in Northern Brooks Range. *U.S. Geological Survey in Alaska: Accomplishments during 1980*. Circular 844, p. 43–46.

Jones, D.L., Silberling, N.J., Berg, H.C., and Plafker, G. 1980. Tectonostratigraphic terrane map of Alaska. *U.S. Geological Survey in Alaska: Accomplishments during 1980*. Circular 844, p. 1–5.

Moore, T.E., Wallace, W.K., Bird, K.J., Karl, S.M., Mull, C.G., and Dillon, J.T. 1994. Geology of Northern Alaska *in* Plafker, G., and Berg, H.C., eds. *The Geology of Alaska*. Boulder, Colorado Geological Society of

America, The Geology of North America, v. G-1, p. 49–140.

Mull, C.G. 1976. Apparent South Vergent Folding and Possible Nappes in Schwatka Mountains. *U.S. Geological Survey in Alaska: Accomplishments during 1976*. Circular 751 B. p. B29–B30.

———— 1979. Nanushuk Group Deposition and Late Mesozoic Structural Evolution of the Central and Western Brooks Range and Arctic Slope. *U.S. Geological Survey in Alaska: Accomplishments during 1979*. Circular 794, p. 5–14.

————, Tailleur, I.L., Mayfield, C.F., and Pessel, E.H. 1975. New Structural and Stratigraphic Interpretations, Central and Western Brooks Range and Arctic Slope.

U.S. Geological Survey in Alaska: Accomplishments during 1975. Circular 733, p. 24–26.

————, ————, ————, Ellersieck, I, and Curtis, S.M. 1982. New Late Paleozoic and Early Mesozoic Stratigraphic Units, Central and Western Brooks Range, Alaska. *American Association of Petroleum Geologists Bulletin* 66:348-362.

National Park Service. 1983. Environmental Overview and Analysis of Mining Effects, Gates of the Arctic National Park and Preserve.

Note: To convert English measurements to metric, go to www.helpwithdiy.com/metric_conversion_calculator.html

C H A P T E R   2 8

Yosemite National Park

EAST CENTRAL CALIFORNIA

Donald F. Palmer   Kent State University

Area: 761,236 acres; 1189 square miles
Established as a National Park: October 1, 1890
Designated a World Heritage Site: 1984
Address: Superintendent, P.O. Box 577, Yosemite
National Park, CA 95389

Phone: 209-372-0200
E-mail: yose_web_manager@nps.gov
Website: www.nps.gov/yose/index.htm

*From deep glacial valleys on the west to the crest
of the Sierra on the east, the park has
magnificent scenery and the highest and most
spectacular waterfalls in the United States.
Batholithic intrusions that make up the greatest
part of the bedrock represent the root zone of a
volcanic-plutonic complex that formed in
Mesozoic time as a function of subduction on
North America's western margin.*

*A series of glacial advances are recorded by
sculpted topography and thick sections
of glacial tills and moraines.*

FIGURE 28.1   Yosemite Falls, on the north wall
of Yosemite, descends 2425 feet in three steps to the
flat-floored valley below. The upper falls is 1430 feet
high. The lower fall is 320 feet and the intermediate
cascades between the main falls are 675 feet. Photo by
Ronald F. Hacker.

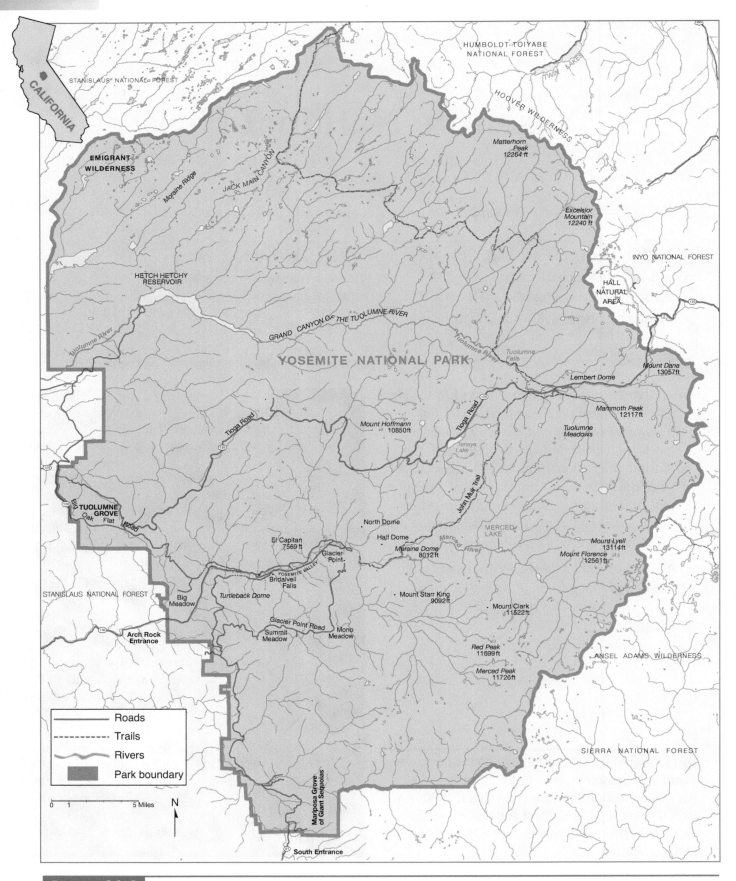

FIGURE 28.2 Map of Yosemite National Park, California.

Yosemite National Park is located approximately in the center of the Sierra Nevada, the largest continuous mountain range in the United States. Early settlers and travelers found the range to be an impediment to communication and progress, so the region was of little interest until the discovery of gold in the foothills of the Sierra Nevada in 1848. The resulting gold rush brought miners and traders into direct competition with the Miwok and Paiute Indian tribes that had lived for many decades in the region between the foothills and the valleys east of the Sierra Nevada.

It was this confrontation that led to the opening of Yosemite Valley and the high Sierra to tourists and nature lovers. In 1851, the Mariposa Battalion, under the direction of Major James Savage, entered the valley from the western end in pursuit of a group of perhaps 200 Indians of the Ahwaneechee tribe led by their chief, Tenaya. Major Savage's plan was to force the Indians, who had been responsible for a number of raids against trading posts in the area, out of their stronghold and to convince them to give themselves up to the Indian commissioners. While the majority of the battalion thought primarily of the immediate military objectives, the company physician, Dr. Lafayette Bunnell, was nearly immobilized by the grandeur of the valley before him. It was Bunnell who suggested that the valley be named Yosemite, an anglicization of *Uzumati,* the name of the Indian tribe, meaning "grizzly bear." Bunnell, whose impressions were later recorded in his book *The Discovery of the Yosemite,* also named many of the major topographic features and lakes of the area. His enthusiasm for the scenery was infectious among the company and their reports on their return to the foothill towns interested many in the region. By 1855, the area had become tamed enough so that James Hutchings of San Francisco could visit the valley with three others, including the artist Thomas Ayres. Ayres had been sketching landscapes since he had arrived in northern California in 1849, and Hutchings felt that the artist's talents would be well employed in bringing to the public the first accurate pictures of the wonders of Yosemite Valley. Ayres' drawing of the Yosemite Falls was published in the fall of 1855, and in July 1856, the first issue of *Hutchings' California Magazine* presented four drawings of Yosemite by Ayres.

Ayres returned to the valley in 1856 and visited the high country near Tuolumne Meadows. His sketches and personal accounts were published throughout the country and a show of his drawings was given in New York City. Other landscape artists, notably Albert Bierstadt, Thomas Hill, and William Keith, were to follow in the next decade.

In 1859, Hutchings brought to Yosemite its first photographer, Charles Leander Weed, whose photographs were shown in San Francisco in September of that year. Weed was followed soon by the great photographer Carleton Watkins, who had the technical mastery and the artistic sensibility to make Yosemite come to life for those who had not seen it. It is impossible to overestimate the importance of the early artists in establishing Yosemite as the first wilderness park.

The education of the public, through the combined efforts of artists, explorers, scientists, and entrepreneurs, aided the political efforts of those who sought to preserve the area. In June 1864, Congress granted Yosemite Valley and the Mariposa Grove of Big Trees to the state of California in a proclamation signed by President Lincoln. Thus, Yosemite became the first public park under the administration of a state.

Galen Clark, who had campaigned for the preservation of the area, was appointed the first guardian of the park. Clark's management recognized the two-fold mission that still confronts the National Park Service today: to protect and preserve the wilderness and at the same time to allow reasonable access for public recreation and enjoyment. Clark worked to make the valley more pleasant and accessible to tourists. In 1878, in order to reduce the swampy conditions in parts of the valley, he dynamited part of a recessional moraine that dammed the Merced River. This resulted in fewer mosquitoes and a dryer, more hospitable valley floor of pleasant meadows. He cleared forest litter from some areas in order to allow clear pathways through the groves of Big Trees. Hiking and horse trails were constructed and maintained, and a variety of access roads were built into the valley and high country.

John Muir, the Scottish-born naturalist and conservationist, came to California as a young man in 1868 and made the state his home. He wrote many articles about Yosemite and campaigned tirelessly for the expansion of the park beyond the valley itself. He persuaded numerous influential people to visit the park. Ralph Waldo Emerson, then in his late sixties, spent ten days with Muir, camping and traveling on horseback throughout the Yosemite region in 1871. Emerson's admiration for Yosemite was conveyed to others in the eastern United States. Years later in 1903, President

Theodore Roosevelt spent three days with Muir in Yosemite.

Largely through Muir's efforts, Yosemite National Park was established around the original Yosemite Grant. In 1905, the state of California returned its original grant to the federal government in order to unify the administration of the park.

Wagon roads were laid out in the park in 1878; before that, travel by horseback and on foot were the only methods of getting around in the park. In 1914, the first year in which automobiles were admitted, 127 cars passed through the area. In recent years, with over 2 million people visiting the park annually, automobile travel has had to be restricted because of traffic problems and pollution.

While the aim of the National Park Service is to preserve the wilderness of Yosemite, the effects of man's development of the region were far-reaching. The introduction of domesticated animals and plants in the 1860s led to changes that were largely reversible, but the actions of the first guardians of the park brought about changes that were sometimes unexpected and often irreversible. When the early explorers and tourists first entered the Valley, they beheld grand vistas across large open meadows. Over 750 acres of meadows were measured on the Valley floor in the 1860s. Since then, thick forest growth has invaded the original meadows, decreasing the number of unobstructed views of the cliffs and waterfalls. This stark change in the vegetative cover of the valley floor was caused in part by the aim of preservation, which kept people from clearing areas and cutting smaller trees for firewood, as the Indians had done, and in part by the actions of Galen Clark, who sought to make the Valley more hospitable. In blasting out the recessional moraine to reduce the swampiness of parts of Yosemite Valley, Clark accelerated the spread of the forests into the newly dried soils of the meadows. Today, the approximately 340 acres of meadow can be maintained only by judicious clearing of trees and shrubs.

A more permanent change was effected in part of the park by the construction of a dam across the Tuolumne River. The dam resulted from the needs of the rapidly growing city of San Francisco for water and electric power as it was recovering from the great earthquake of 1906. Interests focused on the Hetch Hetchy Valley as an "ideal" reservoir site. This valley was similar to Yosemite Valley with cliffs over 2000 feet high and with great waterfalls (fig. 28.3). In spite of opposition by conservation groups, Congress gave permission for the flooding of the Hetch Hetchy Valley in 1913. The dam, completed in 1923 and enlarged in 1938, flooded scenery that once rivaled Yosemite Valley in grandeur and beauty. The loss of this area is acutely felt today, for the population pressures exerted on Yosemite Valley could have been alleviated by the existence of an alternate valley within the park.

Geologic Features

Within Yosemite National Park, a complex of igneous rocks of varying composition was intruded over a period of more than 100 million years. These batholithic rocks were intruded into older metasedimentary and metavolcanic rocks now preserved as isolated masses at the eastern and western margins of the batholiths. Because the granites, granodiorites, quartz diorites, and monzonites that make up the batholiths in this area have different properties, the rocks have responded differently to geologic processes since their emplacement. Thus the character of the igneous structures influenced, and to a large extent controlled, the development of the magnificent and varied scenery that we see in the park today.

Finally, the morphology of the surface has been sculpted by glaciers that steepened the river valleys and modified the upper surfaces of the Sierra Nevada by erosion and deposition. Perhaps nowhere else in North America is there a better place to see the effects of glaciation on an igneous rock terrane.

Bedrock Geology

The bedrock geology of the Yosemite region developed geologically over a period of over half a billion years, evolving from a stable continental margin in the Precambrian and early Paleozoic to an active convergent margin in the mid-Paleozoic and into Cenozoic time. Because of its long complex geologic history, rocks of the region may have been derived from continental, island-arc, or oceanic sources and may have reached their present positions after much lateral motion toward or parallel to the old continental margin of North America. In the northern part of California, the evidence shows the development of a late Proterozoic to early Paleozoic continental margin recorded in rocks that formed on the abyssal plains of the ocean, on sea mounts, and on the continental rise and continental

FIGURE 28.3 Hetch Hetchy Valley was dammed after pressure from the public for a water supply for San Francisco. In 1913 Congress gave permission to flood the valley in spite of opposition from conservation groups. The dam was completed in 1923 and enlarged in 1938. It destroyed a miniature Yosemite Valley that had 2,000 foot high cliffs and great waterfalls. The photograph shows how it looks today. Photo by Michael Little.

slopes. A Pre-Late Devonian collision placed these rocks onto the continental margin in an event that may be correlative with the Antler orogeny in central Nevada. Later, a Late Devonian to Permian island arc assemblage developed and was incorporated into the continent of North America by the Middle Triassic. Not all of these events are recorded in the rocks within Yosemite National Park, and where pre-batholithic rocks do occur, they are metamorphosed and often difficult to correlate and date. Nonetheless, geologists working within the park have identified many of the earliest rocks that attest to the history of the region.

The oldest rocks in the park are metamorphosed units occurring on its western and eastern margins and as isolated masses suspended within the batholithic intrusions. These rocks fall into two main groups. The first—and older—group consists of Proterozoic and Lower Cambrian miogeosynclinal rocks related to quartzite and schists deposited on the old continental margin that is now to the southeast in and around Death Valley (Chapter 48). These rocks were deposited in shallow water and were derived largely from continental sources. They included limestones, sandstones, and shales that have been metamorphosed to marble, quartzite, and slate. This first group, which appears to have been offset along a large strike-slip fault, now crops out in the Snow Lake pendant in the Emigrant Wilderness in the northern part of Yosemite and in other isolated pendants in the central part of the park.

The second group consists of the Paleozoic rocks that mark the record of the convergent plate motions that occurred from the Middle Paleozoic to Early

Mesozoic. These rocks crop out extensively on the western side of the park and include the Shoo Fly Complex and the later Calaveras Complex. The Shoo Fly Complex consists of schists and gneisses, representing the metamorphosed sedimentary rocks derived largely from continental sources during Proterozoic and Early Paleozoic time, and intrusive granite of Late Devonian age. The younger Calaveras Complex is a mixture or mélange of rocks derived largely from oceanic sources and includes shale, siltstone, and chert, which form a matrix for inclusions of mafic igneous rocks, serpentinite, and other units representative of disrupted island arc volcanics and oceanic crust. Although these two major complexes developed at different times, they were juxtaposed along a major thrust by Early Triassic time.

Along the crest of the Sierra near Mt. Dana, on the eastern boundary of the park, an apparently different group of Permian metavolcanic rocks overlap an earlier Paleozoic section of metasedimentary units. In almost all cases, the age determinations have been questioned because of later metamorphism that has destroyed fossils and reset the radiometric ages to some extent.

The Mesozoic rocks that crop out within the park occur mostly on the eastern margin along the crest of the Sierra. They occur as elongate roof pendants parallel to the mountain range. This sequence of Mesozoic rocks is exposed at the top of Mt. Dana as a series of volcanic tuffs and flows interbedded with sandstones of volcanic origin. These rocks are Jurassic in age and may be the volcanic expressions of the Jurassic intrusions that are the early stages of the Sierra Nevada Batholith. North of Tioga Pass, the eastern entrance to the park, a third section of metavolcanic rocks suggests nearly continuous igneous activity from Triassic to early Jurassic time (225 to 185 million years before present). In the southeastern part of the park an early Jurassic metasedimentary and metavolcanic sequence has been found that correlates with that of the Boyden Cave sequence near Kings Canyon National Park (see Chapter 49).

Beyond the western boundary of the park and west of the Calaveras Complex lies a series of rocks that throw considerable light on the evolution of the Yosemite region. Here thick sections of island-arc volcanic and deformed ultramafic rocks attest to the existence of a subduction zone immediately west of the Yosemite region from late Paleozoic through early Mesozoic time.

Mesozoic Batholiths

The tectonic events occurring west of the Yosemite region led to the fusion of deep crustal material and the intrusion of the Sierra Nevada Batholith. In the park, dozens of plutons of varying sizes, ages, and compositions were emplaced. Ages of the plutons tend to fall into two groups with a gap of time between them. The first group of dates, ranging from 210 to 150 million years before present (mybp), indicates that a period of plutonism occurred beginning in late Triassic and continuing throughout Jurassic time. This period of activity was intimately associated with the volcanic activity discussed previously. The second group marks intrusions over a period of 50 million years and represents the main batholithic event in the Yosemite region. In Yosemite Valley, the early Cretaceous intrusive phase lasted from roughly 120 to 100 mybp. The main late Cretaceous phase occurred from about 95 to 80 mybp. In Yosemite Valley, minor episodes occurred between these two. The southeastern part of the park, however, is underlain by major plutons and associated metavolcanic rocks that date from 100 to 95 mybp. The distribution of ages shows that major episodes of plutonism were often separated by large intervals of time. During early Cretaceous time, the Jurassic and older sedimentary rocks were intruded by a sequence of magmas. In Yosemite Valley this sequence is called the Western Intrusive Suite, the members of which have been generally named for prominent exposures (table 28.1). The sequence has been worked out based on the rules of crosscutting relationships and the inclusion of fragments of the earlier intrusions in later plutons. The most plentiful, or widespread, of this group is the El Capitan Granite, which crops out over a large area.

The next sequence within Yosemite Valley is known as the Minor Intrusive Suite. These intrusions are known to come after the early intrusive group because the Minor Intrusives cut the older intrusions. One of the most notable of the Minor Intrusives is the diorite of the "Map of North America," which is crudely shaped like a map of North America and appears as a dark mass on the light-gray cliff face of El Capitan (fig. 28.4). Some thin, light-colored pegmatite dikes, younger than the diorite, cut across the maplike intrusion and further demonstrate the order of intrusion in the area.

The youngest of the Cretaceous intrusions are the rocks of the Tuolumne Intrusive Suite (table 28.1). The

FIGURE 28.4   The face of El Capitan is made up of massive El Capitan Granite with few through-going joints. The dark irregular mass in the center of the cliff is called "the Map of North America" because of its shape. This mass is part of a complex of diorite dike intrusions that invaded the plutons of the earlier Western Intrusive Suite. The very thin white lines crossing both the diorite and El Capitan Granite are later pegmatite dikes. Photo by Jan McGuire.

earliest members of the suite crop out in Yosemite Valley; but the youngest members (Cathedral Peak Granite and Johnson Granite Porphyry) crop out north and east of the valley toward Tuolumne Meadows. This suite of intrusions is a nested group of plutons that were intruded as parts of a single magmatic system over a relatively short period of time. The term "nested" refers to the concentric zonation that developed as later intrusions were emplaced into the cores of the earlier plutons. The Tuolumne Suite shows a gradual change in the composition and texture of the rocks throughout time. The earliest and outermost rocks are richer in the mafic minerals hornblende and biotite and have a higher percentage of the sodium-calcium feldspar, plagioclase. The younger, central units are poorer in the mafic minerals and contain more of the potassium feldspar, orthoclase.

Texturally, the earlier rocks are *equigranular,* meaning that the different minerals are approximately the same size throughout the rock. On the other hand, the central intrusives in the suite are *porphyritic,* a texture in which large crystals occur in a matrix of smaller grains. Porphyritic textures develop in two stages. First the magma cools and crystallizes slowly at depth, leading to large crystals suspended in the magma. Later, the partially crystallized magma is remobilized and emplaced at a higher, cooler level in the crust. At this level, the magma solidifies more rapidly into the finer grained matrix around the early formed crystals.

The combination of chemical and textural changes in the Tuolumne Intrusive Suite is consistent with the evolution of a single magmatic system brought about by *fractional crystallization.* Fractional crystallization occurs as minerals that crystallize at higher temperature, such as hornblende, biotite, and plagioclase, are separated from the magma. The residual liquid, therefore, is changed in its composition, and rocks that crystallize later from the residual magma will be lighter colored and may be richer in potassium feldspar and quartz.

An alternative hypothesis for the origin of the intrusive suites is that they represent sequential formation of magmas by melting of rock at deep crustal levels. By this hypothesis, as the region cooled, only lower temperature magmas could be mobilized and intruded. Whatever the explanation, the existence of the intrusive suites does demonstrate the episodic nature of igneous activity in a region that was hot and active over many millions of years.

Late Cenozoic Volcanic Activity

Most of the late Cenozoic volcanics are outside park boundaries, mainly to the east in the vicinity of Mono Craters. However, some volcanic and intrusive activity occurred in the northern and eastern parts of the park. North of the Grand Canyon of the Tuolumne River, lava flows of intermediate (andesite) composition as well as mudflows of volcanic ash (lahars) have been found. On the flanks of Mt. Dana, dikes and plugs occur, some of which appear to have been intruded along old fault zones. These isolated bodies are remnants of extensive activity that occurred throughout the Sierra. The evidence shows that rhyolite ash flows covered much of the northern part of the range about 30 mybp. Ten million years ago andesitic and basaltic lavas were extruded during part of the uplift of the range. Finally, some remnants of flows and intrusions, which date at 3.5 mybp, are found in the eastern part of the park.

Joint Systems and Landform Development

The surficial processes that have shaped the features of this geologically young mountain range are weathering, stream erosion, mass wasting, and glaciation. The unique and distinctive character of the topography, however, was determined by preexisting structures within the rocks. In Yosemite National Park, the types and systems of jointing of the various granitic rocks have been a key factor in the development of the landforms.

Joints are planar surfaces in rocks along which separation occurs. The motion of the two blocks of rock on separate sides of the joint are away from one another. Joints differ from faults in that faults have motion of the rock parallel to the surface of the fracture. The joints that occur in the batholithic rocks in Yosemite are formed by pressure release associated with the uplift of the range and the unloading of the area by erosion of the rocks that once overlay the intrusions.

Joints usually occur in sets of variable spacing. Large continuous joints separated by considerable distances are termed "master joints." They act to control drainage patterns and occasionally lead to parallel cliff faces. Lesser joint sets involve smaller joints, which are not as continuous, but usually occur more closely together.

In some areas, massive, unjointed rocks form large cliffs such as El Capitan, which rises 3604 feet above the valley and is made up of massive granitic rock that is highly resistant to weathering and erosion because it has almost no joints or fractures. Close by, however, are intricately jointed granitic rocks that form castle-like landforms of striking appearance.

Intersecting joint sets occur in many of the Yosemite granites, with joints oriented in two or more different directions. Where the joints are fairly closely spaced, pillars, columns, and pinnacles have been formed by erosional processes. Examples in Yosemite Valley are Watkins Pinnacle, Split Pinnacle, Washington Column, and Cathedral Spires. A larger feature that is bounded by more widely separated joints is the Three Brothers, with its asymmetrical, "gabled-roof" summits. At a number of locations, rock with widely spaced vertical joints occurs beside rock with close, intersecting joints. An example is Cathedral Rocks (massive features) next to Cathedral Spires (columns and pinnacles).

Sheet jointing, or *exfoliation,* is a type of physical (mechanical) weathering that is related to vertical pressure release from unloading. As the great weight of rock overlying a batholith is removed by erosion, the upper part of the batholith is able to expand slightly because of the reduced confining pressure. Cracks, or sheet joints, may develop parallel to the outer surface as a result of expansion. As the rock is exposed, mechanical weathering causes the thin layers of rock to spall off or break loose in concentric slabs from the granite mass, a process usually aided by gravity (mass wasting). In massive granites devoid of much vertical jointing, *exfoliation domes* have developed. Exfoliation of concentric shells of rock can be readily observed at Sentinel Dome, Turtleback Dome, or North Dome (fig. 28.5)

Half Dome developed due to both vertical jointing and exfoliation, with some help from glacial activity. This famous landform was never a whole dome that was cut in two by a tremendous glacier, as it may appear; in fact, the "other half" was never there. The steep face of Half Dome (fig. 28.6) began to take shape as thin slabs of rock between parallel vertical joints were loosened by weathering and mass wasting and fell away, layer by layer—a process that is continuing at the present time. The orientation of the face of Half Dome is parallel to a line of cliffs south of the Merced River, thus it may represent one of a set of "master joints" in this region of the Sierra. Half Dome was modified in the Pleistocene when the Tenaya Glacier moved down the valley, reaching within 500 feet of the top of Half Dome. The glacier cleared away the exfoliation debris

and talus; and because it was flowing in a direction parallel to the trend of the vertical joints, it removed more sheets of rock by plucking, thus leaving the precipitous face that we see today. The rounded back of Half Dome developed more gradually as curved shells of rock were exfoliated (fig. 28.7).

In some of the massive granites, exfoliation produced arches on the steep side of the domes. Royal Arches near the head of Yosemite Valley is the classic example. These are not "see-through" arches, but are, rather, a series of recessed arches set back into the main arch that curves out over the dome's base, a thousand feet below the top of the overhang. Here the Tenaya Glacier probably removed the lower part of the concentric shells of rock, leaving the upper parts unsupported.

FIGURE 28.5   View of Lambert Dome shows the effects of exfoliation. The curved lines in the lower part of the dome mark the edges of exfoliation slabs. National Park Service photo.

FIGURE 28.6   The planar north face of Half Dome is defined by a nearly vertical set of joints, along which rock was easily removed. The stepped character of the jointed face can be seen in the photograph as straight or broken white lines such as those on the upper left or lower center of the cliff. The dark vertical lines are water drip lines. The crest of Half Dome is 8852 feet and the steepest planar part of the upper face is nearly 2000 feet high. Photo by Sherry Schmidt.

FIGURE 28.7   The view of Half Dome from Glacier Point shows the exfoliated top and south side of Yosemite's most famous landmark. Joints that intersected the horizontal sheeting controlled the vertical north face. Photo by Ronald F. Hacker.

Effects of Glacial Erosion

Some domes such as Liberty Cap, Lembert Dome, and Mount Broderick were actually overridden by glaciers. These domes took on the characteristic profile of giant *roches moutonnées*. They have a smooth, rounded, and polished side where the glacier approached and overrode the resistant outcrop, and a steep, often precipitous face on the side where the glacier flowed away from the dome. On the down flow side the glacier was able to pluck slabs of rock from the face by refrozen meltwater and by ice that worked its way into joints in the outcrop.

Abrasion by rocks and gravel carried at the bottom of the glaciers was severe, and glacial polish occurs on many of the surfaces in Yosemite. The comparative freshness and sheen of the polish shows that weathering has not yet affected it very much. Polished surfaces tend to roughen due to long-term weathering so that no polish remains on glacial pavement scoured by the earliest glaciers. Thus the polish can be helpful in determining the extent of the various stages of glaciation.

Glacial erratics, some quite sizable, are a common sight in the park. Glacier Monument, which is 1500 feet high, has erratics on top that were brought by the ice from an area 12 miles to the east. Erratics, along with patches of till, are found mostly on the uplands above the valleys. Morainal remnants are numerous in the park. The ones that are deeply weathered and dissected are associated with earlier glacial stages. Nunataks, such as Half Dome and El Capitan, provide clues as to the size and extent of glaciers. (*Nunataks* are mountain peaks that projected above the glaciers and were not overridden by ice.)

In the high basins where the glaciers originated, cirque lakes formed following the retreat of the ice. Some still exist but others have silted in, becoming subalpine marshes and meadows.

Glacial Lake Yosemite, which was impounded by a moraine near El Capitan, was about 5.5 miles long when the ice was melting back. The lake rapidly filled with sediment and disappeared, leaving the swampy meadows found by the first explorers. Seismic surveys show that the bedrock is 2000 feet below the floor of the valley near Glacier Point. The glacial sediments filling the valley include thick sections of glacial outwash and till and at least six separate layers of lake sediments. This sedimentary succession shows the complexity of the glacial history of Yosemite Valley. It suggests that early glaciations gouged out the valley floor and then a long sequence of glacial advances and retreats yielded glacial and interglacial sediments that refilled and smoothed the valley floor.

FIGURE 28.8   The Royal Arches, near the head of Yosemite Valley, are the result of exfoliation in massive granite. The inset arches are set back into the main arch. Photo by Ronald F. Hacker.

The floors of Little Yosemite and Tenaya Canyons are covered by similar deposits of drift and outwash. Most of the moraines that were left in Yosemite Valley by retreating glaciers were washed out by swiftly flowing meltwater or were covered over by lake deposits. The destruction and burial of moraines makes them difficult to correlate; therefore, age dating of glacial advances is obscure. Furthermore, the scattered occurrence of moraines in isolated valleys and in highland basins contributes to the uncertainty in the details of glacial history.

Waterfalls

The jointing of the bedrock, the stream erosion associated with the uplift, and Pleistocene valley glaciers all influenced the development and location of Yosemite's spectacular waterfalls. Most of the highest waterfalls in the country are in Yosemite National Park. Among the highest are:

Yosemite Falls—2425 feet (fig. 28.1).

Ribbon Fall—1612 feet.

Wapama Falls (Hetch Hetchy Valley)—1500 feet.

Staircase Falls—1300 feet.

Tueeulala Falls (Hetch Hetchy Valley)—1000 feet.

Bridalveil Fall—620 feet (fig. 28.10).

Nevada Fall—594 feet.

Illilouette Falls—370 feet.

Vernal Fall—317 feet.

Yosemite, Ribbon, Wapama, Tueeculala, Bridalveil, and Illilouette Falls all issue from hanging valleys that existed before glaciation and were heightened by glacial action. The regional uplift and tilting that raised the Sierra Nevada Range to its present height disrupted the earlier drainage systems of the whole area and caused the main streams, such as the Merced River, to flow generally parallel to each other in a southwesterly direction. In the later stages, the uplift was so rapid that the increased stream gradients allowed these main streams to erode their valleys at an accelerated rate. This erosion was much faster than that of the tributary streams, which flowed parallel to the range with gradients that were little changed by the uplift. By the time the Ice Age began, the tributary streams of Yosemite were already entering the main valley in steep cascades, tens to hundreds of feet above the floor of the main valley. When the glaciers moved into the valleys, the erosive action of the ice straightened and deepened

FIGURE 28.9   View of Granite Lake near Tioga Pass, located on the east side of Yosemite National Park. Photo by Sherry Schmidt.

FIGURE 28.10   View of Yosemite Valley looking east. The Mariposa battalion first saw the valley from this vantage point (now the Wawona Tunnel parking lot). The present morphology of the valley has evolved through uplift and stream erosion, followed by steepening of the valley walls by glaciation. This left a series of hanging valleys such as that of Bridalveil Creek, which joins the Merced River after the 620 foot drop of Bridalveil Falls. Photo by Ronald F. Hacker.

the main valley, making its walls even steeper and accentuating the "hanging" relationship of the tributary valleys. Many of the cascades then became free-leaping falls.

Nevada Fall and Vernal Fall have a somewhat different geomorphic history. They descend over giant glacial steps created as ice moved down valley, plucking out sections of closely jointed bedrock between strips of massive granite and forming a giant glacial staircase. The action in this case was very similar to that which steepened the downflow faces of overridden domes such as Lembert Dome.

Geologic History

The Yosemite region developed as a result of a long, complex series of events that affected the entire Sierra Nevada range and, indeed, the whole western United States. Those events led to the formation of the rocks of the area and affected the evolution of landforms in the Sierra.

1. Plate collision and the volcanic-plutonic complex.

The early geological evolution of the region has been largely obscured by the great batholiths of Jurassic and Cretaceous time. To understand the early history of the region, geologists have had to look to the areas west and east of the batholiths and at the isolated masses of metamorphosed rock in the roof pendants of the plutons.

The rocks around the Yosemite region show that the area was close to the continental margin during much of the Paleozoic. Late in Paleozoic time, a subduction system developed that accreted slabs of material onto the western edge of the continent. East of the Sierra, thick sections of Precambrian and lower Paleozoic sedimentary rock occur that are representative of sedimentation on the continental margin.

As subduction continued into the Mesozoic Era, volcanic and plutonic activity began to be important within the Sierra Nevada province. Thick sections of oceanic crust and upper mantle were accreted onto the western coast of North America in what are now the

FIGURE 28.11 View of Mount Dana from Dana Plateau. The weathered rock in the foreground is called *felsenmeer*. Frost-shattered blocks and coarse, angular fragments have accumulated below the summit in an area of well jointed bedrock. The snowfield on Mount Dana was a source of snow for the Pleistocene Tuolumne glacier. Photo by Sherry Schmidt.

foothills of the Sierra Nevada. In late Jurassic time, regional metamorphism and structural deformation marked the Nevadan orogeny. The rocks were later remetamorphosed and (in many places) obliterated by the intrusions of the Cretaceous batholith.

Within the Sierra, a great volcanic plutonic complex developed as a result of the subduction. Geochemical studies suggest that the magmas were derived in part from the down-going slab and in part from continental crustal material, that was melted and assimilated by the magmas coming from greater depth. The depth at which the batholiths were emplaced has been estimated to be about six miles, although those closely related to volcanic rocks may have been shallower. As the activity continued, the subduction center moved to the west, and the center of batholithic intrusion shifted slightly to the east. This latter shift is seen

in the "younging" of the batholiths towards the crest of the Sierra in Yosemite, Sequoia, and Kings Canyon National Parks (see also Chapter 49).

2. Uplift and erosion.

The main batholithic phase ended 80 million years ago and was followed by a period of uplift and extensive erosion during late Cretaceous and early Cenozoic time. Cenozoic volcanic activity covered much of the northern Sierra. The fact that the remnants of these once extensive lavas and ash flows are directly over the Cretaceous plutons shows that by the middle of Cenozoic time, erosion had stripped away most of the metamorphic rocks into which the batholith was emplaced. Further uplift and erosion continued episodically throughout the late Cenozoic. The region was alternately one of steep and of subdued topography.

Table 28.1	Geologic Column, Yosemite National Park

Time Units			Rock Units		Geologic Events
Era	Period	Epoch	Hetch Hetchy Valley	Yosemite Valley	
Cenozoic	Quaternary	Holocene	Glacial deposits, alluvium, landslides, rockslides, talus		Erosion and deposition by glacial, fluvial, and mass wasting process
		Pleistocene	Glacial tills, moraines, etc.; some early deposits buried or destroyed by later ice advances		Severe erosion during repeated glacial advances. Sculpturing of giant staircases, hanging valleys, basins, etc.
	Tertiary — Neogene / Paleogene	Pliocene Miocene	Mudflows (lahars) Andesite lava		Volcanism (northern part of park) Series of uplifts with faulting and tilting of Sierra Nevada block.
			//////////////////////////////////	//////	
Mesozoic	Cretaceous		*Tuolumne Intrusive Suite*		
			Aplite Yosemite Creek Granite Mt. Gibson Diorite	Johnson Granite Porphyry Cathedral Peak Granite	Major intrusions into country rock and igneous rock of the composite batholith Dikes, veins
			Half Dome Quartz Monzonite Sentinel Granodiorite		
			Minor Intrusive Suite		
				Diorite of "Map of North America" Quartz-mica diorite Bridalveil Granite Leaning Tower Quartz Monzonite	Minor intrusions of composite batholith
			Western Intrusive Suite		
			Ten Lakes Alaskite Rancheria Mt. Granodiorite Double Rock and Mt. Hoffman Granodiorites	Taft Granite	Major intrusions into country rock and emplacement of composite batholith
			El Capitan Granite Tamarack Creek Quartz Diorite	Granodiorite of the Gateway Arch Rock Granite Diorite of the rockslides	
	? — Jurassic		South Fork of Tuolumne River Quartz Diorite Diorite	? Mariposa Slates	Deformation, metamorphism, beginning of intrusions Interbedded sediment and volcanics, faulting
	Triassic				
Paleozoic			Metasediments and metavolcanics (mostly outside park; some preserved in roof pendants) includes Shoo Fly Complex Proterozoic–Early Paleozoic and younger Caloveros Complex.		Prolonged sedimentation followed by deformation, metamorphism, erosion

Sources: Modified after Smith, 1962; Peck, Wahrhaftig, and Clark, 1966; Harbaugh, 1975.

The latest period of uplift involved the development of the asymmetric mountain range we see today. Tilting increased with successive episodes of uplift so that by the Pleistocene Epoch, the eastern side of the range was a very steep, fault-bounded face, and the western slope had become steeper than it was originally. The range is geologically young, and historic earthquake activity and faulting show that the mountains are still rising.

3. Development of the Merced River valley.

As the regional slope to the west developed due to uplift and tilting, the Merced River (and the other main streams) began to flow toward the west and southwest. During this time, the Merced valley lay between rolling hills several hundred feet high. With the gradient fairly low, the Merced meandered back and forth, forming a flood plain. As uplift continued, the gradient increased, causing the river to incise its channel in the broad valley floor.

Late in the Cenozoic Era, the continued tilting of the Sierra block began to have a major effect on the landscape. Valley walls became steeper as the incised channel of the Merced became deeper. On the other hand, the tributary streams, which were oriented more nearly parallel to the mountain range, did not experience the increased gradients that affected the westward-flowing streams. Because of their lower gradients, the tributaries on the upland were not able to cut their valleys as rapidly as were the main streams. Thus hanging valleys began to develop, and the water in the tributaries flowed down steep cascades into the narrow V-shaped gorge of the main valley.

In Yosemite Valley where tributaries such as Bridalveil Creek flowed on resistant, unjointed granite, erosion was slow and the streams cut down slowly. As a result, renewed or continued uplift accelerated downcutting of the Merced River valley, and waterfalls began to develop on the tributaries flowing over resistant rock.

Major faults forming on the east side of the range permitted the block to rise even higher, and Owens Valley was formed as the result of the western side of the block that formed the mountain dropping down. By the onset of the Ice Age, the highest part of the Sierra had reached an elevation of at least 14,000 feet.

4. Pleistocene glaciation.

Details of the glacial history of Yosemite National Park have been difficult to work out because of the problems of dating and correlating successive moraines from one isolated valley or basin to another. Chief among these problems has been the correlation of glacial sediments from the western to the eastern side of the Sierra Nevada ranges. There is still considerable disagreement, but four major periods of glaciation with numerous advances and retreats have been defined in the area. From oldest to youngest these stages are called: pre-Tahoe, Tahoe, Tenaya, and Tioga. The last three stages have been correlated with the Wisconsinan glaciation, which was the last major advance of continental glaciers in North America. The dates of the advances are only approximately known. However, the Tioga glacial stage is thought to have begun about 30,000 years ago, to have reached its maximum perhaps 20,000 years ago, and to have ended 10,000 years ago. The timing of the maximum is hard to determine in part because there appear to have been many fluctuations in the position of the ice margins during the glacial maximum. The age and definition of Tenaya glaciation is very much in question, but the age of the Tahoe glacial stage is generally agreed to have reached a maximum about 70,000 years ago.

The last advance of the Tioga glaciation was the least extensive, and during this period ice did not fill Yosemite Valley. The pre-Tahoe advance, on the other hand, inundated virtually all of the landmarks around the valley. The advances of the Tenaya and much of the early Tiogan stages did fill the valley as far as Bridalveil Fall. It was following these advances that the most recent of the several Lake Yosemites was formed. The glaciers that advanced, retreated, and readvanced during the earlier stages lasted the longest and were the most erosive. The great Tuolumne Glacier, for example, was fed by the snowfields of such mountains as Dana, Gibbs, Lyell, and Maclure. Miles wide, it spread over the uplands and then flowed around Mount Hoffman and divided into two trunk glaciers. Snows were so abundant that glacial ice accumulated in the high areas faster than the trunk glaciers could carry it away. One filled Big Tuolumne Canyon and Hetch Hetchy Valley. The other crossed a 500-foot divide that separated the Tuolumne and Merced basins and then flowed through Tenaya Canyon into the Yosemite Valley. Branch glaciers filled and coalesced until finally only the highest peaks and domes, such as El Capitan, Eagle Peak, Sentinel Dome, and Half Dome, stuck up through the ice. In the narrow canyons and gorges, spurs and projections were truncated by the moving ice, which also gouged out the bedrock of the valley floors.

The 2000-foot-thick glaciers of Tenaya Canyon and Little Yosemite joined together and filled the main Yosemite Valley. Tenaya Glacier moved parallel to one set of bedrock joints, carving and deepening a canyon with a gently sloping floor. However, the ice flowing down Little Yosemite Valley, which is underlain by massive granite with widely spaced joints, created the giant staircase over which Vernal and Nevada Falls now descend.

FIGURE 28.12 There are three giant sequoia tree groves in Yosemite National Park. The largest is the Mariposa Grove with 200 trees, the smaller Tuolumne Grove (25 trees) and the Merced Grove (20 trees). After the last retreat of the glaciers the trees were able to reestablish themselves on the western slope of the Sierra. They grow on sandy soils derived from the weathering of granite. Photo by Ronald F. Hacker.

5. Deglaciation and the present landscape.

Glaciers are great excavators; but when the active digging stops and the melting processes take over, the basins and troughs that have been scooped out make convenient receptacles for water, glacial debris, and sediment. Yosemite Valley was filled in stages as glaciation waned and glacial advances became less extensive. Streams are still filling in the basins and lakes, as well as eroding their valleys. Exfoliation of domes and arches and active mass wasting continue. Landslides dammed Tenaya Creek near the mouth of the canyon, forming Mirror Lake. Near the inlet a delta is building up in the shallow water. Rockslides (possibly triggered by earthquakes) have formed long talus slopes in the valleys.

Many small glaciers exist in the park; the two largest are Lyell and Maclure Glaciers. These produce glacial features today on a small scale. However, the present active glaciers are not remnants of the earlier glaciers. They are new glaciers formed during the latest of a number of Holocene glacial advances. These "Little Ice Age" glaciers have advanced and retreated several times in the last 3000 years.

Probably between 3000 and 4000 years ago, the giant sequoia trees reestablished themselves in scattered groves on the western slopes of the Sierra. Once widespread in temperate North America, the sequoias and redwoods were nearly exterminated by the multiple glaciations of the Pleistocene. Since in their native stands, giant sequoias grow only on sandy loams (or loamy sands), the soil derived from granitic rocks is favorable to their establishment. The sequoia groves in the park are Mariposa (200 trees), Tuolumne (25 trees), and Merced (20 trees) (fig. 28.12).

Geologic Map and Cross Section

Bateman, P.C., Kistler, R.W., Peck, D.L., and Busacca, A. 1983. Geologic Map of the Tuolumne Meadows Quadrangle, Yosemite National Park, California. U.S. Geological Survey Map GQ 1570.

California Division of Mines and Geology. 1967. *Geologic Atlas of California*, Mariposa sheet.

Calkins, Frank C. and others. 1985. Bedrock Geologic Map of Yosemite Valley, Yosemite National Park, California. U.S. Geological Survey Map I-1639.

Huber, N.K.; Bateman, P.C.; and Wahrhaftig, C. 1989. Geologic Map of Yosemite Valley, Yosemite National Park, Central Sierra Nevada, California. U.S. Geological Survey, Miscellaneous Investigations Series, I-2149.

Kistler, Ronald W. 1966. Geologic Map of the Mono Craters Quadrangle Mono and Tuolumne Counties California. U.S. Geological Survey Map GQ-462.

Kistler, Ronald W. 1973. Geologic Map of the Hetch Hetchy Reservoir Quadrangle, Yosemite National Park, California. U.S. Geological Survey Map GQ1112.

Krauskopf, Konrad B. 1985. Geologic Map of the Mariposa Quadrangle, Mariposa and Madera Counties, California. U.S. Geological Survey Map GQ-1586.

Peck, Dallas L. 1980. Geologic Map of the Merced Peak Quadrangle, Central Sierra Nevada, California. U.S. Geological Survey Map GQ-1531.

Sources

Bateman, P.C. 1992. Plutonism in the Central Part of the Sierra Nevada Batholith, California. U.S. Geological Survey Professional Paper 1483, 186 p.

California Division of Mines and Geology. 1962. *Geologic Guide to the Merced Canyon and Yosemite Valley, California.* California Division of Mines and Geology Bulletin 182.

Ditton, R.P.; and McHenry, D.E. 1973. *Yosemite Road Guide. Yosemite National Park, California.* Yosemite Natural History Association.

Guyton, B. 1991. Muir, Whitney, and the Origin of Yosemite Valley. *California Geology* 44: (12) 275–283.

Harbaugh, J.W. 1975. *Field Guide to Northern California,* K/H Geology Field Guide Series. Dubuque, Iowa: Kendall/Hunt Publishing Company.

Heady, Harold F., and Zinke, Paul J. 1978. Vegetational Changes in Yosemite Valley. National Park Service Occasional Paper Number Five.

Hill, Mary. 1975. *Geology of the Sierra Nevada.* Berkeley: University of California Press.

Huber, N. King. 1987. *The Geologic Story of Yosemite National Park.* U.S. Geological Survey Bulletin 1595. 64 p.

Lahren, M.M. 1991. Snow Lake Pendant, Yosemite-Emigrant Wilderness, Evidence for a Major Strike-slip Fault within the Sierra Nevada, *California Geology* 44: (12) 267–274.

Matthes, F.E. 1930. Geologic History of Yosemite Valley. U.S. Geological Survey Professional Paper 160 (includes map).

———. 1950. *The Incomparable Valley*, ed. F. Fryxell. Berkeley: University of California Press.

———. 1958. *The Story of the Yosemite Valley.* U.S. Geological Survey. Text on back of topographic map.

Muir, John. 1961 reprint (originally published in 1894). *The Mountains of California.* Natural History Books.

Peck, D.L.; Wahrhaftig, C.; and Clark, L.D. 1966. Field Trip Guide to Yosemite Valley and Sierra Nevada Batholith. In *Geology of Northern California,* ed. E.H. Bailey, pp. 487–502. California Division of Mines and Geology Bulletin 190.

Robertson, David. 1984. *West of Eden: A History of Art and Literature of Yosemite.* Yosemite Natural History Association and Wilderness Press, 174 p.

Russell, C.P. 1968. *100 Years in Yosemite.* Yosemite Natural History Association.

Smith, A.R. 1962. Petrography of Six Intrusive Units in the Yosemite Valley Area, California. California Division of Mines and Geology Special Report 91.

Note: To convert English measurements to metric, go to www.helpwithdiy.com/metric_conversion_calculator.html

North Cascades National Park

NORTH CENTRAL WASHINGTON

Area: 504,781 acres; 789 square miles

Established as a National Park: October 2, 1968

Address: 810 State Route 20, Sedro Woolley, WA, 98284-1239

Phone: 360-856-5700

E-mail: NOCA_Interpretation@nps.gov

Website: www.nps.gov/noca/index.htm

The North Cascades Range supports over 300 valley and cirque glaciers within North Cascades National Park. High, jagged peaks intercept moisture-laden winds that nourish glaciers, waterfalls, lakes, streams, and lush forests.

Deep glacial troughs and glacially dissected uplands were left by the greater glaciers of the Ice Age; frost-split rocks tower above active alpine glaciers of the present.

Exceedingly complex igneous and metamorphic rocks and geologic structures suggest a tempestuous tectonic history involving plate collisions, subduction, accreted terranes, uplift, and volcanism.

FIGURE 29.1   "Sawtooth"skyline of the Picket Range. Frost action and mass wasting have emphasized foliation patterns of the intensely metamorphosed rock exposed in the cliffs. Glaciation has formed the horns and arêtes. National Park Service photo.

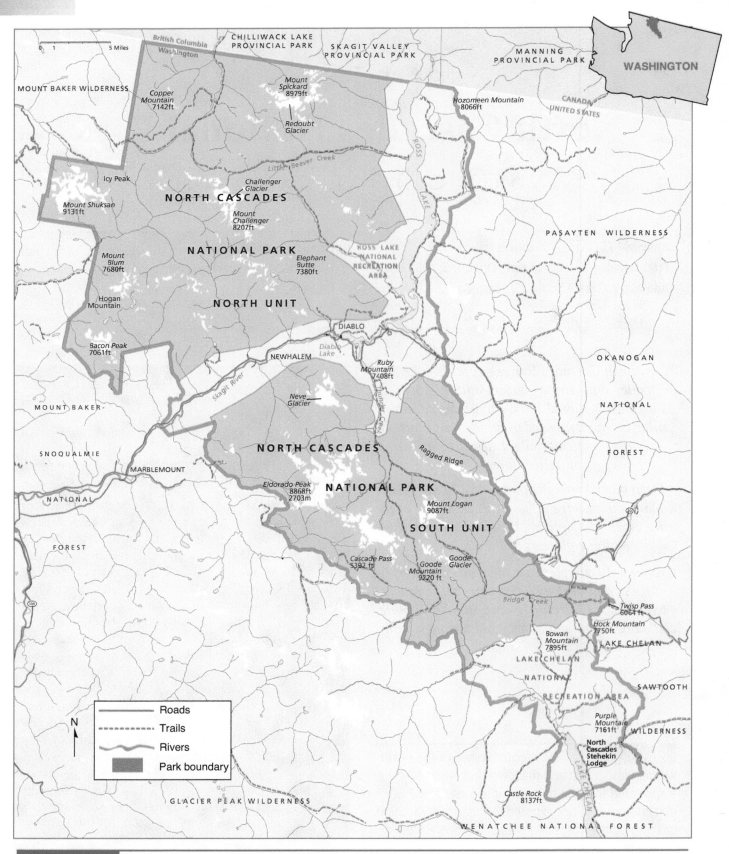

FIGURE 29.2   North Cascades National Park, Washington.

Geographic Setting

A tribal legend explaining the Cascades Range illustrates the Indians' awareness of their natural world. When the earth was young, the story goes, the Cascades region was flat. Rain never fell, but forests and plants drew the moisture they needed from underground water that kept coming to the surface. After a while, water stopped coming up in eastern Washington. In desperation, the tribe sent a delegation west to Ocean to plead with him to send rain. Ocean agreed to help and sent his children, Clouds and Rain, eastward with a good supply of water. All went well for a while, but the thirsty Indians would not let Clouds and Rain go back home. The Indians kept digging large pits for Clouds and Rain to fill with more water. Ocean promised the Indians that if they would let Clouds and Rain come home, he would send water regularly, but still the Indians refused to let Ocean's children go. Ocean then appealed to the Great Spirit to punish the Indians. So the Great Spirit scooped up a huge mass of dirt and rock and molded it into the Cascades. Ocean quickly flowed into the ditch where the earth had been, and this made Puget Sound. East of the Cascades the land dried up and has remained fairly dry to this day. The only water left to the Indians was what was in the pits, the largest pit being Lake Chelan. (Lake Chelan, in the National Recreation Area south of the park, is 55 miles long, steep-sided, narrow, and deep.)

Presumably, the Indians were struck by the contrast between the mist, fog, rain, and snow typical of the western slopes of the range and the dryness of the land on the east side. They appear to have understood the principle of the *rain-shadow* effect of a high mountain range that lies in the path of prevailing winds that are moisture-laden from having passed over water.

Hunters and trappers were the first non-Indians to explore the North Cascades, and they were followed by miners. Before the 1860s, gold, copper, molybdenum, and zinc were mined on a small scale. Today mining is not an important enterprise, although some commercial mining has been carried on in the region.

The prospectors who looked for "strikes" in the North Cascades, especially near the Canadian border, must have endured great hardships. The same could be said for the geologic field parties who did the original reconnaissance work in these mountains. Travel was by foot, with pack mules carrying equipment and supplies. Even today, the only road that crosses the park is Washington Highway 20, which goes through Ross Lake National Recreation Area between North Unit and South Unit. Most geologists backpack their supplies when working in remote areas, although mining companies and the U.S. Geological Survey sometimes use helicopters for supplying field parties. Under the best of conditions, field work in high areas is difficult. Some snow falls during summer months. Rain and mist reduce visibility much of the time. Below the treeline, the forest cover is heavy. But for people who appreciate wilderness beauty, hiking or backpacking in the North Cascades is rewarding.

North Cascades National Park was created to preserve the remaining wilderness areas of the range and protect the watershed. Following the recommendations of a study team, the park was established in 1968 with four units placed under National Park Service jurisdiction: (1) North Unit, including the Picket Range Wilderness; (2) Ross Lake National Recreation Area, comprising a hydroelectric dam and reservoir holding the Skagit River drainage and separating North and South Units; (3) South Unit, including Eldorado Peaks Wilderness; and (4) Lake Chelan National Recreation area.

Geologic Features

Water, ice, and jagged peaks are the scenic elements for which the North Cascades are called "the American Alps." Adding interest to the skyline—although they are not inside national park boundaries—are two volcanoes of the Cascade volcanic chain; namely, Mount Baker and Glacier Peak. Both are several thousand feet higher than the ranges of the North Cascades, which average around 5000 feet in elevation south of the park and rise to more than 8000 feet within the park.

The active Cascades glaciers within and around the park make up this country's largest concentration of glaciers outside Alaska. Also in this wild and rugged region are many spectacular arêtes, horns, cirques, and deep glacial troughs that were sculptured by Pleistocene glaciers. Why do these mountains of only moderate height bear so much ice? The answer is water—water as rain, fog, mist, snow, and more snow, all adding up to the average annual precipitation on the western side of the range of 110 inches. A winter's accumulation of 46 feet of snow is not unusual, and four to five feet of snow may pile up in a single storm.

In the North Cascades, where the snowline on the western slopes is between 6500 and 7500 feet in elevation, glaciers are more numerous and extend lower than in Waterton-Glacier International Peace Park (Chapter 26), which is at the same latitude and has higher peaks. The difference is because Waterton-Glacier is several hundred miles farther inland from the moisture source, the Pacific Ocean.

In the *rain-shadow zone,* east of the North Cascades, annual precipitation rates drop sharply to only 35 inches, for example, around Lake Chelan, and a mere 12 inches in the Pasayten Wilderness. The prevailing winds, carrying moisture from the Pacific Ocean, are forced upward over the North Cascades. Expanding and cooling, the rising air loses moisture. Descending air on the eastern side of the Cascades compresses and warms, thus holding moisture and causing high evaporation. Therefore, the downwind side of the range receives scant rainfall.

Development of the North Cascades

The North Cascades Range is a region of great scenic beauty that is also geologically and geographically complicated and challenging. The spectacular Cascade mountains are carved into uparched bedrock from a part of the earth's crust that was uplifted along a north-south axis late in Cenozoic time. Intrusive and volcanic activity accompanied uplift. Yet the mountains we see represent only the latest and relatively mild event in a region that has been in the midst of intense geologic activity since Precambrian time. The record of orogenies involving severe compression and crustal shortening is exposed in rocks throughout the North Cascades.

FIGURE 29.3   View of Mount Shuksan (located in the western section of North Cascades National Park) as seen from Ptarmigan Ridge on Mount Baker, Mount Baker National Forest. Photo by Douglas Fowler.

The latest uplift may be largely a reflection of isostatic adjustments.

In the North Cascades we have in close proximity very old rocks and some of the youngest intrusive and extrusive rocks in North America. Mount Baker and Glacier Peak, the two volcanoes just outside park boundaries and only about 60 miles apart, were formed from Quaternary lavas that rose through many layers of much older rocks.

Glacier Peak, a relatively inaccessible cone in the middle of the range (in Glacier Peak Wilderness), is built on top of a high ridge of pre-Cretaceous metamorphic rocks (slate, phyllite, and greenschist) between two plutons of quite different age. Andesite lavas and, later, dacite lavas that make up the volcano have no apparent relation to the older rocks that compose the mountains of the range. Late in the Pleistocene Epoch and in Holocene time, the dacitic lavas produced a great deal of ash and pumice because this type of lava (more silica-rich than andesite) tends to erupt explosively. Glacier Peak's most recent ash eruption was only about 200 years ago (Beget 1982).

Mount Baker, in the foothills near the western edge of the range, has a much larger cone (elevation, 10,775 feet) that was built on top of older sedimentary and volcanic rocks. In shape and construction, Mount Baker is a typical *composite volcano*. Volcanoes of this type are long-lasting and tend to construct large, lofty cones, built up over time by alternating layers of andesite lava and pyroclastics. Because of emissions of steam and sulfurous ash for several months in 1975, Mount Baker (along with other Cascade volcanoes, Chapters 35, 36, 37) is being continuously monitored.

North Cascade Rocks and Structures

Distinct structural blocks, or *tectonostratigraphic terranes,* can be identified from west to east across the North Cascades Range. Steeply dipping north or northwest fault systems, approximately parallel with the trend of the range, separate the terranes (fig. 29.4). Within each block, the rocks are fairly consistent in age and type; but the rocks in each block differ markedly from the rocks in adjacent blocks. Field interpretation of the fault systems shows the presence of thrust faults and strike-slip faults that resulted from compressive and lateral movements. Younger vertical faults have been mapped within blocks.

The terranes arrived in the North Cascades region as microplates in late Mesozoic and early Tertiary time. First as eastward-moving and later as northward-moving, the plates were driven against older, complex

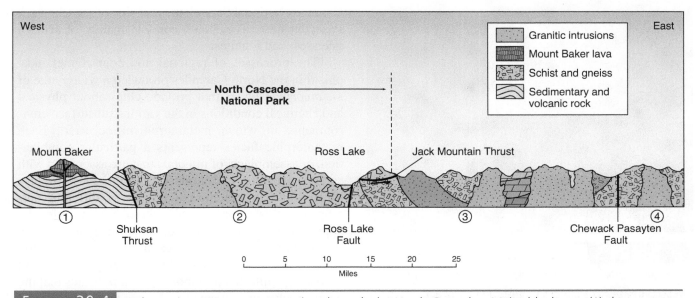

FIGURE 29.4   Schematic west-east cross section through the North Cascades. Major blocks are (1) the western foothills—folded sedimentary and volcanic rocks; bounded on the east by the Shuksan fault; (2) Chilliwack Batholith intruding metamorphosed schists and gneisses; (3) the Methow graben, a down-faulted block between Ross Lake and Chewack Pasayten faults; (4) the eastern highlands. Adapted from Bates McKee, *Cascadia* (fig. 7.3) © 1972 McGraw-Hill Book Company. Used by permission.

rocks (already deformed by past orogenies) that lay along the continental margin. The incoming rocks as well as the rocks already there were profoundly affected by the intensity of the collisions. The fault systems between the terranes and the old continental margin represent boundaries or junctions where vertical and horizontal movements of tremendous force occurred, along with oblique subduction, folding, metamorphism, volcanic activity, and the emplacement of plutons.

For convenience, the rocks across the range from west to east are described in four major zones of rock

FIGURE 29.5   Ross Lake has been carved out of the Skagit Gneiss along the Ross Lake Fault. The left (west) bank of Ross Lake is where the Chilliwack Batholith has intruded the Mesozoic-aged gneisses (Skagit) and schists (Cascade River). The right (east) side belongs to the Methow graben, which contains sedimentary and volcanic rocks. National Park Service photo.

distribution (fig. 29.4). Most of North Cascades National Park is within the second zone, where the Chilliwack Batholith intruded gneisses and schists. Small areas of the park extend into the first zone, on the west, and the third zone, on the east. The four rock distribution zones are described briefly below, but the emphasis throughout the rest of the chapter is on the rock units exposed within park boundaries.

Zone 1 is the western foothills region, bounded on the east by the Shuksan thrust fault. Mount Baker, a young volcano, rises from a platform of folded Late Paleozoic and Mesozoic sedimentary and volcanic rocks in this zone.

In Zone 2, the composite Chilliwack Batholith lies between the Shuksan fault and the Ross Lake fault. The granitic magmas of the batholith intruded older, severely deformed and metamorphosed schists and gneisses.

In Zone 3, the Methow graben, a downdropped fault block, lies between the Ross Lake fault and the Chewack Pasayten fault. The Methow graben consists of Upper Mesozoic sedimentary beds that have been intruded by Cenozoic granites. Near the western edge of the block, the Jack Mountain thrust, a wedge of rock forming the tops of several peaks in the park, rests on top of younger Upper Mesozoic beds. The park's eastern boundary is in this zone.

Zone 4, the eastern highlands, is outside North Cascades National Park on the east side of the Chewack Pasayten fault. This block consists mainly of schists, gneisses, and granites.

The processes of regional and contact metamorphism in the North Cascades produced a wide range of *metamorphic facies* that provide clues about physical and chemical conditions in the various subsurface environments in which metamorphism occurred. Each metamorphic facies represents a particular combination, or assemblage, of minerals that are associated with certain pressure and temperature relationships. By identifying the metamorphic facies of different outcrops of schists, for example, geologists can infer the approximate depth at which metamorphism took place. In a metamorphic facies series, a greenschist suggests an environment of relatively low temperatures and pressures, while an amphibolite schist indicates that the parent rock was subjected to higher temperatures and pressures at a greater depth.

Under some conditions of metamorphism, migmatites, which are interlayed or intermixed rocks, partly igneous and partly metamorphic, are produced. The

Box 29.1 **How Do Glaciers Increase the Hazards of Cascade Volcanoes?**

The glacier-covered summits of Mount Baker and other Cascade volcanoes greatly increase the hazards of any eruption because of the possibility of a *lahar,* or volcanic mudflow, being triggered when snow and ice, suddenly transformed to water and steam, mix with hot ash and other debris, making a slurry. A lahar roars down a mountain valley destroying everything in its path. When thermal activity on Mount Baker increases, the level of Baker Lake, a reservoir at the foot of the mountain, has to be lowered because U.S. Geological Survey geologists fear that a mudflow might cause the dam to break, which would devastate communities farther down the river valley. In 1980, a lahar set off by an explosive eruption at Mount St. Helens (170 miles to the south, air distance) was not so destructive as it might have been because reservoir levels had been lowered when earthquake swarms and steam eruptions indicated that the volcano might erupt.

Composite volcanoes, like those in the Cascades, sometimes have explosive eruptions, depending upon the amount of gas in the lava and its *viscosity,* or resistance to flow. The viscosity of a lava controls the ease or difficulty with which the gas escapes into the atmosphere. Composite cones are built up by lava that is predominantly intermediate, or andesitic, in composition. Andesitic lava is ordinarily more viscous than basaltic lava; but if the temperature of an andesite lava is extremely hot, then it becomes more fluid, or nonviscous, and is able to flow easily from a crater or vent. If a vent becomes clogged and gas pressure builds up, then an explosion occurs that ejects pyroclastic debris, clouds of ash, and lava. Eruptions of composite volcanoes tend to be intermittent, with periods of violent activity followed by years of apparent dormancy.

Sudden release of intense heat is the immediate danger when a volcanic explosion occurs. All living things on the volcano's flanks may be incinerated by the blast. Glaciers and snowfields turn to hot water and steam that mix with ash, forming dangerous mudflows. Water storage reservoirs that have been constructed on the lower slopes may serve as catch basins for mudflows—providing enough warning has been given so that the water can be drawn down. Whether or not the dams can hold is uncertain. A large mudflow could overtop a dam and endanger communities downstream. Hot falling ash carried downwind may start fires over a wide area, contaminate water supplies, collapse roofs, render highways impassable, clog storm drains, endanger aircraft, and damage crops.

Can eruptions be predicted? The monitoring program of the U.S. Geological Survey, in cooperation with state surveys, seeks to interpret data and issue warnings based on several types of indicators: (1) increases in thermal activity; (2) increases or changes in seismic activity; (3) gravity and magnetic data; (4) remote sensing and infrared studies; (5) gas and temperature monitoring; and (6) precise measurements of changes, such as bulging or tilting, in a volcano's outward conformation.

In the Cascades, volcanic eruptions have been infrequent in terms of human life spans. Before Mount St. Helens erupted, many residents did not believe the scientists who warned of potential hazards. However, the 1980 explosion and subsequent less intense eruptions made believers out of skeptics. Major evacuations are really the only adjustment that human beings can make when a volcano becomes threatening.

North Cascades migmatites were probably formed at great depth during regional metamorphism as very hot water, bearing mineral ions, was forced at high pressure through planes of weakness (foliation) in the metamorphosed rocks, changing them to granite along

the paths where the water traveled. The well known "Skagit Gneiss," a striking migmatized schist, crops out in roadcuts along the highway (Washington 20) that crosses the range between North and South Units. Veins and dikes of pegmatites and other igneous rocks swirl, crosscut, and interfinger the banded gneiss in colorful, intricate networks (Misch, 1987). At the Diablo Overlook, where large specimens of North Cascades rocks are displayed in a "rock garden," the following "recipe" appears on a plaque identifying Skagit Gneiss:

Place schist and assorted granitic rocks deep underground.

Allow time to metamorphose at somewhat below melting point. Be patient!

Add more granitic rocks, hot liquids. Increase pressure.

Stew and shake for a few million years.

Allow to cool, and let rise for ten million years.

Result: The very complex, layered, folded, and intruded rock formation visible in the highway cuts nearby.

The North Cascades are not rich in mineral deposits, but some gold, copper, and silver grains were carried by mineral-bearing hydrothermal solutions into quartz veins during cooling of the intrusions. Some of the gold was subsequently freed by weathering and erosion from its quartz matrix and washed into stream gravels. Since gold has a high specific gravity, it tends to sink to the bottom of a stream and the grains ("gold dust") and nuggets become concentrated in sand bars and outwash. Such an accumulation is called a *placer deposit*. Before the filling of the Ross Lake reservoir, placer gold was taken from glacial outwash and stream gravels in Ruby Arm, but this site is now under water.

Glacial Features

From the air, looking down over North Cascades National Park, one sees an awesome terrain—snowfields from peak to peak; ice spilling out of cirques and tumbling down valleys; sharp, jagged ridges that extend for miles; steep-sided horns; precipitous mountain slopes; deep U-shaped troughs; and bubbling cascades of meltwater falling into wooded valleys. High country and low country are inextricably mixed. Some of the names that prospectors gave to the rugged terrain near the 49th parallel (the U.S.-Canadian boundary) evoke images of difficulties they must have had in trying to get around in this region on foot—names like Devil's Dome, Terror Ridge, Fury Ridge, and Ragged Ridge (fig. 29.6).

Excavation by Pleistocene glaciers provided some access into these formidable mountains. Washington Highway 20 crosses the North Cascades by following large glacial troughs through the Ross Lake National Recreation Area and the national forest on the east side

FIGURE 29.6   Arêtes and horns on the serrate crest of the ice-sculptured Picket Range (North Unit). Taken from the north. National Park Service photo.

FIGURE 29.7 A roche moutonnée near Klawatti Glacier. Roche moutonnées are elongate knobs or hillocks of bedrock that have been smoothed and scoured by moving ice on the upglacier side so that the rock is gently inclined and rounded. On the downglacier side the surface is steep and jagged because of glacial plucking. The blunt steep end points in the direction that the glacier is traveling. National Park Service photo.

of the park. The Skagit River and its tributary, Granite Creek, now drain the deep, steep-walled valleys that were excavated by an enormous, branching trunk glacier that flowed from the uplands far down into the Puget Sound lowland. A well known trail, once used by Indians, prospectors, and pioneers, and today by backpackers, is another way into the range. The trail goes from fiordlike Lake Chelan on the southeast side of the range, up the glacial valley of the Stehekin River, across Cascade Pass (a glacial col), down the Cascade River, and then down the Skagit River.

Many examples of alpine erosional features are in the park. Above Cascade Pass (elevation 5392 feet) towers Eldorado Peak, 8868 feet high, a sharply eroded glacial horn rising above active glaciers. Part of the Boston Glacier, longest glacier in the park, is on Eldorado. Numerous active glaciers are in the vicinities of Mounts Spickard, Challenger, and Shuksan, Bacon Peak, and Colonial Peak. The two volcanoes outside the park, Mount Baker and Glacier Peak, have extensive glaciers. Glaciologists have pointed out that the Cascade volcanoes—even though they are higher and tend to support more glaciers—show the effects of gla-

cial sculpturing less than the nonvolcanic peaks of the North Cascades. The inference is that since volcanic eruptions continued during and after the Pleistocene, the volcanoes were able to keep "repairing" their cones.

Mount Shuksan (fig. 29.8), near Mount Baker, is a glacial horn with glacial cirques indenting its flanks. The Picket Range, with its thin, serrated ridge, was named for its resemblance to a picket fence extending along the skyline. Mount Johannesberg, at the headwaters of the Skagit River, is another knife-edged glacial arête. To the south, paralleling Lake Chelan, is the aptly named Sawtooth Ridge. Three Fools are glacial horns on the crest of a nearly vertical headwall that drops thousands of feet to the cirque floor.

Of the glacial troughs, Lake Chelan (in Lake Chelan National Recreational Area and Wenatchee National Forest) is the most remarkable. The bottom of Lake Chelan lies 400 feet below sea level, and its 1500-foot depth makes it one of the nation's deepest lakes. The Chelan trough was excavated by an 80-mile-long glacier that came down the preglacial valley of the Stehekin River from the uplands above Cascade Pass. Many of the flat-floored glacial troughs, such as Ruth

Mount Shuksan (9131 feet) as seen from the south is on the crest of an uplifted fault block of highly metamorphosed schist and gneiss. National Park Service photo.

Creek valley and Nooksack valley, contain braided streams because of the low gradient and high sediment load. Hanging valleys, most of them containing waterfalls, are numerous throughout the park. The most spectacular is a 1500-foot waterfall from the cirque on Bacon Peak that holds Green Lake. Ann Lake and Rainy Lake are also examples of cirque lakes.

Except for moraines on valley floors and sidewalls, some outwash deposits, and similar accumulations, glacial depositional features in the park are less significant than erosional landforms. Most of the glacial debris was moved southward by the Cordilleran ice sheet that pushed down from Canada. Erratic boulders, however, from British Columbia were left sitting on some of the lower summits that were scoured by the ice sheet.

Geologic History

l. The oldest rocks, the Yellow Aster Complex.

The term "complex" is appropriate because the early geologic history of the Yellow Aster rocks, which consist of crystalline basement material, has been obscured by repeated igneous and tectonic activity, as well as long erosion. The Yellow Aster Complex is thought to be old continental crust, clearly pre-Devonian in age and probably made up of orogenically deformed early Paleozoic and Precambrian sedimentary and igneous rocks. Zircons from the Yellow Aster Complex have been dated as older than 1650 million years. Yellow Aster rocks crop out in the western foothills (Zone 1 in fig. 29.4).

2. Early and middle Paleozoic erosion.

As earlier tectonic activity decreased in intensity, long periods of erosion produced unconformities.

3. Middle and late Paleozoic deposition and volcanism.

West of the park, the metamorphosed rocks of the Chilliwack Group (not to be confused with the much younger Chilliwack batholith) were deposited as alternating volcanics and sedimentary beds, mostly in shallow water. In some places the Chilliwack beds are associated with Mesozoic clastics (conglomerates, slates, etc.), suggesting that deposition was more or less continuous in ocean trenches between plates from the end of the Paleozoic Era into Mesozoic time. An episode of metamorphism is dated, on the basis of zircons, at 415 million years in age (Silurian). Another date, derived from zircons in metamorphosed sedimentary rocks, places the Marblemount belt in early Triassic time, 220 million years ago.

4. Mesozoic deposition and volcanism.

Part of the Ross Lake glacial trough, Mount Hozomeen, and the eastern boundary of North Unit are within a belt of rocks of the Hozomeen Group that are believed to be Triassic (possibly Permian) in age. However, most of the thick sequences of Mesozoic volcanic and sedimentary rocks are outside park boundaries in the western foothills and the eastern highlands. The rock units are typical ocean trench deposits that accumulated as microplates drifted toward the continental margin.

5. Late Mesozoic to early Tertiary docking of tectonostratigraphic terranes, with accompanying subduction, faulting, metamorphism, and mountain building.

Geological activity in the North Cascades region increased in complexity and intensity as microplates were squeezed between the converging Pacific and North American plates and "stitched" onto ancient complex rocks of the continent. During the 100 million years since the onset of these geologic events, segments of continental and oceanic rocks of various types and origins were buried and uplifted; heated and cooled; folded, crushed, contorted, and shattered; intruded by granitic magmas; covered by lavas and ashfalls; and intricately combined to produce a "collage" of interesting and puzzling "metamorphic suites" and associated rock units.

The rocks of the Skagit Metamorphic Suite and the Shuksan Metamorphic Suite (Zone 2, fig. 29.4) are prominently exposed in North Cascades National Park. The term *suite,* as used here, means an assemblage of associated plutonic or metamorphic rock units. (In a stratigraphic sense, a suite is comparable to a group.)

During the movement of the Shuksan thrust fault on the west side of the range, rocks were sheared, shattered, and metamorphosed, producing the complex structures of the Shuksan Metamorphic Suite, which trends north-northwest in the western foothills and along the western edge of the park. The unit consists of the Darrington Phyllite (parent rock, shales) overlain by the Shuksan Greenschist (parent rock, submarine basalt flows). Unusual blueschists occur with the greenschists. The schists were altered in a low-temperature, high-pressure environment (i.e., a greenschist metamorphic facies). Tectonic slices of ancient Yellow Aster rock were brought up from deep within the crust, along with slices of ultramafic rock (largely olivine) that may have been derived from upper mantle rock, just below the earth's crust).

The Skagit Metamorphic Suite, which is older than the Shuksan Metamorphic Suite, dominates the main part of the range. The Skagit rock units show varying degrees of metamorphism from greenschist metamorphic facies to amphibolite facies; that is, from low temperatures with high pressures to deep burial with conditions of high temperatures and high pressures. The Cascade River Schist (part of the Skagit Metamorphic Suite) crops out mainly in the western part of the park's South Unit. The principal component of the Skagit Metamorphic Suite, the Skagit Gneiss, crops out in both North Unit and South Unit and is the migmatized equivalent of the Cascade River Schist.

Most of the rock units in the Skagit Metamorphic Suite are part of very old continental crust that drifted across the Pacific for millions of years before colliding with North America. Zircon dates of 2000 to 1452 million years ago have been derived from the Skagit Gneiss. Several granitic plutons were emplaced before, during, and after the arrival of the microplate and were themselves metamorphosed or remetamorphosed. The Eldorado Gneiss, for example, consisting of metamorphosed granitic rock, has yielded a zircon date of 92 million years.

On the east the Ross Lake fault separates the Skagit Gneiss from a long, thin belt of metamorphosed rocks. In this belt is the Elijah Ridge Schist (probably derived from the North Creek Volcanics), overlain by the Jack Mountain Phyllite. The Jack Mountain thrust fault moved older rock units eastward over younger beds (fig. 29.4). Parts of these fault zones have been destroyed by younger intrusions.

6. Late Mesozoic and early Cenozoic deposition; minor deformation and continuing uplift.

In late Cretaceous and early Tertiary time, highlands (precursors of the present mountains) that were elevated by the Cretaceous orogeny became a source of sediment for deposits in marine embayments on the west and east sides of the range. Then, continental beds (mixed sedimentary and volcanic) were laid down over

Table 29.1		Geologic Column, North Cascades National Park	
Time Units		**Rock Units (W = west side, E = east side)**	**Geologic Events**
Era	**Period**		
Cenozoic	Quaternary	Glacial and alluvial deposits Cascade extrusives (W) (Mount Baker, Glacier Peak)	Fraser glaciation (latest of successive major episodes) Severe glacial erosion Construction of composite volcanoes
	Tertiary	Tertiary volcanics (W and E) Chuckanut Formation (W)	Uparching along north-south axis, block faulting, erosion, sedimentation, volcanism
		Late Cretaceous and Tertiary plutons and volcanics, including Chilliwack Composite Batholith and the Eastern Intrusive Belt	Oblique subduction, thrust faulting, compression, metamorphism, migmatization, vertical and lateral fault movements
Mesozoic	Cretaceous	Shuksan Metamorphic Suite (W) (including Shuksan Greenschist and Darrington Phyllite) Jack Mt. Phyllite (E) Eldorado Gneiss Skagit Metamorphic Schist (including Skagit Gneiss/Cascade River Schist)	Docking of terranes Multiple intrusions (batholiths, stocks, dikes, veins, mineralization)
	Triassic	Elijah Ridge Schists (E) (derived from North Creek Volcanics) Hozomeen Group (E)	
Paleozoic	Upper	Chilliwack Group (W) not exposed in park	Accumulation of sediments and volcanics; several orogenies
		Major unconformity	Uplift and erosion
	Precambrian	Yellow Aster Complex (W) (Pre-Devonian crystalline basement)	Highly deformed cratonic rocks

Sources: Misch, 1966; McKee, 1972; Brown, 1987.

the marine units. Patches of these beds, such as the Chuckanut Formation and Tertiary volcanics, exist in the park, but mainly the sequences are found outside park boundaries.

7. **Multiple Cenozoic intrusions, including the Chilliwack Composite Batholith and the Eastern Intrusive Belt.**

Emplacement of a large, composite batholith that along with associated older metamorphics forms the core of the Cascade Range began with the intrusion of stocks having a diorite or gabbroic composition and was followed by a series of major intrusions, most of which were quartz diorite. Later in the sequence they

tended to be granodiorite. Next, stocks of very light-colored granitic magma pushed up into the plutons placed earlier. Small intrusions of granitic composition continued through the rest of the Tertiary.

An Eastern Intrusive Belt, emplaced east of the Ross Lake fault, includes the Ruby Creek Heterogeneous Plutonic Belt, the Black Peak batholith, the Goldenhorn batholith (comprising the only true granite in the North Cascades), and the Perry Creek intrusives of granodiorite and quartz diorite.

8. **Late Tertiary block-faulting and uplift.**

Uplift along a north-south axis was greater toward the north, with maximum uplift occurring in the area of the 49th parallel. Concurrent and subsequent erosion exposed the older metamorphic rocks and deep-seated intrusions in the North Cascades.

9. Pleistocene and Holocene composite volcanoes.

The continuing subduction of oceanic crust and sediments far below the North Cascade range generated lavas that rose through thousands of feet of metamorphic and igneous rocks to form the composite volcanoes, Mount Baker and Glacier Peak. Eruptions in the Cascade volcanic chain have continued into the present (Chapters 35, 36, and 37). Frequent earthquakes in central and western Washington are another indication of ongoing tectonic activity. Movement of the oceanic crust (the small Juan de Fuca plate) has been determined as 35 mm per year of northeasterly motion, relative to North America. Calculations based on seismic data suggest that the subducting slab under Puget Sound has a dip of 9° (Tabor 1986).

10. Pleistocene and Holocene glaciations.

Glaciers have been present in the North Cascades since before the beginning of the Pleistocene, and have been extensive during most of this time. Again and again, glaciers in the mountains of western Canada grew until they formed a Cordilleran ice sheet that moved down over northern Washington, expanding and at times overrunning the alpine glaciers of the North Cascades. The latter enlarged into piedmont glaciers, accumulated great ice fields, and eventually merged with the ice sheet from the north. During warmer interglacial periods, the process was reversed as the ice sheet melted back, leaving glaciers occupying the valleys where they are found today.

In the high mountains, evidence of early and middle Pleistocene glaciations was destroyed by later glacial advances. Fraser glaciation, the last major advance of the Cordilleran ice sheet, began about 25,000 years ago and attained its maximum stand about 50 miles south of Seattle between 15,000 and 13,500 years ago. The ice sheet's recession was followed by a readvance into northern Washington about 11,000 years ago, after which it disappeared. Since that time, the North Cascades glaciers have advanced and retreated many times in response to climatic changes. Significant advances occurred in Neoglacial time, between 3500 and 2000 years ago and about 1000 years ago. In the last 800 years, the glaciers have made a number of minor advances and retreats.

FIGURE 29.9   A meltwater stream has cut through the gravelly end moraine of the Nooksack cirque. National Park Service photo.

FIGURE 29.10   The Stehekin River Valley (South Unit) has the typical U-shape of a valley that has been glaciated. During the melt season, remnant glaciers and snowfields among the peaks furnish much needed water to ranches and communities in central Washington. National Park Service photo.

The Cascade glaciers are an important water resource in the Pacific Northwest because they store ice and snow in the winter season and release meltwater in summer when it is needed for irrigation and other purposes.

Geologic Map and Cross Section

Brown, E.H., Blackwell, D.L. Christenson, B.W., Frasse, F.I., Haugerud, R.A., Jones, J.T., Leiggi, P.A., Reller, G.J., Rady, P.M., Sevigny, J.H., Silverberg, D.S., Smith, M.T., Sondergaard, J.N., and Ziegler, C.B. 1986. Geologic Map of the Northwest Cascades. Geological Society of America Map and Chart Series.

Staatz, M.H., et al. 1972. Geology and Mineral Resources of the Northern Part of North Cascades National Park, Washington. U.S. Geological Survey Bulletin 1359. Plate I.

Sources

Alt, D.D. and Hyndman, D.W. 1984. *Roadside Geology of Washington.* Missoula, Montana: Mountain Press. 282 p.

Alt, D.D. 1995. *Northwest Exposures, a Geologic Story of the Northwest.* Missoula, Montana: Mountain Press. 443 p.

Beget, J.E. 1982. Recent Volcanic Activity at Glacier Peak. *Science* 215:1389–90, March 12.

Brown, E.H. 1987. Structural Geology and Accretionary History of the Northwest Cascades System, Washington and British Columbia. *Geological Society of America Bulletin* 99(2):201–214.

Crandell, D.R. 1965. The Glacial History of Western Washington and Oregon. In H.E. Wright, Jr., and D.G. Frey, editors. *The Quaternary of the United States* (VII INQUA Congress). Princeton, New Jersey: Princeton University Press. p. 341–354.

Malone, S.D., and Frank, D. 1975. Increased Heat Emission from Mount Baker, Washington, *EOS Transactions.* American Geophysical Union. p. 679–85.

Mattinson, J.M. 1972. Ages of Zircons from the North Cascade Mountains, Washington. *Geological Society of America Bulletin* 83:3769–3784.

McKee, B. 1972. *Cascadia.* New York: McGraw-Hill Book Company.

Misch, P. 1966. Tectonic Evolution of Northern Cascades of Washington State. H. Gunning, editor. *A Symposium on the Tectonic History and Mineral Deposits of the Western Cordillera*, Special Volume 8. Canadian Institute of Mining and Metallurgy.

Misch, P. 1987. The Type Location of the Skagit Gneiss, North Cascades, Washington. Centennial Field Guide, Cordilleran Section. Hill, M.L. ed. Geological Society of America. v. 1, p. 393–399.

National Park Service. 1985. *North Cascades, a Guide to the National Park Complex, Washington*, Handbook 131, 112 p.

Staatz, M.H., et al. 1972. Geology and Mineral Resources of the Northern Part of North Cascades National Park, Washington. U.S. Geological Survey Bulletin 1359.

Tabor, J.J. 1986. Crustal Structure of the Washington Continental Margin from Refraction Data. *Bulletin of the Seismological Society of America* 76:1011–1024.

Tabor, J.J. and Smith, S.W. 1985. Seismicity and Focal Mechanisms Associated with the Subduction of the Juan de Fuca Plate Beneath the Olympic Peninsula, Washington. *Bulletin of the Seismological Society of America* 75:237–249.

Tabor, R.W., and Crowder, D.F. 1969. On Batholiths and Volcanoes—Intrusion and Eruption of Late Cenozoic Magmas in the Glacier Peak Area, North Cascades. U.S. Geological Survey Professional Paper 604.

CHAPTER 30

Olympic National Park

NORTHWEST WASHINGTON

Area: 922,651 acres; 1442 square miles
Proclaimed a National Monument: March 2, 1909
Established as a National Park: June 28,1938
Designated a Biosphere Reserve: 1976
Designated a World Heritage Site: 1981

Address: 600 East Park Avenue, Port Angeles, WA 98362-6798
Phone: 360-565-3130
E-mail: OLYM_Visitor_Center@nps.gov
Website: www.nps.gov/olym/index.htm

*The highest annual precipitation rates in conterminous United States nourish on high slopes some 60 active glaciers and support a temperate rain forest on the lower western slopes.
By contrast, a rain shadow keeps northeastern slopes dry.*

Very young sedimentary rocks, intermixed with submarine lavas, were scraped off the ocean floor by plate convergence and pushed up into a chaotic, mangled heap on the Olympic Peninsula.

The western shores are exposed to surf as violent and stormy as anywhere in the world.

FIGURE 30.1 A view of Mount Olympus with Hoh Glacier on its summit. Glaciers actively erode resistant sandstones and conglomerates of the Olympic massif. Photo by Gerald Morrison.

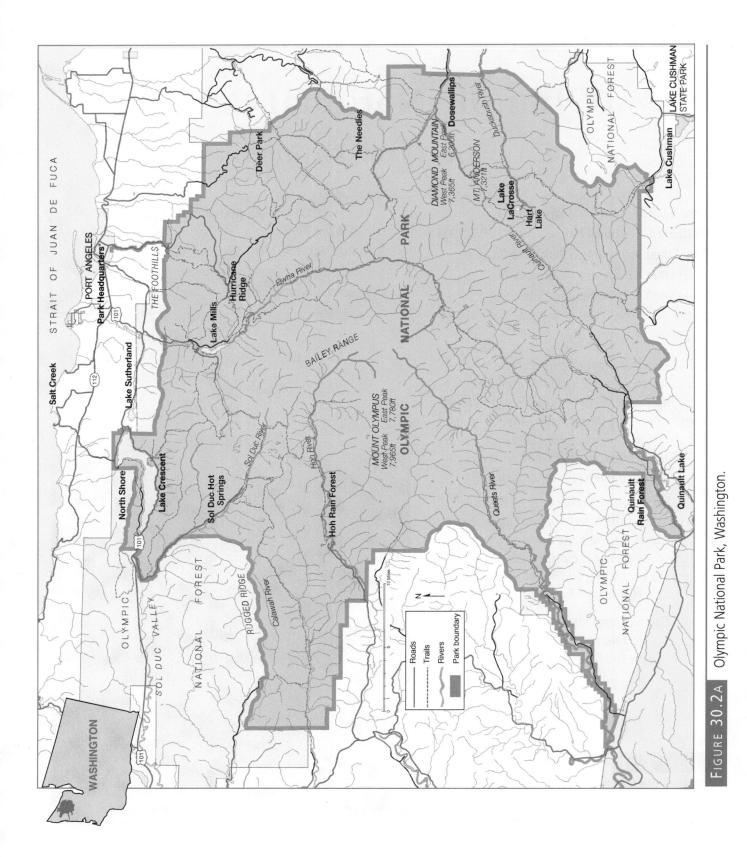

FIGURE 30.2A Olympic National Park, Washington.

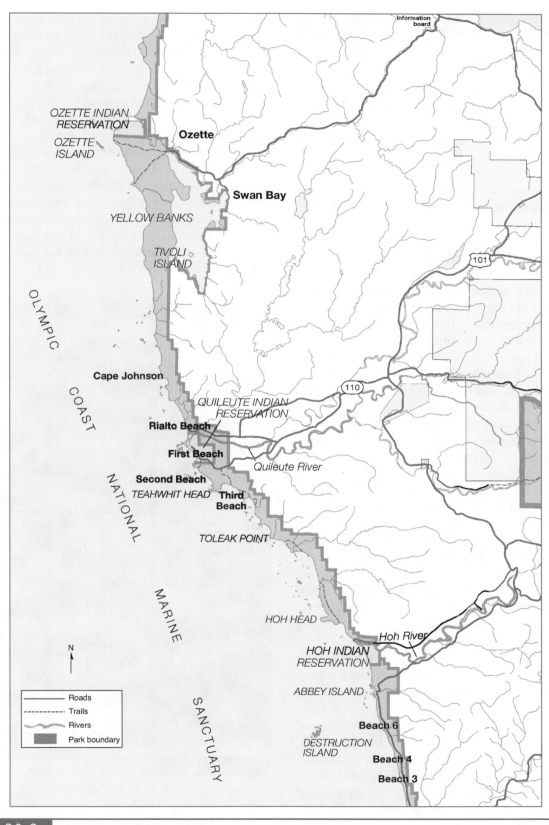

FIGURE 30.2B Olympic National Park, Washington.

Geographic Setting

The natural features of Olympic National Park are remarkable in so many ways it is difficult not to use too many superlatives when discussing the scenery, the wildlife, and the geology. Originally, the land was set aside for the Roosevelt elk. A naturalist, C. Hart Merriam, discovered the Olympic elk and named the new species for Theodore Roosevelt in 1897. In 1909, just before leaving office, President Roosevelt signed the proclamation that created Olympic National Monument. President Franklin D. Roosevelt, 29 years later, approved legislation that established Olympic National Park and protected a much larger area.

Preservation of the wilderness habitat of the elk meant that some of the Olympic rain forest was also saved. Among the unique and valuable trees are towering Douglas fir of record size, Sitka spruce, giant western hemlock, and red cedar. Some trees have diameters of 12 feet or more; some attain heights of 300 feet; many are between 100 and 200 feet tall. Beneath the umbrella of the giant trees, ferns, mosses, and smaller plants grow in profusion on the forest floor (fig. 30.3).

Higher up are forests of thin-spired, subalpine firs that fringe alpine meadows. Above the treeline are bare ledges, glaciers, snowfields, and the peaks of the massif.

The Olympic Peninsula, with its snow-capped mountains looming up, was a landfall for ships exploring the Pacific in the seventeenth and eighteenth centuries. Captain Robert Gray (for whom Gray's Harbor is named) was the first American trader to explore the northwest coast. During his second voyage, from 1790-93, he met Captain George Vancouver, who had been sent from England to chart the coast. Near Destruction Island, the two captains exchanged information about the coastal waters and harbors. In subsequent years, American and English fur traders established trading posts along the Washington and Oregon coast.

The Makah Indians in the region were greatly admired for their courage because they hunted whales in dugout canoes using spears and harpoons with the points fashioned from stone. A few years ago at Cape Avala in the Ozette Indian Reservation, a hillside of clay slumped down, exposing a series of Makah villages that had been built on the site at different levels. Apparently when one village was covered by a slump or slide, the Makahs built a new village on the ruins of the old. Many artifacts, such as paddles, baskets, weapons, and carvings were preserved by the clay. Anthropologists from Washington State University and

Makah descendants living in the area have been excavating the site and cataloguing the artifacts; many are now on display at a tribal museum in the Makah Indian Reservation, adjacent to the northern end of the coastal strip belonging to Olympic National Park.

The interior of the Olympic Peninsula remained an unexplored wilderness for many years after the northern coast and the Puget Sound area were settled. On published maps, the central mountainous area was left blank as late as 1890. Highway maps today show roads skirting the perimeter of the Peninsula but none crossing Olympic National Park from east to west or from north to south. A few paved dead-end roads lead to points of interest, such as the visitor centers at Hoh Rain Forest and Hurricane Ridge, but only trails provide access to the park's interior.

FIGURE 30.3   In Quinault Rain Forest, moss covers many of the trees. Precipitation in the area averages round 150 inches per year. The trees include Douglas fir, sitka spruce, western hemlock, and red cedar. Photo by Gerald Morrison.

A satellite photo of the Olympic Peninsula shows ridges in disarray, appearing to trend in every direction, with few passes through the mountains. Drainage is radial from the high area near Mount Olympus, the highest peak (elevation 7965 feet). Streams rush down narrow valleys between steep interfluves. Getting from one drainage to another is difficult, particularly where glaciers block the way. Below the treeline, dense forests impede one's passage.

The first official exploration of the mountains was led by Lieutenant Joseph O'Neill, U.S. Army, in 1885. His party reached the top of Hurricane Ridge above the Elwha River, which flows north to the coast near Port Angeles.

An expedition sponsored by the *Seattle Press* set out in December 1889 to cross the Olympic Range. The five men in the party had planned to go up the Elwha River by boat, but that proved to be impossible. Next, the sledges they built to carry their supplies bogged down in heavy snow. Then they tried mules, but one fell off a cliff and the other dropped from exhaustion. So they backpacked. With their supplies nearly gone after 14 weeks of climbing and searching for a divide that would get them to a southward-flowing stream, they found a river going to the southwest. Following this stream down, they met a hunting party, who took them by canoe to Lake Quinault. From there, they hiked to the Pacific shore and walked south along the beach to Aberdeen. The trip from Port Angeles to Aberdeen took six months! What they had proved was that the mountains could be crossed. They had marked a trail of sorts and had named many previously unknown peaks and rivers.

In the summer of 1890, Lieutenant O'Neill led a scientific expedition into the Olympic Mountains. In the party were a naturalist-cartographer and a botanist, as well as soldiers to assist them. With great effort, the party constructed a mule trail up the North Fork of the Skokomish River and down the Quinault. They mapped the South Fork of the Skokomish; the Wynoochee River, which joins the Chehalis River; the Humptulips, which empties into Grays Harbor; and several minor streams. Finally, they traveled down the Queets River and reached the Pacific at the southern tip of the park's coastal section.

The Making of Geologic Maps

The geology of the Olympics remained obscure for a long time, even after much of the topographic mapping had been done. Undertaking field work in the rugged and densely forested terrain was difficult; moreover, the geologic structures and stratigraphic relationships are extremely complex. A geologist, who was with a U.S. Geological Survey party in the rain forests, described field work as "going sideways through the trees day after day, soaking wet." Another geologist in an early field party, who collected hand specimens, measured bed thickness, and took strike and dip of beds, recalls "crawling up creek beds through the icy-cold water, next to the bank and beneath overhanging bushes, because that was the only place where bedrock was exposed!" Field garb in those days consisted of heavy, long-sleeved shirts; "tin pants" (duckcloth impregnated with paraffin); hobnail boots; and leather gloves—the latter worn as protection against "devils club," a bushy plant with poisonous barbs—very painful if touched. By this kind of arduous, close-to-the-ground work, hiking and climbing for months at a time, geologists accomplished the task of collecting field data necessary for drafting geologic maps of the Olympic Peninsula.

In making a geologic map, all outcrops visited are marked on a base map, and the type of rock, the strike and dip, and other pertinent observations are entered in a field notebook. Later, field notes are collated; and hand specimens are examined for fossils, bedding characteristics, size of grains, color, and the like. Eventually patterns of rock distribution are developed and their relative positions are studied. Using geologic maps based on field data and laboratory studies, together with scientific knowledge about nearby areas, geologists develop a model for the geologic history of a region. Such a model is by no means finite, for it is revised and refined from time to time as new data and new knowledge become available.

Geologic Features

From early on, geologists have realized that the rocks of the Olympic Peninsula are not old, but only in recent years has it been shown how very young some of them are. For example, we might look at sea stacks (erosional remnants) of Miocene rock—only about 15 million years old—on the western Olympic shore and be awestruck by the fact that some beds are vertical and some are upside down. What force since Miocene time could have stood on end or overturned beds that had been deposited on the sea floor in a horizontal position?

Not far away, at Point of Arches in the Makah Indian Reservation, the oldest rocks of the peninsula jut out into the ocean. These are pre-Cenozoic outcrops of gabbro, at least 80 million years older than the oldest rocks in the Olympic Mountains.

A Plate Tectonics Explanation

What seems evident from the data available is that most of the sedimentary rocks of the Olympics were deposited on basaltic oceanic crust, although some were laid down on partly submerged continental rock that may have been on the leading edge of the North American plate. Another possibility is that the continental rock formed in another location (perhaps to the south) and rode on the oceanic crust as a microcontinent or terrane.

When an eastward-moving segment of oceanic crust, called the Juan de Fuca plate, encountered the westward-moving North American plate, the heavier basaltic oceanic crust was subducted or overridden by the lighter continental crust (fig. 30.4). In the process, during a period of several million years, accumulations of marine sedimentary rock were scraped off the plate, folded, compressed, thrust downward or squeezed upward, fractured, partially metamorphosed, sliced,

and uplifted. The creation of the Olympic Mountains was one result of this powerful tectonic activity. Farther east the effects can be inferred from the upwarping of the North Cascade Range (Chapter 29) and the construction of the northern Cascade volcanoes (Chapters 35, 36).

While the sediments were still accumulating on the ocean floor, great quantities of basaltic submarine lava began welling up and spreading over or interfingering with the sedimentary layers. In some places lava flows piled up so high that offshore islands may have been formed. Most of the flows cooled quickly under water as pillow lavas or brecciated rocks, but some solidified more slowly in dikes and sills between sedimentary beds or in previously erupted lavas (fig. 30.5). The chemical composition of the basalt indicates that the lava rose from very deep in the earth and that it had not been mixed with surface or marine rocks. But the basalt soon was altered as the lavas were subjected to the same forces of deformation as the sedimentary rocks.

Between eruptions of pillow lava, during quiet intervals, a few beds of fossiliferous red limestone were deposited on the ocean floor. The fossils in the rock are Foraminifera (one-celled protozoa that secrete calcite), whose tiny shells can barely be seen with a magnifying glass. The red color of the rock is from iron leached out

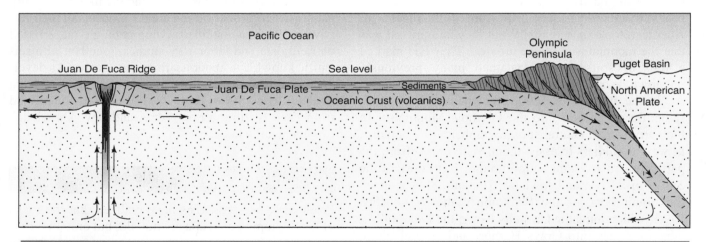

FIGURE 30.4   How plate convergence may have produced the complex structures of the Olympic Peninsula. Oceanic crust, bearing marine sediments, moved eastward for millions of years from the Juan de Fuca ridge, a Pacific spreading center. When heavier oceanic crust met less dense continental crust of the North American plate, the oceanic plate was subducted. Some sedimentary material was dragged down on the oceanic plate; some was scraped off and piled up against the North American plate, enlarging the continent by accretion. From Rau 1980, *Washington Coastal Geology between the Hoh and Quillayute Rivers*, Bulletin 72, State of Washington Department of Natural Resources.

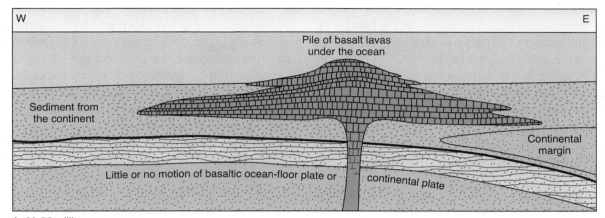

A 30-55 million years ago

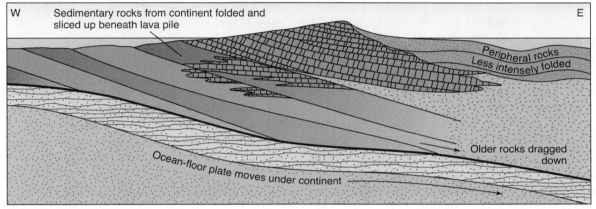

B 12-30 million years ago

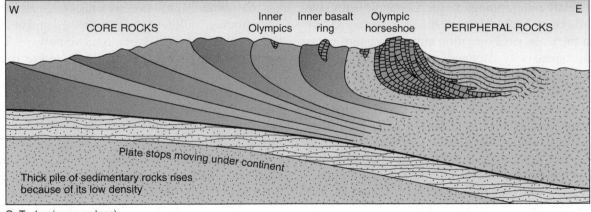

C Today (more or less)

FIGURE 30.5 How the Olympic Mountains developed. *A.* Accumulatoin of basaltic lavas interbedded with sediments, on the ocean floor. *B.* Subduction of the oceanic plate, causing deformation of sedimentary and volcanic rock layers. *C.* Isostatic uplifting of sedimentary rocks due to their lower density. From R.W. Tabor, *Geology of Olympic National Park Hiking and Motorist Guide.* © 1987 Pacific Northwest National Parks and Forests Association. Third printing.

from cooling lava by hot sea water. Associated with the iron are manganese and copper. Some copper and manganese ores were mined early in this century, but the deposits were too small to be of much commercial value. The upper North Fork of the Skokomish River, an area where the red limestones are associated with pillow lavas, was explored by prospectors and miners. A claim staked near Black and White Lakes (named for a brand of whiskey) yielded over a hundred tons of manganese ore that was carried out by mule train during World War I.

Because people were interested in mining prospects, the basalts of the Olympic Peninsula were mapped first. Geologists found that most of the basalts cropped out in an arcuate or horseshoe pattern, partly ringing the peninsula, with the arc portion of the horseshoe on the east side and the arms extending westward. This rock unit, named the Crescent Formation (for exposures at Crescent Bay), is also referred to as the "basaltic horseshoe." The pulloffs and switchbacks on Hurricane Ridge Road, which goes from Park Headquarters to an overlook at about the 6000-foot level, display excellent outcrops of the horseshoe basalts and sedimentary beds, with the oldest rocks at the summit of Hurricane Ridge and the youngest at the base (fig. 30.6).

Since the fossil Foraminifera in the red interbedded limestone were dated as Eocene in age (Rau 1964), the associated lavas were also presumed to be Eocene. No older bedrock has been found on the peninsula except for the small outcropping of pre-Cenozoic intrusive rock on Point of Arches. In fact, the sedimentary layers between the Crescent Formation and the western coast become progressively younger from east to west.

The plate tectonic explanation is that some basalts peeled off and were wedged up against the older rocks of the continent as the ocean floor was underthrust and bent downward (fig. 30.5). Rocks that had been deposited seaward of the basalt pile were crammed accordion-like against the more rigid arch of basalt, and some beds were dragged down by the subducting plate. All the layers were severely deformed, fractured, and sliced by the process, as well as partially metamorphosed by heat and pressure. The younger rocks farther west were also deformed, but they were not so contorted and mashed as the beds jammed inside the basaltic horseshoe. Microfossils (mainly Foraminifera) are more common in the younger beds. The lack of fossils

in the older rocks is probably due to destruction of the shells by severe deformation and recrystallization.

Study of the fossil Foraminifera helped geologists to better understand the structural and stratigraphic relationships of the Olympic sedimentary rocks, but no fossils existed in the lavas. However, the study of *paleomagnetism*, was helpful in discerning age relationships and interpreting structures (box 30.1).

In the last decade, scientists from Oregon State University and the Pacific Marine Environmental Laboratory of the National Oceanic and Atmospheric Administration (NOAA) have discovered and studied new underwater volcanoes on the Juan de Fuca Ridge, southwest of the Olympic Peninsula, 300 miles off the coast. These volcanoes, several more than 100 feet high and a half-mile wide, indicate that the ridge on the sea floor is actively spreading.

Faults and Hot Springs

During the convergence and subduction of plates, the rock layers in the Olympics that had been spread over a large expanse of sea floor were severely sliced up, stacked in a thick pile, turned on edge, and curved into arcuate belts that are roughly concentric with the basaltic horseshoe. Faults separate the rock layers from each other and from the basaltic horseshoe. The most intense disruption and metamorphic alterations occurred in the rocks in the center of the bend of the horseshoe (Tabor 1987a).

As time went on, isostatic adjustments, or "rebounding," allowed the compressed and at least partially subducted rock layers to move upward. The sedimentary rocks, being less dense, tended to rise higher than the heavier basalt. Eventually, as fault movement continued, the Olympic core rocks rose to their present height.

The Calawah fault zone, a major fault on the Olympic Peninsula, follows a wide curving pattern within the basaltic horseshoe and shows evidence of strike-slip motion. The fault zone of sheared and shattered rock extends westward from Crystal Ridge (west of the Elwha River) to the Soleduck River and continues out to the ocean. To the east, the Calawah fault zone bends southward; it parallels the inner curve of the Crescent Formation, and then interlinks with a group of smaller faults and slate belts, called the "southern fault zone" (fig. 30.6). In combination, the two fault zones

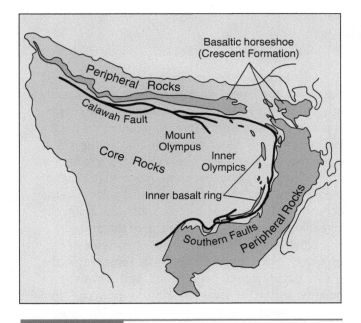

Locations of dominant rock groups and major faults on the Olympic Peninsula. From R.W. Tabor, *Geology of Olympic National Park: Hiking and Motorist Guide.* © 1987 Pacific Northwest National Parks and Forests Association. Third printing.

clearly mark the separation of the simpler structure of the peripheral rocks from the highly deformed, complex rocks of the inner core.

The hot springs in the park, which are located on or close to the fault zones, are not considered a sign of recent or current fault activity. Active faulting probably took place millions of years ago when the rocks now at the surface were deeply buried. The water that issues from the hot springs is not noticeably different in chemical composition from surface water in the vicinity. The most likely explanation for the existence of the hot springs is that ground water percolates deep into the earth, is heated by the high temperatures at depth, and then rises to the surface via the easiest, most direct route, which is through the shattered rock of the fault zones.

The hot water in Olympic Hot Springs issues from rock in the Calawah fault zone in the northern part of the park. The Sol Duc Hot Springs (south of the Soleduck River) do not flow directly from the Calawah fault zone, but reach the surface by way of fractures in sandstone beds adjacent to the fault. Elk Lick, which used to be a hot spring but is now a seep or boggy area

used by elk as a wallow, is in the southeastern part of the park near a branch of the Calawah fault zone.

Turbidites, Tectonic Mélanges, and Piercement Structures

So striking is the appearance of some of the mélange rocks, a hiker is apt to stop and ask, "What's *that?*" To geologists, however, the outcrops may provide significant clues when they are studying a locality and attempting to reconstruct past geologic events.

Turbidites are rocks (or sediments) that were deposited in a marine environment by settling out from muddy water that flowed along a sloping ocean bottom. Such a sediment-laden flow is called a *turbidity current*. Storm waves may start them, or an earthquake that causes undersea sliding, or perhaps a flooding river pouring quantities of sediment into the ocean. As a turbidity current slows down, the sediments tend to sort or grade themselves by size so that the coarsest particles drop first and the finest are deposited last. Turbidite sequences are usually deposited one on top of the other, each sequence grading from coarse to fine, then to coarse again, and so on. When consolidated, the beds become siltstones, sandstones, and conglomerates.

In Olympic National Park, turbidite beds are well exposed in vertical and overturned positions on the western shoreline (fig. 30.7) and in the high mountains. Some have been fractured, as well as folded and overturned. The fractures have been filled with white veins of calcite.

Since tectonic mélanges are associated with intense deformation and shearing, their presence in the Olympic rock sequences is not surprising. By definition, a *tectonic mélange* is a body of rock (large enough to be mapped) that includes fragments and blocks of numerous sizes and kinds of rock embedded in a generally sheared matrix of more malleable rock, such as a claystone or shale. Some of the Olympic mélange blocks are as large as houses. The park's western beaches are the best areas to see the mélange rocks. Since moisture causes the clay minerals of the matrix to swell, this part of the rock becomes crumbly, making it susceptible to wave erosion and separable from the resistant inclusions. Locally these rocks (which are part of the Hoh rock assemblage) are called "smell rocks" or "smell muds" because of their strong petroleum odor. Microfossils in the mélange siltstones indicate that the depositional environment was a deep-water ocean basin beyond the continental shelf (Rau, 1987).

| Box 30.1 | **Paleomagnetism in Iron-Bearing Rocks** |

Many igneous rocks contain iron-bearing minerals, such as magnetite, that recorded the strength and direction of the earth's magnetic field *at the time* they crystallized in a cooling lava flow. The atoms in the crystals responded to the earth's magnetic field by "pointing" toward the location of the north magnetic pole. Unless the rocks are remelted, they permanently retain the magnetic alignment of the time of their crystallization, and this time can be ascertained by techniques of radiometric dating (Chapter 22).

The polarity of the present day places the magnetic north pole close to the geographic North Pole. This is what we call *normal polarity.* However, paleomagnetic studies of ocean floor basalts and basalt flows stacked up on land have revealed that some lavas cooled with a magnetic orientation that is directly opposite to the earth's present magnetic alignment. This means that time after time during the geologic past *magnetic reversals* occurred when the magnetic north pole and the south pole exchanged positions so that the magnetic lines of force ran in the opposite direction (i.e., from north to south) in *reversed polarity,* instead of from south to north, the normal polarity of the earth's magnetic field today.

Scientists have determined that the youngest basalts of the ocean floor are those most recently erupted from a spreading center between two oceanic plates. Outward from a spreading center, the basalts get progressively older on either side as the plates move like two conveyor belts taking the rocks farther and farther away from the point of eruption. Times of magnetic reversal show up in a banded pattern in the ocean-floor basalts, each side (outward from the spreading center) is a mirror image of the other.

When it was discovered that the patterns of progressively older magnetic reversals recorded in the ocean basalts could be matched with reversals of polarity shown in lavas on land, a powerful tool became available to geologists for determining ages and structural relationships of Olympic rocks. The scientific basis for the concept that oceanic crust plunged beneath the North American plate near the present Olympic coast is derived from study of magnetic patterns in basalts on land and on the ocean floor.

Piercement structures are associated with tectonic mélange rocks. Under conditions of high pressures, certain rocks become "plastic," or able to flow; they are squeezed upward into overlying rocks, doming them or even breaking through them. In this way, older rocks may be forced up through younger beds; the result is called a piercement structure.

The Shaping of the Olympic Landscape

Running water. All the while that great masses of rock from the ocean floor were being pushed higher and higher, the agents and processes of erosion were actively shaping the surface of the newly risen land. Moisture-bearing clouds from the Pacific discharge rain and snow copiously on the western slopes. Precipitation averages 140 to 150 inches now and was probably at least that much in the past.

Water running off in every direction from the central massif initiated the radial drainage, and swiftly descending streams cut steep, narrow valleys. The differing hardness of rock types modified the original radial drainage. The basalt ridges tend to be higher and sharper in relief than the less resistant sedimentary beds surrounding the basalt. Since the sandstone beds are harder than the shale beds, the tributaries in the latter tend to cut more deeply than the tributaries that flow over sandstone. As a result, some of the streams in the deeper valleys have captured the drainage of the streams on more resistant rocks.

Also affecting how the radial drainage evolves is the contrast between the copious moisture on the western slopes and the amount that falls in the rain-shadow zone on the eastern slopes. In the city of Sequim, for example, outside the northeast corner of the park, precipitation is only 16 to 18 inches per year, or a little more than one-tenth of what falls on the western slopes. Tributaries on the east side of the Olympic Mountains have steeper gradients, are less numerous, and carry smaller volumes of water than those on the west side.

Glacial ice. The Olympic landscape shows evidence of glaciation in the past; but at the present time, glacial ice, while significant, is secondary to running water as an agent of erosion. What is most remarkable about existing Olympic glaciers is their low elevation on the western slopes due to the abundance of snow that keeps them well nourished. Above the snowline, at around 6000 feet, winter snows last through the summer season. Some of the glaciers extend down their valleys to as low as 4500 feet. The Hoh glacier, descending three miles, is the longest.

Largest in size is the Blue Glacier, which has been closely monitored by glaciologists since 1938. A University of Washington glacier research station, built close to the Blue Glacier in 1957-58, collects data continuously as part of a study to discern relationships between regional climate, accumulation and evaporation of snow and glacial ice, melting rates, and velocity of ice flow. Measurements show that in recent years the Blue Glacier has grown slightly.

Moraines, U-shaped valleys, cirques, rock-basin lakes, erratics, and other glacial features beyond the limits of the existing valley glaciers attest to the extensive alpine glaciation of Pleistocene time. Because of rapid weathering and erosion in the Olympics, some of these landforms are not so long-lasting as glacial features in drier climates.

Rock-basin lakes tend to have short lives in the high Olympics. As a lake silts in, vegetation spreads over the shallows. Soon the lake is a bog; and then it is a meadow, with perhaps a stream winding through it. If the basin is below the treeline, the forest takes over the open space.

Till and glacial outwash overlie the floors of the glacially excavated valleys. One of the most beautiful glacial troughs in the park is the flat-floored Enchanted Valley, lying between steep mountain walls that lead up to hanging glaciers on either side of Mount Anderson. This is the headwaters region of the Quinault River. Farther down, just south of the park boundary, is Lake Quinault (in the Quinault Indian Reservation), dammed by a glacial moraine.

Part of the Olympic Peninsula was covered by advances of continental ice. Granite and gneiss pebbles

FIGURE 30.7 Turbidite beds along Hurricane Ridge Road. Turbidites are formed when rocks or sediments are deposited in a marine environment by settling out from muddy water flowing along a sloping ocean floor. Photo by Gerald Morrison.

in the stream gravels on the north and east slopes are among the indications of continental glaciation. Presumably, the pebbles were dumped by invasions of Cordilleran ice from British Columbia, because no granite or gneiss bedrock is exposed in the Olympic Peninsula. Also foreign to the peninsula are some granitic and metamorphic boulders and cobbles dropped by icebergs in fiordlike lakes that at one time filled valleys on the north and east sides of the mountains. The Cordilleran ice sheet dammed these valleys after the alpine glaciers that carved them melted back. Icebergs broke off ("calved") from the waning ice sheet, floated up the lakes, and melted, dropping the rock debris.

Processes of downslope movement. Mass wasting was most effective during times of glacial retreat when freshly excavated valley slopes, unprotected by vegetation, were left in an unstable condition. Sometimes a ridge collapsed when it was no longer supported by glacial ice. Rock units with tilted beds are particularly susceptible to avalanching (fig. 30.8).

A recent rockslide occurred near the terminus of the Hoh Glacier, which has retreated more than a mile since the early 1800s. Cracks formed on the ridge crest, and the unsupported valley wall gave way; this left a fresh scarp on the valley side slope and a pile of rock debris on the moraine below the glacier. Throughout the park, rocks on oversteepened cliffs or cirque walls fall or slide down.

Some lakes in the park backed up behind natural dams that resulted from landsliding. Three examples are Jefferson Lake, Elk Lake, and Lena Lake in the southeastern part of the park. On the Elwha River, in the fairly recent past, a large rockslide dammed a narrow canyon and formed a lake. In 1967 the dam gave way, releasing the impounded lake water, which washed out a trail bridge and spread gravel over the flood plain downriver. The river had caused the rockslide in the first place by undercutting the base of an unstable slope.

A great landslide mass (that probably came down soon after the last Cordilleran ice sheet retreated) separates Lake Crescent, the largest lake in the park, from Lake Sutherland. The two lakes were originally one

FIGURE 30.8 View from Hurricane Ridge, looking southwest towards Mount Olympus. The forest cover, supported by ample precipitation, helps to stabilize the slopes. Photo by Douglas Fowler.

body of water that occupied a long glacial trough drained by Indian Creek, which flows eastward into the Elwha River. The landslide, which was caused by the collapse of the valley side slopes near Mount Storm King, dammed the valley and cut off Lake Sutherland. Lake Crescent cut a new outlet, forming the Lyre River, which flows north into the Juan de Fuca Strait. Indians may have witnessed the catastrophic landslide because an Indian legend describes the spirit of Mount Storm King becoming so angry over tribal warfare going on in his domain that he threw down great rocks that blocked the valley and put an end to the fighting.

The Olympic National Park Coastal Strip

Between the west coast of Washington and the islands of Japan are 6000 miles of open ocean. Impelled by westerly and northwesterly winds, waves travel unimpeded across this vast stretch of water. Throughout the year, a great deal of wave energy is available to erode the beaches and cliffs and transport beach material. During winter storms, waves 15 to 20 feet high beat on the unprotected shore for days at a time. Comparison of coastal surveys reveals that the Washington Pacific coast near Hogsback retreated 225 feet in 60 years. At this rate, 375 feet of shore are being eroded back every 100 years!

The contrast between the high rate of coastal retreat on the Olympic coast and the relatively low rate of erosion of the coast of Acadia National Park (Chapter 24) is remarkable. The North Atlantic waves that beat on unprotected parts of the northern Maine shore are not significantly less powerful than the North Pacific waves. What makes the difference is the character of the bedrock. The contrast we see is between the effects of wave erosion on very old, highly resistant, igneous and metamorphic rocks in Acadia National Park and wave action on very young, relatively nonresistant sedimentary rocks on the Olympic coast.

The coastal section of Olympic National Park is a narrow strip that extends about 40 miles from the mouth of the Queets River on the south to Cape Alava and the Ozette Indian Reservation on the north. Two small Indian reservations are enclaves in the coastal strip: the Hoh Indian Reservation beside the Hoh River and the Quillayute Indian Reservation at the mouth of the Quillayute River. A coastal highway (Route 101) follows the shore from Queets to the Hoh Indian Reservation. The rest of the park's coastal section is a wilderness area.

In the last 8 million years or so (since early in Pliocene time), the Olympic coast has been at times higher and at times lower than present sea level. Some wave-cut platforms are several hundred feet higher than sea level today, and some old wave erosional surfaces are covered by alluvial deposits that once extended to a shoreline hundreds of feet to the west. Because sea level fluctuated considerably, the shoreline has also shifted eastward and westward a number of times. The raising and lowering of sea level was partly due to Pleistocene glaciation that drew moisture from the oceans when the glaciers were enlarging and then released water as they melted back. But tectonic warping and minor uplift of this coast was also occurring during the Pleistocene, making interpretation of fea-

FIGURE 30.9   Coho salmon run in the Sol Duc River, which rises in Olympic National Park. Photo by Gerald Morrison.

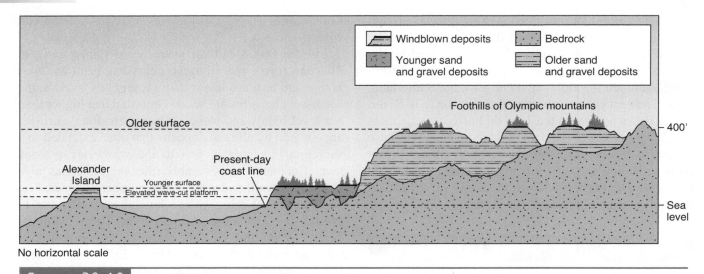

No horizontal scale

FIGURE 30.10   Simplified cross section showing uplift, sea level changes, deposition, and erosion during Quaternary time on this tectonically active coast. The position of the shoreline has changed several times, to elevations both higher and lower than at present, due to changes in sea level associated with Pleistocene glacial episodes as well as tectonic uplift of the Olympic Peninsula. The older land surface (top), built by glacial and fluvial sediments, extended seaward for many miles beyond the present shoreline. As the coast rose, sediments were eroded and a younger wave-cut surface developed. This surface was, in turn, covered by a thinner series of sediments that extended westward of the present shoreline. Wave action, following uplift, has cut back the shoreline. Alexander Island is a remnant of the former mainland. From Rau 1980, *Washington Coastal Geology between the Hoh and Quillayute Rivers,* Bulletin 72, State of Washington Department of Natural Resources.

tures of emergence and submergence difficult (fig. 30.10).

Rapid erosion also alters the position of the shoreline. Cliff retreat produces a widening, submerged, wave-cut platform that extends several hundred feet out from the water's edge. Stacks and small islands of more resistant material that were recently part of the mainland now sit on the wave-cut platform surrounded by water. Sea arches collapse and become stacks. Caves in cliffs are eroded into sea arches (figs. 30.11, 30.12).

As cliffs are undercut and landsliding occurs, the forests that grow right to the edge of the bluff begin to sag and slump. On the north side of the Hoh River mouth is a "drunken forest" where live trees are attempting to bend and grow vertically again after being tilted by a slide. Other drunken forests can be seen along the coast.

Beaches are littered with dead trees that have fallen and great logs that have floated down the rivers to the ocean and then washed ashore. When storms come, these logs act as battering rams hurled against the rocks by the waves.

The beaches between headlands tend to be narrow and have fairly steep slopes. In some areas, the steep cliffs come right down to the water. Nevertheless, the Olympic beaches have been the most convenient "pathway" from one place to another along the coast since prehistoric times because of the difficulty of traveling inland through the thick vegetation and the rain forests. Today beach hikers can follow well marked trails over the cliffs between stretches of beach. However, the warning still holds that hikers must guard against becoming trapped on a narrow beach at high tide, because most of the cliffs are too steep to climb. The beaches must also be avoided when heavy surf or storm waves are running.

Most of the rocks and cliffs in the park's coastal section are in the Hoh rock assemblage of Miocene age. The Kalaloch Rocks, offshore from the Kalaloch Ranger Station, are the southernmost exposure of the Hoh turbidites along the coast. They are mainly sandstones and conglomerates, but they are called "graywackes" because they contain fragments and grains of other rocks. Most of the beach sand is derived

FIGURE 30.11   Sea stacks can be found along the coastal section of Olympic National Park. National Park Service photo.

FIGURE 30.12   The sea stacks in the La Push area are rocks of the Hoh Formation. The Indians referred to it as "smell rock" because of their strong petroleum odor. Photo by Gerald Morrison.

Box 30.2	How Waves Do Geologic Work

Wind-driven water waves erode, transport, and deposit material in a narrow coastal zone where land, water, and air meet. The source of wave energy is wind. Energy is transferred from the moving air into wave shapes that move through the water. The water itself does not move very far; rather, it is the waves, impelled by the wind, that transmit the wind's energy, releasing it when they strike land. The size of waves is a function of (1) wind velocity, (2) the length of time that the wind blows in the same direction, and (3) the *fetch,* or distance, of open ocean over which the wind blows.

Great waves may be several tens of feet high and several hundred feet long. Once wave sets develop in an open ocean, they may move across hundreds of miles of sea before they break against a coast, their power and size unaffected by lighter or changeable local winds.

When a deep-water wave begins to "feel" a shallowing bottom near shore, its lower part begins to flatten due to drag, while the upper part becomes shorter and steeper. Within a few seconds, the top of the wave falls forward, becoming a curling, breaking wave, or *breaker,* that slams against a cliff or rushes up a beach. When forward motion stops, the water runs back down the beach and slides under the curl of the next breaker. This upwash and backwash (called "swish and swash") breaks up and abrades rock and sediment, sorting coarser fragments from finer particles and rolling them back and forth. Coarser sediment tends to accumulate along the top of the beach while finer materials tend to stay in the *surf zone* (the strip of shore where the waves break and roll up the beach or beat against a cliff).

The surf zone shifts up and down the beach twice a day as tides rise and fall. Thus normal wave erosion is most intense on the *foreshore,* the area between mean low-water level and mean high-water level. Storm waves attack any part of the shore they can reach with devastating effect. Sizable rocks may be picked up and tossed high into the air, often shattering as they hit. Tons of sand may be removed from a beach in a single storm.

Most beach sediment is only temporarily at rest. Sand is moved along beaches by *beach* drift. When waves come in obliquely, as they often do, instead of head on, sand is carried up the foreshore at an angle. But the grains of sand roll back perpendicularly, under the influence of gravity, progressing along the beach in a series of looping movements.

Sometimes water "piles up" just outside the surf zone and starts moving in a *longshore current* carrying sand parallel to the shore. When the current slows down or is deflected by a headland, the sand drops and may form a bar or *spit* (a fingerlike extension of a beach, with the far end terminating in open water). This process is called *longshore drift.* Sediment entering the ocean at the mouth of a river may encounter a longshore current and be carried up (or down) the shore, along with beach sediment. Meanwhile, the river has to keep changing its outlet to get around the sandbar or spit built by the longshore current. Records show that the Quillayute River entered the Pacific at several different locations from the late 1800s until 1931, when the U.S. Army Corps of Engineers constructed a harbor at La Push and stabilized the mouth of the river with dikes.

from these rocks or from sediment washed down from the mountains. Blocks of tectonic mélange Hoh rocks form many of the stacks and other strange-shaped erosional remnants sticking up in the water close to shore. The blocks are jumbled masses of resistant basaltic breccias, conglomerate, sandstone, and siltstone. Much of the original matrix of softer rock has been removed by weathering and erosion. Some of the mélange rocks, such as those in the bluffs at Jefferson Cove, are thought to be part of piercement structures.

Overlying the Hoh rocks along the shore are Pleistocene deposits of sand, gravel, and silt. Most of the deposits are alluvial and derived from glacial outwash that poured out of the mountains as the valley glaciers retreated. The youngest Pleistocene deposits consist of fine, wind-deposited silt, called *loess,* which

Table 30.1	Geologic Column, Olympic National Park			

Era	Period	Epochs	Rock Units	Geologic Events
Cenozoic	Quaternary	Holocene	Glacial drift, fluvial and landslide deposits, talus Beach materials	Glaciation Erosion and deposition by streams, mass wasting, and wave action
			Unconformity	Unconformity
		Pleistocene	Younger glaciofluvial and loess Older glaciofluvial deposits	Wind deposition on and near coast Fluctuation of sea level; uplift and warping Repeated episodes of alpine and continental glaciation
	Tertiary (Neogene)		*Unconformity*	Unconformity
		Pliocene	Quinault and Quillayute Formations (western Olympic coast; not exposed in park) Largely marine sedimentary rocks; slightly tilted.	Marine deposition in coastal areas Folding and upwarp
		—?—	*Unconformity*	Unconformity
		Miocene	Hoh rock assemblage (western Olympic coast), late Eocene to Miocene. Steeply tilted, overturned sedimentary rocks; turbidites; tectonic mélange rocks	Faulting and uplift Plate convergence and subduction accompanied by intense folding, shearing, slicing, partial metamorphism
	Tertiary (Paleogene)	Oligocene	Core rocks (central and western) Eocene to Miocene highly folded sandstones, shales; some highly disrupted	Basaltic horseshoe (Crescent Formation); Eocene to Miocene peripheral sedimentary rocks	More or lesss continuous marine sedimentation
		Eocene	Inner Olympic rocks Eocene to Oligocene severely deformed sedimentary rocks and basalts		Submarine eruption of lavas
		Paleocene	*Unconformity*	Unconformity
			Pre-Cenozoic rocks of the old continent (not exposed in park)	

Sources: Modified after Rau, 1980; Tabor, 1987a.

tends to form bluffs because the silt grains, being coherent, cling together. The loess at Quateata headland (First Beach) stands directly on an old wave-cut platform of bedrock and gravel; and at Taylor Point (Third Beach), the wind-blown deposits rest on older alluvial sediment. The loess accumulated near the end of the Pleistocene Epoch when the ice was melting back rapidly but sea level was still quite low. Large areas of land, unprotected by vegetation, were uncovered and left exposed to unusually high winds. Fine surface sediment, picked up by the winds, was redeposited over the coastal area.

The beaches between the headlands are made up of sand, gravel, and cobbles of various sizes and compositions. The sediments are derived from the erosion of local bedrock, from glacial drift, and from material brought down by rivers from the Olympic Mountains. Here and there patches of red sand have accumulated. The red sand grains are tiny garnet crystals that have weathered out of erratics or been transported in glacial debris and then eroded by streams. Being heavier than quartz grains, the garnet crystals tend to be concentrated in stream beds and beach terraces. Presumably the red sands have been transported, concentrated, re-eroded, and re-concentrated several times on their journey to the shore from their places of origin.

Geologic History

1. **The Crescent Formation and the peripheral rocks.**

In early Eocene time the lavas of the Crescent Formation began to pile up on the ocean floor, probably not far from the mainland. Eruptions continued into middle Eocene time, even as subduction of the oceanic plate proceeded. As the plate moved down, the thick mass of basalt acted as a buttress, protecting the sandstones, shales, and conglomerates of the peripheral rocks deposited between the accumulated lavas and the continental margin, so that they were only moderately deformed. The sedimentary peripheral rocks are Eocene to Upper Miocene in age (figs. 30.13, 30.14).

2. **Eocene to Oligocene rocks of the inner Olympics.**

While protecting the peripheral rocks, the Crescent Formation resisted the eastward movement of the sedimentary rocks that were deposited on the seaward side of the lavas. Therefore, the marine basalts, sandstones,

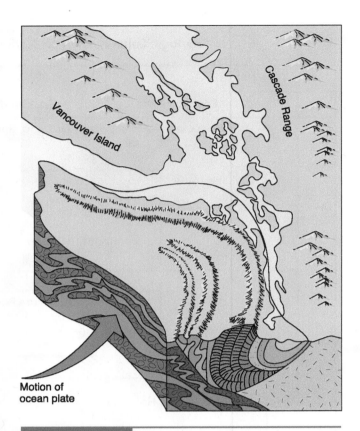

FIGURE 30.13   Opposing movement of oceanic and continental plates jammed the rocks that make up the Olympic Mountains into a tight arc between Vancouver Island and the Cascade Range. From R.W. Tabor, *Geology of Olympic National Park: Hiking and Motorist Guide.* © 1987 Pacific Northwest National Parks and Forests Association. Third printing.

phyllites, slates, and conglomerates within the arc of the horseshoe are more thoroughly metamorphosed, foliated, and recrystallized than the younger beds to the west. All the innermost rocks have been highly disrupted by slicing, shearing, and fracturing.

3. **The core rocks from the inner Olympics westward.**

The inner Olympic rocks and the core rocks are grouped in "lithic assemblages," rather than formations, because boundaries and specific rock types are not well defined. As is evident on a geologic map, the lithic assemblages tend to have an arcuate pattern, similar to that of the Crescent Formation. The core rocks range in age from Eocene to Oligocene; the beds closer to the inner Olympics rocks are more intensely folded and disrupted. Many of the core rock units show a well developed slaty cleavage, which indicates some degree of metamorphism. They consist of sandstones and shales with minor occurrences of conglomerate and basalt (fig. 30.5).

Mount Olympus and the other high peaks in the center of the peninsula are in what is called the "Western Olympic lithic assemblage," a generally younger, less deformed group primarily made up of thick-bedded sandstone, with elongated patches of slate, phyllite, foliated sandstone, and semischist.

4. The rocks of the Olympic Peninsula western coast.

The greater part of the rocks on the Pacific coast are in the Hoh rock assemblage, which is mostly Miocene in age. (Some Eocene volcanics interbedded with siltstones are exposed at Point Grenville, south of the park's coastal section.)

5. The Quinault and Quillayute Formations (not exposed in the park).

The Quinault rocks, Pliocene in age and largely of marine origin, are separated from the Hoh rock assemblage by an unconformity. Quinault outcrops can be seen in the Quinault Indian Reservation, just south of the park's coastal section. For about 40 miles northward along the coast from the outcrops, the Quinault Formation extends seaward, underlying Quaternary sediments on the continental shelf. The formation is of interest because it is regarded as a potential petroleum reservoir. The Quinault beds, which are only mildly tilted, were deposited after the Hoh rocks had been raised above sea level, eroded, and again partly covered by the ocean.

The Quillayute Formation, which is similar to the Quinault Formation, crops out in the Quillayute River valley (upstream from the coast) in a few places where Pleistocene deposits have been eroded away.

FIGURE 30.14 The Sol Duc Hot Springs, south of the Sol Duc River, do not flow directly from the Calawah fault zone as many of the hot springs in this area do. Instead they reach the surface by way of the fractures in sandstone beds adjacent to the fault. Photo by Gerald Morrison.

6. Pleistocene and Holocene glaciation, and sea level changes.

Alpine glaciers have probably existed on the highest mountains of the Olympic Peninsula since the onset of the Pleistocene and have advanced and retreated many times. At their maximum extent, the largest valley glaciers on the west side reached as far as the present coast and perhaps beyond, although parts of this coastal region were left unglaciated.

The far northwest coast and all of the north and east sides of the peninsula were heavily glaciated, because the great Cordilleran ice sheet from British Columbia encountered the Olympic glaciers as it moved southward and then divided into two lobes. One lobe filled the Puget Sound lowland to south of Tacoma, while the other spread over the Juan de Fuca Strait and the northern coast. At least four times, possibly six, the ice sheet advanced and retreated. Cordilleran ice on the northern side of the peninsula, at least 3500 feet thick, scraped the weak peripheral rocks on the outer side of the basaltic horseshoe, destroyed the preglacial topography, and lowered relief along the northern coast.

During warmer interglacial periods, when the glaciers melted back, great quantities of glaciofluvial deposits were laid down over the lowlands and coastal areas. The Pleistocene sands and gravels on the west side and part of the south side are better preserved, because they were not reworked (except in the glacial valleys) by later glaciations. Exposed on the Pacific coastal region are several sequences of Pleistocene and Holocene deposits. The oldest ones, which are semi-consolidated, originally extended for a considerable distance farther to the west than now, but they have been cut back by wave erosion.

In recent years Olympic glaciers have shown slight increases in volume. Stream erosion at present is dominant over glacial erosion; mass wasting is also significant. In the coastal section, wave erosion is unusually rapid.

Sea level has fluctuated on the Olympic Peninsula, during and since Pleistocene time. Some fluctuations have been due to depression by glacial ice and subsequent release by melting; some changes have been the result of tectonic uplift and warping.

Tectonic subsidence, resulting from great earthquakes in Holocene time, may have been the cause of coastal lowlands suddenly dropping several feet. Studies of intertidal muds in the Puget Sound area have revealed that at least six times in the past 7000 years, marshy lowlands have dropped and then been buried by sand, presumably washed in by tsunamis. The earthquakes were apparently generated in the Cascadia subduction zone between the Juan de Fuca Ridge and the Olympic coast (Atwater, 1987).

Geologic Map and Cross Section

Rau, W.W. 1975. Geologic Map of the Destruction Island and Taholah Quadrangles, Washington. Washington Division of Geology and Earth Resources Geologic Map GM-13.

————. 1979. Geologic Map of the Vicinity of the Lower Bogachiel and Hoh River Valleys and the Washington Coast. Washington Division of Geology and Earth Resources Geologic Map GM-24.

Tabor, R.W., and Cady, W.M. 1978. Geologic Map of the Olympic Peninsula, Washington. U.S. Geological Survey Miscellaneous Investigations Series Map I-994.

Sources

Alt, D.D., and Hyndman, D.W. 1995. *Northwest Exposures, a Geologic Story of the Northwest.* Missoula, Montana: Mountain Press. 443 p.

Atwater, B.F. 1987. Evidence for Great Holocene Earthquakes Along the Outer Coast of Washington State. *Science* 236: (4804) 942–944, May 22.

Crandell, D.R. 1965. The Glacial History of Western Washington and Oregon. In *The Quaternary of the United States*, VII INQUA Congress, H.E. Wright, Jr., and D.G. Frey, editors, pp. 341–8. Princeton, New Jersey: Princeton University Press.

Rau, W.W. 1964. *Foraminifera from the Northern Olympic Peninsula, Washington.* U.S. Geological Survey Professional Paper 374-G.

————. 1973. *Geology of the Washington Coast Between Point Grenville and the Hoh River.* Washington Division of Geology and Earth Resources Bulletin 66.

————. 1977. General Geology of the Southern Olympic Coast. In *Geological Excursions in the Pacific Northwest,* E.H. Brown and R.C. Ellis, editors, p. 63–83. Bellingham, Washington: Department of Geology, Western Washington University.

————. 1980. *Washington Coastal Geology between the Hoh and Quillayute Rivers.* Washington Division of Geology and Earth Resources Bulletin 72.

————. 1987. Mélange Rocks of Washington's Olympic Coast. Geological Society of America Centennial Field Guide, Cordilleran section, Hill, M.L. ed. v.1, p. 373–376.

Scharf, J.G., and Wilkerson, J. n.d. *Geologic Guide to the Hurricane Ridge Area, Olympic National Park.* Port Angeles, Washington: Pacific Northwest National Parks and Forests Association. 21 p.

Snavely, P.D., and Wagner, H.C. 1963. Tertiary Geologic History of Western Oregon and Washington. Washington Division of Geology and Earth Resources, Report of Investigation 22.

CHAPTER 31

Glacier Bay National Park and Preserve

SOUTHEASTERN ALASKA

Area: 3,283,168 acres; 5130 square miles

Proclaimed a National Monument: February 26, 1925

Established as a National Park and Preserve:
December 2, 1980

Designated a Biosphere Reserve: 1986

Address: P.O. Box 140, Gustavus, AK, 99826-0140

Phone: 907-697-2230

E-mail: GLBA_Administration@nps.gov

Website: www.nps.gov/glba/index.htm

Icefields, valley glaciers, tidewater glaciers, deep fiords, high coastal mountains.
Glaciers advancing; glaciers retreating.
Glaciated landscapes — outwash plains, drumlins, eskers.
Reforestation after retreat of ice.
Stormy outer coast, protected inner shorelines.
Active tectonism, uplift, earthquakes.
Exotic terranes.

FIGURE 31.1 The crevassed terminus of Margerie Glacier (named after a naval officer), towers above Tarr Inlet at the northern end of Glacier Bay. Chunks of ice, large and small, are "calved" from the ice front. Photo by Ann G. Harris.

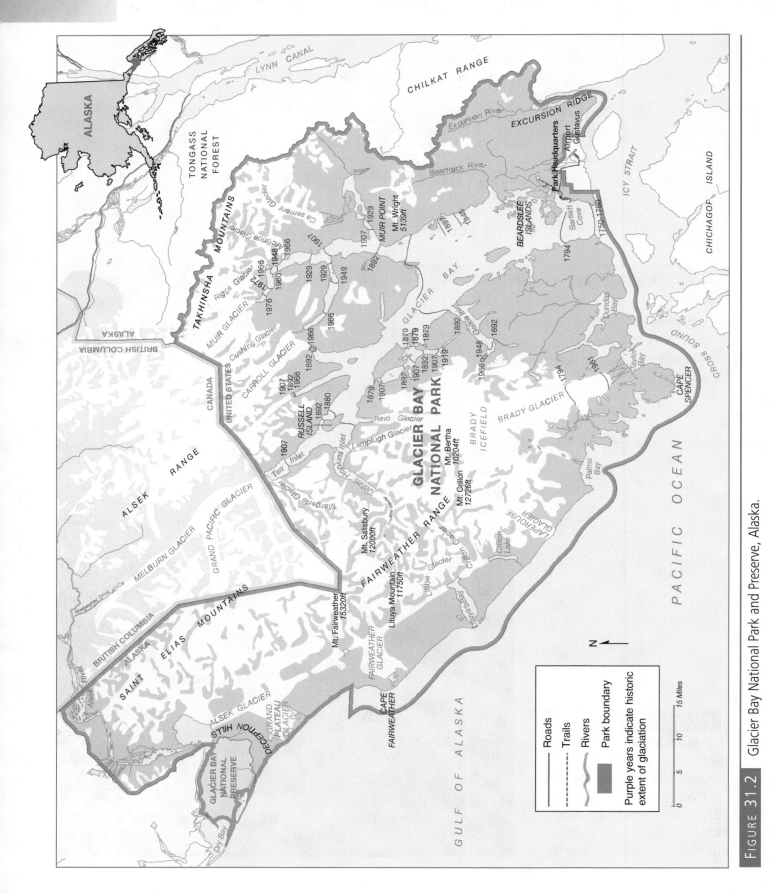

FIGURE 31.2 Glacier Bay National Park and Preserve, Alaska.

Geographic Setting

In southeastern Alaska, at the upper end of the Alaska panhandle, Glacier Bay National Park and Preserve occupies a wild and rugged region of ocean, bay, and inlets; lofty, snow-capped peaks; fiords, glaciers, and rain forests. The park/preserve is on the northern side of Cross Sound and Icy Strait, a waterway between the Lynn Canal and the Gulf of Alaska. The Alexander Archipelago extends to the south. Heading north, the park/preserve's western boundary is drawn a few miles offshore, parallel to the ocean coast up to Cape Fairweather, where it turns into the shoreline and continues on to Dry Bay and the Tongass National Forest. British Columbia lies on the northern border of the park/preserve. Juneau, Alaska's capital city, is about 50 miles east of Glacier Bay on the east side of the Inland Passage. Neither Juneau nor Glacier Bay National Park and Preserve are accessible by road. Transportation is by airplane, ship, or small boat.

Glacier Bay proper is a large, branching fiord, surrounded by mountain ranges and occupying a central lowland in the park/preserve. The huge glaciers that completely covered Glacier Bay 200 years ago have retreated. The snout of Grand Pacific Glacier, which was the largest and longest, is now approximately on the international boundary, some 65 miles from the mouth of Glacier Bay.

The settlement of Gustavus on Icy Strait, at the mouth of Glacier Bay, is both airport and seaport for the park/preserve with park headquarters nearby on Bartlett Cove. During the tourist season, National Park Service concessionaires operate a lodge at Bartlett Cove and run small-boat trips daily up through Glacier Bay to the snouts of tidewater glaciers, such as Johns Hopkins Glacier (fig. 31.3), which descends from the

FIGURE 31.3 An aerial photograph of John Hopkins (right) and Gillman Glaciers (left) located off of Tarr Inlet. It is easy to tell that John Hopkins is a piedmont glacier because of all of the medial moraines that can be seen in the glacier. John Hopkins Glacier calves so much ice that it is seldom possible to get any closer than two miles. National Park Service, Glacier Bay National Park photo.

Fairweather Range and "calves" (breaks off and releases icebergs) into an arm of the bay (fig. 31.4). Cruise ships and planes bring thousands of visitors to the park/preserve every summer.

The mountain ranges in Glacier Bay National Park and Preserve extend north and northwest (across a small wedge of British Columbia) to connect with the ranges in Kluane National Park, Yukon Territory, and those in Wrangell-St. Elias National Park and Preserve (Chapter 32). The coastlines of Wrangell-St. Elias National Park and Preserve and Glacier Bay National Park and Preserve are separated by a narrow stretch of shore in the vicinity of Yakutat Bay, where the Alaska panhandle joins the main part of Alaska.

Mount Fairweather (elevation 15,300 feet), the highest peak in southeastern Alaska, is at the northern end of the Fairweather Range, a south-trending extension of the St. Elias Range. Mount Fairweather and its neighbor, Mount Quincy Adams (elevation 13,560 feet), mark adjacent corners on the international boundary between Alaska and British Columbia. East of the park/preserve, also near the international boundary, is Klondike Gold Rush National Historical Park at Skagway, Alaska.

Glaciers flourish in the region because storms sweeping in from the Gulf of Alaska dump rain and snow on the high coastal mountains the year round. Precipitation is estimated to be as high as 180 inches

FIGURE 31.4　A small iceberg, showing the annual bands of debris, floats past Lamplugh glacier located in the John Hopkins Inlet of the Park. Photo by Ann G. Harris.

per year midway up the slopes along the western side of the Fairweather Range. The ocean coast is cold, foggy, and virtually uninhabited, with the temperature seldom rising above 75° F. Strong winds from the gulf blow inland. Heavy ocean waves beat against the shore almost constantly. The combination of high mountains, a prolific source of moisture, cool temperatures, and many stormy or overcast days is what brings about active glaciers and an extensive rain forest along the outer coast of the park/preserve.

At the present time only one tidewater glacier calves directly from the park/preserve into the Gulf of Alaska. This is LaPerouse Glacier, between Lituya Bay and Icy Point. Grand Plateau Glacier (south of Dry Bay) descends to sea level but terminates in a bay.

High tides are characteristic of the inland water-ways, with a tidal range of up to 20 feet in many bays and inlets. The high tides are due to a combination of high latitudes and the constricting effect of bays and narrow channels. Tidal currents run at velocities of 5 to 7 miles per hour in the passages, even higher in Lituya Bay. Navigation is made more difficult in the foggy channels of the inner bay by the presence of rocks, floating ice, and icebergs.

When Russian explorers accompanying Vitus Bering reached the Glacier Bay area in the summer of 1741, the bay was blocked by a solid front of glacial ice. Probably they did not suspect the bay's existence. However, the Tlingit Indians, the native inhabitants, knew the bay. Recorded in their legends are stories about prehistoric retreats and advances of glaciers. Before the glaciers advanced and they had to move southward, the Tlingits camped on the islands and shores of Glacier Bay to fish and gather plants. Many years later, as glaciers retreated, the Tlingits resumed their traditional activities throughout the Glacier Bay area.

The Tlingits were a hardy people who fashioned cedar canoes that they navigated with great skill through the treacherous tides and rough waters of the region. The Russian sailors did not have that skill and knowledge. A landing party of 11 men in a small boat rowed up an inlet from Icy Strait and were either lost, wrecked, or possibly killed by the natives. The Russian captain sailed off, after several days of waiting, to rejoin Bering.

Captain James Cook, a British naval officer, named Cross Sound, which he discovered on Holy Cross Day in 1778. He described it as a bay appearing to branch into several arms, the largest of which turned north; but glacial ice blocked further entry. Cook was searching for the fabled Northwest Passage across North America. He sailed on up the coast and named Mount Fairweather. Perhaps he saw it on a rare sunny day.

A French explorer, Comte de LaPerouse, put into Lituya Bay in 1786 and spent nearly a month there after his longboat was swamped. He became acquainted with the Tlingits who were living there and "purchased" Cenotaph Island from one of the chiefs. Mt. LaPerouse and LaPerouse Glacier are named for him.

In 1794 Captain George Vancouver, who had served under Captain Cook on the earlier voyage, returned to Glacier Bay with his own expedition. He again explored Cross Sound, sailed into Taylor Bay, discovered the Brady Glacier, and found glacial ice at tidewater a few miles inside the opening of Glacier Bay. Yet only 85 years later, when the intrepid John Muir came to study the glaciers, the ice had retreated 48 miles up the bay!

Aside from Russian fur traders and a few prospectors, few outsiders had visited the Glacier Bay region before Alaska was purchased from Imperial Russia in 1867. But John Muir, who made several visits to the area, aroused the interest of both scientists and venturesome tourists. He moved about in a small boat, made temporary camps along the shores, journeyed to the glacier snouts, and climbed up on top of the glaciers and on some of the ridges and peaks above the ice. In his dispatches to California newspapers, he reported his observation of glaciers calving and retreating.

Commercial steamships began bringing tourists to Glacier Bay to view glaciers in action, and parades of scientists came to study the glaciers' behavior and the reclamation by vegetation of recently uncovered land. Many of the place names on topographic features honor these early scientists, especially the geologists. Some names were assigned by Professor H.F. Reid, who mapped glaciers and made careful measurements in the 1890s. Some names are those of eminent glaciologists who visited Glacier Bay during a field trip sponsored by the 12th International Geological Congress in 1912. All agreed that glacial rates of retreat in Glacier Bay are the fastest recorded anywhere on earth.

The glacial phenomena, rugged mountains, the great natural beauty of the region, the marine mammals, birds and other wildlife, and the plant successions all inspired a strong movement to protect Glacier Bay. Because of intense lobbying by individuals and commercial interests, shortly after the national monument was proclaimed, Congress passed legislation permitting mining in the area. Over the years extensive prospecting has been done, and several small mining operations have been carried on without much profit. A test drilling program verified the presence of a sizable nickel-copper deposit beneath the Brady Icefield, and claims have been staked out by a mining company. The mining interests believe that ore can be extracted without serious degradation of the environment, but some scientists claim that the environment is too fragile to permit this kind of exploitation.

Geologic Features

Structurally Controlled Topography

The topography of Glacier Bay Park and Preserve has a pronounced "grain" inasmuch as most of the mountain ranges, glacial valleys, and waterways trend northwest to southeast. Beginning with the Gulf of Alaska coastline on the west and going inland, the major features that follow this trend are the Alsek Glacier; St. Elias and Fairweather Ranges; Brady Icefield and Brady Glacier; Glacier Bay and its extending arms, Tarr and Muir Inlets; the Beartrack and Excursion Rivers; the Takinsha Mountains; and—just outside the park/preserve's eastern boundary—the Chilkat Range (fig. 31.5). Between these major topographic features are a number of smaller landforms that parallel the main trend, such as the Carroll Glacier.

Two major features with an east-west orientation that cut across the regional trend are the Alsek River at the northern edge of the park/preserve and the Cross Sound/Icy Strait waterway marking the southern boundary. The Alsek River flows south from Yukon

FIGURE 31.5   In the Chilkat Range many of the glaciers have retreated toward the heads of their valleys, away from the sea. The meltwater streams still enter the sea. Photo by Ann G. Harris.

Territory, then turns west and flows from British Columbia across the park/preserve to empty into Dry Bay. The Alsek is of interest to geologists because it is a transverse stream, and it is the only stream that crosses the St. Elias Range. A few smaller landforms, such as Lituya Bay (opening on the Gulf of Alaska) and Geikie Inlet (on the west side of Glacier Bay) also have an approximate east-west trend. One has only to glance at these features on a map to conclude that the topography of Glacier Bay Park and Preserve is structurally controlled.

The long, northwest-southeast-trending valleys follow fault lines that are zones of less resistance to weathering and erosion. Branching faults and faults that cut across the fault blocks probably underlie the prominent transverse features. Weathering and preglacial stream erosion deepened the fault zones; valley glaciers scoured and enlarged them. These valleys are magnificent, large-scale examples of how structural features control the locations and orientations of drainage lines and ice flow. Only when the ice became deep enough to bury the divides between stream valleys, as with the Fairweather Glacier, did ice tend to flow across the dominant valley trends. Later as glaciers retreated, flow directions again were confined by the controlling structural patterns.

Three major faults trending northwest-southeast in the Glacier Bay area show up on the geologic map of Alaska as part of major fault systems that extend southward along the Canadian and United States west coast and northward into the Alaska Range, joining the Denali fault system (Chapter 34). The Border Ranges fault, for example, which includes a segment in the park/preserve called the Fairweather fault, runs from the St. Elias Range southward to the Alexander Archipelago. It is described as a late Mesozoic plate boundary where an entire rock complex became welded to the continental margin by widespread deformation, metamorphism, and plutonism during late Cretaceous and Tertiary time.

Recent studies show the faults as boundaries between tectonostratigraphic terranes in southeastern Alaska (Brew and Ford, 1982). (A *tectonostratigraphic terrane* is a fault-bounded region characterized by rocks and geologic history differing markedly from those of neighboring regions.) Since the southeast Alaska terranes are believed to have been "rafted" into their present position rather than formed on the site, they are referred to as *exotic,* or *accreted ter-*

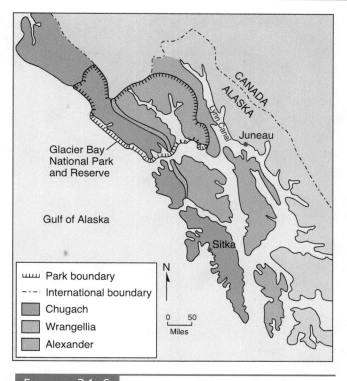

FIGURE 31.6 Simplified representation of major tectonostratigraphic terranes in Glacier Bay National Park and Preserve and vicinity. After Brew and Ford 1982.

ranes. Complicating our understanding of these features is evidence of overthrusting brought about by compression plus considerable lateral displacement caused by impinging and pivoting tectonic plates in the area (fig. 31.6).

Glacial Features

Since many of the valley glaciers in the Glacier Bay area come down to sea level with their snouts extending out into salt water, they are called *tidewater glaciers.* They may float free if the water is deep enough, or they may ground themselves on inlet floors. The irregular forward movement of brittle ice at the snouts and the action of waves and tides cause the ice to fracture. Large blocks and crumbling masses of ice break or fall continually from glacier snouts, a process called *calving* that produces icebergs of all sizes and floating, shattered pieces of ice called "growlers." The crumbling and falling of ice creates enormous waves and splashes, accompanied by loud booming and cracking.

FIGURE 31.7   Margarie Glacier calving. Because about half of the glacier is underwater, ice chunks sometimes calve from the bottom of the glacier instead of from the top. As a result, an iceberg may suddenly come bobbing up without warning. It is for this reason that cruise ships are required to stay a mile away from the glacier front. National Park Service, Glacier Bay National Park photo.

When ice hits the water, it goes under the surface, then bobs up again, bouncing and flopping around (fig. 31.7). The calving process stirs up the water, bringing food to the surface and attracting birds that shriek as they swirl and dip over the water. This dramatic and noisy display continues to enthrall visitors to Glacier Bay Park and Preserve.

Glacial erosion in the park/preserve has tended to produce two quite different mountain shapes. The higher peaks, such as Mt. Fairweather and Mt. Salisbury, are horn-shaped due to dissection by cirques and frost action. At lower elevations, especially alongside inlets and narrow bays, ridges and peaks have smooth, rounded, domelike outlines. Abrasion and scouring by overriding glacial ice have left these bedrock masses bare of soil and vegetation. Smaller knolls that are oriented

parallel to the direction of ice flow are termed *roches moutonnées. Nunataks* are isolated, rounded bedrock knobs or peaks protruding through an icefield or a glacier. A splendid example of such a feature is The Nunatak, an isolated knob left standing in Muir Inlet after the ice surrounding it had retreated.

Glacial deposits, whether laid down by advancing or retreating glaciers, take many forms in the park/preserve. Among subglacial deposits are *drumlins*—oval, elongated hills or mounds of till with the steeper slope at the up-ice end and a tapering, streamlined "tail" of till at the down-ice end; crevasse fillings; cross-valley ridges of till; *crag-and-tail* features—like roches moutonnées, that are smooth and scoured, while the lee end is steep and hackly from glacier plucking, at the stoss end; and *eskers*—narrow, sinuous, steepsided ridges made up of layers of sand and gravel deposited by meltwater in channels on top of ice or in ice tunnels in a stagnant glacier. Moraines and outwash plains along with temporary lakes and iceblock holes called kettles are found on lowlands near present and former glacier snouts.

Studies of Glacial Advances and Retreats

When John Muir first came into Glacier Bay in a canoe with Tlingit guides, he studied the area around their first encampment (near Geikie Inlet) with growing excitement. Around the small rocky beach, he saw a landscape being born. Everything was new and raw, and all the signs indicated that ice had only recently left the campsite. A few sparse mosses were the only living plants, but stumps of sizable trees stood among the wastes of sand and gravel. Muir thought these were petrified and was surprised to see the Indians using them for firewood. Actually the stumps were the remains of forests that had been overwhelmed by an earlier ice advance.

On this first trip, Muir and his party reached the tip of Russell Island, which was where the glacier front was in 1879. Muir predicted that an island and a large fiord would soon be exposed, for he noted that the sea was undermining the glacial ice. Muir was a keen observer. On his next expedition in 1880, he deduced that in Taylor Bay (which opens into Cross Sound) a large glacier was advancing, in contrast to the other glaciers that they had visited, all of which were retreating! By the summer of 1890, steamships were bringing supplies and tourists to Glacier Bay, and Muir established a camp at what is now called Muir Point. There a geologist, Professor H.F. Reid, leading a scientific party, joined Muir. The rest of the summer was spent studying, mapping, and exploring the area around Muir Glacier. Reid was the first to plot the positions of glacial ice using a theodolite and plane table. W.O. Field, Jr., over a span of nearly forty years, mapped the termini of glaciers in the Glacier Bay area and recorded their rates of retreat (or advance). More recently, geologists from the Institute of Polar Studies at the Ohio State University continued Field's work by carrying on detailed studies of glacial deposits and glacial retreats and advances in the fiords and on the slopes of Glacier Bay National Park and Preserve.

Basically the aim of scientists is (1) to measure and survey the melting back of glacial ice and from this determine rates of retreat; (2) to search for evidence that might explain the rapid changes observed; (3) to find out how subglacial landforms are created and how lakes appear and disappear; (4) to study the return of plants and animals to recently uncovered, barren landscapes; and (5) to measure and study Glacier Bay's exceedingly rapid rate of *glacial rebound*. As melting releases the weight of ice, the shorelands and islands rise and expand in size.

In the last two centuries, the glaciers in Glacier Bay have retreated about 15 times faster than most glaciers elsewhere in the world. Rates of retreat per year in Glacier Bay have been from 600 to 1100 feet horizontally, and ice surfaces have been lowered at rates of around 12 vertical feet per year. The rapid unloading of glacial ice has caused the land surface to rise at rates greater than anywhere else in southeast Alaska. At Bartlett Cove the land rises at a rate of about an inch and a half per year, and farther up in the bay, where ice left more recently, the rebound rate is even higher. The creation of new beaches, tidal zones, and shoals makes marine charts short-lived! Tectonic activity in the Glacier Bay area may be a contributing factor in these rebound rates.

A few glaciers in the park/preserve have been advancing. Sometimes this has been due to "surging" caused by excess water within the glacier that acts as a lubricant between ice and bedrock, permitting faster sliding as well as rapid movement along ice fractures. In other cases, glacial advances may have occurred because of high snowfall, snow avalanching, increased debris cover, or perhaps changes in weather patterns that result in less runoff and melting.

No clear-cut, simple explanations have been developed to explain the extreme fluctuations and rapid movement of the glaciers in the Glacier Bay area. Several factors are probably involved, with individual glaciers responding differently to each factor and often at different rates. It appears, for example, that the Brady Icefield has been much more stable than glacial ice on the east side of Glacier Bay. Because of its elevation and nearness to the coast, Brady Icefield has a steady supply of snow, and it is also fed by large glaciers descending from the Fairweather Range.

The large-magnitude changes that occur in the eastern part of the park/preserve and the Takhinsha Mountains may be due to generally rising mean annual temperatures since the 1700s and may also be the result of a partial rain shadow effect produced by the

Fairweather Range. Less snow accumulates east of the bay. Another factor is elevation of the source area. Advancing glaciers generally come down from snowfields at higher elevations that receive and retain more snow.

The tidewater glaciers are the most variable in their behavior with some advancing, some retreating, and some changing very little. In general, snouts of glaciers that discharge in deeper water or tidal estuaries fluctuate more than those on dry land or those that rest on shoals.

Since southern Alaska is a seismically active area, earthquakes "shake down" snow accumulations in the high areas, resulting in increased pressure from the added load all down the line. An earthquake may also devastate a tidewater glacier. This happened to Muir Glacier in 1899 when an earthquake broke up the terminus and caused the glacier to begin a rapid retreat.

Glacier Bay became choked with a mass of floating ice within a few hours, and for several years ships could not get into Muir Inlet.

Reforestation. Muir and the scientists who followed him were fascinated by the way in which plant zones followed glacial retreat. Within a few years after melting ice uncovers bare rock, hardy species such as mosses, dryas, dwarf fireweed, horsetails, and lichens take root. Alder and willow follow within 60 years. Sitka spruce begins to grow in the young forest and eventually dominates the stand, but as time passes, is displaced by western hemlock—except on relatively sterile gravels and exposed beaches (figs. 31.8a, b, c, d).

Between 2500 and 5000 feet in elevation, the species of the arctic alpine zone flourish, and rich alpine meadows appear. At higher elevations, vegetation shrinks to heath saxifrage and crustal lichens (fig. 31.7).

FIGURE 31.8A   Glacial barrens with dryas, a hardy species of plant that is one of the first plants to take root, along with moss, lichen, dwarf fireweed, and horsetails. National Park Service, Glacier Bay National Park photo.

Bedrock

Geological mapping shows that bedrock exposed by ice retreat is remarkably diverse, a partial confirmation of the theory that Glacier Bay Park and Preserve is made up of exotic terranes. One finds carbonate, volcanic, metamorphic, and gabbroic rocks in no particularly logical arrangement. Recognizable sedimentary rocks range in age from early Silurian through late Tertiary, about 20,000 to 30,000 feet of Paleozoic rock and over 12,000 feet of Tertiary beds. Detailed mapping in the area east of Mt. Fairweather shows a typical assemblage containing, in ascending order: Willoughby Limestone, 5000 feet; Tidal Formation, 10,400 feet; Upper Tidal Formation, 2100 feet; Pyramid Peak Limestone, 2300 feet; Pendleton Formation, 2000 feet; Black Cap Limestone, 4500 feet. Plutons associated with the Coast Range batholith of Mesozoic age intrude these stratified rocks. (The Coast Range batholith is a large-scale pluton of multiple intrusions that crops out from Washington State to Alaska.) Metamorphism of low to intermediate intensity has altered some of the rocks. Numerous small and marginal mineral deposits are scattered throughout the park/preserve.

As described earlier, several northwest/southeast-trending faults cut through the rocks of the park/preserve. The faults divide and rejoin; some lie offshore; others pass beneath the Brady Icefield and Glacier Bay. Similar fault structures are beneath the Lynn Canal, a natural waterway close to the park's eastern boundary.

Field geology in the Glacier Bay area has always been challenging. Geologists must make rain-soaked climbs in fog and drizzle and cross uneven, blocky stretches of snow and ice to reach critical outcrops. Transportation is usually by small boat, but in recent years the U.S. Geological Survey has used helicopters for transporting scientists and supplies from a floating barge to mapping sites.

Geologic History

1. **Possible origin of terranes in Precambrian and Paleozoic time.**

Fossil and paleomagnetic evidence, although far from complete, seems to indicate that some of the terrane rocks were formed in the Southern Hemisphere.

FIGURE 31.8B   Within 60 years after the glacier retreats an emerging forest develops: first are the cottonwoods, followed by alder, willow, and spruce. National Park Service, Glacier Bay National Park photo.

The terranes may have been islands, seamounts, microcontinents, or pieces broken off from larger land masses. At this point we can only speculate about the route of their "journey," but it must have taken many millions of years. What is now the eastern side of the plate

FIGURE 31.8c   In the line of succession of plants the alder and willow trees are crowded out by the sitka spruce. Finally the western hemlock crowds out the sitka spruce and is the last tree in the line of succession; hence it is considered a climax forest. National Park Service, Glacier Bay National Park photo.

boundary was probably located somewhere to the south and east of the present position. Indications are that the region had long been a continental margin.

2. **Mesozoic and Cenozoic accretion.**

The terranes probably "docked" from middle Mesozoic through middle Tertiary time and were pushed with great force against the North American plate. The three terranes that are the largest and most recognizable are (from west to east) Chugach, Wrangellia, and Alexander (fig. 31.8). The long, prominent, roughly parallel faults are the *sutures,* or junctions, between these microplates. As the Pacific plate pushed against the North American plate, thrusting occurred from the southwest toward the northeast.

Strike-slip faulting moved the southwest block toward the northwest (in relation to the northeast block) as the Pacific plate rotated counterclockwise while pressed against the western edge of North America. Because oceanic crust was subducted as the plates converged, folding and reverse faulting occurred and plutons of granitic, dioritic, and gabbroic composition were emplaced, some as late as Miocene time. Regional and contact metamorphism, along with deposition of minerals, also took place in the converging zone (Plafker et al. 1994).

3. **Late Tertiary and Quaternary tectonic activity.**

The forces that drove exotic terranes against southeastern Alaska are still active and are causing the mountains to rise. Major earthquakes continue to occur. In 1958 an earthquake registering 8 on the Richter scale caused millions of tons of rock to break loose from the rock walls enclosing Lituya Glacier. A great rockslide tore off the glacier snout, and the chaotic mass of rock gravity and ice crashed into Lituya Bay. A giant wave, 1740 feet high, surged over the ridge, fell back, and then swept down to the ocean. Waves sloshed back and forth in Lituya Bay for 20 minutes. When it was over, all the trees in a wide swath along the sides of the bay were gone. Tlingit legends warn of similar earthquakes and destructive waves at Lituya Bay in the prehistoric past.

4. Pleistocene glaciation.

In this part of Alaska only sketchy evidence for pre-Wisconsinan glaciation exists. However, thick tillites have been found in offshore drilling in the Gulf of Alaska. Much of this glacial detritus is believed to be late Tertiary and early Pleistocene in age. Apparently the Fairweather Range was being uplifted throughout Pleistocene time with uplift continuing into Holocene time. (Estimated rate of uplift on the outer coast is now 0.7 cm per year.) As the mountains rose, precipitation on the slopes increased and glaciers enlarged. Possibly the Fairweather fault, which skirts the range front,

FIGURE 31.8D   When the glaciers return they bury the forest. Upon the retreat the cycle begins over again, but this time the old tree stumps are uncovered by erosion, which indicate that this has happened before. National Park Service, Glacier Bay National Park photo.

Table 31.1			Simplified Geologic Column, Glacier Bay National Park and Preserve	
Time Units			**Rock Units**	**Geologic Events**
Era	**Period**	**Epoch**		

Cenozoic	Quaternary	Holocene	Neoglacial drift accumulations	Rapid glacial rebound
				Rapid rates of glacial retreat
				Numerous episodes of glacial advance and retreat (since late Tertiary time)
				Ongoing tectonic activities (since Mesozoic time):
				    Strong earthquakes
				    Horizontal displacement with Pacific side sliding northwest
				Vertical uplift
				Thrust faulting
				Deformation by compression
				Subduction with plutonism
				Regional metamorphism

		Pleistocene	Wisconsinan drift Thick older drift in offshore marine sections	
	?			

Mesozoic			Wide variety of rock types (Silurian to late Tertiary in age):
			    Carbonates
			    Granite, diorite, granodiorite, gabbro, lavas
			    Marble, schist, gneiss, quartzite, hornfels, granulite
				Accretion (Jurassic through Eocene time) of at least three exotic terranes: Alexander, Wrangellia, Chugach

Paleozoic	Silurian		Willoughby Limestone	

Sources: Beikman, 1980; Brew and Ford, 1982, 1986; Mackevett, 1971; Nokleberg, James, and Silberling, 1985; Rossman, 1959, 1960; Seitz, 1959.

blocked or split most of the glaciers, keeping all but the largest away from the coast. Evidence of middle and late Pleistocene glaciation is missing in many of the outer coast areas.

Wisconsinan glaciation dominated the region for thousands of years, attaining maximum extent between 21,000 and 15,000 years before the present. Ice sheets accumulated in three areas: (1) the Brady Icefield, west of Glacier Bay (fig. 31.9); (2) the Takinsha Icefield along the northeast edge of the park/preserve; and (3) on the northern highlands along the international boundary. Ice in the northern highlands may have coalesced into an intermontane ice sheet or a piedmont glacier.

5. Neoglacial advances during Holocene time.

In late Pleistocene and Holocene time, a combination of cooler climate and the long-continued uplift of the Fairweather Range increased snowfall markedly and brought on glacial advances. Apparently the main flow within the piedmont ice sheet moved southward over Glacier Bay, turned seaward, and passed through Cross Sound. Holocene warming came slowly to Glacier Bay, but about 10,000 years ago, ice retreated

FIGURE 31.9   The Brady Ice Field feeds most of the glaciers in the park. Only the peaks of the buried mountains beneath the ice field show through; they are called nunataks. This is what Glacier Bay looked like when Captain James Cook sailed by in 1778. National Park Service Park, Glacier Bay National Park photo.

and the land rose. The trend reversed some 4000 to 5000 years ago, and icefields again expanded. Rejuvenated glaciers overwhelmed forests, buried the stumps in till and drift, and deposited till and outwash in lowlands and valleys. Glaciers continued to grow until about 500 years ago, at which time retreat began. The retreat picked up speed between A.D. 1660 and 1760 and has maintained a fast rate of meltback since then, surpassing rates measured elsewhere.

Geologic Map and Cross-Section

Beckman, Helen M. 1980. Biologic Map of Alaska, U.S. Geological Survey.

Sources

Anderson, P.S., Goldthwait, R.P., and McKenzie, G.D. (editors). 1986. *Observed Processes of Glacial Deposition in Glacier Bay, Alaska.* Columbus, Ohio: Institute of Polar Studies, Ohio State University. Miscellaneous publication 263. 167 p.

Beikman, H.M. (editor). 1980. Geologic Map of Alaska (scale 1:2,500,000) U.S. Geological Survey.

Boehm, W.D. 1975. *Glacier Bay.* Anchorage: Alaska Northwest Publishing. 134 p.

Brew, D.A., and Ford, A.B. 1982. Tectonostratigraphic Terranes in the Coast Plutonic-metamorphic Complex, Southeastern Alaska. In *The United States Geological Survey in Alaska: Accomplishments during 1982.* Circular 939, p. 90–93.

————. 1986. Southeastern Tectonostratigraphic Terranes Revisited. *Bulletin American Association of Petroleum Geologists*, 69:657.

Clarke, J.M. 1979. *The Life and Adventures of John Muir.* San Francisco: Sierra Club. 326 p.

Field, W.O. Jr. 1942. Glacial Studies in Alaska. *Geographical Review*, 32:154–155.

————. 1947. Glacier Recession in Muir Inlet, Glacier Bay, Alaska. *Geographical Review*, 37:369–399.

Haselton, G.M. 1966. Glacial Geology of Muir Inlet, Southeast Alaska. Institute of Polar Studies, Ohio State University, Report 18, 34 p.

————. 1979. Some Glaciogenic Landforms in Glacier Bay National Monument, Southeastern Alaska. *Moraines and Varves* (Ch. Schluchter, editor), Proceedings of an INQUA Symposium on Genesis and Lithology of Quaternary Deposits, Zurich, September, 1978, p. 197–205.

MacKevett, E.M., Brew, D.A., Hawley, C.C., Huff, L.C. and Smith, J.G. 1971. Mineral Resources in Glacier Bay National Monument, Alaska (geologic map included). U.S. Geological Survey Professional Paper 632, p. 1–89.

National Park Service. 1983. *Glacier Bay: A Guide to Glacier Bay National Park and Preserve*, Alaska. National Park Handbook 123. 128 p.

Nokleberg, W.J., James, D.L., and Silberling, N.J. 1985. Origin and Tectonic Evolution of the Maclaren and Wrangellia Terranes, Eastern Alaska Range, Alaska. *Geological Society of America Bulletin*, 96:1251–1270.

Plafker, G., Moore, J.C., and Winkler, G.R. 1994. Geology of the Southern Alaska Margin in Plafker, G., and Berg, H.C. eds. *The Geology of Alaska*. Boulder, Colorado: Geological Society of America. v. G-2, p. 389–449.

Seitz, J.F. 1959. *Geology of Geikie Inlet Area, Glacier Bay, Alaska*. U.S. Geological Survey Bulletin 1058C, p. 61–119.

To Wonder
Ranger Kevin Richards

We walked across cold cobbles
the sky drizzles and clouds gather
I worry about barnacles
under my boots
but have nowhere else to step
seaweed laces the tideline
like an amber necklace
you find a rock
encrusted with garnets
how they shine even in
this half-light
little crystals perfectly formed
held in the center of your palm
we muse over kelp-draped boulders
about how the garnets
got to be there
maybe wisdom
is not in the knowing after all
but in wondering

From *The Fairweather*, 1993, Glacier Bay National Park and Preserve, National Park Service.

Note: To convert English measurements to metric, go to www.helpwithdiy.com/metric_conversion_calculator.html

C H A P T E R   3 2

Wrangell-St. Elias National Park and Preserve

SOUTH CENTRAL AND SOUTHEASTERN ALASKA

Monte D. Wilson   Boise State University

Area: 13,188,325 acres; 20,707 square miles	**Address:** 106.8 Richardson Hwy, P.O. Box 439, Copper Center, AK 99588
Proclaimed a National Monument: December 1, 1978	
Established as a National Park and Preserve: December 2, 1980	**Phone:** 907-822-5234
	E-mail: wrst.interpretation@nps.gov
Designated a World Heritage Site: 1979	**Website:** www.nps.gov/wrst/index.htm

In this largest unit of the National Park System is North America's largest assemblage of peaks above 16,000 feet. Mount St. Elias, elevation 18,008 feet, is the second highest peak in the United States. Spanning three of Alaska's four climatic zones, the park/preserve encompasses diverse and spectacular landscapes: rugged coasts, surging tidewater glaciers, the continent's largest piedmont glacier, high icefields on massive ranges, valley glaciers, wild rivers in deep canyons, expanses of tundra, a rich concentration of wildlife.

Major faults bound tectonosratigraphic terranes. Earthquakes and volcanoes are signs of ongoing tectonic activity.

Old mining works recall days when riches in copper, silver, and gold were taken out of the mountain.

FIGURE 32.1   Mt. St. Elias with a height of 18,008 feet towers over the landscape as seen from Icy Bay. Photograph by Ann G. Harris.

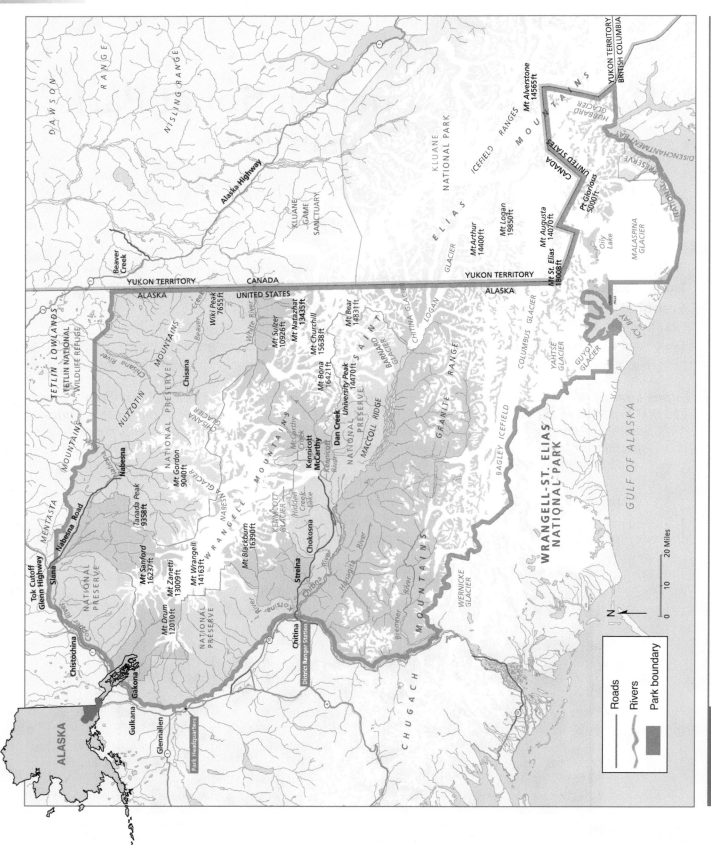

FIGURE 32.2   Wrangell-St. Elias National Park and Preserve, south central and southeastern Alaska.

Geographic Setting

Wrangell-St. Elias National Park and Preserve is by far the largest unit in the National Park System; it is larger than several of the smaller states. In conjunction with the adjacent Kluane (kloo-ah-nee) National Park in the Yukon Territory of Canada, this is the world's largest parkland and is, in fact, much larger than the combined size of all national parks in the other 49 states and American Samoa.

Creation of a national monument in the Wrangell-St. Elias Mountains and adjacent areas was recommended in 1938 by Ernest Gruening, who was then director of U.S. Territories and later served as Alaska's governor and U.S. Senator. Forty years after Gruening's recommendation, national monument status was proclaimed by President Jimmy Carter. In 1979, the United Nations Educational, Scientific, and Cultural Organization (UNESCO) designated Wrangell-St. Elias National Monument and Kluane National Park as a World Heritage Site because of the area's natural and cultural resources.

Wrangell-St. Elias was established as a National Park and Preserve in 1980, also under President Carter's administration, but due in large part to the support and effort of Interior Secretary Cecil Andrus. The area extends from Yakutat Bay and Malaspina Glacier on Alaska's south coast north to the upper Tanana River Valley in the interior and from the Canadian border west to the Copper River.

Wrangell-St. Elias National Park and Preserve displays a great diversity of spectacular landscapes ranging from rugged coastlines to high mountain ranges and includes deep canyons, large glaciers, tundra areas and whitewater rivers. The National Weather Service recognizes four climatic zones in Alaska: Maritime, Transitional, Continental, and Arctic (fig. 32.3). Wrangell-St. Elias lands span three of these zones, missing only the Arctic zone. The Gulf of Alaska coast and the adjacent Chugach and St. Elias Mountains are in the moist Maritime climate. Along the coast, mean annual precipitation exceeds 130 inches and moderate temperatures prevail, ranging from winter lows around 0° F to summer highs in the 70s. The St. Elias Mountains are the world's highest coastal range; the land rises in only 15 miles from sea level at Icy Bay to an elevation of 18,008 feet at the peak of Mt. St. Elias. This uniquely steep topographic gradient causes heavy precipitation to fall from moist maritime air masses and

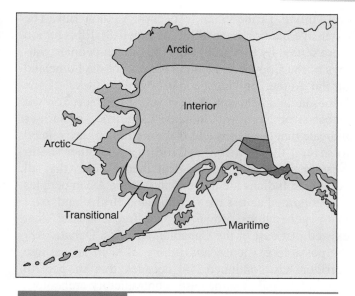

FIGURE 32.3   Climatic zones of Alaska. Wrangell-St. Elias National Park and Preserve, the darkest area, spans three of the state's four climactic zones.

produces average annual snowfalls of more than 600 inches (50 feet) at higher elevations. These conditions have resulted in the largest assemblage of glaciers in North America.

Most of the Copper River basin, inland from the coast, is in the Transitional climate zone. Precipitation at various elevations in this area ranges from about 10 to 25 inches per year, mostly delivered as snow. The Transitional climatic zone has colder winters and warmer summers than the Maritime zone. The Continental climatic zone, north of the Wrangell and St. Elias Ranges, lies in the rain shadow of these high mountains. Villages just north of the park/preserve are Northway, Alaska, and Snag, Yukon Territory; both have mean annual precipitation of 8 to 10 inches and temperatures that range from –70° F to the high 80s.

Four distinct groups of Indians currently occupy the Wrangell-St. Elias region, in areas established by their ancestors during prehistoric times. Most anthropologists agree that all of these people descended from Asian hunters who migrated across the Bering Strait between 8000 and 12,000 years ago near the close of the last great ice age.

In the coastal zone the Eyak tribe settled the Copper River Delta area and people from the Tlingit (pronounce the first syllable halfway between "cling" and "thling") tribe came to occupy Yakutat Bay. The extensive interior areas of the Wrangell-St. Elias region were settled by two groups of Athapascan peoples; valleys of the Copper and Chitna Rivers are the homeland of the Ahtna, or "ice people," while valleys of the Nabesna and Chisana Rivers are homeland for the Nabesna or Northway Indians. Linguistic similarities indicate that the Ahtna and Nabesna Indians are related to several Indian tribes of California and other southwestern states. Anthropological theory holds that all American Indians are descendants of the Asian peoples who migrated across the Bering Land Bridge and, over many generations, down an "ice-free corridor" that existed just east of the Canadian Rockies. Debate over the possible existence and timing of such an ice-free corridor has continued for decades due in part to incomplete and in some cases contradictory archeological and geological evidence.

The abundance of both marine and land resources, as well as the mild coastal climate, enabled the Eyak and Tlingit Indians to meet life's basic needs relatively easily. This left time for them to develop elaborate social customs and beautiful art forms, including the totem poles and cedar houses of the Tlingit. Subsistence in the interior was more difficult due to the rigorous climate and limited diversity of food resources; famine, particularly in late winter and early spring, was an occasional occurrence.

High-grade copper deposits in the southeastern part of the Wrangell Mountains were exploited by Ahtna Indians living along the Copper and Chitna Rivers. Some of these people became skilled at using heat to work the copper into sharp projectile points and cutting implements. Copper items were traded among native groups for at least 1400 years prior to contact with whites. In the spring of 1885, Lieutenant Henry T. Allen of the U.S. Army found Chief Nicolai's Copper River Indians, part of the Ahtna group, using bullets of a copper-silver alloy produced from their own mines. Chief Nicolai showed Lt. Allen the location of a rich vein of copper that was being worked by his people.

The Wrangell-St. Elias Park and Preserve contains many significant prehistoric and historical archeological sites. Those of Athapascan origin are most numerous. Others are from the Eyak and Tlingit Indians as well as from the Chugach Eskimos, who visited the area occasionally.

The earliest recorded entry of Europeans into Alaska was by Vitus Bering, a Dane, who in 1741 explored this area for Czar Peter the Great of Russia. Bering named Cape St. Elias (on Kayak Island) in honor of the saint upon whose feast day the cape was first sighted. He also recorded sightings of the rugged, glaciated mountains, including the prominent peak that came to be called St. Elias.

Over the next several decades, this part of the Gulf of Alaska coastline was visited and claimed, in part or total, by explorers from several nations. In 1778, Captain James Cook visited Yakutat Bay, named several topographic features, and claimed these regions for Great Britain. Captain Salvador Fidalgo named Valdez Bay in 1790 and claimed that area for Spain. Another Spanish land claim was made by the Italian Captain Alejandro Malaspina who in 1791 sailed into Icy Bay; although disappointed in finding that the bay did not lead to the elusive Northwest Passage, he nevertheless had a rock monument built and laid claim to the port for his employer, the King of Spain.

It was the Russians, however, who, by establishing permanent settlements and a local government in the area, were able to hold their land claims. In 1795, Aleksandr Baranov, the Russian governor of Alaska, had a convict labor colony established on Yakutat Bay to engage in agriculture and shipbuilding. After ten years of conflict with local Indians, the town, called "Gloria of Russia," was razed and 40 of its inhabitants were killed by a Tlingit war party. Russian fur traders also traveled into the interior along the Copper River, but hostilities between Ahtna Indians and Russians led to bloodshed in that area as well.

In 1867, William H. Seward, President Andrew Johnson's Secretary of State, arranged the United States purchase of Alaska (then called Russian America) for $7,000,000, or about two cents per acre. The unpopular purchase, called "Seward's folly," ended formal Russian influence in Alaska.

Lt. Allen's 1885 expedition was the first American reconnaissance in the area. He explored the Upper Copper River Basin, the Chitna River Basin, and the Wrangell Mountains before going north over the divide to the Tanana River Valley. In addition to establishing friendly relations with Chief Nicolai and his Copper River group of Ahtna, Allen named the village of Chitna and the Chitistone River (both using the Indian word "chiti," meaning copper) plus Mt. Blackburn, Mt. Drum, and Mt. Sanford.

In 1898, following the discovery of rich placer gold deposits along Canada's Klondike River, the Copper

River Exploring Expedition was organized under command of the U.S. Army. Included in the expedition were two U.S. Geological Survey geologists, Oscar Rohn and T.E. Schrader. Rohn named the Chitistone Limestone and Nicolai Greenstone in the Chitna River drainage. He pointed out similarities to the Lake Superior Greenstones that suggested rich mineral potential. In 1899 the Nicolai mine was located (with assistance from Chief Nicolai) and in 1900 the incredibly rich Bonanza Mine was located eight miles west on a steep ridge, just east of the lower Kennicott Glacier. The mines did not go into full production until 1911, when the completion of a railroad allowed ore to be hauled from the mining town of Kennicott, via a route along the Chitna and Copper Rivers, to the port of Cordova. When the high-grade ore was worked out in 1938, mines in the Kennicott-McCarthy area shut down. By that time they had produced over a billion pounds of copper plus more than nine million ounces of silver, which was recovered as a by-product. For many years there were plans to use the old railroad bed as the base for a road to link Cordova with Alaska's highway system at Chitna. These never realized plans were kept active until the 1964 earthquake destroyed one span of the bridge across the Copper River.

On the north side of the Wrangell Mountains, rich lode deposits (veins in bedrock) of gold were worked in the 1920s and 1930s at both Nabesna and Chisana (pronounced "shoe-shana"). Prior to 1932, most supplies were hauled to these mines by horses in the summer, but dog teams delivered mail in the winter. A particularly accomplished individual of this era was Harry Boyden, who worked as a packer and meat hunter for the mining companies and for several years carried the mail by dog team from McCarthy over the Wrangell Mountains and across glaciers to Nabesna. In 1931 the contract for mail and small freight pieces was taken over by Bob Reeve, who came to be known as the "Glacier Pilot."

Within Wrangell-St. Elias National Park and Preserve, three of the numerous sites associated with the mining industry—one each at Kennicott, McCarthy, and Nabesna—are in the National Register of Historic Sites. In 1995, there were 348 patented lode claims, 99 patented placer claims, and 11 patented mill site claims within park/preserve boundaries. In order for a mining claim to become patented, government regulations concerning discovery and development of mineral resources must be met; this usually requires the invest-ment of much time, effort, and money with no guarantee of return. On patented claims, the land and mineral rights have been transferred from government to private ownership.

The Wrangell and St. Elias Ranges include more high mountain peaks than any other part of North America. Mt. Logan, located only 15 miles outside the United States border, is Canada's highest point at 19,850 feet, and second only to Mt. McKinley at 20,306, as the highest peak in North America. Mt. St. Elias with its summit of 18,008 feet situated on the United States-Canada border, is the second highest peak in both countries. Mountain climbers have challenged these peaks for over a century. Climbing the high peaks in these ranges is extremely difficult, partly because of the area's inaccessibility and very rugged topography, but also because the combination of high latitude and high altitudes produces a rigorous climate that supports extensive snow fields and many glaciers.

Beginning in 1886, several attempts were made to climb Mt. St. Elias before it was conquered in 1897 by an Italian expedition under the leadership of His Royal Highness, the Duke of Abruzzi. Mt. Blackburn (16,390 feet) was first climbed in 1912 by Dora Keen and her party of Americans. In 1925 the summit of Mt. Logan was reached by the MacCarthy-Lambart party of Canadians. Aircraft support of mountain-climbing parties in this region began in 1937 when Bob Reeve, the Glacier Pilot, landed scientists Bradford Washburn and Robert Bates at the 8500-foot level to begin the first ascent of Mt. Luciana (17,147 feet). The following year, Washburn and Terris Moore were the first to climb the volcanic cone of Mt. Sanford (16,237 feet). The second successful climb of Mt. St. Elias, and the first by an American party, was accomplished in 1946 by a Harvard Mountaineering Club expedition led by glaciologist Maynard Miller. Beginning in 1949 many climbing expeditions to the area made first ascents of major peaks; McArthur Peak (14,400 feet), the last unclimbed peak higher than 14,000 feet, was finally climbed in 1961.

Geologic Features

Bedrock Sequences

The northernmost part of Wrangell-St. Elias National Park includes a small portion of the Yukon-Tanana terrane. That terrane consists of varied metamorphic and

Box 32.1	Plate Tectonics and Accreted Terranes

Different parts of the Wrangell-St. Elias area have drastically different types of bedrock and geological structures. A thousand geologists could work for a thousand years in this area, but unless they understood plate tectonics (particularly the accreted terranes aspect of plate tectonics) they would probably not reach any generally acceptable interpretation of regional geologic history. It seems that the different portions of the Wrangell-St. Elias area can best be explained as being accreted terranes.

Geologists use the word *terrain* for its common meaning of general landscape characteristics or "lay of the land." Pronounced the same, but spelled differently, the term *terrane* refers to a block of crustal rock (having surface dimensions of tens to a few thousand kilometers) with a unique sequence of rock units that have resulted from its particular geologic history. Because of its history, one terrane might have massive granites at its base overlain by a thick sequence of Paleozoic carbonate sedimentary rocks with warm-climate fossils and an uppermost zone of basalt flows interbedded with terrestrial sediments. Another terrane might consist of oceanic crust at the base overlain by Mesozoic island-arc basalts and associated sediments, which are then capped by a sequence of clastic sediments containing coldwater fossils.

Finding such dissimilar rock sequences on adjacent ridges in southern Alaska leaves the field geologist with a very difficult problem in geologic history so long as it is assumed that this area has always shared the general geologic history of the North American continent. However, such major differences in bedrock can easily be accounted for if it is assumed that the various blocks, or terranes, originated in different parts of the globe and had completely different geologic histories until they "drifted" via the mechanism of plate tectonics to collide with, and accrete to, North America. Such terranes apparently existed either as "microplates" or as islands of continental materials within the upper part of an oceanic plate moving toward, and subducting beneath, the North American continental plate. When they collided with North America, the terranes were "scraped off" and attached to the main continental mass. It now appears that a wide strip of western North America from Alaska to California is made up of accreted terranes, including all, or portions of, several national parks.* The concept of accreted terranes is relatively new, and it is generally agreed that there is a lack of detailed understanding of the mechanism. Lacking also are adequate factual data derived from geological mapping, fossil collections, and absolute ages.

Even in its immature state, however, the concept of plate tectonics and accreted terranes helps us to understand the rock sequences, volcanoes, and earthquakes of Wrangell-St. Elias. Among the terranes in

---

*Further discussions of accreted (tectonostratigraphic) terranes are in Chapter 24, Acadia National Park; Chapter 29, North Cascades National Park; Chapter 31, Glacier Bay National Park and Preserve; Chapter 33, Kenai Fjords National Park; Chapter 34, Denali National Park and Preserve; Chapter 38, Katmai National Park and Preserve; Chapter 39, Lake Clark National Park and Preserve; Chapter 50, Channel Islands National Park; Chapter 51, Redwood National Park; Chapter 54, Shenandoah National Park; and Chapter 55, Great Smoky Mountains National Park.

---

igneous rocks, with the oldest being Precambrian in age. Future geologic mapping may show the Yukon-Tanana terrane to be a composite of several smaller microplates.

The Wrangellia terrane has been studied in more detail than most accreted terranes, so its major rock units are well known. A thick assemblage of island-arc volcanic rocks of late Paleozoic age forms the base of the Wrangellia terrane. This volcanic unit is overlain by fossil-bearing limestones, shales, and cherts of Permian age. The lowest Triassic rocks are cherty shales; resting upon them is a very thick sequence of greenstones that were metamorphosed from basalts of both submarine and subaerial origin. The Nicolai Greenstone is of particular importance within this sequence because it contains deposits of copper similar to those of Precambrian

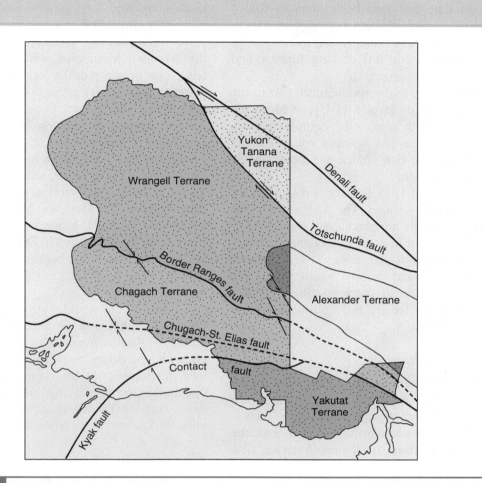

FIGURE 32.4   Major faults and accreted terranes in Wrangell-St. Elias National Park and Preserve.

Wrangell-St. Elias National Park and Preserve are: Yukon-Tanana, Wrangellia, Chugach, Alexander, and Yakutat (fig. 32.4). Paleomagnetism and fossil evidence indicate that some of these terranes originated in equatorial regions. Boundaries between terranes are partially delineated by major faults trending northwest-southeast, such as the Denali fault, Totschunda fault, Border Ranges fault, and St. Elias fault.

age in Isle Royale National Park (Chapter 23). Uppermost Triassic rocks are limestones and argillites of marine origin. Jurassic and Cretaceous rocks of the Wrangellia terrane are mostly sandstones with some shale and conglomerate. They are overlain by Lower Tertiary volcanics, which, in turn, are partially covered by Upper Tertiary and Quaternary lavas that issued from volcanoes of the Wrangell Mountains.

A small part of the Alexander terrane, consisting of Paleozoic marbles, lies within Wrangell-St. Elias National Park and Preserve, just on the Alaska side of the international boundary. Other rock types are present in the more extensive Canadian portion of the terrane.

The Chugach terrane, which forms a belt across Southern Alaska, makes up most of the southern part of Wrangell-St. Elias National Park and Preserve. Much of the Chugach terrane consists of Upper Cretaceous graywackes and slate that have been intensely folded and weakly metamorphosed. Interspersed within these rocks are discontinuous assemblages that include chert, gabbro, pillow basalts, and tuff of Late Jurassic and Early Cretaceous marine origin.

The Yakutat terrane, in the southeastern part of the park/preserve, is made up largely of Upper Mesozoic graywacke and shale with interleaved, complexly faulted units of chert, argillite, and volcanic rocks. Eocene age basalts and organic-rich shales are the youngest rocks of the Yakutat block.

Surficial deposits of Late Tertiary and Quaternary age are found on all the terranes. The most voluminous of these surficial deposits are glacial and fluvial materials, both of which occur in great variety and are widely distributed throughout the area. Other surficial deposits include beach and shallow marine sediments along the coast, widespread mass wasting materials (especially from rock glaciers and solifluction), and two fields of sand dunes.

Earthquakes and Volcanoes

Wrangell-St. Elias National Park and Preserve is located in one of the world's most active zones with regard to plate tectonics. In this region, the Pacific Plate is being subducted beneath North America and the Yakutat terrane is accreting to North America. The associated volcanic belts, active faults, frequent earthquakes, and high mountains are expected results of the high heat flows and stresses in crustal rocks of this plate tectonic setting. The major zone of volcanoes is in the Wrangell Mountains, with Mt. Wrangell (14,163 feet), one of the world's largest andesitic volcanoes, being the park's only active cone. Its last reported eruption was in 1930. Carl Benson of the University of Alaska reported an increase in heat flow in the early and middle 1980s near the summit of Mt. Wrangell.

Quaternary lavas make up several other spectacular, but presently inactive, volcanoes in the park, such as Mt. Bona (16,421 feet) in the St. Elias Range, Mt. Blackburn (16,390 feet), and Mt. Sanford (16,237 feet), both in the Wrangell Mountains. Another inactive volcano in the Wrangells is Mt. Drum (12,010 feet), which is of particular interest because of three large thermal springs, known as the Klawsi Mud Volcanoes, on its western flank.

Some of the world's strongest earthquakes have occurred in this area. In September 1899, two earthquakes of Richter magnitudes 8.3 and 8.6 were recorded with epicenters near Yakutat. The several earthquakes with magnitudes in the 7 and 8 range that have occurred in the area during this century give ample evidence that the Yakutat microplate is actively colliding with, and accreting to, North America.

Mountains and Canyons

From north to south, the major ranges in the park/preserve are the Mentasta, Nutzotin, Wrangell, St. Elias, and Chugach Mountains. Although these ranges have separate names, not all of the boundaries between them are obvious, and all of these mountains are parts of larger regional features. The Mentasta and Nutzotin Mountains, along the northern margin of the park/preserve, are actually part of the more extensive Alaska Range. Skolai Pass is the boundary between the Wrangell and St. Elias Mountains, which together make a southeast-trending, continuous range. The Chugach Mountains run approximately east-west and merge with the larger Wrangell-St. Elias Range near the park's southeastern corner. Nine of the 16 highest peaks in the United States, including Mt. St. Elias, the second highest, are in Wrangell-St. Elias National Park. Mt. Logan and many of Canada's highest peaks are in the adjacent Kluane National Park.

Deep valleys and canyons cut the high mountains in several places. Particularly spectacular in this regard are the sheer walls and plunging waterfalls of

FIGURE 32.5 An unnamed valley glacier is joining a larger piedmont glacier in Wrangall-St. Elias National Park. Note the lateral moraine along the side of the main glacier. National Park Service photo.

Chitistone and Nizina Canyons, both of which are much larger than Yosemite Valley.

Glaciers and Rivers

Relatively warm waters of the Japanese current wash the shores of Southern Alaska. Air masses moving over these warm currents take up water vapor until they become completely saturated. As these air masses move inland, the high mountains force them to rise. While the air is rising, it cools and releases precipitation, mostly as snow. Much of the snow is converted to ice with the result that the Wrangell, St. Elias, and Chugach Ranges have the largest concentration of glaciers in North America. Bagley Icefield, which runs along the crest of the Chugach Range for about 150 miles, is the largest subpolar icefield in North America. It covers most of the core of the Chugach Mountains and nourishes dozens of valley glaciers that drain down both sides of the range.

Malaspina Glacier, in the southernmost part of the park/preserve, is the largest piedmont glacier in North America. It is nearly 40 miles across and is larger in area than the state of Rhode Island. The word *piedmont* means "foot of the mountain." The term is applied to more or less flat-lying glaciers that form by the coalescence of two or more alpine glaciers where they flow beyond a steep slope. Several large alpine glaciers drain southward off the St. Elias Range to nourish the Malaspina Glacier, which covers most of the coastal plain between Yakutat Bay and Icy Bay.

A particularly interesting aspect of the glaciers in this area is their high level of activity. Many of them sustain high rates of flow. Additionally, many of the glaciers in this area tend to undergo periodic surges. These surges are times of rapid advance that last from a few weeks to a year or more and are usually followed by rapid meltback. Bering Glacier, a large (2300-square-mile) piedmont glacier, just outside the southwest boundary of the park, has surged a least five times this century, including a six-mile advance between October, 1993, and September, 1994. Variegated Glacier surges every 20 years, but most surging glaciers do not have an obvious periodicity.

At several places in the park/preserve, glaciers tend to create glacier-dammed lakes on an almost regular basis. Kennicott Glacier blocks a stream drainage to form Hidden Lake; each summer, raised water levels due to snowmelt cause the lake to overtop the glacier and break out, the result being intense flooding on the Kennicott River. Oily Lake and Malaspina Lake are

Not all of the mountains are covered with glaciers in Wrangall-St. Elias National Park. National Park Service photo.

also glacially dammed, but their outburst floods are not always annual.

In 1986, a particularly dramatic glacial damming event that occurred on the margin of the park/preserve got a great deal of public attention. Hubbard Glacier (fig. 32.7), which flows into Yakutat Bay at the southeastern corner of the park, has an unusually high continuous velocity of about 32 feet per day in some parts of the glacier. This large alpine glacier, after advancing for months, blocked off the 30-mile-long Russell Fjord on May 29, 1986. The ice dam transformed Russell Fjord from an arm of the sea into a salty lake that began to rise as it received runoff from many streams. Public attention was focused on the possible starvation of many marine mammals, mostly seals and porpoises, trapped in the lake by the ice dam. Over the next few months, Lake Russell accumulated 3.3 cubic miles of fresh water and rose to a height of about 82 feet above sea level. Local residents were concerned about two possible hazards: (1) the potential for sudden failure of the ice dam with ensuing flooding and iceberg damage at Yakutat; (2) the possibility that the ice dam might remain stable and cause Russell Lake to rise and overflow its southern end, which would damage the salmon spawning beds of the Situk River, a very important economic resource in the area. Early in the morning of October 8, the dam failed and the lake drained out in only 36 hours. The

FIGURE 32.7   Hubbard Glacier which flows into Yakutat Bay, located at the southeastern corner of the park, flows at a velocity of about 33 feet per day in some parts of the glacier. On May 29, 1986, it blocked off the 30-mile-long Russell Fjord. The ice dam transformed Russell Fjord from an arm of the sea to a salty lake that began to rise as the result of streams entering the lake. Marine animals, mostly seals and porpoises, were trapped behind the ice dam and in danger of starving. The ice dam broke on October 8, 1986 and the animals were freed. Bones and mummies found in the area indicate that this has happened many times in the past. Photo by Douglas Price.

tremendous flooding and enormous icebergs brought some changes to the shoreline and bottom of the bay, but no major damage to the town of Yakutat.

The Copper River and its tributaries, especially the Chitna River, form the largest stream drainage system in the park/preserve. Waters in this system run generally southward into the Gulf of Alaska. The Nabesna, Chisana and White Rivers, which drain the north slopes of the ranges, flow north and westward through the Yukon River system into the Bering Sea. Rivers in this area tend to have large annual fluctuations. They have very high discharge rates during summer runoff from melting snow and glaciers, but very low discharge rates during the frozen times of mid- and late winter.

Melting glaciers form the headwaters of all major streams in the park/preserve. The great volume of sediment in glacial meltwaters causes the streams to carry large amounts of silt during the summer. Because of their muddy waters, the streams do not support large populations of resident fish, but do serve as migration routes between the ocean and spawning beds, and also between various clearwater lakes and tributaries. Three streams, the Tebay River, Hanagita River, and Beaver Creek, that do not start from melting glaciers have clear water and larger resident fish populations. The park/preserve offers excellent whitewater boating and fishing opportunities. Information on river hazards and fragile resources, such as salmon-spawning beds and swan-nesting areas, can be obtained from the National Park Service.

Periglacial and Permafrost Features

Wrangell-St. Elias National Park and Preserve lies mostly within the discontinuous permafrost zone. Areas at high elevations or in the northern part of the park/preserve are underlain by permafrost. However, permafrost is absent or spotty at lower elevations, particularly in the southern part of the park/preserve along the Copper and Chitna River Valleys and near the coast. Areas of intermediate elevations tend to have permafrost where there is a layer of sphagnum moss or other protection from summer heat. Distribution of permafrost and thickness of the active zone (the surface layer that is frozen in winter and thawed in summer) is a primary control of vegetation in such areas. For example, in the Copper and Chitna River Valleys, stands of "scrubby" black spruce usually indicate permafrost at shallow depths.

Periglacial features in Wrangell-St. Elias are similar to those of Denali National Park and Preserve (Chapter 34). Soils in periglacial areas tend to be thin, discontinuous, and coarse-grained. In such cold temperatures, chemical weathering of minerals is extremely slow so the breakdown of rocks is accomplished by mechanical weathering—primarily by frost action. Silt- and sand-size particles are produced in abundance by frost wedging, but this process cannot produce finer clay-sized particles. This is why cold-climate soils are usually sandy and silty, but have little clay. Bacterial decay of organic material is also slow in cold climates. Because of this, humus and peat have accumulated to considerable thicknesses at some favorable localities in Wrangell-St. Elias Park and Preserve.

Geologic History

1. **Complex metamorphic rocks of Paleozoic and possibly Precambrian age in the Yukon-Tanana Terrane.**

The oldest rocks in Wrangell-St. Elias National Park and Preserve are exposed north of the Denali fault in the Yukon-Tanana Terrane. Foliated metamorphic rocks such as quartz-mica schist and gneiss are common; within the nonfoliated metamorphic rocks, quartzite is common and marble occurs in a few places. Metamorphosed rhyolite and plutonic rocks are also abundant in this terrane. Based on limited field studies and radiometric dating, it appears that the metamorphic rocks developed either from sediments deposited dur-

ing Precambrian time, or from sediments that were derived by erosion of Precambrian rocks during the Paleozoic Era. These marine deposits accumulated either at the outer edge of the North American continental shelf or in a marine basin far removed from the continent. In any case the sediments were lithified, and by mid-Paleozoic they had been deformed, metamorphosed, and intruded by granite plutons.

2. **Mid-Paleozoic accumulation of carbonate sediments on the Alexander Terrane.**

During the Devonian Period, carbonate-rich muds were deposited on the flanks of a volcanic island arc located far from the North American continent. Lithification changed these muds into limestone; subsequent metamorphism transformed the limestone into marble. The island arc was transported thousands of kilometers by crustal plate movements until, much later, it accreted to North America as the Alexander Terrane. Most of the Alexander Terrane occurs in Canada as an elongate area approximately parallel to the west coast. Although the portion of the Alexander Terrane within Wrangell-St. Elias National Park con-

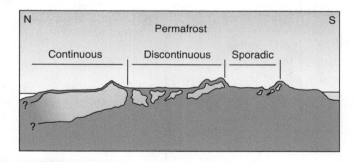

FIGURE 32.8   Diagrammatic north-south cross section through Alaska showing permafrost zones. In the continuous zone, permafrost extends to depths exceeding a thousand meters; the active layer (which thaws in summer) ranges from 4-20 inches. In the discontinuous zone, the active layer ranges from 10 inches to 10 feet in thickness while maximum permafrost depth is usually less than 109 yards and taliks (nonfrozen areas) exist. Within the sporadic zone, permafrost exists only in particularly favorable spots. The southern part of Wrangell-St. Elias National Park and Preserve is in the zone of no permafrost; much of the park preserve is in the sporadic zone, while the northern part lies in the zone of discontinuous permafrost.

Table 32.1			Simplified Geologic Column, Wrangell-St. Elias National Park and Preserve	
Time Units			**Rock Units**	**Geologic Events**
Era	**Period**	**Epoch**		
Cenozoic	Quaternary	Holocene	Holocene drift, alluvium, colluvium and marine sediments	Holocene glaciation, fluvial action, and (W) volcanism
		Pleistocene	Wisconsinan drift	Accretion of Yakutat Terrane, ongoing faulting, earthquakes (W) Quaternary volcanism
	Tertiary	Pliocene	Older drift	Numerous glacial advances and retreats since Late Tertiary
		Miocene		Ongoing tectonic activities, especially faulting and uplift
		Oligocene		
		Eocene	(W) Andesite	(W,Y) Volcanism
			(Y) Basalt and shale	
		Paleocene		
Mesozoic	Cretaceous		(Y) Graywacke and shale (C) Graywacke and slate (C,W) Clastic sediments (C) Pillow basalt	Accretion of four terranes during Jurassic through Eocene: (Y-T), (A), (W) and (C)
	Jurassic Triassic		(W) Limestones and shale, Nicolai Greenstone, cherry argillite	
Paleozoic	Permian Pennsylvanian		(W) Limestones, shale, argillite, and chert (W) Volcanic breccia, water-laid tuffs, minor limestone	(W) Island-arc volcanism (Pennsylvanian-Triasssic) with on-going marine sedimentation
	Mississippian Devonian Silurian Ordovician Cambrian		(A) Marble (Y-T) Wide variety of metamorphic rocks Precambrian to Mississippian: schist, gneiss, quartzite, marble, and plutons	
	Precambrian			

Explanation: (A) Alexander Terrane; (C) Chugach Terrane: (W) Wrangell Terrane (Y) Yakutat Terrane (Y-T) Yukon-Tanana Terrane.

Sources: Burns, 1983; Csejtey, et al., 1982; Jones, et al., 1982; Silberling and Jones, 1984.

tains primarily Devonian marble, other parts include diverse marine volcanics and sediments that have undergone varying degrees of metamorphism.

3. Late Paleozoic through early Mesozoic volcanism and sedimentation in the Wrangell Terrane.

An arc of volcanic islands developed in Wrangellia early in the Pennsylvanian Period. After producing lavas for many millions of years, the volcanic arc became inactive. Cooling, subsidence, and accumulation of sediments followed, continuing into the Triassic Period. An episode of rift-zone volcanism occurred during the middle part of the Triassic.

Rocks of the Wrangell Terrane reflect this complex geologic history. The basal rock unit consists of Upper Paleozoic volcanic breccias, lava flows, and associated clastic sedimentary rocks. As plate movements transported the arc toward North America and volcanic activity ceased, the rocks slowly cooled, became more dense, and subsided. This subsidence caused the older rocks to be covered first by shallow and then progressively deeper marine waters in which were deposited a sequence of limestones, shales, argillites, and chert beds. By mid-Triassic time, the rifting of the Wrangell Terrane produced thousands of meters of pillow basalts and lava flows, which built up into an oceanic platform and later an island. In time, as volcanic activity ceased, a second sequence of cooling, subsidence and sediment accumulation occurred on Wrangellia.

4. Mid-Mesozoic to early Cenozoic accretion of exotic terranes to western North America.

Absolute dates for accretion of the several exotic terranes in the Wrangell-St. Elias area have not yet been determined; however, the chronological sequence can be interpreted. The first microplate to accrete must have been the Yukon-Tanana because it adjoins the "original" North American continent. This "docking" of the Yukon-Tanana Terrane was followed by accretion of the Wrangell, Alexander, and Chugach Terranes. Actually there is good evidence that these three microplates collided and joined together as a composite terrane long before they accreted to North America as a single large unit early in the Cenozoic Era.

As the microplates moved together and then collided with North America, there was a considerable amount of volcanic and tectonic activity. Evidence of submarine volcanism is preserved in Jurassic pillow basalts of the Chugach Terrane, whereas andesite flows indicate Eocene terrestrial volcanism on the Wrangell Terrane.

The accretion of terranes onto the North American continent, which occurred over tens of millions of years, brought about a great deal of uplift, folding, faulting, and metamorphism of rock units in the region. This tectonic activity produced the basic structures of the several mountain ranges in the region.

5. Late Cenozoic and ongoing accretion of the Yakutat Terrane.

Details of the origin and early history of the Yakutat Terrane are not known. The terrane is composed of both continental and oceanic rocks, some as old as late Mesozoic, and it originated thousands of kilometers from its current position (fig. 32.9).

Collision of the Yakutat block with North America began in Pliocene time and is still going on. The Yakutat Terrane occurs in the southern portion of Wrangell-St. Elias National Park and Preserve. Forces generated by its accretion are responsible for uplift of the Chugach and St. Elias Ranges to their present elevations.

The Yakutat Terrane is still actively accreting. This is one of the reasons why there are so many earth-

quakes in the region, including several very strong earthquakes in the last century.

6. Quaternary volcanism.

Some of the most impressive peaks in the park/preserve owe their present forms to volcanic activity in the Quaternary Period. Within the St. Elias Range, Mt. Bona is the highest peak built up by Quaternary volcanism. Great outpourings of andesite lavas in the Wrangell Mountains produced Mt. Sanford, Mt.

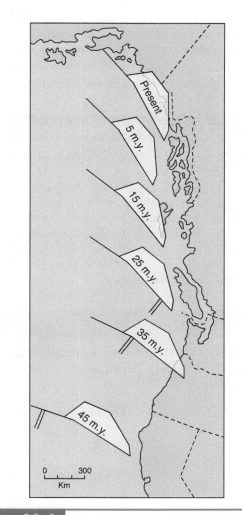

FIGURE 32.9   Travel path of the Yakutat block with respect to North America. The double lines indicate the approximate position of a spreading center; the single line extending westward from the Yakutat block is a magnetic anomaly interpreted as a plate boundary. The viewer should keep in mind that both North America and the Yakutat block were and are moving. Modified from Burns 1983.

Blackburn, and Mt. Drum. Mt. Wrangell, which also formed from Quaternary volcanism, has the distinction of being the only active volcano in the park/preserve.

7. Late Cenozoic glaciation.

Since late in the Tertiary Period there have been several major episodes of glacial advance and retreat in the Wrangell-St. Elias region. Most evidence of earlier glaciations was destroyed by subsequent glacial advances, so a detailed history has not been worked out.

Rates of glacial advance and retreat in this region are rapid. Historic records document glacial recession of tens of kilometers within several decades. The residents of Yakutat understand the possibilities for rapid glacial advance, and this causes them to be quite concerned about how Hubbard Glacier might affect their future.

Geologic Map and Cross Section

Beikman, Helen M. 1980. Geologic Map of Alaska, U.S. Geological Survey.

Howell, D.G., and others. 1985. Preliminary Tectono-stratigraphic Terrane Map of the Circumpacific Region. American Association of Petroleum Geologists, Tulsa.

Silberling, N.J. and Jones, D. L. 1984. Lithotectonic Terrane Map of Alaska (West of the 141st Meridian), U.S. Geological Survey Open-File Report 84-523 Part A, map and 12-page explanation.

Sources

Burns, Terry R. 1983. Model for the Origin of the Yakutat Block, an Accreting Terrane in the Northern Gulf of Alaska, *Geology* 11: 718–721.

Csejety, Bela, Jr., and others, 1982, The Cenozoic Denali Fault System and the Cretaceous Accretionary Development of Southern Alaska. *Journal of Geophysical Research* 87:(135) 3741–3754.

Eliot, John L. 1987. Glaciers on the Move. *National Geographic* 171:104–119.

Janson, Lone E. 1975. *The Copper Spike*. Edmonds, WA: Alaska Northwest Publishing Co. 175 p.

Jones, David L., and others. 1982. The Growth of Western North America, *Scientific American* 247:(5) 70–84.

———. 1982. Character, Distribution, and Tectonic Significance of Accretionary Terranes in the Central Alaska Range, *Journal of Geophysical Research* 87:(135) 3709–3717.

Motyka, R.J., and Benson, C.S. 1983. Fluctuations of Heat Flow from the North Crater, Mt. Wrangell, Alaska, EOS, 64:(9) 90.

Plafker, G. and Berg, H.C. eds. 1994. *The Geology of Alaska*. Boulder, Colorado: Geological Society of America, The Geology of North America, v. G-1, 1066 p.

Richter, D., Rosenkrans, D., and Steigerwald, M. 1995. Guide to Volcanoes of the Western Wrangell Mountains. U.S. Geological Survey Bulletin 2072, 46 p.

Sharp, R.P. 1942. Soil Structures in the Saint Elias Range, Yukon Territory. *Journal of Geomorphology* V:(4) 274–301.

Wrangell-St. Elias, International Mountain Wilderness, 1981. *Alaska Geographic* 8:(1) 144 p.

Note: To convert English measurements to metric, go to www.helpwithdiy.com/metric_conversion_calculator.html

Kenai Fjords National Park

SOUTH CENTRAL ALASKA

Donald S. Follows   Interpretive Geologist, Anchorage, AK

Area: 669,541 acres; 1046 square miles

Proclaimed a National Monument: December 1, 1978

Established as a National Park: December 2, 1980

Address: P.O. Box 1727, Seward, AK 99664

Phone: 907-224-3175

E-mail: KEFJ_Superintendent@nps.gov

Website: www.nps.gov/kefj/index.htm

The geologic story of Kenai Fjords National Park is centered in the ice-capped southern Kenai Mountains, where dynamic interplay between a mile-high icefield with outward radiating glacial tongues and the fjords below creates spectacular scenery sculptured by earthquakes, ocean and ice.

FIGURE 33.1   Harding Icefield is the main source of ice for the glaciers in Kenai Fjords National Park. Snow-covered granite nunataks jut up through the icefield surface. National Park Service photo taken by Keith Trexler.

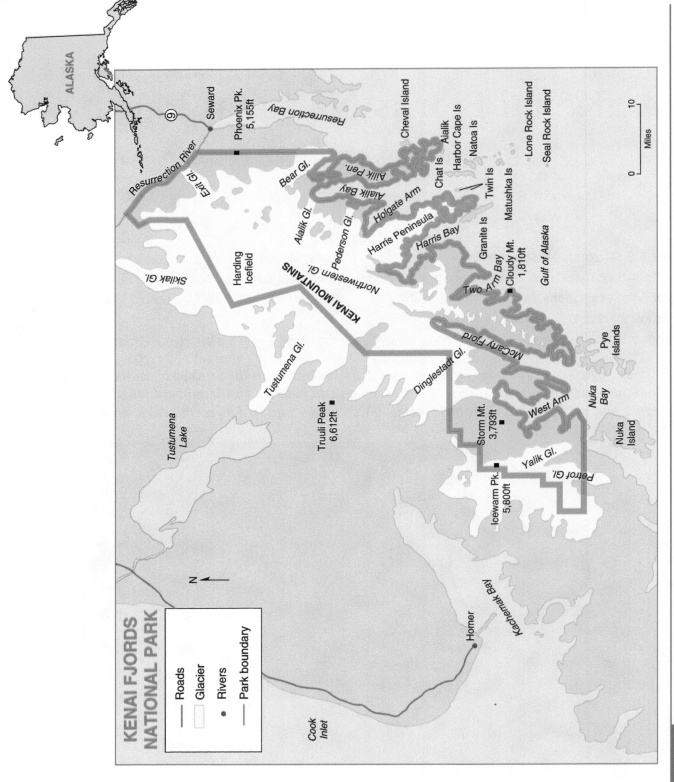

Figure 33.2 Kenai Fjords National Park, Alaska.

Geographic Setting

Kenai Fjords National Park was established by Congress in 1980 to recognize the 300-square-mile Harding Icefield, with its associated glacial system of at least 38 glaciers, and to protect representative wildlife populations, onshore and offshore, that are attracted to emerging and deglaciated land. The Kenai National Wildfire Refuge complements the national park ecosystem along its west and northwest boundaries.

The major portion of the national park is located in the mountains above the wild south coast of the Kenai Peninsula. The area known as "Kenai Fjords" includes an indented, drowned coastline from approximately Port Dick on the west to Resurrection Bay near the port of Seward. Part of this coast is in the national park, part is in the Alaska National Maritime Refuge, and a small part is privately owned. The state of Alaska also owns some of the coastal lands. Although only 80 airline miles from Anchorage, the park is isolated and undeveloped.

Kenai Fjords National Park celebrates a true "geoecosystem" where living and nonliving components of the environment respond to the gradual retreat of the glaciers. Each year tens of thousands of breeding seabirds and marine mammals colonize embayments and offshore islands to feed from nutrient-rich ocean waters upwelling from the continental shelf along the Kenai Fjords coast. Fractured granite islands, steep cliffs, and sea stacks create ideal habitats for colonial-nesting seabirds. Steller sea lions haul out on wide rock benches at the tideline, while harbor seals doze on drifting ice floes. Dall porpoise, sea otters, and humpback, minke, and killer whales ply the productive waters of the fjords.* Ling cod, black bass, halibut, salmon, red snapper, shrimp, and crab are harvested by both commercial and recreational fishermen.

Farther back in the fjords, bald eagles soar over black bear, moose, coyote, and mountain goat foraging thinly vegetated slopes and coves between the icefield and the ocean shore. River otter play along the short, swift streams that connect ocean to mountainside. Dominating the large embayments are seven tidewater glaciers that calve into the ocean fjords.

Although glaciers are retreating, the climatic conditions that create a dumping ground high in the moun-

---

*The spelling of "fjord" as a common noun as well as a proper noun in this chapter conforms to preferred usage in Kenai Fjords National Park.

tains for massive snowfalls have changed little since the Ice Age. As relatively warm air masses rise from the ocean and cool over the Kenai Mountains, the moisture falls as rain and snow. Average precipitation is estimated at 200 inches per year, including 500 to 1000 inches of snow before compaction. Two to four times as much precipitation falls as snow on the Kenai Mountains than as rain along the fjord coast.

On the Kenai coast, the ocean is seldom calm. Storm waves and heavy swells beat against the sea stacks, arches, and cliffs. Safe landings for boats are few on this outer fringe. Summer temperatures range from 40° F to 60° F, but freezing weather occurs on the icefield in even the warmest months.

The Harding Icefield rests atop the southern half of the Kenai Mountains, the range that forms the backbone of the Kenai Peninsula. Rising 3000 to 6000 feet above sea level, the Kenai Mountains are lithologically and structurally part of the Kenai-Chugach Mountains, an arcuate coastal range that bends eastward at the top of Kenai Peninsula and stretches alongside the Gulf of Alaska. To the southwest, many miles from the Kenai Peninsula, the same mountains rise from the ocean to form the spine of Kodiak Island.

Changes in relief are abrupt in the southern Kenai coastal area where elevations of a few feet are close (horizontally) to features 5000 feet above sea level. In McCarty Fjord, a 6300-foot peak is less than two miles from tidewater, while the ridge along the fjord is nearly 5000 feet in elevation. Farther north and west, the 6612-foot summit of Truuli Peak ranks as the highest point on the entire Kenai Peninsula. ("Truuli" was the Kenatzie Indian name for the Kenai Mountains.)

As glaciers continue to retreat, uncovering the higher ridges and summits, the rectangular, blocklike appearance of the Kenai Mountains becomes more evident; but over half the park's area is still covered by glaciers. The thickness of the icefield over ancient preglacial stream valleys may be more than a thousand feet. When Pleistocene ice began to accumulate, the glaciers followed the previously eroded stream courses, scouring the mountain valleys into U-shaped troughs and lowering passes across the trend of the range.

The geoecosystem of Kenai Fjords National Park extends several miles beyond its present shoreline out onto the continental shelf. The 300-foot depth contour on the ocean bottom seems to correlate with glacial features and deposits left during the advance of valley glaciers before the Holocene rise in sea level (and/or coastal subsidence) that flooded the troughs.

FIGURE 33.3A The Harding Icefield feeds the Holgate Glacier, an active tidewater glacier that ends in an ice cliff from which small icebergs break off. The bluish color seen in the ice is due to the ice's density caused by most of the air having been forced out of the ice. Photo by Ann G. Harris.

FIGURE 33.3B From a distance, the Holgate Glacier does not seem very large, but a photograph taken with a ship in front of the glacier gives a better idea of its size. Photo by Ann G. Harris.

Geological investigations of the southern Kenai Mountains began early in this century, but the lack of major economic discoveries, coupled with the ruggedness and extreme remoteness of the outer coast, soon terminated general interest. In 1909, when geologists first explored the south coast, it was virtually uninhabited from English Bay to Seward and has remained that way until the present day, except for a few hardy gold miners in Nuka Bay. During a National Park Service field reconnaissance of some 600 nautical miles of south coast shoreline in 1976, the investigators did not encounter a single inhabitant.

Although Captain Cook in 1778 and Captain Vancouver in 1794 sailed past these shores, the rugged, uncharted coast, rough weather, and treacherous ice floes kept the explorers from putting in. Later on, native and Russian fur traders regularly used Portage Pass at the northern terminus of the Kenai Mountains for access to and from Prince William Sound. The portage avoided the long, difficult passage around the wild south coast of the Kenai Peninsula.

The park's interior was even less well explored than the coast. It is believed that the first party to cross the entire Harding Icefield did so in 1968 after climbing Truuli Peak at the southwest edge and dropping off the icefield at Exit Glacier in the northeast corner.

In 1970, a local entrepreneur set up a flightseeing and snowmobile tour on Harding Icefield, complete with an overnight cabin in which fly-in visitors could take shelter. While the interest level was high and tourists were attracted from all over the world, the continual burial of the cabin by snow made trips increasingly difficult. Because the Harding Icefield receives much more snow than is melted during the summer months, it became impossible to maintain the cabin on each season's new snow layer. After the second winter, the cabin was buried too deeply to be dug out again and had to be abandoned.

Geologic History and Features

1. Oldest rocks, Mesozoic in age.

Great gaps exist in the geologic record, making the regional sequence of events difficult to discern. No Paleozoic rocks have been identified in the park. The oldest rocks in the region crop out around Nuka Bay and also northwest of the park in a highly deformed mélange belt that parallels the Border Ranges fault system along the western Kenai-Chugach Mountains. (A *mélange* is a

mappable body of rock characterized by a lack of internal continuity of contacts or beds.) Included in the mixture are blocks and fragments of all sizes, both exotic and native, embedded in a fragmented matrix of finer-grained material. Known as the McHugh Complex, these rocks are believed to be Triassic to Cretaceous in age. Blocks of white limestone in the McHugh Complex that are exposed along the east wall of Petrof Glacier contain fossils found elsewhere only in Afghanistan and China. Their presence reveals the magnitude of plate motions that formed the McHugh Complex (Bradley and Donlevy, 1995).

2. The flysch rocks of the Valdez Group.

A *flysch* is a thinly bedded sequence of marine sedimentary rocks consisting of sandy shales, marls, muds, conglomerates, graywackes, etc. Upper Cretaceous flysch rocks of the Valdez Group are the dominant rock series in the park. Their lack of fossils has for years challenged geologists trying to date the sequence. The bulk of the Kenai Mountains is composed of this great series of dark-colored, metamorphosed, marine sediments that consist generally of a monotonous series of rhythmically interbedded sandstones, siltstones, and minor mudstones and pebble conglomerates that have been intensely folded and regionally metamorphosed. The sequence forms a flysch (turbidite) unit several thousand feet thick (fig. 33.4).

The distribution of rock types in the region is broadly parallel to the major faults that border the southern Kenai Mountains to the east and west. The Border Ranges fault, lying west of Kenai Fjords National Park, marks the topographic break between mountains and lowlands and is believed to represent a "seam" joining together distinctly different bodies of rock of regional size, called *terranes* (tectonostratigraphic terranes). In south central and southeastern Alaska the terranes that lie in a great bend along the Gulf of Alaska, are considered *exotic terranes;* that is, the rocks formed in another location far to the southwest and were subsequently "rafted" to their present site, possibly late in Cretaceous and early Tertiary time. The McHugh Complex and the Valdez Group are part of a belt of rocks that extends from southeast Alaska for hundreds of miles along the Gulf of Alaska coast to the Shumagin Islands in the Aleutians.

3. Accretion and tectonic activity early in Tertiary time.

To the east of the park, in Prince William Sound, rocks of the Orca Group were deformed and accreted in

Paleogene time along the south side of the Valdez Group. Both the Valdez and Orca Groups were partially subducted and underwent severe regional deformation. In some areas, intense folding and faulting shortened and compressed the rock section by as much as 50 percent or more!

Recent petrographic studies of the Orca and Valdez rock sequences across Prince William Sound suggest that there is no terrane boundary or break between these two groups that separates discrete packages of rocks with distinct geologic histories. Instead, the Orca and Valdez probably form a single package derived from a common source during Late Cretaceous into Eocene time (Dumoulin, 1987).

4. Tertiary uplift, erosion, and intrusions.

By the beginning of Tertiary time, practically all of what is now Alaska had emerged from the sea. During Eocene time, vast quantities of sand and mud were removed by streams from highlands bordering Cook Inlet and the Susitna lowlands west of the park and from the rising Kenai Mountains as well. Small plutons, dikes, and the Harding Icefield granitic batholith, all Eocene in age, were emplaced in the deformed rocks of the southern Kenai Mountains. Contacts are sharp and steeply dipping where the batholith intruded slate and sandstone beds of the Valdez Group. Most of the batholith is buried beneath the Harding Icefield, but the granite crops out in

Table 33.1			Geologic Column, Kenai Fjords National Park	
Time Units				
Era	**Period**	**Epoch**	**Rock Units**	**Geologic Events**
Cenozoic	Quaternary	Holocene	Alluvium, glacial drift	Coastal subsidence; subduction Earthquakes, sporadic faulting At least two glacial advances
		Pleistocene	Wisconsinan drift	Wisconsinan and earlier glaciations; Cook Inlet and Prince William Sound covered by ice sheets; erosion and scouring
	Tertiary	Pliocene		Uplift of Kenai Mountains; increased erosion
		Miocene		Earthquakes Accretion of Yakutat block (eastern Gulf of Alaska) Uplift, erosion
		Oligocene	Granitic intrusives	Emplacement of Harding Icefield batholith Subduction; regional metamorphism
		Eocene		Accretion of composite Chugach-Prince William Terrane (central Alaska)
		Paleocene	Orca Group (dark gray flysch rocks, turbidites)	Thrust faulting Strike-slip faulting
Mesozoic	Upper Cretaceous		Valdez Group (metasedimentary flysch—sandstone, siltstone, conglomerate; major unit in park) ?	
	Jurassic		McHugh Complex (mélange of siltstone, sandstone, conglomerate, tuff, pillow basalt, chert, limestone, argillite)	Deformation along subduction zone
	Triassic			

Sources: Plafker et al., 1977; Nilsen, 1981; Dumoulin, 1987; Bradley and Donlevy, 1995.

FIGURE 33.4   Much of the rugged coast of the Kenai Peninsula consists of slates and graywackes of the Valdez Group. Photo by Ann G. Harris.

cirques, glacial troughs, and in the isolated peaks jutting up through the ice, called *nunataks* (fig. 33.1). The Aialik and Harris Peninsulas in the park are part of the batholith, as are the offshore islands, except for Nuka Island just outside the park's southwest corner. The Pye Islands, across Nuka Bay from Nuka Island, are a southern extension of the batholith. (Nuka Island, the Pye Islands, and several other islands offshore from Kenai Fjords are in the Alaska National Maritime Refuge, south of the national park.)

A zone of thermally altered granitic rocks that weather to a reddish-brown color is exposed in the inner part of Northwestern Fjord, near the head of Harris Bay. Also associated with the batholith are felsic dikes composed of plagioclase feldspar with minor amounts of muscovite, quartz, and orthoclase. Where these light-colored rocks intrude the dark metamorphic rocks of the Valdez Group, a pattern of black-and-white stripes is formed on cliff walls.

The granite islands (Cheval, Chat, Chiswell, Lone Rock, Seal Rocks, Matushka, Granite, etc.) offshore from the park and in the Alaska Maritime National Wildlife Refuge are hauling and nesting places for marine mammals and seabirds. During extensive wildlife surveys, National Park Service teams have found that coastal seabirds have an affinity for granitic rocks, perhaps because the fractured nature of these rocks allows underground nesting sites. However, exposures of pillow basalt on Cape Resurrection provide rookery sites for a large population of kittiwakes. By contrast, Nuka Island, composed of a large outcrop of graywacke, is relatively lacking in seabird colonies.

From Eocene through Pliocene time, no sedimentary record was left in the park region, although erosive processes continued to carry detritus from the uplands to the sea. At that time, the general distribution of mountain mass and lowlands was about the same as what can be seen today. The Kenai-Chugach Mountains had been raised to approximately their present elevations. However, the landscape did not look like the glaciated uplands we see in the Kenai Mountains today. Because of the relatively rapid uplift that occurred in Tertiary time, the preglacial streams cut steep, narrow slot valleys that followed weak planes and fractures in the rocks. The mountain range had a somewhat more massive, blocky appearance.

5. The onset of glaciation.

In most of Alaska, the Ice Age began earlier, was more intense, and lasted longer than in the lands to the south. As world temperatures cooled, the snowline on mountains crept down to lower elevations. Small glaciers expanded, grinding and gouging their way down valleys. Eventually they filled stream courses brim to brim and overtopped ridges.

The Kenai-Chugach Mountains became the nourishing grounds and source area for a multitude of alpine glaciers that gradually coalesced into one massive icecap. At the time of the glaciers' maximum extent, ice was over 3000 feet thick in many valleys. Only the highest peaks escaped the effects of glacial abrasion and scouring. Prince William Sound and Cook Inlet were entirely covered by expanding glacial systems.

Throughout the many glacial stages in Pleistocene time, the high Kenai Mountains were source and supply of glacial ice as storms continued to bring snow to the uplands. The great icefield, which then encompassed the entire backbone of the southern Kenai Mountains, remained relatively intact while its long glacial arms advanced and retreated through the valleys below, swelling or shrinking like a giant glacial octopus. These climatically controlled advances and retreats of the ice edge in the valleys and over the lowlands sculpted the landforms and magnificent scenery that we see today.

6. Naptowne Glaciation and Neoglacial time.

The Naptowne episode, the last major glaciation of southern Alaska, took place 14,000 to 10,000 years ago. The Harding Icefield expanded to the north and west, completely filling the Resurrection and Kenai River valleys. Glacial ice then flowed into the Kenai lowlands to merge with a large piedmont glacier occupying the western flank of the Kenai Mountains. The glaciofluvial deposits left as the ice melted back about 10,000 years ago provided a habitat of great diversity that today burgeons with thriving life forms of many kinds. This geoecosystem is now protected in the Kenai National Wildlife Refuge west of the national park.

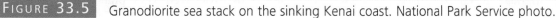

FIGURE 33.5 Granodiorite sea stack on the sinking Kenai coast. National Park Service photo.

FIGURE 33.6 Three-Hole Arch in Aialik Bay shows the effects of intense wave action. Beaches are lacking on this drowned coast. Photo by Ann G. Harris.

The advance of Neoglacial ice began about 4000 years ago and ended about the middle of the nineteenth century. Since that time glaciers in southern Alaska have retreated fairly rapidly. In Kenai Fjords National Park, rocks very recently uncovered by the melting back of ice are barren of vegetation; in other places land exposed a little longer to natural processes is slowly becoming habitable for new generations of plants and animals.

7. Ongoing tectonism and a look ahead.

The boundaries of Kenai Fjords National Park were drawn to include the icefield and its glaciated forelands, the deep-water fjords, drowned cirques, igneous intrusions, and an *active tectonic zone along an axis of*

maximum subsidence. Geologic evidence suggests that for at least the past 900 years and perhaps for as long as 1360 years, the coastal region within this tectonic zone has been sinking into the ocean. The axis of subsidence has been located roughly along the crest of the coastal mountains. During the 1964 Good Friday earthquake (which affected the entire south central coast of Alaska), tectonic movement dropped the Kenai Fjords coastline 7 to 9 vertical feet!

Tectonic movements associated with the 1964 quake were but one pulse in an ongoing trend of regional deformation that may have begun in late Pliocene time. More than half of Alaska is seismically active at present. Quaternary tectonism can be detected by the presence of faulted deposits, uplifted glacial terraces, offsets of stream courses, and altered marine terraces. As the Kenai coast is submerged by ongoing geologic processes, glacial cirques become half-moon bays; peaks that once topped coastal mountains are now islands, battered by storm waves (fig. 33.5).

Kenai Fjords National Park with its glaciers, deep fjords, rain forests, wildlife, and sinking coast is a place of subtle beauty as wild and mysterious as can be found in Alaska. For life forms, clinging to the raw and broken edges of an ice age, this park preserves an ever-challenging environment.

Geologic Map and Cross Section

Bradley, D. and Donlevy, T. 1995 (draft). Geologic Map of Kenai Fjords National Park and Vicinity. U.S. Geological Survey.

Sources

Capps, S.R., 1940. The Geology of the Alaska Railroad Region. U.S. Geological Survey Bulletin 907. p. 51–64.

Dumoulin, J.A. 1987. Sandstone Composition of the Valdez and Orca Groups, Prince William Sound, Alaska. U.S. Geological Survey Bulletin 1774.

Field, W.O., Editor. 1975. *Mountain Glaciers of the Northern Hemisphere,* v. 2. New York: American Geographical Society.

Follows, Donald S. 1978. Glacial Landscapes and Geology of the Harding Icefield Complex, Southern Kenai Mountains, Alaska, in Landscapes of Alaska Seminar. Alaska Resource Library, Anchorage.

Grant, U.S. and Higgins, D.F. 1913. Coastal Glaciers of Prince William Sound and Kenai Peninsula, Alaska. U.S. Geological Survey Bulletin 526. p. 52–56.

Jones, D.L., and Pessagnon, E.A. Jr. 1977. A Cretaceous Accretionary Flysch and Mélange Along the Gulf of Alaska Margin. U.S. Geological Survey in Alaska: Accomplishments during 1976. p. B1–B112.

Karlstrom, T.N.V., 1964. Quaternary Geology of the Kenai Lowland and Glacial History of the Cook Inlet Region, Alaska. U.S. Geological Survey Professional Paper 443. p. 56–57.

Lethcoe, J.R. 1989. *An Observer's Guide to the Geology of Prince William Sound.* Valdez, Alaska: Prince William Sound Books.

Miller, D.W. 1984. *A Guide to Alaska's Kenai Fjords.* Seward, Alaska: Wilderness Images.

———. 1986. The Kenai Fjords. Where Mountains Meet the Sea, Alaska's Gulf Coast. *Alaska Geographic*, v. 13, no. 1. p. 63–81

Nilsen, T.H. 1981. Accretion Model for the Cretaceous Chugach Terrane, Southern Alaska. U.S. Geological Survey in Alaska: 1981 programs. p. 93–97.

Péwé, T.L. 1975. *Quaternary Geology of Alaska.* U.S. Geological Survey Professional Paper 835. p. 31–34.

Plafker, G. 1969. *Tectonics of the March 27, 1964, Alaska Earthquake. U.S.* Geological Survey Professional Paper 543-1. p. I-1/I-74.

Rice, B. 1987. Changes in the Harding Icefield, Kenai Peninsula, Alaska Unpublished M.S. thesis, University of Alaska, Fairbanks.

Note: To convert English measurements to metric, go to www.helpwithdiy.com/metric_conversion_calculator.html

Denali National Park and Preserve

S O U T H   C E N T R A L   A L A S K A

Phillip Brease   Park Geologist, NPS

Area: 6,500,000 acres; 10,156 square miles

Established as a National Park: February 26, 1917

Separate Denali National Monument Proclaimed:
December 1, 1978

Both Incorporated into and Established as Denali National Park and Preserve: December 2, 1980

Designated a Biosphere Reserve: 1976

Address: P.O. Box 9, Denali Park, AK 99755

Phone: 907-683-2294

E-mail: denali_info@nps.gov

Website: www.nps.gov/dena/index.htm

Large glaciers, 20 to 30 miles long, descend more than 15,000 vertical feet from the high peaks of the Alaska Range to the flats below.

Mount McKinley, North America's highest mountain, which rises to 20,320 feet from its base at the 2000-foot level, has the greatest vertical relief of any mountain in the world.

Plate collisions, accretion of terranes, and subduction, pile up crustal material and push the McKinley massif higher. As the mountains rise, vast amounts of sediment, eroded by grinding ice and swift streams, choke valleys and spread over lowlands.

FIGURE 34.1   North side of Mount McKinley as seen from the Park Road, which skirts the base of the massif for some 90 miles between Denali National Park and Preserve Headquarters and Kantishna. Cotton grass (beside thaw lake) and other tundra plants thrive on glacial outwash and alluvium in areas of discontinuous permafrost. Photo by Douglas Price.

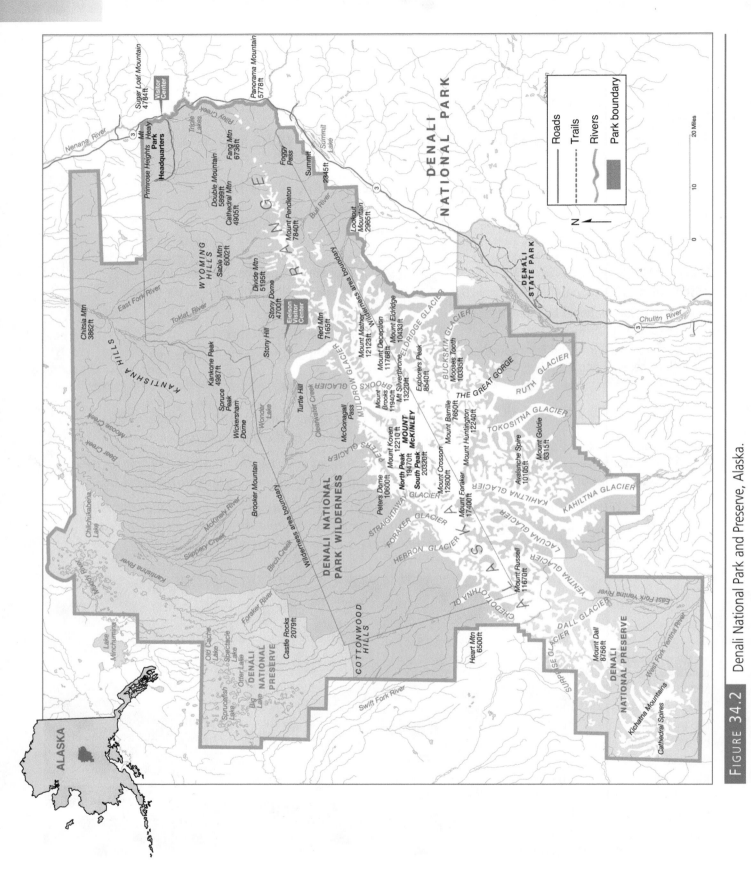

FIGURE 34.2   Denali National Park and Preserve, Alaska.

Geographic Setting

The Athabascan name for Mount McKinley is Denali, "the High One." (This is the name most Alaskans prefer, although McKinley remains the official name.) Although Mount McKinley's two peaks are often shrouded in clouds in summer, its great bulk dominates the landscape. When the clouds part, revealing the magnificent massif, the summit, South Peak (elevation 20,320 feet), towers over the lower North Peak (elevation 19,470) and over all the mountains of the Alaska Range.* On clear days, McKinley's majestic, snow-covered summit can be seen from ships in the Gulf of Alaska. From the north, its height is even more impressive. The steep north face of Mount McKinley rises 15,000 feet from its base. Everything in Denali seems to be on a grand scale (fig. 34.1).

The Alaska Range is a curved mountain chain about 600 miles long that forms a topographic barrier and drainage divide between the coastal lowlands around Cook Inlet and the Yukon lowlands of the Alaskan interior. Denali National Park and Preserve lies like a saddle across the central and highest part of the range. In the massif section, where the Alaska Range is about 60 miles wide, the mountains trend in a generally northeast-southwest alignment. Nearly 20 peaks in this part of the range are over 10,000 feet in elevation; elsewhere in the Alaska Range, the mountains are mostly between 7000 and 9000 feet high. Mount McKinley and its lofty, white-crowned companions in the park/preserve—Mounts Mather (12,050 feet), Silverthrone (13,169 feet), Crosson (12,772 feet), Hunter (14,580 feet), Foraker (17,395 feet), and Russell (11,676 feet)—are challenged only by the highest peaks in Wrangell-St. Elias National Park and Preserve (Chapter 32), some 200 miles (air distance) to the east.

The lands added in 1980 to create Denali National Park and Preserve surround the original McKinley National Park on three sides and ensure the protection of the entire massif, its glaciers, and its ecosystem. The Kichatna Mountains on the southern flanks of the mas-

sif, including the geologically unique Cathedral Spires, are now in the park/preserve. Eight major glacial troughs radiate from Cathedral Spires headwalls that rise nearly a mile from their bases. The historic Kantishna mining district on the northwest side of the original park has also been placed under National Park Service protection along with migration routes and wintering and calving areas of caribou.

Mount McKinley was named in 1896 by a gold prospector in honor of the 25th President of the United States, who was elected in that year. In 1898, George H. Eldridge, a geologist, and Robert Muldrow, a topographer, were sent by the U.S. Geological Survey to explore the southern approaches to the range. Eldridge and Muldrow determined the map location and calculated the elevation of Mount McKinley with an accuracy close to currently accepted figures, which was an amazing achievement considering the difficulties of mapping a region without (at that time) topographic control points. The two men also plotted the route for the Alaska Railroad, built 25 years later from Seward to Fairbanks. The railroad, which skirts the eastern boundary of the park/preserve, is a principal means of access to the Denali area.

Lt. Joseph S. Herron led a U.S. Army expedition in 1899 that was the first to cross the Alaska Range from south to north. They discovered and named Mount Foraker, the second highest peak of the Alaska Range, the mountain that Native Americans call "Denali's Wife."

The first expedition to reach the base of Mount McKinley was led by Alfred H. Brooks, a pioneer geologist who took a U.S. Geological Survey party to the northern side of the Alaska Range in 1902. Brooks climbed a shoulder of Mount McKinley but was forced to turn back at about the 7500-foot level because of icy slopes and steep cliffs. He placed a note in a cartridge shell in a rock cairn at the spot. The note and shell were recovered in 1954 by geologists of the U.S. Geological Survey and are now on display in the Park Museum.

The scaling of Mount McKinley, to reach either of the two peaks (two miles apart with a difference in elevation of more than 800 feet), is a hazardous undertaking that requires many hours of continuous ice climbing. Less rock is exposed and the snow and ice are thicker and more treacherous than on most other high mountains. Only experienced climbers who have National Park Service permission may attempt an ascent. Climbing back down the mountain is equally dangerous.

---

*University of Alaska-Anchorage scientists, using gravity meters and satellites, have determined Mount McKinley's elevation as 20,306. This figure is 14 feet lower than the elevation of 20,320 feet calculated by Bradford Washburn in 1954. The USGS accepted the elevation of 20,320 feet as correct.

Weather on the mountain is unpredictable. On a clear, bright day a blizzard may come up within 15 minutes, or dense fog and freezing mists may suddenly overtake climbers. Temperatures as low as –90°F and winds as high as 100 miles per hour have been recorded. "Normal" temperatures are around 0° F, with winds 25 to 30 miles per hour. Even in July, the best month to attempt a summit climb, conditions are often perilous. A website displaying hourly temperatures, and wind speed and direction can be found at www.denali.gi.alaska.edu.

In 1903 Judge James Wickersham and a party of four from Fairbanks made the first attempt to reach the top of Mount McKinley. The route they chose was by way of Peters Glacier to the north face and a sheer escarpment, called Wickersham Wall, which is almost impossible to climb. The highest elevation they reached was 10,000 feet. A map of the Kantishna Hills that they filed with a mining claim after their return was the impetus for a gold rush. Rich placer diggings were discovered in the Kantishna mining district. Later in life, when he was the territorial delegate to Congress, Judge Wickersham introduced the bill that provided for the establishment of Mount McKinley National Park.

In 1910, four miners in the Kantishna district, calling themselves the Sourdough Party, boasted that they could get the job done. Having prospected in the foothills, the miners knew the approaches to McKinley well and had already decided that the Muldrow Glacier on the north side offered the best access to the summit. They discovered McGonagall Pass, leading to the upper Muldrow Glacier, and established a base camp on a ridge at 11,000 feet. Two miners reached the top of North Peak, believing it to be the higher of the two summits. Once there, they realized their mistake, but planted a spruce pole to mark their achievement. They descended North Peak to the saddle, and started toward South Peak but turned back because of bad weather.

The Muldrow Glacier route the Sourdoughs used is still a principal route by which climbers approach Mount McKinley's summit. The West Buttress route, which is shorter and safer and ascends by way of Kahiltna Glacier, was pioneered in 1951 by the explorer-scientist Bradford Washburn, who led a number of expeditions into the high areas of the McKinley massif. In 1947, his wife, Barbara Washburn, became the first woman to climb to Mount McKinley's summit. The Stuck-Karstens Expedition, a party of four men, achieved the honor in 1913 of making the first success-

ful ascent of Mount McKinley's South Peak. Harry Karstens later became the first superintendent of Mount McKinley National Park in 1917.

Visitors can reach Denali National Park and Preserve via the Alaska Railroad, by Denali Highway (#8), George Parks Highway (#3), and by private plane and charter aircraft (fig. 34.3). Private vehicles are not allowed on the park roads (except to get to the campgrounds), but shuttle buses operate daily during the summer season. During the winter, visitor facilities are closed, but the park/preserve is open to skiers and dogsledders. Winter seasons, drier and less cloudy than summer months, last for about half the year, but snow

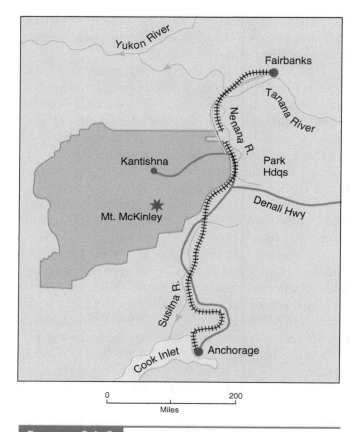

FIGURE 34.3 Sketch map showing access to the park/preserve and principal drainage ways. The Alaska Railroad and the paved Anchorage-Fairbanks highway follow essentially the same route and are open year-round. The Denali highway, approaching from Tok Junction (southeast), is closed in winter. The Alaska Range forms a drainage divide between north-flowing streams to the Nenana and Yukon Rivers and south-flowing streams that join the Susitna River.

may fall in any month. Rivers are frozen from late October to late April. Summer temperatures may range from 40° F to 80° F. January has less than four hours of light each day, but in June, skies may be brilliantly colored at midnight.

Geologic Features

Denali National Park and Preserve consists of three rock provinces that occupy east-west bands through the park, and generally represent the oldest to youngest rocks from north to south (see geologic map, fig. 34.4). The largest province takes up most of the northern half of the park, where the oldest, most highly altered marine and volcanic rocks underlie smaller pods or veneers of Quaternary and Tertiary sediments. A smaller group of slightly younger, and slightly less altered, marine sediments can be found along the Alaska Range crest, where in places, they are pierced and/or covered by much younger granitic and volcanic rocks. Younger still are the majority of the rocks in the southern third of the park, where the great Mount McKinley massif, and other plutonic rocks, intrude shallow marine sediments.

These rock provinces which to some extent represent *tectonostratigraphic terranes* (fault-bound stratigraphic packages with different depositional histories from adjacent packages), are separated by the Denali fault system. This is a series of major crustal fractures that arc through most of the southern third of the state. The fractures extend from Canada northwesterly into Alaska, then southwesterly out into Bristol Bay and the Bering Sea (see fig. 34.5). A portion of the Hines Creek strand of the fault system is located nearly along the park road for the first 30 miles of its course from the park entrance to the Teklanika River area. It separates the oldest Yukon-Tanana terrane from Mesozoic Pingston/McKinley terranes, and bisects portions of the Paleozoic Farewell terrane. The McKinley strand of the fault system crosses the main north-south highway between Fairbanks and Anchorage just north of the town of Cantwell, and continues westerly into the park on the south side of the Alaska Range some 50 miles to Anderson Pass, where it then crosses the range to the north side to continue to arc around Mount McKinley and on to the southwestern regions of the park and beyond. The McKinley strand separates the Farewell terrane from the Late Mesozoic-Early Tertiary Kahiltna terrane in the eastern portion of the park, while it splits the Farewell in the western parts of the park (fig. 34.6).

The park, as is the case for most of Alaska, has a long history of collision tectonics and accretionary terrane accumulation, which has resulted in nearly continuous orogeny and varying conditions of alteration (metamorphism) to nearly all rocks. In general, rock units have, in places, been folded, faulted, and altered to the *greenschist facies* (rocks of low temperature, low pressure metamorphism), often resulting in homogeneous lithologies, as well as effectively blending faults and other contacts into oblivion. This alteration condition is further complicated by the intrusion of (and contact alteration by) some of the largest igneous plutons in the world, such as that of the Mount McKinley massif. Additionally, Denali National Park geology is only mapped at a reconnaissance level because of the vast size of the Alaskan state, and because detailed mapping, often driven by mineral or oil resource concerns, is limited in national parks, where economic mineral or energy exploration is not encouraged.

Yukon-Tanana Terrane

Visitors using ground transportation (car, bus, or train) are most likely to first set foot on or see the oldest rocks in the park (called *basement rocks* because they lie beneath all others) when they disembark from the train station or stop at the visitors' center. The marbled rusty brown, buff to gray and white rocks underlying the park entrance area are referred to as the *Yukon-Tanana Crystalline* rocks. These are a series of metamorphic (fig. 34.6) rocks, including quartz mica schist, quartzite, phyllite, slate, marble and greenstone, which originated as shallow ocean sediments ranging from 400 million to one billion or more years ago. At least four periods of regional metamorphism and *tectonism* (folding and faulting) have thoroughly altered this sedimentary and volcanic basement rock to a greenschist facies.

The Yukon-Tanana crystalline rocks are found throughout most of the Alaskan interior, extending as far north as Fairbanks, and easterly into the Yukon Territory of Canada. In the park, these rocks are exposed in an east-west band along the northern boundary of the park, but are best seen near the Savage River, just beyond park headquarters, the Kantishna Hills at the west end of the park road and in the rugged Nenana Canyon just outside the park entrance (see fig. 34.7).

The Geology in the Area of Denali National Park and Preserve

Adapted by P. Brease, 2002, from: Geologic Map of Alaska — Helen Beikman, 1980
Albers equal area projection

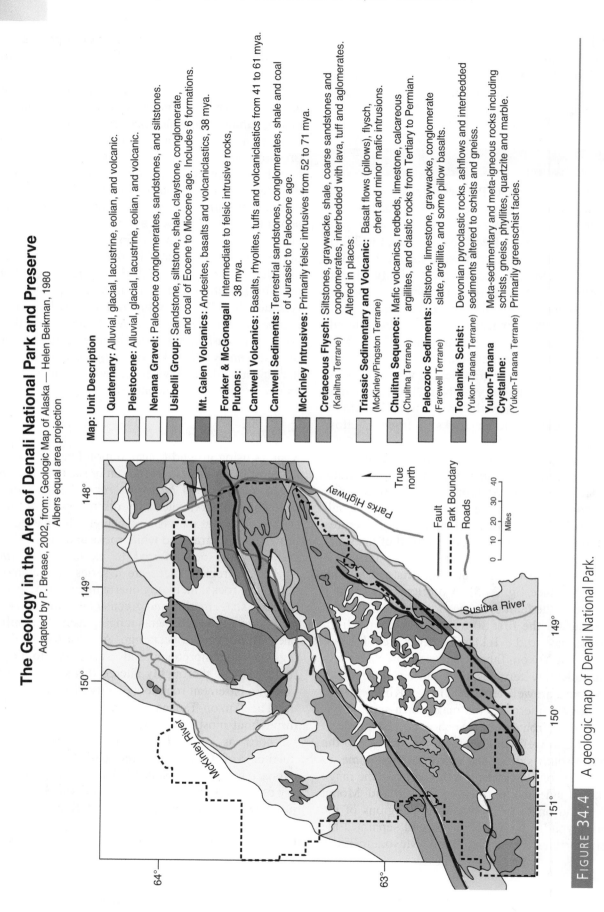

Map: Unit Description

Quaternary: Alluvial, glacial, lacustrine, eolian, and volcanic.

Pleistocene: Alluvial, glacial, lacustrine, eolian, and volcanic.

Nenana Gravel: Paleocene conglomerates, sandstones, and siltstones.

Usibelli Group: Sandstone, siltstone, shale, claystone, conglomerate, and coal of Eocene to Miocene age. Includes 6 formations.

Mt. Galen Volcanics: Andesites, basalts and volcaniclastics, 38 mya.

Foraker & McGonagall Intermediate to felsic intrusive rocks,
Plutons: 38 mya.

Cantwell Volcanics: Basalts, rhyolites, tuffs and volcaniclastics from 41 to 61 mya.

Cantwell Sediments: Terrestrial sandstones, conglomerates, shale and coal of Jurassic to Paleocene age.

McKinley Intrusives: Primarily felsic intrusives from 52 to 71 mya.

Cretaceous Flysch: Siltstones, graywacke, shale, coarse sandstones and
(Kahiltna Terrane) conglomerates, interbedded with lava, tuff and aglomerates. Altered in places.

Triassic Sedimentary and Volcanic: Basalt flows (pillows), flysch,
(McKinley/Pingston Terrane) chert and minor mafic intrusions.

Chulitna Sequence: Mafic volcanics, redbeds, limestone, calcareous
(Chulitna Terrane) argillites, and clastic rocks from Tertiary to Permian.

Paleozoic Sediments: Siltstone, limestone, graywacke, conglomerate
(Farewell Terrane) slate, argillite, and some pillow basalts.

Totalanika Schist: Devonian pyroclastic rocks, ashflows and interbedded
(Yukon-Tanana Terrane) sediments altered to schists and gneiss.

Yukon-Tanana Meta-sedimentary and meta-igneous rocks including
Crystalline: schists, gneiss, phyllites, quartzite and marble.
(Yukon-Tanana Terrane) Primarily greenschist facies.

True north

Parks Highway

Susitna River

McKinley River

Fault
Park Boundary
Roads

0 10 20 30 40
Miles

FIGURE 34.4   A geologic map of Denali National Park.

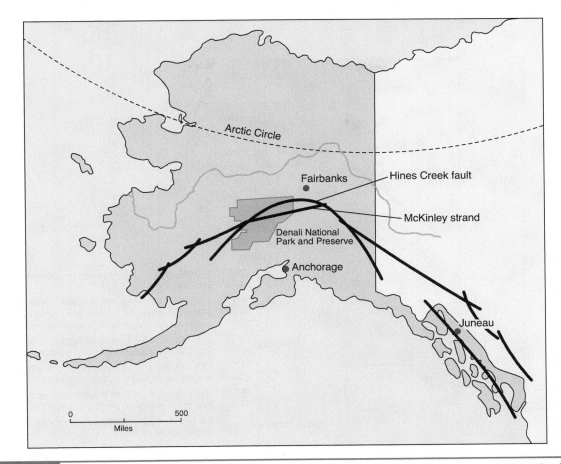

FIGURE 34.5 The Denali fault system arches across southern Alaska. Principal movement along the fault system has been horizontal, with the south side moving west and northwest relative to the north and northeast side. Vertical movement has also occurred on several fault segments. Both types of movement still go on.

Two rock units of similar type and appearance, but of slightly younger age, are the Totatlanika Schist and the Wyoming Hills Sequence. The Totatlanika Schist, made up of schists, gneiss, phyllites, marbles, and metavolcanic rocks, is located in patches along the northern boundary, and intermittently as portions of the buttress foothills on the northwest side of the park. The Wyoming Hills Sequence, a group of quartzites, phyllites, marbles, chert, and limestone, are found a few miles north of the park road between the Teklanika and Toklat Rivers. The Totalanika consists of substantial amounts of submarine volcanic ash flows which, interbedded with marine sediments, have been altered to greenschist facies, and are described as thrust sheets, often overlying the Yukon-Tanana Crystalline rocks. All three rock units are lumped into the Yukon-Tanana terrane, due to a long, similar history of alteration, are similar in appearance to the Yukon-Tanana Crystalline rocks.

Farewell Terrane

Although the dominantly granitic Mount McKinley steals the show in the Alaska Range, many of the high mountain peaks within the park are of Paleozoic (250–500 mya) marine origin, and come from a wide range of depositional environments, from deep ocean basins to slope and shelf or nearshore platforms. These rocks were formerly subdivided into smaller terranes, including a principally basinal sequence (Dillinger terrane), platform facies (Nixon Fork terrane), and a slope sequence (Mystic terrane). Most researchers have now lumped them into the Farewell terrane (fig. 34.6).

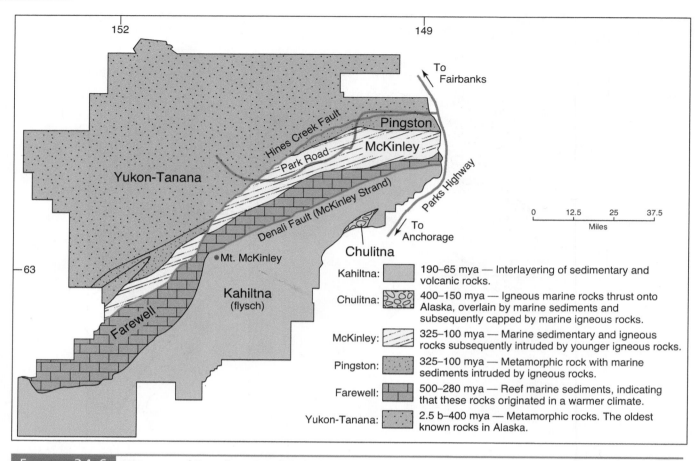

FIGURE 34.6   Terranes of Denali National Park.

The oldest and generally lowest Dillinger sequence, consists of limestones, mudstones, shales, and chert, which range in age from Silurian to Early Ordovician (about 410–500 mya), as determined by poorly preserved fossil *graptolites* (extinct planktonic organisms) and a few corals. These rocks make up an extensive area to the west and south of the park, but are mostly found as scattered outcrops in the southwest corner of the park.

The slightly younger part of the terrane is a collection of sandstones, siltstones, limestones, shales, and cherts that can be found in a northeast-southwest band on both sides along the McKinley Strand of the Denali fault (fig. 34.5). These rocks make up some of the peaks visible from the park road including Scott Peak, Mount Pendleton, and other high ridges in the headwaters of the Toklat and Teklanika River areas.

Where not altered by the intrusion of the McKinley or Foraker plutons, the rocks contain abundant marine fossils of many types, suggesting a very warm, tropical depositional environment of about 380 million years ago. East-west trending massive limestone lenses found within the unit suggest that these rocks were an ancient reef complex, which resembles the present-day reef system off the coast of Bermuda. Limestone outcrops in the southwest corner of the park near a place called Shellabarger Pass (fig. 34.8), contain a brachiopod (called *Myrospirifer breasei*), the only "new" species of any living thing, plant, or animal, existing or extinct, within the boundaries of the park.

Some marine shales, limestones, chert, and pillow basalt are overlain in some locations by terrestrial flysch-like sediments including sandstones, conglomerates, and lesser amounts of chert and limestone. A portion of the terrestrial sediments, informally called the "Mount Dall Conglomerate," are minimally 4921 feet thick, and contain plant fossils (ferns and broadleafs) within shales and siltstones, which provide a Middle Pennsylvanian to Permian (250 to 323 million years

FIGURE 34.7   Nenana Canyon. The Nenana River runs north along the eastern edge of the park, cutting a canyon through the Alaska Range and northern foothills. The buff colored rocks in this picture are schists, while the darker brown rocks are basalts of the Yukon-Tanana crystalline terrane, the basement complex of Denali National Park. The park's highway (leading north to Fairbanks) is on the right side of the photograph. National Park Service photo by Phil Brease.

old) depositional age. Among the Permian fossils found within the Mount Dall Conglomerate are specimens of a particular broadleaf plant (*Zamiopteris sp.*) that are the only known Permian flora in Alaska, and are the only known Permian flora specimens in North America, north of the Dunkard Basin of Western Pennsylvania. Similar *Zamiopterids* are only found in a Siberian setting referred to as Angaraland, of Permian age.

Pingston, McKinley, and Chulitna Terranes

Two terranes, the Pingston and the McKinley, are found as smaller east-west slivers on the north side of the Denali fault, and consist of a similar series of Mesozoic marine sediments including marble, chert, limestone, shale and sandstone, interbedded with flows and intrud-ed by gabbro, diabase, and diorite. The basalt flows are frequently *pillowed* (cooled under water), suggesting an island arc or ocean/seamount deposition. Fossil *bivalves* (marine clams) and *radiolarians* (plankton), in the sedimentary interbeds, are dated Triassic in age.

In the eastern portion of the park, rocks of the McKinley terrane are found in the upper Sanctuary—upper Windy Creek areas, and in the northern head-walls of Mount Pendleton and the Polychrome Glacier vicinities. In the upper Sanctuary, the volcanic unit is accessible in the middle section of Refuge Valley, a popular destination for backpacking in the park. In the western portions of the park, the Pingston is displayed north of the park road after Toklat River, and in the road cut along Eielson Bluffs, where pillow basalts are visible (fig. 34.9).

Perhaps the most exotic terrane located in and near Denali National Park and Preserve is a marine sediment and volcanic series that represents a small package of rocks like no others found in the state of Alaska, or perhaps anywhere else. These rocks, referred to as the Chulitna terrane, consist of thrust fault slivers of ocean crust type volcanic rocks that are interbedded with deep water marine sedimentary rocks, and are known as an *ophiolite* sequence. The ophiolite rocks (serpentine, basalt, chert, and gabbro) are suspected to represent fragments of oceanic crust and upper mantle material. These rocks are interbedded with deep-water sedimentary rocks (limestone and cherts), and overlain by island arc volcanics (volcanic conglomerates) and continental sediments (sandstones, argillites, tuffs, redbeds, and basalts). These rocks are referred to as the Chulitna Sequence, and their depositional history as a terrane suggests a South Pacific origin, some possible docking time at the ancestral seacoast of Idaho, and then final (or current) arrival in Alaska.

The rocks are found just within the boundary in the southeast corner of the park, south of Broad Pass, and form the higher peaks of the west wall of Chulitna valley. The unit strikes roughly northeast-southwest between the West Fork Chulitna River on the north, and the terminus of the Eldridge Glacier on the south. The greatest bulk of exposures of the "Chulitna Sequence" are not in the national park, but are instead within Denali State Park just to the southeast of Denali (figure 34.10).

Younger "Overlap" Rocks

The Kahiltna terrane is a *flysch sequence* (rapid marine foredeep deposition) consisting of a great variety of marine sedimentary rock types in nearly all conditions of *induration* (hardness) and deformation, including portions near the McKinley or Foraker plutons that are intensely metamorphosed. The dominant rock types, in order of frequency, are dark grey to black argillites,

FIGURE 34.8   Located in the far southwestern corner of the park, Shellabarger Pass is an area of 400 million year old ocean reef limestones that have yielded numerous marine fossils. National Park Service photo by Phil Brease.

PILLOW BASALTS

FIGURE 34.9   Hot, molten lava erupted into an ocean basin from near-shore volcanoes about 200 million years ago to become "pillow basalts." Pillows form when molten lava flows into water and separates into blobs (similar to those seen in a lava lamp), covered with a thin crust of crystalline rock created by the cooling effect of seawater (cooling rinds). The original pillows can be seen in this photo as light brown spheres separated by the grayish-brown rock of the cooling rinds. This rock is part of the McKinley terrane and is exposed along the Denali National Park road at Eielson Bluffs and milepost 67. National Park Service photo by Phil Brease.

coarse-, to finer-grained lithic (contains fragments of previously formed rocks), graywacke, pebble conglomerates, shales, and minor amounts or interbeds of dark-grey to black chert, limestones, sandstones, and silt-stones. Near the local intrusives, the sedimentary rocks have been altered to slates, argillites, phyllites, horn-fels, schists, and gneisses.

The Kahiltna terrane is considered by most to be an *autochthonous* (locally derived) sequence, representing a closing marine basin during the last accretionary event in the late Cretaceous time (approximately 100 mya). Marine fossils, particularly *radiolarians* (tiny, plankton-like sea creatures) found within the flysch, provide a Jurassic-Cretaceous age for these rocks. The Kahiltna flysch can be found as a large unit, principally on the south side of the Denali fault, and appears to be the primary country rock that has been intruded by the McKinley granites. Climbers ascending Mount McKinley are likely to encounter the altered form of the Kahiltna rocks, as the summit cap of the northern peak is a large *roof pendant* (overlying country rock) of the flysch (fig. 34.11). Similarly, the southern, or true

summit, also has scattered blocks of flysch at the "top of the world."

Overlying the flysch, possibly somewhat conformably, is a series of generally coarse-grained, mostly nonmarine conglomerates, sandstones, siltstones, mudstones, and coal known as the *Cantwell Sediments*. These coarse sediments were laid down in a depositional trough (Cantwell Basin) that existed on the north side of the Alaska Range at least as early as the late Cretaceous time based on fossil flora and *palynologic* (fossil pollen) evidence. Small amounts of limestone in the lower portion of the unit records a late Cretaceous *transgression* (seawater invasion inland), which probably represents the last of the marine deposition within the park area.

The bedding, lithology, and coarseness of the Cantwell sediments suggests high energy, wet, alluvial fan, braided stream, and some *lacustrine* (lake) deposition. These sediments indicate the origin was from local terranes north and south of the Cantwell Basin, and structural interpretations of the progressively steeper dipping strata toward the southern boundary of the unit

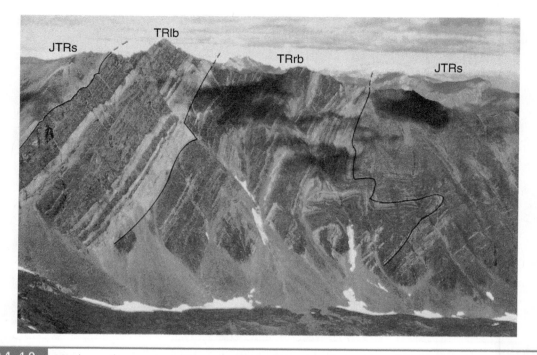

FIGURE 34.10   Perhaps the most exotic geological terrane in Alaska, the Chulitna Sequence, purportedly originated as marine sedimentary rock in a South Pacific oceanic environment. It then drifted eastward to "dock" near present-day Idaho, and finally moved north to arrive in central Alaska near the southern boundary of Denali National Park and Preserve. Certain rock types within this sequence, called redbeds (the red rocks visible in the center of this picture labeled TRrb), are found only in Idaho and at this location in Alaska. The redbeds of Idaho and Alaska were probably once connected (deposited in the same place) and are evidence of the long distance these rocks have traveled. U.S. Geologic Survey photo by David Jones.

suggest northward thrusting, probably associated with northward accretionary tectonics in southern Alaska. Look for the Cantwell rocks along the park road, from the park entrance (south side ridges) to Toklat River. They are well exposed in the park road cuts just west of the Polychrome rest stop (fig. 34.12).

Following a period of increased orogeny in Cretaceous time, *magmatic activity* (emplacement or eruption of magma) produced widely scattered plutons, stocks, and smaller structures in the Farewell and Kahiltna terranes, mostly to the south of the present range axis. These igneous rocks cooled at a depth beneath the ancestral Alaska Range, and would not become lofty, rugged, visible mountains until a new period of uplift in the late Tertiary. Collectively, these crystalline rocks are called the McKinley Intrusive Sequence. The sequence involves at least seven separate plutons and several minor stocks. The tallest mountain in North America, Mount McKinley, is chief among the intrusions, and the true summit, or South Peak is made up of biotite granite, while the upper 1000 feet of the North Peak of Mount McKinley consists of the earlier mentioned Cretaceous flysch (fig. 34.11). Age dating of the McKinley Intrusive Sequence rocks by radioactive methods (potassium-argon) has resulted in age ranges of 52.3 to 65.6 million years old, with the South Peak of Mount McKinley dated at 56.7 million years old.

Beside forming large and small intrusions, magma also erupted on and near the surface, creating the Cantwell Volcanics, a series of basalt and rhyolite flows and ash deposits that blanketed the Cantwell sediments and other rocks to the north. Radiometric ages of the Cantwell Volcanics range from 41 to 61 million years old, with a median age of 56 mya. They can be found as an east-west belt from the Sanctuary to the Toklat River and south of the park road. These are the rocks that make up Cathedral Mountain, Double Mountain, and the colorful Polychrome Pass area (fig. 34.13). Flow direction data suggest that the source area for

FIGURE 34.11 North/South Summits of Mount McKinley. The high point on the left side of the picture is the true summit at 20,320 feet, while the north peak caps out at 19,470 feet. Although the bulk of Mount McKinley is made up of 56 million year old granitic rock, the upper 1000 feet of the north peak is a cap (roof pendant) of 100 million year old flysch (shallow marine rock) sequence. National Park Service photo by Phil Brease.

FIGURE 34.12 East Fork Anticline. Folded and overturned Cantwell sedimentary beds exposed in the bank of the East Fork Toklat River, one mile downstream from the park road. The Cantwell Formation has undergone considerable folding and faulting since its deposition in Cretaceous to Paleocene time. National Park Service photo by Phil Brease.

these rocks was to the southwest of the unit, somewhere in the vicinity of Mount McKinley.

Magmatic activity in the Alaska Range vicinity was renewed during the Oligocene to Eocene Epochs, resulting in two major plutons crystallizing in the central and western portions of the park. The Foraker pluton, a biotite-hornblende granodiorite, is intermittently exposed in a 325 square mile area, making up the bulk of Mount Foraker (17,395 feet), and portions of Mount Russell (11,676 feet), as well as many other unnamed high peaks and ridges on the south side of the Denali fault. Located slightly to the east of the Foraker is the McGonagall pluton, a 120 square mile surface exposure of similar granodiorite located on the north side of the Denali fault. Both plutons have radiometric dates of approximately 38 million years, and some authors consider them to be parts of the same intrusion that was split or displaced by right-lateral (strike-slip) movement on the Denali fault (see section on Denali fault).

The Mount Galen Volcanics, made up of andesite, basalt, breccia, and tuffs, intruded the McKinley terrane and erupted at several vents, blanketing a large area (10 square miles) with andesitic ash falls and flows. One such vent has been mapped with the park boundary on Thorofare Mountain, and material from this location has been dated at 38 million years, the same age as the McGonagall and Foraker plutons. These volcanic rocks can be found at the far western end of Eielson Bluffs, on Mount Galen, and the mountains south of Thorofare Pass. Evidence of the Mount Galen eruptive products indicates short periods of violent eruptions.

In mid-Tertiary time (15 to 30 million years ago), subsiding basins developed north of the present-day Alaska Range, which accumulated sediments primarily from highlands to the north and alternated between swampy lake and high-energy river deposition to create the Usibelli Group. This unit contains four members (Grubstake, Lignite, Suntrana, and Healy Formations) that are made up of interbedded fresh water sandstones, claystones, conglomerates, and coal. Significant ash deposits are reported in a member called the Grubstake Formation, in which petrified tree trunks are found standing in five-foot thick volcanic ash beds in locations outside the park. The Usibelli Group can be viewed in the park as scanty outcrops of sandstone and thin coal primarily on the Teklanika River Bluffs across the river from the Teklanika Campground, and on the

FIGURE 34.13   At the Polychrome Pass rest stop, 41 to 61 million year old basalt (dark rock) and rhyolite (light rock) make up the interbedded lava flows that gave Polychrome Pass its name. National Park Service photo by Phil Brease.

FIGURE 34.14   Located just north of the peak near Healy, Alaska, the Suntrana area is the type section for the Usebelli Group, a collection of five formations that are the source for coal in the area. The remains of historic mining structures are visible in the foreground. Surface and underground mining have taken place since the early 1900s. Here a collection of Miocene to Eocene sandstones, claystones, conglomerates, and coal overlie the late Paleozoic to Precambrain Yukon-Tanana rocks, representing an unconformity in the neighborhood of 425 million years. National Park Service photo by Phil Brease.

south side of the road at Sable Pass. This formation is best viewed at and around the town of Healy, where the formation is mined for coal (fig. 34.14).

Uplift began or renewed on the southern margin of the Alaska Range in the late Pliocene Epoch (about 6 million years ago), and the uplift spread northerly to lift the "Outer Range" as well. Streams began to flow northerly once again from a new Alaska Range into the Healy basin area and points farther north.

These streams and rivers carried great amounts of sediment, and deposited onto, as well as further eroded, the Usibelli Formation. The new outer range uplift, combined with rapid erosion, created alluvial fans and bajadas (depositional cones of gravel) on the north and south sides of the outer range. The deposition partially filled the Cantwell trough and Healy basin area, and is most prominent against the north side of the outer range. This material (sands and gravels) became the poorly consolidated sandstones and conglomerates of the Nenana Gravel, which can be found within the park where the park road traverses it from milepost 17 to the Sanctuary Campground and again in Highway Pass. Nenana Gravel can also be found as a bajada on the northern boundary, east and west of the Teklanika River.

Faults and Earthquakes

The Denali Fault system, which has a northward-convex surface trace more than 720 miles long across the entire width of the state of Alaska, has been the subject of geoscience controversy since it was first described in the 1930s by U.S. Geological Survey geologist Stephan Capps (see fig. 34.5 for fault location). Evidence for the fault includes intermittent scarp traces and sag areas (visible from space), juxtaposed lithologies and unrelated strata, a few offset glaciers and streams, and some fault gouge deposits (fig. 34.15). Most investigators, usually focusing on small specific segments, have described the fault as a major strike-slip feature with as much as several hundred miles of Cenozoic right-lateral offset, with some vertical movement as well.

Within the park, perhaps one most noteworthy of these right lateral interpretations, the Foraker and McGonagall plutons, are described as a single plutonic body that was originally emplaced more centrally on each side of the Denali Fault, and was later split or displaced some 19 miles right-laterally in the last 38 million years. A later investigation of the same segment revealed that the McGonagall pluton has an intrusive contact at its boundary near the Denali fault, and coupled with new chemical and petrological evidence, structural interpretations, and visible slip features, the fault in this area demonstrates more vertical or oblique displacement, rather than a large degree of horizontal strike-slip (see fig. 34.16).

In recent time, attempts have been made to determine active movement on the Denali Fault. A trilateration network (high precision survey) established across the McKinley Strand, east-west for 30 miles near Cantwell by the U.S. Geologic Survey in 1975, found no significant movement in the 14 year period of measurement. In a current, ongoing effort using survey grade global position system (GPS), surface stations measured along the highway between Fairbanks and Talkeetna have shown a 0.24 to 0.36 inch per year westerly migration rate in a six-year study period for stations north and south of the fault (McKinley Strand). The rates of westerly migration of GPS stations are slightly greater just south of the fault (fig. 34.17).

Until recently, earthquake, or seismic activity, along the fault has been inconclusive regarding the nature or frequency of movement on the Denali fault. Denali Park is within the high risk seismic zone, and numerous earthquakes are felt in the vicinity each year, mostly from near or beneath Mount McKinley. In figure 34.17, all seismic events of magnitude 3 and greater are shown for the first six months of 1994 (January 1, 1994 to May

FIGURE 34.15 Bull Divide—Denali Fault Line Scarp. Linear features, such as this fault-line trace in the Bull River basin on the south slope of the Alaska Range, can be found in places along the McKinley Strand of the Denali Fault System. The fault is not well understood, and investigators frequently debate what type of fault it is. Most think it is a right-lateral strike/slip fault. Movement that has taken place along the fault in recent geologic times is from 10 to over 100 miles. National Park Service photo by Phil Brease.

FIGURE 34.16   Gunsight Pass, the low saddle on the right edge of the photo, is where the Denali fault (McKinley Strand) courses through from the Muldrow Glacier (foreground and left side of photo) to the Peters Glacier valley (out of sight behind the saddle). The granitic rocks of the McGonagall pluton are visible in an irregular (intrusive) contact with marine sediments of suspected Farewell terrane. A normal fault "footwall" is in evidence in the marine sediments just left of the intrusive contact. National Park Service photo by Phil Brease.

30, 1994). Some 600 seismic events occur within the park per year. It can also be noted that linear seismic activity along the Denali fault (diagonal yellow line) is minimal in this view, except where it is associated with the uplift of Mount McKinley (center mass of dots).

Linear seismic patterns on the fault along most of its trace through the state have been surprisingly minimal.

However, on October 23, 2002, an earthquake at magnitude 6.7 occurred on or near the Denali fault just 30 miles east of the park, followed 11 days later by a main shock at magnitude 7.9 on November 3, 2002. This was the largest earthquake to occur in North America in 2002, and the largest to hit interior Alaska in recorded history. The main shock happened 55 miles east of the park on the Denali fault. The *focal mechanisms* (the direction or geometry of seismic energy) of both foreshock and main shock have been resolved as dominantly right-lateral strike-slip, thus clearly suggesting that the Denali fault is active and demonstrates right-lateral strike-slip in some areas. The fault movement history at many locations still remains poorly

understood, and additional investigations are needed to fully unravel the nature, timing, and historic movement of the overall Denali fault system.

Glaciers and Permafrost

Glacial features dominate much of the southern half of Alaska and once covered most of the area to the north, almost to Fairbanks. Glaciers cover about 5 percent of the state today, and about 16 percent of the 6 million acres of Denali Park. At least 5, and possibly 7 major glacial advances have been identified on the north and south sides of the Alaska Range within and adjacent to Denali National Park. Principal ice erosional and depositional features are visible in most of the river valleys on the north side along the park road, on the east side along the Nenana River, and on the south side in the various headwater streams of the Susitna River.

Along the park road, U-shaped canyons and glacial terraces are common in the Savage, Sanctuary, Teklanika, Toklat, Thorofare, and McKinley Rivers.

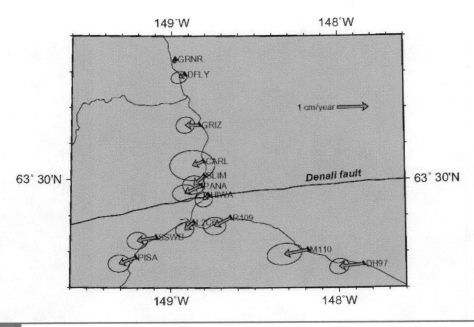

FIGURE 34.17 GPS Movement. Global Position Stations (GPS) used to evaluate surface crustal motion in the Interior of Alaska. Triangles mark starting locations, and arrows show degree and direction of movement in a six-year period. Movement rates average .24 inches to .35 inches per year for stations near, and 22 miles north of, the Denali Fault (McKinley Strand). Figure courtesy of Hilary Fletcher, Geophysical Institute, University of Alaska.

Terminal or recessional moraines are visible at Riley Creek (just south of the railroad trestle), at Sanctuary (easterly road cut just before the Ranger station), and on the McKinley River a few miles downstream from Wonder Lake. Some of the best *kame terraces* (ice and meltwater deposition) are found in the Nenana River valley just north of the park in the town of Healy, where 14 levels of depositional or erosional terraces have been recognized.

Perhaps most notable of glacier deposition are the *erratics* (rock fragment carried some distance from source) found on glacial terraces and ridge tops in many places throughout the park. The headquarters erratics, which are house-sized boulders of granite, are prominent on the skyline of the ridge to the south as seen from the park road when climbing "government hill" and traveling by the park headquarters offices. These erratics are from the Yanert Valley some 30 miles to the east of their present lofty location (fig. 34.19).

The more extensive glaciers and snowfields are on the southeastern side of the range because moisture-bearing winds from the Gulf of Alaska drop more snow on this side. The five largest south-facing glaciers are

the Yentna (20 miles long), Kahiltna (30 miles), Tokositna (23 miles), Ruth (31 miles), and Eldridge (30 miles).

On the drier north side, except for the Muldrow Glacier, which is 32 miles long, glaciers are smaller and shorter. The Muldrow Glacier's length is due to the fact that it follows the northeast-trending McKinley Strand of the Denali fault to Anderson Pass, then bends sharply and flows northwestward down to the lowlands. Twice in the last hundred years the Muldrow glacier has "surged" (moved forward for a short time at a greatly increased rate of speed), flowing over lower stagnant ice and making a jumble of broken ice-blocks. Higher up, ice ledges on the granite walls were left hanging when sections of the glacier collapsed and sank. A build-up of water between the bottom of the glacier and the bedrock channel floats the ice (by hydrostatic pressure), which causes surging.

The glaciers of the Alaska Range carry enormous amounts of rock debris on, in, and beneath the ice as they move downslope. Loose fragments and blocks that fall from cliffs or are scraped from rock walls accumulate as low ridges of till that ride along on the

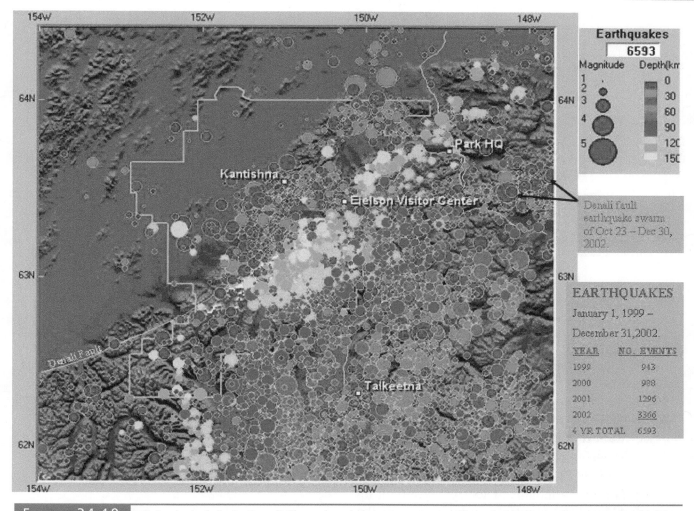

Earthquakes

6593

Magnitude	Depth(km
1	0
2	30
3	60
4	90
5	120
	150

Denali fault earthquake swarm of Oct 23 – Dec 30, 2002.

EARTHQUAKES

January 1, 1999 – December 31, 2002.

YEAR	NO. EVENTS
1999	943
2000	988
2001	1296
2002	3366
4 YR TOTAL	6593

FIGURE 34.18 Seismic Activity in Denali-The colored dots represent earthquakes greater than magnitude 2 in Denali vicinity during the first half of 1994 (January 1-May 30, 1994). The dot sizes represent the magnitude of each event, while the color of the dots indicates how deep the focus was. The Park (yellow stepped boundary) gets about 600 seismic events per year, most of them concentrated around the base of Mount McKinley (center of figure). These events suggest that rapid uplift of Mount McKinley is an ongoing process. From "Seismic-Eruption" program-copyright, Alan Jones, State University of New York and Binghamton.

edges of the glaciers. These are *lateral moraines.* Where tributary glaciers merge with each other or the trunk glacier, adjacent lateral moraines join to form *medial moraines* that are also carried down on the surface of the moving ice. Some of the large trunk glaciers have accumulated lateral and medial moraines that, seen from above, look like lane markers on a superhighway (fig. 34.20).

From the snouts of the glaciers, braided meltwater streams, heavily loaded with rock debris, continually shift and intertwine their channels over the valley floors. New channels open up as old channels are filled. Valley trains build up as streams drop quantities of poorly sorted sediment. *Valley trains* are long, narrow accumulations of glacial outwash, confined by valley walls. In some places coalescing

FIGURE 34.19   Headquarters Erratics. This house-sized granite boulder (notice the person below on the right for scale) is located on a ridge overlooking Denali Park Headquarters. Plucked from bedrock, it was carried here from 36 miles to the southeast by an ice advance called the Brown Glaciation. The Brown Glaciation may have been the maximum ice advance in central Alaska during the Pleistocene Epoch (10,000 to 2 million years ago). However, even this maximum expansion of glacier ice did not cover the interior basin of Alaska (between the Alaska Range and the Brooks Range), which remained an arid grassy steppe that was a corridor for migration of plants, animals, and humans between Asia and North America. National Park Service photo by Phil Brease.

outwash fans at the mouths of valleys have built up *outwash plains* over lowlands on both sides of the range.

Sometimes when glaciers are retreating and melting is rapid, blocks or masses of ice buried by till and outwash may persist for many years far beyond a glacier's terminus. When the ice finally does melt, the sediment slumps in, forming depressions called *kettles.* When filled with water, the depressions become *kettle lakes.* Wonder Lake, near Kantishna and the end of the park road, is the result both of glacial erosion and ice-block melting. About 4 miles long and 280 feet deep, the lake occupies a depression scooped from weak rock by a glacier flowing off a resistant ridge. After the glacier retreated and the ice melted, Wonder Lake was created.

Old moraines and kettles on the outwash plain are indications that the glaciers extended to lower altitudes in the past. Also, the fact that the present glaciers are "underfit" in their valleys shows that many of the glacial features that we see in the park/preserve today are the result of more extensive glacial action in the past.

Permafrost and Periglacial Activity

During colder Pleistocene climates, all the ground in the park/preserve area became solidly frozen. On the north side of the range, much of the ground is still frozen because average annual temperatures have not risen enough for melting to get below the thin topsoil. Permanently frozen ground, or *permafrost,* is *discontinuous* in the Denali area. The mean surface ground temperature, –30° F to –41° F (a few degrees below freezing), has allowed some thinning of the permafrost and a great deal of local variability. (North of the Arctic Circle permafrost is *continuous* and generally more than 500 feet in thickness.)

Permafrost is covered by an active layer that freezes and thaws seasonally. The depth of the active layer is controlled by variable factors, such as vegetation, surface material, topography, amount of moisture, and so on. In the Denali area, the active layer ranges from less than an inch to 10 feet or more, depending upon local conditions. Thicknesses of permafrost below the active layer have been measured between 30 and 100 feet at various locations around the Alaska Range.

FIGURE 34.20   Ruth Glacier, flowing south from the McKinley Massif. Lateral and medial moraines from tributary glaciers form a wide band of medial moraines being carried downslope on Mount Denali. Photo by Ann G. Harris.

FIGURE 34.21   Horseshoe Lake, a mile north of Park Headquarters, is in a lowland area close to the Nenana River. Details of the land surface are constantly altered by periglacial activities and by streams working over glacial till and outwash. National Park Service photo by V. Watson.

The *periglacial milieu* is the active layer above the permafrost where freeze-thaw oscillations are more or less continual. Water changing back and forth between its liquid and solid states produces periglacial phenomena. In the active layer, water is available (and usually plentiful) because it cannot seep down into the impermeable permafrost below. (Voids and pores in permafrost are filled with ice.)

Simple frost action—a process of mechanical weathering that breaks up rock—is not usually dominant in temperate regions. In subpolar climates, however, the freeze-thaw cycles are predominant over other processes most of the time. Periglacial effects are noticeable around the margins of melting glaciers, on slopes, in river valleys, and on the tundra plains. Under periglacial conditions, *frost splitting* of wet rocks produces pinnacles like Cathedral Spires in massive, resistant rocks and thick accumulations of talus in fissured rocks. Sometimes talus masses in draws or gullies retain moisture, freeze, and become *rock glaciers* that move slowly downslope.

Frost heaving disrupts unconsolidated material, both sorted and unsorted. Soil volume increases irregularly when it freezes and then decreases irregularly when thawing occurs. Thus the surface of the soil keeps heaving up and falling back a little differently each time. As this process goes on, grains, pebbles, and cobbles on a flat surface tend to rearrange themselves in polygonal patterns and strips. Underneath the surface, lenses of coarse and fine material slump and sag chaotically.

On slopes, the active layer may become saturated during summer melt seasons due to the interaction of rain, meltwater, and low evaporation. The saturated soil (usually carrying a cover of vegetation) slides and slumps down a slope, exposing fresh permafrost underneath. This process is called *solifluction,* or soil flow. A stand of white spruce growing on a lower slope of Mount McKinley is called the Drunken Forest because the oddly leaning trees appear to be staggering as lobes of sliding soil slowly carry them down the mountainside. Marshy flats south of the park road also have areas of drunken trees leaning in all directions as the thawed soil layer slides over the permafrost.

Many of the shallow ponds in the park/preserve, like those in the tundra plains northwest of the range, are *thaw lakes* and *cave-in lakes* (fig. 34.1). They form where sunwarmed water has melted basins in the underlying permafrost. The lakes gradually deepen from summer to summer, as warmth penetrates the frozen ground, and they enlarge as the rims collapse.

In winter, the freeze-thaw cycles slow and then become suspended, but ice, having 10 percent greater volume than water, continues to exert pressure. Thermal expansion and contraction at depth can cause cracks to develop in the ice-cemented permafrost. During the melt season, water gets into the cracks, forming veins of ice called *ice wedges,* that enlarge with successive seasons of freezing and thawing. In periglacial regions, excavations or landslides sometimes reveal very large ice wedges that have been buried in sand and gravel for centuries.

Periglacial activity and the presence of permafrost cause engineering problems for highways, railroads, landing fields, and buildings because structural failures tend to occur whenever the natural thermal regime is disturbed. Uneven thawing may cause roadbeds and foundations to tilt or collapse.

Geologic History

1. Precambrian and early Paleozoic marine deposition.

Near the equator sediment was deposited in shallow, fluctuating seas along the western emergent coastline of a newly separated North America from Gondwanaland. The proto-Pacific ocean plate subducted beneath the Pangean continent, and onshore volcanic eruptions or island arcs contributed volcanic flows and intrusions to the Yukon-Tanana terrane. (fig. 34.22).

2. Mid-Paleozoic rifting, with carbonate shelf and basin development.

Tropical climates at low latitudes developed abundant marine life along the western coastline of ancestral North America. A portion of the western coastal range resulting from the Antler Orogeny rifted away from the proto-Canada and Alaska, creating the Cache Creek Seaway, and enhancing the erosion of volcanic and other sedimentary material into shallow seas. As the North American continent migrated northerly, the Farewell basin developed, and barrier reef flora flourished in the future location of the park (fig. 34.23).

3. Late Paleozoic Seaway and basin closure and terrane accretion.

Northeast migration of the proto-Pacific plate continued closing the Cache Creek Seaway, and pushing up over the McKinley/Pingston and Farewell compos-

Table 34.1

Geologic Column, Denali National Park and Preserve

Era	Period		MYA	North of Denali Fault	South of Denali Fault	Geologic Events
Cenozoic	Quaternary	Holocene	0.1	Cantwell Ash Bed (3700) ybp (years before present)		Unknown southerly volcanic source.
				GLACIAL ADVANCE (Maximum Extent)		
				Nenena-McKinley River Area	**Sustina River Area**	
		Pleistocene	2	Carlo Creek readvance (8000 ybp) / Riley Creek-Wonder Lake (9-25,000 ybp) / Healy-McLeod Creek (65-75,000 ybp) / Dry-Lignite Creek (125,000 ybp) / Bear Creek (125-150,000 ybp) / Browne (150,000+ ybp) / Teklanika (Pleistocene-Pliocene?)	Alaskan (5500 ybp) / Naptown (5500-45,000 ybp) / Knik (50,000-75,000 ybp) / Eklutna (90,000-110,000 ybp) / Sustina	Most of Alaska Range under ice prior to 150,000 ybp. / Teklanika may have extended as far north as Nenana.
	Tertiary	Pliocene	12	**Nenana Gravel** (Pliocene-Miocene) / Poorly indurated conglomerates, sandstone, claystone and lignite. Glacial-fluvial origin.	**Kenai Group** (Pliocene-Miocene) / 1) **Sterling Formation:** Poor to well indurated sandstones and conglomerates. / —— UNCONFORMITY —— / 2) **Tyonik Formation:** Sandstone and conglomerates with siltstone, claystone, coal and volcanic ash.	Nenana Gravel is a bajada from uplifting Alaska Range.
		Miocene	26	**Usibelli Group** (Miocene-Eocene) / 1) **Grubstake Formation:** Claystone and conglomerate / 2) **Lignite Formation:** Sandstone, claystone and coal. / 3) **Suntrana Formation:** Sandstone and coal. / 4) **Healy Creek Formation:** Sandstone and conglomerate. / —— Unconformity ——	—— Unconformity-Disconformity ——	Deformation and thrust faulting creates new Alaska Range emergence. / Subsiding basin deposition north and south of uplifting Alaska Range.
		Oligocene	37	**Mount Galen Volcanics** (38 ma) / Andesite and basalt flows, breccias and tuffs / Ash falls and pyroclastic flows are common	**McGonagall and Foraker Plutons** (38 ma) / Biotite and hornblende granodiorite, with some sulfide mineralization.	Possible same magma source @ 38 ma.
		Eocene	53	**Cantwell Volcanics** (41-61 ma) / Basalt, rhyolite, andesite, obsidian, in flows, intrusions, and pyroclastic deposits	**McKinley Intrusive Sequence** (52-57 ma) / **McKinley:** Biotite granite, granodiorite, (56 ma) / **Kahiltna:** Biotite and muscovite granite and granodiorite / **Ruth:** Biotite and muscovite granite	Possible same magma source @ 56 ma.
		Paleocene	65	**Cantwell Sediments** (Paleocene-Cret) / Conglomerate, sandstone, siltstone, shale, thin coal beds and tuff layers . . . some flora in shales	**Cathedral:** Granite and granodiorite. / **Composite:** Granite-monzonite-lamprophyre	Emergent Alaska Range weather into Cantwell trough.
Mesozoic	Cretaceous		136	—————— Unconformity ——————		Accretion and overriding of Talkeetna Superterrane closes Susitna Basin and initiates period of orogeny.
				CRETACEOUS FLYSCH (Kahiltna Terrane)		
				Jet black slates interbedded with thin sandstones, massive coarse graywacke, siltstone, argillites, chert and some breccia. Varies in alteration related to tectonics and igneous proximity.		
				MESOZOIC SEDIMENTARY/VOLCANIC (Pingston, McKinley, Chulitna Terranes)		
				Calcareous shales, argillites, sandstones and sandy limestones, some graywackes and cherts.		
	Jurassic		190	**Nikolai Greenstones:** Submarine basalt flows (pillow basalts), volcanclastic and cherts. Overlies Triassic conglomerates and sandstones, siltstones, argillites and thin interbeds of limestone.	**Chulitna Sequence: (Chulitna terrane)** limestones, tuffs, basalts, crystal tuff, gabbros, serpentines and red beds. / **Mount Dall:** Terrestrial conglomerates and fossil shales.	Chulitna sediments and ophiolites of possible South Pacific origin.
	Triassic		225			
Paleozoic	Permian		280	**UNDIVIDED PALEOZOIC MARINE ROCKS (Farewell terrane)** / Turbidites (sandstones, argillites, and dolomitic limestones), cherts, pillow basalts, shales and massive fossiliferous limestones.		Mount Dall plant fossils suggest possible Siberian origin.
	Pennsylvanian					
	Mississippian		345	**TOTATLANIKA SCHIST** (Early Mississippian-Dev) [Yukon-Tanana terrane]		Totatlanika originated as submarine ash flows on continental margin.
	Devonian		395	Metasediments and meta-volcanclastic schists, marble, phyllite, and gneiss		
				"Wyoming Hills Sequence" / Quartzose and carbonaceous phyllites, and slates, quartzite, marble, limestone and chert.	**"Dillinger River Sediments"** / Interbedded lime mudstones and shale, with sandstones, limestones and graywackes. Some mafic dike intrusions.	
	Silurian					
	Ordovician		590	—————— FAULT AND UNCONFORMABLE CONTACTS ——————		
	Cambrian			**YUKON-TANANA CRYSTALLINE SEQUENCE (Yukon-Tanana terrane)**		Originated as shallow to fluctuating seas south of an ancient North American continent.
	Precambrain		600+	Original deposition as quartz rich shallow marine sediments with interbedded volcanics and limestone. Multiple periods of alteration creates sericite schist, quartzite, graphitic and calcareous phylites and marbles. Sequence locally includes Keevy Peak Formation and Spruce Creek Sequence. Formerly known as "Birch Creek Schist"		

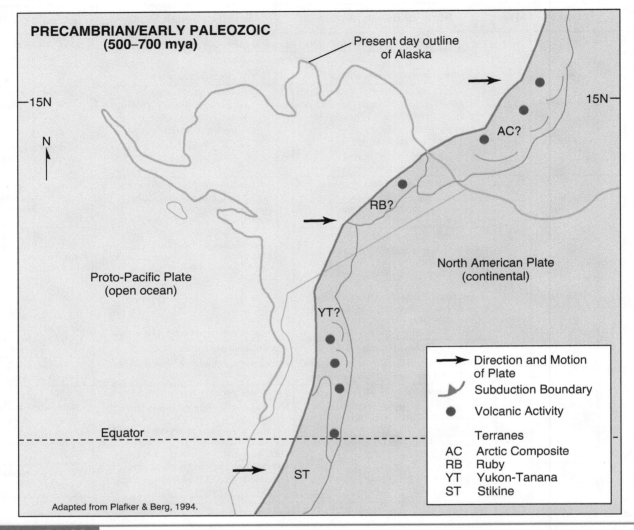

PRECAMBRIAN/EARLY PALEOZOIC
(500–700 mya)

FIGURE 34.22   Precambrian/Early Paleozoic Alaska (500–700 ma).

ite terranes onto the Yukon-Tanana backstop. Ancestral Alaska (and North America) rotated slightly counterclockwise, migrated northerly, and were back in union with Gondwanaland (fig. 34.24).

4. Late Mesozoic arrival of Talkeetna superterrane and Kahiltna basin closure.

Gondwanaland and Pangea/Laurasia separated the ancestral North America (part of Laurasia) migrated to place Alaska some 70°–80° north latitude. Continued northeast migration of the Pacific (Farallon) plate drove a major offshore island arc referred to as the "Talkeetna Superterrane" into the southern Alaska coastline. Sediments were rapidly shed from North American continental sources and the approaching terrane, and caught up in the closing of the Kahiltna basin to become the Kahiltna flysch. Possibly exotic micro terranes, such as the Chulitna, were caught in the closure as well (fig. 34.25).

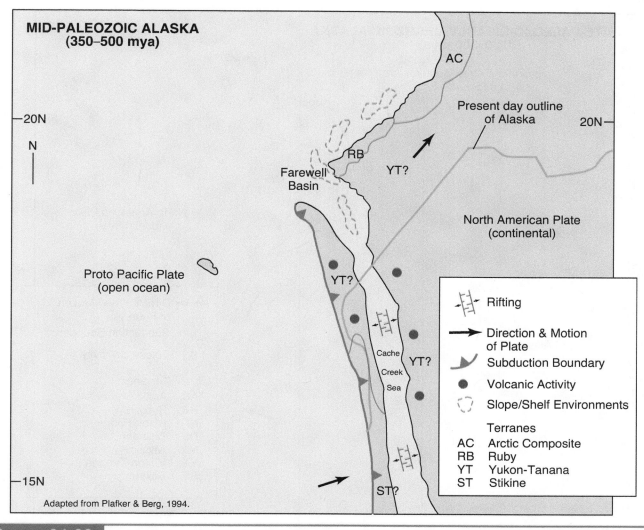

MID-PALEOZOIC ALASKA
(350–500 mya)

AC

Present day outline
of Alaska

20N

N

RB

Farewell
Basin

YT?

20N

North American Plate
(continental)

Proto Pacific Plate
(open ocean)

YT?

Cache

Creek

Sea

YT?

Rifting

Direction & Motion
of Plate

Subduction Boundary

Volcanic Activity

Slope/Shelf Environments

Terranes
AC Arctic Composite
RB Ruby
YT Yukon-Tanana
ST Stikine

15N

ST?

Adapted from Plafker & Berg, 1994.

FIGURE 34.23   Mid-Paleozoic Alaska (350–500 ma).

**5. Paleocene final terrane assemblies and onset of
plutonic and volcanic activity.**

Alaska underwent an increased period of orogeny
from the final north-south squeeze from the northerly
thrusting Talkeetna terrane and the southerly push
from the spreading Arctic Basin. The subduction of
the Farallon/Kula plates beneath a newly assembled
Alaska enhanced magmatic activity. Uplift of the
ancestral Alaska Range began; simultaneous erosion
of this range into basins on the north and south sides
of the range axis became the Cantwell Sediments. The
Mount McKinley intrusives crystallized beneath the
Alaska Range, while volcanic eruptions (Cantwell
Volcanics) overlapped the sediments. A second round
of magmatic activity crystallized to produce the
Foraker/McGonagall plutons, and the Mount Galen
volcanics. Denali essentially was in final terrane
assemblage (fig. 34.26).

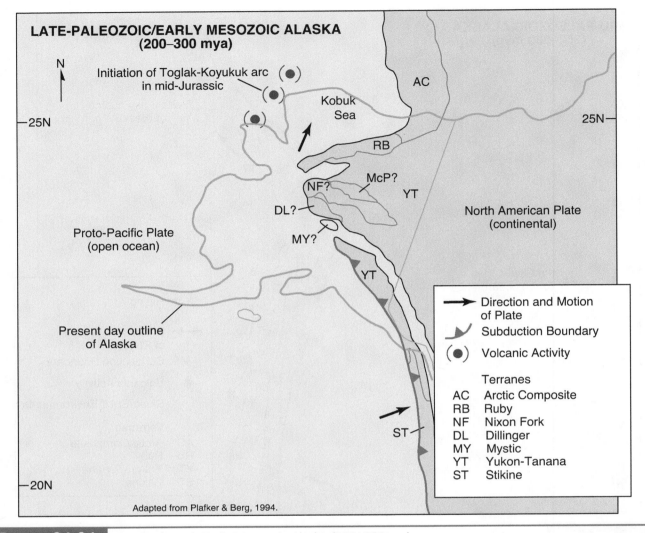

**LATE-PALEOZOIC/EARLY MESOZOIC ALASKA
(200–300 mya)**

N

Initiation of Toglak-Koyukuk arc
in mid-Jurassic

Kobuk
Sea

AC

25N

RB

McP?

NF?

YT

DL?

North American Plate
(continental)

Proto-Pacific Plate
(open ocean)

MY?

YT

Present day outline
of Alaska

→ Direction and Motion
of Plate

↘ Subduction Boundary

(●) Volcanic Activity

Terranes
AC Arctic Composite
RB Ruby
NF Nixon Fork
DL Dillinger
MY Mystic
YT Yukon-Tanana
ST Stikine

ST

20N

Adapted from Plafker & Berg, 1994.

FIGURE 34.24   Late Paleozoic/Early Mesozoic Alaska (200–300 ma).

6. Neogene sedimentation and uplift.

A period of increased warming and precipitation increases erosion of the ancestral Alaska Range, and subsiding basins on the north and south side of this range developed. Southerly flowing, high-energy streams provided silicic materials that deposited cyclically with periods of swampy warm basin environments to produce the Usibelli Group. Renewed thrust faulting and uplift initiated a period of rapid uplift, creating the current Alaska Range, the outside range, and associated alluvial fans and bajadas that are associated with the Nenana Gravel.

7. Pleistocene glaciation.

Major glaciation enveloped the Alaska Range about 150,000 to 200,000 years before present, covering most of the mountains and extending an ice sheet out into the southerly Cook Inlet area. Lobes of ice also extended into the northern interior, at least as far north as the town of Nenana. At least five or six major retreat and advance cycles carved U-shaped valleys, shaved off some mountain tops, and produced much of the glacial/fluvial deposition seen in the valleys today.

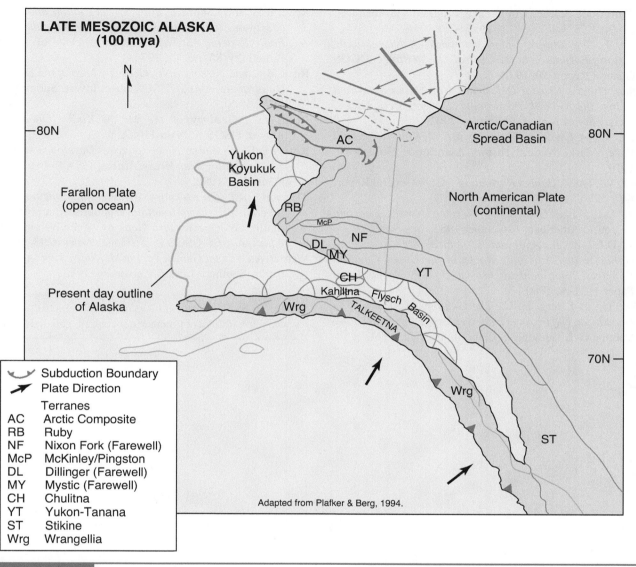

LATE MESOZOIC ALASKA
(100 mya)

N

—80N   80N—

Yukon
Koyukuk
Basin

AC

Arctic/Canadian
Spread Basin

Farallon Plate
(open ocean)

RB

North American Plate
(continental)

McP

NF

DL
MY

YT

CH

Present day outline
of Alaska

Kahiltna

Wrg   Flysch   Basin

TALKEETNA

70N—

Wrg

ST

Adapted from Plafker & Berg, 1994.

⌣ Subduction Boundary
↗ Plate Direction

Terranes
AC   Arctic Composite
RB   Ruby
NF   Nixon Fork (Farewell)
McP   McKinley/Pingston
DL   Dillinger (Farewell)
MY   Mystic (Farewell)
CH   Chulitna
YT   Yukon-Tanana
ST   Stikine
Wrg   Wrangellia

FIGURE 34.25   Late Mesozoic Alaska (100 ma).

Sources

Brooks, A., *Geography and Geology of Alaska.* USGS Professional Paper #45, 1906.

Brooks, A., *The Mt. McKinley Region, Alaska.* USGS Professional Paper #70, 1911.

Bundtzen, T.K., *A History of Mining in Kantishna Hills.* Alaska Journal, pp. 151-161, Spring 1978.

Bundtzen, T.K., *The Geology and Mineral Deposits of the Kantishna Hills.* MS Thesis, University of Alaska, 1981.

Capps, S.R., *The Eastern Portion of Mt. McKinley N.P.* USGS Bulletin 836-D, 1932.

Collier, M., *The Geology of Denali National Park.* Alaska Natural History Association, 1989.

Coney, P., et al, *Cordilleran Suspect Terranes,* Nature, v. 288, 1980.

Csejtey, B., et al, *Geology and Geochronology of the Healy Quadrangle, Alaska,* USGS Misc. Investig. Series, Map I-1961, 1992.

Csejtey, B. et al, *Requiem for the Denali Fault of Alaska,* American Association for the Advancement of Science,

50th Arctic Science Conference, Program and Abstracts, 1999.

Decker, J., *The Mount Galen Volcanics; A New Middle Tertiary Formation in the Central Alaska Range.* DGGS General Report 50, 1978.

Fletcher, Hillary, *Crustal Deformation in Alaska Measured Using the Global Positioning System*, Phd Thesis, University of Alaska, Fairbanks, 2002.

Gilbert, W., *A Geologic Guide to Mt. McKinley National Park.* Alaska Natural History Association Guidebook, 1979.

Gilbert, W., *The Teklanika Formation* DGGS General Report #47, 1976.

Jones, D.L., *The Growth of Western North America.* Scientific American, November 1982.

Jones, D.L., et al, *Age and Structural Significance of Ophiolite and Adjoining Rocks in the Upper Chulitna District, South-central Alaska.* USGS Professional Paper 1121-A, 1980.

Plafker, G., and Berg, H. C., *The Geology of Alaska—The Decade of North American Geology Project Series—* Volume G-1, Geologic Society of America, 1994.

Reed, B.L., and Lanphere, M.A., *Offset Plutons and History of Movement along the McKinley Segment of the Denali Fault System, Alaska:* Geologic Society of America Bulletin, v. 85, no. 12, 1974.

Reed, B.L, and Nelson, S.W., *Geologic Map of the Talkeetna Quadrangle, Alaska.* USGS Misc. Invest. Series Map I-1174, 1980.

Reed, J.C., *Geology of the Mt. McKinley Quadrangle, Alaska.* USGS Bulletin 1108A, 1961.

Ritter, D.F., *Complex River Terrace Development in the Nenana Valley near Healy, Alaska.* GSA Bulletin, v. 93, p 346-356, 1982.

Thorson, R., *Quarternary Glacier Expansions from America's Highest Mountain.* Unpublished report, 1980.

Wahrhaftig, W. *Quarternary Geology of the Nenana River Valley, Alaska.* USGS Professional Paper #293, 1958.

Werner, A., *Glacial Geology of the McKinley River Area.* MS thesis, Southern Illinois University, 1982.

Note: To convert English measurements to metric, go to www.helpwithdiy.com/metric_conversion_calculator.html

PART IV

Volcanic Features
and Volcanic Activity

FIGURE PIV.1   The Hilina fault system on the south slope of Kilauea Volcano. The series of fault blocks lead in step-like fashion down the slope and continue under water. On land they appear as steep cliffs and under the ocean as undersea landslides. The rift zones erupt when lava forces its way into them causing the volcano flanks to become unstable, thus helping create the faults. Photo by Ann G. Harris.

Volcanoes, wherever they are, dominate the surrounding landscape. Volcanic eruptions, like earthquakes, are compelling events to human beings. To geologists, a volcano erupting is "doing its thing." To monitor an eruption, a geologist may drop everything and hurry to observe the activity. As much as any tourist, a geologist regards a volcanic eruption as awe-inspiring, magnificent, and perhaps even frightening; but primarily, geologists see an eruption as a fascinating scientific opportunity. Volcanoes and eruptions are significant geologically far beyond their contribution to landscapes.

Volcanic Features Protected in the National Park System

Examples of volcanism in the national park system are remarkable combinations of uniqueness, scientific merit, and landscape beauty. The two volcanoes close to North Cascades National Park (Chapter 29) and the volcanoes in Wrangell-St. Elias National Park (Chapter 32) were described in Part III. Of the nine national parks in Part IV, Hawaii Volcanoes National Park has had the most eruptions during recent years, but all nine have had volcanic activity of one sort or another during Holocene time. Moreover, eruptions have occurred within the last few centuries in all but one of these national parks. The exception, Yellowstone National Park, has notable geothermal features; more active volcanism may resume in the future.

Listed below are national monuments that highlight volcanic features protected in the national park system:

- **Aniakchak National Monument and Preserve,** Alaska, has an immense caldera and an active volcano (fig. PIV.2).
- In **Bandelier National Monument,** New Mexico, 13th-century cliff houses were dug into volcanic tuff.
- A young, symmetrical cinder cone rises in **Capulin Volcano National Monument,** New Mexico.
- **Chiricahua National Monument,** Arizona, displays a maze of spires and pedestals eroded from volcanic rock.
- **Craters of the Moon National Monument,** Idaho, is a desolate landscape of cones, craters, and/or tunnels.
- **El Malpais National Monument,** New Mexico, is a spectacular volcanic area with spatter cones, a 17-mile-long lava tube, and ice caves.

- In **Devils Postpile National Monument,** California, tall, pipelike basalt columns form imposing cliffs.
- The nation's first national monument, **Devils Tower,** Wyoming, an 865-foot pillar of columnar rock, is a remnant of an intrusive body that resembles a volcanic rock because of columnar jointing.
- **Lava Beds National Monument,** California, has hardened lava flows, lava tubes, and cinder cones.
- **Pinnacles National Monument,** California, was a Miocene volcanic field that was split in two and its parts separated by movement of the San Andreas fault.
- The colorful summit crater of **Sunset Crater Volcanic National Monument,** Arizona, was formed about 900 years ago.

Mount St. Helens in Washington erupted violently in May 1980, destroying a large part of its beautiful cone. Since that time, quieter eruptive activities have built a dome in its crater. In 1982, Congressional legislation established the 110,000-acre **Mount St. Helens National Volcanic Monument** in order to protect and preserve this evolving volcanic feature and its surroundings. The monument is administered by the U.S. National Forest System.

Volcanism and Extrusive Rocks

The variety of volcanic landscapes and landforms in nature suggests that the igneous rocks that formed them must have originated from different sources in many different ways. *Extrusive igneous rock* (which includes all types of volcanic rock) is the surface manifestation of hot magma that has risen through the crust. (*Magma* is mobile, liquid rock material generated within the earth. Magma that hardens in the crust below the surface becomes *intrusive igneous rock.*) What produces the many different kinds of igneous rocks, whether intrusive or extrusive, is mainly a matter of a magma's chemistry and temperature and the crustal conditions it encounters as it works its way upward. If it reaches the surface before it solidifies, magma then becomes lava, which may erupt quietly as hot liquid or be ejected violently as *pyroclastic debris,* or *tephra.* Pyroclastic material (tephra) includes fragments of all sizes and shapes—pumice, ash, cinders, bombs and blocks—that are thrown out from a volcanic vent during an explosive eruption.

FIGURE PIV.2   A new volcano rises in the huge Aniakchak caldera that exploded and collapsed less than a thousand years ago. **Aniakchak National Monument and Preserve** is on the Alaska Peninsula, about 130 miles southwest of Katmai National Park and Preserve (Chapter 38). Both Aniakchak and Katmai are on a volcanic arc that marks the plate boundary where the Pacific plate is being subducted beneath the North American plate. National Park Service photo.

Characteristics of Lava

Why do some lavas erupt explosively while others are relatively "quiet"? The amount of gas in lava and whether or not gas can easily escape into the atmosphere are the key factors; a lava's *viscosity,* or resistance to flow, determines how gases are released. A highly viscous lava with a large volume of gas trying to escape creates a violently explosive eruption. A less viscous lava that releases gas as it rises and flows out may not explode at all. Most of the gas released during eruptions is water vapor in the form of steam, with small amounts of hydrogen sulfide, carbon dioxide, and hydrochloride acid also given off.

The degree of viscosity depends on a lava's chemical composition, the amount of water it contains, and its temperature. Silica (SiO_2) is the major component in all volcanic rocks, with the amount ranging from about 45 percent to 75 percent of the total weight. Lavas with high percentages of silica (65% or more) have high viscosity and tend to solidify at much lower temperatures (roughly 1292° F). Silica-poor lavas, which solidify at around 2732° F, have low viscosity and usually erupt as very hot fluid, sometimes with the consistency of molasses. The wide range of combinations possible accounts for differences in the behavior of lavas, as well as for the many varieties of extrusive rock.

How Lavas and Volcanic Rocks Are Classified

Silica-poor volcanic rocks and their parent lavas (around 50 percent silica by weight plus oxides of aluminum, calcium, magnesium, and iron) are considered *mafic*. Because they are composed of mostly dark green or greenish black minerals, mafic extrusive rocks tend to be dark in color. Basalt is the most common rock in this group. Rocks and lavas that are silica-rich (65 percent silica or more, with aluminum, sodium, and potassium oxides plus very small amounts of the oxides of calcium, magnesium, and iron) are classified as *felsic*. The most abundant volcanic rock with a felsic composition is rhyolite, which is generally light-colored because it contains mixtures of white, pink, light gray, and pale green minerals. *Intermediate extrusive rocks* are those that are transitional in chemical composition between mafic and felsic. *Andesite,* which typically has a dark to medium grayish color, is the most common intermediate volcanic rock.

Types of Volcanoes and Volcanic Belts

Not surprisingly, the kinds of lavas that reach the surface and the ways in which they erupt have a great deal to do with the shape and size of the volcanic features and landforms that are produced. Lavas that are predominantly mafic tend to construct *shield volcanoes,* which have broad, convex-upward cones, built up by layers of basaltic lava flows. Kilauea, in Hawaii Volcanoes National Park (Chapter 40), is a shield volcano built up from the ocean floor.

A composite cone, such as Mt. Rainier (Chapter 35), tends to have a slightly concave profile and is constructed of alternating layers of mainly andesitic lavas and pyroclastics.

Small, steep, conical hills, formed by the accumulation of ash, cinders, and other pyroclasts, are called *cinder cones.* Usually of basaltic or andesitic composition, cinder cones may be the result of local concentrations of gas that cause pyroclasts to erupt through a volcanic vent. A typical example is Cinder Cone in Lassen Volcanic National Park (Chapter 37).

Very viscous, felsic lavas (dacite or rhyolite) may solidify in or just above a volcanic vent, forming a steep-sided *volcanic dome.* This is a less common type of volcano but is apt to be hazardous due to its explosive potential. Some of the most destructive volcanic explosions in recorded history have been the result of

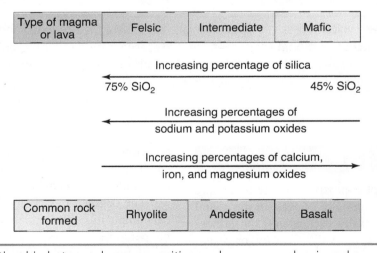

 FIGURE PIV.3   Relationship between lava composition and common volcanic rocks.

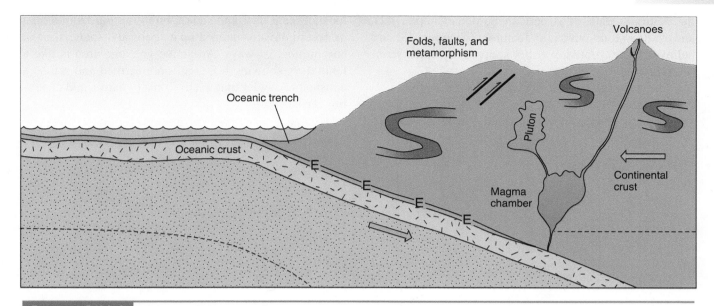

FIGURE PIV.4   An oceanic plate and a continental plate may come together, forming a plate convergence boundary. The more bouyant continental crust overrides the denser oceanic crust, which bends down as it slides under continental crust. Heat generated by friction as plates grind against each other and intense pressure at depth cause oceanic crust, plus marine sediments and scraped-off continental material, to melt, creating magma. Within the continental crust above, rocks subjected to the compression of the converging plates become folded and undergo varying degrees of metamorphism. Some of the magma works its way up through the continental crust and may erupt in volcanoes. Other magma cools within the crust, forming plutons. The "E's" in the subduction zone represent deep-focus earthquakes occurring along the top of the descending crustal slab.

the eruption of volcanic domes. Lassen Peak (Chapter 37) is an unusually large volcanic dome.

Volcanoes have a tendency to change their eruptive patterns over time. Some start out with explosive eruptions of silica-rich lavas and then change to a long period of fluid, basaltic eruptions. Others build andesitic cones of lava and ash and then blow their tops off when rhyolitic lavas rise to the surface. Since volcanologists don't have much specific information about the chemical mixtures in the magmas that are "cooking" below the crust, making predictions is more of an art than a science; but progress has been made in recent years as more data are collected and observations made.

Most volcanoes are located in belts that parallel the boundaries of tectonic plates. When plates are separat-

ing and pulling apart at a *spreading center*, the rifting allows hot lavas (mainly basaltic) to reach the surface. The Icelandic volcanoes in the Atlantic Ocean emanate from the rift along the mid-oceanic ridge. The volcanoes along the North American west coast (northern California to Alaska) are part of the circum-Pacific volcanic belt and are related to the subduction of oceanic crust beneath the continent (fig. PIV.4). The heat and pressure resulting from plate convergence causes rocks to melt, producing lava that works its way to the surface (box 35.1, Chapter 35).

Volcanic activity *not* related to plate boundaries may be due to the presence of a *hot spot* at depth that causes molten rock material to rise to the surface. The Hawaiian volcanoes (Chapters 40, 41), the volcanoes in the National Park of American Samoa (Chapter 42), and Yellowstone volcanic features (Chapter 43) are believed to be located over hot spots in the upper mantle, far below the surface.

The national parks included in Part IV display volcanic features that are easy to recognize. Each has a story to tell of how it was formed—in hot clouds of incandescent dust and steam, bubbling out from a lake of boiling lava, squeezed up as gobs and spatters, or in any number of ways. An active volcanic area is like a laboratory showing how rocks are formed and is also a constant reminder that we live on a dynamic and changing earth.

CHAPTER 35

Mount Rainier National Park

CASCADE RANGE, WASHINGTON

Area: 235,613 acres; 368 square miles
Established as a National Park: March 2, 1899
Address: Tahoma Woods, Star Route, Ashford, WA
  98304-9751

Phone: 360-569-2211 ext. 3314
E-mail: MORAInfo@nps.gov
Website: www.nps.gov/mora/index.htm

*Mount Rainier, a composite volcano, is the highest mountain in the Cascade Range. Explosive
eruptions in Holocene time destroyed the summit crater and upper part of the cone.
Steaming hot mudflows (lahars) have roared down the volcano flanks
in the recent geologic past and may reoccur in the future.*

*Glacial erosion has modified the volcano's surface, creating beautiful alpine scenery.
Downslope from ice and snow, subalpine meadows give way to dense forests.*

FIGURE 35.1   East side of Mount Rainier. Gibraltar Rock, Steamboat Prow, and Willis Wall are erosional
remnants of the volcano's cone when the summit was at its maximum elevation, some 75,000 years ago. Glaciers
on the east side descend to lower elevations because they are shaded from the afternoon sun. Emmons, the longest
glacier in the park, extends into the White River Valley. Photo by Alan Jacobs.

Geographic Setting

At the beginning of this century, when Mount Rainier (elevation, 14,410 feet) was believed to be the highest mountain in the United States, the name "Columbia Crest" was given to what was thought to be the highest point on the crater's rim. More recent surveys have shown Register Rock to be the highest point at the summit. The grand old title of Columbia Crest remains, although the distinction of being "the highest" has long since passed to loftier peaks.

Few mountains anywhere dominate their surroundings as does Mount Rainier. Viewed from Puget Sound on the west, the volcano appears to rise from the sea. And from any side, Rainier's cone towers over other peaks on the skyline. Yet, several of these lesser peaks around the base of Mount Rainier are over 6000 feet high.

To the Indians who inhabited the region for thousands of years before nonnative people arrived, the mountain was Tahoma, or "Snow Mountain." Out of respect and awe, Indians hunted on the lower slopes only, not daring to approach the summit. A myth of the Nisqually Indians personifies Mount Rainier as a Monster Spirit, who sometimes spewed out her venom in great torrents. This was the Indians' explanation for occasional volcanic mudflows (lahars) that roared down the mountain, filling valleys and overspreading lowlands. Indian encampments were destroyed by prehistoric mudflows, and some of the inhabitants perished.

Mount Rainier impressed Captain George Vancouver of the British Navy when he sailed his ship into Puget Sound in 1792 on his mission to explore the North Pacific coast. He named the mountain for his friend and superior officer, Rear Admiral Peter Rainier, who never saw the great volcano that bears his name.

Mount Rainier, a difficult climb under the best of circumstances, challenged early explorers. The first officially recorded attempt to reach the summit of Rainier was made by Lieutenant A.V. Kautz and his party in 1857. Kautz, Dr. R.O. Craig, four soldiers, and a Nisqually Indian guide almost made it to the top but were turned back by high winds and violent storms. Their approach was from the south and up the Kautz Glacier.

Thirteen years later, in August 1870, General Hazard Stevens and P.B. Van Trump carefully planned an approach to the summit up a narrow arête, the

Cowlitz Cleaver, that separates the snowfields of Cowlitz and Nisqually Glaciers and leads to Gibraltar Rock, which is close to the top (fig. 35.1). When Sluiskin, the Indian guide, realized that Stevens and Van Trump really meant to get to the top, he refused to accompany them because he believed that an evil spirit lived in a lake of fire at the summit. He promised to wait three days and asked for a letter absolving him of blame for their deaths if they did not come back. The two climbers did reach the top but had to spend the night there because of darkness. They thought they would freeze to death until they discovered ice caves and steaming, foul-smelling vents on the summit. They spent an uncomfortable night roasting by the vents and shivering in the caves, but survived and descended the next day. When the anxious Sluiskin saw the two ice-encrusted figures emerging from the fog that had enveloped the mountaintop, he fled in terror, believing them to be ghosts. After they convinced him that they were alive, the party returned to the settlements. The Stevens route to the summit via Gibraltar is now considered the safest approach.

James Longmire, an early settler in the area, assisted Samuel F. Emmons of the U.S. Geological Survey in the early geologic reconnaissance of Mount Rainier and the surrounding region. Longmire himself finally climbed the mountain in 1883. On that trip he discovered mineral springs on the southwest side, and he later built a resort hotel at the site. The Park Headquarters and Visitor Center are now close to Longmire's springs, where the first settlement in the park area was located.

Considerable disagreement arose in the late 1880s over what to call this mountain area that was fast developing into a park—Tahoma, Tacoma, or Mount Rainier. The last name won out, and in 1899 the Mount Rainier area became the fifth national park to be established.

Geologic Features

From Mount Rainier's broad summit, glaciers radiate in all directions. During the volcano's existence of less than a million years, fire and ice (volcanism and glaciation) have continually interacted in molding, altering, and reshaping its form and appearance (fig. 35.3).

Matthes (1928) described Rainier's cone as "resembling an enormous tree stump with spreading base and irregularly broken top." Yet, lofty and massive as the summit is now, it was, as recently as 6000 years

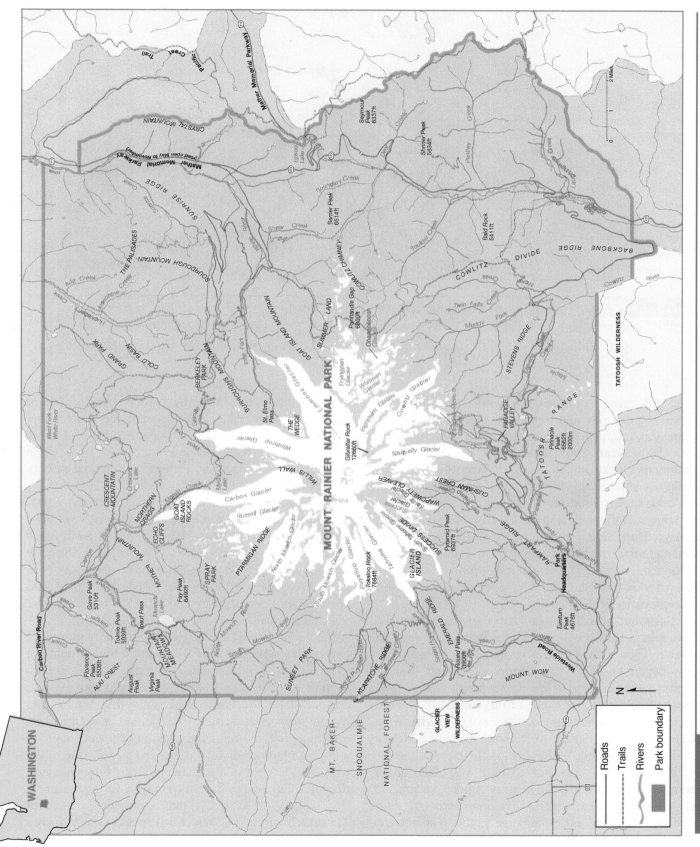

FIGURE 35.2   Mount Rainier National Park, Washington.

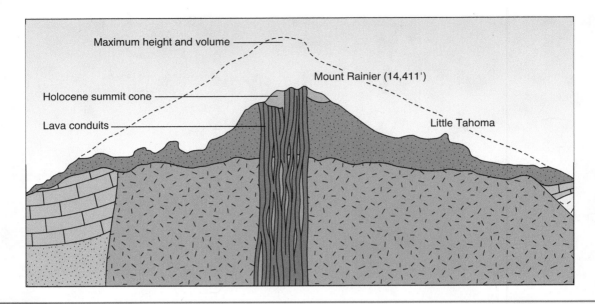

FIGURE 35.3 Cross section of Mount Rainier, with an approximate profile of the volcano's maximum size above its present outline. Modified from *Fire Mountains of the West, The Cascade and Mono Lake Volcanoes* by Stephen L. Harris. Copyright © 1988 by Mountain Press Publishing Company, Missoula, MT. Used by permission.

ago, more than a thousand feet higher than today. A series of eruptions and explosions began about 6600 years ago, culminating in an extremely violent explosion 5700 years ago that destroyed the summit. Enormous blocks of rock and glacial ice were tossed up in the air and landed miles away. Great avalanches of ice and rock roared down the mountain. Burning, steaming mudflows plunged with great speed down the valleys. Dark clouds of ash and steam blotted out the sun, and torrential rains began to fall. The Indians' fear of the mountain god Tahoma is not surprising since stories of this disaster must have been passed down through many generations.

Snow and glacial ice again accumulated and the glaciers rebuilt themselves, beginning again their work of erosion. Then, a mere 2500 years ago, lava flows and ash falls began to repair the summit of Mount Rainier. Within the old *caldera* (a large basin produced by explosion and collapse) a new cone rose. Later, lava opened another smaller crater just to the east of the earlier vent. The two craters intersect at Register Rock, the highest point of the volcano. Within historic time (which in the Pacific Northwest does not go back very far), minor events in the late nineteenth century (smoke, steam, and ash emissions) were reported by observers; but since then the mountain has been relatively quiet. The volcano

is continuously monitored by seismographs placed at various locations and elevations; infrared images that record variations in summit temperatures; and other observations, direct and indirect. Data from these sources have shown recent increases in thermal activity and some movement of magma at depth.

Mount Rainier differs from other Cascade volcanoes in that about 90 percent of its eruptions have been in the form of lava flows. Having only a tenth of its bulk made up of pyroclastics *(tephra)* is abnormally low for a composite cone. Moreover, some of those pyroclastic layers were originally lava that was blasted apart by exploding gases and later reconsolidated as volcanic breccia.

By contrast, Mount St. Helens has had a more violent history with few lava flows and a high volume of pyroclastics. In fact, much of the ash and pumice on Rainier's slopes came from St. Helens during explosive episodes like the 1980 explosions that littered the region with pyroclastics. (*Pumice* is a frothy rock resulting from the sudden explosion of lava containing a large amount of dissolved gas and steam. Due to the explosive release of pressure, tiny gas bubbles cause the solidified lava to have a spongy or frothy texture.)

Around 4900 B.C., the explosion of Mount Mazama in Oregon (Chapter 36) spread several inches

Box 35.1	The Cascade Volcanoes, a Part of the Pacific "Fire Zone"

Mount Rainier is one of more than a dozen geologically young volcanoes with composite cones (stratovolcanoes) that are perched on older rocks of the Cascade Range. In Chapter 28, North Cascades National Park, you learned something of the geologic history of the Cascade Range and the development of two of these volcanoes, Mount Baker and Glacier Peak. In this chapter and the two that follow—36, Crater Lake National Park and 37, Lassen Volcanic National Park—we take the view that in the Cascade region, times of general quietness are merely lulls during a half million years of fairly intense volcanic activity.

The Cascade volcanoes are aligned in a north-south direction, roughly paralleling the Pacific coastline for about 500 miles, from northern California to southern British Columbia (fig. 35.4). They are a small segment of the circum-Pacific volcanic belt, or "Ring of Fire," that encircles the Pacific Ocean. This major volcanic belt marks converging plate boundaries where oceanic crust is subducted beneath overriding continental crust or an island arc.

The volcanoes of major volcanic belts are almost entirely composite cones and andesitic in rock composition. Composite cones are so named because they are constructed of alternating layers of lava flows and pyroclastic eruptions (ash falls, etc.) The Cascade volcanoes are typical composite cones in that the prevailing rock type is andesite and each volcano has grown, over many thousands of years of time, while brief eruptive episodes (lasting years or decades) were interspersed with long periods of quiescence.

Andesite is classified as an intermediate igneous rock because it is compositionally between the light-colored, silica-rich rocks (e.g., rhyolite) and the dark, basaltic rocks that are relatively deficient in silica. Andesite, commonly dark gray or greenish black in color, is made up of approximately equal amounts of light-colored minerals (plagioclase feldspars) and dark minerals (hornblende, olivine, pyroxenes). To geologists studying the Cascade volcanoes, variations in chemical composition, texture, and structure of the lavas and rocks are important inasmuch as very minor differences often enable investigators to identify and date lava flows and ash falls. Volcanic emissions are like fingerprints; that is, no two are exactly alike when samples are examined under a microscope and analyzed chemically.

Although andesite is clearly the predominant rock of the Cascade volcanoes, it is not the exclusive rock type; for at some time in its history, each volcano has had basalt and rhyolite eruptions. Several of the more explosive volcanoes have had numerous dacitic eruptions, dacite being a light andesite with a higher percentage of quartz (i.e., closer to rhyolite). Lavas containing higher percentages of silica are more viscous and tend to erupt more explosively.

The oceanic crust that is being subducted is largely basalt. Where does the andesite come from? According to plate tectonic theory, the basalt originated from submarine eruptions along a spreading zone in the Pacific and has moved like a conveyor belt away from the oceanic ridge to the far boundary of the plate. Along the Pacific Northwest coastline, where the westward-moving North American plate is moving against a segment of eastward-traveling oceanic crust, the less dense continental crust overrides the denser oceanic crust. The oceanic crust, which is relatively young (10 million years old) is subducted beneath North America at a rate of approximately 1.58 inches per year (fig. PIV.4). The subducted slab of basalt, along with some of the overlying sediment, descends at an angle to progressively greater depths, becoming hotter and hotter as it goes down—partly due to the geothermal gradient and partly because of the friction of the slab grinding past the overlying crust. (The *geothermal gradient* is the rate of temperature increase associated with increasing depth beneath the surface of the earth.) At some critical depth, the temperature is high enough so that some of the rock melts and forms magma. As pockets of magma coalesce, they tend to rise—perhaps as large blobs, shouldering aside rocks of the continental crust, or perhaps ascending rapidly through fissures and erupting on the surface.

Why is this rising magma andesitic? Several lines of speculation have been suggested. Perhaps a silica-

(continued)

Box 35.1	The Cascade Volcanoes, a Part of the Pacific "Fire Zone" (continued)

deficient basaltic melt is formed along the subduction zone and is contaminated by silica-rich rocks of the continental crust to produce an intermediate magma; or perhaps only part of the basaltic crust beneath the subduction zone melts, giving a melt higher in silica than the solid residue of oceanic crust remaining in the earth's interior. Or, possibly silica-rich sediment and sedimentary rock carried downward by the descending slab are incorporated into the molten basalt to form andesitic magma.

The line of Cascade volcanoes is believed to be located approximately above the zone of melting deep within the crust. It is supplied by magma rising from this zone and erupting at the surface as lava. Does it follow, then, that all the Cascade volcanoes have had a similar history? The fact is that they are more like a large family of brothers and sisters. They share a common background and have similar characteristics, but each grew and developed in a highly individual way.

Is there a potential for hazards other than volcanic eruptions in the Cascade region? Specifically, what are the estimations of seismic risk? Historically, the Pacific Northwest coastal area has had few large-scale earthquakes, but severe earthquakes accompanied by tsunamis have occurred in prehistoric time in the region. Seismologists suspect that the Cascade subduction zone may be storing strain energy that may be released by catastrophic earthquakes in the future. Any severe shaking of the Puget Sound or Willamette Valley area would cause destructive tsunamis along the Washington and Oregon coasts.

of ash over Mount Rainier and left a blanket of ash on the entire Pacific Northwest, parts of Canada, and even western Montana. Ash that erupted before and during early glacial stages is difficult to trace or date with accuracy because processes of glacial erosion and deposition have left very few undisturbed patches.

The postglacial pumice and ash beds on Mount Rainier are identified by letter; i.e., the fall from Mount Mazama is called Pumice Layer O. The fall of maximum thickness, 20 inches, is Pumice Y from Mount St. Helens. Pumice Layer X, which erupted from Mount Rainier itself between 100 and 150 years ago, has a maximum thickness of only one inch. More recently, the 1980 eruption of Mount St. Helens deposited fresh ash on Rainier's slopes.

Glacial Features

The cone of Mount Rainier, from summit to base, has been deeply furrowed and scarred by longcontinued glacial erosion. Remnants of the volcano's outer layers show on the mountain flanks as sharp crags and ridges (the "cleavers") that jut through the ice and separate the radiating glaciers descending from the snowy summit. One of these residual rock masses, Mount Tahoma, is described by Matthes (1928) as "a sharp, triangular tooth on the east flank" of Rainier that rises to an elevation of over 11,000 feet. "In its steep, ice-carved walls one may trace ascending volcanic strata aggregating 2000 feet in thickness that point upward to the place of their origin, the former summit of the mountain" (more than a thousand feet higher than the present top).

The caldera rim left by the explosion that destroyed the summit has been breached and the rock crumbled by overriding ice cascades. Even the new cone is beginning to show minor indications of erosion by glacial ice. Several hundred feet of snow fill the two young craters. Beneath the snow are the steam caves. An extensive network of tunnels is in the east crater. The west crater has a smaller cave system but contains a pool of meltwater at the present time under a canopy of ice. If only a small increase in vent temperature occurred for any length of time, lakes might form on the snowpack surface in the craters. This has happened from time to time in the past.

Storm clouds from the Pacific Ocean, borne by prevailing westerlies, drop a large part of their load at the lower and middle elevations on Mount Rainier, giving these zones more precipitation than the summit.

Box 35.1	The Cascade Volcanoes, a Part of the Pacific "Fire Zone" (continued)

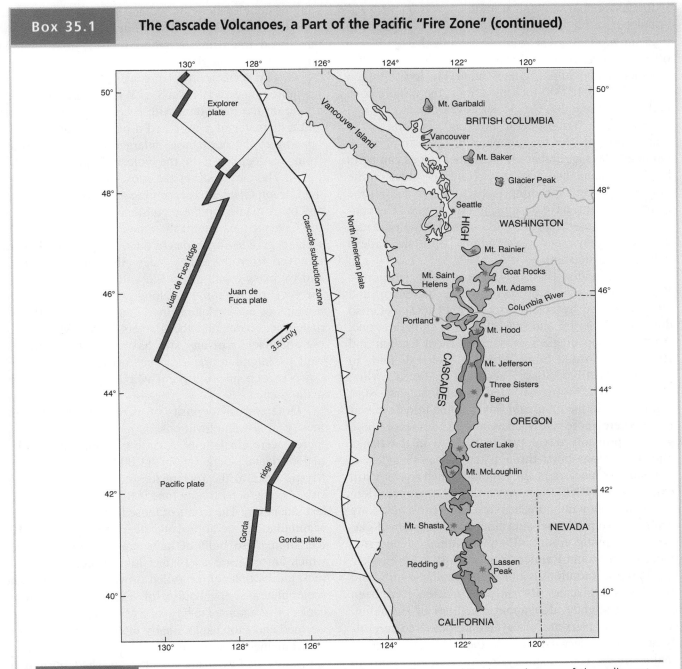

FIGURE 35.4   Plate tectonic elements—the Gorda, Juan de Fuca, and Explorer plates—of the adjacent Pacific basin are pulled down into the subduction zone and underneath the North American plate. The arrow indicates the movement and direction of convergence between the Juan de Fuca and North American plates. The Cascade volcanoes are aligned at the surface over a melt zone at depth. Older volcanic rocks are shown in lighter stippled areas; darker stippled areas locate younger volcanic rocks. Christiansen, R.L. and Yeats, R.S. 1992. Post-Laramide geology of the U.S. Cordilleran region (Chapter 7) in Burchfiel, B.C., Lipman, P.W. and Zoback, M.L., eds. *The Cordilleran Orogen: Conterminous U.S.* Boulder, Colorado: Geological Society of America. *The Geology of North America,* v. G-3. p.347. Reprinted with permission.

Estimated annual snowfall at middle levels on the western slopes of the mountain exceeds 100 feet. Added to high snowfall are great volumes of snow that are blown off the summit or come down in avalanches. And since summer temperatures remain relatively low between the 8000- and 10,000-foot elevations, the middle zone is the most favorable for glacier development.

Rainier's "atmospheric layering" is familiar to those who live near the mountain or come to climb it. On many days a climber on the higher ridges can be in bright sunshine and look back and see thick white clouds covering everything below; valleys and all the lower mountains have disappeared from view. The converse is true (as climbers should be so cautioned); sudden storms may envelop the summit while the valleys below are in sunshine.

The glaciers that originate at the summit—Kautz, Tahoma, Winthrop, Emmons, Ingraham, and Nisqually—are "refueled" after they get 4000 feet or so down from the top and reach the high snowfall zone. Another group of glaciers—among them Carbon and Russell, North and South Mowich with Edmunds in between, Puyallup, South Tahoma, Paradise, Cowlitz, Ohanapecosh, Fryingpan, and several others—start from basins in the mountain flanks at the middle elevations where more snow accumulates. Some of these glaciers probably developed in hollows in the flanks after explosions burst through the volcano's sides. A few smaller glaciers in the Mount Rainier system still exist at around 6000 feet on north-facing slopes. Here the winter snowfalls are heaviest, and snowfields are protected from summer sunshine. In all, 26 named glaciers, occupying around 40 square miles of land surface, are in Mount Rainier National Park.

Periodic variations in climate affect the volume and length of the glaciers. Recently the glaciers have been advancing slightly after about a century of shrinking. The larger glaciers are surveyed and measured carefully at regular intervals, because they are the source of major streams (the Nisqually, Puyallup, White, Carbon, and Cowlitz Rivers) that supply hydroelectric power to the region. Emmons Glacier, largest in the park, supplies the White River. Flow in these rivers tends to be low in winter and high in the summer melt season. The effect of meltwater on the streams is perhaps most striking near the headwaters. In the early morning a small stream issuing from a glacier can be crossed easily on foot; but by midafternoon on a sunny summer day this same stream is a torrent of milky-white water,

banging cobbles along its bed and far too swift and dangerous to wade across. The white color is due to the pulverized fragments or grains of *rock flour* scoured and abraded by moving glaciers.

Glacial erosion on a large scale is most evident in precipitous headwalls, such as Willis Wall, the largest cirque on the mountain, with a headwall of 3600 feet and a diameter of a mile and a half. This great basin, opening to the north, was enlarged by glacial erosion from "a deep gash in the volcano's side" (Matthes 1928). The cirque stores the snow and ice that enables the Carbon Glacier to descend to an altitude lower than any other glacier on the mountain.

Conspicuous lateral moraines descending with the glaciers give us some idea of the quantity of rock being removed from the valley sidewalls, both by abrasion and by mass wasting. On the valley sidewalls, above the rock debris moving along with the ice, are the remains of earlier lateral moraines built when the climate was colder and the glaciers larger. Most of the present glaciers of any size have medial moraines as well as lateral moraines. The medial moraines trail down from a junction point where a branch joins the main glacier.

Drift and till deposits of various ages are found mostly at lower elevations in moraines built during more severe glaciations. A large moraine from the last major glacial stage (about 11,000 years ago) dams Mystic Lake at the foot of the mountain on the north side. *Rock glaciers,* now inactive, were also left by the last glaciation. The fronts of these large talus and debris accumulations are abrupt and steep and from 25 to 100 feet high. Also built of talus are *protalus ramparts,* which are ridges 5 to 15 feet high of coarse, angular rock pieces that fell or rolled down from heights and accumulated at the bottom of large perennial snowbanks piled against the base of a cliff. When the climate warmed and snowbanks finally melted, a protalus rampart remained, standing apart from the toe of the slope.

Valley trains, made up of outwash deposited by meltwater streams, cover the floors of the narrow valleys that descend from the ends of the glaciers. Valley trains that come down from Mount Rainier have been modified by ash falls and mudflows. Some valley trains open out onto outwash plains at lower elevations.

Lovely cirque lakes, like Tipsoo Lake, Crescent Lake, Mowich Lake, and others, can be seen in Mount Rainier National Park. Some tributary streams that flow in hanging valleys have waterfalls. Christine Falls and

FIGURE 35.5   Mount Rainier as seen from the air. It is considered the next volcano that is likely to erupt in the lower 48 states. Photo by Ann G. Harris.

Comet Falls are examples. Ice caves, formed by meltwater, are usually visible at the bottom of Paradise, Carbon, and several other glaciers (fig. 35.6).

Mass Wasting Processes on Mount Rainier

Rockfalls, rockslides, avalanches, debris flows, mudflows, and other processes of downslope movement, singly or in combination, operate more or less continually in the Mount Rainier environment. The interaction of processes of mass wasting, weathering, volcanism, and glaciation has created a powerful group of mechanisms for eroding, transporting, and depositing debris far and wide. Their potential for seriously disrupting the environment is a matter of grave concern to volcano-watchers. Consider the following items.

On almost any clear day, those who visit the Paradise or Sunset overlooks (or other vantage points) can see dust clouds almost hourly rising from small rockfalls and avalanches near the summit. And scanning the snowfields of major glaciers with telescope or field glasses, one sees fresh, dark debris that has fallen since the last snow came, perhaps the previous night or within the past few days.

Geologists have mapped more than 55 lahars (flows of pyroclasts, volcanic debris, mud, etc., usually mixed with hot water) that have flowed down Mount Rainier in Holocene (postglacial) time. The longest and largest of these, the Osceola Mudflow (about 5700 years ago) ran about 70 miles to the northwest, altered the course of the White River, and filled in an arm of Puget Sound.

In October 1947, within a span of about 8 hours following heavy rains, successive mudflows roared down Kautz Creek to the Nisqually River, leaving behind several cubic miles of muddy debris in the

FIGURE **35.6**   Inside the Paradise Glacier Ice Cave that is located on the south side of Mount Rainier. The deep blue color of the ice is due to the fact that most of the air has been forced out, making the ice extremely dense; therefore, it absorbs every wave length of visible light except blue, which it reflects. Photo by Alan Jacobs.

Kautz valley. Witnesses reported that large trees and boulders 13 feet in diameter were borne along by the mudflows, which had the consistency of wet cement (fig. 35.8B).

In December 1963, a series of rockfalls and avalanches, apparently triggered by a steam explosion, fell in quick succession from the north face of Little Tahoma Peak. Plummeting downward, this great volume of broken rock struck the Emmons Glacier with great force and then bounced and flew down the White River valley. While dropping some 6200 feet in altitude, the avalanche traveled nearly four miles from its source. Some of the boulders that were transported are as large as buildings.

In the 1990s, repeated (and continuing) outburst floods from Tahoma and South Tahoma Glaciers caused closing of the Westside Road and Tahoma Creek trail.

These examples are by no means unusual events in the geologic story of Mount Rainier. Mixtures of ash, pumice, fragmented bedrock, and solid blocks blanket the mountain flanks and are found as thick fill in the valleys that lead away from the cone. The action of hot water and steam on the lavas and ashfalls at the summit and on the slopes causes any consolidated material to break up both mechanically and chemically, producing quantities of debris for mass wasting processes, glaciers, and streams to carry away.

Glaciers are efficient at breaking and grinding up less resistant volcanic rock. On Mount Rainier active glaciers oversteepen slopes, expose fresh surfaces, and pulverize loose fragments. Continual freeze-and-thaw action at high altitudes on rock unprotected by soil or vegetation tends to break up and shatter the rock. Loosened fragments fall off cliffs and roll down slopes. Tremors or even slight explosions shake loose more rocks.

The fact that andesite, the dominant rock, is more susceptible to chemical weathering than, say, granite is significant. The feldspars and ferromagnesian minerals that make up andesite tend to weather out as clay minerals and soluble salts, especially in the presence of hot water, which hastens the chemical action. Given the availability of great quantities of loose (or easily loosened) material, the dynamics of slides and flows can be explained as follows: (1) Precipitation and runoff on Mount Rainier are high. Rocks and debris are water-soaked much of the time. Material that is wet tends to move, slip, and slide faster than dry material. (2) Slopes at high altitudes are generally steep and unprotected by vegetation. Heavy snow accumulations tend to set off avalanching and slides. (3) Eruptions drop ash and pumice on snow accumulations. Mix in steam and hot water and all at once you have a mudflow. Add melting glacial ice and rock debris, and you have a disaster roaring down valleys (fig. 35.7).

The distinction between debris flows and mudflows is gradational; that is, a *mudflow* is a variety of *debris flow* in which the solids are at least 50 percent sand, silt, ash, and clay. The water content critically affects velocity and mobility. By gaining water along the way, a mudflow becomes more of a slurry; and if it runs into a good-sized river, then that river becomes an overloaded stream that floods lowlands and deposits the mountain's debris on flood plains and, in the Pacific Northwest, in harbors.

According to estimates, lahars and mudflows travel at speeds of about 50 miles per hour. This doesn't allow much time for people and vehicles to get out of the way. However, authorities at Mount Rainier National Park monitor conditions daily and have set up early warning systems and evacuation plans. Campgrounds have been relocated on higher ground. The park would be closed

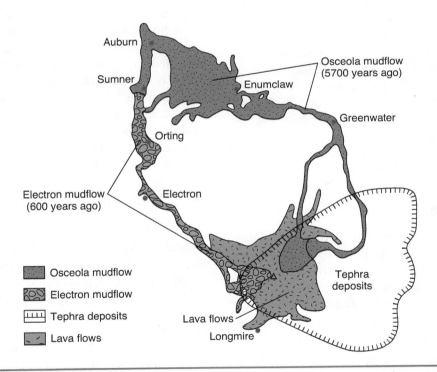

FIGURE 35.7 Tephra deposits, lava flows and mudflows from Mount Rainier in Holocene time. River valleys radiating from the north side of the volcano appear to be at greatest risk. Modified from *Fire Mountains of the West, The Cascade and Mono Lake Volcanoes* by Stephen L. Harris. Copyright © 1988 by Mountain Press Publishing Company, Missoula, MT. Used by permission.

immediately if thermal or seismic data indicated any possibility of an eruption. Meanwhile, visitors are able to enjoy the park and its many natural wonders without being overly apprehensive.

Geologic History

1. Eocene lowland.

In the early part of the Cenozoic Era, the area now occupied by the Cascades was a coastal plain. Over several million years, sand, clay, and organic swamp deposits accumulated in subsiding tidal marshes. These sediments became a 10,000-foot thick sequence of sandstone, shale, and coal (some of which has been mined) known as the Puget Group. Arkose of the Puget Group interfingers with the younger Ohanapecosh Formation near the park's western boundary, but is not exposed in the park.

2. Eocene volcanism, continuing into Oligocene time.

About 40 million years ago, volcanic activity began as a local manifestation of large-scale tectonic activities that affected western North America from late Mesozoic time on into the present. The eruptions marked the beginning of subduction of oceanic crust. Evidence for this volcanism, which occurred millions of years before Mount Rainier's cone was built, is in the Ohanapecosh Formation, overlying the Puget Group. The Ohanapecosh Formation contains breccia from volcanic explosions and material from lava flows, mudflows, and ash falls, all of which accumulated in shallow water. Sinking continued to keep pace with the build up of deposits on the ocean floor.

3. Oligocene uplift.

As convergence of plates continued, uplift accompanied by compression raised the Puget and Ohanapecosh rocks above sea level and gently folded them during early Oligocene time. During this period, a warm, wet climate prevailed, resulting in deep weathering and extensive erosion. A thick red soil developed, and streams carved gullies between low hills.

4. Oligocene volcanism; Stevens Ridge Formation.

Later in the Oligocene, perhaps 25 or 30 million years ago, volcanism began again. Ash falls and flows covered the area. Violent, exploding clouds of fine pumice erupted. As these glowing avalanches settled, ash particles with still molten edges bonded together to form a *welded tuff*. Ash flows filled the valleys, and eventually a plateau was built. This rock unit became the Stevens Ridge Formation. Outcrops of the lightgray welded tuff, containing darker inclusions of angular pumice fragments, can be seen in the southern part of the park along the highway through Stevens Canyon.

5. Continued volcanism in Miocene time; Fifes Peak Formation.

The Fifes Peak Formation shows that the character of the volcanism was changing. Many flows of both andesite and basalt lava, ranging from 150 to 500 feet thick, piled up to a total thickness of 2500 feet. Rocks of the Fifes Peak Formation make up many of the cliffs and peaks in the northwestern part of the Mount Rainier National Park.

6. Miocene deformation and uplift; intrusion of granodiorite.

As Fifes Peak volcanism ceased, the region was again uplifted and gently folded. (The already folded, underlying Ohanapecosh Formation was compressed into tighter folds.) Some faults, with displacements of thousands of feet, resulted from the compression.

A body of magma welled up through the crust. A small amount erupted as lava, but most of the magma solidified underground, forming the Tatoosh pluton, which is composed largely of coarse-grained granodiorite having a "salt-and-pepper" appearance (i.e., light-colored and speckled with large dark mineral grains). Dikes and sills are associated with the pluton, and some contact metamorphism occurred in the country rock. The Tatoosh pluton has been dated radiometrically as being about 12 million years old.

7. Pliocene uplift and erosion of the Cascade Range.

The Cascade Range continued to rise, and deep erosion carved the region into peaks and ridges separated by stream-carved valleys. The rocks overlying the Tatoosh pluton were eroded, and the pluton was "unroofed." The granodiorite of the pluton is exposed in the Carbon and White River valleys and part of the upper Nisqually River valley. Plutonic rocks also form the central part of the craggy Tatoosh Range in the southern section of the park, near the base of Mount Rainier.

8. The building of Mount Rainier, a composite cone; repeated episodes of Pleistocene glaciation.

The climate was already very cold and the Tatoosh Mountains were undergoing glaciation when construc-

Table 35.1			Generalized Geologic Column, Mount Rainier National Park	

Time Units			**Rock Units**	**Geologic Events**
Era	**Period**	**Epoch**		
Cenozoic	Quaternary	Holocene	Surficial deposits: avalanches, lahars, ash, pumice, glacial and alluvial material Andesite lava: pumice, tuff	Mudflows, debris flows, ash falls Growth and retreat of modern glaciers New cone constructed in old caldera Osceola mudflow and lahars Explosive destruction of old summit
		Pleistocene	Ash, pumice, lahar deposits Interbedded andesite lavas and breccias Glacial deposits throughout epoch Andesite lavas Intracanyon lavas	Building of composite cone to maximum height Glacial and interglacial stages Beginning of Mount Rainier Filling of valleys by lavas
	Tertiary — Neogene	Pliocene		Beginning of glaciation Uplift and erosion of Cascade Range
		Miocene	Tatoosh Granodiorite - Fifes Peak Formation	Intrusion of Tatoosh pluton with associated volcanics dikes, sills Andesite eruptions; basalt lava flows; pyroclastics
	Tertiary — Paleogene		— ? — ? — ? — ? —	
		Oligocene	Stevens Ridge Formation	Extensive ash flows, ash falls; welded tuff
			/////////////////////////	Unconformity Folding, erosion; weathering of red soil
			Ohanapecosh Formation	Andesite lava eruptions; volcanic breccias, mudflows, ash falls deposited in shallow water and on land
			— ? — ? — ? — ? —	
		Eocene	Puget Group (not exposed in park) — ? — ? — ? — ? —	Sand, clay, swamp deposits; coal beds

Source: Modified after Fiske et al., 1963; Crandell, 1969.

tion of the cone began with a series of relatively fluid lava flows early in the Pleistocene Epoch. The lavas flowed into stream-carved canyons, eventually filling them. Continued eruptions built a low, broad volcano on top of the mountainous terrain. Mudflow and glacial deposits were interlayered between lava flows, indicating that glaciation was ongoing throughout this time. Then a large number of short lava streams from a central crater, interspersed with ejections of breccia and ash, began constructing the cone, which rose to a height of perhaps 15,500 to 16,000 feet. A plug of solidified magma filled the central vent, and dikes radiated from the center, like spokes in a wheel. The peak probably attained its greatest height about 75,000 years ago.

9. Erosion of the cone and destruction of the summit.

Glacial erosion steepened the mountain flanks, and by the end of the Pleistocene, perhaps a fourth of the

FIGURE 35.8A   Sign in campground gives a warning to campers. Photograph by Gerald Morrison.

FIGURE 35.8B   In October 1947, within a span of eight hours following heavy rains, a series of mudflows roared down Kautz Creek to the Nisqually River, filling the valley with debris. Photo by Alan Jacobs.

FIGURE 35.9   Crevasse system of Nisqually Glacier located on the south side of Mount Rainier. The crevasses are bowed in the center because the ice can move faster when it is ice sliding over ice and is not slowed down by abrasion, as it is along the sides. Photo by Alan Jacobs.

mountain's bulk had been removed. Steam and occasional eruptions at the summit prevented glaciers from eroding at the top of the mountain, so the summit remained high.

Beginning about 6600 years ago, a series of violent explosions blasted away an enormous portion of the eastern side of the mountain and, later, the summit itself. It was probably the last and greatest explosion that caused the catastrophic Osceola Mudflow that traveled as far as Puget Sound. Rainier's lowered summit area became a caldera about two miles across, the highest points on the rim barely 14,000 feet above sea level. Glacial ice filled the caldera and eroded the rim and caldera basin.

10. Reconstruction of the summit by lava flows and pumice.

Reconstruction began about 2500 years ago when pyroclastic eruptions, followed by lava flows, built a new cone within the caldera. In time, eruptions shifted somewhat, creating a double crater for this "volcano upon a volcano." The mountain probably erupted at least once between 1820 and 1854 while Mount St.

Helens was having a series of pyroclastic eruptions. Minor eruptions may have occurred again on Mount Rainier in the 1870s and 1880s. Fumaroles (small vents emitting vapor and gas) and occasional steam explosions on the flanks of Mount Rainier indicate continued high heat flow. *Microseisms* (very small earth tremors) suggest that eruptions may begin again.

Geologic Map and Cross Section

Fiske, R.S., Hopson, C.A., and Waters, A.C. 1964. Geologic Map and Section, Mount Rainier National Park, Washington. U.S. Geological Survey Misc. Geological Investigations. Map I-432.

Sources

Christiansen, R.L.; Yeats, R.S., et al. 1992. Post-Laramide Geology of the U.S. Cordilleran Region in Burchfiel, B.C.; Lipman, P.W.; and Zoback, M.L., eds. *The Cordilleran Orogen*, Conterminous U.S. Boulder, Colorado Geological Society of America, v.G-3, p. 261-406.

Crandell, D.R. 1969. Surficial Geology of Mount Rainier National Park, Washington. U.S. Geological Survey Bulletin 1288 (text plus map in folder).

————. 1969. The Geologic Story of Mount Rainier, Washington. U.S. Geological Survey Bulletin 1292.

————. 1971. Postglacial Lahars from Mount Rainier Volcano, Washington. U.S. Geological Survey Professional Paper 677.

Crandell, D.R. and Mullineaux, D.R. 1967. Volcanic Hazards at Mount Rainier, Washington. U.S. Geological Survey Bulletin 1238.

Fiske, R.S.; Hopson, C.A.; and Waters, A.C. 1963. Geology of Mount Rainier National Park, Washington. U.S. Geological Survey Professional Paper 444.

Harris, S.L. 1988. *Fire Mountains of the West: The Cascade and Mono Lake Volcanoes.* Missoula, Montana: Mountain Press Publishing Co.

Heaton, T.H. and Hartzell, S.H. 1987. Earthquake Hazards on the Cascadia Subduction Zone. *Science* 236: 162-168, April 10.

Hopson, C.A., Waters, A.C., Bender, V.R., and Rubin, M. 1962. The Latest Eruptions from Mount Rainier Volcano. *Journal of Geology* 70:635-647.

Matthes, F.E. 1928. Mount Rainier and Its Glaciers. National Park Service.

————. 1955 (revised) Mount Rainier National Park, Washington. Text on back of quadrangle map. U.S. Geological Survey.

Sigafoos, R.S., and Hendricks, E.L. 1972. Recent Activity of Glaciers, Mount Rainier, Washington. U.S. Geological Survey Professional Paper 387-B.

Note: To convert English measurements to metric, go to www.helpwithdiy.com/metric_conversion_calculator.html

C H A P T E R   3 6

Crater Lake National Park

SOUTHWEST OREGON

Area: 183,224 acres; 287 square miles

Established as a National Park: May 22, 1902

Address: P.O. Box 7, Crater Lake, OR 97604

Phone: 541-594-3100

E-mail: CRLA_Information_Requests@nps.gov

Website: www.nps.gov/crla/index.htm

Crater Lake, 21 square miles in size and over 1900 feet deep, lies in the caldera of a collapsed volcano. Mazama, one of the composite cones in the High Cascades, exploded in climactic eruptions nearly 7000 years ago.

In multicolored walls, encircling the lake and rising 500 to 2000 feet from the water, are the records of intense volcanic and glacial activity during Mazama's construction and destruction.

Pumice pinnacles, a pumice "desert," plug domes, satellite craters, cinder cones, U-shaped valleys, and more are in this remarkable national park.

FIGURE 36.1 Looking north over Crater Lake and Wizard Island. Llao Rock juts out from the caldera rim. National Park Service photo by C. L. Summer.

. . . The color of the water is ultramarine, bordered by turquoise along the shores. Set in majestic cliffs, it is a most attractive natural jewel ...

—*J.S. Diller, U.S. Geological Survey, 1911.*

If you have seen Crater Lake, you can imagine how astonishing and mysterious it must have seemed to early visitors to Oregon Territory. John Wesley Hillman, a young prospector, came upon the place quite by accident in 1853 as he was following his mule up a ridge in search of a lost mine. When the mule suddenly stopped on the brink of a cliff, Hillman looked down over the edge and saw below the indigo water of the lake in its great caldera, a good five miles from rim to rim. He called it "Deep Blue Lake." The spot where he stood, on the southwest side of the rim, is now Discovery Point. Later, settlers in the region decided "Crater Lake" was the ideal name. This name has been retained, despite assertions by teachers, geologists, and park rangers that it should be called "Caldera Lake"!

The lake's intense blue color is due to the purity of the water and the great depth of the caldera. Crater Lake's officially measured depth is 1932 feet, which makes it the deepest lake in the United States. The water, which has accumulated from rain and snow, contains little dissolved or suspended material. Measurements of visibility in this transparent water have reached depths of over 325 feet. The lake was devoid of aquatic life until some 60 years ago when fish were introduced. Freshwater shrimp had to be brought in first so that when the lake was stocked with trout, the fish would have something to eat.

Except for minor seasonal changes, the level of the water remains fairly constant from year to year; the intake from precipitation balances loss from seepage and evaporation. Springs on the outer slopes of the caldera are fed by local ground water, not by subterranean discharge from Crater Lake. Summer temperatures of the lake water average 55° F; winter temperatures approach freezing, but the lake seldom freezes over.

The seasonal and annual variability of Crater Lake make it an extremely complex and dynamic system. Monitoring programs involving consistent data collection have not been in place long enough for scientists to understand fully how the lake's ecosystem works, although considerable progress has been made since the Limnological Study (funded by Congress) began in the

early 1980s. Nevertheless, except for the consequences of fish introductions, long-term changes caused by human activities could not be separated from changes caused by natural phenomena.

For example, during a period in the 1970s, when the lake's clarity appeared to decrease, higher than expected nitrate levels were found in water samples, especially from one of the springs in the caldera wall. As a precaution, a septic system's drain field was moved well away from the caldera rim. Since then nitrate levels have not changed significantly in the lake or the spring, but in 1994, researchers recorded the highest clarity readings to date in the lake. Since chemical analyses did not show a connection between the original drain field and the spring in question, it may be that the water in that spring has a natural source for the slightly higher nitrate readings. The main source of nitrogen in the lake water, about 90 percent, comes from the atmosphere.

As a part of the study, a submersible was brought in to explore the lake floor. Helicopters landed the 7000-pound submersible and more than 20,000 pounds of equipment on Wizard Island. Scientists who cruised the lake bottom in the submersible observed hydrothermal vents where they discovered blue saline pools, bacteria mats, and fields of moss (Weingrod 1995).

An Oregonian, Judge William Gladstone Steel, was instrumental in the movement to preserve the Crater Lake region as a national park. As a boy in Kansas, he had read about a mysterious, deep blue lake in Oregon. When he was 18 years old, his family moved to Oregon, and soon young Steel began searching for the lake, asking all the oldtimers where it was. Finally, in 1885, after 13 years of searching, he stood on the caldera's rim and looked down at Crater Lake. Then and there, he decided that it must be protected and preserved.

Geologist Clarence Dutton helped Judge Steel organize a U.S. Geological Survey party in order to gather scientific data to support their hopes for a national park. In 1886, members of the party carried in a 26-foot boat that they lowered into the lake. By making soundings with a pipe and piano wire in different parts of the lake, the men obtained a depth reading quite close to the official depth of today (1932 feet). Meanwhile, a topographer surveyed the caldera rim and vicinity and prepared the first map of Crater Lake. Steel gave names to many of the features.

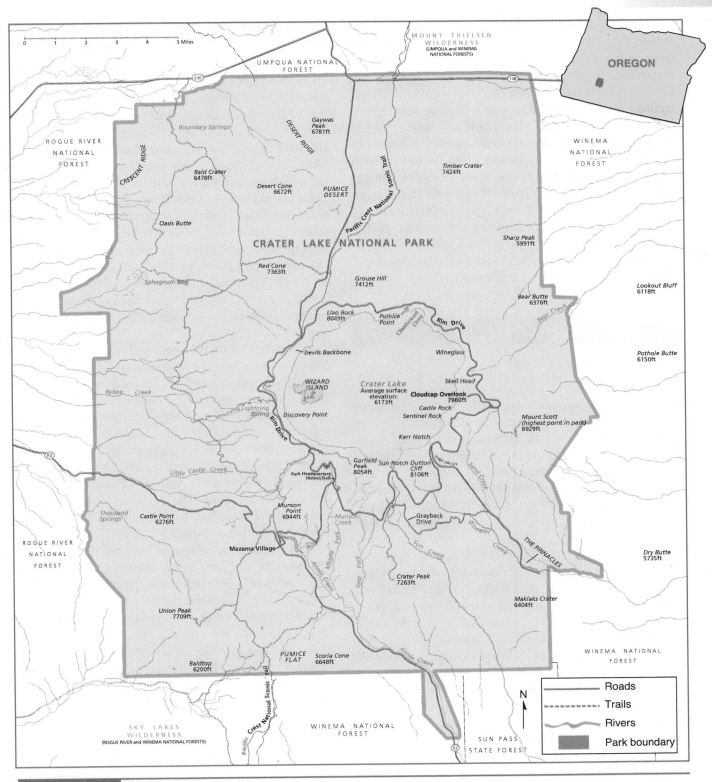

FIGURE 36.2   Crater Lake National Park, Oregon.

Highways into the park lead to a scenic drive around the caldera rim. Probably the best vantage point to see many features of geologic interest is the top of Mount Scott, reached by a trail from Rim Drive. Looking down one can see both volcanic and glacial landforms around the caldera, as well as an incomparable view of the lake. To the north and south rise the High Cascades, topped by snowcapped volcanoes. To the west is "a sea of peaks and ridges"—the Western Cascades. Beyond them, the more distant Klamath-Siskiyou Mountains are seen. To the east stretches an immense platform of basalt (the Columbia River basalts) that goes farther than one can see into Idaho. The lava plateau is cut by canyons, and a few steep-scarped, tilted, fault-block mountain ranges break the flatness of the surface in the distance.

Geologic Features

Mazama's Volcanic Rocks

Basalt, andesite, trachyte, and *rhyolite* are compositional varieties of dense-textured, extrusive igneous rocks, ranging from the darker basalts (higher in iron and magnesium, lower in silica) to light-colored rhyolite (lower in iron and magnesium, higher in silica).

Dacite, a compositional variety containing principally plagioclase and quartz, is difficult to distinguish from rhyolite without the use of petrographic techniques. Much of the rock at Crater Lake is described as *rhyodacite.*

Pumice is a textural term for a frothlike volcanic glass that cooled from liquid lava during eruption. A typical pumice contains so much trapped air that it can float. The composition is usually that of rhyolite or trachyte.

Tuff is a general term for consolidated rocks made up of pyroclastic fragments of various sizes. A *tuff breccia,* for example, consists of about half ash and half larger fragments. In a *welded tuff,* glass shards, hot ash, pumice, and cinders have been fused by hot gases and the weight and heat of overlying materials. Uneven compaction causes laminations and banding. The composition is highly siliceous.

Scoria refers to cinders or a cindery crust that has a vesicular texture (full of holes). Escaping volcanic gases during the eruption and solidification of lava produce vesicular texture. Scoria, which is usually darker and heavier than pumice, is andesitic or basaltic in composition.

Volcanic Features

The large caldera occupied by Crater Lake plus the encircling flanks and ridges are about all that is left of once-mighty Mazama. Its early history was similar to that of the other Cascade volcanoes. Like its contemporaries—Mount Rainier (Chapter 35) and Lassen (Chapter 37)—Mazama grew during the Pleistocene, while alpine glaciers advanced and retreated on its slopes. From its foundation rocks at the base (between 5000 and 6000 feet in elevation), Mazama rose another 6000 feet to an elevation of approximately 11,000 feet above sea level. The cone was broader and less steep than Shasta's or Rainier's because the subsidiary vents and overlapping cones gave the volcano a bulky, irregular shape. The present caldera rim is 7000 to 8000 feet above sea level, and geologists have estimated that the original cone probably rose 2500 to 3500 feet above that.

Associated with the construction and destruction of Mazama are a variety of extrusive and intrusive features. Thick pumice and ash deposits, relics of the violent explosive eruptions, lie on the flanks of the mountain, relatively fresh and only mildly weathered by the elements since they fell. Some of the deposits are welded tuffs produced by fiery clouds and ash flows. The Pumice Desert, which stretches away to the north from the rim, is a very thick blanket of ash that is mostly devoid of vegetation. The pyroclastic material is too loose to hold water and too siliceous to develop soil or support much plant growth.

The Pinnacles, an area of tall spires about 5 miles southeast of the rim, are erosional remnants of tuff deposits, 200 to 300 feet thick, ejected in flaming ash flows during the climactic eruptions. Steam clouds rose for years after the hot pyroclastics blanketed the slopes and filled the canyons. Where steam was localized, as in fumaroles, particles became cemented together in the shape of pinnacles, or spires, and were more resistant to erosion. Thin, hollow tubes, where gas escaped, still go down through the centers of some of the pinnacles (fig. 36.3).

Associated Composite Cones

Mount Bailey and Mount Thielsen, both a short distance north of the park, probably became inactive early in the Pleistocene, perhaps around the time that Mazama started up. Mount Scott, highest point in the park (8926 feet), developed as a parasitic cone from

Mazama's eastern base. Mount Scott's very steep cone is made up of layers of andesite lava, closely related chemically to some of the earlier flows of Mazama. Evidently, Mount Scott became extinct before the end of the Pleistocene. A large cirque on the northwest side indicates glacial erosion that was not "repaired" by postglacial volcanic activity. Not much is left of Union Peak in the southwest corner of the park except its volcanic plug, an erosional remnant of lava that solidified to resistant rock in the neck of the volcano.

Shield Volcanoes and Cinder Cones

Basalt lavas and andesitic basalts built a number of satellite cones and craters around Mazama. Crater Peak formed as a shield volcano of andesite and basalt flows and later erupted pyroclastics of andesite and dacite. Another shield volcano, Timber Crater, in the northeast corner of the park, is also made up of basalt and basaltic andesite flows, but its shield is capped by two cinder cones. None of the cinder cones are more than a thousand feet high from base to summit. Examples are Desert Cone, Bald Crater, Red Cone, and Diller Cone. Volcanic *bombs* (fragments of lava, or ejecta, usually rounded, ropey, or spiral, that hardened while in the air) lie on the cinder cones in myriad sizes and shapes. Some of the cinder cones are aligned on what seems to have been a fissure system tapping a magma chamber.

Wizard Island, the large cinder cone in Crater Lake, resembles a wizard's hat, for which it was named. It formed in the caldera from eruptions after Mazama's collapse. The cone rises nearly 800 feet above lake level. Its well developed crater is 400 feet across and about 90 feet deep (fig. 36.1).

Dikes and Other Features of the Caldera Walls

Most of the large andesite dikes exposed in the park are old conduits and radial fissure fillings that probably served as feeders for Mazama's flank eruptions. Devil's Backbone, which runs up the cliffs of Crater Lake from the water's edge to the west rim, is the most conspicuous dike (fig. 36.4). A similar dike forms a high vertical wall immediately beneath the Watchman and merges upward into the Watchman lava flow (western rim). Phantom Ship, a much older eroded dike or conduit remnant, sticks out of the water near the lake's south shore (fig. 36.5B).

FIGURE **36.3** The Pinnacles, some five miles southeast of the caldera rim, were fused by hot gases escaping through many feet of ash after the climatic eruption over 6,000 years ago. Erosion has removed the thick cover of loose ash, leaving the resistant pillars standing in the deep canyon of Sand Creek. On the wall in the background the lower light area of the welded tuff has a rhyodacite composition, while the upper darker area has an andesitic or basaltic composition. National Park Service photo.

thought to have been formed by a lava flow that filled a glacial trough; but more recent evidence indicates that the flow was emplaced in an explosion crater (Bacon, 1983).

Feeder dikes of Hillman Peak are exposed in the remaining half of the old conduit that was laid bare by explosions that removed the eastern half of its cone (fig. 36.6). Hillman Peak, highest spot on the rim, was one of the parasite cones of Mazama and now reveals a nearly perfect cross section of its interior due to the blasting out of the caldera.

The unusual colors displayed in the caldera walls—hues of yellow, buff, brown, orange, and faint purple—are the result of hydrothermal weathering of the volcanic rocks by ascending ground water and chemically charged gases. Caves in the caldera walls, mostly along the water level, are the result of the erosion of lavas or mudflows between lava flows. Some caves, also eroded from lava, are outside the rim.

Dacite Domes

Dacite flows that began erupting toward the end of Mazama's development tended to become thicker and more viscous. Some earlier dacite flows are overlain by andesite or by glacial drift. The later, more viscous dacites formed domes. Some, such as Lookout, Pothole, and Dry Buttes, are isolated domes. Others, like the group south and east of Mount Scott, are in clusters.

Glacial Features

Between eruptions, when the mountain was quiet, glaciers rebuilt themselves and carved out deep valleys on the flanks of the volcano. When eruptive activities resumed, the glacial troughs provided convenient channels for hot lavas, ash flows, lahars, and glowing avalanches. The culminating eruptions and explosions truncated, or "beheaded," glacial valleys that extended from the high slopes to elevations lower than the present caldera rim. Kerr Valley and Sun Notch on the southeast side are spectacular U-shaped troughs beheaded by the explosions and now exposed in the precipitous caldera wall. Also exposed in the cliffs rising from the lake are horizons of glacial debris interspersed with layers of lava and pumice. Grooves of glacial striations come to the very brink of the cliffs; where layers of andesite jut out slightly from the cliff

FIGURE 36.4   The Devil's Backbone, a large andesite dike exposed by erosion of the soft material in the caldera's west wall. National Park Service photo.

Dacite dikes, several of which are clustered near Llao Rock, are smaller and less prominent. Llao Rock (fig. 36.1) was formed by a massive rhyodacite flow that came from near the bottom of a satellite cone that exploded, not long before Mazama's climactic eruptions. Because of its shape, Llao Rock was earlier

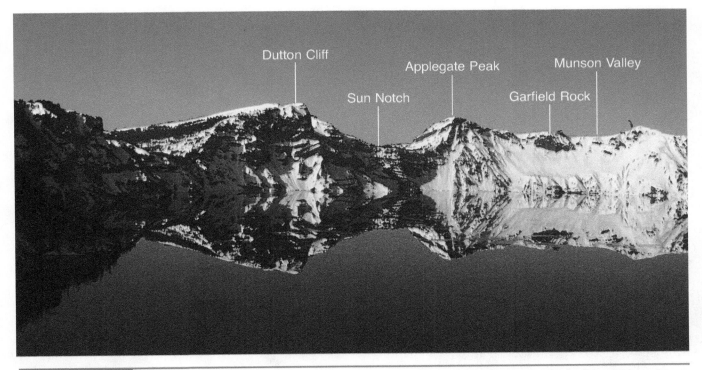

FIGURE 36.5A Sun Notch and Kerr Notch, U-shaped glacial troughs beheaded by the explosion of Mazama, are separated by Dutton Cliffs. National Park Service photo.

FIGURE 36.5B The Phantom Ship is a remnant of a volcano that existed before Mount Mazama. It was covered by the younger volcano and was not exposed until the caldera formed. National Park Service photo.

FIGURE 36.6   Hillman Peak is 8100 feet high on the caldera's west rim. Pumice deposits and blocks of lava blanket the slope between the peak and Rim Drive. National Park Service photo.

walls, more striations can be discerned on the angular tops. This indicates that after a lava flow hardened, glaciers had time to abrade the surface (fig. 36.7).

A large glacier on Mazama flowed down Munson Valley on the southwestern flank by the present location of park headquarters. Munson Glacier went on into Annie Creek Valley and down beyond the park's southern boundary, perhaps as far as Fort Klamath. Meltwater streams from the extensive glaciers of the southern and eastern slopes carried large quantities of outwash into the Klamath Marsh. On the west side, several glaciers joined and went down the valley of the Rogue River; while on the north slopes, the Mazama glaciers joined those of Mounts Bailey and Thielsen.

Moraines are of many ages and types. Some are almost completely buried by pumice and by the ash of the culminating explosions. The youngest moraines were deposited after the explosions had ceased, indicating an interval of colder climate sometime within the past 6000 years. No glaciers exist in Crater Lake National Park at the present time, but heavy snows fall each winter, about 50 feet from September through May.

Geologic History

1. **Development of the Western Cascades, middle and late Tertiary time.**

Along the western part of the Cascade Range lies a belt of older lavas and pyroclastics that make up the Western Cascades. These mountains were folded and deeply eroded long before the Cascade volcanoes began to erupt. Both the Western Cascades and the younger High Cascades are the result of the continuing subduction of oceanic crust beneath the westward-moving North American plate (box 35.1). None of the Tertiary volcanics are exposed in Crater Lake National Park; however, some of these rocks may underlie the base of Mazama.

The grooves on these rocks perched on the caldera rim are glacial striations that were incised by glaciers on Mount Mazama long before the climatic eruptions. National Park Service photo.

Table 36.1	Geologic Column, Crater Lake National Park

Time Units			Rock Units	Geologic Events
Era	Period	Epoch		
Cenozoic	Quaternary	Holocene	Andesitic ash, lava Rhyodacite lava Ash-flow deposits Intracaldera debris Wineglass Welded Tuff Air-fall tephra, pumice Rhyodacite lava and pumice	Cinder cones within caldera: Wizard Island, Merriam Cone Lava platform in caldera Minor glacial advance Climactic eruptions and collapse of caldera Cleetwood, Grouse Hill, Llao Rock, and Redcloud flows
		Pleistocene	Dacitic and rhyodacitic volcanics interbedded with glacial deposits, debris flows, reworked ash, etc. Older volcanic rocks: andesite, rhyolite, dacite	Early, middle, and late Wisconsinan lava flows and ash falls building up Mazama cone; parasitic cones, flank eruptions Intense alpine glaciation concurrent with volcanism throughout epoch Construction of composite cones near Mazama: Union Peak, Mounts Bailey, Thielsen
	Tertiary	Pliocene	Western Cascade volcanics	Buildup of lava plateau; uplift Western Cascade volcanoes Regional uplift

Sources: Bacon, 1983; Cranson, 1982.

2. **Miocene uplift and Pliocene volcanism; building of the High Cascades.**

As the volcanic mountain system was uplifted, it became a barrier to the moisture-carrying winds from the Pacific that dropped rain and snow on the slopes. Glaciers had probably begun to form at the higher elevations by the end of Miocene time.

The High Cascades evolved as a north-south belt east of the Western Cascades, with younger volcanics overlapping older Tertiary rocks. Pliocene lava flows, mostly basalts and andesites, built up lava plateaus and the older volcanoes near Crater Lake, such as Mounts Bailey and Thielsen and Union Peak.

3. **Construction of Mazama in middle and late Pleistocene time.**

The Mazama cone began to grow on top of Pliocene and Pleistocene lavas approximately 75,000 years ago. Mazama's summit may have attained about the same elevation as Lassen, Hood, and Baker—around 11,000 feet—but never grew as high as Rainier and Shasta. Like the other Cascade volcanoes, Mazama was built up by mostly andesite lava flows and pyroclastic eruptions. The volcano had a number of parasitic cones and numerous flank eruptions. Mount Scott is the largest surviving satellite cone.

4. **Pleistocene glaciations before, during, and after volcanic eruptions.**

Throughout much of its history, Mazama was capped by snow and ice. In the caldera wall, layers of glacial deposits are interbedded with lavas down to and below the lake surface. From this record, we can see that glaciers advanced and retreated many times, because of climatic variations and because of destruction by volcanic activity.

When Mazama was at its maximum height, high peaks and ridges stuck up through the glacial ice. The absence of till at the top of Hillman Peak and Watchman suggests that these peaks were never entirely covered by ice. Hillman Peak probably acted as a "cleaver," or arête, similar to Mount Rainier's cleavers that divide glaciers coming down from the summit (Chapter 35).

Long periods of quiescence followed times of active eruptions. Bands of soil containing charred vegetation at the top of till deposits, later covered by lava flows and pumice, record such quiet times when reforestation of the slopes could take place.

It seems likely that the time of maximum glaciation was near the end of the andesitic eruptions and after the early dacitic flows that erupted mainly from fissures along the north side of the mountain. This activity may be the reason why glacial erosion was more severe on the south side. The large U-shaped valleys on the south side, Munson, Kerr, and Sun, were deeply excavated during this time. Glacial ice was probably a thousand feet thick in the troughs.

5. **Quiet ending of Pleistocene time; differentiation of magma.**

After Mazama had reached its maximum height, large-scale volcanic activity on the main cone ceased for about 40,000 years. In the interim, the Wisconsinan stage of glaciation came to a close. The discovery of carbonized stumps of trees that had been growing in till in a glacial col near Discovery Point implies that a long period of quiescence preceded the last catastrophic explosions. The pumice that buried the trees can be correlated with those explosions. The presence of trees on lower slopes also suggests that at the time of the explosions, the glaciers had receded to higher elevations on the mountain.

Several miles below the surface of the High Cascades, magma collected and became concentrated in the magma reservoir while the volcanoes were relatively quiet. Mazama's magmatic system was reorganized and strengthened. Gradually a kind of chemical zoning took place in the magma chamber, with layers of lighter, more siliceous magma rising nearer the top and the denser, more mafic magma sinking toward the bottom.

6. **The beginning of the end of Mazama.**

During the last few hundred years before the climactic eruptions, several vents along the northern slopes of Mazama began to "leak out" volcanic material. These dacitic eruptions of lava and pumice included the Llao Rock, Redcloud, and Cleetwood flows plus several others. During the brief, but very intense, Cleetwood flow, which directly preceded the culminating explosions, quantities of hot tephra were ejected, and blocks of old andesite from inside the summit were thrown out with the pumice.

7. **Mazama's climactic eruptions 6845 ± 50 years ago.**

The single-vent phase. The initial tremendous explosion was sudden, thunderous, and violent. A huge column of fiery, gaseous pumice shot 5 to 10 miles high in the air at nearly twice the velocity of sound (fig. 36.8A). The column may have been a mile in diameter and perhaps "spouted" for more than a day. As it began to collapse, glowing avalanches sped down from the vent. Some became the intensely welded tuff of the Wineglass (northeast part of the caldera rim) (fig. 36.8B). The tephra ejected in the initial explosion was almost entirely rhyodacitic in composition.

On the slopes and around the base of Mazama, many feet of ash and pumice fragments of various sizes accumulated. Finer ash particles were carried many miles by the prevailing winds, principally toward the east and north. Deposits of various thicknesses fell over a considerable range, landing as far away as Alberta and Saskatchewan to the north, Nevada and northern California to the south; Idaho and Montana to the east, and even to the northwest corner of Yellowstone National Park in Wyoming (fig. 36.9).

Radiocarbon dating of the Mazama ash layer places these eruptions at about 6845 years ago, with a plus-or-minus factor of 50 years. However, some pollen studies

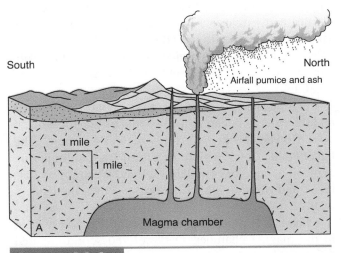

FIGURE 36.8A   A great eruption column burst from a vent on the north side of Mazama. Falling ash spread to the northwest for hundreds of miles.

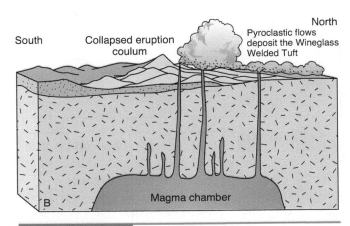

FIGURE 36.8B   As the column collapsed, fiery pyroclastic flows sped downslope.

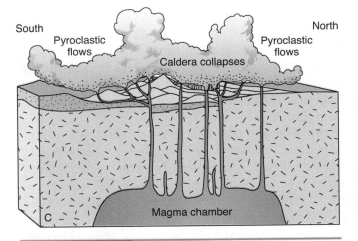

FIGURE 36.8C   The whole summit area caved in on itself, new vents opened, and more pyroclastic flows roared down the slopes on all sides of Mazama.

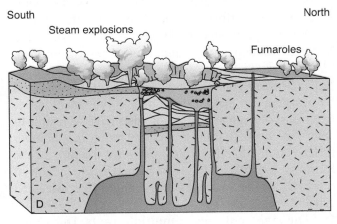

FIGURE 36.8D   Dust-filled steam clouds boiled up from the caldera; choking jets of gas rose from canyons clogged with ash; a toxic volcanic haze hung over the incinerated surface of the land.

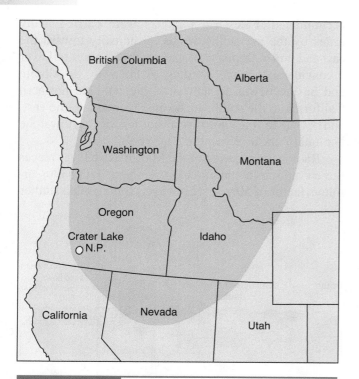

FIGURE 36.9 The darker area shown on the map was blanketed by varying thicknesses of Mazama ash from the climactic eruption. Dust and ash, borne on prevailing winds, also fell farther to the east and northeast, far beyond the Pacific Northwest. Modified after Williams 1942.

have yielded dates that nearly pinpoint the eruption. Dates from a bog in western Montana, in which pollen is associated with Mazama ash, place the ash fall within a maximum time "window" of only three years around the postulated 6845 years ago. Wherever a Mazama ash deposit is found, it provides a significant date of reference for geologists and archaeologists. Indian artifacts (mats, points, scrapers, etc.), preserved by Mazama ash, have been found at sites many miles to the east of Crater Lake. Geologists have been able to determine average rates of sedimentation in a number of lakes in the Pacific Northwest since the time of the Mazama ash falls. They can also tell which glacial moraines in the Cascades are younger than Mazama ash and which ones are older.

The ring-vent phase: multiple vents. Due to the release of pressure and the rapid extrusion of large volumes of tephra, the cone's upper section could no longer support its own weight. The whole summit area fractured and collapsed back into the volcano's center. A series of large glowing avalanches boiled out from vents all around the collapsing crater and roared down the mountain flanks, branching into the glacial valleys and spreading destruction everywhere. The whole mountain was set ablaze as trees and anything else combustible were incinerated. One fiery ash flow traveled 40 miles down the Rogue River valley. Another went north in the valley between Mounts Bailey and Theilsen, crossing Diamond Lake and then dumping pumice bombs into the valley of the North Umpqua River. Flows that went east carried large boulders of pumice many miles across the flats east of the volcano. To the south, some flows ended in the Klamath Marsh. Floating chunks of pumice, resembling styrofoam, were later washed down into the Klamath Lakes (fig. 36.8C).

During the ring-vent phase, as collapse continued and the caldera began to form, some of the hot ash and lava flowed back into the volcano's center. The Cleetwood flow, which preceded the culminating explosion, had not completely cooled when it was flattened and rifted by backflow oozing down the caldera wall.

The last eruptions apparently came from very deep in the magma chamber. Dark-colored scoria covered some of the light-colored pumice of the earlier avalanches. On the north side, dark scoria created Pumice Desert. Light and dark pumice layers are especially noticeable in the Pinnacles.

After the eruptions had ended, the air was full of choking, acrid volcanic smog and dust. Heavy rains fell and clouds of steam and fine dust continued to billow upward. Pumice-filled canyons remained hot and smoking for years. When the dust finally settled and the air cleared, only a great pit was left where the cone had been. The caldera was 5 miles wide and 4000 feet deep (fig. 36.8D).

How much of Mazama's previous bulk disappeared due to explosion and collapse? According to estimates based on a number of measurements, approximately 15 cubic miles of brecciated rock and pyroclastic material were engulfed by the collapse of the cone into the caldera. The volume of ejected and erupted material was even greater. In all, about 25 cubic miles of tephra were thrown out by the ash falls and ash flows.

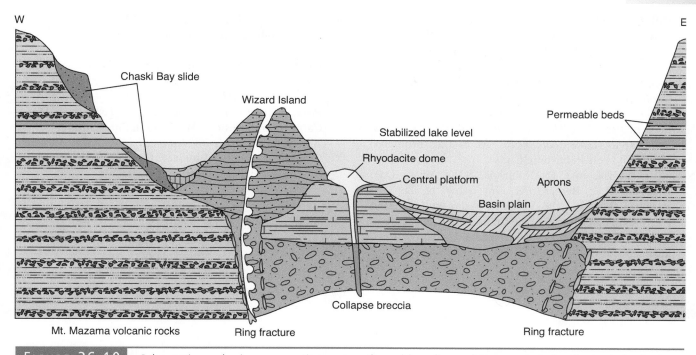

W   E

FIGURE 36.10 Schematic geologic cross section across the caldera floor of the present-day Crater Lake (not drawn to scale). From C.H. Nelson et al. 1994. Geological Society of America Bulletin, p. 701. © Geological Society of America. Used by permission.

8. Postcollapse activity.

Volcanic features younger than the climactic eruptions developed only within the caldera and were by no means extensive. Lava oozed out over the caldera floor, constructing Wizard Island, Merriam Cone, smaller cones, the central platform, and, eventually, a rhyodacite dome. A variety of sediments accumulated on the caldera floor, for the most part during the first 2000 years after the great explosion. As the caldera cooled down, rain and snow collected, forming a lake. Large-scale wall failures around the rim formed avalanche debris fans; sheetwash moved loose materials into sediment aprons on the lakebed. Turbidites spread finer sediments over the basin plain. Steam vents, along with hydrothermal springs, were active during this time. As the caldera continued to be quiescent for many years, very fine sediments, consisting of airborne silt, ash, mudflows, and siliceous ooze, covered the older features in the lakebed (fig. 36.10).

Around the caldera, the walls were attacked by the forces of erosion—landsliding, mud flows, freezing, and thawing. A brief recurrence of small glaciers at higher elevations left drift patches on top of pumice. Running water reestablished a radial drainage pattern as it tried to clear out the debris-choked valleys. After a time, dense forests grew around the base and up the slopes, hiding some of the scars.

Geologic Map and Cross Section

Williams, H. 1942. *The Geology of Crater Lake National Park, Oregon* (with geologic map). Washington, D.C.: Carnegie Institution of Washington Publication 540.

Sources

Alt, D.D. and Hyndman, D.W. 1978. *Roadside Geology of Oregon,* Missoula, Montana: Mountain Press Publishing Co.

Bacon, C.R. 1983. Eruptive History of Mount Mazama and Crater Lake Caldera, Cascade Range, U.S.A. *Journal of Volcanology and Geothermal Research* 18:57–115.

————. 1987. Mount Mazama and Crater Lake Caldera, Oregon. Centennial Field Guide, Cordilleran Section,

Hill, M. ed. v. 1. Boulder, Colorado: Geological Society of America. p. 301-306.

———. 1988. Crater Lake National Park, Oregon; Map and text on back of Map U.S. Geological Survey.

Cranson, K.R. 1982. *Crater Lake, Gem of the Cascades, the Geological Story of Crater Lake National Park,* 2nd edition. Lansing, Michigan: KRC Press.

Harris, S. L. 1988. *Fire Mountains of the West: the Cascade and Mono Lake Volcanoes.* Missoula, Montana: Mountain Press Publishing Co.

Nelson, C.H.; Bacon, C.R.; Robinson, S.W.; Adam, D.P.; Bradbury, J.P.; Barber, J.H., Jr.; Schwartz, D.; and Vagenas, G. 1994. The Volcanic, Sedimentologic, and Paleolimnologic History of the Crater Lake Caldera Floor, Oregon: Evidence for Small Caldera Evolution. Geological Society of America Bulletin 106:684–704, May.

Orr, E.L.; Orr, W.N.; and Baldwin, E.M. 1992. *Geology of Oregon,* 4th ed. Dubuque, Iowa: Kendall/Hunt Publishing Company. p. 141–166.

Weingrod, C. 1995. A Commitment to Clarity. *National Parks* 69:42–47, July/August.

Williams, H. 1942. *The Geology of Crater Lake National Park, Oregon* (with geologic map). Washington, D.C.: Carnegie Institution of Washington Publication 540.

———. 1976. *The Ancient Volcanoes of Oregon,* 7th edition. Condon Lectures. Eugene, Oregon: Oregon State System of Higher Education.

Note: To convert English measurements to metric, go to www.helpwithdiy.com/metric_conversion_calculator.html

C H A P T E R   3 7

Lassen Volcanic National Park

NORTHERN CALIFORNIA

Area: 106,372 acres; 166 square miles
Proclaimed a National Monument: May 6, 1907
Established as a National Park: August 9, 1916
Address: P.O. Box 100, Mineral, CA, 96063

Phone: 530-595-4444
E-mail: LAVO_Information@nps.gov
Website: www.nps.gov/lavo/index.htm

Lassen Peak, the world's largest volcanic dome, was constructed in the ruins of a composite cone that collapsed toward the end of Pleistocene time. Lassen has not erupted for about 80 years; but hot springs, steaming fumaroles, mud pots, and sulfurous vents indicate volcanic activity at depth. A wide range of volcanic phenomena is in the park, including very young lava beds and an active cinder cone.

Glacial erosion and deposition modified some of the volcanic features.

FIGURE 37.1   Bumpass Hell, an active thermal area, has bubbling mud pots, boiling springs, steam vents, and the pervasive odor of sulfur. Photo by Michael Little.

When President Theodore Roosevelt proclaimed Lassen Peak and Cinder Cone National Monuments in May 1907, the generally held assumption was that these little known areas in the northern California wilderness contained interesting volcanic phenomena, obviously extinct, but worthy of preservation for future generations to study and enjoy. The Indians who lived in the area had told settlers that the mountain was not "dead." The Indians said that Lassen Peak was full of fire and water, and that one day the mountain would blow itself to pieces.

Seven years after the monument was proclaimed, in May 1914, Lassen Peak awoke with a roar, opening up a new crater and emitting lava and ash. After several hundred intermittent eruptions of varying length and severity between 1914 and 1921, the volcano quieted down again. It was during the period of activity, after some impressive outbursts, that Lassen Volcanic National Park was established on August 9, 1916.

Lassen Peak, the southernmost volcano in the Cascade Range, is near the northern end of the Sacramento Valley. The 10,457-foot high mountain was a landmark for immigrants coming to the Sacramento Valley in the mid-1800s. The peak was named for Peter Lassen, a Danish blacksmith who settled in Northern California in the 1830s. As a sideline, Lassen guided parties of immigrants to the Sacramento Valley. In later years, California-bound wagon trains followed Nobles Emigrant Trail, parts of which can be seen in the park today as deep ruts and wagon tracks (fig. 37.2).

Although Nobles Emigrant Trail winds along the base of Cinder Cone in the northeast corner of the park, apparently only a few miners, homesteaders, and travelers witnessed the most recent eruptions of ash, steam, and fragmental material from Cinder Cone in 1850-51. Observers reported that fire on the top of the cone could be seen for many nights from points as far distant as even a hundred miles away. Population in the region was still sparse in 1914 and 1915, when the more violent Lassen eruptions occurred. Some property damage resulted from the explosions and mudflows, but people were able to get out of the way in time. It was, perhaps, sheer luck that enabled the photographers and observers who recorded the events at close range to survive with only minor injuries.

At the recommendation of the U.S. Geological Survey, the Visitor Center and accommodations near Manzanita Lake were closed in 1974. These facilities would be in the path of an avalanche from Chaos Crags should movement of the unstable slopes be triggered by seismic activity or renewed volcanism.

Geologic Features

Unlike Shasta (80 miles to the north) and the other major volcanoes of the Cascade Range, Lassen Peak is not an andesitic composite cone. Lassen, which is composed of dacite, is a volcanic dome, the largest in the world. A *volcanic dome* is characterized by an upheaved, plug-like conduit filling that forms when viscous lava is forced through a vent onto the earth's surface. Like toothpaste squeezed out of a tube, the viscous lava is unable to flow freely, so it builds a bulbous dome over its vent.

The Lassen dome is an offspring of a once mighty composite volcano, called Tehama, more than a thousand feet higher than Lassen, with a diameter of 11 to 15 miles at its base. Late in Pleistocene time, Tehama (also known as Brokeoff Volcano) collapsed, forming a caldera more than two miles in diameter. Preceding its collapse, vast quantities of fluid dacite streamed from vents on Tehama's northeast flank. Subsequently, several small dacite domes rose (Eagle Peak, Vulcan's Castle, Mount Helen, and Bumpass Mountain) in the area just south of Lassen Peak, perhaps at the same time that the Lassen dome was beginning to form. The outpouring of so much lava probably weakened Tehama's structure and may have caused its collapse.

About 18,000 years ago, the volcanic dome that became Lassen Peak began pushing up through a vent in the pre-collapse dacite flows on Tehama's northern flank. The huge dome's rise appears to have taken place with great force and in a remarkably short time. The quantities of talus that surround and partially bury Lassen's dome were produced by the fracturing of older rock as the plug of dacite forced its way to the surface.

How Eruptive Activity Is Related to Type of Lava

Dacite, the rock type of the Lassen dome, is intermediate in composition between rhyolite and andesite. Because dacite has a higher percentage of silica than andesite, it tends to flow less easily, hence its propensity for constructing domes. Dacite is associated with more explosive eruptions because gases are not able to bubble out freely through the viscous magma. Gas is trapped until enough pressure builds up to blast a way out.

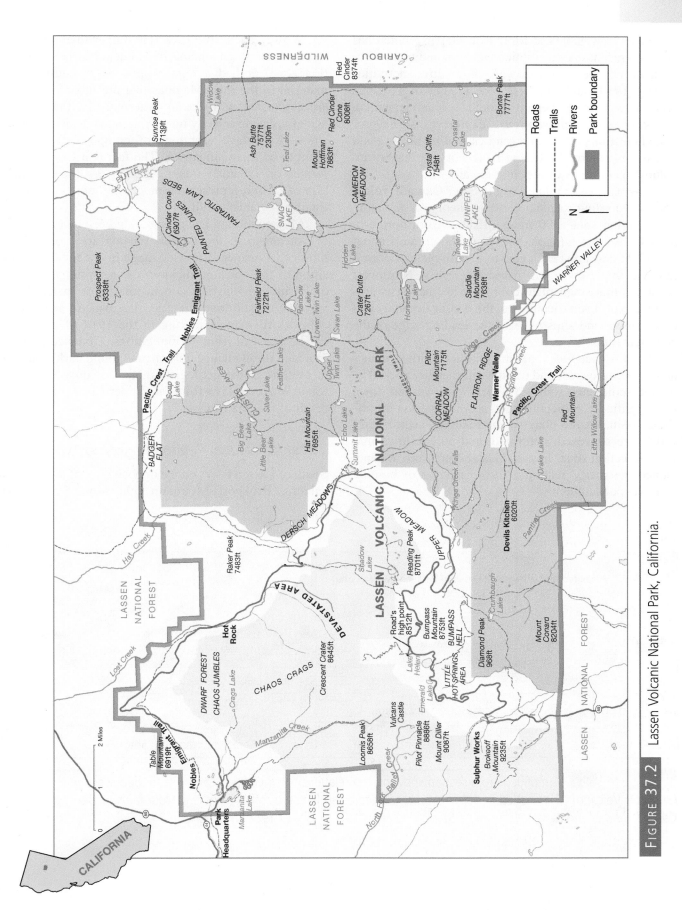

FIGURE 37.2 Lassen Volcanic National Park, California.

Volcano-watchers at Mount St. Helens, which has also had recent dacite eruptions, have noted many similarities between its renewal of eruptive activities (spring 1980) and the first year (1914) of Lassen's eruptions. Activity began at both volcanoes with a series of *phreatic eruptions* (explosive, steam-propelled ejections of ash and mud caused by heating and expansion of ground water) that shot ash and steam columns thousands of feet into the air.

How is Lassen related to the other Cascade volcanoes? According to plate tectonic theory, a spreading center, lying some 300 miles offshore from northernmost California, Oregon, and Washington, adds about an inch of new crust each year to a small plate of oceanic crust, thus moving it toward the Pacific Northwest coast. As this small plate collides with the westward-moving North American plate, the denser oceanic crust bends down and slips beneath the leading edge of the continental plate.

The Cascade volcanoes are products of the subduction of oceanic crust and marine sedimentary rock. The marine sediments, some of which were washed down from uplifted continental crust, accumulated in an offshore basin. As the plates came together, sedimentary beds were squeezed, deformed, and pulled down into the subduction zone. Intense heat and pressure at depth melted the rock, creating a mixed magma of basaltic and felsic constituents.

Why is dacite the rock type rather than andesite? Perhaps the magma came from deeper in the subduction zone, or had a high proportion of sedimentary material in the mix. Magmas that are more silica-rich are associated with dacite. Alternatively, the relative age of volcanic activity may explain the dacitic composition. Late-stage eruptions of composite volcanoes seem to be more silica-rich. The fact that Lassen Peak formed on Tehama's flanks suggests that the construction of Lassen represents a much later episode in continuing volcanic activity.

Magma mixing has presumably been characteristic of volcanism in the Lassen-Tehama area throughout most of its history. The wide range in chemical composition of the volcanic rocks in the park suggests disequilibrium of magmas due to mixing of materials of various crustal and mantle origins. At least two episodes of magma mixing in the 1914-21 eruptions can be deduced from accumulations of banded pumice in which flows change color from light to dark from top to bottom. Some geologists think that quartz crystals found in recent lava flows at Cinder Cone may be the result of magma mixing.

Volcanic Features Related Mainly to Tehama

South of Lassen Peak, on the opposite rim of the ancient caldera, Brokeoff Mountain (elevation, 9235 feet)

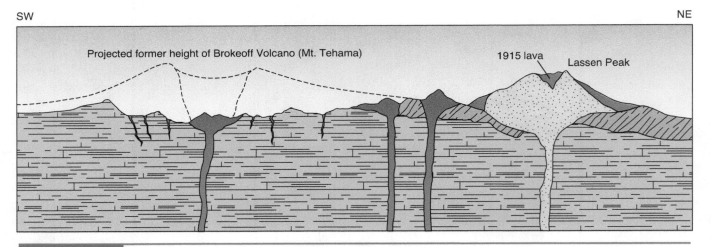

SW   NE

Projected former height of Brokeoff Volcano (Mt. Tehama)   1915 lava   Lassen Peak

FIGURE 37.3   Simplified cross section showing the relationship of Lassen Peak to its predecessor Tehama (Brokeoff Volcano). The dashed line shows Tehama's projected former profile. From *Fire Mountains of the West, The Cascade and Mono Lake Volcanoes* by Stephen L. Harris. Copyright © 1988 by Mountain Press Publishing Company, Missoula, MT. Reprinted by permission.

stands as an erosional remnant. Brokeoff Mountain is the second highest peak in the park. Other prominent remnant peaks, outlining the rim of the old caldera, are Mount Diller, Pilot Pinnacle, and Mount Conrad. Lava pinnacles, once part of the caldera, show up on the skyline. Much of this severe erosion was due to repeated glacial advances that followed Tehama's collapse.

Thermal features in the southwest corner of the park, in or close to the old caldera, are associated with Tehama. Sulphur Works, with its hot springs, fumaroles, and strong, rotten egg odor (indicating the presence of hydrogen sulfide), is thought to be at the center of the former cone. Between Sulphur Works and Little Hot Springs Valley (an area of hydrothermal activity) is Diamond Point, an old conduit through which lava moved upward during the construction of Tehama.

As a result of hydrothermal alteration, many of the rocks at the Sulphur Works and Little Hot Springs Valley—which were formerly hard, gray-green andesite lavas—have been transformed into bright-colored clays. Other rocks are rapidly decaying and have become yellow, buff, and red in color from iron oxides. The hot water coming from the fissures contains both sulphurous acid and sulfuric acid, which react with the minerals of the original rock to produce clays and iron oxides. Nearly pure opal (not gem quality) forms where the acidity of the hot water is very high (fig. 37.4).

Just beyond Little Hot Springs Valley is Bumpass Hell, a 16-acre tract of boiling springs, steaming sulfur vapors, and bubbling mud pots that are associated with old caldera fissures. Bumpass Hell is named for a cowboy who worked in the area in the 1860s. When he broke through the crust over a mud pot and scalded his feet, he told friends that he had been "in hell." Later, when he was escorting a newspaper editor through the area, Bumpass broke through the crust again. This time he was so badly burned that he lost a leg. The superheated steam in the springs at Bumpass Hell and the fumaroles are an indication of the late stages of volcanism in the Tehama area (fig. 37.1).

Devils Kitchen is another area of hot sulfurous springs, located in Warner Valley about seven miles southeast of Lassen. The springs are so strongly acidic that holes and pits have been eaten into the bedrock. Note on the map (fig. 37.2) that the hot springs and thermal features are clustered along a northwest-southeast trend and appear to be associated with an inactive fault area.

The Lassen Eruptions

The volcanic dome of Lassen Peak, with its surrounding talus, represents about a cubic mile of material (fig. 37.5). This is a large summit area for a dome, but its bulk is small compared to the composite cone of Mount Rainier (Chapter 35). A feature of Lassen Peak was its development of summit craters from which pyroclastics and lava were ejected, as happened in the phreatic eruptions in 1914. Most domes do not have summit craters because the explosions are emitted from the base of the plug. Chaos Crags, for example, is a complex dome on Lassen's northwest flank that is made up of several dacite plugs, but it does not have a crater.

In the crater bowl on Lassen's summit, a pre-1914 vent and a 1914-15 vent were filled and covered by a large dacite lava flow in May 1915, and a third summit crater, or vent, was opened. A fourth crater was blasted out of the northwest corner of the summit by a series of steam and ash explosions in April, May, and June of 1917 (fig. 37.6). These were the last of the major eruptions of the 1914-1921 period of activity. Slight fumarolic activity at the summit (steam escaping from small fissures) has persisted to the present.

The dacite flow on May 19, 1915, poured down the western side of Lassen about 1000 feet before it hardened. Observers some 20 miles away to the west and south reported that beginning about 9 o'clock that night the mountain was "boiling over" for two hours. The photographer B.F. Loomis and the postmistress of nearby Manton had described a fire at the summit a few days earlier and then two days before the eruption, a large black mass had begun to rise at the top. This "black mass" was the plug, or column of lava, that was pushing up through the conduit, breaking the rock, and opening a new vent.

Hot lava and ash flowed down the steep northeast side onto packed snow left from the winter, and quickly combined with the snow to form a thick, pasty lahar that slid down the slope, burying a milewide swath of forest in its path, filling Lost Creek, then gliding up over a divide and down into Hat Creek. In the valleys north of Lassen, fences, bridges, farm buildings, and fields were totally devastated. Fortunately, homesteaders were awakened by the roar of the oncoming mudflow, just in time to flee to high ground and safety. The worst was yet to come.

A few days later, on May 22, 1915, at 4:30 P.M., a violent explosion shot a great cloud of ash thousands of

<figure>FIGURE 37.4 The sulfur works area is located over the old vent system of ancient Mount Tehama. Gasses from the Fumaroles sublimates on the surrounding rocks. Photo by Sherry Schmidt.</figure>

feet into the air. Photographs of the event, referred to as the Great Eruption, show an enormous cloud in the shape of a mushroom over Lassen Peak. Prevailing westerly winds carried fine ash particles for many miles to the east. Much of the force of the blast was deflected downward as a glowing avalanche swept down the northeast slope, snapped off many more trees (some with trunks 6 feet in diameter), and threw them many feet from their stumps. Hot ash—mixed with melting snow—reactivated the mudflow and the whole scalding mass sped down the same valleys, split in front of Mount Raker, and rushed by, the two branches extending beyond the park's northern boundary. Smaller mudflows streamed down the north and northwest slopes.

The tract of greatest destruction is a wide swath about four miles long, called the Devastated Area, which marks the path of the mud and ash flows that roared down the mountain during the May 19 and 22 eruptions. Visitors who drive through the park can see that scars on the landscape are being covered by natural reforestation, but the trees in the Devastated Area

FIGURE 37.5 Lassen Peak from Kings Creek Meadow. U.S. Geologic Survey photo.

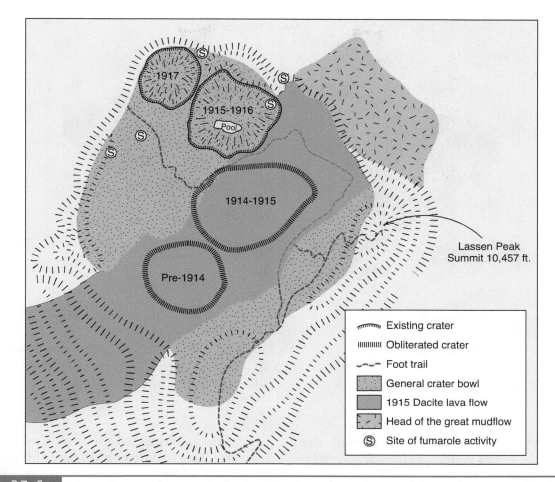

Existing crater
Obliterated crater
Foot trail
General crater bowl
1915 Dacite lava flow
Head of the great mudflow
Ⓢ Site of fumarole activity

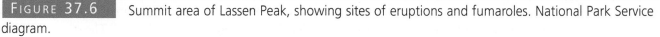

FIGURE 37.6 Summit area of Lassen Peak, showing sites of eruptions and fumaroles. National Park Service diagram.

are sparse and stunted. The siliceous ash soil is lacking in nutrients and too porous to hold sufficient water for normal tree growth.

Chaos Crags (north of Lassen Peak) is younger than the other volcanic domes. It was formed by four or more dacite domes less than 1200 years ago. Chaos Jumbles, a rock avalanche that extends toward the northwest corner of the park from the base of Chaos Crags, was set off by the explosion and partial collapse of one of the domes about 300 years ago. Avalanching transported large angular blocks a mile or more from their source. At speeds of around 100 miles per hour, the rock debris apparently traveled some of the distance on a cushion of compressed air that reduced friction (a phenomenon called "air-layer lubrication"). The damming of Manzanita Creek by the avalanche formed Lake

Manzanita. Soil development and regrowth of trees has been slower on the blocky landslide debris of Chaos Jumbles than in the Devastated Area. The unstable Chaos Crags slopes are monitored carefully because the loose rock may start moving again.

Inactive Shield Volcanoes

Four shield-type volcanoes, built up by successive flows of fluid andesite, are located in the park. These volcanoes are Raker Peak, north of Lassen; Prospect Peak in the extreme northeast corner; Red Mountain, next to the south-central boundary; and Mount Harkness in the southeast corner. All are between 7000 and 8400 feet in elevation. They have been modified by glacial erosion and by ash falls from Lassen

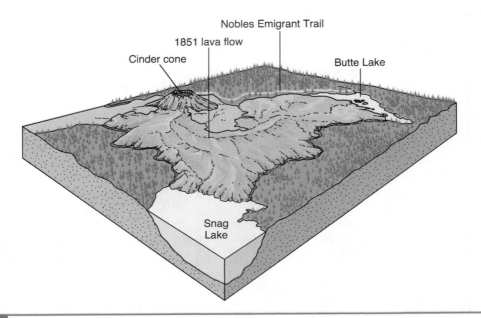

Nobles Emigrant Trail

1851 lava flow

Cinder cone

Butte Lake

Snag
Lake

FIGURE 37.7   In 1851 (and prior to that time), basaltic lava erupted from vents near the base of Cinder Cone. Flows dammed drainageways, creating Snag Lake and then Butte Lake. Water from Snag Lake seeps through sediments under the lavas and keeps Butte Lake from drying up. Nobles Emigrant Trail skirts Butte Lake and follows the edge of the lava flow around Cinder Cone. Drawing made from a National Park Service photo by C. Peters.

Peak and Cinder Cone. Although apparently constructed entirely of lava without interbedded pyroclastic deposits, each of the four shield volcanoes is topped by a cinder cone rising from a central summit crater. The cinder cones and perhaps the top layers of lava appear to be postglacial in age. On Raker Peak, a large dacite dome has pushed up on the southwest side, forming a line of remarkable dacite cliffs with columnar structure.

Cinder Cone

Located about 10 miles northeast of Lassen Peak, Cinder Cone has been built up during the past few centuries by a succession of ash eruptions. It rises as a symmetrical cone from the flatlands, 750 feet high and about a half-mile in diameter. A cinder sheet of varying thickness surrounds the cone. Two circular craters are at the top. After cinder eruptions had built the cone, several lava flows oozed out from the base of the volcano and spread northeast and southeast, damming creeks that feed Butte Lake at the north and Snag Lake at the south. The area of the young, unweathered flows is called Fantastic Lava Beds because of their blocky, jagged, rough surface.

An unusual feature of the basaltic rock that cooled from the young lava flows at the base of Cinder Cone is the presence of tiny, sparkling quartz crystals that can be seen with the unaided eye. Quartz basalt, a rare combination, is considered a "hybrid rock." The quartz crystals are partially assimilated *xenocrysts,* defined as "foreign" crystals in a body of igneous rock. They may have been picked up by the basalt as it rose toward the surface, or they may have been produced by magma mixing.

Glacial Features

In Lassen Volcanic National Park, glacial debris and volcanic debris are found mixed together and reworked in mudflows, lahars, landslides, and piled up in a few moraines. Some moraines and other glacial features were buried by tephra or covered by lava flows.

Some of the volcanoes, such as Red Mountain, Raker Peak, and Lassen Peak, which was a center of ice radiation, were more severely modified by glacial erosion than others. Small valley glaciers persisted at the higher altitudes and in protected locations well into post-Pleistocene time. Lakes formed in some of the cirque basins. Two examples are Emerald Lake and

Lake Helen. The latter was named in honor of Helen Tanner Brodt, whose ascent of Lassen Peak in 1864 was the first by a woman. Summit Lake, where campgrounds are located, is dammed by a moraine. Other lakes of glacial origin include Feather, Silver, Big Bear, Little Bear, Swan, Echo, and Twin Lakes.

Geologic History

1. The Pre-Cenozoic rocks.

Inasmuch as Paleozoic and Mesozoic rock units, intruded by granite, crop out only 15 miles southeast of Lassen Volcanic National Park, it seems reasonable to assume that Sierra Nevada basement rocks extend northward under the park. Moreover, similar rocks of the same age make up the Klamath Mountains that rise west of Lassen and extend northward up the coast into Oregon. Gravity studies have shown that the lighter (less dense) volcanic rocks occupy a deep basin and that very far below them are the basement rocks. Presumably, rocks linking the Sierra Nevada and the Klamath Mountains are buried beneath the Southern Cascade volcanoes and the great lava plateau on which they stand.

2. Cenozoic uplift and the beginning of volcanism.

As the region began to be uplifted, and tilted westward, extensive volcanism was generated by a subduction zone at depth. Huge lahars poured down stream channels on the gentle western slopes of the Sierra Nevada. These became the Tuscan Formation of Pliocene age. Although the Tuscan rocks are not exposed in Lassen Volcanic National Park, they lie just below the surface in a number of places.

3. Willow Lake basalt flows in Pliocene time.

Beginning in the southern portion of the park, basaltic flows poured out of fissures and vents and covered an increasingly wide area. These flows, and many subsequent outpourings, built up the lava plateau.

4. Extensive andesite flows in Pliocene and Pleistocene time.

The basalt flows were covered over by a thick series of very fluid andesite flows called the Jupiter lavas, The Twin Lakes lavas of black porphyritic andesite flooded a large area at about the same time. The Twin Lake andesites, although much older than the quartz basalt lavas at Cinder Cone, are also rich in xenocrysts of quartz. The Flatiron andesites erupted and spread over the southwestern part of the park. By this time the landscape had become a fairly flat lava plain stretching over a wide area. No pyroclastic debris had accumulated.

A renewal of basalt flows (the Eastern basalts) built low hills along the park's eastern boundary, which were later eroded to a rugged topography. With the advent of pyroclastic eruptions, some cones began to appear in the northern part of the park.

5. Development of Tehama (Brokeoff Volcano) during Pleistocene glaciation.

In the southwestern corner of the park, Tehama became larger and larger from successive andesite flows, interbedded toward the top of the cone with increasing amounts of tuff and breccia. The great composite volcano became an impressive feature above the lava plain, rising to over 11,000 feet.

6. Eruption of the four shield-type volcanoes.

Eruptions of fluid andesite formed Raker and Prospect Peaks, Red Mountain, and Mount Harkness during the late stages of Tehama's development. These shield-type volcanoes rose at the four corners of the Central Plateau area.

7. Late Pleistocene destruction of Tehama; eruption of Pre-Lassen dacites.

Near the present location of Lassen Peak, on the northeast flank of Tehama, (which had been inactive for 400,000 years), a new vent opened and extruded flows of fluid, black, glassy dacite, 1500 feet in thickness. These flows hardened as columnar lavas that now encircle the base of Lassen Peak. Tehama's cone collapsed in on itself, forming the huge caldera. This was well before Lassen Peak formed.

The pre-Lassen dacites had erupted during an interglacial period. When the glaciers returned, they enlarged the caldera by eroding the soft, decaying rocks of the old volcano.

8. Construction of Lassen Peak about 11,000 years ago.

A pause in volcanic activity followed Tehama's demise. Then, close to the end of the Pleistocene Epoch and after an ice sheet had melted away, a large volcanic plug of partly solid, partly viscous, dacite lava began to rise through the northeastern vent on the old cone's flank. The rock enclosing the vent shattered and cracked as the lava kept on rising. This produced an

Table 37.1	Geologic Column, Lassen Volcanic National Park

Time Units			Rock Units	Geologic Events
Era	Period	Epoch		
Cenozoic	Quaternary	Holocene	Dacite; pyroclastics, lavas, mudflows	Lassen Peak eruptions, 1914-1921
			Quartz basalt flows Pyroclastics, cinders	Eruptions from base of Cinder Cone, 1570-1871 Construction of Cinder Cone
				Chaos Jumbles produced by avalanching from Chaos Crags 300 years ago
			Dacite breccia	Eruption of Chaos Crags, 1000-1200 years ago Construction of Lassen Peak on flank of Tehama 11,000 years ago Eruption of smaller domes, south side of Lassen Peak
			Dacite domes	
		?	Pre-Lassen fluid dacite flows	Columnar dacite at base of Lassen Peak Erosion of Tehama caldera Decline and collapse of Tehama
		Pleistocene	Fluid andesite lavas; some basalt flows	Construction of shield volcanoes
			Glacial deposits interbedded with volcanics	Episodes of glaciation before, during, and after volcanic eruptions
			Andesite flows and pyroclastics	Construction of composite volcano, Mount Tehama
			Eastern basalts and pyroclastics	Lava ridge; later eroded to rugged hills on east side of park
	Tertiary	?	Flatiron Andesite Juniper Lake Andesite; Twin Lakes Andesite	Construction of Central Plateau and lava plain
		Pliocene	Willow Lake Basalt	Flowed across southern portion of park
			Tuscan Formation (not exposed in park)	Pyroclastics, lahars

Source: Modified after Macdonald 1963, 1965; Kane, 1980; Harris, 1988.

extensive cover of talus on the steep slopes of the emerging volcanic dome. Within a comparatively brief time, Lassen Peak attained its present height (fig. 37.8). Because the ice sheet had gone, the talus accumulations are relatively undisturbed. Later, valley glaciers reformed and eroded shallow cirques on the north and

northeast sides of the Peak. Small moraines mark the extent of these glaciers.

Smaller, satellite domes of dacite pushed up, forming Chaos Crags north of Lassen Peak. Tuff and pumice were thrown out by pyroclastic explosions. The most recent of these explosive eruptions caused

FIGURE 37.8 Development of Lassen Peak. *A.* Dome of viscous dacite lava rises on the rim of the eroded Tehama, toward the end of Pleistocene time. *B.* As the dome kept pushing up, some of the lava overflowed. *C.* Talus blocks from the plug dome and the fractured vent accumulated on the slopes. During a late glacial stage, small glaciers reformed on the north side of Lassen. From *Fire Mountains of the West, The Cascade and Mono Lake Volcanoes* by Stephen L. Harris. Copyright © 1988 by Mountain Press Publishing Company, Missoula, MT. Reprinted by permission.

the rock avalanche that created Chaos Jumbles about 300 years ago.

9. Construction of Cinder Cone by pyroclastic eruptions.

Cinder Cone began several centuries ago with violent pyroclastic eruptions that spread ash over 30 square miles. After the cone was built, quartz basalt flows erupted from vents at the base. The most recent pyroclastic activity occurred in 1851 (fig. 37.7).

10. Eruptive activity of Lassen Peak, 1914–1921.

Before the eruptions began, the pre-1914 crater contained a fairly deep lake. This crater was totally destroyed and the depression filled by the explosions and subsequent lava flows of the active period. Two craters at the summit can be seen today.

Nearly 400 outbursts, a few major but most of them minor, were recorded by observers during the eruptive period. Scientific data collected at that time and subsequent studies of Lassen's activity have been helpful to geologists who study the Cascade volcanoes and try to anticipate what they will do next. Monitoring of volcanic features in the park by seismometers, tiltmeters, and inclinometers is carried on by the U.S. Geological Survey in cooperation with the National Park Service.

Geologic Map and Cross Section

Macdonald, G.A. 1963. Geology of the Manzanita Lake quadrangle, California. U.S. Geological Survey map GQ 248.
———. 1964. Geology of the Prospect Peak quadrangle, California. U.S. Geological Survey map GQ 345.

Sources

Amesbury, R. 1967. *Nobles' Emigrant Trail*. Privately published by R. Amesbury. (Distributed by Loomis Museum Association) 42 p.
Bumpass Hell Nature Trail, Lassen Volcanic National Park. 1985 revision. Loomis Museum Association leaflet.
Cinder Cone Nature Trail, Lassen Volcanic National Park. 1984. Loomis Museum Association leaflet.
Crandell, D.R. 1972. Glaciation Near Lassen Peak, Northern California. U.S. Geological Survey Professional Paper 800-C pp. C179-Cl88.
Harbaugh, J.W. 1975. *Northern California,* K/H Geology Field Guide Series. Dubuque, Iowa: Kendall/Hunt Publishing Company. p. 86–109.
Harris, S.L. 1988. *Fire Mountains of the West: The Cascade and Mono Craters Volcanoes*. Missoula, Montana: Mountain Press Publishing. 320 p.
Heiken, G. and Eichelberger, J.C. 1980. Eruption of Chaos Crags, Lassen Volcanic National Park. *Journal of Volcanology and Geothermal Research,* 7:443–448.

Kane, P.S. 1980. *Through Vulcan's Eye, the Geology and Geomorphology of Lassen Volcanic National Park.* Loomis Museum Association. 118 p.

——— 1982. Pleistocene Glaciation, Lassen Volcanic National Park. *California Geology* 35 (5):95–105, May.

Loomis, B.F. 1971 (3rd revised edition). *Eruptions of Lassen Peak: Pictorial History of Lassen.* Loomis Museum Association. 96 p.

Macdonald, G.A. and Katsura, T. 1965. Eruption of Lassen Peak, Cascade Range, California, in 1915: Example of Mixed Magmas. *Geological Society of America Bulletin,* 76:475–482.

Schultz, P.E. 1983 revision. *Road Guide to Lassen Volcanic National Park.* Loomis Museum Association. 40 p.

Williams, H. 1929. The Volcanic Domes of Lassen Peak and Vicinity, California. *American Journal of Science,* 18:313-330.

——— 1931. The Dacites of Lassen Peak and Vicinity. *American Journal of Science,* 22:385–403.

Note: To convert English measurements to metric, go to www.helpwithdiy.com/metric_conversion_calculator.html

Katmai National Park and Preserve

SOUTHWEST ALASKA

Area: 4,090,000 acres; 8809 square miles

Proclaimed a National Monument: September 24,1918

Established as a National Park and Preserve:
  December 2, 1980

Address: P.O. Box 7 #1 King Salmon Hall,
  King Salmon, AK 99613

Phone: 907-246-3305

E-mail: KATM_Visitor_Information@nps.gov

Website: www.nps.gov/katm/index.htm

Katmai's active volcanoes are in the Alaskan-Aleutian volcanic belt. The violent eruption of Novarupta in 1912, created the ash-filled Valley of Ten Thousand Smokes.

Powerful waves attack an emergent coast of fiords, cliffs, and waterfalls.

Wildlife abounds in an ecosystem of mountains, glaciers, lakes, streams, forests, tundra, and marshes. The Alagnak Wild River, rising in a northern lake in the preserve, teems with salmon.

FIGURE 38.1   Mount Griggs was named after the botanist Robert F. Griggs from Ohio who made several trips to Katmai to study the plant regeneration in the Valley of Ten Thousand Smokes. National Park Service photo.

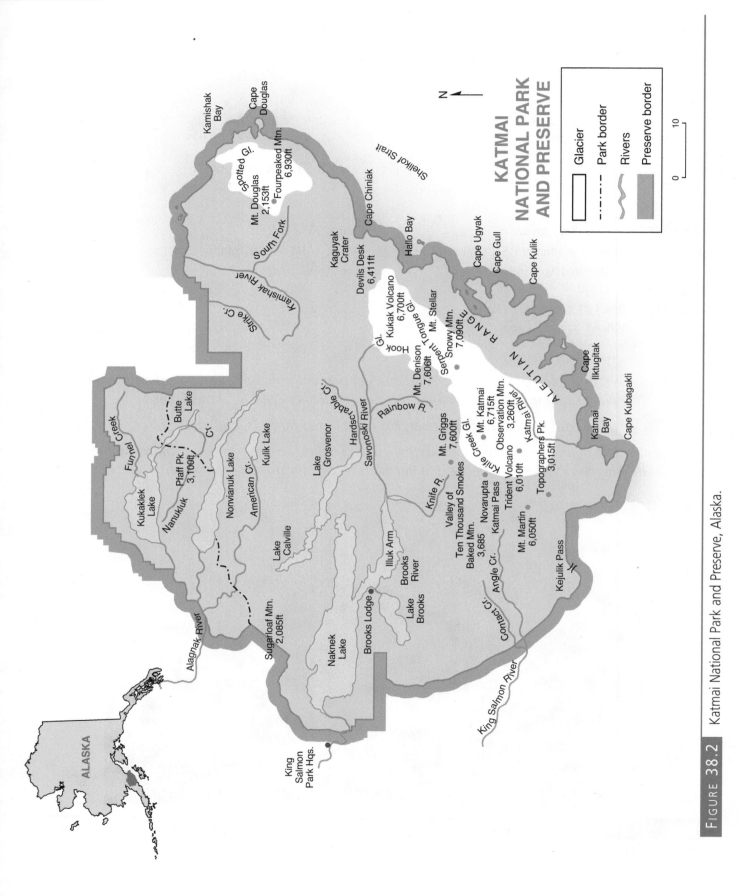

FIGURE 38.2   Katmai National Park and Preserve, Alaska.

Geographic Setting

Located in the northern part of the Alaska Peninsula, overlooking the Shelikof Strait, Katmai National Park and Preserve is 290 air miles southwest of Anchorage. The park/preserve extends along the coast from Kamishak Bay south to the Becharof National Wildlife Refuge and westward nearly to the community of King Salmon. To the north, at the head of the peninsula, is Lake Clark National Park and Preserve (Chapter 39). The active volcanoes in the two parks are part of the Alaskan-Aleutian volcanic belt (fig 38.3).

The Katmai volcanoes, like those in the Lake Clark region, trend from northeast to southwest, parallel to the eastern coast of the Alaska Peninsula. In this part of the peninsula, the volcanoes rise over 7000 feet from the shores of Shelikof Strait. Inland from the coast, the mountains of the Aleutian Range have elevations of 3000 to 5000 feet; and the whole range, including the volcanoes, is about 100 miles wide from east to west. Several hundred miles down the peninsula, the range breaks up into the Aleutian Island chain, which stretches in an arcuate pattern westward for another thousand miles out into the North Pacific. Throughout its length, the Aleutian Range has 60 volcanoes, of which 47 are known to have been active since 1760. In the spring and summer of 1986, Augustine Volcano, which rises from the floor of Kamishak Bay (between Katmai National

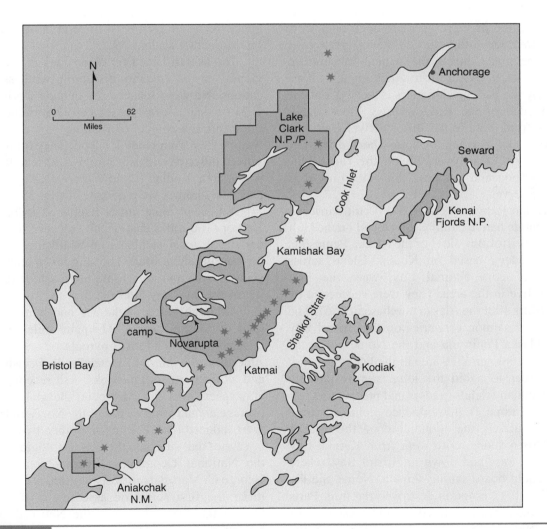

FIGURE 38.3 Active volcanoes of the upper Alaska Peninsula. The line of volcanoes marks the inner edge of an ongoing subduction zone. In this part of the Alaska-Aleutian volcanic arc, the magma that builds the volcanoes rises through continental crust, then erupts as lavas of variable composition. Modified after Fierstein 1984.

Park and Preserve and Lake Clark National Park and Preserve), erupted large quantities of ash (Chapter 39).

The Katmai climate is wet and stormy. From whatever direction the wind blows, moisture-laden winds are forced to rise over the mountainous Alaska Peninsula. Light rains can last for days, and strong winds and gusts ("williwaws") frequently sweep the area. In summer, skies are clear perhaps 20 percent of the time. However, summer temperatures (averaging 50° to 60° F) are mild by Alaska standards.

The world's largest predatory land mammals, the Alaska brown bears, thrive in the wild, remote Katmai area. The huge size of these Ice Age survivors, weighing from 600 to 1500 pounds, is attributed to fish oils in their rich diet. The bears feed on the red salmon that spawn in the lakes and streams of the Alaska Peninsula.

Facilities in the park/preserve are open only from June to September, although charter flights over Katmai can be made during the winter. In summer, access to Katmai is by air from Anchorage to King Salmon, where the park headquarters is located, outside the western edge of the park/preserve. Float planes ferry visitors from there to the Brooks River Camp on the shore of Lake Naknek. A four-wheel drive bus makes daily trips to an overlook on the rim of the Valley of Ten Thousand Smokes. This is the only road in the park/preserve.

The Aleutian Range was such a difficult barrier to cross that separate native cultures developed on the Gulf of Alaska and Bristol Bay sides of the Alaska Peninsula. Russian fur traders, based on Kodiak Island across Shelikof Strait from Katmai Bay, were the first Europeans to live in the area. They were succeeded by American traders after the Alaska purchase in 1867. The establishment of salmon canneries on the Bristol Bay side of the Alaska Peninsula and the Nome gold rush brought more people into the area in the late nineteenth century. In order to avoid the long, stormy passage around the Aleutian Islands, traders and prospectors followed the old Katmai Trail, used since prehistoric time as a shortcut across the mountains of the Alaska Peninsula. This arduous route went from Katmai Bay through Katmai Pass and down to Bristol Bay, where prospectors could board sailing ships to Nome and the goldfields. The 1912 eruption destroyed the trail. Parts of the route have been restored on the west side of Katmai Pass and can be used by backpackers, but the section from Katmai Pass to Katmai Bay has not been opened because of heavy brush, braided streams, quicksand, quickmud, and frequent slides.

The 1912 Eruption

Apparently no lives were lost during this catastrophic event, probably due to the sparse population and the fact that a week-long series of earthquakes, prior to the explosion, prompted natives in mountain villages to flee to coastal areas on either side of the Alaska Peninsula.

Around noon of June 6, 1912, people on the northern and eastern sides of the Gulf of Alaska knew that an extraordinary volcanic event had occurred because they heard loud booms and saw in the distance great clouds of ash. Around midnight of that day, a severe earthquake occurred, estimated to be at least magnitude 6.4 on the Richter scale.

The ashfall lasted for three days in the immediate vicinity of the volcano, disrupting wireless communications. No word reached the outside world until June 9. More than a week passed before a rescue party, paddling through chunks of floating pumice, could reach Katmai Bay from Kodiak Island. They found the native village deserted and ash from 3 to 15 feet deep over the whole area. Kodiak Island was covered by a foot of ash and many homes were destroyed (fig. 38.4).

A U.S. revenue cutter happened to be in Kodiak Harbor (100 miles east of the volcano) on June 6. The crew took local residents on board in complete darkness during the ashfall. The ship lay at anchor until the skies cleared and it was safe to go ashore on the morning of the third day after the first explosions.

Although much of the ash and pumice fell in the Gulf of Alaska, over 3000 square miles of land were covered by about a foot of pyroclastic debris. Ash fell in Juneau on the Alaska Panhandle, in British Columbia, and as far north as Fairbanks. Ash reaching the upper atmosphere was carried around the globe, causing the average annual temperature of the Northern Hemisphere to be reduced by 1.8° F for more than two years.

Later that summer, the U.S. Geological Survey and the National Geographic Society sent a geologist, George C. Martin, from Washington, D.C. to Katmai to make the first scientific assessment of the disaster. During the next few years, the National Geographic Society sponsored four more scientific expeditions to the eruption area. All but one were led by Robert F.

Griggs, a botanist, who originally went to devastated Kodiak Island to study the effects of the eruption on plant regeneration. In the course of his first three expeditions to the mainland (1915, 1916, and 1917), Griggs explored, under difficult and often dangerous field conditions, the calderas of Katmai and Novarupta and the Valley of Ten Thousand Smokes. After Katmai was proclaimed a national monument, the society sponsored two subsequent expeditions in 1919 and 1930, both led by Dr. Griggs. His illustrated account of his scientific adventures (Griggs 1922) vividly portrays the Katmai scene. Griggs and his fellow scientists accumulated a remarkable body of data and observations. Their work, and studies done since, have enabled scientists to reconstruct a fairly complete account of what happened during the Katmai eruptions and how they came about.

Geologic Features

The Katmai-Novarupta Connection

Katmai Volcano, which was probably about 7500 feet in elevation before its summit collapsed, and its lower neighbor, Novarupta (elevation, 2757 feet), were connected in the eruptive sequence in a rather unusual way. The first visitors to the area thought that the Katmai vent was the main source of the approximately 4 to 6 cubic miles of volcanic debris ejected during the eruptions. The entire top of Katmai had collapsed, leaving in place a great caldera. Not until 1953 was it determined that all the ash flows and pumice had been ejected from Novarupta's vent rather than from the Katmai outlet. Why, then, did Katmai collapse in on itself?

The earthquakes and fault movements preceding the eruptions may have allowed magma from Katmai to flow laterally through fissures and connect with conduits containing magmas of differing chemical composition rising beneath Novarupta. As the magma masses mingled, they frothed and exploded, ejecting from Novarupta great clouds of fine, white rhyolitic pumice, followed by dacitic ash, and finally andesitic scoria. Fissures opened in the smaller volcano, releasing colossal pyroclastic flows that roared northwestward down the valley of the Ukak River. The incandescent mass devastated everything in its path, buried the valley floor, and covered 40 square miles with ash as much as 700 feet deep. When Griggs named this area the Valley

of Ten Thousand Smokes in 1916, countless fumaroles of steam, some a thousand feet high, still issued from the plain of ash. At the present time, only a few active vents remain.

As the magma chamber emptied, the unsupported summit of Katmai collapsed, forming a caldera three miles long, two miles wide, and 3700 feet deep from the rim down. Only minor explosive eruptions have occurred in these volcanoes since 1912, the most recent activity on Katmai being in 1968. The post-eruption elevation of the rim of Katmai is 6715 feet (fig. 38.5).

Crater Lake formed in the Katmai caldera not long after the initial cooling. In recent years the lake level has risen, but the lake probably won't run over the caldera rim because water seeps through the porous pyroclastics making up the caldera walls.

The Valley of Ten Thousand Smokes

On his second expedition in 1916, Robert F. Griggs went up the Katmai River valley. On their way to Katmai Pass, he and his companions made the first ascent of Katmai to the caldera rim and looked down at the lake, more than 1000 feet below. They saw hundreds of steam jets shooting into the air, roaring "like a great locomotive when the safety valve let go," as Griggs described it.

Wanting to press on to the Ukak River valley, they continued toward Katmai Pass. But when Griggs crossed the divide, he saw a most awesome sight. Thousands of columns of steam rose 500 feet and more from the ash-filled valley below. Escaping steam hissed and roared, and the sulfurous odor was overwhelming. Brightly colored encrustations sparkled around the spouting vents. The river was no more, but hot water bubbled up and ran off in innumerable rivulets. He knew that in this desolate place, which he named the Valley of Ten Thousand Smokes, no living thing could have survived the eruptions.

Scientists who later studied the Valley of Ten Thousand Smokes found that the heat is "rootless"; that is, the fumaroles, escaping gases, steam jets, and hot water were generated by the gradually cooling ash-flow deposits rather than by magmatic heat at some depth below the surface. Fumarolic activity in the valley had died out by the 1930s. Near the Novarupta caldera, however, a few fumaroles still give off steam as groundwater percolates down to the hot rocks below the volcanic vent (fig. 38.6).

FIGURE 38.4 A succession of photographs showing the aftereffects of the 1912 eruption. National Park Service photos by C.G. Martin.

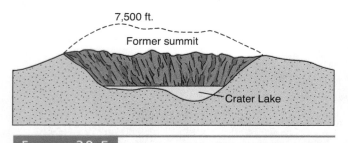

FIGURE 38.5   Cross section of Katmai after the summit collapsed in 1912. The dashed line shows the volcano's projected height before the eruption occurred (after Griggs 1922).

The Valley of Ten Thousand Smokes is still barren of vegetation; but new growth now covers much of the ash-fall layers outside the valley. Clumps of grass and low plants struggle to survive in sheltered crannies in the ash-flows.

Tephra in the valley consists of fine ash, *lapilli* (small fragments), and *volcanic bombs* (larger fragments). Massive pyroclastic flow deposits fill the valley. Air-fall deposits cover the ash flow deposits and are spread over the region. The chemistry of the ejecta shows that the magma was of mixed composition—40 percent rhyolite, 50 percent dacite, and 10 percent andesite.

Novarupta

In 1917, Griggs discovered what he recognized as a new caldera west of Katmai. He named the feature Novarupta, meaning "newly erupted." Novarupta's pre-1912 configuration is unknown, but Griggs hypothesized that a crater did not exist at that location before 1912. The explosion of Novarupta produced a shallower, smaller caldera, less impressive than Katmai's broad, steep-walled caldera. It is not surprising that the early investigators were convinced that most of the ejected pumice came from Katmai.

In later years, detailed mapping and study of ash deposits by Garniss Curtis, Howel Williams, and other geologists show that the ash layers thicken away from Katmai and toward Novarupta, despite changes in wind direction that occurred during the eruptions. Clearly, Novarupta was the true outlet for the tephra thrown out in the 1912 explosions and eruptions.

The eruption probably began with deafening booms and roars as a great high plume of gas and ash shot up into the atmosphere, like a jet stream from the nozzle of a giant fire hose. It "played" for perhaps 60

hours. Curtis (1968) has estimated the rate of expulsion of ejecta during this phase as 35 billion cubic feet per hour. As the plume began to subside, pyroclastic flows, or glowing avalanches, swept down the slopes, and falling ash blanketed all the surrounding terrain. Some of the ash and pumice welded into dense rock due to great heat and pressure from overburden. The mixing of the magmas also produced blocks of "banded pumice" with striking black and white swirls.

In a late phase of the 1912 eruption, a rhyolitic dome was pushed up from the floor of Novarupta's collapsed caldera, and possibly a dacite dome emerged on the floor of Katmai's caldera (fig. 38.7).

Related Volcanoes and Eruptions

Kaguyak, about 40 miles northeast of Katmai, has a caldera and a lake, but it has been inactive in historic time. Fourpeaked Mountain, farther to the northeast, has had no volcanic activity in historic time and is almost completely covered by an ice field and glaciers. Neighboring Mt. Douglas, also ice-covered except for its 7063-foot peak, steams intermittently. Between Kaguyak and Katmai are four inactive volcanoes—Devils Desk, Stellar, Denison, and Snowy—and one, Kukak, that is fumarolic. Mount Griggs, which rises above the Valley of Ten Thousand Smokes, across from Novarupta, has been inactive in historic time but has two sulfurous fumarole fields (*solfataras*). Trident, Mageik, and Mount Martin, which trend southwest from Katmai, have been active in historic time. Trident has had a number of small eruptions. The most recent was in 1974 when a dome formed. Between 1953 and 1968, several flows of blocky andesitic lava moved down Trident's flanks, covering 1912 ash. Trident's neighbor, Mageik, had its most recent ash eruption in 1953, simultaneously with Trident. Another feature of Mageik is the "boiling lake" in its crater. The warm, acidic water in the crater bubbles vigorously due to escaping gases. Mount Martin steams intermittently and had a small ash eruption in 1951.

Glacial Features

The active glaciers in the park/preserve, all of which are on the flanks of the coastal volcanoes, are nourished by moisture-bearing winds and protected from melting by the nearly continual cloud cover. Serpent Tongue glacier, between Mt. Denison and Snowy Mountain, is the largest glacier in the park/preserve. The Knife

FIGURE 38.6   A few fumaroles near Novarupta caldera still give off steam as groundwater percolates down to the hot rocks below the volcanic vent. The photograph was taken in July 1953. Note the helicopter for scale. National Park Service photo.

Creek glaciers, which flow down Katmai's western slopes, are so well insulated by their cover of ash that they have changed little since the 1912 ejecta blanketed them. They look more like mud slides than glaciers.

In the foothills on the western side of the Aleutian Range, a half-dozen large, elongated lakes and many small lakes occupy glacial troughs, basins and cirques excavated by the mightier glaciers of the Pleistocene Epoch. Moraines dam most of the lakes. Naknek Lake, which is the largest, was for a time connected to Bristol Bay. Kukaklek Lake, in the preserve (northwest side of the park), is the source of the Alagnak Wild River, noted for whitewater rafting and sport fishing.

Geologic History

1. Paleozoic rocks.

North of Puale Bay, about five miles south of the park/preserve's southern boundary, limestones bearing mid-Permian fossils have been found, interbedded with volcanic flows and breccias. The limestones are part of a 4000-foot sequence of sedimentary and volcanic rocks.

2. Mesozoic rocks.

Thick sequences of Jurassic and Lower Cretaceous rocks, consisting of clastic and volcanic sediments, make up the main mass of this part of the Aleutian Range. In early to middle Jurassic time, the accumulation of more than 15,000 feet of such material suggests the long-continued presence of an area of subsidence, probably an oceanic trench. Uplifting and rapid erosion of coastal areas to the northwest brought arkosic sediments, which are interbedded with a considerable volume of volcanic debris, presumably from an island arc. These rocks make up the Shelikof Formation and the overlying Naknek Formation that are widespread throughout the region. The Cretaceous Kaguyak Formation is frequently found overlying the Shelikof and the Naknek, sometimes conformably and some-

During the Katmai-Novarupta eruption of 1912, ash and pumice ejected from the Novarupta vent buried a 40 square mile area with up to 700 feet of the welded tuff that formed from the hot debris. Streams are gradually cutting through this material. National Park Service photo.

times unconformably. Fossils are abundant in these Jurassic and Cretaceous formations.

Granitic plutons, part of the Aleutian batholith, began to intrude the sedimentary rocks early in Jurassic time. Plutonic activity continued in a series of pulses through most of the period. Late in Cretaceous and early in Tertiary time, tectonostratigraphic terranes, rafted by a subducting oceanic plate, docked in the area, causing intermittent folding, uplift, and volcanism.

3. Cenozoic rocks.

In the Katmai area, mainly in the southern coastal section of the park/preserve, nonmarine Eocene beds, including coal deposits, are exposed. The coal was probably laid down in coastal swamps. Lava flows are interbedded with the sedimentary rocks. Small intrusions of quartz diorite and quartz monzonite, with radiometric ages ranging from early to middle Tertiary time, have cut the Eocene rocks. Younger Tertiary rocks

are absent in the park/preserve and were probably stripped off by erosion.

Convergence of plates and subduction of oceanic crust continues. A major fault system, the Bruin Bay fault, which trends northeast-southwest down the Aleutian Range, probably represents a suture, or "seam," between the Peninsular terrane on the southeast side and older rocks on the northwest side. Strike-slip displacement has moved rocks on the southeastern side of the fault to the southwest. Vertical movement has also occurred.

4. Quaternary volcanic activity.

Fifteen young volcanoes rise from a volcanic pile on top of Tertiary rocks in Katmai National Park and Preserve. Some cones are presently inactive, but six have erupted since 1912. Studies of Holocene ash layers have revealed at least 12 major volcanic events that preceded the 1912 eruption. The oldest ash layer that

Table 38.1			Generalized Geologic Column, Katmai National Park and Preserve	
Time Units			**Rock Units**	**Geologic Events**
Era	**Period**	**Epoch**		
Cenozoic	Quaternary	Holocene	Volcanic rocks: basalt, andesite, rhyolite, dacite, pumice, welded tuff, etc.	1912 eruption, with ash falls, ash flows, explosions
		Pleistocene		Volcanic activity preceding and following climactic eruption, ranging from steam eruptions, minor ash falls, lava flows to violent eruptions
			Glacial drift	Glacial advances and retreats from late Tertiary to Holocene
	Tertiary		Volcanic rocks Small plutons	Buildup of cones Subduction, deformation, uplift Lava flows; intrusive activity Docking of terranes
		Eocene	Nonmarine sediments; coal beds; volcanics	Erosion of highlands Compaction of coal in swamps and coastal marshes
Mesozoic	Cretaceous		Kaguyak Formation	Intermittent volcanism Fossiliferous marine sediments Ongoing subduction
	Jurassic		Naknek Formation Shelikof Formation Granitic plutons	Accumulation of marine sediments in subsiding basin Emplacement of Aleutian batholith
Paleozoic	Permian		Lava flows, breccias, fossiliferous limestones (adjacent to south edge of park/preserve)	Erosion

Sources: Keller and Reiser, 1959; Buck, 1965.

has been identified and dated fell in 5400 B.C. Other ash layers in the marine sediments of the Gulf of Alaska carry the record back 30,000 years.

With more than 6 cubic miles of ash and pumice discharged, the 1912 Novarupta eruption was the most voluminous of this century in the United States. The volume of ejecta thrown out by Katmai was about 12 times that of Mount St. Helens in 1980. Steam plumes, heat, and occasional eruptions from the more active volcanoes in the park/preserve, plus frequent earthquakes, are indications that we can expect more volcanic events to occur in Katmai National Park and Preserve in the future.

5. **Quaternary glaciation.**

Wisconsinan glaciation in Katmai National Park and Preserve was extensive and severe. Evidences of earlier glaciations were destroyed. In some places, glacial

deposits and volcanic deposits are interbedded, indicating that glaciers were melted by eruptions and then reformed. Holocene glaciers advanced and retreated as patterns of climate changed. The most recent major advance ended about 4000 B.C. However, the fact that two small glaciers reformed on the inner walls of Katmai's caldera since 1912 provides evidence that at high elevations, given sufficient precipitation, glaciers continue to thrive in Katmai National Park and Preserve.

Geologic Map and Cross Section

Buck, C.A. 1965. *Geology of the Alaska Peninsula: Island Arc and Continental Margin* (with geologic map). Geological Society of America Memoir 99.

Keller, A.S. and Reiser, H.N. 1959. *Geology of the Mt. Katmai Area, Alaska* (with geologic map). U.S. Geological Survey Bulletin 1058G, p. 261–298.

Sources

Curtis, G.H. 1968. The Stratigraphy of the Ejecta from the 1912 Eruption of Mt. Katmai and Novarupta, Alaska. In *Studies in Volcanology: A Memoir in Honor of Howel Williams*. Geological Society of America Memoir 116, p. 153–210.

Fierstein, J. 1984. *The Valley of Ten Thousand Smokes, Katmai National Park and Preserve*. Anchorage, Alaska: Alaska Natural History Association. 16 p.

Griggs, R.F. 1922. *The Valley of Ten Thousand Smokes*. National Geographic Society. 340 p.

Hildreth, W. 1983. The Compositionally Zoned Eruption of 1912 in the Valley of Ten Thousand Smokes, Katmai National Park, Alaska. In S. Aramaki and S. Kuchivo (editors). Arc Volcanism, *Journal of Volcanology and Geothermal Research* 18:1–56.

Hildreth, W., and Fierstein, J. 1987. Valley of Ten Thousand Smokes, Katmai National Park, Alaska. Centennial Field Guide, Cordilleran Section, Geological Society of America, v. 1, p. 425–432.

Hildreth, W., Fierstein, J.E., Grunder, A., and Jager, L. 1981. The 1912 Eruption in the Valley of Ten Thousand Smokes, Katmai National Park: A Summary of the Stratigraphy and Petrology of the Ejecta. U.S. Geological Survey Circular 868, p. 37–39.

Montague, R.W. et al., editors. 1975. *Exploring Katmai National Monument and the Valley of Ten Thousand Smokes*. Anchorage, Alaska: Alaska Travel Publications. 276 p.

Snyder, G.L. 1954. Eruption of Trident Volcano, Katmai National Monument, Alaska, February-June 1953. U.S. Geological Survey Circular 318. p. 1–7.

Wallman, P.C., Pollard, D.D., Hildreth, W., and Eichelberger, J.C. 1990. New Structural Limits on Magma Chamber Locations at the Valley of Ten Thousand Smokes, Katmai National Park. *Geology* 18:1200–1244.

Note: To convert English measurements to metric, go to www.helpwithdiy.com/metric_conversion_calculator.html

Lake Clark National Park and Preserve

SOUTH CENTRAL ALASKA

Area: 4,044,132 acres; 6319 square miles

Proclaimed a National Monument:
   December 1, 1978

Established as a National Park and Preserve:
   December 2, 1980

Address: 4230 University Drive, Suite 311, Anchorage, AK 99508

Phone: 907-781-2218

E-Mail: LACL_Visitor_Information@nps.gov

Website: www.nps.gov/lacl/index.htm

Two active composite volcanoes, over 10,000 feet high, overtop the rugged Chigmit Mountains in the southern Alaska Range; two more cones rise to the north and south of the park/preserve. The volcanoes are part of a belt that trends southwestward through the Aleutians. Earthquakes and eruptions signify ongoing tectonic activity.

Lake Clark, 42 miles long, occupies a deep glacial trough, eroded on a fault zone. Dozens of active glaciers and hundreds of waterfalls line the walls of major valleys. Fiords open into Cook Inlet in the coastal section.

Three streams in the park/preserve are protected in the National Wild and Scenic Rivers System.

FIGURE 39.1 Air view, looking northeast, near the north end of Lake Clark. Lake Clark drains into Iliamna Lake, south of the park/preserve, and eventually Bristol Bay on the west side of the Alaskan Peninsula. National Park Service photo.

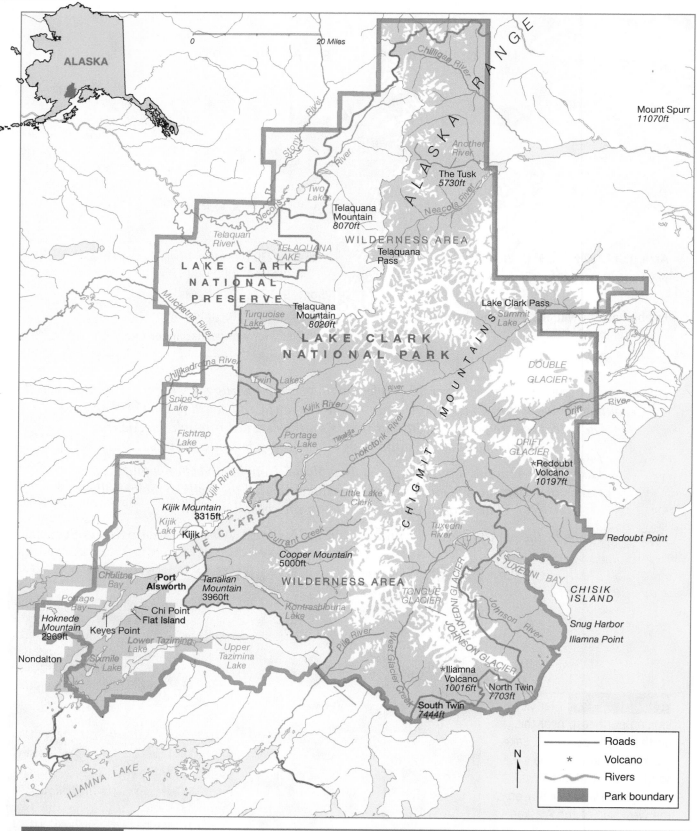

FIGURE 39.2   Lake Clark National Park and Preserve, Alaska.

Geographic Setting

Lake Clark National Park and Preserve, which extends inland from the western shore of Cook Inlet, occupies a rugged, mountainous area southwest of Anchorage. The park/preserve is predominantly wilderness in character. Here at the head of the Alaska Peninsula, the granitic rocks of the Chigmit Mountains link the Alaska Range, curving to the north and east, with the Aleutian Range, which stretches to the south and west. Four distinct physiographic regions, from east to west, give the park/preserve its diversity in ecosystems, scenery, and climate: (1) the coastal strip along Cook Inlet; (2) the abruptly ascending Chigmit Mountains, topped by Iliamna and Redoubt Volcanoes; (3) the foothills of the Chigmit Mountains' western flank, descending among lakes and wild rivers to (4) a boreal forest and a tundra plain.

Maritime elements influence the coastal climate, but a continental climate prevails throughout most of the interior. June through August temperatures on the coast average between 50° and 65° F. Around 60 inches of precipitation falls annually on the eastern slopes. Summers in the interior of the park/preserve are warmer and drier, with temperatures occasionally as high as 80° F; but interior winters are long and severe. Temperatures may plummet to –40° F. Along the coast, winters are more moderate. Strong winds (especially in mountain passes), sudden storms, and subfreezing temperatures may occur at any time of year throughout the park/preserve.

The diversity and richness of plant and animal life reflect the widely differing—but generally favorable—natural conditions within the region's ecosystems. Large mammals like brown and black bears, caribou, moose, and Dall sheep forage in uplands and lowlands along with wolves, lynx, foxes, and smaller creatures. The lakes and streams that drain west and then south into Bristol Bay are important spawning grounds for several species of salmon and trout. On the eastern flanks of the range, rivers cascade to the sea, and coastal cliffs provide rookeries for thousands of puffins, cormorants, and other seabirds. Fish abound in the coastal waters; seals and whales are often seen offshore.

Despite the severe climate and the difficulties in getting around, humans have lived in the Lake Clark region for more than 10,000 years. Migrating clans found sheltered camp sites and an abundance of fish and game. Rock paintings, done with red hematite mixed with animal fat, were discovered in a rock shelter at the head of Tuxedni Bay. The paintings, which are of great antiquity, are similar to some found in Siberia.

Relics of European origin found in old campsites reveal that Russian fur traders and missionary priests had made contact with Native Alaskans by the 1700s. The Russians controlled the fur trade until the Americans took over in the 1870s following the Alaska Purchase. In 1897, John W. Clark, an agent of the Alaska Commercial Company on Bristol Bay, led an expedition into the interior. He followed upriver systems to a large inland lake, later named Lake Clark in his honor. Native Alaskans of Eskimo and Indian descent still live in small settlements in the Lake Clark area.

The salmon industry attracted settlers to the region early in the 20th century. Prospectors and miners looked for gold and other minerals in the Chigmit Mountains without much success, although a few patented claims are still held. The first organized geological investigations were not undertaken until the 1920s, when Stephen R. Capps led U.S. Geological Survey expeditions into this area. He described Lake Clark, fed by hundreds of waterfalls from its rimming mountains, as "one of the most beautiful bodies of water in the world."

No highway access to the park/preserve exists. When weather conditions permit, visitors to the Lake Clark region come by small aircraft, mainly from Anchorage. Sometimes gusting winds in Lake Clark Pass and Merrill Pass are too hazardous for small planes to fly into the interior of the park/ preserve. Aside from field headquarters at Port Alsworth on Lake Clark, no National Park Service facilities are available in the park/preserve. A few licensed operators offer cabins, guiding, and outfitting services; primitive camping is allowed in certain areas. Sport fishing and float trips are permitted in the Chilikadrotna Wild River, which drains Twin Lakes; the Mulchatna Wild River, which flows out of Turquoise Lake; and the Tlikakila Wild River, which flows from Summit Lake (near Lake Clark Pass) southwestward until it empties into the northern part of Lake Clark.

Geologic Features

Even though glaciers, both past and present, have shaped the park/preserve landscape, it is the four active volcanoes that dominate the scene. Except for Augustine (4131 feet), which rises from the floor of

Cook Inlet, the volcanoes are several thousand feet higher than the Chigmit Mountains. The other volcanoes are Mr. Spurr (11,070 feet), northeast of the park/preserve boundary; Redoubt Volcano (10,197 feet), north of Tuxedni Bay; and Iliamna Volcano (10,016 feet), north of Chinitna Bay. Augustine Volcano, occupying Augustine Island, is south of the park/preserve. All four volcanoes have had violent eruptions and all have erupted recently:

- **Mt. Spurr.** A composite volcano with a summit dome, an explosion amphitheater, and a parasitic cone, Mt. Spurr's historic eruptions have been mainly pyroclastic flows. In 1992, the volcano erupted three times and is subject to frequent earthquakes, rockfalls, and slides. Because the volcano is too dangerous for scientists to observe closely, an eight-legged robot, built by NASA researchers, was brought in by helicopter in the summer of 1994 and set down on the rim. Guided by remote control, the robot climbed more than 650 feet down into the crater, walked around inside, took video and still pictures, recorded temperatures and sampled gases emitted by fumaroles. The mission was highly successful, but on the climb back up, the robot fell and was unable to get up. A helicopter succeeded in plucking the robot from inside the crater and set it down safely at the foot of the mountain.
- **Redoubt** is a composite volcano that has many Holocene lahars on its flanks and unstable slopes. Most recently, it began erupting with a roar in December, 1989; and more than 25 eruptions of lava and ash followed throughout 1990. A number of small domes have formed on the summit since the eruptions began.
- **Iliamna,** a strongly andesitic composite volcano has frequent landslides and is deeply eroded. Its last eruption was in 1947, but it has a large fumarolic area near the summit.
- **Augustine** has nested summit domes surrounded by pyroclastic debris, lahars, and debris avalanches. The volcano erupted in August, 1986, pouring our molten andesite from a vent at the base of a summit dome.

Why Are the Volcanoes Here?

The Lake Clark region volcanoes are a major link in the Alaska volcanic belt that extends from Mount Spurr

eastward to Wrangell-St. Elias National Park and Preserve (Chapter 32) and, in the other direction, southwestward to Buldir Island in the western Aleutians. Part of the chain, and closely related, are the volcanoes in Katmai National Park and Preserve (Chapter 38), some 50 to 75 miles (air distance) southwest of Lake Clark. The volcanic belt marks a subduction zone at depth that has been active for millions of years. The recent eruptions represent only the latest phase in a history of tectonic collision and volcanism since late in Mesozoic time and throughout the Cenozoic Era. In the Aleutian trench, parallel to and in front of the 1600-mile arc of volcanoes, oceanic crust is being subducted beneath the North American plate at an average rate of 2.6 inches per year. The subduction zone of the Aleutian trench lies about 30 miles beneath the surface of the Kenai Peninsula (Chapter 33), but steepens abruptly at the western edge of Cook Inlet to a depth of more than 65 miles. Beneath Redoubt and Iliamna Volcanoes, the depth of the subducting plate is greater than 100 miles. This steep, rapid descent causes frequent, strong earthquakes. Since 1972, there have been 13 with a Richter magnitude of 5 to 6. Intense heat, generated by the subduction process, melts rock at depth, forming magma, which subsequently rises, erupts, and builds volcanoes.

Volcano Hazards in the Lake Clark-Cook Inlet Area

Most of the eruptions of the Lake Clark volcanoes are explosive and pyroclastic, involving ash flows, ash falls, lava flows, and lahars (mud flows). The volcano-watchers of the U.S. Geological Survey, basing their conclusions on studies of ash deposits over the past 10,000 years, warn that inhabitants of the Cook Inlet region can expect a measurable ash flow on an average of every 30 years. In recent years eruptions of ash and lava have occurred at a more frequent rate.

The hazard potential of ash falls depends on local winds and upper-air conditions. But the presence of the mountains aligned near the coast tends to channel windblown ash up or down Cook Inlet or eastward over the Kenai Peninsula. During the 1986 Augustine eruptions, and the more recent Redoubt and Spurr eruptions, aviation flights to and from Anchorage were canceled or diverted for several days at a time because of wind-carried ash. Ash in the atmosphere also causes temporary radar blackouts and reduces the electrical generating capacity of turbines.

Tsunamis, caused by sea-floor movement or avalanching (above or below the waterline), are a major potential hazard in the coastal regions. In 1883, Augustine's most violent eruption in historic time produced a tsunami, due to avalanching, that crossed Cook Inlet and inundated Port Graham on the tip of the Kenai Peninsula. No lives were lost, but fishing boats and homes were wrecked.

The presence of glaciers on the slopes of active volcanoes is cause for concern. Because of their height and topographic position (in the path of moisture-carrying winds from the Gulf of Alaska), Mount Spurr, Iliamna, and Redoubt all have sizable glaciers.

In prehistoric time, *lahars* (volcanic mudflows that are pasty masses of snow, ice, volcanic debris, rocks, vegetation, etc.) have roared down valleys on both Redoubt and Iliamna. Some reached Cook Inlet. Lahars are associated with eruptions. On the other hand, *glacier outburst floods* may or may not be triggered by volcanic activity. An outburst flood is a sudden, catastrophic release of glacial meltwater, which may be due to the melting of a drainage channel, the buoyant lifting of ice by meltwater, or an increase in subglacial heat emanating from a volcano's interior. A sudden release of steam from a vent may also trigger a glacier outburst flood.

The Redoubt eruptions that took place between 1965 and 1968 were the cause of several outburst floods. Heat from explosive ash flows melted and broke up the glaciers. Huge blocks of ice and uprooted trees were carried down the Drift River valley. Up to six feet of water backed up behind the ice at the river's mouth, where it meets Redoubt Bay. A Canadian seismic crew (prospecting for oil) had to be hastily evacuated. Again, during Redoubt's 1990 eruptive activity, the Drift River became choked with ice and sediment. This caused flooding at the Drift River oil terminal, which was recently built on the coast outside the park/preserve. Operations at the terminal were severely disrupted.

Glacial Features

Iliamna's glaciers are larger and more numerous than those of Redoubt. Tuxedni Glacier, which descends 16 miles from Iliamna's cone to within a few miles of Tuxedni Bay, is the longest glacier in Lake Clark National Park and Preserve.

Innumerable glaciers in the Chigmit and Neacola Ranges continue to sculpt the mountain landscape of the park/preserve. Most of these glaciers have been in retreat for several hundred years. The large glaciers on the volcanoes, which form at higher elevations and receive more precipitation, are not retreating at the present time. A nearly continuous cloud cover during summer months keeps the glaciers near the coast from melting back.

The legacy of the more extensive glaciers of the past is the spectacular scenery of today. Lake Clark lies in a glacially eroded trough of massive proportions. The lake, 42 miles in length and 1 to 4 miles wide, covers an area of 110 square miles. Its maximum depth is 860 feet. Its location on the Lake Clark fault indicates structural control. Glacial ice, the excavating agent, moved along the "path of least resistance," where rock was already broken up by faulting (fig. 39.1). Several other lakes in or just outside the park/preserve lie in fault troughs that in the past conveyed glacial ice as far as Bristol Bay. Morainal dams, dumped by retreating glaciers, now separate these lakes from each other and from the ocean. Some small lakes occupy kettles in glacial drift. *Kettles* are slumped-in depressions located where a buried ice-block has melted.

In the foothills region west of the Chigmit Mountains, jewel-like lakes fill scoured basins left after glaciers had melted back. The turquoise-green color of the lakes comes from the fine, glacial powder (rock flour) that washes down from glaciers in the mountains. Some examples (from north to south) are Two Lakes, Telaquana, Turquoise, Twin Lakes, Portage, Lachbund, Kontrashibuna, and Tazimina.

The fiords, cliffs, marine terraces, lowlands, and other features in the coastal part of the park/preserve show the effects of changing sea levels (both emergence and submergence) in Pleistocene and Holocene time. In this region of recent glaciation and active tectonism, both glacial rebound and tectonic movements of the sea floor are important factors in sea level rises and drops that have occurred.

Landsliding and Rockfalls

Mass wasting processes are actively modifying the landscape in the park/preserve. Much of the volcanic rock is weak and forms unstable slopes. Valley walls and cliffs, oversteepened by glacial excavation, tend to fall or slide downslope, especially when loose rock and soil are saturated by water. Sometimes earthquakes trigger landslides. Landslide deposits, some fresh, some stabilized and tree covered, are numerous.

On slopes covered with tundra, solifluction lobes are conspicuous features. (*Solifluction* is the slow creep of moist, fine-grained sediment caused by more or less continual freezing and thawing.)

Rocks and Structures

A park naturalist described the Chigmit Mountains as a "frenzy of peaks thrown up by colliding mountain waves." Simply put, granitic rocks, emplaced in older sedimentary rocks, were uplifted and faulted, covered by lavas and pyroclastic debris, and unroofed by erosion. The combination of intrusive, metamorphic, volcanic, and sedimentary rocks, all of which have been disarranged by tectonic activity, presents a complex puzzle.

Here, as in the other national parks located along the Gulf of Alaska, the concept of tectonostratigraphic terranes helps to explain the juxtaposition of discrete, seemingly unrelated crustal blocks, bounded by major faults (sutures). In the Lake Clark region, the Peninsula terrane is the predominant terrane, as it is also in Katmai National Park and Preserve to the south (Chapter 38).

Major faults are the Castle Mountain fault and the Bruin Bay fault (fig. 39.4). Both faults trend northeast-southwest, but the Bruin Bay fault, which separates two subsections of the Peninsular terrane, branches south from the Castle Mountain fault and skirts the coastline, east of Redoubt and Iliamna Volcanoes. The Castle Mountain fault passes on the west side of the volcanoes through the trough occupied, in part, by Lake Clark. High-angle reverse movement has occurred on the

FIGURE 39.3 Because of glaciation many of the valleys have the typical U-shape produced as the glaciers pluck, abrade and remodel the original V-shaped stream valleys. National Park Service photo.

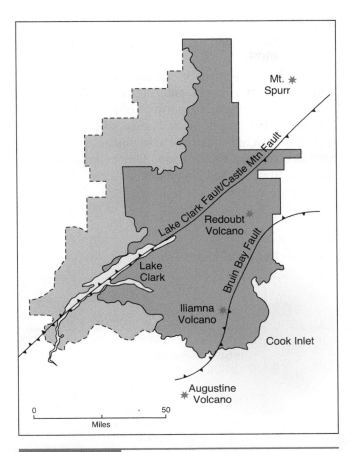

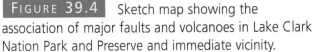

FIGURE 39.4 Sketch map showing the association of major faults and volcanoes in Lake Clark Nation Park and Preserve and immediate vicinity.

Bruin Bay fault, which continues down the coast into Katmai National Park and Preserve. Horizontal displacement on the Castle Mountain fault has been about eight miles and has occurred fairly recently. This is shown by the fact that the displacement is the same across rocks ranging in age from Paleozoic through Tertiary. A topographically prominent glacial trench follows the segment of the Castle Mountain fault (commonly referred to as the Lake Clark fault) as it runs through the park/preserve and extends for many miles northeastward beyond Lake Clark. The fault trace and glacial trench leave the park/preserve after going through Lake Clark Pass. Shear zones and foliation in the granitic and metamorphic bedrock along the fault occupy a belt from 60 feet to as much as 725 feet wide. Tlikakila Wild River, which flows along the fault zone

before emptying into the northern end of Lake Clark, is closely flanked by perpendicular cliffs and alpine glaciers descending from 10,000-foot-high, snow-capped mountains. Fault structures control the flow of a number of other rivers and glaciers in the park/ preserve.

Geologic History

1. Jurassic intrusions.

Granitic magma, in a series of pulses (from 180 to 160 million years ago), was emplaced as a huge batholithic mass beneath a chain of volcanoes. The batholith was 800 miles long and 10 to 20 miles wide. The country rock, beds of Triassic age and older, were metamorphosed into marbles, quartzites, schists, and argillites. Andesitic lavas erupted, forming composite volcanoes. The Jurassic granitoids now form a major part of the Chigmit Mountains.

2. Uplift and erosion.

As the solidified batholith was uplifted, thousands of feet of volcanic debris and underlying granitic rocks were stripped off, transported by streams, and redeposited, mostly in an ocean. At Fossil Point, north of Tuxedni Bay, Jurassic marine fossils have been found, mainly ammonites and pelecypods. The area between Tuxedni Bay and Iniskin Bay (south of the park) has one of the most complete sequences of Jurassic rocks in the United States. These rocks, east of the Bruin Bay fault, are a mixture of volcanic breccia, tuff, siltstone, and graywacke.

3. Late Cretaceous and early Tertiary intrusions and volcanism.

Between 60 and 80 million years ago, coarse-grained igneous rocks were emplaced north of Iliamna Lake and west of the Jurassic batholith. A cluster of volcanoes built up in the Iliamna Lake area. The volcanic rocks in the western part of the park/preserve are early Tertiary (50 to 60 million years) in age. Thousands of feet of ash and welded tuff were explosively erupted from volcanic vents.

4. Accretion of terranes.

At the end of Mesozoic time and early in the Tertiary Period, terrane blocks, or microplates, that had "drifted" in from the south began to collide with the stable Alaska edge on the North American craton. Some of the material was subducted but most of these plates were "sutured" to Alaska, with major faults marking the plate boundaries. The Peninsular terrane is the most prominent block in the Katmai-Lake Clark area.

5. **Middle Tertiary volcanism; construction of the present volcanoes, mainly in Quaternary time.**

A period of quiescence followed the early Tertiary volcanism. Volcanic activity resumed in middle Tertiary time (about 35 million years ago) and has con-

tinued into the present. The volcano belt, and presumably the plate boundary, shifted to approximately the present position.

The four volcanoes in the Lake Clark region, Mt. Spurr, Redoubt, Iliamna, and Augustine, all developed comparatively recently on top of older granitic rocks. The explosive eruptions of highly siliceous ash and lava that has characterized these volcanoes in historic time is expected to continue. Dome building, especially in Redoubt and Augustine, is a typical eruptive pattern.

Table 39.1			**Generalized Geologic Column, Lake Clark National Park and Preserve**	
Time Units				
Era	Period	Epoch	**Rock Units**	**Geologic Events**
Cenozoic	Tertiary	Holocene	Alluvial deposits, ash, glacial drift; landslide debris, talus, etc.	Glacial advances and retreats, stream erosion, mass wasting Ongoing volcanism, earthquakes, subduction
		Pleistocene	Glacial deposits Andesites, breccias, pyroclastics	Several episodes of severe glaciation Construction of composite cones
	Tertiary	Late	Volcanic rocks, mostly andesitic flows	Beginning of glaciation Accumulation of volcanic rocks, pyroclastic debris Subduction, uplift, tilting
		Early	Conglomerates, sandstones, siltstones Basaltic and andesitic flows Plutonic intrusions	Accumulation of marine and nonmarine sediments Uplift, widespread volcanism Subduction
Mesozoic	Cretaceous		Regional unconformity ～～～～～	Arrival of tectonostratigraphic terranes ～
	Jurassic	Late	Naknek Formation Chitnitna Formation Tuxedni Group	Accumulation of thousands of feet of fossiliferous marine sediments, coarse- and fine-grained, mostly clastics Uplift
		Middle	Quartz diorites, quarz monzonites, granodiorites, and other granitic rocks	Emplacement of Aleutian Range batholith
		Early	Talkeetna Formation	Andesitic tuffs, flows, breccias, mixed with clastic sediments
	Triassic		Undivided metamorphic rocks	Complexly folded marbles, quartzites, etc.

Sources: Detterman and Hartsock, 1966; Detterman and Reed, 1973, 1980; Péwé, 1975; Reed et al., 1983.

FIGURE 39.5   This is the land of fire and ice, with its many volcanoes and glaciers. National Park Service photo.

6. Pleistocene glaciation.

Wisconsinan glacial advances, which were extensive in this region, destroyed most of the evidence of earlier glacial stages. Many moraines that confine or separate lakes were left by ice of Wisconsinan age. At their greatest extent, valley glaciers coalesced to form large lobes and piedmont glaciers. Glacial ice dammed Cook Inlet and extended to the Kenai Peninsula.

7. Neoglacial advance in Holocene time.

Glaciers in the park/preserve began readvancing around 4500 to 4800 years ago, and were still advancing up until about 200 years ago. Some glaciers have retreated about a quarter of a mile in the past 25 years. In the late 1930s, Lake Clark Pass was occupied by a glacier about five miles wide, but only parts of this former ice field remain.

Sources

de Laguna, F. 1974. Magical Cave Paintings from Prehistoric Alaska. In M. Sherwood, editor. *The Cook Inlet Collection*. Anchorage, Alaska: Alaska. Northwest Publishing Company. p. 1–7.

Detterman, R.F., and Hartsock, J.K. 1966. *Geology of the Iniskin-Tuxedni Region, Alaska*. U.S. Geological Survey Professional Paper 512. p. 1–78.

Detterman, R.F., and Reed, B.L. 1973. Surficial Deposits of the Iliamna Quadrangle, Alaska. U.S. Geological Survey Bulletin 1368A, p. Al–A64.

———. 1980. Stratigraphy, Structure, and Economic Geology of the Iliamna Quadrangle, Alaska. U.S. Geological Survey Bulletin 1368B, p. B1–B86.

Hirshman, F., editor. 1986. Lake Clark and Iliamna Country. *Alaska Geographic* 13 (4).

National Park Service. n.d. Geology of Lake Clark (unpublished).

Miller, T.P., and Richter, D.H. 1994. Quaternary Volcanism in the Alaska Peninsula and Wrangell Mountains., Alaska; in Plafker, G., and Berg, H.C., eds. *The Geology of Alaska.* Boulder, Colorado: Geological Society of America, *The Geology of North America,* v. G-1, p. 759–779.

Péwé, T.L. 1975. *Quaternary Geology of Alaska.* U.S. Geological Survey Professional Paper 835.

Reed, B.L., Miesch, A.T., and Lanphere, M.A. 1983. Plutonic Rocks of Jurassic Age in the Alaska-Aleutian Range Batholith: Chemical Variations and Polarities. *Geological Society of America Bulletin* 94:1232–1240.

Note: To convert English measurements to metric, go to www.helpwithdiy.com/metric_conversion_calculator.html

Hawaii Volcanoes National Park

ISLAND OF HAWAII, HAWAII

Area: 299,177 acres; 358 square miles

Established as a National Park: August 1, 1916

Redesignated Hawaii Volcanoes National Park:
September 22, 1961

Designated a Biosphere Reserve: 1980

Designated a World Heritage Site: 1987

Address: P.O. Box 52, Hawaii National Park, HI, 96718

Phone: 808-985-6000

E-mail: HAVO_Interpretation@nps.gov

Website: www.nps.gov/havo/index.htm

Magma, melted by a "hot spot," rises from reservoirs beneath the ocean floor and builds Hawaiian volcanoes. Cones attain elevations above sea level of more than 10,000 feet. As oceanic crust passes slowly over the hot spot, old volcanoes become dormant; new ones form and grow from the ocean depths.

Frequent, "gentle" eruptions are characteristic of volcanic activity on Hawaii Island; but lava may squirt in high fountains from volcanic vents, or cascade down slopes in rivers of incandescent rock. Observers have recorded frequent eruptions of Mauna Loa and Kilauea since the early 1800s. A new volcano, Loihi, is being born in the ocean south of Hawaii.

Earthquakes and precise measurements on the volcanoes yield data that allow forecasts of eruptive behavior.

FIGURE 40.1 Lava from Pu'u' O'o covered most of the highway, but a few sections were left to show where it had been. Photo by Ann G. Harris.

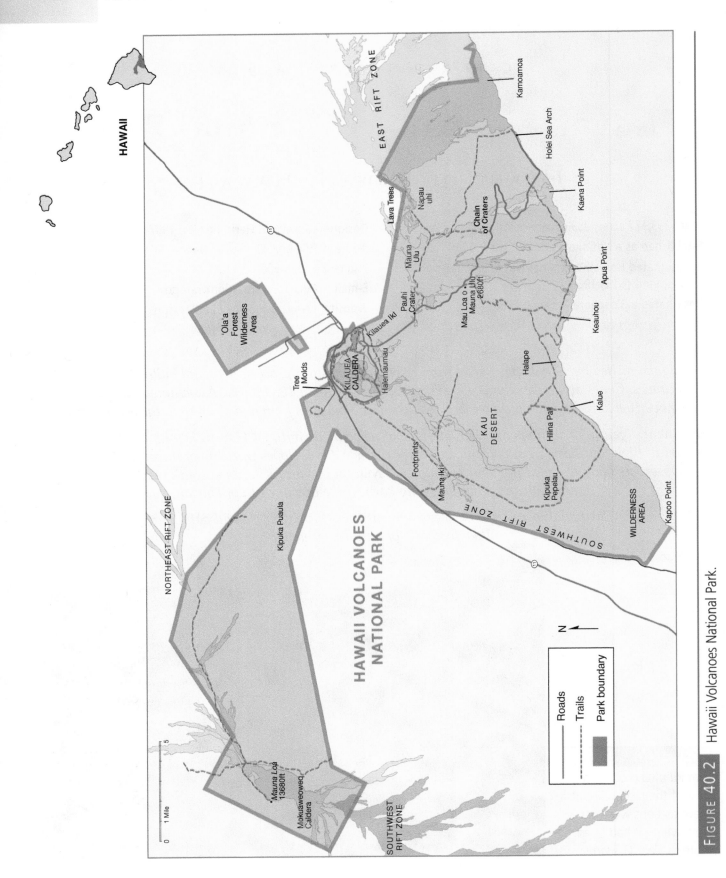

FIGURE 40.2 Hawaii Volcanoes National Park.

Box 40.1	The Geography and Geology of the Hawaiian Archipelago

The Hawaiian Archipelago in the Pacific Ocean is a group of shoals, atolls, and islands, stretching 1700 miles on a northwest-southeast linear trend, and a little over 2400 miles west of Los Angeles. The southeast part of the chain, made up of eight main islands and numerous islets, includes: Hawaii, at the southeast end, largest island and location of Hawaii Volcanoes National Park; Maui, the second largest island and site of Haleakala National Park (Chapter 41); Kaho'olawe; Lana'i; Moloka'i; Oahu, where Honolulu, the state capital, is located; Kaua'i; and Ni'ihau.

Extending toward the northwest from Ni'ihau are the small, mostly uninhabited atolls and shoals of the Hawaiian Islands National Wildlife Refuge. The Midway Islands and Kure, at the far end of the chain, are some 1100 miles northwest of Honolulu. The state of Hawaii includes all the islands and reefs in the chain from Hawaii to Kure, except the Midway Islands, which are administered by the U.S. Navy. Ka Lae (South Cape) at the southern tip of Hawaii is the southernmost point in the United States.

Climate. The subtropical climate is pleasant and mild. Air temperatures range from below freezing at high elevations to the high 90s on the lee coasts. Because the Hawaiian Islands lie in the path of the northeast trade winds, rainfall is heavy on the windward sides of the islands. Occasionally in winter, winds from the south or southwest ("kona winds") bring heavy storms and may dump as much as 24 inches of rain in four hours. Maximum precipitation on the islands occurs between 2000 and 6000 feet above sea level on windward slopes (i.e., facing northeast), but rain shadow zones reduce precipitation on some islands. Annual rainfall ranges from about 10 inches on lee slopes to about 450 inches in the wettest belts. Stream densities are high and stream erosion is intense on the windward sides of the islands.

Geologic Development. The islands in this long chain consist mostly of inactive volcanoes built of basaltic lavas that punched through the crust beneath the ocean floor. Repeated eruptions of thin, fluid lava flows (2 to 3 feet thick) brought the volcanoes up to the surface from depths of some 2.5 to 4 miles below sea level in less than a million years. Above water, their evolution as volcanic islands has followed a consistent pattern. Some reached greater heights than others; but regardless of how high above sea level the volcanoes grew, each one evolved into an "alkalic cap stage," which means that the basaltic lavas became poorer in silica and richer in sodium and potassium. Eruptions were less frequent and more explosive during this stage and lava flows became more viscous.

As eruptive activity waned, erosion cut deep valleys into the soft rock of the volcano slopes. On the older islands, Ni'ihau, Kaua'i, Oahu, Moloka'i, Maui, and Kaho'olawe, erosion was interrupted by renewed volcanism of even more alkalic and explosive lavas. Diamond Head on Oahu, for example, is the result of a period of renewed volcanism. As each volcanic island aged and was eroded, it sank as oceanic crust beneath adjusted to the rapid growth of newer volcanoes in the chain. Coral reefs, like those on Midway, encircled the older volcanoes. With continued subsidence, waves eroded the tops of extinct volcanoes, and fringing reefs became atolls surrounding lagoons.

Radiogenic dates derived from rock samples taken at various places along the Hawaiian chain confirm that the volcanoes get progressively younger toward the southeast. Ages of principal lava flows, in millions of years, beginning with Kaua'i (the oldest) are 5.8 to 3.4; Oahu, 3.5 to 2.3; Moloka'i, 1.5 to 1.3; Maui, 1.3 to 0.8; and Hawaii, less than 0.7 and still growing (Clague 1989).

A seamount named Loihi, about 20 miles south of Hawaii Island, is expected to be the next island in the chain. Loihi has risen 12,000 feet from the ocean floor and grown to within 3000 feet of water level. Surveys show its shape as similar to Mauna Loa and Kilauea, with a flat summit area, a caldera, and two probable rift zones.

What caused the Hawaiian chain to form this linear pattern of older to younger volcanoes across the ocean floor? Since the volcanoes are erupting near the middle of the Pacific plate, more than 2000 miles from

(continued)

| Box 40.1 | The Geography and Geology of the Hawaiian Archipelago (continued) |

the nearest plate boundary, the volcanism cannot be associated with tectonic activity occurring at plate margins. Moreover, the oceanic crust in that part of the Pacific is much older (Cretaceous) than the lava flows that constructed the Hawaiian volcanoes. Most of them are younger in age than Miocene.

The most likely hypothesis suggests the existence of an immobile hot spot, a thermal high at depth, perhaps as much as 200 miles across, that melts rock material below the overriding Pacific plate. The magma accumulates in pockets that bulge up and eventually break through a stretched or weakened place in the crust. Hot lava then rises rapidly through conduits and fissures to the ocean floor. The release of great quantities of basaltic lava builds up shield volcanoes that rise above water level in about 300,000 years. As the Pacific plate drifts slowly northwestward, successive volcanoes form over the hot spot, are rafted away on the moving plate, and become dormant as they are cut off from the source of fresh magma. The hot spot theory thus explains the pattern of a volcanic chain with active volcanoes at the southeast end and older volcanoes becoming inactive or extinct toward the northwest.

Polynesian voyagers from the South Pacific settled the Hawaiian Islands. These courageous, seafaring people began arriving at the islands about 2000 years ago, their final migration being around 750 A.D. Volcanoes, believed to be the abodes of deities, were central to Polynesian culture and religion. Pele, the goddess of all volcanoes, lived in Halema'uma'u, the fire pit in Kilauea caldera on Hawaii.

Captain James Cook discovered the islands in 1778 on a voyage of exploration for the British Admiralty. From Captain Cook on, visitors to the islands have been fascinated by the Hawaiian volcanoes. Travelers from all over the world came to see the amazing spectacle of Halema'uma'u, the boiling lake of lava; among them was Mark Twain, who said that, by comparison, Vesuvius was a "soup kettle"!

James Dwight Dana, a young geologist attached to a U.S. Navy expedition commanded by Lieutenant Wilkes, conducted the first scientific investigation of the Hawaii volcanoes in 1840-41. Dana recognized the trend of decreasing age of the volcanoes from northwest to southeast because he observed differences in the degree of erosion that had taken place on the islands since eruptions had occurred. He theorized that a great fissure zone, from which lava erupted, might have opened on the ocean floor.

The movement to protect and preserve the land around the volcanoes on Maui and Hawaii Islands was led by L.A. Thurston, editor of the *Honolulu Advertiser*. As founder of the Hawaiian Volcano Research Association, Thurston was also active in helping to establish the Hawaii Volcano Observatory. It was formally organized in 1911 under the auspices of the Massachusetts Institute of Technology and the Carnegie Institute of Washington. Professor T.A. Jagger, a geologist and volcanologist at Massachusetts Institute of Technology, chose Kilauea as the site for the proposed observatory because of the constancy of the volcano's activity and because investigators could get close to active lava without exposing themselves to too much danger. Continuous recording of seismic and volcanic activity has been carried on at the Hawaii Volcano Observatory since 1912 under the successive sponsorship of the Massachusetts Institute of Technology, the U.S. Weather Bureau, the National Park Service, and (since 1956) the U.S. Geological Survey. The Observatory buildings are located in the park on the rim of Kilauea Crater, overlooking Halema'uma'u.

Included in the park is the summit area of Mauna Loa, a strip of land between Mauna Loa and Kilauea, and most of the southern flank of Kilauea, from the caldera to the coast (fig. 40.2). Mauna Loa and Kilauea make ideal natural laboratories for studying volcanic activity. They erupt frequently but seldom explosively; they emit enormous amounts of lava; and they have large, relatively simple structures.

Geologic Features

Five shield volcanoes make up the island of Hawaii, called "Big Island." Its area is 4030 square miles. Mauna

FIGURE 40.3   Kilauea Caldera with Halemaumau Pit Crater in the background. Steam is coming from the molten lava deep in the volcanic neck. Photo by Ann G. Harris.

Loa and Kilauea are both active volcanoes; Kohala, at the northern end of the island, and Mauna Kea, on Kohala's south side, have been inactive in historic time. Haulalai, on the west coast of Big Island, erupted last in 1801. Mauna Kea, the highest cone (elevation 13,784 feet), is also the world's highest peak on an island. At 13,680 feet, Mauna Loa is a close second in elevation.

Approaching the island of Hawaii, whether on board ship or in a plane, travelers see the summits of Mauna Loa and Mauna Kea from great distances if the weather is clear. The closer one gets, the more impressive their bulks appear as they rise from the ocean. Yet much more of their great size is hidden beneath the water; the ocean floor on which the volcanoes stand is 18,000 feet below. Thus Mauna Loa's true height is 32,000 feet, or more than 6 miles! Kilauea, whose highest point is 4090 feet above sea level, is nearly 10,000 feet lower than Mauna Loa and Mauna Kea. All of the rock above sea level on Big Island has been erupted during the past million years, which gives us an idea of the rapidity with which this tremendous volume of lava has been extruded.

Because Mauna Loa and Kilauea have been active during historic time, many changes in their features have been documented. During the past century and a half, Mauna Loa has produced a lava flow on the average of every seven years and has exhibited minor eruptive activity in the years between. Its shield-shaped

dome is about 70 miles long and 30 miles wide. Even more impressive is its volume of about 10,000 cubic miles, which makes it the world's largest active volcano. At the summit, the Moku'aweoweo caldera coalesces with adjacent pit craters. Formed by collapse rather than explosion, the caldera has changed its size, shape, and depth frequently during historic time as lava flows have spread out over the caldera floor or breached the rim.

Rising from the southeastern flank of Mauna Loa is Kilauea, which has a shield dome about 50 miles long and 14 miles wide. Its caldera is three miles long and two miles wide (fig. 40.3). Within the caldera is the fire pit (active crater) of Halema'uma'u that from time to time contains a lake of liquid lava. A *lava lake* forms when magma rises within the volcano and floods the pit in the floor of the caldera. Sometimes a lava lake lasts for years, bubbling and glowing (fig. 40.4).[*]

---

[*]In *Roughing It,* Mark Twain describes in awed terms seeing the lava lake after dark: "The greater part of the vast floor of the . . . [caldera] under us was as black as ink and apparently smooth and level; but over a mile square of it was ringed and streaked and striped with a thousand branching streams of liquid fire! . . . Imagine a coal-black sky shivered into a tangled network of angry fire." Mark Twain visited Halema'uma'u in 1866. The fiery "branching streams" he described were lava fountains.

If gas pressure raises the level of the molten rock high enough, lava spills out over the rim of the fire pit and spreads over the caldera floor. More often, lava breaks out through the flanks of the volcano above or below sea level. As lava pours out of the flanks, the level of the lake falls and the lake may disappear, perhaps for years at a time. If the whole plumbing system of the volcano is drained, then the caldera floor collapses. After a time, more magma rises into the reservoir, pushes up through the conduit, and the whole process—with modifications—may be repeated.

Volcanic Rocks That Form from the Lavas

The Hawaiian basalts are of two general types, or suites—*tholeiitic* and *alkalic,* both believed to be derived from the upper mantle, but perhaps from different levels. The tholeiitic basalts make up the bulk of the shields (99%). The alkalic basalts (1%) are found mainly on the caps of the shields. The predominant phenocrysts in both types of Hawaiian basalts are olivine. Tholeiitic basalts commonly contain low- and high-calcium pyroxenes, both poor in aluminum. Pyroxene phenocrysts are rare in these rocks. Alkalic basalts contain only one pyroxene, a high-calcium and moderately high-aluminum variety that occurs commonly as phenocrysts as well as groundmass grains.

Types of Lava and Lava Features

Since most of the Hawaiian lavas are fluid, they flow from vents relatively quietly and congeal with a smooth, billowy or ropy surface called *pahoehoe* (pah-hoay-hoay) (fig. 40.5). Typically, lava erupts in a series of pulses that build up layers of pahoehoe, one on top of the other. Sometimes a flow hardening as pahoehoe changes to another common type of congealing termed *aa* (ah-ah), which has a rough, jagged, or clinkery surface. The basalt in the interior of aa flows is dense and thick with many stretched vesicles left by escaping gas. The front of the flow moves along as a red-hot, clinkery wall. Massive aa flows may show columnar jointing. The change from a pahoehoe flow to aa involves the initial gas content of the lava and changes in lava viscosity. An aa flow never reverts to pahoehoe, but a fresh pahoehoe flow may be erupted on top of an earlier aa flow.

When a lava lake is active, or when a flank vent is first erupting, gases bubbling up from depth produce *lava fountains* that may spout from a few inches to 30 feet in height, rising and falling on the hot lava surface (fig. 40.6). Occasionally lava fountains shoot up much higher, as happened at Kilauea Iki in 1959 when a lava fountain reached a height of 1900 feet, the highest ever observed. Pu'u 'O'o, which began erupting in 1983,

FIGURE 40.4   Inside of Pu'u 'O'o crater showing the lava lake and its crust. Photo by Ann G. Harris.

FIGURE 40.5 A pahoehoe lava flow is a basaltic flow having a ropy or billowy surface. It forms when a crust forms over the flow; then the flow moves again, wrinkling the crust. Photo by Ann G. Harris.

delighted volcano-watchers with its exceptionally high and frequent fountaining.

During high fountaining, *thread-lace scoria,* or *reticulate,* which is a light, feathery type of pumice, may form and be carried many miles downwind. Tiny strands and droplets of fountaining lava, also caught by the wind, harden in the air and fall. These glassy pellets and strings, called *Pele's tears* and *Pele's hair,* are fragile and short-lived.

Lava tubes and *lava tunnels* beneath the congealing surface of a pahoehoe flow convey fresh, hot lava out to the moving front (fig. 40.7). Later, when the whole mass has cooled, the tunnels and tubes remain. A lava tube serves as a large conduit for the movement of lava down a slope far from the vent. The tube's outer crust acts as an insulator, keeping lava from cooling. The Thurston lava tube on the east side of Kilauea, which is 20 feet high and 22 feet wide in places, was formed by

the crusting over of a pahoehoe flow from one of the Twin Craters. Some lava tubes contain *lava stalactites,* produced as liquid lava hardened while dripping from the tube roof.

Tree molds and *lava trees* may form in pahoehoe lava when a flow inundates a forest. Hot lava, chilled by its initial contact with a tree, either sticks up against the bark, hardening quickly, or flows around the trunk. After the charred tree rots away, the lava "trees" are left standing. Tree molds, resembling empty pails in the flow, show where tree trunks stood. Tree molds are numerous in Hawaii Volcanoes National Park and in Lava Tree State Park.

Mounds or hillocks on the surface of pahoehoe lava are called *tumuli* (singular, *tumulus*). They tend to form parallel to the flow front in places where lava is confined, especially in craters. Solid crust at the top of the flow is dragged along and domed up by the faster-

The fire pit Kilauea Iki had a lava lake in it in the 1950s. In 1959 there was a spectacular eruption, with numerous fountains erupting in the lava lake. National Park Service photo.

moving fluid lava underneath. Sometimes fresh lava dribbles or spatters out, hardening into *driblet spires* of fantastic shapes.

Lava balls are accretionary features characteristic of aa flows and range in size from a few inches to 10 feet or more. They form when a fragment of solidified lava rolls along, enlarging as it picks up sticky lava. (Children in snowy regions make snowballs large in much the same way by rolling them in sticky snow until they're too big to push.)

A Hawaiian term, *kipuka,* meaning "opening," refers to island-like areas of older land surrounded by more recently erupted lava. Kipukas, ranging in size from a few square feet to several square miles, are usually the result of irregularities in the land surface. A kipuka surface may be higher or lower than the level of the surrounding lava, the latter case being more common. New lava is sterile and barren of vegetation, but some kipukas have weathered long enough to have soil that can support plant life, making them islands of

greenery in a landscape of bare volcanic rock. Several kipukas are in the park. The best known is Kipuka Puaulu, or Bird Park, northwest of Hawaii Volcano Observatory.

Ash Explosions and Pyroclastics

Kilauea has been an ash producer, probably because more moisture is trapped on its windward slopes. Ash explosions are commonly associated with the sudden mixing of ground (phreatic) water and hot lava or magma. Violent explosions have occurred when water drained down into newly emptied vents or conduits and reached magma. The result has been a *phreatomagmatic explosion.* In 1790 such an explosion killed a band of Hawaiian warriors who were marching across the Ka'u desert (a rain shadow zone on the southwest slope of Kilauea). Despite years of weathering, the warriors' footprints are still faintly visible on an ash surface close to the park's western boundary.

FIGURE 40.7   Lava tubes and lava tunnels beneath the congealing surface of a pahoehoe flow convey fresh, hot lava out to the moving front. Occasionally a section of the roof will fall in, creating a "skylight" through which the flowing lava can be seen. Photo by Ann G. Harris.

Phreatic explosions occur any time that hot lava encounters ground water and turns it to steam. The most explosive steam explosions in historic time occurred during the 1924 eruptions from Kilauea when large blocks of old lava (that can be seen scattered around the summit area) were ejected from the volcano's main vent (fig. 40.8). More typically, phreatic explosions produce ash, cinder cones, and pyroclastics of various sizes, especially at or near the shore where moisture is plentiful. Most cones, such as those along Kilauea's east rift zone, are a combination of cinders and spatters.

Rift Zones on the Volcanoes

Hawaiian shield volcanoes are built up not only by summit eruptions but also by eruptions from *rift zones* that radiate from the summit calderas and extend down into the ocean. Their orientation is influenced by the buttressing effects of neighboring volcanoes and by gravitational stresses. The rift zones are scored at the surface by many cracks, some of which have served as vents for erupting lava. Below the surface, dikes of congealed lava fill fissures. Each upwelling of magma from depth causes swelling of the summit and flanks, which creates new stresses on the rift zones and opens new cracks and fissures or reopens old ones. An eruption relieves pressure in the magma reservoir, and the volcano then deflates.

Mauna Loa and Kilauea each have two principal rift zones (fig. 40.9). Mauna Loa's main growth has been along the two principal rift zones that extend from the caldera to the southwest and to the northeast. Kilauea's south flank is pushing into the sea, growing a few inches per year due to repeated intrusions of magma into the southwest and east rift zones. Since it began erupting in 1983 from the east rift zone, Pu'u 'O'o has built its cone

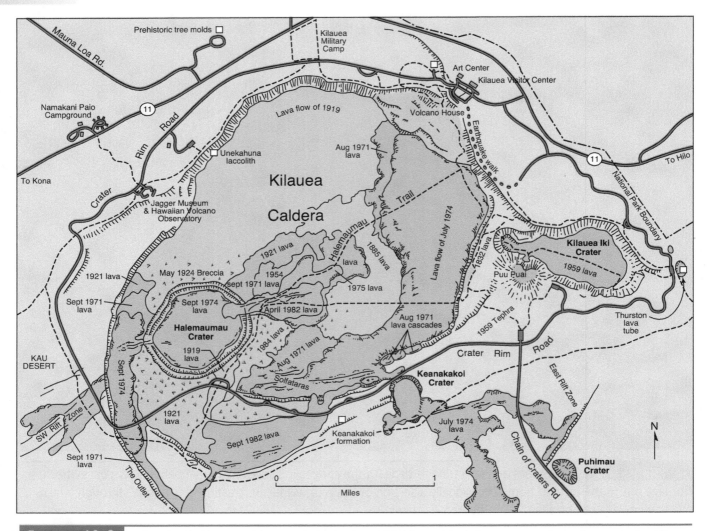

FIGURE 40.8 Sketch of Kilauea caldera and vicinity in 1987. Heavy black lines are roads; trails are dashed lines. Historic flows are outlined and dated. Ridges in some flows represent eruptive fissures with spatter ramparts. Breccia symbols scattered around Halemaumau show the distribution of pyroclasts from the May 1924 explosion. From Hazlett 1987, Geological Society of America Centennial field guide, Cordilleran section, v. 1, p. 17. © Geological Society of America; used by permission.

to a height of more than 830 feet, making it the highest volcanic landform that has grown and been monitored by scientists in historic time. In 1986, after the Pu'u 'O'o conduit ruptured, lava erupted to the east of the cone from a new vent and began to build up a shield. Flows from the vent, named Kupaianaha, formed a system of lava tubes that carried fresh lava over a fault scarp and down to the ocean (fig. 40.10).

Fault Systems and Landslides

Rift zone eruptions tend to make the volcano flanks out of balance or unstable. Earthquakes accompanied by sudden movements along fault lines release the accumulated strain. A fault system between Mauna Loa and Kilauea appears to act as an "expansion joint" between the two volcanoes. A jolting 6.6 earthquake on this fault system in November 1983 damaged nearby roads, trails, and some buildings. Along Mauna Loa's southeastern base, step faulting, by which the mountain side has moved downward in response to gravity, has produced a series of fault scarps (called "palis") facing southeast.

Similar step faulting takes place from time to time in the Hilina fault system along Kilauea's southern slope. Here the fault blocks form a series of steps down

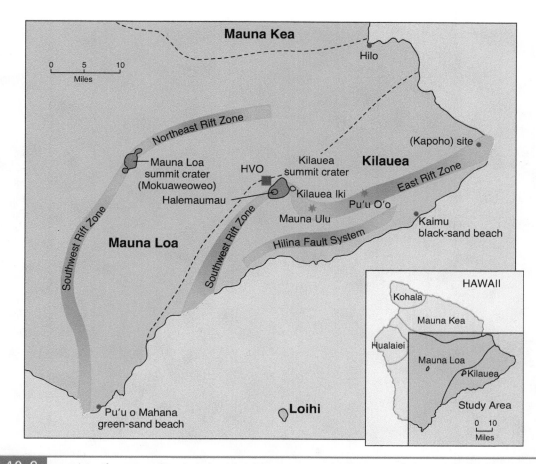

FIGURE 40.9   Major rift zones associated with Mauna Loa and Kilauea and the Hilina fault system. Note underwater location of Loihi offshore. From Tilling, Heliker, and Wright 1987.

to the coast, with the trend being parallel to the shore. Measurements after a magnitude 7.2 earthquake in November 1975 revealed that some rocks along the Hilina fault had dropped 11 feet and shifted southward as much as 24 feet (Tilling, 1987). The fault scarps of the Hilina fault system show up on Kilauea's south flank as steep cliffs overlooking the sea. Lava flows have cascaded down over the fault scarps into the ocean.

The U.S. Geological Survey recently completed imaging of the sea floor around the Hawaiian Islands by means of a sidescan sonar device (Geologic Long Range Inclined Asdic, or "GLORIA"). The sonar monitor revealed many undersea landslides consisting of giant slumps that had slid gradually off the volcanoes into the sea. However, in some places, the slides were debris avalanches that had given way explosively, leaving on the volcano flanks headwalls of great "amphitheaters," as huge sections of volcanic material dropped to the sea bottom. Geologists believe that the major fault systems on the Big Island are related to the upper parts of the giant landslides hidden below sea level. At present, the south side of the Big Island is slumping toward the sea while the East Rift Zone is migrating southward. Scientists at the Hawaiian Volcano Observatory continue to carefully monitor the area.

Submarine and Shore Features

Lava flows that reach the ocean surface produce cinders and ash when the hot lava mixes with the sea water. Cinder cones and benches are formed when lava enters the ocean. Waves erode and transport volcanic sediment, depositing "black sands" on beaches. Since the source of supply is limited and some sand is washed out to sea each year, the black sands may last only two or three centuries.

Older sea cliffs and sea arches undergo wave erosion and are retreating, but in some places along the

FIGURE 40.10   A lava flow derived from Pu'u 'O'o entering the ocean. Note how the cliff is gradually breaking away, forming terraces. Much of the lava entering the ocean is chilled instantly, forming obsidian, a natural black volcanic glass. This is the composition of the sand of the black beaches. Photo by Ann G. Harris.

shore, recent lava flows protect the cliffs and have enlarged them.

When fluid lava erupts under water or flows from a tube that opens in the ocean, *pillow structures* may form. With water removing heat rapidly, the lava quickly crusts over and solidifies as a pillowlike or tongue-like protrusion. More lava squeezes out (like toothpaste from a tube) and stacks another "pillow" on top of the first. The process is repeated many times as the lava flow advances under water. Because the lava is under pressure from the weight of the overlying water, explosive interactions seldom occur between the cold ocean water and the hot lava. In recent years scuba-diving scientists have observed and filmed the forming of pillow structures during underwater eruptions of lava near Big Island.

Geologic History

1. **Initial eruptions on the sea floor, probably early in Pleistocene time.**

The age of the rock at the base of the volcanic pile has not been determined. Putting down drill holes in order to get core samples for age determination is a difficult and costly operation. Moreover, in some locations, drill bits might hit magma that would melt them.

2. **Construction above sea level of Hawaii Island.**

Five shield volcanoes, each with its own magma reservoir but all having the same magma source in the upper mantle, built up the island during Pleistocene time. Kilauea, the youngest volcano, is still enlarging

FIGURE 40.11   A Moorish Idol swims among the coral on a reef that hasn't been destroyed by the lava flows on the Kona side of the Island outside of park boundaries. Photo by James Ellashek.

at the southeastern end of the island. South of Hawaii, Loihi seamount is rising. It is regarded as the newest expression of the hot spot beneath the Pacific plate. Using a submersible, serviced by a research vessel, scientists monitor Loihi's development and collect samples of vent fluids, gases, and lavas. Loihi is expected to reach the ocean surface in 60,000 years, more or less.

3. The development of Mauna Loa.

Rock units. Mauna Loa's rocks are mainly olivine basalts. The oldest that are exposed are the lavas of the Ninole Formation that make up the core of the mountain. Ninole rocks crop out in low hills on Mauna Loa's southeast slope. Deep canyons were cut in these lavas during an erosional period, producing an unconformity between Ninole beds and the overlying lavas of the

Kahuku Formation. The Kahuku lavas partially filled the canyons and built up the slopes of Mauna Loa. Pahala ash, 5 to 50 feet thick, mantles the Kahuku rocks. Most of the Pahala ash in this area came from Kilauea and was probably the result of violent phreatomagmatic explosions that accompanied the collapse of Kilauea's caldera. Much of the Pahala ash has been covered by younger lavas of the Ka'u Formation. The fairly fresh and relatively unweathered rocks of the Ka'u include lavas from eruptions during historic time. Mauna Loa had probably attained its present height by the end of the Pleistocene Epoch.

Eruptions in historic time. Mauna Loa's eruptions since early in the nineteenth century have been mainly of two types: (1) summit eruptions without flank flows and (2) the more typical flows from the rift zones that may be accompanied by brief summit eruptions. Eruptions gen-

Box 40.2	**Forecasting Eruptions**

Crisscrossing the island of Hawaii is a network of scientific instruments for monitoring the volcanoes. Collecting and analyzing data and making direct observations are the responsibility of volcanologists at the Hawaii Volcano Observatory and the University of Hawaii. Tied in with this network are seismic stations and investigators in other parts of Hawaii, Japan, New Zealand, and many locations throughout the world.

What are changes or signs that scientists look at? What combinations of factors indicate an impending eruption?

1. Periodicity. What is the eruptive history of the volcano? How long are its quiet periods? Such information is useful in a general way, although eruptive patterns of volcanoes may change unexpectedly.
2. Changes in the behavior of fumaroles and in the output, composition, and temperature of gas. Steam and gas emissions are affected by atmospheric temperatures and humidity, but observations in this category are useful. On Kilauea, for example, an increase in the hydrogen content of gas emissions usually signals the onset of an eruption.
3. Shifts in the orientation of the volcano's magnetic field and reduction of magnetic attraction of rocks due to increasing heat. Changes in magnetism are difficult to interpret, but they appear to indicate movement of magma within the interior of a volcano. Often an eruption does follow such movement, but sometimes a volcano quiets down without an extrusive event occurring.
4. Infrared photographs and remote sensing. Increases in heat and changes in the locations of hot areas show up when these techniques are used. Any eruptive activity is invariably preceded by an increase in heat on the volcano's surface.
5. Geoelectrical changes. Variations in electrical currents in a volcano have been observed just before explosions or eruptions, suggesting that geoelectricity may be useful as an indicator.
6. Tumescence and tilting. Kilauea and Mauna Loa have histories of swelling and tilting before eruptions, but occasionally swelling goes down without an eruption occurring. Land deformation on any active volcano is considered an ominous sign. The cause of bulging, rising of the summit, and tilting is assumed to be magma filling the volcano reservoir and rising in the conduits. Tiltmeters

erally begin with harmonic tremors (swarms of little earthquake tremors) and the release of clouds of steam and gas. Lava fountains may shoot up many feet in Moku'aweoweo, the summit caldera. Strands of Pele's hair are blown out and away by uprushing winds. During a summit eruption, lava pours from cracks in the caldera wall and spreads over the floor. Pit craters may develop or enlarge in and around the caldera.

When eruptions begin in the rift zones, lava and hot rising gases shoot up, making a "curtain of fire" along the length of a fissure opening. Clots, blobs of pumice, and threads of volcanic glass harden in the air and then fall glistening to the ground. During the first few days, very fluid pahoehoe pours down over the slopes like a slow-moving river (10 to 25 miles per hour). Later flows may change to aa.

The 1859 eruption, which lasted for 10 months, produced a 33-mile flow, the longest recorded in historic time. Several eruptions from the northeast rift zone have endangered the city of Hilo and its harbor on the eastern coast of the island. The closest call was a pahoehoe flow 29 miles long in 1881 that stopped on Hilo's outskirts.

A major eruption of Mauna Loa took place in June 1950. Lava first began erupting near the caldera at the top of the southwest rift zone. In minutes the erupting fissures had opened up the southwest rift zone for several miles down the slope. Within 24 hours, lava fountains blazed in a curtain of fire down the mountainside for 8 miles. On both sides of the rift ridge, rivers of lava poured down the slopes. Three flows reached the ocean. The volume of lava (19.37 cubic miles) that was

on Kilauea and Mauna Loa continuously monitor land surface deformation as well as migration of the center of a tumescent area.

7. Changes in seismicity. From earliest times, people who lived near volcanoes associated earthquakes with eruptions. On Hawaii Island, certain kinds of earth tremors and rumblings were called "Pele's thunder." These were regarded as warning signals of her anger.

Earthquakes are frequent on Big Island; many of them are so slight that they are imperceptible to residents. Most of the earthquakes are shallow, local, and indicate a shifting around of rocks in response to stresses in the volcanic structures. The Big Island's most severe earthquake in historic time, estimated to have had a magnitude of 8, happened in 1868 after days of strong shocks. The series of quakes triggered mud avalanches and a 30-foot drop of the island's south flank. The southern coastline subsided as much as seven feet and a destructive tsunami washed in over the shore. Both Mauna Loa and Kilauea erupted shortly after the earthquakes (Parks 1993). On Maui and the other Hawaiian Islands, earthquakes are less common and are probably caused by isostatic adjustments of the oceanic crust to the load of overlying volcanoes.

Seismic data collected on Hawaii have been studied for many years by scientists seeking clues as to the imminence of eruptions. Several significant patterns have shown up, but they must be combined with other observations to be meaningful. Earthquake "swarms," consisting of thousands of little earthquakes, are recorded on seismographs before eruptions occur on Mauna Loa and Kilauea. These quakes, called *harmonic tremors,* are probably the result of minor faulting that accompanies tumescence or the inflation and pushing up of the volcano top. Fissures and troughs may develop along the rift zones. After the magma has drained and the summit starts to shrink, quakes begin again, sometimes stronger than those that accompany the swelling. Evidence from tiltmeters may confirm whether the volcano is shrinking or possibly swelling again. This is important in Hawaii where eruptive periods last for weeks, months, or years.

Forecasts regarding eruptions are necessarily based on several lines of evidence and are expressed in probabilistic terms. Nevertheless, even though anticipating volcano behavior remains an inexact science, forecasts can be made that can help to prevent loss of lives when dangerous areas need to be evacuated.

extruded during the three weeks of the 1950 eruption equaled the amount of lava that erupted in the 10-month eruption of 1859. In terms of volume, these two eruptions are the largest that have been recorded on the island of Hawaii. Although Mauna Loa has been quiet most of the time since 1950, more eruptions are expected to occur.

4. The development of Kilauea.

Rock units. Eruptions of Hilina Formation lavas from the southeast flank of Mauna Loa began building Kilauea in about the middle of Pleistocene time. The Hilina lavas are presumed to be roughly equivalent in age to the Kahuku Formation on Mauna Loa. Hilina basalts are exposed in the fault scarps along the southern coast. Flows of both pahoehoe and aa can be distinguished.

Pahala ash overlies the Hilina Formation. Above the Pahala ash are rocks of the younger Puna Formation. Puna lavas were erupting at about the same time as the Ka'u lavas on Mauna Loa, and the flows interfinger in the area between the two volcanoes. Puna flows were erupted largely from vents in Kilauea's caldera and from the rift zones. Spatter and cinder cones formed on both the southwestern and eastern rift zones. Those on the eastern rift zone were more numerous.

Eruptions in historic time. The types of activity have been (1) flank flows, more often than not preceded by summit activity or eruption; (2) shield construction; (3) rising and falling of lava lakes; (4) brief summit eruptions; and (5) violent explosive eruptions (fig. 40.8).

The lava lake in the fire pit on Kilauea's summit apparently filled not long after the violent explosion of

Table 40.1	Geologic Column, Hawaii Volcanoes National Park				

Time Units			Rock Units and Geologic Events	
Era	Period	Epoch	Mauna Loa	Kilauea
Cenozoic	Quaternary	Holocene — "Historic" / "Prehistoric" — ?	*Younger Ka'u Formation* Summit eruptions, 1975, 1984 Numerous summit and rift zone eruptions, 1832–1950 Devastating earthquake and landslide, 1868 Continuing shield construction by flows of aa and pahoehoe Enlargement of caldera by pit craters *Older Ka'u Formation* Shield construction by aa and pahoehoe; cinder cones, spatter cones, etc. Development and collapse of Moku'aweoweo caldera; pit craters *Kahuku Formation*, capped by *Pahala Ash* Building up of shield by thick accumulations of aa and pahoehoe	*Younger Puna Formation* Construction of satellite cones Mauna Ulu, Kilauea Iki, Pu'u 'O'o, Kupaianaha since 1968 Numerous summit and east and west rift zone eruptions, 1790 to present Continuing shield construction by flows of aa and pahoehoe Explosive eruptions; 1790, 1924 Appearance, disappearance, reappearance of lava lakes; enlargement of caldera *Older Puna Formation* Shield construction by aa and pahoehoe flows; ash falls, cinder cones, spatter cones Development and collapse of Halema'uma'u in summit caldera; pit craters *Hilina Formation*, capped by Pahala ash Beginning of shield construction by many flows, both aa and pahoehoe, building up thick accumulations; some ash layers, indicating past explosive eruptions
		Pleistocene	Great erosional unconformity Fluctuations at sea level and submergence; intense erosion of shield; canyon cutting	
			Ninole Formation Construction of Ninole volcano Volcanic "basement" rocks (not exposed at surface); age unknown	Volcanic "basement" rocks (not exposed at surface); age unknown

Sources: Modified after Macdonald et al., 1983; Tilling et al., 1987.

although some sinking, followed by refilling, occurred from time to time after lava had been drained by flank eruptions below sea level. The lava lake was at a high level at the beginning of 1924 when a series of earthquakes began. They first occurred at the summit and then shifted outward along the eastern rift zone, suggesting that magma was moving in that direction. Meanwhile, the level of the lake dropped several hundred feet. The explosions began in May and lasted for about three weeks. Ash-filled clouds of steam and gas shot up. Sulfur fumes killed vegetation over a wide area. Blocks of caldera rock were thrown a mile into the

air. Local earthquakes, lightning, and heavy downpours of rain accompanied the explosions. Afterwards Halema'uma'u had increased in diameter from 500 feet to over 3000 feet, and the floor of the caldera had sunk 11 feet. Even the caldera rim was lowered a few feet. What caused the explosions and collapse was ground water pouring into the hot vents as lava drained away. Instantaneous boiling resulted, and then the violent escape of the steam. The explosions were not magmatic since no fresh lava was ejected with volcanic debris.

Since 1924, the lava lake that had bubbled and flamed for so many years has refilled partially, but only

FIGURE 40.12   Pu'u 'O'o is a satellite cone on the flank of Kilauea Volcano; the most recent lava flows have come from this vent. Photo by Ann G. Harris.

for very brief intervals. The fire pit is quiet except for wisps of steam. Summit eruptions have been short and infrequent. Flank eruptions, on the other hand, have been numerous and sustained. One summit eruption, however, was spectacular. In 1959, Kilauea Iki, a large pit crater close to the caldera, had a major eruption. The lava fountains that shot up were the highest on record, and cinders and pumice devastated a nearby forest.

Before the lava lake disappeared, the southwest rift zone was more active than the eastern rift zone. Mauna Iki is a secondary lava cone southwest of the caldera that was built by an eruption in 1920. Fresh lava flowed out near Mauna Iki in 1971.

The east rift zone has been exceedingly active for many years. From the caldera, the east rift zone goes southeastward and then takes a sharp bend almost due east to Cape Kumukahi and on out into the ocean. The Chain of Craters, on the upper part of the eastern rift zone, used to be a row of 13 pit craters coming down from the caldera. A highway went through the area from the Crater Rim road southeastward to the coast. The road was buried by lava in 1969 when an entirely new vent opened up and Mauna Ulu, a satellite cone, began to grow. Mauna Ulu continued to erupt until 1974, and many of its flows reached the ocean. Next to Mauna Ulu is the Alae shield, which was built up over an old pit crater that was 440 feet deep. Other nearby craters were

partly filled or buried, and some reopened and began pouring out lava or filling up and forming lava lakes.

The following year, in 1960, lava erupted 28 miles away from the summit near Cape Kumukahi at the extreme eastern tip of the island. Lava flowing from the eastern rift zone destroyed the village of Kapoho and created some 500 acres of new land where the flow went into the ocean. (The question of who owns new land has been in litigation.)

The Pu'u 'O'o eruptive series, which has been continuously active since 1983, has become the longest and largest eruption in Kilauea's history (fig. 40.12). Flows from Pu'u 'O'o and its associated vent Kupaianaha have constructed a new shield and added new land to the island where lava has entered the ocean. The coastal community of Kalapana, which lay in the path of the flows, was wiped out by repeated lava flows. More than 189 homes and businesses were destroyed, as well as a National Park Visitor Center, archeological sites, roads, trails, and other structures.

5. Glaciation.

A small icecap existed on Mauna Kea, Hawaii's highest volcano, during the Pleistocene Epoch. Glaciers were also present on Mauna Loa at the same time, but all evidence has been destroyed or buried by recent lava flows. In today's climate, snow falls every winter on Mauna Kea and Mauna Loa, but does not last long.

6. Changes in sea level.

Old shorelines, some higher than the ocean's present stand and some lower, testify to the fluctuating ocean level around the Hawaiian Islands during their comparatively brief geologic history. The most significant long-term trend is land submergence. As the volcanic pile has gotten higher and heavier, the sea floor has tended to sink. However, older sea levels are not consistent from place to place because local tilting and warping have occurred. Thus tectonic changes have affected the relative position of land and ocean. The islands' shorelines were also affected by worldwide changes in sea level that occurred during Pleistocene time.

Geologic Map and Cross Section

Macdonald, G.A. 1971. Geologic Map of the Mauna Loa Quadrangle, Hawaii. U.S. Geological Survey Map GQ-897.

Peterson, D.W. 1967. Geologic Map of the Kilauea Crater Quadrangle, Hawaii. U.S. Geological Survey Map GQ 667.

Sources

Abbott, A.T. and Peterson, F.L. 1983, 2nd edition. *Volcanoes in the Sea: The Geology of Hawaii.* Honolulu: University of Hawaii Press.

Clague, D.A., Dalrymple, G.B., Wright, T.L., Klein, F.W., Kayanagi, R.Y., Decker, R.W., and Thomas, D.M. 1989. The Hawaiian-Emperor Chain (chapter 12) in Winterer, E.L., Hussong, D.M., and Decker, R.W. eds. *The Eastern Pacific Ocean and Hawaii.* Boulder, Colorado: The Geology of North America, v. N. p. 187-287.

Decker, B., and Decker, R.W. 1987, 2nd edition. *Road Guide to Hawaii Volcanoes National Park.* Mariposa, California: Double Decker Press.

Decker, R.W., Wright, T.L., and Stauffer, P.H. (editors). 1987. *Volcanism in Hawaii.* U.S. Geological Survey Professional Paper 1350, 2 vol. 1667 p.

Griggs, J.D., Takahashi, J.T., and Wright, T.L. 1986. Volcano Monitoring at the U.S. Geological Survey's Hawaiian Volcano Observatory, *Earthquakes and Volcanoes* 18 (1):3-71. U.S. Geological Survey.

Hazlett, R.W. 1987. Kilauea Caldera and Adjoining Rift Zones. Geological Society of America Centennial Field Guide, Cordilleran section, v. 1, p. 15–20.

———— 1990. *Geological Field Guide, Kilauea Volcano.* Hawaii Natural History Association.

Heliker, C. 1990. *Volcanic and Seismic Hazards on the Island of Hawaii.* U.S. Geological Survey General Interest Publication.

Herbert, D.; and Bardossi, F. 1968. *Kilauea: Case History of a Volcano,* New York: Harper & Row Publishers.

Macdonald, G.A. 1972. *Volcanoes.* Englewood Cliffs, New Jersey: Prentice-Hall.

———— and Hubbard, D.H. 1989, 9th edition (revised by Jon W. Erickson). *Volcanoes of the National Parks in Hawaii.* Hawaii Natural History Association and National Park Service.

Parks, N. 1993. The Fragile Volcano. *Earth* 2(6): 42–49.

Stearns, H.T. 1985. *Geology of the State of Hawaii.* Palo Alto, California: Pacific Books.

———— 1978, 2nd edition. *Road Guide to Points of Geologic Interest in the Hawaiian Islands.* Palo Alto, California: Pacific Press.

Macdonald, G.A. 1946. Geology and Groundwater Resources of the Island of Hawaii. Hawaii Division of Hydrography Bulletin 9.

Swanson, D.A., and Peterson, D.W. 1972. Partial Draining and Crustal Subsidence of Alae Lava Lake, Kilauea Volcano, Hawaii. U.S. Geological Survey Professional Paper 800-C.

Tilling, R.I, Heliker, C., and Wright, T.L. 1987. *Eruptions of Hawaiian Volcanoes: Past, Present, and Future.* U.S. Geological Survey, General Interest Publication. 55 p.

Note: To convert English measurements to metric, go to www.helpwithdiy.com/metric_conversion_calculator.html

Haleakala National Park

ISLAND OF MAUI, HAWAII

Area: 28,099 acres; 44 square miles
Established as a National Park: August 1, 1916
Redesignated Haleakala National Park:
   September 13, 1960
Designated a Biosphere Reserve: 1980

Address: P.O. Box 369, Makawao, Maui, HI 96768
Phone: 808-572-4400
E-mail: HALE_Interpretation@nps.gov
Website: www.nps.gov/hale/index/htm

Spectacularly eroded summit "crater" of 10,023-foot-high Haleakala Volcano.
Youngest lava flow from Haleakala erupted c. 1790.

The island of Maui is made up of two old volcanoes. The taller of the two, with its mid-slopes
wrapped in clouds and its lower levels lost in the haze above the sea, often seems to be floating.

— "The Maui Scene" (Kyselka and Lanterman 1980)

FIGURE 41.1   Haleakala Crater, a little over seven miles long, two miles wide, and a half mile deep, displays volcanic features of widely different ages and has undergone extensive erosion. The "crater" is actually two erosional valleys that joined; the original crater was between 2000 feet and 3000 feet higher. Photo by Ann G. Harris.

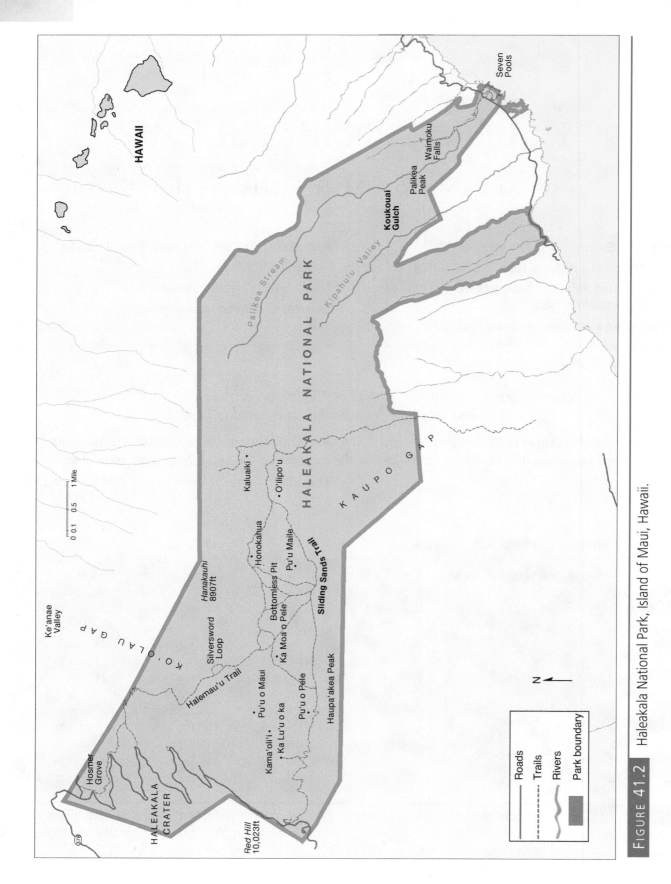

FIGURE 41.2   Haleakala National Park, Island of Maui, Hawaii.

Haleakala National Park is on Maui, the next island northwest of Hawaii Island, where Hawaii Volcanoes National Park is located. (See box 40.1 in Chapter 40.) Maui was a religious and cultural center for the Polynesian people who settled the Hawaiian Islands. In the Hawaiian language, *Haleakala* (ha-lay-ah-ka-lah) means "House of the Sun."

Lahaina, a port on the west side of Maui and the site of the first nonnative settlement in the Hawaiian Islands, was an important whaling center in the nineteenth century. Humpback whales spend the winter months in the channel waters between Maui and the neighboring islands of Lana'i, Kaho'olawe, and Moloka'i.

Most of the park is contained within Haleakala Crater; but in 1969 the Kipahulu Valley was added, which connects the crater section with the ocean. This narrow corridor was set aside so that the fragile ecosystem of Kipahulu Valley, with its rare and endangered species of plants and birds, could be protected. At least 25 bird species that lived on Maui before 1900 became extinct after mosquitoes carrying bird malaria arrived on a ship from Central America. The forested slopes of the Kipahulu Valley are not open to the public but are kept in a natural state for scientific study.

"Climatic layering" on Maui produces unique weather patterns and tends to control the prevalence of, or lack of, vegetation. Maximum precipitation is greatest on northeastern slopes between the 2000- and 6000-foot levels, where the trade winds drop more than 300 inches of rain per year. Precipitation, clouds, and fog often envelop the middle-level slopes of Haleakala and are responsible for the luxuriant growth in the rain forests. The Kipahulu Valley is a notable example. Above 6000 feet and on the leeward slopes, the environment tends to be semiarid. From the air, one sees heavy forests on the lower and middle flanks of the volcano and barrenness at the higher elevations around the summit and crater. Layering of vegetation is also due to the extreme porosity of the volcanic rocks in the crater. Rain water drains away so fast that not much soil develops and vegetation is sparse. Runoff moves loose volcanic debris down the inner slopes of the crater and into the stream valleys.

Among the rocks in the desertlike crater and on the rim grows the tall silversword, a rare, yuccalike plant, native to Hawaii, that has a mass of silvery, saberlike leaves. At maturity, the plant produces hundreds of purple flowers and then dies after the seeds ripen. A hundred years ago, the silversword covered acres of the crater floor, but nearly became extinct before coming under National Park Service protection. Also protected is a flock of nene (naynay), or Hawaiian geese, which were close to extinction early in the century. The nene is Hawaii's state bird.

Geologic Features

Maui's two eroded volcanoes, Haleakala and West Maui, were once part of a large volcanic massif made up of six major volcanoes. When sea level was lower, during the coldest times of the Pleistocene Epoch, the six volcanoes formed one island. Now there are four islands—Maui, Kaho'olawe, Lana'i, and Moloka'i—separated by channels.

Maui's volcanoes are older and more deeply eroded than the great volcanic shield complex making up the island of Hawaii (Chapter 40). In map view, Maui's appearance is like an asymmetrical dumbbell, with the smaller West Maui volcano on one side and the larger East Maui volcano, Haleakala, on the other side. Joining the two volcanoes is a low isthmus constructed of lavas from Haleakala and sandy sediments that accumulated in dunes during the Ice Age and have since been lithified. A pumping station located on the isthmus supplies Maui's water, which is drawn from a freshwater lens beneath Haleakala and the isthmus. The fresh water floats on top of the dense saltwater that saturates the entire base of the island.

West Maui is 5788 feet high and 19 miles across. By contrast, on East Maui, the rim of Haleakala Crater, rises to over 10,000 feet and the volcano is about 33 miles across. Maui's total area is 728 square miles. The highest point on the rim, Red Hill (elevation 10,023 feet), is one of three prominent cinder cones at the summit of Haleakala (fig 41.3). The National Park Service maintains geological exhibits at overlooks on the rim. Like the island of Hawaii, the summit of Haleakala has three radiating rifts: one going almost north, another east, and the third toward the southwest. The Ke'anae Valley follows the north rift from the crater out through the Ko'olau Gap and down to the ocean. The Kaupo Gap drains the south part of the summit depression; the Kipahulu Valley does not start in the crater, but descends to the sea along the east rift. Throughout Haleakala's history, many eruptions emanated from the rifts; and, to a certain extent, their structure influenced the development of drainage and erosion on the volcano.

The so-called crater, the dominant topographic feature of the island, covers 19 square miles and is several thousand feet deep. Its inward-facing slopes are much steeper than its outer flanks. Because it looks like a crater or caldera, this impressive landform will probably always be called "Haleakala Crater," even though by geologic definition it is neither a crater nor a caldera (fig. 41.1).

Described by Hackett (1987) as "one of the largest erosional depressions on earth," its development must have required several hundred thousand years of mass wasting and erosion. Haleakala had a real crater originally and was once much higher. But after the volcano became inactive, rapid stream erosion removed the youngest lava flows, ash, and cindery material. Because rainfall is heaviest on north and east-facing slopes, due to the trade winds, streams draining those areas were able to erode faster and capture the headwaters of lesser streams. Ground water was also a significant factor in erosion and capture. Several master valleys developed; two of them—by cutting headward into the mountain's core—finally merged, creating a huge depression on the summit. A low dividing ridge across the floor of this caldera-like feature separated the two drainage basins, the Kaupo Gap on the southeast and Ke' olau Gap toward the northeast.

Volcanic activity resumed after a long erosional period. The old rift zones of the volcano reopened. Lava pouring from fissures on the sides of the surrounding walls and from the crater floor flowed down the eroded stream valleys. An enormous mudflow, along with cinder blocks and lava, went through the Kaupo Gap and down to the sea. Cinder cones, constructed of lava, pumice, cinders, blocks, ash, and

FIGURE 41.3 The Re'anan Valley follows the north trending rift from the crater out through Ko'olau Gap and down to the ocean. The highest point on the rim of Haleakala is Red Hill at 10,023 feet; with an elevation that high a cloud layer forms in the crater almost every morning and is burned off in the afternoon. National Park Service photo.

Small cinder cones erupted on the floor of Haleakala Crater when old rift zones reopened long after the volcano's summit had been deeply eroded. In the background is the south rim. National Park Service photo.

obsidian, built up on the floor of the depression, with some cones rising as much as 600 feet in height (fig. 41.4). Sizable volcanic bombs, formed from molten fragments that cooled before striking the ground are strewn about. Bizarre spatter cones formed, such as Pele's Pigpen and Bottomless Pit, the latter named for its deep vent. Lava tubes like Holua tube were left when pahoehoe lava cooled on the outside while fresh, hot lava on the inside drained away. Bubble-shaped caves in the lava are the result of gases expanding and lava hardening over the bubble instead of collapsing.

One of the striking features of Haleakala Crater is that the rocks in the steep walls and in many of the fea-

tures of the floor are multicolored. One reason for this is that when the sun's rays strike the rocks at different angles during the day, the colors vary as the sun's position changes. Also, volcanic gases have left varicolored crusts on the rocks due to precipitation and sublimation. Chemical weathering of the basaltic rocks adds still more colors to the landscape as iron-bearing minerals are oxidized and hydrated. The prevailing warm temperatures and frequent mists and rains intensify the effects of weathering processes on these easily weathered rocks.

Most of the streams on Maui flow only during and right after rains. These are *ephemeral streams*. On the windward slopes, however, streams cascading down over volcanic cliffs and ledges have created beautiful pools and waterfalls that are cut in the valley-filling lavas. Hikers and swimmers who visit the "Seven Pools" of 'Ohe'o Gulch in the lower Kipahulu Valley (fig 41.5) on the eastern shore of Maui, can see interesting lava outcrops, birds, and flowers. (Actually, the number of pools—usually *more* than seven—depends on the amount of water in 'Ohe'o Stream. After heavy rains, the stream is a torrent, and no pools can be seen.) In the rugged sea cliffs along the shore, older and younger lavas are exposed, as well as partially buried cinder cones and other volcanic features.

FIGURE 41.5   The "Seven Pools" of 'Ohe'o Gulch in the lower Kipahulu Valley on the eastern shore of Maui has many interesting lava outcrops where plunge pools have formed as waterfalls spill over the ledges. It is a favorite location for hikers and swimmers. Photo by Ann G. Harris.

Table 41.1			Generalized Geologic Column, Haleakala National Park	

Time Units			Rock Units	Geologic Events
Era	Period	Epoch		
Cenozoic	Quaternary	"Historic time" / Holocene		Small lava flows c. 1790
		?	Hana Formation	Valley-filling lavas, crater lavas, ash, cinder cones, spatter cones
		Pleistocene	Mudflow debris	Kaupo mudflow
			Accumulations of sediment	Landsliding, canyon-cutting; development of amphitheaters and Haleakala Crater by stream action, frost action, mass wasting
			Great erosional unconformity	Fluctuation of sea level and submergence
	? ?	?	Kula formation	Cinder cones, ash; alkalic cap construction
			Kumu'iliahi Formation	Extensive aa and pahoehoe flows
	Tertiary / Neogene	Pliocene	Honomanu Formation	Build up of shield accumulations of pahoehoe and aa flows
			Volcanic "basement" rocks (not exposed at surface)	Lava flows building up from the ocean floor

Sources: Bacon, 1983; Cranson, 1982.

Geologic History

1. Construction of the volcanic massif.

In late Pliocene time, the shields of six volcanoes were built up above sea level from a basaltic lava foundation on the ocean floor. As the volcanoes continued to erupt, they coalesced as a great volcanic complex.

2. Eruption of Honomanu Formation lavas on East Maui.

In earliest Pleistocene time, flows of the Honomanu Formation, many thousands of feet thick, constructed the bulk of the East Maui volcano. West Maui, which was older, was built up by similar outpourings. Rocks of both volcanoes consisted of tholeiitic basalts and olivine basalts that hardened from pahoehoe and aa flows. Honomanu rocks above sea level were buried by the flows of later lavas and are now exposed only in the cliffs along the northeast shore of the island and in the walls of the deeper stream valleys.

3. Eruption of Kumu'iliahi Formation lavas in Pleistocene time.

At the base of the south wall of Haleakala Crater (north and northeast of Kumu'iliahi Peak) and in a small knob at the east edge of the Ko' olau Gap, are thin-bedded pahoehoe and aa lavas formerly assigned to the Honomanu Formation but now regarded as a discrete rock unit on the basis of chemical composition. Kumu'iliahi rocks are too alkalic to be considered tholeiitic, but are less alkalic than the overlying rocks of the Kula Formation.

4. The Pleistocene Kula Formation.

Kula lavas erupted along the two rift zones that trend to the southwest and southeast from the summit of Haleakala. The third rift zone that extends northward

FIGURE 41.6   Sliding sands trail (to the right) leads down and across the "crater" floor. The trail has received its name because much of it consists of loose pyroclastic debris dust, ash and cinder-sized particles. National Park Service photo.

from the summit did not reactivate. The new outpourings of lava buried the shield formed by the Honomanu lavas and formed a thick cap (2500 feet) over the summit, raising the top of the volcano some 2000 to 3000 feet above its present elevation. The rocks of the Kula Formation include alkalic basalts, but hawaiite,* an andesitic rock, is the predominant type.

The Kula flows, mainly aa, contain interbedded layers of ash and pyroclastics. Since many of the eruptions were explosive, large cinder cones formed with most of

them astride the volcano's rift zones. Lenses of stream gravel between some Kula lavas indicate that quiet intervals separated eruptive periods. Exposed in the walls of Haleakala Crater are a few dikes of Kula age that parallel the rift zones. The dike rocks, which have a higher proportion of feldspar minerals, tend to be more resistant than the surrounding lavas. The ancient Hawaiians made adzes from an even more resistant, fine-grained, igneous rock exposed in a few intrusive masses emplaced in the west wall of the crater.

5. **The great submergence; weathering and erosion.**

While volcanism subsided, sea level rose until at times the ocean met the shore at levels more than 100

---

*Not to be confused with the mineral hawaiite, a gem variety of olivine.

feet higher than at present. The isthmus between East and West Maui was under water. Throughout the Hawaiian Islands, extensive beach deposits and deltas are graded to high sea stands, which were probably the result of warmer climatic intervals with melting of Pleistocene glaciers. Stream erosion and younger lava flows have obliterated most of the erosional and depositional features produced by waves on Maui.

When cooler climates prevailed during the Pleistocene, more rain and snow than usual fell on Maui. Glaciers probably did not develop; but frost action may have been severe, and snow accumulated in winter. Stream erosion and mass wasting accelerated, and erosion was intense. The height of Haleakala was reduced, perhaps by more than 2000 feet. The south slope of Haleakala became steeper and shorter because seaward slip faults triggered huge landslides on that side of the volcano. On the northeast flanks, where precipitation is heavier due to the trade winds, the larger stream valleys cut downward several thousand feet. Because of fluctuations of sea level, the mouths of some stream valleys were scoured out below present sea level and later partially filled with gravel and mudflow debris. In the summit area, intensive headward erosion, downcutting, and the rapid removal of rock and debris produced large, amphitheater-shaped basins at the heads of major stream valleys. The headward widening that formed the amphitheaterlike valleys at the summit, narrowing as they descend to the sea, is characteristic of Hawaiian stream erosion.

6. Development of Haleakala Crater.

Harold T. Stearns, who for many years was District Geologist for the U.S. Geological Survey in the Hawaiian Islands, developed the theory that the caldera-like depression at the summit of Haleakala was mainly the result of coalescence of amphitheaters of the Ke'anae and Kaupo Valleys. The large size of Haleakala Crater is due to the valley heads being offset, instead of joining in a straight line across the top of the mountain. Throughout the crater area, evidence is clear that a great deal of erosion took place, although some of the erosional features have been buried under younger volcanic lavas and mudflows.

7. Late Pleistocene to Holocene volcanic activity.

The lava flows, ash falls, and cinder cones of the Hana Formation that erupted after the erosional period are conspicuous on the floor of the crater, on its west-ern rim, and on the dry, unforested southwest flanks of the volcano. The surface of the most recent flows in the crater area and on the southwest rift zone is fresh and only slightly weathered, indicating that at least some eruptions occurred a few hundred years ago. Hana rocks are alkalic basalt and, like the Kula rocks, are distinctly lighter in color than the dark tholeiitic basalts that built the primitive shield.

Many of the Hana flows traveled much farther from their vents than Kula lavas did; this suggests that, in general, the Hana lavas were more fluid. Some Hana lavas went down the Ke'anae and Kaupo valleys to the ocean, and some erosional valleys were nearly filled. After the Kipahulu Valley filled with Hana lava, the stream cut a new gorge along the north side of the filled valley; and then fresh Hana lava partly refilled the canyon again. During one eruptive episode, an enormous mudflow went down the Kaupo Valley and reached the ocean.

8. Volcanic activity during historic time.

Carbon-14 ages for charcoal associated with unweathered Hana lavas from the southwest rift zone of Haleakala have been determined as A.D. 1070 (±170 years), A.D. 1310 (±140 years), and A.D. 1370 (±120 years). The most recent eruption from the southwest rift zone took place about 1790. At the present time, Haleakala is considered inactive. Earthquakes that occur from time to time on Maui are probably due to subsidence of the crust from the weight of the volcanoes.

Geologic Map and Cross Section

Macdonald, G.A. 1978. Geologic Map of the Crater Section of Halaeakala National Park, Maui, Hawaii (with accompanying summary). U.S. Geological Survey Map 1-1088.

Sources

Culliney, J.L. 1988. *Islands in a Far Sea*. San Francisco: Sierra Club Books. 410 p.

Daws, G. 1989, 3rd edition. *Hawaii, the Islands of Life*. The Nature Conservancy of Hawaii. Honolulu: Signature Publishing. 156 p.

Decker, R.W., Wright, T.L., and Stauffer, P.H., editors. 1987. *Volcanism in Hawaii*. U.S. Geological Survey Professional Paper 1350, 2 vol., 1667 p.

Hackett, W.R. 1987. Haleakala Crater, Maui, Hawaii. Centennial Field Guide, Cordilleran section, Geological Society of America, v. 1, p. 11-14.

Kyselka, W. and Lanterman, R. 1980. *Maui, How It Came to Be*. Honolulu: University of Hawaii Press. 159 p.

Macdonald, G.A. and Hubbard, D.H. 1989, 9th edition, revised by Jon W. Erickson. *Volcanoes of the National Parks in Hawaii*. Hawaii Natural History Association and National Park Service. 65 p.

Macdonald, G.A. and Abbott, A.T. 1983, 2nd edition. *Volcanoes in the Sea*. Honolulu: University of Hawaii Press.

Reber, G. 1959. Age of Lava Flows on Haleakala, Hawaii. *Geological Society of America Bulletin* 70:1245-6, September.

Stearns, H.T. 1942. Origin of Haleakala Crater, Island of Maui, Hawaii. *Geological Society of America Bulletin* 53:1–14, January.

———. 1985. *Geology of the State of Hawaii*. Palo Alto, California: Pacific Books.

Note: To convert English measurements to metric, go to www.helpwithdiy.com/metric_conversion_calculator.html

National Park of American Samoa

CENTRAL SOUTH PACIFIC

Area: Parts of three volcanic islands: 10,520 acres, including 420 acres offshore

Authorized as a National Park: October 31, 1988

Officially Established as a National Park: 1993

Address: Superintendent, National Park of American Samoa, Pago Pago, American Samoa, AS 96799

Phone: 011-684-633-7082

E-mail: NPSA_Administration@nps.gov

Website: www.nps.gov/npsa/index.htm

These National Park units on three different volcanic islands are leased from the American Samoan government and native villages. The island of Tutuila is mostly rainforest, mountains, and beaches. Ofu has a mountain background, a coral sand beach and fringing coral reef. Ta'u has some of the tallest sea cliffs in the world, around 3,000 feet high.

FIGURE 42.1 Coral sands on Ofu Beach. In the distance, the inactive volcano on Olosega Island rises more than 2,000 feet above the ocean. National Park Service photo.

Geographic Setting

The Samoa islands are south of the equator and just east of the International Date Line. The islands are about 2,600 miles southwest of Hawaii and 1,800 miles northeast of New Zealand (fig. 42.2). The Polynesians of the eastern Samoa islands; that is, American Samoa, have had close ties with Americans since the 1870s. United States officials negotiated treaties in 1872 and 1878 with Samoan chiefs for trading rights and the use of the landlocked, deep-water harbor at Pago Pago (PAHNG-oh PAHNG-oh), one of the best natural harbors in the Pacific. The U.S. Navy used the site for a coaling and repair station and, during World War II, as a staging area. Western Samoa (the western islands of the archipelago), which had been under the protection of New Zealand, became independent in 1962.

America Samoa, an unincorporated territory of the United States, has its own government with an elected territorial governor, a bicameral legislature, and an independent judiciary. In the National Park of American Samoa, land tenure and jurisdiction are different from those in other parks.

Within authorized park boundaries, all lands are village or communally owned and controlled. In 1993 a 50-year lease was signed after five years of negotiation with nine chiefs in village councils and the American Samoan government for $419,000 per year. The federal government has management authority and exercises it to protect natural and cultural resources, as well as provide for visitor use. The villagers continue to carry on their traditional subsistence and gathering activities as long as they do not endanger natural or cultural resources that the government is required to protect.

The word Samoa is derived from the native Samoan words *sa ia Moa* (sacred to Moa/center of the earth). This refers to the legend that "the rocks cried to the earth, and the earth became pregnant." The god of the rocks, named Salevao, noticed movement in the center of the earth (*moa*). When the child was born he was named *sa ia Moa* after the place from which he came. Salevao promised that everything that grew would be sacred to Moa (*sa ia Moa*). Therefore, the earth and rocks were called sa ia Moa or SAMOA.

Polynesians arrived in Samoa as early as 1000 BC and spread from there to other Pacific islands. Samoans, the people of Polynesia's oldest culture, are well tuned to their island environment, managing it communally and holding it to be precious. The national park ethic of preserving environmental and cultural values fits well and naturally with the traditional Samoan way of life, which is called "fa'asamoa."

American Samoa, as a political unit, consists of five islands, five islets, and two atolls (fig. 42.3). The most populated island in American Samoa is Tutuila (too-too-EE-lah) with 95 percent of the total population. The territory time is one hour earlier than Hawaii. Temperatures range from a high of 18° C (90° F) in December, the hottest month, to a low of 10° C (75° F) in August. Annual rainfall averages 190 inches, the rainy season being from December to March. The humidity averages 80 percent but is a little higher in the summer months. Because the islands trend in the same direction as the prevailing southeast trade winds, the difference between the leeward (north and west sides) and windward (south and east sides) is not so great as in Hawaii. Pago Pago harbor (south of the park) has the highest yearly precipitation of close to 195 inches because the high ridge of Rainmaker Mountain at 1,718 feet, which is also a National Natural Landmark, intercepts precipitation.

Hurricanes periodically strike these islands. Three recent ones were quite destructive; Tusi (1987), Ofa (1990), and Val (1991). Of the three, Val was the worst. In 1998 El Nino produced a drought.

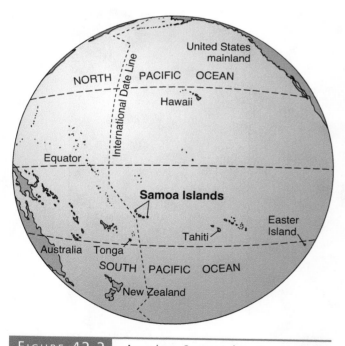

FIGURE 42.2 American Samoa's location in the Pacific is below the equator and the east of the International Date Line.

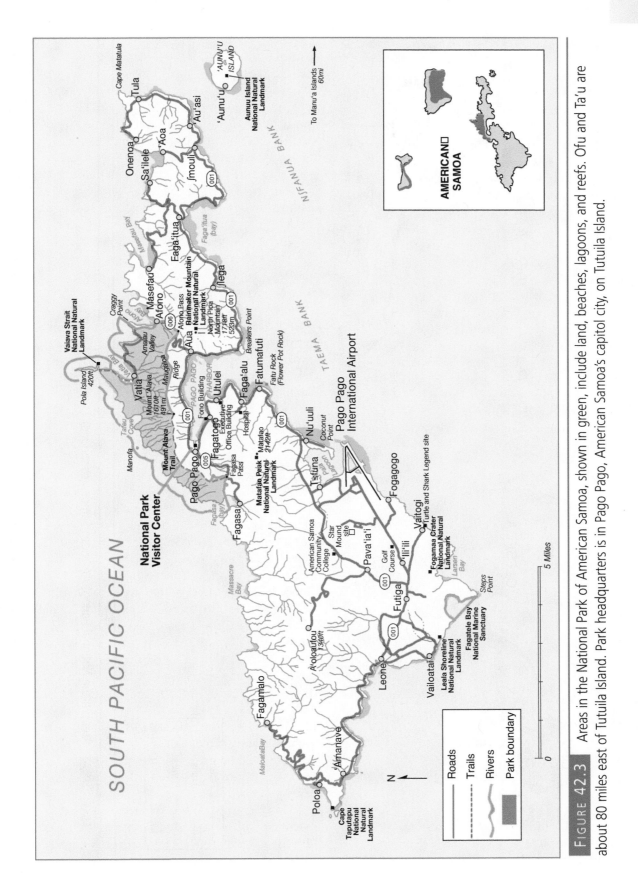

(continued)

FIGURE 42.3   Areas in the National Park of American Samoa, shown in green, include land, beaches, lagoons, and reefs. Ofu and Ta'u are about 80 miles east of Tutuila Island. Park headquarters is in Pago Pago, American Samoa's capitol city, on Tutuila Island.

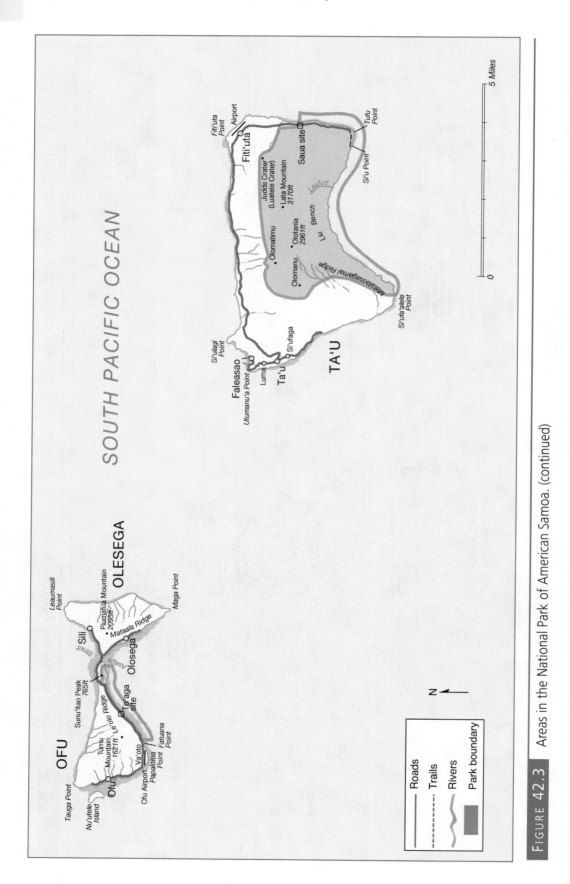

F I G U R E 42.3 Areas in the National Park of American Samoa. (continued)

Only 10 percent of the land is arable, supporting subsistence agriculture on young, shallow volcanic soils. Mountain slopes—nearly everywhere—are too steep (40 to 70 percent) for the growing of crops. The forest is a mixed species paleotropical (old world) rainforest, the only one in the National Park system.

The National Park Units

The three islands that have parklands and protected offshore waters are Tutuila, Ta'u (tah-OO) and Ofu (OH-foo). All three islands have drowned barrier reefs offshore. Because of weathering and leaching, the soils are low in silica, and high in iron and aluminum, and therefore have low fertility. They are classified as latosols. Wildlife is abundant, with about 900 species of fish, 200 species of coral, 35 species of birds, and 3 species of mammals (all bats).

Ta'u Unit. This is the largest park unit with 190 acres of land and 300 acres) offshore. The entire island is 15 square miles and 3,051 feet in elevation. The park's land is located on the south side of the island.

The park's flora is mostly undisturbed rain forest that includes coastal lowlands, montane, and cloud forest communities. On a rare clear day, the view from the cloud forest—largest in Samoa—toward the ocean is spectacular. Sea cliffs drop in two or three steps, from the top of Lata Mountain—which at 3,170 feet is American Samoa's highest peak—down to the rugged coastline (fig 42.4). Laufuti stream, the only perennial stream on the island, has the spectacular Laufuti Falls, which pours down from the cliff more than 1,000 feet. Thriving here are rare plants, tropical birds, two species of large flying fox (fruit bats) with wingspans up to three feet.

Ofu Unit. Nine miles across open water from Ta'u is the island of Ofu. The park portion consists of a beautiful

FIGURE 42.4   On Ta'u Island's south coast, sea cliffs—some of the tallest in the world—rise over 3000 feet, almost to the summit of Lata Mountains, American Samoa's highest peak. National Park Service photo taken by Doug Cuillard.

A marine scene of coral and fish off Ofu Island. National Park Service photo by Doug Cuillard.

coral sand beach and one of the best examples of a healthy coral reef in the South Pacific. It is an excellent site for snorkeling (fig. 42.5). Ofu is separated from Olosega by a 656 foot wide sea channel but is connected by a bridge. Both islands are remnants of a large shield volcano. The island of Ofu is a complex of volcanic cones, buried by lava from two merging flows, and remnants of a large volcano that is deeply eroded and has collapsed into a caldera. Lava flows and dikes can be seen in the walls, along with thick pyroclastic beds. The basic structure is a shield with parasitic cinder cones. Geomorphically, it is still in the shield building stage.

The reef supports hundreds of species of fish, coral, and other marine life. Endangered hawksbill sea turtles and green sea turtles also find refuge in these waters.

Tutuila Unit. This park area is located north of the capital of American Samoa, Pago Pago. The unit consists of about 2,670 acres of land and 450 acres of ocean. The size of Tutuila Island is approximately 48 square miles, with an elevation of 2132.5 feet.

Narrow irregular ridges form the backbone of the island of Tutuila. These ridges are older beds that have been deeply eroded. The park unit is typical of the rest of the island; i.e., closely spaced streams that are deeply eroding the volcanic slopes made up of basalt. The amount of precipitation and warm temperatures have formed a *laterite* soil. Most of the island has no craters because of substantial weathering, with the exception of the Leone Peninsula, formed about 70,000 years ago. Also found on the island are *drowned valleys*, small coastal flats, and volcanic intrusions such as Rainmaker Mountain, which is a *trachyte plug*, 1,718 feet high and outside of the park boundaries. It is a National Natural Landmark, one of seven in American Samoa.

Tutuila has a cyclic history of erosion followed by subsidence. Pago Pago harbor is a drowned volcanic caldera south of the park. At Vaisa Point on the west side of Tafeu Cove, the small streams enter the sea in hanging valleys because the wave action has eroded the cliffs faster than the streams can downcut. Maugaloa Ridge separates the park southward from the rest of the island.

This unit includes an undisturbed expanse of rainforest in the north central section of the island, as well as offshore waters and a fringing reef. Coastal lowland, montane, and ridge communities exist in the rainforest. Sea cliffs rimming the coast provide safe rookeries for thousands of sea birds.

In Amalau Cove (between Vatia Bay and Afona Bay) *pillow lavas* can be seen along the shore (fig. 42.6). The pillow lavas indicate extrusion under water. The beach at the head of the bay is a mixture of basaltic boulders and cobbles, with large chunks of coral that have been rounded by wave action. Because of the shallow slope, vigorous wave action, and size and shape of the rocks, they can be heard rubbing against one another and making a noise as the waves move back and forth. They are called the "singing rocks." Vaiava Strait is another National Natural Landmark. It separates Pola Tai Island (Cock's Comb), which is 420 feet high from Tutuila Island and includes Matalia Point, Cockscomb Point and Polauta Ridge (fig. 42.7). The strait's shape is

FIGURE 42.6   Pillow lavas on the shore of Amalau Cove on Tutuila Island. Pillow lavas are submarine extrusions of basaltic lavas that solidify as pillow-shaped masses closely fitted together. National Park Service photo by Doug Cuillard.

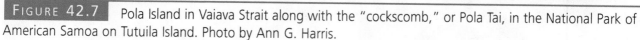

FIGURE 42.7   Pola Island in Vaiava Strait along with the "cockscomb," or Pola Tai, in the National Park of American Samoa on Tutuila Island. Photo by Ann G. Harris.

due to the fact that it is one of the trachyte intrusions. Mount 'Alava at 1610 feet is part of the caldera rim of Pago Pago volcano. A rugged trail leads to the summit of Mount 'Alava. The beginning of the three-mile trail starts at Fagasa Pass, about two miles west of Pago Pago. Hikers should allow about three hours for the climb up and about two hours for the hike down.

The island as a whole has a rugged terrain and a deeply embayed coastline showing erosion and subsidence. Pillow lavas and a flat plain between Nu'uuli and Leone (outside of the park) were built on a submerged reef by lava and tuff of Holocene age.

Geologic Features

The volcanoes of Samoa developed on the Pacific plate over a *hot spot* similar to the hot spot beneath the Hawaiian Islands. In 1985, J.H. Natland and D.L. Turner, using evidence from potassium/argon dating, field observations, and petrologic data, concluded that the Samoan shield volcanoes formed in age progression from west to east just as the rest of the Pacific Ocean

island chains. What created confusion was the fact that when the western islands entered the post-erosional stage of volcanic activity, there was enough eruption of lava to basically cover most of the island. The islands to the east, i.e., Ofu and Ta'u, are still in the shield building stage.

Additional evidence is Vailulu'u Seamount (a *seamount* is an underwater volcano), which eventually will become the next Samoan island. It is on the easternmost end of the Samoan swell, which is the leading edge of the hot spot track. Originally discovered in mid-1973, confirmation was established in 1999. It has risen 14,764 feet from the sea floor. Vailulu'u means "the sacred sprinkling of rain."

The summit of the seamount has three peaks along the crater rim. The summit crater is 1312 feet deep and 1.24 miles wide. The summit peaks extend down the slopes of Vailulu'u into three rift zones of the seamount. Radiometric dating of the lavas show dates of 5 to 50 years. Underwater studies have shown large-scale mass wasting of the sides.

New measurements show that in this area the Pacific plate, which moves in a northwest direction, is

FIGURE 42.8 Pago Pago harbor, viewed from the top of Mount Alava. A collapsed caldera, cut by ring dikes and modified by erosion, has formed this deep-water harbor. National Park Service photo.

the speediest tectonic plate on the globe. The Pacific plate is sliding down into the mantle beneath the Indo-Australian plate at the rate of almost 9.5 inches a year.

The main islands in American Samoa form a short *aseismic ridge*, that is, a submarine ridge that is a fragment of continental crust and not subject to earthquakes. It runs from the northwest to the southeast; therefore Tutuila, the largest island, is also the oldest, probably Pliocene in age. Age dates range from 1.24 to 1.4 my for the island. The younger islands are probably Holocene in age. The islands are not single volcanoes but are made up of intergrown and superimposed cones of *shield* volcanoes (flattened domes, built by flows of very fluid basalt). Tutuila, with four eroded shield volcanoes and one volcano of Holocene age, occupies the top of the volcanic pile that rises about 15,000 feet from the ocean floor. Eruptions have been largely basaltic and fluid, but a greater variety of igneous rocks have been found on Tutuila than on the other islands that are younger. Most lavas were of the *pahoehoe* type, but *cinder cones, tuff cones,* and remnants of *calderas* can

be found. Pahoehoe lava consists of smooth ropy flows. Cinder cones are made up of pyroclastic (material solidified in mid-air) debris, while tuff cones are consolidated pyroclastic debris. Pago Pago harbor is a caldera that has been breached by stream erosion and "drowned" by rising sea levels and slow subsidence (fig. 42.8).

Although the volcanoes of American Samoa have been inactive for some time, the hot spot beneath the moving tectonic plate still gives indications of its presence. A low bulge of prehistoric volcanic origin on the south side of Tutuila (the Tafuna Plain), has a surface so young that streams have not eroded it. A submarine eruption occurred off Olosega (southeast of Ofu) in 1866; in 1973, a submarine eruption was detected east of American Samoa, by an underwater listening system (SOFAR).[*]

---

*Smithsonian Institution, 1981 *Volcanoes of the World*, pg. 46, Hutchinson & Ross.

In the warm, humid, rainy climate of Samoa, the local basalt disintegrates and decomposes rapidly. Patterns of radial drainage on the mountains are typical. Erosion has cut deeply into the cones, leaving peaks, narrow ridges, and multiple, closely spaced stream valleys. Landslides leave scars on the steep slopes. Debris is swept down slopes and chokes lower stream valleys. Extensive dike complexes, i.e., Masefau Dikes, Fagasa Dikes, and Afano Dikes, display what were once "feeders" rising from volcanic depths.

Resistant trachyte plugs (fine-grained equivalent of syenite, similar to granite but much less quartz) make up the high peaks of Rainmaker and Matafao on Tutuila. Along exposed shores, waves have battered the rocks, eroding high sea cliffs and sheltered coves.

Limited drilling has identified bedded calcareous sandstone and siltstone, rather than solid coral within the fringing reefs of Tutuila. The sediments that make up these deposits were washed down from the mountain slopes into the water and gradually buried by coral growth as sea levels rose and subsidence occurred. Some coastal areas have traces of even higher shorelines of late Pleistocene and Holocene age.

The islands of American Samoa are located along the crest of a submarine ridge, called the Samoan Ridge, that is essentially perpendicular to the Tonga Trench, at approximately S 70° E. This ridge is probably the surface expression of a large sea-floor fault, along which the volcanoes erupted and formed islands. The Tonga-Kermadec oceanic trench is 2300 miles long, running from New Zealand (fig. 42.9) to the Samoan Islands. Here Pacific oceanic crust is subducting under flooded, shallow continental crust of the Indo-Australian plate. The island arcs of the southwest Pacific, with their andesitic volcanoes that developed along the continental side of the trench, are not directly related to the basaltic Samoan volcanoes despite their proximity. However, recent studies (Farley, 1995) suggest that fluids brought up from the mantle by the Samoan hot spot are geochemically similar to pelagic (ocean bottom) sediments that have been drawn down into the Tonga Trench and "recycled."

Beside the basaltic lava flows, another type of volcanic rock formed in the coastal areas, on both hills and offshore islands. Lava rising to the surface comes in contact with surface water, causing huge, powerful, explosions, such as the ones seen as lava flows enter the ocean today in Hawaii. The result is a mixture or layer of ash, volcanic rock, and pieces of coral called *tuff.* This is the origin of the tuff cones found on the islands of Tutuila, Ofu and Ta'u.

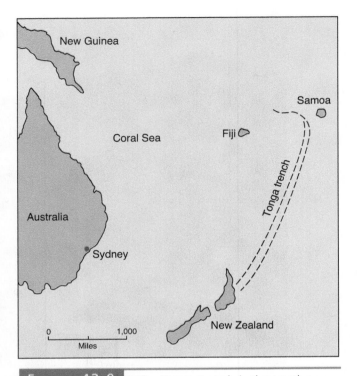

FIGURE 42.9   The Tonga trench is the northern section of this long oceanic trench, which is over six miles deep. The Pacific plate moves almost directly westward here before it bends down and is pulled into the trench.

Geologic History

New geologic techniques have been useful in studying the great pile of lavas and ash that have accumulated on the Pacific ocean floor and built up the Samoan archipelago. Radiometric dating methods have given absolute ages for rocks. Geochemical and petrologic analyses have allowed investigators to describe and identify units one from another. The geologic column (table 42.1) with its relative dates and formation names is the result of such methods.

1. **According to Stearns (1944), prior to 1.4 million years ago the intrusion from a volcanic rift zone formed the Masefau Dike complex.**

Intrusion of basaltic dikes that were either vesicular or occasionally *amygdoidal* (filled in vesicles) from a rift zone on the ocean floor to form either a shield volcano or layers of lava. Much of the lava that was intruded has been *brecciated,* that is broken into angular fragments; some of it may have been the broken material that fell back into the magma chamber. The thickness of this complex is about 197 feet.

Table 42.1				Geologic Column, National Park of American Samoa			

Eon	Era	Period	Epoch	Formation	Member	Stage	Geologic Events
Phanerozoic	Cenozoic	Quaternary	Holocene			Post Caldera	Mass wasting in the form of talus accumulation at base of cliffs and landslides. Deposition of sediments by water. Wave action creates beaches of basalt, coral and sands on all islands. Formation of Vailulu'u seamount.
							∼∼∼∼∼ UNCONFORMITY ∼∼∼∼∼
				Faleasao and Fitiuta Tuff			About 30,000 years ago the formation of several tuff cones in the northwest corner of Ta'u. Additional flows erupted at a different time in northeast corner of Ta'u.
							∼∼∼∼∼ UNCONFORMITY ∼∼∼∼∼
				Aunu'u Tuff Leone Basalts			About 70,000 years ago the formation of a tuff cone southeast of Tutuila Island forming Aunu'u island. Extrusion of Leone Volcanics from several vents creating the Leone peninsula on western half of Tutuila Island.
							∼∼∼∼∼ UNCONFORMITY ∼∼∼∼∼
			Pleistocene				Fluctuation of sea level because of world-wide glaciation. Development of fringing and barrier reefs, first on Tutuila Island (oldest) and later on Ofu' and Ta'u Islands after they formed.
				Lautele Formation		Shield Building	Formation of lava cone from flows along a fault on the northwest and northeast corners of Ta'u Island; lava pools in crater.
				Tunao Formation			Olivine basalts mixed with volcanic ash and tuffs from satellite cones in an east-west direction along a rift zone on Ta'u Island.
				Trachyte Plugs / Lata Formation	Post-Caldera		Formation of tuff cones inside and outside of the caldera on Ta'u Island.
					Intra-Caldera		Mixture of shales and olivine basalts. Flows emitted from rift zones that trend northeast and northwest about 500,000 years ago on Ta'u Island.
						Post Caldera	∼∼∼∼∼ MAJOR UNCONFORMITY ∼∼∼∼∼
							Intrusion of trachyte plugs 1.3 ± 0.03 mya into Olomoana, Alofau, Pago, and Taputa volcanoes about 100,000 years after they formed on Tutuila island.
				Late Formation		Shield Building	A series of lava flows forming Ta'u Island as a shield volcano.
				Nuu Tuffs			Shallow submarine eruptions forming pyroclastic material on Ofu Island.
				Taufanua Formation			Olivine basalts forming a shield volcano and a complex of volcanic cones that were buried. Still forming Ofu Island.
				Asago Formation			Mostly breccia forming the base of Ofu Island (and Olosega Island).
		Tertiary	Pliocene	Taputapu Volcanics		Post Caldera	Thin-bended olivine basalts forming over rift zone in a dome structure. Tuffs interbedded with lavas, cinder cones, westernmost volcano on Tutuila Island.
				Pago Extra-Caldera Volcanics	Upper Member		Upper member massive basaltic andesites and andesite lavas interbedded with tuff and cinder layers on Tutuila Island.
					Lower Member		Lower member a thin-bedded olivine basalt associated with dikes, tuffs, and breccia of earlier rocks. Bed dip away from caldera on Tutuila Island.
				Pago Intra-Caldera Volcanics	Upper Member		Flows of porphyritic basalts, andesite lavas, tuffs, breccias and cinder cone. Filling of caldera on Tutuila Island.
					Lower Member	Shield Building	Formation of subaerial portion of shield volcano between 1.54 and 1.28 million years ago. Formation of caldera 1.27 ± 0.02 million years ago on Tutuila Island.
				Alofau Volcanics			Second shield volcano forming over a rift zone. Olivine basalt flows, tuffs, dikes and breccia to the west of Olomoana volcano. Caldera formed on Tutuila Island.
				Olomoana Volcanics			Initial eruptions forming Tutuila Island. A shield volcano forming about 1.4 million years ago with olivine basalts, andesitic basalts, tuff and cinder cones.
							∼∼∼∼∼ UNCONFORMITY ∼∼∼∼∼
				Masefau Dike Complex			Intrusion of basaltic dikes from a rift zone on the ocean floor into lava layers that were brecciated by the intrusion. Erosion.

Sources: Modified after Keating, 1992; Natland, 1980; Whistler, 2002; Amreson et al., 1982.

2. **Around 1.4 million years ago the shield building stage began and four volcanoes formed nearly contemporaneously during the Pliocene Epoch, creating the island of Tutuila.**

Tutuila is the oldest island of American Samoa and the largest. The island has a rugged terrain and a deeply embayed coastline, showing evidence of erosion and subsidence. Going from east to west, four volcanoes formed at almost the same time on Tutuila Island. An overlap of each succeeding volcano indicates that they formed in a series at slightly different times. The Olomoana shield volcano, which formed first, consists of mostly olivine basalts that were eventually capped with andesitic basalts. Within the cone are exposures of many layers of glassy tuffs. Several cinder cones on its flanks are visible. The thickness of these beds is 1,057 feet. Weathering of the beds shows only a few gullies that are not very long and have not been deeply eroded because they are close to base level (the lowest point to which a stream can be eroded).

The second volcano, the Alofau, developed over a rift zone with an N 70° E trend. It has a composition of olivine basalt, and an almost mile wide caldera developed. Several hundred dikes have been exposed. Some of them are poryphyritic; that is, the bubble holes, or vesicles, have been filled with secondary material called amygdules. The caldera decreases in height to the southwest and part of it is submerged offshore. The thickness of these beds is 3,156 feet. Only a limited canyon developed on the flanks of the caldera.

The Pago volcano is located in the center of Tutuila Island. The subaerial portion was constructed between 1.54 and 1.28 million years ago. The caldera, which is about 6 miles long and 3 miles wide, formed 1.27 ± 0.02 million years ago. The southeast portion of the caldera forms Pago Pago harbor. The northwest rim, Maugaloa Ridge, forms the southern boundary of the park. The Pago volcanic series is divided into the Intra-Caldera and Extra-Caldera units. The thickness of these beds is 1,053 feet. The Pago Intra-Caldera unit partly fills the wide caldera and consists of massive poryphyritic basaltic and andesitic lava flows. There are also associated cinder cones, tuffs, and breccias, plus a few thin lenses of interbedded gravel. This unit has poorly graded deep valleys. The canyons are headed by amphitheaters. The Pago Extra-Caldera unit has a lower and upper member. The lower member is prim-

itive olivine basalt and is thin bedded. Associated with it are thin dikes, glassy tuff, and glassy tuffs with fragments of previously formed rocks. The beds dip away from the caldera at about a 25° angle. The upper member is a mixture of massive basaltic andesites and andesite lava flows, interbedded with glassy tuffs and scattered cinder layers, up to 492 feet. They rest conformably on the lower member. The erosion of this type of rock has formed deeply incised canyons and entrenched gullies. The canyon fill is highly laterized. Being derived from basalts, it is rich in oxides of iron and aluminum and is a typical product of weathering in tropical climates.

The last volcano is Taputapu, the westernmost volcano; its lavas appear to overlap the Pago volcanics. It is a shield volcano that developed over a rift zone that trends N 75° E. The olivine basalts are thin bedded and interbedded with glassy tuffs and cinder cones. The beds are 1,614 feet thick. The soils from the basalts have become deeply laterized. Isolated canyon ridges have formed and the canyons are moderately incised. These four volcanoes have undergone the first three stages of submarine origin, producing a dome or shield, and then the formation of a caldera, as Stearns had proposed.

3. **The formation of Ofu Island (and Olosega Island) from a submarine shield volcano during the Pliocene Epoch. Formation of modern fringing reefs.**

Most of the shield volcanoes in Samoa formed from volcanic centers. Examples of this are Ofu-Olosega and Ta'u plus a submerged bank west of Ofu-Olosega. The two islands of Olosega to the east and Ofu Island to the west are separated by a 656 foot sea channel. The national park unit is located on the eastern tip of Ofu Island. Ofu is much younger than Tutuila Island, as is illustrated by the fact that it is not so highly eroded as Tutuila. There are no offshore banks that are wide or were left over from the drowning of the reefs during late Pleistocene time. In their place are shallow sea cliffs that lack reefs at the edge of the Holocene flows. There are some reefs, but they occur on the older flows as is seen in the national park on Ofu. They are either fringing reefs or very narrow barrier reefs. These islands did not develop until Tutuila volcanoes were extinct and underwent extensive erosion before their reefs were drowned during Pleistocene time.

The Asaga Formation (the oldest unit that is Pliocene in age) is mostly a breccia; that is, angular fragments that are cemented together. It probably formed as the result of magma forcing its way up to the surface. This formation and the reefs are found in the park. Ofu and Tutuila were once part of a large shield volcano (Tuafanua Formation) with its center most likely north of the two islands, according to evidence. The Tuafanua units are divided into the ponded flows, and flows that are mixed with pyroclastic debris. A series of older tuff and cinder cones are scattered over the island, such as the cinder cone at Tauga Point (northwest Ofu), a tuff cone at the beach at Samo'i (the western end) plus a composite cone that is exposed by To'aga (southeastern Ofu) in a cliff. They have been buried by lava from two merging flows.

Eventually the crater of the large shield volcano collapsed and formed a caldera. The collapse exposed dikes and lava flows.

When lava flows come in contact with water, there is a tremendous explosion of steam, lava, and coral producing tuff. One example of this is a tuff cone just off the west coast of Ofu that makes up the island of Nu'utele. This formation, known as the Nuu Tuff, is also found on the western side of Ofu. These pyroclastic beds are deeply weathered, forming laterized fragmental volcanics. Additional flows covered some of this material.

4. **Intrusion of trachyte plugs during the post-caldera stage into all of the volcanoes approximately 1.3 ± 0.03 mya, during Pliocene and early Pleistocene Epochs. This was followed by a time of major erosion, known as the first erosional stage, producing a great erosional unconformity.**

About 100,000 years after the four main volcanoes formed, more or less contemporaneously, there was a series of intrusions of dense, cream colored trachyte plugs, some dikes, and also some crater filling. This is the fourth stage proposed by Stearns. Only one trachyte plug (called the Lefulufulua plug) intruded into the Olomoana Volcano. The Lelia plug intruded the Alofau volcano. Things were more active with the Pago volcano. The lower layer has three trachyte plugs, the Matafao (forming Matafao Peak, a National Natural Landmark), and Papataele and Pioa (part of Rainmaker Mountain). The upper layer has five; the Pioa (rest of Rainmaker Mountain, Matafao, Ta'u, Vatia and Afono plugs).

By using the submarine profile of the western Samoan Islands, it has been estimated that the volcanoes were originally a little over a mile high as compared to Matafas Peak, the highest mountain on Tutuila Island today, which is only 2,142 feet high.

The trachyte plugs are dense and very resistant to erosion. This is the type of rock that forms the sharp peaks, such as the tip of Vasia Point. There are two more trachyte plugs in the park; one is along the Mount 'Alava Trail, about a mile and a quarter from the beginning of the trail on the west end of the park land. The other trachyte plug is about a mile east of Fagasa Bay. There are also outcrops of either cinders or tuff throughout the park. Long-term erosion has removed any older volcanic cones that once existed on the flanks of Pago volcano.

5. **The formation of Ta'u Island from a remnant of a shield volcano about 500,000 years ago.**

The history of Ta'u is very similar to that of Tutuila, but it is the youngest of the park islands. The island is a remnant of a shield volcano that collapsed in Holocene time. It consists of many thin lava flows and some tuff beds. The Lata Formation of Pliocene age is the building stage of this volcano. The Pleistocene age Lata Formation represents the Intra-Caldera Member and Post-Caldera Member. The presence of a mixture of shale and olivine basalt suggests a fluctuation of sea level producing shales between eruptions. After the formation of the caldera, numerous tuff/cinder cones formed inside and outside the caldera on the northern flanks, making up the Tunoa Formation, a mixture of ash, tuff, and olivine basalt.

The Lautele Formation is a mixture of ponded lavas and tuffs that formed at the northwest and northeast corners of Ta'u. The Faleasao Tuff (northwest corner) was deposited about 30,000 years ago, making it the most recent volcanism observed on the island. This island is in the shield building stage and is on the leading end of the island chain. Erosion and landslides dissect the terrain.

The highest sea cliff is 3000 feet and is on the north side of Ta'u, with 328 to 656 foot cliffs on the west side. There is a prominent terrace at Si'ufa'alele Point in the southwest corner of the park that is 10 to 15 feet high. These terraces consist of sand and coral shingles. Most of the beaches are less than 98.5 feet wide. They are protected by offshore fringing reefs and the fact that there is a low tidal range. The beaches contain mostly fine material consisting of sand to gravel sized grains. Local springs and seeps flow from the lava layers into the beach deposits or lava on top of tuff beds. The fringing reefs around the island are a maximum distance of 689 feet from shore. They have surge channels in the reef front, which is made up of colonies of coral and algae. The surge channels are up to 26 feet wide and 15 feet deep. The reef flat consists of calcareous sands and coral and coralline algae. An inland escarpment called Liu Bench has a displacement at about 1476 feet. New studies have shown it to be a mass wasting feature and not the north side of a caldera as originally thought. If this bench slumps into the ocean, it may create a tsunami (seismic sea wave) that could be strong enough to destroy Fiji.

6. **During the Pleistocene Epoch a lowering of sea level, stream erosion of the land, and formation of fringing reefs on a wave cut platform. A later rise in sea level produced drowned valleys.**

Studies of Tutuila have indicated a profile of a mature erosional surface, fringing reefs, a drowned reef platform, and a barrier reef at 190 to 236 feet below sea level. This shelf has been tilted towards the southeast. There are benches and sea cut caves at 20 feet and 10 feet above high tide. The 20 foot bench is very noticeable because of the erosion of the lavas around the island, which form cliffs from 33 to 328 feet high. Off Tutuila's coastline the barrier reefs and fringing reefs merge, forming a narrow structure that is only 35 feet wide or is sometimes missing, especially along the southern coast. This structure is missing from around Ofu/Olosega and Ta'u. On Tutuila a modern fringing reef in Vatia Bay is located at the end of the paved section of Route 006. The rest of the shoreline of the park at this location is a drowned barrier reef at about 50 feet deep. There is a broad reef shelf at about 98 feet and the shelf edge then breaks at 131 to 164 feet and plunges to the submarine portion of the volcano.

The islands Ofu/Olosega and Ta'u have merged benches that are presently being eroded. They are about 3 feet to almost 7 feet above high tide. On Ofu Island, 295 foot high sea cliffs have formed around the island with talus slopes at their base. Wave action has modified the slopes. Within the park boundaries on Ofu is Sunu'itao Peak, which is 765 feet high.

The rise in sea level produced a series of drowned valleys. Examples on Tutuila are Afone, Fagaituo, Fagasa, Nua Seetga, Pago Pago, and Vatia Bays.

7. **The formation of Aunuu Island during the Holocene time at or about the same time as the Leone Basalts.**

The island, a National Natural Landmark located off the southeast coast of Tutuila, was formed by submarine volcanic activity. The tuffs accumulated to the thickness of 656 feet. The eastern half of the small island is a tuff cone. The western half was swampy because of an impermeable layer of weathered tuff, but it has been drained and taro is presently raised on it.

8. **Formation of the Leone Volcanics from several locations (0.07 mya) in the Holocene time during the post-erosional volcanism stage.**

The Leone eruptions built an almost flat plain from a fissure that ran between Fagamaa Crater to Ototele Peak about 70,000 years ago. A cone chain marks the fissure. Emitted were olivine basalts, in the form of pahoehoe flows, and pyroclastic tuffs. The tuffs veneered much of the area and the flows covered the submerged reefs. A cone about 246 feet high was formed at the source of the Tafuna flow (the Cinder Member). The Stony Ash Member formed another cone 197 feet high. Three craters near Steps Point ejected tuff that covers the volcanic ash. An erosional unconformity separates the overlying Fagamaa Tuff from the Fagatele Tuff. Because they are so young, geologically speaking, they are weathered the least. The drainage is still in the process of developing; hence the drainage patterns are not well defined. Soils are just starting to form. Associated with the tuffs are cinder and ash cones. Many of the flows are pahoehoe or ropy lava flows. Stearns (1944) suggested that a fissure from Otolele Peak about 4 miles long opened across the coral reef with fire fountains developing a cone chain. The trade winds blew much of the pyroclastic debris to the northwest on Tutuila Island and filled both the valley and the bay, producing the plain. The thickness of the beds is about 197 feet.

9. **Modern deposition of unconsolidated alluvium.**

Alluvium is unconsolidated material that has been transported by running water and downward movement to the valley bottoms. Talus consists of the rocks and boulders that accumulate at the bottom of the cliffs and steep slopes. Talus slopes are found on the west coast of Ta'u and south coast of Ofu. Because of the long duration of weathering, high humidity, and precipitation, the basaltic soils have been leached and have high concentrations of aluminum and iron, with low concentrations of silica. Even though the soil supports dense vegetation, it has relatively low fertility. Along the beaches are coral gravels mixed with calcareous sands and shells. In a few places the rocks on the beach have been cemented by calcite.

The tropical climate is conducive to both physical and chemical weathering; hence soils are very deeply weathered. Also affecting the soils are the fracturing and faulting of the original bedrock, and physical erosion, which make slopes susceptible to landsliding during tropical storms and hurricanes. For example, in October of 1979 after several days of heavy rain, seventy landslides were triggered and four people were killed. There is evidence of submarine landsliding in the geologic past, that may have caused tsunamis, which damaged nearby land areas.

10. Formation of the Vailulu'u seamount.

Vailulu'u seamount, located east of Ta'u, is a future Samoan Island. Radiometric dating of the lava flows indicates that the flows range in age from 5 to 50 years. Located on the easternmost end of the Samoan swell is the leading edge of the hot spot track. The seamount has risen 14,764 feet from the sea floor. Studying this seamount can give geologists insight as to how the Samoan Islands have formed.

Geologic Map and Cross Section

Keating, B.H. 1992. *The Geology of the Samoan Islands, Geology and Offshore Mineral Resources of the Central Pacific Basin,* Circum-Pacific Council for Energy And Mineral Resources, *Earth Science Series,* v. 14, N.Y. Springer-Verlag, p. 127–178.

Sources

Amerson Jr., A.B., Whistler, W.A., Schwaner, T.D. 1982. (*Wildlife and Wildlife Habitat Of American Samoa I.* *Environment and Ecology*) Department of the Interior, U.S. Fish and Wildlife Service, Washington, D.C. p. 8–14.

Farley, K.A. 1995. *Rapid Cycling of Subducted Sediments into the Samoan Mantle Plume,* Geology v. 23, no. 6: June, p. 531–534.

Keating, B.H. 1992. *The Geology of the Samoan Islands, Geology and Offshore Mineral Resources of the Central Pacific Basin.* Circum-Pacific Council for Energy and Mineral Resources, Earth Science Series, v. 14, N.Y. Springer-Verlag, p. 127–178.

National Geographic Society, 1955. *The Earth's Fractured Surface.* Multi-colored map 1:48,000,000, with accompanying text.

National Park of America Samoa Official Map and Guide, reprint 1998, National Park Service.

National Park Service-American Samoa Government, 1988. *National Park Feasibility Study, American Samoa,* p. 138.

National Park Service, n.d. *The National Park of American Samoa "America's Newest and Least Known National Park"* unpublished material, National Park Service n.d., National Natural Landmarks, p. 3

Natland, J.H. 1980. The Progression of Volcanism in the Samoan Linear Volcanic Chain, *American Journal of Science,* v. 280-A, p. 709–735.

Stearns, H.T. 1944. Geology of the Samoan Islands, Bulletin Geological Society of America v. 55, p. 1279–1332.

Stearns, H.T. 1975. *American Samoa* in Fairbridge, R.W. ed. Encyclopedia of World Regional Geology, Part 1, pg. 1; Hutchinson & Ross.

Proposal request from Oceans Ventures Fund, n.d. *Volcanic Hazards Map for Ta'u Island* unpublished 6 p.

Whistler, W.A. 2002. (manuscript) The Tropical Rainforest of Samoa Isle Botanica Publication, Honolulu, Hawaii, Chapter 1.

Wright, D.J., Donahue, B.T., & Naar, D.F. 2001. Sea Floor Mapping & GIS Coordination at America's Remotest National Marine Sanctuary (American Samoa) in Wright D.J. (ed.) Undersea With GIS, ESRI Press, Redlands, California, 32 p.

Note: To convert English measurements to metric, go to www.helpwithdiy.com/metric_conversion_calculator.html

Yellowstone National Park

NORTHWEST WYOMING, EASTERN IDAHO, SOUTHERN MONTANA

Area: 2,219,791 acres; 3468 square miles

Established as a National Park: March 1, 1872

Designated a Biosphere Reserve: 1976

Designated a World Heritage Site: 1978

Address: P.O. Box 168, Yellowstone National Park, WY 82190-0168

Phone: 307-344-7381

E-mail: yell_visitor_services@nps.gov

Website: www.nps.gov/yell/index.htm

Over 100,000 geysers, hot springs, mud pots, and colorful travertine terraces make Yellowstone the largest, most spectacular, and most complex thermal area on earth. A "hot spot" below the continental crust is believed to be the heat source.

Volcanic breccia, ash, and lava have built up the Yellowstone Plateau, over a mile above sea level. At least three explosive eruptions in the last two million years have developed giant calderas. Geophysical indicators suggest that eruptive activity may resume in the future.

Lakes, waterfalls, cirques, high mountain meadows, petrified trees, Obsidian Cliff, and the Grand Canyon of the Yellowstone River, are some of the wonders of this first national park.

FIGURE 43.1 Castle Geyser, in Upper Geyser Basin, erupts water and steam. The cone building up around the vent is made of geyserite, an opaline silica. Photo by Sherry Schmidt.

Yellowstone and the National Park Image

As the oldest and best known national park, Yellowstone symbolizes to Americans all that is awe-inspiring, beautiful, and wonderful about our national parks. Old Faithful has fascinated millions of visitors. Every day during the summer, thousands of people gather to watch its eruptions. Parking lots fill, empty, and then refill as this world-famous geyser puts on performance after performance. Mammoth Hot Springs, its great colorful terraces ever growing and changing, impresses visitors with its size. In the Grand Canyon of the Yellowstone—a deep gash in the Plateau—are the strange "yellow rocks" for which the park was named. Black bear, grizzlies, bighorn sheep, moose, elk, mule deer, bison, and wolves (newly reintroduced) roam in the forests and glades. The administrative structure developed by the National Park Service at Yellowstone National Park has served as a model for national parks throughout this country and the world.

No serious, long-term damage seems to have resulted from the series of fires that devastated sections of Yellowstone in the dry summer of 1988. At that time, many years had passed since the last great fires, and the whole Yellowstone area was in a vulnerable condition. Not only was the weather unusually dry, but in places, trees had been downed by wind storms and insect infestations, and trees and brush had filled in previously open areas. Burnable material was abundant once the fires started, whether due to lightning or human carelessness. After autumn rains and snows ended the fire danger, scientists studied its effects on the Yellowstone ecosystem. Some mudslides and slumps occurred on slippery or steep slopes due to the loss of binding vegetation, but landslides are a natural part of Yellowstone processes because many slopes are unstable. The fires did open up old vistas and some clearings where archeological and geological sites of interest were found. Meanwhile, in the years since the fire, healthy new growth has taken hold in burned-over sections. National Park Service managers hope that in the future, smaller, controlled fires will prevent a buildup of dangerous conditions.

Discovery and Exploration

Indians hunted and fished in the Yellowstone region for thousands of years. They made tools and weapons for their own use and for bartering from the plentiful obsid-ian. Large piles of obsidian chips near Obsidian Cliff and the finding of arrowpoints made from Yellowstone obsidian in the Great Plains and as far away as the Mississippi Valley are evidence of the importance of the obsidian trade.

John Colter, a member of the Lewis and Clark expedition, was probably the first non-Indian to visit Yellowstone. He left the expedition on the return trip in 1806 to join a party of trappers. The following year, alone and on foot, he crossed the Wind River Mountains and the Teton Range and explored the Yellowstone region, making contact with Indian tribes. Although wounded in a battle between Crow and Blackfoot Indians, he managed to make his way to a trading post at the mouth of the Bighorn River (south central Montana). There his description of a place of "fire and brimstone" was dismissed as the result of delirium.

Little was known about the region until the famous mountain man, Jim Bridger, after returning from an 1857 expedition, began spreading tales of spouting water, boiling springs, yellow rock, and a mountain of glass. Since Bridger was known as a "spinner of yarns," his reports were regarded as preposterous. Nevertheless, Bridger's descriptions aroused the interest of influential men, among them the geologist and western explorer, F.V. Hayden. In 1859, he and an Army surveyor, W.F. Raynolds, began a two-year reconnaissance of the Upper Missouri region with Bridger as the party guide. They reached the approaches to Yellowstone but were unable to get through the mountain passes because of very heavy snows. The outbreak of the Civil War put a stop to their explorations. Hayden, who was a medical doctor before becoming a geologist, served as a battlefield surgeon with Union troops. Eleven years passed before he was able to fulfill his dream of making a scientific study of Yellowstone.

Meanwhile, intrigued by the reports of miners who had traveled up the Yellowstone River in an unsuccessful search for gold, a group of Montanans organized the Washburn-Langford-Doane expedition. Headed by Henry Washburn, the surveyor-general of Montana Territory, the party included Nathaniel P. Langford (who later became known as "National Park" Langford). An Army detachment, commanded by Lt. Gustavus Doane, accompanied them. They spent nearly a month exploring the Yellowstone region, naming features, and collecting specimens. As they were gath-

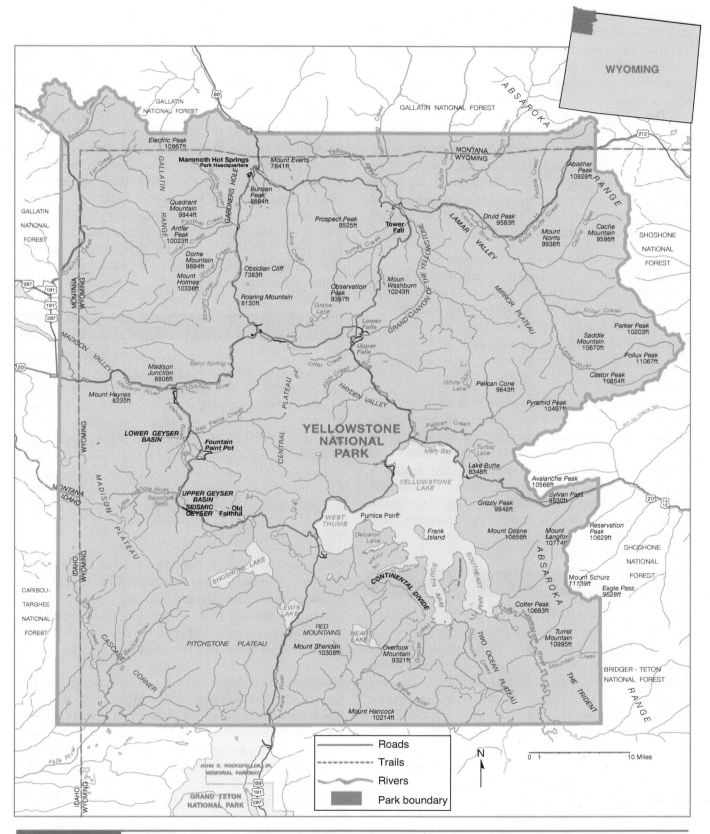

FIGURE 43.2   Yellowstone National Park, Wyoming, Idaho, and Montana.

ered around a campfire on their last night in the area, they talked about what should be done with this strange and marvelous region. Out of this discussion came the idea of making it a preserve under the protection of the government and not open to settlement or mining. All agreed to work toward getting the land set aside.

The next year, in 1871, F.V. Hayden led a large, government-sponsored expedition into Yellowstone. His comprehensive report, supplemented by the excellent photographs of W.H. Jackson and paintings by Thomas Moran, helped to convince the Congress to withdraw this unique area from public sale. On March 1, 1872, President Grant signed a bill authorizing Yellowstone National Park. The new law provided that

".. . a tract of land in the territories of Montana and Wyoming, lying near the headwaters of the Yellowstone River is hereby reserved and withdrawn from settlement, occupancy, and sale under the laws of the United States, and dedicated and set apart as a public park or pleasuring ground for the benefit and enjoyment of the people. . . ."

N.P. Langford, the first superintendent of the park, served for five years without pay, as did several succeeding superintendents. Congress did not appropriate funds to implement what was a noble intention. In 1886, the responsibility for protecting and managing the park was given to the Army. After the National Park Service was organized in 1916, a corps of rangers, under the leadership of a civilian superintendent, took over the park's care.

Geographic Setting

Yellowstone National Park is astride the Continental Divide, the topographic ridgepole of North America that separates Atlantic and Pacific drainages. The Divide enters the park near the southeast corner, crosses between Yellowstone Lake and Shoshone Lake, and then goes into Idaho south of the West Yellowstone entrance to the park (fig. 43.3).

The Yellowstone River rises south of the park, collects water from tributaries as it flows northward across the park, and joins the Missouri River in Montana. The Missouri, in turn, joins the Mississippi, which empties into the Gulf of Mexico. The drainage from the southwestern third of the park finds its way into the Snake River, which passes through Grand Teton National Park

(Chapter 44) and then flows west to join the Columbia River on its way to the Pacific Ocean. Clarks Fork of the Yellowstone River, which rises just inside Yellowstone National Park, was designated by Congress in 1990 as a Wild and Scenic River for its first 21 miles.

Yellowstone National Park is on a high plateau, averaging about 8000 feet in elevation and nearly surrounded by ranges of the Middle Rockies, 10,000 to nearly 14,000 feet high. To the southwest, the plateau slopes gently down to the Snake River Plains. The ranges that encircle the Yellowstone Plateau are (beginning on the west and going clockwise) the Madison Range; to the north, the Gallatin Range and the Beartooth Mountains; on the east, the Absaroka Mountains; and to the south, the Teton Range and Grand Teton National Park.

Geologic Features

Geologically, as well as topographically, the two national parks in the Middle Rockies—Yellowstone and Grand Teton (Chapter 44)—are related to Rocky Mountain National Park (Chapter 25) in the Southern Rockies and to Waterton-Glacier International Peace Park (Chapter 26) in the Northern Rockies. The ranges of the Rocky Mountain chain evolved through the same time span of geologic history; were subjected to Laramide tectonic forces of great intensity; and have rocks and structures of similar age.

The Surrounding Ranges

Many features of the Yellowstone landscape are associated with volcanic events that took place in relatively recent geologic time; but long before Cenozoic volcanism began, the structures of the ranges were evolving. Thus the story begins not on the Yellowstone Plateau but in the mountains that rise above it.

Madison and Gallatin Ranges. These ranges, which extend into the northwest section of the park, are faultblock mountains with north-south river valleys following downfaulted blocks. The Gallatin River flows north between the two ranges. The Madison River crosses the park's western boundary and heads north along the western flank of the Madison Range. The Yellowstone River outlines the eastern margin of the Gallatin Range.

Like many other Rocky Mountain ranges, the Madison and the Gallatin have a core of Precambrian

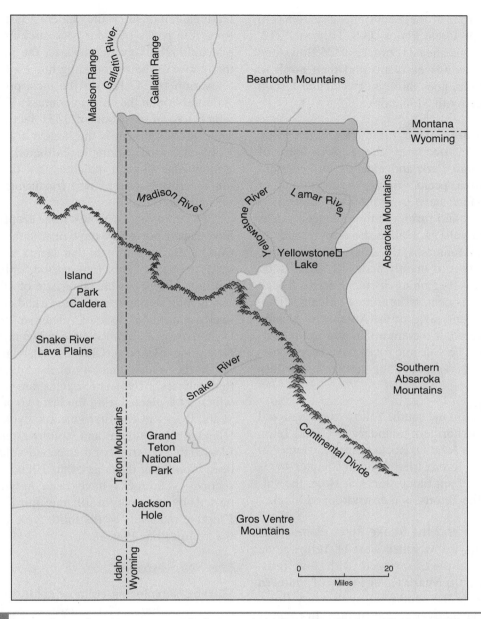

FIGURE 43.3 Locations of major topographic features in Yellowstone National Park and vicinity.

granites and gneisses with flanking Paleozoic and Mesozoic sedimentary rocks, some of which are overlain by volcanics. The northwest-southeast structural trend of the two ranges is characteristic of the general trend of the Rocky Mountains. Typical, also, is evidence of pulses of Laramide mountain-building and uplift. The present mountains are large blocks bounded by post-Laramide faults that in Cenozoic time cut obliquely across older structural trends in a nearly north-south direction.

The Beartooth Mountains. Uplift and vertical faulting have raised the Precambrian granites and gneisses of the Beartooth massif high into the air. From its southern tip in Yellowstone National Park, the massif rises to Montana's highest point, Granite Peak, elevation 12,799 feet. At the top of the range are a few outliers of a once extensive sedimentary cover, Beartooth Butte being an example. Other Paleozoic rocks, mainly limestones, lap up on the flanks. U-shaped valleys, 3000 to 4000 feet deep, dissect the glaciated uplands.

The bold eastern front of the Beartooths rises 4000 to 5000 feet above the Great Plains. U.S. Highway 212, which leaves the northeast corner of Yellowstone National Park and winds eastward and then north to Red Lodge, Montana, goes through spectacular terrain in crossing the Beartooth Mountains.

Absaroka Mountains. This large range, trending northwest-southeast, occupies the eastern part of Yellowstone National Park and extends into western Wyoming. Absaroka peaks, rising to elevations of 10,000 to 12,000 feet, have been intensely eroded by streams and glaciers and form a jumble of steep ridges above fairly broad valleys. Unlike the other ranges of the Middle Rocky Mountains, the Absarokas are made up of over 10,000 feet of mainly andesitic, pyroclastic volcanics. Eruptions beginning in the Eocene Epoch came from numerous centers in a large volcanic field. The tuffs, breccias, and lavas of the Absaroka Volcanic Supergroup overlie older metamorphic and sedimentary rock sequences. The volcanic rocks have been somewhat uplifted and warped, but are only mildly deformed.

The Teton Range. On the south, Yellowstone National Park is joined to the northern boundary of Grand Teton National Park by the John D. Rockefeller, Jr. Memorial Parkway. The Teton Range (described in Chapter 44) is a tilted and uplifted fault block. Jackson Hole, the valley at the foot of the Tetons, is a downdropped block.

Island Park Caldera and the Snake River Plains. The Island Park caldera, a few miles west of Yellowstone National Park, is a plateau ringed with low hills. Beyond the caldera, the Snake River Plains of southern Idaho slope gently to the southwest, with many flows of young basaltic lava covering the surface. In *Craters of the Moon National Monument* (about 120 air miles southwest of Yellowstone), some basalt flows erupted as recently as 2000 years ago. The volcanism of the Island Park caldera and the Snake River Plains is an integral part of the volcanic history of the Yellowstone Plateau.

The Yellowstone Plateau

A huge collapsed caldera, nearly filled with thousands of cubic miles of young volcanic rocks and volcanic debris, occupies the major part of the Yellowstone Plateau. Geyser basins, hot springs, and other thermal features demonstrate the caldera's lasting heat. Except for a few peaks of older rock sticking up through the plateau's relatively flat surface, the rocks are younger than those of the surrounding ranges. Mount Washburn (elevation 10,243 feet) is the most prominent summit. A depression in the caldera contains Yellowstone Lake, which has an elevation of 7331 feet. With an area of 139 square miles, it is the largest high-altitude lake in North America. Sporadic earthquake activity in the Yellowstone Plateau indicates that faults are responding to crustal stresses and fracturing associated with volcanic activity.

The voluminous Yellowstone eruptions, which may have lasted but a few days or months, were separated by long, quiet intervals that lasted hundreds of thousands of years. This long-period, episodic pattern of volcanic activity is characteristic of silicic magmas in continental settings and is in marked contrast to a more nearly continuous style of volcanism typical of basaltic systems, especially those in oceanic regions such as the Hawaiian volcanoes (Chapters 40, 41).

In the volcanic history of the Yellowstone Plateau, three phases, or volcanic cycles, are recognized, all of which took place during the last two and a half million years. Rocks of the first volcanic cycle are exposed in Island Park caldera and southwestern Yellowstone. Identifiable outcrops of the second volcanic cycle have been found near the edge of the Island Park caldera, but second cycle rocks have been largely destroyed or buried by the output of the third volcanic cycle. Eruptive materials of the third cycle make up most of the Yellowstone Plateau.

The Yellowstone Caldera

Mazama's explosion-collapse caldera (Chapter 36) has an area of barely 20 square miles. What exists at Yellowstone is a multiple-eruption caldera with an area of about 1500 square miles! The volume of magma erupted at Yellowstone probably amounts to 50 times the output of Mazama. By means of extensive field investigations and with the aid of satellite imagery, the outlines of Yellowstone caldera have been accurately determined. About 40 miles long and 30 miles wide, it reaches from the Madison River (near Madison Junction) to the eastern shore of Yellowstone Lake and from Lewis Falls north to the foot of Mount Washburn.

Studies of the predominantly rhyolitic rocks of the three volcanic cycles in the Island Park–Yellowstone

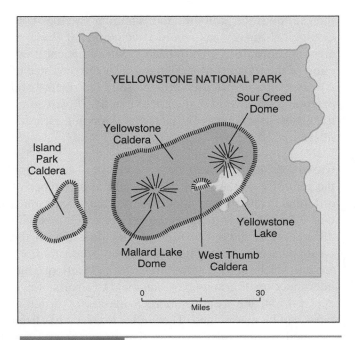

YELLOWSTONE NATIONAL PARK

Sour Creed Dome

Yellowstone Caldera

Island Park Caldera

Mallard Lake Dome

West Thumb Caldera

Yellowstone Lake

0 ——————— 30
Miles

FIGURE 43.4   Location and extent of Island Park and Yellowstone calderas, both Pleistocene in age, but Island Park is older. Two currently expanding domes are within the Yellowstone caldera. The small crater in West Thumb of Yellowstone Lake is late Pleistocene.

area suggest that during each episode great volumes of lava rose to the surface and erupted, building up thousands of feet of ash, tuff, breccia, and layers of lava (fig. 43.5). Small lava eruptions—related to the third cycle, but occurring after the Yellowstone caldera was formed—persisted until about 70,000 years ago.

Any explanation for such great volumes of extrusive material requires the presence of a fairly large reservoir of magma, presumably originating in the upper crust and extending fairly close to the surface. As magma was withdrawn by eruptions through fractures and weak zones, the roof rocks over the chamber stretched and thinned, finally collapsing from their own weight into the liquid magma below. Since no identifiable roof rocks or old volcanic cones have been found, we can assume that the missing surface material was either incorporated into the melt or ejected with ash. The cataclysmic explosions were of a magnitude not known in recorded history.

For thousands of years following the climactic eruption, lava continued to flow periodically until the tremendous depression was nearly filled. Then, about 150,000 years ago, the floor of the Yellowstone caldera began to bulge up again. This doming exposed rocks and structures that have enabled geologists to put

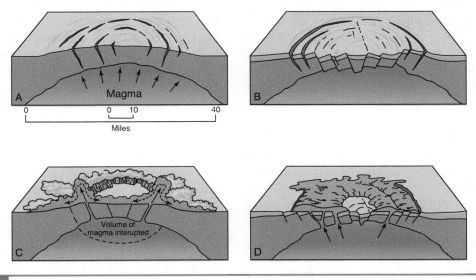

FIGURE 43.5   Development of Yellowstone caldera during the most recent volcanic cycle, 600,000 years ago. *A.* Upwelling magma arches roof rocks, causing concentric ring fractures that extend down toward the magma chamber. *B.* When ring fractures tap the magma chamber, gases explode as pressure is released, triggering catastrophic pyroclastic flows. Borne by prevailing winds, ash particles spread over thousands of square miles. *C.* Roof rocks collapse, forming an immense caldera. *D.* Magma, depleted of gas, flows out from the ring fractures over the caldera floor. From W. H. Parsons, *Middle Rockies and Yellowstone.* © 1978 Kendall/Hunt Publishing Company. Used by permission.

together a sequence of events to account for this remarkable feature and also to develop hypotheses regarding future volcanic activities in Yellowstone.

Yellowstone's Next Eruption?

Heat flow. The earth's crust gets warmer with depth. The rate of heat increase with depth is called the *geothermal gradient.* Heat energy tends to move from hotter areas to cooler areas. Despite the fact that rocks are poor conductors, the earth's internal heat moves from great depths toward the surface. Amounts of heat flow can be measured both at the surface and in drill holes. Heat flow values taken from a number of points on the Yellowstone Plateau are abnormally high, more than 30 times the average global heat flow. In the bottom of wells drilled by the U.S. Geological Survey, the average temperature is over 400° F at depths of 500 to 600 feet below the plateau surface.

Curie depths. Most rocks lose their magnetism when they are heated beyond the *Curie point,* a threshold temperature of about 560° C. Thus information about the temperature of rocks below the surface can be inferred from the magnetic properties of rock. Rock under Yellowstone reaches the Curie point at a remarkably shallow depth. Using geophysical techniques, investigators have determined that beneath the surface of the Yellowstone Plateau, Curie depth is about 6 miles down. Furthermore, in some spots under the caldera, Curie depth is less than 4 miles (Smith and Christiansen, 1980). By contrast, Curie depths under other parts of the continent are between 10 and 20 miles below the surface.

Electrical conductivity. At higher temperatures, earth materials tend to be good conductors of electricity. Since data obtained by geophysical techniques show that electrical conductivity at Curie depth under Yellowstone is good, the patterns of magnetic and electrical data are in agreement.

Seismic wave velocities. Seismologists have found that earthquake waves increase or decrease in speed as they travel through the rocks of the crust, depending upon the density of the materials they encounter. Seismic P waves, for example, travel faster through the high-density basalts of oceanic crust and then slow down when passing through granitic continental crust, which has lower density. Seismograph monitoring in the Yellowstone caldera and vicinity by the U.S. Geological Survey has recorded velocities of P waves arriving from distant earthquakes. Mathematical modeling of the travel-time delays suggests the presence of a body of low-density material close to the surface and extending downward to the upper mantle. (P waves are the fastest seismic waves that travel through the earth's interior.)

Gravity data. A *gravity survey* consists of measuring the intensity of the gravitational field at a number of different locations and then plotting the results on a map. The object is to associate gravity variations with differences in the distribution of rock densities. If the crust consists of homogeneous materials, the variations are due solely to differences in surface elevation. Thus interpretation of gravity data can yield two sorts of evidence. First, the location of rock masses of different types can be determined. Then, after a survey has been repeated at intervals, perhaps five or ten years, any changes in elevation that have occurred can be calculated and the results compared.

Gravity measurements derived from 900 locations on the Yellowstone caldera and the surrounding region have indicated the presence of low-density material, both under the caldera and also in the subvolcanic basement, with the latter showing even lower readings than the material nearer the surface. What this implies is that the presumed shallow chamber contains magma that is high in silica, and that the silica content in rocks *above* the chamber is less high, perhaps because the silica has been somewhat dispersed. Rhyolites, which are plentiful in Yellowstone, are high in silica and have a density of about 2.5 to 2.6. Mafic rocks, such as basalts, which are less common in Yellowstone, have somewhat higher densities of around 3.0. Gravity data have also confirmed geodetic surveys that indicate the caldera area is rising at the geologically rapid rate of a little less than half an inch per year.

Depth of focus and earthquake frequency. The *focus* is the point within the earth that is the center of an earthquake. It is the initial rupture point of the earthquake from which seismic waves emanate. The *epicenter* is a point on the earth's surface *directly above* the focus of an earthquake. A seismologist can pinpoint the location of an earthquake and the depth of focus by comparing the seismic wave data recorded on the seismograph at his station with similar information from other stations and then analyzing the results.

With one earthquake of over magnitude 7 and seven magnitude 6 earthquakes, Yellowstone National Park and adjacent parts of Idaho have been the most seismically active region in the Northern and Middle Rocky Mountains in historic time. The strongest earthquake was the Hebgen Lake quake (7.1 Richter scale) in August 1959, which caused extensive landsliding (fig. 43.6). The epicenter was just outside the northwest corner of the park. Monitoring of many earthquakes in the Yellowstone area (most of which have registered between 1 and 6 on the Richter scale) has revealed that the average depth of focus is around 9 miles, which is considered shallow. This means that the rocks are solid to that depth because only solid rocks can fracture and produce earthquake waves. *Earthquake swarms* (clusters of minor earthquakes) occurring in Yellowstone suggest movement of magma below the surface.

Ages of the Caldera rocks.

U.S. Geological Survey scientists have ascertained absolute ages for volcanic rocks from the three volcanic cycles using the potassium-argon method of radiometric dating. Combining the dates with information from stratigraphic studies and geologic mapping produced the following sequence. The first volcanic cycle began about 2.2 million years ago in the Island Park area and lasted until about

FIGURE 43.6   The Hebgen Lake landslide, that created Earthquake Lake, occurred on August 19, 1959 and was triggered by a 7.1 (Richter Scale) earthquake. The highly jointed Mississippian-aged Madison Limestone formation was tilted on edge and acted as a retaining wall for the very badly weathered Precambrian-aged gneisses and schists. The collapse of the Madison Limestone caused the weathered material to come rushing downhill with such a velocity that it crossed the Madison River, damming it, and went halfway up the opposite side of the valley before it slid down to the valley floor. Photograph by Ann G. Harris.

1.6 million years ago. The climactic eruptions and caldera collapse occurred about 2 million years ago. Climactic eruptions of the second volcanic cycle, which also began in the Island Park area, were about 1.3 million years ago. Overall, a smaller volume of material was erupted. The third volcanic cycle, which was marked by a shift in volcanic activity from the Island Park area to the center of the Yellowstone Plateau, began about 1.2 million years ago. Climactic explosions about 630,000 years ago produced the present Yellowstone caldera.

Composition of the lavas.

While most of the volcanic rocks produced by eruptions from the Yellowstone caldera are rhyolitic, some basalt flows were extruded late in the third volcanic cycle. Basaltic lava flows also followed the rhyolitic flows of the first and second volcanic cycles. The pattern seems to be one of extensive rhyolitic volcanism, with the center slowly moving from southwest to northeast, followed periodically by episodes of basaltic volcanism along the same general trend.

The eruption of rhyolitic lavas and then basaltic lavas from the same vents with no intermediate lavas (e.g., andesites) appearing in the sequence calls for a special explanation. The composite volcanoes described in Chapters 35 through 39 changed the composition of their lavas as eruptions ran their course, but the transition from one type of lava to another was usually more gradual.

Chemically and mineralogically, the two rock types are quite different. Rhyolite is high in silica and low in iron and magnesium, while basalt is high in iron and magnesium and low in silica. The predominant minerals in rhyolite are potassium feldspar and quartz; in basalt, they are usually calcium plagioclase feldspar, olivine, augite, and hornblende. Minerals that form basalt crystallize out from a melt or magma at considerably higher temperatures than do the minerals that form rhyolite.

It seems logical to theorize that at Yellowstone two different source materials melted in order to form two distinct liquids, one rhyolitic and the other basaltic. Basaltic magma may be formed by the partial melting of the upper mantle, while rhyolitic magma may be derived by the partial melting of metamorphic rocks (such as gneisses and schists that are high in silica) in the earth's lower crust. Long-continued uplift and extension, or stretching, of the crust might cause upward displacement of the mantle. This, in turn, causes reduction of confining pressure that allows materials

with low melting points to be liquefied. As heat radiates upward in the crust, additional high-silica rocks are incorporated into the melt because they tend to melt first. In this way, a high-silica magma, having lower density and greater viscosity and being somewhat more buoyant, begins to accumulate in the upper crust. In the late phase of a volcanic cycle, after rhyolitic magmas have been ejected, molten basalt from the upper mantle begins to rise under conditions of reduced pressure.

Expanding domes.

According to geophysical and geochemical evidence, a body of very hot, low-density, high-viscosity material, capable of generating and extruding tremendous volumes of volcanic material, has existed beneath the Yellowstone region for the past several million years. Within the Yellowstone caldera, two small resurgent domes (Sour Creek dome and Mallard Lake dome) are bulging up the plateau center at a faster rate—a little more than one-half inch per year—than the general uplift going on in the Yellowstone caldera.

Scientists do not yet have enough information to make specific predictions, but past volcanism, contemporary uplift with high uplift rates, length of time between volcanic eruptions, earthquake swarms, recent volcanism in the caldera, and the presence of a shallow heat source in the upper crust all point to a future of continuing volcanic activity in the Yellowstone Plateau (fig. 43.7). The very high heat flow, along with the anomalous behavior of seismic waves, implies that shallow hot bodies, probably containing a large volume of fluid in the form of steam, water, and partially molten rock, are present underneath the plateau (Smith and Braile 1984, p. 108).

What form might future volcanic activity take? At present, there is no way of knowing whether small eruptions of lava may occur around the margins of the Yellowstone Plateau, or medium-size eruptions within the caldera, or perhaps even a major explosion and flow of ash. However, any renewal of volcanism will undoubtedly be preceded by such warning signs as localized earthquakes, variations in the velocity of seismic waves, increasing gas emissions and doming or other deformation. All these phenomena can be monitored and continuously evaluated.

The Hot Spot Theory

What is the source of the heat that drives the Yellowstone thermal and volcanic activity? And why have the eruptive centers shifted over time along a

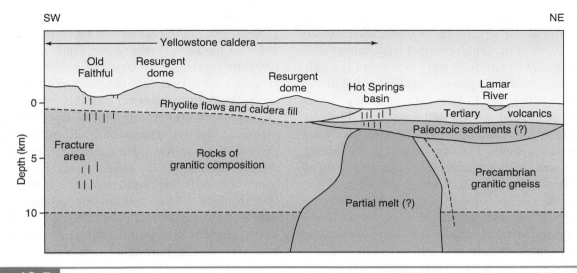

SW   NE

FIGURE 43.7   Simplified southwest-northeast cross section through the Yellowstone caldera and the Lamar valley, based on seismic and other geophysical interpretations. Modified after Smith and Braile 1984.

trend from southwest to northeast? Geologists theorize that a hot spot in the upper mantle, comparable to the one that produced the Hawaiian Island chain (Chapter 40), may be the cause of volcanic episodes in the Yellowstone region. However, the Pacific hot spot lies under thin oceanic crust, while thick continental crust overlies the Yellowstone hot spot. As explained earlier, eruptive patterns in the two regions have developed along profoundly different lines. Although high heat flow keeps the geyser basins and hot springs active between eruptive cycles, major eruptions in Yellowstone have been more episodic than eruptions in the Hawaiian chain, which have tended to be more or less continuous.

At present, the North American tectonic plate is moving 1.77 inches per year in a southwestern direction over the hot spot. Since the assumption is that the hot spot has remained stationary, the centers of volcanic activity have shifted northeastward as the crust drifts toward the southwest. In general, the ages of the volcanic rocks tend to fit this model. The older rocks are in the vicinity of Island Park and younger rocks are in the central part of the Yellowstone Plateau. The young basaltic rocks of the Snake River Plains, which largely covered earlier rhyolitic eruptions, may represent partial melting of upper mantle material that occurred after the emptying of rhyolitic magma from an older chamber. Geologists have speculated that within a few hundred thousand years, if the trend continues, a fourth caldera may form in the northeast corner of the Yellowstone National Park.

Hot Springs and Geysers

For observing hydrothermal features, no place in the world can compare to Yellowstone, which has about 62 percent of all known geysers, plus thousands of hot springs, mud pots, and fumaroles. A *hot spring* is, by definition, any thermal spring whose temperature is above that of the human body. A *geyser* is an intermittent hot spring that regularly or irregularly erupts jets of hot water and steam, as a result of ground water being heated by hot rock, creating steam under conditions preventing free circulation. Confined steam builds up pressure underground until it attains enough force to be forcibly ejected at the surface.

Hot springs, which occur in many parts of the world, were well known to Europeans who settled North America. But none of the early western explorers had ever seen a geyser before coming to Yellowstone and had doubtless never heard of Geyser in Iceland, the namesake for all the world's geysers. Geyser fields are rare, and are located in remote areas, such as Iceland, New Zealand, and Indonesia, places visited by few nineteenth-century Americans and Europeans. To come upon an area in the wilderness where great jets of steam and scalding water suddenly spouted from the ground, shooting 100—even 200—feet into the air, only to stop and then, inexplicably, begin again, must have been unnerving, as well as awesome, to the mountain men.

The source of the geysers' heat and the origin and operation of their "plumbing systems" were subjects of scientific speculation for more than a hundred years, but a generally accepted theory of how they work has been developed. For geysers to function, three conditions must be present: (1) a powerful heat source, usually volcanic; (2) a plentiful supply of water; (3) a fairly watertight plumbing system, capable of holding a large volume of water and withstanding the tremendous pressure of steam and superheated water (fig. 43.8).

Geysers do not have open underground chambers for accumulating water. The tube or vent at the surface leads down to a shallow plumbing system consisting of intricate conduits that connect small openings in the rock and lenses of porous sediment capable of storing water. Normally, several geysers in a geyser basin are supplied by an interconnecting fracture system. The passageways in the system are lined with *sinter,* which is a chemical sedimentary rock, mainly siliceous, deposited as a hard encrustation by precipitation. Very hot water, deep underground, dissolves silica from the rhyolitic rocks and carries it through the system to the surface, depositing precipitated silica along the way. The sinter deposits help to keep the system nearly watertight and tend to produce constrictions that prevent free circulation of water. Most of the world's geyser fields are in areas where rhyolite is abundant.

Rainwater and meltwater from snow infiltrate the ground, percolating slowly down through cracks and pores in the rock. Some water seeps into the geyser plumbing systems; but much of it goes several thousand feet lower, where it is heated by contact with the hot rocks over the magma chamber to a temperature of more than 400° F (fig. 43.9). Due to hydrostatic pressure, the water cannot boil; but it begins to expand and rise toward the surface. Steam bubbles, rapidly forming in the rising water, heat the cooler water that is filling the lower part of the plumbing system. This filling and mixing process continues until the system is full of water ready to boil over. The average boiling point at the surface of the geyser basins in Yellowstone (elevation about 7400 feet) is about 199° F (93° C).

Steaming water flows, instead of spouting, from the hot springs in the geyser basins, because the hot springs have larger openings and less restricted underground circulation. In the geysers, rapidly rising steam bubbles choke or clog constricted passageways, and the weight of the water above creates a pressure-cooker effect. As the water temperature rises, it reaches a critical point, which varies from geyser to geyser. Upward

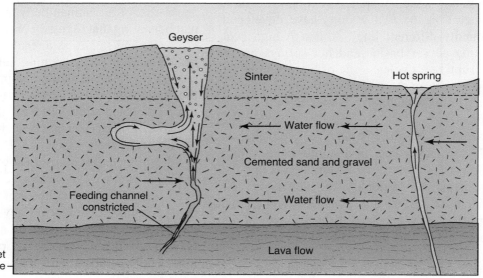

FIGURE 43.8   How geysers and hot springs work. Circulating ground water is heated at depth under pressure. Water boiling in the geyser system produces steam bubbles, causing expansion and overflow at the surface. With pressure suddenly reduced, water trapped in side chambers flashes into steam. The resulting "explosion" clears the upper part of the geyser tubes and shoots water and steam into the air. An eruption is not repeated until enough water seeps into the system and becomes again superheated, filling all tubes and side chambers. In hot springs, where boiling water is not constricted, it rises through the tubes and flows out continuously.

pressure of steam bubbles forces out some of the water; this lowers the pressure enough so that the water below flashes to vapor. The vapor then blasts out of the vent with such force that the whole system is nearly emptied as boiling water and steam shoot into the air. After that, the whole cycle begins again.

Because the silica in solution precipitates both above and below ground as the water temperature drops, sinter may build up enough to seal off diversionary passages in the plumbing system, thus changing a geyser's eruptive behavior. Some geysers build large cones or mounds of sinter around the vents. Hot springs may have fantastically shaped mounds of sinter over which water flows out. Most of the Yellowstone sinter is light colored, but some is tinted various hues.

The Geyser Basins.

Upper Basin has the highest concentration of geysers and hot springs, the most famous being Old Faithful. Its eruptions are watched, photographed, and cheered by the thousands of visitors who come to view the spectacle. Old Faithful overflows for a few minutes at the beginning of the eruption, as if to clear its throat. Then with a roar, the mighty stream shoots up 100 to 180 feet into the air. (Average height of its eruptions is about 130 feet.) After a few minutes, the top of the jet slowly drops and then

sinks down. Some puffs of steam drift out, and Old Faithful settles back to fill again.

Old Faithful plays as high as it did when first described, but average intervals between eruptions have lengthened after each of the last three major regional earthquakes. This is probably due to shifts in circulation of hot water away from the geyser. Nevertheless, Old Faithful's predictability remains unique, even though its timing is not so precise as many visitors think. During an eruption, Old Faithful ejects from 3700 to 8400 gallons of water, most of which drains into the Firehole River.

Castle Geyser (fig. 43.1), in the Upper Geyser Basin, is about 1400 feet northwest of Old Faithful. Castle's imposing cone, with a circumference of about 120 feet, stands at the top of a large mound that rises about 40 feet above the Firehole River. This geyser's eruptive behavior has been highly variable, with some periods of great irregularity and, more recently, a pattern of steam eruptions about every nine hours. Most eruptions have been of the steam phase type although the geyser has had periods of continually splashing water over the rim like a giant boiling pot (Marler 1971).

In Lower Basin, south of Upper Basin, the groups of geysers and hot springs are more scattered. The

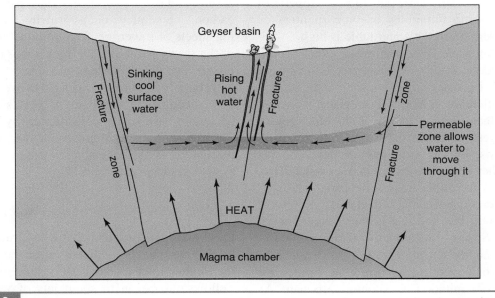

FIGURE 43.9 Relationship between a geyser or hot springs basin and a magma chamber at depth. Groundwater, infiltrating from the surface, may percolate down to a depth of more than a mile before it becomes very hot and then begins to rise. From W.H. Parsons, *Middle Rockies and Yellowstone.* © 1978 Kendall/Hunt Publishing Company. Used by permission.

FIGURE 43.10 Hot water riverlets drain into the Firehole river. The primitive algae that lives in the water gives the riverlets different colors because each species lives in a different temperature range and has a different color. Photo by Ann G. Harris.

Fountain Paint Pots, a cluster of multicolored mud pots, are in this basin amid a large geyser complex. *Mud pots* are hot springs that contain boiling mud and dissolved compounds that hot water brings up from the rocks below. Mud pots do not have enough water to flush out sediment as it accumulates (fig. 43.11).

Fumaroles, or steam vents, puff in both Upper and Lower Basins. Some fumaroles become spouters or even small geysers when the water table is high.

Norris Basin, near the northwest edge of the Yellowstone caldera, is the hottest geyser basin in the park, presumably because magma is closest to the surface here. Another difference in Norris Basin is that the water is acidic rather than alkaline, as it is in the other geyser basins. The largest geyser in the world, Steamboat, is located in Norris, but its eruptions are few and far between. Six other geyser basins are in Yellowstone, but several are in less accessible areas of the park.

A quite different thermal feature is in the northwest corner of the park, but not in a geyser basin. This is Mammoth Hot Springs, a mountain of travertine terraces covering nearly a square mile. *Travertine* is a dense, massive or concretionary limestone formed by rapid chemical precipitation of calcium carbonate. At Mammoth Hot Springs, the buildup of travertine deposits by precipitation is much more rapid than the accumulation of sinter deposits around the hot springs and geysers in the geyser basins (fig. 43.12).

Thick beds of marine limestone underlie Mammoth Hot Springs. Since limestone is highly soluble, the rising warm water (cooler than waters in the geyser basins) dissolves and brings to the surface enormous quantities of calcium carbonate—over 2 tons per day in solution. As warm, mineralized water flows from the springs and down over the terraces, particles of calcium carbonate precipitate out as growing icicles and spreading sheets of travertine. The millions of algae that live in the warm pools tint the rock shades of brown, red, orange, and green. Springs and terraces change rapidly. New outlets form as old ones close over; active springs may become inactive in a year's time. Terrace Mountain, still under construction by Mammoth Hot Springs, continues to grow, change, and enlarge. At present, it is the largest carbonate-depositing spring system in the world.

Seismic Activity and the Thermal Features

Earthquakes disrupt and change hydrothermal systems in Yellowstone. Immediately after the main shock of the Hegben Lake quake in 1959, hundreds of springs in Yellowstone began to spout jets of water. Clepsydra Geyser erupted wildly for three years after the quake and then became dormant. Sapphire Pool, which used to be a placid spring with occasional eruptions a few feet high, began erupting violently and voluminously.

Hundreds of tons of sinter broke up and washed away before Sapphire's eruptive power gradually declined. Fumaroles opened near Firehole River, and mud pots spouted. Seismic Geyser in Upper Basin is the direct result of a crack in the earth that opened during the 1959 quake. Old Faithful's eruptive intervals have lengthened and been less regular since the Borah Peak earthquake (magnitude 7.3) in 1983.

Glacial Features

Because violent eruptive events and later glaciations buried or destroyed most of the older glacial features in Yellowstone National Park, those that remain in the area belong to the Pinedale substage of Wisconsinan time, which ended in Yellowstone about 8000 years ago. Although winters on the high Yellowstone Plateau are severe, and snowfalls are heavy, no glaciers are present-ly on the plateau. At higher elevations in the encircling mountains, small glaciers assigned to Neoglacial time (post-Wisconsinan) are moderately active.

In the colder and wetter Pleistocene climates, valley glaciers started high in the mountains. As they grew and flowed down to lower elevations, they coalesced to form piedmont glaciers. The Lamar Glacier, which originated in the Beartooth Mountains, flowed down from the northeast; the Upper Yellowstone glacier, advancing from the southeast, was fed by valley glaciers in the southern part of the Absaroka Range. Small icecaps covered the plateau and the foothills during prolonged episodes. Mount Washburn and other high peaks stuck up above the ice as nunataks. At its greatest extent, the Pinedale icecap flowed radially outward from about the present location of Yellowstone Lake. To the south and southwest, ice flowed over the Continental Divide and down into Jackson Hole (Chapter 44).

FIGURE 43.11   Black Dragons Cauldron, a constantly boiling mud pot in Lower Geyser Basin. Photo by Edward Mooney.

FIGURE 43.12   Mammoth Hot Springs covers nearly one square mile and is a mountain of travertine terraces. They exist because limestone, rather than the volcanic rock that lies under the majority of the park, underlie the area. The warm springs dissolve the limestone as the water rises, and the travertine is precipitated out as the water cools on the surface. The algae living in the hot water color the terraces that are still forming. The portion of the terrace that is dull looking is no longer active. Photo by Douglas Fowler.

A large glacial erratic of Precambrian gneiss, left by a Pinedale glacier, gives some idea of its transporting power. The erratic, weighing more than 500 tons and measuring about 24 x 20 x 18 feet, stands near Inspiration Point, a short distance from the western rim of the Grand Canyon of the Yellowstone River. To get from its source to its present location, ice must have carried the erratic at least 15 miles (fig. 43.13).

The Grand Canyon of the Yellowstone River

The Yellowstone River, which rises in the Absaroka Range southeast of the park, carries runoff from the eastern side of the Continental Divide. The river flows into the southern end of Yellowstone Lake, out through the northern end, and then northward into Montana. A principal tributary is the Lamar River that drains the northeastern part of the park.

The Grand Canyon of the Yellowstone is a narrow, steep-walled canyon that evolved through several stages of cutting and filling during the past half a million years. Ash and volcanic debris from eruptions repeatedly filled the river valley, but the early Yellowstone River cut down its channel again after each episode. Then advances of Bull Lake glaciers (early Wisconsinan time) blocked the Lamar-Yellowstone drainage system. Water backed up, creating a long lake in the canyon, and layers of lakebed sediment accumulated. After glacial retreat, rapid erosion cleared the canyon, but the Pinedale icecap overrode the canyon and again blocked the system. About 12,000 years ago, the last ice in the canyon melted away. Since that time, the Yellowstone River has eroded most of the volcanic debris and the glacial and lakebed sediment from the canyon.

The canyon walls are cut mainly in rhyolitic lava and tuff. Intense chemical weathering from hot ground water and hot springs has changed the color of the rock from an unweathered grayish black to a startling array of reds, browns, and yellows, producing the strange "yellow rocks" for which Yellowstone was named. Near the bottom of the canyon, where hot springs bubble out and fumaroles emit steam, the processes of hydrothermal alteration soften and dissolve the rock, changing its composition and physical properties.

At the head of the deepest and most spectacular part of the gorge is Lower Falls, 308 feet high (fig. 43.14). Here the river flows over the edge of an unaltered, resistant rhyolite flow that erodes more slowly than the tuff and softened lava below. Above Lower Falls is Upper Falls, which is 109 feet high. The Grand Canyon of the Yellowstone ends near Tower Falls, where Tower Creek flows over the rim of a hanging valley and drops 132 feet into the Yellowstone River below.

Columnar Jointing and Obsidian

Prominent columnar jointing is displayed in basaltic lava flows in Yellowstone National Park, especially in the walls of the Grand Canyon of the Yellowstone and at Sheepeaters Cliff along the Gardner River. The characteristic parallel columns of *columnar jointing,* polygonal in cross section, form by contraction as a lava flow cools. The rock tends to divide into columnar segments approximately perpendicular to the cooling surface (fig. 43.15).

The Yellowstone obsidian, which was of such value to the Indians, is a dark volcanic glass. Obsidian Cliff, along the road from Norris to Mammoth Hot Springs, is presumed to be the "Glass Mountain" that Jim Bridger talked about. *Obsidian,* or natural glass, forms when lava (usually rhyolitic) cools and hardens before mineral grains have had time to crystallize.

Buried Forests

At Specimen Ridge, on the south side of the Lamar River valley in the northeast corner of the park, petrified tree trunks (some with root systems preserved) stick up through volcanic material in lifelike positions. Erosion of volcanic material on the slopes has exposed

FIGURE 43.13 A glacial erratic that was transported by ice during the Pinedale glaciation. It is located near the Grand Canyon of the Yellowstone River. Photo by Ann G. Harris.

these fossil forests that were buried long ago by Eocene ash flows and lahars (mudflows). The trees were lithified as silica-enriched ground water infiltrated and preserved the woody structure (fig. 43.16).

FIGURE 43.14 Lower Falls. 308 feet high, and the Grand Canyon of the Yellowstone River. At Lower Falls, the river drops from a ledge of resistant rhyolite. The canyon is cut in the young lavas and pyroclastics of the central part of the Yellowstone Plateau. Hot water, rising through fractures, accelerates chemical weathering processes, softening rock in the canyon walls and changing the color to shades of yellow and orange. Photo by Diane Ludwig.

Another, smaller buried forest is at the tip of the Gallatin Range where it extends into the northwest corner of Yellowstone.

At Specimen Ridge, 20 or more successive forests (the exact number is uncertain) were buried, one above the other from bottom to top, during Eocene time. Perhaps as much as 2000 years passed between burials, as again and again violent eruptions from the Absaroka volcanic field sent ash flows and mud rolling down slopes, smashing and devastating the forests. The tree varieties most commonly identified are sycamore, walnut, magnolia, chestnut, oak, redwood, maple, and dogwood. Eocene climate throughout this time remained mild and humid, probably like the climate of southeastern United States today.

Geologic History

1. Precambrian erosion, sedimentation, and tectonism.

In the northern part of the park, in the Gallatin and Beartooth Ranges, outcrops of contorted schists and gneisses represent fragments of the geologic history of Precambrian time. Some have been dated as over 2.7 billion years old. The schists were probably volcanics, sandstones, and shales deposited in basins in primitive crust. The gneisses may have been derived from granitic intrusions. Presumably, the rock materials have been "recycled" several times by the processes of erosion, deposition, metamorphism, and tectonism.

2. Paleozoic and Mesozoic deposition.

Over 40 separate rock formations of Paleozoic and Mesozoic age have been identified in Yellowstone National Park. They crop out across the northern section of the park in the Gallatin and Beartooth Ranges, as well as in the uplands in the southern margin of the park. Many of the Mesozoic and Paleozoic beds were stripped off during uplift or lie buried under many feet of volcanic rock. Little correlation exists between the northern and southern sedimentary rock units in the Yellowstone region.

Predominantly marine, the rocks are an assortment of conglomerates, sandstones, shales, and limestones. Approximately 3000 feet of Paleozoic rocks and 6000 feet of Mesozoic rocks were deposited, for the most part in shallow seas that repeatedly advanced and withdrew over what is now western North America. Beds of terrestrial deposits on beaches and flood plains are

FIGURE 43.15   At Tower Junction in the northern part of the park basaltic lava flows with prominent columnar jointing are visible between the thicker rhyolite flows. The columnar jointing formed as the lava cooled and contracted. Intersecting joints produced the hexagonal columns. Photo by Edward Mooney.

interspersed in the Mesozoic formations. For several hundred million years, this was a relatively quiet region of low elevation and low relief.

3. **Laramide orogeny in late Cretaceous and early Tertiary time; faulting and erosion.**

The mountain-building, compression, folding, and rapid increase in sedimentation that began in late Cretaceous and Paleocene time was the result of plate convergence to the west. The subducted Pacific plate underthrust the North American plate for hundreds of miles, shortening the crust and raising up the Rocky Mountain chain. Tectonic activity continued for some 30 million years.

In south central Yellowstone, Paleozoic and Mesozoic beds were tightly folded into anticlines (trending northwest-southeast), separated by synclines and faults. Movement along one reverse fault in this

area was locally more than 10,000 feet. Reverse faulting of this magnitude also occurred in the northern part of Yellowstone. Fault blocks, created by the tremendous stresses, were raised, dropped, or tilted differentially. Intensified erosion was an immediate result of the uplift. Sediment stripped from the heights was deposited in nearby downwarped areas and basins. The Harebell Formation, for example, in south central Yellowstone, of latest Cretaceous age, received more than 8000 feet of sediment.

4. **Eocene volcanism; the Absaroka volcanic field.**

Tremendous outpourings and explosions of andesitic lava and pyroclastics built up the Absaroka and Washburn Ranges, as well as part of the Gallatins, during the Eocene Epoch. Most of the Yellowstone area was at least partially covered by volcanic debris. The eruptions of viscous lava, ash, and breccia came from

FIGURE 43.16 Petrified Sequoia tree stump. Trees that grew in Eocene time were buried by volcanic breccia during an eruption and subsequently lithified by silica-enriched, percolating ground water. The silica that filled the wood cells is more resistant to weathering and erosion than the breccia that originally completely covered the stumps. Photo by Ann G. Harris.

large vents, one of which was in Mount Washburn. Composite thickness of the Absaroka Volcanic Supergroup is estimated to be over 12,000 feet. Intrusive equivalents (having the same age and composition) of the Absaroka volcanics were emplaced as stocks, dikes, and sills. Interbedded with the volcanics are sedimentary beds of reworked ash and volcanic debris. Many deposits show the effects of extensive mass wasting, such as landsliding and slump. As eruptions subsided, large volumes of volcanic sediment were eroded and transported out of the Yellowstone area.

5. Late Tertiary (Pliocene) uplift, faulting, and erosion.

Regional uplift raised Yellowstone and the Middle Rockies several thousand feet. Mountain blocks were further elevated by faulting, while downthrown blocks became basins. The Gallatin Range was uplifted to its present height. To the south, the Teton Range was raised and Jackson Hole dropped down (Chapter 44). With these tectonic events, the landscape of the Yellowstone region began to assume its present framework.

Uplift rejuvenated the streams, which began cutting deep canyons and removing more of the Absaroka volcanic rocks. Remnants exposed in the park, such as the Mount Washburn eruptive center, represent only a small amount of the great volume of Absaroka lavas that buried the Yellowstone surface in Eocene time. About 40 million years elapsed before volcanic activity—unrelated to the Absaroka episodes—resumed in Yellowstone.

6. Pleistocene rhyolitic eruptions; volcanic cycles and caldera collapse.

The largely rhyolitic rocks that built up the Yellowstone Plateau during the three Pleistocene volcanic cycles are designated the Yellowstone Group (fig. 43.17). The first volcanic cycle produced a large dome, quantities of welded tuff, and eventually a collapsed caldera, approximately 30 by 50 miles in size and several thousand feet deep. The Huckleberry Ridge Tuff is associated with the explosive eruptions that occurred as the caldera collapsed, about 2 million years ago.

The second volcanic cycle, similar in pattern but shorter in duration, was not so widespread or productive of lava as the first cycle. The second cycle reached its climax about 1.3 million years ago with the explo-

sive eruptions that formed the Mesa Falls Tuff. Again, the caldera collapsed.

About 1.2 million years ago, the third, and most recent, volcanic cycle began with a definite shift in activity from the Island Park area to the Yellowstone Plateau. Very likely, ground water, plentifully supplied by rain and melting snow, played a role in triggering the catastrophic explosions of about 630,000 years ago, during which more than 1000 cubic kilometers of magma were expelled. Gaseous steam, suddenly released, propelled enormous fiery clouds of ash high into the air. Foaming lava poured from the vents, dev-

Table 43.1	Generalized Geologic Column, Yellowstone National Park				

Time Units			Rock Units		Geologic Events
Era	Period	Epoch			
Cenozoic	Quaternary	Holocene	Alluvial, landslide, glacial, and hot spring deposits		Earthquakes; landsliding Geothermal activity, hot springs, geysers Neoglaciation Regional uplift; doming in center of Yellowstone River Pinedale glaciation Erosion of Grand Canyon of Yellowstone River
		Pleistocene	Yellowstone Group (about 20 formations and members)* (fig. 43.17)	? — — — (Plateau rhyolites and basalts) (Lava Creek Tuff) (Mesa Falls Tuff not exposed in park) (Huckleberry Ridge Tuff)	Post-caldera lava flows; Obsidian Cliff Bull Lake glaciation Pre-Bull Lake glaciation Climactic eruptions and caldera collapse of third volcanic cycle; rhyolitic ash falls, mudflows, welded tuff, breccias Second volcanic cycle; Island Park caldera and western part of Yellowstone Plateau First volcanic cycle; caldera collapse in Island Park area; ash falls, pyroclastics, breccias, etc.; rhyolitic eruptions
	Tertiary — Neogene	Pliocene / Miocene	/////////		Erosion of ranges, canyon-cutting, sedimentation in basins
	Tertiary — Paleogene	Oligocene			Regional uplift; block-faulting
		Eocene	Absaroka Volcanic Supergroup* (9 formations)		Repeated eruptions of andesitic pyroclastics and lavas; burial of fossil forests
		Paleocene			Basaltic breccias and flows Post-orogenic faulting, uplift and erosion
Paleozoic/ Mesozoic	(Cretaceous)		Over 40 formations, some in northern and some in southern sections of park*		Laramide orogeny; intense folding and thrust-faulting; uplifting of Rocky Moutains
					Extensive deposition of marine and nonmarine sedimentary rocks as seas advanced and retreated over downwarped area
Precambrian			/////////		Major angular unconformity
			Gneisses and schists		Orogenic cycles; intrusions, metamorphism; sedimentation and erosion

*For details of the stratigraphic record, see Keefer (1971, p. 9–11), Parsons (1978, p. 214), Christiansen and Blank (1972, p. B6), Love and Keefer (1975, p. D9), Smedes and Prostka (1972, p. C7).

astating the surroundings. Borne on prevailing winds, ash blanketed the midcontinent area. Collapse of the chamber roof produced a caldera 45 by 25 miles in extent and a half a mile deep. Among the features of the Yellowstone landscape formed at this time were the welded tuff deposits of the Lava Creek Tuff, the depression holding Yellowstone Lake, and the blurred but still visible caldera rim.

A resurgence of fairly strong volcanic activity took place about 160,000 years ago in the depression now occupied by the West Thumb of Yellowstone Lake. After dome expansion, explosive eruptions and collapse left a small caldera.

For thousands of years after the collapse of the main caldera, viscous lava periodically poured out over the caldera floor, sometimes in flows 1000 feet thick. Most flows stayed within the caldera, but a few spilled over the rim. Lava occasionally came up through the ring fissures. A prominent, small flow of this last type is the one exposed at Obsidian Cliff. Minor basalt flows also oozed out.

7. Glaciation and canyon-cutting.

Toward the end of Wisconsinan glaciation, the level of Yellowstone Lake was approximately 140 feet higher than it is now. In some places in the park, glacial drift and alluvium are interbedded with volcanic debris, indicating that eruptions occurred when glaciers were in existence.

After the disappearance of Pinedale glaciers during the postglacial optimum (between 6000 and 2000 B.C.), the Yellowstone River cleared out its Grand Canyon. The processes of weathering, mass wasting, and stream erosion continue to modify this remarkable feature.

8. Post-Pleistocene uplift and erosion; geothermal activity.

The Yellowstone Plateau and the caldera have undergone uplifting in Holocene time. The caldera's more rapid rise is continuing at the present time. Two small resurgent domes in the caldera may presage a resumption of volcanic activity. Geysers and hot springs have prob-

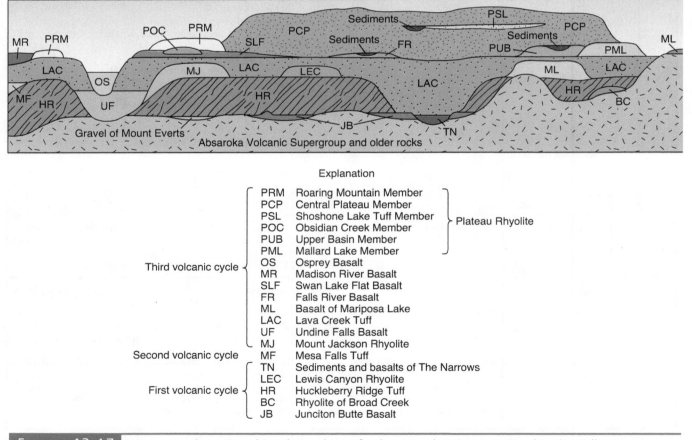

	Explanation	
	PRM	Roaring Mountain Member
	PCP	Central Plateau Member
	PSL	Shoshone Lake Tuff Member
	POC	Obsidian Creek Member
	PUB	Upper Basin Member
	PML	Mallard Lake Member
	OS	Osprey Basalt
Third volcanic cycle	MR	Madison River Basalt
	SLF	Swan Lake Flat Basalt
	FR	Falls River Basalt
	ML	Basalt of Mariposa Lake
	LAC	Lava Creek Tuff
	UF	Undine Falls Basalt
	MJ	Mount Jackson Rhyolite
Second volcanic cycle	MF	Mesa Falls Tuff
	TN	Sediments and basalts of The Narrows
	LEC	Lewis Canyon Rhyolite
First volcanic cycle	HR	Huckleberry Ridge Tuff
	BC	Rhyolite of Broad Creek
	JB	Junciton Butte Basalt

Plateau Rhyolite: PRM, PCP, PSL, POC, PUB, PML

FIGURE 43.17   Names and stratigraphic relationships of volcanic rock units recognized in the Yellowstone caldera area. From Christiansen and Blank 1972, p. B7.

FIGURE 43.18   Conglomerate from an ancient streambed overlain by airborne volcanic tuff. Photo by Earl Harris.

ably been active on the Yellowstone Plateau ever since the end of Pinedale glaciation, and they may have been intermittently active during interglacial periods. Location of the hydrothermal features is believed to be related to faults and fractures in the bedrock.

Despite the fact that the surface of Yellowstone Lake freezes over in winter and stays close to the freezing mark in summer, much higher temperatures exist with increasing depths in the lake. In recent years, using a remote-controlled submersible, scientists have discovered bacterial mats (similar to those found in Crater Lake, Chapter 36) in deep, dark, hot places and around vents on the lake floor. These bacteria "eat" chemicals dissolved in rising geothermal waters. In turn, creeping worms feed on the bacteria. The lake floor is fairly smooth outside the rim of the caldera, but within the caldera are tall cliffs, deep chasms, talus slopes, fumaroles, and oozing, near-boiling muck. The lake bottom appears to be geologically active and undergoing constant change, apparently because of movement of molten rock beneath the caldera (Milstein 1990).

Rates of heat flow on the Yellowstone Plateau, the frequency of shallow earthquakes, and the inferred presence of a body of hot, low-density material below the surface all point toward the possibility of eventual new eruptions. Investigators have noted that a major caldera has formed in the Yellowstone region about every 600,000 to 900,000 years in the past two million years. Inasmuch as the present Yellowstone caldera collapsed about 630,000 years ago, we can speculate that a new one may be in preparation.

Table 43.2	Comparison of estimated volumes of recent explosive eruptions with estimated volumes of Yellowstone volcanic explosions. Note that the Yellowstone eruptions were much greater in volume (20 times or more) than any modern volcanic explosive eruptions.		

Volcano	Date	Volume of ejecta (cubic miles)
St. Helens	1980	0.744 mi³
Lassen	1914–21	0.806
Katmai	1912	4.34
Krakatoa	1885	11.16
Mazama	6845 B.P.	46.5
Tambora	1815	93
Yellowstone/Island Park		
First cycle	2.2 mya	2500
Second cycle	1.2 mya	280
Third cycle	0.6 mya	1000
	Yellowstone total	3780

Sources: Williams and McBirney 1979; Decker and Decker 1981, in Smith and Braile 1984.

Geologic Map and Cross Section

U.S. Geological Survey. 1972. Geologic Map of Yellowstone National Park. Miscellaneous Geological Investigations, Map I-711.

U.S. Geological Survey. 1972. Surficial Geologic Map of Yellowstone National Park. Miscellaneous Geological Investigations, Map I-710.

Sources

Barger, K.E. 1978. Geology and Thermal History of Mammoth Hot Springs, Yellowstone National Park, Wyoming. U.S. Geological Survey Bulletin 1444, 55 p and map.

Boyd, R, D. 1961. Welded Tuffs and Flows in the Rhyolite Plateau of Yellowstone Park, Wyoming. Geological Society of America Bulletin 72:387–436.

Bryan, T.S. 1979. The Geysers of Yellowstone. Boulder, Colorado: Colorado Associated University Press.

Christiansen, R.L. 1984. Yellowstone Magmatic Evolution: Its Bearing on Understanding Large Volume Explosive Volcanism, in National Research Council 1984. Explosive Volcanism: Inception, Evolution, and Hazards, Studies in Geophysics, Series II. Washington, D.C.: National Academy Press. p. 85–95.

——— and Blank, H.R. 1972. Volcanic Stratigraphy of the Quaternary Rhyolite Plateau in Yellowstone National Park. U.S. Geological Survey Professional Paper 729-B.

——— and Hutchinson, R.A. 1987. Rhyolite-basalt Volcanism of the Yellowstone Plateau and Hydrothermal Activity of Yellowstone National Park, Wyoming; Geological Society of America Centennial Field Guide, Rocky Mountain Section, v.2, p. 165–172.

Eversman, S. and Carr, N. 1992. Yellowstone Ecology: A Road Guide. Missoula, Montana: Mountain Press Publishing Company, 242 p.

Fritz, W.J. 1982. Geology of the Lamar River Formation, Northeast Yellowstone National Park. 33rd Annual Field Conference, Wyoming Geological Association Guidebook. p. 73–101.

———. 1985. Roadside Geology of the Yellowstone Country. Missoula, Montana: Mountain Press Publishing Company, 144 p.

Hammond, A.L. 1980. The Yellowstone Bulge. *Science 80* 1:68–73.

Keefer, W.R. 1971. The Geologic Story of Yellowstone National Park. U.S. Geological Survey Bulletin 1347. 7th printing 1987. Yellowstone Library and Museum Association.

Love, J.D., and Keefer, W.R. 1975. Geology of Sedimentary Rocks in Southern Yellowstone National Park. U.S. Geological Survey Professional Paper 729D.

Marler, G.D. 1971. Studies of Geysers and Hot Springs Along Firehole River, Yellowstone National Park. Yellowstone Library and Museum Association.

Milstein, M. 1990. Discoveries in the Deep. *National Parks* 64 (3–4): 29–33.

Parsons, W.H. 1978. *Middle Rockies and Yellowstone,* K/H Geology Field Guide Series. Dubuque, Iowa: Kendall/Hunt Publishing Company.

Pierce, K.L. 1979. History and Dynamics of Glaciation in the Northern Yellowstone Park Area. U.S. Geological Survey Professional Paper 729-F.

Smedes, H.W., and Prostka, H.J. 1972. Stratigraphic Framework of the Absaroka Volcanic Supergroup in the Yellowstone National Park Region. U.S. Geological Survey Professional Paper 729-C.

Smith, R.B., and Braile, L.W. 1984. Crustal Structure and Evolution of an Explosive Silicic Volcanic System at Yellowstone National Park, in National Research Council 1984. *Explosive Volcanism: Inception, Evolution, and Hazards,* Studies in geophysics, Series II. Washington, D.C.: National Academy Press. p. 96–109.

——— and Christiansen, R.L. 1980. Yellowstone Park as a Window on the Earth's Interior. *Scientific American* 242:104–117.

U.S. Geological Survey. 1977. *Geysers.* 23 p.

White, D.E.; Fournier, R.O.; Muttler, L.J.P.; and Truesdell, A.H. 1975. Physical Results of Research Drilling in Thermal Areas of Yellowstone National Park, Wyoming. U.S. Geological Survey Professional Paper 892.

PART V

Landscapes and Structures in Areas of Complex Mountains

FIGURE PV.1   In Great Smoky Mountains National Park, in the high ranges to the Southern Appalachians, erosion on ancient rocks and structures has created a mountain landscape of gently rounded hills. Photo by David B. Hacker.

Mountains are challenging to climbers and to geologists. The rocks exposed, the shapes and elevations of summits, and the character of the land surrounding mountain ranges all give clues about the earth's internal processes—crustal movements and tectonic forces—that created mountains and placed them where they are today. The results of external processes show up on mountains, too. Weathering, mass wasting, and the agents of erosion have altered the rocks and the scenery.

Geologists like to do field work in mountains. They find more different kinds of rocks and geologic structures than are ordinarily visible in flatlands. But deciphering the nature and processes of mountain-building can be difficult. A vital outcrop may be covered by glacial snow and ice or may be inaccessible. Avalanches may have buried geologic structures or erosion may have removed them.

Mountains are seldom isolated. They are generally components of a mountain mass, or *mountain range,* consisting of a succession of mountains or mountain ridges. *Major mountain belts,* which are chains or cordilleras consisting of many mountain ranges, tend to be more than a thousand miles in length, but are usually only a few hundred miles wide. The Appalachians, which extend from Newfoundland south into Georgia, are an example. Acadia National Park (Chapter 24) along with Shenandoah and Great Smoky Mountains National Parks (Chapters 54 and 55 in this section) are in the Appalachians.

Rimming the western side of the continent—from the Aleutian Islands in the Bering Sea southward into Mexico—are the multiple ranges of the North American Cordillera. Within its domain are the majority of American national parks, a dozen Canadian national parks, and about a score of national parks in Mexico and Central America. The North American Cordillera, which is the dominant geological component of western North America, has been evolving since the middle of the Paleozoic Era. Ever since that time, the Pacific tectonic plate has been intermittently in contact with the western margin of the North American plate. At the present time, the North American plate moves one to two inches a year in a westward direction, while the Pacific plate slowly rotates against the North American western margin in a

counter-clockwise direction. This creates "strike-slip" structures and causes great discordance between rock units in the two plates.

The plate collisions that have occurred might be described as a more or less continuous interplay of constructional and destructional processes, both local and continental in scale. In some places great outpourings of lava buried older rocks and structures. The magnificent Cordilleran landscapes that we see today in western North America are the result of erosional processes operating on a dynamic framework of structures of all types composed of a great variety of rocks. Geologists who study Cordilleran history sometimes call the ranges "the enigmatic mountains" because the more data are collected, the more complexity becomes apparent.

Characteristics of Major Mountain Belts

Mountain ranges are the result of severe deformation of crustal material plus igneous activity in a particular area. Mountain belts are produced along plate margins by plate convergence or collision with accompanying tectonic activities on a greater and more prolonged scale. In general, three stages occur in the evolution of mountain belts: (1) an accumulation stage; (2) an orogenic stage; and (3) a crustal extension, block-faulting, and uplift stage with prolonged erosion.

In an *accumulation stage,* great thicknesses of sedimentary and volcanic material are deposited in a marine environment. Sediments that accumulate between converging oceanic and continental plates are compressed and pulled down or scraped off when the oceanic plate is subducted. Over time, the squeezed and folded sedimentary rocks may be elevated and eventually exposed by erosion in mountainsides. For example, in Redwood National Park (Chapter 51), which is located in the Northern California Coast Ranges, a greenish rock that is a contorted mixture (mélange) of deep-water, shallow-water, land-derived, and volcanic sediments is conspicuous in road cuts and cliffs along the shore.

The *orogenic* ("mountain-building") *stage* begins while accumulation of sediments is still going on. The convergence or collision of plates causes the layered rocks to be tightly folded and then broken by thrust faults. Deeply buried rock layers are metamorphosed by heat and pressure; still deeper in the crust or upper mantle, magma forms batholiths that may supply volcanoes or eventually cool below the surface as plutons. The plunging folds in Hot Springs National Park, Arkansas (Chapter 52) are evidence of orogenic activity involving compression and deformation due to plate collision (fig. PV.2 A,B).

The *crustal extension, block-faulting,* and *uplift stage* is probably a reflection of isostatic adjustments following the orogenic stage. Continental crust, being less dense, tends to ride higher on the mantle than oceanic crust. Mountain-building causes a mountain belt to be out of equilibrium in relation to the *craton* (the continent's stable interior). The continental crust of the mountain belt, greatly thickened by accumulation of sedimentary rocks and intrusions of igneous rock, begins a long slow rise as erosion strips rock from slopes. Normal faults may develop due to tensional (horizontal) forces pulling apart the rising crust so that fault-block mountains are created. Block-faulting may also result when vertical movements in the crust are unequal. When a mountain mass "floats" upward in order to regain isostatic balance, tilting may occur, or some blocks may rise faster than others. In Great Basin National Park (Chapter 45), block-faulted mountain ranges dominate the landscape (fig. PV.3).

Continents enlarge as major mountain belts evolve along continental margins. Some growth is due to *tectonic accretion,* or addition of tectonostratigraphic terranes to a continental margin. As the terranes are caught

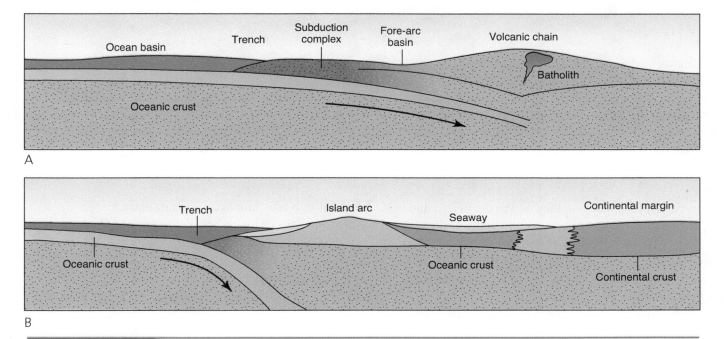

Schematic representations of two ways in which subduction may occur along a plate boundary. *A.* In this convergence of an oceanic plate and a continental plate, many miles of oceanic crust are disappearing beneath an on-thrusting subduction complex. Sediments accumulate in the shallow trench and the fore-arc basin; hot magma rises from the subduction zone; young mountain ranges, including a volcanic chain, build up parallel to the continental margin. *B.* Here a deep trench has formed between converging oceanic plates. Oceanic crust is subducted beneath an island arc, which is separated from the continental margin by a seaway. Sediments accumulate in the trench and the seaway. Volcanoes erupt along the island arc. The island arc may become a tectonostratigraphic terrane, jammed against the continental margin, with crushed and deformed oceanic crust and sediments in the seaway slivered between.

FIGURE PV.3   West side of the snow-capped Snake Range, Great Basin National Park, Nevada, and Spring Valley. The Snake Range is an upthrown, tilted fault block. Spring Valley is a basin formed by a down-dropped block. Photo by Ronald F. Hacker.

in crustal compression, they add to the height and bulk of coastal mountain ranges. While many of the terranes that have been welded to western North America are relatively young in terms of geologic time, terranes added to North America's eastern margin are older and are associated with earlier (Paleozoic and Precambrian) orogenies.

CHAPTER 44

Grand Teton National Park

NORTHWEST WYOMING

Area: 309,994 acres; 484 square miles
Established as a National Park: February 26, 1929
Address: P.O. Box 170, Moose, WY 83012-0170

Phone: 307-739-3300
E-mail: GRTE_Info@nps.gov
Website: www.nps.gov/grte/index.htm

Silvery gray, ice-sculptured peaks—distinctive landmarks for Indians, pioneers, and pilots—rise more than a mile above sagebrush flats. The Tetons, youngest mountains in the Rocky Mountain system, are made up of some of the continent's oldest rocks.

Tectonic forces are still raising the upthrown Teton Range and tilting the downdropped Jackson Hole block. A few modern glaciers cling to cirque basins. Rock-basin lakes and alpine meadows nestle among the peaks; moraine-dammed lakes lie along the fault trace at the foot of the range.

The Teton Range and Jackson Hole are part of the Greater Yellowstone ecosystem.

FIGURE 44.1 The Teton Range, the Snake River, and part of Jackson Hole in Grand Teton National Park. The Snake River, flowing south away from the viewer, has cut terraces in outwash deposits left by Pleistocene glaciers. Photo by Jan McGuire.

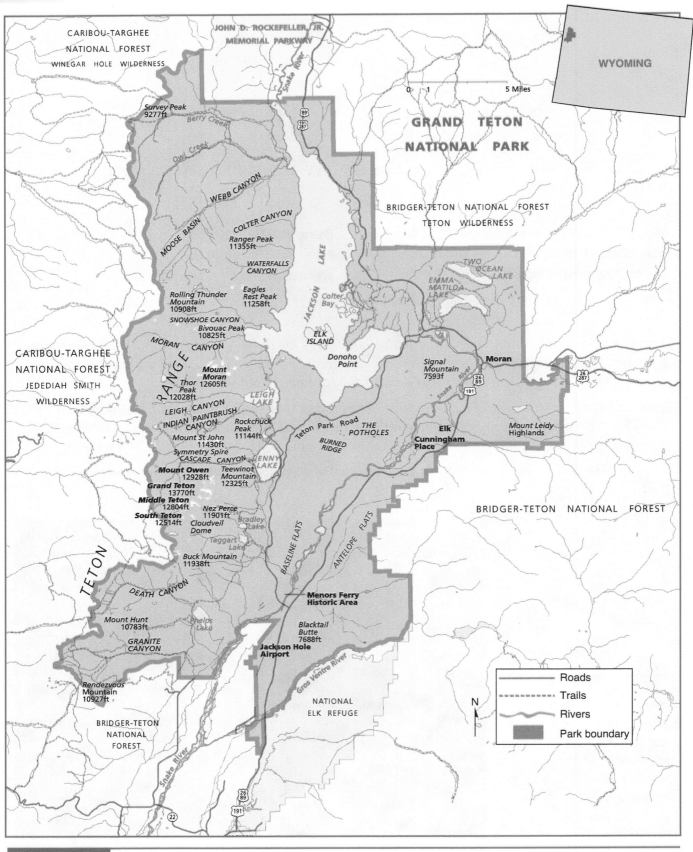

FIGURE 44.2 Grand Teton National Park, Wyoming.

Small bands of early hunters, drifting down from the Northern Rockies, camped beside Jackson Lake some 12,000 years ago. While following the seasons and the migrations of animals, hunters and gatherers looked for the distinctive skyline of the Teton Range, just as we today scan the horizon, waiting for those dramatic peaks to come into view, whatever our mode of travel. Grand Teton and its companion peaks, rising without foothills from the flats, are like a beacon. Once seen, their outline against the sky is never forgotten. Who has stood on the floor of Jackson Hole, face tilted upward, and regarded the formidable east face of the Teton Range without a sense of awe?

Several routes of travel and trade passed through Jackson Hole. The Indians could follow the Snake River north into Yellowstone for obsidian, or go south to the Colorado Plateau or east to the Great Plains. Teton Pass, at the southern end of the range, was a shortcut to the Pacific Northwest. Indian tribes knew the area well and most of the time regarded the valley as neutral ground.

To the trappers and traders of the eighteenth and early nineteenth centuries, a deep valley ringed by high mountains was a "hole," likely to be a safe place where they could "hole up." Jackson Hole, the favorite valley of trapper David Jackson, was named for him in 1829 by his fur trade partner. Early French voyageurs gave the name *Les Trois Tetons* to the three peaks now called Grand Teton, Middle Teton, and South Teton. A common translation used by the American mountain men was "The Three Paps." These easily distinguished peaks, which can be seen from both western and eastern approaches, were the landmarks that guided the trappers.

F.V. Hayden, the geologist who led the first government-sponsored, surveying and scientific expedition into Yellowstone in the summer of 1871, decided to also map the Teton valley that he had visited in 1860 with the Raynolds expedition. In the spring of 1872, Hayden split his group into two parties. He led one team into Yellowstone from the north. Geologist James Stevenson led the second party from Utah north to Fort Hall in Idaho and then east across Teton Pass into Jackson Hole. The two parties met late in the summer in Yellowstone. With Stevenson was N.P. Langford, who had been appointed superintendent of the newly established Yellowstone National Park; the photographer, William Henry Jackson; the artist-geologist, William Henry Holmes; several other geologists

and topographers, packers, and a cook; and a mountain man as guide.

Among the lasting accomplishments of that summer and fall of 1872 were the remarkable photographs of Teton country taken by William Henry Jackson; the superb drawings of William Henry Holmes, including his famous panorama showing the Teton Range; topographic maps; scientific reports; fossil and rock collections; and plant and animal specimens, including birds and insects. Two more expeditions in 1877 and 1878 completed the mapping of most of Wyoming, Idaho, and Montana, and furnished material for the voluminous Hayden Survey reports covering those years. By the 1880s, this part of the American West was no longer unknown territory to the civilized world.

When Jackson Hole was opened to homesteaders after the completion of the surveys, a few settlers established homes and ranches. However, the growing season was too short for most crops, and the valley was isolated for weeks each winter when snow blocked the passes. The population remained sparse well into the twentieth century, even after Teton County was organized in 1921.

Meanwhile, a movement was begun to protect the timber south of Yellowstone. Pierce Cunningham, a rancher whose cabin is now a historic site, circulated a petition asking that the valley be set aside "for the education and enjoyment of the Nation as a whole." In 1929, part of Teton National Forest was designated Grand Teton National Park. This act by Congress protected the Teton Range and some of the lakes, but most of the floor of Jackson Hole, including Jackson Lake, remained in private hands. Cattlemen strongly opposed proposals to incorporate grazing acreage into the national park.

Nevertheless, interested citizens throughout the country realized that historically, culturally, and geologically, the Teton Range and the valley floor below were two halves of a natural environment that needed to be protected and managed as an integrated unit. John D. Rockefeller, Jr., who believed in this concept, formed a land company in 1927 and began buying up land to be held until such time as it could be administered by the National Park Service. President Franklin Roosevelt took the next step in 1943 by proclaiming an area of 223,000 acres (much of which was already public land) as Jackson Hole National Monument. The Rockefeller land was accepted as an addition to the monument in 1949. Congress established the enlarged area as Grand Teton National Park in 1950, with part of the former Jackson Hole National Monument being absorbed by the National Elk Refuge south of the park. In recognition of Rockefeller's vision and generosity, the scenic highway that goes through Grand Teton National Park and into Yellowstone National Park (Chapter 43) was named John D. Rockefeller, Jr. Memorial Parkway.

Geologic Features

Photographs cannot capture the grand scale of the landforms that the human eye sees in Grand Teton National Park. When you drive along the highway on the floor of Jackson Hole, the massive, gray mountain front looms up almost within reach. Snow clings to the rough-hewn rock heights, and small glaciers hang in high, narrow, steep-walled valleys. In the canyons incising the slopes, waterfalls drop in silvery white lines, easily visible from many points along the highway.

Grand Teton National Park includes the major portion of two great landforms: (1) the Teton Range, an elongate, upfaulted block tilted to the west and about 45 miles long; and (2) Jackson Hole, a narrow, down-dropped fault block, about the same length and 6 to 12 miles wide. The valley's remarkably flat floor varies in elevation from 6000 to 7000 feet. At least seven of the Teton peaks exceed 12,000 feet in elevation; Grand Teton is 13,770 feet high. These tall peaks, concentrated in the 4-mile center of the range, make up the "Cathedral Group," between Avalanche and Cascade Canyons. The precipitous drop from peaks to valley floor might be even greater if not for intense erosion that has crenulated and dissected the mountain range and deposited thousands of feet of sediment that cover the bedrock of the valley floor (fig. 44.1).

Rocks and Structures

The sedimentary rocks that once covered the mountain block have been stripped from the top of the range, exposing Precambrian crystalline rock. Remnants of sedimentary rock flank the western slopes of the range. Similar units beneath Jackson Hole are buried by alluvial and glacial deposits. In geomorphic terms, the mountain features are erosional landforms and the valley features are depositional landforms.

The crystalline rocks. The Precambrian rocks that dominate the Teton Range consist of layered gneisses, granitic gneisses, granites, pegmatites, and diabase dikes. The banded nature of some of the gneisses suggests that they were sedimentary and volcanic rocks that accumulated in a subsiding basin or trench and were subsequently metamorphosed during orogenic activities. The massive gneisses were once granitic intrusions that were later metamorphosed. The gneisses are the oldest of the Precambrian rocks in the park. Strontium-rubidium dating places the age of the granites at about 2.5 billion years, and the black diabase dikes are 1.3 billion years old.

Extending vertically up the face of Mount Moran from base to summit is a black band that is the exposed edge of one of the diabase dikes that cut through the older rocks. The Mount Moran dike is about 150 feet thick and can be traced westward for about 7 miles. *Diabase* is a dense, black, igneous rock studded with light-colored plagioclase crystals (resembling small laths) incorporated with pyroxene crystals.

Some of the gneisses contain dark crystals of garnet set in white "halos," which look like eyes peering out from the rock face. Geologists call these rocks "bright-eyed gneiss." Pods of heavy dark green or black serpentine (soapstone) are scattered throughout the gneisses. Indians carved bowls from this smooth, relatively soft rock. Pebbles of serpentine—if found in stream gravels outside the park—can be polished and used as gems. This so-called "Teton jade" is softer and less lustrous than true jade.

Paleozoic sedimentary rocks. On the western flanks of the Teton Range and at the northern and southern ends of the range, gently dipping sequences of Paleozoic marine rocks top ridge crests and crop out on valley walls. The beds lie unconformably over the Precambrian rocks. Projections updip of the rock layers indicate that they formerly extended eastward clear across the Teton Range but were eroded as the mountain block was uplifted and tilted. Breaks in the rock record, shown by erosional unconformities, indicate many interruptions of deposition. Within the sequence are limestones, dolomites, shales, and sandstones. Careful study of the sedimentary units yields details of source areas, directions of sediment movement, positions of the Paleozoic seas, and evolving marine life forms. Drilling and subsurface geophysical exploration have made possible the identification of the same Paleozoic sequences deeply buried beneath Jackson Hole.

Mesozoic sedimentary rocks. Marine and nonmarine Mesozoic rocks that were deposited on top of the Paleozoic sequence were stripped off the Teton Range, exposing the Precambrian and Paleozoic rocks. Mesozoic rocks crop out in the northern, eastern, and southern parts of the Teton region, outside the park, and are also buried below the Jackson Hole sediments.

Cenozoic rocks and sediments. Tertiary rocks are exposed along the eastern margin of Jackson Hole and at the north end of the Teton Range, but most of the Tertiary units are buried beneath younger deposits that cover the floor of Jackson Hole. The Tertiary sequence of conglomerates, sandstones, shales, volcanic breccias, and tuffs have an aggregate total thickness of about 6 miles! No other region in the United States contains a more complete nonmarine Tertiary section than the Teton region. The rock units reflect the tectonic activity of the Laramide orogeny, Eocene volcanism that produced the Absaroka lavas and ashfalls, and the movement of the fault blocks.

Uplift of the mountains accelerated erosion, producing vast amounts of debris that spread over the downfaulted basin. A resumption of volcanic activity in Oligocene and Miocene time contributed additional material. Then in late Pliocene and Pleistocene time, the great volcanic events occurring in the Yellowstone area (Chapter 43) spread rhyolitic debris and tuff over the northern end of the Teton Range and over Jackson Hole.

All surface materials in Jackson Hole, with the exception of outliers of older rock in Signal Mountain and Blacktail Butte, are Pleistocene or Holocene in age. These young deposits, which consist of volcanics, lakebed sediment, glacial drift, and alluvial materials, blanketed the thick sequences of Paleozoic, Mesozoic,

FIGURE 44.3 Terrace along Snake River; riprap has been placed against the bank to slow down the erosion by the Snake River. Photo by Ann G. Harris.

and Tertiary rocks that fill the Jackson Hole basin. Scattered bones, pollens, and snail shells in the lakebed sediment document the presence of Cenozoic plants and animals that lived in the Teton region prior to the present.

Faults and Fault Blocks

Tensional forces produced the Teton fault system, split apart the two large crustal blocks, and tilted them toward the west, with the mountain block hinging up and the eastside block sliding down. Separating the huge blocks is the 40-mile-long trace of the fault, which runs along the base of the Teton Range, approximately where the slope breaks, between the mountain front and the western side of Jackson Hole. In structural terms, the Teton fault is a steeply dipping *normal fault*. By definition, the hanging wall of a normal fault has moved down in relation to the footwall. The Teton

fault dips to the east. Therefore, the hanging wall side, the Jackson Hole block, moved downward in relation to the Teton Range block, which moved up (fig. 44.5).

How did geologists know the fault was there? How did they determine when the movement occurred? How much did the blocks move and in what direction? What do geologists know about the rate of movement, and what do they think caused this displacement of crustal units? Most of the geologic evidence was found in the rocks that were briefly described previously.

1. Geologists suspect faulting when rocks of quite different types and ages are next to each other and the stratigraphic and geomorphic conditions make an unconformity unlikely. Beneath the floor of Jackson Hole, the Precambrian crystalline rocks of the Teton Range butt up against much younger rocks. Although the fault contact is obscured by glacial and stream deposits, geologists can locate

FIGURE 44.4   Sedimentary rocks still cap the western part of the Teton fault block. These Lower Paleozoic rocks sit uncomfortably on top of the Archean gneisses and granites that make up the core of the Teton Range. Photo by Douglas Fowler.

the fault system by drilling and seismic prospecting, and by projecting data collected from isolated outcrops.

2. The absence of foothills is significant. This indicates that the mountains were not the result of erosion.

3. The relatively straight and smooth eastern front of the Teton Range is suggestive of faulting. Not many geologic processes other than faulting produce linear features. Because the mountain front has been eroded and cut by canyons, it does not represent the original fault scarp, but does suggest the approximate slope and location of the scarp. A few small, triangular facets, remnants of the original fault scarp, are spaced along the base of the range.

4. The most likely cause of the asymmetry of the range is faulting and tilting.

5. Small, fresh fault scarps in the vicinity of the Teton fault system are regarded as evidence of postglacial faulting and tilting.

With the presence of the Teton fault inductively confirmed, the next step is to determine which side moved up and how much movement occurred by matching rock types (stratigraphic correlation) across the fault. In matching one side with the other, the best marker is the contact between the Paleozoic rocks and the underlying Precambrian rocks. On the Teton block, we project the westward-dipping Precambrian-Paleozoic contact toward the east until it intersects the steeply dipping plane of the fault. On the Jackson Hole block, the Paleozoic-Precambrian contact can be located far below the surface by deep drilling and by geophysical techniques. The matching of contacts shows that the western block appears to have moved up relative to the eastern block a distance of roughly 30,000 feet (fig. 44.5).

The shape of the Teton block suggests that the Teton fault dips steeply eastward. Data from geophysical studies support this conclusion. Therefore, the Jackson Hole block (the hanging wall) moved downward in relation to the Teton block (the footwall), making the Teton fault a normal fault.

According to the principle of cross-cutting relationships, a fault is younger than the youngest rock that it cuts and older than the oldest rocks that are *not* displaced or cut by fault movement. Applying this rule to the Teton fault, we find that it truncated the youngest sedimentary rocks in the basin, that it began about 9

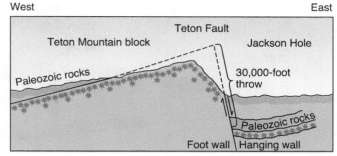

FIGURE 44.5 Simplified cross section of the Teton fault system, showing displacement of Paleozoic beds, the amount of "throw," and a projection of volume of rock removed by erosion. (Buck Mountain fault not shown.)

million years ago, and that it is still active. Most of the movement, which was spasmodic rather than continuous, occurred late in Tertiary time before Pleistocene glaciation. The average rate of movement (derived by dividing the amount of "throw" by the number of years) was about one foot for every 300 years. The continued tilting of the Jackson Hole block affects stream drainage and causes slopes to be susceptible to landsliding, sometimes triggered by earthquakes.

Several other faults have significantly affected the Teton landscape. The highest mountains in the central part of the Teton Range were elevated by a reverse fault that is older than the Teton fault. The Buck Mountain fault, a reverse fault of moderate displacement, can be traced for a distance of about 10 miles in a north-south direction on the western side of the Teton summit area (figs. 44.6, 44.7). (A *reverse fault* is one where the hanging wall moves up in relation to the footwall.) West of the Buck Mountain fault trace, the Paleozoic sedimentary rocks are bent sharply upward, and on the narrow, uplifted block on the east side of the fault are the dominant peaks of the range—the Three Tetons and several others—in a fairly straight alignment. A logical assumption is that the Buck Mountain reverse fault was a result of orogenic activity that preceded the normal Teton fault of late Tertiary time. Thus the crest area that had been raised by reverse faulting was later elevated by normal faulting to its present height.

On the east side of the Jackson Hole block is a nearly vertical normal fault with the west side upthrown and the east side dropped down. Other faults

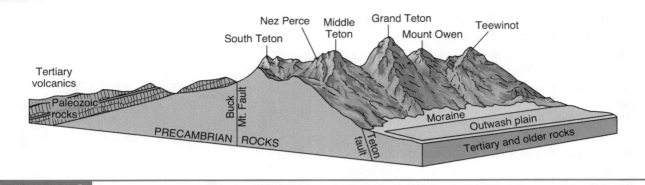

FIGURE 44.6 Block diagram of the crest of the Teton Range, bounded on the east by the Teton fault and by the Buck Mountain fault on the west. Outwash and moraines obscure the base of the fault scarp. On the western slopes are remnants of Paleozoic rocks that once lay unconformably over the Precambrian crystalline rocks. After Horberg 1938; reproduced by permission of Augustana Library Publications.

have been mapped at both the northern and southern ends of the valley.

The ultimate cause of the faulting in the Teton area is related to the geologic history of the surrounding region. During the Laramide orogeny, which began at the end of Cretaceous time, whole regions of western North America were folded, uplifted, and subjected to reverse faulting and thrust faulting on a grand scale. After compressive forces subsided, crustal extension and uplift produced many fault-block ranges and basins.

How Surface Processes Modified the Teton Landscape

Faulting has been the most significant geologic process at work in developing the landscape in Grand Teton National Park. But as the structurally controlled mountain framework rose, moisture-bearing clouds were forced to release rain in order to surmount the heights. Greater precipitation at the high elevations fed snow-fields and increased the volume and velocity of streams. Running water and glacial ice, with the help of weathering and mass wasting processes, set about sharpening peaks, excavating canyons, and sculpturing the range. Sediments soon covered the floor of Jackson Hole.

The Teton Range is a world-famous example of *alpine topography*. The Teton skyline, as seen from below, is a series of concave curves passing from peak to peak with lower passes in between. The serrated ridge, conical horns, and great cirques are evidence of glacial erosion. The higher peaks in the center of the range are the focus of this grand scenic display.

Despite the fact that the Precambrian rocks are extremely hard and resistant to erosion, the high peaks, which remained above the ice as glaciers grew and shrank, were badly shattered and cracked by frost action on all exposed surfaces. Below the rough, jagged rock of the peaks are smooth ledges that have been scoured and scraped by ice. Shallow, ice-eroded depressions in the bedrock are occupied by small lakes. Cirques and rock basins in the glacial troughs also contain lakes.

The large cirques and groups of cirques that head the major canyons are like amphitheaters with sheer walls hundreds of feet high. The canyons, or glacial troughs, dissect the eastern face of the range. Excavated several thousand feet deep by mass-wasting and ice, they have broad floors and sheer, oversteepened walls, forming the U-shape typical of glacial valleys. Tributary glaciers scoured hanging valleys, now occupied by waterfalls. Some canyons have glacial staircases with falls and cascades rushing down the slopes.

Several times during the Pleistocene, valley glaciers became so large that they coalesced and spilled out over the floor of Jackson Hole. Here they joined great tongues of ice from the north and east that filled the basin. Moraines built at the mouths of the canyons mark the extent of the last Pleistocene glaciation. These moraines now hold a series of lakes (Leigh, String, Jenny, Bradley, Taggart, and Phelps) close against the foot of the mountains. Several of the lakes are surprisingly deep, such as Jenny Lake (226 feet) and Leigh Lake (250 feet). Meadows in the moraines have replaced former, more shallow lakes that have been filled in or have drained away. Large glacial boulders,

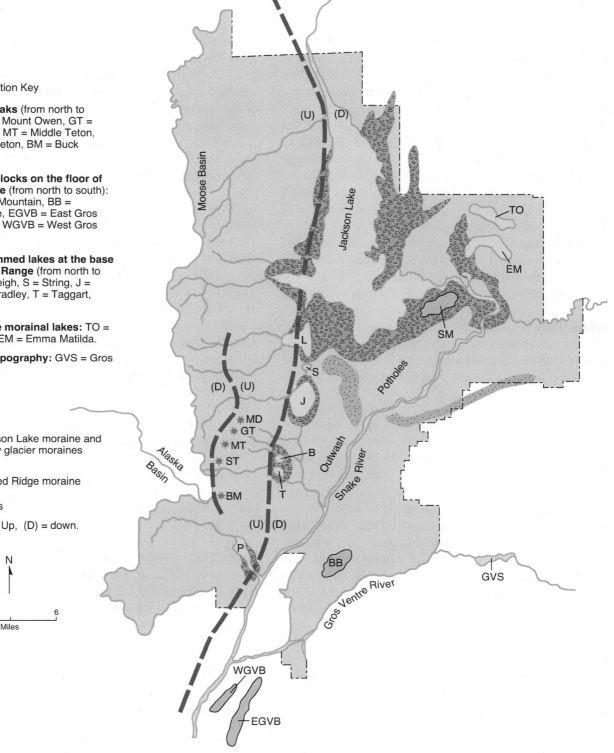

Location Key

Mountain peaks (from north to south): MO = Mount Owen, GT = Grand Teton, MT = Middle Teton, ST = South Teton, BM = Buck Mountain.

Small fault blocks on the floor of Jackson Hole (from north to south): SM = Signal Mountain, BB = Bucktail Butte, EGVB = East Gros Ventre Butte, WGVB = West Gros Ventre Butte.

Moraine-dammed lakes at the base of the Teton Range (from north to south): L = Leigh, S = String, J = Jenny, B = Bradley, T = Taggart, P = Phelps.

Pre-Pinedale morainal lakes: TO = Two Ocean, EM = Emma Matilda.

Landslide topography: GVS = Gros Ventre Slide.

Jackson Lake moraine and valley glacier moraines

Burned Ridge moraine

Faults

(U) = Up, (D) = down.

N

0   6
Miles

FIGURE 44.7   Location map showing major faults and geologic features. The moraines shown are Pinedale in age. The Jackson Lake recessional moraine holds in Jackson Lake. Burned Ridge is a terminal moraine. The Potholes are in glacial outwash that extends to the south over Jackson Hole.

carried down by the ice, dot the moraines and rim the lakes.

Jackson Lake, the largest body of water in the valley, occupies the trough of the last major glacier from the north that combined with ice from the northern canyons and covered part of Jackson Hole. A large recessional moraine that was built across the valley floor holds in Jackson Lake. South of this moraine is a pitted outwash plain, called the Potholes. The small depressions, some filled with water, are *kettleholes* that were left when stagnant, debris-covered ice blocks melted. Farther down the valley, the Burned Ridge moraine marks the terminus of the last glacial advance. Beyond this moraine, a thick outwash plain, built of glacial gravel, extends for some 10 miles down the valley, its surface sloping gently southward (fig. 44.7).

During times of glacial advance, more debris is put into the meltwater streams than they can carry, and they rapidly aggrade their beds by building up thick accumulations of sand and gravel. When glaciers retreat, streams are swollen with meltwater but carry less debris. They begin cutting trenches into the accumulated outwash, thereby deepening their channels and producing terraces. Terrace tops are built up during glacial advances; terrace faces are cut during times of glacial retreat.

Taming the Snake River. In Jackson Hole, the Snake River has cut several levels of prominent terraces (fig. 44.1) that can be related to moraines of various ages. The amount of work the Snake has done resorting and redepositing unconsolidated material is impressive. With its rapid changes in volume, high velocity, and shifting channels, the Snake deserved the name "Mad River" given it by the mountain men. Even today, despite the damming of the Jackson Lake outlet and the building of levees, the Snake is occasionally difficult to control. Most of the time, however, the Snake River twists around gravelly bars and quietly reworks loose material in its braided channel.

From its source in the Teton Wilderness, the Snake River flows into Jackson Lake, goes out the southeast end, and then runs on south through Jackson Hole, collecting runoff from the front slopes of the Teton Range and from the highlands on the east side of the valley. The tributaries deliver sediment to the valley floor, and much of it eventually reaches the Snake. The streams from the less steep mountains on the east tend to build alluvial fans where they enter the valley and then flow directly west, some of them joining the Snake almost at right angles. By contrast, the tributaries on the west side of the valley have much steeper gradients but do not develop extensive alluvial fans. Instead, the streams leave the canyons, flow east a short distance, and then bend south, flowing along the mountain front and joining the Snake at the south end of Jackson Hole. Several of the streams flow into interconnected morainal lakes at the base of the Teton Range on their way to the Snake River. This "lopsided drainage," which is due to the tilting of the Jackson Hole block, is another indication of tectonic control of topography.

Mass wasting. Great quantities of broken rock and sediment were carried down the mountain sides by glaciers, meltwater, and landslides during Pleistocene time, and the dumping of rock debris in Jackson Hole did not stop with the retreat of the large-scale glaciers. The work of water and ice goes on, aided by the force of gravity, although only small glaciers in shaded locations now occupy cirques and the upper parts of glacial troughs. Talus accumulates at the bottom of cliffs and slopes, but rests there only temporarily. The journey of rock debris downslope can be fast or slow or by fits and starts. Some talus, mixed with water, snow, and ice, flows down slowly in rock glaciers. Active rock glaciers can be seen in Granite Canyon (southern part of the park) and in a north-trending valley near Eagles Rest (a peak in the northern part of the range). Some talus moves by the very slow process of creep. Some is brought down rapidly by snow avalanches, especially in late winter when avalanches are frequent.

Vegetation patterns. In Jackson Hole, vegetation patterns reflect in an interesting way the general locations of glacial and alluvial deposits. The evergreen forests that grow on the mountain slopes below treeline extend out onto the valley floor, but only along the moraines (fig. 44.8). On the outwash plains and where outwash deposits have buried moraines, only sagebrush, coarse grasses, and a few other hardy plants can survive. Lodgepole pines grow well on the moraines because the unsorted glacial debris is more compact and retains moisture better. Also, the soil has a higher clay content and is richer in minerals that plants need for growth. The coarse, quartzite gravels of the out-

FIGURE 44.8   Mount Moran and the outwash-covered floor of Jackson Hole. A prominent diabase dike on Mount Moran extends from the summit to Falling Ice Glacier. Tree-covered moraine at the base of the range is Burned Ridge, terminus of the last major advance of Pinedale glaciation. Photo by Douglas Fowler.

wash plains, on the other hand, are dry and infertile. Moisture rapidly drains away to the nearest streambed. Along streambeds, below the parched and barren outwash plains, aspens, cottonwoods, and willows thrive on the available moisture and nutrients. Some irregular patches and lenses of loess, blown into the valley between glacial episodes, provide the only good soil in Jackson Hole (except for small amounts of silt). Badgers and coyotes dig their burrows where loess crops out on banks because loess is easy digging and does not collapse.

The Gros Ventre landslide. In June 1925, after an unusually wet spring season, a large rockslide moved several cubic miles of rock and debris down Sheep Mountain and across the Gros Ventre River valley in about a minute's time. This occurred about three miles upstream from the southeastern boundary of Grand Teton National Park. The slide dammed the river, backing up a lake 200 feet deep and five miles long. What caused the slide was a combination of saturation of sedimentary rock layers on Sheep Mountain and the undercutting of the slope by the Gros Ventre River that left the base of the slope unsupported. Triggered by minor earth tremors, a rain-soaked slab of Tensleep Sandstone (of Pennsylvanian age) broke loose and slid downdip over soft shale underneath. Two years later the debris dam blocking the river gave way due to the pressure of high water from melting snow. The resulting flood washed out the town of Kelly five miles downstream, and six people drowned. Today, many years later, the jumbled landslide topography in the valley and a landslide scar 200 feet above the river on Sheep Mountain are still clearly visible.

Geologic History

1. Accumulation of sedimentary and volcanic rocks in a Precambrian trough.

The oldest rocks were deposited about 2.8 billion years ago as sandstones, limestones, and shales in a marine environment. Volcanic rocks, probably from an island arc, were interbedded with sedimentary rocks.

2. Precambrian orogenies and intrusions.

The sedimentary and volcanic rocks were intensely folded and metamorphosed. The gneisses that resulted were intruded by granites that were also metamorphosed. Subsequently, lighter-colored granites and pegmatite dikes were emplaced.

3. Intrusion of diabase dikes late in Precambrian time.

About 1.3 billion years ago, black diabase dikes, ranging from 5 to 200 feet thick, cut through the older Precambrian rocks. Long erosion ensued.

4. Deposition of Paleozoic sedimentary rocks.

About nine major formations, representing all periods of the Paleozoic Era except the Silurian, were deposited in Paleozoic oceans that covered the Teton region. The sedimentary units have a composite thickness of about 4000 feet and include sandstones, shales, limestones, and dolomites. Deposits are discontinuous, but no significant deformation occurred. The best fossil record is in the Paleozoic rocks of the Alaska Basin, just outside the park's western boundary. The limestones and dolomites contain brachiopods, trilobites, bryozoans, and corals that once grew in Paleozoic seas. Remnants of the Paleozoic rocks are exposed in the park, but more complete sequences are found outside park boundaries on the north, south, and west (fig. 44.4).

5. Mesozoic deposition.

Although Mesozoic rocks lie conformably over the Paleozoic rock units, the depositional environments of the Mesozoic Era were more varied, shifting from marine to transitional to continental, and alternating back and forth as crustal conditions changed in the Teton region. In general, sedimentation became more rapid throughout the era, especially toward the close. More than 15 formations with a composite thickness of between 10,000 and 15,000 feet accumulated during Mesozoic time. The Cretaceous units, which are the most numerous and have the greatest thickness, record the last transition from marine to continental rocks. Much of the sediment came from the mountains rising to the west that preceded the mountain-building that was soon to take place in the Teton region. Beds of bentonite between marine sands and shales represent ashfalls from western volcanoes that drifted down over a shallow sea. Landslides caused by the swelling of wet bentonite clay sometimes block access roads in the northern part of the park.

Coal accumulated in swamps as the last Cretaceous sea withdrew. Coal beds crop out in abandoned mine workings near the park's eastern margin. A broad, general uplift and the beginning of folding brought the Mesozoic to a close.

6. Laramide orogeny, beginning late in the Cretaceous Period.

The compressive forces of the Laramide orogeny initiated crustal buckling along a northwest-southeast trend and broad uplift throughout the region. From the Targhee uplift, northwest of the Teton region, rapidly eroding rivers brought large quantities of sand, gravel, and quartzite cobbles, which were spread south and east over the Teton region. These sediments made up the 5000 feet of conglomerates and sandstones of the Harebell Formation (northern and northeastern parts of the park).

Mountains in the region were raised higher as the orogeny intensified with more folding and faulting. The climax came early in Eocene time as great thrust faults and reverse faults produced mountain blocks and basins. The Buck Mountain fault probably formed at this time.

7. Tertiary nonmarine sedimentation.

In Jackson Hole and the immediate vicinity, vast quantities of clastic sediment that came from uplifted adjacent areas spread over the land. The Pinyon Conglomerate, which overlies the Harebell Formation, is of Paleocene age. On top of coal beds and claystone at the base of the Pinyon are thick zones of well rounded quartzite pebbles derived from the Targhee uplift, like those in the Harebell Formation. The quartzite fragments increase in size toward the northwest and many of them show percussion scars from being pounded during transport by swift mountain streams with steep gradients (Love and Reed 1971). The nature

Table 44.1	Generalized Geologic Column, Grand Teton National Park

Time Units

Era	Period	Epoch	Rock Units	Geologic Events
Cenozoic	Quaternary	Holocene	Alluvium, talus, landslide deposits; minor glacial deposits	Weathering, erosion, mass wasting, Neoglaciation Continued faulting and seismic activity
		Pleistocene	Moraine and outwash deposits Loess deposits (interglacial) Outwash deposits	Pinedale glaciation (morainal lakes, outwash plain, terrace cutting, etc.) Bull Lake glaciation Buffalo glaciation, extensive glacial erosion and deposition Preglacial lakes in Jackson Hole Volcanic eruptions in southern Jackson Hole
	?—?	?	Bivouac Formation	Welded tuff eruptions from Yellowstone area
	Tertiary / Neogene	Pliocene	Teewinot Formation Conglomerates, tuff	Teewinot Lake on subsiding Jackson Hole block; lakebed deposits, ashfalls Development of Teton fault system; uplift of Teton Range Crustal extension
	Tertiary / Paleogene	Miocene / Oligocene	Volcanic conglomerates, tuff, sandstones, etc.	Volcanic eruptions in Jackson Hole; basin filling
		Eocene	Volcanic tuffs, breccias, conglomerates, sandstones, claystone, shale, coalbeds	Extensive volcanism in the Absarokas
		Paleocene	Pinyon Conglomerate	Buck Mountain fault, uplift Laramide orogeny Sediment accumulation
Mesozoic	Cretaceous		Harebell Formation	Extensive nonmarine sedimentation
	Jurassic		About 15 nonmarine and marine formations in Jackson Hole region (for details, see p. 80, Love and Reed 1971)	Upwarps and subsidence, intermittent shallow sea; marine, transitional, and nonmarine sedimentation
	Triassic			
Paleozoic			About 9 major formations, limestones, dolomites, shales, sandstones (for details, see p. 68, Love and Reed 1971)	Intermittent deposition in shallow seas between episodes of upwarp and erosion
Precambrian time			Diabase dikes Light-colored granites, pegmatites Dark granites Gneisses	Uplift and erosion Intrusions Orogenies Metamorphism Accumulation of marine and volcanic sediment in crustal trough

Source: Modified after Love and Reed, 1971; Behrendt et al., 1968.

of the upwarp and subsidence of that time can be inferred from the rock record, which demonstrates how tectonic activities controlled sedimentation. Uplift was rapid in the mountain areas that supplied the coarse sediment. At the same time, Jackson Hole was subsiding more and more, thus providing a site for thick accumulations of eroded sediment.

8. Volcanic activity in Tertiary time.

As the intensity of Laramide mountain-building diminished in Eocene time, volcanoes began erupting in the Yellowstone-Absaroka area, producing the tremendous volume of volcanic rocks that make up the Absaroka Volcanic Supergroup. Tuff from these eruptions accumulated in Jackson Hole. In Oligocene and Miocene time, volcanoes on the eastern margin of Jackson Hole produced additional volcanic debris that filled basins and buried uplands.

9. The Pliocene Teewinot Lake in Jackson Hole.

Movement on a fault that trended east-west across the floor of Jackson Hole, near the present south boundary of the park, impounded the first large freshwater lake in the valley. The water in the Teewinot Lake, dammed by the fault scarp, backed up toward the north. For about 5 million years, sediment (including tuff from nearby volcanoes) accumulated in the lake, which persisted because the rate of subsidence kept up with the rate of sedimentation. The presence of fossil snails, clams, and small mammal bones in the sediment indicates that the lake remained shallow.

10. Pliocene and Pleistocene volcanics.

Rhyolitic lava, spewing from volcanoes in Yellowstone, flowed down into the northern end of Teewinot Lake and solidified as obsidian, which has been dated by the potassium-argon method as 9 million years old. (Indians found the obsidian many years later and used it for cutting tools, spears, and arrowpoints.) After the Teewinot Lake dried up, glowing avalanches poured down from Yellowstone, spreading welded tuff over the old lake bed. The Bivouac Formation, exposed on the flanks of Signal Mountain, was the result of one of these catastrophic eruptions. The welded tuff juts out

 FIGURE 44.9   The Teton Range, as seen from Schwabher landing. Photo by Jan McGuire.

as a ledge on the mountainside with conglomerates above and below. A series of volcanic eruptions also occurred in Pleistocene time, near the southern boundary of the park. Remnants of the volcanic rocks cap East and West Gros Ventre Buttes.

11. Teton block-faulting over the last 9 million years.

The long-continued upfaulting that elevated the Teton Range and tilted it westward initiated the erosion of the Mesozoic and Paleozoic rocks that once covered the summits. The range still stands high, however, because of ongoing faulting and the resistance of the Precambrian rocks. In Jackson Hole, sedimentation keeps up, more or less, with subsidence. Local faulting, upwarping, and downwarping caused the impounding of two successive preglacial lakes in the south and southwestern part of the valley. East and West Gros Ventre Buttes, as well as Blacktail Butte and Signal Mountain, are small fault blocks that stick up above the flat surface of Jackson Hole.

12. Pleistocene glaciation.

Cenozoic climates, which were initially subtropical and humid, gradually cooled. By Pliocene time, the Teton region had about the same climate as today. Pollen preserved in the Teewinot lakebed sediment came from plants and trees such as fir, spruce, pine, and sage, not unlike those that now grow in the park. With the onset of much colder Pleistocene climates, glacial ice accumulated in the Teton Range, in the Absaroka Mountains to the north, and in the Wind River Range and the Gros Ventre Mountains to the east. At least three times during the Pleistocene, glaciers from these highlands flowed down over Jackson Hole.

During Buffalo glaciation, the first well documented invasion of ice, glaciers from the north converged and flowed south along the eastern front of the Teton Range. Glaciers from the east flowed down the Gros Ventre River valley and joined the main ice stream near the Gros Ventre Campground. Signal Mountain and the other three buttes in southern Jackson Hole were overridden and abraded by a 2000-foot-thick mass of ice. At its maximum extent, the ice flowed down into eastern Idaho. Neither of the two later glaciations reached that far. After the ice melted, Jackson Hole was a wasteland. All accumulated soil had been scraped or washed away, and the floor of the valley was covered with quartzite boulders and cobbles. The parts of the valley not affected by the later glaciations are still in this barren state;

that is, lacking in soil and too stony for all but the hardiest plant varieties to grow.

Bull Lake, the second glaciation, was less extensive. Some moraines and outwash deposits of Bull Lake glaciation were covered by wind-deposited loess after the glaciers melted back and before vegetation was reestablished. Later glaciation scraped off or covered much of the loess.

The glacial features associated with Pinedale glaciation, especially those in Jackson Hole, are familiar elements of the park landscape because this was the last major glacial episode. Beginning about 25,000 years ago and lasting until about 10,000 years ago, Pinedale glaciers left behind most of the lake basins, moraines, and gravelly outwash plains that we see today, although Pinedale glaciation had less volume than its predecessors. The terminal moraine of Pinedale ice is the Burned Ridge moraine, a series of low, hummocky, tree-covered ridges that slant southward across the middle part of Jackson Hole. The Potholes area, described earlier, is between Burned Ridge moraine and the recessional moraine that dams Jackson Lake. The string of morainal lakes at the foot of the range are the result of Pinedale valley glaciers. The basins of Two-Ocean Lake and Emma Matilda Lake, however, were left by Bull Lake glaciation.

13. Holocene and historic time.

Pinedale glaciers probably melted completely at the end of the Pleistocene, but small Neoglaciers formed in the Teton Range during episodes of colder climate. These glaciers have not done much eroding, but they are picturesque and are regarded with caution and respect by mountain-climbers.

By building a dam at the south end of Jackson Lake, engineers have continued the work begun by Pinedale glaciers. The purpose of the dam is to prevent excessive raising and lowering of the lake level and to keep the Snake River from going on a rampage every spring. Sometimes in dry seasons, the needs of the Idaho ranchers downstream for more water from the Snake conflicts with the needs of people in the park who want the lake level kept high for recreation purposes.

South of Jackson Lake the Snake River keeps trying to cut a channel more to the west because of continued tilting of the Jackson Hole block. Engineers keep reinforcing levees in order to force the river to remain where it is.

Although severe earthquakes have not occurred in the valley since it was settled, minor tremors occur from time to time. Geologists who monitor the Teton fault system think that a major shock is possible, and perhaps probable. Meanwhile, rockslides, earthflows, and avalanches interfere from time to time with human activities.

A few years ago, divers in Jenny Lake discovered about a dozen trees—one of them a 70-foot tall Engelman spruce—standing upright in water, between 35 and 80 feet deep, on the lake bottom. Carbon-14 dates derived from samples of the wood revealed an age of 600 ± 100 years. J.D. Love (1987) and other scientists have speculated as to the significance of this phenomenon. Are some or none of the trees rooted? Did they slide into the lake while standing, or did they assume an upright position later? If they are rooted, then the lake bottom has dropped at least 80 feet in the last 600 years, possibly during an earthquake. If none of the trees are rooted, where did they slide from? Nearby canyons do not appear to be locations from which a large enough landslide or snow avalanche could have descended; that is, one that could have carried the trees down into the lake. In the Grand Teton region, as we have seen, continual change is to be expected.

Geologic Map and Cross Section

Love, J.D.; Reed, J.C., Jr.; Christiansen, R.L.; and Stacy, J.R. 1972. Geologic Block Diagram and Tectonic History of the Teton Region, Wyoming and Idaho. U.S. Geological Survey Miscellaneous Investigations, Map I-730.

Love, J.D.; Reed, J.C., Jr.; Christiansen, R.L.; and Stacy, J.R. 1992. Geologic Map of Grand Teton National Park, Teton County, Wyoming, U.S. Geological Survey Miscellaneous Investigations, Map. I-2031.

Sources

Behrendt, J.C.; Tibbetts, B.L.; Bonini, W.E.; Laven, P.M.; Love, J.D.; and Reed, J.C. 1968. A Geophysical Study in Grand Teton National Park and Vicinity, Teton County, Wyoming. U.S. Geological Survey Professional Paper 516E.

Fryxell, F. 1938. *The Tetons, Interpretations of a Mountain Landscape*. Reprinted by Grand Teton Natural History Association.

Horberg, L. 1938. *The Structural Geology and Physiography of the Teton Pass Area, Wyoming*. Publication No. 16. Rock Island, Illinois: Augustana Library Publications.

Love, J.D.; and Reed, J.C., Jr. 1971. (reprint 1989). *Creation of the Teton Landscape* (with geologic map). Grand Teton Natural History Association.

———. and Love, J.M. 1983. Road Log, Jackson to Dinwoody and Return. Public Information Circular 20, Geological Survey of Wyoming. 32 p.

Love, J.D. 1987. Teton Mountain Front, Wyoming. Geological Society of America Centennial Field Guide, Rocky Mountain section, v.2, p. 173–178.

National Park Service. 1984. *Grand Teton, a Guide to Grand Teton National Park, Wyoming*. Handbook 122. 96 p.

Parsons, W.H. 1978. *Middle Rockies and Yellowstone*, K/H Geology Field Guide Series. Dubuque, Iowa: Kendall/Hunt Publishing Company.

Note: To convert English measurements to metric, go to www.helpwithdiy.com/metric_conversion_calculator.html

Great Basin National Park

EASTERN NEVADA

Area: 77,100 acres; 120 square miles **Proclaimed a National Monument:** 　January 24, 1922 **Enlarged and Established as a National Park:** 　October 27, 1986	**Address:** 100 Great Basin National Park, Baker, NV 　89311 **Phone:** 775-234-7331 **E-mail:** grba_interpretation@nps.gov **Website:** www.nps.gov/grba/index.htm

High above the desert floor are snowy peaks, cirques, alpine meadows, and cirque lakes. Five ecologic zones exist between the crest of the range and the desert floor.

In Lehman Caves, decorated galleries and passages extend deep into the mountain flank within a wedge of Cambrian limestone.

A 75-foot-high natural arch stands like a sentinel in a creek valley.

The Snake Range is an example of a metamorphic core complex.

FIGURE 45.1　A small glacier clings to Wheeler Peak (elevation 13,063 feet) high above the desert floor. The bedrock is resistant quartzite of Cambrian age. Photo by Sherry Schmidt.

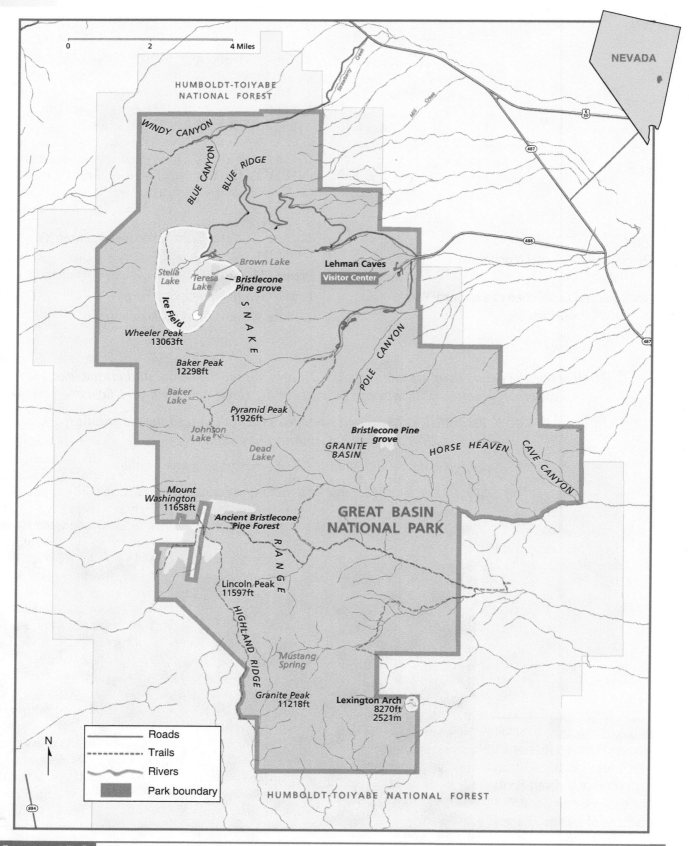

FIGURE 45.2 Great Basin National Park, Nevada.

Geographic Setting

Great Basin National Park lies on the Southern Snake Range in eastern Nevada. This is a sparsely populated area near the Utah border, just south of U.S. Highways 6 and 50. The small village of Baker is the closest community.

As a descriptive term, "Great Basin" applies to a large area encompassing most of Nevada, a sizable portion of western Utah, and small slices of Wyoming, Idaho, Oregon, and California. *Drainage divides* (the high areas separating drainage basins) delineate the boundaries of this geographical region characterized by interior drainage. No streams from the Great Basin flow to the ocean. All surface runoff and streams flow into saline lakes and marshes or gradually sink into sand and gravel in the beds of disappearing streams. The Humboldt River and the Carson Sink in Nevada are examples of interior drainage.

The physiographic province called Basin and Range includes most of the Great Basin but is delineated by geologic structure rather than by drainage divides. Block-faulted mountains and sediment-filled basins characterize the province, which extends farther south than the Great Basin, to include more of southern California, the southern half of Arizona and New Mexico, the western corner of Texas, and adjoining areas of Mexico.

Great Basin National Park, its boundaries well within both these overlapping regions, displays the notable characteristics of both areas as well as some distinctive and unusual features of its own. Recent past and present climatic conditions in the park region have been responsible for many of the landscape features.

The semiarid climate of the Great Basin is under the dominance of a rain shadow created by the Sierra Nevada, some 350 miles to the west in California. Moisture-laden air masses coming in from the Pacific Ocean are forced to rise into much cooler air in order to pass over the high ridges of the Sierra. Cooling reduces the ability of the clouds to hold moisture, causing rain and snow to fall on the windward side of the ranges. Descending on the lee side, the air warms and picks up moisture. This deprives the land below of rain, and surface moisture quickly evaporates. The water table is discontinuous and usually far below the surface. Local precipitation is strongly influenced by topography. Whereas 4 inches of rain per year may fall in the lower basins, more than 30 inches may drop annually on the higher mountain summits, and much of it is in the form of snow in winter. In summer, rain falls as infrequent but torrential showers of short duration. These conditions have prevailed in historic time, but during the pluvial episodes of the Pleistocene Epoch, (not long ago in terms of geologic time), the climate was wetter.

During Pleistocene time, glaciers accumulated on Wheeler Peak and the other high summits of the Snake Range. Late Pleistocene snowlines are estimated to have been at about the 11,800-foot level. *Pluvial lakes* (lakes formed during cooler, wetter glacial climates) occupied Spring Valley and Snake Valley, the basins on either side of the Snake Range. During some of that time, the lake in Snake Valley was the westernmost arm of Lake Bonneville, a large pluvial lake that spread over eastern Nevada, Utah, and southern Idaho. (Today, Great Salt Lake represents what's left of Lake Bonneville.) Dry saltpans on the flats of Spring Valley and Snake Valley remind us of the pluvial lakes that sparkled there before the waters dried up.

In Great Basin National Park, between the desert floor and the high peaks of the Snake Range, displayed in about 8000 feet of relief, are five ecosystems existing in a desert region in temperate latitudes. Controlled at their respective levels by temperature and moisture, the five zones are, from bottom to top: (1) sagebrush and pinyon/juniper of the desert; (2) manzanita and mountain mahogany; (3) aspen and Douglas fir; (4) Engelmann spruce and pine; and (5) hardy alpine plants above the treeline. In the highest zone, at the edge of the treeline, are the centuries-old, gnarled and twisted bristlecone pines surviving in an ancient forest, now under strict protection (fig. 45.3). On Wheeler Peak, enough precipitation falls to support two small lakes, an alpine meadow, small snowfields and a glacier.

At 13,063 feet, Wheeler Peak is the second highest mountain in Nevada. It was named for Lieutenant George Montague Wheeler, a topographer in the U.S. Army Corps of Engineers, who led expeditions to the Western Territories from about 1869 to 1879. G.K. Gilbert, one of the giants of American geology, was a member of Wheeler's party and wrote the first geologic report on the Great Basin region.

Some of the early routes to California crossed this part of Nevada. The journey was difficult because of the ups and downs of the ranges and basins as well as the scarcity of water and grasslands for sustaining draft animals and stock. The Pony Express trail, the Overland Telegraph, and the Overland Stage route all

passed along the northern edge of the Snake Range. This was a direct way for those who could travel fast.

During the latter part of the nineteenth century, mining camps and widely scattered ranches were started in the region, with boom-and-bust activities in mining continuing to the present time. An enterprising miner, Absalom S. Lehman, who came to the area in 1867, failed to strike it rich on his mining claims but was alert to the heavy demand for food in the mining camps. He cleared a site on the lower slopes of the east face of Wheeler Peak and began to farm. A stream coming down the mountain and a spring supplied water for crops, stock, and an orchard. Soon he was supplying meat, dairy products, eggs, and garden produce to the mining camps, and the venture became ever more successful as time went on.

In 1885 Lehman discovered an opening in the rocks above his ranch that, to his amazement, led into a sizable underground chamber. He lowered himself by rope into the cave and began exploring the passages with a lantern. After announcing his discovery through the local newspaper, he guided over 800 people through the passages in the first year. Lehman continued to escort visitors through the caves that bear his name for the rest of his life (fig. 45.7). In 1922, the caves and a square mile of land above ground became Lehman Caves National Monument. The protected area was enlarged and established as Great Basin National Park in 1986. Included in the park are Wheeler Peak, the neighboring peaks of the Southern Snake Range, and Lexington Arch, a 75-foot-high natural arch near the southern boundary. A scenic drive winds up to the 10,000-foot elevation on Wheeler Peak. Prospecting is not permitted in the park, but unpatented mining claims may be worked, provided a plan of operation has been approved by the National Park Service. None of the claims are producing at the present time.

Geologic Features

The Rocks

At least 14,000 feet of marine sedimentary rocks, ranging in age from Cambrian through Pennsylvanian, are exposed in Great Basin National Park. Wheeler Peak owes its topographic dominance to the superior resistance of the thick Prospect Mountain Quartzite of Cambrian age.

Carbonates are the dominant rock type, with older rock units thicker than those near the top of the section. Because of complex faulting, stratigraphic sections of the Paleozoic rocks are incomplete, and precise correlations are difficult. Cutting through the sedimentary rocks are much younger *stocks* (small plutons) that form resistant peaks and ridges. Lapping up against the Paleozoic rocks are younger nonmarine units—coarse volcanic clastics, lavas, ash, and widespread alluvial deposits.

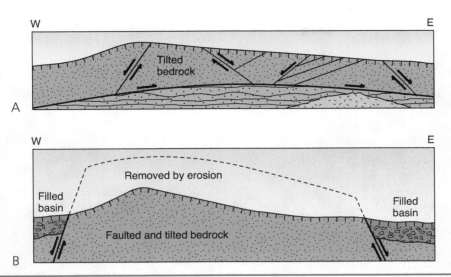

Simplified cross sections of the Snake Range. *A.* As the core of the range rose, thick sequences of rock slid along décollement faults that generally followed bedding planes. This faulting took place at least six miles below the surface. Cambrian beds are below the fault zone; younger Cambrian and Ordovician beds are above. The pluton is Cretaceous in age. Crustal stretching (extension) broke the rocks above the fault into blocks separated by normal faults. *B.* Faulting tilted the Snake Range toward the east. Erosion stripped rock material from the upthrown and tilted block and deposited fill in the basins. The lower diagram has a larger horizontal scale than the top diagram; the top diagram extends deeper into the crust than does the lower one. Adapted from Drewes 1958 and Whitehead 1969.

The two stocks in the range have been identified as coarse-grained quartz monzonite and coarse-grained, porphyritic quartz monzonite. They are close in age and probably came from the same magma chamber. The stock with the porphyritic texture is unusual in that it contains potassium feldspar phenocrysts, 1/4 to 1/2 inch long, set in a coarse-grained matrix of plagioclase, quartz, biotite and muscovite. In most igneous rocks, the potassium feldspar minerals form later in the crystallization sequence. The crystal sizes in the porphyritic stock probably represent unusual cooling conditions at the time the plutons were emplaced, presumably during late Cretaceous and early Tertiary time.

Ore bodies. For over a century, mining has been a principal industry in Nevada. Hillsides are pockmarked with prospect pits. A number of these sites have yielded commercial amounts of metallic ores. Copper, lead, gold, and silver have been mined in eastern Nevada; beryllium and tungsten have also been reported. Most of the ore bodies were deposited in the interval from late Cretaceous time to late in the Tertiary and are presumed to be related to the emplacement of plutons.

The Metamorphic Core Complex and Block-faulting

Great Basin National Park extends for some 20 miles along the crest of the southern end of the Snake Range, a north-south-trending *fault-block mountain range* about 150 miles in length. This type of feature is a crustal unit formed by uplift along normal (or sometimes vertical) faults. Tensional forces, pulling apart the uplifting crust, cause normal faults to divide the crust into structural blocks of different elevations and orientations. As is typical of fault blocks, the Snake Range is asymmetrical and, in this case, steeper on the western side. The block has been tilted toward the east and southeast. The amount of vertical displacement is difficult to estimate because the faults bounding the range are buried by gravels. However, the mountain block has probably moved up more than 9000 feet. Numerous high-angle normal faults within the range have produced a series of slices and smaller blocks of rock that are unevenly tilted relative to each other. On the eastern side, fault movement was about half as much as on the western flanks.

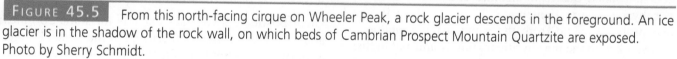

FIGURE 45.5   From this north-facing cirque on Wheeler Peak, a rock glacier descends in the foreground. An ice glacier is in the shadow of the rock wall, on which beds of Cambrian Prospect Mountain Quartzite are exposed. Photo by Sherry Schmidt.

Deep erosion has gullied the mountain slopes and deroofed the plutons, exposing igneous rock on summits and slopes. Several thousand feet of roof rocks have been removed and washed downslope, blanketing the basins on either side with great quantities of fill (fig. 45.4 A,B).

Complicating the geology of the Snake Range are numerous nearly horizontal faults that moved upper slabs eastward and northeastward relative to lower slabs. Detailed mapping in the southern part of the range has identified at least six discrete fault surfaces where younger beds have been pushed over older beds. This type of bedding-plane faulting is called *décollement*. In one area, for example, formations younger than middle Cambrian lie on the upper surface of the décollement and are themselves broken by numerous

younger (Mesozoic) horizontal faults, as well as by even younger (Tertiary) block faulting.

The rock layers and fault slices have been bent into broad folds that trend in several directions. These structures may be associated with uplift or plutonic intrusions, or both, rather than with compression.

This type of structure, characterized by décollement faulting, is called a *metamorphic core complex*. In this case, a core of low-grade metamorphic rocks, Precambrian and Cambrian in age, was uplifted, while a cover of unmetamorphosed Paleozoic rocks fractured and slid toward the northeast. Late Mesozoic or early Tertiary crustal extension in the Snake Range has been estimated as causing over 30 miles of displacement, measured by use of a separation of the Prospect Mountain Quartzite. This low-angle faulting in the region preceded the high-angle normal faulting that produced the typical basin-and-range structures.

Glaciation

The erosive action of successive Pleistocene cirque glaciers gave Wheeler Peak its distinctive, hornlike profile against the sky. Neighboring peaks also display cirques, and the highest summits of the range have snowfields year-round, a remarkable sight when one is traveling across the hot desert floor in midsummer. Two episodes of glaciation left moraines on the upper slopes at about the 8100-foot elevation for the older Lehman Creek moraine and at 9500 feet for the Dead Lake moraine.

A north-facing cirque on the summit of Wheeler Peak is above the snowline and holds a small glacier. Snow in this north-facing cirque is protected from direct sunlight by the shadow of the peak and is also in a good position to catch snow blown off the summit by prevailing westerly winds (fig. 45.5). Because no other north-facing cirque in the central Great Basin region is high enough to intersect the snowline, the glacier on Wheeler Peak is the only one existing in the area at the present time (Osborn, 1990).

An active rock glacier nestles in a draw on Wheeler Peak's northeast flank. A *rock glacier* (a type of mass wasting) is a slowly moving accumulation of talus, ice, snow, and water that creeps and slides down a mountain slope, usually only a few inches per day. Lovely alpine meadows and several small cirque lakes in the summit areas of the Snake Range remind us of the Ice Age by preserving a habitat for alpine wildflowers and other cold-acclimated life forms.

FIGURE 45.6   Aspen and pines grow on the flanks of Wheeler Peak. Photo by Sherry Schmidt.

Lehman Caves

Part way up on Wheeler Peak's eastern flank, behind the Visitor Center, is the entrance to Lehman Caves (fig. 45.7), a marbly limestone solution cavern known for the diversity of its cave decorations. The cave passages are in a thick, fairly uniform formation, the Pole Canyon Limestone, Middle Cambrian in age, that has been faulted upward as a large, wedge-shaped unit of bedrock. Although the beds dip steeply, the 10,000 feet of cave passages are nearly horizontal. This indicates that the cave passages were formed late in the history of the range, *after* most of the tilting and uplift had occurred and most likely within the last three or four million years. Hence base level and the water table had more control over the attitude of the cave system than the bedding planes in the limestone. However, the size and location of chambers and passages were influenced by fractures and bedding planes in the limestone.

The Pole Canyon Limestone is alternately pale-gray and white and (for the most part) massive. When exposed at the surface in this area, the rock is a cliff-former. The Pole Canyon Limestone underwent low-grade metamorphism during an earlier time, which made it purer, denser, and whiter in color than it was originally. During the process, fossils were mostly obliterated, and carbonate grains were dissolved and redeposited on adjacent grains, thus compacting most of the intergranular spaces. The Pole Canyon Limestone is considered a low-grade marble.

Many beautiful and varied dripstone features (speleothems) are displayed in the underground rooms and passages (fig. 45.8). Besides the usual stalactites, stalagmites, and columns, visitors can see "soda straws," scallops on sidewalls, shields (called "parachutes"), draperies, cave popcorn, and forms called *helictites,* which are twiglike decorations that grow in

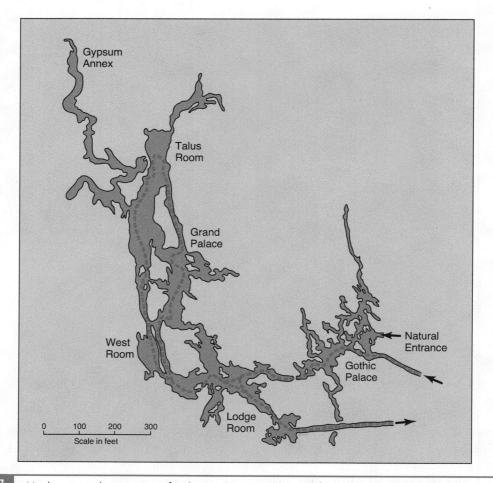

FIGURE 45.7   Underground passages of Lehman Caves. Adapted from National Park Service.

FIGURE 45.8 Lehman Caves is partway up on the eastern flank of Wheeler Peak. Because the cave formed after the bedrock was tilted, the cave passages are nearly horizontal. Photo by David B. Hacker.

FIGURE 45.9 Stalactites hanging from the ceiling follow the joint system. Large stalagmites are on the floor of Lehman Caves in Great Basin National Park.

curlicues and spirals out from the walls. Aragonite crystals and delicate gypsum deposits form lacy decorations in a few chambers. (How speleothems form is explained in Part II, in Chapters 14, 15, and 16.)

The Talus Room, a large chamber partly filled with broken rock and rubble, reveals a record of earthquakes, possibly 10,000 to 30,000 years ago, that brought down the entire ceiling of this 280-foot room.

Earthquakes in Nevada, which occur from time to time, indicate that the region is tectonically active.

Some of the processes that formed Lehman Caves are still going on today although probably at a slower rate. Most of the hollowing out of the passages took place during times of higher precipitation, presumably during the pluvial episodes of Pleistocene time. Ground water (and meltwater) percolated downward through the fractured rocks of Wheeler Peak. At times, all the rock openings and cracks were filled above the present level of the cavern floor. Flowing and eddying water dissolved, rounded, and smoothed the walls, ceilings, and floors of the passages. The fastest moving and largest volume of ground water was probably dissolving rock just below the water table.

When the climate became drier, the water table dropped. Then air entered the cave passages, evaporation took place, and dripstone began to precipitate. Some of the cave features may have undergone several cycles of solution-deposition-solution-redeposition as the water table rose and fell with changes in climate.

At present the year-round temperature inside the caves is about 50° F and the humidity is in the high 90s, making the environment exceedingly pleasant for visitors who come in from the desert on a hot day.

Lexington Arch

This imposing natural arch, 75 feet high, is the result of weathering and erosion of limestone. The arch stands near the base of Granite Peak in the southern part of Great Basin National Park. A trail from Lexington Creek leads up to Lexington Arch. Geologists surmise

Table 45.1	Simplified Geologic Column, Great Basin National Park			
Time Units				
Era	**Period**	**Epoch**	**Rock Units**	**Geologic Events**
Cenozoic	Quaternary	Holocene	Basin fill, talus slopes, sand dunes Speleothems Rock glaciers	Increasing aridity Erosion/deposition of speleothems Mass wasting
Cenozoic	Quaternary	Pleistocene	Glaciofluvial and lakebed deposits	Expanding and shrinking pluvial lakes Glacial cirque erosion Morainal deposition Episodes of Wisconsinan glaciation Solution erosion of cave passages Rising and falling water table Forming of speleothems
Cenozoic	Tertiary	Oligocene	Basaltic lavas, ash Tuffs, conglomerates, andesitic lavas	Development of basin-and-range topography by faulting and uplift Volcanism (two phases)
Mesozoic	? Cretaceous		Older conglomerates Rhyolite porphyry dikes, sills; mafic sills Quartz monzonite, quartz monzonite porphyry	Dikes and sills intruded Emplacement of stocks Décollement faulting; uplift of metamorphic core complex
Paleozoic	Ordovician to Pennsylvanian (?)		10 or more formations (limestones, dolomites, chert, shales, sandstones, quartzites)	Long-continued deposition of marine sediments in continental shelf, near-shore, and shore environments
Paleozoic	? Upper Cambrian ?		Pogonip Group Corset Spring Shale Johns Wash Limestone Lincoln Peak Formation	
Paleozoic	Middle Cambrian		Pole Canyon Limestone Pioche Shale	
Paleozoic	Lower Cambrian		Prospect Mt. Quartzite	

Source: Modified from Drewes, 1958; Whitebread, 1969.

that the arch is a remnant of a cave passage that became exposed by erosion.

Geologic History

1. Accumulation of sedimentary rocks, Cambrian to Pennsylvanian Periods.

Throughout most of Paleozoic time, this region lay along the edge of a continental landmass. The older rocks, including the Cambrian strata, accumulated in a continental shelf environment. Seas transgressed and receded numerous times across the region. Later in the era, from middle to upper Paleozoic time, complex geologic events took place with increasing frequency. Subduction occurred during orogenies.

2. Late Mesozoic and early Tertiary faulting.

Evidence of geologic events in early and middle Mesozoic time is missing from the geologic record; but late in Mesozoic time, the region was affected by repeated episodes of faulting caused by the upward movement of a metamorphic core complex. This tectonic activity, which was related to the Laramide orogeny and accompanied by plutonism, continued through Tertiary time. In the Snake Range, as described earlier, stocks were emplaced and gravity sliding moved blocks eastward. Although the stocks cut the faults, the intrusions are quite close in age to the faulting. Some low-angle faults slid along bedding planes in the shales.

One might logically think that the low-grade metamorphism evident in the Pole Canyon Limestone of the

---

Box 45.1   Crustal Extension in the Great Basin

The Great Basin section of the U.S. Cordillera displays a significant part of the large-magnitude intercontinental extension that occurred during Cenozoic time. The well exposed block-faulted valleys and mountains and the metamorphic core complex in the Snake Range are among the results of this major extensional event. The development of the Rocky Mountains and the other great ranges of the Cordillera was the result of compression and crustal thickening; but here in the Great Basin (and in certain other western areas), as compressional forces subsided, extension and crustal thinning took place.

As the upper crustal rocks become thinner and more stretched, subsidence and eventual burial will probably occur; and perhaps, in some millions of years, rifting will allow the sea to move in over the land. Meanwhile, a thick layer of less dense continental crust, beneath the upper crustal rocks, prevents the Great Basin from sinking below sea level.

High regional heat flow from a bulging upper mantle may be causing the rocks of the upper crust to stretch and thin. Low-velocity seismic waves indicate a shallow depth to the mantle. This shallowness allows basaltic magma to move toward the surface from deeper zones, resulting in some eruptions. High seismicity is another characteristic of the area, with earthquakes occurring from time to time.

The complex metamorphism, uplift, and faulting that are caused by the extensional forces take place in the top seven to ten miles of the crust. At greater depths, fluid behavior and lateral flow of rock materials adjust to the vertical movements and the magmatic activity. The interaction and strains of intense pressure and traction from below are expressed in the surface features that we see, as well as in subsurface features that can be inferred from geophysical mapping and other evidence.

The metamorphic core complexes are characterized by large, shallow-dipping normal faults (detachments or décollements) where low-grade metamorphic rocks are often in juxtaposition with intensely metamorphosed older rocks from the middle crust. The latter are broken and crushed gneisses (called "tectonites") that have been forced upward. Estimates of movement along the décollement faults run in tens of miles, with rates of movement being about the same magnitude as velocities of plate movements (Wernicke 1992).

Later stages of extension show the breaking up of décollement sheets, as well as fracturing of surrounding stable blocks. Geologists have speculated that this action has produced the typical, shallow, block-faulted basin-and-range structures that resemble a series of giant tilted dominoes. Driving west from Great Basin National Park on U.S. Highway 50, one goes up and down over the ranges and across the basins for many miles.

Snake Range would have been caused by contact metamorphism from the plutons. However, field evidence indicates that the metamorphism preceded the horizontal faulting, inasmuch as unaltered rocks are thrust over metamorphosed rocks and the plutons are younger than the faults. Perhaps the rocks were metamorphosed during a Paleozoic orogeny. Mafic dikes and rhyolitic dikes and sills intruded the Snake Range following faulting and emplacement of the stocks.

3. Middle Tertiary rocks.

Andesitic volcanism was extensive in the Great Basin area during this time. Conglomerates, ash flows, and tuffs accumulated in the Snake Range. A unit of tuff has been dated by the potassium-argon method as 31 million years ± 1.6 million (Oligocene). The volcanic activity was accompanied by irregular uplifts and deep erosion.

4. Block faulting, Pliocene to present.

The entire region was uplifted and broken by many high-angle normal faults that produced the tilted mountain ranges and basins. A second phase of volcanism, typically basaltic in composition, accompanied block faulting. The geometry of normal faulting, which is produced by tensional stresses, indicates crustal extension; that is, a thinning and stretching of the crust. Fault movements were both continuous and intermittent, with many individual fault surfaces on both sides of the range (box 45.1).

5. Effects of Quaternary climatic changes.

The summits of the Snake Range were eroded by alpine glaciers during episodes of Wisconsinan glaciation. Pluvial lakes occupied the basins on either side of the range. A valuable legacy of Pleistocene time are the reservoirs of ground water stored in the basin fill beneath the lakebeds. Lehman Caves was formed largely during Pleistocene time, and dripstone was deposited in the chambers and passages. Extensive weathering, downslope movement, and erosion have continued throughout Quaternary time and quantities of fill have accumulated around the base of the range. Snow collects in sheltered hollows at the summits. Sand dunes pile up on the flats below.

Geologic Map and Cross Section

Whitebread, D.H. 1969. Geologic Map of the Wheeler Peak and Garrison Quadrangles, Nevada and Utah. U.S. Geological Survey Miscellaneous Geological Investigations, Map I–578, with accompanying text.

Sources

Bartley, J.M., and Wernicke, B.P. 1984. The Snake Range Décollement Interpreted as a Major Extensional Shear Zone. *Tectonics*, 3: (6) 647–657, November.

Drewes, H. 1958. Structural Geology of the Southern Snake Range, Nevada. *Geological Society of America Bulletin*, 69:221–240.

———. and Palmer, A.R. 1957. Cambrian Rocks of the Southern Snake Range, Nevada. *American Association Petroleum Geologists Bulletin*, 41:104–120.

Fiero, Bill. 1986. *Geology of the Great Basin*. Reno: University of Nevada Press

Halladay, O.J., and Peacock, V.L. 1972. *The Lehman Caves Story*. Baker, Nevada: Lehman Caves Natural History Association.

McPhee, John. 1981. *Basin and Range*. New York: Farrar, Straus, Giroux.

Misch, P., and Hazzard, J.C. 1962. Stratigraphy and Metamorphism of Late Precambrian Rocks in Central Northeastern Nevada and Adjacent Utah. *American Association Petroleum Geologists Bulletin*, 46:289–343.

Osborn, G. 1990. Wheeler Peak Cirque and Glacier/Rock Glacier. Unpublished report prepared for the Great Basin Natural History Association. 59 p.

Stewart, J.H. 1971. Basin and Range Structure: A System of Horsts and Grabens Produced by Deep-seated Extension. *Geological Society of America Bulletin*, 82:1019–1044.

Wallace, R.E. 1975. Basin and Range Province, in Fairbridge, R.W., editor. *The Encyclopedia of World Regional Geology, Part I: Western Hemisphere*, p. 541–548. Stoudsburg, Pennsylvania: Dowden Hutchinson and Ross.

Wernicke, B. 1992. Cenozoic Extensional Tectonics of the U.S. Cordillera, *in* Burchfiel, B.C.; Lipman, P.W.; and Zoback, M.L. eds. The Cordilleran Orogen: Conterminous U.S. Boulder, Colorado: Geological Society of America, The Geology of North America, v. 6–3, p. 553–581.

Note: To convert English measurements to metric, go to www.helpwithdiy.com/metric_conversion_calculator.html

Saguaro National Park

SOUTHEASTERN ARIZONA

Area: 88,000 acres; 138 square miles

Proclaimed a National Monument: March 1, 1933

Established as a National Park: October 31, 1994

Address: Saguaro National Park—Headquarters and Rincon Mountain District, 3693 South Old Spanish Trail, Tucson, AZ 85730-5601

Phone: 520-733-5153

-or-

Saguaro National Park—Tucson Mountain District, 2700 North Kinney Road, Tucson, AZ 85743

Phone: 520-733-5158

E-mail: sago_information@nps.gov

Website: www.nps.gov.sagu/index.htm

Giant saguaro cacti may reach a height of 50 feel in the park's cactus forest.

The structural characteristics of the Rincon Mountains make them a "showpiece" of metamorphic core complexes.

Eroded fault-blocks make up the Tucson Mountains.

FIGURE 46.1   In Saguaro East, the cactus forest grows on rocky slopes below the ridges. Photo by Ronald F. Hacker.

Saguaro National Park contains two separate districts: Saguaro East, located about 20 miles east of Tucson, and Saguaro West, 18 miles west of Tucson. On the valley floor and foothill slopes grow the giant saguaro cacti. This cactus variety grows nowhere else in the world but in portions of the Sonoran Desert, a large arid region that extends from southern Arizona and the southeast corner of California for several hundred miles south into Mexico. The two park districts, on either side of the Tucson Basin, were set aside to preserve and protect the life and landscape of this especially beautiful portion of the Sonoran Desert.

Here where summer temperatures at lower elevations commonly climb above 100° F, less than 12 inches of rain may fall in a typical year. Winters may have wet spells, with precipitation often in the form of snow; but months may pass between rainy seasons without a drop of rain. Yet, despite the harsh conditions, life forms endure and even thrive. Cacti and other desert plants (fig. 46.2), as well as animals, reptiles, insects, and birds have evolved and adapted in the desert environment.

Elevation determines climatic zones that support the different plant communities in the park. Saguaros do not grow above an elevation of 4500 feet; but on higher mountain sides of the two districts, where it is cooler and moister, scrub oaks grow, then mixed oak and pine. On the peaks and ridges of the Rincon Mountains, at elevations between 7000 and 8600 feet, mainly on north-facing slopes, are stands of ponderosa pine and Douglas fir. The summit of Mica Mountain, highest peak in the Rincons, is 8666 feet above sea level.

The saguaro, for which the park is named, is a long-lived, slow-growing plant that starts from abundant, shiny black seeds that are the size of a period on this page. Only a few seeds germinate and grow to maturity. At 15 years, a young saguaro may be a foot tall; by 30 years, saguaros begin to flower and produce fruit. A 75-year-old saguaro usually has sprouted paired arms (branches) that grow straight out from the trunk and then curve upward. The 50-foot giant trees are 150 years old or more and may weigh over eight tons! Saguaros have the ability to swell as they absorb in their fleshy interior large quantities of water after a cloudburst or a heavy rain. In the long dry intervals, their circumference decreases, and pleats become more pronounced as the water is used up.

During late Pleistocene pluvial climates, large mammals roamed in the Tucson Basin but died out or migrated as the climate became warmer and drier. Early groups of hunter-gatherers (Hohokams) settled in the basin nearly 2000 years ago and practiced agriculture. Mortars in Box Canyon (Saguaro East)—conical holes ground into the gneiss bedrock—show where ancient women pulverized mesquite pods. The mortars are half a foot to a foot deep. Hohokam pictographs have been found on sheltered rocks (fig. 46.3). These early people abandoned their sites about 500 years ago. The modern O'odham Indians (also called Papagos), who came into the Tucson Basin later, used saguaro fruit and seeds for food. The strong woody ribs were made into fences and shelters. Spaniards displaced the Indians. Later miners and ranchers came. Prospect holes, where miners searched for silver, gold, and copper, are scattered over the landscape. Heavy use, cutting down of trees, and overgrazing denuded large parts of the cactus forest. Since the park districts were set aside and open grazing was stopped, plant and animal communities have slowly begun to recover although natural hazards have caused some setbacks. Some saguaros, especially older or weaker specimens, have been lost to severe freezes, lightning-caused fires, high winds, or prolonged drought.

Geology of Saguaro East

Both sections of Saguaro National Park are in the Southern Basin and Range region, but geologically, the two districts are markedly different. The higher Rincon Mountain District, east of Tucson, displays a three-humped metamorphic core complex; while the Tucson Mountain Range, west of Tucson, consists of fault-block structures. The city of Tucson occupies the relatively flat-floored structural basin between the two mountain areas. Geologists regard the Rincon Mountains (including the Santa Catalina and Tortolita Mountains) as the "showpiece" of metamorphic core complexes (Davis, 1987).

A *metamorphic core complex* (see also Great Basin National Park, chapter 45) is a domal uplift of intensely deformed metamorphic and plutonic rocks overlain by a tectonically detached, less metamorphosed cover. The inbetween zone, where slipping and sliding movements and the most severe deformation occurred, is called a *décollement* (fig. 46.5). These complicated pet-

FIGURE 46.2 The east section has a greater variety of cacti, such as barrel and prickly pear, and, of course, saguaro. National Park Service photo.

FIGURE 46.3 In the east section pictographs left by the Hohokam Indians can be seen on rocks that have been sheltered. Photo by Ronald F. Hacker.

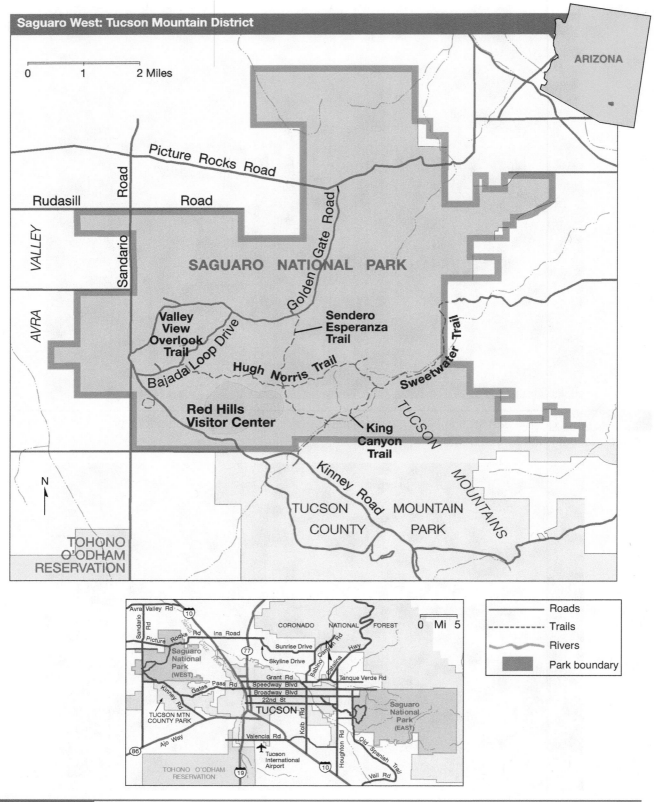

FIGURE 46.4A *Above:* Saguaro National Park (West), Arizona. *Below:* Both park regions shown relative to Tucson.

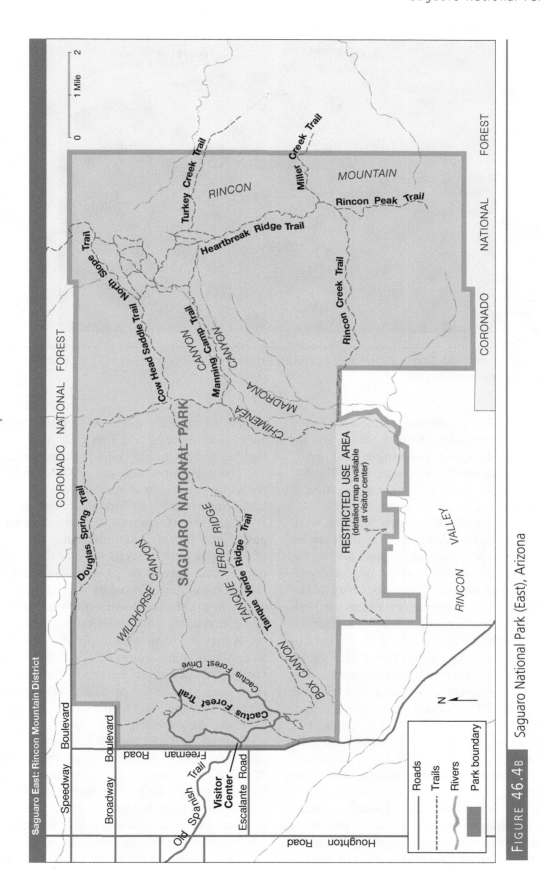

Saguaro East: Rincon Mountain District

RINCON

MOUNTAIN

Turkey Creek Trail

Miller Creek Trail

Rincon Peak Trail

Heartbreak Ridge Trail

North Slope Trail

Rincon Creek Trail

CORONADO NATIONAL FOREST

Cow Head Saddle Trail

CANYON Trail

Manning Camp Trail

MADRONA CANYON

Douglas Spring Trail

SAGUARO NATIONAL PARK

CHIMENEA CANYON

RESTRICTED USE AREA
(detailed map available at visitor center)

WILDHORSE CANYON

TANQUE VERDE RIDGE

Tanque Verde Ridge Trail

RINCON VALLEY

CORONADO NATIONAL FOREST

BOX CANYON

Cactus Forest Drive

Cactus Forest Trail

Speedway Boulevard

Broadway Boulevard

Freeman Road

Old Spanish Trail

Visitor Center

Escalante Road

Houghton Road

N

Roads
Trails
Rivers
Park boundary

FIGURE 46.4B   Saguaro National Park (East), Arizona

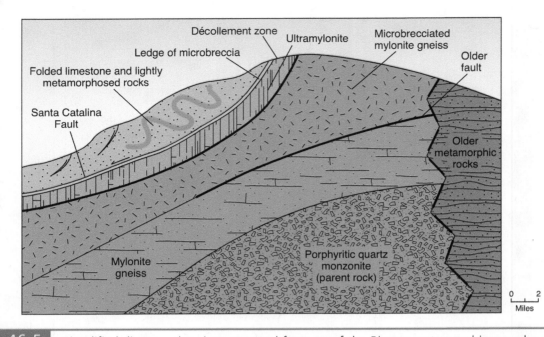

FIGURE 46.5   Simplified diagram showing structural features of the Rincon metamorphic complex. After Nations and Stump 1981; Davis 1987.

rographic and structural assemblages, which are found in the interior of southwestern United States, are the result of extensional tectonics involving crustal uplifting and crustal stretching that began early in Tertiary time, following a long period of intense compression of western North America.

In order to understand how such features were formed, geologists make detailed examinations of rocks and outcrops exposed in the canyons and on the slopes. Field work is followed by petrographic and geochemical study of rock samples in the laboratory. In the Rincon Mountains, the rocks (with few exceptions) fall within the general term "gneiss," with the Catalina gneiss being the most common rock that visitors see in the Rincons and along Tanque Verde Ridge. This gneiss is a coarse-grained mylonite, Precambrian in age, that is strongly banded or layered. (*Mylonites* are massive metamorphic rocks that have been so severely sheared, pulverized, and rolled by overthrusting, pressure, and heat that the original grain sizes and shapes have been thoroughly altered.) Before the Catalina gneiss was mylonitized, it was a porphyritic quartz monzonite resembling granite.

Ultramylonites, very dense rocks that are the result of such severe crushing that primary mineral structures are obliterated, are exposed in a shear zone that can be seen in small washes or on low ridges. In some places,

patches of mylonite gneiss have been "microbrecciated," i.e., crushed to very fine grain sizes by prolonged faulting and fracturing under high pressures.

These dynamic changes probably occurred when the rocks were six to seven miles deep within the earth. Later, the rocks were brought closer to the surface. When the area that is now southern Arizona was being pulled apart in early and mid-Tertiary time, the thinning of the upper crust relieved some of the weight burdening the deep crustal rocks so that they began to rise. Deeper rocks moved more by ductile flow, while those nearer to the surface underwent brittle faulting. It was this domal uplift that formed the Rincon Mountains.

A massive shear zone, more than a mile thick, and a gently dipping normal fault—the Santa Catalina fault—stretches along the top of the plug of mylonitized rock. This zone of extensive movement is the décollement (fig. 46.5). Above the décollement, capping the metamorphic core complex, are the cover rocks that include folded limestones and weakly cemented gravels. Ranging in age from Precambrian to Miocene, the cover rocks were moved laterally by shearing and faulting more than six miles in a west-southwest direction. The principal foliation of the metamorphic rock masses below the fault show the same direction. The amount of dip of the fault decreases away from the center of the cover. The fault surfaces

FIGURE 46.6   The Javelina rocks have been severely fractured. National Park Service photo.

of the décollement are smooth and display slickensides. Younger normal faults, caused by crustal extension, cut the cover rocks.

A loop drive, called the Cactus Forest Drive, that begins at the Visitor Center, passes through or close to numerous areas of geologic significance. The Santa Catalina fault, for example, runs north-south in this area, roughly bisecting the loop drive. Visitors can examine exposures of ultramylonite gneiss, microbrecciated mylonite gneiss, beautifully layered Catalina mylonite gneiss, and rocks such as folded limestones penetrated by calcite veins.

Geology of Saguaro West

Although different, the geologic features of the Tucson Mountain District are as complex as those of the Rincons. The Tucson Mountain Range, which is about 20 miles long and 7 miles wide, is typical of the Basin and Range Province and consists of a faulted, east-tilted wedge of chiefly Paleozoic and Mesozoic sedimentary rocks. The ancient marine limestones, exposed in the Twin Hills area, probably underlie the mountain block and much of southern Arizona as well.

Precambrian granites, gneisses, and schists make up the highest summit, Wasson Peak (4687 feet), as well as nearby Amole Peak. Late Cretaceous and Tertiary volcanic rocks, such as rhyolite, andesite, and basalt, were extruded during numerous episodes and blanketed older rocks. Basalt flows from a volcanic plug that cut through the older rocks spread over much of the area. Cat Mountain Rhyolite forms the topmost layer of the Tucson Mountains.

Sheets of sandstone, mud, and tuff, washed down by streams, accumulated between lava flows. The Recreation Redbeds, exposed at the Red Hills Visitor Center consist of red sandstone colored by hematite (fig. 46.7). The Amole Arkose, a coarse rock with angular fragments, makes up a large portion of the Tucson Mountains. The fact that very few fossils have been found in the redbeds or the arkose has made correlation of the rocks difficult.

The Tucson Mountain Chaos, east of the Visitor Center and overlying the redbeds, is a puzzling hodge-podge of rocks containing boulders and house-sized blocks of various older rocks, mixed with lava flows, ash, and fault debris.

As the stretching and thinning of the crust began in Tertiary time, the rocks that had been deposited broke into large blocks that tilted and slid past each other, forming the down-dropped basins and uplifted ranges that we see today. Many of the rocks were metamorphosed by the intense pressures of tectonic pulling and twisting as blocks were uplifted or dropped thousands of feet.

Spanish settlers in the Tucson Basin roasted great quantities of the Paleozoic and Mesozoic carbonate rocks in lime kilns, remains of which are seen in the park. Miners found pockets of ore containing silver and copper at contacts between igneous intrusions and the limestones. That is why the slopes of the Tucson Mountains are spotted with dozens of prospect holes from old mining claims.

Geologic History—Tucson Mountains

1. **Prior to 1.75 billion years ago (early Proterozoic) deposition of marine sediments offshore. Formation of the Precambrian X Pinal Schist.**

Graywackes and shale were deposited in a deep-water marine environment. Intrusion of granites was accompanied by regional low-grade metamorphism (greenschist facies) during deformation. Some of the rocks had been changed to bluish-gray phyllites and gray meta-graywackes, but others were a meta-rhyolite porphyry, quartzites and meta-basalts. The formation is

FIGURE 46.7   The Recreation Redbeds are sandstones that have been colored red with hematite. National Park Service photo.

Table 46.1	Geologic Column of Saguaro National Park, East and West Units

Time Units				Rock Units			Geologic History (T=Tuscon Mts.) (R=Rincon Mts.)
Eon	Era	Period	Epoch	Group	Formation East (Tuscon Mix)	Formation West (Rincon Mix)	
Phanerozoic	Cenozoic	Quaternary	Holocene		Basin Sediments	Gravel and Sand	Erosion of mountains, formation of bajadas (T) and pediments (R). Tucson Basin has been filled with 7000 feet of sediment.
			Pleistocene			*Erosion*	
		Tertiary	Pliocene (Neogene)				Detachment fault slid the Tucson Mountain Caldera off of Catalina Foothills Granite 53 miles to the west 30-17 mya (T).
			Miocene (Neogene)			*Erosion*	
			Oligocene (Paleogene)		Mid-Tertiary Volcanics	Pantano Formation	Lava and tuffs ejected in a series of flows. Dacile lava in northern part of range, rhyolite tuffs and basaltic lavas on east side about 30-20 mya (T). Mostly limestone formed that contains large clasts, rhyolite intrusions (R).
			Eocene (Paleogene)		Amole & Watson Igneous Intrusions	Catalina Granite Caldera	Pulling apart of the crust in an east-west direction, igneous intrusions into tilted beds (T). Formation of a volcano from a granite pluton, eventual collapse of crater to form a caldera (R).
			Paleocene (Paleogene)			*Erosion*	Second metamorphism of Precambrian Metamorphic Core Complex (Catalina Gneiss) during the Laramide Revolution along with the overlying Paleozoic beds (R).
	Mesozoic	Early Cretaceous		Bisbee	Cat Mountain Tuff Tuscon Mountain Chaos	*Erosion*	Caldera-Cycle Volcanics. Caldera collapses in a trap door fashion. The series forming a caldera-filling breccia. Some fragments are up to 1642 ft in diameter creating a mega-breccia. A series of ash flow tuffs, some were so hot that they became welded (T).
					Amole Arkose		Deposition in streams and lakes, mostly arkose but also has limestones and conglomerates. Conglomerate contains fragments of Paleozoic Naco Group limestones, Recreation Red Beds and some arkose that formed on caldera floor. Imbricate thrust sheet (T).
		Late Jurassic			Recreation Red Beds	*Erosion*	Sedimentary, volcaniclastic and volcanic sequence. Volcanic breccia contains fragments more than 1 1/2 feet. Tuff is mixed with the volcanic breccia, unconformable contact with Amole Arkose. Red color contains piedmontite (T).
	Paleozoic	Permian		Naco	Rain Valley Formation	*Erosion*	Deposition of sandstone, limestone, and dolomite (T). Deposition and erosion of beds (R).
					Concha Limestone		Fossiliferous limestones deposited in a sea. Contains productio brachiopods (T), deposition then later removed by erosion (R).
					Sherror Formation		Quartzite, dolomite and sandstone deposited in a sea (T). Deposition then later removed by erosion (R).
		Mississippian			Escabrosa Formation	*Erosion*	Marine deposit of limestone, crinoid fragments and chert nodules (T). Deposition and later removal by erosion (R).
		Devonian			Martin Formation		Ordovician and Siluran beds removed by erosion. Limestones contain corals and brachiopods; (T) Deposition and erosion (R).
		Cambrian			Abrigo Formation	*Erosion*	Deposited over an old erosion surface. Contains trilobites (T). Deposition and erosion (R).
					Bolsa Quartzite	*Erosion*	Marine deposition. Contains stromatolites (T). Deposition and removal by erosion (R).
					Intrusion of Andesite Porphyry	*Erosion*	Intrusion of andesite poxphyry between 1450–1250 mya (T).
Proterozoic		Precambrian X				Quartz Monzonite Porphyry granite	Intrusion 1,660–1,450 mya. First metamorphism of beds that will eventually become Catalina Gneiss.
		Precambrian X			Pinal Schist	Pinal Schist	Formed 1.75–1.50 mya during mountain building.

Source: After Kring, Lapman, Hill, Chronic, McColley et. al., 2002.

known as the Pinal Schist even though very little schist is present. It has been estimated that 32,808 feet of sediments accumulated before deformation.

The Pinal Schist formed between 1.75 and 1.50 billion years ago during an episode of mountain-building, early in Proterozoic time that eventually resulted in the enlargement of the southwestern portion of the North American continent, followed during the Mazatzal orogeny by the formation of a mountain range. That can be seen just north of the entrance to the eastern Rincon Mountain section of Saguaro National Park. The Pinal Schist, which is considered the basement rock in this region, underlies much of the Tucson region.

2. **The intrusion of quartz monzonite and poryphyritic granite plutons 1660 million and 1450 million of years ago, during Precambrian X time.**

None of the intrusions are exposed within park boundaries but can be found in nearby regions. Veins and dikes of younger igneous rocks of varying compositions have intruded the older rocks.

3. **After 1450 mya and before 1250 mya, mountains from various tectonic episodes were removed by erosion. Intrusion of andesite.**

Erosion of the Mazatzal Mountains and other features that had formed during Precambrian X underwent erosion, forming a vast area of low relief, with occasional hills and valleys. Life had evolved only to the stage of tiny microbes. An andesite porphyry, dated at 1,440 mya, is exposed in the eastern Rincon section along Cactus Forest Drive.

4. **Transgression of the oceans onto the land, burying the roots of the former mountains with Lower Cambrian sediments.**

Over 3,281 feet of sediments, such as conglomerates, sandstones, and limestones, were deposited. Before the beginning of the Paleozoic Era, they were removed by erosion from the Tucson area. The Middle Cambrian-aged Bolsa Quartzite marks the beginning of the Paleozoic beds in this region.

This formation, which is about 540 million years old, is found over much of southern Arizona and is a mixture of quartzite, sandstone, and minor amounts of conglomerate and shale. The color of the rock ranges from gray to purplish-gray to brownish-gray. The bed-

ding is from medium to thick. Above the Bolsa Quartzite is the Abrigo Formation, a mixture of brownish-gray shales, fine-grained to medium-grained sandstone, and a thin-bedded limestone. Fossil trilobites help to date these beds that were deposited in vast seas that covered an old erosional surface that was relatively stable.

5. **Erosion of the region during the Ordovician and Silurian Periods, then deposition in the Late Devonian time of the Martin Formation, followed by the deposition of the Mississippian Escabrosa Limestone.**

The Martin Formation is primarily a dolomite (a calcium-magnesium carbonate), but also contains some shale, limestone, and sandstone. The dolomite portion is coarse-grained and brown, medium-bedded with some chert. The limestone is gray and the shale is reddish brown. Fossil corals and brachiopods are found in these beds.

The medium-gray Escarbrosa Limestone ranges from a thin to thick-bedded formation. It is bioclastic (made up of animal fragments) consisting of crinoid (sea lilies) disks and stems that are 0.2-0.4 inches in diameter. Much of the formation consists of large chert nodules.

6. **During Pennsylvanian time and the beginning of the Permian Period, erosion occurred. Deposition began during Early Permian time of the Naco Group that consisted of three formations in this region.**

Pennsylvanian-aged beds deposited in the Tucson area were removed by erosion. During Early Permian time the Naco Group was deposited, reflecting various environments. They are the Scherrer Formation, Concha Limestone, and Rain Valley Formation.

The Scherrer Formation is a mixture of sandstone, quartzite with minor amounts of dolomite, and siltstone. The sandstone and quartzite range in color from a pinkish-gray to a light-gray and are fine-grained. The siltstone is a thick layer and reddish-gray in color, while the dolomite is thin-bedded and light-gray in color. The younger Concha Limestone contains brachiopods (that help to date the formation) and chert, and is a fine-grained, medium-gray limestone. The Rain Valley formation is a mixture of limestone, dolomite, and sandstone and is sometimes missing from the geologic column in this region.

7. During the Mesozoic Era, several episodes of volcanic activity occurred; the first was during the Jurassic Period and the second during the Cretaceous Period.

About 150 mya volcanoes became active. Materials expelled by the volcanoes were mixed with stream deposited sediments, forming the Jurassic-aged Recreation Redbeds (fig 46.7). (Hematite, an iron oxide mineral, is the reason for the red color.) The Recreation Redbeds can be found at the Red Hills and Brown Mountain (along Kinney Road). The beds are a mixture of sandstone, siltstone, volcanic conglomerate, and tuff. Some of the sandstones and siltstone are cross-stratified because the water current changed directions between deposition of each layer. The volcanic conglomerates have some large fragments/clasts that are more than 1.5 feet in diameter. The clasts are mainly volcanic andesites with some quartz and quartzites. Also includ-ed are shale, limestone, granite, arkose, and redbed fragments. The conglomerate and volcanic tuffs are interbedded. Within the tuff is a reddish-black or reddish-brown mineral called piedmontite or piemon-tite. The Brown Mountains (located southwest of the park) used to be referred to as the Piedmontite Hills. Using the K-Ar radiometric method of dating, the igneous dikes that are cross-cutting the Recreation Redbeds, yield a date of 155 my. Another estimate comes from an andesite bed that intruded and deformed the redbeds in the Brown Mountains (northern end), giving a date of 159 my, which is Upper Jurassic time.

After the volcanic activity ceased, a period of quiet and erosion followed. By Cretaceous time, the streams deposited many layers of sandstone and some limestone along their banks and in lakes. The formation is called the Amole Arkose (Amole is Spanish for "grinding stone"). An arkose sandstone is indicative of rapid ero-sion and rapid deposition. As this type of environment is

FIGURE 46.8   A natural spring. National Park Service photo.

not conductive to the preservation of fossils, very few are found. The Amole Arkose contains pebble-, cobble-, and boulder-conglomerates. The different types of sediments were sorted by size according to the depositional environment, The Amole Arkose is dominated by siltstone and silty mudstone besides the arkose. These beds, which were derived from both the Paleozoic carbonates and the underlying Recreation Rebeds, vary in color and frequently are laminated and thin-bedded. The conglomerate made up of cobbles was deposited in alluvial fans at the base of the mountains that ringed the Amole basin. The streams running across the alluvial fans transported the fine-grained material (clay, silt and sand) deeper into the basin. Braided streams received the sand-sized particles, and the silt and clay were deposited in the deltas that formed as the streams entered the lake. The very fine silts and clays settled out in the deeper quiet water of either the lake or stream channels. An unconformable contact of the Amole Arkose is between the Rain Valley Formation (Permian) and Recreation Redbeds (Jurassic). The lack of fossils and usable radioactive minerals makes it difficult to date these beds. The Amole Arkose is considered a member of the Bisbee Group, which places it as Early Cretaceous, that is about 100 million years old.

8. **Collapse of a Late Cretaceous volcano 70 million years ago, forming a caldera. During the collapse, an ash-flow cooled into a welded tuff named the Cat Mountain Tuff.**

To the west, a subduction zone had formed, producing a series of volcanoes fed by the magma that formed as one of the plates was subducted. Among these volcanoes was one that was located at the site of the present Catalina Mountains, located just northwest of the Tucson Mountains.

A huge explosion produced a large collapse caldera that was 12 by 15 miles in size. The caldera was formed when the chamber of the volcano was emptied as an ash flow tuff, the Cat Mountain Tuff, was ejected. It has a rhyolitic composition. At first there was a series of ash flows and after most of the gases were lost, the ash changed to lava. The lava flows came from small vents that were located on the rim of the caldera, and flowed to the bottom of the caldera. The Tucson Mountains are the remnants of these flows, derived from the bottom of the caldera that formed during the Cretaceous Period (70 mya). The Cat Mountain Tuff forms the crest of the Tucson Mountains. The flows are easily distinguished by their red color and flattened pumice fragments. The fragments were flattened by the weight of the overlying flows. The flows also contain glass shards and fragments of older rock. This type of flow is called a *nuée ardente,* or fiery cloud. The heat of the flow welded the tuff together.

About five miles below was the Catalina Granitic Pluton, which was the source of the magma that was ejected from the volcano. This pluton cooled for 40 million years. As it cooled, it heated the rocks above and weakened them.

During Pliocene-Miocene time, the subduction zone became a strike-slip zone. As a result, the forces applied to the surrounding rocks caused the Tucson Mountain Caldera to become detached from the rocks below by a low-angle normal fault referred to as a *detachment fault.* Over a span of 13 million years, the caldera slid many miles to the southwest to its present location.

Fifteen million years ago, during the Miocene Epoch, additional forces continued to pull apart and stretch the land between the volcanic Tucson and the granitic Catalina Mountains. Erosion removed the top of the present-day Tucson Mountains and *isostatic rebound* uplifted the granitic pluton that we call today the Catalina Mountains. The land between them is the Tucson Basin. This section of land between them was tilted while it subsided, creating a type of topography called *basin and range.*

The Catalina Mountains and Tucson Mountains underwent erosion from Miocene time to the present and the basin has been filled with sediment. Not all of the rocks eroded evenly; for example, outside of the park to the east, rocks on the bottom of the caldera have now been exposed by erosion.

9. **Possible gravitational collapse of an oversteepened caldera wall producing the Tucson Mountain Chaos during the Tertiary Period.**

The Tucson Mountain Chaos is a mixture of Precambrian schists, sedimentary rocks that are Paleozoic and Mesozoic, and Tertiary volcanic rocks, all imbedded in a tuff and sandstone matrix. Many of the blocks are about the size of a house.

Some of the studies of the younger ash flows indicate that the megabreccia (a breccia in which the blocks are really huge) was produced when the steep walls of the caldera collapsed after the caldera floor subsided. These walls were up to 3168 feet high, and some of the blocks were up to 1600 feet in diameter.

The Cat Mountain Tuff and Tucson Mountain Chaos are intermixed as alternate layers, with the Tucson Mountain Chaos found predominately in the lower part of the sequence. The thickness of these beds increases from south to north, which suggests that the caldera floor may have dipped down like a trap door. The hinge would have been to the south. Hence the beds are at least 13,100 to 16,400 feet at the north end of the caldera where they ponded and only 330 to 660 feet thick at the south end. Volcanic dust and ash was blown into the air, and depending upon the direction of the prevailing winds, deposited over much of the western landscape. The remaining material in the magma chamber (about 90% of the total) solidified into a granitic pluton.

10. **Magmatic intrusions and lava flows after the collapse of the caldera and the emplacement of the megabreccia and tuffs during Late Cretaceous time.**

A series of lava flows was extruded from new vents that formed on the floor of the caldera. Because most of the gases had been lost in the tuff flows, the lava was more like the lavas of the Hawaii today. They were more siliceous than the basaltic flows and much more viscous, hence they formed flows up to 500 feet thick. The first flows were near the Old Yuma Mine located just south of Picture Rocks Road by the eastern border. The lava flowed to the northwest as far as Panther Peak, located by the northern boundary of the park. Interfingering of this flow with volcaniclastic sediments created a 8200 feet sequence by Panther Peak. The activity shifted to the southeastern part of the caldera. Small vents extruded flows of intermediate composition, dacite tuffs and andesite and dacite lava flows. These flows were very viscous.

After the flows a large granitic pluton called the Amole Pluton intruded on the northwest side of the caldera along a ring fracture. As the pluton intruded the various beds, they were uplifted and tilted, forming a resurgent dome (there are two of them in Yellowstone National Park). The magma chamber that produced the dome is found primarily in the western portion of the park covering approximately 140 square miles. Erosion exposed the top of the Amole Peak. It

FIGURE 46.9 Mountain view with waterfall. National Park Service photo.

showed that the pluton was zoned. The upper part was a granodiorite and the interior was of granite. Evidence indicates that it was probably formed slightly after the Cat Mountain Tuffs, since it has intruded some of the flows. Two other bodies rose elsewhere outside of the park boundaries (Saginaw Hills and Sedimentary Hills).

A series of ring dikes that were silica rich, named the Silver Lily dike swarm, crosscut the entire complex. These light colored dikes are up to 66 feet wide and some can be traced a distance of approximately 19,700 feet. Outside of the park, irregular intrusions and dikes with an andesitic composition were intruded from the magma chamber, primarily in the Twin Hills area.

There is evidence of an extensive system of circulating hydrothermal solutions beneath the surface. They altered the sedimentary and volcanic rocks and deposited copper sulfides and other ores in the aureole zone around the plutons. This was the end of the post-collapse sequence of rocks. The crystallization of these rocks probably took about a million years.

For a period of 40 to 45 million years the region underwent erosion. The dikes are more resistant to erosion than the sedimentary rocks, so they form topographically high ribs of rock.

11. During mid-Tertiary time, volcanic activity resumed with a series of smaller eruptions.

In the northern part of the Tucson Mountain range, a dacite lava field (Safford Dacite) was extruded onto the landscape. In the eastern side of the range basaltic lavas and tuffs with a rhyolitic composition formed, while several domes of rhyolite or dacite rose on the south side. This took place around 20 to 30 million years ago.

Some remnants of this field can be seen as rhyolite to dacite lava flows, volcanic vent spatter, and volcanic vent intrusions, all intermixed with volcaniclastic sediments. The thickness of the flows is 82 to 328 feet. Safford Peak, located in the northern part of the park, is a vent intrusion with four lava flows, rather than the one or two flows that are found in most areas. Panther Peak, to the west of Safford Peak, has a nearby vent intrusion called Glassy Spatter. Other flows outside the park boundaries form landmarks, such as Sentinel Peak, that faces Tucson.

12. Cenozoic sediments were deposited in association with the mid-Tertiary volcanics.

These sediments are only about 100 feet in thickness; they are volcaniclastics with identifiable fragments of mid-Tertiary volcanics. These beds have accumulated in just the last couple of million years. They are a mixture of colluvial deposits that slumped down due to gravity. They range in size from silts to boulder-sized fragments. They are undergoing erosion and further dissection at the present time.

Geologic History—Rincon Mountains

1. About 2 billion years ago, the Precambrian granitic basement complex evolved.

The basement rock underwent low-grade regional metamorphism, forming the "Pinal Schist." The metamorphic rocks were intruded by more than one generation of intermediate to granitic plutons. A diabase sill intruded sedimentary rocks that had been deposited later and had not been metamorphosed.

2. During the Paleozoic Era, a huge inland sea depositing sandstone, shale, and limestone covered the region.

Undisturbed sediments of sandstone, shale, and limestone were deposited in the shallow sea. These sediments were the Middle Cambrian Bolsa Quartzite; Middle and Upper Cambrian Abrigo Formation; Upper Devonian-aged Martin Formation; the Mississippian-aged Escabrosa Limestone; Upper and Middle Pennsylvanian Horquilla Limestone; Upper Pennsylvanian and Lower Permian Earp Formation; Lower Permian (from oldest to youngest) Colina Limestone, Epitaph Dolomite, Scherrer Formation, Concha Limestone and Rain Valley Formation. Most of these formations were removed by erosion in this region but can be seen in adjacent areas or were incorporated as a portion of younger formations such as the Pantano Formation that is found in this area. The Pantano beds were deposited during Pennsylvanian-Permian times as limy sediments or as thin-bedded clastics that are preserved as an unconformable contact between the fault blocks that contain them and the gneiss-granite contact.

3. During the Mesozoic Era, deposition of sedimentary beds and igneous intrusions.

The sedimentary beds were deposited in an environment of shallow seas or deposited on land. The beds then underwent intrusions ranging in composition from gabbro, an igneous rock with a basic composition, to granite with an acidic composition. Some of the magma reached the surface in some areas and was extruded as lava flows. The Bisbee Group was deposited in Early Cretaceous time and is a mixture of siltstones, shales,

FIGURE 46.10   Hells Gate, Happy Valley. National Park Service photo.

sandstones, and arkoses. These beds were metamorphosed during the Laramide Revolution and converted to quartzites (sandstones), phyllites, schists, and gneisses (depending upon the degree of metamorphism). The intrusion of the granites may have helped in the metamorphism or even been responsible for it. There is a high probability that the intrusion and metamorphism of the beds may have taken place during the second period of metamorphism of the Catalina Gneiss.

4. **During Lower Tertiary time magmatic activity occurred. Formation of the Metamorphic Core Complex. The term Catalina Gneiss is a general term for the type of terrain in this region.**

After the Sevier and Laramide orogenies, the crust thickened in this region. At the same time that the igneous activity was taking place, the Metamorphic Core Complex was forming around 40 to 25 mya during the Lower Eocene and Oligocene Epochs. Near the entrance of the park is the Loop Drive. A clockwise tour around the Loop Drive shows progression in the transformation of the deformed rocks. First the Precambrian and Tertiary quartz monzonites formed the *mylonites* and *ultramylonites* (pulverized rocks formed by faulting and metamorphism that is active.

Secondly, this was followed by *microbreccias* derived from the mylonite and ultramylonite that had been altered by chlorite (a green mineral). This was followed by the altered chloritic microbreccias that had been changed into a *fine-grained microbreccia* known as cataclasite and ultracataclasite. These are crushed and granulated rocks on a microscopic level. Next is the *detachment fault* (or *décollement*); this is the separation plane that is on the bottom of the fault from the overlying cover of noncataclastic, nonmylonitic rocks above the fault; and some folded Paleozoic cover rocks. (See fig. 46.5.)

The general term Catalina Gneiss includes a coarse-grained biotite quartz monzonite augen gneiss; medium- to coarse-grained, garnetiferous-muscovite quartz, monzonitic augen gneiss; mylonitic schist; and abundant pegmatite and aplite. These rocks can be seen between Rincon Peak and Mica Mountain in the eastern part of the park, especially along Tanque Verde Ridge. Evidence indicates the two periods of metamorphism had affected this unit.

5. **During Mid-Tertiary time an orogeny occurred 30 to 20 million years ago, stretching the crust in an east-west direction.**

The Catalina detachment fault, a low angle normal fault, is responsible for bringing the metamorphic core complex to the surface for the Rincon and Santa Catalina Mountains. This took place during the Oligocene to Lower Miocene Epochs. The lower plate rocks have the composition of mylonitic injection gneiss in which the banding is strongly compositional; it shows signs of brittle deformation. It is highly fractured and faulted in many places. The composition of the upper plate of the detachment fault consists of Paleozoic carbonates. Metamorphic core complexes are a mixture of metamorphic and igneous rocks forming the core of the mountains; some of the rocks have been highly deformed.

The western portion of the Rincon Mountain section of Saguaro National Park is divided by a northeast-trending fault. East of the fault is the Tanque Verde Ridge, which consists of the Catalina Gneiss. This makes up the main portion of the Rincon Mountains. On the west side of this fault is a pediment (sloping rock surface at the base of the mountain) that has been carved into granite and schist. The pediment has been subdivided into three blocks by two vertical faults that are parallel to each other and perpendicular to the northeast fault.

6. **The stretching of the crust produced a series of faults. Gravels and limestones moved down next to the newly formed gneiss.**

Both an Oligocene rhyolitic agglomerate (volcanic breccia) and the Oligocene-Miocene Pantano Formation, which is a mixture in size from a megabreccia to claystone, can be seen in the park on Loop Drive, unconformably overlying the Pinal Schist. The Pantano is moderately well sorted and cemented. The pebbles and cobbles are subrounded. The material was derived locally from Tertiary volcanics. The formation is a mixture of sedimentary rocks (Paleozoic and Mesozoic in age) and granodiorites (Precambrian).

Something triggered the downhill movement of the blocks, perhaps further tilting, as they slid down slope until the base of the gneiss ridge stopped the movement.

The gravels were made up of fragments of the core rocks; the Catalina Gneiss and the original cover rocks. These beds can be seen along Scenic Drive

7. **Additional movement along the major faults caused some of the basement crust to be uplifted as much as 30 miles, including the gneiss.**

The movement of the major faults continued to uplift and tilt the rocks in the area until a total of 30 miles had been displaced. This may have taken place around Pliocene time. Some of these are the rocks that formed the pediment. The Pantano sediments were eroded from the southerly raised block. At the central block (at Observation Hill), it was exposed and tilted to the southwest. The northern block had dropped and the sediments remained undisturbed. Isostatic adjustment occurred after the overlying sediments were removed by erosion, causing the gneiss to rise, forming a broad, dome-like mountain. This mountain has three structures that trend in a northeast-southwest direction. They are from north to south, the Tanque Verde Antiform, Rincon Valley Synform, and Rincon Peak Antiform. They are located essentially from Rincon Peak to just north of Mica Mountain, both of which are located within the park boundaries. An antiform is a fold in which the limbs close upward in bedrock of an unknown stratigraphic sequence. A synform is just the reverse because the limbs close downwards.

8. **The Catalina Fault passes both through the foothills of the Catalina Mountains and around the southern and western edges of the Rincon Mountains.**

In the park the major fault, the Catalina Fault, is a low angle normal fault that bisects the Scenic Drive area. The fault runs in a north-south direction and dips to the southwest. The fault may run beneath Tucson and the Tucson Mountains, where the western portion of Saguaro National Park is located.

9. **A pediment at the base of the mountain is covered with alluvium derived from the erosion of the Catalina Gneiss.**

Pediments are sloping rock surfaces at the base of a mountain, covered with a thin veneer of gravel. They form in arid or semi-arid climates. The pediment of the Rincon Mountains is covered with alluvium derived from the Catalina Gneiss.

The dissected remains of the pediment blocks that have been uplifted and the main mountain mass are gradually being leveled by erosion.

Geologic Map and Cross Section

Drewes, Harald. 1978. Geologic Map and Sections of the Rincon Valley Quadrangle, Pima County, Arizona. U. S. Geological Survey Miscellaneous Investigative Series Map I-997.

Kamilli, R.J. and Richard, S.M., Editors, 1998. Geologic Highway Map of Arizona, Arizona Geological Society and Arizona Geological Survey, 1 sheet containing text and figures, Scale 1:1,000,000 Tucson, Arizona.

Richard, S.M., Reynolds, S.J., Spencer, J.E., and Pearthree, R.A. (compilers) 2000. Geologic Map of Arizona, Arizona Geological Survey, Map-35.

Wilson, E.D., Moore, R.T., and O'Hare, R.T. 1960. Geologic Map of Pima and Santa Cruz Counties, Arizona, Prepared by the Arizona Bureau of Mines, University of Arizona, Tucson.

Sources

Arizona Volcanic Mountains: The Tucson Mountains n.d. http://www.geo.arizona.edu/geos256/azgeology/tucsonmts.html

Chronic, H. 1983. *Roadside Geology of Arizona*. Missoula, Montana: Mountain Press Publishing Co. 321 p.

Davis, G.H., 1978. Twenty-ninth Field Conference, November 9-11. Third Day Road Log from Tucson to Colossal Cave and Saguaro National Monument, New Mexico Geological Society in cooperation with the Arizona Geological Society. Pgs. 77–87.

Davis, G.H. 1987. Saguaro National Monument, Arizona: Outstanding Display of Structural Characteristics of Metamorphic Core Complexes. Geological Society of America Centennial Field Guide, Cordilleran Section, v.1 p. 35–40.

Ellwood, B.B. 1996. *Geology and American's National Park Areas*. Prentice Hall pgs. 261-262.

Friends of Saguaro National Park- About Us: About Saguaro National Park n.d. http://www.friendsofsaguaro.org/geology.html 5 pgs.

Gehrels, G.E. and Spencer, J.E. Eds. 1990. *Geological Excursions Through the Sonoran Desert Region,* Arizona and Sonora Arizona Geological Survey Special Paper 7.

Geology-3: A Mountain: *The Forming of Tucson* n.d. http://www.geo.arizona.edu/geos256/azgeology/amtn/geo_3.html.

Hargrave, n.d. ASU (Arizona State University) Geology 510 (Advanced Structural Geology Class) http://activetectonics.la.asu.edu/advstruct/Students/hargrave/text.html 5 p.

Kamilli, R.J. & Richard, S.M. (Editors) 1998. *Geologic Highway Map of Arizona*. Geological Society & Arizona Geological Survey. 1 sheet containing text & figures, scale 1:1,000,000. Tucson, Arizona.

Kiver, E.P. & Harris, D.V. 1999. *Geology of U.S. Parklands*. John Wiley & Sons, Inc.

Kring, D.A., 2000. *Desert Heat-Volcanic Fire: The Geologic History of the Tucson Mountains and Southern Arizona*. Arizona Geological Society Digest 21 p. 103.

Lipman, P.W., 1994. U.S. Geological Survey Research on Mineral Resources, Part B—Guide Book For Fieldtrips, Ninth V.E. McKelvey Forum on Mineral and Energy Resources. U.S. Geological Survey Circular 1103-B p. 62–69.

Lipman, P.W., 1994. Tuscon Mountain Caldera—*A Cretaceous Ash-Flow in Southern Arizona*. U.S. Geological Survey Circular 1103-B p. 89-102

Mayo, E.B. 1968. *Arizona Geological Society Southern Arizona Guide Book III,* Prepared For Field Trip Held in Conjunction with U.S. Geologoical Survey Cordilleran Section 64th Annual Meeting, April 11–13, 1968, Tucson Arizona (edited by S. R.Titley) p. 155–170.

McColly, R.A., 1991. The Geology of the Western Portion of the Saguaro National Monument, *Arizona Geological Survey Digest*, Volume IV, p. 87–91.

Nabhan, G.P. and Huey, H.H. 1986. *Saguaro—a View of Saguaro National Monument and the Tucson Basin*. Southwest Parks and Monuments Assoc. 74 p.

Naruk, S.J. and Bykerk-Kaufman, A. 1990. Field Trip Guidebook 86th Annual Meeting Cordilleran Section Geological Society of America, Tucson, Arizona. *Late Cretaceous and Tertiary Deformation of the Santa Catalina Metamorphic Core Complex, Arizona* p. 41–50.

National Park Service, 2003. Saguaro National Park http:www.nps.gov/sagu/ 1 p.

Nations, D. and Stump, E. 1981. *Geology of Arizona*. Dubuque, Iowa: Kendall/Hunt Publishing Co, 221 p.

Paquette, A. 1998. *The Forming of Tucson,* Geology-3: A Mountain http://www.geo.arizona.edu./geos256/azgeology/amtn/geo-3.

Rystrum, V.I. (Compiler) n.d. *Metamorphic Core Complexes*. http://www.colorada.edu/GeolSci/Resources/wus tectonics/core 13 p.

Shakel, D. 1996. Geology at Pima Community College, *Geology of Tucson and Vicinity.* http://www.azstarnet. com/~dshakel/glgtus01.html 8 p.

The American Southwest: Arizona Guide Saguaro National Park n.d. 2 p. http://www.americansouthwest.net/ aruziba/saguaro/national_park.html.

Thorman, C.H. and Drewes, H.,1981. Rincon Area, Pima County, *Arizona Studies Related to Wilderness,* Rincon Wilderness Study Area, Arizona. U.S. Geological Survey Bulletin-1500 p. 8–30.

Wernicke, B. 1992. Cenozoic Extension Tectonic of the U.S. Cordillera, in Burchfiel, B.C.; Lipman, P.W.; and Zoback, M.L. eds. The Cordilleran Orogen: Conterminous; U.S.; Boulder, Colorado: Geological Society of America, the Geology of North America, v. G-3, p. 553–581.

Note: To convert English measurements to metric, go to www.helpwithdiy.com/metric_conversion_calculator.html

Joshua Tree National Park

SOUTHERN CALIFORNIA

D.D. Trent   Citrus College

Area: 794,300 acres; 1239 square miles

Proclaimed a National Monument: 1936

Established as a National Park: October 31, 1994

Designated a Biosphere Reserve: 1984

Address: 74485 National Park Drive, Twentynine Palms, CA 92277-3597

Phone: 760-367-7511

E-mail: JOTR_Info@nps.gov

Website: www.nps.gov/jotr/index.htm

Desert landscape with up-faulted mountains and down-dropped basins. Ancient (Proterozoic) and Mesozoic rocks in odd and curious shapes due to prolonged weathering and erosion. Oases sited along fault traces. Named for the magnificent stands of Joshua Trees that are unique to the Mojave Desert.

FIGURE 47.1   Inselbergs at Hidden Valley Campground display the rectangular joint system that is prevalent in the White Tank monzogranite. Photo by D. D. Trent.

Joshua Tree National Park, one of the nation's newest national parks, preserves a typical area of California desert landscape that includes parts of two deserts; the lower elevation Colorado Desert and the higher elevation Mojave Desert. The first-known inhabitants were the Pinto Basin people, an early culture whose artifacts have been found along the shorelines of an ancient lake that once occupied the Pinto Basin in the park's eastern Wilderness Area. The dating of these artifacts suggests that the Pinto Basin people lived here from about 7000 to 5000 years ago. More recent native American populations included the hunters and gatherers who are the ancestors of the modern Cahuilla, Chemehuevi, and Serano people. The Oasis of Mara, at the site of the Oasis Visitor Center and Park Headquarters at Twentynine Palms, was the home of Chemehuevis until the early 1900s.

In the late nineteenth century, the Oasis of Mara was a popular watering stop for miners on their way to and from the gold mines in the surrounding region. Today, within the park boundaries, are the remains of more than 2000 abandoned mines and mining prospects. Among the more productive mines were the Lost Horse, Silver Bell, El Dorado, and Desert Queen. Estimates of the gold production from these mines range from $40,000 to $40,000,000.

Cattle raising went on at about the same time as the mining activity as cattlemen found the grasses in the high desert suitable for their stock. Grazing continued within what is now the park until 1945.

Beginning in the early 1920s, homesteaders began taking up land in the Twentynine Palms area. The reasons for this were the availability of water at the Oasis of Mara and the desert climate. Disabled veterans of World War I were encouraged to settle in the area in the hope that the dry air would help cure their ailments. Many of the health-seekers found that an out-of-door desert lifestyle did indeed have therapeutic value. Gradually the area became more popular, bringing new housing, more roads, an influx of land developers, and cactus poachers.

A dedicated lady from Pasadena, Mrs. Minerva H. Hoyt, who had a passion for the desert, agonized over the removal of cacti and other desert plants from the Joshua Tree area to the backyard rock gardens of Los Angeles. Her efforts to protect the desert environment culminated in the creation of Joshua Tree National Monument, proclaimed by President Franklin D. Roosevelt in 1936. The national park was established 58 years later in 1994.

Location and Geography

Joshua Tree National Park is on the eastern end of the broad mountainous belt called the Transverse Ranges, that stretch from Point Arguello, 50 miles west of Santa Barbara, eastward for nearly 300 miles to the Eagle Mountains in the Mojave Desert (fig. 47.2).

The park region includes several distinct mountain ranges, the Little San Bernardino, Cottonwood, Hexie, Pinto, Eagle, and Coxcomb Mountains. Both the southern and the northern margins of the park are marked by steep escarpments that rise abruptly from the lower desert areas. Elevations within the park range from 1,000 feet in Pinto Basin to 5,814 feet at the summit of Quail Mountain. Valleys lying between the mountain ranges are of two types: (1) structural basins formed by the down-dropping of a block between two faults and (2) erosional valleys. Pleasant Valley, between the Little San Bernardino and Hexie Mountains, is an example of the structural type; Queen Valley, in the central part of the park, is an example of the eroded type.

Climate

The climate of the high desert of the Joshua Tree region is that of a mid-latitude desert with relatively moderate temperatures. For example, the average temperature at Twentynine Palms, elevation 1960 feet, is only 67.3°F and at Hidden Valley Campground, 4200 feet, the average temperature is about 7° to 12° F cooler. Two factors cause eastern California to be a desert: (1) the rain shadow effect produced by the high mountains on the west, and (2) the existence during summer months of a semi-permanent high-pressure air mass, the Hawaiian High, which builds up over the northeastern Pacific Ocean and blocks the passage of frontal storm systems over California. Occasionally, during the summer and fall, the Hawaiian High weakens and moist air from the Gulf of Mexico slips into the region across Arizona, bringing thunderstorms. For this reason, August has the highest rainfall (table 47.1) which, curiously enough, is usually the driest month for the more humid regions of the state.

The Hawaiian High usually dissipates during the winter months and southern California is subjected to an average of four or five frontal storms that originate in the northeastern Pacific. Consequently, it is in December and January that the desert's second rainy season occurs (table 47.1). The average rainfall at

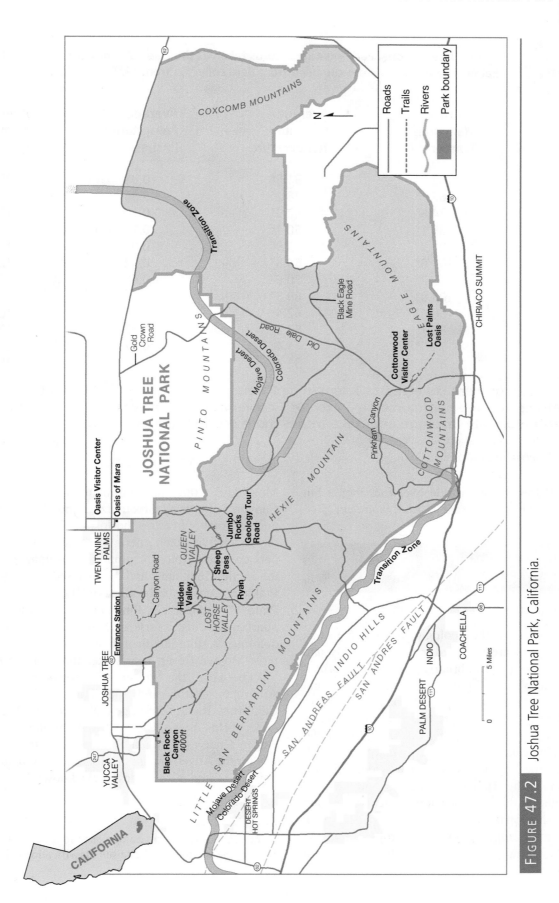

Table 47.1	Joshua Tree National Park weather records taken at the Oasis Visitor Center at Twentynine Palms. The averages were compiled from data collected from 1936 through 1991.			
Month	Average Maximum Temperature	Average Minimum Temperature	Average Precipitation (inches)	Average Humidity (percent)
January	62.8°	31.5°	0.38	28.2
February	67.7°	38.2°	0.35	24.1
March	74.8°	42.9°	0.30	19.2
April	83.1°	50.1°	0.10	16.5
May	90.3°	55.9°	0.06	14.0
June	100.6°	64.8°	0.02	12.3
July	105.2°	70.6°	0.62	17.0
August	103.0°	69.5°	0.68	20.8
September	96.4°	62.4°	0.31	16.4
October	85.7°	52.6°	0.32	19.5
November	71.5°	41.4°	0.27	25.7
December	62.3°	31.6°	0.46	28.4

Weather readings were collected at Twentynine Palms at an elevation of 1960 feet. Temperatures average approximately 7° to 10° F cooler at higher elevations in the park. Higher elevations also average about 3.5 inches more precipitation annually.

Source: U.S. National Park Service, 1992.

Twentynine Palms is only a little over four inches but at higher elevations the average rainfall is greater.

Types of Rock Exposed in the Joshua Tree Region

Metamorphic Rocks

The earliest events in the geologic history of Joshua Tree National Park are recorded in rocks of early and middle Proterozoic (Precambrian) age that were formed by the metamorphism of preexisting sedimentary and igneous rocks. These rocks, formerly named the Pinto Gneiss, are now recognized as fragments of a widespread metamorphic complex that was caught up in the Mesozoic tectonic arc along the Pacific margin of North America, became fragmented, and was widely distributed throughout the Transverse Ranges and vicinity. Four sub-units of this complex are recognized within Joshua Tree National Park: the Joshua Tree Augen Gneiss, a granitic augen gneiss that crops out in the Chuckwalla, central Eagle, and south-central Pinto

Mountains; the metasedimentary suite of Placer Canyon, composed of quartzite and dolomite, which unconformably overlies the Joshua Tree Augen Gneiss; the distinctive augen gneiss of Monument Mountain, a dark colored porphyritic granodiorite-monzogranite,[1] in the Hexie Mountains; and the metasedimentary suite of Pinkham Canyon in the Chuckwalla, Eagle, Hexie, and Pinto Mountains. The Pinkham Canyon rocks include quartzite, schist, very fine-grained granofels, and dolomite, a suite identical to strata in the northeasternmost Mojave Desert near Baker that offers a tentative link between the Proterozoic rocks of the North American craton and the Transverse Ranges of California. Radiometric age dating of these Proterozoic rocks yields ages of 1.65 to 1.70 billion ybp for the Augen Gneiss of Joshua Tree and 1.65 to 1.68 billion ybp for the Augen Gneiss of Monument Mountain, making these some of the oldest rocks known in California.

---

[1] The igneous rock terminology used in this chapter follows the modified Streckeisen classification (1973).

Igneous Rocks

At least five different major plutons, ranging in age from middle Proterozoic to Cretaceous, have intruded the metamorphic complex described above (fig. 47.3). The oldest are a succession of intrusions of the igneous protoliths (parent rocks) of foliated metamorphosed hornblende gabbro, diorite, and amphibolite, laminated granodioritic to monzogranitic orthogneiss, and various leucocratic granite orthogneiss (commonly garnetiferous) that intruded the metasedimentary rocks of Pinkham Canyon. The amphibolite and the leucocratic granitic orthogneiss have yielded isotopic ages of 1.71 billion ybp and 1.68 billion ybp respectively. A younger Proterozoic suite of plutonic rocks is an anorthosite-syenite intrusive complex in the southeastern section of the park, which yields a radiometric age of about 1.2 billion ybp.

The Triassic and Cretaceous plutons in Joshua Tree National Park, in common with the granitic rocks of the Sierra Nevada, the Peninsular Ranges, the Klamath Mountains, and the White-Inyo Mountains, are believed to have been generated in an oceanic-continental convergence zone. Examples of the intrusive contacts of these rocks with the Proterozoic gneiss country rocks are well exposed along the trail to Fortynine Palms Oasis and along the east side of Lost Horse Valley (fig. 47.4).

The oldest Mesozoic intrusion, the Twentynine Palms megacrystic quartz monzonite, consists of a matrix of small mineral grains that encloses large phenocrysts of potassium feldspar that attain lengths up to two inches. The pluton is of Triassic age, yielding a preliminary radiometric age of 245 million ybp. It is part of a widespread belt of Permo-Triassic plutonic rocks exposed in southern California. This belt of rocks is significant because their intrusion signals the onset of an ocean–continent tectonic convergence and subduction plutonism along the continental margin. The

FIGURE 47.3 Proterozoic gneiss on the trail to the Lost Horse Mine. Photo by Anthony Belfast.

FIGURE 47.4   The contact between the White Tank monzogranite and Proterozoic Gneiss along the east side of Lost Horse Valley. Photo by D.D. Trent.

Twentynine Palms pluton crops out along the trail to Fortynine Palms Oasis and along the arroyo on the east side of Indian Cove campground.

The principal plutons of Cretaceous age include the Queen Mountain Monzogranite, the White Tank Monzogranite, and the Oasis Monzogranite. These rocks appear to be part of the late Cretaceous intrusive events recognized in the eastern Mojave Desert, Peninsular Ranges, and the Sierra Nevada.

The oldest of the Cretaceous plutons in Joshua Tree National Park is the Queen Mountain Monzogranite. It is coarse-grained, consisting of plagioclase, potassium feldspar, quartz, and either biotite or hornblende. The Queen Mountain has yielded a radiometric age date of 104 million ybp.

The light-colored Cretaceous White Tank Monzogranite predominates in the more accessible parts of the park. The White Tank pluton resembles the Queen Mountain monzogranite but differs by being finer-grained, and by containing small amounts of biotite and/or muscovite but no hornblende. What are perhaps the most scenic areas of the park underlain by the White Tank monzogranite are Indian Cove, the Wonderland of Rocks, Jumbo Rocks, White Tank, and Lost Horse Valley (fig. 47.5).

The youngest of the Cretaceous plutons, the Oasis Monzogranite, is a garnet-muscovite-bearing pluton exposed in the area around Fortynine Palms Oasis. The garnets are blood-red and small, but large enough, nevertheless, to be visible without magnification. The muscovite grains impart a distinct glitter to the rock on sunny days.

In addition to the large monzogranite and quartz monzonite plutons already described, there are smaller masses of a similar rock, granodiorite, and small dark plutons of Jurassic age named the Gold Park Diorite. Cutting across all of these rock masses, and thus being younger in age, are dikes of felsite, aplite, pegmatite, andesite, and diorite. Pegmatite dikes in the park consist mainly of quartz and potassium feldspar with a

FIGURE 47.5 Rock sheeting in the White Tank monzogranite forms domelike landforms at Barker Dam in the Wonderland of Rocks. Photo by D.D. Trent.

composition close to that of granite. Making them distinctive is the very large size attained by the mineral grains, often 3.8 to 4 inches long.

Even younger than these dikes are veins of milky quartz that, over the years, have been prospected for gold. The quartz is sometimes stained reddish brown from the weathering of pyrite (fool's gold). Pyrite is a common mineral in quartz veins and is sometimes associated with gold or other valuable minerals. Chemical alteration of the pyrite produces reddish iron oxides that stain the rocks and serve prospectors as clues that gold, silver, copper, lead, or other important ores may be present.

Basalt occurs at three places within the easily accessible parts of the park: (1) near Pinto Basin, where the basalt probably originated as extrusive flows, (2) at Malapai Hill on the Geology Tour Road (figs. 47.6A,B),

and in the Lost Horse Mountains. These exposures show much in common with other basalt bodies in the eastern San Bernardino Mountains and the Mojave Desert that have been age-dated at between 8 to 10 million ybp. In addition to basalt, another mafic rock, lherzolite, occurs as inclusions within the basalt at Malapai Hill and in the Lost Horse Mountains. Lherzolite, is an olivine-rich peridotite that is derived from the mantle; thus, the basalt has risen some 30 to 50 miles in order to carry the inclusions to the surface.

Structural Geology

Faults

Joshua Tree National Park is surrounded by active or recently active faults. The Pinto Mountain fault, trend-

FIGURE 47.6A   Malapai Hill, an eroded endogenous volcanic dome, rises above the pediment of southern Queen Valley. The Hexie Mountains, behind Malapai Hill, expose the contact between Proterozoic gneiss (dark rocks) and the White Tank monzogranite. Photo by D.D. Trent.

FIGURE 47.6B   Malapai Hill from Lost Horse Mountain. Photo by D.D. Trent.

Table 47.2			**Geologic Column, Joshua Tree National Park**		
Time Units					
Era	**Period**	**Epoch**	**Rock Units**		**Geologic Events**
Cenozoic	Quaternary	Holocene	Alluvium, talus, playa lake sediments, dune sands		Weathering, erosion, mass wasting
Cenozoic	Quaternary	Pleistocene	Basaltic eruptions, playa lake sediments, alluvial fans		Volcanism, faulting, uplift, weathering, erosion, mass wasting
Cenozoic	Tertiary		Basaltic eruptions, alluvial fans		Volcanism, faulting, uplift, weathering, erosion, mass wasting
Mesozoic	Cretaceous		Oasis Monzogranite White Tank Monzogranite Queen Mountain Monzogranite		Intrusions and orogeny
Mesozoic	Jurassic		Gold Park Diorite Mafic and felsic dikes		Orogeny?
Mesozoic	Triassic		Twentynine Palms megacrystic quartz monzonite		Intrusion
Paleo–zoic			(no record)		
Proterozoic	Z		(no record)		
Proterozoic	Y		Anorthosite-syenite complex (1.2 bybp)		
Proterozoic	X		**Eagle Mountians Assemblage** Includes the metasedimentary suite of Placer Canyon and granitic augen gneiss of Joshua Tree 1.70–1.65 bybp; complex suites of meta-sedimentary and meta-igneous rocks	**Hexie Mountains Assemblage** Augen Gneiss of Monument Mountain (1.68–1.65 bybp); metamorphosed grabbo, diorite, and amphibolite (1.71–1.4 bybp); meta-sedimentary suite of Pinkham Canyon; suite of meta-sedimentary and metamorphic rocks	Tectonic episodes involving metamorphism and plutonism, probably resulting from plate interactions along the western edge of the North American landmass

ing nearly east-west along the north side of the Pinto Mountains, is one of the most prominent. The fault zone is followed closely by Twentynine Palms Highway (State Highway 62) between Morongo Valley and Twentynine Palms.

Between Morongo Valley and Yucca Valley, the fault is marked by side hill benches, triangular faceted spurs, and a probable left-lateral stream offset. Quaternary basin-fill buries much of the geomorphic evidence of the fault from Yucca Valley to Copper Mountain, but just west of Copper Mountain the fault is marked by a line of vegetation. A prominent escarpment is formed by the main trace of the fault along a 1.2 mile-long shutterridge at Copper Mountain. (A *shutterridge* is formed by displacement on a fault traversing ridge-and-valley topography. The displaced part of a ridge "shuts in" an adjacent canyon.) In Twentynine Palms, the fault at the Oasis of Mara, immediately west of the Oasis Visitor Center, is marked by a line of vegetation about 1.5 miles long

FIGURE **47.7** Columnar jointing, northwest corner of Malapai Hill. Photo by Walter Stephens/D.D. Trent.

and by a scarp about a half-mile long and three to six feet high (fig 47.8).

The Blue Cut fault extends east-west through the Little San Bernardino Mountains, about a half-mile south of Keys View, under Pleasant Valley and into the Pinto Basin (fig. 47.9). The Blue Cut fault branches from the Dillon fault, which is even farther south and trends southeastward through the Little San Bernardino

Mountains. The Blue Cut and Pinto Mountain faults are both left-lateral faults. They appear to belong to a system of faults, all about the same age, that include the many north-northwest-trending right-lateral faults of the Mojave Desert. Included in this fault system are the Johnson Valley, Camp Rock, and Emerson northwest-trending right-lateral faults that ruptured in the 1992 Landers earthquake (M = 7.5), the epicenter of which was about 22 miles northwest of Twentynine Palms. This right-lateral fault set appears to extend southeastward from the central Mojave Desert into the western part of Joshua Tree National Park in the region of Black Rock Canyon campground.

South of the Dillon and Blue Cut faults lies the San Andreas fault zone. The trace of the San Andreas fault is clearly visible from Keys View (fig. 47.10). Along this portion of the San Andreas, the fault divides into two main branches, the Banning and Mission Creek faults. The traces of these faults are marked by the Indio Hills, an uplifted block wedged between the faults, and by a number of palm oases that are aligned along the faults.

In addition to the major faults are many minor faults throughout the region of the park. Such fault zones are often important in localizing springs. Movement by faults causes impervious zones composed of pulverized rock fragments that form subsurface barriers and may force ground water to rise. The oasis at Cottonwood Springs, for example, appears to be localized by a fault zone that has provided the fissures along which ground water reaches the surface. The Oasis of Mara at the Twentynine Palms Visitor Center has been formed in a similar manner along the Pinto Mountain fault.

Joints

These small fissures cutting rocks may occur in sets of parallel joints and in systems of two or more intersecting sets. The White Tank monzogranite has a system of rectangular joints that is primarily responsible for the spectacular landforms in the park. One set, oriented roughly horizontally, results from the erosional removal of the overburden stress of many miles of rock that once overlay the monzogranite. These joints, sometimes called lift joints, cause exfoliation. Lift joints (or pressure release joints) are due to expansion from the release of overburden stress, somewhat analogous to that of a seat cushion resuming its shape after a person sitting on it arises. Where vertical joints are lacking or widely spaced, lift joints (rock sheeting) form domelike landforms (fig. 47.5).

FIGURE 47.8   Fault scarp of the Pinto Mountain Fault at Oasis of Mara and Twenty Nine Palms Oasis. Photo by D.D. Trent.

FIGURE 47.9   Blue Cut Fault (southern branch) near Cholla Cactus Garden. Photo by D.D. Trent.

FIGURE 47.10   The northern end of the Salton Trough as seen from Keys View in Joshua Tree National Park. The high peak in the distance is Mount San Jacinto, 10,786 feet in elevation. Palm Springs and Cathedral City are at the base of the mountains. The Indio Hills, running from left to right in the middle distance, mark an uplifted block wedged between two branches of the San Andreas fault. Photo by D.D. Trent.

Another set of joints is oriented vertically, roughly paralleling the contact of the White Tank monzogranite with its surrounding rocks. The third set is also vertical but cuts the second set at high angles. The resulting system of joints tends to develop rectangular blocks. Especially good examples of the joint system may be seen at Jumbo Rocks, Wonderland of Rocks, and Split Rock (fig. 47.5).

Sculpturing the Landscape

Weathering

Perhaps the most impressive aspect of the landforms at Joshua Tree National Park is the strange and picturesque shapes assumed by the bold granitic rock masses at the Jumbo Rocks, Wonderland of Rocks, Split Rock, and elsewhere. The sculpturing of these rock masses is the result of the combined action of rock jointing, and chemical and physical (mechanical) weathering. The combination of these processes results in spheroidal weathering,

the peeling-off of thin concentric shells of rock that form spherical rock masses. Spheroidal weathering results from slight pressures that have been built up in the outer portions of the rock from chemical decomposition of the aluminum silicate minerals into clay minerals. For example, when potassium feldspar comes into contact with hydrogen ions and water, the following chemical reaction, called hydrolysis, takes place:

$$\text{potassium feldspar} + \text{hydrogen ions} + \text{water} \rightarrow \text{kaolinite} + \text{potassium ions} + \text{silica}$$

$$2KAlSi_3O_8 + 2H^+ + H_2O \rightarrow Al_2Si_2O_5(OH)_4 + 2K^+ + 4SiO_2$$

Kaolinite, a clay mineral formed by hydrolysis in an arid environment, occupies a greater volume than the original feldspar. The expansion is especially great at the edges and corners of the jointed granitic rocks resulting in the jointed blocks of rock losing their sharp edges and corners to eventually assume rounded or spheroidal shapes.

The stresses, in addition to popping off thin shells of rock, cause the mineral grains of the rock to disintegrate physically and form a loose mineral soil called *grus*. Furthermore, frost-wedging and root-wedging also contribute to the breakdown of rocks by physical action (Chapters 3 and 54).

Subsoil Weathering

The concave hollows or pits that pockmark granitic rock surfaces in Joshua Tree National Park are called *tafoni*. Although a rather common phenomenon, the process explaining the formation of tafoni is not clearly understood. The most common explanations include hydrolysis and hydration, thermal differences due to freezing and thawing or insolation, recrystallization of salts on exposed rock faces, and wind erosion.

In many areas of tafoni, nearly all of the declivities are aligned parallel to the outcrop-soil contact. This implies that the tafoni may have originated several centimeters beneath the surface of the soil by a two-step process of chemical and mechanical weathering. The first step involves the episodic wetting and drying that caused local chemical weathering of bedrock beneath the soil surface. In the second step, climate change and soil erosion forced removal of the soil cover, exposing the chemically altered sites that became pits when exposed to the atmosphere. Because of lower evaporation rates in the declivities, they weathered more rapidly than the surrounding rock, enlarging and sometimes merging into larger pits, with Skull Rock being an especially good example (fig. 47.12).

Nearly vertical surfaces, commonly on the shady sides of rock outcroppings throughout the park, have been formed by subsoil weathering. The action of moisture trapped in the soil at the base of the vertical surface

FIGURE 47.11   Lost Horse Valley from Lost Horse Ridge. In the valley the Lost Horse mine was one of the more productive gold mines during the late nineteenth century. Within the park boundaries are the remains of more than 2000 abandoned mines and mining prospects. Photo by D.D. Trent.

FIGURE 47.12   Skull Rock at Jumbo Rocks Campground illustrates cavernous weathering and undercutting by subsoil notching. Photo by D.D. Trent.

causes undercutting and accounts for many of the steep cliff faces because the process of wearing back and rounding off higher on the cliff cannot keep up with the undercutting at the base.

Erosion. Of the dynamic processes that carry away surficial rock material, running water, even in arid environments, is by far the most important erosional agent. Wind action is important in the desert, but the long-range effects of the wind are small when compared to the action of water.

Erosional and weathering processes presently operating in the arid climate conditions in the region of Joshua Tree National Park are not entirely responsible for the spectacular sculpturing of the rocks. The present Joshua Tree landscape, and that of much of the Mojave Desert, is essentially a collection of *relict* features inherited from earlier times of higher rainfall and lower temperatures. Thus, the desert landscape we see now is a "fossil" landscape. For example, Fortynine Palms Canyon could not have formed in the present rainfall regime. Such deep canyons must be attributed to former pluvial conditions during an epoch when the area

of the southwestern United States received much greater precipitation than at present, when evaporation was considerably less, and the mean annual temperature was several degrees cooler.

Landforms of the Desert

The landforms encountered in Joshua Tree National Park are typical of those found in the arid portions of the southwestern United States:

1. *Arroyos,* or dry washes—deep, flat-floored stream courses that contain water only a few hours or perhaps a few days each year;
2. *Playas*—lakes that may contain water a few weeks a year during the rainy season;
3. *Alluvial fans*—fan-shaped deposits of sediment formed at the base of mountains in arid regions;
4. *Bajadas*—broad sloping aprons of rock debris that form by the coalescing of several alluvial fans;
5. *Pediments*—gently sloping bedrock surfaces that are erosional surfaces carved along the base of desert mountains.

Pediments are a curious desert landform typical of the southwestern United States and many other desert regions. Superficially, pediments look like bajadas (depositional features) rather than products of the bedrock erosion. The slopes of pediments are slight, from 1/2° to about 6°, and they are usually carved on homogeneous crystalline rock, such as granite. Pediments may be covered with a thin mantle of gravel, but if overlain by more than ten feet of gravel cover, the resulting landform is considered depositional and is called a bajada. In order to determine whether a gently sloping desert surface is a pediment or a bajada, the observer must look at thickness of the gravel veneer, exposed along the drainage channels.

Apparently a pediment is formed by the retreat of a mountain front, leaving an extensive planed bedrock surface that marks the path of the retreating foot of the slope. Rill wash, sheetfloods, winds, and lateral planation by streams tend to sweep the pediment clean of debris except for local accumulations of alluvium or gravel.

A pediment may be seen at Malapai Hill (Stop 6 on the Geology Tour Road[2]). Large expanses of bare granite pavement and bold dikes weathered out of the granite are exposed on the surface of the pediment (fig. 47.13).

Some investigators regard pediments as the only true desert landforms that can be attributed solely to arid conditions operating at present. Others regard pediments as features that have evolved in a sequential manner over a period of many years. At issue are the relative roles of past and present processes in explaining the development of these arid region landforms.

The origin of pediments may be closely linked to the origin of another characteristic desert landform, *inselbergs,* prominent, steep-sided residual hills and mountains rising abruptly from erosional plains (figs. 47.14 and 47.15). Studies in Uganda conclude that inselbergs are residuals of deep chemical weathering during the more humid environments of the late Tertiary

---

[2]Road guide available at Park Visitor Center.

FIGURE 47.13 Pediment (eroded on the White Tank monzogranite) from White Tank Campground. Photo by D.D. Trent.

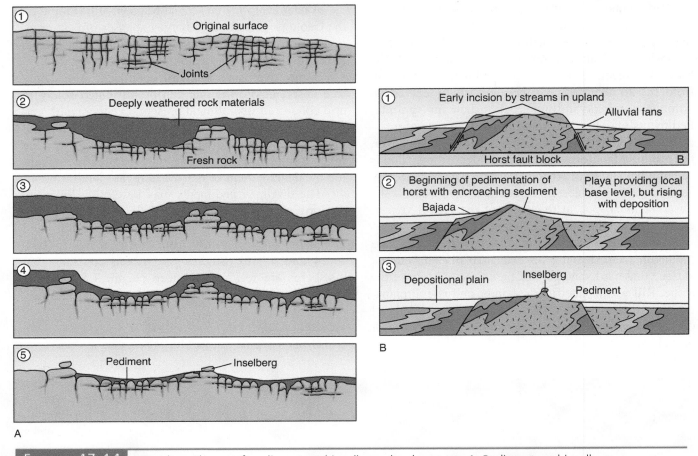

FIGURE 47.14 Two hypotheses of pediment and inselberg development. *A.* Pediment and inselberg development in Uganda (*after Ollier, 1975*): (1) Subsurface jointing in the original substrate. (2-4) Deep and complete weathering of the rock with closely spaced joints, but unconsumed rectangular blocks in regions of widely spaced joints. (5) Removal of weathered rock leaves pediments and inselberg remnants. *B.* Pediment and inselberg development in the southwestern United States resulting from a combination of deep weathering of a horst upland, stream erosion, and rising base levels in the adjacent down-faulted basins. After Garner 1984; Bradshaw and others 1978.

and Quaternary Periods (fig. 47.14A). Inasmuch as subsurface weathering is more intense in areas of closely spaced jointing, but less so in areas of widely spaced joints, pediments form by the removal of the deeply weathered rock materials, leaving behind the sparsely jointed rock residuals as inselbergs.

The origin of inselbergs in Uganda is not totally applicable to the deserts of the southwestern United States where, unlike Uganda, tectonism has been active for millions of years and continues up to the present. Tectonism has created fault-block mountain ranges and down-dropped basins. The internal drainage of the basins results in the gradual filling of the basins with rock debris derived from the adjacent uplands, which

cause a slow rise of local base levels. Stream erosion, accompanied by rising base levels, is important in forming pediments in the Mojave Desert (figure 47.14B).

Seven to nine million years ago the landscape of the Mojave Desert was one of rolling hills covered with a mantle of well vegetated soil rather than the raw, rugged bedrock slopes that we see today. The climate and the amount of plant cover at the time must have been similar to that in today's semiarid mountains of Riverside and San Diego County. The hills contained cores of relatively unjointed, massive granite, whereas the intervening valleys were developed in areas of more intensively jointed, easily weathered and eroded rock. This more humid time was a period of net soil forma-

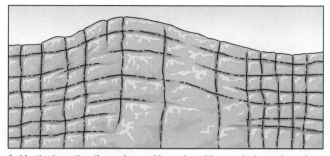

A. Vertical section through granitic rocks with a varied spacing of joints some 20 million years ago.

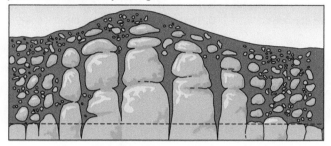

B. During the Pliocene Epoch, after a period of sub-humid climate and decomposition of the rock by groundwater that percolated downward along joints to the water table. Rotted and decomposed rock is shown in black.

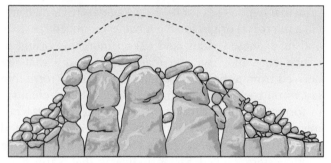

C. Boulder-mantled slopes developed in the past few tens of thousands of years of the Pleistocene Epoch by the removal of the decomposed rock under arid conditions. Present-day examples: along the Fortynine Palms Oasis trail and along the highway between the town of Joshua Tree and Hidden Valley Campground.

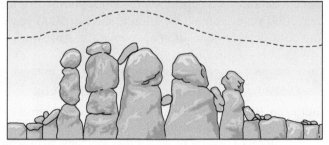

D. The present. In higher elevations with longer exposure to conditions of arid weathering, the boulder mantle has been largely decomposed, leaving steep-sided bold outcrops rising abruptly above the surrounding surface. A thin veneer of grus covers the horizontal surface. Examples are at Hidden Valley, Caprock, Ryan Campground, and Jumbo Rocks.

FIGURE 47.15   Diagram illustrating the formation of inselbergs at Joshua Tree National Park. From Trent 1984.

How arid region landforms differ from landforms in humid regions:

1. The internal drainage basins in deserts provide base levels of erosion that may lie well *above,* or even *below,* sea level. In humid regions, however, the ocean surface provides the base level of erosion.

2. Base levels of erosion in most deserts, and clearly in the Mojave Desert, are constantly *rising* as the products of erosion accumulate within the internal basins; in humid regions, the ocean provides a relatively constant base level.

3. Products of erosion in humid regions are carried great distances, eventually to the ocean. But erosion products in the desert are carried only short distances, resulting in the conspicuous accumulation of loose depris in the form of sand dunes, talus, alluvial fans, and bajadas.

---

tion, with more soil forming than eroding. As the climate dried, however, vegetation receded and erosion increased, which stripped the residual soils from the steeper hillsides, leaving behind the subangular and spherical granite boulders that were too heavy for wind or water to transport. These boulders had been subsurface joint blocks that had become rounded and isolated by chemical decomposition along the joint planes. Today these residual granite boulders, or corestones, form *boulder-mantled slopes,* which may be seen along the road between the city of Joshua Tree and Hidden Valley Campground (figure 47.16). Eventually, boulder mantles crumble into grus leaving only inselbergs, the cores of former hills and ridges, such as those that form the spectacular prominences at Hidden Valley, Cap Rock, Jumbo Rocks, and along the Geology Tour Road (fig. 47.11). The presence of these masses of undecomposed rock is evidence that the renewal of boulder mantles by present-day weathering processes is not taking place. Thus, the granitic landscape of Joshua Tree National Park, and elsewhere in the Mojave Desert, may be thought of as a relict landscape that has evolved over a time span of several million years.

Evidence for this interpretation for the origin of inselbergs and boulder-mantled slopes comes from the Black Mountain along State Highway 247 at Pipes Wash, about 22 miles west-northwest of Twenynine Palms. Here a reddish iron-oxide and calcite-rich soil, and boulders in a weathered ancient soil matrix, have been preserved beneath basalt lava flows with radio-

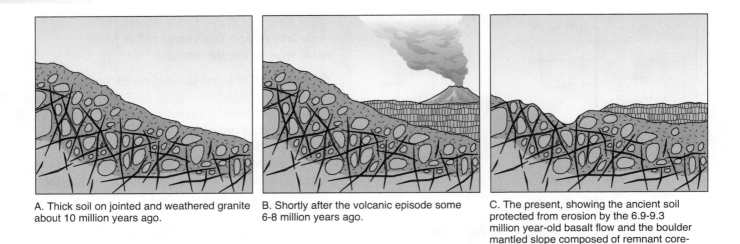

A. Thick soil on jointed and weathered granite about 10 million years ago.

B. Shortly after the volcanic episode some 6-8 million years ago.

C. The present, showing the ancient soil protected from erosion by the 6.9-9.3 million year-old basalt flow and the boulder mantled slope composed of remnant corestones that formed beneath the ancient soil.

FIGURE 47.16   Diagrammatic sketch of geologic relations at Black Mountain along highway 247 north of Yucca Valley. After Trent and Hazlett 2002; age date from Morton 1985.

metric ages of 6.9 to 9.3 million years[3]. Similar reddish soils form today in warm regions under the cover of dense plant growth where the average rainfall exceeds 10 inches annually. Continuity between these buried relict soils and associated debris, and the boulder-mantled slopes, establishes the boulder mantles as relict features inherited from a time of semiarid, deep chemical weathering, most likely about 9 million years ago in the late Teritary Period (fig. 47.16). Presumably, the boulder-mantled slopes in Joshua Tree National Park are of approximately the same age as those at nearby Black Mountain.

The Final Polish

Nearly all of the rock surfaces in the region of the park show some degree of *desert varnish,* a thin patina of insoluble clay, plus iron and manganese oxides. In some cases the surface impregnation of varnish is deep enough into the partially decomposed rock that it binds the material together and produces a dark-brown, metallic-looking rind called "case hardening." The Proterozoic gneiss and the monzogranite cropping out at Indian Cove and along the Fortynine Palms Oasis Trail reveal especially good examples of desert varnish.

Varnish is not unique to the desert, but it is best revealed there. Varnish forms today wherever water seeps onto rock surfaces. In humid regions, it forms in tunnels and along railroad cuts; in the southwest, it

forms where there are seeps along canyon walls (Chapters 5 and 8). Apparently water is needed to transport the iron and manganese onto rock surfaces. The principal hypotheses for the origin of desert varnish are: (1) a microbial origin in which bacteria concentrate iron and manganese oxides, and (2) an inorganic origin in which clay and iron and manganese oxides that are derived from airborne dust and other sources form thin layers on rock surfaces. Regardless of the mechanism of formation, the varnish formed long ago when the climate was different from that of today. Abundant and archeological evidence from the Old World and the southwestern United States indicates that varnish on today's dry surfaces must have been deposited more than 2000 years ago. Examination of the pyramids and other stone monuments in Egypt indicate that there has been essentially no deposition of desert varnish for the last 2000 years, some deposition in the last 5000 years, but considerable deposition on even older stoneworks.

Geologic Map and Cross Section

Bortugno, E.J., and Spittler, T.E., 1986, Geologic Map of the San Bernadino Quadrangle, Regional Map Series: California Division of Mines and Geology, map scale 1:250,000.

Dibblee, T.W., Jr., 1967. Geologic Map of the Joshua Tree Quadrangle, San Bernardino and Riverside Counties, California: U.S. Geological Survey Miscellaneous Geologic Investigation Map I-516, scale 1:62,500.

——— 1968. Geologic Map of the Twentynine Palms Quadrangle, San Bernardino and Riverside Counties, California: U.S. Geological Survey Miscellaneous Geologic Investigation Map I-561, scale 1:62,500.

Sources

Allen, C.C., 1978. Desert Varnish of the Sonoran Desert—Optical and Electron Probe Micro Analysis. *Journal of Geology*, v. 86, n. 6, p. 743–752.

Axelrod, D.I., 1958. *Evolution of the Macro-Tertiary Paleobotany*: Carnegie Institution of Washington, Publication 590, Washington, D.C., 323 p.

Bradshaw, M.J., Abbott, A.J., and Gelthorpe, A.P., 1978. *The Earth's Changing Surface*. Halsted Press, John Wiley and Sons, 336 p.

Brand, J.H., and Anderson, J.L., 1982. Mesozoic Alkalic Monzogranites and Peraluminous Adamelites of the Joshua Tree National Monument, Southern California (abstract): Geological Society of America Abstracts with Programs. v. 14, n. 4, p. 151–152.

Decker, Barbara, and Decker, Robert, 1994. *Road Guide to Joshua Tree National Park*. Double Decker Press, 48 p.

Ernst, W.G., 1981. Summary of the Geotectonic Development of California, *in* Ernst, W.G., editor, *The Geotectonic Development of California*. Prentice-Hall, Inc., p. 601–613.

Garner, H.F., 1974. *The Origin of Landscapes—A Synthesis of Geomorphology*. Oxford University Press. 734 p.

Hopson, R. Forrest, 1998. Quaternary Geology and Neotectonics of the Pinto Mountain Fault, Mojave Desert, Southern California. *California Geology*, v. 51, n. 6, Nov.–Dec, p. 3–63.

Hughes, III, R. O., Evenson, E. G., Gosse, J. C., and Harrington, C., 1995. Tafoni, Genesis in Arid Climates; Paleoclimatological Implications [abs.], Abstracts with Programs, Geological Society of America 30th Annual Northeastern Sectional meeting, p. 56.

Morton, D.M., 1985 *cited* in Reynolds, R.E., Busing, A.V., and Beratan, K.K., 1992, Old Routes to the Colorado: The 1992 Mojave Desert Quaternary Research Center Trip *in* Reynolds, R.E., editor, *Old Routes to the Colorado*: Redlands, California, San Bernadino County Museum Association, Special Publication 92-1, p. 5–27.

Oberlander, T.M., 1972. Morphogenesis of Granite Boulder Slopes in the Mojave Desert, California. *Journal of Geology*, v. 80, n. 12, p. 1–20.

——— and Dorn, R.I., 1981. Microbial Origin of Desert Varnish: *Science*. v. 213, p. 1245–1247.

Ollier, C.D., 1975, *Weathering*. Oliver and Boyd, London, 384 p.

Potter, R.M., and Rossman, G.R., 1977. Desert Varnish: The Importance of Clay Minerals. *Science*, v. 196, p. 1446–1448.

Powell, R.E., 1993. Balanced Palinspastic Reconstruction of Pre-late Cenozoic Paleogeology, Southern California: Geologic and Kinematic Constraints on Evolution of the San Andreas Fault System, *in* Powell, R. E., R. J. Weldon II, and Jonathan C. Matti, Memoir 178, The San Andreas Fault System: Displacement, Palispastic Reconstruction, and Geologic Evolution. Geological Society of America, p. 1–106.

Streckeisen, A.L., 1973. Classification and Nomenclature Recommended by the I.U.G.S. Subcommision on Systematics. *Geotimes*, v. 18, p. 26–30.

Trent, D.D., and Hazlett, Richard W., 2002, Geology of Joshua Tree National Park. Joshua Tree National Park Association, Twentynine Palms, CA. 65 p.

Wooden, J.L., Tosdal, R.M., Howard, K.A., Powell, R.E., Matti, J.C., Barth, A.P., 1994. Mesozoic Intrusive History of Parts of the Eastern Transverse Ranges, California: Preliminary U-Pb Zircon Results (abstract). Geological Society of America Abstracts with Programs, v. 26, n. 2, p. A104–105.

Note: To convert English measurements to metric, go to www.helpwithdiy.com/metric_conversion_calculator.html

Death Valley National Park

EASTERN CALIFORNIA AND SOUTHWESTERN NEVADA

Lauren A. Wright   Pennsylvania State University
and Marli B. Miller   University of Oregon

Area: 3,299,840 acres; 5156 square miles	**Address:** P.O. Box 579, Death Valley, CA 92328
Proclaimed a National Monument: February 11, 1933	**Phone:** 760-786-3200
Established as a National Park: October 31, 1994	**E-mail:** DEVA_Superintendent@nps.gov
Designated a Biosphere Reserve: 1984	**Website:** www.nps.gov/deva/index.htm

The hottest, driest, lowest place in North America, is the floor of Death Valley.
Ancient rocks and young rocks, in complex exposures, record a variety of faults and displacements.
Snow-topped peaks rise abruptly thousands of feet from the valley floor, which is below sea level.
Displayed in the barren landscape are salt pans and playas, young volcanic craters, large alluvial
fans, imposing fault scarps, sand dunes, wineglass canyons, turtlebacks, abandoned mines, and more.

FIGURE 48.1   Double rainbow and central Funeral Mountains. Photo by Marli B. Miller.

Geographic Setting and Human History

True to its reputation, the floor of Death Valley is, indeed, the hottest, driest, and lowest land in the United States. Summer temperatures frequently exceed 120° F and the mean annual precipitation is only about 1.5 inches. But, more importantly, Death Valley National Park encloses one of the world's most spectacular desert landscapes and a mountain range that remains snow-capped during most winter seasons. Even on the floor of the valley, temperatures occasionally drop below freezing.

Death Valley lies in the Mojave Desert of southeastern California and is arid because it lies in the rain shadow of several more westerly mountain ranges, including the Sierra Nevada. It also contains the low point of a large region of interior drainage in the southwestern part of the Great Basin. The valley floor, which extends to a depth of −282 feet near Badwater (fig. 48.2), lies between north- to northwest-trending mountain ranges (fig. 48.3). The Panamint Mountains, west of Central Death Valley, being the highest, culminate in Telescope Peak at an elevation of 11,049 feet. The relief between there and the valley floor is one of the greatest obtained within the conterminous United States.

Prehistoric people occupied the Death Valley region for at least 10,000 years. They lived at lower elevations during the winters and at higher elevations in the summers, although at times of milder climatic conditions they could live on the valley floor throughout the year. They obtained water from the region's numerous springs and collected mesquite beans and pinyon nuts as dietary staples.

In 1849, the first people of European descent wandered into Death Valley, having lost their way to the newly discovered gold fields. For the next 80 years, mining dominated the thoughts of most visitors to the region, as prospectors located deposits of gold, silver, copper, lead, and, beginning in the 1880s, borate minerals and talc. The mining of metals was hindered by the extreme aridity. Only the Keane Wonder gold mine and the gold mines at Rhyolite and Harrisburg produced ore of significant value. Both borate and talc mining, however, proved profitable. During the 1880s, borax was recovered from evaporite deposits scraped from the salt pan of the valley floor. In the early 1900s, large deposits of borate minerals were discovered in the Miocene strata of Furnace Creek Wash. By the late 1920s, the Death Valley region had become the world's leading source of borax. Although mining continues in the vicinity of Death Valley National Park today, tourism is the region's principal industry. More than one million persons, including large numbers from other countries, enter the Visitor Center each year.

Shaping of the Present Landscape: Extensional Tectonics and the "Basin and Range Event"

General Features

First-time visitors to Death Valley National Park invariably and correctly sense that these mountains and valleys have formed quite recently in geologic time. Beginning about 16 million years ago and continuing today, this part of the earth's crust has been broken into a gigantic mosaic of mountain blocks (fig. 48.3) each bounded by major faults. Cenozoic sedimentary and volcanic rocks deposited in the intervening basins during the development of the present landscape provide evidence for how and when it formed and for the changing climates of that interval of time. Older sedimentary formations, deposited 35 to 16 million years ago, can be correlated, from place to place, over much larger areas than the deposits of the later and smaller basins. The older Tertiary formations, when reassembled, thus record a considerably less mountainous terrain than the one we observe today.

Most of the present ranges consist of Proterozoic, Paleozoic and/or Mesozoic formations that predate the forming of these ranges by many millions of years (Table 48.1). In fact, we owe much of our knowledge of the earlier events to the fact that these formations have been brought to view along the tilted faces of the ranges. We observe in the oldest rocks evidence of events that shaped and metamorphosed the crust more than 1.7 million years ago. An even clearer record of ancient marine and fluvial environments and accompanying igneous activity is contained in the extensive exposures of the Pahrump Group, emplaced within the 1.2- to 0.8-billion-year interval of time. Largely because the Pahrump rocks remain in essentially their pristine state and are so well exposed on the barren slopes, they have attracted the attention of geologists the world over. The stark mountain slopes also contain continuous exposures of the later Proterozoic and Paleozoic formations that record the pre-Mesozoic history of the western margin

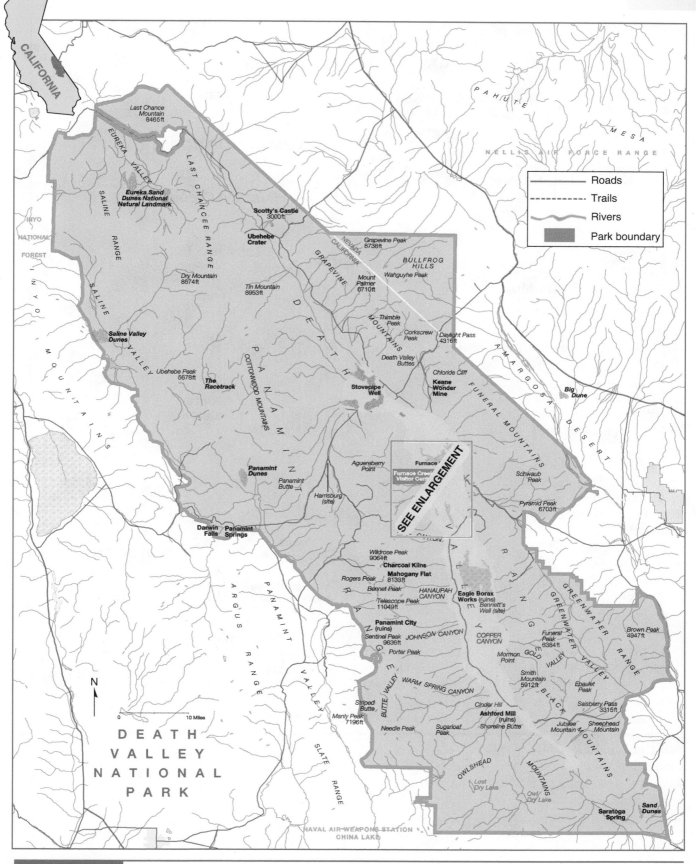

FIGURE 48.2A   Death Valley National Park, California.

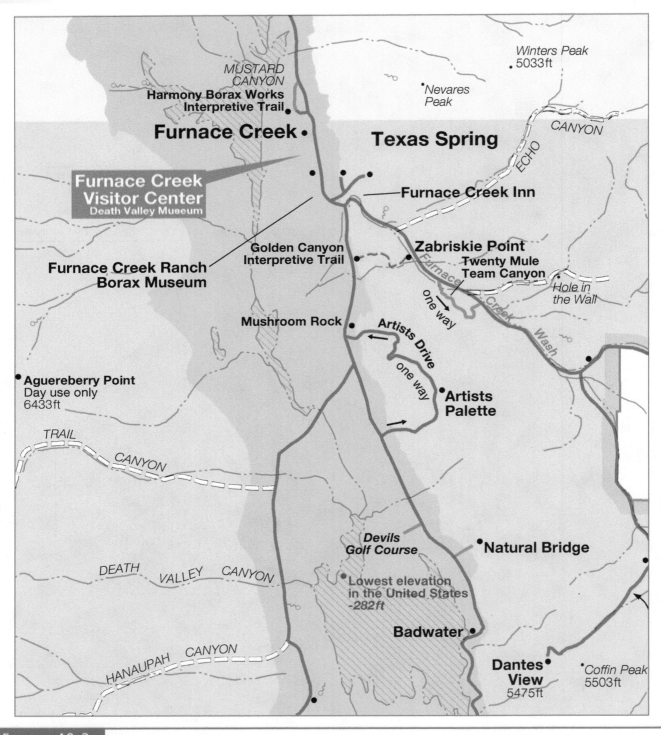

Figure 48.2B   Enlargement.

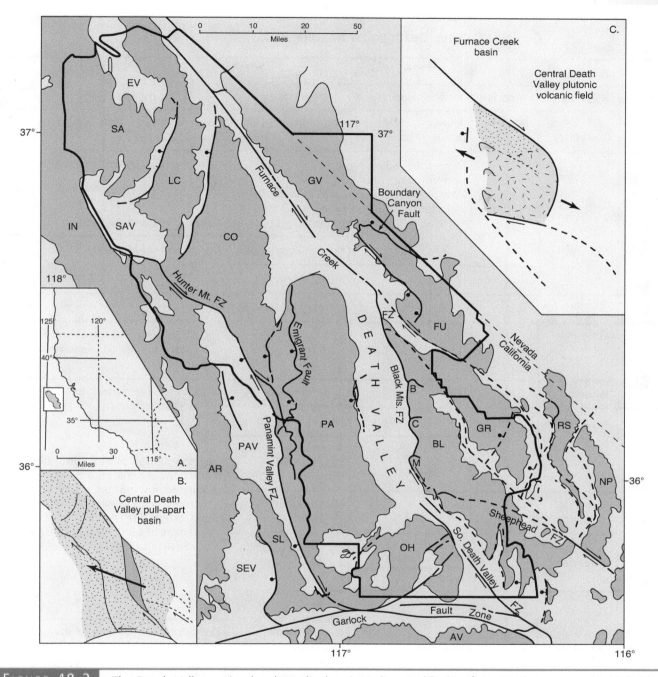

FIGURE 48.3   The Death Valley region has been broken into discrete blocks of crust—the ranges—bounded by major faults, as shown here.

Inset maps: *A.* Location of Death Valley National Park. *B.* Sketch map showing simplified fault pattern—the stippled area of relatively severe extension—between the Furnace Creek fault zone and the Garlock fault zone; also showing the Central Death Valley pull-apart basin between the Southern Death Valley fault zone and the active part of the Furnace Creek fault zone. *C.* Diagram showing an interpretation of the tectonic setting of the Furnace Creek basin and the Central Death Valley plutonic-volcanic field in a rhomboidal area between the Furnace Creek and Sheephead fault zones. **Key to ranges:** AR—Argus Range; AV—Avawatz Mountains; BL—Black Mountains; CO—Cottonwood Mountains; FU—Funeral Mountains; GR—Greenwater Range; GV—Grapevine Mountains; IN—Inyo Mountains; LC—Last Chance Range; NP—Nopah Range; OH—Owlshead Mountains; PT—Panamint Mountains; SA—Saline Mountains; SL—Slate Range. **Key to valleys:** EV—Eureka Valley; PV—Panamint Valley; SAV—Saline Valley; SEV—Searles Valley. **Key to turtlebacks:** B—Badwater turtleback; C—Copper Canyon turtleback; M—Mormon Point turtleback. (The curves in the fault trace along the west edge of the Black Mountain block locate the three turtlebacks.)

| Table 48.1 | | Geologic Column, Death Valley National Park | | |

Time Units			Rock Units		Principal Geologic Events
Era	Period	Epoch	Group	Formation	
Cenozoic	Quaternary	Holocene	Alluvial fans, stream, and playa deposits, dunes		Continued deposition in modern Death Valley
	Tertiary	Pliocene / Miocene	Numerous sedimentary, volcanic, and plutonic units in separate and interconnected basins and igneous fields; includes Artist Drive, Furnance Creek, Funeral, and Nova Formations.		Opening of modern Death Valley / Continuing development of the present ranges and basins / Onset of major extension
				Several Formations	Deposition on relatively subdued terrain
		Oligocene		Titus Canyon	
			~~ Major Unconformity ~~~~~		
Mesozoic	Cretaceous/ Jurassic		Granitic plutons		Thrust faulting and intrusion of plutons related to Sierra Nevada batholith
	Triassic			Butte Valley	Shallow marine deposition ~~~ Unconformity ~~~
Paleozoic	Pennsylvanian			Resting Spring Shale	Development of a long-continuing carbonate bank on a passive continental margin; numerous intervals of emergence, interrupted by deposition of a blanket of sandstone in Middle Ordovician time
	Mississippian			Tin Mountain Limestone / Lost Burro	
	Devonian/ Silurian			Hidden Valley Dolomite	
	Ordovician			Ely Springs Dolomite / Eureka Quartzite	
			Pogonip		
	Cambrian			Nopah / Bonanza King / Carrara / Zabriskie Quartzite / Wood Canyon	Deposition of a wedge of siliciclastic sediment during and immediately following the rifting along a new continental margin
Prototerozoic			Pahrump	Stirling Quartzite / Johnnie / Ibex / Noonday Dolomite	Shallow to deep marine deposition along an incipient continental margin ~~~ Unconformity ~~~
				Kingston Peak / Beck Spring / Crystal Spring	Glacio-marine deposition / Shallow marine deposition / Rapid uplift and erosion
			~~ Major Unconformity ~~~~~		
			Crystalline basement		Regional metamorphism

of North America. So these exposures, too, continue to be of widespread interest.

On the other hand, the contiguous Black Mountains and Greenwater Range, in the eastern part of the park, are underlain largely by bodies of Tertiary igneous rocks, both extrusive and intrusive. These were added to the preexisting crust 12 to 4 million years ago. When the Tertiary record is combined with that of the older rocks preserved in the ranges, geologists can reconstruct an extraordinarily complete record of crustal evolution.

The Faults

In the faults that bound the ranges and basins of the Death Valley region (fig. 48.3), as in the rest of the Basin and Range province, we find visible evidence of an extending crust. The faults are the principal ruptures along which the brittle, upper part of the crust has broken as the great block of the Sierra Nevada has moved westward, away from the west side of the Colorado Plateau. The land between the two has been literally pulled apart. The study of fault patterns generated in this way belongs to a branch of geology called "extensional tectonics."

Classification. The faults that define the ranges and valleys of Death Valley National Park (fig. 48.3) are broadly divisible into three kinds: strike-slip, high-angle normal and low-angle normal. The strike-slip faults, along which movement has been dominantly parallel with the strike of the fault planes, are identified by arrows showing sense of lateral movement. Note that most of these faults strike northwestward and that their southwest sides have moved relatively northwestward, producing a "right-lateral" sense of slip. Movement on the faults identified as normal has been mainly downdip. The large dots in the diagram identify the downthrown sides of both high-angle and low-angle normal faults. The normal faults, by virtue of their geometry, are the simplest expressions of crustal extension; the lower the angle, the greater the opportunity for large-scale extension.

Most students of the structural framework of the Death Valley region view the major range-bounding faults as terminating at depth against essentially horizontal "detachment surfaces." The crust theoretically behaves in a brittle fashion above these surfaces and in a ductile fashion beneath them. Some earth scientists place the principal detachment surface beneath much of the Death Valley region at mid-crustal levels (Serpa et al. 1987); others favor a much shallower depth (Wernicke et al. 1988).

Furnace Creek strike-slip fault zone. The best known of the strike-slip faults exposed within the park boundaries compose the Furnace Creek fault zone (fig. 48.3). These define a linear crustal rupture that extends from the vicinity of Brown Peak (fig. 48.2) northwestward for about 150 miles, including Furnace Creek Wash and the full length of northern Death Valley. A major fault in this zone is particularly obvious along the northeast side of Furnace Creek Wash where it brings Miocene and Pliocene strata of the Furnace Creek Basin into contact with the Proterozoic and Paleozoic formations of the Funeral Mountains (McAllister, 1970). The strike-slip fault truncates the southeast-tilted fault blocks of the Funerals. The Miocene and Pliocene strata, being as much as 15,000 feet thick, also require a maximum vertical displacement of a comparable dimension.

Geologists, beginning with Stewart and his coworkers (1968), have carefully inspected the pre-Cenozoic rocks on both sides of the Furnace Creek fault zone, and have observed features that were once joined and are now separated by movement along the fault zone. In making these matches, they find compelling evidence for displacements measurable in tens of miles.

High-angle normal faulting along the Black Mountains front. The precipitous west face of the Black Mountains is one of the world's most spectacular fault-controlled escarpments. It has been produced by movement on high-angle normal faults along which the Black Mountains are tilting eastward and the floor of Death Valley is dropping downward. Geomorphic features of this escarpment that indicate its recency are discussed below.

The Boundary Canyon low-angle normal fault. The Boundary Canyon fault is the most conspicuous of the low-angle normal faults in the Death Valley region. It is exposed in the northern part of the Funeral Mountains and dips gently northwestward beneath the highly folded and faulted formations of the Grapevine Mountains. Thus the Grapevines lie in the upper plate, and all but a small part of the Funerals lie in the lower plate. In the northern Funerals, the lower plate consists of Proterozoic rocks, including all three formations of the Pahrump Group, the Johnnie Formation and the lower part of the Stirling Quartzite (Table 48.1). The upper plate contains the upper part of the Stirling Quartzite and all of the overlying formations through those of Mississippian age (fig. 48.4).

The Boundary Canyon fault is of special interest in that the rock units of the lower plate have been metamorphosed at temperatures and pressures that characterize mid-crustal levels, whereas the rocks of the upper plate remain essentially unmetamorphosed (fig. 48.4). The components of such a geologic setting are commonly and collectively called a *metamorphic core complex* and require large displacement along the low-angle normal fault that separates the two plates (Chapters 45 and 46).

The Boundary Canyon fault is easily discernible along the west face of the Funeral Mountains east of the highway as one approaches the mouth of Boundary Canyon from the south, and is also exposed on both sides of the lower part of the canyon. In this area, the fault dips gently to the northwest and separates light-colored, unmetamorphosed strata of the middle part of the Stirling Quartzite from drab exposures of strongly metamorphosed and sheared units of the middle and lower parts of the Johnnie Formation (fig. 48.4). Equally metamorphosed rock units of the underlying Pahrump Group are superbly exposed in nearby Monarch Canyon, which drains westward from the crest of the Funeral Mountains.

The Amargosa fault and the Amargosa chaos. Of the extension-related phenomena of the Death Valley region, the best known, most complex, and most controversial is the Amargosa chaos, first described by Noble (1941) and mapped in detail by Wright and Troxel (1984). Noble originally recognized three phas-

FIGURE 48.4   Northeastward view of the Boundary Canyon detachment fault at the northern end of the Funeral Mountains. The fault, which dips northeast to extend beneath the Grapevine Mountains on the skyline, separates underlying dark-colored, highly metamorphosed rocks of the Proterozoic Johnnie Formation and the Pahrump Grounp (foreground) from immediately overlying light-colored and essentially unmetaorphosed rocks of the Proterozoic Stirling Formation. Dark-colored Cambrian rock units form the higher slopes of the upper plate. Photo by Martin Miller.

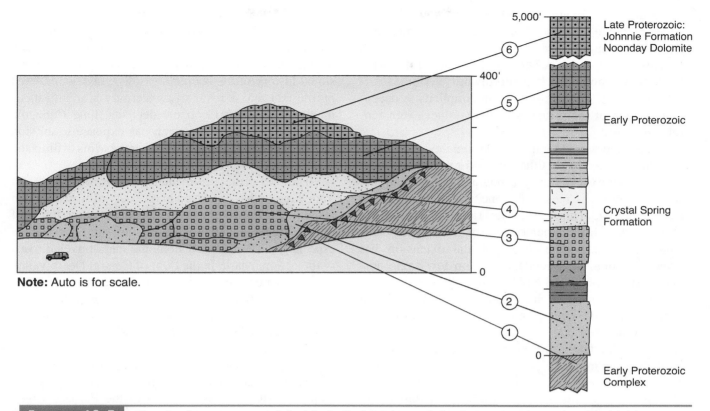

Note: Auto is for scale.

FIGURE 48.5   Sketch of an exposure of the Virgin Spring phase of the Amargosa chaos showing highly attenuated units of the Crystal Spring Formation compared with a columnar section of these units in their natural thicknesses. Modified from Wright and Troxell, 1984.

es of the chaos, but the one that he named the "Virgin Spring phase" is now viewed as true chaos in the sense that he introduced the term. It is exposed in separate localities in the southern part of the Black Mountains from the vicinity of Virgin Spring and Rhodes Washes northward to Gold Valley. An excellent exposure of the chaos lies immediately south of the highway at a point 1.5 miles east of Jubilee Pass.

The general characteristics of this feature are shown in Figure 48.5. In simplest terms, the chaos consists of a mosaic of fault-bounded blocks of Proterozoic and Cambrian formations, arranged in proper stratigraphic order, but highly attenuated to a small fraction of the actual combined thickness of the formations represented. Everywhere the chaos rests, with highly sheared fault contact, upon essentially intact occurrences of the Early Proterozoic crystalline complex. All of the faulting is brittle in nature. In the Gold Valley area, in the central part of the Black Mountains, the chaos is intruded by granitic bodies that predate 10-million-year-old volcanic units. Thus the chaos may be

the oldest extension-related structural feature in the Black Mountains block. Noble originally interpreted the chaos as remnants of a single thrust fault of regional extent and named by him the "Amargosa thrust." It is now ordinarily viewed as composing segments of one or more low-angle normal faults along which the upper plate has been severely extended. The occurrences of the Amargosa chaos have been so disordered by later faulting and folding as to make its original configuration difficult or impossible to accurately reconstruct.

The Ranges

Although the mountain ranges within the boundaries of Death Valley National Park have formed in a framework of interrelated Cenozoic faults, they differ widely in their structural settings and in the principal rock bodies of which they are composed. Most of the ranges of the Death Valley region qualify as fault-block ranges in that they are bounded on one side by a major normal fault and have acquired much of their relief by tilting along that fault. Examples within the park are the Last Chance Range and the Panamint, Cottonwood, and Black Mountains. The Resting Spring and Nopah Ranges, east of the park boundary, also are of this type. Each trends approximately northward and tilts eastward (fig. 48.3). But this apparent simplicity is deceptive as each of these ranges contains other extension-related faults.

The Funeral and Grapevine Mountains differ from the other ranges in that they lie end-to-end forming a single, northwest-trending topographic high parallel with the Furnace Creek fault zone, which bounds it on the southwest. This composite range has extended along normal faults oriented approximately perpendicular to its backbone. Most of the extension has occurred along the Boundary Canyon low-angle normal fault described earlier. Normal faults that dip more steeply are a feature of the southern part of the Funeral Mountains. The Funerals and the Grapevines also have tilted northeastward as a result of a vertical component accompanying lateral movement on the Furnace Creek fault zone. So in this respect, even these ranges qualify as block-fault features.

Most of the ranges of the Death Valley region display the striped outcrops that identify the evenly bedded Proterozoic and Paleozoic formations. The Black Mountains and Greenwater Range stand out from the rest because they are underlain mostly by the less regular bodies of Tertiary plutonic and volcanic rocks. The Owlshead Mountains, in the southern part of the park, are also distinctive in that they are equidimensional in map view, and apparently coextensive with a cluster of granitic plutons of Mesozoic age. The plutons are discontinuously covered with Miocene andesitic flows and are also offset by strike-slip faults. Most of the latter strike northeast and have moved in a left-lateral sense.

The Basins

The Cenozoic sedimentary deposits that have accumulated in the topographic depressions between the ranges of the Death Valley region consist mainly of debris eroded from the high areas and deposited in alluvial fans, ephemeral and perennial lakes, and stream beds. Also included are accumulations of evaporites, principally limestone, gypsum and salt, brought in solution by streams that terminate in the basins. The sedimentary fill is interlayered with extrusive volcanic rocks.

By observing the shapes of these basins and the distribution of the various kinds of sedimentary rocks

that they contain, by dating the extrusive volcanic bodies in the basins, and, on occasion, by employing geophysical methods to detect subsurface features, we can reconstruct the development of the basins and the erosional history of the source areas. The task is hindered by the fact that a cover of Quaternary alluvium hides much of the older Cenozoic deposits. Late Cenozoic faulting and folding, however, has exposed to erosion the pre-Quaternary rocks of several basins within the park boundaries.

The Cenozoic sedimentary basins of the Death Valley region have evolved in a variety of ways, each in response to the interplay of the three major types of faults discussed above. Deposits of three basins are especially well displayed along the main-traveled roads of the national park. In the floor of Central Death Valley sediments are accumulating today in a grabenlike "pull-apart" basin bounded on the east by the zone of normal faults that defines the front of the Black Mountains (figs. 48.3 and 48.7). A thick succession of sedimentary deposits and lava flows, of Middle Miocene to Late Pliocene age, is spectacularly exposed in Furnace Creek Wash. This succession defines the Furnace Creek Basin which, because it lies adjacent to the Furnace Creek strike-slip fault zone, has been called a "strike-slip" basin. In the Emigrant Canyon-Towne Pass area on the northwestern flank of the Panamint Mountains, we observe Late Miocene through Pliocene conglomerates and basaltic to rhyolitic lava flows of the Nova Formation which, in turn, defines the Nova Basin. The Nova Formation has been deposited above a low-angle normal fault called the "Emigrant detachment."

The "pull-apart" basin, of which Central Death Valley is representative, is so named because it involves crust that has been pulled apart between the *en echelon* terminations of two strike-slip fault zones in the manner shown in figure 48.3 (Burchfiel and Stewart, 1966). Although the Furnace Creek fault zone extends well to the southeast of the northern end of Central Death Valley basin, this segment has been less active in Pliocene and Quaternary time than the main, more northwesterly trace of the fault zone. So the fault zone has been viewed as effectively terminating in the vicinity of Furnace Creek Ranch during the development of the Central Death Valley basin. The alluvial fans, salt pan, and lake and fluvial deposits of this basin (Hunt and Mabey, 1966) are described later in this chapter.

Of the Cenozoic basinal deposits exposed within Death Valley National Park, those of the Furnace Creek basin are the most obvious and most admired by visitors. In Furnace Creek Wash and the adjoining northern part of the Black Mountains, we observe the various sedimentary units and basaltic to rhyolitic lava flows that identify the Furnace Creek Basin. These were deposited in Middle Miocene through Pliocene time on crust that lies southwest of a major fault in the Furnace Creek fault zone (McAllister, 1970, 1971, 1973; Cemen et al., 1985) and northwest of another major fault exposed in the vicinity of Badwater. The crust there has moved downward, as well as laterally, to form the northeast margin of the elongate, troughlike Furnace Creek basin. The basinal deposits are about 12,000 feet in maximum estimated thickness in the northwestern part of the basin, but are much thinner in the southeastern part.

The panorama of these deposits viewed from Zabriskie Point is among the most photographed in the entire National Park system. Owing largely to the downdip movement on the frontal fault of the Black Mountains, the sedimentary and extrusive volcanic rocks of this basin have been laid bare and are shown to have been folded onto a broad syncline. The northeast-dipping rock units that compose the Middle and Late Miocene Artist Drive Formation are exposed on both sides of the crest of the Black Mountains. The successively overlying Late Miocene Furnace Creek Formation and Pliocene Funeral Formation coincide with Furnace Creek Wash. The conglomerates of the Furnace Creek Formation, exposed on both sides of the wash, accumulated in alluvial fans that bordered a lake-dominated central part of the basin.

To the Death Valley traveler, the interlayered conglomerates and volcanic rocks of the Nova basin are less accessible than are the deposits of the Furnace Creek basin. They are also more disordered and thus less obviously displayed. This is largely because the basin was fragmented concurrently with the deposition of the Nova Formation and with movement on the underlying Emigrant detachment fault. On the other hand, it is an excellent example of a basin that has evolved in a structural setting of this kind (Hodges et al., 1989).

The Nova Formation is estimated to be more than 6000 feet thick. It is exposed along State Highway 190 both east and west of Towne Pass. The upper part of the Nova is well exposed in Emigrant Canyon on both sides of the road that connects State Highway 190 with Harrisburg Flat and Wildrose Canyon. The conglomerates there, as well as elsewhere in the formation, are derived from the Proterozoic formations exposed in the higher parts of the Panamint Mountains.

Geomorphic Features Related to Active Faulting along the Black Mountains Front

Transition from valley floor to mountain front. The extraordinarily abrupt topographic break between the Black Mountains escarpment and the sedimentary fill of the valley marks the approximate location of the Black Mountains fault zone. Here the mountains rise while the valley floor drops. In most places, relatively small alluvial fans spill out of deep canyons at the fault zone; where there are no canyons, the horizontal deposits of the valley floor meet a wall of rock. These features indicate recent vertical movement because erosion has not had time enough to cut back into the mountain front. Much of the mountain front, therefore, qualifies as a *fault scarp,* an exposed surface produced directly by movement on a fault and essentially unmodified by erosion or weathering. When plotted on a map, the mountain front is shown to be a composite of numerous straight-line segments, each representing a single fault rather than one continuous fault.

The morphology of the Black Mountains escarpment provides important information about the nature of the fault that produced it. When viewed from a distance, such as from the Devil's Golf Course, parts of the mountain front appear incredibly smooth and probably do closely coincide with the actual fault surface. If so, the fault dips westward, the Black Mountains are in the footwall, and the relief is an effect of crustal extension.

Fault scarps. Fault scarps in alluvial fans along the Black Mountains front provide additional and direct evidence that the front is presently active. That they are faults rather than stream-cut features is attested by the observation that they cut across stream-related features. Particularly obvious fault scarps cut the fan immediately south of Badwater (fig. 48.6) and other fans at Mormon Point. The scarp at Badwater is aligned with the spring that feeds the Badwater pond and suggests that the spring is controlled by the fault. Other scarps are very well defined between Furnace Creek Inn and Artist Drive and along the range front south of Natural Bridge.

Faceted spurs. Many west-trending ridges, or spurs, in the Black Mountains end abruptly at about an elevation of about 1500 feet. Below and west of the spurs lies the

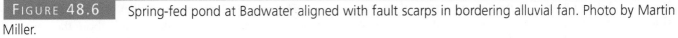

FIGURE 48.6   Spring-fed pond at Badwater aligned with fault scarps in bordering alluvial fan. Photo by Martin Miller.

relatively smooth, steep face of the mountain front which, between the bounding canyons, is triangular in shape. The spur thus appears "faceted," that is, abruptly truncated by the fault zone at the front of the range. Faceted spurs thus indicate that this lower part of the range has been exposed so recently, by the dominantly normal movement along the fault zone, that the spurs remain essentially uneroded.

Wineglass Canyons. On the west side of the Black Mountains escarpment, steep-walled incisions extend eastward into the mountain front. These are the wineglass canyons (fig. 48.7). When viewed from the valley floor, at a right angle to the escarpment, the canyons resemble the cross section of a wineglass. An alluvial fan forms the base of the wineglass; the narrow slot above is the stem; the higher, wider part is the bowl of the glass. The bowls display a rather steep lower part and a flaring, or less steep, upper part. Many of the wineglass canyons have a dryfall at the mouth, at the bottom of the near-vertical stem.

The wineglass stem records the most recent, and still continuing, episode of downdip movement on the frontal fault (i.e., the fault along the base of the mountain). The vertical walls are the result of rapid downcutting by stream action. The bottom of the wineglass bowl marks the former position of the valley floor and indicates a stable interval long enough to permit the canyon to widen by the processes of stream erosion and mass wasting. The less steep, higher part of the bowl is attributable to a still earlier and apparently longer interval of little or no downcutting. The longer period of time would have permitted the canyon to widen to a greater degree than it did during the preceding stable interval.

Particularly good examples of wineglass canyons along the Black Mountains escarpment are at Gower Gulch, immediately south of Golden Canyon, and at Tank Canyon, 0.3 mile south of Natural Bridge. Numerous others can be seen between Badwater and Mormon Point. Titus Canyon along the front of the Grapevine Mountains is still another.

FIGURE 48.7   A wineglass canyon incised in the Black Mountains escarpment. Photo by Martin Miller.

Smaller faults. Much of the bedrock along the Black Mountains front is broken or crushed by movement along the frontal fault zone. Small-scale faults parallel the frontal fault zone and probably formed during movement on it. Where well exposed, these faults display striated surfaces formed by abrasion during slip, the striations being parallel to the direction of movement. The striations on many of these faults trend obliquely and suggest a strong lateral motion during the uplift of the Black Mountains.

Turtleback surfaces. At three places along the west face of the Black Mountains, the mountain front lacks the nearly planar appearance produced by the faceted spurs, and displays the well known and much debated "turtleback" surfaces. One is in the vicinity of Badwater, another immediately south of Copper Canyon (fig. 48.8), and the third at Mormon Point. Each of these features (1) is a convex upward topographic surface, shaped like a turtle shell; (2) curves northeastward into the mountains; and (3) is underlain by a core of the ancient basement complex, mantled by

younger, metamorphosed units, consisting mostly of marble. The mantling rocks have been consistently correlated with the Noonday Dolomite and the Johnnie Formation. Both the basement complex and the overlying younger metamorphic rocks have been broadly folded along northwest plunging axes.

The turtleback surfaces owe their identity to the fact that they are also fault surfaces exposed so recently that they remain essentially uneroded. The faults themselves are excellently exposed along the base of each turtleback surface, where they place the cores of the ancient metamorphic rocks against overlying Cenozoic igneous and sedimentary rocks. Like the other faults along the Black Mountains front, the turtleback faults are normal faults. The turtleback faults differ from the range front faults, however, in that they form surfaces against which many faults in the overlying rocks terminate. In this way the turtleback faults allow the overlying rocks to deform independently of the underlying rocks. The distinctive turtle-shell shapes simply reflect the large folds in the metamorphic rocks below the faults. This control is enhanced by the ten-

Badwater turtleback surface. Photo by Martin Miller.

dency of the turtleback faults to parallel the foliation in the folded metamorphic rocks (Miller, 1991).

The textures and compositions of the metamorphic rocks that mantle the ancient complex below the turtle-back faults indicate that the folding occurred at higher temperatures than exist near the surface. The temperatures were hot enough to cause the rocks to flow instead of fracture. These features provide evidence that these metamorphosed rocks have risen from relatively deep crustal levels along downward projections of the turtle-back faults.

The Black Mountains and the Basin Ranges from Dantes View

From Dantes View at the crest of Black Mountains, nearly 5800 feet above the valley floor, the visitor gains a perspective of Death Valley that is equally instructive but different from the one obtained from below. To the east and west, the landscape consists of alternating, nearly parallel mountain ranges and intervening basins. This basin-and-range topography characterizes much

of the western United States, including all of Nevada, and parts of southern Idaho and southeastern Oregon, western Utah, eastern California, and southern Arizona. Like Death Valley, this topography exemplifies crustal extension, where mountains rose relative to basins along large normal faults.

The perspective from Dantes View also shows the asymmetry of the Black Mountains. Their west flank is steep! It drops more than a mile vertically to the floor of Death Valley, which is only about two horizontal miles away. By contrast, the east flank descends gradually into Greenwater Valley. This asymmetry reflects an eastward tilting of the range as it rises along the Black Mountains fault zone. On the drive down from Dantes View, for example, one can see, to the northeast, tilted volcanic flows along the crest of the Black Mountains.

A careful look across the valley at the Proterozoic and Paleozoic sedimentary rocks of the Panamint Mountains reveals that they also tilt eastward. This tilt is partly a result of normal faulting along the western margin of the range (fig. 48.9). One also observes that the alluvial fans on the west side of the valley are far larger than those on the east side (fig. 48.9). Much of this difference can be explained by the observation that the Panamint Mountains are about twice as high as the Black Mountains and have been more deeply eroded. The difference is probably also attributable to evidence that the floor of the valley is tilting eastward, causing the ends of the fans to be buried in playa sediments (Hunt, 1975).

Lake Manly

At many places in Death Valley we see evidence that this undrained depression was once occupied by a relatively deep and extensive body of saline water known as Lake Manly. Indeed, an actual lake, nonetheless only a few inches deep, does occupy Central Death Valley during seasons of above average rainfall. The most obvious evidence that the lake was once much deeper exists in narrow, horizontal benches, commonly veneered with beach gravels, formed when the lake level was stationary long enough for waves to cut them.

These exist at many localities peripheral to Central Death Valley. They are most numerous and best preserved at Shoreline Butte west of the Ashford Mill site in southern Death Valley. They are also well preserved in the faulted alluvial fans exposed on the northeast side of Mormon Point (fig. 48.10).

As the lake advanced and retreated, gravel bars and accumulations of tufa (calcium carbonate) were deposited along its shores. An excellently preserved gravel bar is cut through by the highway to Beatty, about 2 miles north of Beatty Junction. Nearly horizontal accumulations of tufa, mixed with fragments of the underlying bedrock, cling to the Black Mountains front and are almost continuously in view between Badwater and Mormon Point.

Playa Lakes and the Death Valley Salt Pan

Death Valley National Park contains a number of ephemeral lakes, called playas. Of these, the best known are the salt pan of Central Death Valley and the

FIGURE 48.9   Looking west from Dantes View on the crest of the Black Mountains. The salt pan is below, with the Panamint Mountains rising beyond. Alluvial fans border both sides of the salt pan but those on the west side, along the base of the Panamints, are much larger. Photo by Martin Miller.

FIGURE 48.10 Dissected alluvial fans east of the Mormon Point showing the fault scarps and the horizontal benches cut by Lake Manly. Photo by Martin Collier.

Racetrack playa, although large playas also occupy much of northern Death Valley, northern Panamint Valley, and Saline Valley. During rainy periods these exceedingly flat features may become "flooded" with less than an inch to more than a foot of water. The aridity, however, causes the water to quickly evaporate and leave behind the sediment and dissolved minerals it carried into the playa. Most playas, therefore, are covered by dry, cracked mud commonly associated with evaporites.

The Death Valley salt pan is one of the largest modern salt pans on earth. Although its exact boundaries are poorly defined, it extends from the vicinity of the Ashford mill site northward to the Salt Creek Hills, a distance of about 40 miles. The salt pan is essentially a gigantic, flat sink without a drain. The Amargosa River, which is usually dry, empties into it from the south. It is fed from the north by Salt Creek and from various directions by runoff and spring water originating within the limits of Death Valley. Because this water contains dissolved salts that precipitate as the water evaporates, new salt is continually added to the pan. Much of the salt pan is actually the broad and flat distributary terminus of the Amargosa River. In a strict sense, this part is not a playa, but a low-relief river delta system with alternating channels and flood plain areas. From Dantes View these channels can be seen to extend as far north as Badwater.

Visitors to Death Valley can walk onto the salt pan from nearly anywhere along the Black Mountains front.

At the Devil's Golf Course one can see pinnacles of salt rising above the surface of the pan (fig. 48.11). Polygon-shaped blocks, from one to two meters across, are produced by desiccation of the salt pan. As the water evaporates and the mud beneath the surface dries, cracks develop between the blocks and are then filled with veins of new salt. These ordinarily stand with slight topographic relief above salt within the polygons.

Tire tracks, made by inconsiderate visitors, are visible at many places along the margin of the salt pan. These off-roaders are likely to become mired in the salt mush and require towing. As many of the tracks are more than 20 years old, they graphically record the length of time required for natural processes to heal these scars.

Zonation of the salt pan.

Most of the salts in the salt pan are chlorides, of which halite, ordinary table salt, is the most common. Deposits of sulfate and carbonate evaporites, being less soluble than the salt, precipitate before it does and are distributed along the edges of the pan. Hunt and Mabey (1966) showed that the carbonate and sulfate zones are much wider and better developed on the east side of the salt pan than on the west side.

They reasoned that the difference was due to a gentle eastward tilting of the salt pan along the normal faults of the Black Mountains front that caused the dissolved salt to migrate there.

Underneath the salt pan.

Before Death Valley became a National Monument in 1933, the Pacific Coast Borax Company drilled several exploratory holes into the salt pan in the search for potash (Hunt and Mabey, 1966). These penetrated the valley fill to depths of as much as 1000 feet. Individual drill cores showed that the valley fill ranges in composition from mostly salt, through salt plus other evaporites and clay, to mostly clay with minor proportions of evaporites. The salt-rich parts of the cores thus bear evidence of an arid climate much like that of today, whereas the clay-rich parts indicate less arid times when the basin was occupied by a perennial saline lake from which evaporites were occasionally precipitated. Strong negative gravity anomalies beneath the present salt pan are indicative of the presence of materials of low specific gravity. The anomalies also show that the valley fill is several thousand feet thick and probably includes Late Tertiary as well as Quaternary strata.

FIGURE 48.11   Polygonal blocks about three to six feet across, in the salt pan of Central Death Valley. Photo by Martin Miller.

Young Volcanic Features

Evidence of volcanic eruptions that have occurred within the last one million years can be viewed at two easily accessible localities within the park. At the northeast edge of the Cottonwood Mountains, Ubehebe Crater (fig. 48.12) and several smaller craters formed between 2000 and 3000 years ago. Ubehebe Crater is 450 feet deep, half a mile wide and centrally located in the midst of the other craters. They resulted from numerous explosions, generated when rising basaltic magma contacted groundwater and flashed it to steam. Small explosions that formed the smaller craters were followed by a much larger explosion resulting in the creation of Ubehebe Crater. No lava was extruded, but black cinders and ash, as much as 150 feet thick, blanket an area of about six square miles. Highly colored alluvial deposits underlie the cinders and are exposed in the walls of Ubehebe Crater. The colors are best attributed to oxidation, just prior to the eruptions, by percolating hot ground water.

In southern Death Valley, near its confluence with Wingate Wash, a basaltic cinder cone less than 400,000 years old projects above the valley floor. The cinder cone (fig. 48.13) appears as two red hills, easily visible from the highway southeast of Mormon Point. The hills mark the opposite sides of a cone, originally intact, but now offset several hundred meters by right-lateral movement on a strand of the Southern Death Valley fault zone.

Sand Dunes

Dune fields are widely distributed in Death Valley National Park, but, contrary to a common perception, they underlie a very small fraction of the total area of the park. The most frequented, most photographed and also one of the largest, occupies part of Mesquite Flat north of Stovepipe Wells (fig. 48.14). It borders State Highway 190 on the north and is also accessible from a picnic area at its eastern edge. Equally photogenic, but less accessible dune fields, include those near Saratoga Springs in southern Death Valley, in northern Panamint Valley, and in southern Eureka Valley.

Sand dunes require a steady supply of sand, wind, and a windbreak to bring the sand to rest. Most of the

FIGURE 48.12   Aerial view of Ubehebe and adjacent craters, about a half mile across, with a dark rim of basaltic cinders above colored walls. Photo by Martin Miller.

FIGURE 48.13   Quaternary cinder cone that has been offset (sliced in two) by a fault in the Southern Death Valley fault zone. The straight line beyond the cinder cone is a state highway. Photo by Martin Miller.

sand in dune fields of the park originated as water-transported detritus carried from the mountains to areas intermediate between the edges of the alluvial fans and the silty central parts of the inter-mountain playas. Dunes of the Death Valley area tend to form close to the source areas. They are small in the central part of Death Valley as much of the sand there has become cemented in the salt pan (Hunt and Mabey, 1966).

The Central Death Valley Plutonic-Volcanic Field: a Fault-controlled Igneous Terrane

The Black Mountains and Greenwater Range and the adjoining Furnace Creek Wash, being underlain mostly by Cenozoic igneous and sedimentary rocks, present a varicolored patchwork-like landscape. This landscape differs markedly from the striped forms of the surrounding ranges composed of the uniformly layered Proterozoic and Paleozoic formations. The reason for the localization of the Cenozoic rocks probably lies in the tectonic setting of the rhombic area that contains them. Like the still younger pull-apart basin of Central

Death Valley, this terrane is bounded on the northeast and southwest by the apparent *en echelon* terminations of two right-lateral strike-slip fault zones, the southeastern part of the Furnace Creek fault zone on the northeast and the northwestern part of the Sheephead fault zone on the southwest (fig. 48.3). The igneous bodies that compose the Central Death Valley plutonic-volcanic field are thus theoretically confined to a part of the crust that has been extended more than the surrounding region (Wright et al., 1991). In the early stages of the extension, the Proterozoic rocks that once underlay this area were highly faulted and greatly reduced in thickness as in the Amargosa chaos. These rock bodies are largely hidden beneath the less deformed cover of volcanic and Cenozoic volcanic and sedimentary rocks.

The history of Cenozoic igneous activity within the Central Death Valley region began between 12 and 11 million years ago. Within that interval the composite pluton of the Willow Spring gabbrodiorite was intruded at the site of the Black Mountains and dacitic flows were extruded at the location of the Resting Spring

FIGURE 48.14   Aerial view, looking east, of sand dunes near Stovepipe Wells, with the Grapevine Mountains in the distance. Star dunes, with three or more arms, lie near the lower right of the photo; crescentic dunes, with only two arms, are the elongate dunes on the lower left. Star dunes are apparently the result of variations in the prevailing wind directions. Photo by Martin Miller.

Range, east of the eastern boundary of the park. From 11 to 10 million years before the present, felsic magmas crystallized both as shallow plutons and as lava flows. The felsic plutons are distributed throughout the Central Death Valley area, but most abundantly in the Greenwater Range; the lava flows of that age are exposed in the vicinity of the Sheephead Mountain in the southern part of the field.

The post 10-million-year history of the Central Death Valley plutonic-volcanic field is recorded primarily in extrusive volcanic rocks that eventually covered almost the entire area now occupied by the Black Mountains and Greenwater Range and, in the later stages, much of the adjacent Furnace Creek Basin. These rock units range widely from rhyolitic through andesitic to basaltic in composition. Rhyolitic lava flows and associated air-fall tuffs, comprising the Shoshone Volcanics, the volcanic rocks at Brown Peak, and the Greenwater Volcanics, are the most abundant. Most of the ash-flow tuff is confined to a single formation, the Rhodes Tuff, which is about 9.2 million years old. Felsic volcanism apparently terminated 5 to 6 million years ago following the emplacement of the Greenwater Volcanics. The basaltic and andesitic flows occupy various positions in the volcanic pile. The available radiometrically determined ages suggest a clustering in intervals of 9 to 10, 7 to 8 and 4 to 5 million years before the present.

Springs

Parts of Death Valley have a surprising abundance of spring water. Springs in the area of Furnace Creek Wash, for example, discharge approximately 2000 gallons per minute, easily enough to supply the park village. Elsewhere in Death Valley, particularly along the mountain fronts, springs provide most of the water that supports its diverse and fascinating ecology. Here, as in other areas, springs mark the places where ground water, flowing from higher to lower levels, reaches the surface of the land. Most of the springs in the Death Valley region emanate from fault zones or the toes of alluvial fans (Hunt, 1975). Fault-controlled springs abound along the fronts of the Grapevine, Funeral, and Black Mountains. Some of the more accessible of these are the spring at Badwater, Travertine Spring in Furnace Creek Wash, and the Keane Wonder Spring at the front of the Funeral Mountains just north of the Keane Wonder mine. Klare Spring, along the road down Titus Canyon in the Grapevine Mountains, issues from a fault zone within the mountains.

Water discharges from the toes of alluvial fans where the highly permeable sand and gravel of the fan grades abruptly into the less permeable sand, silt, and clay of the bordering playa. The alluvial fan at the mouth of Furnace Creek is a spectacular example of this kind of spring environment. Numerous spring-fed lines of vegetation originate at the fan-to-playa transition and radiate outward onto the playa. Many other springs, such as Shorty's Well and Bennett's Well, lie at the toes of the fans that slope eastward from the Panamint Range.

The springs that supply water to Salt Creek, on the floor of northern Death Valley, formed differently. North of the springs, ground water moves easily through the permeable alluvium on the floor of the valley. The springs mark the places where the ground water rises upon encountering the impermeable fine-grained strata of the Miocene Furnace Creek Formation. Southward and downstream from there, Salt Creek disappears back into the alluvium.

Most of the water that feeds the fault-controlled springs in the Funeral and Grapevine Mountains originates in the high country of southwestern Nevada. The limestone and dolomite in the Paleozoic formations of the region serve as aquifers. Devil's Hole, a fault-controlled spring on the east edge of the Amargosa Desert and an outlier of Death Valley National Park, provides a "window" into the carbonate aquifer. There, water discharges from a cave in limestone bedrock, but does not escape to the surface. Instead, it forms a pool that is the sole habitat for a rare species of pupfish. Most of the water then continues upon its journey to the springs of Death Valley.

Geologic History

Events That Preceded the Formation of the Present Basins and Ranges

To reconstruct the geologic history that preceded the formation of the present basins and ranges of the Death Valley region, one must theoretically fit these pieces back to where they were before this latest fragmentation began. This procedure involves a restoration of the displacements on the major faults that bound the present ranges. It is not a simple task and markedly different reconstructions have been proposed. But geologists concerned with deciphering the earlier events generally agree that they include the ones shown in Table 48.1 and described in the summary that follows.

1. **The ancient basement and prolonged erosion.**

   The oldest rocks in the Death Valley region are those of an Early Proterozoic crust that served as a basement for the thick accumulations of younger Proterozoic and Paleozoic rocks that underlie most of the ranges of the Death Valley region. This basement consists of a complex of metamorphosed sedimentary and igneous rocks and is characterized by abundant quartz and feldspar. Commonly called a "crystalline complex," it is ordinarily recognizable from a distance by its somber gray color and the nearly featureless nature of most of its exposures. The complex underlies each of the turtleback surfaces along the Black Mountains front. It also is abundantly exposed in the southern part of the Black Mountains and in the Talc and Ibex Hills still farther to the south. The metamorphism has been dated by radiometric methods at about 1.7 billion years bp. This is also the age of similar complexes exposed elsewhere in the southwestern United States. The complex contains belts of pegmatite dikes and widely distributed bodies of granite. A body of this ancient granite in the Panamint Mountains has been dated radiometrically at about 1.4 billion years bp. Except for the intrusion of the granite, little is known of the geologic history of the Death Valley region during the 500-million year interval between the metamorphism of the basement complex 1.7 billion years ago, and the deposition of the lowest beds of the overlying Pahrump Group probably about 1.3 billion years ago. Clearly, however, this was an interval of long-continuing erosion because deep erosion is required to expose a bedrock formed under conditions of higher temperatures and pressures than existed at the surface. Thus, before the basal beds of the overlying Pahrump Group were laid down, large volumes of detritus must have been removed and redeposited, but no one knows where (fig. 48.15).

2. The Pahrump Group: Proterozoic basins and uplands.

The sedimentary rocks of the Pahrump Group are characteristically several thousands of feet thick. The Pahrump Group is composed, in upward succession, of the Crystal Spring Formation, the Beck Spring Dolomite, and the Kingston Peak Formation. These formations underlie much of the west side of the Panamint Mountains. In the northern Funeral Mountains they occupy the northern part of the lower

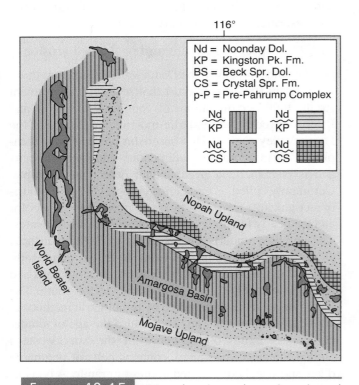

FIGURE 48.15 Map showing paleogeograpic and paleogeologic features of the Death Valley region just before the Noonday Dolomite was deposited. Dark green areas indicate the present exposures of the Pahrump Group. These extend from the Panamint Mountains on the west to the Kingston Range on the east and from the Silurian Hills on the southeast to the northern Funeral Mountains on the north. Linear and dotted patterns indicate the rock units that underlie the Noonday Dolomite. Stipple patterns show high areas from which debris was shed in Kingston Peak time. The original pattern has been distorted by Mesozoic compression and Cenozoic extension. Modified from Wright and Prave, 1992.

plate Boundary Canyon fault. They also are exposed at numerous localities in a belt that extends eastward from the southern Panamints to the east side of the Kingston Range. In the Funeral Mountains and in the central and northern part of the Panamint Mountains, these formations have been highly metamorphosed. Elsewhere they remain but little changed from their original state. Within the park boundaries, the most accessible exposures lie along and east of the front of the Black Mountains, east and southeast of the Ashford Mill site (Wright and Troxel, 1984). All of the Pahrump units described below, as well as the overlying Noonday Dolomite, Ibex Formation, Johnnie Formation, and Stirling Quartzite, can be inspected in a half-day traverse of that locality.

The Pahrump chronicles a succession of events beginning more than a billion years ago when, in the region in and about Death Valley National Park, uplands of the crystalline complex rapidly rose above a shallow sea. The uplands were then dissected and partly buried in eroded sediment ranging from arkosic conglomerate to muddy debris. These strata form the lower part of the Crystal Spring Formation.

The entire seascape was then blanketed by a vast bank of dolomite and limestone growing in shallow water and covered by a laterally continuous algal mat. This deposit forms the middle part of the Crystal Spring Formation. The algae-related and variously shaped columnar structures, known as stromatolites, are abundantly preserved in it. The mat was destroyed by the influx of fine-grained detritus preserved in the siltstones and sandstones in the upper part of the Crystal Spring. This event was followed by the intrusion of sills of diabase above and below the carbonate body. The lowest of the sills extended over hundreds of square miles of this ancient terrane and caused the basal carbonate strata to alter to bodies of commercial talc.

Then the environment reverted to that of a shallow sea and another algae-blanketed carbonate bank, now evidenced in the thinly laminated beds of the Beck Spring Dolomite. These and other Proterozoic algal mats survived on such an extensive scale mainly because animals that feed on algae had not yet evolved.

Again the crust was broken into blocks that formed islands in the Proterozoic sea, and Crystal Spring and Beck Spring rocks were exposed to erosion. Consequently, thick deposits of conglomerates containing abundant clasts derived from these earlier units, along with finer-grained clastic sediments, accumulated in basins between the new or rejuvenated high areas. These strata comprise the Kingston Peak Formation. Some of the conglomerates qualify as *diamictites* in that they are poorly sorted and contain a wide range of particle sizes. Diamictites thus resemble glacial till. Other parts of the Kingston Peak consist of thinly laminated sandstone and siltstone in which clasts of boulder size and larger are embedded. For obvious reasons, these clasts are called "dropstones." The combination of the two textures provides evidence for a glacio-marine environment. Many geologists associate these deposits with a wave of glaciation that swept over the crust of North America 700 to 800 million years ago. Basaltic flows, locally present in this upper part of the Pahrump Group, mark the renewal of mafic magmatism.

3. The Noonday Dolomite, Ibex Formation and Johnnie Formation: foreshadowing the environment of a passive continental margin

The Noonday Dolomite, its deeper water equivalent called the Ibex Formation, and the overlying Johnnie Formation occupy an intermediate stratigraphic position between the Pahrump Group and the thick succession of uppermost Proterozoic and Paleozoic strata that has been long viewed as deposited along an evolving *passive continental margin,* (i.e., a margin associated with rifting). The deposition of the three formations thus theoretically just preceded the splitting of a preexisting continent into two continents separated by a widening expanse of oceanic crust.

The Noonday, which is ordinarily about 1000 feet thick, is yet another algal carbonate unit. Being a cliff-former and colored a distinctive pale yellowish gray, its exposures are particularly easy to recognize, even from miles away. The Noonday and its lateral transition to the Ibex Formation are well exposed above the Pahrump Group at the locality east of the Ashford Mill Site.

There and at other places extending from the Panamint Mountains as far east as the Kingston Range, the contact between the Noonday Dolomite and the underlying Pahrump is an angular unconformity that truncates progressively older units of the Pahrump Group northward. In its most northerly exposures along this belt, the Noonday Dolomite overlies the ancient basement complex. These relationships define a high area in the ancient topography from which much or all of the Pahrump Group was eroded before the Noonday was deposited (fig. 48.16). At more southerly localities, either the Noonday or the Noonday-correlative Ibex Formation lies concordantly upon the Kingston Peak Formation. In the lateral transition from the Noonday Dolomite to the Ibex Formation, the shallow water, algae-related carbonate strata abruptly give way to a succession composed mostly of thinly bedded siltstone and limestone deposited in deeper water in a setting reminiscent of a continental shelf on a small scale.

4. The Stirling Quartzite, Wood Canyon Formation, and Zabriskie Quartzite: the clastic wedge of a developing continental margin

These Late Proterozoic to Early Paleozoic formations (Table 48.1), are widely exposed in the Death Valley region. Within the park, the Wood Canyon is particularly accessible. In this region the formations average about 6000 feet in combined thickness. As they consist mostly of well cemented sandstones and conglomerates, they are more resistant than the underlying, varicolored and shaly Johnnie Formation and the overlying, also varicolored Carrara Formation. The Wood Canyon Formation is particularly accessible west and north of the highway in the vicinity of Hells Gate between the Funeral and Grapevine Mountains. All three formations are well exposed on the north face of Tucki Mountain at the north end of the Panamint Mountains.

By analogy with sedimentary wedges that have accumulated along existing continental margins, geologists view this succession as deposited during the rifting stage of an early margin. They have reasoned that, at that earlier time, the ancient quartz- and feldspar-rich basement complex was split in two and was exposed along the edges of the two developing continents. Thus the Stirling Quartzite, Wood Canyon Formation, and Zabriskie Quartzite are interpreted as debris eroded from the edge of the more easterly continent and deposited as a wedge on the newly rifted crust (fig. 48.16).

These formations also provide the earliest evidence, in the Death Valley region, of complex life forms. They contain the remains of metazoan life forms, particularly trilobites, and also other features, such as tracks and burrows, left by these creatures that rapidly appeared and populated the early oceans. In recent years, fossils of the enigmatic Late Proterozoic Ediacara fauna have been found in the Wood Canyon Formation.

5. The Paleozoic and Early Mesozoic carbonate shelf

The rest of the Paleozoic section in the Death Valley region is dominated by dolomites and limestones and is about 20,000 feet thick (fig. 48.16). These formations constitute the sedimentary record of a long-continuing and slowly subsiding continental shelf. Here they were deposited, with little evidence of disturbance, above the wedge of clastic sediments deposited during the preceding stage of rifting. The carbonate sedimentation was interrupted by numerous periods of emergence and, in Middle Ordovician time, by the deposition of the

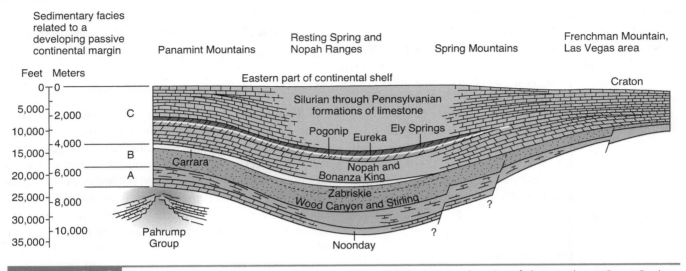

FIGURE 48.16   Stratigraphic cross section of Proterozoic and Paleozoic rock units of the southern Great Basin, including the formations mentioned in this chapter. Simplified from Wright et al., 1981.
A. Upward change from carbonate to mixed carbonate and siliciclastic strata. Earliest evidence of thickening westward away from an incipient continental margin.
B. Siliciclastic strata deposited during and immediately after major crustal rifting to form the margin of the Late Proterozoic and Paleozoic continent. Source of sediments was the ancient basement of the craton.
C. Almost entirely limestone and dolomite recording a slowly subsiding and long-continuing carbonate bank.

Eureka Quartzite. The Eureka is part of a sheet of quartz-rich sand that spread across much of the width of that Paleozoic continent. Carbonate rocks continued to be deposited into Triassic time. But then the sea withdrew and did not return. The oceanic crust, which was theoretically forming as the continental shelf evolved, lay many miles to the west of the site of Death Valley.

Where exposed within the park boundaries, the carbonate succession is about 20,000 feet thick. It underlies large parts of the Panamint, Cottonwood, Grapevine, and Funeral Mountains and is particularly visible in the southern part of the Funeral Mountains. There the Eureka Quartzite can be recognized from a distance as a single, nearly white band, repeated several times by normal faulting. It is underlain by the Pogonip Group, colored various shades of gray, and overlain by the nearly black Ely Springs Dolomite. In Death Valley National Park, the Triassic strata are preserved only in the Butte Valley area at the southern end of the Panamint Mountains.

6. The closing of the Paleozoic ocean in Mesozoic time; thrust faults, and bodies of igneous rock

At numerous localities within Death Valley National Park, we observe that the Proterozoic and Paleozoic rocks have been faulted so that older formations override younger formations. These contacts are the thrust faults that record severe compression of the crust. Together with associated folds and with granitic plutons, also present within the park boundaries, and Mesozoic granitic rock, the thrust faulting marks the end of the passive continental margin and records a foreshortening of the former continental shelf. Features of this kind, together with zones of volcanic rocks, characterize a belt that extends beyond the park area for the full length of North America. The plutons exposed within the park lie along the periphery of a belt composed mostly of igneous bodies and containing the Sierra Nevada batholith to the west. This belt is called the Cordilleran Mesozoic magmatic arc.

In the Death Valley area the foreshortening is expressed mainly by thrust faults that have caused Proterozoic and Cambrian formations to override younger Paleozoic formations. But they are difficult for the short-term visitor to quickly discern amidst the later faults formed while the crust was extending. Among the relatively accessible thrusts are the Schwaub Peak (Wright and Troxel, 1993) and Clery thrusts (Cemen et al. 1985) in the southern part of the Funeral Mountains.

The Mesozoic plutons of Death Valley National Park are of Jurassic and Cretaceous age. They are distributed close to the western boundary of the park, where they underlie most of the Owlshead Mountains, are discontinuously distributed in the western part of the Panamint Mountains, and form the Hunter Mountain batholith at the southern end of the Cottonwood Mountains. Most of the plutons are accessible by unimproved roads.

7. **The missing sedimentary record of Jurassic through Early Tertiary time and the initiation of the Basin and Range event in the Death Valley area**

No sedimentary rocks of Jurassic through Eocene age have been found in the area of Death Valley National Park, although volcanic rocks of probable Jurassic age are exposed in the Butte Valley area of the southern Panamint Mountains. In the national park and the surrounding region, this apparently was a time of uplift and erosion concurrent with and following the thrust faulting and the emplacement of the Mesozoic plutons. During this interval much of the succession of Proterozoic and Paleozoic sedimentary rocks that once covered the Death Valley region was eroded away. In fact, the erosional denudation probably continued over much of the Death Valley region until the inception of the basins and ranges that form the present landscape. Little is known of the river system that must have functioned during this interval, or where the eroded material was deposited. Much of the debris probably was carried into a Cretaceous sea and its environs in the interior of the continent. In the Cottonwood, Grapevine, and Funeral Mountains, however, remnants of this earlier terrain are sealed beneath corresponding remnants of a sedimentary cover deposited in Oligocene and early Miocene time.

8. **The Oligocene and Early Miocene flood plain and the bordering upland**

The Oligocene and early Miocene formations of the Cottonwood, Funeral, and Grapevine Mountains were deposited on a much more subdued landscape than the one we observe today. This conclusion stems from the observation that the contact is characteristically devoid of the major irregularities that would accompany the infilling of an irregular topography. We also note that basal Tertiary beds consistently dip almost as steeply as the underlying Proterozoic and Paleozoic formations of the present ranges. Thus, when rotated back to their original, nearly horizontal positions, these beds are seen as deposited on a broad surface of low relief drained by laterally migrating streams and rivers. The oldest of the Tertiary strata compose the Oligocene Titus Canyon Formation and consist of conglomerates, sandstones, and mudstones. The clasts of the conglomerates are exceptionally well rounded and polished and can be seen in road cuts at Daylight Pass along the

Death Valley to Beatty highway. Clasts of granitic rock point toward a western source area in which Mesozoic plutons were exposed.

In a more southward area that includes most of the Panamint Mountains and the Owlshead and Black Mountains within the national park and also the Resting Spring and Nopah Ranges east of the park boundary, Tertiary rocks older than 14 million years are apparently absent, and the pre-Basin and Range surface is thus obscured. If so, the missing rock units of earlier Tertiary age have either been eroded away or were never deposited. In this area, the 14-million-year and younger Tertiary sedimentary rocks, are part of the evolving Basin and Range terrain.

FIGURE 48.17   A "sliding boulder" on the Racetrack playa. These objects have moved, as yet unobserved, under an unusual combination of weather and surface conditions. The exact nature of these conditions continues to be debated, although strong winds are consistently invoked. Photo by Martin Miller.

Geologic Map and Cross Section

Albee, A.L., Labotka, T.C., Lanphere, M.A., and McDowell, S.D., 1981. U.S. Geological Survey Geologic Quadrangle Map GQ-1532.

Drewes, H., 1963. Geology of the Funeral Peak Quadrangle, California, on the Eastern Flank of Death Valley. U.S. Geological Survey Professional Paper 413, 78 p.

Hall, W.E., 1971. Geology of the Panamint Butte Quadrangle, Inyo County, California. U.S. Geological Survey Bulletin 1299, 67 p.

Jennings, C.W., Burnett, J.L., and Troxel, B.W., 1962. Trona Sheet: California Division of Mines and Geology Geologic Atlas of California.

McAllister, J.F., 1970. Geology of the Furnace Creek Borate Area, Inyo County, California. California Division of Mines Map Sheet 14.

———— 1971. Preliminary Geologic Map of the Funeral Mountains in the Ryan Quadrangle, Inyo County, California. U.S. Geological Survey Open File Map.

———— 1973. Geologic Map of the Amargosa Valley Borate Area—Southeast Continuation of the Furnace Creek Area. U.S. Geological Survey Miscellaneous Geologic Investigations Map 1-782.

Streitz, R., and Stinson, M.C., 1977. Death Valley Sheet: California Division of Mines and Geology Geologic Atlas of California.

Wright, L.A., and Troxel, B.W., 1984. Geology of the North 1/2 Confidence Hills 15′ Quadrangle: California Division of Mines and Geology.

Wright, L.A., and Troxel, B.W., 1993. Geologic Map of the Central and Northern Funeral Mountains and Adjacent Areas, Death Valley Region, Southern California. U.S. Geological Survey Miscellaneous Investigations Series, Map 1-2305.

Wright, L.A., Troxel, B.W., Burchfiel, B.C., Chapman, R., Labotka, T. 1981. Geologic Cross Section from the Sierra Nevada to the Las Vegas Valley, Eastern California to Western Nevada. Geological Society of America Map and Chart Series, MC-28M.

Sources

Burchfiel, B.C., and Stewart, J.H., 1966. "Pull-apart" Origin of the Central Segment of Death Valley, California: Geological Society of America Bulletin, v. 77, p. 439–442.

Cemen, I., Wright, L.A., Drake, R.E., and Johnson, F.C., 1985. Cenozoic Sedimentation and Sequence of Deformational Events at the Southeastern End of the Furnace Creek Strike-slip Fault Zone, Death Valley Region, California, in Biddle, K.T., and Christie-Blick,

Nicholas, eds., Strike-slip Deformation and Basin Formation. Society of Economic Paleontologists and Mineralogists Special Publication, No. 37, p. 127–141.

Collier, Michael, 1990. *An Introduction to the Geology of Death Valley.* Death Valley Natural History Association, 60 p.

Hodges, K.V., McKenna, L.W., Stock, J., Knapp, L., Page, L., Sterniof, K., Silverberg, D., Wust, G., and Walker, J.D., 1989. Evolution of Extensional Basins and Basin and Range Topography West of Death Valley, California: *Tectonics,* v. 8, p. 453–467. Hunt, C.B., and Mabey, D.R., 1966. Stratigraphy and Structure, Death Valley, California. U.S. Geological Survey Professional Paper 494A, 162 p.

Hunt, C.B., 1975. *Death Valley. Geology, Ecology, Archeology.* University of California Press, Berkeley and Los Angeles, California, 234 p.

Miller, M.G., 1991. High-angle Origin of the Currently Low-angle Badwater Turtleback Fault, Death Valley, California. *Geology,* v. 19, p. 372–375.

Noble, L.F., 1941. Structural Features of the Virgin Spring Area, Death Valley, California. Geological Society of America Bulletin: v. 52, p. 941–1000.

Serpa, L., and 6 others, 1988. Structure of the Central Death Valley Pull-apart Basin and Vicinity from COCORP Profiles in the Southern Great Basin: Geological Society of America Bulletin, v. 100, p. 1437–1450.

Stewart, J.H., Albers, J.P., and Poole, F.G., 1968. Summary of Regional Evidence for Right-lateral Displacement in the Western Great Basin: Geological Society of America Bulletin, v. 79, p. 1407–1414.

Wernicke, B., Axen, G.C., and Snow, J.K., 1988. Basin and Range Extensional Tectonics at the Latitude of Las Vegas, Nevada. Geological Society of America Bulletin, v. 100, p. 1738–1757.

Wernicke, B., Snow, J.K., Hodges, K.V., and Walker, J.D., 1993. Structural Constraints on Neogene Tectonism in the Southern Great Basin, in Lahren, M.M., Trexler, J.H., Jr., and Spinosa, C., eds., Crustal Evolution of the Great Basin and the Sierra Nevada. Geological Society of America Field Trip Guidebook, Reno, Nevada, p. 453–479.

Wright, L.A., and Prave, A., 1992. Proterozoic-Early Cambrian Tectonostratigraphic Record of the Death Valley Region, California-Nevada, in Reed, J., et al., eds. Precambrian Rocks of the Conterminous United States. Geological Society of America p. 529–533.

Wright, L.A., Thompson, R.A., Troxel, B.W., Pavlis, T.L., DeWitt, E.H., Otton, J.K., Ellis, M.A., Miller, M.G., and Serpa, L.F., 1991. Cenozoic Magmatic and Tectonic Evolution of the East-central Death Valley Region, California, in Walawender, M.J., and Hanan, Barry B., eds., Geological Excursions in Southern California and Mexico: Guidebook, Geological Society of America, San Diego, California.

C H A P T E R   4 9

Sequoia and Kings Canyon National Parks

EAST CENTRAL CALIFORNIA

Donald F. Palmer   Kent State University

Area: Sequoia: 403,482 acres; 629 square miles. Kings Canyon: 461,901 acres; 722 square miles.

Established as a National Park: Sequoia—September 25, 1890; Kings Canyon—March 4, 1940

Designated a Biosphere Reserve: 1976

Address: Sequoia National Park, 47050 Generals Highway, Three Rivers, CA 93271-9651

Phone: 559-565-3341

E-mail: SEKI_Interpretation@nps.gov

Website: www.nps.gov/seki/index.htm

Sequoia and Kings Canyon National Parks together cover over 1300 square miles and preserve the largest groves of giant sequoias.

The bedrock of the parks is primarily late Cretaceous granitic batholiths with thick sections of early Mesozoic metasedimentary and metavolcanic rocks in roof pendants. The latter rocks record the active subduction system at the western margin of North America from the end of the Paleozoic through the Mesozoic. Uplift of the range occurred episodically with the most recent events beginning in the Pliocene along the Sierra Nevada fault system.

In Pleistocene time, major glacier streams extended to low elevations in both parks.

FIGURE 49.1   The General Sherman Tree is a giant sequoia tree (*Sequoiadendron giganteum*) in Sequoia National Park. The tree is 275 feet high, has a maximum basal diameter of 36 feet, and has an estimated age of 2300 years. Each year enough wood is added to the tree to create a 60 foot tree of normal size. Photo by Douglas Fowler.

Sequoia and Kings Canyon National Parks have grown significantly over a period of 75 years. They began in 1890 when Sequoia and the General Grant Grove were established as national parks, just in time to save their giant sequoia trees from destruction by the lumbering industry. Sequoia National Park was more than doubled in size in 1926 to include the upper drainage basins of the Kaweah and Kern Rivers and the Sierran crest, including Mt. Whitney. In 1940, a large tract of land extending northward from Sequoia was combined with General Grant Grove to form Kings Canyon National Park. Redwood Canyon, between General Grant Grove and Sequoia National Park, has a large grove of sequoias on the flanks of Redwood Mountain and was also included within Kings Canyon National Park. In 1965, the areas of Cedar Grove on the south fork of the Kings River and the Tehipite valley were added to Kings Canyon. The Middle Fork and the South Fork of the Kings River and the upper reaches of the North Fork and South Fork of the Kern River were added to the Wild and Scenic Rivers System by Act of Congress in 1987. The two parks, Kings Canyon and Sequoia, are administered as a single unit.

The major impetus for the founding of the parks was the preservation of the giant sequoias. About 13,000 giant sequoias (*Sequoiadendron giganteum*) grow in scattered patches on the moist western slopes of the Sierra Nevada at elevations between 4000 and 8000 feet. In Giant Forest in Sequoia National Park stands the General Sherman, the largest living tree in the world. The circumference of its trunk is 102.6 feet, and it is almost 275 feet tall. The mature giant sequoias in the parks are between 3000 and 4000 years old.

The giant sequoias and their close relatives, the coast redwoods (*Sequoia sempervirens*), which grow in a narrow belt along the Pacific coast, are the surviving members of a once large family of huge conifers that spread throughout the temperate regions of North America during the Mesozoic Era. Fossilized stumps of trees of the sequoia family have been identified in Yellowstone National Park (Chapter 43) and in Petrified Forest National Park (Chapter 8). The range of the sequoia species became restricted during the Ice Age, which began about 2 million years ago and ended about 10,000 years ago. The Pacific coast species survived Pleistocene glaciations and were able to reestablish themselves in the limited environments in which they now grow (see Chapter 51, Redwood National Park). The mountain ranges paralleling the Pacific coast provide an ideal environment for sequoia species for two reasons: (1) The elevation of the mountains causes the moisture-bearing clouds from the Pacific Ocean to precipitate large amounts of rain and snow; (2) the fresh, sandy soils required by sequoias are derived from the weathering of the crystalline rocks in the mountains.

Although hunters and explorers had found giant sequoias in the Sierra region as early as 1833, A.T. Dowd is credited with the first well publicized discovery of sequoias in 1852 while hunting north of the Yosemite region. The first of Sequoia's Big Trees were found by Hale Tharp in 1858, and those in the General Grant Grove of Kings Canyon National Park were discovered in 1862 by Joseph Thomas. News spread rapidly about the "mammoth trees," and in the 1860s and 1870s claims for land were filed and sawmills were erected. A giant sequoia was felled, cut into segments, and shipped to Philadelphia for the 1876 Centennial Exposition, where it was reassembled and exhibited.

The loggers of the day regarded felling the Big Trees as the ultimate challenge to their skill and bravery. The wood from sequoias is of exceptionally high quality, straight-grained, and naturally resistant to decay. A single tree might yield 300,000 to 400,000 board feet of lumber. Reminders of these early logging days can be seen in Big Stump Basin in Kings Canyon National Park near Generals Highway; where a grove of giant sequoias once flourished, only huge stumps remain.

The exploitation of the area was halted by the action of many individuals who worked for the preservation of the Big Trees and the Sierra Nevada wilderness. Notable among these people is Colonel George Stewart, whose practical efforts resulted in the preservation of Sequoia National Park. But surely the most important person in helping to preserve the Sierran wilderness was the naturalist John Muir. It was Muir who founded the preservationist movement in the United States, and it was his life and writings that formed the basis of argument for the establishment of parks and wilderness preserves in countries throughout the world.

John Muir came to California in 1868 as a young man and settled in the Yosemite Valley. After a thorough study of the Yosemite region, he began in 1872 to explore the area to the south. On foot and often alone, he walked and climbed the entire Sierra between the Tuolumne and Kern Rivers, observing and documenting its natural wonders. In 1873, he met Hale Tharp near where Tharp had first seen the Big Trees; a place Muir named Giant Forest.

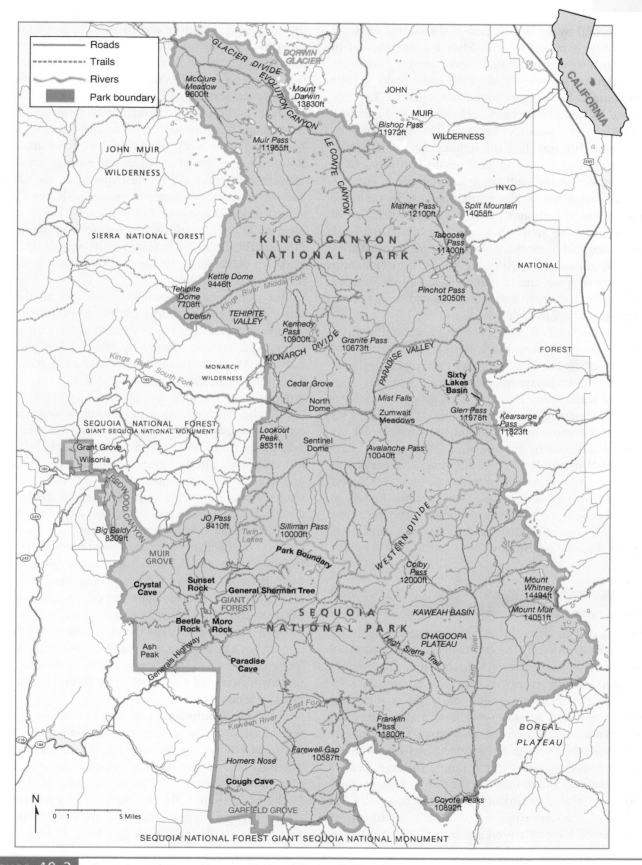

FIGURE 49.2   Sequoia and Kings Canyon National Parks, California.

Appalled by the devastation of the natural environment by loggers and miners, Muir enlisted the aid of politicians, newspaper editors, and other citizens in the campaign to save the Sierra Nevada from further exploitation. He helped found the Sierra Club and became its first president. A tireless naturalist and conservationist, he spent many years of his long and remarkable life traveling throughout the world, studying nature and spreading his message of wilderness preservation for the well-being of all people. The significant expansion of the boundaries of Sequoia and Kings Canyon National Parks was the realization of Muir's dream for the High Sierra. The John Muir Trail, established in his memory, begins in Yosemite Valley and then follows the crest of the Sierra from Tuolumne Meadows in Yosemite to Mount Whitney in Sequoia, winding through three national parks and three national forests on its 218-mile route.

The California Geological Survey, under the direction of geologist Josiah D. Whitney, undertook the mapping of the High Sierra in the 1860s. Their primary mission was to locate occurrences of gold and other minerals with economic value. This group of scientists produced outstanding reports and maps that aided the conservation movement and laid the groundwork for subsequent scientific investigations. The survey party included William Brewer, a botanist who played an important role in the development of soil science; Charles Hoffman, an engineer who became known as the father of modern topographic surveying; and Clarence King, who later became the first director of the U.S. Geological Survey.

Geologic Features

The bedrock of Sequoia and Kings Canyon National Parks is dominated by the late Jurassic and Cretaceous igneous rocks of the Sierra Nevada batholith, which is interspersed with steeply inclined and complexly folded layers of metamorphosed early Mesozoic (late Triassic to early Jurassic) volcanic and sedimentary rock. In a few sites, isolated outcrops of Cenozoic volcanic rock attest to continued igneous activity throughout the period of uplift of the range. In large areas at high elevations and in many of the deeper valleys, thick sections of glacial sediments overlie the bedrock.

The shape and appearance of the landforms in Sequoia and Kings Canyon are strongly influenced by

the batholithic structure and by the weathering characteristics of the intrusions. Because many of the granites are massive and have few joints or fractures, exfoliation processes have shaped many of the exposed bedrock surfaces (fig. 49.3). Exfoliation domes, similar to those in Yosemite National Park (chapter 28), are prominent features in Sequoia and Kings Canyon National Parks. Near the Middle Fork of the Kings River are a group of conspicuous exfoliation domes, including Tehipite and Kettle Domes and the Obelisk. Two large domes on opposite sides of Kings Canyon are Sentinel Dome and North Dome. These domes, over 8500 feet in elevation, are formed in massive granite. Examples of exfoliation in Sequoia National Park are Alla Peak, Beetle Rock, Sunset Rock, and Moro Rock (fig. 49.4).

In well-fractured granites, especially at high altitudes, frost action shatters rock into angular blocks and fragments of various sizes. Vertical cliffs with sharply incised gashes are formed by frost action and mass wasting. At moderate altitudes, where more moisture is available, cascading streamlets have cut slotlike gulches along fractures in canyon walls.

Avalanche chutes are numerous on steep slopes, cirque walls, and canyon walls in the High Sierra. An *avalanche chute* is the steep, smooth track, or trough, left on a slope after an avalanche—carrying ice, rocks, and trees—has descended. Most of the avalanche chutes probably formed during the severe conditions of the Pleistocene; some are still active. Along the crest of the Sierra where parallel, vertical fractures are closely spaced, the avalanche chutes that developed on the fractures give the slope a fluted appearance. On the west flank of Mount Whitney, branching avalanche chutes follow fractures, and some chutes are more than 100 feet deep. In the massive granites (notably on slope faces of the Great Western Divide), large, semicircular, almost perfectly smooth avalanche chutes developed. They have been altered very little since they were repeatedly abraded and polished by sliding ice and rock during Pleistocene time.

At lower altitudes where the vegetation is heavy and moist soil was undisturbed by glaciation, intense chemical weathering has decomposed some of the older granites. This "rotten granite," which is loose and granular, can be seen in roadcuts along the Generals Highway. In some places, blasting was not required during highway construction and roadcuts were excavated through the loose rock by power shovels.

Exfoliation occurs on a series of horizontal joints near Moose Lake, located on the eastern side of Sequoia National Park. National Park Service photo.

Climatic Zones and Precipitation

In many respects, the weathering and erosional processes of the High Sierra are related to climatic zoning, which is fairly well defined on the slopes. The highest parts of Sequoia and Kings Canyon National Parks are barren, windswept, rocky uplands in an Arctic-Alpine zone. Rainfall is limited, and snow accumulates only in scattered patches because of the constant wind. The flat surfaces, such as the Boreal Plateau and the western slope of Mt. Whitney, were never glaciated because snow was blown away before it could accumulate. Relief is subdued in the high uplands. Only the hardiest plants survive in this environment.

In the transitional Subalpine zone, thin patches of soil support hardy shrubs and low plants. Peaks and basins show evidence of past glaciation. The sides of some of the flat uplands are scalloped by small cirques, and small lakes occupy some of the abraded rock basins. Near the treeline, the mountain landscape becomes more irregular. Ridges are serrated with horns and cols. A short distance below the main Sierran crest on the eastern side are the saw-toothed peaks that stand against the skyline so dramatically when viewed from the Owens Valley far below. Prominent cirques are carved out below peaks and arêtes. Many of the avalanche chutes extend down into the forested zones.

In the Canadian life zone, which reaches higher on south-facing slopes because plants receive more sunlight than on north-facing slopes, the firs and lodgepole pines resemble those of forests in more northern lati-

FIGURE 49.4   Moro Rock, an immense exfoliation dome of massive granite, rises to an elevation of 6725 feet above the canyon of the Kaweah River, Sequoia National Park. National Park Service photo.

tudes. Heavy snows cover the ground from November to May. Some of the streams flow in U-shaped valleys that were formerly occupied by glaciers. Mountain meadows and marshes mark the locations of old lake basins that were eroded by glacial ice and are now filled in with sediment.

On the western slopes, precipitation is sufficient to support forests between elevations of 2000 feet and 11,000 feet. The zone of maximum precipitation, approximately between the 7000-foot and the 8000-foot levels, is where the Canadian zone blends into the Transition zone. The lodgepole pines give way to the taller ponderosa pines and the groves of giant sequoias.

Drainage Systems and Divides

The Great Western Divide is a north–south range of rugged mountains located about 10 miles west of the main crest of the Sierra. This high divide separates the westward-flowing Kaweah Basin drainage from the southward-flowing Kern River drainage. The Kaweah River is the southernmost of the major streams that flow generally southwest, following the regional slope of the Sierra Nevada. The western part of Sequoia, which is drained by the Kaweah River, is rugged mountain country, deeply dissected by branching canyons and ridges. The main canyon of the Kaweah at Ash Mountain Park Headquarters is more than 4000 feet deep. Several of the branch canyons, such as East Fork and Middle Fork, are even deeper. However, some plateaulike uplands are found in the Kaweah Basin. Giant Forest occupies one of these gently sloping platforms. The relatively flat surface is apparently due to the resistant nature of the underlying massive granite bedrock that preserves the subdued topography developed prior to the most recent uplift of the range.

The Kern River rises in the upper Kern Basin in the eastern part of Sequoia National Park, between the Great Western Divide and the main crest of the Sierra. The Kern's glacially modified U-shaped valley is incised some 2000 feet in the high upland surface, between the Boreal Plateau on the east and Chagoopa Plateau on the west (fig. 49.5). Tributaries in hanging valleys at right angles to the main stream are notched in the canyon walls. The Kern River's southward course generally follows the trend of a north–south fault in the Sierra block. The fault zone provided a less erosion-resistant course for the stream as the crustal block was uplifted. The canyon becomes shallower as it approaches the southern boundary of the park, and thick glacial and alluvial deposits cover the valley floor.

Many of the topographic features of Kings Canyon National Park are grand in size and simple in form. The relief is very high, and drops of several thousand feet between ridge top and canyon bottom are characteristic of the coarse drainage texture in this part of the Sierra. *Coarse-textured drainage* means that the number of streams and valleys per square mile (or given area) is relatively low. On the massive granites of the tilted Sierra block, only a few major streams have developed. The reason is that as the streams cut deep valleys during the rapid uplift, the larger rivers collected (or captured) the waters of the smaller streams. Moreover, the drainage networks were modified by several cycles of glaciation during uplift. Like the Kaweah drainage in Sequoia, the main tributaries of the Kings drainage have a dendritic pattern and follow the trend of the regional slope.

Pleistocene glaciers scoured some of the domes and drainage divides (fig. 49.6) and excavated U-shaped valleys). The largest and longest of the valley glaciers was the one in Kings Canyon; it was a river of

FIGURE 49.5   A valley that has been glaciated is easy to identify because of its U-shape in contrast to a V-shaped stream valley. Abrasion of the glacier on the valley floor is the main process of widening the valley. Muro Blanco is located along the South Fork of the Kings River in Kings Canyon National Park. National Park Service photo.

FIGURE 49.6   Glacially smoothed and polished granite can be seen at Vidette Meadow in Kings Canyon National Park. National Park Service photo.

ice 44 miles in length that reached a relatively low altitude of 4000 feet on the western Sierra slope. Hanging valleys, some with beautiful waterfalls, such as Roaring River Falls and Mist Falls, once held tributary glaciers that fed the Kings Glacier. Sixty-Lake Basin, near the Sierra Crest, is a glacial staircase. Many small rock-basin lakes dot the broad Monarch Divide (elevation 10,000 to 12,000 feet) that separates the Middle Fork from the South Fork of the Kings River.

The floor of Kings Canyon (South Fork) is from one-fourth to one-third of a mile wide and is covered with outwash and glacial debris. Steep granite walls rise on each side, giving the canyon the U-shape of a glacial trough. By contrast, the lower canyon of the Middle Fork of the Kings River is V-shaped, indicating erosion by stream processes rather than by glacial ice (fig. 49.7). Only the upper reaches of the valley are U-shaped, giving the limit of glacial advance in Pleistocene time.

A few modern glaciers that formed in Neoglacial time, such as the Darwin Glacier, cling to cirques on the mountains along the northeast boundary of Kings Canyon National Park. Their activity is minor compared to the erosion and deposition of their Pleistocene predecessors.

Sierra Nevada Uplift

In broad outline, the geologic story of Sequoia and Kings Canyon National parks is like that of Yosemite (Chapter 28). All three parks lie along the crest and western slopes of the Sierra Nevada, a huge crustal block that has been lifted and tilted westward. Thus, the mountain range is asymmetrical; it has steep slopes on the fault-bounded eastern margin and a more gentle regional slope to the west. Along the eastern foot of the range is the Sierra Nevada normal fault system, which trends north-northwest and marks the break between the bold mountain front and Owens Valley. With an elevation of 14,494 feet, Mount Whitney is the highest summit in the United States outside Alaska (fig. 49.8). The floor of Owens Valley lies nearly 11,000 feet lower. Death Valley National Park (Chapter 48), some 80 miles to the east, contains the lowest point in North America, 282 feet below sea level. These figures give an indication of the magnitude of uplift and downfaulting that has occurred in this part of California. The vertical movement along the Sierra Nevada fault system has been at least 15,000 feet.

Recent earthquakes on the Sierra fault system are evidence that the mountain-building process continues. The largest earthquake in historic time occurred on March 26, 1872, near Lone Pine on the floor of Owens Valley. (The epicenter was about 15 miles east of Mount Whitney.) The quake is estimated to have had a Richter magnitude of at least 8.0. Vertical movement on the fault was as much as 28 feet, and horizontal displacement was about 20 feet. The ground was broken for nearly 100 miles along the fault trace. Nearly all the

FIGURE 49.7 The Middle Fork of the Kings River below Tehipite Valley in King's Canyon National Park. The rounded appearance of the hills has been caused by exfoliation. National Park Service photo.

FIGURE 49.8 With an elevation of 14,494 feet, Mount Whitney is the highest summit in the United States outside of Alaska. National Park Service photo.

buildings in Lone Pine were destroyed, and 27 people were killed (Hill, 1975).

Bedrock Geology

The composite batholith now exposed in the High Sierra is made up of granitic rocks, largely quartz diorites, quartz monzonites, and granodiorites. While the Sierra Nevada batholith consists of intrusions of a wide range of Mesozoic ages, those in the area of Kings Canyon and Sequoia National Parks are predominantly late Cretaceous with very few from the late Jurassic Period. Single intrusions can cover as much as 200 square miles and are often elongated parallel to the trend of the Sierra Nevada range. Many of the individual intrusions have been separated on the basis of age by using the rules of intrusive cross-cutting relationships and the existence of inclusions or xenoliths of earlier rocks in younger ones. Within the parks, a number of intrusive sequences have been identified. Such sequences consist of an associated group of igneous rocks intruded over a relatively short period of time. One such sequence near Giant Forest is the concentrically zoned Sequoia intrusive sequence that was emplaced between 102 and 97 million years before the present. The Sequoia sequence involves a set of four nested intrusions in which the outer zones are more mafic granodiorites surrounding younger intrusions of more silica-rich granite and alaskite (a granite rock having a very low percentage of dark minerals). The existence of these associated intrusions provides geologists with a natural laboratory that confirms the general processes of differentiation in granitic magmas and gives us an idea of the total time an individual magma system may operate and undergo differentiation. *Magmatic differentiation* involves the partial crystallization of an initially homogeneous magma and the development of a residual magma of different chemical composition. By this process, more than one type of igneous rock develops from a common magma. (For more on the intrusive suites of the Sierra Nevada batholith, see Chapter 28, Yosemite National Park.)

Within Sequoia, the oldest intrusions appear as relatively small, sheared bodies of granodiorite on the margins of later plutons or within the fault zone of the Kern River drainage. These have been dated as approximately 160 mybp. Emplacement of the main mass of the batholith appears to have started nearly 120 mybp and to have ended at 80 mybp. The youngest large intrusions occur near the crest of the Sierra with the masses of Paradise and Whitney granodiorite in the south and the LeConte Canyon granodiorite and the alaskite of Evolution Basin in the northern part of Kings Canyon National Park. These were intruded between 86 and 80 mybp.

Roof Pendants

The intrusion of the Sierra Nevada batholith obliterated most of the preexisting rocks in the area. Fortunately, small sections of the country rocks into which the batholiths were emplaced occur between separate granite intrusions as sheets or curtains of metamorphic rock called *roof pendants*. Both Sequoia and Kings Canyon National Parks have roof pendants of varying lithologies and ages (fig. 49.9).

In the northern part of Kings Canyon National Park, the southern part of the Mt. Goddard pendant trends northwest-southeast and separates the youngest intrusions of the batholith on the northeast from slightly older intrusions on the southwest. The Mt. Goddard pendant is composed primarily of early Jurassic metavolcanic rocks, representing the vestiges of a thick volcanic pile of rhyolitic ash flows and tuffs mixed with more mafic lava flows and clastic sedimentary rocks and blocks of limestone. Intrusive breccias crop out along the margins of the pendant; many dikes, both felsic and mafic, are emplaced in the metamorphic rock and the sheared granites associated with the pendant structure. Along the same trend, remnants of this volcanic-sedimentary sequence are seen in the Kaweah Peaks and Mineral King pendants.

The Sequoia Park pendants record a thick section of metamorphosed sedimentary rock, including limestones, quartzite, and calcareous sandstones with lesser amounts of volcanic rock. The relationship between the two sequences may be seen in the Boyden Cave pendant, where the sedimentary sequence of shales, sandstones, and limestones is juxtaposed along the Kings River fault against the kinds of rhyolite ash flows and tuffs found in the eastern pendants.

Just to the west of the two parks, the lower Kings River and Kaweah River pendants record the emplacement of a complex ophiolite mass representing oceanic crust of late Paleozoic to early Triassic age. The roof pendants are all complexly deformed, and geologists do not agree completely upon the stratigraphic sequences within them or upon the original location of the sequences prior to the intrusion of the Sierra Nevada batholith. There is agreement, however, that the mixed volcanic and sedimentary sections represent deposition at the edge of the continent where island-arc volcanics,

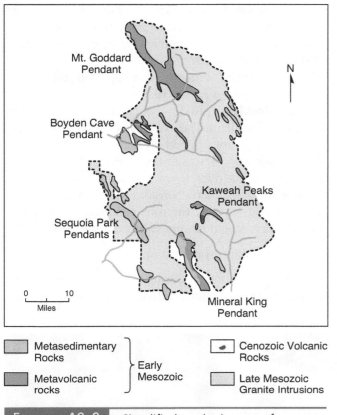

Mt. Goddard
Pendant

Boyden Cave
Pendant

Kaweah Peaks
Pendant

Sequoia Park
Pendants

Mineral King
Pendant

0   10
Miles

Metasedimentary Rocks
Metavolcanic rocks
Early Mesozoic

Cenozoic Volcanic Rocks
Late Mesozoic Granite Intrusions

FIGURE 49.9 Simplified geologic map of Sequoia-Kings Canyon National Parks. The early Mesozoic (late Triassic to early Jurassic) metasedimentary rocks contain both deep-water oceanic and shallow water continental components mixed with volcanic rocks. The Mesozoic volcanic rocks are largely rhyolitic and dacitic ash flows and tuffs with some sedimentary material. The age of the metasedimentary rocks in the roof pendants on the crest of the Sierra on the eastern edge of Kings Canyon National Park is uncertain. Some geologists have suggested that they are Paleozoic in age. After Saleeby and others, 1978.

derived in part from continental crust, were interleaved and mixed with sediments from both oceanic and continental sources.

Caves and Mineralization

In the pendants in the western parts of the parks, marbles have been partially dissolved by ground water to form underground caverns such as Crystal Cave, Clough Cave, Lilburn Cave, Palmer Cave, and Paradise Cave. The metamorphism that changed limestone to

marble destroyed any fossils that may have existed, but the rocks are assigned a late Triassic to early Jurassic age, based on fossils in other nearby pendants. As these carbonate rocks were being raised above sea level, groundwater began percolating along fractures and bedding planes, enlarging openings by removing more of the material. When the openings were raised above the water table by continued uplift, ground water then began to deposit as well as to dissolve calcite. Typical dripstone features formed on the ceilings, walls, and floors of the underground chambers. (Cave features and solution processes are explained in Chapters 14, 15, and 16.)

The mines in Redwood Canyon and along the North Fork of the Kaweah River are in *tactite,* a complex calcareous metamorphic rock formed as a result of hydrothermal activity and contact metamorphism. During the emplacement of the adjacent plutons, chemically active solutions from the hot magma penetrated the country rock and replaced some of its constituents with new minerals. Contained in the tactite are garnet, pyroxene, quartz, the tungsten ore scheelite, and the copper ores chalcopyrite and bornite.

Geologic History

1. Paleozoic deposition.

The western coast of North America has been active over much of geologic time. During the early Paleozoic Era, the coast was part of a passive continental margin and not much is known about the marine sedimentary rocks that were deposited in Paleozoic seas that spread over the region where the Sierra Nevada now stands. Some Paleozoic rocks crop out of the range, and a few scattered patches have been found in the mountains some distance north of Sequoia and Kings Canyon National Parks. In Devonian to Mississippian time, the plate convergence began that led to the Antler orogeny in Nevada. Subduction along the continental margin led to the emplacement of oceanic crust of late Paleozoic to Triassic age along the foothill faults of the Sierra Nevada.

2. Mesozoic deposition and mountain-building.

The Mesozoic history of the two parks may be divided into two parts. The first episode is recorded in the metamorphosed volcanic and marine sedimentary rocks of Triassic and early Jurassic age that are preserved as roof pendants and inclusions in the Sierra Nevada batholith. The second episode tells of the plutonic intrusion of the batholith largely in Cretaceous time.

The rocks of the roof pendants demonstrate that a subduction system existed off the western coast of North America early in Mesozoic time. A volcanic island-arc environment developed adjacent to the old continental shelf, and the subduction system very likely caused large-scale lateral transport and tectonic juxtaposition of the volcanic and sedimentary terranes. Much of the extreme deformation seen in the roof pendants comes from this period.

Toward the end of the Jurassic Period and the beginning of Cretaceous time, the subduction system had moved farther to the west. Igneous activity, resulting in the emplacement of plutons in the Cretaceous batholith, occurred as a series of pulses. The older, early Cretaceous plutons are found in the western parts of the range; and the younger, late Cretaceous granites

are near the crest. In general, the older intrusions are more mafic in composition (quartz diorites, for example), and the younger granites are more felsic (granodiorites, quartz monzonites).

The Nevadan orogeny that uplifted the Sierra Nevada is regarded as the first wave of the Laramide mountain-building episodes that created the great North American Cordillera. The extensive erosion that accompanied the orogeny stripped off many thousands of feet of Paleozoic and Mesozoic rocks. Estimates are that a layer of material over seven miles thick was stripped from the top of the Sierra Nevada. The uncovering of the batholith also exposed the roof pendants of metamorphosed volcanic and marine sedimentary rocks.

3. Cenozoic faulting and uplift.

Erosion continued during the beginning of the Cenozoic Era, but tectonic activity had slowed. Then the

FIGURE 49.10   There are several caves in Sequoia-Kings Canyon Parks. Lilburn Cave in Kings Canyon National Park has a lake. National Park Service photo.

FIGURE 49.11 Granite inclusions in the metamorphic schists formed during the early Mesozoic time by a process called magmatic differentiation. This feature is located in Kings Canyon National Park. The intense pressure has caused the granite to separate from the rest of the rock. National Park Service photo.

Sierra Nevada began to tilt westward. Faulting continued as the mountain block rose higher and higher, and the range began to assume its present position and form. The most recent faulting, uplift, and tilting probably began 2.5 million years ago. Sporadic earthquakes in the southern Sierra are an indication that movement and uplift are continuing. Late in the Cenozoic Era, volcanism occurred in the southern Sierra, but only a few remnants of those lava flows are found within park boundaries (fig. 49.9).

4. Pleistocene and Holocene glaciation.

Episodes of glaciation in the southern Sierra were less severe and less extensive than in the northern Sierra. Pleistocene climates were less harsh in the southern Sierra, and less precipitation was available. Overall, however, the patterns of advance and retreat of ice were similar. Alpine glaciation began early in the Sierra Nevada, probably late in the Pliocene and long before the large continental glaciers accumulated. In Sequoia and Kings Canyon, valley glaciers moved down major stream valleys, enlarging and deepening them. Glaciers were largest and most numerous in the zones of high precipitation on the western slopes. The valley glaciers on the east side of the crest strongly eroded the peaks but did not extend down into the Owens Valley.

At its maximum, the Kern Glacier, the most southerly of the Sierra Nevada trunk glaciers, nearly reached the southern boundary of Sequoia National Park. In the Kaweah Basin, glaciers partly filled many of the converging canyons, but few of them overtopped the divides except in the upper reaches of the basin. Even at their maxima, the largest glaciers in the northeast part of the basin did not extend much below the 5000-foot level.

Glacier systems were more extensive in Kings Canyon National Park, especially toward the north. Valley glaciers occupied the canyons of the Kings River and its tributaries, as well as the upper reaches of the San Joaquin River basin at the northern tip of the park. Glacial erosion was severe.

Three major episodes of glaciation occurred during the Pleistocene Epoch, with substages involving

Table 49.1			Generalized Geologic Column, Sequoia and Kings Canyon Parks	Generalized Geologic Column, Sequoia and Kings Canyon Parks
Time Units			**Rock Units**	**Geologic Events**
Era	**Period**	**Epoch**		
Cenozoic	Quaternary	Holocene	Alluvial, glacial, and mass wasting deposits	Stream erosion; Neoglacial activity; landslides, rockfalls Earthquakes, faulting, continuing uplift
Cenozoic	Quaternary	Pleistocene	Till, moraines, glacial outwash of various ages Avalanche deposits	At least three major glaciations, including substages; intense glacial erosion, avalanche chutes, ice-sculptured landscapes, glacial deposition Faulting, relatively rapid uplift
Cenozoic	Tertiary / Neogene	Pliocene		Probable onset of glaciation Southern Sierra volcanism (outside parks) Canyon-cutting, intense erosion Beginning of faulting and uplift Westward tilting of Sierra Nevada block
Cenozoic	Tertiary / Neogene	Miocene		
Cenozoic	Tertiary / Paleogene	Oligocene		Erosion
Cenozoic	Tertiary / Paleogene	Eocene		Pause in tectonic activity
Cenozoic	Tertiary / Paleogene	Paleocene		Probable beginning of cave development
Mesozoic	Cretaceous		Quartz diorites, granodiorites, quartz monzonites	Emplacement of plutons in composite batholith Metamorphism of country rock Nevadan orogeny
Mesozoic	Jurassic		Limestones, shales, sandstones, interbedded with volcanics	Marine sedimentation and island-arc volcanism
Mesozoic	Triassic			Subduction and plate convergence
Paleozoic			(not exposed in parks)	Long accumulation of sediments in Paleozoic seas

Source: Modified after Hill, 1975; Bateman, 1974; Bateman and Wahrhaftig, 1966.

advances and retreats of ice during these generally colder climatic periods. Till patches and moraines of varying ages mark glacial recessions. The Pleistocene glaciers did not last through the post-Pleistocene climatic optimum.

In Kings Canyon National Park, a few glaciers began occupying the cirques of earlier glaciers during the Little Ice Age (A.D. 1450–1850). These small valley glaciers are in the vicinity of Mount Goddard in the northern part of the park; others are in the Palisades region, just east of the park's northeastern boundary.

Prospects for the Future

More than one hundred years have passed since John Muir began his fight to preserve the California wilderness. Yet today, the national parks and national forests of the High Sierra are seriously threatened by throngs of visitors. The crowds have created problems of overuse that tend to damage the fragile environment, especially in the Alpine and Subalpine life zones. Because Sequoia and Kings Canyon are the national parks closest to the Los Angeles metropolitan area, they are particularly endangered.

Ironically, a recent ten-year study by the National Park Service indicated that the giant sequoias may also be threatened by *overprotection*. Periodic thinning by fires is needed to remove undergrowth and allow seeds to reach the soil and root in the sunlight. A program of "controlled burning" has been started by the National Park Service to help ensure new generations of sequoias. How can we balance the needs of a large urban population that seeks enjoyment of the parklands and, at the same time, protect the park environment from the damaging effects of too many people and vehicles and yet avoid overprotection? For the National Park Service, the answers to these questions involve careful management in the present, long-range planning for the future, and, above all, cooperation and understanding on the part of park visitors.

Geologic Map and Cross Section

Bateman, P.C. 1965. Geologic Map of the Black Cap Mountain Quadrangle, Fresno County, California. U.S. Geological Survey GQ Map 428.

Bateman, P.C., and Moore, J.G. 1965. Geologic Map of the Mt. Goddard Quadrangle, Fresno and Inyo Counties, California. U.S. Geological Survey GQ Map 429.

Jennings, C.W. Geologic Map of California. California Division of Mines and Geology. Sheets covering southern Sierra Nevada.

Moore, J.G. 1978. Geologic Map of the Marion Peak Quadrangle, Fresno County, California. U.S. Geological Survey GQ Map 1399.

———. 1981. Geologic Map of the Mount Whitney Quadrangle, Inyo and Tulare Counties, California. U.S. Geological Survey GQ Map 1545.

Ross, D.C. 1958. Igneous and Metamorphic Rocks of Parts of Sequoia and Kings Canyon National Parks, California. California Division of Mines, Department of Natural Resources Special Report 53.

Sources

Bateman, P.C. 1974. Model for the Origin of Sierra Granites. *California Geology* 27 (1):3–5.

———, and Wahrhaftig, C. 1966. Geology of the Sierra Nevada. In *Geology of Northern California,* ed. E.H. Bailey, p. 107–172. California Division of Mines and Geology Bulletin 190.

Colby, W.E. ed. 1960. *John Muir's Studies in the Sierra.* San Francisco: Sierra Club.

Gray, W.R. 1975. *The Pacific Crest Trail.* Washington, D.C.: National Geographic Society.

Hamilton, W. 1969. Mesozoic California and the Underflow of Pacific Mantle. *Geological Society of America Bulletin* 80:2409–2430.

Hartesvelde, R.J. 1973. *Field Guidebook for Sequoia and Kings Canyon National Parks; the Natural History, Ecology, and Management of the Giant Sequoias.* American Association of Stratigraphic Palynologists, 6th annual meeting, Anaheim, California, October 16–20.

Hill, M. 1975. *Geology of the Sierra Nevada.* Berkeley: University of California Press.

Matthes, F.E. (ed. F. Fryxell). 1965. *Glacial Reconnaissance of Sequoia National Park, California.* U.S. Geological Survey Professional Paper 504-A.

Oakeshott, G.B. 1978. *California's Changing Landscapes.* 2nd ed. New York: McGraw-Hill Book Company.

Oberhansley, F.R. 1974. *Crystal Cave in Sequoia National Park.* Sequoia Natural History Association.

Saleeby, J.B., Goodwin, S.E., Sharp, W.D., and Busby, C.J. 1978. Early Mesozoic Paleotectonic-Paleogeographic

Reconstruction of the Southern Sierra Nevada, in *Mesozoic Paleogeography of the United States*, Society of Economic Paleontologists and Mineralogists, p. 311–336.

Sharp, R.P. 1994. *Geology Field Guide to Southern California*, K/H Geology Field Guide Series. Third ed. Dubuque, Iowa: Kendall/Hunt Publishing Company.

Wheelock, W., and Condon T. 1974. *Climbing Mount Whitney*. Third ed. Glendale, California: La Siesta Press.

Note: To convert English measurements to metric, go to www.helpwithdiy.com/metric_conversion_calculator.html

Channel Islands National Park

OFFSHORE SOUTHERN CALIFORNIA

Donald F. Palmer   Kent State University

Area: 249,354 acres; 390 square miles
Proclaimed a National Monument: April 26, 1938
Redesignated a National Park: March 5, 1980
Designated a Biosphere Reserve: 1976

Address: Channel Islands National Park, 1901 Spinnaker Drive, Ventura, CA 93001
Phone: 805-658-5700
E-mail: chis_interpretation@nps.gov
Website: www.nps.gov/chis/index.htm

The park includes all the Northern Channel Islands — San Miguel, Prince, Santa Rosa, Santa Cruz, and Anacapa Islands — and Santa Barbara Island, one of the Southern Channel Islands.

The Channel Islands have a complex tectonic history and are related geologically to the Transverse Ranges north of the Los Angeles Basin.

Marine terraces on the islands record sea level changes that attest to a recent history of subsidence and uplift. The rise in sea level in Pleistocene time isolated the islands from the mainland and each other and led to their distinctive biological systems. Archeological sites document early human occupation and changes in the plant and animal communities during the last 15,000 years.

FIGURE 50.1   Satellite image of Santa Cruz Island. This island is the largest island in the group and the most rugged. It is 65 miles long and between one and 6.5 miles wide. A major east-west trending fault that separates the north and south mountain ranges was caused by strike-slip movement of the east Santa Cruz Basin fault during the Miocene. The fault is apparent in the aerial running diagonally across the photograph. National Park Service photo.

Geographic Setting

Channel Islands National Park includes six of the nine major islands off the coast of southern California. The Northern Channel Islands, which lie across the Santa Barbara Channel from the California mainland, are geographically distinct from the Southern Channel Islands, which are separated from the mainland by the San Pedro Channel and the Gulf of Catalina. (The Southern Channel Islands not in the national park are San Nicholas, Santa Catalina, and San Clemente.) Three Channel Islands, Santa Barbara (from the southern group), Anacapa and San Miguel (in the northern group), were set aside as a national monument in 1938. The addition of Santa Cruz, Santa Rosa, and Prince Islands in 1980 and the creation of the national park expanded this haven for endangered species of plants and wildlife, many of which are unique to the islands. Also protected are rookeries for sea birds, fur seals, elephant seals, sea lions, and other marine animals. Santa Cruz, for example, contains nine endangered plant species, and the endangered California brown pelican. San Miguel has the only marine mammal rookery in the world inhabited by six species of seals and sea lions. The endangered island fox, much smaller than mainland species, is found on four of the islands. Santa Barbara has a treelike sunflower of the genus *Coreopsis*.

Before the park was established, the fragile ecosystems of the islands and their surrounding waters were threatened by such human activities as ranching, trapping, oil spills, offshore oil exploration and residential growth and development.

Under the present management plan for Channel Islands National Park, primitive camping is permitted only on Santa Barbara and Anacapa Islands. The remaining islands may be visited for one day only with permission from the National park service or property owners. Tourists on San Miguel must be accompanied at all times by park service personnel. Sensitive areas, defined as habitats of endangered plant or animal species, are closed to the public.

The Channel Islands have generally cool, foggy summers and rainy winters (about 12 inches of rain annually). In winter, during the migration season, the islands are some of the best areas along the California coast for watching whales. Species migrating in these waters include the humpback, blue, finback, and sei whales.

Early inhabitants. The Channel Islands have a long history of prehistoric cultures. Definite human skeletal remains have been dated as 11,000 bp (before present), which was approximately the end of Pleistocene time. The climate then was more moist; the drier climate of today began around 7500 years ago.

When sea levels were lower during late Pleistocene time, the Channel Islands occupied a larger land area as an offshore extension of the Transverse Ranges of Southern California. Pleistocene mammoths that had migrated to the islands across the shorter stretch of open water existing then were cut off from the mainland when sea level rose, completely isolating the islands. Once isolated, the mammoths evolved into a pigmy variety, only four feet high at the shoulder. Roasting pits with charred dwarf mammoth bones have been found on the islands, and Indian hunters may have been responsible for the mammoths' extinction.

By 3000 years ago, the seafaring Canalino Indians lived in the region. They used tar from oil seeps on San Miguel and the mainland coast as caulking material for their canoes, for mending bowls, and as adhesive for attaching arrow points and other weapon heads to shafts. It was the Canalino Indians who greeted Rodriguez Cabrillo when he stopped at San Miguel in 1542. Cabrillo, a Portuguese mercenary in the service of Spain, led an expedition to find a sea route to China. After sailing north along the California coast beyond San Francisco, Cabrillo returned to San Miguel to spend the winter. He died on the island in January of 1543.

Diseases introduced by Europeans reduced the native population of the islands in the seventeenth and eighteenth centuries. The Canalino Indians abandoned Santa Rosa after a severe earthquake in 1812, and the last Indian was removed from Santa Cruz in 1813.

In the 1870s and 1880s, the city of Santa Barbara on the California coast (fig. 50.2A) was a health resort because of the numerous natural seeps of tar, oil, and gas along the shore. Natural oil slicks that spread over the coastal waters were believed to purify the air, making infectious diseases and chronic illnesses less prevalent. Most of the seeps emanated from the Monterey Shale, which crops out in some of the Channel Islands as well as on the mainland. San Miguel has two oil seeps, one near Castle Rock, off the island's northwest shore, and the other near Tyler Bight, off the south shore. Another seep has been burning at the eastern end of Santa Cruz Island near Chinese Harbor since before written records existed.

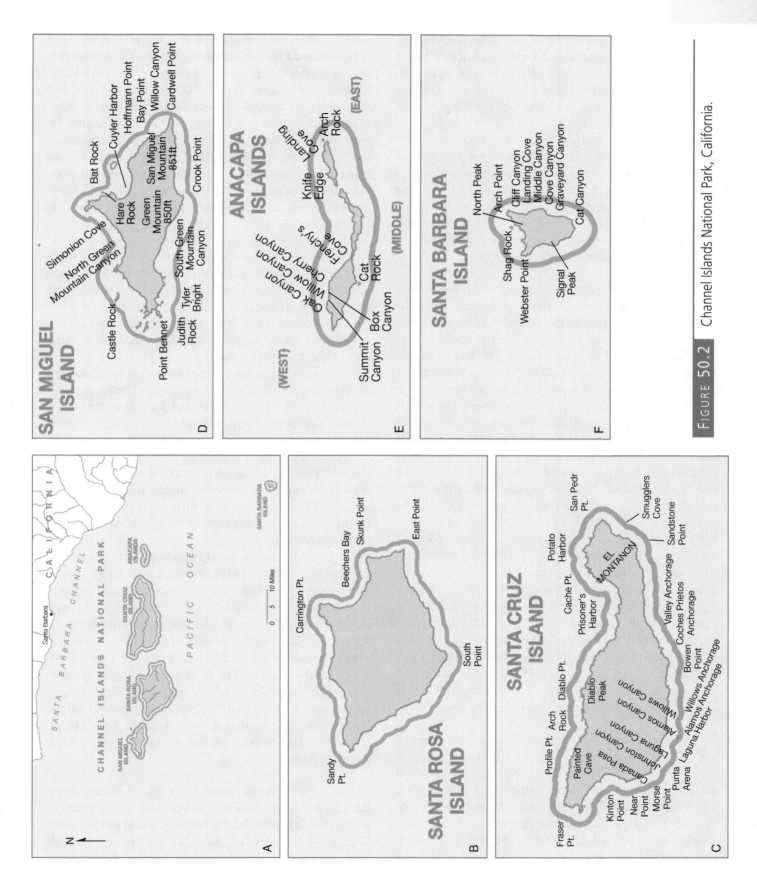

Figure 50.2   Channel Islands National Park, California.

Ever since the beginning of European settlements, the Channel Islands have had separate histories based on the different ownerships and types of land use. These varying patterns of land use are still affecting the natural environments of the islands in the park.

Santa Rosa Island. In 1843 Santa Rosa was awarded to the Carrillo family as a land grant from the government of Mexico, and for more than fifty years, family members maintained a sheep ranch. W.L. Vail and J.V. Vickers bought Santa Rosa in 1902 and converted the ranch to a cattle operation. The Vail and Vickers Company still operate the ranch, although the island was purchased by the federal government in 1987 (fig. 50.2B).

Santa Cruz Island. This island, largest in the group, supported a population of some 3000 Indians when the Spaniards first visited it. The island was named Santa Cruz ("holy cross") in 1769 by a Spanish naval officer after Indians found an iron crucifix that had been lost by the expedition's chaplain. The Mexican government had a penal colony on Santa Cruz for a short time in 1829, but long enough to give Prisoners Harbor its name. Sheep and other domestic animals were first brought over by an English company that acquired the island in 1859. Ten years later the island was sold to Justinian Caire, who operated a sheep and cattle ranch and produced wine. His descendants still own a 6000-acre sheep ranch on the eastern end of Santa Cruz. The rest of the island (about nine-tenths of the total acreage) is owned by the Santa Cruz Island Company. Most of the sheep have been removed and those that remain are confined by fences. Cattle grazing is restricted to parts of the island that are less subject to damage by erosion.

A University of California research facility, the Channel Islands Field Station (with an office located on the University of California-Santa Barbara campus), was established in the 1960s in cooperation with the Santa Cruz Island Company. Eventually the entire island will be owned and managed by the Nature Conservancy in cooperation with the National Park Service (fig. 50.2C).

San Miguel Island. The federal government acquired San Miguel when Mexico ceded California territory to the United States in 1848. George Nidever, who brought sheep, cattle, pigs, and horses to the island in the 1850s, was the first leaseholder. The adobe house that he constructed from local clay still stands. Prior to the introduction of domestic animals, the island had a luxuriant ground cover, but a combination of overgrazing and

repeated droughts destroyed most of the vegetation. Sand dunes, formed by the prevailing northwest winds, prevented regrowth. The severe drought of 1863–64 killed 85 percent of Nidever's stock (5000 sheep, 180 cattle, and 30 horses). An 1870 drought caused the remaining sheep to dig as deeply as two or three feet into the ground for roots. What the sheep didn't kill, the drifting sand dunes did. Subsequent droughts in 1877, 1893–94, and 1923–24 changed the landscape to near-desert conditions supporting only a few grasses and wildflowers. The U.S. Navy owns the island and administers it in cooperation with the National Park Service (fig. 50.2D).

Anacapa Island. Anacapa, which is actually three islets (East, Middle, and West) linked by sand bars, was acquired by the United States from Mexico at the same time as San Miguel. In the late nineteenth and early twentieth centuries, a few sheep were pastured on the island, but all were removed when Anacapa became part of the national monument in 1938.

The name Anacapa is from an Indian word, *eneepah*, which means "deception, or mirage," because under certain conditions of weather and light, a distortion or mirage causes Anacapa to appear to be larger than it actually is.

Arch Rock, located at the eastern tip of East Island, is a well known Anacapa landmark. James Whistler, the nineteenth-century American painter, made an etching of this 40-foot high sea arch when, as a young man, he worked as a map engraver for the U.S. Coast and Geodetic Survey.

An automated 600,000-candlepower beam and foghorn were installed in the lighthouse on East Island in 1966. Prior to that time, Coast Guard personnel manned the lighthouse, which was constructed in 1932. In spite of the warning system, wrecks and groundings occur on Anacapa's rocky shores, usually because of dense fog. A tuna clipper sank off Middle Island in 1949, near a commercial fishing camp. The most famous shipwreck was that of the paddle-wheel steamer "Winfield Scott," which sank in 1853. All 250 passengers got off safely, but spent several anxious days on Anacapa before being rescued. What remains of the wreck lies offshore in about 20 feet of water (fig. 50.2E).

Santa Barbara Island. Sometimes referred to as "Santa Barbara Rock" because of its small size (approximately a square mile), this island is the only one of the Southern Channel Islands to be included in Channel Islands National Park. Archaeological evidence indicates occasional extended visits here by prehistoric people but no permanent camps, probably because of lack of fresh water. A few seeps, like one in Indian Water Cave, are the only natural sources of fresh water.

Originally the island was well covered with vegetation and supported a number of bird and shore species and small animals. However, a century or more of human occupation drastically changed the island's natural environment. Many of the changes are irreversible (fig. 50.2F).

Environment and Organisms

Clearly, Channel Islands National Park is an example of how misuse and overuse of land can change an environment forever. Nevertheless, the park does provide one of the best marine wildlife sanctuaries in the United States, as well as preserving local habitats for some endangered species.

The Channel Islands are located in a transitional region between northern and southern marine fauna governed by the juxtaposition of the south-flowing, cold-water California current on the west and a warm-water countercurrent from the south that flows closer to the mainland. For this reason the westernmost islands—San Miguel, Prince, and San Nicholas (not in the park)—support mainly northern cold-water marine fauna; the easternmost islands—San Clemente and Santa Catalina (outside the park) and Santa Barbara—support primarily southern species. Anacapa, Santa Cruz, and Santa Rosa, which lie between the two currents, have a mixture of southern and northern species. Waters surrounding all the islands are rich in marine algae and kelp, the food supply for marine organisms of many types. About 20 species of whales and porpoises abound in the area, along with six species of sea lions and harbor seals and many species of sea birds.

Most of the islands have sage scrub and grasses along the shorelines, with oak woodlands mixed in on the larger islands. Each island has endemic species and subspecies of plants and animals that live only on one island or, in some cases, adjacent islands; examples are unique snails, deer mice, island foxes, and many kinds of plants. Both native and feral animals live on the islands.

The changes in the character of the Channel Islands are due to a number of factors that can be summarized as follows:

1. The introduction of cats eliminated several species of sea birds.

2. Sheep, rabbits, goats, and other domestic animals reduced or eliminated many native plants. Animals that reverted to the wild have altered the environment and in some instances caused the disappearance of native species.
3. Clearing of land for farming removed or forced out much of the original vegetation. Native species are left only in isolated, protected areas.
4. Fires, both deliberate and accidental, burned most of the vegetation and destroyed so much of the island's habitat that the Santa Barbara song sparrow became extinct.
5. The planting of grains brought about overpopulation of native deer mice who, in turn, decimated land mollusks and sea bird eggs.
6. The introduction of the "ice plant" *(Mesembryanthemum crystallinum)*, a fleshy-leaved plant, led to the crowding out of native species.
7. The combined effects of burning, exotic plants, and domestic animals caused considerable erosion in some areas.

Among threats still endangering the environments of these peaceful islands are: (1) overharvesting of marine resources, such as anchovies (main food of the California brown pelican) and kelp; (2) overgrazing; (3) oil spills from passing tankers; (4) development of offshore oil and gas; (5) sonic booms that disturb seal and bird rookeries; (6) and the continued introduction of exotic species.

Geologic and Other Natural Features of the Channel Islands

If the ocean water were to disappear so that only land could be seen, the islands would stand as high ridges on the continental shelf, separated by a series of basins of varying depths. Although the islands are now in close proximity to one another, they do have distinct geologic characters and histories. Perhaps nowhere is this independence as clearly demonstrated as in the elevated beach ridges and marine terraces on the islands that indicate how intermittent uplift of the islands and lowering of sea level have occurred in the recent geologic past. The elevations along the terraces change from place to place, indicating differential uplift within individual islands; and the completely different terrace levels on some islands reveal independent histories with different deformational episodes and varying rates of uplift.

San Miguel Island

San Miguel, 14 square miles in area, is the northern flank of a faulted anticline trending northwest. Ages of the bedrock range from Cretaceous to Miocene. The rocks are mostly sedimentary, and the beds dip monoclinally to the northeast.

Several levels of beach ridges encircle the island. Due to overgrazing, slopes are severely gullied and the western two-thirds of the island is covered by sand dunes. Rainwater trapped by the dunes is the source of fresh water springs that make San Miguel one of the few Channel Islands with potable water, although poor tasting. Some of the dunes are being stabilized by regrowth of vegetation and by cementation due to ground-water action.

Dune rock, which is an *eolianite* made up largely of dune sand, is formed from wind-deposited clastic sediment that has been cemented below the surface by ground water precipitation of calcite.

On San Miguel, the dune rock is interbedded with layers of soil that represent periods when conditions were moist enough to permit plant growth and soil formation by weathering of bedrock. Intervening drier periods brought about the resumption of dune growth and the dunes buried the vegetation and soils.

Since most of San Miguel's land area is a 400–500-foot high plateau, relief is low and the island is more subject to wind erosion. Prevailing northwest winds have blown away most of the topsoil, exposing caliche and eolianite at the surface. *Caliche* is formed mainly by capillary action drawing mineral-rich ground water to the surface where it evaporates in the wind and sun, leaving crusts of soluble calcium salts, mixed with the sand and silt (fig. 50.3). Caliche also tends to form in a layer just below the surface when not enough moisture is available to flush the calcium salts from the soil. Both surface and subsurface caliche deposits occur over most of the island. Concretions of caliche formed as nodules, tubes and casts around roots and tree trunks that were covered by the sand dunes. A caliche-encased log, 30 feet long and over two feet in diameter, was found on the island. Trunk casts of other logs have also been uncovered by erosion.

Case-hardened caliche is common on the northern and western parts of the island. Here the caliche deposits were exposed to rain and sea spray so that calcium salts on the surface were dissolved and reprecipitated many times, making the crusts thick and hard. Slabs and large chunks of caliche are scattered over the ground, and the sand dunes have become semi-indurated.

The "elevator rocks" of San Miguel are a pebble-and-cobble pavement on old marine terraces that have been raised above sea level and then buried by sediment washed down from still higher terraces or slopes. The finer grains in the sediment are blown away by wind or washed away by water, leaving the coarser gravel and cobbles as an "armor" on the surface.

Many paleontological and over 500 archaeological sites are on the island. A major paleontological find was a large accumulation of bones of the pigmy mammoth *(Mummuthus exilis),* which was found at Running Springs (near Dry Lake on the west side of the island). Another pigmy mammoth site is about three miles west of Cardwell Point on the southeast shore. The bones at Running Springs have been well preserved by calcareous tufa deposited around the springs, which have a high dissolved load of calcite.

Carbon-14 dating of mammoth remains yields ages such as 15,630 ± 460 and 16,520 ± 150 years. Burned and calcined mammoth bones, charcoal made of cypress, mammoth tusks, and teeth all suggest that prehistoric people were present at the sites and that they hunted the pigmy mammoth. There is geologic evidence of overgrazing by the pigmy mammoths, periodic forest fires set by hunters or by lightning, and a 20-year drought cycle.

San Miguel is noted for its unique plant and animal species, now under the protection of the National Park Service. The Harbor seal, California sea lion, Northern elephant seal, Steller sea lion, Northern fur seal, and the Guadalupe fur seal (nonbreeding at this rookery) frequent the seal rookery at Point Bennett on the western tip of San Miguel. Eight kinds of plants are found only on San Miguel.

Prince Island

Tiny Prince Island (also called Gull Island) lies just off the northeast shore of San Miguel. Prince Island's highest elevation is only 296 feet, and its dimensions are about 1000 by 1500 feet. Thin, shallow soils mantle the rocky slopes of this foggy, windswept piece of land. The bedrock consists of Oligocene dacite intrusives, flows, and clastics that are part of the Upper Member of the San Miguel Volcanics. The island is a major sea bird rookery.

Santa Rosa Island

Santa Rosa, 84 square miles in extent, is the second largest island in Channel Islands National Park, and has about 45 miles of shoreline, and is approximately 15 miles long and 10 miles wide. Geologically, the island is a series of fault anticlines and synclines. The Santa Rosa fault trends east-west across the island. Movement along this fault, producing both vertical and horizontal displacement, has caused distinct differences in topography between the northern and southern parts of the island. The northern half of the island, the down-thrown side, has broad, plateau-like terraces into which streams have cut steep canyons. High bluffs face the northern shore. White sandy beaches stretch along the northern and western coast. Most of the beaches are flanked by dunes up to 400 feet high. South of the Santa Rosa fault, the land is higher and more rugged. The highest point (1589 feet) is on a mountainous ridge that follows the east-west trend of the fault. The shales, sandstones, and volcanic rocks on the south side of the island are more resistant to erosion than the shales (Monterey Formation) exposed on the north side.

Many archaeological and paleontological sites have been located on Santa Rosa, including some with remains of pigmy mammoths. A few sites have been studied, but much archeological work still needs to be done. A rare subspecies of Channel Islands fox is found only on this island, as well as an endemic spotted skunk. The Torrey pine grows only here and in Torrey Pines State Park in San Diego County on the mainland. Twenty-six other plants that do not grow on the mainland have been identified on Santa Rosa. The profuse kelp beds along the shore are a refuge for sea otters and a breeding ground for the California sea lion.

Santa Cruz Island

With 96 square miles of land surface, Santa Cruz is the largest Channel Island. It has 65 miles of shoreline, is 24 miles long, and between one and 6.5 miles in width. With its steep shoreline, pocket beaches, and mountains, Santa Cruz is the most rugged of the islands. Sea caves (such as Painted Cave) have been eroded in the cliffs along the shore by wave action. Unlike the other Northern Channel Islands, which have relatively few safe anchorages, Santa Cruz's irregular shoreline has a number of protected coves, especially on the south side.

A major east-west-trending fault cuts across Santa Cruz, separating northern and southern mountain ranges, the principal rock types, and the island's drainage networks. The floor of the intervening valley, eroded along the fault zone by streams, is about the only flat land on the island. Most of the Central Valley is drained by a stream that flows into Prisoners Harbor on the northeastern side of the island. A smaller stream

FIGURE 50.3   The prevailing northwest winds have blown away most of the topsoil, exposing caliche and eolianite. Caliche is formed by capillary action drawing up mineral-rich ground water (mostly calcium salts) toward the surface where the salts precipitate from solution. The "towers" are the roots and tree trunks of past vegetation. Eolianite is dune rock cemented by the caliche. National Park Service photo.

FIGURE 50.4   There are numerous sea caves and arches on Santa Cruz Island. They are created when refracting waves erode a tunnel through both sides of a headland. Photo by Jeffery Dick.

drains to the west. The fault zone is the result of strike-slip movement along the East Santa Cruz Basin fault that brought together the northern and southern halves of the island during Miocene time (fig. 50.1).

In the northern mountain range, Diablo Peak, elevation over 2400 feet, is the highest point. Sharp ridges and steep canyons leading down to the coast dissect the northern range. The less rugged southern mountain range rises 1523 feet above sea level. Broader canyons on the south side, with gentler slopes, provide access to the island's interior.

Some of the sedimentary rocks on Santa Cruz are a continuation of those in neighboring Santa Rosa; but the geology on Santa Cruz is more complex, with exposures of numerous rock types, including volcanic and metamorphic rocks of various ages. Prehistoric inhabitants of the Channel Islands found excellent raw materials for stone tools among this rich variety. Especially prized was a chert of high quality in beds of the Monterey Formation, cropping out on the eastern end of the island. Points and tools chipped from this chert were bartered for goods from mainland Indians. Some

of the volcanic rocks on the island were also used for chipping and grinding weapons and tools.

Among the endemic plants and animals on Santa Cruz are the great island pine, the Santa Cruz pine, spotted skunk, a species of salamander, and the dwarf fox. About 500 native plant species and subspecies grow on the island, 7 of which are endemic, and 31 of which are not found on the mainland. Resident bird species number 44 and include the endangered California brown pelican. Kelp beds surrounding the island provide food for seal and bird rookeries, as well as for fish, shellfish, and other organisms.

Anacapa Island

The three Anacapa islets are segments of a volcanic ridge made up of lava flows, breccias, ash, and cinders erupted from the ocean floor in Miocene time (fig. 50.5). Many of the lava flows have amygdules of olivine and chalcedony. Cobbles and pebbles weathered from the flows are scattered over the beach on Frenchys Cove, one of the few places on Anacapa where boats can land. Sheer cliffs and rocks make up most of the shoreline. The cliffs are particularly high and almost vertical from the ridge crest to the water along the island's south side. Slopes on the north side of the ridge are somewhat less steep.

The highest point on the volcanic ridge, 930 feet, is Vela Peak on West Island. The ridge gradually lowers to 250 feet above sea level, which is the highest point on East Island. From west to east, Anacapa is 5 miles long, but only about one-half mile wide. The largest islet, West Island, has about half of Anacapa's total acreage of a little over one square mile.

Wave-erosional features, such as sea arches, sea caves, stacks, black sand and cobbles, wave-cut platforms, surge channels, and blowholes, are prominent on Anacapa's shore. The sea caves, arches, and surge channels tend to form in bedrock fracture zones. A well developed surge channel and blowhole is near Cat Rock on the south side of West Island. When a wave swell moves into the narrow channel, air that is trapped and compressed within the fissures forces out sea water explosively, spraying the surrounding area.

Raised beaches and marine terraces on parts of Anacapa are indications of changing sea levels in the fairly recent geologic past. The presence of a marine wave-cut terrace at a depth of 240 feet below sea level near East Anacapa Island records the Pleistocene lowering of sea level and perhaps some tectonic activity. While Anacapa Island does not have large faults or folds, it was almost certainly uplifted as part of an episode of faulting and folding along bounding faults that are off the coast.

Santa Barbara Island

Santa Barbara is related lithologically to Anacapa, since both islands are almost entirely volcanic material, primarily basalt. Geologic features of Santa Barbara include numerous faults that cut the bedrock units but are overlain by Pleistocene deposits on the marine terraces; also present are volcanic cinders, pillow basalts mixed with sedimentary marine beds, eolianites, sea caves, landslides, and evidence of churning and slickensides in the soil, caused by differential expansion and contraction of soil materials.

Overgrazing and burning on Santa Barbara have caused severe erosional problems. Cliffs and slopes are unstable and subject to landsliding, especially on the island's north side. Another area of severe erosion is near the base of the eastern side of Signal Peak (elevation 635 feet), the highest point on the island.

FIGURE 50.5 An aerial view of the three Anacapa Islets that are segments of a volcanic ridge made up of lava flows, breccias, ash, and cinders erupted from the ocean floor in Miocene time. From east to west the island is five miles long and one-half mile long. Raised beaches and marine terraces indicate changing sea levels. Anacapa's name is derived from an Indian word *eneepah,* which means "deception" or "mirage" because under certain conditions of weather and light it appears to be larger than it really is. National Park Service photo.

FIGURE 50.6   An aerial view of Santa Barbara Island that is related lithologically to Anacapa Island as they both are almost entirely volcanic, mostly basalt. Numerous faults cut the bedrock units, but are overlain by Pleistocene deposits on the marine terraces. There is a problem with landsliding, especially on the north side. Santa Barbara is a small (one square mile) but important island for the protection of certain plants and animals. National Park Service photo.

Marine terraces, cut by canyons, are prominent on Santa Barbara's northern and eastern sides (fig. 50.6). The highest Pleistocene sea stands probably submerged the entire island. Faunal remains and numerous archeological sites have been found in the terrace gravels. Prehistoric people must have camped often on the island to fish and hunt when canoeing between the Southern and Northern Channel Islands.

Although Santa Barbara is a small island, it is important for the protection of certain native plants and animals: the giant coreopsis (*Coreopsis gigantea*), which attains a height of eight feet and has bright yellow, daisylike flowers in the spring; the liveforever *(Crassulaceae)* and other plants not found on the mainland; an endemic night lizard and certain species of land snails; rookeries of the California sea lion, elephant seal, harbor seal, and species of sea birds; and the habitat of a few endemic mammals, such as a small bat and deer mice.

Geologic History

Channel Islands National Park is located on the west side of the boundary zone between the North Ameri-

can plate and the Pacific plate.[*] The lateral movement along the San Andreas fault system that has been occurring for several million years in this boundary zone has acted to: (1) open the Gulf of California by separating the Baja Peninsula from the mainland of Mexico; (2) move Baja, California, and that part of California west of the San Andreas fault to the northwest; and (3) cause large-scale displacements and rotations of crustal blocks near the California borderland. Tectonic changes of such magnitude, when superimposed upon an already complex geologic history, have made interpretation of the geologic history of the Channel Islands and adjacent mainland very difficult. Continued movement along the San Andreas fault system will lead to greater displacement and perhaps further rotation of the Channel Islands as time goes on.

The Channel Islands are geologically associated with the Transverse Ranges section of the California

---

[*]The national parks in the Sierra Nevada—Yosemite (Chapter 28), Sequoia and Kings Canyon (Chapter 49)—are in the part of central California that is *east* of the San Andreas fault system and the plate boundary.

Table 50.1	Geologic Column, Channel Islands National Park

Time Units			Rock Units	Geologic Events
Era	Period	Epoch	Formation	
Cenozoic	Quaternary	Holocene	Alluvium, stream, and landslide deposits	Continued uplift and sea level changes
		Pleistocene	Terrace gravels	Repeated uplift and development of marine terraces
			Middle Anchorage	Sedimentation of conglomerate
			Alluvium	
			Potato Harbor Formation	Marine transgression, late volcanic activity
	Tertiary (Neogene)	Miocene	Monterey and Blanca Formations	Uplift, faulting, and exposure of the basement complex associated with development of the Transverse ranges, movement and rotation due to plate tectonics.
			Santa Cruz Island Volcanics	Rapid accumulation of volcanic lavas, breccias, and pyroclastic debris. Erosion and reworking of older sediments and deposition under increasing depths of water.
	Tertiary (Paleogene)	Oligocene	San Miguel clastic and volcanic units San Onofre Breccia	
			Rincon Formation	Deposition of marine and nonmarine clastic sediments mixed with volcanics. Uplift in source areas to the north and east.
			Vaqueros Formation	
			Sespe Formation	
		Eocene	Cozy Dell Formation	Inception of coarse clastic sedimentation due to uplifts in surrounding area. Deep water sedimentation
			South Point and Jolla Vieja Formations	
			Cañada Formation	
		Paleocene	Pozo Formation	Continuous deposition in a continental slope environment with a gradually transgressive sea
Mesozoic	Cretaceous (Upper)		Jalama Formation	Marine sedimentation and the Laramide orogeny
	Jurassic		Alamos Pluton Willows Plutonic Complex Santa Cruz Schist	Volcanic activity and sedimentation, intrusion of plutons and regional metamorphism

Source: Modified after Weaver, 1969; Howell, 1976.

Note: 〜〜〜〜〜〜〜〜 erosional events.

Coast Ranges, which are included in the Pacific Border Province. In the Transverse Ranges, the basins, ranges, and most other structures trend east-west or "transverse" to the dominant northwest-southeast trend of the adjoining areas (the California Coast Ranges, the Sierra Nevada, and the Great Valley). The aligned peaks of the Northern Channel Islands are, in fact, the tops of a partially submerged, westward extension of the Santa Monica Mountains on the mainland. The Santa Barbara Channel (north and east of the islands) is the submerged extension of the Ventura Basin, a broad valley north of and approximately parallel to the Santa Monica Mountains (fig. 50.7).

The geologic history of the Channel Islands is told by four kinds of evidence: (1) the sedimentary rocks of the area that reveal the depositional and uplift history of the islands and the surrounding region; (2) igneous rocks and metamorphism and deformation of preexisting rocks, showing the effects of tectonic processes; (3) paleomagnetic data that demonstrate wholesale clockwise rotation of the Channel Islands and adjacent Transverse Ranges relative to the rest of California, which occurred in post-Miocene time; and (4) Pleistocene fossils and topographic features that help to explain recent history.

1. Mesozoic volcanic activity, intrusion, and metamorphism.

The earliest rocks in Channel Islands National Park are a mixture of metamorphosed volcanic and sedimentary rocks called the Santa Cruz Schist. They formed over 160 million years ago as a volcanic pile made up of basaltic and rhyolitic lavas interlayered with clastic sediments of volcanic origin. Later intrusions by the Willows Plutonic Complex and the Alamos Pluton occurred at about 160 and 140 million years ago, respectively. Then in Jurassic and Cretaceous time, the area underwent regional (dynamic) metamorphism that recrystallized and deformed the rocks.

2. Deposition and deformation during the Cretaceous and early Tertiary Periods.

The Cretaceous Period ushered in a marine environment with a tropic ocean that extended north of the Channel Islands area. Over 6300 feet of shales, siltstones, and sandstones were deposited as the Jalama Formation. Deposition continued into the early Tertiary and is marked by coarsening of some sediments as a function of uplift of the adjacent continent and the nearness of the source terrain. During early Tertiary time, 5000 feet of clastic sediments consisting of cobbles, sands, and shales were deposited to form the Pozo, Cañada, Jolla Vieja, South Point, and Cozy Dell Formations. These formations were deposited as turbidites, the sediment coming from a continental mass to the south and southwest of the present islands. (*Turbidites,* which tend to be graded and well sorted, are deposited by a dense, swift, bottom flowing ocean current; i.e., a *turbidity current.*)

During the Oligocene Epoch, terrestrial sediments of the Sespe, Alegria, Vaqueros, Rincon Formations were deposited to a thickness of 3400 feet over a wide

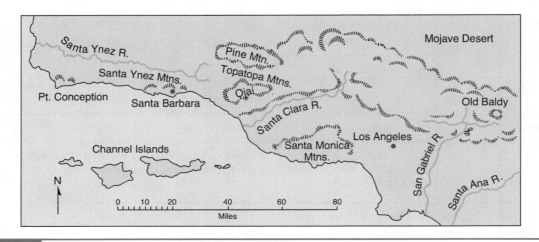

FIGURE 50.7   The Northern Channel Islands shown in relation to the east-west trending Santa Monica Mountains and other Transverse Ranges. Modified from R.P. Sharp, Southern California © 1976 Kendall/Hunt Publishing Company. Used by permission.

area from Santa Cruz Island across the Santa Barbara Channel to the Santa Monica Mountains.

Sedimentation in early Tertiary time was controlled by the Laramide orogeny, which caused a general uplift throughout western North America, and affected specific ranges such as the Santa Ynez Mountains (one of the Transverse Ranges) and other ranges in southern California.

3. Late Oligocene-Early Miocene deformation and continued sedimentation.

Deformation at the end of the Oligocene involved folding and faulting and was followed by renewed volcanic activity in the early Miocene. Volcanic activity was initiated by mafic flows and intrusions and was followed by more felsic flows, breccias, and pyroclastic debris. This 1400 feet of volcanic rock is known as the Blanca Formation, which, in turn, was eroded along with earlier sedimentary rocks and the basement complex. The resulting coarse clastics were deposited as the Beecher Member of the Monterey Formation. Most of the Monterey Formation is a siliceous deposit formed on the continental shelf in warm seas.

4. Clockwise rotation of the Channel Islands and adjacent Transverse Ranges.

Paleomagnetic evidence from Miocene igneous rocks shows that the Channel Islands were rotated between 70° and 80° clockwise since Miocene time. This rotation, which affected much of the Transverse Ranges, appears to be due to "rolling" of the Transverse Range block between the moving north Pacific and North American plates during late Tertiary time. This shearing also created the major offsets along the San Andreas and parallel faults in southern California. The sequence of these events as reconstructed from the paleomagnetic evidence is shown in figure 50.8. The rotation does much to explain the sources of sediment and the location of the seas described earlier. The ocean to the north of the Channel Islands and the south and southwest sources of Tertiary sediments are rotated, upon reconstruction to pre-Miocene time, to a western ocean and eastern sources of sediment, such as we find today.

5. Emergence of the Channel Islands during early Pleistocene time.

Uplift of the Transverse Ranges and Channel Islands formed an area of high relief. Initial uplift formed a major land extension called the Cabrillo Peninsula that stretched from the Santa Monica

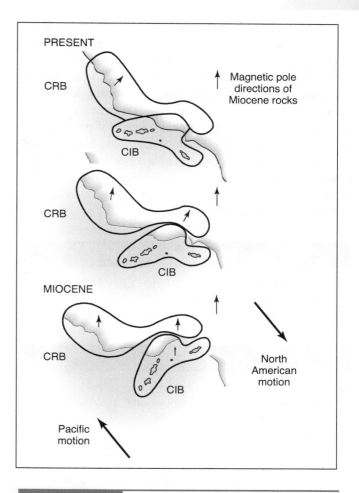

FIGURE 50.8 Rotation of the Channel Islands from Miocene to the present. The diagram shows a schematic representation of the rotations of the Channel Islands Block (CIB) and the Coast Range Block (CRB) relative to the rest of North America. Evidence for rotation is found in the paleomagnetic directions to the pole found in Miocene igneous rocks (shown by black arrows on the different blocks) which point in different directions at the present time. These directions were parallel in Miocene time when the igneous rocks cooled in the Miocene magnetic field (bottom of figure). Subsequent rotation in a clockwise direction led to the situation seen at present (top of figure). The rotation was caused by the relative motion of the Pacific and North American plates (long open arrows), which effectively "rolled" both the Channel Islands and Coast Range Blocks as small fragments caught between the two larger plates. In the figure, the borders of the blocks are approximate, and the land areas (stippled line) are given as a reference only.

Mountains out along the present Channel Islands. It was along this peninsula that Pleistocene land mammals migrated, only to become isolated when the land bridge between the Channel Islands and the mainland subsided. Marine terraces found at various levels on San Miguel Island, and as high as 1800 feet on Santa Cruz Island, attest to the dynamic uplift of parts of the island chain during the Pleistocene Epoch. Such uplifts and areas of subsidence have caused the isolation of the Channel Islands, led to the distinctive biological heritage of many of the islands, and produced the impressive topographic features of today.

Geologic Map and Cross Section

Feroy, D.E., Detking, P. and Reufro, H.B. 1968. Geologic Highway Map, Number 3, Pacific Southwest Region. The American Association of Petroleum Geologist.

Sources

Ernst, W.G., ed. 1981. *The Geotectonic Development of California*, Rubey v. 1. Englewood Cliffs, New Jersey: Prentice-Hall.

Glassow, M. 1977. An Archaeological Overview of the Northern Channel Islands, California, Including Santa Barbara Island. Prepared for the Western Archeological Center, National Park Service, Tucson, Arizona. National Technical Information Service No. 298855.

Hall, C.A., Jr. 1981. San Luis Obispo Transform Fault and Middle Miocene Rotation of the Western Transverse Ranges, California. *Journal of Geophysical Research* 86:1015–1031.

Howell, D.G., ed. 1976. *Aspects of the Geologic History of the California Continental Borderland*. Pacific Section, American Association of Petroleum Geologists, Miscellaneous Publication 24. (Reports by various investigators on the basement rocks, stratigraphy and paleontology, sedimentology, and tectonics presented at the California Borderland Symposium, Santa Cruz Island, 1975.)

Howorth, P.C. 1982. *Channel Islands, the Story Behind the Scenery*. Las Vegas, Nevada: KC Publications.

Johnson, D.L. 1967. *Caliche on the Channel Islands.* Mineral Information Service, Natural Resources Program, Naval Undersea Center, San Diego, California.

Kamerling, M.J., and Luyenduk, B.P. 1979. Tectonic Rotations of the Santa Monica Mountains Region, Western Transverse Ranges, California, Suggested by Paleomagnetic Vectors, *Geological Society of America Bulletin* 90:331–337.

Luyenduk, B.P., Kamerling, M.J., and Terres, R. 1980. Geometric Model for Neogene Crustal Rotations in Southern California, *Geological Society of America Bulletin* 91:211–217.

Power, D.M., ed. 1980. The California Islands: Proceedings of a Multidisciplinary Symposium. Santa Barbara, California: Santa Barbara Museum of Natural History.

Resource Management Workshop, Channel Islands, November 1–3, 1978. Mimeographed sheets.

Sharp, R.P. 1976, *Field Guide to Southern California*, K/H Geology Field Guide Series. Dubuque, Iowa: Kendall/Hunt Publishing Company.

Weaver, D.W. 1969. *Geology of the Northern Channel Islands*. Special Publication. Pacific Sections, American Association of Petroleum Geologists; Society of Economic Paleontologists and Mineralogists.

Note: To convert English measurements to metric, go to www.helpwithdiy.com/metric_conversion_calculator.html

C H A P T E R   5 1

Redwood National Park

NORTHERN CALIFORNIA

Lisa A. Rossbacher   Southern Polytechnic State University

Area: 110,232 acres; 172 square miles

Established as a National Park: October 2, 1968

Designated a World Heritage Site: 1980

Designated a Biosphere Reserve: 1983

Address: 1111 Second Street, Crescent City, CA 95531

Phone: 707-464-6101

E-mail: REDW_Information@nps.gov

Website: www.nps.gov/redw/index.htm

In coastal redwood forests along the northern California shore are the world's tallest trees, hundreds of years old. On this dynamic and scenic coast, sea cliffs, marine terraces, landslides, exposures of the Franciscan rock assemblage, and earthquakes are evidence of recent geologic change.

FIGURE 51.1   The tallest trees on earth are the coast redwoods (*Sequoia sempervirens*), which may grow to 367 feet in height and may have a width of 22 feet at the base. They grow only in a narrow belt along the northern California coast. They can live up to 2,000 years but most of them live only 500-700 years. National Park Service photo.

Father Juan Crespi, a Franciscan missionary on a 1769 expedition to the region wrote about the *palo Colorado* or "red trees." The redwoods were officially named in 1823, when A.B. Lambert designated them *Taxodium sempervirens,* or evergreen cypress. Later, a German botanist realized that these trees did not belong to the bald cypress group, and he renamed them *Sequoia sempervirens.* Both the coast redwood and the giant sequoia of the Sierra belong to the family Taxodiaceae, or swamp cypress family.

In a sense, the stately grandeur of the redwoods has been a factor in their destruction. The attributes that make them so impressive and beautiful in the forest are the same qualities that make them attractive for construction uses; their bark and wood are resistant to decay, fire, and insects. Humans are the trees' only real enemy. Cutting the trees is the ultimate destruction, but even the stress of people walking on their shallow root systems can bruise them.

This part of northern California was dense with redwoods when the explorer Jedediah Smith led a group into the forests in 1828. The pioneers and early settlers cut many trees in the 1880s for buildings in San Francisco and northern California towns. Much of this building was the result of the discovery of gold in northern and central California. Later the advent of motorized equipment made more remote areas accessible to loggers, and more trees were felled.

In order to avert the total destruction of the trees, the Save-the-Redwoods League was established in 1918 to promote the creation of Redwood National Park; it took 50 years to accomplish the League's goal. Feasibility studies were made in the 1920s and 1930s, but no legislative action was taken. In 1946, the Douglas Bill proposed the establishment of a national memorial forest to include 2.5 million acres, about 180,000 of which would be administered according to National Park Service regulations. The Douglas Bill never passed.

Aided by funds from the National Geographic Society, the National Park Service surveyed redwood areas in 1963 and 1964. This study revealed that of the original forest of nearly 2 million acres, only 15 percent remained—and only 50,000 acres (2.5 percent) were protected in state parks! If the logging rate continued unchecked, all unprotected trees would disappear before the year 2000. The study also found a grove of the earth's tallest trees, including one 367 feet high, on private—and unprotected—land.

Redwood National Park was finally established by Congress in 1968, guaranteeing protection of these endangered trees. The national park links three California state parks; from north to south, these are Jedediah Smith, Del Norte, and Prairie Creek State Parks. In addition to virgin redwood groves, the park includes stands of maple, cedar, and oak, private homes and small businesses, some cutover land, and more than 40 miles of coastline. Some of the most impressive redwood stands are located in the three adjacent state parks. The state parks and the national park have some cooperative programs but are administered as separate units.

Establishment of Redwood National Park assured protection of the surviving trees from logging, but other problems remained. Logging operations outside park boundaries threatened to cause serious erosion of Redwood Creek watershed, which includes the tallest trees. Other effects of logging include soil compaction, loss of topsoil, elimination of shade and ground cover, and changes in small drainage basins. To help solve these problems, Congress extended the park boundaries in 1978 by increasing the area by over 80 percent in order to protect a total of 106,000 acres.

The added park area is primarily cutover lands, including 300 miles of logging roads, 3000 miles of skid trails, and countless charred stumps marking former locations of stately redwoods. The Park Service is involved in a watershed rehabilitation program with a long-range goal of speeding recovery of the natural systems in the park area. Since restoration began in 1978, efforts have been concentrated on the 30,000 acres that were logged before the park's establishment. Erosion control priorities are aimed at logging haul roads, tractor-logged hillslopes, active landslides, and natural prairies that were gullied because of road construction. Techniques of slope stabilization and revegetation being developed in the cutover parts of Redwood National Park may aid in reclamation of damaged land in other parts of the country.

The Tall Trees

The redwoods include the tallest trees on earth; many of them are over 200 feet tall, and some stand more than 300 feet high. The tallest coast redwood was discovered by the National Geographic Society in 1963, near the southern end of Redwood National Park in

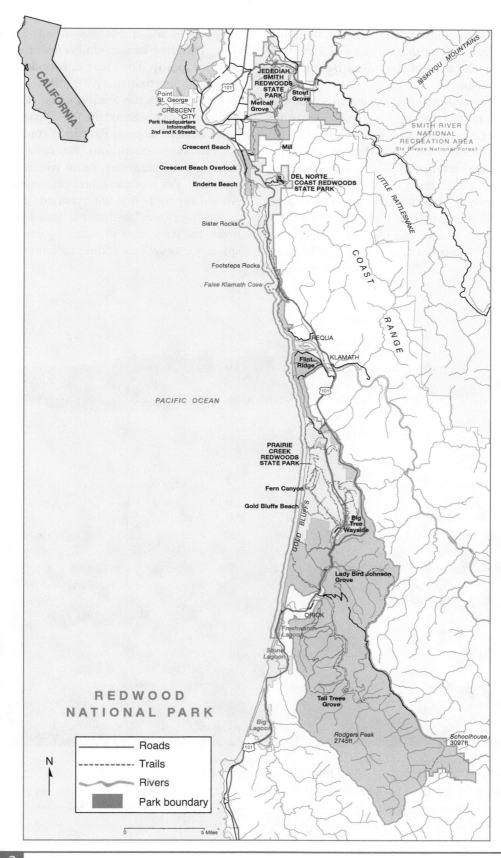

CALIFORNIA

SISKIYOU MOUNTAINS

JEDEDIAH
SMITH
REDWOODS
STATE
PARK
Stout
Grove

Point
St. George
CRESCENT
CITY
Park Headquarters
Information
2nd and K Streets

Metcalf
Grove

Crescent Beach

Mill

SMITH RIVER
NATIONAL
RECREATION AREA
Six Rivers National Forest

Crescent Beach Overlook

Enderts Beach

DEL NORTE
COAST REDWOODS
STATE PARK

LITTLE RATTLESNAKE

Sister Rocks

COAST

Footsteps Rocks

False Klamath Cove

RANGE

REQUA

Flint
Ridge
KLAMATH

PACIFIC OCEAN

PRAIRIE
CREEK
REDWOODS
STATE PARK

Fern Canyon

Gold Bluffs Beach

Big
Tree
Wayside

GOLD BLUFFS

Lady Bird Johnson
Grove

ORICK

Freshwater
Lagoon

Stone
Lagoon

REDWOOD
NATIONAL PARK

Tall Trees
Grove

Big
Lagoon

Rodgers Peak
2745ft

Schoolhouse
3097ft

N

	Roads
---	Trails
~~	Rivers
▓	Park boundary

0   5 Miles

FIGURE 51.2   Redwood National Park, California.

what is now Tall Trees Grove. In 1965, this huge tree measured 367.8 feet high, with a trunk 14 feet in diameter near its base. The ages of the tallest trees may be up to 2000 years or older.

The trees are called redwood because the color of their bark is a dull reddish brown. The bark is dense and fibrous, and may be as much as a foot thick on a very old tree. The leaves are usually less than an inch long, sharply pointed, and evergreen. Redwood cones are egg-shaped and about 1.5 inches in diameter.

The coast redwood, like its cousin the giant sequoia (chapter 49), grew over much of North America when dinosaurs were alive. Pleistocene glaciation, beginning about 2 million years ago and ending only 10,000 years

ago, caused major changes in soil and climate conditions, resulting in a much narrower distribution of the Sequoia family. The coast redwoods are now restricted to a band approximately 30 miles wide and 500 miles in length along the Pacific coast. Fossils of redwood trees have been found in the Dakotas, Montana, and Yellowstone National Park in Wyoming (chapter 43).

Favorable conditions for redwood trees include moderate temperatures, rainy winters (40 to 80 inches per year), dry summers, and year-round fog (fig. 51.3). An old saying is that the redwoods grow "as far inland as the fog flows," and that is about 30 miles here. Fog limits the trees' loss of water by evaporation and transpiration, as well as adding moisture to the soil.

FIGURE 51.3   The fog is important to the coast redwoods, because the fog reduces the evapotranspiration. Other factors that help the trees to survive are the temperate climate, with average temperatures of 45°–61° F, the rich river bottom soils (glacial soil has too much clay and is not rich enough to grow the trees), the 60-140 inches of precipitation from February to May, the amount of tannin in the bark that causes resistance to insects, and the thick bark that protects them from fire. National Park Service photo.

FIGURE 51.4 Where the forest meets the beach is an area of very harsh conditions. Most trees cannot tolerate the salt air. The one exception is the hardy Sitka spruce (*Picea sitchensis*); hence it dominates the most exposed coastal areas, such as Crescent Beach, Freshwater Lagoon Spit, and Gold Bluffs Beach. National Park Service photo.

Geologic History and Features

Redwood National Park is entirely within the Coast Ranges of California, which extend from the Oregon border south to Santa Barbara. This region has provided some important insights into the theory of plate tectonics. The northern Coast Ranges lie east of the San Andreas fault zone, which crosses the coast and goes offshore about 50 miles south of the park. The moist climate that makes this region so favorable for redwoods and other vegetation also makes the geology very difficult to see; thick soils and dense forests result in few exposures. However, the sea cliffs provide an excellent view of the geology for anyone willing to brave the rocky beach (fig. 51.4).

1. **Deep-water, continental, and volcanic deposits in a subducting trough: accumulation of Jurassic-Cretaceous Franciscan rocks.**

The oldest rocks found in the northern Coast Ranges are part of the Franciscan rock assemblage, which is described as a *mélange*, a term reflecting the mixture of diverse materials it contains. These rocks are probably the product of deep-water sedimentation on the continental slope and on oceanic crust in deeper trenches offshore. The sands and muds included submarine volcanics, radiolarian oozes (mud composed of tiny siliceous animal shells), and sediments eroded from land. This mixture of sediments—carried along on an oceanic plate—was subducted beneath the continental margin about 100 million years ago (mid- to late

Cretaceous time). Tectonic activity probably occurred at the same time as deposition, so the unconsolidated sediments were severely disturbed. Some outcrops of the Franciscan have rocks so contorted that they look as if they had been stirred with a spoon. To complicate the situation further, ultramafic rocks intruded the Franciscan beds during and after subduction.

After the subduction ended, the complex Franciscan assemblage was uplifted and exposed at the surface. The rocks' characteristic minerals reflect the low-temperature and high-pressure conditions of their origin.

The Franciscan assemblage is composed primarily of grayish green graywackes. These sandstones contain relatively high proportions of plagioclase feldspar, as well as quartz; a chlorite matrix gives the Franciscan its distinctive greenish color. Some of the volcanics were altered to serpentinite by metamorphism, and these metavolcanics are also green.

The Franciscan is a thick and widespread rock unit in the Coast Ranges. A conservative estimate of its volume is 350,000 cubic miles—which could cover all of the lower 48 states to a depth of 600 feet!

Table 51.1			Generalized Geologic Column, Redwood National Park	
Time Units			**Rock Units**	**Geologic Events**
Era	**Period**	**Epoch**		
Cenozoic	Quaternary	Holocene	Fluvial, alluvial, and mass-wasting deposits	Coastal erosion and spit deposition; accelerated erosion on land; landsliding; continued faulting and uplift
		Pleistocene		Continuing uplift and erosion; formation of marine terraces; fluvial deposits (middle to late Pleistocene)
	Tertiary — Neogene	Pliocene	Stream gravels	Coast Range tectonism extending to the south; thick stream gravels deposited in valleys (especially south of park)
		Miocene	Marine sediments, gravels	Basin forming and sedimentation in Coast Ranges; some volcanism, predominantly submarine
	Tertiary — Paleogene	Oligocene / Eocene / Paleocene	Marine sediments, some quartzose sandstone	Decreasing depositional rates as seaways narrow and become smaller; weathered clays and clean sandstones indicate nearly tropical climate in Eocene. Thrusting over Franciscan rocks.
Mesozoic	Cretaceous / Jurassic		Franciscan rock assemblage (sandstone with quartz, plagioclase feldspar, and chlorite mica), shale, limestone, radiolarian cherts, ultrabasic intrusions	Extensive deposition of marine sediments (over 50,000 feet), followed and accompanied by subduction beneath continental margin, tectonic mixing, and intrusion of ultrabasic peridotite. Metamorphism of sediments and older rocks.
	Triassic		(Not exposed in park)	
Paleozoic			(Not exposed in park)	Long accumulation of sediments in Paleozoic seas (No Precambrian or early Paleozoic rocks have been identified in the Coast Ranges; if they exist, they have been severely metamorphosed.)

Sources: Modified after Norris and Webb, 1976; McLaughlin et al., 1994.

FIGURE 51.5   The False Klamath Cove, located in the Del Norte Coast Redwoods State Park portion of Redwoods National Park, was the original mouth of the Klamath River. A meander loop cut through the bluffs, putting the river at its present location. National Park Service photo.

2. Basin-forming and sedimentation in the Miocene.

The sea advanced over the Coast Range area in the early Miocene Epoch, creating a number of small, nearly isolated marine basins. Volcanism, predominantly submarine, also contributed to the accumulation of deposits during this time.

In the northern Coast Ranges, north of San Francisco Bay, the late Miocene time brought further tectonism. Downwarped areas formed basins, defined by the intervening anticlinal arches trending to the northwest. The area where Redwood National Park is now located was surrounded by two of these depressed and flooded embayments, one at Crescent City to the north and the other along the Mad River near Eureka. A few downfaulted remnants of the basin deposits are all that have survived late Cenozoic erosion to show that marine rocks were formerly widespread here.

3. Thick stream gravels in the Pliocene.

The tectonic activity that began in the northern Coast Ranges during the late Miocene time extended southward throughout the Coast Ranges in the Pliocene Epoch. By late Pliocene time, most of the mountains that exist today were exposed as dry land. Thick layers of stream gravels were deposited in valleys, sometimes nearly burying the surrounding hills. Deposition was most extensive in the southern Coast Ranges, but a few patches of gravel have been found north of Eureka. One place where Pliocene-Pleistocene deposits are exposed is along the beach at Gold Bluff, where the sand does contain small flecks of gold.

4. Quaternary erosion and sea-level changes.

Because most of the land surfaces in the northern Coast Ranges have been exposed since the end of the

Pliocene Epoch (about the last 2 million years), the geologic record has been slowly destroyed. Only a few coastal Quaternary deposits cover the Franciscan in this area. The only major Quaternary activity recorded in the northern Coast Ranges is the development of raised marine terraces. Some terrace formation was caused by worldwide sea-level changes due to glaciation. However, most of the terraces in northern California are wave-cut platforms that have been eroded in bedrock and then raised by localized uplift. In some parts of northern California, the uplift rate has been as high as .04 to .16 inches/year. Marine terraces are visible in the Crescent City area at the northern end of the park.

Tectonic activity has been going on in the Pacific, off the northern California coast, at least since the early Cenozoic Era. At Cape Mendocino, about 50 miles south of the park, the San Andreas fault zone goes offshore and merges with the east-west trending Mendocino ridge and fracture zone. Somewhere between the coast and about 50 miles offshore, the Gorda plate is being subducted under the North American plate (fig. 51.6). This subduction zone was responsible for the November 8, 1980, earthquake in this area; essentially the quake was caused by friction between the two plates. The magnitude of the tremor was 7.0, which made it the largest earthquake in California in 28 years! Northern coastal California has had more than 20 earthquakes with magnitudes of 6.0 or greater in this century.

In 1996, major eruptions took place on the Gorda Ridge, 50 miles off the southern Oregon coast, at a depth of two miles. A huge plume of warm water, six miles wide, wafted over fissures in the sea floor, like a smoky cloud. Detected in water samples from the plume were tiny, heat-loving microbes that can survive in water temperatures up to 230° F. Scientists monitored the eruptions using underwater microphones and other equipment. The eruptions are similar to those that occurred in 1993 on the Juan de Fuca Ridge, north of the Gorda Ridge.

5. Landslides and accelerated erosion today.

Franciscan rocks are extremely fractured and sheared, which makes them highly susceptible to mass wasting. Landslides are so common in this area that they cause more downslope movement of material than any other geologic process—including stream action! Water is often an important agent in initiating landslides, and recent studies have shown that rainfall amounts are useful in predicting landslides. Rock fracturing in fault zones also decreases slope stability. Construction on steep slopes is another significant cause of landslides. Each year, millions of dollars worth of property damage is done by landslides in the northern Coast Ranges.

Stream activity is an important geologic process in the park area today. Stream deposits support some of

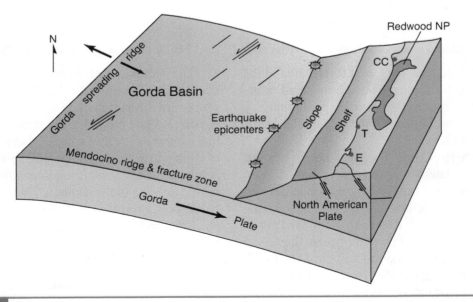

FIGURE 51.6   This generalized block diagram shows the coast and offshore area of northern California. The San Andreas fault swings west and merges with the Mendocino ridge and fracture zone south of Eureka. CC=Crescent City, T=Trinidad, E=Eureka.

FIGURE 51.7   Embayed mouth of the Klamath River, drowned by coastal subsidence. River and tidal currents keep an outlet open through the sandbar. National Park Service photo.

the largest stands of tall trees. Both the Tall Trees Grove on Redwood Creek and Stout Grove on the Smith River grow on alluvial flats. Logging activities in the upstream areas have had numerous effects on downstream reaches. These impacts include accelerated bank erosion, elevated and wider streambeds, higher winter stream discharge, and lower summer discharge.

The coastal region of Redwood National Park has a number of lagoons. These are the drowned mouths of stream valleys that have sand spits separating them from the sea. The lagoons are evidence that this part of the California coastline is sinking, which may be related to subduction offshore.

A noteworthy geologic event in the Crescent City area, at the northern end of the park, was the *tsunami* (seismic sea wave) caused by the 1964 Good Friday earthquake in Alaska (Chapters 31, 33). Sudden

changes in the sea floor, caused by earthquakes or large submarine landslides, can start ocean swells that move across the ocean and hit coastlines with destructive violence. Tsunamis occur with some frequency in Alaska and in the Hawaiian Islands (Chapters 40, 41), but Crescent City is the only place in conterminous United States to have recorded serious damage in historic time from such an event.

Erosion in the area around Redwood National Park has been accelerated by human activities. Extensive lumbering and clear-cutting have nearly eliminated the soil protection in some areas. Gully formation, the loss of rich topsoil, and streams choked with debris are all consequences of using the redwoods as a natural resource. Redwood National Park was established to save the area from further damage, but it will take many years and considerable work to repair the existing damage.

An innovative approach to restoring the whole area—beyond the official park boundaries—involves the establishment of the Park Protection Zone. This 30,000-acre tract, upstream from the park, along Redwood Creek, includes the upper two-thirds of the drainage basin. Because activities upstream have such a powerful impact on downstream reaches within the park, National Park Service resource managers are working with private companies to limit the effects of land use in upstream areas. This philosophy, by recognizing the ways in which land areas within a watershed are connected, is an important step in reclaiming and restoring Redwood National Park.

Geologic Map and Cross Section

Feray, D.E., Detking, P. and Renfro, H.B. 1968 Geologic Highway Map, Number 3, Pacific Southwest Region. The American Association of Petroleum Geologists.

Sources

Alt, D.; and Hyndman, D.W. 1975. *Roadside Geology of Northern California*. Missoula, Montana: Mountain Press Publishing Company.

Bailey, Edgar, ed. 1966. *Geology of Northern California*. California Division of Mines and Geology Bulletin 190.

McLaughlin, R.J.; Sliter, W.V.; Frederiksen, N.O.; Harbert, W.P.; and McCullough, D.S. 1994. Plate Motions Recorded in Tectonostratigraphic Terranes of the Franciscan Complex and Evolution of the Mendocino Triple Junction, Northwestern California. U.S. Geological Survey Bulletin 1997, 60 p.

Norris, R.M. and Webb, R.W. 1976. *Geology of California*. New York: John Wiley & Sons.

Oakeshott, G.B. 1978. *California's Changing Landscapes*. New York: McGraw-Hill Book Company.

Savina, M.E. ed. 1985. *Redwood Country: Field Trip Guidebook*. Northfield, Minnesota: Carleton College and the American Geomorphological Field Group.

Schrepfer, S.R. 1983. *The Fight to Save the Redwoods: A History of Environmental Reform*, 1917–1978. Madison: University of Wisconsin Press.

Note: To convert English measurements to metric, go to www.helpwithdiy.com/metric_conversion_calculator.html

C H A P T E R   5 2

Hot Springs National Park

WEST CENTRAL ARKANSAS

Area: 5,839 acres; 9 square miles

Hot Springs Reservation Set Aside:
April 20, 1832

Redesignated a National Park: March 4, 1921

Address: P.O. Box 1860, Hot Springs, AR 71902

Phone: 501-624-2701

E-mail: HOSP_Interpretation@nps.gov

Website: www.nps.gov/hosp/index.htm

From 750,000 to 950,000 gallons a day flow from 47 thermal springs.

The hydrolic system of the springs developed in intensely metamorphosed, folded, and faulted Paleozoic rocks.

Tufa deposits; novaculite, used for whetstones; and large, tinted quartz crystals are displayed in the park.

FIGURE 52.1   The Display Springs have been left in a natural state so that park visitors can see what the springs looked like originally. The remaining springs have been covered to prevent the loss of gas and contamination. The water of these springs is piped to a central storage area and distributed to the fountains and bathhouses. Photo by Anthony Belfast.

The Indian name for the hot springs was *Tah-ne-co,* meaning "the place of the hot waters," where the Great Spirit's warm breath heated healing waters and vapors. The tribes believed that the springs were meant to be for the benefit of all who were sick. When the foundations for the Quapaw Bathhouse in the park were dug, a narrow winding cave used by Indians that led to the springs was discovered.

Hernando de Soto, in his search for treasure in the New World, visited the hot springs in 1541 shortly before his death. Chronicles of his ill-fated expedition describe his discovery of "large streams of hot water and a small lake" near an Indian village, when he was journeying between the Mississippi River and what is now Oklahoma. Since there are no other hot springs near his route along the Arkansas River, the presumption is that he was in the area of Hot Springs National Park.

In 1804, after the Louisiana Purchase, President Jefferson sent William Dunbar and Dr. George Hunter, a chemist, to explore the Ouachita Mountains and to investigate the hot springs, which Jefferson had heard about. Spending a month in the area, they counted some 70 hot springs and analyzed the water issuing from the rocks. Theirs was the first scientific report to be written about the springs. Other explorers who visited the hot springs were U.S. Army officers Zebulon M. Pike in 1806 and Stephen H. Long in 1818. Two members of Long's expedition, Kearney Swift and Edwin James, the latter a botanist and geologist, studied the springs. They suggested that build-up of *tufa* (calcareous-siliceous deposits around spring openings) (fig. 52.2) blocked the flow from some springs, causing variations from time to time in the total number of springs. In their natural state, the springs flowed from a number of openings in a belt about a quarter of a mile long and several hundred feet wide along the southwest slope of Hot Springs Mountain.

Local residents began putting up cabins, bathhouses, and even hotels to accommodate visitors who sought the benefits of thermal bathing. The natural environment of the hot springs became completely altered. Springs were excavated and lined with masonry to increase and control the flow and then covered to protect the water from contamination. To prevent commercial exploitation and ensure that all people would have access to the springs, President Andrew Jackson persuaded Congress to withdraw from public sale four sections of land (four square miles) around the hot springs, four years before Arkansas became a state. In 1876, a year after a rail line was extended to the area, the city of Hot Springs was incorporated. Part of the city is now surrounded by the park. Today in this

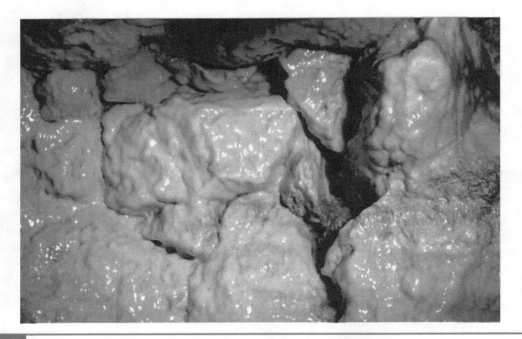

FIGURE 52.2   Tufa is a chemical sedimentary rock composed of calcium carbonate; it will form around the mouth of a hot spring. National Park Service photo.

FIGURE 52.3   The fountain in front of the park office. People will fill their jugs with the water to take home with them so that they may have their daily drink of water. Photo by Ann G. Harris.

famous health resort, five bathhouses offer traditional baths and one is operated as a modern spa. All six are under the supervision of the National Park Service.

The Display Springs have been left in a natural state so that park visitors can see what the springs looked like originally (fig. 52.1). At Tufa Terrace on Arlington Lawn, a hot water display has been developed, using the runoff from the upper springs that created the terrace. Where the thermal water drains down the slope, blue-green algae accumulate and vapor rises from the cascade in cool weather. The remaining springs are sealed to prevent contamination and loss of gas. The combined flow from the springs ranges from a high of around 950,000 gallons a day in winter to 750,000 gallons in summer and fall. The water temperature fluctuates between 95.4° and more than 147° F. Insulated pipes carry water from the springs to large underground reservoirs. Hot spring water is mixed with cooled spring water from the reservoirs to reach desired temperatures. Cooled water is also pumped to drinking fountains and jug fountains (fig. 52.3). The spring water in Hot Springs National Park is of exceptional purity, with no unpleasant taste or odor.

Through the years, the National Park Service has acquired additional land in the region of the springs for the purpose of protecting the infiltration and recharge areas from pollution. The mountains in the park (between 1000 and 1400 feet in elevation) are part of the ZigZag Mountains, a range in the Ouachitas. Music Mountain, in the southwest corner of the park, is the highest point (1405 feet) and is at the center of a great horseshoe-shaped ridge, with West Mountain and Sugarloaf Mountain forming the roughly parallel extended "arms." A few roads and many miles of bridle and hiking trails wind through the woodlands and rugged hill country.

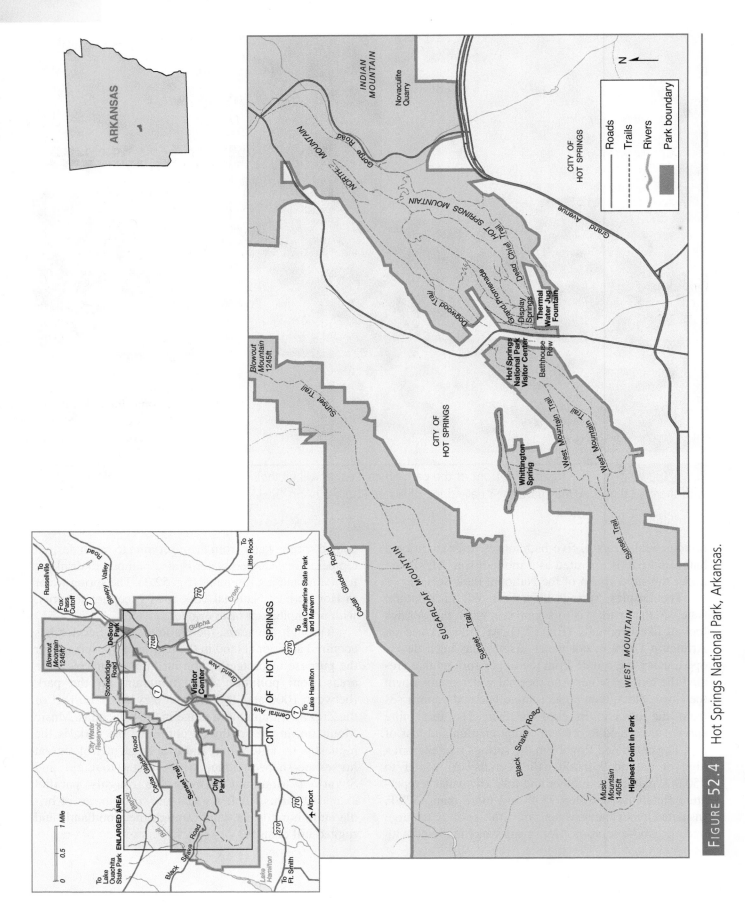

FIGURE 52.4   Hot Springs National Park, Arkansas.

Geologic Features

How Did the Hot Springs Originate?

Hot Springs National Park is in a unique geological setting. Scientists (mainly geologists) have argued over the origin of the water and the heat source. Is rain the source of the water, or is it newly risen from deep in the earth? If the former is the case, how did the water get so hot? Has the water been stored for a long period of time at considerable depth? How and from where has it moved to its point of issue? Since the mid-1800s, numerous investigations have been made, and hypotheses both simple and complex have been proposed.

The modern explanation, which appears fairly conclusive, draws on several lines of evidence. The seasonal variations in flow—higher in wet seasons and lower in dry periods—suggest rainwater as the most likely source of supply for the hot springs. Reinforcing this idea is the fact that the chemical composition of the hot springs water is similar to that of the local ground water, cold spring water, and well water—after allowances are made for the higher content of dissolved silica contained in the hot water. (Hot water can hold greater concentrations of dissolved compounds than cold water.)

Bedinger and his co-investigators (1979) used mathematical models to test several conceptual models of the hot springs flow system. They concluded that the "geochemical data, flow measurements, and geologic structure of the region support the concept that virtually all the hot springs water is of local, meteoric [i.e., from rain] origin." They further deduced that:

> Recharge to the hot springs artesian flow system is by infiltration in the outcrop areas of the Bigfork Chert and the Arkansas Novaculite. The water moves slowly to depth where it is heated by contact with rocks of high temperature. Highly permeable zones, related to jointing or faulting, collect the heated water in the aquifer and provide avenues for the water to travel rapidly to the surface (fig. 52.5 A,B).

The Paleozoic sedimentary rocks of the park, which collect the rainwater and from which the hot springs emerge, have been deformed and metamorphosed by episodes of intense folding, faulting, and thrusting. Many complex geologic structures have resulted. The ZigZag Mountains basically owe their topographic expression to the resistant ridges of the Arkansas Novaculite, which crop out in a zigzag pattern because of the erosion of tightly compressed plunging folds. Weaker, less resistant rocks underlie the valley floors. The landscape in Hot Springs National Park is a good example of control by geologic structure (fig. 52.6).

The main collectors of rainwater are the permeable ridge crests of novaculite (locally intensely fractured) and the Bigfork Chert, which forms lower ridges and rounded knobs. The shale beds below the permeable rocks tend to be impermeable. On Hot Springs Mountain, between the novaculite on the ridge and the

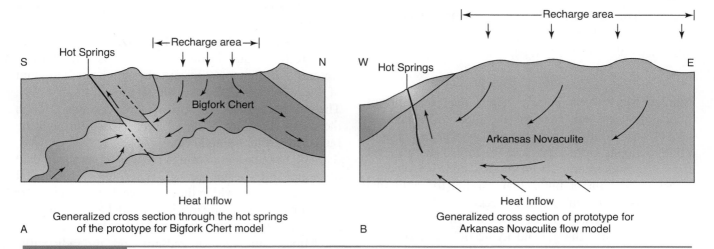

FIGURE 52.5 Cross-sectional models illustrating directions of ground water flow in Hot Springs National Park. (from Bedinger et al. 1979). *A.* North-south cross section approximately through the center of Bigfork Chert recharge area, about two miles from the hot springs. Details of folds and faults not shown. *B.* East-west cross section of the Arkansas Novaculite recharge area, less than a mile from the hot springs. From Bedinger et al., 1979.

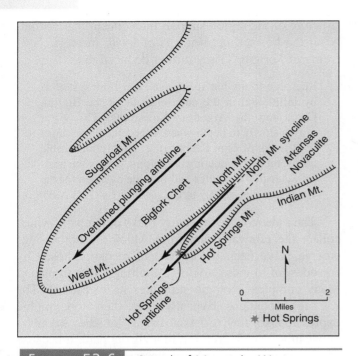

Sugarloaf Mountain, West Mountain, North Mountain, Hot Springs Mountain, and Indian Mountain illustrate zigzag topography produced by erosion of plunging folds. Arrows show the northeast-southwest-trending folds plunging to the southwest. All the folds shown have overturned southeast limbs, which means that both limbs dip northwest. Reverse faults (not shown) that cut the folds trend parallel to them and dip steeply to the northwest. Water that supplies the hot springs infiltrates recharge areas of Bigfork Chert and Arkansas Novaculite, both of which crop out in the middle of anticlines. Simplified from geologic map in Bedinger, et al. 1979.

Stanley Shale beds in the valley, the Hot Springs Sandstone crops out. This massive, quartzitic sandstone (which structurally is near the axis of a plunging anticline) contains large joints and fractures. The water of the hot springs emerges from the sandstone between the traces of two thrust faults that parallel the anticlinal axis. This structural situation accounts for the hot springs' alignment in a belt along the mountain slope. We can visualize the general hydrologic and geothermal flow system as follows (fig. 52.7):

1. In the park's higher elevations among the ridges and higher valleys, rainwater infiltrates recharge areas on the Arkansas Novaculite and the Bigfork Chert.
2. Cold ground water works its way very slowly to depths more than a mile below the surface where, with increasing geothermal gradient, the rocks are very hot.
3. As cold water sinks, heated water rises rapidly, following paths of least resistance—in this case, the permeable fault zones that bring the water close to the surface.
4. From the fault zones, water moves along joints in the Hot Springs Sandstone aquifer; while still very hot, it flows out from openings at the surface. Since the recharge area is at a higher elevation than the discharge area, the springs are artesian; that is, the water has built up a hydrostatic head due to artesian pressure and it flows without pumping.

Estimates as to the age of the hot springs water and the travel time involved in the journey from the surface to depth and back to the surface have been derived from tritium and carbon-14 analyses. *Tritium* (H_3), a naturally occurring, radioactive isotope of hydrogen, is formed by cosmic rays in the upper atmosphere and enters the earth's hydrologic cycle via precipitation. Tritium's half-life is about 12 1/4 years. *Tritium dating*, involving measuring the rate of radioactive decay in water samples, is useful for calculating the age of a sample and also for tracing subsurface movements and velocities of ground water. The carbon-14 dating of ground water is calculated from dissolved soil gas (CO_2) with corrective adjustments made for carbonates dissolved from rock through which the water travels. The results from many carefully collected and analyzed water samples indicate that water in the hot springs is a mixture of water that is about 4400 years old with a small amount of water less than 20 years old (Bedinger et al. 1979). The inference is that movement of water down through the rocks takes much longer than its rise through the fault zones after being heated.

The Arkansas hot springs are less mineralized than the water in most hot springs. What this means is that the rocks through which the Arkansas hot springs water travels have fewer mineral ions available for solution. The dissolved constituents are listed in table 52.1.

Minute amounts of radium and its radioactive byproduct, radon gas, are present in the hot springs water and have been studied by a number of investigators. The source of the radium is not known, but its

FIGURE 52.7   A geologic cross section illustrating the movement of the ground water that produces the hot springs and how it is recharged. The recharge to the hot springs artesian flow system is by infiltration in the outcrop areas of the Bigfork Chert and the Arkansas Novaculite. Water slowly percolates downward, where it is heated by the high temperature of the rock and then is forced to the surface. The age range of the water is as little as 20 years or as long as over 5,000 years. National Park Service diagram.

Table 52.1	Chemical Analysis of Hot Springs Water*	
(milligrams per liter)		
Silica (SiO$_2$)		42.0
Calcium (Ca)		45.0
Magnesium (Mg)		4.8
Sodium (Na)		4.0
Potassium (K)		1.5
Bicarbonate (HCO$_3$)		165.0
Sulfate (SO$_4$)		8.0
Chloride (Cl)		1.8
Fluoride (F)		.2
Dissolved oxygen (DO)	3.0	
Free carbon dioxide (CO$_2$)	10.0	
Radioactivity through radon gas emanation**		
0.81 millimicrocurie per liter		

*Analysis by U.S. Geological Survey, 1972.
**Analysis by University of Arkansas.

presence in trace amounts in the ground water of the region is not unusual. Most of the gas in the hot springs is carbon dioxide (CO$_2$) with smaller amounts of dissolved oxygen and nitrogen. Analyses of water samples from the different springs and from samples collected over a period of time have shown slight variations in dissolved constituents; but on the whole, the water has remained fairly uniform and consistent. Besides being odorless, the water is remarkably free of bacteria, viruses, or other impurities.

Rocks of Special Interest in the Park

The novaculite found in Hot Springs National Park and vicinity is a relatively uncommon rock. *Novaculite* is a dense, hard, even-textured, light-colored, siliceous sedimentary rock that resembles chert (fig. 52.8). Novaculite is *cryptocrystalline* (i.e., having crystals too small to be distinguished under a microscope); but unlike chert, true novaculite is made up of more microcrystalline quartz than chalcedony. Novaculite was formerly thought to be the result of primary deposition of

FIGURE 52.8A   The Arkansas Novaculite makes up part of West Mountain. Note the tilting of the beds as it is part of an overturned plunging anticline. Novaculite is a dense, hard, even-textured, light-colored, siliceous sedimentary rock resembling chert. It is used for whetstone. National Park Service photo.

FIGURE 52.8B   A close up of novaculite. National Park Service photo.

Table 52.2	Geologic Column, Hot Springs National Park			

Time Units			Rock Units	Geologic Events
Era	Period	Epoch		
Cenozoic	Quaternary	Holocene	Alluvium	Erosion and alluviation
		Pleistocene	Terrace deposits	Development of hot springs hydrologic/ geothermal system Pluvial climate
	Tertiary		/////////	Long erosion Major unconformity End of uplift
Mesozoic	Cretaceous		/////////	Syenite stocks intruded close to park; small dikes in park
	Jurassic			Long erosion, major unconformity
	Triassic			Uplift, faulting, extension
Paleozoic	Permian		/////////	Ouachita orogeny (plate collision, thrust faulting, intense deformation, uplift)
	Pennsylvanian			
			(not exposed in park)	Shales, sandstones deposited in closing seaway
	Mississippian		(Formation) Stanley Shale	(Members) Chickasaw Creek Shale Hatton Tuff Hot Springs Sandstone	Flysch deposits in deep trough between closing plates Ash falls Beginning of flysch facies: turbidites, conglomerates, etc. Unconformity
	—— ? ——		////////	////////	
	Devonian		Arkansas Novaculite	Upper	Siliceous shales, ash falls
				Middle	
				Lower	
	Silurian		Missouri Mountain Shale		"Starved basin" deposition
			Blaylock Sandstone		
	——?——	(late)	Polk Creek Shale		Slow accumulation in deep-water, stable basin
			Bigfork Chert		
	Ordovician	(middle)	Womble Shale		

Source: Modified after Bedinger et al., 1979; Viele, 1979.

silica; but in its type occurrence in the Lower Paleozoic rocks of the Ouachita Mountains (the region in which the park is located), true novaculite appears to be a thermally metamorphosed, bedded chert (Keller et al. 1977). The parent rock, a bedded chert, was recrystallized by heat (and to a lesser extent by confining pressure) and thus altered to novaculite.

In some areas outside the Hot Springs National Park, novaculite beds are quarried for use as whetstones, and some specimens that are richly colored are suitable for lapidary work. The Indians prized the rock for making arrowheads and stone tools. Old quarries, opened by Indians, are in the northeast corner of the park on Indian Mountain. The largest quarry is 150 feet in diameter and 25 feet deep. The Indians' method of breaking up and excavating this hard, dense rock was to build a fire to get the rock as hot as possible and then throw cold water onto it. The process fractured the rock into small pieces; but it was dangerous. Sometimes cold dousing of hot rock caused an explosion that filled the air with sharp projectiles of novaculite.

The Hot Springs region is also noted for beautiful quartz crystals of many shapes and sizes. Some crystals are two to three feet long. Specimens must not be collected within park boundaries, but may be purchased at mines and rockshops in the vicinity.

Tufa deposits surround the Display Springs and can be seen at other locations where spring water flowed out naturally in the past. Some of the tufa was precipitated by blue-green algae, the only form of life that can exist in water that hot.

Geologic History

1. Ordovician to Mississippian deposition.

Sediments accumulated very slowly in deep, quiet ocean water. (This ancient ocean basin lay between tectonic plates that eventually became North and South America.) Fine silts, organic material, siliceous particles, and precipitated silica made up the sediment that became shales, chert, and novaculite in a sedimentary sequence characteristic of "starved basin" deposition. Such a sequence implies that the ocean basin remained stable for a long period of time with the volume of sediment gradually decreasing,

Beds in the sequence now exposed in or close to Hot Springs National Park are the Womble Shale, Bigfork Chert (fig. 52.9), and Polk Creek Shale of Ordovician age; the Silurian Blaylock Sandstone and Missouri Mountain Shale; and the Arkansas Novaculite,

Figure 52.9   The Big Fork Chert has been folded and faulted during the uplifting of the area. Chert is a hard, fine-grained sedimentary rock formed almost entirely of silica. National Park Service photo.

FIGURE 52.10   A large boulder of the Stanley Shale, a flysch formation that marks the beginning of the subduction of the American plate. The rock belongs to the lower end of the Stanley Shale that is a conglomerate containing novaculite pebbles, the lighter colored fragments shown here. Photo by Anthony Belfast.

which is Devonian and Lower Mississippian in age. The rocks are not highly fossiliferous, although a few graptolites (extinct colonial marine organisms) have been found in the Polk Creek Shale.

2. Plate convergence in Mississippian time.

The appearance of interbedded siliceous shales in the uppermost part of the Arkansas Novaculite is an indication that the ocean was beginning to close. Siliceous ash, blown from volcanoes on the south side of the narrowing ocean and deposited on the sea floor, was probably the principal source of sediment for these beds.

As continental plates converged, the seafloor margin of the southward-moving North American plate plunged under the leading edge of the north-moving southern plate and was subducted. The Stanley Shale, a flysch formation, marks the beginning of this tectonic event (fig. 52.10). A *flysch* is a marine sedimentary facies that is the result of accelerated accumulation of sediment and rapidly changing conditions of deposition. Typically a flysch is deposited in an elongated deep-water trough between approaching plates and consists of thinly bedded, graded deposits of marls and sandy shales interbedded with conglomerates, graywackes, and the like. Beds that show evidence of deposition by fastmoving turbidity currents (i.e., turbidites) are characteristic.

The Hot Springs Sandstone, the lower member of the Stanley Shale, contains novaculite pebbles in conglomerate layers and shows some interbedding of thin black shale. It is from the Hot Springs Sandstone outcrops on Hot Springs Mountain that the springs emerge.

The Hatton Tuff (of volcanic origin) and the dark shales and graywackes of the upper part of the Stanley Shale are less resistant beds that underlie the wide valleys in the Hot Springs region.

3. The Ouachita orogeny in late Pennsylvanian and Permian time.

The Ouachita orogeny was part of the tectonic activity that formed the supercontinent Pangea, which existed in late Paleozoic time (300 to 200 mya). Caught between colliding continental plates, the Paleozoic marine beds were severely deformed by intense folding, thrust-faulting, and uplifting. As the plates locked, some thrusting was to the north, and some was to the south. Plunging anticlines and synclines, some of them overturned, produced the complex mountain structures of the Ouachita Mountains.

4. Continuing uplift and erosion during Mesozoic time.

The geologic history of the Ouachita Mountains in postorogenic time is somewhat obscure. Most probably, isostatic adjustments were made as crustal blocks rose differentially in order to attain equilibrium. A gradual uplift continued in the Ouachitas until early in the Tertiary Period. The local faults that control the flow of the hot springs water may have developed during uplift. In Cretaceous time, intrusions of syenite (an intermediate igneous rock) were emplaced near the park, with a few dikes extending into the park. Erosion was rapid as streams stripped off weaker beds, exposing more resistant rock layers and revealing fold patterns, as in the ZigZag Mountains.

5. Development of the hot springs hydrologic system.

The hot springs hydrologic system is probably older than the oldest water now carried in the system. Pluvial climates of the Pleistocene, when moisture was plentiful, may have brought about the development of the hot springs. Flow will probably continue as long as the regional climate is temperate and humid. Only a slight lessening of the average flow has been observed since monitoring of the system began. The main concern of National Park Service officials is to protect the recharge areas, maintain the quality of the local ground water, and prevent contamination of the hot springs so that people can continue to enjoy and benefit from these unique thermal waters and the beautiful surroundings.

Geologic Map and Cross Section

Bedinger, M.S.; Pearson, F.J., Jr.; Reed, J.E.; Sniegocki, R.T.; and Stone, C.G. 1979. *The Waters of Hot Springs National Park, Arkansas—Their Nature and Origin.* U.S. Geological Survey Professional Paper 1044-C.

Viele, G.W. 1979. Geologic Map and Cross Section, Eastern Ouachita Mountains, Arkansas. Map and Chart Series MC-28F. Plate Margins Working Group, U.S. Geodynamics Committee. Geological Society of America.

Sources

Bedinger, M.S., et al. 1979. *The Waters of Hot Springs National Park, Arkansas—Their Nature and Origin.*

———. 1974. *Valley of the vapors: Hot Springs National Park.* Eastern National Park and Monument Association.

Hanor, J.S. 1980. *Fire in the Folded Rocks: Geology of Hot Springs National Park.* Eastern National Park and Monument Association.

Keller, W.D., Viele, G.W., and Johnson, C.H. 1977. Texture of Arkansas Novaculite Indicates Thermally Induced Metamorphism. *Journal of Sedimentary Petrology* 47:834–843.

McFarland, J.D. III. 1988. Geologic Features at Hot Springs, Arkansas. Centennial Field Guide, Geological Society of America, South Central Section, v. 4, p. 263–264.

Stone, C.G., ed. 1977. *Symposium on the Geology of the Ouachita Mountains*, vols. 1, 2. Arkansas Geological Commission.

Stone, C.G.; Hale, B, R.; and Viele, G.W. 1973. *Guidebook to the Geology of the Ouachita Mountains, Arkansas.* Arkansas Geological Commission.

Viele, G.W. 1989. The Ouachita Orogenic Belt, in Hatcher, R.D. Jr.; Thomas, W.A.; and Viele, G.W. edsp. The Appalachian-Ouachita Orogen in the U.S. Boulder, Colorado: Geological Society of America, The Geology of North America, v. F-2, p. 555–561.

Viele, G.W. and Thomas, W.A. 1989. Tectonic Synthesis of the Ouachita Orogenic Belt, in Hatcher, R.D. Jr.; Thomas, W.A.; and Viele, G.W. eds. The Appalachian-Ouachita Orogen in the U.S. Boulder Colorado: Geological Society of America, The Geology of North America, v. F-2, p. 695–728.

Note: To convert English measurements to metric, go to www.helpwithdiy.com/metric_conversion_calculator.html

Big Bend National Park

SOUTHWEST TEXAS

Area: 801,163 acres; 1252 square miles

Established: June 12, 1944

Designated a Biosphere Reserve: 1976

Address: P.O. Box 129, Big Bend National Park, TX 79834

Phone: 915-477-2251

E-mail: BIBE_Info@nps.gov

Website: www.nps.gov/bibe/index.htm

One of the great rivers of the continent, the Rio Grande, loops deeply to the south, forming the "Big Bend" in the border between Texas and Mexico. Three spectacular canyons, cut in uplifted Cretaceous strata by the river, reveal much of the region's geologic history.

The Chisos Mountains display Tertiary lava flows and intrusions amid thick sequences of Cretaceous sedimentary rocks. Superb desert scenery, fantastically eroded rocks, high-angle faults, a breached anticline, unusual fossils, hot springs, old mines, archaeological sites, and a high, dry wilderness ecosystem are some of Big Bend's wonders.

FIGURE 53.1   From Big Bend National Park on the Texas side, visitors can look at the dipping beds of Cretaceous Santa Elena Limestone displayed in the cliffs across the Rio Grande on the Mexican side. Here the river has carved Boquillas Canyon through Sierra del Carmen, a range that extends from Texas southward into Mexico for many miles. Photo by Ann G. Harris

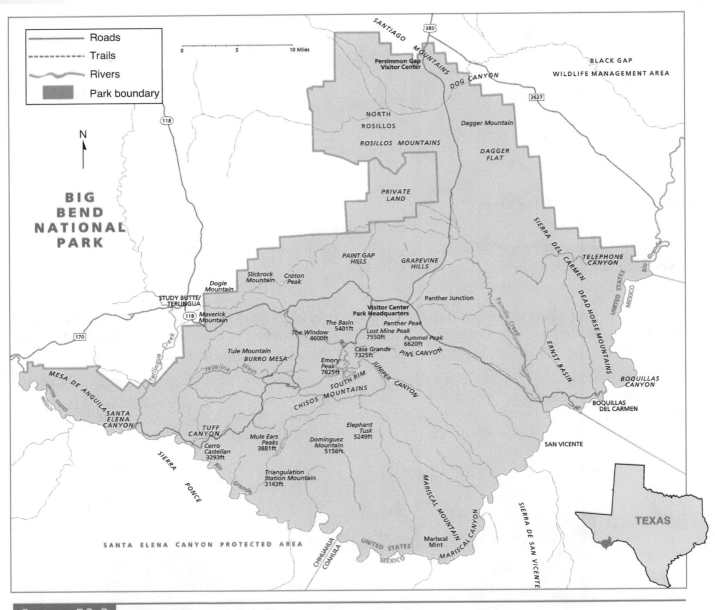

FIGURE 53.2   Big Bend National Park, Texas.

Geographic Setting

The first superintendent of Big Bend National Park was Ross Maxwell, a geologist who worked in Big Bend country for many years. In his guidebook (Maxwell 1968), he relates an Indian legend to show that prehistoric people recognized the rock formations and landforms as unusual and puzzling. According to the story, when the Great Spirit had finished making the earth, the stars, and all living things, he couldn't decide what to do with the stony trash left over. Finally, he piled it all up in a heap and left it. That was how Big Bend country came to be.

Big Bend country is a triangular-shaped region formed where the Rio Grande runs southeast and then northeast in a great loop. At the elbow of the bulge is Big Bend National Park. Mexico lies across the river. Although Big Bend National Park occupies only about a third of what is thought of as Big Bend country in west Texas, it is still one of the larger national parks in the United States. Within its bounds are spectacular canyons of the Rio Grande, the jagged igneous peaks of the Chisos Mountains, barren desert highlands, rare plants and animals.

The Chihuahuan Desert, which is in the Basin and Range Province, extends for many miles both north and south of the international boundary. Because of the generally sparse vegetation, angular landforms and rock units exposed in all their starkness dominate the landscape of this geologically young desert.

Big Bend National Park is far from a settlement of any size. The town of Marathon, 70 miles north of the park, is on the Southern Pacific Railroad as well as the nearest east-west highway. Interstate 10, the main highway across southern Texas, is an additional 50 miles north of Marathon. Big Bend is not the isolated region it once was because a modern highway (US 385) goes from Marathon to Persimmon Gap (the north entrance of the park), and a second highway (US 118) goes north from the western part of the park to Alpine, Texas.

Today's climate is harsh, but 10,000 years ago, nomadic bands hunted elephants, camels, bison, horses, and other big game in Big Bend country. Animals and hunters had drifted south, well beyond the limits of the last great continental glacier to what was then a temperate, well watered region of forests and grasslands. Then, in a relatively brief period, as the climate warmed, the ice sheet melted and dryness spread over the land. Moisture-loving plants died out, except for a few that survived around springs and at high elevations. The Ice Age mammals and the hunters disappeared.

In a later time, tribes who had adapted to desert life camped in sheltered caves along the river while they gathered plants and hunted desert animals. They left hearths, grinding pits, burial sites, petroglyphs, and artifacts in the park area. Chisos Indians from Mexico, a more recent group, were in Big Bend country when the Spaniards came in the sixteenth century, but were driven out by Mescalero Apaches. The Spaniards, later the Mexicans, tried to control the Indians along the Rio Grande, but neither missionaries nor soldiers had much success in subduing the tribes. The Comanches, a nomadic group, began appearing in the region late in the 1700s. The Comanche Trail goes through Big Bend from Persimmon Gap south to the river. Every autumn, Comanches rode from Oklahoma across Texas and into Mexico to raid ranches and steal horses and cattle. Both Apaches and Comanches resisted for many years the incursions of settlers and other intruders into their territory.

After the Mexican War in the 1840s, Major W.H. Emory surveyed the Rio Grande and the Big Bend area for the International Boundary Commission. His 1852 report was the first scientific description of the region. Emory Peak (7835 feet), the highest point in the park, is named for him. The United States War Department sent an expedition to Big Bend to locate the Comanche Trail and try to devise ways of protecting the Texas settlements from Indians. The expedition was also for the purpose of determining the capabilities of camels imported for military transport in the arid southwest. An Army lieutenant named Echols led the caravan along the base of the Santiago Mountains, entering territory that is now Big Bend National Park near Persimmon Gap and Dog Canyon. They crossed the difficult desert terrain and followed Terlingua Creek to the mouth of Santa Elena Canyon on the Rio Grande, a trip that took more than a month. Lt. Echols reported that the twenty camels stood the trip better than the mules and the men. With the beginning of the Civil War, the camel experiment ended. The camels were turned loose in the desert and disappeared.

Mining activities began in Big Bend in the 1890s with the opening of quicksilver (mercury) mines near Terlingua, now a ghost town outside the park's western boundary. Some quicksilver was also produced at the Mariscal mine in the southern part of the park. Maxwell (1968) reports that the quicksilver mined and smelted in the Terlingua district from 1896 until the 1940s amounted to one-fourth the total mercury production in the United States. The quicksilver was extracted from the mineral cinnabar (red mecuric sulfide), which the Indians had used for war paint and for pictographs. A hundred years of prospecting in and around the mountains of the park failed to turn up significant deposits of other ore metals, despite legends of gold and silver from "lost mines" supposedly worked by the Spaniards. However, across the Rio Grande from Boquillas, fluorspar (fluorite) ore, lead, and zinc have been mined in Mexico.

A few hardy ranchers and homesteaders settled in the canyon bottoms along the Rio Grande in the early 1900s. The settlers were able to grow cotton and livestock feed by drawing irrigation water from the river. At the site of some warm springs near the river, a small village and a bathhouse were erected. Invalids came there to soak in tubs of warm water. Nevertheless, Big Bend was still the "Wild West." Gangs of outlaws and Mexican revolutionaries crossed the border to plunder ranches and settlements. U.S. Army outposts were established to protect settlers and patrol the border, but raids persisted until about the time of World War I.

The movement to establish a park in the Big Bend began in 1933 when Texas Canyons State Park was set aside by an act of the Texas legislature, and the process of acquiring additional land was started. Two years later the United States Congress authorized the inclusion of the area in the National Park System, pending

FIGURE 53.3   Casa Grande ("big house") is a
lava-capped mesa with nearly vertical sides and flat top.
The rocks are alternate layers of lava (darker material)
and volcanic ash (lighter material) of the South Rim
Formation. Photo by Ann G. Harris.

completion of land acquisition. Through donations and
state appropriations during the next ten years, the work
went on. Finally in 1944, a large tract of mountains,
desert, and 107 miles of international boundary was
transferred to the federal government, thus establishing
Big Bend National Park.

Rare species of plants, birds, and animals live in
the park, especially in the Chisos Mountains, where
there is a little more moisture at higher elevations. A
few species of colder-climate trees, such as ponderosa
pine, cypress, and aspen, have survived among the
peaks since the Ice Age. When the climate was cooler
and moister, these species grew over a wider area, but
desert vegetation replaced them as conditions became
drier and hotter.

Throughout Big Bend country, rainfall averages
less than 10 inches per year. When rain does come, usually during late summer or early fall, precipitation is
local, frequently in a cloudburst, perhaps as much as an
inch in an hour. At these times, erosion is intense.
Temperatures are warm the year round and very hot in
summer. NEVER hike or drive in Big Bend National
Park without taking along drinking water.

Geologic Features

The Landforms

The locations and shapes of the infinite variety of landforms in Big Bend National Park can be explained in
terms of geologic structure and the degree of resistance
to erosion of specific rock units. For example, the three
great canyons of Big Bend, through which the Rio
Grande flows, are located where geologic structures
have produced mountains. In the southeastern part of
the park, Boquillas Canyon (longest of the three) cuts
across the Sierra del Carmen, a range of mountains that
trends north-south along the park's eastern boundary
(fig. 53.1). Upstream, at the park's southern tip, is
Mariscal Canyon, where the Rio Grande crosses the
structure of Mariscal Mountain. At the western end of
the park, Santa Elena Canyon, the most spectacular of
the gorges, was cut by the river during uplift of Mesa
De Anguila, a fault-block mountain. The principal
bedrock of all these mountains is folded and faulted
Cretaceous limestone.

Seemingly unrelated are the volcanic and plutonic
rocks that make up the Chisos Mountains in the center
of Big Bend National Park. The rugged Chisos, which
are several thousand feet higher in elevation than the
features along the Rio Grande, were built by igneous
emplacements and extrusions, Oligocene in age or
younger, are the youngest bedrock in Big Bend.

The oldest rocks in the park are Paleozoic units
exposed in Persimmon Gap at the northern entrance.
Persimmon Gap, an opening in the Santiago Mountains,
is a wind gap left by stream capture that probably took
place early in Cenozoic time.

People have given vivid and colorful names to oddly shaped natural landmarks in Big Bend National Park. Most of these features were produced by the processes of weathering and erosion on rocks of differing resistance. Cowboy Boot, for example, which looks like a giant inverted boot, is a rock pillar eroded along joints in volcanic rock. Two resistant dikes, side by side and looking like a mule's ears sticking up on a mountaintop, are the result of differential erosion. The landform is named Mule Ear Peaks. Elephant Tusk, a steep-sided peak with a pointed top, was eroded from a resistant igneous intrusion. Alsate's Face is an eroded ridgetop that resembles the head in profile of a recumbent Indian gazing at the sky. (Alsate was an Apache chieftain of local renown.) Casa Grande ("big house") is a lava-capped mesa with nearly vertical sides and a flat top (fig 53.3). A small cave eroded in a cliff near the top of Lost Mine Peak is named the Watchman's House. The cave, according to legend, is occupied by the ghost of an Indian who was left to guard a mine once worked by the Spaniards.

A large open valley (elevation 5401 feet), nestled among the peaks of the Chisos Mountains, is an erosional feature, called the Basin. The Window, a narrow gorge cut in the Basin's western rim, opens onto a 75-foot dry fall. The Window is the outlet for all the drainage from the Basin (fig. 53.4). The remnants of two calderas that erupted large volumes of basaltic

FIGURE 53.4   Looking west from the Basin, an erosional feature in the Chisos Mountains, down Oak Canyon and through the Window, a narrow gorge cut in the Basin's western rim. The Window is the outlet for all the drainage from the Basin. Vernon Bailey Peak (on the north) in the Pulliam Mountains consists of resistant igneous rock (a felsic quartz-riebeckite) emplaced in a small stock and subsequently exposed by erosion. The rock is believed to be the intrusive equivalent of the Burro Mesa Riebeckite Rhyolite, which is the youngest eruptive rock in Big Bend National Park and has been dated, as Oligocene or younger. Ward Mountain is to the south. Photo by Douglas Fowler.

lavas and tuff early in Cenozoic time rise above the Basin: Pine Canyon caldera on the east side and the Sierra Quemada caldera on the southwest.

Evidence of deposition is as widespread throughout Big Bend as erosional features, although the sheetwash gravels, filled basins, terrace gravels, and gravel-strewn pediments lack the curious shapes of the erosional landforms. Great quantities of sand and gravel have been washed down from the mountains and deposited in valleys and basins. Deposition was especially heavy in Pleistocene and post-Pleistocene time.

Because of the scarcity of vegetation, Big Bend is a place where many of the features plotted on geologic maps (anticlines, fold axes, fault traces, fault blocks, etc.) can be readily seen. Noteworthy examples are Cow Heaven anticline and the Mariscal anticline.

In learning about such a multiplicity of structural features and landforms, it is helpful to remember that the limestones and other sedimentary units are mainly Cretaceous; the igneous rocks are early Cenozoic in age; and the major structures are related to Laramide tectonic activity in latest Cretaceous and early Cenozoic time. Blockfaulting, extension, and mountain uplift, characteristic of the Basin and Range province, occurred in Tertiary time along with intense erosion and deposition.

The Rio Grande

Throughout most of its 1885-mile length, from southern Colorado to the Gulf of Mexico, the Rio Grande is younger than the geologic structures over which it flows. It developed its course by flowing across basin fills and fault blocks. This is evident in the Big Bend international boundary area, where the Rio Grande is a *transverse stream;* that is, it flows in a crosswise direction over the faults and fault blocks. The river is also *superimposed,* which means that it was let down from higher levels by erosion, cutting through the formations on which the river originally developed and through unconformities below. During downcutting, the Rio Grande and its tributaries removed sand and gravel that once filled the basins and exposed the bedrock of the upfaulted blocks. A major tributary is the Rio Concho, which crosses northeastern Mexico and joins the Rio Grande at Presidio, upstream from Big Bend National Park. Water from the Rio Concho had a significant role in the cutting of the canyons.

Detailed studies of river gravels, terraces, and pediments in the Rio Grande valley (mainly in New Mexico) show the occurrence of alternating periods of *aggradation* (the building up of the land surface by deposition) and *degradation* (the lowering of the land surface and general reduction of relief by stream action). This sort of cut-and-fill also went on in the Big Bend canyons. In some places on less steep sidewalls, pediments cutting across both bedrock and gravel were left high above the present valley floor.

The volume of the river has fluctuated in late Cenozoic time. During the Pleistocene Epoch, the Rio Grande was fed by valley glaciers in the Southern Rockies. Their meltwater greatly increased the river's volume and erosive power. Periods of pluvial climate brought greater precipitation to the whole southwest region; but at other times, the regional climate was even drier than at present. Climatic fluctuations profoundly affect the river's regimen, especially its cycles of flooding, and also cause changes in the processes of weathering, mass wasting, and runoff in the Big Bend region.

In 1978, a 191-mile strip on the American shore of the Rio Grande was placed in the Wild and Scenic Rivers system. The protected area begins in Big Bend National Park and continues downstream to the Terrell-Val Verde County line more than 75 miles from the park's eastern boundary. Except during flood stage, float trips by permit are allowed.

Hot Springs

The Hot Springs area, which has been developed for tourists in southeastern Big Bend National Park, is the best known of many hot springs along the course of the Rio Grande, on the Mexican as well as the Texas side. Three more hot springs, all downstream from Hot Springs, are located in the park. One is in a cane break next to the river, nearly a half-mile downriver; another is adjacent to Boquillas Canyon; the third serves as the water supply for Rio Grande Village. At Hot Springs, which has the highest temperature (106° F), the water is collected in a rectangular pool beside the river. The other springs have lower temperatures—105°, 95°, and 96° F, respectively. The springs maintain fairly constant temperatures and amounts of discharge, and are seemingly unaffected by seasonal changes in weather and rainfall. Discharge rates for the Big Bend hot springs are ordinarily less than 52 gallons per minute, which is not especially high.

Geologists think that the hot springs are remnants of what was a highly active hydrothermal system during late Tertiary time when crustal thinning (extension), involving normal faulting and numerous eruptions, was occurring. The hot springs are related to ground water circulation. In this region of high heat flow, hot rocks are comparatively near the surface. Ground water percolates down through the many faults and fractures in the limestone, is heated at relatively shallow depths, and then rises to the surface.

Vertebrate Fossils

Near park headquarters at Panther Junction, an exhibit shelter displays mammal and dinosaur bones that were discovered in sedimentary rocks in the park. The remains of very ancient, primitive mammals were found in Paleocene beds near the roadside exhibit. These extinct mammals, the oldest that have been found in Texas, were small, forest-dwelling carnivores and herbivores having no direct relationship to modern species. (Such mammals may have coexisted with late Mesozoic dinosaurs, whose fossilized remains have also been found in Big Bend.)

Eocene beds overlying the Paleocene rocks near the exhibit contain the bones of somewhat more advanced mammals. These animals were not direct ancestors of living mammals, but had an anatomical similarity to later species. The remains include the earliest "horse" (*Hyracotherium,* or eohippus), an early cat, pig-like and hippopotamus-like animals, and a rhinoceros-like species. In younger rocks elsewhere in Big Bend, doglike mammals, a rabbit, camels, a small deer, and sheep-like animals have been identified.

Dinosaur fossils in Big Bend have been found in shallow marine and nonmarine deposits, late Cretaceous in age, that accumulated in a warm, swampy, coastal environment. In addition to dinosaur bones, fossils in these rock units include petrified wood, turtle shells, fish scales, shark teeth, and crocodile bones.

By far the most remarkable finds have been bones from pterosaurs (flying reptiles). In 1971, a student at the University of Texas, Douglas Lawson, discovered wing bones (including a large forearm bone) of a giant pterosaur, which has become known as the "Big Bend pterodactyl." Diligent later searches turned up fossil fragments of other individuals of the same species (fig. 53.5). The mechanics of pterosaur flight are puzzling. How were such enormous wings controlled and maneuvered? How did the flying reptiles take off and how did

they land? Did they "ride" updrafts and air currents as hang gliders do? Did the huge creatures hunt over land as well as water? How did they get around when they were on the ground? Paleontologists and aerodynamicists have been constructing working models of giant pterodactyls in an effort to answer some of these questions. (Langston 1981, 1986).

At the end of the Cretaceous Period, the pterosaurs, who had dominated the skies for millions of years, died out—along with all other dinosaurs on earth. Modern birds, who took over the realm of the sky, developed from another form of reptilian life, unrelated to the pterosaurs.

Mining in the Big Bend Area

The quicksilver that was mined for many years in Big Bend country came from cinnabar ore hydrothermally deposited, probably during Tertiary igneous activity. The mineral *cinnabar* (mercuric sulfide) was brought upward in hydrothermal solutions that were then caught in a stratigraphic trap under the impermeable Del Rio Clay and deposited along with calcite in the Santa Elena Limestone, just below the clay. The stratigraphic trap was formed by a faulted anticline. Most of the ore was found within 50 feet of the bottom of the Del Rio Clay, but in some places the ore extended downward as much as 200 feet into the Santa Elena beds. Some of the richest ore was discovered in collapsed sinks and pipes.

In the Terlingua district (outside the park's western boundary), the cinnabar ore occurs in an east-west belt about 16 miles long and 6 miles wide, extending from Study Butte to Lajitas on the Rio Grande. Scattered occurrences of cinnabar have also been reported at sites in the lower Big Bend area. A mine was operated for a few years in the Mariscal Mountain area. Here the ore is associated with hydrothermal activity during the emplacement of sills in the Boquillas Formation.

Cinnabar ore from the Big Bend mines was smelted locally and the recovered quicksilver was shipped out in iron flasks. Obtaining fuel for the furnaces was always a problem. Most of the forests on the lower slopes of the Chisos Mountains were cut down for fuel. Because of the soil erosion that resulted and the arid climate, regrowth of trees was not possible and the slopes have remained barren. After deposits of subbituminous coal were located northeast of the mines, coal was used in the smelters.

FIGURE 53.5   In Cretaceous time, these huge pterosaurs (*Quetzalcoatlus northropi*) soared over what is now Texas. They had a wing span of 36 feet, weighed more than 120 pounds, and were the largest animals that have ever flown. Adapted from a drawing by G. Paul in *Earth Science*, Winter, 1985.

Geologic History

1. Sedimentation and mountain-building in the Paleozoic Era.

The oldest rocks in Big Bend National Park are deformed Paleozoic units exposed in thrust slices at Persimmon Gap. Here in the northernmost section of the park are the Santiago Mountains, which trend northwestward into the Marathon Basin and the Glass Mountains of west Texas. The rocks at Persimmon Gap, which include shales, limestones, chert, and novaculite, are related on the basis of age and lithology to Paleozoic sequences of the Ouachita fold belt in Oklahoma and Arkansas. The novaculite, a dense, light-colored, siliceous rock, is similar to that found in Hot Springs National Park (Chapter 52).

The Paleozoic rocks were deposited in a deep-ocean trough between northern and southern continental plates. When the ocean closed late in Paleozoic time, the continental plates collided, resulting in the mountain-building of the Ouachita orogeny. During the folding and faulting, some of the rock units from the deep ocean basin were thrust over rock units along the northern continental margin.

In Persimmon Gap and in roadcuts of the highway through the gap, some unique exposures of these rocks show the effects of three major orogenies. The Paleozoic rocks in the thrust slices are the result of the Ouachita orogeny. Later, during the Laramide orogeny, these rocks and their Cretaceous cover units were altered by overthrusting. More recently, the rock units have undergone extensional faulting due to Basin and Range deformation (Tauvers and Muehlberger 1988).

Table 53.1			Geologic Column, Big Bend National Park	
Time Units			**Rock Units**	**Geologic Events**
Era	**Periods**	**Epochs**		
Cenozoic	Quaternary	Holocene Pleistocene	Younger gravels and terraces	Erosion and deposition Erosion, deposition, major canyon-cutting
	Tertiary — Neogene	Pliocene Miocene	Older gravels, pediment gravels, older terraces	Uplift, renewed faulting Basin filling
			///////////////////////	Crustal extension and block faulting
	Tertiary — Paleogene	Oligocene - - ? - -	South Rim Formation	Volcanic eruptions Hydrothermal mineralization
		Eocene	Chisos Formation	Igneous intrusions
			Canoe Formation	Block faulting Mixed clastics and ash
		Paleocene	Hannold Hill Formation Black Peak Formation	Nonvolcanic deposits (mammal bones)
Mesozoic	Cretaceous		///////////////////////	Laramide orogeny, folding, thrust-faulting
		(Upper)	Javelina Formation	Nonmarine deposition (dinosaur bones, silicified wood)
			Aguja Formation	
			Pen Formation	
			Boquillas Formation	
		(Lower)	Buda Limestone	
			Del Rio Clay	
			Santa Elena Limestone	Transitional and shallow marine deposition in transgressing and withdrawing Cretaceous seas
			Sue Peaks Formation	
			Del Carmen Limestone	
			Telephone Canyon Formation	
			Glen Rose Limestone	
Paleozoic	(Ordovician to Pennsylvanian)		///////////////////////	
			Deformed shales, sandstones, novaculite, conglomerates, probably Tesnus, Caballos, and Maravillas Formations	Ouchita orogeny
				Shelf deposits; deep-ocean deposits

Source: Modified after Maxwell, 1968.

Except for the outcrops at Persimmon Gap, the Paleozoic rocks of the Big Bend area have been stripped off or buried by much younger rocks. However, in Mexico, across the Rio Grande from Boquillas, some older metamorphic Paleozoic rocks are exposed in the Sierra del Carmen escarpment.

FIGURE 53.6 Vertical cliffs in Santa Elena Canyon. Exposed here from bottom to top are the Glen Rose Formation (beds just above the waterline), the Telegraph Canyon Formation (covered with vegetation), and the Del Carmen limestone, a cliff former. All these formations are found below the Santa Elena Limestone and are Lower Cretaceous in age. Photo by Douglas Fowler.

2. Accumulation of Cretaceous sediments.

The absence of Triassic and Jurassic rocks in the Big Bend sequence probably indicates a long period of erosion. Throughout Cretaceous time, the Big Bend region was part of a coastal plain and a shallow marine shelf along the margin of a continental landmass that sloped toward an ancestral Gulf of Mexico. Unconformities in the sedimentary sequence mark several advances and withdrawals of the ocean during the 90 million years of the Cretaceous Period. The final draining away of seawater occurred when the region was uplifted late in Cretaceous time.

Massive, limey units like the Glen Rose Formation, the Del Carmen Limestone, and the Santa Elena Limestone accumulated during higher sea stands (fig. 53.6). Clastic sediment from the north and northwest spread over the ocean floor when sea level was lower or the land higher. Units interbedded with the limestones include claystone and mudstone (some of it bentonitic), shales, calcareous sandstones, thin sandstone beds, stringers of silt and coal, gypsiferous marls, chalk, cherty limestone, and basal conglomerates. Careful study of these sequences allows geologists to document sedimentary facies changes that show transgressions and withdrawals of the sea (fig. 53.7), episodes of erosion, and a wide variety of sedimentary environments.

Most of the rocks are fossiliferous and contain molluscan remains typical of marine Cretaceous sequences (fig. 53.8). In the transitional and nonmarine units (the Aguja and Javelina Formations) that accumulated late in Cretaceous time are fossilized logs and various remains, including dinosaur bones.

3. Laramide orogeny, late Cretaceous-early Tertiary time.

The Big Bend region is close to the eastern edge of the vast area of western North America that was profoundly affected by the complex and long-lasting tectonic events associated with the Laramide orogeny. Deformation was moderate in Big Bend in comparison to that occurring in regions farther north and west. Compressive forces produced tight folds and overthrusting along the eastern and western margins of the Big Bend and broader anticlinal folds in between. Some of the older folds and structures, such as those at Persimmon Gap, were realigned and refolded with the overlying Cretaceous beds. If you draw a line on a geologic map from Santa Elena Canyon to the northeast corner of Big Bend National Park, you will note that the line crosses more than a dozen folds and several major faults.

FIGURE 53.7   The Santa Elena Limestone was deposited in shallow water and the seas transgressed and regressed on the land. The ripple marks shown on the piece of float in the center of the photograph indicated shallow water deposition. Photo by Ann G. Harris.

On the mountainous topography that developed as a result of tectonic activity, the processes of erosion went to work. Structural trends controlled the patterns of valleys and ridges that evolved. Not much evidence of these landforms can be found in the present landscape of Big Bend because they were destroyed or buried by subsequent geologic events.

4. Paleocene and Eocene nonvolcanic deposition.

Erosion of the Cretaceous rocks and structures produced clastic sediments that accumulated between folds and fault blocks, forming nonvolcanic rock units, Lower Tertiary in age and approximately 3000 feet in thickness. These units lie unconformably over the older Cretaceous rocks. Lithologically the rock units are composed of conglomerates, sandstones (including channel sandstones), and shales. In the Canoe Formation nonvolcanic beds grade into mixtures of clastics, tuff, and ash, interbedded with lava flows, signifying the onset of volcanic activity. Because fairly intense erosion continued on the mountain slopes and on the nonvolcanic sequences that had accumulated in

the valleys, local unconformities cut across all three early Tertiary formations and some of the uppermost Cretaceous units as well.

5. Igneous extrusive and intrusive activity.

In early Tertiary time, after the slackening of Laramide compressive forces, magma began rising into the crust. Molten rock reached the surface and was extruded as lava or pyroclastics. Magma below the surface cooled in dikes, sills, sheets, laccoliths, plugs, and other discrete bodies of irregular shape. Many were covered by lavas and pyroclastics and some may have been feeders for volcanic eruptions. Because most of the intrusions are fine grained, they probably cooled at depths of a few thousand feet or less. Subsequent erosion in Big Bend National Park and adjacent areas has exposed many of the intrusions, which are largest and most abundant in a broad, northwest-trending belt centering in the Chisos Mountains. In terms of composition, the intrusive rocks range from felsic rhyolites (the riebeckites) through intermediate rocks, such as trachytes and andesites, to mafic

FIGURE 53.8   A fossil-bearing layer in the Lower Cretaceous Santa Elena Limestone near Boquillas Canyon. Huge clams (bivalves) have been found up to three feet, but most fossils are much smaller. Photo by Douglas Fowler.

basalts. They range in age from latest Cretaceous to late Oligocene.

Volcanic centers opened in the Big Bend area and from these vents, lavas, and pyroclastics were erupted. In basins and valleys, volcanic rocks are interbedded with sandstone and other clastics. Some beds, formerly continuous, have been interrupted by faulting and erosion. Between eruptions, streams reworked and redeposited loose debris that was later buried by new volcanic material.

The Chisos and South Rim Formations are made up of 3000 to 5000 feet of massive lava flows, flow breccias, conglomerates, tuffaceous sandstones, mudstones, tuffaceous clays, and ash beds. Chisos rocks have been dated as mainly late Eocene in age, and the South Rim Formation as probably Oligocene or younger. Potassium-argon dates taken from basalts in the lower part of the Chisos Formation yield dates of 38 to 42 million years before the present. Dates from riebeckite rhyolite near the top of the South Rim Formation record eruptions that occurred about 30 million years ago.

6. Tertiary block-faulting.

Big Bend country, which is in the easternmost part of the Basin and Range province, has a landscape shaped by the erosion of upfaulted blocks and deposition in downfaulted basins. Toward the close of Laramide tectonism, segments of crustal material that had been contorted and squeezed together by compression began to readjust isostatically. As the crust sagged, thinned, and stretched, tensional forces broke it into blocks that moved up or down (fig. 53.9). In the Big Bend area, evidence based on crosscutting relationships indicates that throughout Tertiary time high-angle faulting occurred intermittently, producing a series of northwesttrending fault blocks. Faulting was most intensive, however, during early and middle Tertiary time.

7. Extensive late Tertiary erosion and deposition.

Unconsolidated sediments, shown as "older gravels" on geologic maps of the Big Bend, are the remnants of a sedimentary cover that was once more wide-

FIGURE 53.9 The fault block mountains of the Sierra del Carmen Range that extends south into Mexico. After being squeezed toward the end of the Laramide orogeny, segments of the crustal material began to readjust isostatically and were stretched and thinned; as a result, tensional forces broke it up into blocks that moved up or down forming high angle faults such as the Sierra del Carmen Mountains. Photo by Douglas Fowler.

spread. Eroded clastic and volcanic debris filled the fault basins and probably covered some of the fault-block ranges. During this time the present drainage system, graded to the Rio Grande, was organized.

8. **Pleistocene and Holocene faulting and erosion.**

Due to complex changes in rainfall and runoff patterns, a large part of the basin fill and other Tertiary deposits were eroded and carried away by swollen Pleistocene streams. Renewed uplift on some of the Tertiary faults elevated the highlands, at the same time increasing the erosive power of the streams. The Rio Grande and its larger tributaries cut canyons and widened and deepened valleys. Landforms referred to as "older terraces" developed at this time, and cliff recession produced pediments that truncated bedrock and older gravels.

In post-Pleistocene time, as the regional climate became increasingly arid, streams continued to erode and deposit but at a slower rate. Young gravels have accumulated in valley bottoms and on narrow flood plains. Pediments, fans, and terraces are being dissected by ongoing erosion.

Geologic Map and Cross Section

Maxwell, R.A., and Lonsdale, J.T. 1966. Geologic Map of the Big Bend National Park, Brewster County, Texas. University of Texas Bureau of Economic Geology.

Sources

Chesser, K. and Estepp, J.D. 1986. Hot springs of Big Bend National Park and Trans-Pecos Texas, in Pause, P.H. and R.G. Spears, editors. 1986. p. 97–104.

Langston, W., Jr. 1981. Pterosaurs. *Scientific American* 244 (2):122-136.

——— 1986. Rebuilding the World's Biggest Flying Creature: The Second Coming of *Quetzalcoatlus Northropi,* in Pause, P.H. and R.G. Spears, editors. 1986. p. 125–128.

Maxwell, R.A. 1968. *The Big Bend of the Rio Grande,* a Guide to the Rocks, Landscape, Geologic History, and Settlers of the Area of Big Bend National Park. Guidebook no. 7 (geologic maps in pocket). Austin, Texas: University of Texas Bureau of Economic Geology.

———; and Dietrich, J.W. 1965. *Geology of the Big Bend Area, Texas;* Field Trip Guidebook with Road Log and Papers on the Natural History of the Area. West Texas Geological Society Publication 65-51.

———; Lonsdale, J.T.; Hazzard, R.T.; and Wilson, J.A. 1967. *Geology of the Big Bend National Park, Texas.* Publication no. 6711. Austin, Texas: University of Texas Bureau of Economic Geology.

National Park Service. 1983. *Big Bend.* Handbook 119. 128 p.

Pause, P.H. and Spears, R.G. editors. 1986, *Geology of the Big Bend Area and Solitario Dome, Texas.* West Texas Geological Society 1986 Field Trip Guidebook, publication 86-82 (geologic maps in pocket). 277 p.

Price, J.G. and Henry, C.D. 1988. Dikes in Big Bend National Park; Petrologic and Tectonic Significance. Geological Society of America, South-central Section, Centennial Field Guide, v. 4, p. 435–440.

Schiebout, J. 1986. Big Bend: A Crossroads in the Beginning of the Age of Mammals, in Pause, P.H. and R.G. Spears, editors, 1986. p. 129–134.

Spearing, D. 1991. *Roadside Geology of Texas.* Missoula, Montana: Mountain Press Publishing, p. 291–321.

Tauvers, P.R. and Meuhlberger, W.R. 1988. Persimmon Gap in Big Bend National Park, Texas; Ouachita Facies and Cretaceous Cover Deformed in a Laramide Overthrust. Geological Society of America, South-central Section, Centennial Field Guide, v. 4, p. 417–422.

Tyler, R.C. 1984. *The Big Bend: a History of the Last Texas Frontier.* National Park Service. 288 p.

Note: To convert English measurements to metric, go to www.helpwithdiy.com/metric_conversion_calculator.html

Shenandoah National Park

NORTHWEST VIRGINIA

Area: 196,466 acres; 307 square miles

Established as a National Park: December 26, 1935

Address: 3655 U.S. Highway 211 East, Luray, VA
22835-9036

Phone: 540-999-3500

E-mail: SHEN_Superintendent@nps.gov

Website: www.nps.gov/shen/index.htm

From Skyline Drive, along the crest of the Blue Ridge Mountains, spectacular vistas overlook Shenandoah Valley to the west and the rolling Piedmont to the east. Precambrian and Paleozoic rocks in complex structures record ancient orogenies. Waterfalls, racing streams, high meadows, ridges, and hidden hollows are seen along 500 miles of trails, including a 95-mile segment of the Appalachian Trail.

FIGURE 54.1   Dark Hollow Falls, like other major falls in the park, represents an erosive break between units of Catoctin volcanic flows. National Park Service photo.

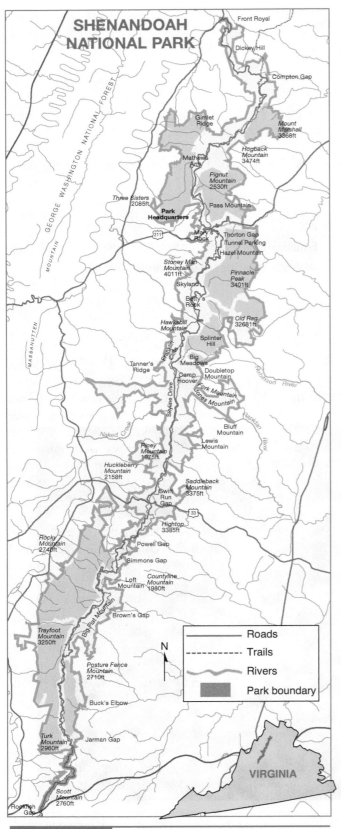

FIGURE 54.2 Shenandoah National Park, Virginia.

Geographic Setting

South of New England, the Appalachian Mountain belt divides into four physiographic provinces that trend south in roughly parallel bands. From east to west, these divisions are the Piedmont (a sloping plateau), the Blue Ridge, the Ridge Valley, and the Appalachian Plateau (fig. 54.3). In the Blue Ridge province, which has the highest elevations in the Appalachian system, are two national parks—Shenandoah in Virginia and Great Smoky Mountains (Chapter 55) in Tennessee and North Carolina.

The Blue Ridge Mountains extend from southern Pennsylvania southwestward. From South Mountain in Pennsylvania, which is a single narrow ridge only about 10 miles wide, the mountains become closely spaced ridges of greater height and width, culminating in a broad bulge of complex ridges that cross North and South Carolina and Tennessee, and dip into Georgia. Great Smoky Mountains National Park occupies the highest part of the southern Blue Ridge province.

Shenandoah National Park, near the northern end of the Blue Ridge, is long and narrow. The park follows the crest of the Blue Ridge for about 80 miles from Front Royal, Virginia (about 72 miles west of Washington, D.C.), to the southern entrance of the park near Waynesboro, Virginia. Elevations in Shenandoah National Park range from around 600 feet at Front Royal to 4051 feet at the top of Hawksbill Mountain.

Skyline Drive, a scenic highway famous for its overlooks, runs the length of Shenandoah National Park and connects with the Blue Ridge Parkway, which continues south to Great Smoky Mountains National Park. The Appalachian Trail goes through both parks as it follows the crest of the Appalachian Mountains from Maine to Georgia.

Due to their greater age, the Blue Ridge Mountains and other ranges of the Southern Appalachians are more rounded and less angular than the mountains of western United States. Much of the bedrock of the central and southern Appalachians is mantled by *residual soil* that accumulated in place through millions of years of weathering. This soil and ample precipitation distributed evenly throughout the year support dense hardwood forests on the mountain slopes. Evergreens that have persisted at higher elevations are probably remnants of coniferous forests that spread southward during cold Pleistocene climates. Glaciers did not reach Virginia, but periglacial conditions that prevailed at high altitudes intensified weathering and erosional

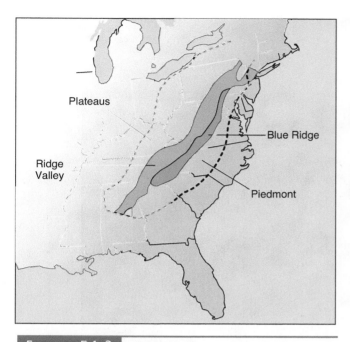

FIGURE 54.3 Physiographic provinces of the Southern Appalachians.

processes. The large boulder fields on many slopes of the Blue Ridge are the result of periglacial activity. Even in today's climate, snow blankets higher levels of the Blue Ridge for part of the winter season. Frost wedging is a significant factor in the weathering of rocks.

Major streams are the Potomac River, which cuts through the Blue Ridge at Harpers Ferry (north of Shenandoah National Park) and the James River, which has carved scenic gorges for its course through the Blue Ridge south of the park. No streams cross the Blue Ridge within the park. To the west, in the Shenandoah Valley, the Shenandoah River flows north, paralleling the Blue Ridge, and joins the Potomac at Harpers Ferry. The Blue Ridge itself is dissected by tributary streams, many with waterfalls, that drain the mountain slopes. Streams have been the dominant agent of erosion in the region during and since the intermittent uplifts of Paleozoic time. Erosion is fairly intense, despite the protective cover of vegetation. During heavy rains, streams that drain logged-over areas (mostly outside the park) become muddied by loose sediment. When storms are prolonged or extremely heavy, mudslides occur on steep or denuded slopes.

Prehistoric gatherers and hunters frequented the Blue Ridge for thousands of years before Europeans came to Virginia. Indians left stone burial mounds, arrowheads, spear points, and other artifacts in Shenandoah National Park. Elk, deer, and bison grazed in the high open meadows and clearings in the forest. Fires that kept the meadows free of trees were started by lightning, or sometimes kindled by Indian hunters.

The English who settled in the Virginia Tidewater did not expand westward for about a hundred years. The Piedmont (the area between the coast and the mountains), the Blue Ridge, and the Shenandoah Valley on the far side remained unsettled and largely unexplored until the early 1700s when settlers began to establish farms and villages on both sides of the Blue Ridge and in the Shenandoah Valley.

Settlements in the Shenandoah Valley and the Piedmont became increasingly prosperous as time went on, but the mountain people who farmed the upland valleys of the Blue Ridge became ever poorer and more isolated. They were limited in their choice of favorable sites not only by the rugged topography but also by the nature of the bedrock. The well watered, fertile soils of the broad summit meadows and the gentler eastern slopes—underlain mostly by granites and basaltic green stones—were more suitable for homesteads than the rocky, steep western ridges that developed thin, infertile soils on resistant quartzites and phyllites. Moreover, on the west side, springs were fewer and intermittent. What good soils there were in the uplands soon became depleted as crops were planted year after year in the same small fields. Erosion was severe, especially after timber cutting removed protective vegetation from the slopes. A disastrous blight, early in the 20th century, exterminated the chestnut trees, which had been a reliable food source as well as one of the region's finest hardwoods.

In the 19th century, the discovery of small deposits of iron, manganese, and copper in the Blue Ridge helped the region for a time; but as ores were worked out, mines were abandoned. Some manganese mining was carried on as late as World War II. In the western foothills of the park and outside the western boundary are more than 40 old mine sites.

As times became ever harder, the mountain people and their unique way of life seemed doomed. But toward the end of the 19th century, the development of mountain resorts high in the Blue Ridge brought money, new people, and a revival of interest in the region's potential. The beauty of the mountains and the benefits of pure mountain air were now accessible to the large populations of the nation's capital and the

Stony Man Mountain (as seen from Skyline Drive), the second highest peak in the park, was given that name because its outline against the sky looks like a reclining, bearded man. National Park Service photo.

cities of the mid-Atlantic seaboard, areas that sweltered in summer heat and humidity. The idea of establishing a national park in the Virginia Blue Ridge attracted support. George Freeman Pollock, who operated Skyland Resort on the mountaintop and controlled some 5000 acres along the crest, became enthusiastic about the project. After Congress authorized the park's acquisition in 1926, he helped to raise over a million dollars for land purchases. With no public lands in the area, all ground for the park had to be purchased from private owners or acquired by donation. President Herbert Hoover gave his land and fishing camp on the Rapidan River. The Virginia legislature appropriated a million dollars for compensation and relocation of landowners. In 1935, a large tract along the mountain crest was turned over to the federal government for the establishment of Shenandoah National Park. Later additions brought the total area to over 300 square miles.

Skyline Drive, with its beautiful panoramas to the east and west, traverses the park along the spine of the Blue Ridge. Most of the land has been allowed to revert to a wilderness state. Several ancient clearings, including Big Meadows (fig. 54.5), are kept open, but new forests have reclaimed the mountain farms, orchards, and logged-over areas. More than three million people visit this exceedingly popular national park each year.

Geologic Features

Bedrock and Geologic Structures

Except in roadcuts, fresh exposures of bedrock are rare in Shenandoah National Park. Outcrops in streams and on bare slopes show evidence of deep weathering. Ledges may be shattered or largely covered by broken and crumbling talus. Weathering and vegetation may have masked bedrock and structures, but their controlling influence on landscape development is plain to see.

Consider Big Meadows, part of a broad summit area on the crest of the Blue Ridge (fig. 54.5). Here is a plateau of low relief, about 3550 feet in elevation and roughly five square miles in extent. Elsewhere along the northern Blue Ridge, the crest is relatively narrow and deeply dissected. Two intermittent swamps, fed by springs and runoff from higher slopes, have long existed in shallow basins in Big Meadows. What lies

FIGURE 54.5   Big Meadows, located south of Franklin Cliffs and Dark Hollow Falls in the central portion of Shenandoah National Park, is part of a broad summit area on the crest of the Blue Ridge. It is a plateau of low relief, about 3350 feet in elevation and approximately five square miles, lying on top of Precambrian lava flows. National Park Service photo.

beneath Big Meadows are at least 1800 feet of ancient lava flows of the Catoctin Formation, originally basaltic but later metamorphosed to greenstones. These resistant greenstones cap the highest ridges and peaks of the park (Hawksbill Mountain, for example). In the Big Meadows area the top of the formation and overlying sedimentary rocks have been removed by erosion, producing the gently rolling topography of this part of the Blue Ridge crest.

The flood basalts of the Catoctin Formation, which erupted close to the end of Proterozoic time, are comparable in magnitude to the Columbia River basalts of Oregon and Washington and the Parana flood basalts of Brazil. Like these younger great flows, the Catoctin lavas probably erupted within a relatively short span of time, about three to five million years. The denuded

masses of Catoctin greenstones, however, tend to be more resistant than the more recent flood basalts.

Most of the waterfalls in the park cascade over cliffs and benches of Catoctin greenstones. Dark Hollow Falls, which drains much of the Big Meadows is an example (fig. 54.1). Prominent cliffs of columnar-jointed lava, such as Franklin Cliffs and Crescent Rocks, can be seen from Skyline Drive. Field investigations have shown that some Catoctin flows that exhibit columnar jointing erupted subaerially, while other flows were subaqueous. Some pillow basalts at Dark Hollow Falls may have resulted from lava flowing into a localized body of water. Thicknesses of flows appear to be remarkably constant, and few unconformities can be observed. Erosion in intervals between flows was only moderately significant. Porphyritic and

FIGURE 54.6 The Catoctin Formation greenstone, exposed here on the peak of Hawksbill Mountain, was metamorphosed from Precambrian flood basalts. Sediments that covered the greenstone have been removed by erosion. National Park Service photo.

amygduloidal basalts are prominent in some zones of the Catoctin Formation.

Amygdules have been filled with quartz, albite, calcite, hematite, and epidote. Large, irregular pods of epidosite (epidote + quartz) are common, especially along flow margins. These pods were probably segregated during metamorphism (Badger 1992).

Some of the Precambrian plutonic rocks that are older than the Catoctin Formation have been cut by greenstone feeder dikes that represent intrusions of molten basalt that did not reach the surface. Because the ancient granitic rocks are even more resistant than

the greenstones, differential erosion has produced deep notches of varying widths where the dikes were emplaced. Along the Ridge Trail on Old Rag Mountain, a flight of natural stairs, has weathered out between vertical granite walls. The stairs on the dike surface weathered from columnar joint blocks.

The Old Rag Granite, exposed at the summit of Old Rag Mountain and elsewhere in the park, is easily recognized by its very light color and coarse grain size. Because this granitic bedrock is very resistant, weathering acts mostly along joints where water and expanding ice can attack the rock. Between joints, large spher-

Box 54.1	An Outline of Appalachian Evolution—Georgia to New York

For two centuries, the Appalachian Mountains have been the inspiration and testing ground for many of the grand ideas and concepts of geology. It was in dealing with the complexities of these ancient mountains that American geologic sciences came of age. From observations in the field, geologists learned the nature of folds and thrust faults in mountain belts and recognized an association of sea-floor rocks with initial stages of mountain-building. Hypotheses were developed to account for orogenies and for variations in landscape development.

Still many questions regarding the origin and development of the Appalachians and other mountain belts remained unanswered. One puzzling fact was that masses of rock exposed in the higher Appalachians appeared to be geologically incongruent (unrelated) to rock units in the foothills and lowlands on either side. Moreover, Paleozoic deposits thickened and became coarser toward the *east,* but where was the great landmass that had supplied the sediments? Had a "lost continent" foundered beneath the waters of the broad Atlantic Ocean? What forces generated the great lateral contractions and the horizontal compressions that produced the folded and faulted rock units in the ranges?

Pre-plate tectonics concepts recognized evidence indicating that tectonic pulses in a series of episodes throughout Paleozoic time were involved in the geologic history of the Appalachians; but the advent of plate tectonic theory, with its concept of moving crustal segments, disposed of the need for fixed landmasses as sediment sources. Instead, continents and continental fragments were visualized as coming together or breaking apart, with intervening oceans closing or opening during the process. Crustal plates may converge in such a way that downbuckling occurs and sediment accumulates in clastic wedges between plates. Colliding landmasses may crumple or one may override the other. Fragments of plates may be sutured onto a larger plate as tectonostratigraphic terranes during an orogeny. Such terranes contain rocks that are usually unrelated to the rocks of neighboring areas, in terms of age, fossil forms, deformational and/or metamorphic history, and paleomagnetic signatures.

Present approaches toward gaining an understanding of the evolution of mountain belts are based on theoretical reconstructions of former positions and movements of tectonic plates during geologic time. In this work, scientists rely on classic field geology (structural geology and stratigraphy) supplemented by interpretations of paleomagnetic data, orientation of deformed rock assemblages, sedimentary facies, distribution of fossil forms, and geophysical techniques, such as deep seismic reflection profiling, originally developed for oil exploration by the petroleum industry.

Although the Appalachian Mountains belt divides into two major segments along its length, the overall structural features exhibit a remarkable parallelism of strike, from Newfoundland to Alabama. The two segments, which have related but somewhat different orogenic histories, are 1) the Newfoundland-New England segment and 2) the Pennsylvania-Alabama segment, which includes the Blue Ridge. Emphasis in this discussion is on the second segment.

1. Middle and Late Proterozoic (Precambrian) time.

By the middle of Proterozoic time, during the Grenville orogeny, a supercontinent had formed, which consisted of an amalgamation of most of the continental crust that existed at that time, including the crust of North America and Africa. The details of sedimentation, deformation, plutonism, and volcanism have been resolved in the gneissic remnants exposed in the core of the Blue Ridge; but some of the history was masked by the Grenville orogeny and later events.

Then, as rifting began late in Proterozoic time, the supercontinent broke up and a seaway opened that became the Iapetus Ocean. (This forerunner of the Atlantic Ocean was named Iapetus for the father of Atlas, the Titan of Greek mythology, for whom the Atlantic Ocean was named.) As rifting continued and landmasses drifted apart, two or more seaways developed, separated by the Piedmont and Avalon terranes, which eventually became attached to the North American craton. Outpourings of flood basalts overspread part of the Shenandoah area; and for millions of years, sediments accumulated on the plate margins, in the oceans, and between the island chains.

(continued)

Box 54.1   **An Outline of Appalachian Evolution—Georgia to New York (continued)**

2. Early Cambrian through Early Ordovician time.

Orogenic activity began again, involving the creation of volcanic arcs, ocean closing, and subduction. Ocean crust was subducted beneath the arcs, which were jammed together. The overriding plate thrust westward onto the continental margin of North America. The weight of this newly arrived plate on the North American margin caused the accumulated Cambrian and Ordovician carbonates to founder and drown, creating a trench with water depths up to 1000 meters (more than half a mile). This trench was later filled with mud and sand washed in from the arriving thrust sheet.

3. Middle Ordovician into Early Silurian time.

The Taconic orogeny involved a volcanic arc-continent convergence, during which oceanic crust and the arc were thrust onto the edge of the continental crust. The Avalon-Carolina composite terrane, which became attached to North America during this time is the largest and most easily recognized terrane in the Appalachians (fig. 54.7).

4. Late Devonian to Early Mississippian time.

The North American and African plates were joined by the Acadian orogeny, which had a more profound effect in the northern segment of the Appalachians than in the southern segment.

5. Late Pennsylvanian to Permian time.

The Alleghany* orogeny brought about the final closing of the proto-Atlantic seaways and the forming of the supercontinent Pangea. Intense compression and thrusting dominated orogenic activity in the

_____

*Spelled *Allegheny* in some localities.

Southern Appalachians. Estimates of crustal shortening in a northwest-southeast direction across the mountain belt from late in Mississippian time through Permian time are greater than 50 percent (approximately 170 miles) in the vicinity of Great Smoky Mountains National Park. Displacements decrease northward and southward from the Great Smokies. In and near Shenandoah National Park, crustal shortening amounted to about 20 percent, or less than 75 miles.

6. Mesozoic time.

A period of rifting began late in the Triassic Period as the crust stretched and thinned, causing block-fault basins to develop, with accompanying volcanism, from North Carolina to the Gaspé Peninsula in Canada. Nonmarine shales and sandstones accumulated in the basins. Dinosaur tracks are preserved in many of these sedimentary layers. As the supercontinent Pangea broke up, the modern Atlantic Ocean opened and widened while the landmasses drifted apart. In the Appalachians, isostatic adjustments enabled less dense continental rocks to rise gradually. The high mountains that resulted were attacked by the processes of erosion. Great thicknesses of rock were stripped off and deposited along the base of the mountains, on the Atlantic Coastal Plain, and in the ocean. The amount of erosion is inferred from metamorphic rocks, now exposed, that in places must have been formed at more than approximately 10 miles below the earth's surface.

7. Cenozoic time.

The North American plate has continued to drift toward the west. Isostatic uplift continued at a subdued rate while fluvial erosion dug ever deeper. Deposits of sediments have spread farther to the east, extending the Atlantic Coastal Plain.

oidal boulders (some as much as 10 feet in diameter) have weathered in place (fig. 54.8).

The resistance of the gneissic Pedlar Formation relates to the character of the bedrock exposures; that is, whether they are massive or strongly foliated. Excellent outcrops of strongly foliated Pedlar gneisses can be seen along Skyline Drive, especially at the Tunnel and Buck Hollow Overlooks (fig. 54.9). The Pedlar and Old Rag rocks are the oldest rocks in Shenandoah National Park.

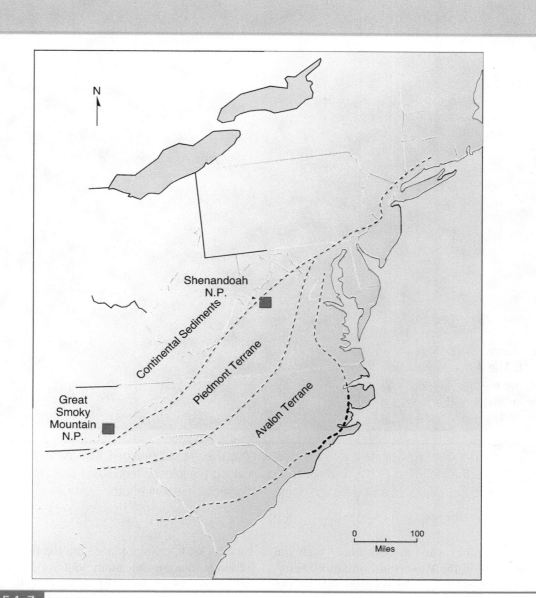

FIGURE 54.7 Major tectonostratigraphic terranes in the Southern Appalachians. The Piedmont terrane was pushed onto the margin of the North American craton probably early in Cambrian time. The Avalon, which extends farther northeast than shown here, arrived during the Taconic orogeny. The continental sediments west of the parks accumulated in the Paleozoic seas that overspread the midcontinent.

In general, the ancient greenstones and granites are the rocks that form the crest and gentler eastern slopes of the Virginia Blue Ridge, the areas favored by the Indians and later by the settlers. On the western side are the Cambrian clastic rocks of the Chilhowee Group that have been metamorphosed to quartzites and phyllites. Steep, sharp ridges, V-shaped hollows, and slopes covered by coarse rock debris characterize the topography formed on these rocks.

FIGURE 54.8   The Old Rag Granite is a highly resistant massive granitic bedrock. Because of the resistance of the granite to the weathering, most of the weathering occurs along the joint system. Over time, frost action and spheroidal weathering has produced an exfoliation dome. National Park Service photo.

The Cambrian rocks vary in resistance, with the quartzites, like those in the Weverton Formation, being more resistant to erosion. On the western side of the Blue Ridge, the thin, sandy, and shaly soils that developed on the rocks support forests, but the trees tend to be less diverse, smaller, and less luxuriant than those that grow in more fertile soil on the plutonic and volcanic rocks of the eastern side. Part of the difference may be due also to lower precipitation and moisture on the western side of the ridge.

Cleavage and Joints

In the Shenandoah region, metamorphism and compressive deformation were so severe that cleavage obscures the original bedding of most fine-grained metasedimentary rocks. Cleavage planes are the dominant and most obvious feature on many outcrops. *Rock cleavage,* which is the tendency of rocks to split along closely spaced, parallel planes, is independent of bedding and is the result of regional metamorphism and deformation. (Rock cleavage is not the same as *mineral cleavage,* which is a physical property of some minerals that break along planes related to molecular structure and crystal faces.) Cleavage and also foliation are pervasive, even in massive plutonic rocks and quartzites.

Joints (bedrock fractures without displacement) are conspicuous in the Blue Ridge, with most outcrops displaying fracture trends (joint sets) in several directions. Joint planes, which are ordinarily less closely spaced than cleavage planes, are caused by tensional deformation and also by pressure release, or unloading.

FIGURE 54.9   The strongly foliated gneiss of the Pedlar Formation can be seen at Mary's Rock Tunnel. It is among the oldest rock in Shenandoah National Park. National Park Service photo.

As weathering processes break up bedrock into boulders, blocks, and rubble, talus accumulations reflect the particular characteristics of the bedding, cleavage, and joint systems of the bedrock source. White angular blocks, for example, pile up at the foot of ledges of massive, resistant quartzite. From a distance, the talus slopes look like fresh rockslides because vegetation cannot grow over them. Soluble minerals available for soil enrichment that may have been present in the original sandstone have long since been removed, and the quartzite is extremely resistant to chemical weathering. Talus slopes of basalt, on the other hand, are difficult to see from a distance because trees have grown over the rocks in soil developed by weathering. Sizes of granite blocks in talus slopes range from small to enormous, depending upon the

fractures in the bedrock source. Below ledges and cliffs of layered rocks, the talus is usually made up of many small blocks of near-uniform size because cleavage planes, bedding, and joints are normally closely spaced.

Folds and Faults

Asymmetric folds, large and small, are visible in some outcrops of Chilhowee Group formations, especially in the southern section of the park. Folds in light-colored, resistant beds of quartzite are easy to trace along hillside slopes. Most folds trend northeast-southwest, parallel to the long axis of the Blue Ridge.

Faults have strongly influenced landform development in Shenandoah National Park. The low- to high-

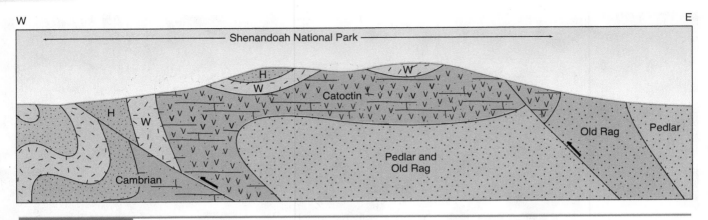

FIGURE 54.10   Simplified geologic cross section of the northern Blue Ridge in the area of Shenandoah National Park. Complex folding and thrust faulting are the result of intense tectonic compression that engulfed the region during much of the Paleozoic Era. Key: W = Weverton Formation; H = Hampton Formation; both are Cambrian in Age. Adapted from Gathright, 1976.

angle thrust faults, which dip generally to the southeast, are the most important structural features in the park, although fault traces are apt to be obscured by talus, gravel, or alluvium. Displacements produced by faulting moved the rock mass of the Blue Ridge over younger Paleozoic rocks (fig. 54.10).

Also significant are steeply inclined strike-slip (transcurrent) faults that trend northwestward across the Blue Ridge. Uneven movement on the two types of faults produced uneven summits on the Blue Ridge and an irregular lobe and embayment pattern of rock distribution along the Shenandoah Valley's eastern edge.

In the massive crystalline rocks, brecciated shear zones reveal some fault traces. Zones of fault breccia that mark thrust faults in the quartzites of the Erwin Formation are especially striking. Red iron oxide cement has filled voids in the fault zone, binding the brecciated quartzite into very resistant rock.

Water Gaps and Wind Gaps

A *water gap*—like the James River gorge, south of Shenandoah National Park—is a deep pass in a mountain ridge, through which a stream flows. A *wind* gap is a former water gap that has been abandoned by the stream that eroded it. Some of the wind gaps in Shenandoah National Park, carved into the resistant rocks of the Blue Ridge by rivers of an earlier time, are more than a thousand feet deep. Transverse valleys of this type are called wind gaps because their shape tends to funnel the summit winds. Typical examples are Compton Gap, Thornton Gap, Swift Run Gap, Powell Gap, Jarman Gap, and Rockfish Gap.

What happened to the streams that eroded the wind gaps? And why, for such a long time, did the rivers flow across the Blue Ridge, a formidable topographic and structural barrier, instead of going around it? A plausible explanation for evolution of drainage on the northern Blue Ridge might be as follows:

1. Streams flowed southeastward to the Atlantic Ocean from a drainage divide in the Appalachian Plateau province to the west. As the streams crossed the Blue Ridge, which was being uplifted, their course and direction were controlled by the regional slope of the land surface.

2. As streams cut down, erosion removed a considerable volume of rock material, creating water gaps that were transverse to the trend of the Blue Ridge structures. In this way, the original dendritic stream pattern was *superimposed* on structures and rock units.

3. Because of lithologic and structural variations in the bedrock, some streams had an advantage in downcutting and maintaining their courses. The Potomac River, for example, eroded its channel across the Blue Ridge north of the park, where the resistant layers of Catoctin lava thin to less than 50 feet. Favored streams and their tributaries were able to establish lower local base levels and erode headward faster than the streams without such advantages.

4. The Shenandoah River, a tributary of the Potomac, cut headward up (south) the Shenandoah Valley, which is underlain by limestone, much more erodible in a humid climate than the crystalline rocks of

the Blue Ridge. Because of its location on a belt of nonresistant limestone, the Shenandoah River was able to *capture* the headwaters of streams flowing across the Blue Ridge between the Potomac and the James Rivers. The water gaps then became wind gaps, usually with two small streams flowing away from the crest of the gaps, one to the east and the other to the west. By the capture of the Blue Ridge tributaries, the drainage in the Shenandoah Valley became *adjusted* to the structure. In other words, *adjustment of drainage* had taken place.

5. Differential resistance of bedrock and structures continues to control the development of the topography. During isostatic upwarping in Cenozoic time, streams have maintained their courses by downcutting and adjusting to rocks and structures.

Geologic History

1. **The oldest rocks, Proterozoic Y, in Shenandoah National Park.**

Radiometric dating procedures have yielded an approximate age of 1100 million years ago for the crystallization and metamorphism of the Pedlar Formation and the Old Rag Granite, which make up the resistant core of the Blue Ridge. The intensely foliated gneisses of the Pedlar Formation were probably intruded by the Old Rag Granite. Both were metamorphosed at depth, under conditions of high temperatures and pressures, during the Grenville orogeny.

2. **Deposition of the Swift Run Formation in late Proterozoic time.**

A discontinuous sequence of conglomerates, sandstones, and pyroclastics was deposited by streams eroding highlands and building up valley floors.

3. **Eruption of the Catoctin Formation lavas near the end of Proterozoic time.**

Plateau basalts, pouring out over the ancient landscape, filled the valleys and eventually covered the eroded mountains. The rifting that allowed the basaltic magma to well up preceded the initial opening of the Iapetus Ocean early in Cambrian time. Ash falls interlayered with the lava flows indicate that some of the eruptions were explosive. The uppermost layer in the volcanic sequence is a purplish slate (representing an extensive ash fall) that marks the boundary between the Catoctin Formation and the overlying Cambrian beds.

4. **Deposition of the Chilhowee Group, Cambrian marine sedimentary rocks.**

Great quantities of clastic (detrital) sediments were washed down from the highlands and deposited in the ocean. The Weverton Formation, which lies unconformably on Catoctin rocks and, in some places, on the Pedlar Formation, is made up of 100 to 500 feet of conglomerates, sandstones, and shales metamorphosed to quartzites and phyllites.

The earliest traces of life forms appear in the quartzites of the Hampton Formation. In these beds are tracks, trails, and *Skolithos* tubes, or burrows, left by a wormlike, marine organism. The Hampton Formation is composed of 1800 to 2200 feet of clastic rocks— thin-bedded sandstones and siltstones (now quartzites and phyllites). In the overlying Erwin Formation, *Skolithos* tubes are more abundant and varied, especially in the quartzites. The resistant, massive beds of Erwin quartzite form sharp peaks, flatirons, and ridges along the western flanks of the Blue Ridge. Most of the manganese and iron deposits in the area have been found in the topmost Erwin beds.

5. **Paleozoic orogenies and docking of tectonostratigraphic terranes.**

In a long series of tectonic episodes throughout Paleozoic time, the Piedmont terrane and a succession of other tectonostratigraphic terranes drifted in from the southeast. The continental margin was overridden by arriving crustal segments. Seaways were compressed and closed. Later, as rifting began again, a new ocean opened. The Taconic orogeny in Ordovician time was marked by the attachment of the composite Avalon-Carolina terrane. The Devonian Acadian orogeny had less effect on the Blue Ridge since its severity was greatest in the Northern Appalachians. Terranes from the east were accreted during the Alleghany orogeny in Permian time, and some were thrust over younger rocks along the continental margin. Folds that are oversteepened to the northwest, thrust faults pushed from southeast to northwest, and cleavage systems oriented to compressive stresses from the southeast were all produced by repeated convergence and collision of tectonic plates.

Table 54.1	Geologic Column, Shenandoah National Park

Time Units			Rock Units		
Era	**Periods**	**Epochs**	**Group**	**Formation**	**Geologic Events**
Cenozoic	Quaternary	Holocene	Alluvium		Stream erosion, mass wasting
		Pleistocene	Periglacial talus		Intensive frost action and stream erosion
	Tertiary	Pliocene Miocene Oligocene Eocene Paleocene			Stream erosion, evolution of modern drainage Isostatic uplift
Mesozoic	Cretaceous Jurassic Triassic		Diabase dikes		Uplift, erosion, sedimentation in basins Tensional rifting and block faulting, intrusion of diabase
Paleozoic	Permian Pennsylvanian				Alleghany orogeny, thrust-faulting, folding, metamorphism, uplift and erosion
	Mississippian Devonian Silurian				Acadian orogeny, uplift and erosion
	Ordovician				Arrival of terranes Taconic orogeny, uplift and erosion Plate rifting—opening of Iapetus Ocean, sedimentation
	Cambrian		Chilhowee Group	Erwin (Antietam)	Quartzites (with *Skolithos* tubes,) interbedded shales
				Hampton (Harpers)	Shales, siltstones, interbedded quartzites
Proterozoic Z (Late Precambrian)				Weverton	Conglomeratic quartzites, shales Rifting, erosion ——————
			Catoctin		Dark greenstones (originally basaltic lavas), dike intrusions, ash, and tuff
			Swift Run		Volcanoclastic deposits
Proterozoic Y			Old Rag Granite		Orogeny and erosion —————— Metamorphosed light-gray plutonic rock
			Pedlar		Metamorphosed greenish gray gneisses

Sources: Gathright, 1976; Reed, 1969.

6. Intrusion of Triassic-Jurassic dikes, erosion, sedimentation, and uplift.

Long, slow uplift during the Mesozoic Era was accompanied by tensional and strike-slip faulting as isostatic adjustments followed the mountain-building of the last Paleozoic orogeny. Active erosion removed great volumes of rock from the uplands and filled the basins. The diabase dikes that rose through fractures in the Blue Ridge bedrock are associated with block-

FIGURE 54.11   Bearfence Mountain at an elevation of 3620 feet is one of the few places in the park that a 360° view is possible. National Park Service photo.

faulting in the mountain belt. Only a few of the dikes are exposed in Shenandoah National Park, most of them in the central and southern sections. McCormick Gap, near the southern entrance, is eroded along the northwest-southeast trend of an extensive diabase dike. The dike rock is less resistant than the Catoctin greenstones that it intruded.

7. **Cenozoic upwarping and development of the modern landscape.**

The upwarping of the Appalachian region in Cenozoic time (which continues at a very slow rate) was due to isostatic compensation of the crust, a result of gradual widening of the Atlantic Ocean.

During Pleistocene time, erosion intensified because of increased frost action and precipitation. The

wide alluvial fans along the foothills, sloping westward to the Shenandoah Valley, were built up in the Pleistocene Epoch. The fans are now being eroded by streams.

Geologic Map and Cross Section

Gathright, T. M., II. 1976. *Geology of the Shenandoah National Park, Virginia* (geologic maps in pocket). Virginia Division of Mineral Resources Bulletin 86.

Sources

Badger, R. 1992. Stratigraphic Characterization and Correlation of Volcanic Flows Within the Catoctin Formation, Central Appalachians. *Southeastern Geology* 32(4):175–195.

Bobyarchich, A. R. 1981. The Eastern Piedmont Fault System and its Relationship to Alleghanian Tectonics in the Southern Appalachians, *Journal of Geology* 89(3): 335–345, May.

Hatcher, R.D. Jr. 1989. Tectonic Synthesis of the U.S. Appalachians; in Hatcher, R.D. Jr.; Thomas, W.A.; and Viele, G.W., eds. *The Appalachian-Ouachita Orogen in the United States*. Boulder, Colorado: Geological Society of America, The Geology of North America, v.F-2, p. 511–535.

Heatwole, H. T. 1981, 2nd edition. *Guide to Skyline Drive and Shenandoah National Park*. Shenandoah Natural History Association 226 p.

Reed, J.C., Jr. 1969. Ancient Lavas in Shenandoah National Park Near Luray, Virginia. U.S. Geological Survey Bulletin 1265.

Note: To convert English measurements to metric, go to www.helpwithdiy.com/metric_conversion_calculator.html

Great Smoky Mountains National Park

TENNESSEE-NORTH CAROLINA BORDER REGION

Area: 520,269 acres; 813 square miles

Established as a National Park:
June 15, 1934

Designated a Biosphere Reserve: 1976

Designated a World Heritage Site: 1983

Address: Great Smoky Mountains National Park, 107 Park Headquarters Road, Gatlinburg, TN 37738

Phone: 865-436-1200

E-mail: grsm_smokies_information@nps.gov

Website: www.nps.gov/grsm/index.htm

The Great Smoky Mountains, the loftiest range east of the Black Hills, are the result of uplift and prolonged erosion. The rounded, resistant summits of the range expose deformed and metamorphosed rocks, Precambrian in age.

Late in Paleozoic time, major thrust faults pushed older rocks on top of much younger rocks, mainly on the west side of the range.

Moist air from the Gulf of Mexico waters dense forests and produces a smoky haze, for which the range is named.

FIGURE 55.1 The rounded appearance of the Great Smoky Mountains is the result of chemical and physical weathering in a temperate, humid climate, and the added breakdown of the bedrock by the abundant vegetation.

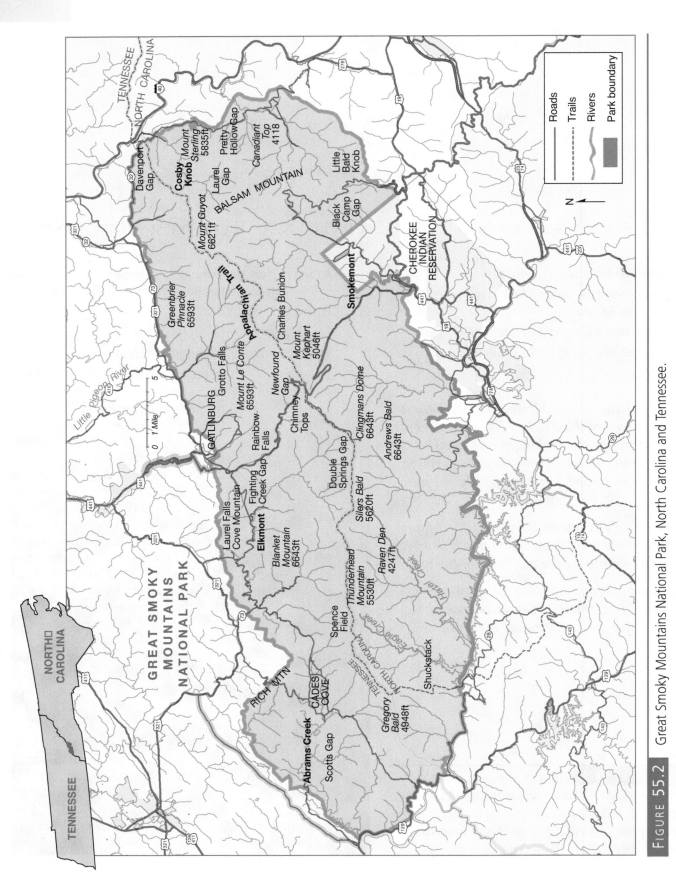

Geographic Setting

Great Smoky Mountains National Park, which straddles the boundary between Tennessee and North Carolina, is in the highest part of the Appalachian Mountains belt. Only Mount Mitchell (6684 feet in elevation) in the Black Mountains northeast of the park is higher than the major peaks of the Great Smokies. Clingmans Dome, highest point in Great Smoky Mountains National Park, is 6642 feet in elevation; the second highest peak, Mount Guyot, is 6621 feet. More than 20 other summits in the Great Smokies are over 6000 feet in height. The two highest mountains in the park were named for T. L. Clingman and Arnold Guyot, the geologists who made the first scientific reconnaissance of the range in the late 1860s.

The crest of the southern Blue Ridge in Great Smoky Mountains National Park is more than 2000 feet higher than the Blue Ridge crest in Shenandoah National Park (Chapter 54), about 400 miles to the north. In Shenandoah, the spine of the Blue Ridge is relatively narrow, with spurs jutting out on either side. In Tennessee, North Carolina, and Georgia—at the southern end of the Blue Ridge physiographic province—the range becomes a sprawling complex of ridges, lacking the distinctive lineation of the mountains to the north. What is locally called "the Blue Ridge" in North Carolina is a prominent frontal scarp that rises from the Piedmont along the eastern edge of the Blue Ridge province.

No rivers cross the Blue Ridge upland south of the Roanoke River in Virginia. The drainage divide follows the crest of the Blue Ridge province. Streams on the eastern slopes and the Piedmont drain to the Atlantic. Most of the water from the southern Blue Ridge flows west or northwest and then south to the Gulf of Mexico.

The southern highlands are humid because warm, wet winds, sweeping northeastward from the Gulf of Mexico, bring moisture to the mountain slopes. Clingmans Dome and the other high summits in the Great Smokies average more than 85 inches of precipitation annually, making these mountains nearly as wet as those of the Pacific Northwest. Elevation determines temperature as well as precipitation. Average summer temperatures on Clingmans Dome range from 51 ° F to 64° F. By contrast, at Park Headquarters in Gatlinburg, Tennessee, elevation 1500 feet, summer temperatures average 59° F to 87° F. Gatlinburg and the valleys in the southern Blue Ridge receive 50 or more inches of rainfall per year.

An abundance of moisture and a thick mantle of residual soil in valleys and foothills has supported for millions of years a luxuriant and diversified plant life, with many species attaining extraordinary size. Some tree and plant species that became established in the region during the colder, wetter climates of the Pleistocene have continued to thrive in the mountain environment.

In the warmer seasons of the year, droplets of mist in the air, combined with terpene vapors (resins and plant oils) rising from the lush vegetation, and wisps of fog are what cause the natural "smoke" that gave the Great Smokies their name. The gently swirling rise and fall of the haze under varying conditions of temperature and light produces an effect of bluish-gray smoke around the summits and in the valleys (fig. 55.3).

Because of its unique ecological resources, Great Smoky Mountains National Park is carefully monitored to determine the impact of air pollution, especially those effects due to acid rain. Trees at higher elevations are especially vulnerable, and a large number are dying or have died. Increased emissions of sulfur dioxide particles from coal- and oil-burning power plants are the main cause of acid rain. The particles are carried into the park region in droplets of water. Unfortunately, haze resulting from pollution has begun to outweigh and mask the natural haze that has characterized the Great Smokies atmosphere. Since the 1950s, visibility in the mountains has declined 30 percent.

Great Smoky Mountains National Park is visited by more than 10 million people annually, partly because a major highway (US 441) crosses the mountains through the park (fig. 55.2). Here, also, is the southern terminus of the scenic Blue Ridge Parkway (Chapter 54). Yet despite the fact that this park has more visitors than any other in the National Park system, about 75 percent of its 800 square miles is maintained as a wilderness. Although two-thirds of the total area had been logged over or burned before coming under national park protection, establishment of Great Smoky National Park saved the largest remaining virgin forest in eastern United States.

The Cherokee Indians, primarily an agricultural people, had developed a settled and advanced culture in the Southern Appalachians, long before the coming of Europeans. The Great Smoky Mountains were the sacred ancestral home of the Cherokee nation. In 1838, the Cherokee were forcibly deported from their ancient homeland and escorted by troops to Oklahoma Territory. Many Indians died during the march, called

FIGURE 55.3   The bluish-gray haze, or "smoke," that has given the Smoky Mountains their name rises and falls over the summits and valleys as weather conditions, hours of daylight, sunshine, and temperature change seasonally. Mist or fog in the air combines with rising vapors of resins and plant oils (terpenes), creating the effect of smoke. Photo by Ronald Hacker.

the "Trail of Tears." Some Cherokee escaped deportation by hiding in the mountains. Eventually the federal government permitted Cherokee who wished to do so to return. About 4500 Cherokee now live in the Qualla Indian Reservation in North Carolina on land adjoining Great Smoky Mountains National Park.

In the late 18th and early 19th centuries, emigrants from England, Scotland, and Ireland began to occupy Cherokee land (legally and illegally) in the foothills and valleys on both the Tennessee and North Carolina sides of the Great Smokies. After the fertile bottom lands in the valleys were taken, families sought farm sites higher on the slopes. A fortunate few found hidden "coves" and "balds." Cove is the Southern Appalachian term for an open, flat, or gently sloping, valley underlain by limestone bedrock and protected from wind by high ridges. (Their geologic origin is discussed later in the chapter.) The balds are treeless openings in the forest on ridges at higher altitudes. Some are grassy, but most are covered by a maze of closely intertwined shrubs, like rhododendron and mountain laurel. The balds may be long-weathered scars of old mudslides or

tracts that were burned over to provide grazing areas.

Cades Cove, first settled in 1818, is maintained as a historical and cultural preserve by the National Park Service (fig. 55.4). In this isolated valley, deep in the Great Smoky Mountains, the pioneers found fertile farm land, produced by weathering of the limestone that underlay the floor of Cades Cove. Today, exhibits, museums, a blacksmith shop, and a rebuilt grist mill tell the story of life as it was in the southern highlands for well over a century—self-sufficient, independent, and shut off from the outside world. The overshot water wheel that powers the grist mill is an example of the dozens of small water wheels that settlers built on mountain creeks for grinding grain. With high gradients and high velocities, the streams were a reliable and convenient energy source.

In the early 1900s, lumber companies acquired large tracts of land in the Great Smoky Mountains. Hardwood forests centuries old were harvested, and mountain streams were used as sluiceways for logs. Splash dams were constructed on creeks so that the water could be released in a flood that washed logs

FIGURE 55.4 View of Cades Cove, looking east-southeast. The trace of the Great Smoky fault is the foreground, between the viewer and the house, and follows around the cove floor along the edge. Across the valley, the Oconaluftee fault lies along the break in the slope between the spurs and the high mountains rising in the background. Rocks above the Oconaluftee fault belong to the Great Smoky Group. Between the Oconaluftee fault and the Great Smoky Group are units of the Snowbird Group. On the cove floor and beneath the Great Smoky fault are Ordovician limestones of the Knox Group. See also cross-section in fig. 55.7. Drawing by Philip B. King (p. 13) in King, P.B., Neuman, R.B., and Hadley, J.B., 1968.

downstream. The Sinks, a series of cascades on the Little River, carried logs by the thousands from the slopes of Clingmans Dome.

Logging, as practiced at that time, was a destructive form of land use. The skidding of logs down slope, as forest cover was removed, caused rapid erosion. Landslides occurred with every heavy rain. Streams were so clogged with sediment that they were unable to sustain aquatic life. Disastrous forest fires, spreading from heaps of dry branches, roared up the slopes. Charlies Bunion, a bare slate knob on the crest of the range, was completely denuded of soil and vegetation by a fire and a subsequent flash flood in the 1920s. Only hardy plants clinging to crevices in the jagged bedrock have grown back.

People in Tennessee and North Carolina who deplored the loss of the forests and the pollution of the streams began to organize a conservation movement for the purpose of establishing a national park in the Smokies. Other Americans, realizing that the mountains were a botanical treasure house, joined in the effort. After Congress authorized the park in 1926, the difficult task of raising money to buy up land from thousands of individual owners began. Citizen groups gathered contributions, and the Tennessee and North Carolina legis-

latures each appropriated two million dollars. John D. Rockefeller, Jr.'s donation of five million dollars, plus a million and a half from the federal government, finally brought the undertaking to successful completion.

Geologic Features

"What geology?" the ranger naturalist said with a chuckle to an enquiring tourist. "We don't have any geology in this park!" What he meant, of course, is that the bedrock geology in Great Smoky Mountains National Park is not obvious. Except in waterfalls, road cuts, and on barren summits, outcrops are scanty; and geologic structures are often obscured by the vegetation. These difficulties, field geologists agree, are "compounded by the enigmatic nature of the rocks themselves" (King, et al. 1968, p. 1).

Some geomorphic aspects of the mountain landscape can be related to geologic processes and climatic factors. Summits and ridges in the park are more rounded than sharp. The outward appearance of these magnificent and imposing mountains is the result of millions of years of weathering and erosion in a humid, predominantly warm to temperate climate.

The alpine glaciers and the great ice sheets that in Pleistocene time altered landscapes of the Northern Appalachians, did not extend to the Southern Appalachians; but the climatic fluctuations profoundly affected the environment of the region. When, year after year, winters were long and severe, plants and animals adapted or perished. Intense frost action attacked the mountain slopes and rocks. The bouldery talus slopes in the steep draws and gullies at higher altitudes probably developed during the Pleistocene Epoch. At lower altitudes, patches of deeply weathered soils may reflect the increased warmth and humidity of interglacial episodes.

The results of mass wasting processes are evident, especially on steep, high slopes where soil cover on bedrock is thin and fragile, or where regrowth, since lumbering ceased, has not become well established. In September 1951, when a thunderstorm dumped four inches of rain in one hour on Mount LeConte, more than 40 slides and avalanches, carrying rocks, trees, and debris, roared down the mountainside. Landslide scars are still visible on the slopes.

Throughout the park, large and small streams, fed by high rainfall, race down slopes through stone-filled channels and plunge over waterfalls. Scenic highways, such as Little River and Laurel Creek Roads, follow winding streams along valleys at lower elevations. Hiking trails, from easy to strenuous, lead to waterfalls that plunge over bedrock ledges (fig. 55.5).

Bedrock and Geologic Structures

The rocks and structures in Great Smoky Mountains National Park are related to those in Shenandoah National Park (Chapter 54). Both regions were pro-

FIGURE 55.5   Waterfalls tumble over resistant ledges of Precambrian Thunderhead Sandstone. *Left:* Laurel Falls, accessible by trail from Little River Road. National Park Service photo. *Right.* Tom Branch Falls is a typical waterfall in the park. The more resistant beds act as the lip of the falls and the less resistant beds are undercut. Photo by © Robert Fletcher.

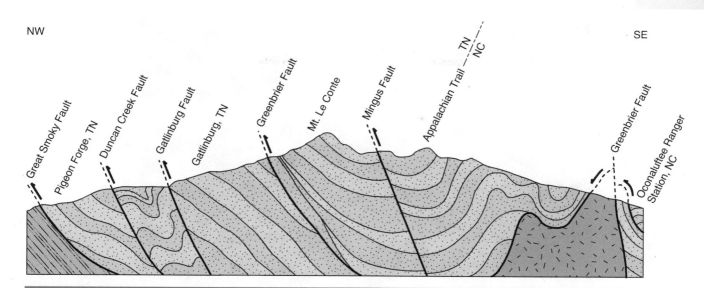

NW   SE

FIGURE 55.6 Northwest-southeast cross section through Great Smoky Mountains National Park. Rocks beneath the Great Smoky fault (*far left*) are Paleozoic. Rocks between Great Smoky fault and Greenbrier fault are Precambrian (Ocoee Supergroup). Rocks beneath the Greenbrier fault (*lower right*) are Precambrian basement complex. The region has undergone intense compression, which is shown by the relative movement along the faults (directions indicated by arrows). From H.L. Moore, *A Roadside Guide to the Geology of the Great Smoky Mountains National Park*. © 1988 The University of Tennessee Press. Used by permission.

foundly affected by the large-scale tectonic events that acted on the Appalachian Mountains belt (box 54.1). A succession of plate collisions and orogenies in Precambrian and Paleozoic times culminated in major thrust faulting near the end of the Paleozoic Era. A significant difference between the two areas is that deformation was more severe and the extent of thrust faulting was considerably greater in the southern section of the Blue Ridge province than in the northern part (fig. 55.6).

In broad terms, three groups of rocks make up the Great Smokies and surrounding foothills. In chronological and roughly geographical order, from southeast to northwest across the mountains, these groups are as follows:

1. The billion-year-old basement complex consists of Proterozoic rocks that have been deeply buried, have undergone repeated episodes of metamorphism, and have been uparched, uplifted, and exposed by erosion.
2. The Ocoee Supergroup is a thick sequence of metamorphosed sedimentary rocks of later Proterozoic age that form the main bulk of the Great Smokies.
3. The Chilhowee Group and other Paleozoic rock units crop out on lower ridges, in the coves, and in the Appalachian Valley, mostly outside the park.

The Basement Complex. In the southeastern foothills, on the North Carolina side of Great Smoky Mountains National Park, are exposures of ancient, deformed crystalline rocks. Deep Creek Campground and the Oconaluftee Visitor Center in the southeast corner of the park are underlain by Proterozoic basement rocks. Some of the layered gneisses and schists were probably formed from sand, silt, and volcanic sediments. The granitic gneisses were plutonic intrusions originally. The basement rocks may be part of the North American craton.

The Ocoee Supergroup. Great thicknesses of Ocoee rocks make up most of the mountain mass in the park and extend for many miles along the mountain trend, both northeast and southwest of the Great Smokies (fig. 55.7). These are the rocks of the great thrust sheets that were pushed northwestward over younger rocks by the Alleghany orogeny late in Paleozoic time. Ocoee rock units consist mainly of many layers of fine- to coarse-grained sandstones. All Ocoee rocks are nonfossiliferous, clastic, marine sedimentary rocks that have been metamorphosed in varying degrees. Most of the beds were probably deposited in a marine trench that existed for millions of years alongside the North American continental margin. Within the Ocoee Supergroup,

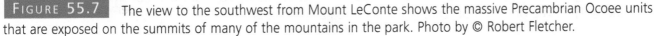

FIGURE 55.7   The view to the southwest from Mount LeConte shows the massive Precambrian Ocoee units that are exposed on the summits of many of the mountains in the park. Photo by © Robert Fletcher.

however, the component groups—the Snowbird Group (which is the oldest), the Great Smoky Group, and the Walden Creek Group—formed in different marine environments. Several unclassified Ocoee sandstone formations are probably transitional between the Snowbird Group and the Great Smoky Group.

The most informative exposures of Snowbird Group rocks, which lie unconformably over the basement complex, crop out outside the park; but Snowbird units make up a belt of rugged foothills along the north side of the Great Smoky Mountains. Snowbird rocks include beds of mostly dark gray argillites, siltstones, sandstones, quartzites, and other sedimentary units.

The formations of the Great Smoky Group, by far the most prominent, the thickest, and the most massive Ocoee units, are exposed on the summits in the park, in numerous roadcuts along the transmountain highway (US 441), and in waterfalls. The fine-grained Elkmont Sandstone is at the base of the Great Smoky Group. The coarse-grained Thunderhead Sandstone, thick-bedded and resistant, forms ledges, cliffs, and the high summits of Clingmans Dome (fig. 55.8) and Mount Guyot. Overlying the Thunderhead Sandstone on Mount LeConte, Chimneys, Sawteeth, and other parts of the crest, are the dark, slaty rocks of the Anakeesta Formation. Craggy pinnacles and steep-sided ridges, eroded from Anakeesta rocks, form prominent landmarks along the Appalachian Trail.

The massive rock units of the Great Smoky Group were initially remarkably homogeneous. Variations in their lithologies are due to the several metamorphic facies that reflect differing degrees of heat and pressure when the rocks were undergoing progressive regional metamorphism. The more intensely metamorphosed rocks (amphibolite facies) are in the southeastern part of the mountains, while rocks of the same formations to the northwest show decreasing grades of metamorphism (greenschist facies). The chemical and mineralogical variations in the rocks of the thrust sheet indicate that on the southeast side metamorphism was probably more prolonged and increasingly intense.

Formations of the Walden Creek Group crop out in the foothills on the northwestern and northern sides of the Great Smokies. Walden Creek rock units are made up mostly of shaly and silty clastics with discontinuous masses of conglomerate and sandstone and a few beds of quartzite, limestone, and dolomite. Walden Creek formations are for the most part intensely folded and

FIGURE 55.8　View from Clingman's Dome, which is capped by the coarse-grained Thunderhead Sandstone, a member of the Great Smoky Group that is Precambrian in age. National Park Service photo.

crumpled or broken into discontinuous lenses because of the stresses produced by the underlying, low-angle Great Smoky thrust fault.

The Paleozoic rocks. Most of the rocks of the Chilhowee Group are exposed outside the park; but exposures close to the park are in English Mountain and Chilhowee Mountain and in roadcuts along the Foothills Parkway that skirts the western and northern boundary of Great Smoky Mountains National Park. Primitive marine fossils (trilobites, brachiopods, etc.) have been found in the oldest rock units of the Chilhowee Group. The Chilhowee beds inside the park are the Lower Ordovician limestones of the Knox Group that underlie the coves, nestled among the western foothills (figs. 55.9, 55.10).

Igneous rocks. A geologic map of the Great Smokies region shows extensive sequences of the dominant sedimentary and metamorphic rocks, but only a very few minor bodies of igneous rocks. (Volcanic and plutonic components in the basement complex rocks have been too greatly altered to be identifiable bodies.) Narrow sills of metamorphosed diorite, Paleozoic in age, crop

out between beds of the Thunderhead Sandstone near the top of Clingmans Dome. Since the sills extend southwestward down the mountain, they were probably the source of ore in old copper mines along Hazel Creek in the southern part of the park.

Folds and Faults

Folds in the layered rocks of Great Smoky Mountains National Park range from broad and open, with long stretches of uniform dip, to small and sharp. In the southeastern area, where metamorphism was more intense, folds of several ages intersect, producing complex relationships. The majority of folds align with the dominant northeast-southwest regional trend.

Three major thrust fault systems dominate the mountain structures of the region: the Greenbrier faults, the Great Smoky faults, and the Gatlinburg faults. The surface traces of the three fault systems trend generally northeast-southwest. Strong compressive forces thrust Late Proterozoic Ocoee rocks as much as a hundred miles northwestward over younger rocks.

Oldest of the three are the faults of the Greenbrier system. Prominent exposures of the faults are in the

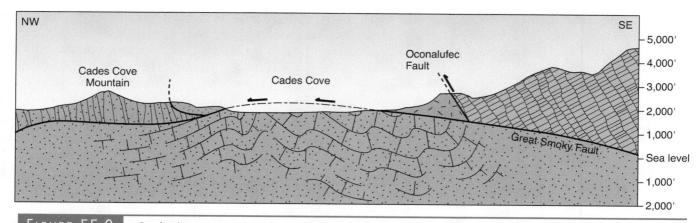

FIGURE 55.9   Geologic cross section through Cades Cove from northwest to southeast. The Ordovician limestones and dolomites of the Knox Group are below the Great Smoky fault; units of the late Precambrian Ocoee Supergroup are above. Erosion during late Cenozoic time exposed the younger rocks below the fault. (The cross section was drawn without vertical exaggeration.) Adapted from Neuman, R.B. and Nelson, W.H. 1965.

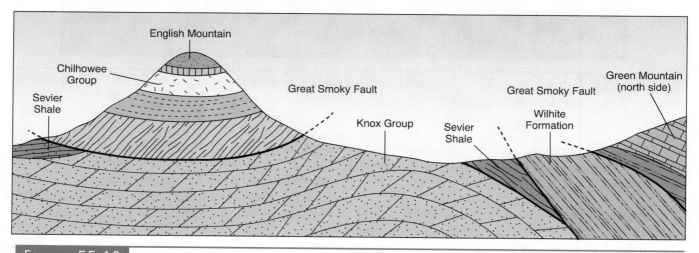

FIGURE 55.10   The structures shown in the diagram are in the northwest section of the park and along the Foothills Parkway. The Great Smoky fault, originally nearly horizontal, has been deformed. Rocks on top of the fault have moved from right to left. Complex folding and faulting produced a syncline under English Mountain and an anticline under the valley between English Mountain and Green Mountain. Weathering and erosion over time on rocks that differed in resistance resulted in a synclinal mountain—English Mountain—and the adjacent anticlinal valley. The Wilhite Formation under Green Mountain is part of the Precambrian Ocoee Supergroup. The Sevier Shale and the Knox Group are Paleozoic in age. Adapted from Moore, H.L., 1988.

southeast boundary area of the park between the Oconaluftee section and the Cherokee Indian Reservation. Greenbrier faults also loop around the northeast corner of the park and extend westward, where they are broken up and obscured by younger faults. The low-angle Greenbrier thrust faults carried rocks of the Great Smoky Group over the underlying rocks of the Snowbird Group, perhaps for a distance of

15 miles. Basement complex rocks in the southeast were also involved in Greenbrier faulting.

The Great Smoky fault system is only one segment of an extensive displacement involving Southern Appalachian structures from Virginia to Alabama. The Great Smoky fault system and associated faults, all low-angle thrusts, moved Ocoee rocks in the Great Smokies area for many miles over younger Paleozoic rocks. Most exposures of the Great Smoky fault system are between the foothills and the Appalachian Valley, west and north of the park, but some are within the park, as, for example, fault segments along the coves (figs. 55.4, 55.9). The main thrust sheet of the Great Smoky fault is split on the western side of the mountains into several fault blocks. The coves—the windows in the thrust sheet—are located in these breaks and openings. The distance from the leading edge of the thrust fault to where the fault surface is exposed in windows represents only a small part of the overall movement of the thrust sheet. Movement along the Great Smoky fault system occurred after the Ocoee rocks had undergone regional metamorphism.

The Gatlinburg faults, believed to be younger than the Great Smoky faults, dip more steeply and cut both Great Smoky and Greenbrier faults in some places. Although most of the Gatlinburg fault system trends east-northeast and lies outside the park's western boundary, the Oconaluftee fault, which is related to the Gatlinburg system, branches off toward the southeast from the vicinity of Meigs Mountain, crosses the crest of the Great Smokies, and goes down the valley of the Oconaluftee River. Gatlinburg faults have had a controlling effect on some streams that have eroded shattered rock along the fault traces. Some fault traces cross from one stream valley to the next through notches in intervening ridges.

How the Coves Developed

Cades Cove, in the western part of the park, is a window in the thrust sheet of Proterozoic Ocoee rocks where younger Paleozoic rocks of the Ordovician Knox Group are exposed (fig. 55.4). (*Window,* when used as a tectonic term, refers to an eroded area of a thrust sheet that displays the rocks beneath the thrust sheet.) Tuckaleechee and Wear Coves, just west of the park boundary, are windows of larger size. The coves, which are fertile, slightly rolling patches of lowland underlain by unmetamorphosed carbonate rocks, are located among the western foothills of the Great Smoky Mountains where the thrust sheet becomes thinner and broken up along its leading edge. Breaks in the thrust sheet became enlarged as runoff from plentiful rains was channeled into the openings while the mountains were being uplifted. Quantities of sediment built alluvial fans and spread over the bottoms. This eventually produced the rounded and roughly oval coves surrounded by high ridges. A few very small, remote coves

that were not cleared by settlers or lumbermen remain in their natural state. Because the limestones beneath the windows in the thrust sheet are less resistant to erosion than the metamorphosed Ocoee rocks, rich soils formed in the protected, well-watered coves. A unique community of plants, the "cove hardwoods" (yellow birches, basswoods, buckeyes, sugar maples), and a profusion of wild flowers flourish in the coves.

Geologic History

1. The basement complex.

The oldest isotopic dates derived from minerals in these rocks indicate that plutonic and metamorphic episodes occurred about 1.1 billion years ago, or Proterozoic Y time. A scattering of later dates suggests that radioactive clocks were reset by recrystallization that took place during Paleozoic orogenies and metamorphism.

2. Deposition of the Ocoee Supergroup.

For millions of years, a developing trench in an ancient ocean received the clastic sediments that make up Ocoee rocks. Fossils have not been found in the rocks, despite diligent searching. Paleogeographers (scientists who try to map the global positions of ancient lands and oceans) speculate that this Precambrian ocean and its adjacent lands lay south of the equator during that time. Changes in grain size and bedding are indications of changing environments of deposition over time; i.e., varying depths of water, ocean currents, types of sediment (pebbly, sandy, muddy), and the like. Approximately 50,000 feet of sediment accumulated in the Ocoee Supergroup. Of this thickness, the Snowbird Group made up about 13,000 feet; the Great Smoky Group, 25,000 (in some places); unclassified formations, 3000 to 4000; and the Walden Creek Group, 8000. The sedimentary processes by which Ocoee units accumulated were modified by tectonic events. Ultimately, tectonic pulses brought about deformation and metamorphism of the rocks.

3. Crustal rifting and deposition of Paleozoic rocks.

As rifts developed in continental and oceanic crust, plates drifted apart and the Iapetus Ocean formed. At its maximum extent, this forerunner of the Atlantic Ocean may have been 1000 miles long and several hundred

Table 55.1			Geologic Column, Great Smoky Mountains National Park		
Time Units					
Eras	**Periods**	**Epochs**	**Rock Units**		**Geologic Events**
Cenozoic	Quaternary	Holocene / Pleistocene	Alluvium / Talus slopes		Stream erosion, mass wasting / Increased frost action, weathering, stream erosion
	Tertiary				Erosion; establishment of modern drainage systems, sporadic uplift
Mesozoic					Rifting; initial opening of Atlantic Ocean, prolonged erosion, uplift, block faulting
Paleozoic	Permian				Alleghany orogeny
	Pennsylvanian				Thrust faulting / Regional metamorphism
	Mississippian		(Outside park)		Deposition / Ocean closing
	Devonian				Acadian orogeny
	Silurian				Erosion
	Ordovician	(Middle)	(Outside park)		Taconic orogeny. Folding and faulting, sedimentation
		(Early)	Knox Group		Uplift and erosion / Carbonates (exposed in coves)
	Cambrian (Early)		Chilhowee Group		Plate rifting; opening of seaways, thick sedimentation
Proterozoic Z	?		Ocoee Supergroup	Walden Creek Group Several formations	Uplift and erosion
					Deformation, metamorphism
				Great Smoky Group Snowbird Group	Thick sedimentary accumulations
Proterozoic Y			Basement complex		Uplift, erosion, orogeny / Highly metamorphosed and deformed igneous and sedimentary rocks

Source: King, Newman, and Hadley, 1968.

miles wide. Sediments accumulated in both deep and shallow parts of the ocean, forming thousands of feet of marine sedimentary rocks.

4. Paleozoic tectonic episodes.

In Middle Ordovician time, plate convergence during the Taconic orogeny involved a volcanic arc and oceanic crust that were thrust onto the continental mar-

gin in the Southern Appalachian region. This event caused the large Avalon-Carolina composite terrane to be attached to the continent, and the uplifted land began to shed clastic sediments.

In the Devonian Period, the Acadian orogeny, which joined the North American and African plates, was more intense in the Northern Appalachians than in the Southern Appalachians. Plutons, generated by tec-

FIGURE 55.11   Charlie's Bunion along the Appalachian Trail was named after one of the early guides who would stop here a while to "rest his bunions." Photo by © Robert Fletcher.

tonic activity, were emplaced in the foothills east of the Blue Ridge.

Late in Paleozoic time, the Alleghany orogeny destroyed what was left of the seaways and created the supercontinent Pangea. Thrust sheets were pushed farther to the northwest, shortening the crust of the mountain belt. The Great Smokies became the crest of a major mountain belt that stretched across Pangea (box 54.1).

5. **Mesozoic isostatic adjustments, erosion, and renewed rifting.**

Although tectonic activity in the Appalachian Mountains belt subsided, isostatic adjustments caused continued uplift of the Great Smoky Mountains. Erosion shaped the ridges and valleys and uncovered many of the older rocks. New rifts opened and the basin of the present Atlantic Ocean began to form. Africa and

Europe drifted away from North America. The climate of the Southern Appalachian region became more temperate as the North American plate moved westward and northward.

6. Continued erosion of the Great Smoky Mountains throughout Cenozoic time.

Sporadic uplift continued into Cenozoic time. Modern drainage patterns developed partly in the direction of regional slopes and partly by differential erosion of rock units. In the Pleistocene Epoch, periglacial processes affected rocks and slopes at higher elevations, although the mountains were not glaciated. Climatic oscillations caused increased erosion and deposition. Some of the unusual plant species found in the Great Smoky Mountains National Park vicinity may have had a much wider range throughout North America prior to the Ice Ages, which put great stress on both plant and animal life. In the Southern Appalachian region, fossil remains have been found of extinct giant mammals, such as mammoths, saber-toothed tigers, and the giant sloth. Their bones have yielded carbon-14 dates of 11,000 to 12,000 years before the present, which means that these creatures managed to survive episodes of cold, harsh weather, probably by sheltering in protected valleys and caves. Why the giant mammals became extinct after the Pleistocene Epoch ended is a scientific mystery.

Geologic Map and Cross Section

King, P.B., Neuman, R. B., and Hadley, J.B. 1968. *Geology of the Great Smoky Mountains National Park, Tennessee and North Carolina* (geologic map and cross section in pocket). U.S. Geological Survey Professional Paper 587. 23 p.

Sources

Hack, J.T. 1989. Geomorphology of the Appalachian Highlands, in Hatcher, R.D., Jr., Thomas, W.A., and Viele, G.W., eds. The Appalachian-Ouachita orogen in the United States. Boulder, Colorado: Geological Society of America, The Geology of North America, v. F-2, p. 459–470.

Hatcher, R.D., Jr., 1989. Tectonic Synthesis of the U.S. Appalachians, in Hatcher, R.D., Jr., Thomas. W.A., and Viele, G.W., eds. The Appalachian-Ouachita orogen in the United States. Boulder, Colorado: Geological Society of America, The Geology of North America, v. F-2, p. 511–535.

Moore, H.L. 1988. *A Roadside Guide to the Geology of the Great Smoky Mountains National Park.* Knoxville: University of Tennessee Press. 192 p.

Neuman, R.B., and Nelson, W.H. 1965. Geology of the Western Great Smoky Mountains, Tennessee. U.S. Geological Survey Professional Paper 349-D, 81 p.

Reed, J.C., Jr. 1987. Precambrian Geology of the U.S.A. *Episodes* 10(4):243–247.

Shelton, N. 1981. *Great Smoky Mountains.* Handbook 112, National Park Handbook Series. National Park Service. 128 p.

Note: To convert English measurements to metric, go to www.helpwithdiy.com/metric_conversion_calculator.html

CHAPTER 56

Black Canyon of the Gunnison National Park

COLORADO

Area: 20,766.14 acres of this 14,135.38 acres are federal and 6,630.76 acres are privately owned. 10,000 acres added on October 21, 1999, as wilderness.

Proclaimed a National Monument: March 2, 1933

Established as a National Park: October 21, 1999

Address: 102 Elk Creek, Gunnison, CO 81230

Phone: 970-641-2337 x205

E-mail: CORE_Info@nps.gov

Website: www.nps.gov/blca/index.htm

The Gunnison River in the Black Canyon is an example of a superimposed stream, with its canyon walls rising more than 2000 feet above the river. The Gunnison has a gradient that averages 95 feet per mile, and the canyon has been eroded into resistant Precambrian rocks, mainly schists and gneisses.

FIGURE 56.1   The view looking toward the east as seen from Tomichi Point suggests how the Black Canyon received its name. Photo by David B. Hacker.

"Several western canyons exceed the Black Canyon in overall size. Some are longer, some are deeper, some are narrower, and a few have walls as steep. But no other canyon in North America combines the depth, narrowness, sheerness, and somber countenance of the Black Canyon of the Gunnison." It is largely because of this unique combination of geologic features that the Black Canyon has been preserved in its wild state.

Geologist Wallace R. Hansen

Geographic Setting

The Black Canyon is located just west of the continental divide and occupies portions of two physiographic provinces, the Southern Rocky Mountains on the east and the Colorado Plateau on the west. Characteristics of both provinces are found within the park, which has a mixture of broad mesas, steep canyons, and complex geology.

In spring, though snow may be gone from the rim, the inner Black Canyon may still have snow and ice in May. Daytime summer temperatures reach as high as 80° F, but nights are cool. Afternoon thunderstorms are common from July to September. In late November or early December, snow generally falls, and drifts often close the North Rim Road in winter.

On the north rim, sedimentary rocks (Paleozoic and younger) overlie the Uncompahgran surface produced when the old highlands were eroded during Pennsylvanian time. The south rim of the canyon exposes the Precambrian bedrock that formed the roots of the Uncompahgre highlands. Here the inner gorge is deepest and most spectacular. A wedge-shaped section of sedimentary Mesozoic rocks is at the head of the canyon, where Tertiary Blue Mesa Tuff forms cliffs that rise to the west on the canyon rim.

The deepest point of the canyon is 2700 feet below the rim. At the Narrows the width is 1150 feet at the rim but only 40 feet at river level. The river is 1725 feet deep at this location. The Black Canyon portion of the Gunnison River is a 53-mile stretch, of which 14 miles are in the park. Because this 14-mile stretch of white water is rated as a Class V, only expert kayakers should attempt the canyon. The problem is extreme gradients, e.g., 360 feet per mile and very difficult portages.

The canyon was named after Captain J.W. Gunnison of the Corps of Topographical Engineers. He was seeking a railroad route through the Rocky Mountains when Paiute Indians killed him on the Sevier River on May 20, 1853. Captain Gunnison and his party bypassed the Black Canyon when searching for a river crossing.

Archaeological evidence suggests that prehistoric Indians lived here. At a much later time the Utes used this area and lived on the rim. Some of the members of the 1873–74 Hayden Expedition (of Yellowstone fame) were the first white men to have seen the canyon. Both the Hayden Expedition and surveying parties for the Denver & Rio Grande Railroad declared that the canyon was "inaccessible."

In 1883, the 53-mile long Black Canyon Gorge fascinated H.C. Wright, who was a part of a surveying team. He observed that the gorge fell 2700 feet into the Gunnison River.

There was no interest in the Gunnison River until the end of the nineteenth century when the Uncompahgre Valley settlers were considering tapping it as a source of water. The first attempt in 1900 to float through the canyon and survey it by the valley settlers was a failure. After a month the five men gave up. The following year two men, William Torrence, from the original railroad surveying team, and Abraham Lincoln Fellows, of the U.S. Geological Survey, learned from the mistakes of the previous trip. They made arrangements to be supplied at various points from the rim and used a rubber mattress for a raft. They were successful and in nine days rafted 33 miles keeping an engineering log. Their information showed that a diversion tunnel was feasible.

In 1905 construction on the tunnel began, but moved very slowly because of violent cascades, unstable rock formations, and weather. When finally finished, the tunnel was 5.8 miles long and could carry sufficient water for the community. President William Howard Taft dedicated the tunnel on September 23, 1909.

In the late 1920s Montrose citizens and surrounding communities and civic groups campaigned to have the area preserved as part of the National Park System. Herbert Hoover proclaimed it as a National Monument on March 2, 1933.

President Clinton signed the bill in October 21, 1999, that created the Black Canyon of the Gunnison National Park. Its 30,000 acres make it the third smallest national park. The National Park Service plans to use funds from Congress and a conservation easement to purchase 3000 acres of private land that abuts the southern boundary of the park to prevent development. As of this writing this has not been done.

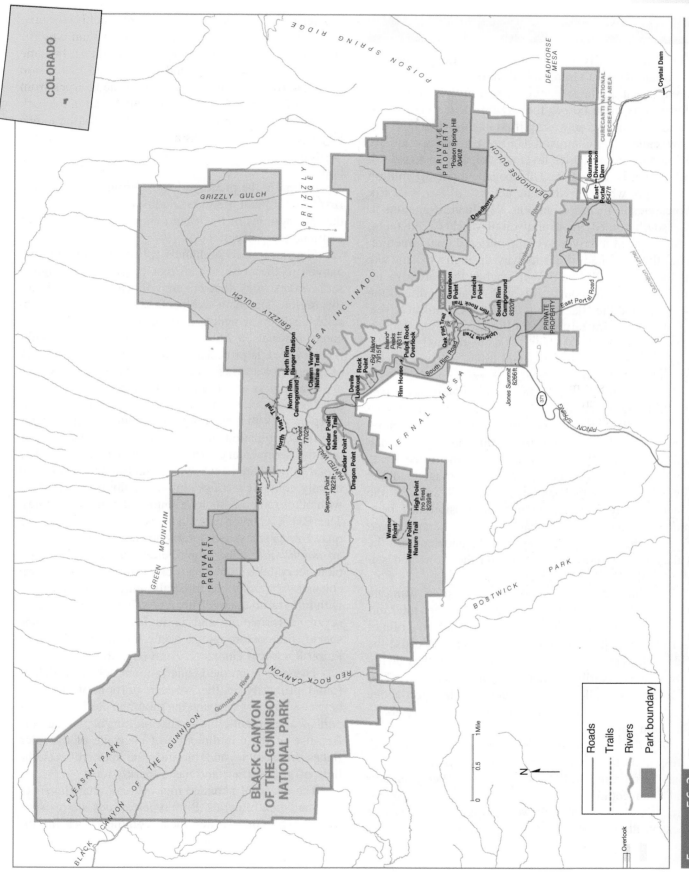

FIGURE 56.2 Black Canyon of the Gunnison National Park, Colorado.

Geologic Features

The Gunnison River flows northwestward, joining the Colorado River at Grand Junction. It is a *superimposed* stream; that is, the structure, which in this case was a dome, was there before the stream. About 60 million years ago this region was uplifted in a movement that is now called the Gunnison Uplift. Erosion eventually leveled the dome. Thick layers of volcanic rock and ash subsequently buried it. This began about 30 million years ago, when the volcanic activity of the West Elk Volcanoes to the north and the San Juan Mountains to the south forced the Gunnison River between them, which placed the river over the buried domal structure. First the river had to cut its way through volcanic ash and sandstone. By the time it reached the harder metamorphic rock the course was well established, and the river created the Black Canyon of the Gunnison River.

In approximately two million years, the canyon downcut over 1700 feet. The watershed for the Gunnison River is 4,000 square miles, which can collect a large amount of runoff. The gradient of the river averages 95 feet per mile, but near the Narrows reaches 480 feet per mile. The rate of erosion of the Gunnison River has been estimated at about an inch per century, or the equivalent to the thickness of a human hair per year. However, because of the three dams upstream that were built for water storage and the diversion tunnel that supplies irrigation water to the Uncompahgre Valley, the rate of erosion has been reduced. The sediment that would be scraping and eroding the sides of the canyon walls and scouring the valley floor is now trapped behind the dams and settling to the floor of the reservoir.

All the side drainages of the canyon are *hanging valleys*. The streams in these side valleys are intermittent. They have water in them only after a heavy rain or during the melting of snow in springtime. Because the side valleys do not have permanent streams, they do not down cut so rapidly as the main valley. Hence the floor of a tributary valley is a considerable elevation above the floor of the main valley and is left hanging above the main valley floor.

As the ancient metamorphic rock was uplifted, stress and strain caused cracks, or *joints,* to develop. They formed in parallel sets of cracks called *joint sets* and intersecting joint sets known as *joint systems.* Cracks expose bedrock to weathering. As the resistant rock weathers more slowly, standing out as ridges, the less resistant rock weathers out more quickly, forming depressions. Such weathering patterns explain why the sides of the canyon have a jagged appearance. In some instances, *monoliths* or islands are created by erosion along the rim, separating a portion of the bedrock from the rest of the rim. The types of joints found in the park are *shear joints,* low angle joints, and exfoliation joints. Shear joints are created as a *stress fracture* that is the result of shearing of one part of a rock past the adjacent part.

Aiding in the process of weathering are *lichen,* the first form of plant life to break down the bedrock. Later the root systems of the more complex plants and trees continue the process.

The flat-topped highlands called *mesas* across the canyon to the north are all that is left of the original sedimentary rock that covered Precambrian rocks of this region. The main sedimentary rocks are Mesozoic-aged sandstones, which, in turn, were covered by volcanic material 30 million years ago. The softer sedimentary rock is protected by the overlying harder rock of igneous lava flows. The harder rock does not weather so rapidly as the softer rock beneath it.

The bedrock of the inner canyon consists of metamorphic rock that accumulated about 1.7 billion years ago. A combination of heat and pressure changed the original sedimentary rock into *gneiss* and *schist*. The *schist* is a foliated rock with a scaly appearance, and usually contains a considerable amount of mica, along with other minerals such as amphiboles and pyroxenes. The Black Canyon area has two types of schist, a quartz-mica schist and hornblende schist also known as an *amphibolite.* The amphibolite (produced by complete recrystallization of a rock during metamorphism) consists of amphibole minerals, plagioclase feldspars with little or no quartz. This type of rock usually shows severe deformation and is highly contorted. Some sparse, thin outcroppings occur on the rim between Pulpit Rock and Gunnison Point (fig. 56.3).

The coarse-grained, quartz-mica schist, which is found primarily in the western portion of the park between Warner Point and the mouth of Red Canyon, is part of a large anticline in that section of the canyon. Because of the one direction of cleavage of the mica flakes (muscovite and biotite), the rock reflects the light off that surface and has a sparkly look.

The *gneiss* is a banded rock with alternate layers of light and dark minerals. Both types of rock are resistant to erosion. Gneiss, the most common rock in the canyon, is quite abundant at the overlooks and in the areas where trails exist. The fine-grained variety can be found at Tomichi Point near the Visitor Center, also at Pulpit Rock Overlook, the Narrows View, Island Peaks View and the Kneeling Camel View. A coarse-grained gneiss can be seen downstream of Warner Point. The layering of the fine-grained gneiss exhibits the relict structure of the original sedimentary rocks. Burial, compaction, and crystallization followed by shearing, granulation, recrystallization, and contortion produced the rock made up of quartz, feldspar, muscovite, and biotite. Accessory minerals are garnet, sillimanite, staurolite, tourmaline and very small amounts of zircon, apatite, and magnetite.

Quartzite (metamorphosed sandstone) has wide distribution in the park and is interbedded with either gneiss or schist. With an increase in feldspar content it grades into gneiss, and with an increase in mica content it grades into schist. The pure individual quartzite layers always contain some feldspar and mica. In spite of the amount of recrystallization the quartzite still shows the details of the original bedding planes. Exposures of this rock can be seen on the north rim about a half a mile east of the North Rim Ranger Station along a primitive road, and also on the slopes north of Gunnison Point on the south rim near the Visitor Center.

Metasediments are sedimentary rocks that have been metamorphosed, but it is still possible to determine what the original rocks were. Some of the primary structural features are preserved, such as the bedding planes. The original rocks were probably impure arkoses, graywackes and sandy shales. Some volcanic rocks may have been mixed in.

Cross-cutting the metamorphic rocks are various igneous rocks, such as granodiorites, quartz monzonites, pegmatites, lamprophyres, and diabases.

The resistance of these rocks to erosion is what has formed the steep parts of the canyon, producing the impressive walls that the observer sees from both rims. The beds are in various forms, such as intrusive sheets, dikes, plutons, and irregular masses. The Painted Wall is mostly gneiss, criss-crossed by pegmatite dikes, as in The Narrows. The lighter- colored dikes stand out in relief because they are more resistant to erosion than the gneiss and schist. The Painted Wall (fig. 56.4) is the highest cliff in Colorado. Nearby Chasm View is totally carved in igneous rock. By volume the various types of quartz monzonites and pegmatites dominate the rocks in the park. In lesser

FIGURE 56.3   Vertical dikes in Precambrian gneiss and schists, as seen from Gunnison Point. Photo by David B. Hacker.

amounts are diabase, aplite, and lamprophyre. All are Precambrian in age.

Granodiorites are similar in composition to granite, with the same amount of quartz but more plagioclase feldspars than potassium feldspars.

Quartz monzonites have been classified under the general heading of "granites" but have less quartz than granite and equal amounts of potassium feldspars and

FIGURE 56.4   The Painted Wall is the most famous view of this canyon. It is sometimes called the Wall with Dragons. Photo by David B. Hacker.

plagioclase feldspars. One of the quartz monzonites, the 1,420 million year old Curecanti Quartz Monzonite, cuts through the lamprophyres indicating that these dark rocks are older than 1,420 million years.

The crenullated, wavy, light-colored rocks that stand out in sharp relief against the darker-colored rocks are *pegmatites*. These pegmatites are actually very coarse granite. They usually form *dikes* (igneous intrusions that cut across the bedding planes of the bedrock that they have intruded). They cut into granites, gneisses, or schists that border the granitic magma chamber. These pegmatites were forced into the darker rock as a liquid from the magma chamber. In the Gunnsion River Canyon the pegmatites are made up of a pink variety of microcline (a potassium feldspar), small amounts of albite (a plagioclase feldspar), the white variety of mica (muscovite), minor amounts of biotite (black mica) and quartz in several forms; one of which is a massive milky vari-

ety, minor amounts of partial clear crystals, and graphic granite (intergrown quartz and albite). Some of the pegmatite dikes have accessory minerals of magnetite or hematite and occasional garnets. Some of the interlocking crystals are spectacular in size. Feldspar crystals can be found 3 to 6 feet across; milky quartz crystals are often several feet long. Books of the muscovite are abundant but are just a few inches in size. Bladed crystals as long as 6 feet of biotite are also present. The injection into the surrounding bedrock cooled the liquid and with the lowered temperature, crystals began to form. The crystals are large because they grew slowly. The longer it takes for the crystals to grow in a *melt* (a solution of ions at a high temperature), the larger the crystals. The pegmatites that consist primarily of quartz, feldspars, and mica are called *simple pegmatites*. In some instances there are ions that for one reason or another do not combine with other materials until the very end of the crystallization process. In this region the minerals are biotite, tourmaline, garnet, magnetite, and occasionally beryl. These are *complex pegmatites* and have the rarer minerals incorporated with the granitic minerals.

The pegmatites are resistant to erosion and stand out as ridges when they occur in the less resistant gneiss and schist as the result of differential erosion. When contrasted with the other igneous rocks, the pegmatites show less difference in the amount of erosion but form striking patterns in color because of the light-colored mineral content. The Painted Wall (fig. 56.4) received its name because of the abundance of the pegmatite intrusions and the colorful patterns they formed.

The pegmatites are exposed in several forms, such as dikes, sills, and irregular-shaped bodies. Gunnison Point is made up of pegmatite, so the rock can be closely examined there. Good views of the pegmatites can be seen from Cedar Point, Chasm View, and Pulpit Rock.

The pegmatites have different intrusive habits when found in different rocks in the park. Dikes cut the sheared or bedded schist in the eastern part of the park, and form vertical swarms across from Pulpit Rock. The dikes in the Vernal Mesa Quartz Monzonite are randomly oriented, dip at low angles to the northeast, and trend northwestward. The differing fracture patterns of the host rocks probably determine the orientation.

Aplites are igneous dikes of quartz and alkali feldspars with a fine-grained sugary texture. In the Black Canyon they are relatively scarce and are gray in color. They also have small amounts of mica and occasionally very small garnets. The aplite dikes are only a few tens of feet long and form straight-walled dikes. Most are only a few inches wide but occasionally they are a few feet wide. Their fine sugary texture distinguishes them from the pegmatites.

Lamprophyres are dark-colored intrusive igneous rocks with a high percentage of biotite, hornblende, and pyroxenes in addition to feldspars. The dark minerals in the dikes are found both as phenocrysts and in the groundmass in contrast to the light-colored pegmatites.

FIGURE 56.5   The granitic dikes show the feldspar crystals that are weathering out. Photo by David B. Hacker.

They can be found in the upper part of Black Canyon and have an unusual geologic occurrence. They are hundreds of feet long but not very wide, ranging from a few inches to a foot or two. They occur at Pioneer Lookout Point. The mineral composition at this location is fine-grained biotite porphyry, dark greenish-gray in color. The matrix is a mixture of microcline, oligoclase, and hornblende. The accessory minerals are quartz, biotite, apatite, zircon, and sphene, all microscopic in size.

A *diabase* is a variety of gabbro, which is the coarse-grained equivalent of basalt. It is also known as dolerite or "trap rock." Its composition is primarily labradorite (a calcic plagioclase feldspar), augite, some olivine and magnetite ("lodestone"). Accessory minerals are orthoclase, quartz, hornblende, biotite, chlorite, iron minerals, and apatite. Lath-shaped plagioclase feldspar crystals are partly or completely included in pyroxene crystals (usually augite). In the Black Canyon they form dikes that are quite long. The longest continuous exposed dike is nearly 8 miles in length and up to 300 feet wide. It is just upstream from Tomichi Point. The full length is unknown because it has been truncated by faults. Other large diabase dikes are scattered throughout the park. Not all of the dikes are long; some are less than a half-mile in length and only a few tens of feet wide. The diabase weathers rapidly because of chemical composition. It is poorly exposed and deeply weathered.

Migmatites, also known as injection gneiss or ribbon gneiss, are *hybrid* rocks that are about half-igneous and half-metamorphic. They formed when any available openings, such as bedding or foliation planes, joints, cracks or fractures were injected with a hot fluid from igneous intrusions. The hot liquids permeate all available spaces, generally with a mixture of quartz, feldspar, and mica-rich material at great depths.

Geologic Structure

The Gunnison Uplift is an upraised, composite, tilted fault block. The limits of the fault block were created during the Precambrian time when the original faulting occurred. Movement during the Uncompahgre and Laramide orogenies readjusted the faults and produced local folds along the faults. The block is flexed into a monocline along the crest of the Gunnison Uplift. As a result, there are large major folds with many minor *isoclinal* folds within. (Isoclinal folds are folds in which both limbs point in the same direction.) The Precambrian rocks of the Black Canyon of the Gunnison River have been folded and faulted repeatedly. Small-scale folds tend to be isoclinal while the large-scale folds have limbs from a few miles to several miles across. The limbs are frequently steep or overturned, especially in the anticlines. High Point, at 8400 feet at the West End of the South Rim Drive, is near the structural crest of the Gunnison Uplift. The topographic high allows visitors to see the geology and structure of the region.

The Dakota Sandstone caps part of the uplift; but along the rim, erosion has removed the sandstone down to the Precambrian bedrock. To the southeast volcanic rocks overlap the Gunnison Uplift. The southern slopes of Vernal and Poverty Mesas (southern side of the rim), coincide with the lines of faulting in the region.

Monoclinal folds have a single flexure, with the beds above and below the fold being parallel. They are formed when previous faults buried by horizontal sediments move again and warp the overlying sediments. The point where the Dakota Sandstone overlies the Ute Indian fault zone in the lower section of the Black Canyon is a good example.

Several fault zones make up the Gunnison Uplift, mainly the Red Rocks Fault, Cimarron Fault (high angle reverse fault), and Ute Indian Fault Zone. The Red Rocks and Cimarron Faults trend in a northwest-southeast direction.

Geologic History

1. **Deposition of impure arkosic sandstone, some conglomerates, graywackes, and sandy shales along with lava flows and volcanic ash during Precambrian time.**

More than 1.7 billion years ago a thick sequence of sediments was deposited on the Precambrian basement rock. Possibly some igneous rocks erupted at the same time. The sedimentary rocks were arkosic sandstones, some conglomerates, graywackes, and sandy shales. The igneous rocks may have varied in composition from granite to volcanic basalt, including lava flows and volcanic ash.

The sediments give a clue as to the environments in which they were deposited. An arkose is a type of sandstone that has 25 percent or more feldspars. The arkosic sandstone indicates that rapid erosion of the highlands occurred with rapid deposition of sediments.

The graywacke is a variety of sandstone that is deposited near volcanic activity as volcanic ash is

Table 56.1				Geologic Column, Black Canyon of the Gunnison National Park
Time Units		**Rock Units**		
Era	**Period**	**Formation**	**Member**	**Geologic Events**
Cenozoic	Quaternary	Modern Alluvium		Alluvium, talus, landslides, formation of canyon about 2 mya by erosion to Precambrian core.
Cenozoic	Tertiary	Alboroto Group		Eruption of San Juan Mountains to the south forming welded tuffs 30 mya. Shifted river over buried Gunnison Uplift. ~~~ UNCONFORMITY ~~~
Cenozoic	Tertiary	West Elk Breccia		Disruption of drainage by lava flows. Mixed lava and gravel layers during the Oligocene Epoch about 32 mya. Pyroclastic flows, lava flows. Formation of a syncline. Unconformity caused by Laramide orogeny 75 mya. ~~~ UNCONFORMITY ~~~
Mesozoic	Cretaceous	Mancos Shale		Last of the extensive shallow seas retreat and erosion for a long time. Gunnison Uplift 75 mya. Muds eroded from mountains to the west and deposited in the sea about 80 mya in a subsiding basin.
Mesozoic	Cretaceous	Dakota Sandstone		Interfingered with overlying Mancos Shale because of fluctuating shorelines. Blanket sand deposited on coastal plains as beaches and in lagoons about 105 mya. ~~~ UNCONFORMITY ~~~
Mesozoic	Cretaceous	Burro Canyon Formation		Erosion of beds. Stream deposits on a broad alluvial flood plain. Some layers contain dinosaur bones and petrified wood.
Mesozoic	Jurassic	Morrison Formation	Brushy Basin Member	Stream deposition of shales mixed with ash from nearby volcanoes. Formation of a basal conglomerate with fragments of chert, shale, and dinosaur bones.
Mesozoic	Jurassic	Morrison Formation	Salt Wash Member	Stream deposition on a coastal plain of white cross-bedded sandstones that are interbedded with fossiliferous shales.
Mesozoic	Jurassic	Wanakah Formation	Junction Creek Sandstone Member	Deposition as sand dunes on dry floor of the former Wanakah Lake. Beginning of Laramide orogeny about 75 mya.
Mesozoic	Jurassic	Wanakah Formation	Middle Member	Deposition in a salty lake of shales, mudstones, sandstone and cherty limestone. Thinning out to the east against the eroded Uncompahgre Highlands.
Mesozoic	Jurassic	Wanakah Formation	Pony Express Limestone Member	Deposition of alternating layers of limestone and gypsum in shallow protected waters. Sediments thinned to the east.
Mesozoic	Jurassic	Entrada Sandstone Formation		Transgression of the sea as beach and intertidal deposition on margin of the seaway, mostly as dunes or desert along the seashore. Basal conglomerate with pebbles of weathered gneiss, quartz, and pegmatites. Extensive erosion wore the Uncompahgre Highlands down to sea level. ~~~ UNCONFORMITY ~~~
Paleozoic	Ordovician Cambrian	Diabase		Erosion of early Paleozoic sediments with the uplift of the Ancestral Rockies (Uncompahgre Uplift). Series of diabase dikes intruded the metamorphic rocks about 500 mya.
Precambrian	Proterozoic Y	Lamprophyres and Pegmatites		Plutons intruded more than 1420 mya. Erosion removed mountains.
Precambrian	Proterozoic Y	Curecanti Quartz Monzonite		Deformation and low-angle faulting during renewed uplift. Intrusion of finer-grained, light-colored plutons along fractures and then spreading laterally 1420 +/– 15 mya.
Precambrian	Proterozoic Y	Vernal Mesa Quartz Monzonite		Intrusion of a series of coarse-grained, dark gray bodies about 1480 +/– 40 mya with pegmatites.
Precambrian	Proterozoic Y	Lamprophyres		Intrusion between 1700 and 1430 mya in two units, one deformed and the other undeformed.
Precambrian	Proterozoic X	Pitts Meadow Granodiorite		Intrusion of granodiorite, pegmatites, and aplite 1730 +/– 190 mya. Regional deformation continued.
Precambrian	Proterozoic X			Mountain building and folding, severe metamorphism, intrusion of pegmatites.
Precambrian	Proterozoic X			Deposition of sedimentary rocks accompanied by lava flows and volcanic ash beds on basement complex.

mixed with the sand grains. The volcanic ash weathers into a mixture of clay, mica, and other minerals and acts as the cementing material of the sand grains.

The sandy shale was deposited in water of moderate depth, close enough to shore that the clay minerals were being mixed with the sand grains.

2. **Mountain building, folding, and the intrusion of the Pitts Meadow Granodiorite (1730 ± 190 mya ago) and metamorphism of sediments. Intrusion of pegmatites, and continued deformation.**

These sediments and igneous material underwent metamorphism and severe folding. The rocks were metamorphosed to the amphibolite facies and eventually to the stage that the rocks are in today. This occurred about the same time as the mountain building and intrusion of the Pitts Meadow Granodiorite around 1730 ± 190 mya. The metasediments reflect the metamorphic changes by injection, metasomatism (high temperature and pressure without melting) and migmatization (mixture of igneous and metamorphic rock). Metamorphic minerals such as staurolite and sillimanite indicate a high degree of metamorphism. However, some primary structural features of sedimentary rocks were preserved, such as bedding planes and possible crossbedding.

The Pitts Meadow granodiorite has been metamorphosed, is gneissic, and well foliated. Alignment of the minerals (lineation) is concordant (parallel to alignment of adjacent rocks). Some of the inclusions in the rock contain sillimanite, indicating a high degree of metamorphism. Some of the tabular bodies appear to have originally been xenoliths (original bedrock incorporated in the batholith) that were smeared out by the flowage of the rock during metamorphism. The development of the wallrock was very complex, suggesting both metasomatism (replacement of original minerals with new minerals with no change in volume) and magmatic intrusion. Much of this took place at the deepest depths of metamorphism of the rocks in place. At the same time, high-grade regional metamorphism was affecting the enclosing rocks. Associated with this process were intrusions of pegmatite and aplite dikes. Many of these pegmatites are quite large. They also vary considerably in age.

The Pitts Meadow Granodiorite is a massive formation and can be seen along the lower Black Canyon and its tributaries. Between 1.7 and 1.43 billion years ago two types of lamprophyre dikes were intruded. One is

not deformed and is straight-sided, and is found in the headward part of the canyon. The other type has been deformed plastically and is weakly to strongly foliated.

These intrusions are porphyritic (have two distinct sizes of crystals) and have chilled borders that suggest the conditions of crystallization were intermediate between intrusive and extrusive conditions.

3. **Intrusion of Vernal Mesa Quartz Monzonite about 1,480 ± 40 mya and intrusion of pegmatites.**

The intrusion of the Vernal Mesa Quartz Monzonite was affected by the folding of the bedrock. The irregular shaped intrusion has been modified by the process of faulting and partially buried by sedimentary rocks. It is sharply discordant locally, which means the contact between the intrusion and the country rock is not parallel to the bedding planes or foliation. Within the national park, the Vernal Mesa Quartz Monzonite is found in the form of several intrusive bodies. The largest body is about 3 square miles in size and forms the walls of the canyon from Rock Point downstream to High Point. A smaller pluton can be seen at the head of Red Rock Canyon.

A quartz monzonite looks similar to granite but has different mineralogy and chemical composition. Granite is composed mostly of quartz, feldspars (microcline or orthoclase), and micas (muscovite and biotite); it contains more of the alkali metals (sodium and potassium) and more quartz. In contrast, the Vernal Mesa Monzonite is a very coarse-grained porphryitic rock containing more calcium and less quartz. The phenocrysts are microcline (a potassium feldspar) ranging from one to two inches long, and the ground mass consists of microcline, quartz, oligoclase (plagioclase feldspar), and biotite. There are also opaque iron minerals, epidote (a metamorphic mineral), sphene, hornblende (an amphibole), apatite, and calcite. The fresh rock is light brownish-gray, and the weathered rock, because of oxidation, has a darker brown color. Some areas are coated with a dark brown to black coating of desert varnish. The orientation of the microcline phenocrysts and the inclusions of the wallrock suggests flowage of the magma in the last stages of the cooling of the pluton. The flowage caused the phenocrysts to rotate and rub against one another so most of them have abraded edges, fractures, or granulated rinds or borders. The parallel orientation of these bodies produced the foliation. The intrusion of the pluton even tore out sections of the country rock consisting of gneiss and schist and incorporated them as xenoliths (means "strange rock" because it is different than the surrounding rock) in the quartz monzonite. The cooling was at a shallower depth than that of the Pitts Meadow intrusion. Many pegmatites intrude the Vernal Mesa Quartz Monzonite.

One of the best places to view the Vernal Mesa Quartz Monzonite is at Chasm View overlook, where there is a good exposure. In a few places in the park, near the Vernal Mesa Quartz Monzonite intrusions, small elongate plutons of a light-colored quartz monzonite differ in composition and appearance. These intrusions are elongate parallel to the foliation or bedding of the surrounding rock. The intrusions themselves have some foliation, which is also parallel to the enclosing rock. These bodies appear to be genetically related to the Vernal Mesa Quartz Monzonite.

4. Intrusion of the Curecanti Quartz Monzonite about 1,420 ± 15 mya and the intrusion of most of the pegmatites.

Many intrusions of pegmatites were injected either during the formation of the Curecanti Quartz Monzonite pluton or shortly afterward. The major pluton of the Curecanti was a large wedge or lens, 3 1/2 miles across from west to east and 2 miles across from north to south. Its roof and floor are exposed in the canyon walls in the upper part of the Black Canyon. There are also scattered minor lens-shaped or irregular thick short dikes or pipes.

A spectacular outcrop of the Curecanti Quartz Monzonite can be seen at the mouth of Curecanti Creek, six miles downstream from the head of Black Canyon. It is a spirelike monolith nearly 800 feet high called Curecanti Needle. The walls of the canyon at this location are almost 1000 feet high.

The plutons of the Curecanti Quartz Monzonite are discordant. The slight foliation in the small plutons indicates that they formed in the last stages of tectonic activity. The large pluton lacks foliation and appears to have formed when there was no tectonic activity. All the plutons lack a chill zone, but they formed in a shallow enough depth that they lifted some of the roof above them. The shape of the intrusion suggests that the monzonite was injected into a set of low-angle fractures in the country rock, and spread laterally. The composition of the Curecanti Quartz Monzonite is quartz, oligoclase, plagioclase, microcline, biotite and muscovite. The large pluton has clear pink garnets. The quartz monzonite is medium-grained and ranges in color from light gray to orange pink. Specimens from the main pluton and smaller ones are identical in appearance. Additional intrusion of lamprophyres and pegmatites occurred with the uplifting of the batholith roof after the Curecanti intrusion. Extensive erosion removed the mountains.

5. During late Cambrian or early Ordovician time, the intrusion of diabase dikes.

During late Cambrian or early Ordovician time diabase dikes were injected into a fracture zone that was trending east-southeast. After the injection, movements along the fracture zones displaced some of them. In the canyon, the diabase is the youngest intrusive igneous rock. It was possible to date the rock radiometrically at 510 ± 60 mya because of the presence of the potassium feldspars. In some of the larger dikes the core of the dike is gabbro, indicating a slow rate of cooling.

6. Sediment in early Paleozoic seas, later removed by erosion.

Early in the Paleozoic Era the landmass that became the North American continent was located on the equator and was covered by shallow seas that deposited a series of sedimentary rocks prior to the Pennsylvanian Period. Sediments accumulated intermittently in the area of the Black Canyon. With the uplifting of the Ancestral Rockies, these sediments were removed by erosion. However, by correlating and reconstructing the paleogeography of the adjacent areas, we can infer the missing sediments in the Black Canyon area. The area to the west of this region, the ancestral Uncompahgre Uplift, affected the Black Canyon of the Gunnison River. The roots of this ancient uplift can be found in the lower section of the Black Canyon, truncated by the Uncompahgran unconformity at the top of Precambrian beds that have been scoured by erosion.

7. Erosion during the Permian and Triassic time, subsidence and encroachment of the seas on the flanks of the ancestral highlands. Deposition beginning in middle Jurassic time of the Entrada Sandstone, Wanakah Formation, possibly the Junction Creek Sandstone Member, and Morrison Formations.

After a long period of erosion, the Uncompahgre Highlands were reduced to a low, featureless westward sloping plain. The terrestrial sediments were deposited on river flood plains, in brackish lakes, and by the wind. Sand dunes that were originally along the shores of the interior sea migrated inland. The Entrada Sandstone was deposited about 170 mya directly on the Precambrian surface. The Entrada Sandstone is exposed in the lower Black Canyon, thins to the east, and finally wedges out within the borders of the park. It can be seen as a thin light-colored band at the northeast rim of the Back Canyon (fig. 56.6). Most of it has been removed by erosion on the southwest rim. At Red Rock Canyon at the western end of the park, the formation is a massive cliff-forming fine-grained red sandstone 85 feet thick. At this location the lower 68 feet are a reddish orange color and the upper 17 feet are a pale yellowish orange color. A basal conglomerate at its base was originally the gravel on the Precambrian bedrock. As the dunes migrated onto the land, the gravels, made up of the Precambrian gneisses, schists, pegmatites, and vein quartz, were buried by the dunes. Part of the Entrada was deposited in a marine environment, possibly along a desert shoreline. The thin bedding planes and oscillation ripple marks indicate shallow water.

The Wanakah Formation consists of a variety of rocks—gypsum, gray cherty limestone, shale of various colors, mudstone, and friable sandstone. Each rock type reflects the environment in which it was deposited, a brackish lake or a landlocked seaway. Like the other formations, it thins to the east, ranging in thickness from 150 feet to nothing, where it rests upon the old erosional Precambrian surface. The Pony Express Limestone Member is found in the western part of the Black Canyon and along the northeast rim.

Outside of the park boundaries, the Junction Creek Sandstone Member crops out, frequently as discontinuous deposits. A crossbedded fine-grained sandstone, pale gray to pink, it was once a series of sand dunes on the dried out floor of the former Wanakah Sea.

8. **During the Jurassic Period deposition of the Morrison Formation and its members.**

Salt Wash and Brushy Basin Members of the famous Morrison Formation (about 155 mya) overlie

FIGURE 56.6   An unconformity with a sedimentary unit, the Entrada Sandstone, above the Precambrian basement complex. Photo by David B. Hacker.

the Wanakah. The Morrison Formation is known for its dinosaur fossils. Unfortunately, it is poorly exposed within the park boundaries, but can be seen in other areas nearby. The Morrison is about 600 feet thick in the Black Canyon area and from a distance looks red, but this is actually surface stain. The formation consists of light gray sandstone, brightly colored shales, and mudstones. The lower Salt Wash Member is a mixture of green, gray, and red shales and mudstones, interbedded with crossbedded sandstones. A few thin freshwater limestones beds are near the base of the Saltwash Member. The thickness ranges from 180 to 250 feet. Differential weathering has caused the sandstone to form ledges, with the shales and mudstones making up slopes. In areas outside of the park the Salt Walsh Member contains uranium. The environment in which the beds of the Morrison Formation formed was a broad alluvial plain with meandering streams. The conglomerates and sandstones accumulated in the channels of the streams; the muds and shales were deposited on the adjacent flood plains. The volcanic ash came from nearby active volcanoes.

9. **During the Cretaceous Period the Burro Canyon Formation was deposited, followed by severe erosion. Deposition in an encroaching sea of the Dakota Sandstone and overlying Mancos Shale on a fluctuating shoreline around 80 mya.**

About 110 mya the fluvial Burro Canyon Formation was deposited. It is about 350 feet thick and is primarily shale or mudstone (a thick-bedded shale), light gray in color. Much of the shale contains bentonite clay derived from the weathering of volcanic ash. The formation also contains gray sandstone, red shale, and a discontinuous conglomerate. The conglomerate in some sections has small pebbles or red and green chert in a matrix of sandstone. It was deposited as a stream and flood plain deposit on a broad alluvial plain. In some areas it contains dinosaur bones and petrified wood. The formation is not easily seen as it forms rubbly slopes covered by brush. Ledges are found only locally.

The Dakota Sandstone is a widespread formation that is found throughout the interior of western United States as blanket sand. This means it is not very thick but covers a large area. It was deposited in a transgressing inland sea on coastal plains, as beaches and in lagoons about 105 mya. In the Black Canyon region, the Dakota Sandstone can be found as rimrock, dip

slopes, and hogbacks. Outside of the park boundary it caps many mesas, and is crossbedded in layers several feet to a few inches thick. In some places coal accumulated in swamps. The depth of the water was shallow, as can be seen by ripple marks, fossilized plant fragments, and worm burrows.

The interfingering of the upper portion of the Dakota Sandstone and lower portion of the overlying Mancos Shale indicates a fluctuating shoreline. The Mancos Shale muds were being eroded from the mountains to the west and deposited on the sea floor about 80 mya. Because it is soft, it is easily eroded and is found in the bottoms of valleys and topographic basins, forming a tan-gray badlands topography that lacks vegetation. The badlands topography is formed because a majority of the shale swells when it gets wet and shrinks when it dries out. This characteristic makes the Mancos susceptible to landsliding. As in other western areas, the Mancos Shale contains calcareous concretions in which cephalopods and bivalves can be found. Gypsum crystals in the Mancos Shale produce the glitter seen on the rock grains.

The sea bottom was slowly subsiding as the Mancos accumulated so the water was basically the same depth. This is similar to what is happening in the Gulf of Mexico today.

10. **The Laramide orogeny about 75 million years ago produced the Gunnison Uplift near the close of the Cretaceous Period.**

Renewed movement along the old Precambrian faults raised the Gunnison Uplift above the rest of the area as monoclinal folds. The Gunnison Uplift was part of the Uncompahgre uplift that was the beginning of the formation of the Rocky Mountains during the Laramide orogeny. The bedrock was bowed as the tilting of the old fault blocks uplifted it and erosion began beveling the crest of the uplift. Once again the Precambrian bedrock was exposed. Some of the bedrock was tilted steeply or overturned.

The uplift caused the inland seas to retreat. Newly created streams transported sediment from the highlands and carried it to various basins. The westward-flowing streams from the newly created Sawatch Range were eventually able to flow unhindered across the Gunnison Uplift because of the removal of the sediments down to the Precambrian core. Only a few isolated hills remained, and they were eventually buried by Tertiary volcanic eruptions.

11. **The formation of early Tertiary lakes such as the Green River Lake outside the park boundaries. A second period of erosion.**

Northwest of the park a huge lake formed in the Uinta-Piceance (pronounced *Pee'-onts*) Basin of the northern Utah-Colorado area. In this lake sediments from distant volcanoes were deposited along with material from nearby highlands that were being stripped by erosion during the middle Eocene Epoch around 50 mya. These deposits are the Green River beds, a mixture of freshwater limestones, evaporites, shales, and sandstones. The beds are known for fish fossils, and the shale layers have so much organic material in them that they became oil shales. They have been stripped from the Black Canyon region but crop out north of the park on Grand Mesa.

A second episode of erosion again flattened the Gunnison Uplift and truncated the Mesozoic sediments that had been tilted by the original uplift. The Precambrian basement complex was exposed once

again as sediments such as the Green River Formation were stripped away. The drainage that was controlled by the structure of the Gunnison Uplift now meandered across the Precambrian rock of uniform hardness.

About 32 mya the ancestral Gunnison River began to establish its course, but this was altered by the nearby West Elk volcanoes to the northeast and the San Juan Mountains located to the south.

12. **Volcanic activity during the Oligocene and Miocene Epochs, covering the region with volcanic debris. Shifting of the ancestral Gunnison River to the south.**

Early in the Oligocene Epoch two nearby volcanic mountain systems, the San Juan and West Elk Mountains (fig. 56.7), blanketed the area with lavas and volcanic breccias (San Juan Formation and West Elk Breccia). By late Oligocene time, the eruptions had changed to a more silicic pyroclastic welded tuff from the San Juan volcanoes. The Oligocene formations are

FIGURE 56.7   Volcanic peaks east of the park in the West Elk area. Photo by David B. Hacker.

the West Elk Breccia (more than 32 mya), Blue Mesa Tuff (more than 28 mya), Dillon Mesa Tuff (more than 28 mya), Sapinero Mesa Tuff (28 mya), Fish Canyon Tuff (27.8 mya) and Carpenter Ridge Tuff (27.5 mya). Many of these tuffs and breccias were hot pyroclastic flows that formed welded tuffs. Last of all was the Miocene Epoch localized extrusion of mafic alkalic lava, called the Hinsdale Formation (18.5 mya). The volcanic activity lasted much longer in the San Juan Mountains than it did in the West Elk Mountains. The pyroclastic tuff from the San Juan Mountains, which is south of the Black Canyon, poured northerly across the Gunnison uplift and reached the flanks of the West Elk Mountains. The volcanic material from the West Elk Mountains had some flows of lava, but the majority of the material appears to be blocks of debris and slurry of mud created when explosions destroyed some of the volcanic cones. Accompanying this outpouring, a syncline developed along the axis of the present day Black Canyon during the Oligocene Epoch. This syncline is evident from the pattern of structure contours plotted at the base of the ash flow tuff volcanic sequence and the plotting of the thickness of the Blue Mesa Tuff. The unloading and doming of the West Elk Mountains, located north of the Black Canyon, most likely caused this syncline. The warping did not stop until after the deposition of the Miocene Hinsdale Formation.

With each eruption, the drainage was disrupted but was restored intermittently as shown by gravels deposited between layers of lava and volcanic ash. The gravels were deposited during outbursts of the river as it broke through volcanic dams. Eventually, about 1.2 mya, the drainage that was flowing westward was forced southward between the two volcanic centers, the West Elk Mountains to the north, and San Juan Mountains to the south.

13. Erosion of the Black Canyon during the Colorado Plateau uplift.

The lower elevations of the region were buried by the successive layers of volcanic material that caused the area to become a sloping plain. The volcanic eruptions ended with the deposition of the Hinsdale Formation, allowing the river to finally establish its course without interruptions. The river was flowing on thick layers of pyroclastic debris and gravels deposited earlier. At this point in time the river was located along the axis of the syncline, which was also across the buried Gunnison Uplift.

Because the volcanic material was soft and easily eroded, a steep-sided valley formed. When the river reached the very resistant Precambrian rock, the cutting of the Black Canyon began. If the Gunnison River had flowed farther north or south, the canyon would have never formed and another broad valley in sedimentary rocks would have formed instead. It probably took about 2 million years for the canyon to be carved. The rate of downcutting of the river of today has been slowed because of the presence of three dams upstream.

Geologic Map and Cross Section

Armbrustmacher, T.J, Barton, H.N., Kulik, D.M., Lee, K., and Brown, S. 1980, Mineral Resources of the Gunnison Gorge Wilderness Study Area, Montrose and Delta Counties, Colorado, U.S. Geological Survey Bulletin 1715-D.

Chronic, H, 1980, Roadside Geology of Colorado, Mountain Press Publishing Company, Missoula, Montana.

Geological Cross Section, Black Canyon of the Gunnison, 1995, Southwest Parks and Monuments Association, Tucson, Arizona.

Sources

Hansen, W.R., 1965, *The Black Canyon of the Gunnison: Today and Yesterday,* U.S. Geological Bulletin 1191, 76 p.

Hansen, W.R., 1981, *Geologic and Physiographic Highlights of the Black Canyon of the Gunnison River and Vicinity, Colorado,* New Mexico Geological Society Guidebook, 32nd Field Conference, Western Slope, Colorado, pgs. 145–154.

Hansen, W.R. 1987, *The Black Canyon of the Gunnison in Depth,* Revised second edition. Southwest Parks & Monuments Association (First published as U.S. Geological Survey Bulletin 1191).

Hansen, W.R. and Peterman, Z.E., 1968, Basement-Rock Geochronology of the Black Canyon of the Gunnison, Colorado, U.S. Geological Survey Professional Paper 600C, pgs C-80 to C-90.

Kiver, E.P., Harris, D.V., 1999, *Geology of U.S. Parklands 5th Edition,* John Wiley & Sons, Inc., New York, p. 387–394.

Note: To convert English measurements to metric, go to www.helpwithdiy.com/metric_conversion_calculator.html

Glossary

aa Lava flows characterized by a spiny, rubbly surface.

abrasion Mechanical wearing, grinding, scraping, or rubbing away of rock by friction or impact. Solid rock particles, transported by ice, waves, running water, wind or gravity, are tools of abrasion.

absolute age Age given in years or some other unit of time.

accreted terrane See *tectonostratigraphic terrane*.

adjustment of drainage, adjusted drainage Condition whereby streams are fitted to structure by headward erosion along outcrops of less resistant rocks.

agate A cryptocrystalline variety of quartz; banded or variegated chalcedony.

aggradation Building up of a land surface by deposition.

alaskite Granitic rock containing only a small percentage of dark minerals.

alkalic basalt A group (or suite) of basalts, undersaturated in silica, containing one variety of pyroxene and a considerable amount of olivine.

alluvial fan Large, fan-shaped pile of sediment that forms where a stream's velocity decreases as it emerges from a steep canyon onto a plain or valley floor.

alluvial terrace Landform, made up of alluvium, produced by dissection or downcutting of a flood plain (see *terrace*).

alluvium Unconsolidated gravel, sand, silt, and clay deposited by streams.

alpine glaciers, mountain glaciers, valley glaciers Glaciers that originate in high cirques or icefields and flow in confined paths from higher elevations to lower elevations.

alpine topography Mountainous, glacially modified topography.

amphiboles Group of common, rock-forming mafic minerals (ex. *hornblende*).

amygdule Gas cavity or vesicle in igneous rock filled with deposits of calcite, quartz, or other secondary minerals.

anastomoses (in caves) Nets of braided, interconnected tubes, commonly confined to a discrete layer of rock.

andesite Fine-grained, extrusive rock of intermediate composition.

angular unconformity An unconformity in which younger rock strata overlie an erosional surface on tilted or folded layered rock; bedding planes of the rock units above and below the eroded surface are not parallel.

anhydrite Mineral composed of anhydrous calcium sulfate, common in evaporite deposits.

anticline An upfold opening downward; a fold with older beds in the center (see *syncline*).

aphanitic texture Describing the texture of an igneous rock in which grains are too small to be seen with the unaided eye.

aquifer Body of rock permeable enough to conduct and yield significant amounts of ground water.

aragonite Mineral with the same composition as *calcite* but belonging to a different crystal system.

arch A natural arch produced by weathering, mass wasting, and erosion.

arête Sharp, serrated ridge that separates abutting cirques.

arkose Variety of sandstone containing quartz grains and at least 25 percent feldspar fragments.

atoll Roughly circular coral reef, composed of low islets that enclose, or nearly enclose, a lagoon.

avalanche chute Track or trough left after ice, rocks, soil, and trees have swept down a steep slope in a slide.

back reef Landward side of a reef, including deposits between reef and mainland.

badlands, badlands topography Intricately dissected topography characterized by a fine drainage network, high drainage density, short, steep slopes, and narrow interfluves. Badlands develop on a surface with little or no vegetative cover in unconsolidated or poorly cemented clays and silts.

bajada Broad, gently sloping, depositional surface *(landform)* that extends from a mountain front to a semiarid or arid basin, sometimes by coalescence of alluvial fans.

balanced rock Large rock resting more or less precariously on its base, formed in place by weathering and erosion.

bar (stream) Ridgelike accumulate of sand and gravel in a stream channel, along banks, or wherever a drop in velocity induces deposition.

barchan (dune) Crescent-shaped dune with convex windward side and horns pointing downwind.

barrier reef Long, narrow reef, parallel to a shoreline and separated from the shore by a lagoon.

basal conglomerate Pebbles or coarse rock fragments that form the bottom stratigraphic unit of a sedimentary series and rest on an erosion surface, thereby marking an unconformity.

basalt Dark, fine-grained, mafic igneous rock composed of calcium-rich feldspar and ferromagnesian minerals.

base level Theoretical limiting surface below which a stream cannot erode.

batholith Large, discordant, plutonic mass with no known floor that has more than 40 square miles in surface exposure.

beach Strip of sediment (sand, shell fragments, cobbles, and/or mud) that extends from low-water line inland to a cliff or zone of permanent vegetation.

beach drifting Lateral movement of sand and pebbles along the foreshore caused by oblique arrival of waves against a coast.

beachrock A sedimentary rock formed in the intertidal zone in tropical and subtropical regions that consists of sand and gravel, commonly including coral debris, cemented with calcium carbonate.

bedding plane Nearly flat surface separating each layer in a sedimentary sequence from the beds above and below.

bentonite Very fine-grained sedimentary rock (composed mainly of montmorillonite and colloidal silica), derived from volcanic debris, tuff, and ash, that absorbs water quickly and swells when wet. When saturated, it slumps or may even flow as a fine muck. After it dries, it hardens again.

bioherm Moundlike mass of rock built by sedentary organisms, such as colonial corals, calcareous algae, mollusks, gastropods, and so on. Detritus, organic and inorganic, has filled voids in the structure.

biota All living organisms (plants, animals) of an area.

blanket sand Sedimentary deposit of fairly uniform thickness and wide distribution, typically a sandstone deposited by a transgressing sea.

blowout Hollow formed by wind sweeping away sand or fine soil; the accumulating material may form a *blowout dune*.

bomb (volcanic) Fragment of lava, or ejecta, rounded, ropy, or spiral in shape, that cooled and hardened while spinning in the air.

boulder A detached rock mass having a diameter greater than 10 inches (at least the size of a volleyball); the largest size of rock fragments regarded as sediment.

boxwork (cave) After a network of intersecting blades of calcite was deposited in cracks in highly fractured limestone, the host rock was removed by solution, leaving the intricately crisscrossed fins of the boxwork on cave walls.

braided stream An interlacing, tangled network of rivulets and branches separated from each other by islands and bars.

breakdown (cave) Debris from collapse of the ceiling or walls of a cave.

breaker Wave that has become so steep that the crest topples forward, moving faster than the main body of the wave.

breccia Coarse-grained clastic rock composed of large, angular, broken rock fragments cemented together in a finer-grained matrix; can be of any composition, origin, or mode of accumulation.

brecciated texture Describing rocks formed from an accumulation of angular fragments or broken and crushed particles.

butte An isolated, steep-sided hill, with or without a flat top, but usually smaller than a mesa.

calcite A rock-forming mineral composed of calcium carbonate ($CaCo_3$); the principal constituent of limestone.

caldera Large, bowl-shaped depression that forms when a volcano cone explodes and collapses.

caliche A rock layer, common in arid regions, composed of crusts of soluble calcium salts mixed with sand and silt; formed mainly by capillary action drawing mineral-rich ground water toward the surface where the salts precipitate from solution.

calving Breaking away of blocks of ice from a floating glacier front, forming icebergs.

canyon passage (caves) Underground stream bed that resembles a chasm.

capture (stream) The natural diversion of a stream's headwaters into the channel of another stream having greater erosional activity and flowing at a lower level.

carbonates Sedimentary rocks that contain more than 50 percent by weight of calcite or other carbonate minerals.

carbonation Type of chemical weathering in which carbon dioxide in air and soil combines with water to form carbonic acid, which reacts with minerals or rocks.

carnelian Translucent red or orange-red variety of chalcedony.

cave (cavern) Naturally formed underground chamber, large enough for a person to enter; develops in carbonate rock by solution processes.

cave earth Residue of clay particles and other insoluble materials found in all caves; may consist of limestone residue or fine sediment washed into a cave form the outside.

cave ice, calcite rafts Thin, rectangular pieces of calcite that form in bodies of standing water in caves.

cave-in lake Shallow body of water in a basin produced by collapse of soil due to unequal thawing of ground ice in regions underlain by permafrost.

cave lining Coating of calcite or silica deposited on cave walls during an interval when cave was filled with water.

cave pearls Concretions of calcite formed in caves when small particles become coated with calcium carbonate.

cave popcorn Nodular, grapelike protrusions formed where water seeps out of pores in the rock walls of a cave.

Cavernous weathering Chemical and physical weathering that loosens rock grains and enlarges hollows and recesses in a cliff face (see *tafoni*).

cementation Process by which clastic sediments become consolidated into compact rock, usually through precipitation of minerals in spaces between sediment grains.

chalcedony A cryptocrystalline variety of quartz; possesses *conchoidal fracture*.

chattermarks Small, curved cracks made by vibratory chipping of ice-embedded rock fragments against a bedrock surface.

chemical weathering See *decomposition*.

chert Hard, fine-grained sedimentary rock formed almost entirely of silica.

cinder cone Relatively small volcano constructed of loose fragments *(tephra)* ejected from a central vent.

cinnabar A heavy, red mineral, mercuric sulfide (HgS), the principal ore of mercury.

circum-Pacific belt The great volcanic belt that rings the Pacific Ocean and marks the location of plate boundaries and major tectonic activity.

cirque Steep-walled, bowl-like hollow carved into a high mountain slope by glacial erosion and mass wasting.

cirque glacier Glacier occupying a cirque basin at the head of a glacial valley; usually a remnant of a larger glacier.

cirque lake Small lake occupying a cirque; has no prominent inlet or outlet, is fed by runoff from the cirque walls and is dammed by a bedrock lip or a small moraine.

clastic (detrital) Describing fragments derived from preexisting rocks or minerals, transported as particles to depositional sites by water, ice, wind, or gravity. When consolidated, such sediment becomes *clastic sedimentary rock*.

clastic dike Body of clastic materials, derived from underlying or overlying beds, that cuts across bedding of a sedimentary formation.

clastic texture Rock texture produced by the lithification of an accumulation of fragments.

clay Sediment made up of particles less than 4 microns in diameter, i.e., smaller than very fine silt grams.

clay minerals Group of finely crystalline, hydrous aluminum silicates, formed chiefly by weathering of silicate minerals (feldspars, pyroxenes, etc.); able to absorb substantial amounts of water.

cleavage (mineral) Property of some minerals to break along planes related to a mineral's crystal structure.

cleavage (rock) Characteristic or tendency of a rock to split along parallel, closely spaced planar surfaces.

closed drainage basin A drainage basin lacking sufficient stream flow to discharge into another basin, river, or the sea; an area of interior drainage.

coarse-textured drainage Descriptive of an area where the number of streams and valleys per unit of area is relatively low.

coast Strip of land of indefinite width that extends from low-tide line inland to the first major change in landform features.

cobble Rock fragment, somewhat rounded by abrasion, larger than a pebble and smaller than a boulder (i.e., between 2.5 to 10 inches in diameter).

col Saddle-shaped depression or pass in a ridge, carved by glacial ice breaking down an arête between cirques.

collapse block (in a cave) Segment of a cave ceiling or wall that has fallen to the cave floor.

collapse breccia Breccia resulting from the collapse of rock overlying an opening, as in a cave.

colloidal dispersion Condition existing when submicroscopic particles carry a slight charge of electricity that causes them to repel each other, keeping the particles in suspension.

column (dripstone)   A cave feature formed as a stalactite growing downward meets and joins a stalagmite growing upward.

columnar structure, columnar jointing   Volcanic rock solidified from lava in parallel, vertical columns, polygonal in cross section; formed as a result of contraction during cooling.

composite volcano   Volcano constructed of alternating layers of pyroclastics and rock solidified from lava flows; also called *stratovolcano*.

compression, compressive force   A force that tends to shorten a body (cf. *tensional force*).

conchoidal fracture   The characteristic of certain minerals or rocks to break along smoothly curved surfaces.

concordant   Having boundaries parallel to bedding or earlier developed planar structures; structurally conformable.

concretions   Hard, compact, rounded masses, or nodules, of mineral matter, of varying size and composition, that have been precipitated out from solution, usually around nucleii of some type (such as fossils).

conglomerate   Coarse-grained clastic sedimentary rock formed by cementation of rounded gravel in a matrix of sand.

contact   Interface between two types or ages of rock.

contact metamorphism   A local process of thermal metamorphism related to the intrusion of magma; affects country rock at or near its contact with the intruded igneous body.

continental crust   The thick granitic crust that underlies the continents.

continental glacier   Ice sheet of considerable thickness covering a substantial part of a continent.

continental shelf   Submarine platform at the edge of a continent, inclined gently seaward.

continuous permafrost   Zone uninterrupted by pockets or patches of unfrozen ground.

converging plate boundary   Boundary between two tectonic plates that are moving toward one another.

coquina   Soft, porous, coarse-grained, off-white limestone made up of loosely aggregated shells and shell fragments.

coral   Cylindrical-shaped marine invertebrate that lives either singly or bound together structurally with other corals in a colony attached to the sea floor; a reef-forming organism.

cordillera   See *mountain belt*.

correlation   Determination of age relationships between rock units (or geologic events) in separate areas.

country rock   Any rock older than and intruded by an igneous body.

crag and tail   Glacially carved, streamlined ridge with a knob (the "crag") of resistant bedrock and an elongate body (the "tail") of less resistant bedrock or till on the lee side.

crater   Saucerlike depression over a vent on a volcanic cone.

craton   A portion of continental crust that has been structurally stable for a prolonged period of time.

creep   Very slow, continuous downslope movement of soil or unconsolidated debris; type of mass wasting.

crevasse   Open fissure in a glacier caused by fracturing of ice during movement.

cross-bedding   A sequence of crisscross, dipping beds deposited by water or wind currents of variable direction.

cross fold   Complex structure produced when a fold intersects a preexisting fold of different orientation.

crust   The thin, outermost layer of rock that makes up the earth's surface.

crustal rebound   See *glacial rebound*.

cryptocrystalline   Having an internal crystalline structure so fine that it cannot be detected by a petrographic microscope.

crystalline (rock)   Referring to a rock consisting wholly of crystals or crystal fragments, such as an igneous rock that cooled from a melt or a metamorphic rock that has undergone recrystallization.

cuesta   An asymmetrical ridge that slopes gently on one side, conforming to the dip of resistant beds, and has on the opposite side a steep scarp or cliff formed by the outcropping of resistant beds.

Curie point   The temperature above which thermal agitation prevents spontaneous magnetic ordering. Specifically, the temperature at which there is a transition from ferromagnetism to paramagnetism. Above the Curie point, rocks do not retain their magnetic properties.

cutbank   Develops on the outside of a stream meander where the bank is being undercut by the stream's lateral migration.

dacite   Fine-grained, extrusive igneous rock, intermediate in composition between rhyolite and andesite.

daughter product   isotope of an element produced by radioactive decay.

debris flow   Mass wasting involving rapid flowage of debris of many sizes.

décollement   Detachment structure of beds showing disharmonic folding and characterized by independent styles of deformation in the rocks above and below the décollement; associated with overthrusting.

decomposition   Weathering processes by which rock is changed due to chemical activity, mainly involving water and atmospheric gases; also called *chemical weathering*.

deflation   Removal by wind of fine sediment particles from a land surface.

degradation   The lowering of a land surface and general reduction of relief by erosion.

dendritic drainage pattern   Drainage pattern of a stream and tributaries that resembles a branching tree or veins in a leaf.

desert pavement   Closely packed surface of wind-polished pebbles and rock fragments, somewhat cemented together by calcium carbonate; tends to form on flat surfaces in arid regions after wind and sheetwash have removed fine particles. Once formed, desert pavement protects underlying sediment from further deflation (wind erosion).

desert varnish   Thin, dark patina that develops on the surface of pebbles, boulders, or bedrock, mainly in desert regions; consists of iron and manganese oxides with some silica.

detrital (detritus)   See *clastic*.

diabase   Dense, black, intrusive igneous rock, with ophitic texture, studded with light-colored plagioclase crystals within pyroxene crystals; frequently found in dikes.

diamictite   Poorly sorted, nonmarine sedimentary rock (such as a *tillite*), containing a wide range of particle sizes.

differential erosion (or weathering)   Erosion (or weathering) that occurs at irregular or varying rates, caused by differences in the resistance and hardness of rocks, minerals, or surface materials.

differentiation (magmatic)   Process by which more than one rock type develops from a single magma.

dike   A tabular, discordant body of intrusive igneous rock (cf. *sill*).

diorite   Coarse-grained, intrusive igneous rock, intermediate in composition; composed of approximately equal amounts of plagioclase feldspar and ferromagnesian minerals.

dip   The vertical angle and direction at which a bedding plane or fault plane is inclined from the horizontal.

disconformity   An unconformity that indicates missing rock layers, although beds above are parallel to those below.

discontinuous permafrost An intermediate zone containing patches of unfrozen ground.

discordant Cutting across bedding or earlier developed planar structures.

disintegration The breaking up of rock into smaller pieces by frost action, abrasion, etc. Also called *physical weathering* or *mechanical weathering*.

displacement Refers to the relative movement of two sides of a fault, as measured in a specific direction.

diverging boundary Boundary separating two tectonic plates that are moving away from each other; a *spreading center*.

dolomite Sedimentary rock composed of the mineral dolomite (calcium-magnesium carbonate), commonly formed by replacement of limestone.

dome pit (cave) Solution-formed vertical passage or high chamber with domed ceiling and vertical walls; also called a *vertical shaft*.

drainage basin Total area drained by a stream and its tributaries.

drainage divide Boundary between adjacent drainage basins.

driblet cone, driblet spire Spire-shaped mound resulting from the splattering of lava ejected from an opening in a lava tube.

dripstone General term for calcite or other deposits formed in caves by dripping water.

drumlin Smooth-surfaced, elongate hill or ridge (the steeper slope at the up-ice end), made up of compacted glacial till.

dune rock Consolidated dune sand (see *eolianite*).

earthquake swarm Series of minor earthquakes occurring in a limited area and time.

ejecta Material, both solid and liquid, thrown out by a volcano. Sizes may range from ash particles to large blocks of rock and volcanic bombs.

endellite Clay mineral formed by alteration of other clay minerals in an acidic, sulfate-rich environment.

end moraine Ridgelike accumulation of till deposited at the lower or outer end of a glacier.

entrainment The process of water or wind currents picking up and transporting sediments.

eolian Referring to wind action.

eolianite Sedimentary rock consisting of clastic material deposited by wind.

eon The longest geologic-time unit, next in magnitude above an era; e.g. the Archean Eon.

ephemeral stream Flows briefly only in direct response to precipitation in the immediate locality (cf. *intermittent stream*).

epicenter The point on the earth's surface directly above the focus of an earthquake.

epoch Unit of geologic time shorter than a *period*; e.g., the Holocene Epoch.

epsomite Mineral consisting of native Epsom salts; occurs in colorless crystals, masses, and encrustations in gypsum mines, limestone caves, or in mineral waters in solution.

era Geologic-time unit next in order in magnitude below an eon and including two or more *periods*; e.g., the Paleozoic Era.

erosion The processes whereby the materials of the earth's crust are loosened, dissolved, or worn away and removed by running water, glacial ice, wind or gravity.

erosion surface A lowered land surface, smoothed by prolonged erosion.

erratic Large rock fragment that has been transported by ice from its place of origin; commonly different in composition from underlying bedrock.

esker Long, narrow, sinuous ridge of stratified, glacial meltwater deposits (sand, gravel) that accumulated in channels in or beneath stagnant ice.

evaporite Sediment that precipitated from a saline solution; usually marine but sometimes nonmarine.

exfoliation, sheet jointing Weathering process by which concentric scales or shells of rock are successively spalled or stripped from the surface of a large rock mass; caused by physical and chemical forces producing differential stresses within the rock, *or* by release of confining pressure (i.e., *pressure-release jointing*).

exfoliation dome Large, rounded landform developed in massive rock, such as granite, by the process of exfoliation (sheet jointing).

extrusive rock Igneous rock that has cooled or solidified at the earth's surface, whether from a lava flow or from fragments of volcanic debris.

faceted spur A ridge or spur with an inverted-V face that was produced by faulting or by the beveling action of streams, waves, or glaciers (see also *truncated spur*).

facies, facies change The term *facies* refers to the characteristics of an assemblage of rock, mineral, or fossil features that reflect a rock's conditions of origin. *Facies change* refers to gradual

lateral or vertical changes in sediments or fossils that indicate an evolving depositional environment (see *metamorphic facies, sedimentary facies*).

fault Fracture in bedrock along which movement has taken place.

fault-block mountain Uplift along normal or vertical faults forms structural mountain blocks or ranges that are products of crustal extension.

feldspar group Includes the earth's most common minerals that make up about 60 percent of the crust. All feldspars contain silicon, aluminum, and oxygen and may contain potassium, sodium, calcium or other elements in varying combinations. Feldspars are components of many types of rock.

felsenmeer, block field Frost-riven blocks and coarse, angular fragments that have accumulated on a summit area.

felsic Describing silica-rich igneous rocks with a relatively high content of potassium and sodium.

ferromagnesian minerals Generally dark, iron-magnesium-bearing minerals, such as olivine, hornblende, biotite, augite; the *mafic* minerals.

fetch The extent of open ocean over the surface of which the wind blows with constant speed and direction, thereby creating a wave system.

fin (ridge) Local term (Colorado Plateau) for a sharp, narrow ridge, usually eroded from sandstone.

fine-textured drainage Descriptive of an area where the number of streams and valleys per unit of area is relatively high.

fiord Glacially eroded coastal inlet occupied by an arm of the ocean.

firn Compacted mass of granular snow, transitional between snow and glacial ice.

firm line Irregular, shifting line marking the highest level at which a glacier's winter snow cover is lost during a melting season.

flood basalt Fluid basaltic lava that erupts repeatedly from fissures and spreads over a vast area.

flood plain That part of a valley floor, periodically covered by flood waters, which leave a blanket of sediment on either side of the stream channel.

flowstone General term for any accumulation in caves of calcium carbonate or other mineral matter formed by flowing water on the walls or floor.

flute (cave) An incised, vertical or nearly vertical groove developed in a cave wall by dripping water.

flysch Marine sedimentary facies that is the result of accelerated accumulation of sediment and rapidly changing conditions of deposition; e.g., alternating beds of shales and sandstone.

focus   The point within the earth where an earthquake rupture starts (cf. *epicenter*).

fold   A bend or warping in bedding or foliation in rock, the result of deformation.

foliated (foliation)   Planar arrangement of textural or structural features in any type of rock, especially in metamorphic rock.

footwall   The underlying side of a fault, especially the wall rock beneath an inclined fault (see *hanging wall*).

fore reef   Seaward-facing section of a reef; normally covered with reef talus.

foreshore   Area between mean low water level and mean high water level; the zone regularly covered and uncovered by rise and fall of tides, and upon which waves usually break.

formation   Mappable, lithologically distinct rock unit, usually made up of one or more beds of sedimentary rock, but may be composed of metamorphic or igneous rock units. Contacts separating a formation from adjacent rock units must be recognizable.

fossil   Any evidence of ancient organisms naturally buried and preserved in rocks or sediments.

fringing reef   Relatively small, linear reef formed parallel to a shoreline and directly attached to or bordering the shore, normally without a lagoon.

frost action   Mechanical weathering processes caused by cycles of freezing and thawing water in pores, cracks, and other openings at or near the surface.

frost heaving   Uneven lifting of rock or soil by expansion of freezing water.

frost splitting, frost wedging   Breaking up of rock or soil due to pressure exerted by freezing of water contained in cracks or fissures.

frostwork (cave)   Bristlelike cluster of aragonite needles deposited on cave popcorn.

fumarole   Volcanic vent from which gases and vapors are emitted.

gabbro   Coarse-grained, mafic igneous rock composed mainly of ferromagnesian minerals and calcium-rich plagioclase feldspar.

geode   Hollow, or partly hollow, globular-shaped body occurring in sedimentary rock that weathers out as a discrete nodule or concretion.

geologic agent, geologic process   Running water, wind, ice, and the like, are *geologic agents* that do geologic work by means of natural *geologic processes,* such as erosion and deposition.

geothermal gradient   Rate of temperature increase with depth of crust. Average geothermal gradient in the crust is about 25° C/ km of depth.

geyser   Hot spring that intermittently ejects jets of hot water and steam.

glacial drift   Unconsolidated rock debris, stratified and unstratified, that has been transported by glaciers or meltwater streams and deposited directly on land or in the sea.

glacial pavement   A bedrock surface that has been planed down, striated, smoothed, and polished by glacial abrasion.

glacial rebound   Upward movement of the earth's crust after the weight of glacial ice that had depressed the crust is removed; crustal rebound.

glacier   Large, long-lasting mass of ice, formed on land by compaction and recrystallization of snow, which moves because of its own weight; forms where winter snowfall exceeds summer melting.

glacieret   Miniature alpine glacier; commonly a remnant of a larger glacier.

glacier outburst, glacier outburst flood, glacier burst   Sudden release of meltwater from a glacier or glacially dammed lake; may be caused by breakout, buoyant lifting of ice, or subglacial thermal or volcanic activity.

glassy (vitreous) texture   Extrusive igneous rock texture similar to broken glass; develops as a result of rapid cooling of lava without visible crystallization into individual mineral grains.

glauconite   Dull-green mineral, iron-potassium silicate of the mica group; its presence in a sediment or sedimentary rock indicates very slow accumulation.

glowing avalanche, glowing cloud   Incandescent, turbulent, swiftly flowing cloud of hot ash, dust, and gas erupted from a volcano. The denser, lower part of the cloud may form an ash flow.

gneiss   Coarse-textured metamorphic rock comprised of bands or lenses of light and dark minerals.

gradient   Degree of inclination or steepness of a surface or slope; ratio of vertical to horizontal, expressed in feet/mile, meters/kilometer, degrees, etc.

granite, granitic rock   Any felsic, coarse-grained, light-colored, intrusive igneous rock containing quartz as an essential component, along with feldspar and mafic minerals.

granodiorite   Coarse-grained plutonic rock composed mainly of quartz, potassium feldspar, and plagioclase, plus mafic minerals in lesser amounts; the intrusive equivalent of *rhyodacite*.

granular disintegration   Type of mechanical weathering consisting of grain-by-grain breakdown of rock masses. Mineral grains tend to separate from one another along their natural contacts to produce coarse sand or gravel.

gravel   Rounded sediment particles coarser than sand.

gravity survey   Measuring direction and intensity of the earth's gravitation field at different locations, using a gravimeter, and plotting the measurements on a map.

ground moraine   Blanket of till deposited by a glacier or released as glacial ice melts.

ground water, underground water   The water beneath the ground surface that fills the cracks, crevices, and pore space of rocks.

group (stratigraphy)   Two or more geographically associated formations with notable features in common.

grus   Loose mineral soil produced *in situ* (in place) by the granular disintegration of granite and granitic rocks.

gypsum   A mineral (hydrous calcium sulfate) which, when lithified, becomes a sedimentary rock, called *rock gypsum*.

gypsum flower, cave flower   Gypsum that occurs as a curved, elongate cave deposit resembling the shape of a flower. (Epsomite also can occur as cave flowers.)

halite   Mineral composed of sodium chloride; referred to as rock salt when lithified.

hanging valley   Smaller valley that terminates abruptly above a main valley or shoreline. A waterfall may develop because the floors of the small valley and the main valley (or shoreline) are discordant in level.

hanging wall   Overlying side of, or surface above, an inclined fault plane (see *footwall*).

hardground   A thin layer on the sea floor in which the sediment is lithified so as to form a hardened surface that commonly serves as a substrate for the growth of organisms.

harmonic tremors   Small earthquakes that often occur before a volcanic eruption.

headward erosion   Uphill growth of a valley by gullying, mass wasting, and sheet erosion.

helictite (cave)   A contorted cave deposit (speleothem) that angles or twists erratically.

high-energy coast Exposed to ocean swell and stormy seas; characterized by average breaker heights of greater than 50 cm.

hogback Long, narrow, sharp-crested ridge formed by differential erosion on the outcropping edges of steeply inclined or highly tilted resistant rocks (cf. *cuesta*).

hoodoo, earth pillar, rock pillar, toadstool Column or pillar of bizarre shape, produced by rainwash and differential weathering of horizontal beds; may be small and delicate or large and massive.

horn, matterhorn Sharp peak formed when cirques cut back into a mountaintop from three or four sides.

hornblende Commonest mineral of the amphibole group; found in both igneous and metamorphic rocks.

hornfels Fine-grained, unfoliated metamorphic rock.

hot spot Deep, persistent heat source in the earth's mantle that causes magma to rise in a plume toward the surface; associated with Hawaiian-type volcanoes.

hot spring, thermal spring Natural spring with water temperature warmer than human body temperature.

hydration Type of chemical weathering in which water combining with minerals in rock causes mineral grains to expand, weakening the rock.

hydromagnesite White, earthy magnesium mineral, sometimes called "moon milk."

hydrothermal metamorphism Alteration of rock by passage through it of hot water carrying mineral ions in solution.

iceberg Block of glacier-derived ice floating in water.

icecap Large glacier, not restricted to a valley, extending in all directions.

ice cave An artificial or natural cave in which ice forms and persists throughout all or most of the year.

ice sheet See *continental glacier*.

ice wedge Large, wedge-shaped, vertical, or inclined sheet, dike, or vein of ice that tapers downward into permafrost; originates by freezing of water in a narrow crack produced by thermal contraction of the permafrost. An ice wedge enlarges with subsequent freezing and thawing.

igneous rock Rock formed from solidification of magma on the earth's surface or within the crust.

ignimbrite Rock formed by consolidation of pyroclastic debris, welded tuff, recrystallized ash flows, etc.

impermeable Rock, soil, or sediment that cannot transmit fluid, such as water or petroleum, under pressure.

incised meander Meander that retains its sinuous curves as it cuts downward below the level at which it originally formed. Renewed downcutting may be the result of crustal uplift.

inclusion Fragment or enclave of rock distinctly different in composition from the enclosing igneous rock.

inselberg Prominent, rounded, smoothed, steepsided residual hill rising abruptly from an erosional plain, normally in hot, dry regions.

intermediate rocks Transitional or intermediate in composition between felsic and mafic igneous rocks.

intermittent stream Flows only at certain times of the year (cf. *ephemeral stream*).

intrusion Upward movement of a magma body into the crust; also, an igneous rock mass so formed within surrounding rock.

intrusive rock Igneous rock that has cooled within the earth's crust.

intrusive sheet Tabular igneous intrusion, usually concordant.

isoclinal fold, isocline Fold having parallel limbs, with both limbs having the same amount and direction of dip.

isostasy State of balance or equilibrium, comparable to floating, between adjacent blocks of crust resting on the mantle's plastic layer.

isostatic adjustment Concept of vertical movement of blocks or sections of the earth's crust in order to attain equilibrium.

jasper Variety of chert, characteristically reddish in color.

joints, joint sets Surface fractures, or partings, in rock without displacement. A joint surface is usually a plane. A group of parallel joints forms a *joint set*.

kaolinite Clay mineral formed by hydrolysis in an arid environment.

karst topography Develops due to erosion of limestone or dolomite (occasionally gypsum) by solution of rock; characterized by closed depressions, sinkholes, caves, underground drainage, and the like.

keratophyre Light-colored volcanic rock, composed predominantly of plagioclase feldspar and quartz; extruded from submarine vents.

kettle, kettlehole Depression in glacial outwash caused by melting of a buried block of stagnant ice.

kettle lake Body of water occupying a kettle in a pitted outwash plain or moraine.

kipuka Islandlike surface area surrounded by a lava flow; highly variable in size (Hawaiian word meaning "opening").

klippe An outlier, or isolated, erosional remnant, of a thrust sheet; normally surrounded by younger rocks.

knob (topographic) Protruding mass of resistant rock; any rounded mound or hillock.

lahar Mudflow on a volcano flank; debris carried in the flow consists of pyroclastics, mud, vegetation, etc.; sometimes triggered by glacial meltwater.

lamina The thinnest recognizable unit or layer of deposition in a sediment or sedimentary rock.

landform Any of the natural features (canyon, stream, arch, mountain, etc.) that make up the earth's surface character.

landscape An assemblage of landforms.

landslide, landsliding General inclusive term for slope failure or downslope movement of rock and debris; may be sudden, fairly rapid, or slow.

lapilli Pyroclastic fragments, with diameters ranging from 2 to 64 mm, thrown out by an erupting volcano.

lateral moraine Low, ridgelike accumulation of till carried along the side of a valley glacier.

lava Magma, solid or molten, that reaches the surface, either through fissures or volcanic vents.

lava ball Accretionary feature that forms when a fragment of solidified lava rolls along, growing in size as it picks up sticky lava.

lava fountain Jet of incandescent lava that shoots into the air as lava is forced out of a vent by hydrostatic pressure and expansion of gas bubbles.

lava lake Lake of lava (usually basaltic) in a volcanic crater or depression; may be solidified or partly solidified.

lava stalactite Iciclelike pendant of lava that hardened while dripping from the roof of a lava tube or lava tunnel.

lava tree Tree mold of lava projecting above the surface (see *tree mold*).

lava tube, lava tunnel Hollow space, cave, or tunnel beneath the surface of a solidified lava flow; formed by withdrawal of molten lava after a crust forms over the flow.

lee (glacier) Indicates the side of a hill, knob, or prominent rock located away from an advancing glacier or ice sheet; facing the downstream side of a glacier and sheltered from severe abrasive action (see *stoss*).

lignite Brownish-black coal, intermediate in coalification between peat and subbituminous coal.

limb   One side of a fold; each limb is usually part of two adjacent folds.

limestone   Sedimentary rock, organic or inorganic in origin and composed mainly of calcite; a carbonate rock.

lithification   Consolidation and conversion of sediment to form sedimentary rock.

loess   Fine-grained deposit of wind-blown dust, composed of angular, unweathered grains of quartz, feldspar, and other minerals; unconsolidated, but may be held together by calcareous cement.

longshore current   Moving mass of water that develops parallel to a shoreline, caused by oblique impingement of waves against a coast. This piles up water, some of which then flows along the shore.

longshore drift   Movement of sediment of longshore currents, parallel to the shore.

low-energy coast   A sheltered coast protected from strong wave action by headlands, islands, or reefs and characterized by average breaker heights of less than 10 cm.

mafic   Describing silica-poor igneous rocks having a relatively high content of magnesium, iron, and calcium.

magma   Mobile, melted rock, containing water and gases under pressure; generated at high temperatures within the lower crust and upper mantle.

magma chamber   Reservoir in the earth's crust occupied by magma.

magnetic field (of the earth)   Region of magnetic forces surrounding the earth.

magnetic poles (of the earth)   The points of emergence at the earth's surface where the strength of the magnetic field is greatest. Magnetic lines of force appear to enter at the negative (presently south) pole and leave at the positive (presently north) pole; also called *geomagnetic poles.*

magnetic reversal   A change in the earth's magnetic field between *normal polarity* and *reverse polarity.*

mantle   The thick rock shell of the earth's interior below the crust and above the core. The upper mantle is believed to be partially plastic in character.

marble   Coarse-grained metamorphic rock made up of interlocking calcite crystals, or recrystallized calcite. Some marble contains dolomite crystals.

marker flow   Lava flow with distinctive characteristics that allow it to serve as a stratigraphic reference or be traced over a distance.

massive (rock)   Describes stratified rock that appears to be without internal structures, such as minor joints and lamination; also applies to igneous rocks (granite, diorite) that may have fairly homogeneous texture over a wide area.

mass wasting   Loosening and downslope movement of soil and rock in response to gravitational stress; rates of movement range from creep (very slow) to very rapid falls and slides.

matrix   Finer-grained material (groundmass) enclosing larger grains or particles of a rock or filling spaces between grains.

meander   A sinuous curve or loop in a stream's course; produced by lateral erosion.

meander scar   Meander that has dried up or filled with sediment and vegetation after abandonment by the stream that formed it.

mechanical (physical) weathering   See *disintegration.*

medial moraine   A long ridge of till riding on top of a glacier; formed by joining of adjacent lateral moraines of merging glaciers moving downslope.

mélange (tectonic)   Body of rock, large enough to be mapped, that includes many fragments and blocks of various sizes and kinds of rock embedded in a sheared matrix of malleable rock, such as claystone or shale.

member (stratigraphy)   Rock unit comprising some portion of a formation.

mesa   Broad, flat-topped hill bounded by cliffs and capped by a resistant rock layer.

metamorphic core complex   Domal uplift of metamorphic and plutonic rock overlain by a tectonically detached, unmetamorphosed cover of younger rock.

metamorphic facies   Group of metamorphic rocks having similar mineral compositions and physical characteristics and hence assumed to have formed under a particular range of pressure and temperature conditions.

metamorphic rock   Rock that has been transformed from preexisting rock into texturally or mineralogically distinct new rock as a result of high temperature, high pressure, or both, but without the rock melting in the process.

mica   Group of rock-forming minerals (muscovite, biotite) that commonly occur in flakes, sheets, shreds, or scales. Micas, which are complex silicates, are characterized by low hardness and perfect basal cleavage.

microbrecciated   Describes rock that has been crushed to a very fine grain size by faulting and fracturing under high pressure.

microplate   Small tectonic plate.

microseisms   Very small earth tremors.

mid-oceanic ridge   A continuous, mostly submarine mountain range, with a central rift valley, extending through the Arctic, Atlantic, Indian, Pacific, and Antarctic Oceans; see *spreading center.*

migmatite   Complex rock composed of mixed igneous and metamorphic material; has a banded and veined appearance.

mineral   A naturally occurring, inorganic, usually crystalline solid with a definite chemical composition and characteristic physical properties that are either uniform or variable within definite limits.

mirabilite   White or yellow monoclinic mineral occurring as a residue from saline lakes, playas, and springs, and as an efflorescence; also called Glauber's salt.

monadnock   Upstanding hill or mountain, rising conspicuously above the general level of an erosion surface; remnant of a former upland.

monocline, monoclinal fold   An open fold (usually in relatively flat-lying beds) that bends, or flexes, from the horizontal in only one direction and has one limb.

montmorillonite   Group of clay minerals derived from the alteration or weathering of calcic feldspars, ferromagnesian minerals, and volcanic glass; common in soils, a typical insoluble residue in cave earth, and the main constituent of bentonite.

moraine   Body of till being carried on or by a glacier or left behind after a glacier has receded.

mountain belt, cordillera, mountain chain   Complex, interconnected series of mountain ranges having a well defined, longitudinal trend; commonly several thousand miles in length and several hundred miles wide.

mountain range   A single mass of mountain ridges, closely related in age and origin.

mudball   Spherical mass of mud or mudstone (may be as large as 8 inches in diameter) in a sedimentary rock; developed by weathering and breakup of clay deposits.

mud cracks   Crudely polygonal pattern formed in drying clay or silt by shrinkage.

mudflow   Debris flow consisting of soil, sand, silt, ash, clay, etc., mixed with enough water so that it moves downslope as a viscous mass.

mud pot   Hot spring choked by dissolved compounds in boiling mud.

mylonite   Hard, chertlike rock, with banded or streaky structure, formed by pulverizing of rocks during faulting or intense dynamic metamorphism; commonly found in shear zones.

natural bridge   Arch-shaped landform, produced by weathering, mass wasting, and erosion, that spans a water course or a dry stream valley.

natural levees   Low ridges of flood-deposited sediment, formed on either side of a stream channel, that thin away from the channel.

nonclastic   Describes a sediment or rock formed by chemical precipitation or organic secretion (see *clastic*).

nonfoliated   Pertaining to metamorphic rocks (or massive igneous rocks) lacking *foliation*.

normal fault   Caused by tensional force, which moves the hanging wall down in relation to the footwall.

normal polarity   Natural remanent magnetization closely parallel to the present ambient magnetic field direction; i.e., the direction of magnetic north in a rock, set when it was formed, is approximately the same as modern polarity.

novaculite   Dense, hard, even-textured, light-colored, siliceous sedimentary rock resembling chert.

nunatak   Isolated bedrock peak, projecting above an icefield or glacier.

obsidian   Volcanic glass, usually black or reddish-brown and rhyolitic in composition; characterized by conchoidal fracture.

oceanic crust   That part of the earth's crust that underlies the ocean basins; predominantly basaltic.

oceanic trench   Deep, narrow submarine trough, parallel to an island arc or continental margin; occurs along a subduction zone.

offset   Horizontal displacement along a fault, measured perpendicular to the interrupted horizon.

olivine   Common, dark-colored, rock-forming ferromagnesian mineral.

ooids, ooliths   Tiny, spherical grains about the size of fish roe, usually composed of calcium carbonate, that form by precipitation around a nucleus (shell fragment, sand grain, etc.) in shallow, wave-agitated water. The grains accumulate as sedimentary layers.

oolite   Sedimentary rock, usually limestone, made up of ooliths, cemented together.

ophiolite   Masses of mafic and ultramafic igneous rocks whose structure and composition identify them as segments of oceanic crust pushed by plate collision into continents; commonly found embedded in mountain belts.

ophitic texture   Igneous rock texture (especially in diabase) in which lath-shaped plagioclase crystals are partially or completely included in pyroxene crystals (typically augite).

ore   Naturally occurring material that can be profitably mined.

original horizontality, principle of Most water-laid sediment is deposited in horizontal or near-horizontal layers closely parallel to the earth's surface.

orogeny   Forming of mountain systems by deformational (compressive) crustal processes, accompanied by metamorphism and igneous activity.

orthoclase feldspar   Light-colored, rock-forming mineral of the alkali feldspar group.

outwash (glacial)   Sand and gravel deposited by meltwater streams in front of or beyond the terminal moraine or margin of a glacier.

outwash plain   Smooth, broad, gently sloping plain formed by deposits of heavily loaded meltwater streams draining a glacier margin.

overbank flow   That part of high water or flood water that rises out of the stream channel, depositing sediment over the flood plain surface.

oxbow lake   Crescent-shaped body of standing water occupying a cutoff or abandoned meander.

oxidation   Uniting of oxygen with another element or compound; a process of chemical weathering.

pahoehoe   Basaltic lava flow having a ropy or billowy surface texture (see *aa*).

paleomagnetism   The record of the earth's ancient magnetic fields.

paleosol   Soil layers of past geologic time buried by younger sediments; usually recognizable in a soil profile.

parabolic dune   Deeply curved dune in a region of abundant sand; horns (arms) point upwind and may be anchored by vegetation.

patch reef   Small, moundlike organic reef, usually part of a larger reef complex.

patterned ground   More or less symmetrical forms circles, polygons, nets, steps and stripes-characteristic of, but not necessarily confined to, surficial material subject to intensive frost action-as in polar and subpolar regions.

pebbles   Small, rounded, waterworn stones, ranging in size between a small pea and a tennis ball (1/6 to 2.5 inches).

pediment   Broad, flat (or gently sloping) bedrock erosional surface that develops at the base of a mountain front, usually in a dry region; typically covered with a thin veneer of gravel.

pedestal, mushroom, rock pedestal Slender rock column or neck topped by a bulge of more resistant rock; formed by undercutting due to weathering and erosion.

pegmatite   Extremely coarse-grained igneous rock, usually granitic in composition and made up mostly of large quartz and feldspar crystals; commonly found around the margins of deep-seated plutons.

Pele's hair, Pele's tears   Natural spun glass, "Pele's hair," tends to stream out as threads of lava from the pear-shaped "tears," which are ejected from a vent in a liquid state and solidify in flight. Lava threads break off and may be borne by wind for some distance.

periglacial   Refers to processes, conditions, areas, climates, and topographic features at the margins of glaciers and ice sheets, where freezing and thawing occur seasonally.

period   Geologic time unit longer than an *epoch;* subdivision of an *era;* e.g., the Cambrian Period.

permafrost   Ground that remains permanently frozen over a period of many years.

permeability   Capacity of a rock to transmit a fluid—water, petroleum, etc. (see *porosity*).

permineralization   Process of fossilization whereby the original hard parts of an organism have had additional mineral material deposited in their pore spaces.

petroglyphs   Prehistoric carvings or drawings on rock.

phaneritic texture   Describing the texture of an igneous rock in which individual components are visible to the unaided eye.

phenocryst   Relatively large, conspicuous crystal in the groundmass of a porphyritic igneous rock.

phreatic eruption   Explosive, steam-propelled ejections of ash and mud caused by heating and expansion of ground water.

phreatic explosion   Explosion of steam, mud, and nonincandescent material; caused by heating and consequent expansion of ground water due to an underlying heat source; more violent than a *phreatic eruption.*

phreatic zone   Subsurface zone (below the water table) in which all rock openings are filled with water; also called the *saturated zone.*

phreatomagmatic explosion   Violent volcanic explosion that extrudes magmatic gases and steam along with pyroclastics; caused by mixing of magma with ground water or ocean water.

phyllite   Fine-grained metamorphic rock in which clay minerals have recrystallized into microscopic micas. A silky sheen on corrugated cleavage surfaces is a characteristic.

physiographic province   Geomorphic region showing in all its parts similar climate and geologic structures and having relief features readily differentiated from those of neighboring regions.

pictograph   Picture-writings or signs scratched or painted on rock, hides, etc.

piedmont   General term specifying a feature at the base of a mountain; a foothill.

piedmont glacier   Thick, continuous sheet of ice resting on land at the base of a mountain range; formed by the spreading out and coalescing of valley glaciers from higher elevations.

piercement structure, diapir   Piercing or rupturing of domed or uplifted rocks by mobile material pushed up by tectonic stresses.

pillar (cave)   Bedrock support in a cavern remaining after removal of surrounding rock by solution.

pillow lava, pillow structure
Submarine extrusions of basaltic lavas that solidify as pillow-shaped masses, closely fitted together. "Pillows" range in size from a few inches to a few feet in diameter.

pingo   Large, blisterlike mound, pushed up by pressure of freezing water in permafrost.

piping   Erosion by percolating water in a layer of subsoil, resulting in caving and in the formation of narrow conduits, tunnels, or "pipes" through which soluble or granular soil material is moved.

pisolite   Sedimentary rock, usually limestone, made up chiefly of small, round accretionary bodies cemented together; similar to *oolite*.

placer   Surficial deposit, usually on a beach or in a stream bed, containing particles of valuable minerals or native metal (e.g., gold) in unusual concentrations due to their higher specific gravities.

plagioclase feldspar   Abundant group of rock-forming minerals containing sodium and/or calcium in addition to aluminum, silicon, and oxygen.

plastic   Describes a substance capable of being molded, bent, or deformed under strain without rupturing (up to the limit of its malleability).

plastic flow zone (glacier)   Bottom part, or sole, of a glacier where ice is molded or bent under stress so that it flows without fracturing.

plateau   Broad, comparatively flat landform of considerable extent, elevated above adjacent countryside; usually bounded by cliffs or an escarpment on at least one side.

plate, tectonic plate   Large, moving slab of rock making up part of the earth's crust.

plate tectonic theory   The concept that intense geologic interactions occur along boundaries between rigid, slowly moving plates that make up the earth's surface.

plucking, quarrying   Glacial erosional process by which freezing of water in cracks and joints loosens and detaches large rock fragments and blocks from bedrock, which are then torn out by advancing ice.

pluton   Any massive body of igneous rock formed at depth in the earth's crust by consolidation of magma.

pluvial climate   Climatic interval characterized by abundant rainfall, as during a transitional or interglacial period. A *pluvial lake* is one that formed during such an interval.

polyp   An individual living coral.

porosity   The percentage of openings (interstices) in the total volume of a rock.

porphyritic texture   Igneous rock texture in which larger crystals (phenocrysts) are set in a finer groundmass, which may be crystalline or glassy.

pothole   A general term for any naturally formed, circular pit or depression in soil or rock; sometimes used as a synonym for *kettle* or *kettlehole,* or, in desert regions, may refer to a "rock tank," in which water collects. Most commonly, a *pothole* designates a cylindrical hollow eroded into the rock of a stream bed by abrasive action of the stream's sediment load.

Precambrian shield   Extensive area of ancient Precambrian plutonic and metamorphic rocks at or near the surface; part of a craton.

pressure-release jointing   Exfoliation that occurs in once deeply buried rocks that have been unloaded by erosion, thus releasing their confining pressure.

primary structures   Features formed during the time of a rock's origin but before its final consolidation.

protalus rampart   Ridge of loose, angular rock fragments that accumulated along the toe of a large perennial snowbank lying at the base of a cliff. When the climate warmed, the snowbank melted, leaving an arcuate ridge of boulders and other coarse debris.

pseudomorph   Mineral whose outward crystal form is that of another mineral species.

pumice   Light-colored, vesicular, glassy rock, commonly of rhyolitic composition and often sufficiently buoyant to float on water; formed during explosive eruptions from lava containing gas and water vapor in large amounts.

pyroclasts, pyroclastic debris or material
Rock fragments formed by volcanic explosion or ejection from a volcanic vent. Rock consolidated from such material is said to have *pyroclastic texture.*

pyroxene group   Dark, rock-forming silicate minerals (e.g., augite) similar in chemical composition to the amphiboles.

quartz   Crystalline silica (SiO_2), an important rock-forming mineral.

quartzite   *a)* Metamorphic rock formed by recrystallization of sandstone or chert during regional metamorphism. *b)* Very hard but unmetamorphosed sandstone, consisting of quartz grains solidly cemented with secondary silica.

quartz monzonite   Coarse-textured granitic rock of varying composition; intrusive equivalent of *rhyodacite.*

quicksand   Bed of fine, smooth sand, saturated with water flowing upward through voids, forming a soft, shifting, semiliquid mobil mass that yields easily to pressure and will not support much weight.

radioactive decay   Spontaneous disintegration or transformation of certain unstable atoms, called the parent material, to daughter products through gain or loss of nuclear particles.

radiometric dating   Methods for determining absolute age in time units of geologic materials, based on nuclear decay of naturally occurring radioactive isotopes; e.g., carbon-14, uranium-lead ratio, etc.

rain-shadow zone   The downwind (lee) side of a mountain range receives little precipitation because clouds drop more rain and snow on the upwind side of the range.

recharge   Addition or replenishment of water in the zone of saturation or in an aquifer.

recrystallization   Development in a rock of new crystalline grains that may or may not have the same composition but are usually larger than the original grains.

reef (organic)   Resistant mound or ridge formed in the ocean by accumulation of calcareous remains of plants and animals and debris.

reef crest   Upper surface of a reef.

reef terrace   Shelflike, eroded surface that may develop at several levels on a reef because of successive episodes of submergence and emergence.

regional metamorphism Metamorphism affecting an extensive area and involving relatively high temperatures and pressures; associated with orogenic belts.

regression (marine) Retreat or withdrawal of a sea from land areas, and the consequent evidence of such withdrawal.

relative time, relative age Refers to a sequence of geologic time determined by organic evolution or superposition (see *absolute age*).

relict Describes a topographic feature developed by erosional or depositional processes no longer operating because of a pronounced change in environmental conditions.

relief Vertical difference in elevation between highest and lowest points on an area of the earth's surface.

remanent magnetism The permanent magnetism induced in *a)* an igneous rock when it cooled from a liquid to a solid state (i.e., past the *Curie point*), *or* induced in *b)* a sedimentary rock by settling of magnetic grains during deposition so that they became aligned with the earth's magnetic field.

residual soil Soil cover formed in place by weathering of underlying bedrock (cf. *transported soil*).

reversed polarity Natural remanent magnetization opposite to the present ambient magnetic field direction. When the rock was formed, the direction of magnetic north was approximately 180° different from what it is now (see *normal polarity*).

reverse fault Fault, caused by compression, in which the hanging-wall block moved up relative to the footwall block (see *normal fault*).

rhyodacite Fine-grained porphyritic rock, intermediate in composition between dacite and rhyolite.

rhyolite Fine-grained, felsic extrusive rock composed chiefly of feldspar and quartz.

rift zone *a)* System of crustal fractures (e.g., the rift valley along the crest of the mid-oceanic ridge) where tensional forces pull plates apart. *b)* On Hawaiian volcanoes, a linear zone of volcanic features associated with underlying lava conduits.

rigid zone (glacier) Upper, brittle part of a glacier; does not flow plastically.

rimstone (cave) Thin, crustlike deposit of calcite forming a ring around an overflowing basin or pool of water in a cave.

ripple marks Small-scale troughs and ridges produced by wind or water currents sweeping over loose sediments.

roche moutonnée Elongate knob or hillock of bedrock that has been smoothed and scoured by moving ice on the upglacier (stoss) side so that the rock is gently inclined and rounded. On the downglacier (lee) side the rock is steep and hackly from glacier plucking or quarrying.

rock Naturally formed, solid aggregate composed of one or more minerals or a mass of mineral matter, sediments, precipitates, or organic matter.

rock-basin lake Small body of water occupying a depression scoured in bedrock by glacial action.

rock cycle Theoretical concept for relating tectonism, erosion, and rock-forming processes to common rock types.

rockfall Sudden free fall of a bedrock segment from a cliff or steep slope; the fastest form of mass wasting.

rock flour Fine, powdery glacial sediment consisting of ground-up rock, produced by abrasion underneath a moving glacier.

rock glacier Mass of poorly sorted, angular boulders and fine materials, cemented by interstitial ice and flowing downslope by means of glacier motion. An inactive or relict rock glacier may descend by *creep* (mass wasting).

rock salt Sedimentary (evaporite) rock composed chiefly of the mineral halite.

rockslide Rapid sliding of a block or mass of bedrock that has become detached from an inclined surface of weakness, such as a bedding plane.

roof pendant Downward projection of *country rock* into a batholith or pluton.

salt anticline Anticlinelike structure formed by the bulging upward of a linear salt core into overlying sedimentary beds.

saltation Sediment transport by which stream currents move sand grains along a channel in short, abrupt leaps.

salt dome Circular piercement structure produced by upward movement of a pipelike salt plug.

sandblasting Abrasion by wind-blown sand.

sandstone Sedimentary rock formed by cementation of sand-sized grains, usually 85 to 90 percent quartz; may be deposited by water or wind.

sand wave A general term used for any very large, subaqueous sand ripple; in effect, it is an underwater sand dune.

savanna Tropical or subtropical grassland, characterized by scattered shrubs, sparse trees, and a pronounced dry season.

scallop (cave) Asymmetrical solution hollow formed by turbulent water flow.

schist Strongly foliated, crystalline metamorphic rock with excellent cleavage; characterized by coarse-grained minerals oriented approximately parallel to each other.

Scleractinia The formal taxonomic name for the group that includes all present-day true corals. It is an order within the class Anthozoa or the phylum Cnidaria.

scoria *a)* Frothy, high vesicular basalt; *b)* clinkers or cinders in a burned coalbed.

sea wall (beach) Long, steep-faced, natural embankment or berm of shingle or boulders, built by powerful storm waves along the higher-water mark.

sediment Unconsolidated, loose, solid fragments or particles that can originate by 1) weathering and erosion of preexisting rocks; 2) chemical precipitation from solution, usually in water; and 3) secretion by organisms.

sedimentary facies Significantly different rock types occupying laterally distinct parts of a layered rock unit, such as a *formation*.

sedimentary rock Rock that has formed from 1) lithification of any type of sediment, 2) precipitation from solution, 3) consolidation of remains of plants and/or animals; or 4) combinations of processes.

seismic waves Oscillations of energy that produce earth vibrations during an earthquake.

selenite Clear, colorless, cleavable variety of gypsum.

shale Fine-grained, clastic sedimentary rock, formed mainly by compaction of clay; splits easily into thin slabs parallel to bedding planes.

shatter zone Area of randomly fissured or cracked rock; may have network of veins filled with mineral deposits.

shield volcano Broadly convex, gently sloping, volcanic cone built by flows of low-viscosity lava.

shingle Coarse, loose, waterworn detritus of various sizes.

shutterridge Formed by displacement on a fault traversing ridge-and-valley topography; the displaced part of the ridge "shuts in" an adjacent canyon.

sill Tabular, concordant body of intrusive igneous rock.

silt Fine-grained sediment intermediate in size between clay particles and sand particles.

siltstone Fine-grained clastic sedimentary rock consisting of compacted silt and mud; does not split like shale.

sinkhole (karst) Closed depression found on land surfaces underlain by carbonate rock; formed by solution or by collapse of a cavern roof.

sinking stream Surface stream that disappears into an underground channel.

sinter Chemical sedimentary rock, mainly siliceous, deposited as a hard encrustation by precipitation.

slate Fine-grained metamorphic rock that splits easily along smooth, parallel cleavage planes.

slough Wet, marshy area; may be a sluggish channel or a backwater.

slump block Mass of rock material that slides downward (slumps), usually with a backward tilting relative to the slope from which it descends.

slope wash, rain wash Movement of sediment downslope by mass wasting assisted by trickling or running water not confined to a channel.

solfatara Fumarole that emits sulfurous gases.

solifluction Downslope flow of water-saturated debris over impermeable material; a type of mass wasting common in permafrost areas.

solution (weathering) Process of chemical weathering by which soluble mineral and rock constituents pass into solution and become *dissolved load.*

spatter cone Low, steep-sided mound formed of globs of lava piled up around a volcanic vent.

speleothem Any secondary mineral deposit formed in a cave by the action of water.

spheroidal weathering Process of chemical weathering that tends to round off a block of rock by loosening and detachment of successive, concentric shells of decayed rock form the block; similar to *exfoliation.*

spilite Green, mafic igneous rock composed predominantly of plagioclase feldspar and chlorite; an altered basalt.

spit Fingerlike extension of a sandy beach, with the far end terminating in open water.

spongework (cave) An entangled complex of irregular, interconnecting, tubular channels or cavities of various sizes in the walls of limestone caves; separated by intricate and perforated partitions and remnants of partitions.

spreading center The crest of the mid-oceanic ridge, where the ocean floor is being pulled apart and new crust is forming; a crustal rift where tensional forces separate tectonic plates.

stack, sea stack Pillarlike erosional remnant left detached from a headland as a wave-eroded coast retreats inland.

stalactite Icicle-shaped, dripstone pendant formed on a cave ceiling.

stalagmite Cone-shaped dripstone mass formed on cave floors, generally directly below a stalactite.

steppe Generally flat, treeless, semiarid region, with short grasses and scattered bushes.

stock Small pluton having an area of surface exposure of less than 40 square miles.

storm surge A rise in water level along an open coast during a storm, caused primarily by onshore wind piling water up against the coast.

stoss Describes the side of a hill, knob, or prominent rock facing the upstream side of an advancing glacier; the side most exposed to abrasive action (see *lee*).

striations Furrows or lines, generally parallel, inscribed on rock surfaces as rock fragments embedded in ice are dragged along by a moving glacier.

strike Compass direction or trend of a line formed by the intersection of an inclined plane (such as the fault plane) with a horizontal plane.

strike-slip fault Fault in which movement is parallel to the strike of the fault surface. Two blocks slide past each other, resulting in lateral displacement.

stromatolite Laminated, calcareous sedimentary structure consisting of the remains of blue-green algae; occurs in various shapes and forms—horizontal, columnar, branching, etc.

subduction The sliding down of the leading edge of an oceanic plate beneath a continental margin or an island arc.

suite Set of geographically related crystalline rocks (igneous and/or metamorphic).

supergroup (stratigraphy) Assemblage of related formations and groups in a geographic locality.

superimposed (stream) Describes a stream or drainage system let down by erosion through rock units on which it originally developed and then maintained its course while cutting across different rock structures lying unconformably beneath.

surf zone Strip of shore where the waves break and roll up the beach or beat against a cliff face (see *foreshore*).

suspension Process of sediment transport in which fine particles are held indefinitely in eddying, turbulent water.

syenite Coarse-grained intrusive rock containing one or more feldspars, less than 50% dark minerals, and little or no quartz (see *trachyte*).

syncline Downfold opening upward; fold with younger beds in the center.

tactite Complex, calcareous metamorphic rock formed by contact metamorphism.

tafoni Granitic or gneissic blocks or boulders pitted and hollowed out by cavernous weathering (sing. *tafone*)

talus Broken rock accumulating at the base of a cliff or slope, most commonly as a result of frost action and mass wasting.

tectonite Any rock whose fabric reflects its deformation.

tectonostratigraphic terrane, accreted terrane, exotic terrane Fault-bounded region made up of rock units having a geologic history different from that of contiguous terranes; considered to be a discrete portion of oceanic or continental material not formed in place.

tension, tensional force Stress that tends to elongate or pull apart a solid body.

tephra (pyroclastics) General term for fragmental volcanic rocks.

terminal moraine End moraine marking the farthest advance of a glacier.

terrace General term for a landform that resembles a stairstep with a flat tread (the top) and a steep riser (the face).

terrain Physical features characterizing a region of the earth's surface; an ecologic environment.

terrane See *tectonostratigraphic terrane.*

texture (rock) General physical appearance of a rock, including size, shape, arrangement of grains, etc.

thaw lake Pool of water formed on a glacier surface by meltwater accumulation; also, cave-in lake in permafrost.

tholeiitic basalt Suite of basalts that contain several varieties of pyroxenes and little or no olivine; oversaturated in silica.

thread-lace scoria, reticulate Formed by bursting of lava bubbles, creating a delicate, three-dimensional network of glass threads.

thrust fault Reverse fault in which the fault plane has a low angle of dip.

tidewater glacier Glacier that reaches the sea; usually terminates in an ice cliff, from which icebergs are discharged.

till Glacial drift consisting of unsorted and unstratified rock debris carried or deposited by a glacier.

time-transgressioe rock unit Sedimentary formation that becomes younger (or older) when traced across country; usually the result of laterally shifting shorelines when a sea is advancing or retreating.

trachyte Fine-grained, alkaline extrusive rock (extrusive equivalent of *syenite*).

traction Process of sediment transport by which water currents or wind roll or slide particles and fragments along a stream bed (in water), or (in air) over a land surface.

transform fault Variety of strike-slip fault along which displacement abruptly stops or changes form; may be associated with a mid-oceanic ridge.

transform plate boundary Boundary where two plates are sliding past each other.

transgression (marine) Spread or extension of the sea over land areas, and the consequent evidence of such advance.

transported soil Weathered surficial material unrelated to bedrock or older soil on which it rests (see *residual soil*).

transverse stream Flows across bedrock structures, such as folds, faults, and fault blocks.

travertine Any porous deposit of calcite formed by rapid precipitation of calcium carbonate from surface water or ground water or by evaporation around the mouth of a spring, especially a hot spring.

tree mold Cylindrical hollow in a lava flow formed by envelopment of a tree by the flow, solidification of lava in contact with the tree, and disappearance of the tree by burning and subsequent removal of charcoal and ash. The mold preserves surficial features of the tree.

tritium Naturally occurring, radioactive hydrogen isotope, formed by cosmic rays in the supper atmosphere; enters the earths' hydrologic cycle via precipitation.

tritium dating Calculation of age in years by measuring the concentration of tritium in water. The method provides a means of tracing subsurface movement of water and determining its velocity.

truncated spur Faceted lower end of a ridge projecting into a glacial valley that has been eroded by glacial ice (see *faceted spur*).

tsunami Seismic sea wave produced by a large-scale, short-duration disturbance of the ocean floor, principally by a submarine earthquake.

tubular passage (cave) Cavern or tunnel with an elliptical or lenticular cross section; formed in the phreatic zone.

tufa Chemical sedimentary rock composed of calcium carbonate; formed by evaporation as an incrustation around the mouth of a hot or cold calcareous spring or seep.

tuff Rock formed by solidification of volcanic dust, ash, or pumice.

tumescence Swelling of a volcano edifice due to rising magma.

tumuli Mounds or hillocks on the surface of pahoehoe lava (sing. *tumulus*).

tundra Treeless, gently undulating plain, commonly underlain by muck and permafrost; characteristic of arctic and subarctic regions.

turbidites Rocks or sediments deposited in a marine environment by settling out from muddy (turbid) water flowing along a sloping ocean bottom.

turbidity current Flowing mass of sediment-laden water (heavier than clear water) that slides downslope along a lake bottom or ocean floor.

type locality Geographic location where a rock unit is situated and from which it takes its name.

ultramafic rock Dark plutonic rock (such as peridotite) composed almost entirely of mafic minerals.

ultramylonite An ultra-crushed, very dense variety of *mylonite* in which primary structures have been obliterated.

unconformity Planar surface between rock units that represents a time break in the geologic record, with the rock unit immediately above being considerably younger than the rock beneath.

uniformitarianism, principle of The assumption that geological processes operating today are the same processes that operated in the geologic past.

U-shaped valley Characteristic cross profile of a glacially excavated trough, i.e., flatfloored with nearly perpendicular side walls.

vadose zone Subsurface zone located above the *phreatic zone* and between the water table and the surface; zone in which rock openings are filled partly with water and partly with air; also called *zone of aeration*.

valley glacier See *alpine glacier*.

valley train Outwash dropped or spread within the valley of a meltwater stream that flows out from a glacier (see *outwash plain*).

ventifact A boulder, cobble, or pebble with flat surfaces worn or polished by abrasion of wind-blown sand.

vesicle Small cavity or void in volcanic rock caused by gas bubbles trapped in solidifying lava.

vesicular texture Igneous rock texture characterized by abundant cavities formed by entrapment of gas bubbles during solidification of lava.

viscosity Resistance to flow. Thin, runny lava has low viscosity; thick, sticky lava has high viscosity.

volcanic dome Steep-sided, pluglike, or spinelike mass of volcanic rock solidified from viscous lava in or immediately above a volcanic vent; also called *plug dome, lava dome*.

volcanic neck, volcanic plug Vertical, pipelike body of rock that represents the conduit to a former volcanic vent; an erosion remnant of a volcanic cone.

water gap Deep, narrow, low-level pass, through which a stream flows, that penetrates to the base of and across a mountain ridge; a narrow gorge or ravine cut through resistant rocks by a stream.

water table The surface within the ground below which rocks are saturated; separates the *phreatic zone* (below) from the *vadose zone* (above).

wave-cut platform, wave-cut terrace Horizontal bench in the surf zone cut by wave erosion on a coast; usually overspread by beach sediment.

weathering Natural processes that alter or break down rock at or near the earth's surface (see *decomposition, disintegration*).

weather(ing) pits Shallow depressions on flat or gently sloping exposures of granitic rocks; attributed to strongly localized solvent action of impounded water.

welded tuff Glass-rich volcanic rock composed of silicic pyroclasts welded together.

wind gap Shallow notch in the crest of a mountain ridge; former water gap, abandoned by the stream that formed it.

window (landform) Hole that appears in a fin (or narrow ridge) as a result of differential weathering; also, an opening under an arch or natural bridge.

window (structure) Eroded area within a thrust sheet that displays rocks beneath the thrust sheet; also called *fenster*.

xenocryst Crystal foreign to the igneous rock body in which it occurs; resembles a phenocryst.

zone of accumulation The portion of a glacier with a perennial snow cover, where more snow accumulates than melts each year.

zone of wastage The portion of a glacier where melting or loss of ice and snow exceeds the accumulation.

References

Bates, R.L. and Jackson, J.A., editors. 1987. Glossary of geology, 3rd edition. Alexandria, Virginia: American Geological Institute. 788 p.

Lapidus, D.F. with D. Coates. 1987. The facts on file dictionary of geology and geophysics. New York: Facts on File Publications. 347 p.

Index